Auto Drive Trains

TECHNOLOGY

by

James E. Duffy
Automotive Writer

Chris Johanson
ASE Certified Master Automobile Technician

South Holland, Illinois
THE GOODHEART-WILLCOX COMPANY, INC.
Publishers

Riverside Community College
Library
4800 Magnolia Avenue
Riverside, CA 92506

TL 260 .D84 1995
Duffy, James E.
Auto drive trains technology

Copyright 1995
by
THE GOODHEART-WILLCOX COMPANY, INC.

All rights reserved. No part of this book may be reproduced, stored in a retrieval system, or transmitted in any form or by any means, electronic, mechanical, photocopying, recording, or otherwise, without the prior written permission of The Goodheart-Willcox Company, Inc. Manufactured in the United States of America.

Library of Congress Catalog Card Number 94-1440
International Standard Book Number 1-56637-028-0

1 2 3 4 5 6 7 8 9 10 95 99 98 97 96 95

Library of Congress Cataloging-in-Publication Data

Duffy, James E.
Auto drive trains technology / by James E. Duffy, Chris Johanson.

p. cm.
Includes index.
ISBN 1-56637-028-0
1. Automobiles–Power trains. I. Johanson, Chris. II. Title.
TL260.D84 1995
629.24–dc20 94-1440
CIP

Introduction

No vehicle can use the power in its engine without a drive train to deliver that power to the wheels and road. This is unlikely to change in the future. Whether the power plants in future vehicles are piston, rotary, or turbine designs that are powered by gasoline, diesel fuel, electricity, or steam, drive train components will always be needed to convert engine power to vehicle movement. Drive trains are used to move automobiles, trucks, farm equipment, construction machinery, and almost every other engine-powered device. Automotive and other transportation service businesses will always need trained and competent drive train technicians.

Auto Drive Trains Technology is designed to help you become that drive train technician. This textbook contains information on every type of drive train component used on modern cars and light trucks. Much of the information presented here will also apply to drive train components used on other vehicles, both on and off road. This book will be useful for those entering the automotive repair field, as well as for experienced technicians who want to upgrade their skills to include the latest drive train developments. Technicians studying for ASE tests and automotive hobbyists will also find **Auto Drive Trains Technology** a useful source of information.

Each chapter begins with learning objectives, which emphasize the important topics you will study. In the body of the text, new technical terms and manufacturer-specific component names are printed in *italics.* Defined terms are printed in ***boldface italics*** at their point of definition. Terms are defined in context. The words are stressed because of their importance and to help you to quickly learn the technical language of a drive train specialist.

The end-of-chapter material includes a summary, key terms, review questions, and supplemental certification-type questions. The summary will help you review the most important concepts covered in the chapter. "Know These Terms" provides a listing of the words and terms defined in the chapter. You should know the meanings of these and be able to use them. The review questions and certification-type questions will help you assess your knowledge of the information in the chapter. The certification-type questions are especially helpful in that they are written in the same format as questions prepared by the *National Institute for Automotive Service Excellence (ASE)* for ASE certification tests.

Auto Drive Trains Technology will help you pass three of the ASE certification tests: *Manual Drive Train and Axles; Automatic Transmission/Transaxle; Electrical Systems.* All of these subjects receive detailed coverage in this textbook. Chapter 28 summarizes the ASE voluntary testing program. If studied carefully, **Auto Drive Trains Technology** will give you valuable drive train information. This information, combined with the hands-on experience that can be received by completing the jobs in the accompanying workbook, will enable you to become a competent and successful drive train technician.

James E. Duffy
Chris Johanson

Contents

SECTION 5–TRANSAXLES, CV AXLES, AND FOUR-WHEEL DRIVE SYSTEMS

SECTION 6–DRIVE TRAIN ELECTRONICS

SECTION 7–CAREERS, ASE CERTIFICATION

Acknowledgements

The authors would like to thank the following companies that helped make this book possible:

Alfa Romeo, Inc.; American Honda Motor Co.; Audi of America, Inc.; BBU, Inc.; British Leyland Motors, Inc.; Caterpillar Inc.; Champion Parts, Inc.; Chrysler Motor Corp.; Dake Div., JSJ Corp.; Dana Corp.; Deere & Co.; Dorman Products; FAG Bearings Corp.; Federal Mogul Corp.; Fel-Pro Inc.; Fiat Motors of North America, Inc.; Fluke Corp.; Ford Motor Co.; General Motors Corp.[†]; Hayden; Land Rover of North America, Inc.; Lexus; Lisle Corp.; Luk, Inc.; Mac Tools, Inc.; Mazda Motors of America, Inc.; Mercedes-Benz of North America, Inc.; Nissan Motor Corp.; Owattona Tool Co., Div, of SPX Corp.; Perfect Circle; Peugeot, Inc.; Porsche Cars North America, Inc.; RAC; Renault USA, Inc.; Saab-Scandia of America, Inc.; Sachs; Saturn Corp.; Snap-on Tools Corp.; L.S. Starrett Co.; Subaru of America, Inc.; Texaco Inc.; TIF Instruments, Inc.; Torsen; Toyota Motor Sales, USA, Inc.; Vaco Products Co.; Volvo of America; ZF Transmissions, Inc.

[†]Portions of materials contained herein have been reprinted with permission of General Motors Corporation, Service Technology Group.

IMPORTANT SAFETY NOTICE

Proper service and repair methods are critical to the safe, reliable operation of automobiles. The procedures described in this book are designed to help you use a manufacturer's service manual. A service manual will give the "how-to" details and specifications needed to do competent work.

This book contains safety precautions that must be followed. Personal injury or part damage can result when basic safety rules are not followed. Also, remember that these cautions are general and do not cover some specialized hazards. Refer to a service manual when in doubt about any service operation!

Cutaway shows the many systems of a modern automobile, one of which is a drive train system. (Ford)

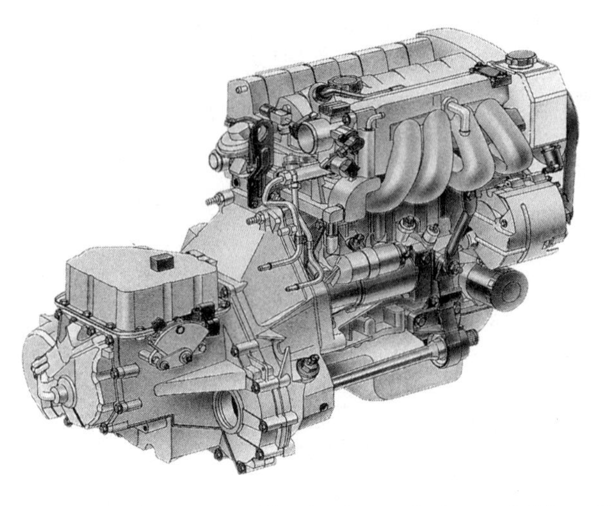

A transaxle, which is a combination transmission and differential assembly, is shown here mounted to a transverse engine.

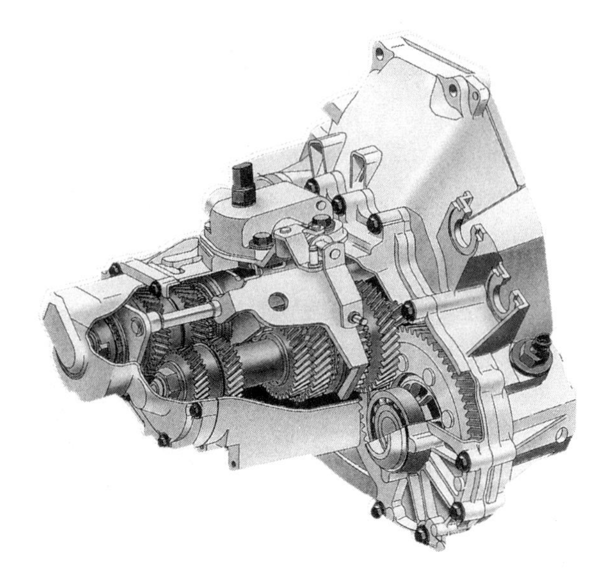

Transmission gearing and differential ring gear and side bearing are shown in this cutaway of a 5-speed manual transaxle. The clutch assembly (not shown) is housed within the bell-shaped housing (right).

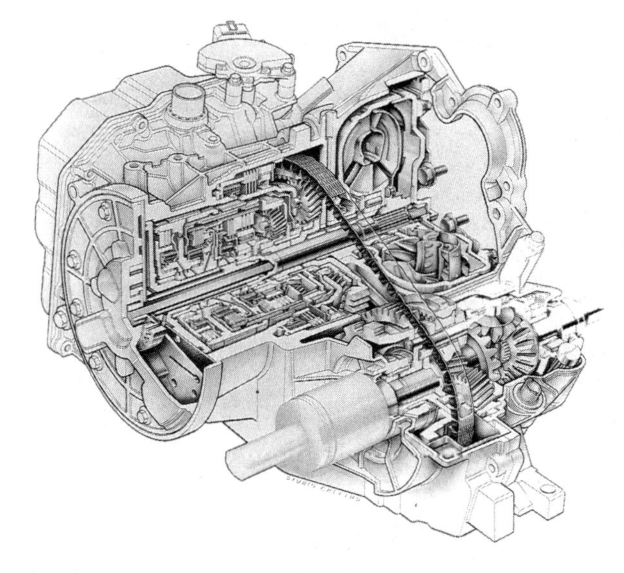

Transmission and differential sections are shown on this Ford CD4E 4-speed automatic transaxle.

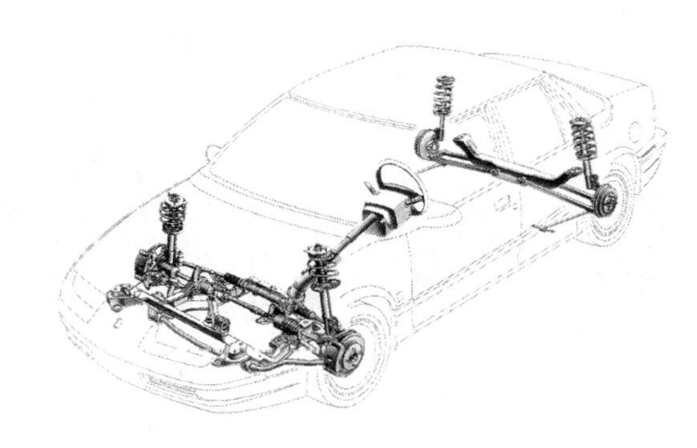

This phantom view shows CV axles on a front-wheel drive car, along with the vehicle suspension system.

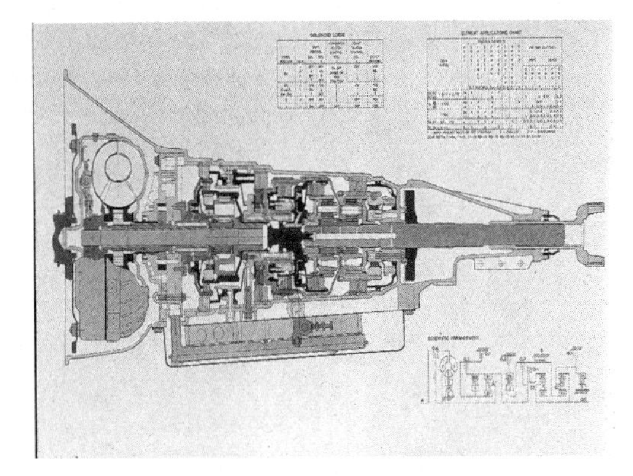

Ford E4OD electronically controlled 4-speed automatic overdrive transmission.

Some of the major components used on vehicle drive trains are represented here. All of these types of components are discussed in detail in this text. (Saturn, Ford)

Chapter 1

Introduction to Drive Trains

After studying this chapter, you will be able to:

- List the basic functions of a drive train.
- Describe the different types of drive trains.
- List the basic parts of a rear-wheel drive system.
- List the basic parts of a front-wheel drive system.
- List the basic parts of a four-wheel drive system.
- Summarize the fundamental purposes for major drive train components.
- Explain the difference between a transmission and a transaxle.

The purpose of this chapter is to present the various vehicle *drive trains* and major drive train parts. The chapter will explain how the drive train relates to engine location and to body and chassis design.

Before you begin to learn the detailed principles and service of drive train components, you should know what components are part of the drive train. You have heard of *gears, clutches, bearings, couplings,* and *drive shafts.* All of these parts relate to a vehicle drive train. This chapter will help you more easily see the place of such parts in the operation of the drive train.

The Drive Train

The vehicle engine, no matter how powerful, is useless if its power cannot be delivered to the wheels. The ***drive train*** is what transmits the engine power to the road. It is that part of the vehicle connecting the engine to the *drive wheels.* The drive train modifies engine power output to match vehicle needs. For example, the drive train provides a way to multiply *torque* (turning force) from the engine when starting off. It also provides a way to reduce engine rpm at higher speeds to improve gas mileage. In addition, the drive train provides a way to select the direction of vehicle travel, or select whether the vehicle moves forward or backs up.

The drive train consists of various mechanical and, sometimes, hydraulic and electronic components that form a path for engine power. This path is referred to as the ***power flow.*** Drive train components are usually arranged in a way that transmits power most efficiently. The arrangement may include, for example, *clutch, transmission, driveline, differential assembly,* and *drive axles.* Two basic ways the parts can be arranged are shown in Figure 1-1. Note that the parts of the drive train are shaded in the illustration.

Figure 1-1A shows a system where the engine is in front, and power is transmitted to the *rear drive axles.* This is a typical *front-engine, rear-wheel drive system.* For many years, this drive train design was almost the only kind used in domestic and imported vehicles.

Figure 1-1B shows a system where the engine is placed directly over the drive axles. This system features a ***transaxle***–a combination of a transmission and a differential assembly. The design can be used to drive either front or rear axles. Transaxles are common on front-engine vehicles. Mid- and rear-engine designs, which use transaxles to drive rear wheels, are less common. Examples of the rear-wheel drive version of this system are the Volkswagen Beetle and the Chevrolet Corvair. Porsche still uses this arrangement on 911 models.

Front-Engine, Rear-Wheel Drive

Figure 1-2 is a further breakdown of a ***front-engine, rear-wheel drive system.*** The placement of the engine in the vehicle is called a ***conventional,*** or ***longitudinal, mounting.*** In this arrangement, the engine is positioned so that the water pump and drive belts are at the front of the vehicle. The crankshaft centerline runs in the lengthwise direction of the vehicle.

The engine feeds power into a ***clutch*** or a ***torque converter.*** The clutch is used on vehicles with *manual transmissions.* The torque converter is used on vehicles

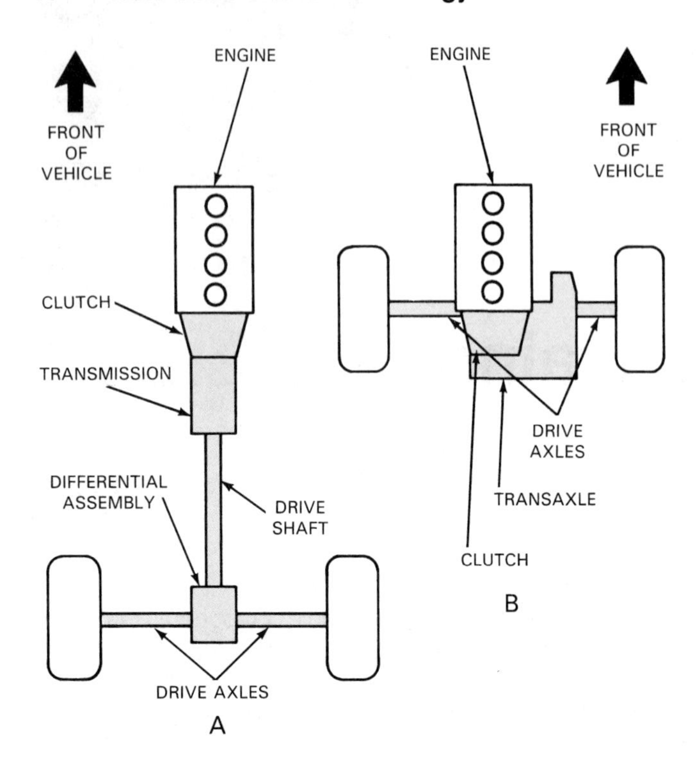

Figure 1-1. Two basic kinds of drive trains. A–This arrangement is found in front-engine, rear-wheel drive vehicles. B–This arrangement is found in some front-wheel drive and some rear-wheel drive vehicles.

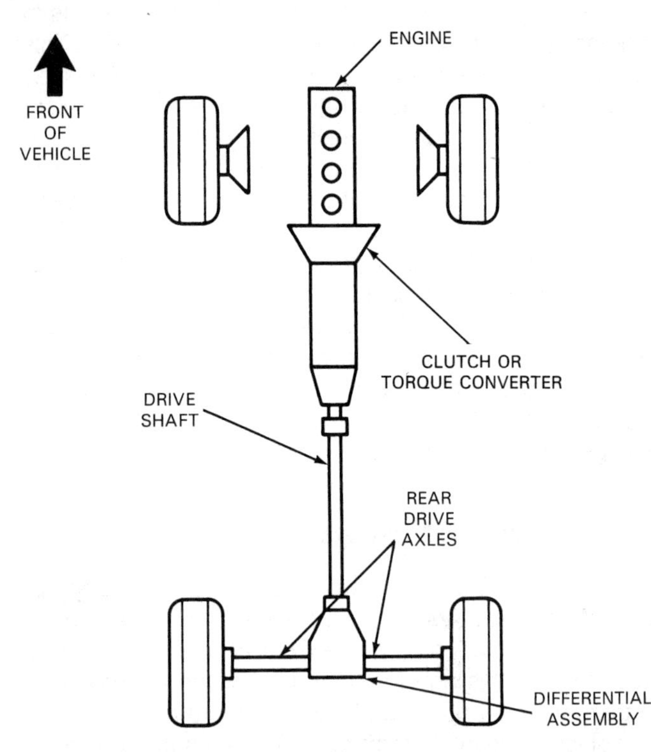

Figure 1-2. Power train layout of a typical rear-wheel drive vehicle. Notice that engine, clutch, transmission, drive shaft, and differential assembly are placed in a straight line. The ring and pinion cause the power flow to take a 90° turn inside the differential assembly to drive the rear wheels.

with *automatic transmissions.* (Though shown separately in Figure 1-2, the torque converter is considered an integral part of the automatic transmission, while the clutch on a manual transmission vehicle is a separate unit.) Both the clutch and torque converter serve as a way to connect or disconnect the flow of power from the engine to the drive wheels. The clutch transmits the power by means of a *friction* coupling, and the torque converter, by means of a *fluid* coupling.

From the clutch, engine power goes directly into the ***transmission.*** Its purpose is to multiply engine torque or reduce engine rpm to match varying *forward* operating conditions. It also provides a way to move a vehicle in the reverse direction. The manual transmission uses gears of different sizes to obtain different gear ratios and achieve these results. The proper gear is selected by the driver on manual transmission vehicles.

Alternately, from the torque converter, engine power continues through the automatic transmission. The automatic transmission drives different combinations of *planetary gears* to obtain different gear ratios, as well as reverse. The proper gear is selected by a hydraulic control system on the automatic transmission.

From the transmission, the power flow goes to the ***drive shaft assembly,*** or ***driveline.*** This assembly consists of a drive shaft, front and rear *universal joints,* and a front *slip yoke.* The driveline is designed to provide a smooth delivery of power to the *rear axle assembly,* while allowing the rear axle to move up and down to compensate for road conditions.

The ***rear axle assembly,*** sometimes called the ***rear end,*** sends power from the driveline to the drive wheels. It is made up of a differential assembly and drive axles, which are encased in a housing that is considered part of the rear axle assembly.

Note that the rear axle assembly is sometimes referred to as the ***final drive.*** This term, which also applies to front-wheel drive vehicles, refers to final stage of a drive train. It includes the combination of the differential *and* drive axles.

The ***differential assembly*** houses a number of gears. They function, in part, to redirect the power flow 90° to drive the rear wheels. This function is served by a *drive pinion gear* and a *ring gear,* commonly referred to as a ***ring and pinion.***

The shaft of the drive pinion gear is connected to the driveline. The gear itself meshes with the ring gear. The rotational axes of the two gears are offset and form a 90° angle, making them *hypoid gears.* The number of teeth in the ring gear compared to the number in the drive pinion gear sets the ***rear axle ratio.*** Higher rear axle ratios provide better acceleration and pulling power. Lower ratios provide better fuel economy.

In addition to redirecting the power flow, the differential assembly allows the drive wheels to revolve at different speeds during vehicle turns. This function is served by two *differential spider gears* and two *differential side gears.* (This will be explained later on in greater detail.) The drive axles are attached to the differential side gears. As these gears rotate, power is transmitted from the rear axle assembly to the rear wheels.

Front-Engine, Front-Wheel Drive

Most ***front-engine, front-wheel drive systems*** appear as in Figure 1-3. Note the engine is installed at a 90° angle to the vehicle. This sideways layout is usually referred to as ***transverse mounting.*** It eliminates the need to change the angle of power flow; therefore, a hypoid ring and pinion is not needed. This is a key difference between drive trains following longitudinally mounted, or *longitudinal,* engines and those following transversely mounted, or *transverse,* engines. A brief description of the latter follows:

Power flows from the engine to the transaxle, flowing first through a clutch in the case of a *manual transaxle.* The transmission, which is part of the transaxle, is located to one side of the engine. It performs the same function as on a rear-wheel drive system. The transaxle transmission sends the power flow directly into the differential assembly, also a part of the transaxle.

Power is split at the transaxle differential. It travels to each wheel through separate axles. These axles are not enclosed. The axles use flexible joints called ***constant-velocity universal joints,*** or ***CV joints.*** CV joints are used because they provide a constant flow of power through angles between drive and driven shafts. The transaxle differential allows each wheel to turn at different speeds during turns.

In some front-wheel drive vehicles, the engine is installed with a longitudinal mounting. See Figure 1-4. The transaxle is similar to that used with a transverse engine; however, it must include a ring and pinion that allows the engine power to make a 90° turn. The other drive train components operate in the same manner as those on a transverse engine.

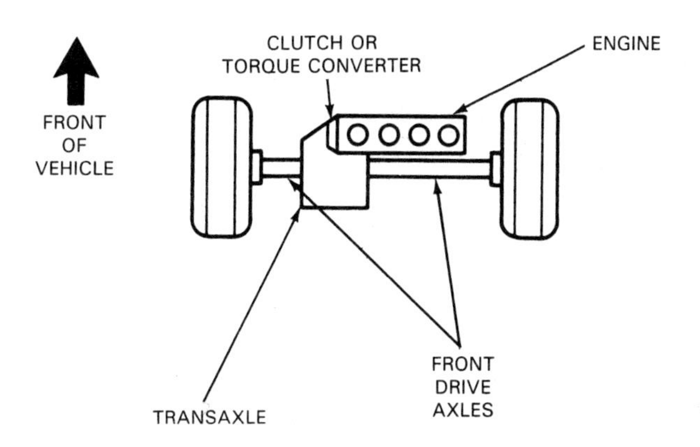

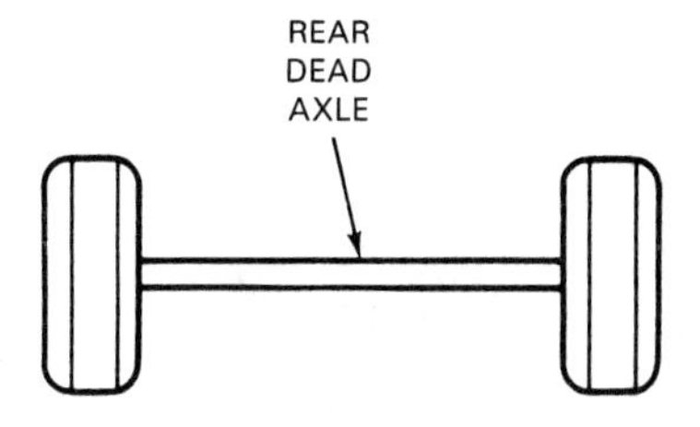

Figure 1-3. Front-wheel drive vehicle with transverse-mounted engine uses a transaxle to transmit power from engine to wheels. The axes of all rotating parts, from engine to wheels, have the same orientation. This makes it unnecessary for the power flow to make a turn at the differential assembly. The design makes a hypoid ring and pinion unnecessary in the differential assembly.

Mid-Engine, Rear-Wheel Drive

The ***mid-engine, rear-wheel drive system*** is illustrated in Figure 1-5. Power flows from the engine (through the clutch in manual transaxles) into the transaxle transmission. From here, power flows directly into the differential section of the transaxle, and then to the drive axles. This design resembles a conventional front-engine system in that the engine is mounted longitudinally. This rear-wheel drive system differs, however, in that it has a transaxle.

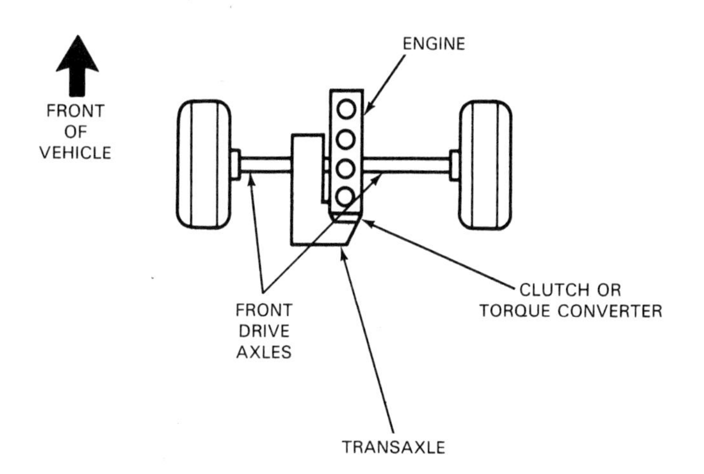

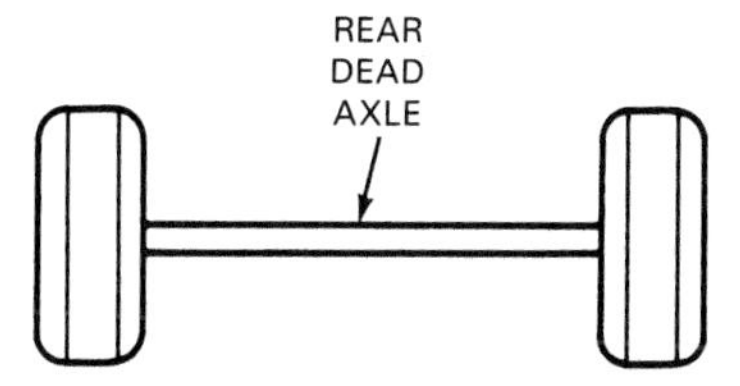

Figure 1-4. This is a front-wheel drive system, but the engine is longitudinally mounted, as it is on a rear-wheel drive vehicle. Although this system uses a transaxle, it has a hypoid ring and pinion.

Rear-Engine, Rear-Wheel Drive

Figure 1-6 shows a ***rear-engine, rear-wheel drive system.*** The engine is transverse mounted. The drive train

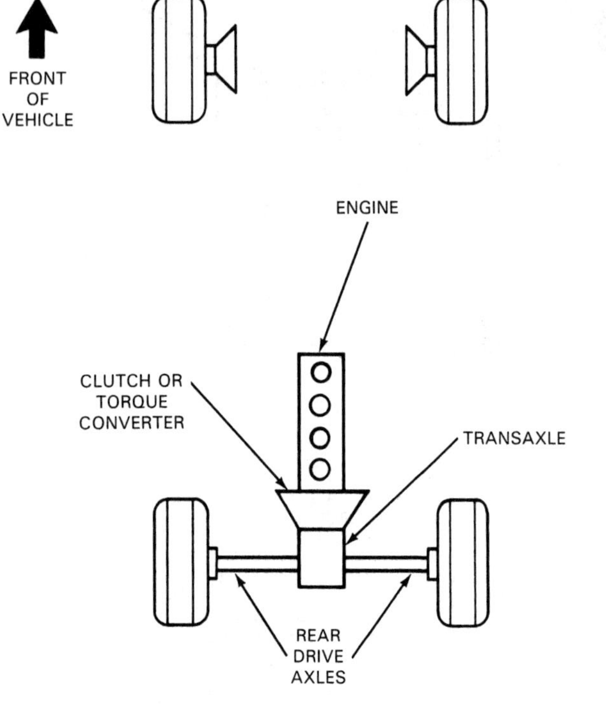

Figure 1-5. A mid-engine, rear-wheel drive system uses a transaxle. The power flow must be diverted 90°.

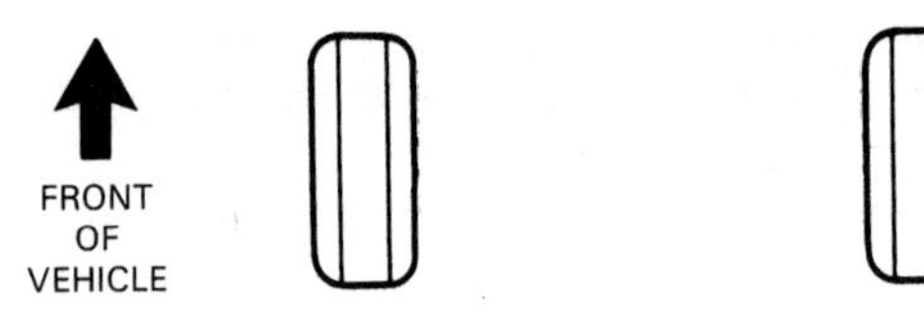

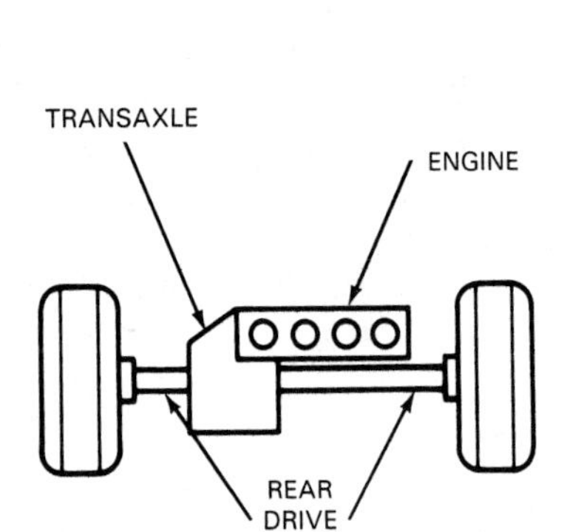

Figure 1-6. The power flow of the transverse-mounted rear-engine, rear-wheel drive system resembles that of a transverse-mounted front-engine, front-wheel drive system. Power flow is from engine to transaxle (through the clutch in manual transaxles) to rear drive axles to rear wheels.

operates in the same manner as that in a transverse front-engine, front-wheel drive vehicle.

On some rear-wheel drive vehicles, the engine is mounted behind the drive train components, Figure 1-7. The transaxle differential section and rear drive axles are mounted in front of the engine. The transmission section of the transaxle sits in front of them. Power flows from the transmission section to the differential section through a hollow shaft. Power flows to the transmission through a shaft that revolves inside of the *transmission output shaft.* Other drive train components operate as they do in the conventional rear-wheel drive system.

Four-Wheel Drive

The ***four-wheel drive system,*** Figure 1-8, somewhat resembles a conventional rear-wheel drive system. In addition, the four-wheel drive system has a front axle assembly, similar to the rear axle assembly. It also has a separate gearbox, called a ***transfer case.*** Through the transfer case, power can be transmitted to front *and* rear axle assemblies through an arrangement of gears. The gear ratio from engine to rear wheels is always equal to the gear

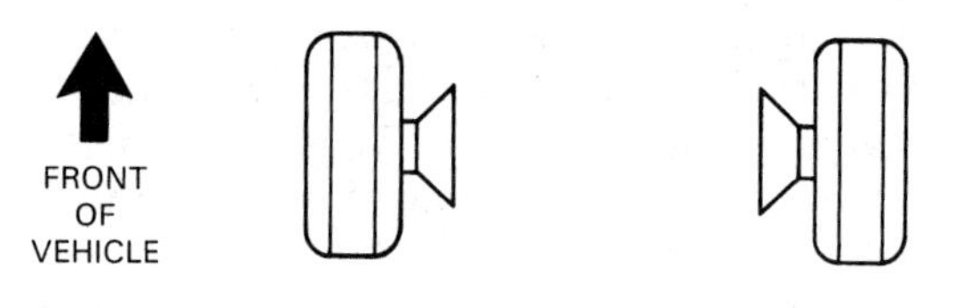

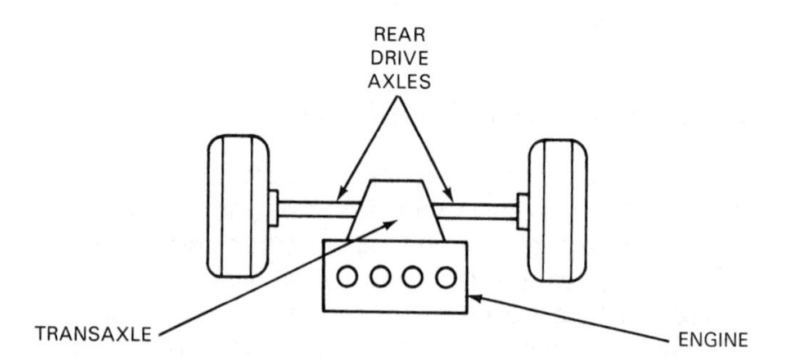

Figure 1-7. Rear-engine, rear-wheel drive system. The engine is mounted at the extreme rear of the vehicle. Power flow is forward to the transaxle. The transaxle differential is often mounted between the engine and transmission section of the transaxle. To transmit power, the input shaft to the transmission is placed inside of a hollow output shaft. Power flows through the input shaft into the transmission. The hollow output shaft transmits power into the transaxle differential.

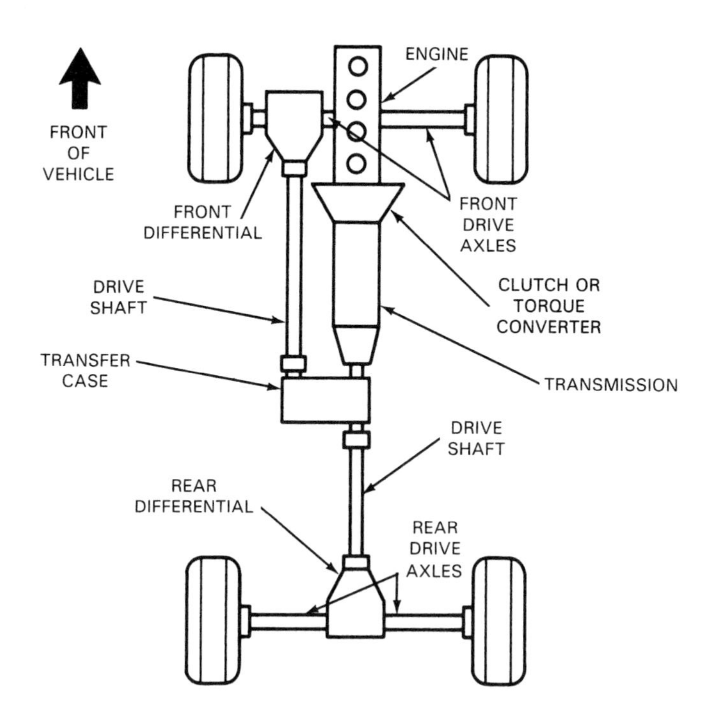

Figure 1-8. In the four-wheel drive system, power may be delivered to both front and rear wheels. The system contains a front drive shaft and final drive, which resemble the rear drive components. The heart of the four-wheel drive system is the transfer case. The driver selects two-wheel or four-wheel drive. In four-wheel drive, the transfer case divides power between front and rear drive shafts. In two-wheel drive, it sends power only to the rear.

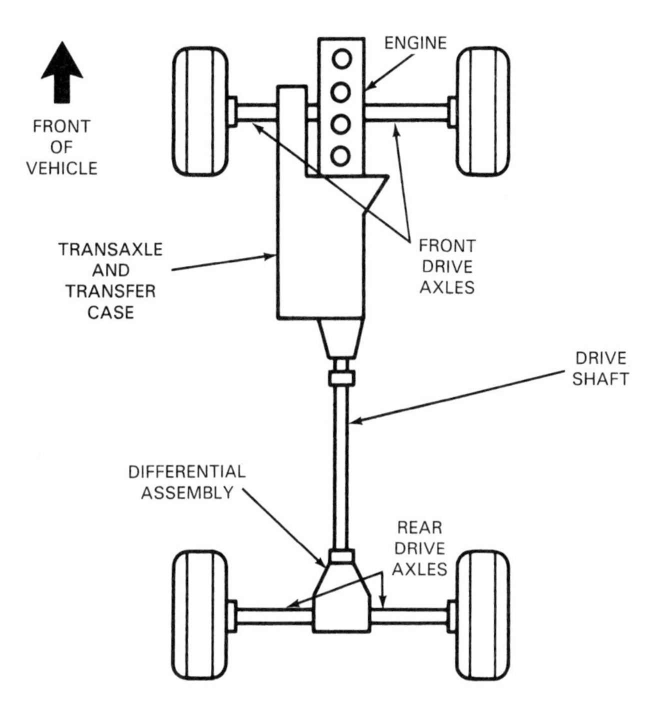

Figure 1-9. Transfer case of all-wheel drive vehicle automatically splits the power flow according to traction requirements. These transfer cases contain an internal *fluid coupling,* which is a device that transfers motion by using a fluid. It allows the front and rear wheels to turn at different speeds.

ratio from engine to front wheels; therefore, front and rear wheels revolve at the same rate.

In a four-wheel drive system, two- and four-wheel drive is selected manually; the vehicle driver decides which drive to use. The ***all-wheel drive system,*** Figure 1-9, a type of four-wheel drive system, is in four-wheel drive at all times. A special internal clutch in the transfer case can divert more or less power to the front wheels depending on road conditions.

Drive Train Components

Certain drive train components are common to all types of drive train layouts. The difference often comes about only in the placement of the components on the various vehicles. Four-wheel drive systems have some additional components. Thus far, all of these parts have been mentioned. Following is a more detailed discussion.

Clutches and Torque Converters

Clutches are always used on vehicles with manual transmissions and transaxles. The clutch is located directly behind the engine. It is almost always mounted on the engine *flywheel.* The clutch connects and disconnects the engine from the other parts of the drive train, Figure 1-10.

Clutches use friction to transmit rotary motion. This is called a *friction coupling.* The presence of friction between two parts in contact with each other will cause, to some degree, resistance of motion between them. At some point, relative motion between the two parts is altogether prevented.

A simple clutch is shown in Figure 1-11. In *normal* position, Figure 1-11A, the two surfaces are pressed together. The clutch is engaged. Rotary motion is transmitted through the disc surfaces. The clutch is disengaged, Figure 1-11B, by depressing a foot pedal connected to a clutch release mechanism. When the *clutch pedal* is released, the clutch reengages.

Torque converters are used with automatic transmissions and transaxles. They provide a hydraulic connection, or *fluid coupling,* between the engine and the other drive train components. The torque converter is installed directly behind the engine, in the front of the transmission.

A simple torque converter is shown in Figure 1-12. The basic construction includes an *impeller,* a *turbine,* and a *stator.* A *converter cover* is attached to the impeller, making the torque converter a sealed assembly. The cover is attached to the engine crankshaft. A brief description follows.

The ***impeller,*** also called a ***pump,*** consists of a set of curved blades, or vanes, attached to the inside of what forms the rear half of the torque converter assembly. As

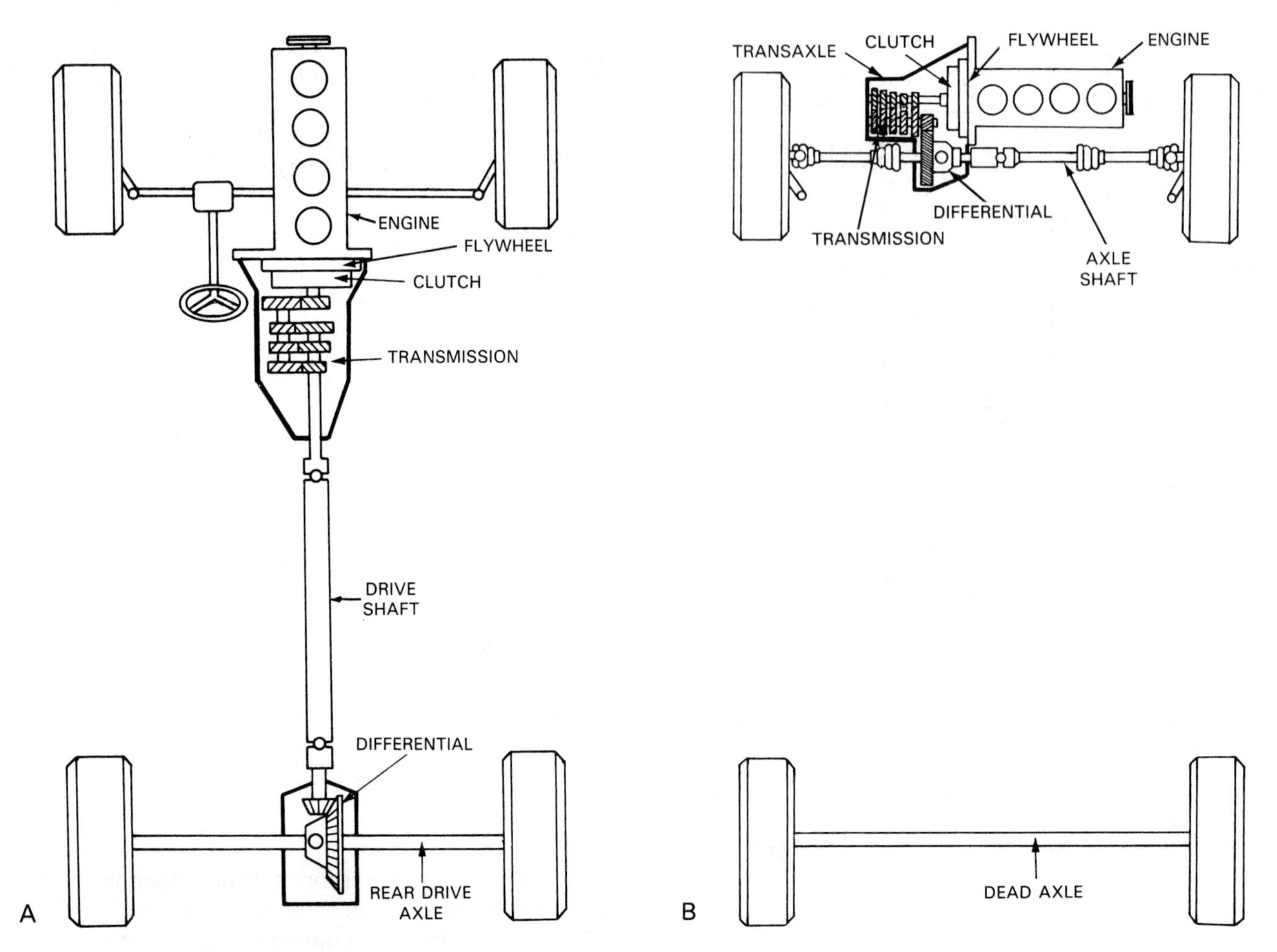

Figure 1-10. Compare the front- and rear-wheel drive systems shown here. The clutch is always mounted directly behind the engine, on the flywheel. Automatic transmissions and automatic transaxles will have a torque converter mounted directly behind the engine, instead of a clutch.

the engine turns, the converter cover and the attached impeller turn. The impeller, then, transfers power to fluid inside the assembly. The rotating fluid causes another set of blades, comprising the ***turbine,*** to revolve. This causes the *transmission input shaft,* which is connected to the turbine, to turn.

The ***stator,*** also called a ***guide wheel,*** consists of a smaller set of curved blades. The curve of the stator blades redirects the fluid, returning it to the impeller in the same direction that the impeller is turning. The stator helps the torque converter multiply power from the engine during takeoff and acceleration.

Torque converters operate whenever the engine is running. They allow the automatic transmission to be left in gear whenever the vehicle is stopped and the engine is running at low speed. In this case, the impeller will not have enough power to turn the turbine and, consequently, the transmission input and output shafts; as a result, the engine does not die. With a manual transmission, the clutch must be physically disengaged, under the same circumstances; otherwise, the engine will stall.

Torque converters are kept filled with fluid by the *automatic transmission oil pump.* Oil is pumped in through the converter rear shaft, which is open. Also, some torque converters have an internal *lockup clutch* that bypasses the fluid portion of the converter to provide direct mechanical drive during highway operation. Fluid leaving the converter passes through the *transmission oil cooler* and then returns to the transmission.

Transmissions

The purpose of the transmission is to match engine power and speed to vehicle conditions. The transmission must provide a *high* gear ratio for takeoff and acceleration and a *low* gear ratio during highway operation. A transmission will have several gear ratios to allow for efficient operation under various conditions. Automatic transmissions can have fewer gears than manual because the torque converter provides some of the torque multiplication.

Transmissions are located behind the clutch in manual transmission vehicles. The automatic transmission mounts

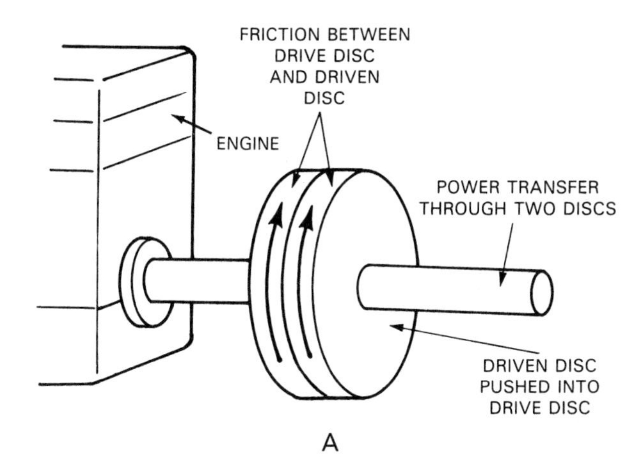

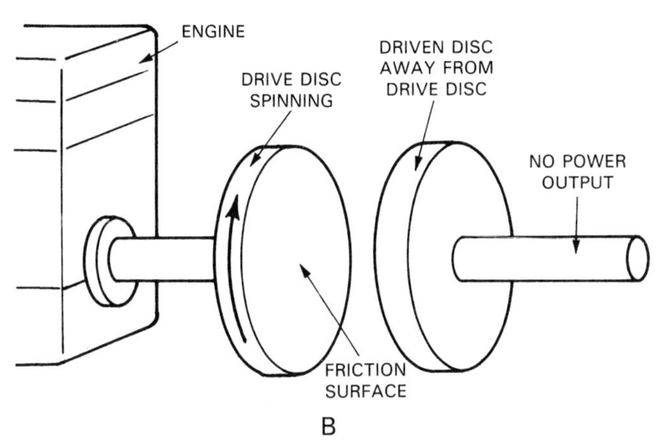

Figure 1-11. Operation of a simple clutch. A–When clutch discs are pressed together, a solid contact is formed. Rotation of the second disc is caused by the rotation of the first disc. B–One disc is spinning. There is no contact between the discs, and rotation is not transmitted.

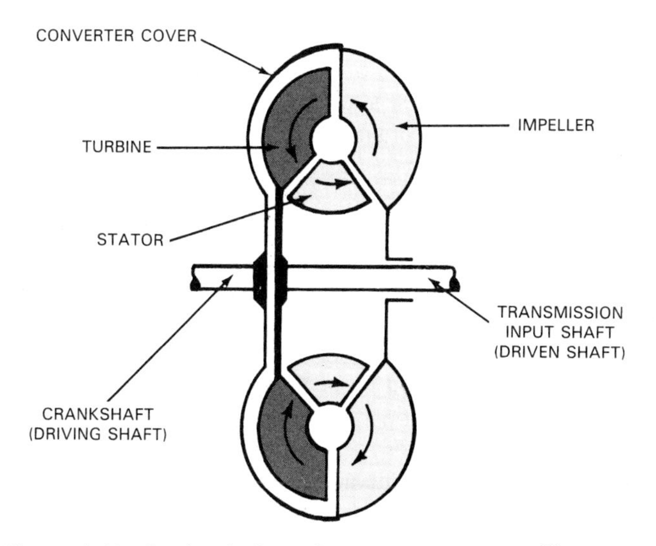

Figure 1-12. Sectional view of a torque converter. The torque converter operates whenever the engine is running. The relative position of the torque converter parts is the same for all converters.

directly behind the engine. Forwardmost in the automatic transmission is the torque converter, which takes the place of a clutch. The transmission output on a front-engine, rear-wheel drive vehicle goes to the drive shaft. See Figure 1-13. On front-engine, front-wheel drive vehicles, a drive shaft is not needed.

Transmissions are divided into two general classes–manual and automatic. In the ***manual,*** or ***standard, transmission,*** Figure 1-14, the gear selection decisions are made by the vehicle operator. Transmission gears and gear *synchronizers* are moved in and out of engagement by *shift forks.* The gears are meshed in different combinations. The driver selects different gear ratios by choosing the combinations needed. The driver must also operate the *clutch pedal* to disengage the clutch when changing gears. Manual transmissions require more skill to operate than automatic transmissions, but they allow more control over vehicle operation. They are usually easier and cheaper to repair.

In the ***automatic transmission,*** the gear selection decisions are made by the transmission automatic control system. Figure 1-15 shows a typical automatic transmission. Automatic transmissions usually use planetary gearsets. Planetary gearsets do not go in and out of engagement. In operation, one of the gears will be locked in place. The remaining, unlocked gears will be driven by engine power and will comprise the input and output. Different gear ratios are achieved, as needed, by alternate combinations of locked and unlocked gears. The gears are operated by holding members called *clutches* and *bands.* The clutches and bands are operated by a hydraulic control system. Late-model automatic transmissions have hydraulic systems that are controlled by *on-board computers.* Vehicles with automatic transmissions are easier to drive than manual. Also, they are more durable for heavy-duty operation, such as trailer towing.

Drive Shaft Assemblies

All front-engine, rear-wheel drive vehicles have a drive shaft. The ***drive shaft,*** or ***propeller shaft,*** is the means of transmitting power from the transmission to the rear axle assembly. It usually consists of a steel tube with a welded or press-fit yoke at each end. See Figure 1-16. The yokes join ***universal joints,*** often called ***U-joints,*** at each end of the shaft. They allow for changes in driveline angles when the rear axle is moving up or down over road surface irregularities. Some drive shafts are two-piece units, with three universal joints.

The front of the drive shaft connects through the U-joint to the ***slip yoke,*** which slips back and forth within the transmission. It allows the drive shaft length to change with rear axle movement. A slip yoke may also be mounted at the rear of the front drive shaft in four-wheel drive vehicles.

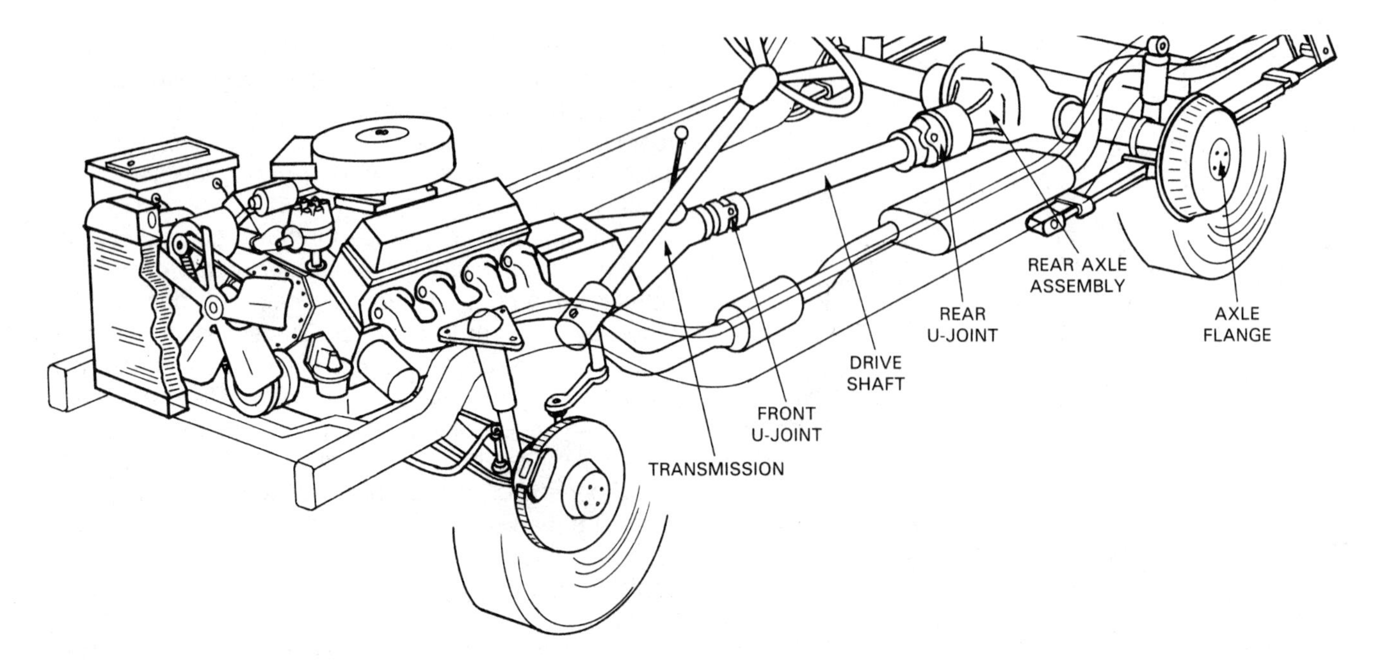

Figure 1-13. This shows how the drive train components of a rear-wheel drive vehicle relate to the other chassis components. The transmission is always directly behind the engine, and the drive shaft connects the transmission to the rear axle assembly.

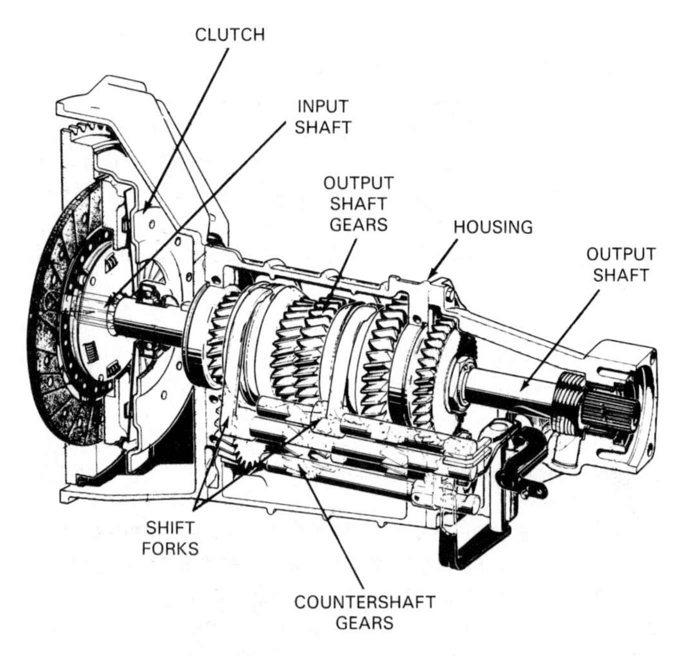

Figure 1-14. The rear-wheel drive system clutch and manual transmission shown here are typical. The components are found on all makes of manual transmissions. (Peugeot)

Rear Axle Assemblies

The job of the rear axle assembly, Figure 1-17, is to transmit power from the drive shaft to the rear wheels. The ***rear axle housing*** contains the differential assembly, drive axles, and related parts. The housing forms a rigid support for the rear of the vehicle. Heavy *gear oil* contained in the housing lubricates the moving parts.

The drive pinion gear is attached to the rear universal joint by the *differential pinion yoke.* The pinion gear is in constant mesh with the ring gear. The pinion, turned by the drive shaft assembly, turns the ring gear. By design, the rpm at the ring and pinion output is reduced from driveline (input) rpm. In this way, the ring and pinion increases engine power. It also changes the power flow 90° so that it can drive the rear wheels. The ring gear has quite a few more teeth than the pinion. Therefore, the pinion must make more revolutions (usually between 2.5 and 4) to turn the ring gear one revolution. Note in Figure 1-18 that the ring and pinion gears are mounted on large bearings. These bearings must be strong to handle the thrust forces created during the transmission of power from ring to pinion.

The rear axle assembly also contains a *differential case.* The case is bolted to the ring gear. The power flow enters the case and passes through a set of small gears before going to the drive axles. The set of gears contains two differential spider gears and two differential side gears. All four gears are meshed together.

The four differential case gears are locked together and rotate as a unit when the vehicle is moving in a straight line. This causes the drive axles to revolve at the same speed. The gears are driven by a shaft that is connected to the differential case and that extends through the two spider gears. The shaft rotates end over end. The four gears do not rotate among themselves when the vehicle is moving straight ahead.

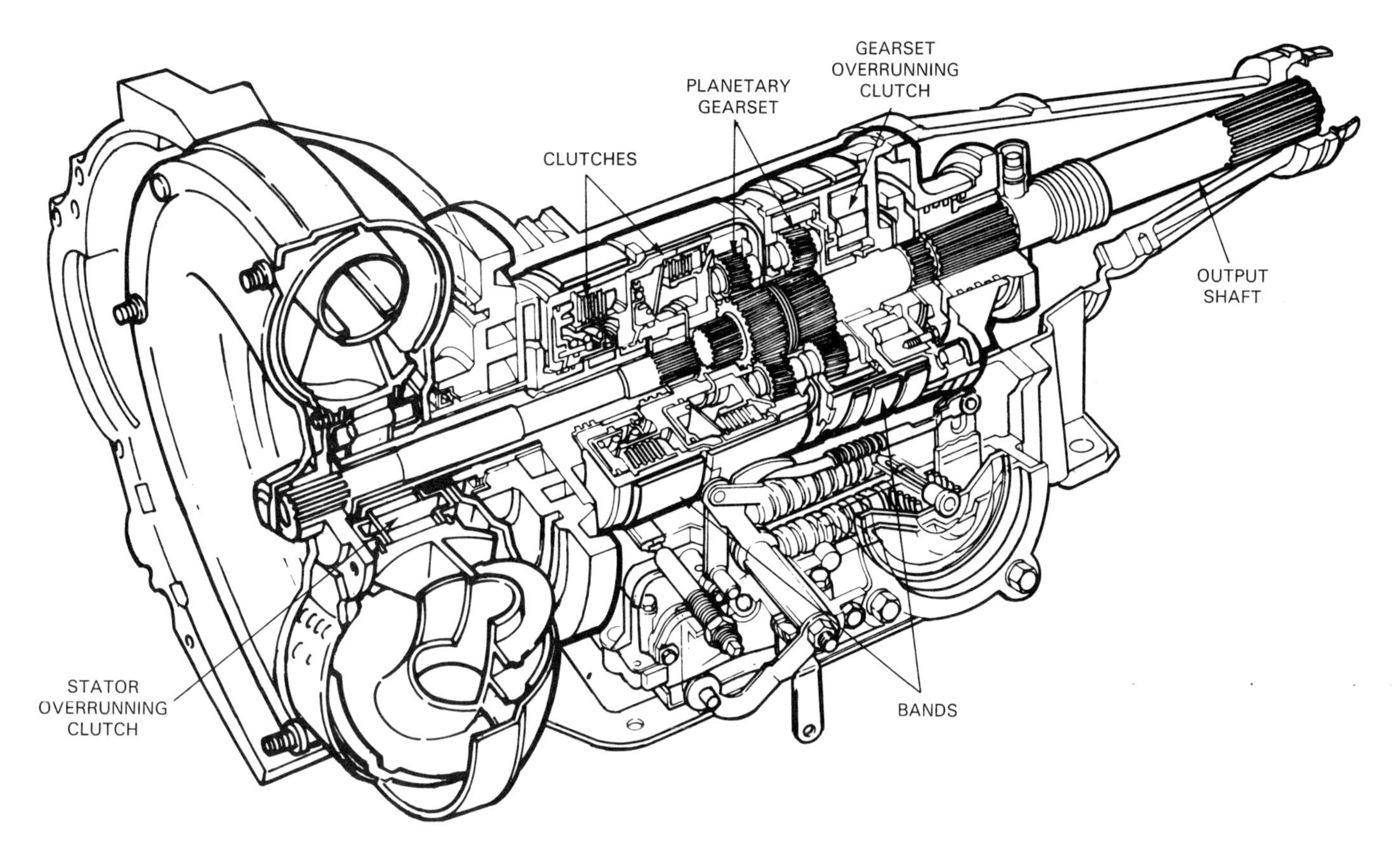

Figure 1-15. The torque converter and automatic transmission shown are typical of those used on front-engine, rear-wheel drive vehicles. The components called out are found on almost all makes of automatic transmissions. (Ford)

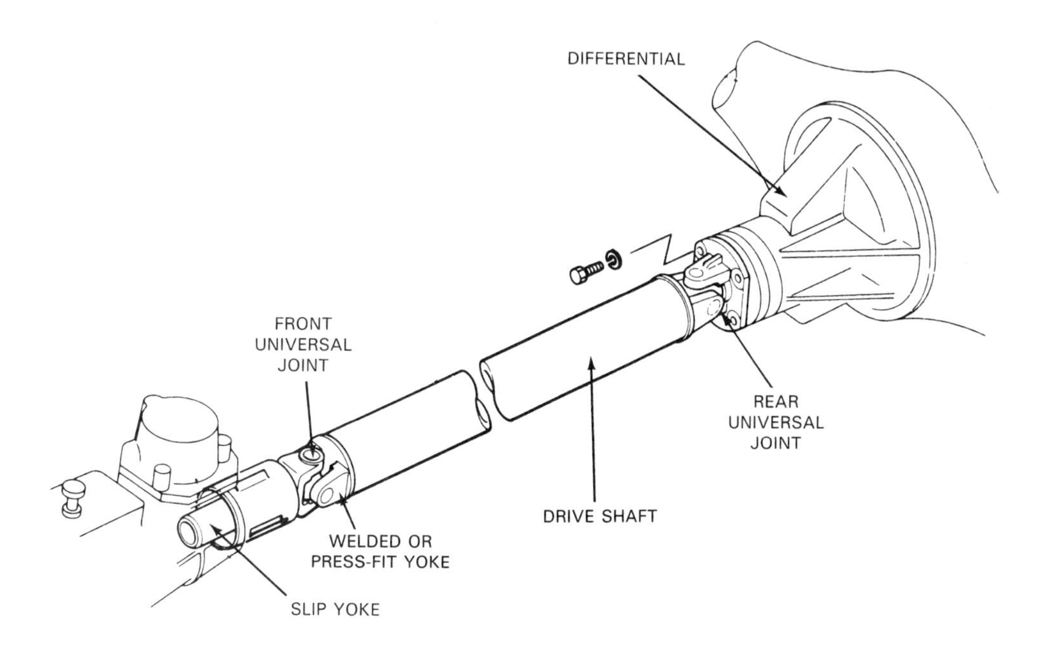

Figure 1-16. A typical rear-wheel drive system drive shaft. The drive shaft used on the front of a four-wheel drive system is similar. Most modern drive shafts have two universal joints. However, you may encounter drive shafts with one or three universal joints.

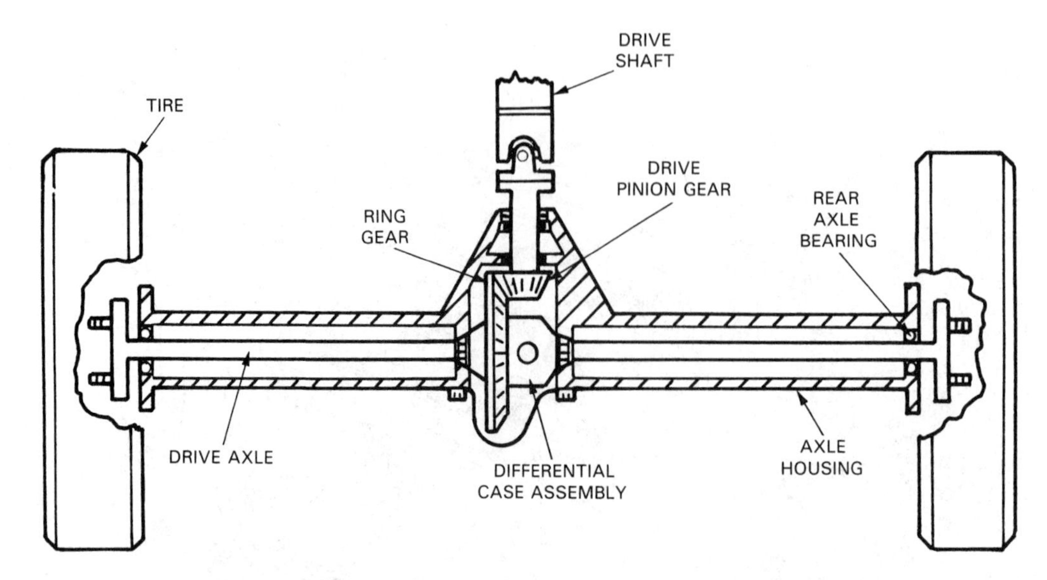

Figure 1-17. A typical rear axle assembly. The drive pinion gear and ring gear cause the power flow to make a 90° turn. The difference in the number of teeth between the ring and pinion is what determines the rear axle ratio. The ring gear is usually bolted to the differential case. The differential case assembly contains a set of four gears, arranged in a way that permits the rear wheels to turn at different speeds. The axles connect the differential assembly to the rear wheels. The entire rear axle assembly is contained in a sealed housing.

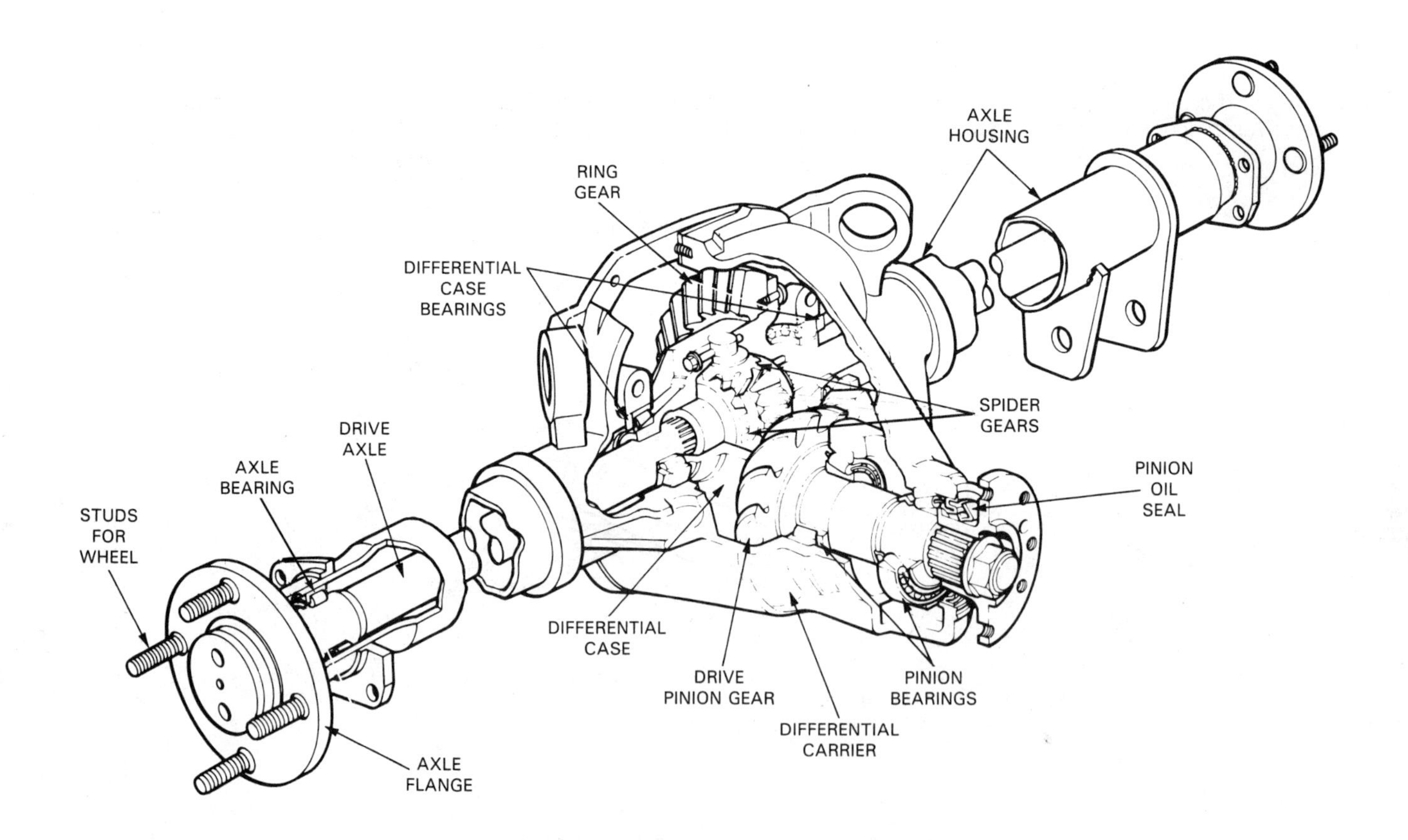

Figure 1-18. A detailed view of a rear axle assembly. Notice that the rear axle parts are supported by large bearings. These bearings are needed to handle the high weight and thrust forces placed on the rear axle assembly during vehicle operation. All parts of the rear axle assembly are under heavy stress. They are lubricated with heavy oil contained in the axle housing. (Ford)

When the vehicle is making a turn, however, the gears move in relation to each other, and the axles can turn at different speeds. This compensates for the difference in *turning radius* of the inner and outer wheels. The inner wheel, with the smaller turning radius, will turn at a slower speed than the outer wheel. This will cause the drive axle of the inner wheel and the related side gear to turn at a slower speed also. The three remaining gears, as a unit, will "walk around" the slower-turning side gear, and the faster-turning wheel will rotate at its own speed.

Sometimes the differential assembly operates when undesired, such as when one wheel is stuck on ice or in snow. When this happens, the wheel without traction spins. The wheel on firm ground stands still as power flow takes the easiest path–that of the spinning wheel. The three differential gears merely walk around the side gear of the wheel with traction, and the vehicle cannot move. To overcome this problem, some differential assemblies are equipped with an internal clutch mechanism. This clutch sends power to the wheel with the most traction or to both wheels, eliminating wheel spinning and allowing the vehicle to move.

The ***rear drive axles*** transmit power from the differential assembly to the rear wheels. Axles of a typical design are straight and have gearlike ridges, or *splines,* at one end. The splined ends are attached to the differential gears, which drive the axles. The outer ends of the axles are flanged. The flanges provide the mounting surface for the wheel rims. The axle outer ends are supported by bearings. Some axles use greased and sealed *axle bearings,* while others lubricate the bearings with differential lubricant. A detailed picture of a rear drive axle is shown in Figure 1-19.

Transaxles

A transaxle is a combination of a transmission and a differential assembly. The transmission and differential are usually contained in a single case. While the fundamentals

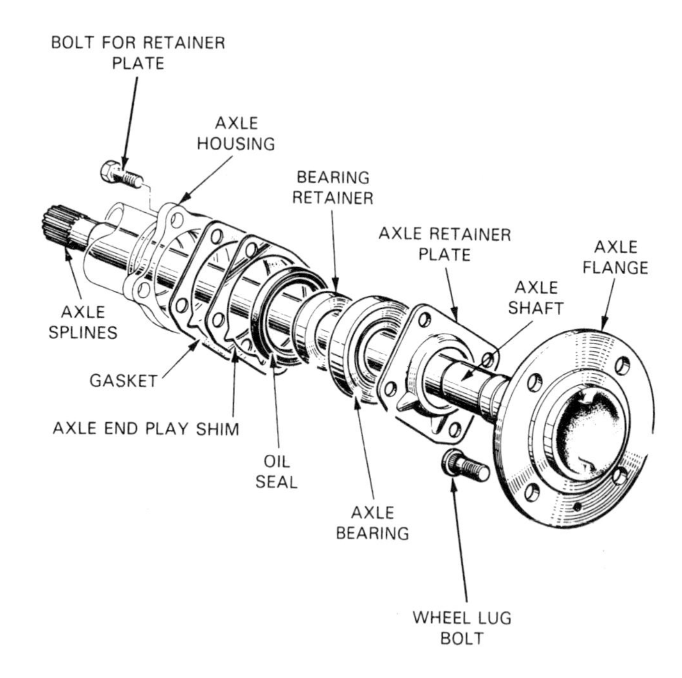

Figure 1-19. Another place where heavy bearings are needed is at the ends of the axle shafts. These bearings support the axle at the wheel. Some axle bearings are *splash lubricated* by the differential oil. Other bearings are permanently packed with grease and are isolated from the differential oil. (Toyota)

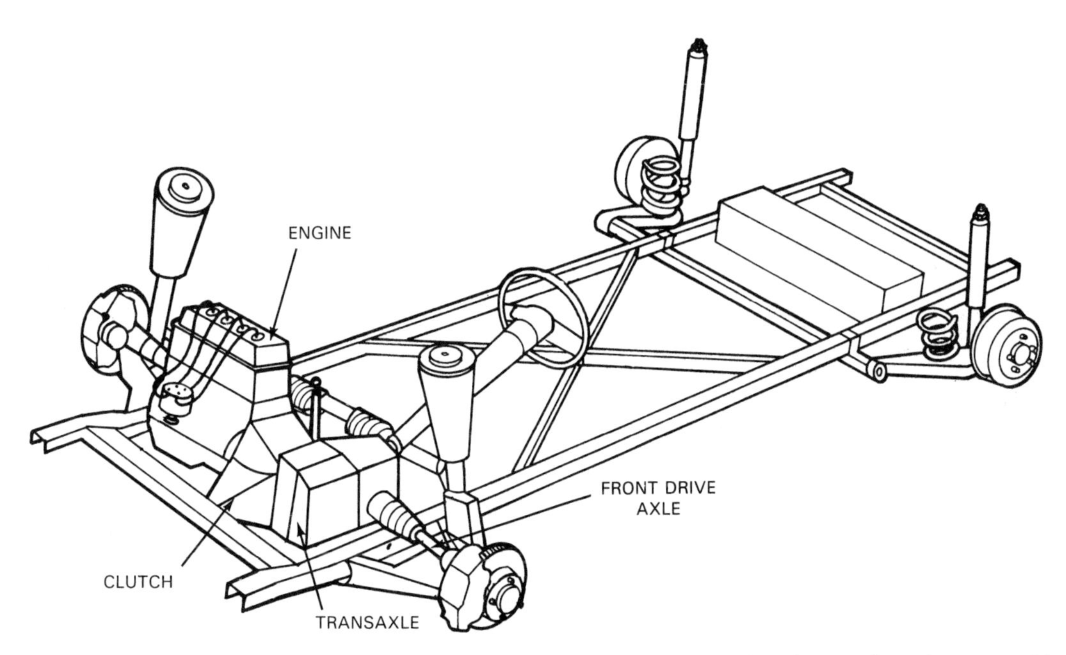

Figure 1-20. This shows how the drive train components of a front-wheel drive vehicle relate to the other parts. Note the relative positions of the engine, transaxle, and drive axles. (Ford)

of the transaxle are similar to those of rear-wheel drive, the component layout is different.

Figure 1-20 shows the placement of the transaxle on a typical front-wheel drive vehicle. Power flows from the engine into the clutch (on manual transaxle vehicles), the transaxle, and then to the drive axles. If the engine is placed transversely, there is no need to divert the angle of power flow, and a conventional hypoid ring and pinion is not needed.

A front-wheel drive system has two drive axles. ***Front drive axles*** are smaller than those used on rear-wheel drive vehicles, and they are usually solid. Most front drive axles use a kind of constant-velocity U-joint called a ***Rzeppa joint.*** This joint is flexible and is, therefore, well suited to a front-wheel drive system. Some longer front axles have a second joint near the transaxle.

Figure 1-21 is a ***manual transaxle.*** Note that the manual transaxle gears resemble those of the manual transmission shown in Figure 1-14. These gears are operated by shift forks similar to those in the rear-wheel drive transmission. Different combinations of gears are selected to obtain the desired gear ratios.

An ***automatic transaxle*** is shown in Figure 1-22. Notice the similarities between the parts of the automatic transaxle and the automatic transmission shown in Figure 1-15.

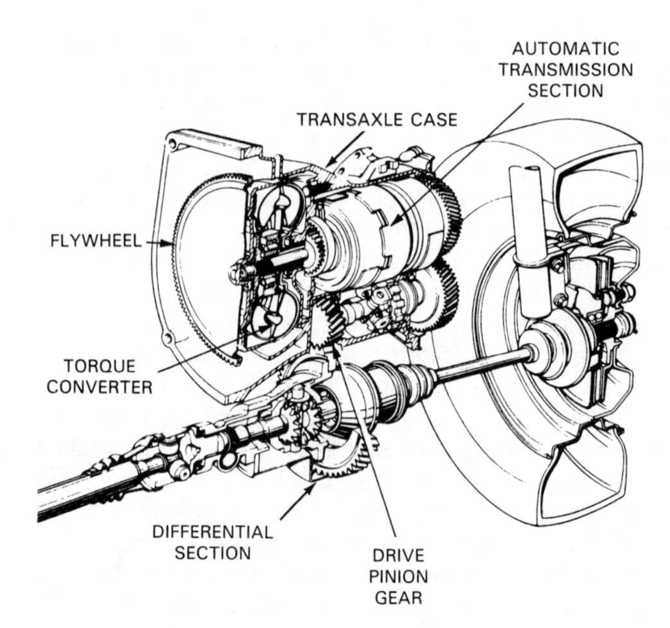

Figure 1-22. The automatic transaxle shown is typical of those used on front-wheel drive vehicles. Note that all automatic transaxles use torque converters. (Chrysler)

Four-Wheel Drive Components

The four-wheel drive system resembles a conventional rear-wheel drive system, with the addition of some extra components. See Figure 1-23. A transfer case sends power to the rear drive shaft and, when desired, to the front drive shaft. The rear drive shaft attaches to a conventional rear axle assembly. The front drive shaft connects to a front axle assembly, which resembles the rear axle. It contains a differential assembly, with a ring and pinion. The front axles drive the front wheels.

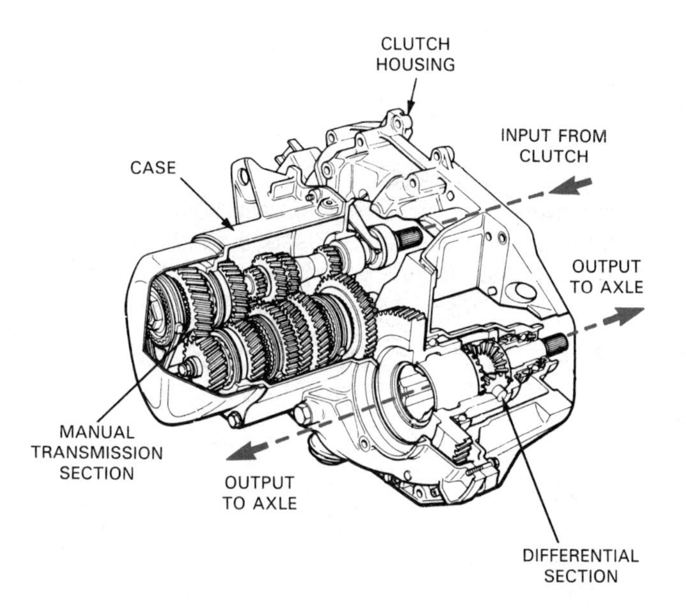

Figure 1-21. Common front-wheel drive manual transaxle. It contains a differential and manually operated transaxle gears. (Renault)

The transfer case contains the mechanism to transmit power to all four wheels or only the rear wheels. The transfer case may also contain extra *reduction gears.* The reduction gears provide a very low gear for heavy towing or off-road use. Some transfer cases are automatic in operation and can vary the amount of power sent to the front wheels, depending on vehicle needs. An automatic transfer case may contain a *central differential* or a *viscous coupling,* in addition to a chain drive and various drive gears.

Summary

The drive train is a vital part of any vehicle. It sends power from the engine to the drive wheels. It may also modify this power so that the vehicle can be more versatile. There are many variations of drive trains. Most drive trains fall into several general classes.

The conventional rear-wheel drive system is used when a front-mounted engine drives the rear wheels. It was the most common kind of drive train for many years. The major drive train components are arranged in a straight line from the engine to the rear of the vehicle. The power flow is diverted 90° at the differential assembly.

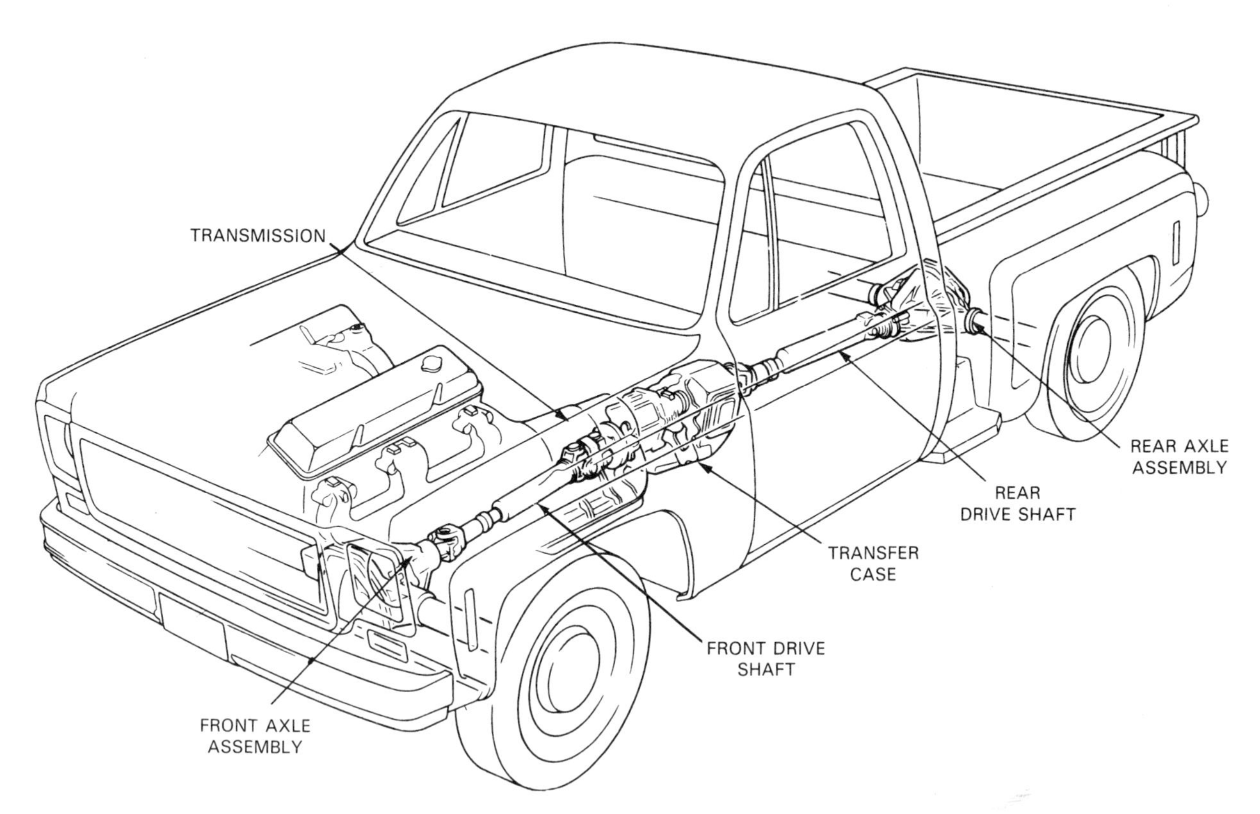

Figure 1-23. Four-wheel drive components. Note the dual drive shafts and that the front axle assembly resembles the rear axle assembly. The four-wheel drive can almost be thought of as two rear axles and two drive shafts tied together by the transfer case. Most major differences between makes of four-wheel drive vehicles occur in the transfer case. (General Motors)

The front engine, front-wheel drive system has become popular. The most common kind of front-wheel drive has a transverse engine. All drive components can be mounted in a straight line, and there is no need to turn the power flow 90° to drive the wheels. The components of the transmission and differential are combined in a single unit called a transaxle.

Rear- and mid-engine vehicles use a transaxle. In some cases, the engine is mounted behind the transaxle.

The four-wheel drive system can apply power to all of the wheels as needed. The transfer case, mounted behind the transmission, divides the power between the front and rear wheels. Some systems are manually operated by the driver, and some systems are automatic.

Drive train components are similar, no matter what type of drive train layout is used.

The clutch and torque converter serve the same purpose–to connect and disconnect the engine from the transmission.

The transmission is used to obtain different gear ratios and to allow the vehicle to be reversed when necessary.

The drive shaft transmits power from the transmission to the rear axle assembly. It uses a series of flexible joints to allow rear axle movement and eliminate vibration.

The rear axle assembly consists of a differential assembly and drive axles. The ring and pinion turns the power flow at right angles to drive the rear wheels and provides the proper gear ratio. The differential assembly also allows the rear wheels to turn at different speeds when making turns. Some rear axle assemblies have internal clutches to provide better traction.

The transaxle combines the functions of the transmission and differential into one unit, usually in a single case. Transaxles are used on front-wheel drive and rear- and mid-engine vehicles. Internal components resemble those used on the front-engine, rear-wheel drive train.

The four-wheel drive transfer case houses a number of gears. These gears can be used to send power to the rear wheels alone, or divide power between the front and rear wheels. Some transfer cases can provide a higher gear ratio when needed. Some have internal fluid couplings and send power to the front wheels only when needed.

Know These Terms

Drive train; Power flow; Transaxle; Front-engine, rear-wheel drive system; Conventional mounting (Longitudinal mounting); Clutch; Torque converter; Transmission; Driveline (Drive shaft assembly); Rear axle assembly (Final drive, or Rear end); Differential assembly; Ring and pinion; Rear axle ratio; Front-engine, front-wheel drive system; Transverse mounting; Constant-velocity universal joint (CV joint); Mid-engine, rear-wheel drive system; Rear-engine, rear-wheel drive system; Four-wheel drive system; Transfer case; All-wheel drive system; Impeller (Pump); Turbine; Stator (Guide wheel); Manual transmission (Standard transmission); Automatic transmission; Drive shaft (Propeller shaft); Universal joint (U-joint); Slip yoke; Rear axle housing; Rear drive axles; Front drive axle; Rzeppa joint; Manual transaxle; Automatic transaxle.

Review Questions–Chapter 1

Please do not write in this text. Place your answers on a separate sheet of paper.

1. What is the basic function of a drive train?
2. The _____ and the _____ _____ serve as ways to connect and disconnect the flow of power from engine to drive wheels.
3. What is the basic function of a rear axle assembly?
4. The _____ _____ _____ is sometimes called the final drive.
5. What is a transfer case?
6. Name the three major parts inside a torque converter.
7. The purpose of a transmission is to:
   a. Reverse a vehicle.
   b. Provide high torque for starting off.
   c. Reduce engine rpm at higher speeds.
   d. All of the above.
8. Some late-model automatic transmissions are controlled by a computer. True or false?
9. A _____ is a combination of a transmission and differential assembly.
10. Why are Rzeppa joints commonly used on front-wheel drive cars?

Certification-Type Questions–Chapter 1

1. Technician A says that modern drive trains consist of various mechanical systems that form a path for engine power. Technician B says that drive trains having manual transmissions usually incorporate hydraulic systems, in addition to electronic systems. Who is right?

(A) A only
(B) B only
(C) Both A & B
(D) Neither A nor B

2. Which of these components allows a vehicle's drive wheels to revolve at different speeds while turning corners?

(A) Engine.
(B) Slip yoke.
(C) Differential assembly.
(D) Torque converter.

3. All of these statements about a clutch are true EXCEPT:

(A) it decides which drive gear to use.
(B) it is located directly behind the engine.
(C) it is almost always mounted on the engine flywheel.
(D) it connects and disconnects the engine from other drive train parts.

4. All of these functions are served by a transmission EXCEPT:

(A) providing a low gear ratio during highway operation.
(B) splitting power at the transaxle differential.
(C) providing a high gear ratio for takeoff and acceleration.
(D) matching engine power and speed to vehicle conditions.

5. Which of these components transmit(s) power from the transmission to the rear axle assembly?

(A) CV joints.
(B) Driveline.
(C) Clutch.
(D) Differential assembly.

6. Which of these components together make up a manual transaxle?
(A) Transmission and torque converter.
(B) Transmission and differential assembly.
(C) Torque converter and differential assembly.
(D) Clutch and differential assembly.

7. All of these components may be found in a front-engine, rear-wheel drive system EXCEPT:
(A) a clutch.
(B) a transaxle.
(C) a transmission.
(D) a torque converter.

8. All of these components are part of a driveline EXCEPT:
(A) a ring and pinion.
(B) a slip yoke.
(C) universal joints.
(D) a drive shaft.

9. A car with an automatic transmission will not move in the reverse direction. Technician A says that the problem is probably due to the slip yoke. Technician B says that the problem is very likely in the hydraulic control system. Who is right?
(A) A only
(B) B only
(C) Both A & B
(D) Neither A nor B

10. Which of these components are not found on a drive train where the vehicle engine is transverse mounted?
(A) U-joints.
(B) CV joints.
(C) Rzeppa joints.
(D) Hypoid gearing.

In addition to the special service tools that are needed to service drive trains, a wide assortment of basic hand tools is essential. Note how many tools are in this set. (MAC Tools)

Chapter 2

Special Service Tools and Safety

After studying this chapter, you will be able to:

- Identify special tools used in drive train service.
- Identify common measuring tools.
- Select the right tool for the job at hand.
- Discuss right and wrong ways to use certain types of tools.
- Cite safety rules that apply to drive train repair.

This chapter will identify and explain special tools and equipment used to service drive trains. It will discuss the ways in which the tools and equipment can be used in specific drive train diagnostic and repair operations. The chapter will give you important information about the safe way to repair drive trains and about general safety.

Drive Train Service Tools

In addition to the normal hand tools used in repair shops, ***special service tools*** are needed to service parts of the drive train. These are tools exclusively designed to service a specific component or else a specific class of components. This contrasts with the usual hand tools, designed to be used on many parts for many purposes. *Spanner wrenches, snap ring pliers, drivers,* and *pullers* are all examples of special service tools.

As mentioned, some special service tools are designed to service a specific component. Many of these have to be obtained through the tool or vehicle manufacturer. These tools are identified in vehicle service manuals, usually in the back of the manual. Figure 2-1 shows some of the special service tools used to service drive train components of rear-wheel drive vehicles. Figure 2-2 shows some of the tools that could be used for front-wheel drive vehicles. The part number appearing with each tool is a coded description of that tool. This number is specified when ordering.

Service manuals also contain information on when to use a tool and how to use it properly. Most service manuals

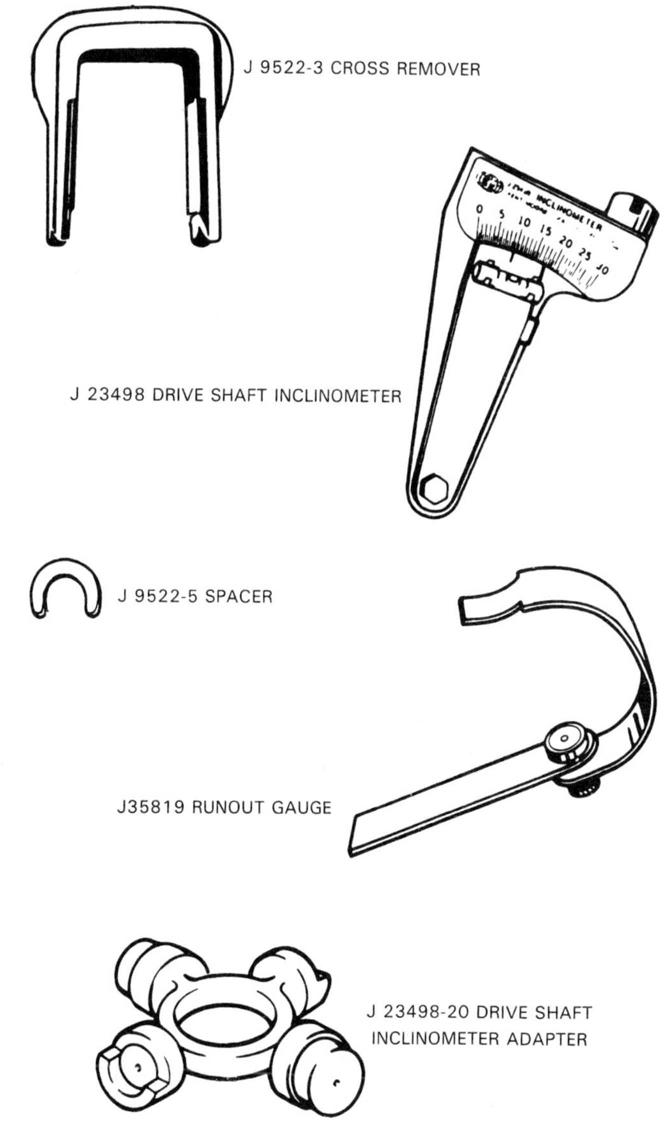

Figure 2-1. Special tools typical of those used to work on a drive shaft and its component parts. (General Motors)

Figure 2-2. These special service tools are used to repair automatic transaxles on some front-wheel drive vehicles. (General Motors)

also contain the address for the manufacturer of the tools and information on how to order tools. Prices of special service tools can be obtained from the tool manufacturer. In some cases, these tools must be ordered through the vehicle manufacturer, as mentioned.

Special Service Wrenches and Sockets

Wrenches are tools used to grip and turn fasteners or other parts. Some wrenches have jaws; others have sockets. Special hand wrenches may have openings to fit a fastener with an unusual head shape. Fasteners with special heads are often used when space problems mean that common tools cannot be used. Sometimes the wrench handle is specially shaped to fit where clearances are tight. Figure 2-3 shows a wrench designed to fit a fastener having both an unusual head and a hard-to-access location.

Figure 2-4 shows another type of wrench. ***Spanner wrenches,*** as they are called, are special service tools used to tighten large parts or to hold parts in place while other tools are used. Most spanner wrenches are one-piece units. Some spanner wrenches, like the one in Figure 2-4A, have lugs that fit into holes in the part. Others, like the spanner wrench shown in Figure 2-4B, have provisions for bolting to the parts that they hold.

In addition to such wrenches, special *sockets* are available that have shapes or depths to fit fasteners that are different from standard or to fit where clearance is a problem. See Figure 2-5. These sockets are designed to fit standard handles and drive sizes, such as 3/8-in. or 1/2-in. ratchets and extensions. They are used the same way as any other socket.

Punches and Drifts

Punches and ***drifts*** are removal and installation tools designed to be struck with a hammer. They are used to remove pins, plugs, and other pressed-in parts from bores. The punch or drift is placed against the part to be removed. The top of the tool is struck with the hammer. A punch and a drift are included among tools shown in Figure 2-6. Note that drifts can be made from discarded steel or brass rods.

Snap Ring Pliers and Picks

Snap ring pliers are special service tools used for removing and installing *snap rings,* or *retainer rings.* As pictured in Figure 2-7, snap ring pliers have special jaws

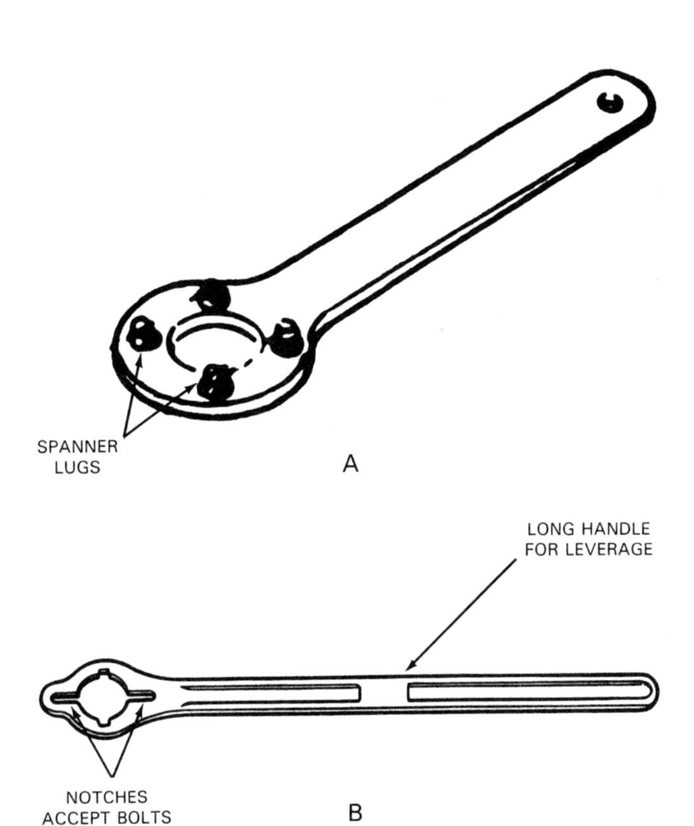

Figure 2-4. Spanner wrenches are used on certain parts of the drive train. A–This spanner wrench has special lugs that fit into holes in the *drive pinion flange.* B–This type of spanner wrench is bolted to the part on which it is used.

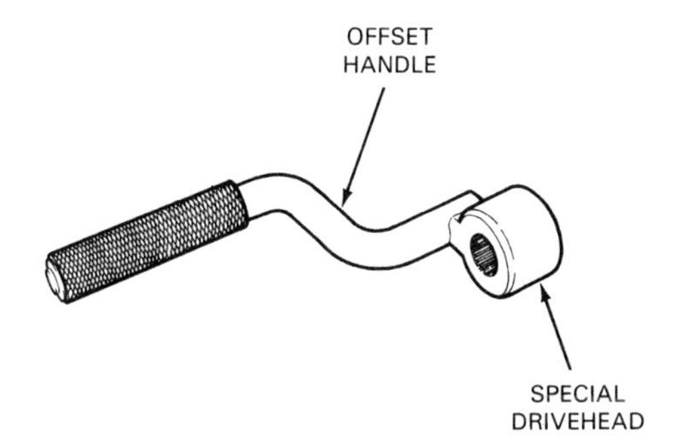

Figure 2-3. A hand wrench for servicing a transmission assembly. Notice the offset handle on this special wrench. The head is shaped to fit an unusual fastener.

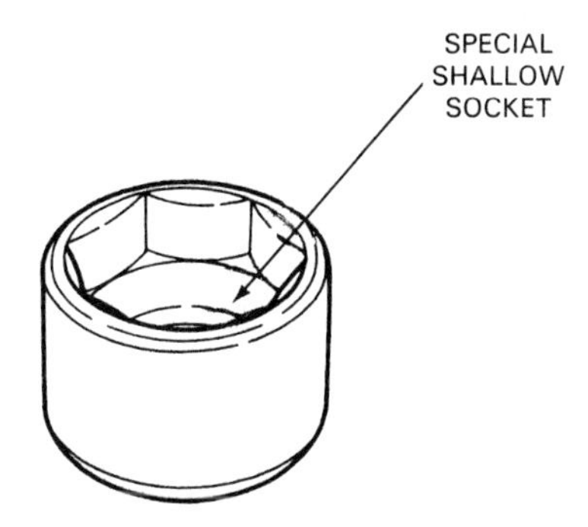

Figure 2-5. This special socket is made for a *drive pinion lock nut.* A socket such as this is used when a standard socket will not fit because of clearance problems.

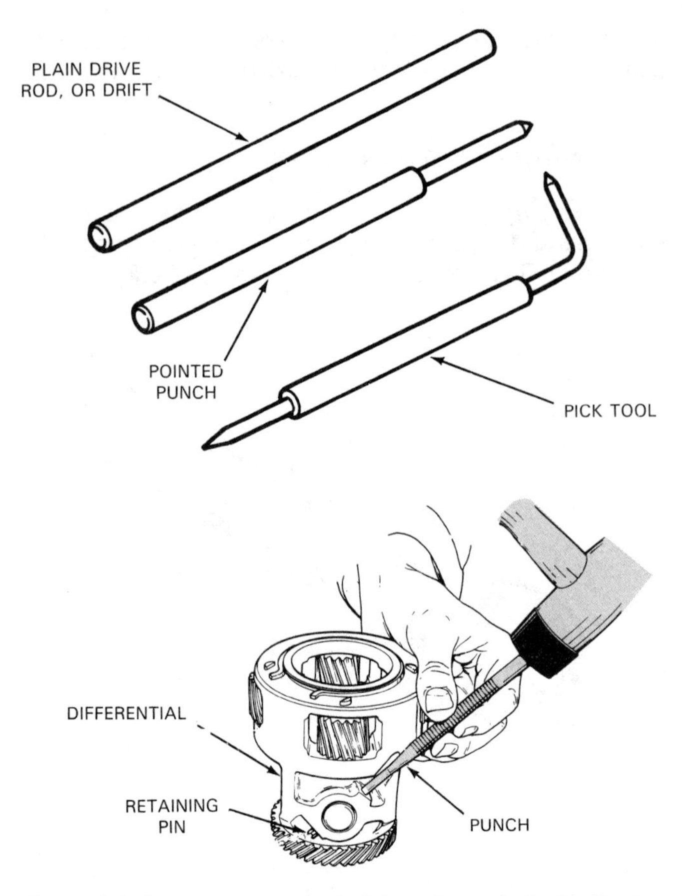

Figure 2-6. An assortment of special service tools. In this illustration, the *pinion shaft retaining pin* is being removed with a hammer and a punch. (General Motors)

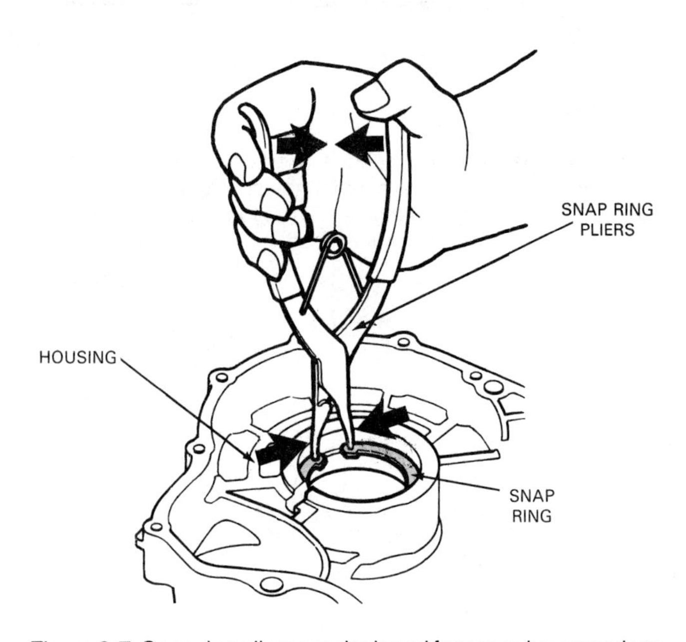

Figure 2-7. Snap ring pliers are designed for removing snap rings. The illustration shows a snap ring being installed in a transmission housing. (Honda)

that fit and grasp the snap ring. Note that eye protection should always be worn when working with snap rings. A flexed ring can shoot into your face with considerable force.

Picks may also be used for removing snap rings. Picks are pointed tools. (Refer back to Figure 2-6.) Ordinary icepicks are often used as picks. Picks are useful for prying on parts and for jobs requiring a small pointed tool. These tools can be used to get under, pry up, and remove snap rings when there is no other way of removing them.

Presses

In many cases, drive train parts are an ***interference fit*** with a mating part. This means that the one part was machined slightly larger than the space of the mating part into which it is fit. A shaft that is driven into a smaller hole is an example. The parts will be pressed together tightly, and they cannot be removed easily. A ***press fit*** is one type of interference fit. In this type of fit, considerable pressure is applied to the mating pieces to get them to fit together.

Often, gears and bearings have an interference fit with the shafts on which they ride. These parts can be easily removed from a shaft with a ***press.*** A press is a piece of equipment consisting essentially of a frame and a ram, used for applying pressure on an assembly. Presses are also used for *installing* gears and bearings on a shaft.

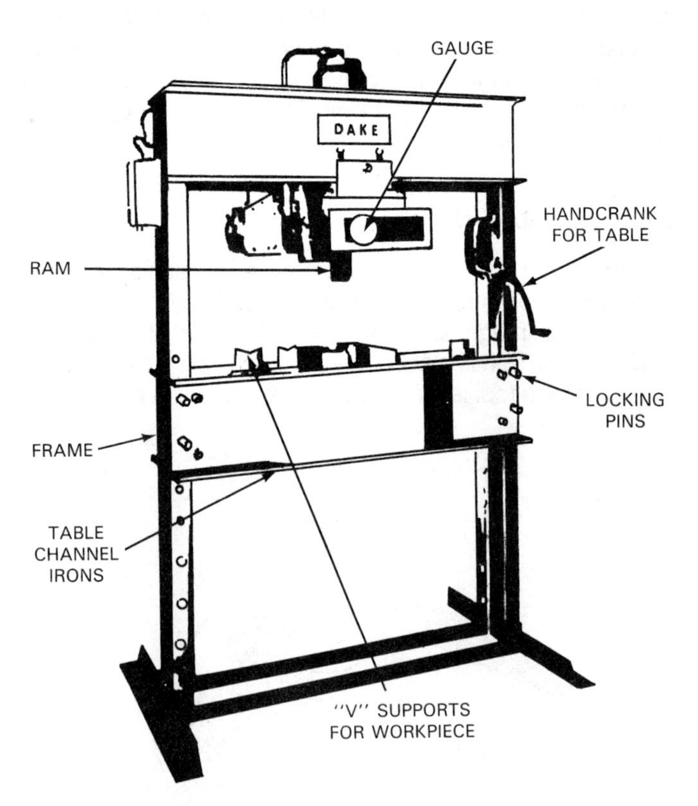

Figure 2-8. The hydraulic press is used to remove many pressed-on parts, such as axle bearings and automatic transmission bushings. (Dake)

The most common press found in repair shops today is the *hydraulic press,* Figure 2-8. This press has a hydraulic ram. Pressure is developed by a hand-operated piston or, on larger presses, by an electrically powered pump.

Special *adapters* are often used to make the pressing operation easier. The adapter shown in Figure 2-9 is used for pulling bearings. The tool fits under the bearing. It provides a firm surface for the bearing to rest on. Adapters are usually adjustable. This permits use with many sizes of gears and bearings.

Another type of press is the mechanical *arbor press.* It is hand operated and is used for lighter-duty jobs. It performs the same function as a hydraulic press, but ram-pressure capability is much less.

Drivers

There are several different types of ***drivers.*** One type is used to install and remove bearings and seals in housings. This driver is driven by a hydraulic press or an arbor press. If a press is not available, a hammer can sometimes be used with the driver, Figure 2-10. When using either method, always wear eye protection. Also, always ensure that the part is properly aligned in the bore.

Before using a driver, always make sure that it is the proper size for a particular job–that it exactly fits the seal or bearing to be installed or removed. The shape of the driver is critical. If the driver does not match, the part will be damaged. See Figure 2-11.

Figure 2-12 shows drivers being used to install and remove *bushings,* which are a type of bearing. In either case, the outside diameter of the driving surface should be slightly less than the bore diameter. This allows the driver to push the bushing into or out of a bore without contacting the bore itself. Drivers for bearings that are to be installed to specified depths have a shoulder that is larger than the housing bore, as shown in Figure 2-12A. The bushing is at the proper depth when the shoulder bottoms on the housing. Figure 2-12B shows a driver used for removing a bushing. This driver does not need or have such a shoulder, for obvious reasons.

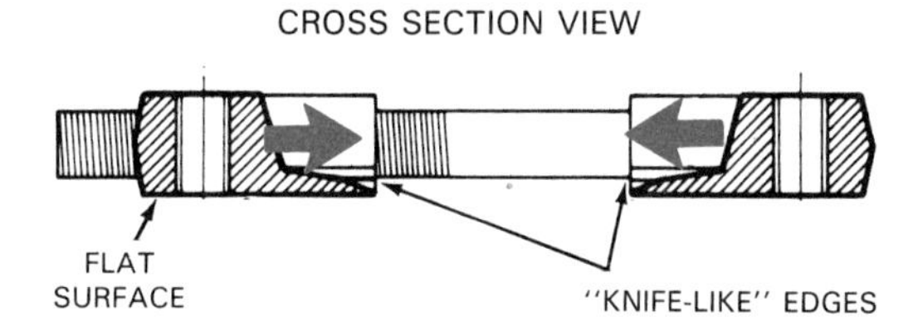

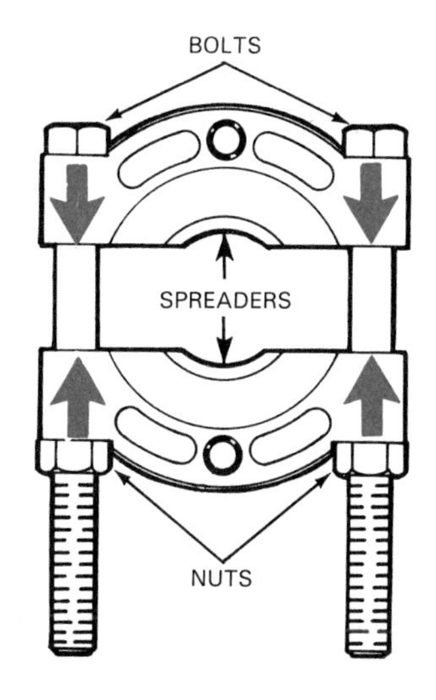

Figure 2-9. Bearing adapters are used to make bearing removal jobs easier and speed the process. The adapter is installed under the bearing.

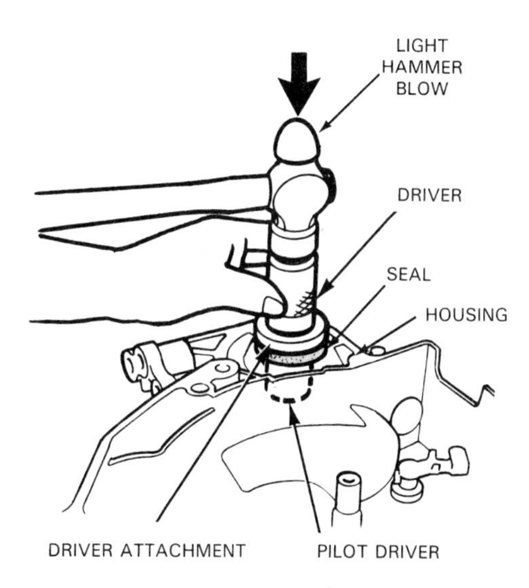

Figure 2-10. A hammer is used on a driver to install the oil seal in a torque converter housing. Note the pilot driver is used to ensure that the seal is driven in straight. It aligns the driver and attachment. (Honda)

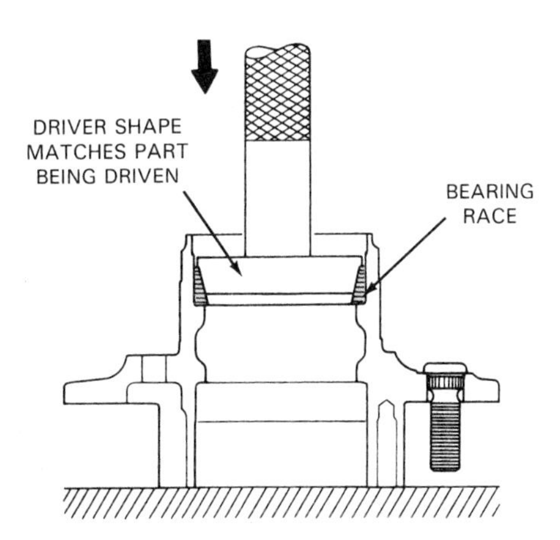

Figure 2-11. When installing a bearing race, check that the driver closely matches the inner shape of the race or damage may result. (Chrysler)

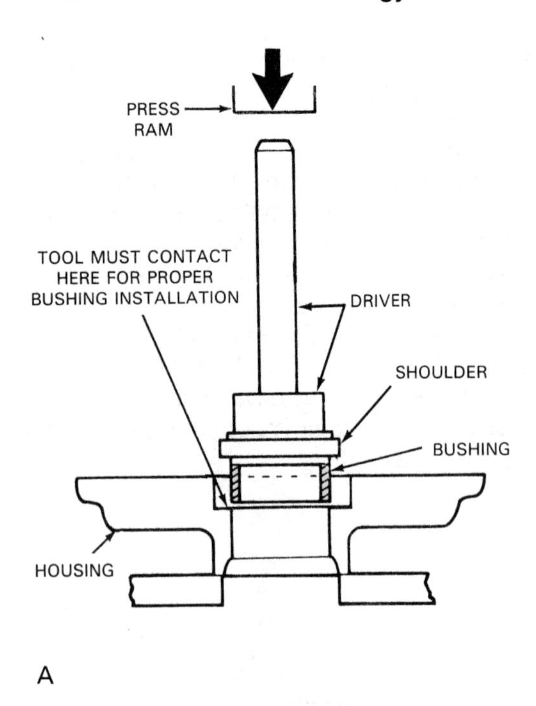

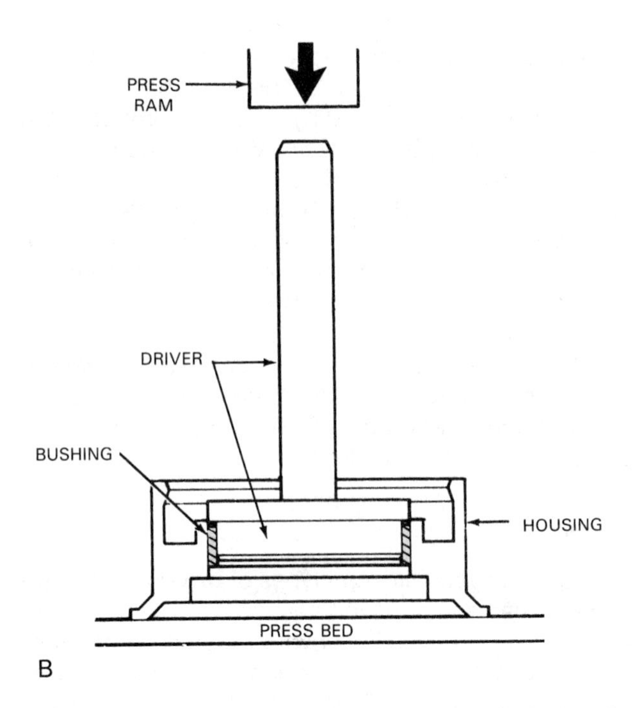

Figure 2-12. Make sure the driver diameter matches the part being installed. A–Driver being used with hydraulic press for installation of a pump bearing, or bushing. B–Driver being used with hydraulic press for removal of the *forward clutch hub bushing.* (Ford)

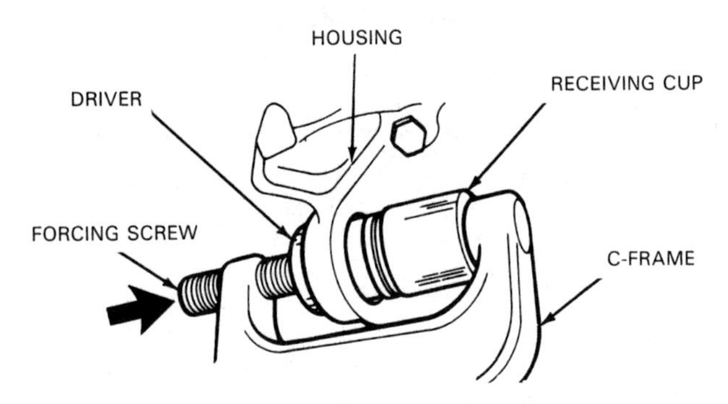

Figure 2-13. The C-frame driver is a special purpose pressing tool. It is used in many places where a large force must be applied in a confined space. (Ford)

Another type of driver is the *C-frame driver,* shown in Figure 2-13. This type is used for installing specific parts, such as is shown in this ball joint installation. Proper adapters are selected before using the C-frame driver. The screw is then tightened to install the part.

Pullers

To remove parts that are pressed together, ***pullers*** are often used. A puller is a device that exerts pressure on such parts to separate them. There are many kinds of pullers.

Figure 2-14 shows some of the pullers that you might use to remove drive train parts. A *wheel puller,* Figure 2-14A, is a tool that uses screw threads to exert pressure. It is used to remove gears, pulleys, or bearings from a central shaft. The puller bolts attach the screw and plate assembly to the part to be removed. The central screw is then tightened to pull the part from the shaft.

The operation of another kind of puller is similar to that of the wheel puller, Figure 2-14B. This kind of puller is usually used to remove seals or other parts that are lightly pressed in. The tool is fastened onto the part by the threaded connection. Tightening of the central screw, then, causes the part to be pulled out. One disadvantage of this puller is that threading the tool into the pressed-in part usually ruins the part.

A *jaw puller,* a third kind of puller, is shown in Figure 2-14C. This *bearing puller,* as it is also called, is used to remove bearings from shafts or housings. It consists of a central block, with a center screw and jaws that pivot from the block. The puller jaws are placed over the bearing to be removed. The jaws are securely tightened in place with bolts or a thumbscrew that draws the jaws in toward the center. The center screw is then tightened to pull the bearing from the shaft.

Another tool used to remove parts is the *slide hammer,* or *impact puller,* Figure 2-14D. Its pulling force comes from impact, instead of screw force. The puller is installed on the part to be removed. This is sometimes done by threading the *guide rod* into an existing threaded hole on the part. Other times, an adapter, attached to the threaded end of the guide rod, is used. Then, the slide, which is weighted, is pulled sharply against the shoulder of the guide stop. The resulting impact is transmitted to the part, and the part is removed from position.

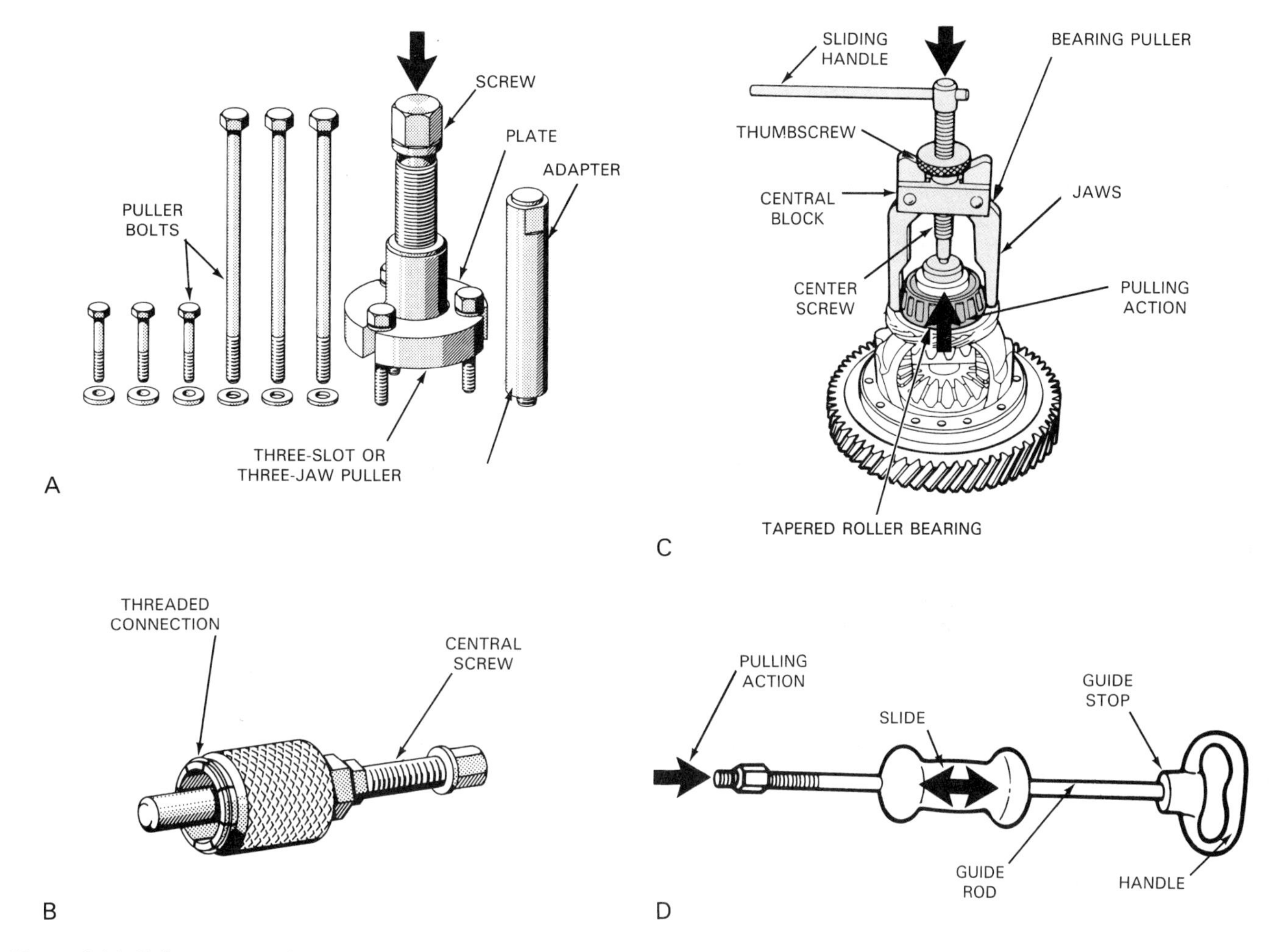

Figure 2-14. Pullers are used to remove many components of the drive train. A–This kind of puller is used to remove pressed-on parts that have threaded holes to accept puller bolts. B–This puller threads into the part to be removed. The part is usually ruined in the pulling process. C–The jaw puller is used to remove many antifriction bearings. It must be carefully installed. D–The slide hammer can be used to remove many kinds of bearings and seals. This tool will accept any number of adapters. (Mercedes-Benz, Honda, Chrysler)

To remove oil seals, a special adapter is sometimes used with the slide hammer. This adapter consists of a sheet metal screw that is brazed or welded onto a nut. The internal threads of the nut match the external threads of the slide hammer guide rod. To remove a seal, a small hole is drilled in the metal portion of the seal. The adapter screw is then threaded tightly into the hole. The guide rod is threaded into the adapter nut. After the adapter is attached to the guide rod, the slide portion of the slide hammer can be moved sharply against the shoulder of the guide stop to remove the seal. If the seal is difficult to remove or the hole becomes stripped, another hole can be drilled on the opposite side of the seal, and the removal process can be repeated.

There are some special precautions that should be taken when using pullers:

- *Always* wear eye protection. Failure to do so could result in irreversible eye damage.
- *Always* install the puller properly. An improperly installed puller could cause injury or damage.
- *Never* apply force through the rolling elements of a ball or roller bearing. The bearing will most likely be ruined, and it may fly apart, causing additional damage and injury.

Spring Compressors

Spring compressors are used to remove and install coil springs. In operation, the tool compresses the spring. This shortens the spring and takes pressure off the parts against which it is seated; thus, the spring and associated parts can be removed.

In drive train service, spring compressors will be used mostly in automatic transmission overhaul. The spring compressor is used to compress the *clutch release spring* of the transmission. This will take the pressure off the spring retainer and snap ring for the *clutch apply piston.* In doing so, the parts can be removed without damage. See Figures 2-15 and 2-16.

WARNING! Use extreme care when removing and installing springs. A loaded spring can cause serious injury if it is released suddenly. Be sure that you are wearing eye protection when working with springs, and be careful!

Seal Protectors

Seal protectors are metal or plastic sleeves. They are used to prevent seal damage when a seal is installed over a shaft with a sharp area, such as a *keyway,* Figure 2-17. Once the seal is in place, the protector can be removed.

Tubing Tools

It is sometimes necessary to repair or replace tubing during drive train service. Tubing that connects the automatic transmission to the transmission *oil cooler* in the vehicle radiator is one example. To do this properly, some special tubing tools are needed. There are considered to be three general types of tubing tools. Two of these are shown in Figure 2-18.

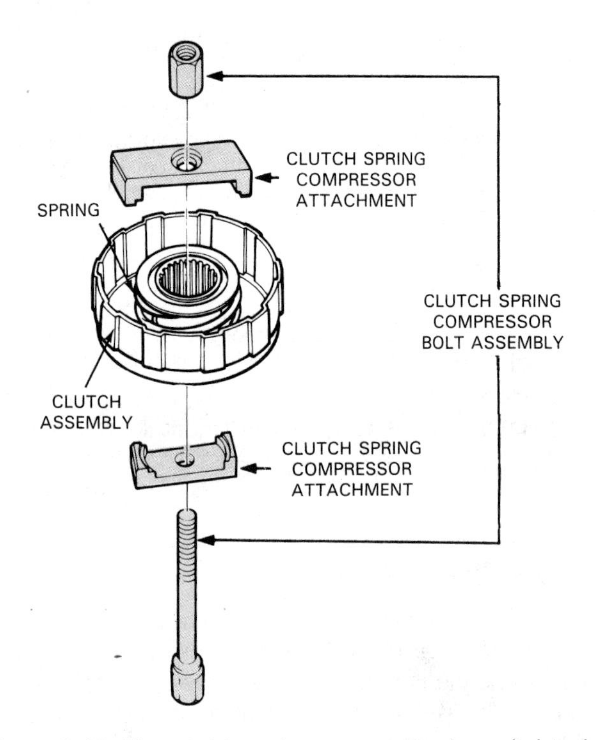

Figure 2-15. The *clutch spring compressor* is a vital tool for repairing automatic transmissions. Automatic transmissions have clutch assemblies that require its use.

Tubing cutters

A ***tubing cutter*** is shown in Figure 2-18A. The tool essentially consists of a cutting wheel and opposing rollers that guide the tool and provide a surface against which to cut.

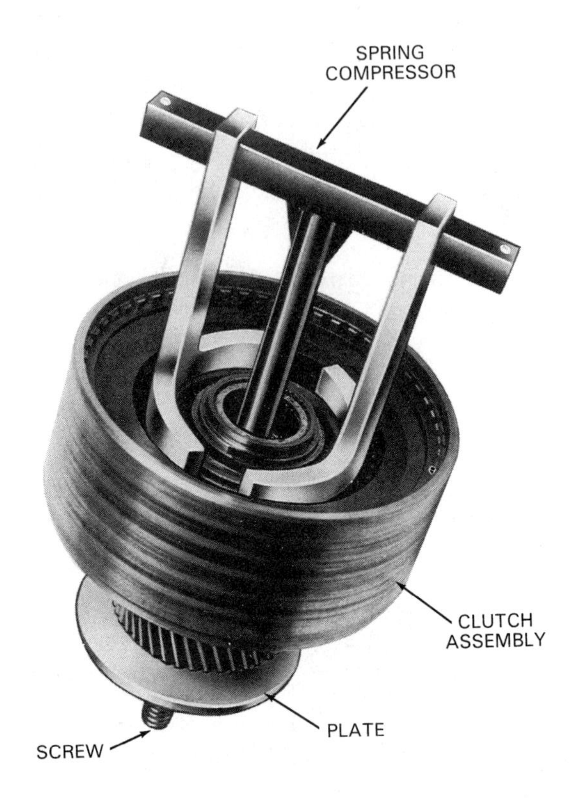

Figure 2-16. A clutch spring compressor is shown in operation. Once the spring retainer is depressed, the retaining snap ring can be removed. Then, the compressor is carefully released and the clutch disassembled. (Owatonna Tool)

Figure 2-17. Using seal protectors prevents damage when seals are installed over shafts. (Owatonna Tool)

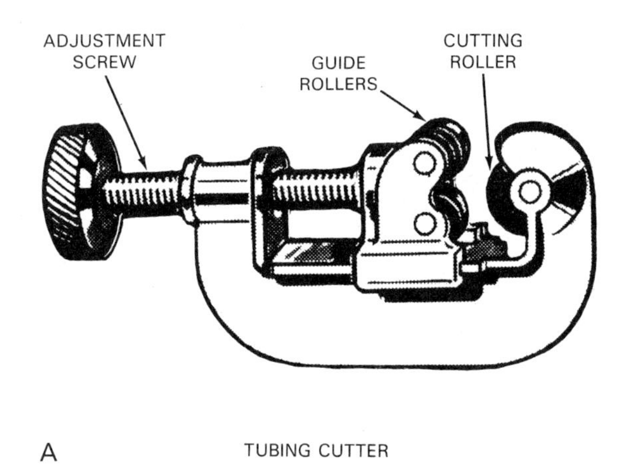

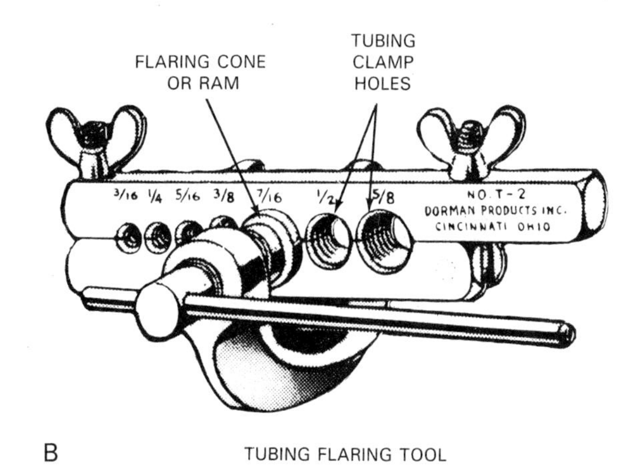

Figure 2-18. Tubing tools are a necessity when transmission cooler lines must be repaired or new lines must be run. A–The tubing cutter is used to make a clean cut in tubing. B–Tubing can be flared easily with a flaring tool. (Dorman Products)

In cutting the tubing, the tubing cutter is gradually tightened, and the tool is rotated completely around the tubing. The cutting wheel should be tightened in small steps to avoid deforming the tubing. After the tubing is cut, it should be *deburred* to remove any sharp projections. Many tubing cutters have a *reamer blade* for this purpose.

Tubing flaring tools

Tubing flaring tools are used to *flare* the tubing, which means to make an enlarged lip at the end of the tubing. Flaring tools consist of a ram and a block that contains different-size holes to match various sizes of tubing. See Figure 2-18B.

To flare the end of a length of tubing, place it in the proper hole, with the recommended amount of tubing exposed. Tighten the tubing firmly in the block. Then, tighten the ram into the tubing until the flare is formed.

Bending tools

If tubing is bent in a more or less sharp angle, it will usually kink at the inside of the bend. To prevent this, special ***bending tools*** should be used. There are several kinds of bending tools, all of which accomplish the job handily when used correctly.

Alignment Tools

Alignment tools are usually used in the clutch and transmission. A *clutch alignment tool,* Figure 2-19, is used to keep the *clutch disc* in alignment as the *clutch cover attaching bolts* are tightened. The tool is sometimes called a *pilot shaft.*

Transmission alignment tools are used to align some transmission internal parts during assembly. These parts are usually moving or are supports for moving parts. Transmission alignment tools will vary.

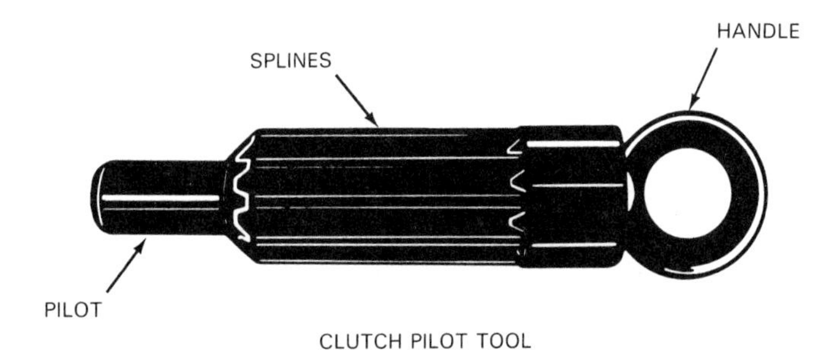

Figure 2-19. A clutch alignment tool holds the clutch disc in alignment as the pressure plate is tightened.

Flywheel Turning and Locking Tools

Often, the engine flywheel must be turned or held stationary so that bolts can be reached to be loosened or tightened. ***Flywheel turners*** move the flywheel by engaging two or more teeth on the flywheel ring gear. The handle is long enough to provide good mechanical advantage, which enables the flywheel to be turned easily. See Figure 2-20.

A ***flywheel holder*** is used for keeping the flywheel from turning. It has teeth that engage the teeth on the flywheel ring gear. The other end of the tool is bolted to the engine block or to another stationary part of the drive train. See Figure 2-21.

Organizing Trays

An ***organizing tray,*** Figure 2-22, can be useful when you are working on components with many small, similar parts. The tray will provide a way to organize parts so that they can be reassembled in the same position as they were originally. Organizing trays are handy when working on automatic transmission *valve bodies,* for instance. They can also help avoid the accidental switching of identical

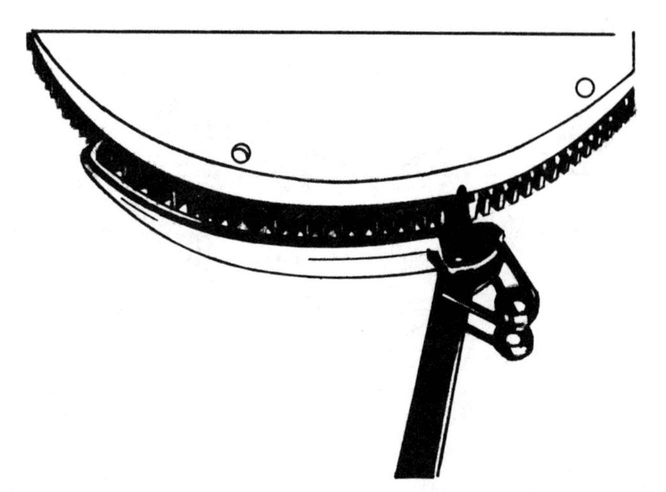

Figure 2-20. Flywheel turners are useful when the flywheel must be turned by small amounts, such as when converter or pressure plate bolts are being installed.

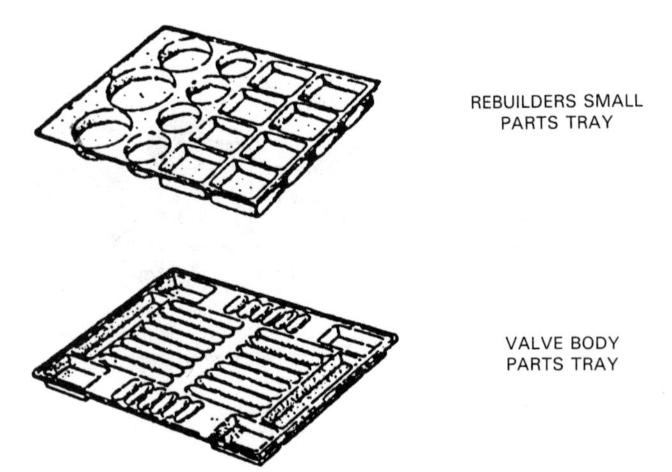

Figure 2-22. Parts trays are useful when disassembling components with many small parts, such as transmissions, transaxles, and rear axle assemblies. (Hayden)

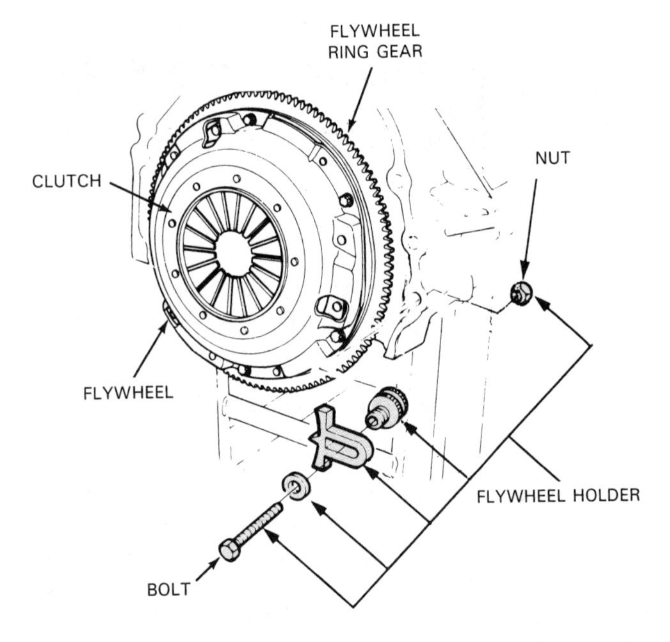

Figure 2-21. Flywheel holders are useful when the flywheel must be held stationary, such as when converter, flywheel, or pressure plate bolts are tightened or loosened. (Honda)

parts, which must be returned to their original assembly. (Though such parts look identical, they will have worn differently; thus, they should not be switched.)

Drain Pans

If oil is allowed to drip on the floor during repairs, it will create a safety hazard. Further, if tracked onto carpets and other floor treatments, oil can cause damage. ***Drain pans*** are handy for catching oil that drips from assemblies as they are drained or disassembled, Figure 2-23.

Screw Extractors

Screw extractors are used to remove broken screws or other broken fasteners from parts. Screw extractors are hardened shafts, designed to grip a broken fastener from the inside. To use this kind of extractor, first, drill a hole in the broken fastener. (This hole must be the proper size for the extractor to be used, and it must be exactly in the center of the shaft.) Then, lightly tap the extractor into the hole and back it out with a wrench. The fastener will unscrew in the process. This is shown in Figure 2-24.

Other tools are available for extracting fasteners from parts. These are designed to grip the fastener from the outside. They can be used only if the broken fastener extends from the part. A special *stud puller* is an example. Sometimes, vise grip pliers can be used for this purpose.

Drive Train Lifting and Holding Equipment

Most drive train components are large, heavy, and awkward. To assist in removing and servicing them, some special equipment is needed, as described below.

Holding fixtures

Holding fixtures, like the one shown in Figure 2-25, are handy for holding large components, such as transmissions, rear axle assemblies, and transaxles. Some holding fixtures are made to be used with only one component, while others can be used with any kind of transmission or transaxle.

Transmission jacks

Transmissions and transaxles are large, heavy assemblies. They can cause serious injury if they are allowed to fall. Lifting them into place during installation can also

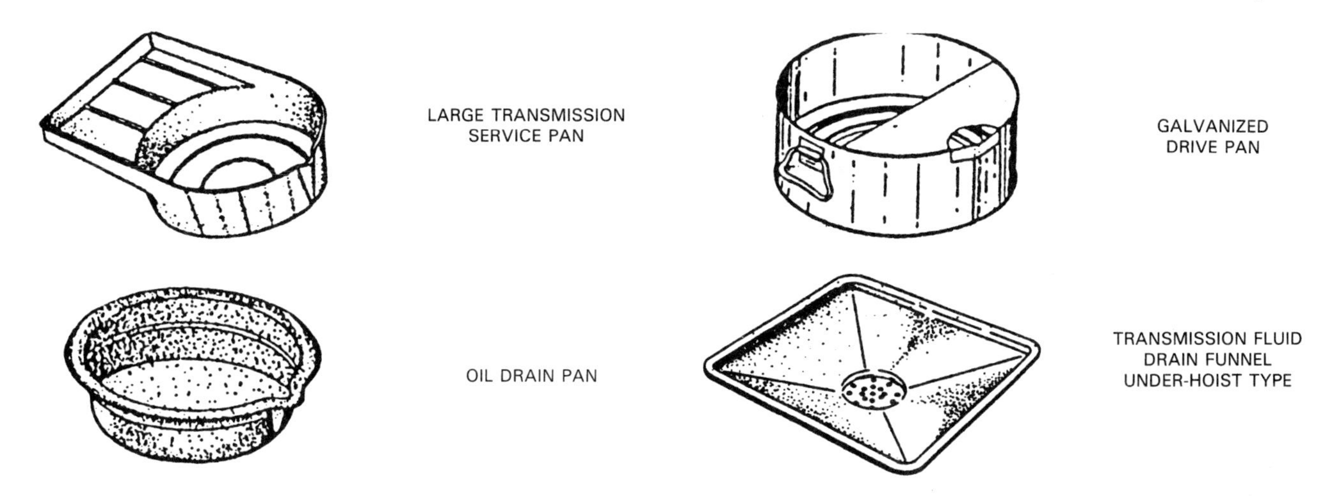

Figure 2-23. The drain pans shown here are useful for catching oil.

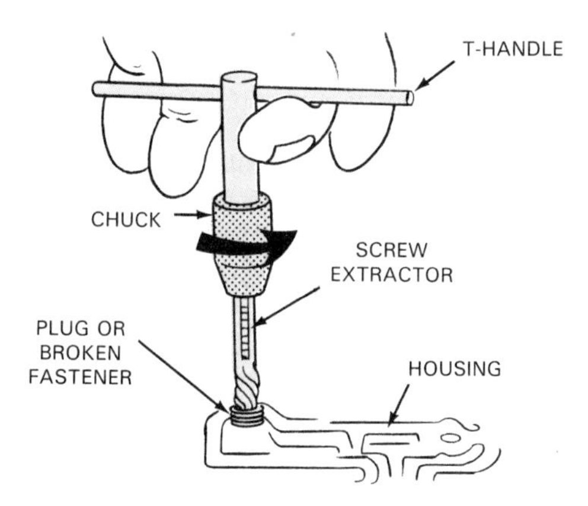

Figure 2-24. A screw extractor is used for removing broken fasteners from parts. A hole must be drilled in the fastener before the kind of extractor pictured can be used. (General Motors)

pose problems. To prevent this, ***transmission jacks*** should be used to remove and install transmissions. See Figures 2-26 and 2-27.

Drive Train Cleaning Equipment

Drive train parts are located under the vehicle. As a result, they collect dirt from the road. The dirt mixes with leaking oil. Heavy grease deposits result. Internal drive train parts usually require cleaning because of the heavy oils used. Further, where there is internal damage, heavy debris and metal deposits will be present. For these reasons, special cleaning equipment is needed. Different types of cleaning equipment are presented in the next few paragraphs.

Line and converter flushers

In many cases, by the time that an automatic transmission requires overhaul, the transmission fluid has become contaminated with debris. The fluid has been badly overheated, which caused the fluid to break down and nonmetallic materials to deteriorate. Varnish and carbon have built up throughout the transmission as a result of the fluid. A defective transmission oil cooler may have caused the fluid to become contaminated with water and antifreeze. The transmission parts must be thoroughly cleaned. Special equipment is needed to clean and flush the oil cooler and lines and the torque converter.

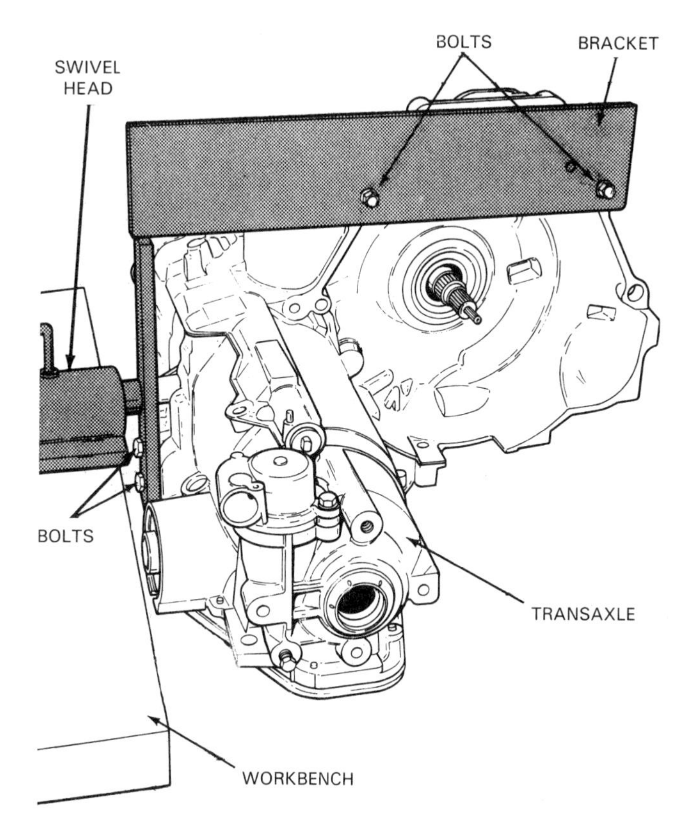

Figure 2-25. A holding fixture simplifies the job of overhauling a transaxle. (General Motors)

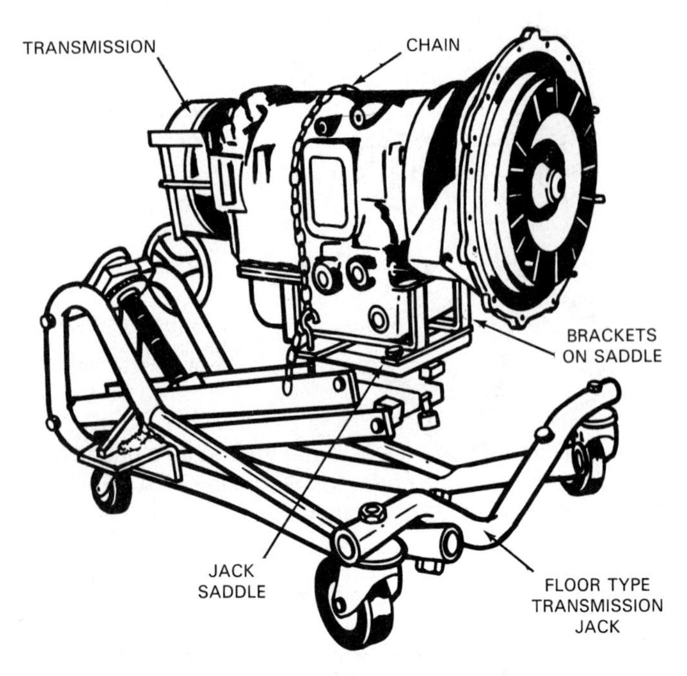

Figure 2-26. A transmission jack, like the under-the-vehicle model shown, makes transmission removal much easier. (General Motors)

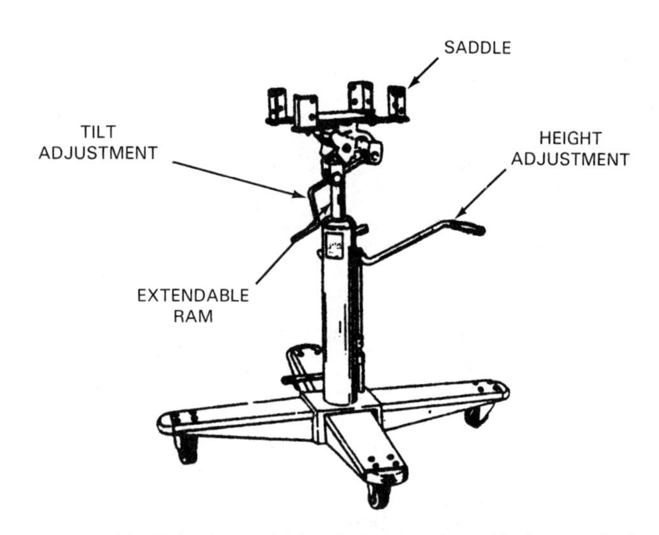

Figure 2-27. This transmission jack is designed to be used when the vehicle is on a hoist. (Hayden)

The ***oil cooler and line flusher,*** Figure 2-28, uses solvent to remove contaminants from the cooler lines. Once the proper connections are made, solvent is forced through the lines, cleaning and flushing them in the process. A final blast of compressed air removes the solvent, and the cooler and lines are ready for reuse.

The ***converter flusher,*** Figure 2-29, is used to remove debris from inside of the automatic transmission torque converter. It is a special device that moves a pulsating flow of solvent in and out of the converter. In addition, the

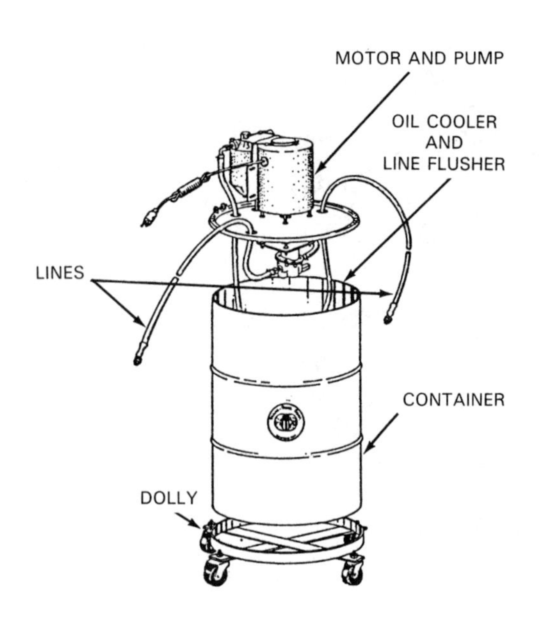

Figure 2-28. Automatic transmission cooler lines often become filled with sludge, antifreeze, or metal. They must be cleaned when a transmission is overhauled. Cooler lines can be flushed easily with this pressurized equipment. (Hayden)

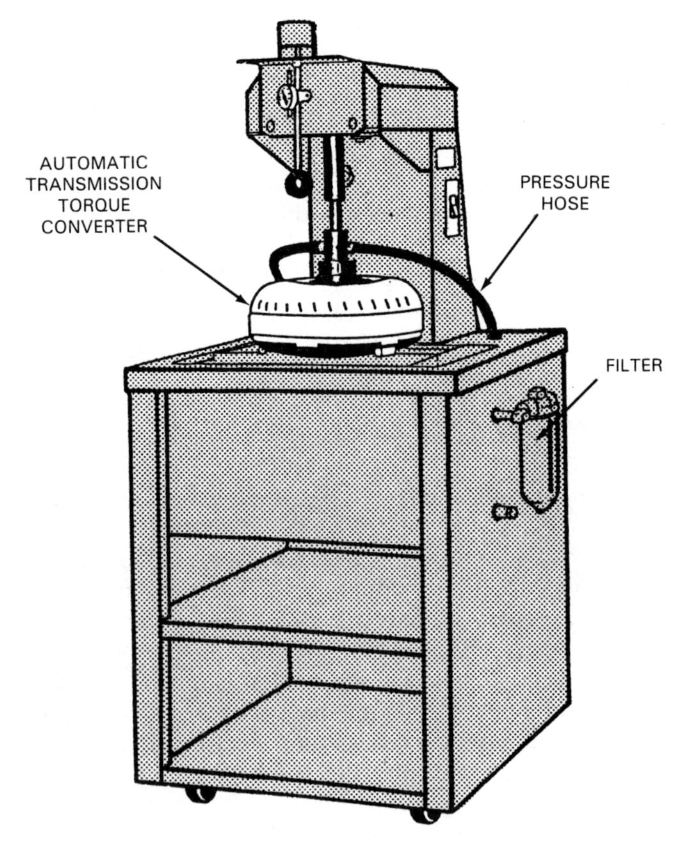

Figure 2-29. To avoid the expense of replacement, many shops clean sealed torque converters with a converter flusher. (Hayden)

converter flusher rotates the converter to further agitate the solvent inside of the converter. This agitation and flow through the converter removes varnish and debris. The flushing action is usually accomplished in a few minutes.

Parts washers

After you have removed all of the parts from an assembly, everything should be cleaned. Problems can be hard to see when a part is covered with oil, grease, or carbon deposits. Closer part inspection can be done during and after part cleaning. ***Parts washers*** are often used for cleaning. Different types are available, depending on the part construction and type of material. See Figure 2-30.

One type of parts washer is the *cold-soak tank.* It can be thought of as a sink that runs cleaning solvent instead of water. The solvent aids in dirt and grease removal. Most parts washers have a pump to recirculate the solvent. Parts are left to soak in the solvent. When necessary, brushes and scrapers are used for cleaning. Since the solvent is much harsher than water, you should always protect your hands and face when using it.

Another more powerful kind of parts washer is called an *immersion cleaner,* or a *hot tank.* The part to be cleaned is placed in the tank, which usually is filled with strong corrosive chemicals. Cleaning is done by an automatic process.

Disposal of vehicle maintenance products

Certain products involved in vehicle maintenance, such as rust removers and parts cleaners and degreasers, may contain hazardous materials. Such products may contain toxic chemicals or strong acid or alkaline solutions, for example. They must be disposed of properly. Petroleum products used for vehicle operation, such as automatic transmission fluid and gear oil, are also examples of materials that require special handling. If you generate hazardous waste, you might be subject to regulations that govern the disposal of these waste products. In particular, any business that maintains or repairs vehicles, heavy equipment, or farm equipment is subject to the requirements of the Resourse Conservation and Recovery Act. Check with your state's hazardous waste management agency or Regional EPA (Environmental Protection Agency) office for more information.

Drive Train Measuring Tools

Many drive train components require special measuring tools in the reassembly process and, afterwards, to ensure that they have been put together correctly. ***Gauges*** are used for this purpose. A gauge is an instrument used to measure something. In drive train service, *pressure gauges, size gauges,* and *angle gauges* are often used. Another gauge commonly used measures torque. This gauge is an integral part of a *torque wrench.*

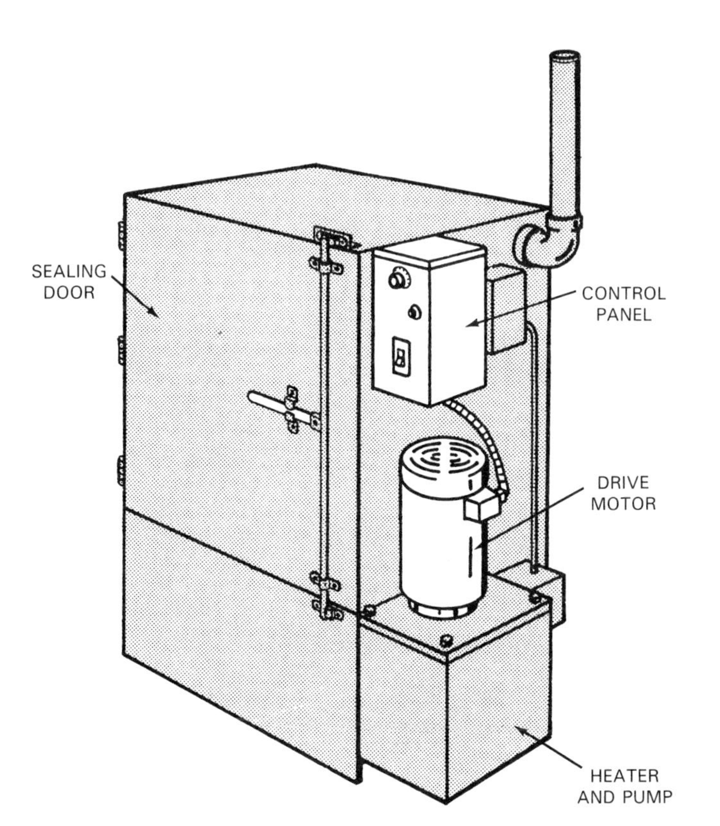

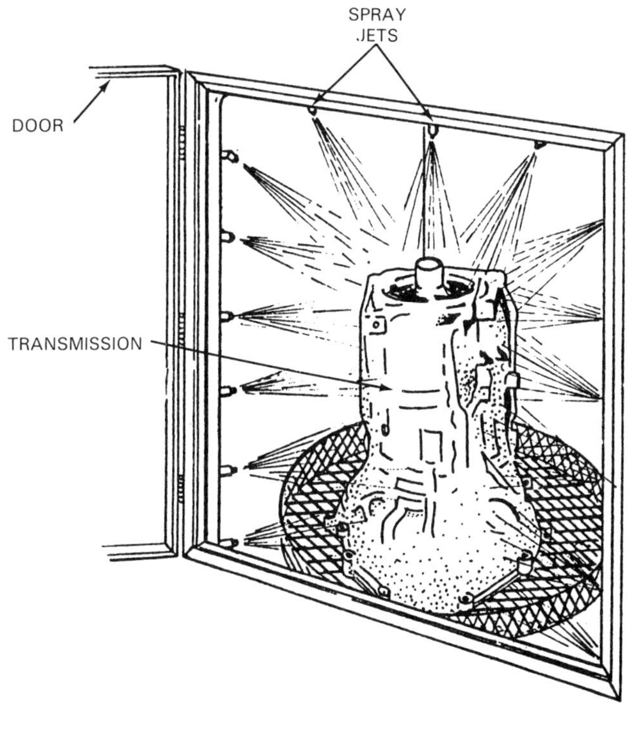

Figure 2-30. This parts washer will clean the interior and exterior of large parts with spray action. (Hayden)

Pressure gauges

Pressure gauges measure pressure of a fluid. Most automotive gauges use the ***gauge pressure*** scale. This means that normal *atmospheric pressure* (14.7 psi [101 kPa] at sea level) is chosen as a zero reference pressure. These automotive gauges, then, measure pressure relative to that of the surrounding atmosphere. The zero reference is denoted 0 psi*g*, where the *g* stands for *gauge;* it is equivalent to 14.7 psi (101 kPa). In many cases, this distinction for gauge pressure is unimportant, and just *psi* is used.

Pressure gauges are used in the troubleshooting of automatic transmissions and transaxles. They allow the technician to precisely measure internal oil pressure. All automatic transmissions have at least one pressure connection, called a *tap,* or *plug,* where the gauge can be attached. The gauge is installed at the tap, and the engine is started. Transmission pressures can then be checked for each of the transmission operating positions. The actual pressures are checked against the specifications given in the manufacturer's service manual. Any variation from the specified pressures means that there is a problem in the transmission hydraulic system. See Figure 2-31.

A ***vacuum gauge*** is also shown installed in Figure 2-31. This type of pressure gauge is used to measure a ***vacuum,*** which is pressure below normal atmospheric pressure. Thus, vacuum gauges measure pressures *below* 0 psig (0 kPag). Vacuum pressure is often given in terms of *inches (millimeters) of mercury (in. Hg vacuum or mm Hg vacuum).* On the vacuum pressure scale, 0 in. Hg (0 mm Hg) is equivalent to 0 psig. A vacuum pressure of 2 in. Hg (52 mm Hg) is approximately equal to -1 psig, and 10 in. Hg (259 mm Hg) equals about -5 psig.

Vacuum gauges are useful for measuring engine manifold vacuum when diagnosing an automatic transmission that has a *vacuum modulator.* The vacuum gauge is attached to the intake manifold, and the engine is started. Lower than normal vacuum indicates an engine problem that could affect transmission operation. A vacuum gauge can also be connected to the modulator line at the modulator. If the vacuum is much less at the modulator than at the intake manifold, the modulator line is plugged or is leaking.

Size gauges

Size gauges are linear measurement devices. They are used for measuring part dimensions and clearances–the space between two parts. An example of one type of size gauge is the *micrometer,* shown in Figure 2-32. This

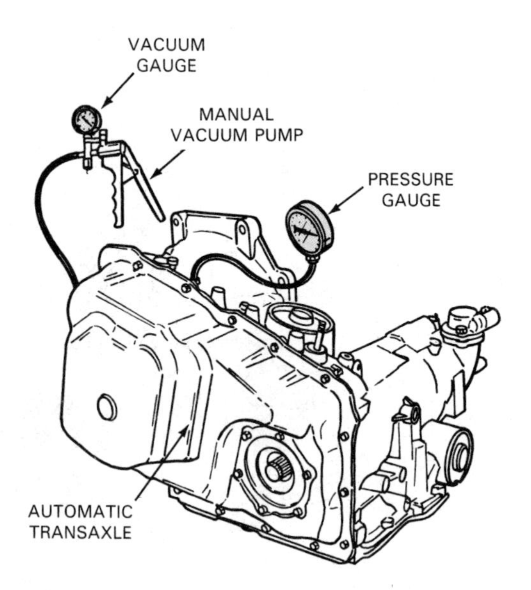

Figure 2-31. Pressure gauge used here to check internal transmission pressures. This is an aid to locating internal fluid leaks or other problems. Manual vacuum pump is used to simulate vacuum condition. Vacuum pressure is indicated on vacuum gauge.

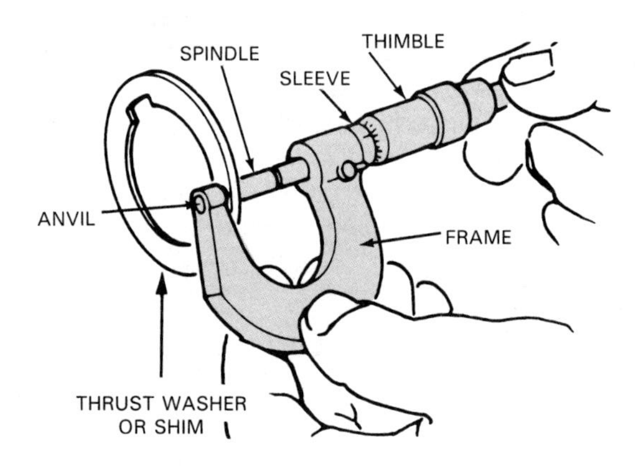

PART NUMBER	THICKNESS
90441–PL5–000	4.00 mm (0.157 in.)
90442–PL5–000	4.05 mm (0.159 in.)
90443–PL5–000	4.10 mm (0.161 in.)
90444–PL5–000	4.15 mm (0.163 in.)
90445–PL5–000	4.20 mm (0.165 in.)
90446–PL5–000	4.25 mm (0.167 in.)
90447–PL5–000	4.30 mm (0.169 in.)
90448–PL5–000	4.35 mm (0.171 in.)
90449–PL5–000	4.40 mm (0.173 in.)

Figure 2-32. The micrometer is used on machined parts to check that parts fall within tolerance. It is shown being used to check a thrust washer for proper thickness. (Honda)

type of gauge has a graduated scale from which very accurate measurements are read. Other size gauges give measurements by choosing the *best fit* from a set of gauges that range in size. *Feeler gauges,* consisting of strips of metal made to exact thicknesses, are an example.

Size gauges are often used prior to assembly in order to determine what clearances of mating parts will be after they are assembled. This eliminates the problem of assembling and reassembling parts to make adjustments.

Outside micrometers. In general, a ***micrometer (mike)*** is a measuring device that can make very accurate measurements. Refer to the ***outside micrometer*** pictured in Figure 2-32. Note that this micrometer has a *frame, sleeve, thimble, spindle,* and *anvil.* The opening between the anvil and spindle can be made larger or smaller by turning the thimble. The size of the opening can be read on the sleeve and thimble. Micrometers are available in *inch-based* and *metric-based* versions.

To use an outside micrometer, place the part to be measured between the anvil and spindle. Figure 2-32 demonstrates this with a transmission thrust washer. Close the mike until it tightens lightly against the part. The part should be able to move between the anvil and spindle with a light drag. Do not *over*tighten the micrometer. This will distort the frame and cause inaccurate readings.

The basic steps of reading a micrometer are similar for both inch-based and metric-based micrometers. The following is an explanation of reading the inch-based micrometer, Figure 2-33. With the part in place, read the greatest numbered division visible on the sleeve. These numbers are tenths of an inch (0.1"). Read the fractional increment visible on the sleeve. Each division is 25 thousandths of an inch (0.025"). Add these readings together and write them down. Then, read the number on the thimble that matches the center line on the sleeve. Each division is 1 thousandth of an inch (0.001"). To obtain the final reading, add the sleeve and thimble readings together.

Note that a fourth decimal place is added to the final reading if there is an auxiliary scale, known as a *vernier,* on the sleeve of the mike. This figure, which will be in ten-thousandths of an inch, will be from *the* vernier line that exactly aligns with one of the lines on the thimble.

The following is an explanation of reading the metric-based micrometer, Figure 2-34. With the part in place, read the greatest division visible above the center line of the sleeve. Each division is 1 millimeter. Add 0.5 millimeter if a following division mark is visible below the center line. (On some mikes, 1 mm divisions fall below the center line, and 0.5 mm divisions fall above.) Write the number down. Add to this the number on the thimble that matches the center line on the sleeve. Each division is 1 hundredth of a millimeter (0.01 mm).

Hole gauges and telescoping gauges. These tools, which are shown in Figure 2-35, are used to make accurate measurements of the inside diameter of bushings, and other internal surfaces. The ***hole gauge,*** Figure 2-35A, is used to measure small diameter holes or openings. The gauge is inserted into the hole and adjusted to fit. Then, it is removed and measured with an outside micrometer. The ***telescoping gauge,*** Figure 2-35B, is generally used to measure larger openings. The gauge is placed in the hole, properly tightened, and locked. It is then withdrawn from the hole and measured with a micrometer. The micrometer reading is the size of the opening.

Inside micrometers. An ***inside micrometer,*** Figure 2-36, is used for measuring larger inside diameters or other part openings. The instrument is placed within the bore or other opening and adjusted to fit. It is read in the same manner as an outside mike.

Depth micrometers. A ***depth micrometer*** is similar to the outside mike discussed above, but it is used to measure the depth of openings in machined surfaces. Figure 2-37 shows a depth micrometer being used to check the depth of a bearing bore in a transmission case.

Calipers. Instruments called ***calipers*** make up a whole class of tools used for external or internal measurements. *Outside* and *inside calipers,* Figure 2-38, are essentially a pair of movable legs. In measuring, the legs are adjusted to fit the dimension in question. The caliper can

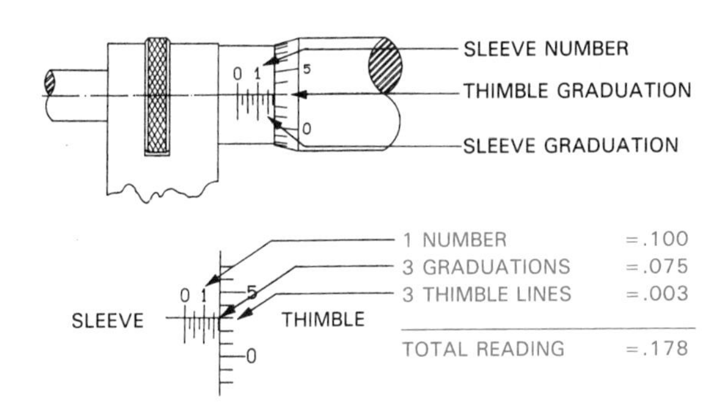

Figure 2-33. Study basic steps for reading a micrometer graduated in thousandths of an inch. Read number, then sleeve graduations, then thimble. Add these three values to obtain reading. Total reading above is 0.178 inches. (Starrett)

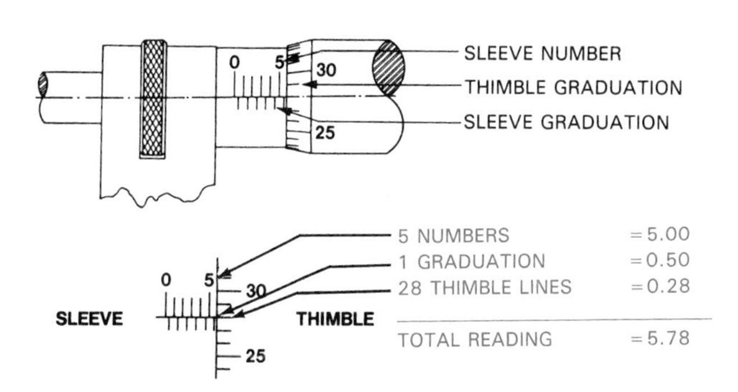

Figure 2-34. Metric micrometer is read like an inch-based micrometer. However, note metric values for sleeve and thimble. Total reading above is 5.78 millimeters. (Starrett)

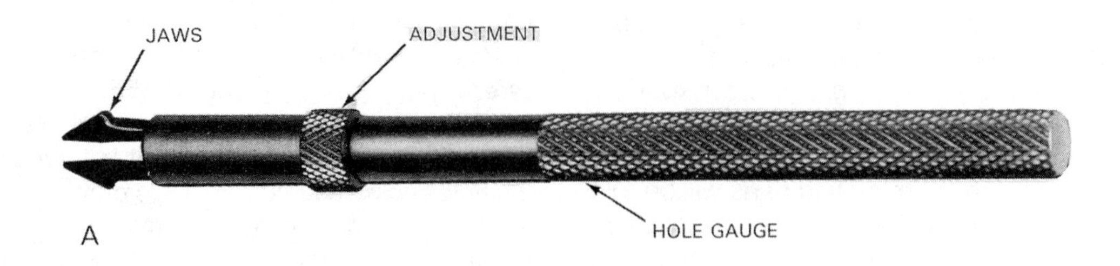

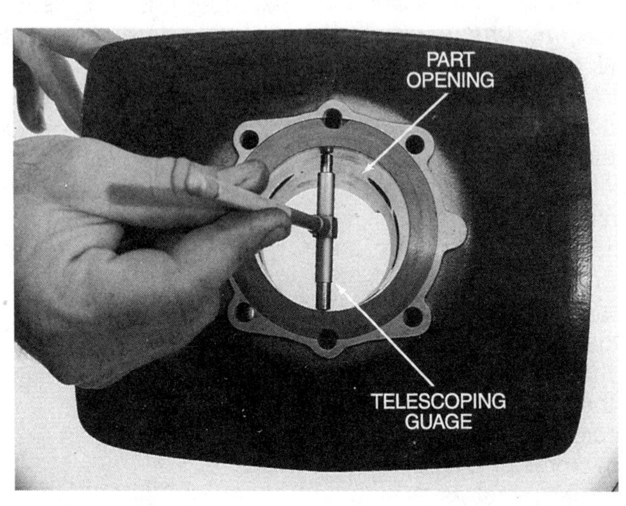

Figure 2-35. Tools used to measure diameters of small holes and openings. A–Hole gauge. B–Telescoping gauge. (Vaco)

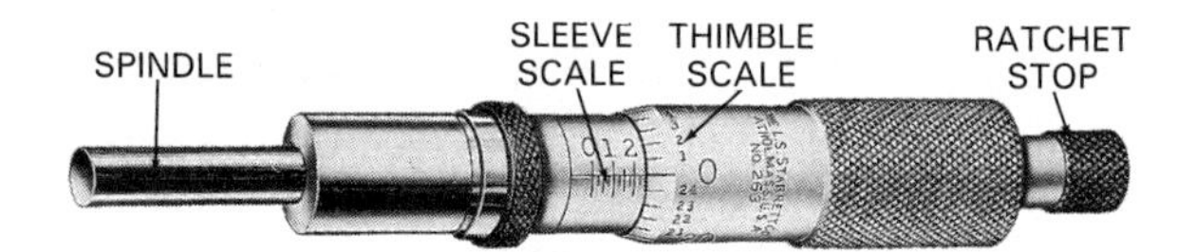

Figure 2-36. Inside micrometers measure larger holes and openings. (Starrett)

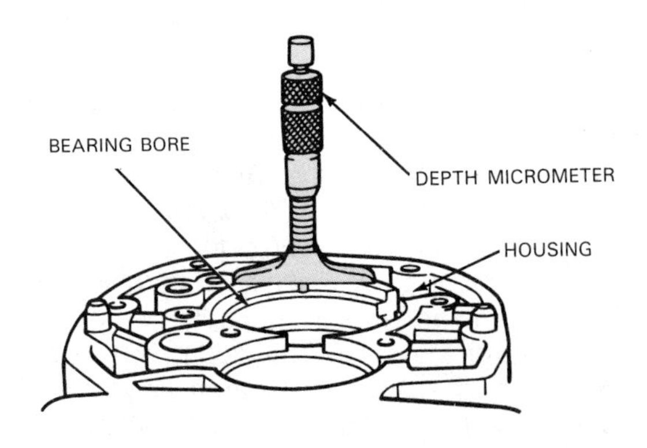

Figure 2-37. The depth micrometer measures depth of bores and depressions in machined parts. The graduated scale (not shown) of the depth micrometer is read in the same manner as the outside micrometer. (Ford)

then be laid over a rule, and the span, measured. Calipers of this type are less accurate than micrometers.

The *vernier caliper* is a type of *sliding caliper.* It is an accurate tool, often used for measuring inside and outside diameters. This device is a ruled instrument having a set of jaws–one fixed, one movable. The movable jaw has a sliding scale, or vernier, that is part of it. A component is placed within the jaws, and the movable jaw is adjusted to fit. The sliding scale moves with the adjustable jaw, as they are a unit. Figure 2-39 shows a typical use of a vernier caliper. Note that the caliper shown has an additional set of jaws for making inside measurements.

Vernier calipers come in inch-based and metric-based versions. Readings can be obtained to 0.001 in. and 0.02 mm, respectively. Both types are read in more or less the same way. The measurement marked by 0 on the vernier scale is read off the fixed scale. To this, the amount on the vernier is added. This amount will be read from the vernier line that coincides with a line on the fixed scale. Only one vernier line will line up exactly with a line from the fixed scale.

A *dial caliper* is another type of sliding caliper. With this type, the vernier is replaced by a dial gauge. Otherwise, the two instruments look and operate pretty much the same. The dial gauge reads directly to 0.001 in. or 0.02

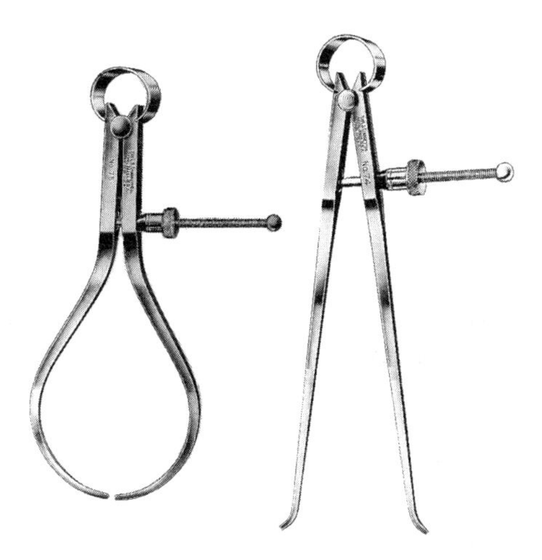

Figure 2-38. Outside and inside calipers used for making rough measurements on part openings. (Starrett)

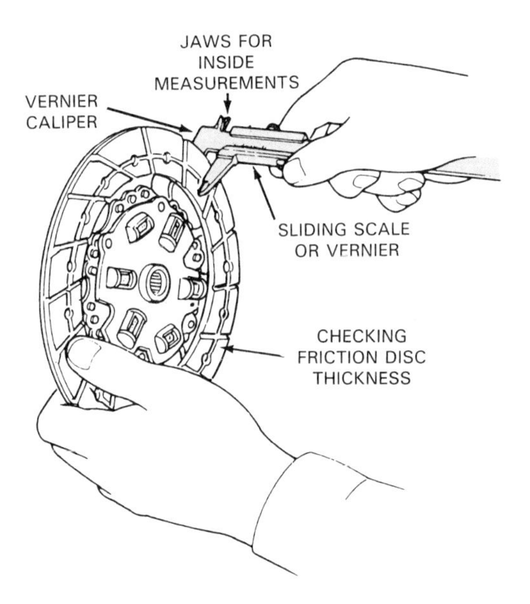

Figure 2-39. A vernier caliper being used to check the thickness of a clutch disc. (Honda)

mm. It makes precision measurements easier to read. Both vernier and dial calipers can be used and read quickly.

Feeler gauges. A ***feeler gauge*** is used to measure spaces, or gaps, between two surfaces. A *flat feeler gauge* is made of a thin strip of steel or brass. A *wire feeler gauge* is made of a wire of specified diameter. Feeler gauges are usually sold in sets. A set will contain a number of gauges of different diameters or thicknesses. See Figure 2-40.

A feeler gauge is used by inserting it in the gap to be measured. Different thicknesses are tried. The correct gauge will drag slightly when pulled between the parts. The clearance, then, will be given by the size written on the gauge.

Figure 2-41 shows a ***straightedge*** being used, along with a feeler gauge, to check a pressure plate surface for warping. Straightedges are used to check for warping of part surfaces. These hardened-steel bars are manufactured so that one edge is very straight. They serve as the standard against which to compare. The tool is placed directly on the part to be checked, and the part surface is checked in relation to the straightedge. If a feeler gauge over a certain size can be inserted between the part and straightedge, the part is excessively warped.

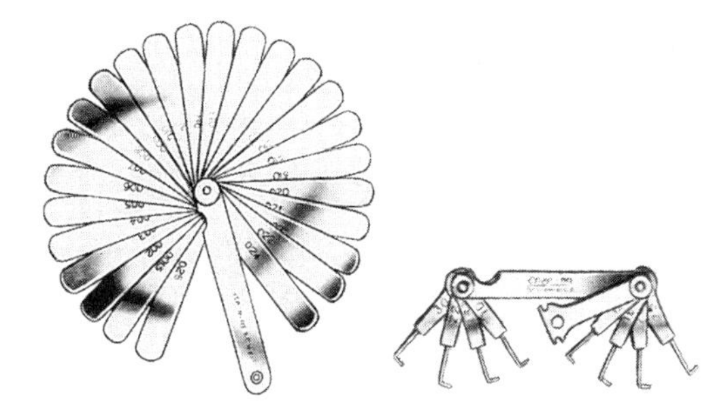

Figure 2-40. Flat and wire feeler gauges, useful for checking small clearances.

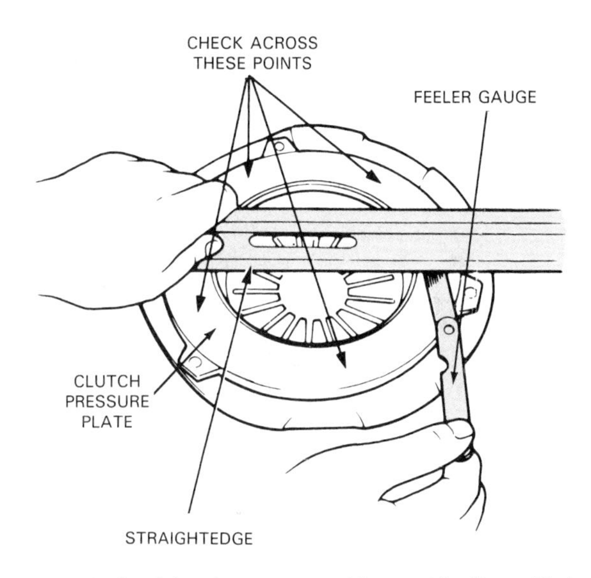

Figure 2-41. Straightedges are used in combination with feeler gauges to check surfaces for flatness. The straightedge is placed on the part, and various feeler gauges are inserted. If too large of a feeler gauge can be inserted, the part is excessively warped.

Dial indicators. A ***dial indicator*** is a gauge that is used on moving drive train parts to measure small amounts of movement. These instruments are frequently used to check gear *backlash,* or clearance between teeth, shaft *end play,* cam lobe lift, and other similar kinds of part movements. The gauge, which has a dial face, is operated by a plunger. Typical scale graduations are in thousandths of an inch, or in hundredths of a millimeter, but finer divisions are available. The gauge can be *zeroed* (adjusted to zero) by turning the dial face assembly.

The dial indicator is mounted on or near the component to be checked. The contact point, or *plunger* is positioned so that it contacts the movable portion of the component with a slight pressure. The gauge assembly should be secured so that only the plunger can move. Once the gauge is firmly in place, it should be zeroed. This is done by turning the dial face. The movable portion is then moved back and forth or is rotated, and a movement amount will be indicated on the dial face, Figure 2-42.

Angle gauges

Angle gauges are special devices that measure angles between two components or *inclinations*–angles relative to a plane of reference. Typically, they are used on driveline and drive axle components to determine the angles between them. This is done to correctly set angles for universal joints; otherwise, if an angle is too sharp, speed fluctuations and vibration problems can occur. The *inclinometer,* Figure 2-43, is an example of an angle gauge.

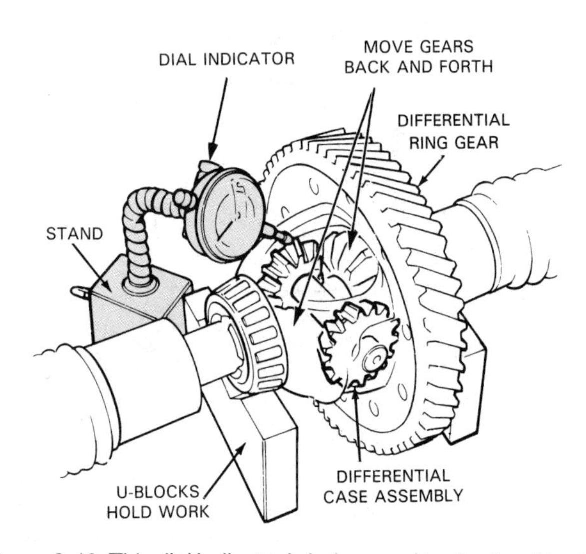

Figure 2-42. This dial indicator is being used to check a drive train assembly for improper clearance. If the dial indicator is installed and adjusted properly, the part can be checked to a thousandth of an inch or better. (Honda)

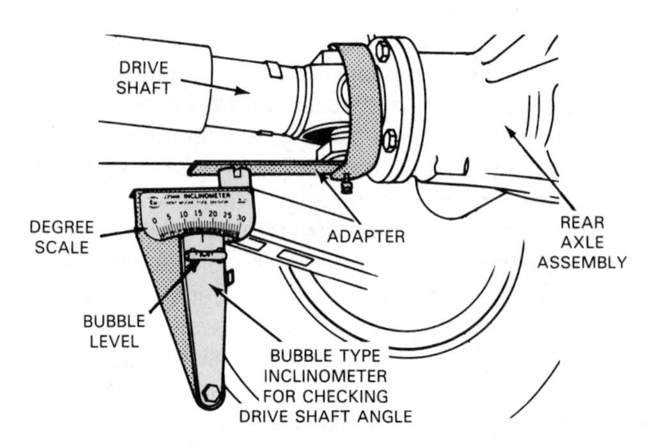

Figure 2-43. An inclinometer is used here to check drive shaft angle. Note that several types of inclinometers are made. (General Motors)

Torque wrenches

A ***torque wrench*** is a combination wrench and measuring device. It is used to apply a *measured* turning force, or torque, to a threaded fastener. The amount of torque is indicated on the tool while the force is applied, Figure 2-44. Torque wrench scales usually read in *foot-pounds (ft.-lb.)* or in metric units of *newton-meters (N-m).*

In Figure 2-44A, the torque wrench is used to tighten lug nuts to a specified value. When the proper value is read on the scale, the nut is as tight as it should be. In Figure 2-44B, the wrench is used to verify the amount of preload (mild pressure) on an installed bearing. In this procedure, enough torque is applied to the tightened fastener to cause it to move. Reading the scale at this point will give the

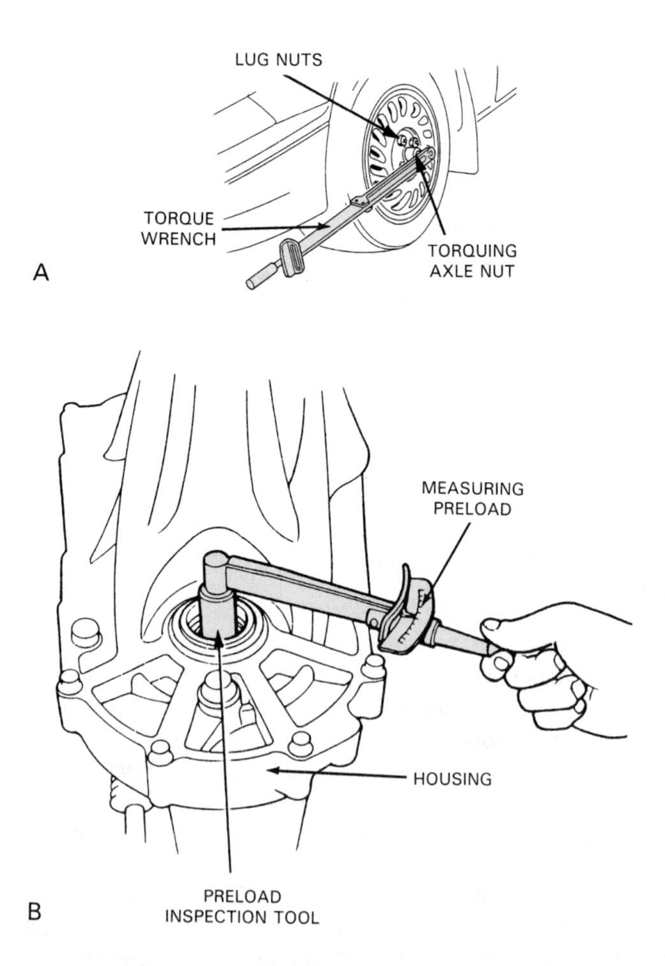

Figure 2-44. Torque wrenches shown in two different applications. A–The wrench is being used to tighten a fastener to the proper torque value. B–The torque wrench is being used to check a bearing for the amount of preload to determine if it is properly adjusted. (Honda)

BOLT TORQUE

BOLT SIZE	GRADE 5 N·m	GRADE 5 ft-lbs (in-lbs)	GRADE 8 N·m	GRADE 8 ft-lbs (in-lbs)
1/4-20	11	(95)	14	(125)
1/4-28	11	(95)	17	(150)
5/16-18	23	(200)	31	(270)
5/16-24	27	20	34	25
3/8-16	41	30	54	40
3/8-24	48	35	61	45
7/16-14	68	50	88	65
7/16-20	75	55	95	70
1/2-13	102	75	136	100
1/2-20	115	85	149	110
9/16-12	142	105	183	135
9/16-18	156	115	203	150
5/8-11	203	150	264	195
5/8-18	217	160	285	210
3/4-16	237	175	305	225

A

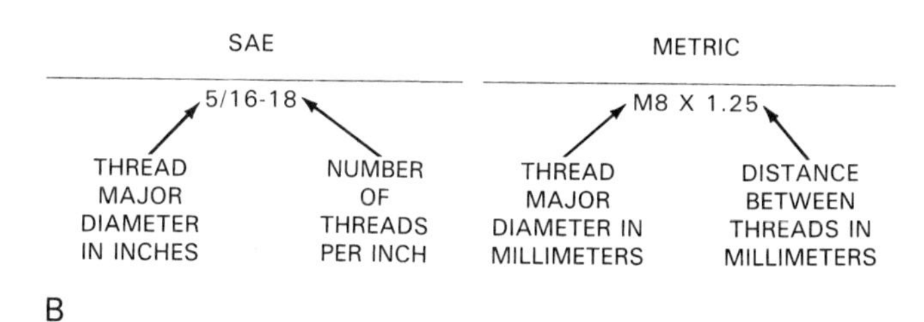

B

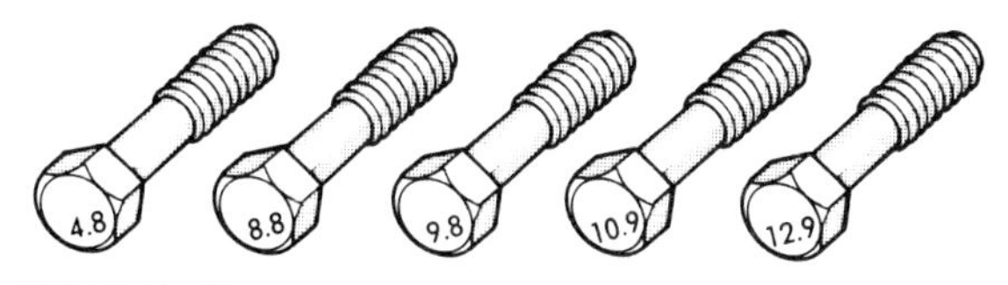

C

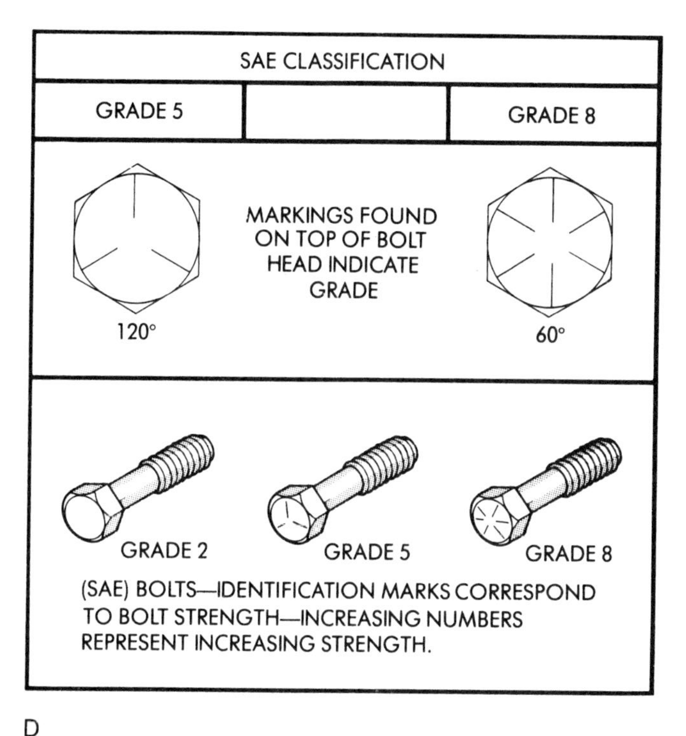

D

Figure 2-45. Torque references should be consulted when torque requirements of a particular bolt is unknown. A–This table shows the maximum torque values for various sizes and grades of threads. B–This reference explains how to interpret SAE (inch) and metric thread notations. C–Strength of metric bolts may be determined by identification class number on head of bolt. D–Strength of SAE bolts may be determined by identification markings embossed on bolt head. (Chrysler)

torque on the bolt, which equates to the amount of preload on the bearing.

Every threaded fastener has a torque value. Some parts of the vehicle, such as the engine and drive train, require very precise torquing procedures. These procedures and the torque values are given in the vehicle service manual. If the specifications are not available, standard torque references, Figure 2-45, must be used to determine the proper torque.

Service Manuals

Although they are seldom thought of as such, ***service manuals*** could qualify as tools. The vehicle service manual provides the specifications and the step-by-step instructions needed to properly service a particular vehicle. Figure 2-46 shows some typical specifications for major drive train parts that are found in service manuals. Also found in service manuals are the location of serial numbers to aid in obtaining parts. See Figure 2-47. The service manual can provide valuable information to the technician, especially when in doubt about something.

Safety

You should always work safely and carefully. A moment of carelessness or inattention can result in a serious injury or even death. Even when no one is hurt, an accident can cause severe damage to vehicles and to shop equipment. Accidents almost always result in lost time, which

Clutch — Section 12

	MEASUREMENT	STANDARD (NEW)	SERVICE LIMIT
Clutch pedal	Pedal height Stroke Pedal play Disengagement height	213 (8.39) to floor 140—150 (5.5—5.9) 15—20 (0.59—0.79) 70 (2.76) min.to floor	— — — —
Clutch release arm	Free play at arm	3.0—4.0 (0.12—0.16)	—
Flywheel	Clutch surface runout	0.05 (0.002) max.	0.15 (0.006)
Clutch disc	Rivet head depth Surface runout Radial play in spline at circumference (200ϕ) Thickness	1.3 (0.05) min. 0.8 (0.03) max. 0.7—2.1 (0.028—0.083) 8.1—8.8 (0.32—0.35)	0.2 (0.008) 1.0 (0.04) 4.0 (0.157) 5.7 (0.224)
Clutch release bearing holder	I.D. Holder-to-guide sleeve clearance	31.00—31.15 (1.220—1.226) 0.050—0.239 (0.002—0.009)	31.2 (1.228) 0.28 (0.011)
Clutch cover	Uneveness of diaphragm spring	0.8 (0.03) max.	1.0 (0.04)

Manual Transmission — Section 13

	MEASUREMENT		STANDARD (NEW)	SERVICE LIMIT
Transmission oil	Capacity ℓ (US.qt., Imp.qt.)		1.8 (1.9, 1.6) at oil change 1.9 (2.0, 1.7) at assembly	
Mainshaft	End play Diameter of needle bearing contact area Diameter of third gear contact area Diameter of 4th, 5th gear contact area Diameter of ball bearing contact area Runout		0.11—0.18 (0.004—0.007) 25.977—25.990 (1.0227—1.0232) 33.984—34.000 (1.3380—1.3386) 26.980—26.993 (1.0622—1.0627) 21.987—22.000 (0.8656—0.8661) 0.02 (0.0008) max.	Adjust with shim 25.92 (1.020) 33.93 (1.336) 26.93 (1.060) 21.93 (0.863) 0.05 (0.002)
Mainshaft thrid and fourth gears	I.D. End play Thickness 	 3rd 4th 3rd 4th	39.009—39.025 (1.5358—1.5364) 0.06—0.21 (0.0024—0.0083) 0.06—0.19 (0.0024—0.0075) 30.22—30.27 (1.1898—1.1917) 30.12—30.17 (1.1858—1.1878)	39.07 (1.538) 0.33 (0.013) 0.31 (0.012) 30.15 (1.187) 30.05 (1.183)
Mainshaft fifth gear	I.D. End play Thickness		37.009—37.025 (1.4570—1.4577) 0.06—0.19 (0.0024—0.0075) 28.42—28.47 (1.1189—1.1209)	37.07 (1.459) 0.31 (0.012) 28.35 (1.116)
Countershaft	End play Diameter of needle bearing contact area Diameter of ball bearing contact area Diameter of low gear contact area Runout		0.17—0.38 (0.0067—0.0150) 30.000—30.015 (1.1811—1.817) 24.980—24.993 (0.9835—0.9840) 35.984—36.000 (1.4167—1.4173) 0.02 (0.0008) max.	0.53 (0.021) 29.95 (1.179) 24.93 (0.981) 35.93 (1.415) 0.05 (0.002)
Countershaft low gear	I.D. End play Thickness		41.009—41.025 (1.6145—1.6152) 0.03—0.10 (0.0012—0.0039) 29.41—29.44 (1.1579—1.1591)	41.07 (1.617) 0.22 (0.009) 29.36 (1.156)
Countershaft second gear	I.D. End play Thickness		44.009—44.025 (1.7326—1.7333) 0.03—0.11 (0.0012—0.0043) 29.92—29.97 (1.1780—1.1799)	44.07 (1.735) 0.23 (0.009) 29.85 (1.175)

Figure 2-46. Note typical drive train specs that can be found in service manuals. These specifications are very important when repairing drive train components.

means lost money. Sometimes, an injury can cause someone to be out of work for many months. In some cases, the injured technician may not be able to resume his or her job. The automotive technician can never be too safe or know too many safety rules. Safety may be thought of as a tool–one that you must use whenever you work on a vehicle. Throughout this course in Auto Drive Trains Technology and throughout your career, make safe practices your most-used and most-valuable tool. Your life depends on it!

Drive Train Safety

Injuries that occur when working on drive trains commonly result from dropping heavy objects and from contacting moving drive train parts. Improper use of transmission jacks or improper securing of components can cause a heavy component to fall, crushing a body part. Trying to lift heavy parts can cause strains or more serious injuries. A moving drive shaft or front drive axle can strike you, or it can grab at loose clothing, pulling you into the

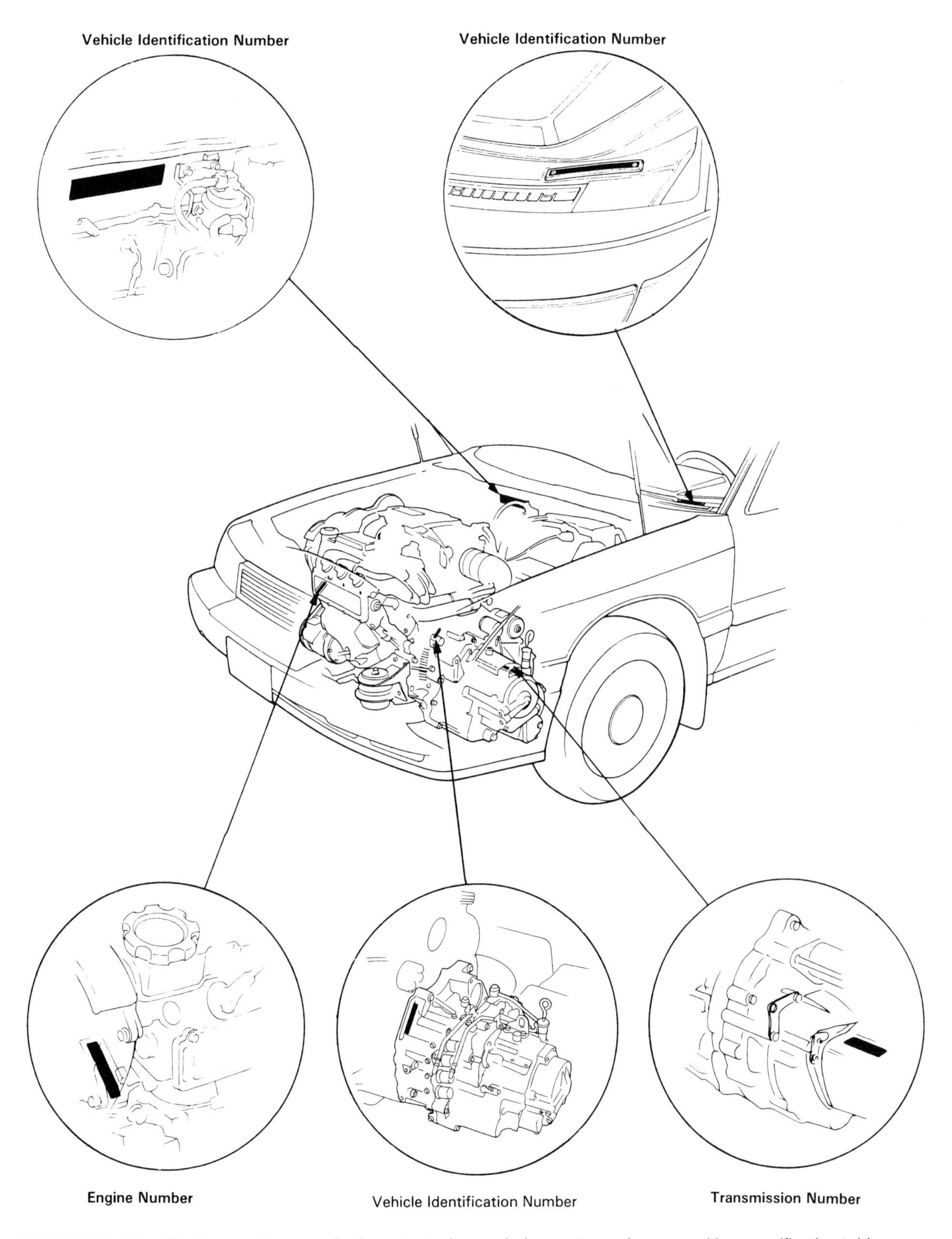

Figure 2-47. Vehicle identification numbers can be important when ordering parts or when consulting specification tables.

moving part. In addition, many other accidents can happen: vehicles falling from hoists or jackstands; gasoline or oil fires; burns from hot engine oil or coolant, and other incidents.

You should always observe safety rules when working around drive trains. Some general drive train safety rules are:

- *Always* wear eye protection when working around moving drive train parts and when using cleaning solvents and compressed air. Examples of eye protection are shown in Figure 2-48.
- Do *not* wear loose clothing or jewelry when working around moving drive train parts. Long hair should be tied up or otherwise secured. Loose items or jewelry could get caught in the moving parts. Rings, watches, or other metallic items could cause electrical burns if they come in contact with electrical terminals.
- *Protect* your hands and arms when using cleaning solvents; wear gloves.
- *Avoid* breathing asbestos dust from clutch facings. Remove dust with an approved vacuum collection system and clean clutch housings or components with a liquid cleaner to avoid creating dust.
- *Use caution* when removing or installing heavy components, such as transmissions, transaxles, and rear axle assemblies. Use proper lifting equipment to avoid strains and injury.
- *Always* disconnect the battery before working on the vehicle. This will prevent injury should someone accidentally try to crank the engine. This rule is especially important when working around the vehicle flywheel or ring gear.
- When using compressed air to clean parts or for other purposes, *always* direct the air stream away from yourself and other people. *Never* use compressed air to dust your clothes. Do *not* spin bearings with compressed air.
- *Use extreme caution* when using hammers, pullers, or hydraulic presses. Parts can break or come apart, causing metal fragments to fly.
- Do *not* operate front drive axles with the wheels hanging. High speed operation at high axle angles will cause the CV joints to fly apart.
- *Always* use extreme caution around hot engines, transmissions, exhaust system parts, fluids, and cooling systems.

General Safety

Some of the safety rules just given are relevant not only when working on drive trains but anytime a vehicle is serviced. In addition to the drive train safety rules given above, you should remember and observe all general safety rules. These rules should be adhered to at all times by all parties concerned.

- No smoking! Gasoline vapor can ignite easily and cause a fire. Diesel fuel, motor oil, transmission fluid, brake fluid, and many other substances in the shop are also fire hazards.
- Keep all work areas clean and organized. One of the major causes of accidents and shop fires is allowing old parts, dirt, and oil to accumulate in the shop. A clean shop also makes it easier to locate needed tools and supplies.
- *Be careful* when working with electric power tools. Make sure that the tools are properly grounded. Ensure that you are not standing in water and, in general, that your feet are well-insulated from ground. In addition, make sure that no other part of your body is directly touching (through noninsulated contact) any other grounded medium or object, such as an electrical box.
- Before using pneumatic tools, such as impact wrenches or air chisels, *make sure* that all attachments are securely attached to the tool. Make sure that you are familiar with the tool's operation.
- *Make sure* that the work area is well lighted and ventilated. Always turn on ventilation fans when running any engine in the shop. Use drop lights when working under any vehicle.

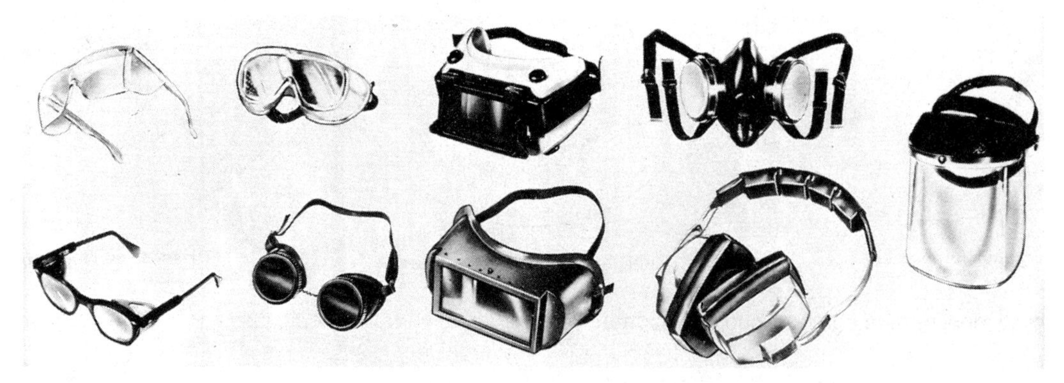

Figure 2-48. Wear approved eye protection when needed, such as when using compressed air or solvents. Shown here are numerous safety articles worn for protection. (Snap-on Tools)

- Work in an organized, professional manner. Haphazard or careless work habits and horseplay in the shop result in accidents and sloppy repair jobs.
- Learn to use vehicle hoists, jacks, and jackstands safely. Most fatal injuries in repair shops are caused by improper use of vehicle lifting and holding equipment. Figure 2-49 shows some typical jack and lift points on one make of vehicle. Figure 2-50 shows the proper use of a *hydraulic jack* and jackstands on some vehicles. This is critical to prevent accidents or death.

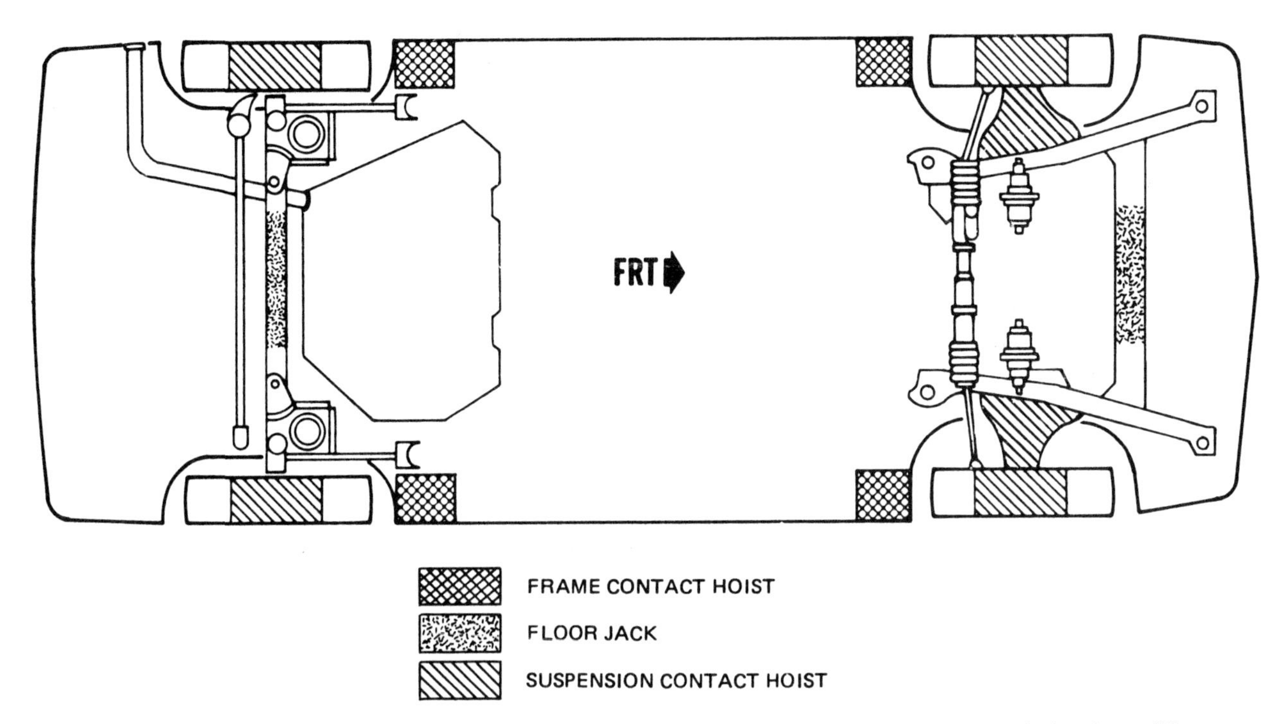

Figure 2-49. Specific locations for placement of frame contact hoist, floor jack, and suspension contact hoist–three different ways to lift the vehicle–are given. (General Motors)

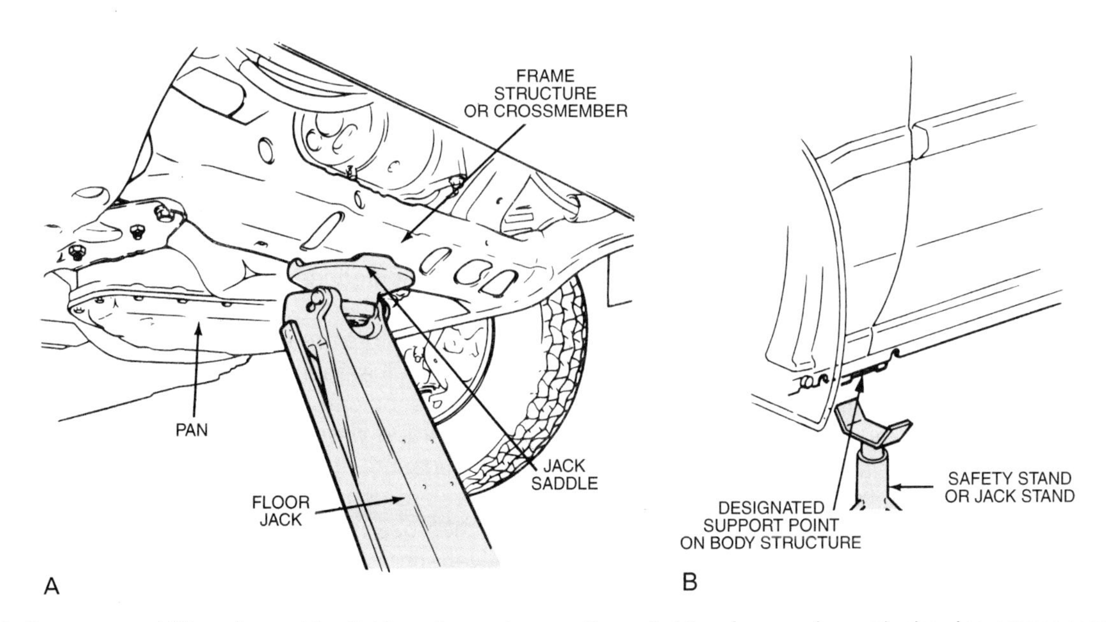

Figure 2-50. Proper use of lift equipment is vital to safe repair operations. A–View from underneath showing proper contact point on this vehicle for floor jack. B–View showing proper placement of jackstands on this vehicle. (General Motors, Honda)

- *Be sure* that you are thoroughly familiar with *any* piece of shop equipment before using it, including understanding its operation *and* proper application.
- Use common sense. *Always* think of the consequences of any action before you perform it.
- Report unsafe conditions and practices to your instructor or supervisor.

Summary

Special service tools are often needed to repair drive train parts. These tools are made specifically for troubleshooting or repairing a particular component or a group of similar components.

There are many special service tools, most of which can be purchased from independent parts suppliers. Other tools must be purchased from special tool manufacturers, listed in vehicle service manuals.

Special wrenches are designed to turn fasteners that have odd head shapes or that are positioned so that they cannot be reached with the more common hand wrenches.

Punches and drifts are special service tools for removing small pins or shafts. They are designed to be used with a hammer. Picks and snap ring pliers are designed to remove snap rings or other small parts.

Hydraulic presses are useful for removing or installing press-fit parts. Various adapters must be used during disassembly and reassembly of bearings to securely hold them in place. Presses should always be used very carefully.

Drivers are used to install bearings or seals in housings. They must be used carefully so that the part is installed correctly. Some drivers are used in combination with a hydraulic press, while others are designed to be struck with a hammer.

Pullers are used to remove pressed together parts, such as bushings, bearings, or seals. These pullers operate by applying pressure through screw threads or by the impact of a weight sliding on a shaft.

Spring compressors are used to compress coil springs, which removes pressure from snap rings so that they can be easily removed. Seal protectors are used to prevent seal damage when installing seals over shafts.

Special tools are required to repair or make tubing connections. Cutting tools cut the tubing in the most efficient manner. Flaring tools are used to flare the tubing ends. Bending tools are used to bend tubing without kinking it.

Alignment tools are used to keep drive train parts in alignment. An example is the clutch disc alignment tool. Flywheel locking and turning tools are used to turn or lock the flywheel, to gain access to fasteners or to lock them thoroughly.

Organizing trays are used to keep small parts in proper order. They are useful when working on drive train components containing many small parts. Drain pans will prevent a potentially dangerous mess by catching oil that would otherwise end up on the shop floor.

Screw extractors are used to remove broken fasteners from parts. Some extractors are designed to grip the fastener from the outside. Others are used internally by drilling a hole in the broken fastener.

Holding fixtures are useful for holding large heavy components, such as transmissions and transaxles. Some holding fixtures can be used on only one kind of transmission, while others are designed to fit many types of transmissions and transaxles.

Transmission jacks make transmission removal and replacement much easier. Some jacks are used under the vehicle, while others are used when the vehicle is raised on a hoist. To prevent injury and accidents, they should always be used if available.

Drive train components collect large amounts of dirt, sludge, and metal particles, and they must be carefully cleaned. Line flushers, converter flushers, and parts cleaners are necessary to thoroughly clean drive train parts. They should be used according to manufacturer directions.

There are many gauges available. Size gauges are used to measure a size or clearance. Angle gauges measure angles. Pressure gauges are used to diagnose automatic transmissions. They measure internal transmission oil pressures. Vacuum gauges measure intake manifold vacuum to diagnose vacuum modulator problems.

Micrometers are able to make very precise size measurements. To accurately use a micrometer, you must proceed very carefully. The micrometer must be carefully tightened on the part. The micrometer must also be carefully read. Micrometer reading varies for inch-based and metric-based micrometers.

Hole and telescoping gauges are used to make very accurate measurements of inside diameters. A depth micrometer is used to measure the depth of openings in machined surfaces. Calipers have movable legs or jaws to gauge the size of openings.

Feeler gauges are made in different thicknesses. They are used to gauge openings, or gaps. They come in flat and wire varieties. Straightedges are flat bars, used to gauge the straightness of parts that they are placed against. Straightedges are often used in combination with feeler gauges.

Dial indicators are used for moving drive train parts to measure small amounts of movement. They are easy to use but must be carefully installed, adjusted, and read.

The torque wrench is used to apply torque to a part and to measure this torque at the same time. Torque wrenches are usually used to tighten fasteners. They are sometimes used to measure bearing preload. There are many varieties of torque wrenches. Always consult factory tightening specifications before tightening a part.

Service manuals must be used to obtain tightening values and sequences, repair procedures, and other specifications. The proper service manual must be used to properly repair a drive train.

Safety is very important. A moment of carelessness can result in an accident that can change your entire life. You can be cut, bruised, burned, or hurt in many ways. Equipment and vehicles can also be damaged. Always follow the safety rules given in this chapter.

Know These Terms

Special service tool; Spanner wrench; Punch; Drift; Snap ring plier; Pick; Interference fit; Press fit; Press; Driver; Puller; Spring compressor; Seal protector; Tubing cutter; Tubing flaring tool; Bending tool; Alignment tool; Flywheel turner; Flywheel holder; Organizing tray; Drain pan; Screw extractor; Holding fixture; Transmission jack; Oil cooler and line flusher; Converter flusher; Parts washer; Gauge; Pressure gauge; Gauge pressure; Vacuum; Vacuum gauge; Size gauge; Micrometer; Outside micrometer; Hole gauge; Telescoping gauge; Inside micrometer; Depth micrometer; Caliper; Feeler gauge; Straightedge; Dial indicator; Angle gauge; Torque wrench; Preload; Service manual.

Review Questions–Chapter 2

Please do not write in this text. Place your answers on a separate sheet of paper.

1. Special service tools are designed to be used on many parts for many purposes. True or false?
2. _____ _____ are special tools used to tighten large parts or to hold a part while another tool is used.
3. The shape of a driver for a particular job is critical. True or false?
4. A press fit is an example of:
   a. A clearance fit.
   b. A shrink fit.
   c. An interference fit.
   d. A running fit.
5. List three safety rules to follow when using pullers.
6. How do you use a tubing flaring tool?
7. Why might you use a flywheel turning tool?
8. With today's small transaxles, transmission jacks are seldom used. True or false?
9. How would you clean out the inside of a torque converter?
10. List 10 points of safety to be aware of and observe when servicing drive trains. Cite reasons for observing these rules.

Certification-Type Questions–Chapter 2

1. All of these statements about spanner wrenches are true EXCEPT:
- **(A)** they are sometimes used to tighten large fasteners.
- **(B)** they are sometimes used to hold parts as they are tightened by another wrench.
- **(C)** they are designed to fit standard 3/8-in. or 1/2-in. ratchets.
- **(D)** some are bolted to the part that they hold.

2. Technician A says that a punch should be placed against the part to be removed before striking the punch with a hammer. Technician B says that both punches and chisels can be struck with a hammer. Who is right?
- **(A)** A only
- **(B)** B only
- **(C)** Both A & B
- **(D)** Neither A nor B

3. Technician A says that snap ring pliers can be used for both removing and installing snap rings. Technician B says that picks should never be used to remove snap rings or retainers. Who is right?
- **(A)** A only
- **(B)** B only
- **(C)** Both A & B
- **(D)** Neither A nor B

4. All of these can be installed by the use of drivers EXCEPT:
- **(A)** seals.
- **(B)** O-rings.
- **(C)** bearings.
- **(D)** bushings.

5. Technician A says that some pullers remove parts by the use of screw force or impact. Technician B says that some pullers remove parts by the use of hydraulic pressure. Who is right?

(A) A only
(B) B only
(C) Both A & B
(D) Neither A nor B

6. In drive train service, the spring compressor is usually used in automatic transmission service to remove:

(A) diaphragm springs.
(B) clutch release springs.
(C) leaf springs.
(D) torsion springs.

7. Technician A says that special tools should be used to cut, flare, and bend tubing. Technician B says that using a tubing bender will cause the tubing to kink. Who is right?

(A) A only
(B) B only
(C) Both A & B
(D) Neither A nor B

8. Technician A says that flywheel turners are bolted to the engine or another stationary part of the drive train. Technician B says that flywheel turners engage the teeth on the flywheel ring gear. Who is right?

(A) A only
(B) B only
(C) Both A & B
(D) Neither A nor B

9. Cold-soak tanks and hot tanks are types of:

(A) parts washers.
(B) converter flushers.
(C) oil cooler flushers.
(D) drain pans.

10. A vacuum gauge is a type of pressure gauge that is most commonly used to measure:

(A) automatic transmission pressures.
(B) intake manifold vacuum.
(C) bolt torque.
(D) coil spring release tension.

11. All of these are types of size gauges EXCEPT:

(A) micrometers.
(B) calipers.
(C) torque wrenches.
(D) dial indicators.

12. A flywheel is being checked for excessive warping. Technician A says that a feeler gauge and straightedge can be used to make this check. Technician B says that a dial indicator can be used to make this check. Who is right?

(A) A only
(B) B only
(C) Both A & B
(D) Neither A nor B

13. Technician A says that some threaded fasteners do not have torque values. Technician B says that an experienced technician does not need to use a torque wrench. Who is right?

(A) A only
(B) B only
(C) Both A & B
(D) Neither A nor B

14. Technician A says that loose clothing should not be worn when working around moving drive train parts. Technician B says that new shop equipment can best be understood by trying to use it before reading the instruction manual. Who is right?

(A) A only
(B) B only
(C) Both A & B
(D) Neither A nor B

Chapter 3

Gears, Chains, and Bearings

After studying this chapter, you will be able to:

- Identify certain basic parts of a gear.
- Discuss gear reduction and overdrive.
- Name and describe types of gears found on drive trains.
- Calculate gear ratio.
- Describe the construction and explain the operation of a planetary gearset.
- Summarize basic methods of gear removal and inspection.
- Discuss common gear problems.
- Describe the construction and explain the operation of a chain drive.
- Summarize general methods of chain drive lubrication and inspection.
- Discuss aspects of the different types of bearings.
- Summarize fundamental methods for servicing bearings.

All vehicle drive trains, no matter how complex, are made up of simple parts. Parts such as *gears, chains,* and *bearings* are the building blocks of the drive train. They can be arranged in many ways to create durable and efficient devices that can transmit and modify engine power as needed.

Engine power is transmitted by the action of gears. Engine torque is multiplied and rpm is reduced through gears. Vehicle direction is reversed through gears. Increasingly, these actions are accomplished by chains and sprockets. Simple gears and chains are arranged into some complex drive train mechanisms, but their basic design does not vary.

All shafts and other rotating parts of the drive train revolve on bearings. Without bearings, the moving parts of the drive train would encounter too much friction, resulting in wasted engine power and premature *wearing* of parts. Drive trains use many kinds of bearings, but they all serve the same purpose: to reduce friction and improve efficiency.

In this chapter, you will learn how gears, chains, and bearings function. You must understand how they operate before you begin to study more complex drive train parts. In later chapters, you will see how these simple parts are combined in transmissions, transaxles, and rear axle assemblies.

The inspection and service procedures covered in this chapter should always be performed when drive train components such as transmissions or differentials are overhauled. Learning to perform these tasks *now* will allow you to concentrate on overall service procedures as they are presented in later chapters.

Gear Drives

A simple ***gear*** is a toothed wheel. In operation, gears engage other gears or mechanical parts for the purpose of transmitting power. Such a system is referred to as a ***gear drive.*** The gear that transmits the power is called the *drive gear;* the one that receives the power is called the *driven gear.* Gears provide *positive* (non-slipping) power transmission. Gears that are engaged are said to be *in mesh.* Gears can be used to increase or decrease turning speeds. They are often used to increase torque. Another use is maintaining proper *timing* between two rotating parts, such as an engine crankshaft and camshaft.

A gear that is used in a drive train is usually made of a high-strength steel. In its manufacture, a gear *blank*–a steel disk without any features–is produced from relatively soft steel. The gear teeth are cut on a special machine. The most common type of gear-making machine is a *gear shaper.* Once the teeth are cut, the parts of the gear that will contact other moving parts are ground and polished to a smooth finish. The finished gear is hardened by a heat-treating process. The process makes a durable gear and allows for a precision fit.

Gears that are used in light service are often made of brass, aluminum, fiber material, or plastic. These materials are cheaper, and they permit quieter operation. Some examples are speedometer drive gears, engine distributor gears, and the internal gears found in many electric motors.

Types of Gears

There are a number of different types of gears available. Many of these are commonly used on automotive drive trains. Types of gears include *spur gears, helical gears, herringbone gears,* and *bevel gears.* Additional types include *hypoid gears, worm gears, rack and pinion gears,* and *planetary gearsets.* Following is a description of each.

Spur gears

Spur gears are a type of gear used to connect parallel shafts. The teeth on spur gears are always cut straight across, or *axially.* In other words, they run parallel to the axis of rotation. Spur gears are found most often in manual transmissions where gears must be manually moved in and out of engagement. The straight cut of the teeth allows a smooth engagement and disengagement. Spur gears are inexpensive to manufacture. A disadvantage to these gears is that they become noisy if the clearance between mating gears becomes too large. Typical spur gears are shown in Figure 3-1.

Helical gears

Helical gears are similar to spur gears. They, too, are used to connect parallel shafts. However, the teeth of helical gears are cut at an angle, called a *helix angle,* across the gear surface. A helical gear is shown in Figure 3-2. Helical gears operate more smoothly and quietly than spur gears. The engagement of individual teeth is a gradual sliding motion, and several teeth are engaged at once. Helical gears are often used for the forward gears of manual transmissions, where high gear speeds make smooth, quiet operation all the more important. The sliding action of the teeth causes more friction than encountered by spur gear teeth. As a result, helical gears must be kept well lubricated.

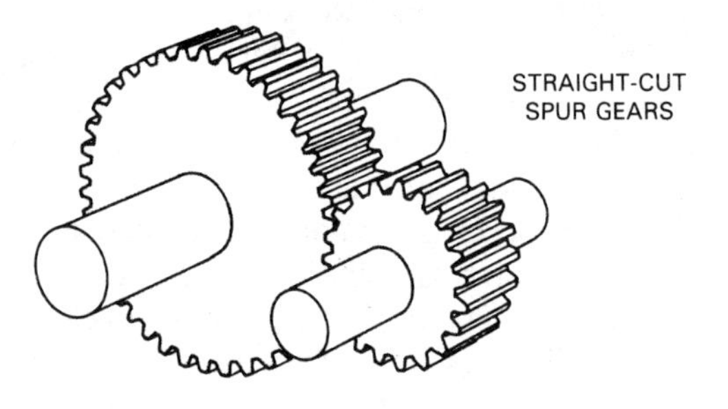

Figure 3-1. The simplest gear found in automotive drive trains is the spur gear. Spur gear teeth are cut axially, or parallel to the gear axis of rotation. These gears are strong, but they may become noisy if the clearance between the teeth of the mating gears becomes too great. (Deere & Co.)

A thrusting action occurs along the shaft whenever power is being transmitted through a helical gear. This is due to the helix angle. See Figure 3-3. This action is called ***end thrust***. Two meshing helical gears will try to push each other in opposite directions. If end thrust is too large, the gears will disengage. *Thrust washers* or *thrust bearings* are often used to control end thrust.

Herringbone and double-helical gears

The ***herringbone gear*** is a variation of a helical gear. It is sometimes used to overcome the end thrust problems of helical gears. A typical herringbone gear is shown in Figure 3-4. The thrust forces of the two sides of the

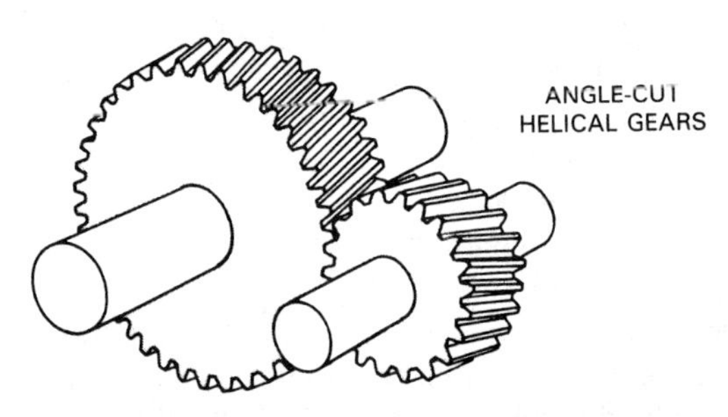

Figure 3-2. Helical gear teeth are cut at an angle. Helical gears are used in many places, including transmissions and final drives. Helical gears provide smoother and quieter power transmission than do spur gears because the meshing of teeth is more gradual. (Deere & Co.)

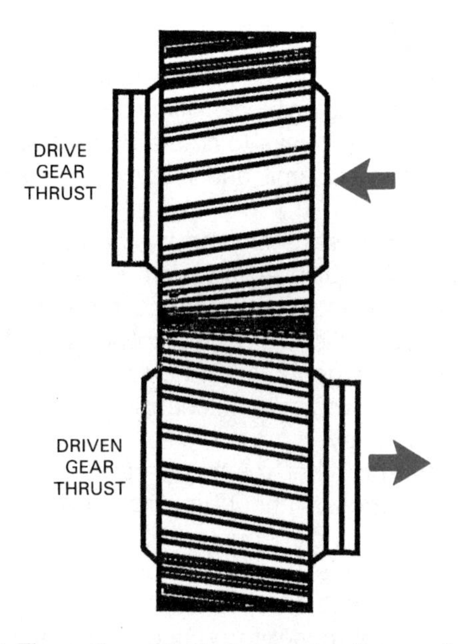

Figure 3-3. The action of helical gears tends to push the gears away from each other. This puts pressure on the gears and on the shaft on which they ride. The pressure must be controlled by other parts in the drive train, or gear movement will be excessive.

herringbone gear cancel each other. The herringbone gear operates more like a spur gear.

A variation of the herringbone gear is called a ***double-helical gear.*** This gear differs in that the opposing teeth have a space between them. Herringbone gears and double-helical gears are expensive to machine, and they do not work well at high speeds. For these reasons, they are rare in automotive drive trains.

Bevel gears

When axes of rotation are not parallel, ***bevel gears,*** Figure 3-5, are used. Bevel gears connect shafts that run at angles. In vehicle drive trains, the angle is usually 90°, but bevel gears can be used to achieve any angle. The bevel gear resembles a *truncated cone,* or a cone with the top cut off. Bevel gears are used, for example, in the differential assembly.

Spiral bevel gears

A variation of bevel gear is the ***spiral bevel gear,*** shown in Figure 3-6. Whereas bevel gear teeth are straight, spiral bevel gear teeth are curved. The spiral bevel design results in smoother power transfer, since the teeth engage gradually, and since parts of more than one tooth are always engaged. The sliding action of spiral bevel teeth means that lubrication is even more critical than for straight bevel teeth.

Hypoid gears

Hypoid gears are shown in Figure 3-7. These gears, like bevel gears, are used in applications where power flow must be diverted by some angle. They are similar to spiral bevel gears. In the case of bevel gears, however, shaft *axial centerlines* (rotational axes) intersect. With hypoid gears, this is not the case. The one gear is set at a lower or higher point in relation to the other, and the centerlines

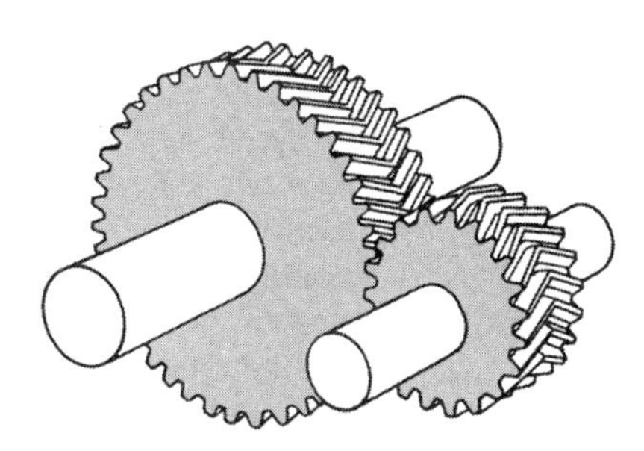

Figure 3-4. Herringbone gears consist of two mating helical gears, each angled in opposite directions. The design of this gear eliminates some of the problems experienced by other gears–end thrust in helical gears, for example. However, these gears are expensive, and they do not work well at high speeds. They are seldom used on vehicle drive trains. (Deere & Co.)

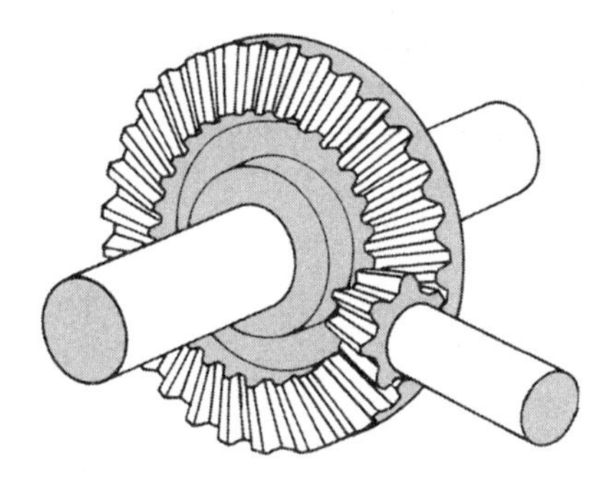

Figure 3-5. Bevel gears can be used to change the course of power flow by some angle. The angle is usually 90° in vehicle drive trains. (Deere & Co.)

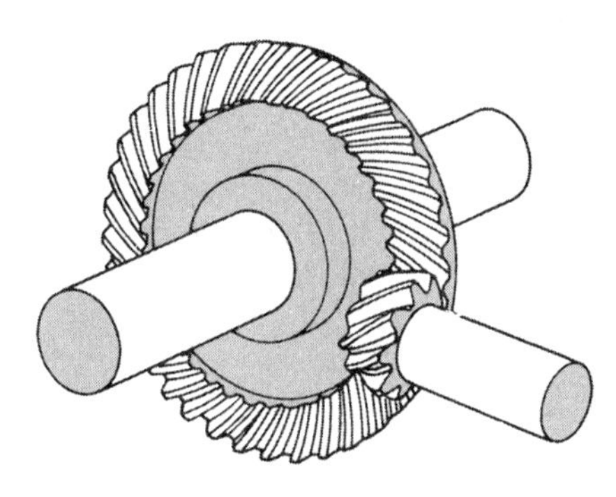

Figure 3-6. A spiral bevel gear has curved, in addition to beveled, gear teeth to allow a gradual meshing of teeth. This gear combines the angle change feature of the bevel gear with the smooth operation feature of the helical gear. (Deere & Co.)

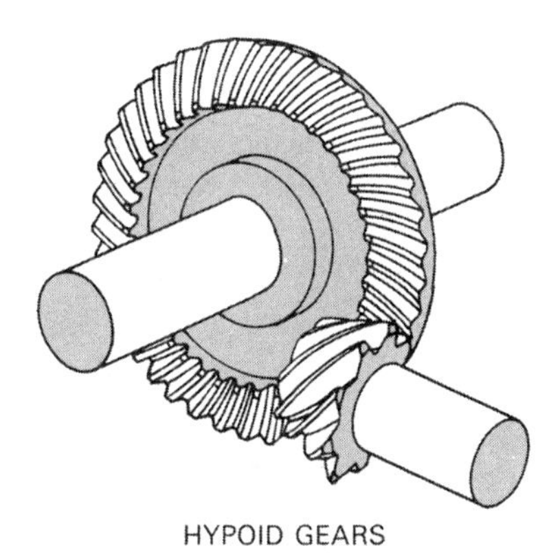

Figure 3-7. The hypoid gear is a variation of the bevel gear. The hypoid system is used when the gear shaft axes are at different levels. Hypoid gears are used extensively in rear-wheel drive vehicles. The drive gear can be placed lower in the rear axle housing to allow the drive shaft hump to be smaller. (Deere & Co.)

of the two do not meet. The most common use of hypoid gears is in the rear axle assembly. The ring and pinion of the differential assembly are hypoid gears.

Worm gears

Worm gears, Figure 3-8, also connect shafts that run at an angle. When worm gears are used, the angle is always 90°. A worm gearset has two major components–the *worm* and the *worm wheel.* Worm gears are often used when a large speed reduction is needed. Except for the speedometer drive, they are seldom used on drive trains. Outside of the drive train, they may be found on power takeoffs, used to operate vehicle-mounted winches and other equipment.

Rack and pinion gears

Rack and pinion gears, Figure 3-9, are usually associated with the vehicle steering system. The *rack* is a flat bar with teeth cut into it. The *pinion* is a small gear. The rack and pinion converts rotary motion into *linear* (straight line) motion when the pinion drives the rack. If the rack drives the pinion, linear motion is converted into rotary motion.

Planetary gearsets

Planetary gearsets are gear assemblies used in automatic transmissions and in manual transmissions with *overdrive units.* They provide *gear reduction* and *direct drive.* They also provide overdrive.

A planetary gearset, as shown in Figure 3-10, consists of three types of gears: a ***sun gear*** in the center; ***planet gears,*** or ***planet pinions,*** surrounding and meshed with the sun gear; a ***ring,*** or ***annulus, gear*** meshed with the planet gears. The planet gears are attached to a ***planet,*** or ***planet-pinion, carrier,*** yet they are free to rotate.

The unique feature of a planetary gearset is that all gears are in constant mesh. In driving power through the gearset, one of the three gear types–sun, ring, or planet–will be held stationary; the other two will rotate, serving as input and output. Different combinations of input, output, and stationary are used to achieve different effects–for example, an increase or decrease of speed. Another example is neutral; a system will be in neutral when all three gears are free.

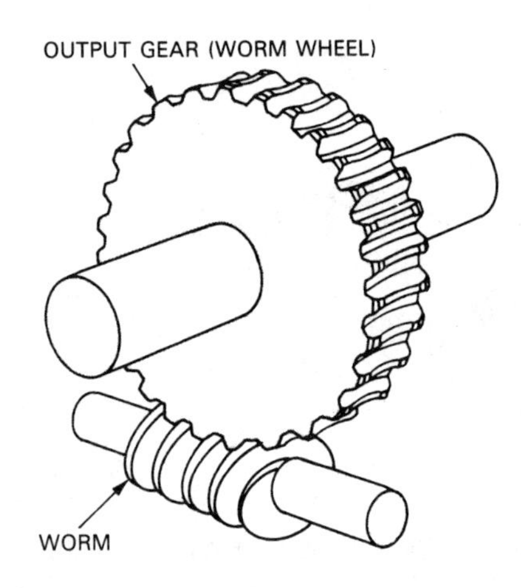

Figure 3-8. The worm gear is used when a large speed reduction is needed and the gear shafts do not have to be on the same plane. These gears are seldom used on most drive trains, except for the speedometer drive. (Deere & Co.)

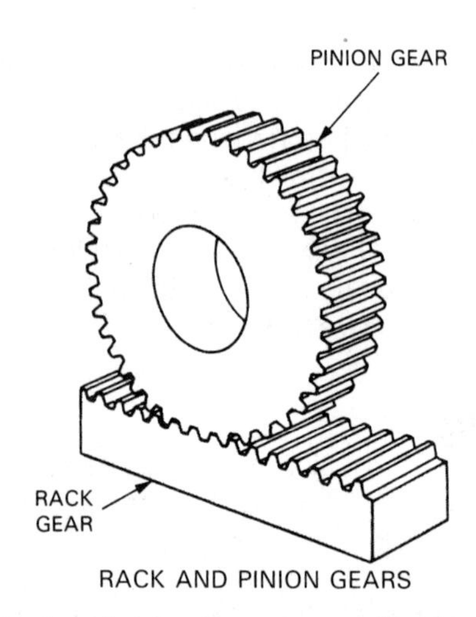

Figure 3-9. Rack and pinion gears are used in steering systems. The pinion gear is actually a spur gear. The spur gear teeth mate with matching teeth on the rack. (Deere & Co.)

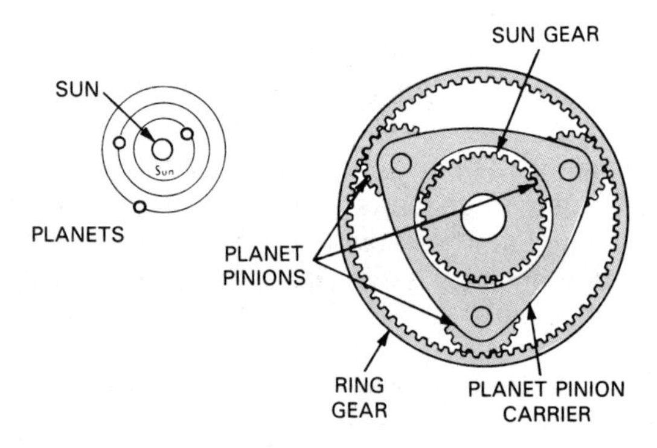

Figure 3-10. A planetary gearset somewhat resembles our solar system. Planetary gears are in constant mesh. Gears rotate or are held stationary in different combinations to achieve different gear ratios. A single gearset can be used to provide reduction, direct drive, overdrive, and reverse. (Deere & Co.)

Gear Nomenclature

There are terms to describe almost every possible dimension of a gear. Design drafters and machinists involved in gear manufacturing are very familiar with these terms. Automotive technicians do not need to know every term. This is because servicing drive train gears usually means replacing the *damaged* gear with an exact duplicate of the original. However, to aid in obtaining the proper replacement gear, the technician should know a few common gear terms. Some of these are given in Figure 3-11.

The contact surfaces of gear teeth are called ***faces.*** The width of the tooth measured parallel to the gear axis is the ***face width.*** The base of the gear tooth is called the ***root.*** The distance from the *outside diameter* of the gear to the *root diameter* is the ***whole depth.*** Stated another way, the whole depth is the distance from the top of the tooth to the root. The face width and whole depth should be measured and the number of teeth should be counted when obtaining replacement gears.

An imaginary circle around the gear located at about midway between the outside diameter and the root diameter is the ***pitch circle.*** Pitch circles of two mating gears are tangent to one another. ***Circular thickness*** is the thickness of a tooth measured between its faces and along the pitch circle. The ***working depth*** is the overlap measured between outside diameters of the two gears. Finally, there must always be some space between the root of one tooth and the top of the mating tooth. This space is referred to as ***clearance.***

Gear backlash

Gear backlash is a term the technician will commonly use. Backlash, in gears, is the amount by which the space between neighboring gear teeth exceeds circular thickness of a mating gear tooth. The difference is figured at the pitch circles. In general, the greater the amount is, the more play there will be between two meshing gears. The degree of backlash must be correct for quiet gear operation and long gear life. On some gear trains, the backlash can be corrected by adjusting the positions of the gears. On others, the gears must be replaced to obtain the proper backlash. Some gears, by design, have less backlash problems than others. Herringbone gears, for example, are better in this respect than spur gears.

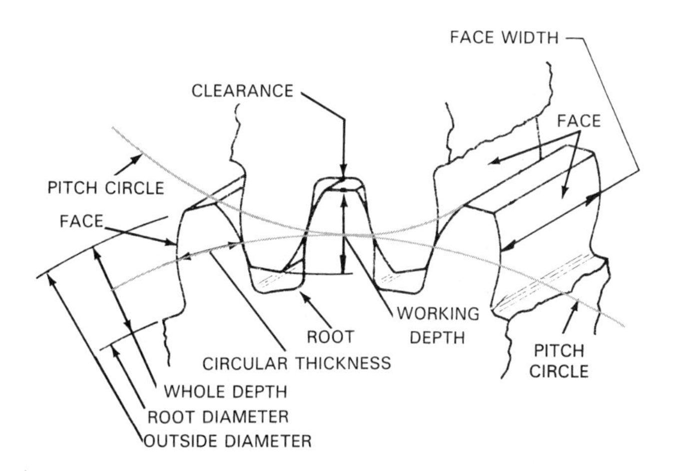

Figure 3-11. Basic gear terms should be learned to help understand gear operation. Gear faces are the contact surfaces of gear teeth. The faces carry the load of the gears. Depth and width of the teeth are critical for proper gear operation. Backlash is the clearance at the pitch circle between faces of two mating teeth.

Gear Ratio

You are probably familiar with the term ***gear ratio.*** It is the speed relationship between two gears. The gear ratio shows the difference in the number of teeth between two gears. Mathematically, it is the ratio of the number of teeth on the driven gear to the number of teeth on the drive gear. This is the same as saying it is the number of teeth on the driven gear *divided by* the number of teeth on the drive gear.

Figure 3-12 is an example of gear ratio. In the illustration, the driven gear has 90 teeth. The gear that drives it, or the drive gear, has 30 teeth. The ratio between the two gears is found to be: 90/30 (90 divided by 30), which equals 3/1, which equals 3:1 (read as "3 to 1").

The gear ratio relates the speed of the drive gear to the speed of the driven gear. A 3:1 gear ratio is an example of what is called ***gear reduction.*** A 3:1 gear reduction means that it takes three revolutions of a drive gear to turn the driven gear through one whole revolution. Refer to Figure 3-12A. A set of gears with a ratio of 3:1 is considered a *reduction gearset.* This is because the speed at the output shaft will be less than the speed of the input.

While a reduction gearset reduces speed, it has the opposite effect on torque; the reduction gearset multiplies torque. For example, with a 3:1 reduction gearset, 1 ft.-lb. (1.35 N-m) of torque on the drive gear generates 3 ft.-lb. (4.05 N-m) on the driven gear. Refer to Figure 3-12B.

To sum up these mathematical relationships:

$$\text{gear ratio} = \frac{\text{number of teeth on driven gear}}{\text{number of teeth on drive gear}}$$

$$= \frac{\text{rpm of drive gear}}{\text{rpm of driven gear}}$$

$$= \frac{\text{ft.-lb. on driven gear}}{\text{ft.-lb. on drive gear}}$$

A vehicle requires different gear ratios for different operating conditions. For instance, internal combustion engines develop only minimal torque when engine speed is low. To overcome this problem, reduction gearsets are used in the transmission. A typical reduction in low gear is about 3:1. For example, a 900-rpm engine speed would be reduced at the transmission output to 300 rpm. At the same time speed is being reduced, torque is being multiplied. If

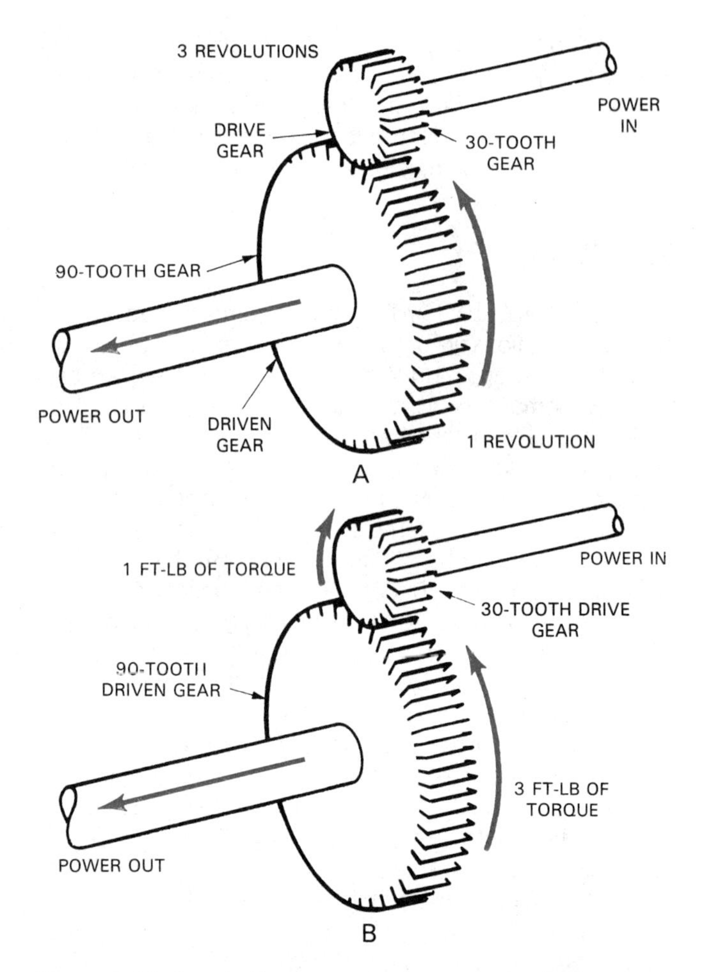

Figure 3-12. A reduction gearset arrangement, which reduces speed while increasing torque, is shown. A–The output shaft completes one revolution for every three of the input shaft. B–A torque of 1 ft.-lb. (1.35 N-m) on the input shaft translates to 3 ft.-lb. (4.05 N-m) on the output shaft.

50 ft.-lb. (67.5 N-m) of torque enters the gears, 150 ft.-lb. (202.5 N-m) leaves. The reduction gearset, then, serves as a torque-multiplying device to get the vehicle moving at low speeds.

As vehicle speed increases, less torque is needed to keep the vehicle moving, due to its *inertia* (resistance to change in state of motion). Also, the engine is turning faster and is producing more power. The vehicle can then be shifted into a gear with less reduction, such as 2:1. As an example, at the transmission output, an engine speed of 1500 rpm will be reduced to 750 rpm, and an engine output of 75 ft.-lb. (101.25 N-m) will be increased to 150 ft.-lb. (202.5 N-m).

When the vehicle is moving at highway speeds, gear reduction is no longer needed or desired. The vehicle is shifted into high gear, which is direct drive. ***Direct drive,*** in which gears are not actually used to transmit power, has a gear ratio of 1:1. It allows the engine to run at reduced speed while maintaining vehicle speed. With this gear ratio, there is no change in either speed or torque between engine output and transmission output.

For greater fuel economy at higher speeds, ***overdrive*** is often employed. Most *overdrive gearsets* have a ratio of around 1:1.5 (sometimes written as 0.66:1). This means that an engine speed of 1000 rpm will exit the transmission at 1500 rpm. With this type of ratio, there is a speed increase through the gearset. This allows the engine to run more slowly than a higher gear ratio would permit at the same speed of travel. Since the engine is running more slowly, it will use less fuel and last longer.

Reverse Gear

Simple gears can be arranged to reverse direction. In the normal action of two mating gears, the drive gear and driven gear rotate in opposite directions. A third gear will restore the original direction of rotation, which has been reversed by the second gear. A manual transmission, for instance, employs a ***countershaft gear assembly*** between input and output shafts. The assembly consists of a shaft and several gears, which rotate as a unit. See Figure 3-13.

To change directions, a vehicle has a *reverse gear,* which is part of the transmission. Figure 3-13 shows reverse gear engaged. Note the use of the ***reverse idler gear,*** or simply, the ***idler gear,*** turning the output shaft in the same direction as the countershaft gear assembly, which is opposite to the direction of the input shaft. The reverse idler gear does not change the gear ratio.

Gear Removal

It helps to know a little bit about the ways gears are installed before discussing how to remove them. Gears are joined or fit onto their shafts in several different ways. Commonly, gears are a press fit on the shaft. In contrast, many gears slip over the shafts on which they ride. Such gears are held in position by other parts of the drive train or by snap rings. If the gear must turn with its shaft, it is held by a *key* that fits into matching slots in the gear and shaft. These slots are called *keyways.* (Keys and keyways are covered later in this chapter.) In many cases, a gear and shaft are a solid unit, machined from a single casting. If the gear teeth are damaged, the entire part must be replaced.

There are a couple of ways to go about removing gears from their shafts. The method depends upon the way the gear is held onto its shaft. Gears that are held in place by any kind of retainer will usually slide off the shaft once the retainer is removed. Removing press-fit gears requires the use of a gear puller or a press–either mechanical or hydraulic. The gear is pulled or pressed off. In some cases, the gear can be driven off, but this must be carefully done to prevent damage or injury. A typical gear removal operation using a press and adapter is shown in Figure 3-14.

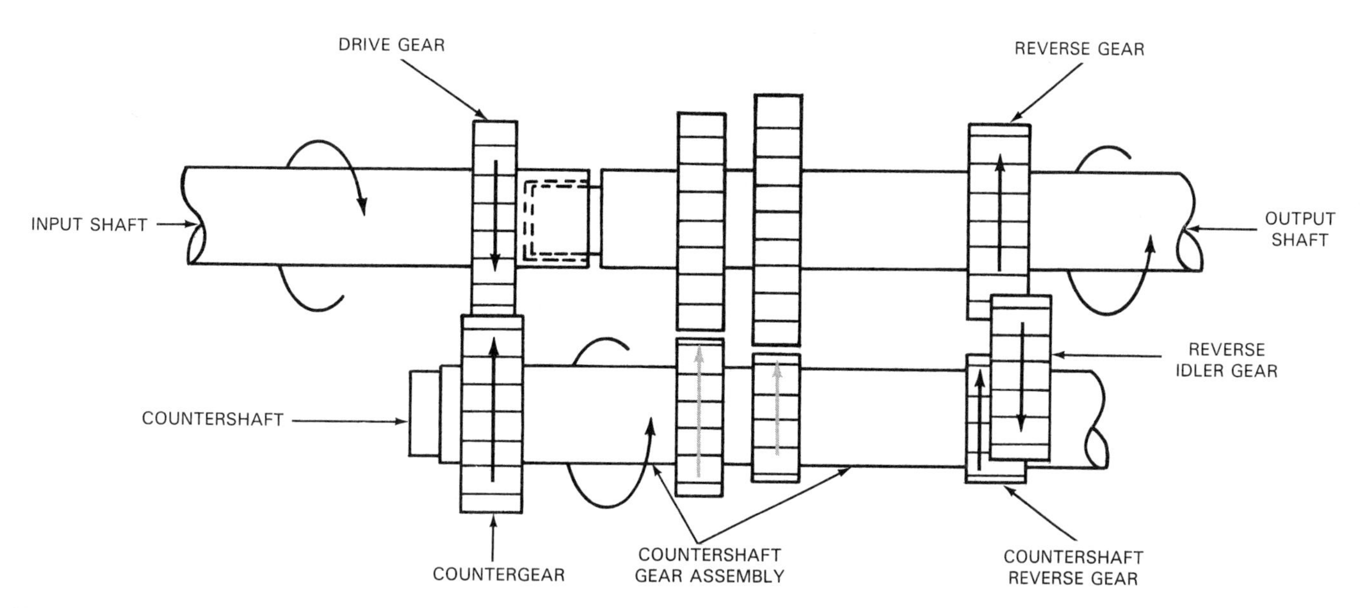

Figure 3-13. Simplified drawing of a transmission with reverse gears engaged. Countergear causes countershaft gear assembly to rotate opposite to direction of input shaft. Countershaft reverse gear meshes with reverse idler gear, which meshes with reverse gear of output shaft. Reverse gears rotate in the same direction because of the reverse idler gear. Output shaft rotation is opposite of input shaft.

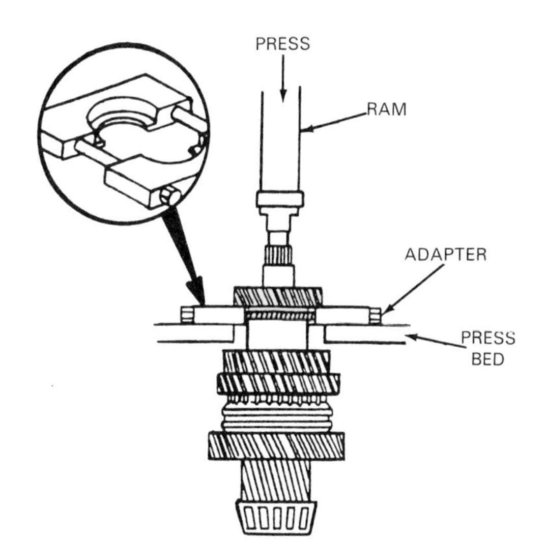

Figure 3-14. A press is shown in the process of removing a gear from a manual transaxle output shaft. The press pushes down on the shaft. The gear is restrained by the adapter plate, and as a result, it comes off of the shaft. Other gears remain in place. (General Motors)

Gear Inspection

Gears should be inspected whenever there is evidence of gear problems or when they are removed from the vehicle for any reason. Some of the problems to look for are covered in these next few paragraphs.

Wear is erosion of a surface caused by relative motion between two parts. Gears are subject to wear, and, given time, a gear will wear out from use. In most automotive drive trains, the various gears are made to be durable. These gears will operate satisfactorily for many miles. Some gears, such as those of the rear axle assembly, will typically last for the life of the vehicle.

Gear wear can be responsible for gear ***failure***–the total breakdown in structure so that it can no longer fulfill its purpose. A gear with normal wear from high mileage will appear worn but in good condition. A gear removed from a well-maintained vehicle with very high mileage may show considerable wear. In either case, the faces will show an even wear pattern and will appear smooth and shiny. They should be light gray in color.

At some point, gear wear may become so extreme that the teeth are too thin to support the loads that are placed on them. Such gears usually develop excessive backlash, which increases the shock loads on the teeth when they are first engaged. When this happens, a gear tooth may develop a crack and break. The broken tooth falls to the bottom of the gear housing; however, the gear may operate for a while longer. This increases the shock loads on the remaining teeth, causing other teeth to break. Sooner or later, the gear will fail completely.

Early gear failure is usually caused by a drive train defect, lack of lubrication, or abuse. The most common cause of gear tooth failure on a low mileage vehicle is abuse of the drive train. Placing heavy loads on a vehicle not designed to carry them can overstress the teeth and cause gear failure. Loading the gears too rapidly will cause shock loads, which can overload the teeth and break the gear. On a manual transmission vehicle, this may happen when the clutch is engaged abruptly with the engine running at high speed. On an automatic transmission vehicle, placing the car in gear at high engine speeds can break gear teeth.

Figure 3-15 shows some common types of gear ***damage.*** These are conditions short of failure. Left uncorrected, these problems of gear damage will lead to gear failure. A common gear problem is ***pitting.*** Pitting is evidenced by small holes, or pits, in the gear teeth. The pits reduce the gear contact area. If the cause of the problem is not corrected, the pits can become larger and more plentiful. Pitting is usually caused by abrasive particles or corrosion. These can be attributed to dirt or metal debris or water in the lubricant.

Severe pitting can lead to ***spalling,*** a condition in which sections of the gear teeth flake or chip off. Spalling, like pitting, can also be caused by lack of lubricant or severe gear loads.

Another cause of gear failure is ***peening,*** which appears as indentations in the gear faces. Peening is usually caused by severe loads on the gear teeth or by large particles of foreign matter in the lubricant.

Scoring is another cause of gear failure. It appears as scratches on gear teeth faces, caused by improper machining or lack of lubrication. If the scoring is not corrected, it will lead to spalling or a condition called ***welding.*** Welding occurs when mating gear teeth overheat, melt together, and then pull apart as the gear turns. This will quickly destroy the gear.

Sometimes lack of lubrication will cause the gears to overheat. Overheating will undo the heat treating applied in manufacturing to harden the gears. The gears become relatively soft, and they soon wear out. Overheated gear teeth usually turn a different color from the undamaged part of the gear. This color can vary anywhere from yellow to deep blue.

Chain Drives

A ***chain drive*** consists of a series of links, or a ***chain,*** and two or more gears called ***sprockets.*** The chain serves to transmit motion between the two sprockets. The links form a chain, or *flex belt.* Chain drives have a positive driving action, similar to that of two meshing gears. The sprocket teeth match the openings in the chain links and mesh with them to transmit power. One sprocket transmits power. It is called the *drive sprocket.* The other one receives the power and is called the *driven sprocket.*

Chain drives transmit power in a straight line between the two sprockets. This means of power transmission is compact. For this reason, chain drives are often used to connect torque converters to the transmission mechanical system in front-wheel drive vehicles where space is limited. Chain drives are found not only in transaxles, but in four-wheel drive transfer cases. These will be explained in later chapters. Figure 3-16 is an example of a chain drive assembly.

Types of Chains

There are two types of chains used on modern vehicles. The more common is the ***roller chain,*** Figure 3-17. Roller chain is so called because it employs rollers that rotate on *drive pins.* The rollers contact the sprocket teeth. The spaces between the rollers match the size of the teeth. The rollers provide a rolling contact, which reduces friction between the chain and sprocket.

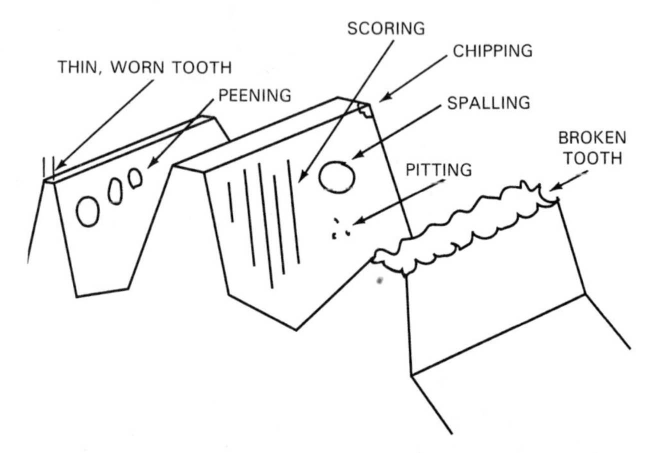

Figure 3-15. Excessive gear wear and some other problems leading to complete gear failure. Note that, in reality, gear teeth are likely to show just one kind of damage.

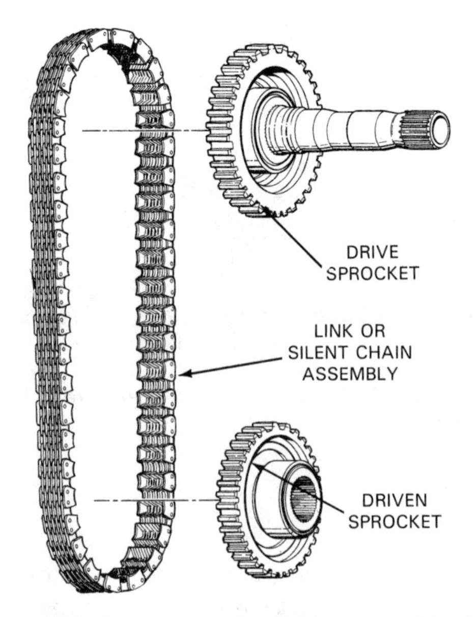

Figure 3-16. This is a typical chain drive assembly. The major parts of the chain drive are the chain and drive sprockets. Every system has at least two sprockets. (Ford)

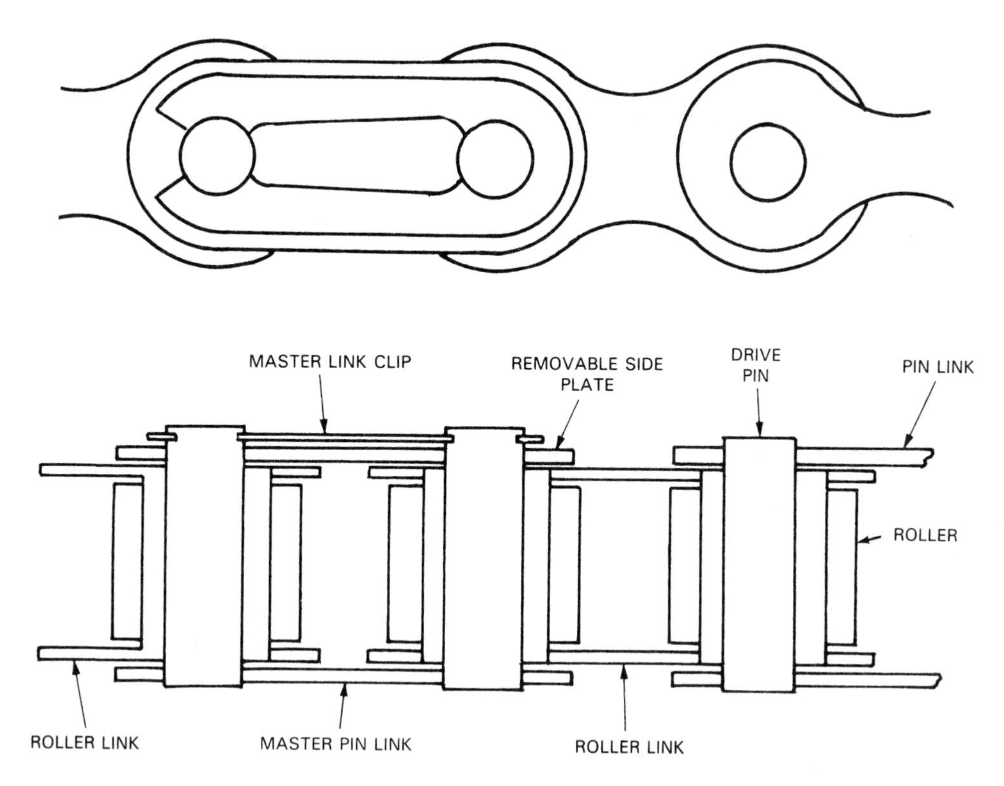

Figure 3-17. Roller chains are made up of pin links and roller links. The rollers mesh with the sprocket teeth, providing positive drive.

Roller chain can be thought of as alternating *pin links* and *roller links.* (Refer to Figure 3-17.) Roller links are coupled together with pin links. An enlarged view of a pin link is shown in Figure 3-18. Many chains have a master pin link, which is held together with a clip. The clip can be removed to separate the chain for easy replacement. Chain links vary in design, but all serve the same function.

The ***silent chain,*** Figure 3-19, is made up of a series of flat metal links, which provide quiet operation. These links have teeth that engage matching teeth in the sprockets. The links are placed side by side. This creates a solid contact for engaging the sprocket. They are held in alignment by pins that pass completely through the chain. This is opposite from the roller chain, wherein the pins (or rollers) engage the sprockets, and the links perform the task of alignment. The construction of the silent chain allows some internal movement of the chain parts. This reduces vibration and noise and partially compensates for wear and sprocket misalignment.

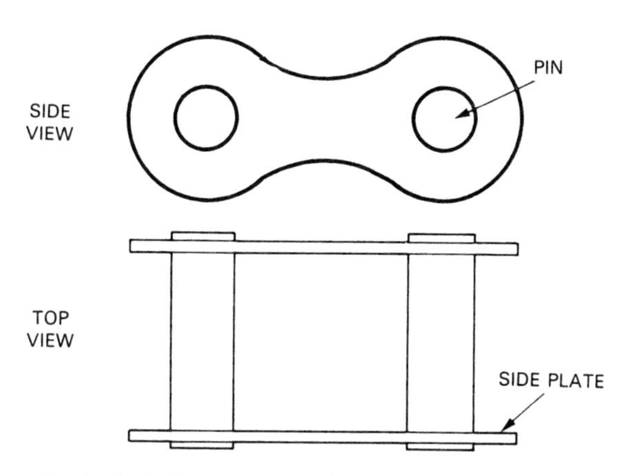

Figure 3-18. A pin link consists of two side plates riveted to two pins.

Chain Tensioners

Chains eventually wear. This can cause the chain to lose its tension. To prevent this, many chain drives have a ***chain tensioner*** to take up the slack. Most chain tensioners are spring-loaded devices similar to that shown in Figure 3-20. They maintain constant pressure on the return, or *nondrive,* side of the chain.

One drawback of a *spring-loaded chain tensioner* is that as the chain wears, the tensioner exerts less tension on it. The spring extends further to take up the slack, and, eventually, the chain is too worn for the tensioner to have any effect. In most cases, the chain will become noisy before it fails completely.

Another type of tensioner is the *hydraulic chain tensioner.* This type uses system oil pressure to keep constant tension on the chain, even as parts wear. It is not as common as the spring-loaded tensioner.

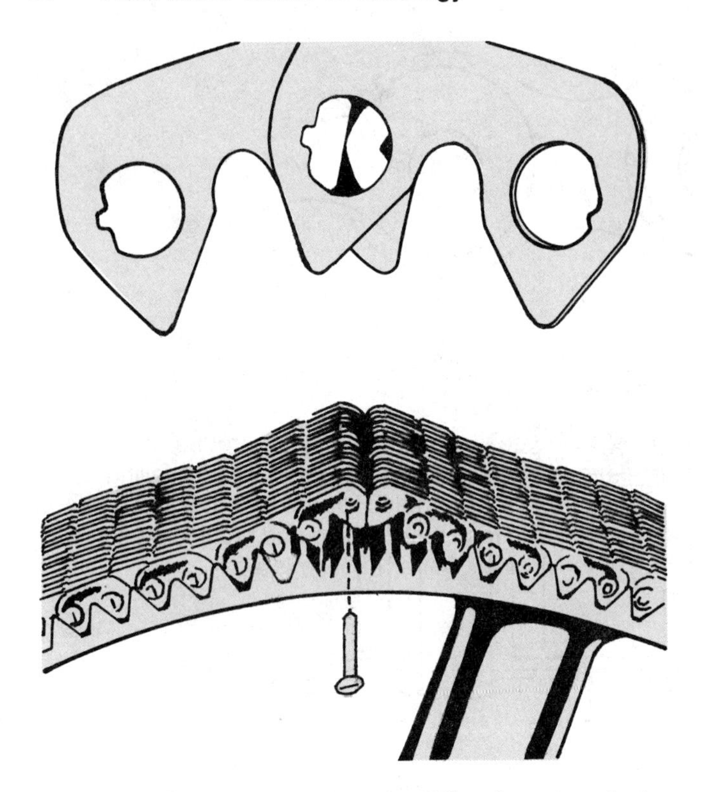

Figure 3-19. Silent chain construction differs from that of other chains. On a silent chain, the links are a series of toothed leaves that extend the entire width of the chain. The links engage the sprocket. They are held in place by pins. The pins keep the links in alignment. This is just the opposite of the roller chain.

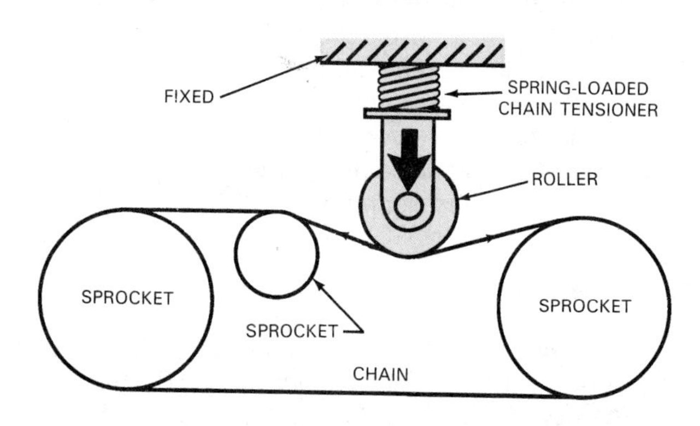

Figure 3-20. Like any mechanical part, chains and sprockets will wear. Wear makes the chain loose. A loose chain will be noisy and may jump off the sprockets. To keep the chain from becoming too loose, a chain tensioner is often used. Most tensioners are spring loaded. A spring is compressed against the chain, and this puts tension on the chain, taking up the slack.

Chain Drive Ratio

To find chain drive ratio, divide the number of teeth on the driven sprocket by the number of teeth on the drive sprocket. Refer to Figure 3-21 for an example. The example finds the ratio for a final drive that employs a chain drive.

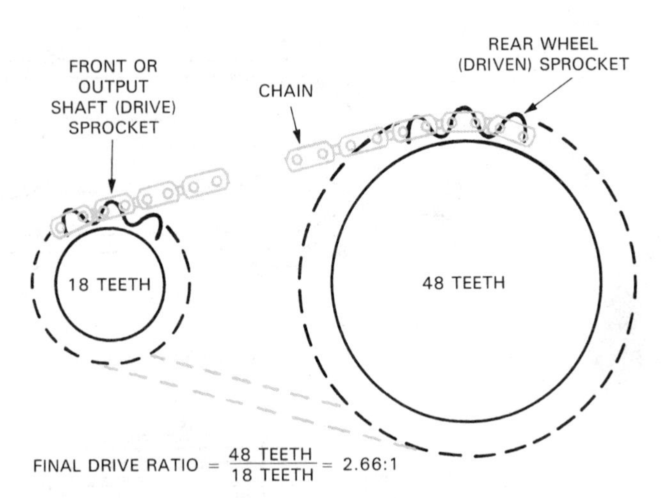

Figure 3-21. Chain drive ratios are figured in the same way that gear ratios are. The number of teeth on the driven sprocket is divided by the number of teeth on the drive sprocket. The resulting figure is the ratio. The length of the chain has no effect on the ratio.

In another example, assume that a driven sprocket has 30 teeth and the drive sprocket has 10 teeth. Dividing 30 by 10 gives 3. The gear ratio in this example is 3:1. This is a reduction ratio because for every 3 revolutions of the input, the output completes only 1 revolution. In other words, the speed of the output is less than the speed of the input.

If the driven sprocket had 10 teeth, and the drive sprocket had 30 teeth, the gear ratio would be 1:3. This would be an overdrive ratio because the speed of the output would be greater than the speed of the input.

If the drive and driven sprockets have an equal number of teeth, the ratio is 1:1. The input and output turn together. There is no multiplication of torque or increase of speed. Note that the size of the chain has no effect on the ratio.

Chain Lubrication

Even the most efficient chains require lubrication. One advantage of chain drives is that they are usually enclosed. This way, they are kept lubricated, as well as protected from dirt and moisture.

There must always be some provision for chain lubrication. The chains used in vehicle drive trains are usually designed so that they pass through a pool of oil, which collects in the bottom of the chain case. In other cases, oil is sprayed on the chain through a calibrated oil hole.

Chain Inspection

A chain drive should be inspected whenever there is evidence of chain problems or when the chain is removed from the vehicle for any reason. Chain and sprocket wear usually shows up as a loose chain. A loose chain is hard on sprocket teeth. If the chain is loose, there is a very good chance that the sprocket is damaged; therefore, always be sure that you check both.

Sometimes a chain will become so loose that it jumps off of the sprockets. Usually the chain will become noisy before this happens. A loose chain that is still in place will have too much slack, so there will be too much play in the chain. This will be more noticeable on the side that is not under tension. If the chain is removed from the sprockets, lay it sideways, as shown in Figure 3-22. Any noticeable sagging means that the chain is worn out and should be replaced.

The sprockets should be checked for excessive wear or cracks. The tooth faces of roller chain sprockets may appear as hooked teeth. In extreme cases, even these hooks will be worn away. Sprocket problems often resemble gear problems. The tooth faces of silent chain sprockets will often show a series of indentations where the chain makes contact.

A worn or damaged sprocket should be replaced. The chain may be reused, in this case, depending on its condition. If a chain needs to be replaced, the sprockets should always be replaced along with it, as a set. Otherwise, the old sprocket may damage the new chain.

A problem encountered in chain drives is misalignment of sprockets. When chains are used, there is little room for misalignment, which is one of the disadvantages of chain drives. Misalignment will cause excessive chain and sprocket wear and noise. In extreme cases, the chain will jump out of engagement with the sprockets.

For proper alignment, sprocket shafts must be parallel. Check distances between shafts on both front and back sides of the sprockets; they should be equal. Also, for proper alignment, the two sprockets must be in line with each other, or fall in the same plane. This can be checked by connecting a straight edge across the flat surfaces of the sprockets.

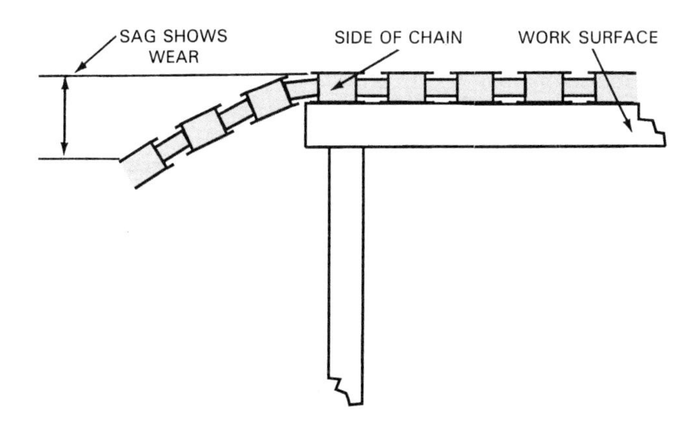

Figure 3-22. Chains that are in place should be checked for looseness. A removed chain may be checked on a workbench by laying the chain on its side and projecting it off the edge. A chain that sags even slightly should be replaced.

Bearings

When two objects rub together, resistance to movement arises between them. This resistance is friction. Bearings are used to reduce friction. This section will discuss how bearings accomplish this task, as well as types of bearings and inspection and service procedures, but first, a few more words about friction.

Friction in Bearings

Friction is a part of everyday life. Often, it is useful. Friction between your shoes and the ground, an example of *static friction,* allows you to walk without falling down. Friction holds assembled parts together. It prevents threaded parts from unthreading and provides the holding power for press-fit parts. Friction is what makes a brake system work.

There are times when friction can be a handicap. Friction between a part that is supposed to turn freely and a stationary part will turn power into heat and wear. Friction can be a severe problem in the drive train, where the rotating parts are moving at high speeds. In addition, drive train action often puts side loads on a shaft, subjecting one side of a stationary part to more friction.

Bearings are devices used to reduce friction between rotating and stationary parts. Further, they guide and support rotating parts, to prevent damage from misalignment or excessive clearances. The relationship between a rotating shaft, a stationary housing, and a bearing is shown in Figure 3-23.

Types of Bearings

There are two major classes of bearings–*plain* and *antifriction.* Figure 3-24 illustrates the two types. The ***plain bearing,*** or ***friction bearing,*** as it is sometimes called, is

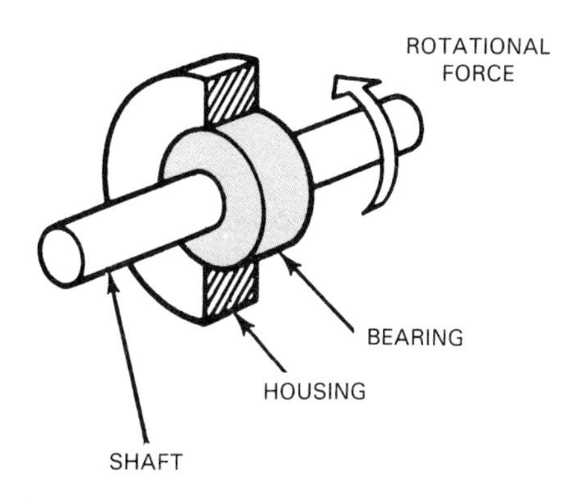

Figure 3-23. Note the relationship between rotating shaft, stationary housing, and bearing. The bearing ensures the shaft is properly aligned with the housing. It also serves as a buffer. It incurs the wear, rather than the shaft or housing, making servicing easier and cheaper. (Federal Mogul)

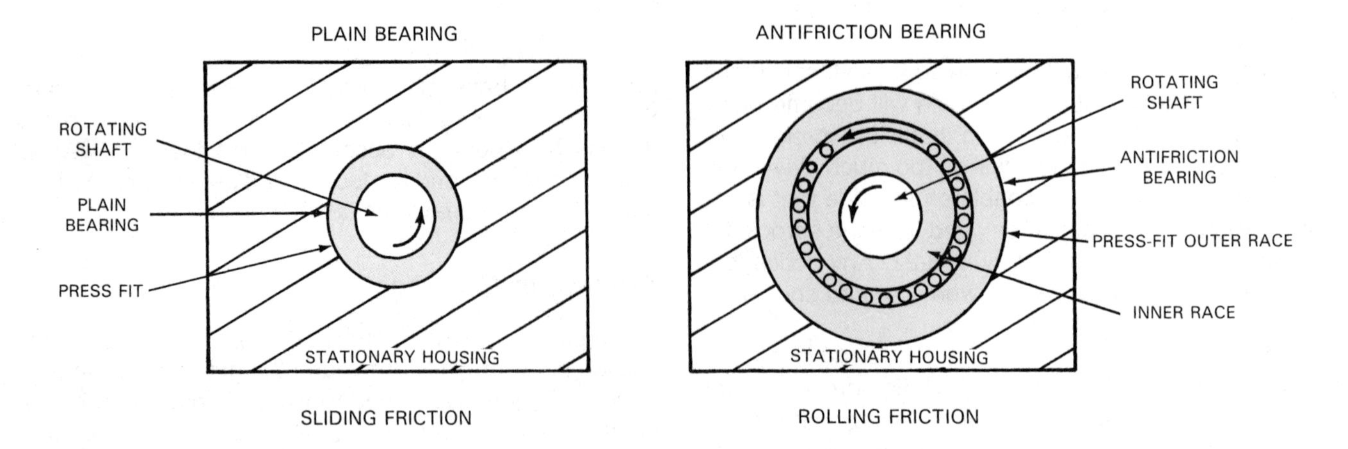

Figure 3-24. The two main types of bearings are shown. Note that sliding friction is seen in the plain bearing, while rolling friction is seen in the antifriction bearing.

pressed into place and does not move. A rotating shaft slides around on the plain bearing surface. The ***antifriction bearing*** uses an assembly of balls or rollers contained within a housing, where they are aligned and free to roll. The bearing is pressed into place; however, the one side (inner or outer) is free to move with the rotating member because of the rolling elements. This changes *sliding,* or *kinetic, friction* between the mating surfaces to *rolling friction.* A detailed discussion about each type of bearing follows.

Plain Bearings

Plain bearings are used for supporting rotating shafts. A rotating shaft slides around the supporting surface of the bearing, which is stationary. These bearings have a one-piece or two-piece construction. *Sleeves* and *bushings* are one-piece plain bearings. Bushings are small sleeves. Two-piece bearings are called *insert bearings.*

Plain bearing material

Friction bearings are composed of many materials. Most bearings used in transmissions are made of a steel shell with a coating of a soft metal, such as copper, brass, lead, or aluminum. Sometimes *babbit metal,* which is an alloy of tin or lead, copper, and antimony, is used for the coating.

Plain *pilot bearings* in the clutch assembly, used to support transmission input shafts, and some bearings used in transmissions, are solid brass or bronze. One thing is true about any plain bearing, no matter what its composition–it is always designed to wear before the shaft metal. This is done so that the bearing, which is relatively cheap, will wear out instead of the expensive shaft.

Plain bearing lubrication

A rotating shaft can contribute to friction through two types of loads–*radial* and *axial.* ***Radial loads*** are perpendicular to the axis of rotation. ***Axial,*** or ***thrust, loads*** are parallel to the axis of the shaft. To reduce friction from these loads, the bearing must be lubricated.

For proper bearing lubrication, the shaft and bearing must be separated entirely by a film of lubricant, Figure 3-25. When enough lubricant is present, the sliding action takes place in the lubricant between the shaft and the bearing; the shaft *rides* on the film.

In order to accommodate the bearing lubricant, a clearance is needed between the shaft and bearing. An ***oil clearance,*** Figure 3-26, must allow the lubricant to enter and circulate properly. If the clearance is too tight, the lubricant cannot form an adequate surface between the mating parts. If it is too loose, the lubricant will leak out too quickly to maintain the surface separation. The shaft and

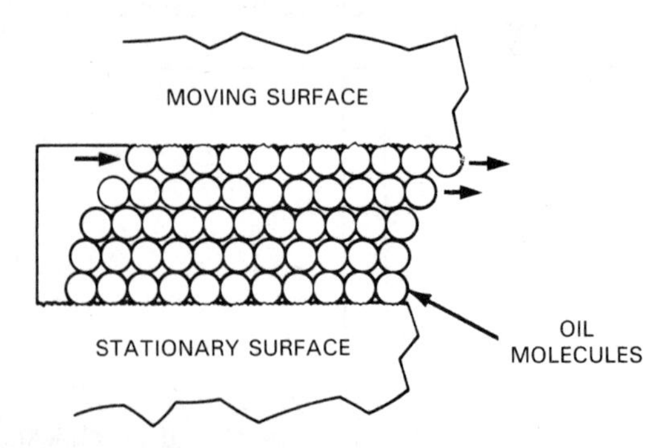

Figure 3-25. A film of oil between two surfaces can act as a bearing. Oil molecules serve as tiny ball bearings. The oil film is what makes the use of friction bearings feasible. (Federal Mogul)

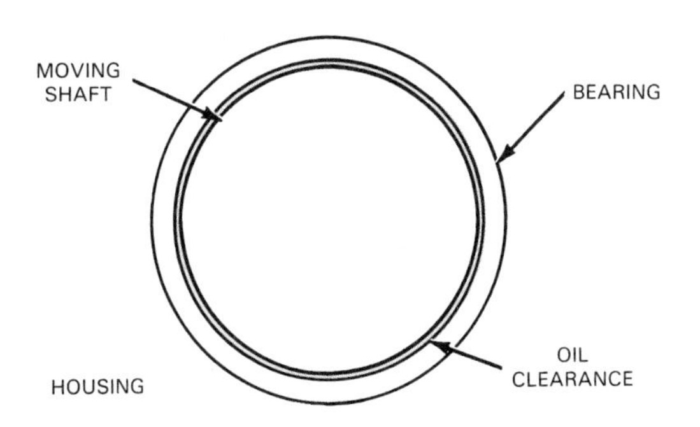

Figure 3-26. The oil clearance between a bearing and a shaft is usually a few thousandths of an inch. This clearance must be loose enough to allow the oil to reach the bearing and shaft surfaces. However, it must not be so loose that the oil leaks out too quickly. (Federal Mogul)

Figure 3-27. An assortment of plain bearings, showing grooves and slots for lubrication. (Federal Mogul)

bearing are said to have a ***running fit*** when the oil clearance is sufficient to enable parts to turn freely and receive proper lubrication.

The lubricant itself may be a type of oil, supplied by pressure, splash, or immersion. It may be a heavy grease, applied periodically. Plain bearings used in automatic transmissions are often supplied with oil from the transmission oil pump. Bearings used in standard transmissions are lubricated by splash from the turning gears. Plain pilot bearings in the clutch assembly are greased only when the clutch is serviced.

Most plain bearings will have slots and grooves for lubricant. The grooves usually run the entire length of the bearing surface. They allow the lubricant to be picked up and evenly distributed around a moving shaft. Figure 3-27 shows a number of plain bearings. Notice the grooves and slots in these bearings.

Plain bearing removal

Bushings and sleeves, plain bearings of one-piece construction, are press fit into position. There are a couple of ways to go about removing them. The best way is to use a puller made for the particular bearing. The puller is secured to the bearing. Applying pressure will remove the bearing from the housing.

A different method of bearing removal can be used when the bearing housing is a *blind hole*–that is, open on one side, blocked off on the other. An example of this is the clutch pilot bearing. Pack or fill the area behind the bearing with heavy grease. Insert a shaft that is of the same diameter as the bearing opening into the bearing. Put on eye protection and strike the exposed end of the shaft with a hammer. The hammer force will be transferred into the grease. The region behind the bore becomes pressurized. The pressure will be applied to the bearing, and the bearing will be forced from the bore.

A *cape chisel* or a *round-nose cape chisel* is an ideal tool for removing small bushings where both ends are accessible. A *pin punch* is another tool that can be used. Where applicable, these tools can be used to remove sleeves and insert bearings as well. Always use care to avoid damaging the bearing housing.

Plain bearing inspection

Whenever a drive train part is disassembled, the bearings should be closely inspected. Clean the bearings thoroughly before inspection. Use a solvent that will not damage or coat the bearings.

During inspection, check for wear and damage. A normally worn bearing will have smooth contact surfaces. Bearings that are to be reused should be checked for proper size. Often a bearing will look good, but will be worn excessively. Bearing clearances often measure only up to a few thousandths of an inch, or several hundredths of a millimeter.

One quick way to check a plain bearing is to place it onto the shaft on which it rides. If the fit feels loose, the bearing, and possibly the shaft, is worn. This method depends on your knowing what a loose bearing fit feels like. It is hard to tell without experience.

If you are not sure, check actual dimension of bearing inside diameter using a caliper or other such instrument. If the bearing has excessive clearance in one area only, it has been subjected to heavy loads on one side. Check that the shaft is not bent or misaligned. Any bearing with excessive clearance should be replaced.

In many cases, when the bearing is replaced, the shaft might also have to be replaced. Do not assume that replacing the bearing will solve the problem. Proper operation of the assembly depends on both the bearing and the shaft. Always check the shaft where it mates with the bearing. If there is no obvious damage, use a micrometer or sliding caliper to check for proper diameter and taper. Check the shaft in several places.

Another problem to look for is scoring of the bearing. Scoring will appear as scratches or ridges partially or totally inscribed about the bearing surface. Scoring is usually caused by lack of lubrication or by abrasive particles in the lubricant. If the bearing is badly scored, the shaft is probably also damaged.

A bearing that has been in use for some time may have pitting or even large craters in its surface. Pitting starts with dirt or water in the lubricant. Over time, large pits, or craters can develop.

Sometimes, a missing ground strap between the engine and car body can cause craters. The engine electrical components will ground through the drive train. This electrical action will cause metal transfer, leading to pitting.

Sometimes, a plain bearing will be operated without lubrication or under other abusive conditions. This can lead to severe bearing damage. Bearings may overheat, spin in their housing, or weld themselves to the shaft. A sign of a bearing that has been subjected to overheating is a change in the material's color. Indications of spun or welded bearings will usually be obvious. In many cases, the shaft or housing must also be replaced or repaired.

Plain bearing installation

Bearing drivers are made to closely fit into a bearing. The bearing will bottom against a slightly larger area on the driver. Pressure on the driver is transmitted to the bearing, and the bearing is forced into its housing.

Bearings should be installed with the proper-size driver, preferably, on a hydraulic press. If a press is not available, some bearings can be hammered into place, using a driver to avoid part damage. Strike the driver with a hammer. Do *not* strike the bearing with the hammer. Also, attempt hammering only if the proper-size driver is available.

The bearing should be installed carefully. If the bearing has any oil supply holes, line them up with the oil holes in the housing for the bearing. Make sure that the bearing enters the housing bore squarely and does not become cocked. A poorly installed bearing will quickly destroy itself and the shaft that it supports.

After the bearing is installed, additional service may be required. Some bearings will require *reaming* and *honing,* to widen and smooth the bearing inside diameter to fit the mating shaft. Some bearings must be *staked.* This involves taking a punch to the bearing surface, which produces a raised dimple on the side of the bearing against the housing. This action locks the bearing in place and keeps it from spinning or slipping out of position. After reaming, honing, or staking, the bearing should be thoroughly cleaned.

Replacement bearings should be checked to ensure that they have the proper clearance with the shaft. Before reassembly, that is, before inserting the shaft into the bearing, lightly lubricate the bearing surface. Use the same kind of lubricant that the bearing will have during normal operation. This will ensure adequate lubrication the first time a load is placed on the bearing. If the bearing is only lubricated during service (a clutch pilot bearing, for example), be sure to apply the proper amount of grease before reassembly.

Antifriction Bearings

Antifriction bearings contain rolling elements that operate within a housing made up of one or two pieces of metal. The rolling elements can be balls or rollers. In general, the pieces of the housing are called ***races.*** For conical antifriction bearings, the races are sometimes called *cones.* For long life, all of these parts are made of heat-treated, high-strength steel.

A typical antifriction bearing, with inner race, outer race, and rolling element, is shown in Figure 3-28. Note that antifriction bearings do not always have both an inner and outer race to contain the rolling elements. Sometimes, one of these is omitted. Then the elements are in direct contact with the mating surface.

In addition to rolling elements and races, many antifriction bearings contain a *cage* to keep the rolling elements in position. Also, some bearings are prelubricated, and these have seals to keep the lubricant in and to keep dirt and moisture out. The placement of seals is shown in Figure 3-29.

Antifriction bearings may be pressed or slipped into position–onto a shaft, for example, or into a stationary housing. Frequently, the bearing is pressed onto the shaft, and then the shaft and bearing are slipped into place in the stationary housing. In some cases, such as rear axle pinion gear bearings, the bearing is pressed into the stationary housing.

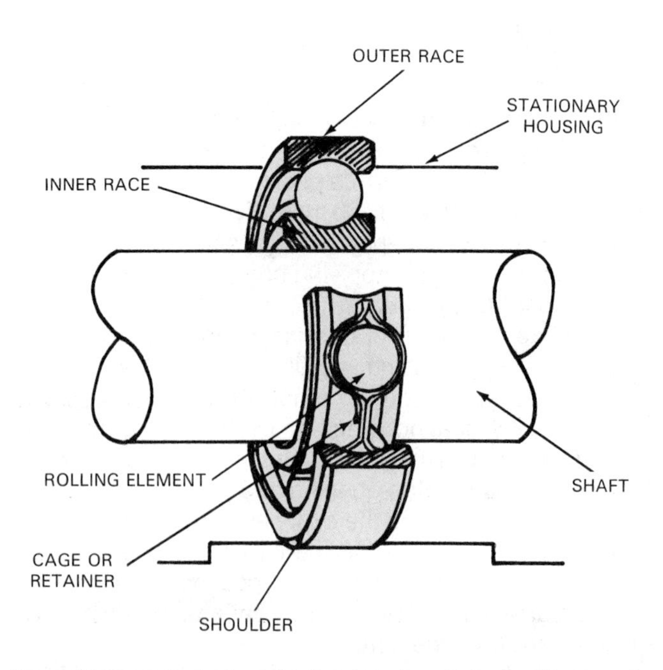

Figure 3-28. A typical antifriction bearing. Note the three major parts–inner race, outer race, and rolling element.

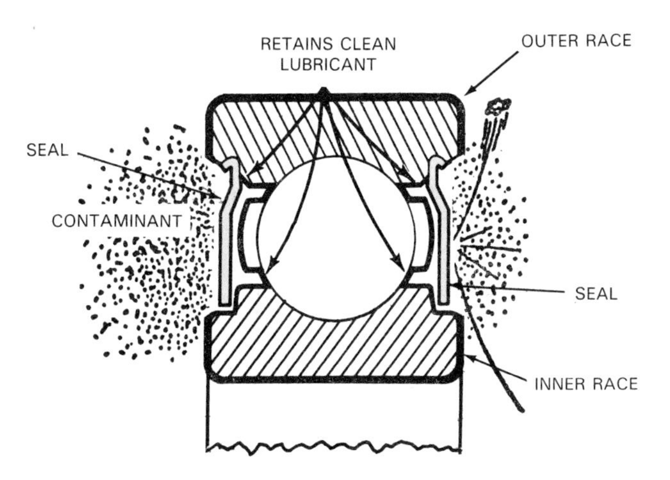

Figure 3-29. Some bearings have a seal, as shown in this partial sectional view of a ball bearing. The seal serves to keep bearing lubricant in and dirt and moisture out. Bearings of a transmission, transaxle, or rear axle assembly will not have a seal.

The advantage of the antifriction bearing is that it uses a rolling motion, rather than a sliding motion like the plain bearing. The rolling motion produces less friction. The antifriction bearing is usually used where rotating parts are highly stressed or where it is difficult to supply adequate amounts of lubricant. The antifriction bearing is more efficient than the plain bearing, but it is more expensive. Sometimes, however, size, clearance problems, or the back-and-forth movement of the shaft prevent the use of an antifriction bearing.

Types of antifriction bearings

The two major types of antifriction bearings are *ball bearings* and *roller bearings,* Figure 3-30. Ball bearings are used for clutch *throwout bearings,* which are used for clutch release. They are also used for pilot bearings and in some transmissions and transaxles. Roller bearings of various types are used as pilot bearings and also in transmissions and rear axle assemblies. *Tapered roller bearings, needle bearings,* and *thrust bearings* are all variations of the roller bearing.

Ball bearings. Although they vary in size, ball bearings used in automotive drive trains are all of the same basic design; however, methods of lubrication vary. Clutch throwout bearings and some *rear axle bearings* are greased by the manufacturer and are not intended to be regreased. Ball bearings used in other parts of the drive train are lubricated by oil splashed from other moving parts.

Roller bearings. As mentioned, there are several types of roller bearings. ***Straight roller bearings*** are used for axle bearings in rear axle assemblies, Figure 3-31. They are also used in manual and automatic transmissions. This type of bearing usually comes as a one-piece unit.

A ***tapered roller bearing*** is shown in Figure 3-32. Rollers of this bearing assembly are tapered. The outside

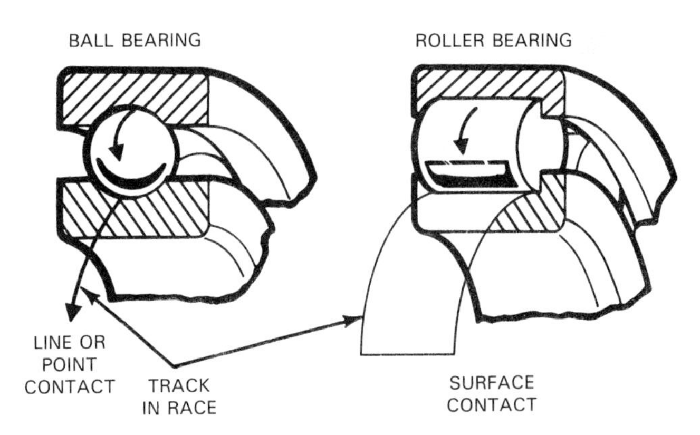

Figure 3-30. Two types of antifriction bearing–ball and roller–are shown here. (Federal Mogul)

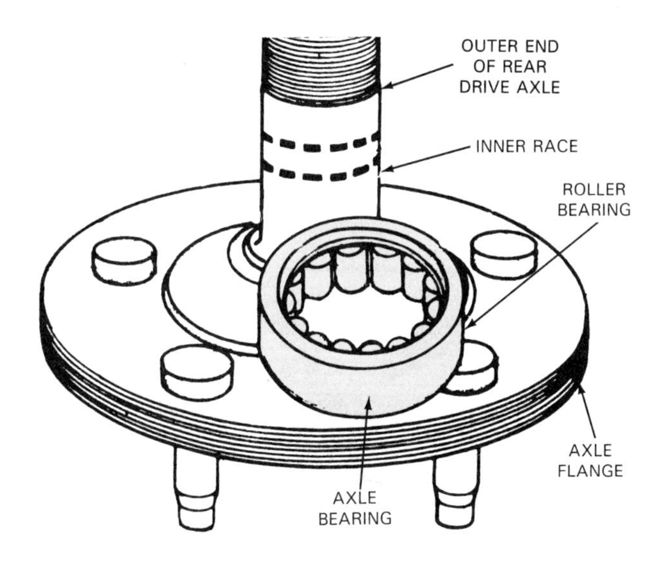

Figure 3-31. An example of a straight roller bearing. Notice that the bearing shown here has only an outer race. The axle shaft is functioning as the inner race. The rolling elements fit right up against the shaft. (Ford)

diameter of the inner race (cone) and the inside diameter of the outer race (cup) are both tapered, also, to fit the rollers. The assembled bearing, however, including its inside diameter, is cylindrically shaped. Tapered roller bearings are useful for heavy loads. They are used in rear axles and also in front drive axles.

Needle bearings, Figure 3-33, are another type of roller bearing. Needle bearings perform the same function as other roller bearings, but their diameter is much smaller in relation to their length. Needle bearings have tiny rollers that resemble needles, hence, the name *needle bearing.*

Thrust bearings, Figure 3-34, are flat, disklike bearings that resemble washers. Thrust bearings are made up of needle rollers. The rollers are arranged *radially.* This means they radiate out along imaginary lines projected from a center point. This kind of bearing is sometimes called a ***Torrington bearing.***

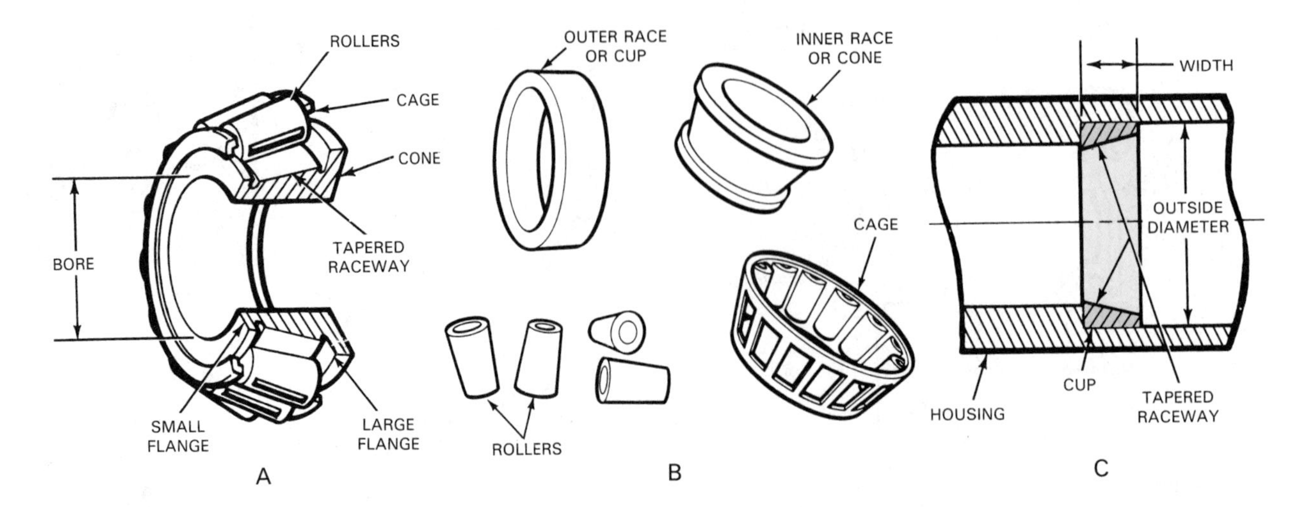

Figure 3-32. Tapered roller bearing. A–This sectional view of an assembled tapered roller bearing shows the bearing with the outer race removed. B–This shows various parts of a tapered roller bearing. C–This sectional view shows the outer race, or cup, installed in a housing.

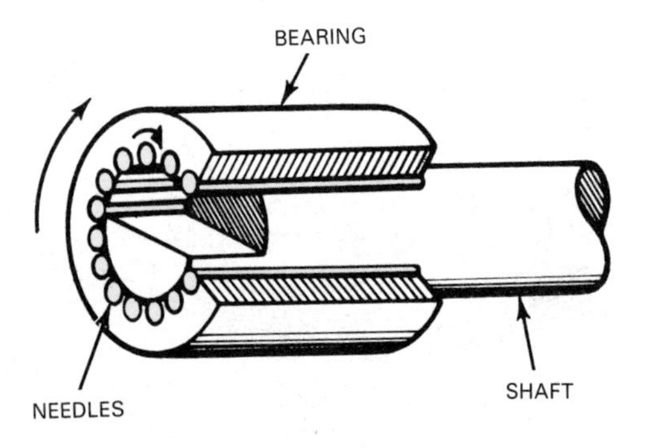

Figure 3-33. The rollers of a typical needle bearing are longer and thinner than those used with other roller bearings. Note that the bearing shown has only a single race.

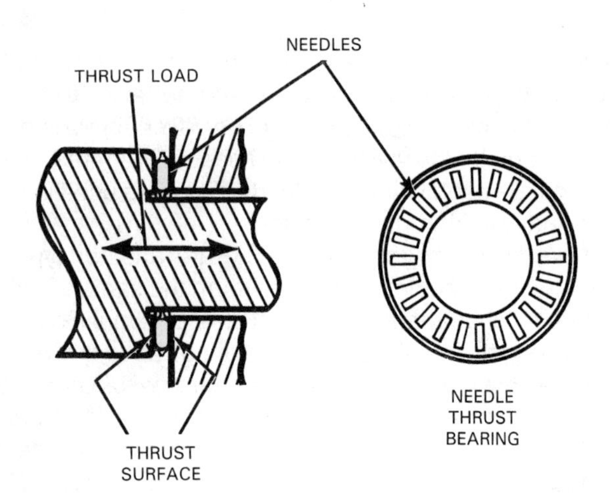

Figure 3-34. Note a thrust bearing using needle rollers. The needles are arranged radially; their axial centerlines radiate outward from a center point. (Federal Mogul)

Bearing load factors

Antifriction bearings can be subjected to radial loads and thrust loads. Figure 3-35 shows these. Applied radial loads are directed perpendicular to a shaft's rotational axis. Thrust loading is directed parallel to the axis. Thrust loading is caused by the back and forth movement of the shaft. The design of helical and bevel gears inherently creates thrust loads, which must be compensated for in bearing design. In some cases, such as at the pinion shaft of a rear axle assembly, the bearings are *preloaded* to reduce the deflection of parts and minimize thrust variation. Most bearing loads are *combination loads,* which are the result of both thrust and radial loads.

Ball and roller bearings are affected differently by loads. The ball bearing absorbs both thrust loads and radial loads on the ball surface. The straight roller bearing absorbs radial loads on the rolling surface, and minor thrust loads are absorbed on the sides of the rollers. The tapered roller bearing is designed to absorb both radial loads and thrust loads on the roller surfaces.

Applied loads will not be perfectly distributed about the bearing. Several factors can cause loading to be greater on one side or another. Gravity is a more obvious factor. The manufacturing tolerances of gears and shafts–the fact that they are not perfectly round or symmetrical–is another. Helical gears, too, by their design, will cause loading to be greater on one side. These conditions will cause the bearing elements to be alternately loaded and unloaded as the shaft rotates. Figure 3-36 depicts this action. Loading intensity is illustrated by the shading of the bearings–the darkest being the point of greatest intensity. In conclusion, the bearing must be manufactured to withstand not only heavy loads, but also to withstand the effects of *changing* loads.

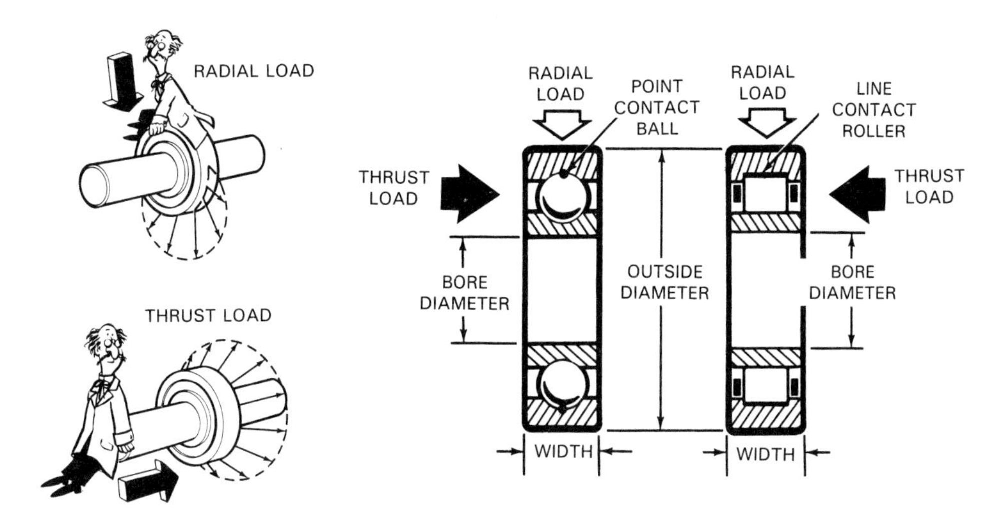

Figure 3-35. Two types of loads placed on a bearing are radial and thrust. The two loads acting upon a bearing at the same time may be referred to singly as a combination load. Most drive train bearings are subjected to combination loads. (Federal Mogul)

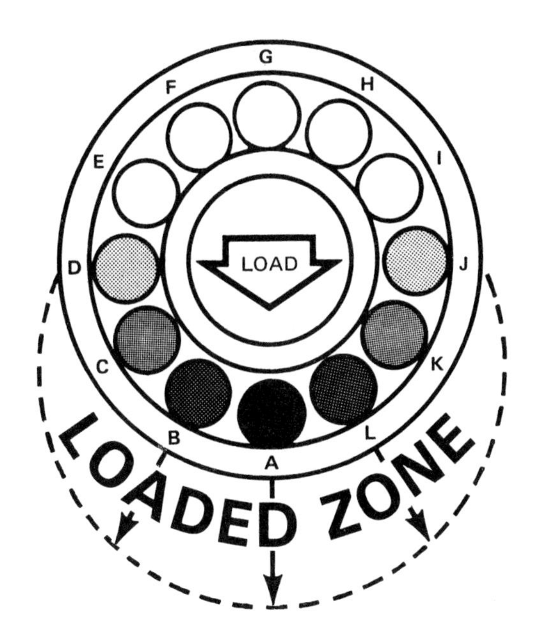

Figure 3-36. Complicating the bearing loading process is the fact that bearing loads vary about the bearing, according to gravity and shaft tolerances. Bearing rollers are alternately loaded and unloaded as the shaft rotates. (Federal Mogul)

Bearing servicing precautions

Upcoming paragraphs cover servicing aspects of antifriction bearings. This includes antifriction bearing removal, inspection, and installation. Before you study these aspects, there are some special precautions that you must be aware of and take whenever servicing antifriction bearings. They are as follows:

- *Always* wear eye protection when removing or installing a bearing.
- Do *not* apply excessive force to a bearing or shaft during removal or installation.
- Do *not* apply any force through the rolling elements of an antifriction bearing. Only apply force to the press-fit side of a bearing. The bearing could be damaged or could burst apart and cause injury.
- *Never* spin a bearing with compressed air. The rolling elements could be thrown out with explosive force and severely hurt you or someone else. Mechanics have been killed by using compressed air to spin bearings.

Antifriction bearing removal

Some bearings are held to shafts by snap rings or by large nuts that thread onto the shaft. In these cases, bearing removal is simple. The snap ring or nut is removed, and the bearing is slid from the shaft. In other cases, the bearing is pressed on, and special methods must be used to remove it.

The best way to remove a pressed-on antifriction bearing from a shaft that has not been removed from its installation is to use a bearing puller. Some pullers are made for a particular bearing; however, others are universal designs that can be used on almost any bearing. These pullers have adjustable jaws that can be tightened around the bearing. Using a puller, the bearing can be forced from its shaft.

An ideal way to take a bearing off a shaft that is removed from a vehicle is to use a hydraulic press. A support is placed under the bearing, and pressure is applied to the shaft to press the shaft from the bearing. Remember! Pressure should only be applied to the *press-fit* race to avoid applying force through the rolling elements, which could damage the balls, rollers, or races. Refer to Figure 3-37 for examples of proper and improper bearing removal.

Some bearings are press fit into a housing instead of around a shaft. A brass drift can be used to drive the race

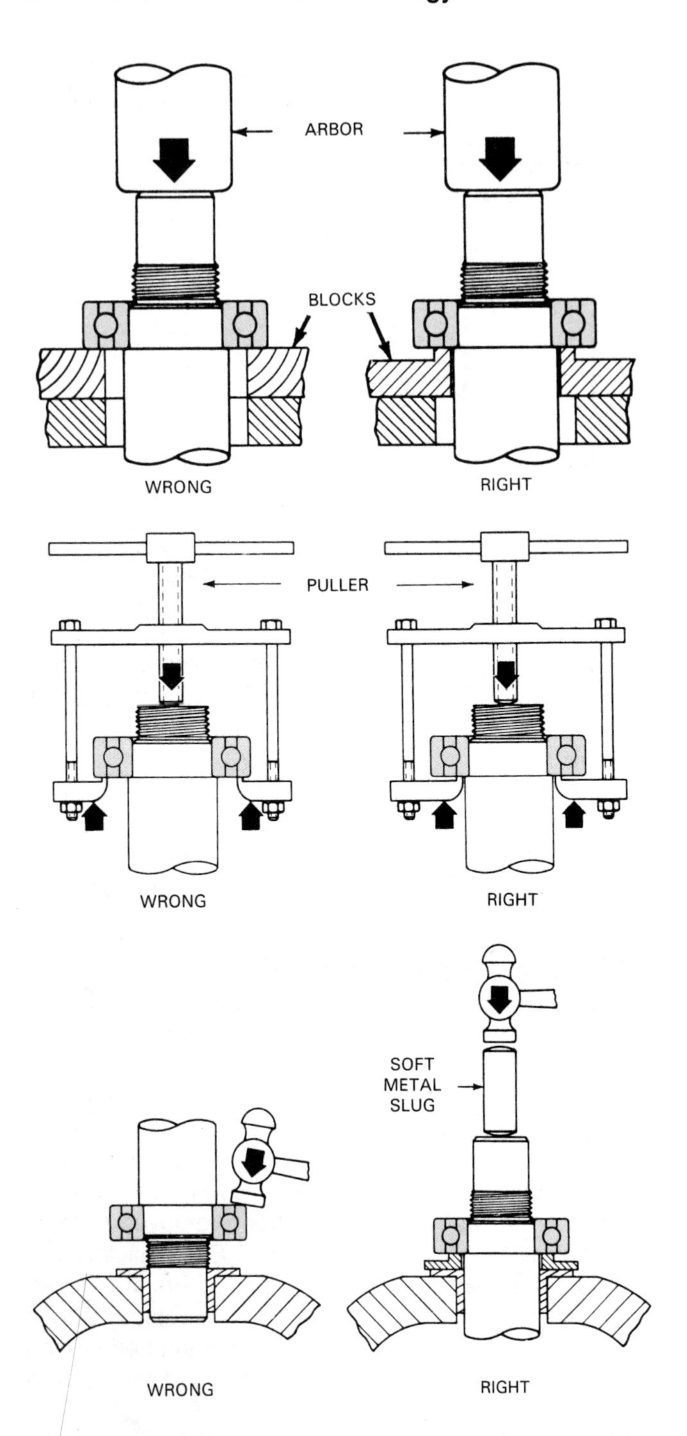

Figure 3-37. Different methods of bearing removal. Do you see the difference between right and wrong ways of removal in each drawing? (FAG Bearings)

from the housing. Look for a notch cut into the shoulder of the outer race and carefully apply the drift to it to remove the bearing.

Sometimes, a *bearing retainer,* which is a force-fit metal ring or plate, must be removed before removing a bearing. This would apply to some rear axle bearings. The retainer is removed with a chisel or torch–carefully, to avoid damaging the shaft. In some cases, it is possible to cut the bearing from the shaft with a cutting torch. The torch must be used with caution to prevent overheating or otherwise damaging the shaft. If there is any way to avoid using a torch, do so.

WARNING! Wear eye protection when cutting off a retainer plate.

Antifriction bearing inspection

Bearings should be thoroughly cleaned before inspection. Often, what appears to be a bearing defect is dirt that collected on the bearing during disassembly. At other times, actual bearing defects are hidden by dirt or sludge. If you are cleaning a bearing with compressed air, always hold both races so that the bearing does not spin. Remember! Never spin bearings with compressed air.

After cleaning, rotate the bearing by hand as shown in Figure 3-38. Note if there are any places where you feel roughness or binding. If you find any problems, replace the bearing.

Visually inspect the bearing for signs of wear and damage, including conditions of scoring, overheating, and corrosion. Bearings may show signs of extreme wear or damage due to problems of overloading and lack of lubrication. Spalling and *brinneling,* Figure 3-39, are indications of these problems. Check for a bent or otherwise damaged cage. Cage damage is usually caused by careless installation.

Identifying replacement bearings

It is very difficult to match bearings by a visual inspection. It is, therefore, important to make sure that the replacement bearing you get is the bearing that you need. If the old bearing has a part number, Figure 3-40, compare it with the number of the replacement bearing. If the old

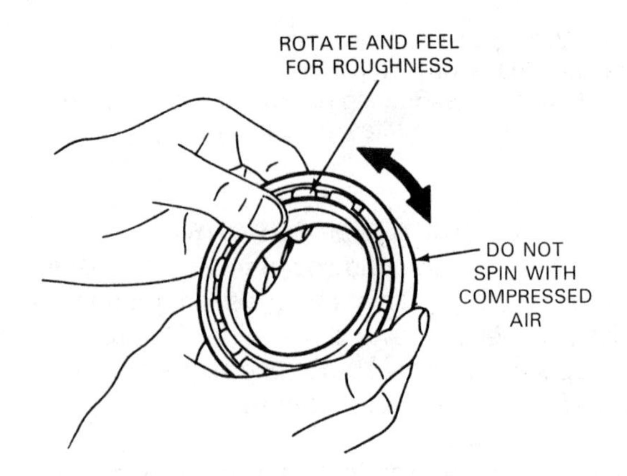

Figure 3-38. Hold clean bearing like this during inspection and rotating outer race by hand. Do not use compressed air to inspect. An unloaded bearing spun at high speeds can come apart with explosive force! (Nissan)

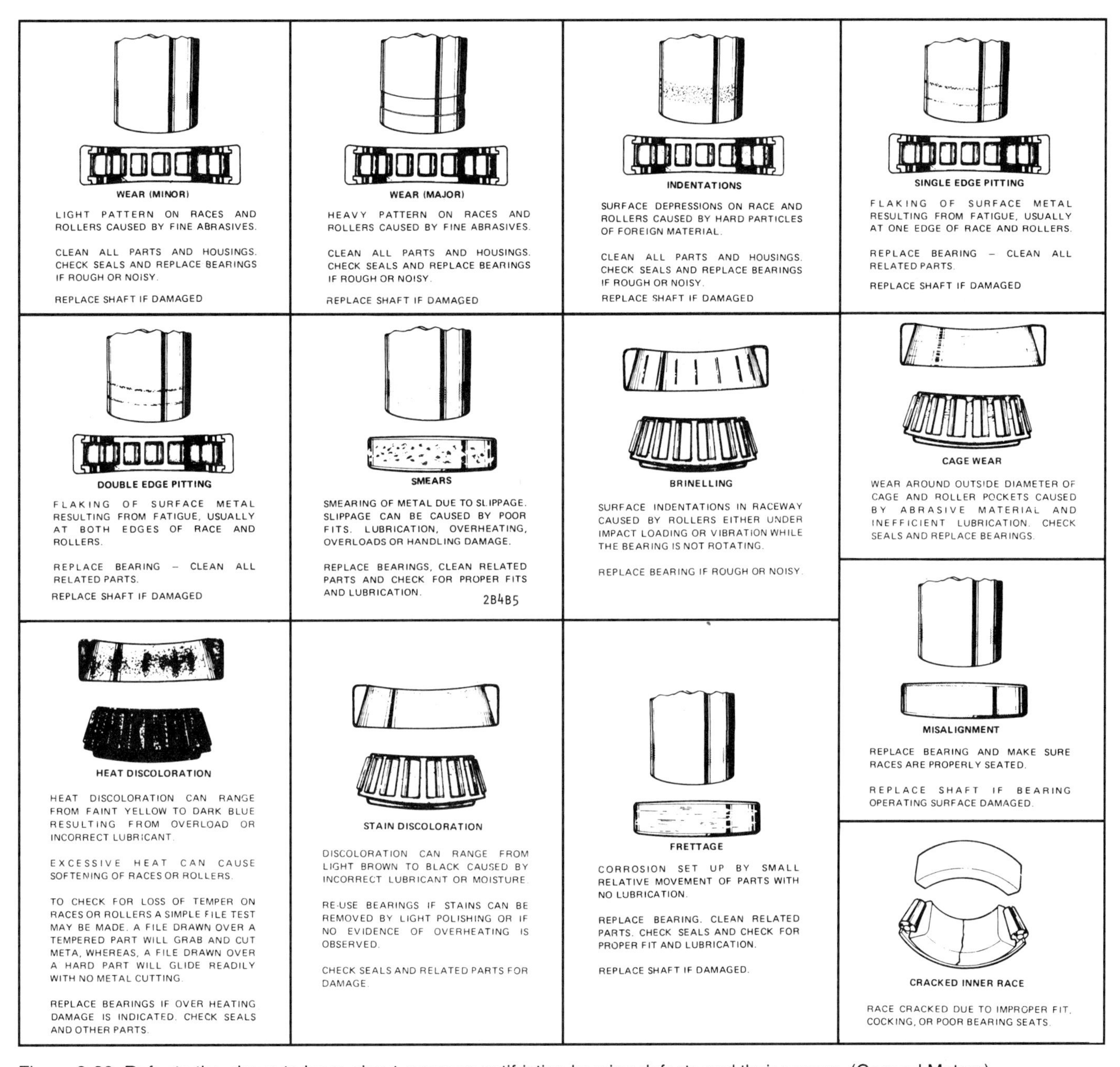

Figure 3-39. Refer to the above to learn about common antifriction bearing defects and their causes. (General Motors)

bearing does not have a number, or the number cannot be read, consult an appropriate parts catalog to ensure that the bearing is correct for the make and model of the vehicle that you are working on. In some cases, you may need to obtain the serial number of the vehicle or of a particular component (transmission, rear axle, etc.).

Storing bearings

Cleanliness is vital to the successful operation of antifriction bearings. Bearings should be left in their original wrappings until just before installation. If a bearing must be greased before installation (this usually applies to bearings of front drive wheels, clutch throwout bearings, etc.), pack the bearing in a manner similar to that shown in Figure 3-41. The grease must reach all of the *voids,* or spaces, in the cage. Be careful not to drop any bearings; since the hardened steel is brittle, it may chip or develop a shock crack.

Antifriction bearing installation

The best way to press a bearing on is with a bearing driver and a hydraulic press. A driver of the proper diameter

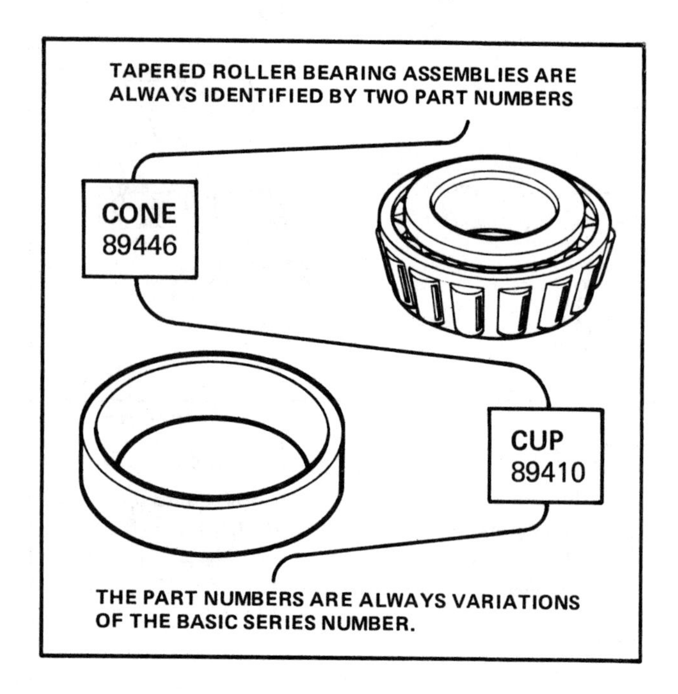

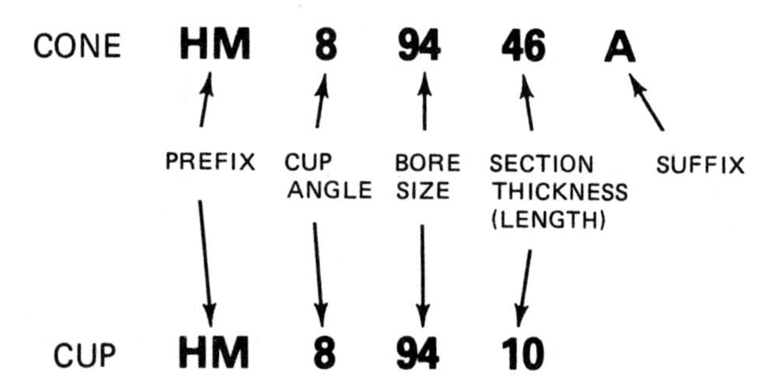

Figure 3-40. Always try to identify the part number of the original bearing before ordering replacements. An example is shown here. If the number is missing, you will need to gather information about the vehicle to obtain a replacement part. (Federal Mogul)

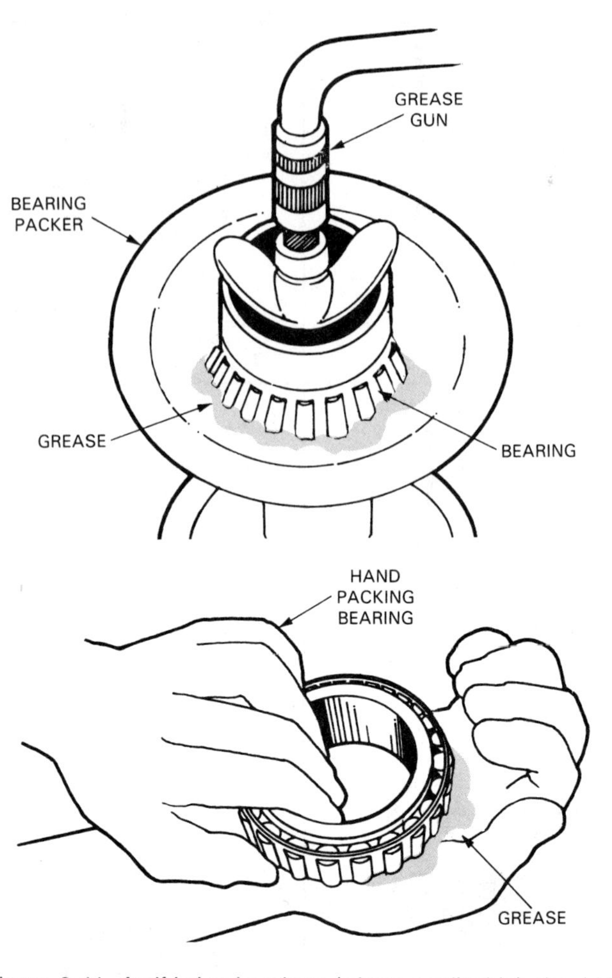

Figure 3-41. Antifriction bearings (where applicable) should be carefully packed with grease as shown here. Bearings can be packed with a machine made for the purpose or by hand.

is selected to fit the bearing. Pressure on the driver from the hydraulic press is transferred to the bearing, and the bearing is forced onto the shaft.

If a hydraulic press is not available or cannot be used, some antifriction bearings can be driven into place by tapping on the race that is to be press fit. Remember not to hammer on the free race when installing a bearing. This will cause damage to the rolling elements and races and can cause injury.

When press fitting a bearing, watch closely to ensure that it is not being cocked or pressed in too far. If the shaft has a shoulder, the bearing should firmly bottom on the shoulder. Also, make sure that dirt or other debris is removed from the bearing before installation. Refer to Figure 3-42 for examples of proper and improper bearing installation.

If a shaft is too large, a bearing can be heated and expanded somewhat. This might enable the bearing to be dropped into place over the shaft. If you decide to use this method, make sure that you do not overheat the bearing. Doing so will destroy the properties of the metal and ruin the bearing.

Just as a bearing can be expanded by heating, it can also be shrunk by cooling. Therefore, if a bore is too small for a bearing, you can try chilling the bearing in a freezer. This may shrink it just enough to allow installation into the bore.

After the bearing is installed, check it for proper operation and alignment. Some bearings, like pinion and front axle bearings, must be preloaded, or adjusted so they are under mild pressure. This is done to prevent bearing looseness. When bearings must be preloaded, follow the manufacturer's procedures exactly. Bearing preloading is discussed again in *Manual Transmission* and *Rear Axle Assembly* chapters herein.

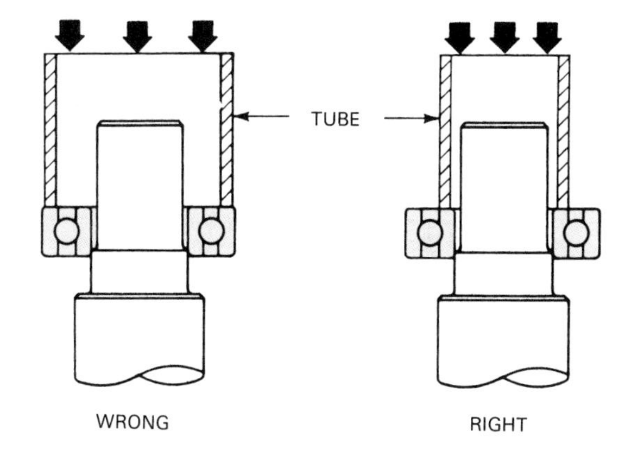

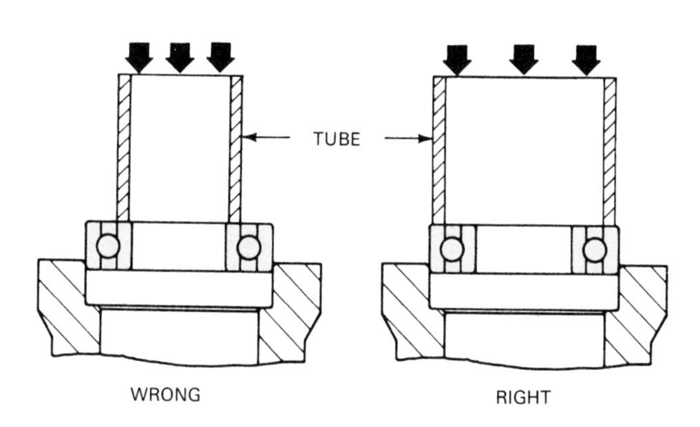

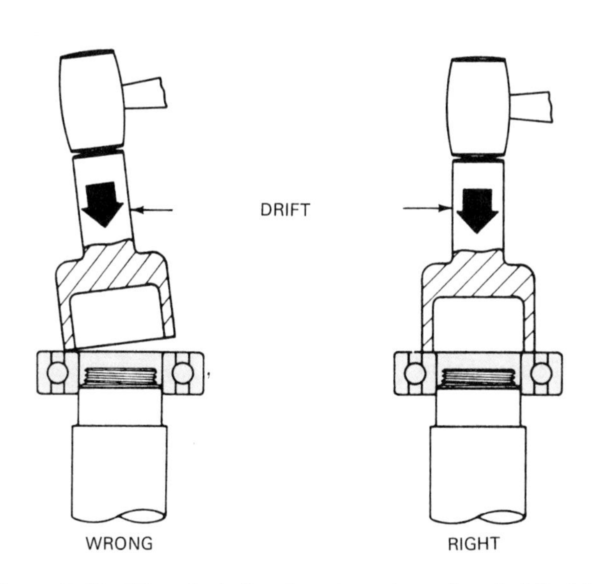

Figure 3-42. When installing bearings, keep in mind that force should never be applied through the rolling elements. Also, when pressing a bearing onto a shaft or into a housing, always make sure that the bearing is going on or in straight. Analyze these diagrams showing different methods of installing bearings. Do you see the difference between right and wrong ways of installation in each drawing? (FAG Bearings)

Related Components

Many components are used with gears and bearings in the drive train. Some of these are explained below.

Thrust Washers

Thrust washers are used between the end of a rotating shaft and a stationary housing or between two rotating shafts. They prevent excessive axial, or back-and-forth, shaft movement by limiting the amount of ***end clearance,*** or the distance between the two parts. This type of shaft movement is often called ***end play.***

Figure 3-43 is a typical thrust washer. Like plain bearings, thrust washers are made of a metal that is purposely softer than the parts against which they operate. Most thrust washers have a key or tab extension. This is to lock the washer to one component, so that all wear occurs on the side contacting the other component. The grooves in the thrust washer allow lubricant to flow between the contact surfaces of the washer and rotating part. Note that Torrington bearings are often used as thrust washers.

Most thrust washers are replaced when end clearance becomes too large. Thrust washers for some drive train units are available in different thicknesses. This way, improper end play in the unit can be corrected by selecting a thicker or thinner washer.

Retainer Plates

Retainer plates are used to hold gears and bearings in position. They are attached to a stationary housing with bolts, machine screws, or snap rings. Most retainer plates have a center hole for a shaft to pass through.

Retainer plates are used on most manual transmissions. The *front bearing retainer* is an example. They are also used on some drive axles. They attach to the ends of the axle housing. The plate keeps the bearing and axle from moving outward. In other words, it keeps them in the axle housing. See Figure 3-44.

Collars

Collars are another type of retainer. These are steel rings used to hold gears or bearings on shafts, as shown in Figure 3-45. Collars are usually used where space is too limited to allow a different kind of retainer. They are usually a press fit.

Collars are most often removed with a chisel, which is used to split the part open. There are other ways to remove a collar, too, but they usually expand and distort the collar. A collar, in any case, should never be reused. If it is not destroyed in the removal process, it will most likely be distorted and not fit properly.

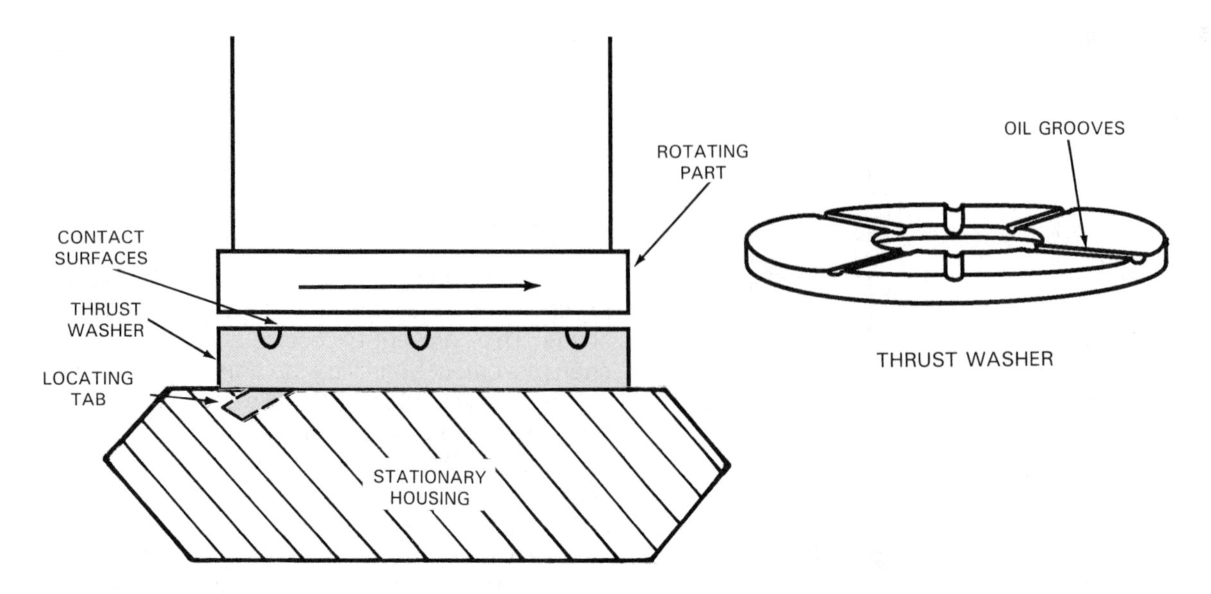

Figure 3-43. Thrust washers are installed between two parts where there is relative rotational motion between the two. Thrust washers keep back-and-forth movement of shafts to a minimum. This is a concern when using helical gears, which tend to cause this sort of movement. Thrust washers also reduce thrust caused by friction and wear.

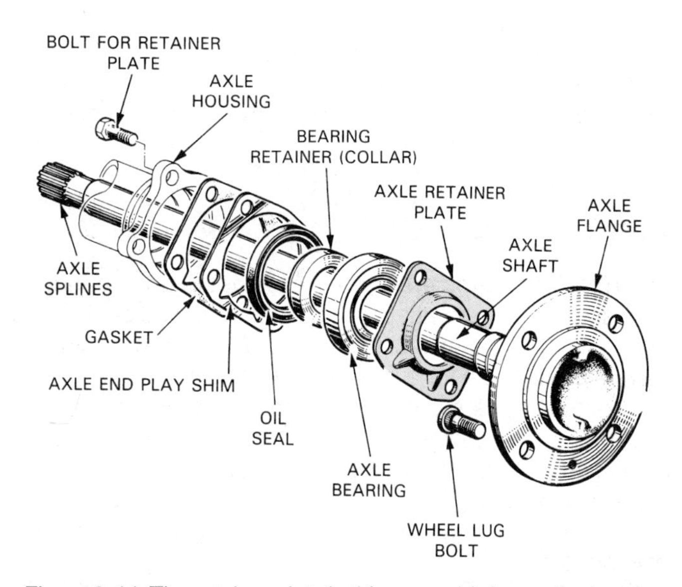

Figure 3-44. The retainer plate in this assembly keeps the bearing and axle from coming out of the axle housing. (Toyota)

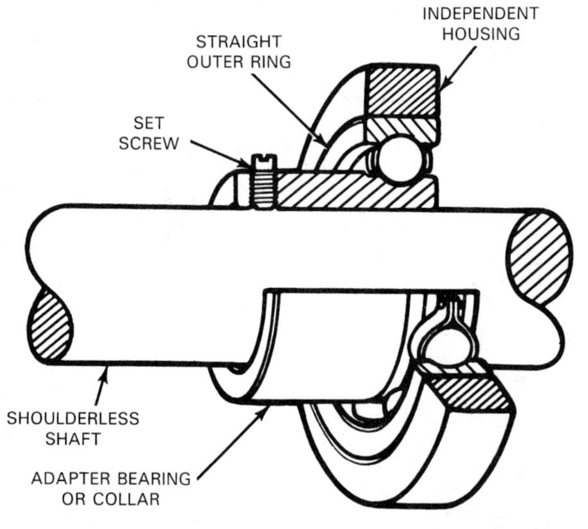

Figure 3-45. Collars are generally used to retain bearings on shafts, as shown here. They are also sometimes used to retain gears. (Federal Mogul)

Splines

A ***spline*** is a type of slot that is cut into a shaft or other part. The slot runs *longitudinally,* or lengthwise to the axis of rotation. A spline may be straight, running parallel to the axis of the shaft or part, or it may be a spiral, running at an angle. A ***splined shaft*** will have many equally spaced, parallel splines. Splines can be internal or external. Internal splines of one component are designed to mate with external splines of another component. The two components, then, rotate as a unit.

Splines serve the important purpose of allowing two parts to move back and forth in relation to each other, while allowing power to be transferred through them. They are used on the *sliding* gears of manual transmissions and on slip yokes of drive shaft assemblies. They are used on clutch friction discs to drive manual transmission input shafts and on torque converter turbines to drive automatic transmission input shafts. They are used on other drive train parts as well. Splines are shown on the end of the axle in Figure 3-44. This is a typical splined shaft.

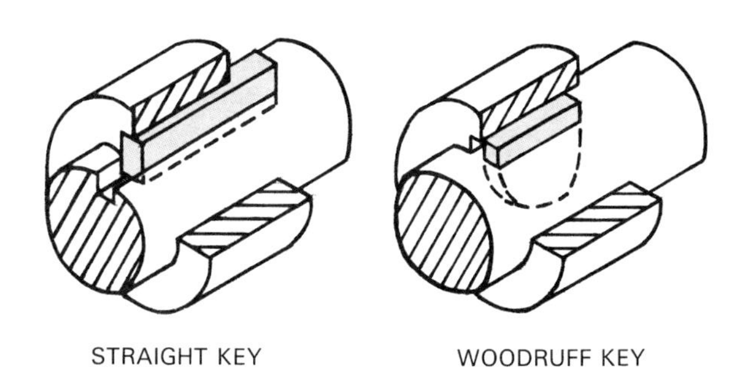

Figure 3-46. Keys are used to prevent rotation between a machine part and a shaft.

Keys and Keyways

To keep a shaft and machine part, such as a gear or pulley, from rotating in relation to each other, keys and keyways are often used. Slots, or ***keyways,*** are cut in the two mating parts, and a ***key*** is inserted in the slots. Keys and keyways come in many variations. The most common for automotive use are the *straight key* and the *Woodruff key,* Figure 3-46. Sometimes, a single external spline is cut into a part and used as a key. A matching slot is cut into the mating part. The spline and slot are locked in place, and the two parts rotate as a unit.

Summary

Vehicle drive trains are made up of many simple parts such as gears, chains, and bearings. Engine speed can be increased or decreased at the output of a gear or chain drive. A vehicle's direction can also be reversed by these parts. Friction can be reduced through the use of bearings.

Simple gears are toothed wheels that engage one another for the purpose of transmitting power. Speed and torque are increased (or decreased) through drive train gears. To improve fuel economy at higher vehicle speeds, overdrive gears are used.

Several classes of gears are used in vehicle drive trains. Spur gears are used to connect parallel shafts. The teeth on spur gears are straight and parallel with the shaft. Spur gears are sometimes used in manual transmissions. Spur gears are the simplest kind of gear. They can be noisy if the clearance is too large.

Helical gears have teeth that are cut at an angle. Helical gears operate more smoothly and quietly than spur gears. The angled cut of helical teeth puts a sideways thrust on the gears and shafts. This is often called end thrust.

Herringbone gears overcome some of the thrust and noise problems of helical and spur gears. Herringbone gears are sometimes called double-helical gears. These gears are seldom used in drive trains.

Bevel gears connect shafts that run at an angle. The angle of shafts used with drive train bevel gears is usually 90°. Similar to the bevel gear is the spiral bevel. Bevel gear teeth are straight, and spiral bevel gear teeth are curved.

Hypoid gears are similar to bevel gears, but the shaft centerlines do not intersect as they do with bevel gears. Otherwise, hypoid gears resemble spiral bevel gears.

Worm gears connect shafts that run at a 90° angle. Worm gears are useful when a large speed reduction is needed. They are seldom used on vehicle drive trains.

Rack and pinion gears consist of a flat bar with teeth cut into it (rack) and a small gear (pinion). Rack and pinion gears are usually used on the vehicle steering system.

A planetary gearset consists of a sun gear in the center, planet gears surrounding and meshed with the sun gear, and a ring gear meshed with the planet gears. The planet gears are attached to a planet carrier but are free to revolve within it. All planetary gearset gears are constantly in mesh with each other. To change gear ratios, different parts of the gearset are driven or held stationary. Planetary gearsets are used in overdrives and automatic transmissions.

Gear ratio is the speed relationship between gears. It can be summed up as the ratio between the number of teeth in any two meshing gears. A vehicle requires different gear ratios for different operating conditions. A vehicle must also have a reverse gear. Two or more gears can be arranged into a gear train that will provide an output shaft rotation that is opposite to engine rotation.

Given time, gears will wear out. Normal gear wear will occur after many miles. Low mileage gear failures are usually caused by abuse, overloading, or poor maintenance. Common tooth failure conditions are pitting, spalling, peening, scoring, and welding. Gear teeth may also overheat and lose their strength. Overheated gears wear out quickly.

Many gears are a slip fit with their shafts. Slip fit gears are held in position by other parts of the drive train or by snap rings. If the gear must turn with its shaft, it is held by a key or by splines. Sometimes the gear and shaft are machined from a single casting. Some gears are pressed onto the shaft that they ride on. Removing these gears requires a gear puller or a hydraulic press.

Chain drives are often used where space is limited or to transmit power between a large distance. Every chain drive assembly consists of a chain and at least two sprockets. Some systems use more than two sprockets.

There are two major classes of drive chains. The roller chain consists of a series of pins, usually encased in rollers, that are held in position by links. The pins or rollers engage the sprocket teeth. The silent chain consists of a series of leaves that are toothed to engage the sprocket teeth. The leaves are held in position by alignment pins.

Drive chain wear is compensated for by automatic tensioners that place a calibrated load on the chain. These tensioners are usually spring loaded.

Drive chains can be checked for wear on or off the vehicle. A loose chain should always be replaced with a new one. If the sprockets show any tooth wear, they should be replaced. The chain and sprockets are usually replaced at the same time.

Bearings are used to reduce friction between moving and stationary parts. Bearings also guide and support rotating parts. The two major classes of bearings are friction and antifriction. Plain bearings are often pressed into a housing. They have no moving parts.

A rotating shaft slides around the surface of a plain bearing. Friction is reduced by a film of lubricant between the shaft and bearing. The lubricant may be supplied by pressure or splash. Some bearings are permanently lubricated by grease. Most plain bearings have slots to aid in oil distribution.

Plain bearings should always be checked for wear and proper size during any kind of service. Worn or damaged bearings should be replaced. In many cases, the shaft that rides on the bearing should also be replaced. Replacement bearings should be installed with a driver. Oil holes should be aligned with the matching holes in the housing before installation.

Antifriction bearings contain balls or rollers that rotate between races. The balls and rollers change sliding friction into rolling friction.

Ball bearings are used in clutches, transmissions, and transaxles. Roller bearings are used in transmissions and rear axles. Tapered roller, needle, and thrust bearings are all varieties of roller bearing.

Bearing loads usually change with shaft rotation, so that the bearing is alternately loaded and unloaded. This makes the use of high quality bearings critical.

The best way to remove an antifriction bearing is with a special puller or a hydraulic press. *Always* wear eye protection when removing a press-fit bearing. *Never* spin an antifriction bearing with compressed air.

Carefully clean and inspect antifriction bearings. A bearing with any kind of defect must be replaced. Always make sure that you obtain the proper replacement bearing. New bearings should be left in their original wrappings until needed.

Antifriction bearings should be installed with the proper bearing driver. A hydraulic press is preferred. Never put driving pressure through the rolling elements, as this will damage the bearing and may cause injury. After installing the bearing, check that it rolls freely, with no binding.

Thrust washers are used to prevent excessive back-and-forth shaft movement. Retainer plates are devices used to hold gears and bearings in position. Collars are used to hold gears or bearings on shafts. Splines allow two rotating parts to move backward and forward in relation to each other while power is transferred through them. The internal splines on one component are designed to mate with external splines on another component. Keys and keyways keep two parts rotating with each other. The key is placed in two slots cut in the mating parts.

Know These Terms

Gear; Gear drive; Spur gear; Helical gear; End thrust; Herringbone gear; Double-helical gear; Bevel gear; Spiral bevel gear; Hypoid gear; Worm gear; Rack and pinion gear; Planetary gearset; Sun gear; Planet gear (Planet pinion); Ring gear (Annulus gear); Planet carrier (Planet-pinion carrier); Face; Face width; Root; Whole depth; Pitch circle; Circular thickness; Working depth; Clearance; Gear backlash; Gear ratio; Gear reduction; Direct drive; Overdrive; Countershaft gear assembly; Reverse idler gear (Idler gear); Wear; Failure; Damage; Pitting; Spalling; Peening; Scoring; Welding; Chain drive; Chain, Sprocket; Roller chain; Silent chain; Chain tensioner; Bearing; Plain bearing (Friction bearing); Antifriction bearing; Radial load; Axial load (Thrust load); Oil clearance (Running fit); Race; Ball bearing; Roller bearing; Straight roller bearing; Tapered roller bearing; Needle bearing; Thrust bearing (Torrington bearing); Brinneling; Thrust washer; End clearance (End play); Retainer plate; Collar; Spline; Splined shaft; Key; Keyway.

Review Questions–Chapter 3

Please do not write in this text. Place your answers on a separate sheet of paper.

1. Some gear drives are capable of providing an increase in speed relative to the drive input. True or false?
2. Which of the following statements regarding gear manufacture is *not* true?
   a. Gear blanks are made from a relatively hard steel.
   b. The faces of a gear are polished to a smooth finish.
   c. Gears are hardened in a heat-treating process.
   d. None of the above.
3. List and explain six types of gears.
4. Gear _____ is the amount of clearance or play between two meshing gears.
5. Describe what happens when gear teeth wear thin.
6. A failure condition of gears evidenced by flaking and chipping off of metal is known as _____.
7. A failure condition of gears evidenced by scratches on the tooth faces is known as _____.
8. Peening in gears may be caused by severe loads or by large foreign particles in the lubricant. True or false?
9. A drive sprocket has 10 teeth, and the driven sprocket has 40 teeth. What is the drive ratio?
10. A _____ _____ can be used to take up slack in a loose chain of a chain drive.
11. Chain drives are found in transaxles and in four-wheel drive transfer cases. True or false?
12. How do roller chains and silent chains differ?

13. What should you look for when inspecting a chain?
14. Sleeves, bushings, and insert bearings are all types of ______ ______.
15. What is color change in a bearing a sign of?

Certification-Type Questions–Chapter 3

1. **All of these are building blocks of the typical drive train EXCEPT:**
   (A) gears drives.
   (B) chain drives.
   (C) bearings.
   (D) belt drives.

2. **All of these are achieved through the action of gears EXCEPT:**
   (A) friction reduction.
   (B) torque multiplication.
   (C) power transmission.
   (D) direction reversal.

3. **All of these would result if bearings were not used in the drive train EXCEPT:**
   (A) excessive friction between moving parts.
   (B) wasted engine power.
   (C) increased turning speeds.
   (D) premature wearing of parts.

4. **Technician A says that gears used in light-service applications can be made of plastic. Technician B says that gears used in light-service applications are generally made of high-strength steel. Who is right?**
   (A) A only
   (B) B only
   (C) Both A & B
   (D) Neither A nor B

5. **All of these gears are used to connect parallel shafts EXCEPT:**
   (A) spur gears.
   (B) bevel gears.
   (C)herringbone gears.
   (D) helical gears.

6. **All of these are parts of a planetary gearset EXCEPT:**
   (A) sun gears.
   (B) spur gears.
   (C) planet gears.
   (D) ring gears.

7. **All of these are common gear problems EXCEPT:**
   (A) pitting.
   (B) scoring.
   (C) splining.
   (D) spalling.

8. **Chain drives transmit power in a straight line between:**
   (A) pulleys.
   (B) keyways.
   (C) CV joints.
   (D) sprockets.

9. **Technician A says that chain tensioners are designed to take up slack in chains. Technician B says that chain tensioners maintain pressure on the drive side of the chain. Who is right?**
   (A) A only
   (B) B only
   (C) Both A & B
   (D) Neither A nor B

10. When finding chain drive ratio, Technician A says that the number of teeth on the driven sprocket should be added to the number of teeth on the drive sprocket. Technician B says that the number of teeth on the driven sprocket should be divided by the number of teeth on the drive sprocket. Who is right?

(A) A only
(B) B only
(C) Both A & B
(D) Neither A nor B

11. Technician A says that it is not necessary to replace sprockets when a chain is replaced. Technician B says that it is not necessary to replace a chain when its sprockets are replaced. Who is right?

(A) A only
(B) B only
(C) Both A & B
(D) Neither A nor B

12. All of these functions are served by bearings EXCEPT:

(A) reducing friction between parts.
(B) guiding and supporting rotating parts.
(C) preventing damage from misalignment.
(D) transmitting power.

13. The two major classes of bearings are:

(A) plain and ring bearings.
(B) radial and axial bearings.
(C) plain and antifriction bearings.
(D) roller and antifriction bearings.

14. All of these are used as a coating for friction bearings except:

(A) brass.
(B) copper.
(C) chrome.
(D) lead.

15. Technician A says that if the oil clearance is too tight, the oil cannot form an adequate surface between mating parts. Technician B says that a loose oil clearance will allow oil to leak out too quickly to maintain surface separation. Who is right?

(A) A only
(B) B only
(C) Both A & B
(D) Neither A nor B

16. To increase service life, all parts of an antifriction bearing are made of:

(A) copper.
(B) low-carbon steel.
(C) high-strength steel.
(D) babbitt metal.

17. All of these are roller bearing variations EXCEPT:

(A) needle bearings.
(B) thrust bearings.
(C) ball bearings.
(D) Torrington bearings.

18. Technician A says that it is acceptable to apply a light force through the rolling elements of an antifriction bearing. Technician B says that it is acceptable to spin the rolling elements of an antifriction bearing with compressed air. Who is right?

(A) A only
(B) B only
(C) Both A & B
(D) Neither A nor B

19. Technician A says that an antifriction bearing can be heated to facilitate installation. Technician B says that an antifriction bearing can be cooled to facilitate installation. Who is right?

(A) A only
(B) B only
(C) Both A & B
(D) Neither A nor B

20. Thrust washers are used:

(A) to hold chain links.
(B) around bevel gears.
(C) to connect two sprockets.
(D) between the end of a rotating shaft and its housing.

21. Technician A says that thrust washers have grooves that allow oil to flow between the washers and the rotating parts. Technician B says that thrust washers are generally replaced when end clearance becomes too large. Who is right?

(A) A only
(B) B only
(C) Both A & B
(D) Neither A nor B

22. All of these are used to attach a retainer plate to a stationary housing EXCEPT:

(A) cotter pins.
(B) machine screws.
(C) bolts.
(D) snap rings.

23. All of these statements about collars are true EXCEPT:

(A) they are steel rings.
(B) they are a type of retainer.
(C) they should never be reused.
(D) they keep a shaft from rotating.

24. Technician A says that splines are used on sliding gears of manual transmissions. Technician B says that splines are used on clutch friction discs and on torque converter turbines. Who is right?

(A) A only
(B) B only
(C) Both A & B
(D) Neither A nor B

25. Which of these is one of the most common keys for automotive use?

(A) Spur key.
(B) Wood key.
(C) Woodruff key.
(D) Torrington key.

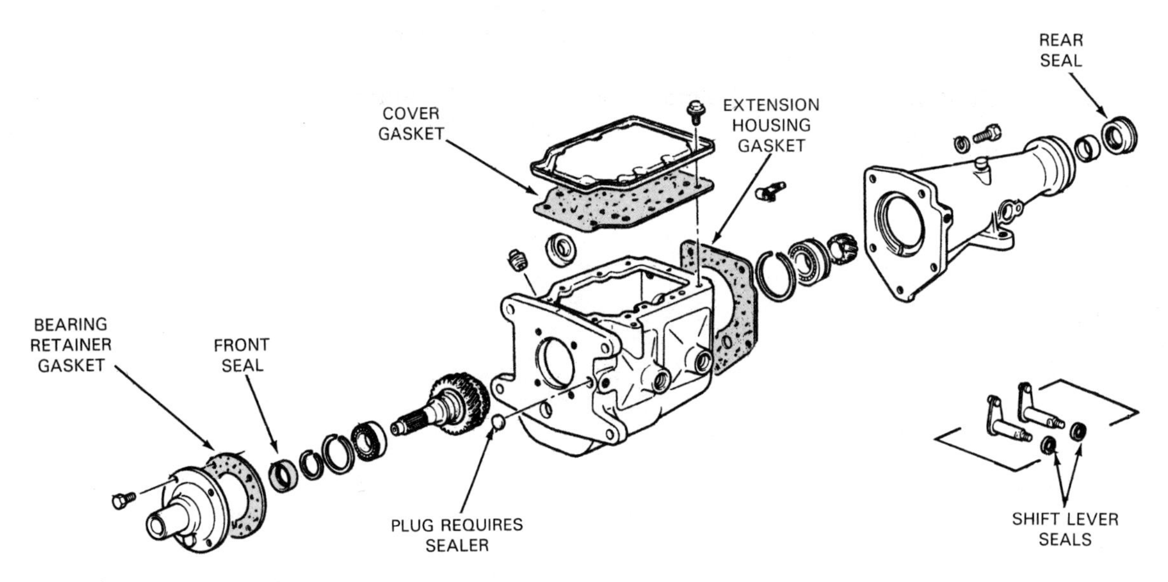

Note placement of gaskets and seals in this exploded view showing part of a three-speed manual transmission.

Chapter 4

Sealing Materials, Fasteners, and Lubricants

After studying this chapter, you will be able to:

- Discuss the types of seals found in a drive train.
- Point out typical problems to look for when replacing seals.
- Explain proper methods of removing and installing seals.
- Discuss the importance of proper gasket installation.
- Describe how sealants are used in drive train repair.
- Summarize the different lubricants used in a drive train.

Drive train components must have adequate *lubricant* to operate properly. This means there must be protection against leaks, which could reduce lubricant supply. Components must also be protected from damaging dirt and moisture. This chapter identifies the ways that lubricants are retained in major drive train components and the ways dirt and moisture are kept out. In addition, this chapter identifies different types of *fasteners* and major classes of drive train lubricants and tells where they are used.

This chapter will prepare you to perform some basic procedures required to successfully overhaul transmissions, transaxles, and other front and rear axle components. Later chapters will be referring to descriptions and procedures covered here.

Seals

Seals are devices that keep fluids, such as lubricant, within a component or desired space. They also keep contaminants, such as air, dirt, and water, from getting into such places. Seals may be used for one or both of these purposes. In some applications, they are used to separate two different fluids. Seals are usually used with rotating shafts or with shafts that slide back and forth axially. Special types of seals are sometimes used to seal stationary parts. Some seals provide controlled leakage of fluid to allow some fluid for lubrication of moving parts. These are called *nonpositive seals.*

Types of Seals

Seals are found throughout the drive train. They are used in transmissions, transaxles, and rear axle assemblies to seal rotating shafts. Stationary seals are used to seal many nonrotating parts in transmissions, transaxles, and drive shafts. There are several different types of seals. The type that is used depends upon the specific application. Seals commonly found in drive trains include *lip seals, ring seals, seal rings,* and *boots.*

Lip seals

Lip seals are used on rotating shafts in the drive train. A typical lip seal is shown in Figure 4-1. The casing is a rigid support for the other seal components. It is often made from stamped steel. The sealing element is designed to contact the rotating shaft. Most sealing elements are made from oil-resistant, synthetic rubber. The contact area is made in the form of a lip. A steel coil spring, often called a *garter spring,* or a *garter,* is wrapped around the sealing element. It places pressure on the sealing element to hold it tightly against the shaft. Some lip seals do not have a spring.

Figure 4-2 shows typical sealing action of a lip seal. Spring pressure plus *internal pressure*–fluid pressure in the seal *cavity*–cause the sealing element to be pressed tightly to the shaft. An automatic transmission *front pump seal* typifies this condition. If there were not pressure behind the seal, as would be the case for the transmission *rear seal,* the garter spring alone would still keep the sealing element in contact with the shaft.

A thin oil film (Figure 4-2) at the sealing lip reduces friction and provides additional sealing. If the seal is properly sized and installed, the oil film does not leak past the lip. Note that some lip seals have a felt *wiper* on the outside of the seal to catch small amounts of lubricant that get past the lip.

Different conditions require that different kinds of lip seals be used. Some of the variations are shown in Figure 4-3.

Figure 4-3A shows some of the ways that the garter spring seal can be made. Notice the various lip constructions and placements of the garter. The design of the lip and garter spring depends on the special sealing needs of certain shafts and in certain applications. Variations in pressure, shaft end play, and exposure to outside elements affect the design of the seal.

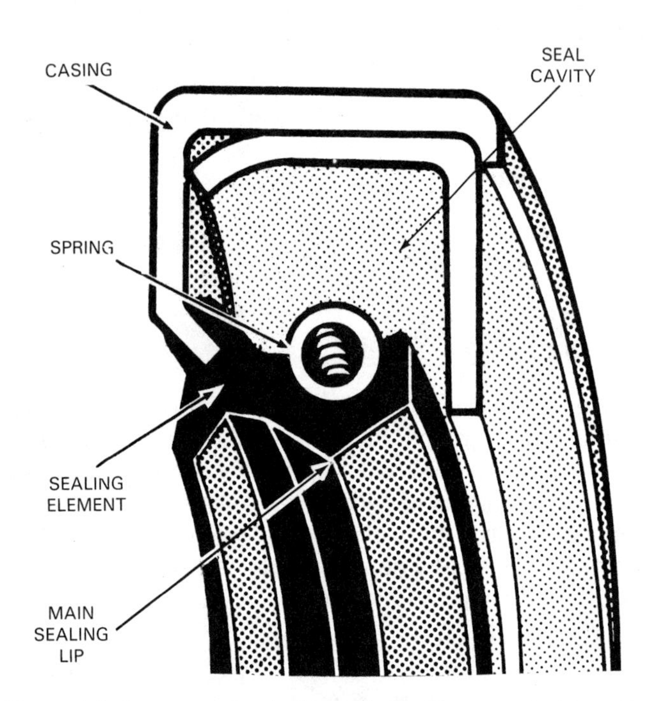

Figure 4-1. Components of a lip seal. *Casing* acts to hold seal in position when installed. It is usually press fit into a stationary housing. *Sealing element* has the lip that contacts the shaft. *Main sealing lip* is placed nearer, or toward, fluid to be retained. Optional *spring* puts pressure on sealing element to assure contact with the shaft at all times. Note that the shape of the components may vary. (Ford)

Figure 4-3B shows a variation of the lip seal. This type uses a *finger spring,* which is a cone-shaped series of *fingers* that place pressure on the sealing lip. The function of the finger spring is similar to the garter spring; however, the action of the finger spring is such that it causes less pressure to be placed on the shaft and lip. This reduces seal and shaft wear; however, some sacrifice is made in sealing ability. This type of seal is used in lower-pressure systems or for sealing components lubricated with grease.

Figure 4-3C illustrates lip seals that have no springs. These seals are used when the main requirement is to hold back contaminants. They may also be used in places where *internal* fluid pressure will force the sealing element against the shaft. These seals would not be appropriate for applications where they would be subjected to *external* pressure. Most seals under external pressure require a spring to insure a good seal.

Figure 4-3D shows seals that have more than one lip. These seals are used where external pressures are high or where seals are exposed to harsh conditions. The outer lip stops oil that has gotten past the inner lip, or main sealing lip, and it protects the inner lip from damage. *Double lip seals* are often used on drive train parts where there is exposure to the outside atmosphere. Examples include rear axle *pinion oil seals* and some transmission rear, or output shaft, seals.

A variation of the double lip seal has lips pointing away from each other, as shown in Figure 4-3E. These seals are used in some transaxles, when it is necessary to keep transmission fluid and differential gear oil separated.

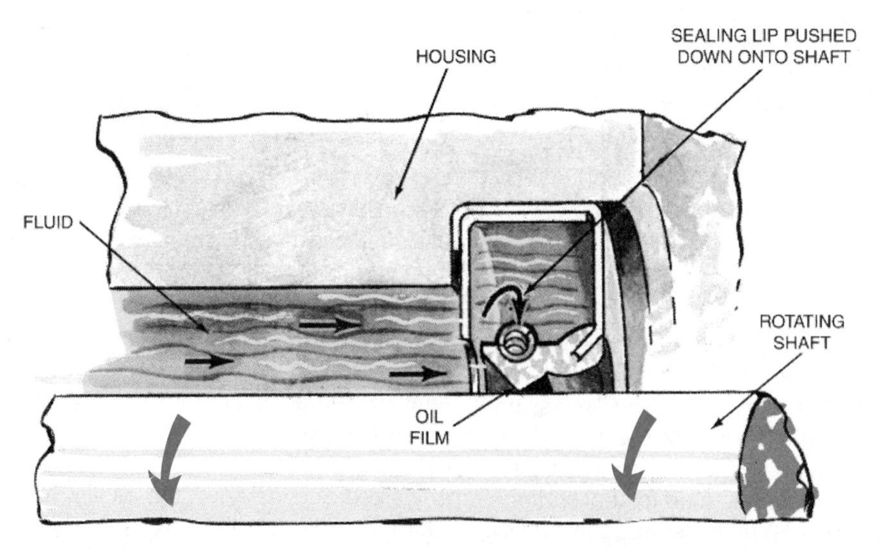

Figure 4-2. Note how a lip seal works. Pressure helps the spring hold the lip against the shaft. A film of oil between the lip and shaft helps to seal and reduces wear. Slight leakage at the lip is usually acceptable. (Caterpillar)

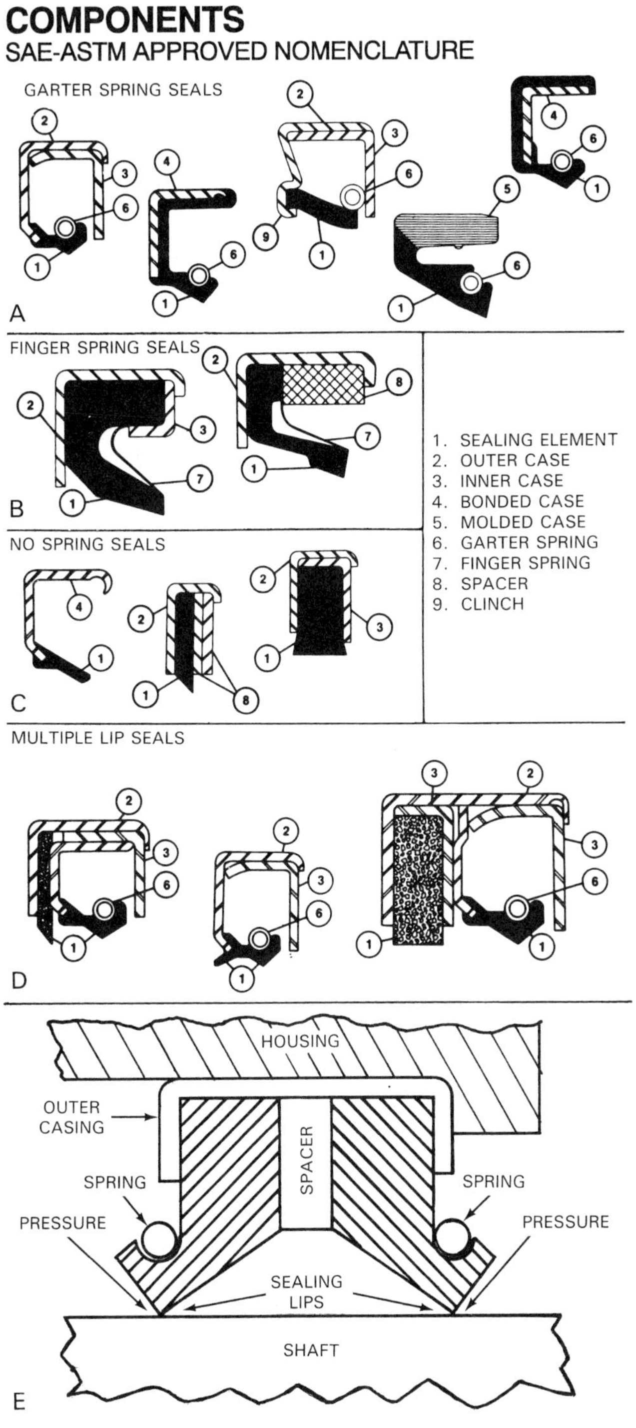

COMPONENTS
SAE-ASTM APPROVED NOMENCLATURE
GARTER SPRING SEALS
A
FINGER SPRING SEALS
B
NO SPRING SEALS
C
MULTIPLE LIP SEALS
D
1. SEALING ELEMENT
2. OUTER CASE
3. INNER CASE
4. BONDED CASE
5. MOLDED CASE
6. GARTER SPRING
7. FINGER SPRING
8. SPACER
9. CLINCH
HOUSING
OUTER CASING
SPACER
SPRING
SPRING
PRESSURE
PRESSURE
SEALING LIPS
SHAFT
E

Figure 4-3. Study different kinds of lip seals used in drive trains. A–Design of sealing lip and garter spring depends on usage. B–This variation of lip seal employs a finger spring. C–Some lip seals require no springs. Fluid pressure may force sealing element against shaft. D–Some seals may have two or more lips. These seals are often used to seal exposed drive train parts. E–Lips point away from each other on this double lip seal. These seals are used where it is necessary to keep two fluids from mixing together. (Federal Mogul)

Some seals are protected by a ***dust shield,*** Figure 4-4. Dust shields are used in places where the seal would be exposed to large amounts of dirt and water. Examples are the front wheel bearing assemblies on front-wheel drive vehicles and the transmission and pinion oil seals of rear-wheel drive vehicles. A massive assault from these elements would overcome the seal, and the seal would quickly be destroyed, allowing lubricant to leak out. The dust shield acts as a deflector or baffle to keep most of the dirt and water away from the seal. The seal can then handle the small amount of material that does get through to it. Dust shields are very effective in helping to reduce the occurrence of seal failure.

DUST SHIELD
BEARING AND SEAL HOUSING

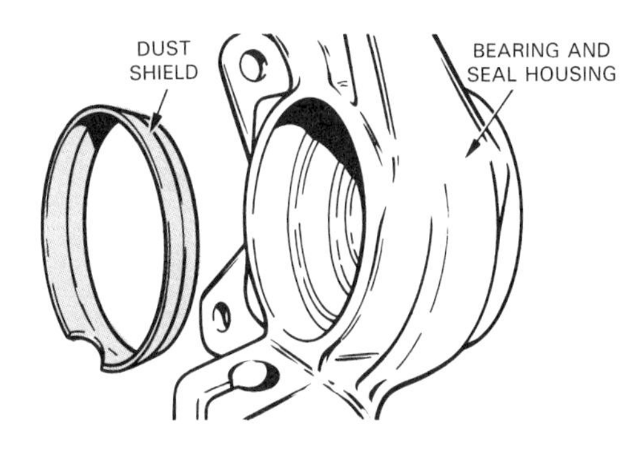

Figure 4-4. A dust shield keeps dirt and water away from sealing elements. This reduces seal failures in parts of drive train exposed to excessive dirt and water. (Ford)

O RING
D RING
SQUARE-RING

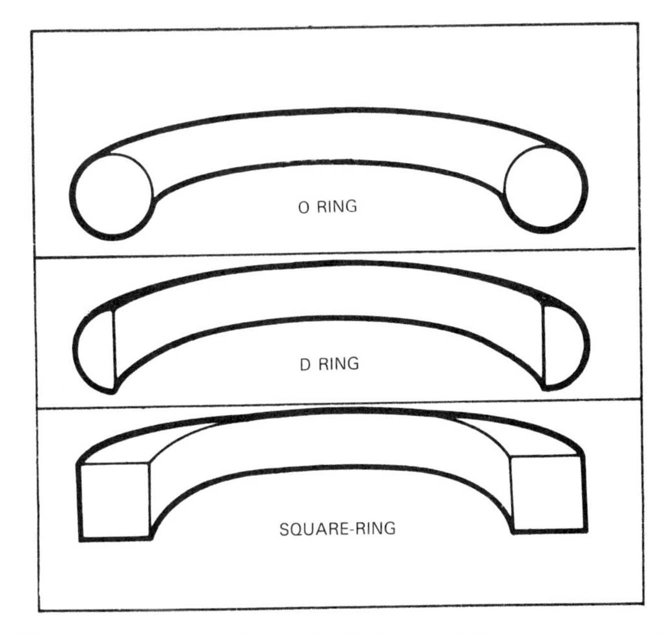

Figure 4-5. Types of ring seals. O-rings and D-rings are almost always made of synthetic rubber. Square-rings may be made of rubber, metal, or teflon. These rings are all designed for specific applications. (Ford)

Ring seals

Ring seals are used to seal stationary parts or hydraulic pistons that slide in their bores. The ring seal fits into a shallow groove cut into one or both parts. Ring seals can be round, half-round, or square in cross section. See Figure 4-5. In drive trains, the round seal, or ***O-ring,*** is most common. Most O-rings have a small diameter through the cross section, as compared to the *annulus,* or ring. O-rings are used with both stationary and moving parts, while *D-rings* and *square-rings,* as they are called, are usually used with moving parts.

Stationary ring seals fit into nonmoving parts and are found in many parts of the drive train. They are often called ***static seals.*** They are intended to form a perfect seal. Ring seals used on hydraulic pistons are found inside of automatic transmissions–on the clutch apply piston, for example. These are called ***dynamic seals.*** They are intended to allow slight fluid leakage, just as piston rings in an engine are expected to allow some *blow-by* of compressed gases. See Figure 4-6, which compares the two kinds.

Seal rings

Seal rings (not to be confused with ring seals) are metal or teflon rings. They act as seals for oil pressure that must be directed to or through passageways in rotating shafts or drums. Seal rings are nonpositive seals. They are designed to allow slight leakage for lubrication of moving parts. The rings are installed in grooves machined in the shaft. They are commonly used in automatic transmissions.

Metal seal rings resemble small piston rings. Some metal rings are locking types, with small latch tabs to hold the ring together. Other metal rings are the straight, or butt, type. This type depends on the size of the surrounding bore for a tight fit. See Figure 4-7.

Teflon seal rings are one-piece rings made of teflon and plastic. Teflon seals resemble square-cut ring seals (square-rings), although teflon seal rings are usually made in bright colors. They are often used in original installations because they allow lower-cost assembly.

Both types of seal rings seal in hydraulic fluid to maintain pressure. They do this by forming a mechanical seal between the moving parts. The outside diameter of the ring seats against the inner surface, or bore, of a rotating part. The side of the ring, moved by pressure, seats against one side of the shaft groove. This prevents practically all fluid leakage. See Figure 4-8.

The seal ring should rotate with the bore. If not, the bore surface will wear, allowing pressure loss. (The side of the ring will rotate against the shaft groove, and the groove will incur slight wear.) Note that seal rings used to seal *servo pistons* and *accumulators* (discussed in Chapter 10) do not rotate.

Metal seal rings can be removed without being damaged, although most manufacturers recommend replacing the rings when a transmission is overhauled. Teflon seal rings cannot be removed without being destroyed. In many cases, the original teflon rings can be replaced with metal rings.

Boots

A ***boot*** is a pleated, or accordian-like, flexible cover made of synthetic rubber or plastic. It is used, for example, to enclose constant-velocity U-joints used on front-wheel drive vehicles. The ends of the boot are fastened by straps or clamps–one end around the drive axle, one end around

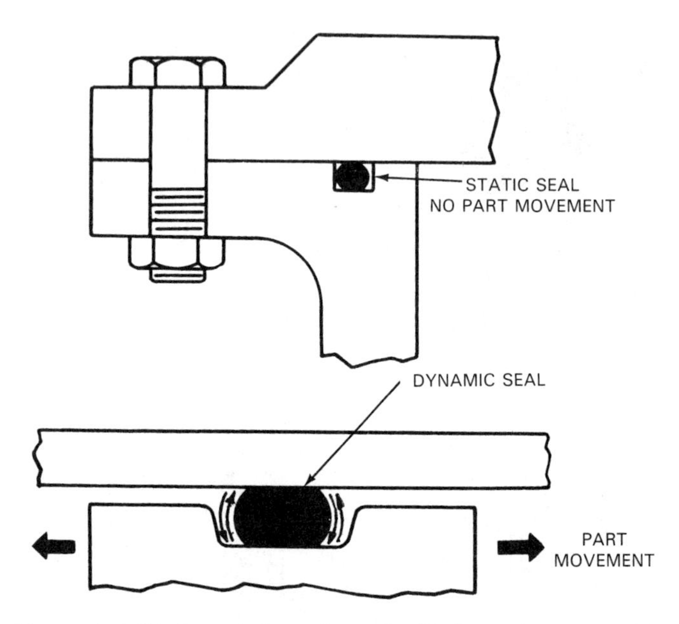

Figure 4-6. Static and dynamic seals. Static seals are used to seal parts that will not move in relation to each other. Dynamic seals seal where there is relative motion between two parts. (Deere & Co.)

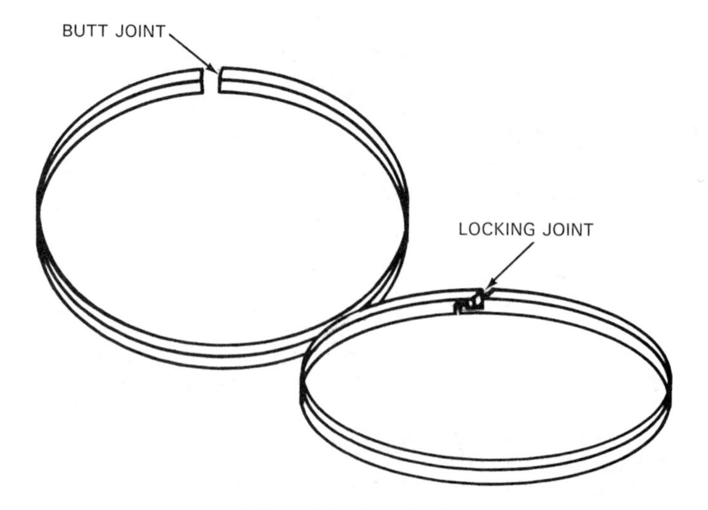

Figure 4-7. Seal rings with locking and butt joints are shown here. Be careful how you handle these as they are very brittle.

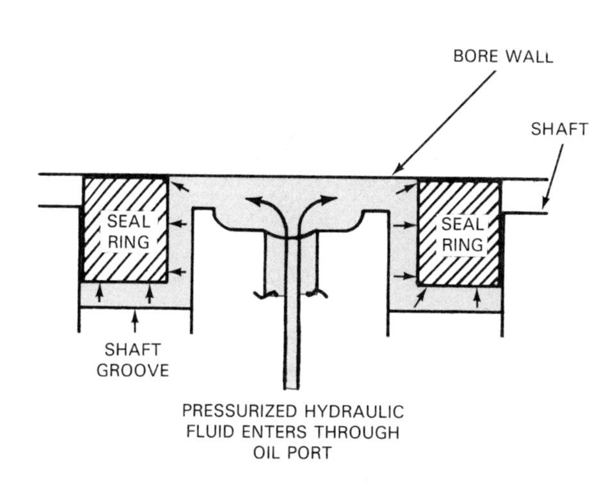

Figure 4-8. This is a partial view of a shaft in a bore. Seal rings are installed in the shaft grooves. Pressurized hydraulic fluid keeps seal rings tight against the side of their grooves.

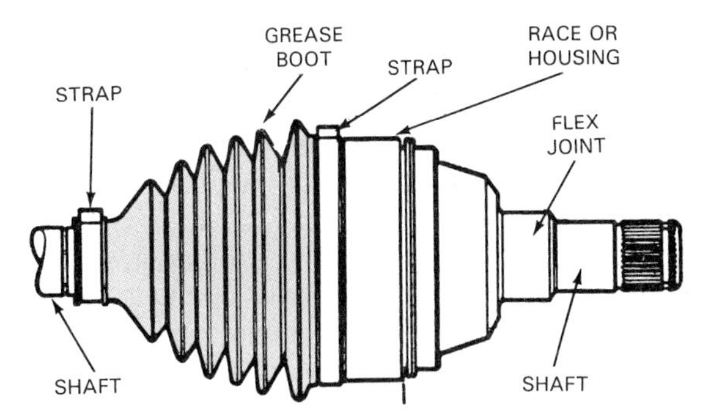

Figure 4-9. Boots are designed to retain grease and keep dust out of parts, such as constant-velocity U-joints. Boots such as these are usually found on front-wheel drive vehicles. (General Motors)

the CV joint yoke or outer race. The boot allows angular movement, while keeping out dirt and retaining grease. Look at Figure 4-9.

Seal Service

A leaking seal should be replaced when it is discovered. When a seal leaks *out,* a component will eventually run out of lubricant and be destroyed. The leaking lubricant often causes messy, slippery floors, and it may leak on and damage other vehicle components. A seal that leaks *in* will allow lubricant to become contaminated, which may result in damage to components.

Seal leaks

Almost all seal problems show up first as leaks. Seal leaks can be caused by several problems. The most common problem is a worn or damaged seal. All seals will wear out, given time, and it is not unusual for a seal in a vehicle with high mileage to leak. This is usually caused by wear and hardening of the sealing element. Sometimes, the leak occurs because a lip seal garter spring becomes weak. Figure 4-10 points out areas of a lip seal that commonly wear. Ring seals usually become hard and shrink. When this happens, they no longer conform to the grooves that they are sealing. A new ring seal usually solves the problem.

Sometimes, the shaft or housing for the seal wears out, causing a leak. When replacing a lip seal having many miles of service, always check the shaft for signs of wear. The shaft tends to wear where the sealing lip contacts it. See Figure 4-11. If this area is not worn, the standard replacement seal will work well. Many times, however, the shaft will be so worn that the standard replacement will not

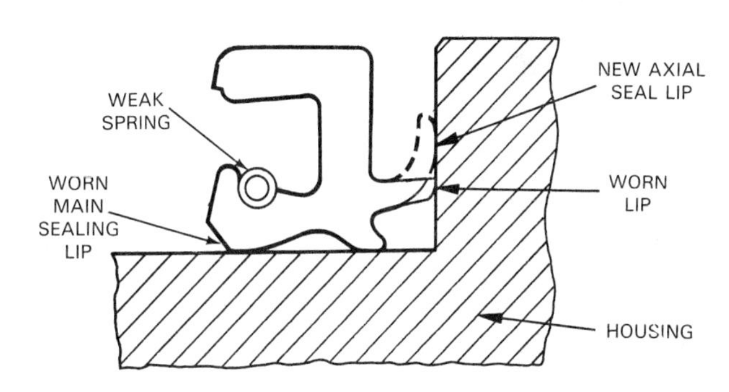

Figure 4-10. Depicted are conditions of wear and damage that can affect performance of a lip seal. Any one of these problems can cause a leak. (Ford)

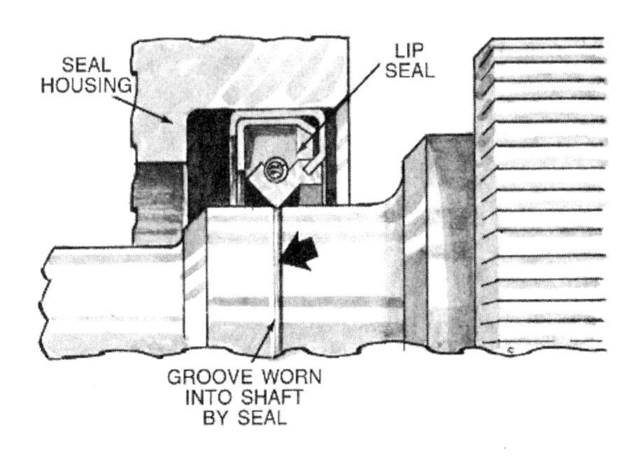

Figure 4-11. A groove worn in a shaft can cause a leak. A new seal alone may not solve this problem. In many instances, you will find that the shaft must be replaced or a sleeve must be installed. (Caterpillar)

correct the problem. In most of these instances, the shaft must be replaced.

In some cases, however, the problem can be corrected by using a replacement seal that will contact the shaft to either side of the contact area of the original seal. Sometimes, on more expensive parts, the shaft can be repaired by pressing a sleeve on over the worn shaft area. A seal is then fitted to the sleeve. The sleeve adds to the diameter of the shaft, so a larger lip diameter will be required to fit the sleeve.

When checking a shaft for signs of wear, check for excessive play in the shaft. If there is too much *lateral,* or side-to-side, movement, the seal cannot form a good sealing surface. The shaft or shaft bearings should be replaced. If a seal seems to have failed at low service mileage, suspect shaft or housing problems; check them out. Also, check for excessive dirt or metal particles, which will cause wear sooner than normal.

One of the most common causes of seal leakage following a repair is using the wrong seal. The wrong size lip seal will leak almost immediately. An improperly sized ring seal may seal for a few minutes or hours, but it will eventually leak. A seal that is not the right size should never be used. See Figure 4-12.

The other most common cause of seal leakage after a repair job is improper installation, evidenced by a cocked or distorted seal. See Figure 4-13. O-rings that are improperly installed will leak immediately. A lip seal that is not perfectly aligned with the shaft will not make good contact with the shaft. The seal will begin leaking, wear unevenly, and possibly damage the shaft.

Shaft or housing damage attributed to other causes can damage a seal and cause it to leak. Examples of bore and shaft damage are shown in Figure 4-14. Scratches and dents from hammering out an old seal, grooves, sharp keyway edges, corrosion, and improper machining can all cause seal leakage. A loose or bent shaft will cause seal distortion and leaks, even with a new seal. All of these problems should be looked for and eliminated before installing a new seal.

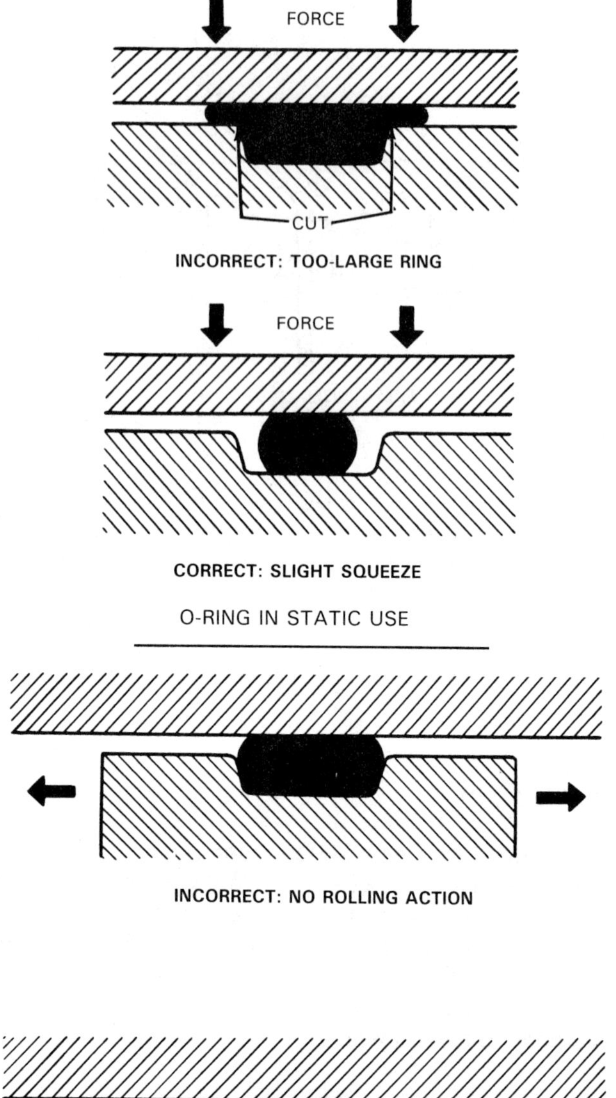

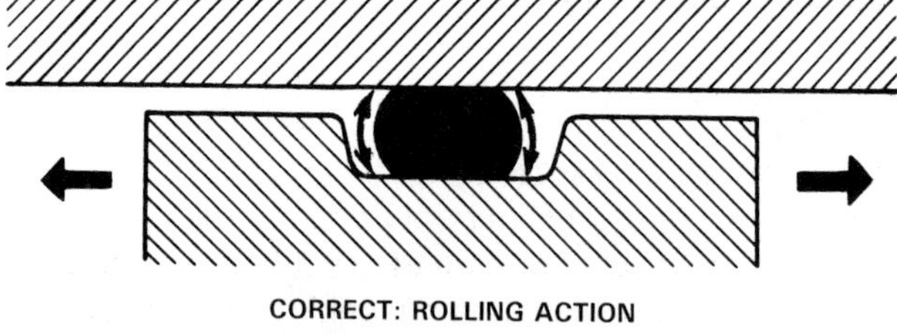

Figure 4-12. Proper sizing of the O-ring is important. Note what happens when proper and improper sizes of static and dynamic seals are used. (Deere & Co.)

Seal removal

The following precautions should be observed during any seal removal operation:

- Whenever you are using a seal puller or driving the seal from its housing, *always* wear eye protection.
- Do *not* damage the seal bore or the shaft.
- Check all related parts for damage that could result in a seal leak.

Seals should be removed carefully. Most static O-ring seals are compressed between two stationary drive train parts. Some O-rings are held by a retainer plate. The plate or component is disassembled, and the O-ring can be lifted or pried out of its groove. Dynamic O-rings are removed by removing the piston from its cylinder to expose the O-ring. The O-ring is then carefully pried from its groove.

Most lip seals are pressed into a bore that is part of a drive train component. A special tool, such as that shown in Figure 4-15, should be used to pry out the old seal. Alternate pressure on the seal from side to side. This will help the seal come out easily, with minimum damage to the bore.

Another way to remove the seal is to use a slide hammer with a special puller jaw. The jaws are pushed through the seal and then expanded. In lieu of this, drill a small hole on the side of the seal in the metal casing. Thread a screw attached to a slide hammer into the hole. Using the slide hammer, then, pull the seal from the bore.

To remove the seal evenly and prevent it from becoming wedged, drill two holes across from each other. The slide hammer can then be alternated to remove the seal.

If you can reach the seal from behind, you can drive it out of its bore with a seal driver or a hammer and a soft metal drift. Be careful not to damage the housing!

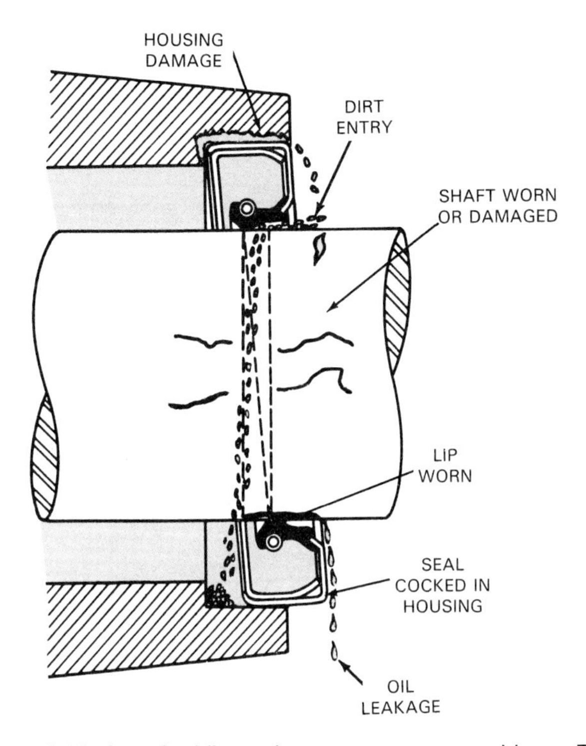

Figure 4-13. A cocked lip seal can cause many problems. The most obvious problem is leaking, but a cocked seal can also cause shaft and housing damage. The housing will usually be damaged when the seal is installed. The shaft will wear because of the uneven pressure placed on it by the lip. (Deere & Co.)

Once the old seal has been removed, check the housing, shaft, and all other related parts. If leakage was a problem, these parts may provide clues as to the cause. Correct any problems before installing a new seal.

If a standard replacement seal is to be used, check its part number against the old seal's to ensure that they match. If they do not, it could be that a new number has since been assigned to the particular seal. Otherwise, it is possible that the old seal was the wrong one. If there was a problem with the original seal, this could well have been the cause.

Seal installation

Two general points should be made in regard to seal installation prior to discussing specifics. First, a seal should not be removed from its wrappings until ready to be installed. This will help keep the replacement seal clean.

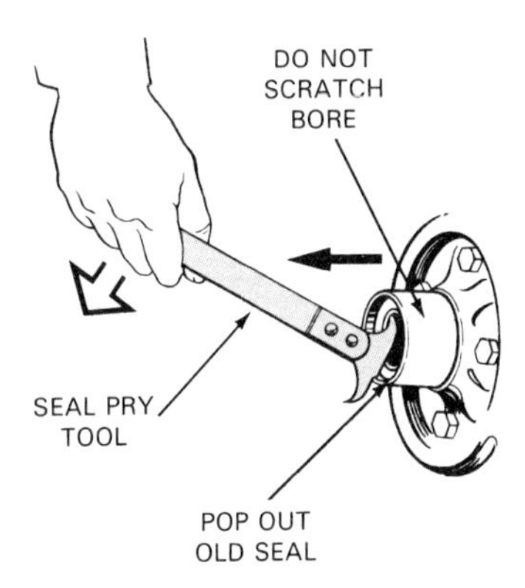

Figure 4-15. The best way to remove a seal is with a special *seal puller.* If a puller is not available, seals can be removed by several other methods. Always be careful not to damage the seal housing during removal. (Lisle)

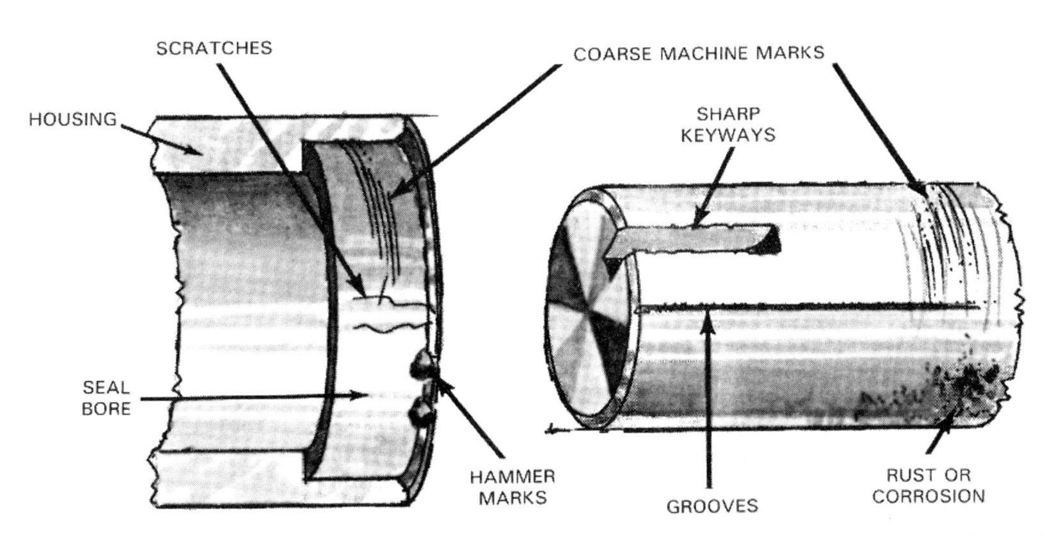

Figure 4-14. Defects in the shaft and housing can ruin a seal during installation. Sharp edges on shafts can cut the seal as it is slipped into place. Housing defects can damage the seal casing. (Caterpillar)

Also, the seal contact surfaces should be coated with lubricant prior to assembly, Figure 4-16. For this, you should use the same kind of lubricant used in the part or assembly.

If a ring seal is to be installed, lubricate the groove. Carefully install the ring in the groove. If an O-ring is being used between two stationary parts (a static seal), make sure that it fits properly in the groove before reassembling the component. If the O-ring is the dynamic type, check it for proper size around its piston and in its bore.

If installing a lip seal, lubricate the shaft as well as the sealing lip. Again, use the correct lubricant. If the seal has a metal casing and it is not precoated by the manufacturer, coat the casing lightly with a *nonhardening sealer,* Figure 4-17.

If the seal will be installed into a housing bore with the shaft already installed, make sure that the shaft is free from burrs or scratches that could damage the seal. If necessary, polish the shaft with *crocus cloth,* which is an abrasive cloth that has very fine particles of iron oxide. In addition, to protect a seal from drilled holes, splines, keyways, square shaft ends, and other sharp spots, a temporary collar may be formed over the spot with a piece of smooth, heavy paper; otherwise, a plastic sleeve may be placed over the spot.

Slide the seal over the shaft until it reaches the housing, making sure the seal is oriented in the proper direction. Drive the seal in with a driver that closely matches the outside diameter of the seal. Drive it in squarely, until it is properly seated. After installation, check to make sure that the garter spring (if used) did not come loose and that there was no damage from the driving action. Figure 4-18 shows a seal driver being used to install a seal in a bore.

In an emergency, a large socket or pipe can be used as a driver, if it is the right size. It is also possible to install a seal with a hammer, if done carefully. The seal should be lightly tapped in. Alternate hammer taps from side to side, until the seal is seated. Failing to do it this way will cause the seal to become cocked and bent.

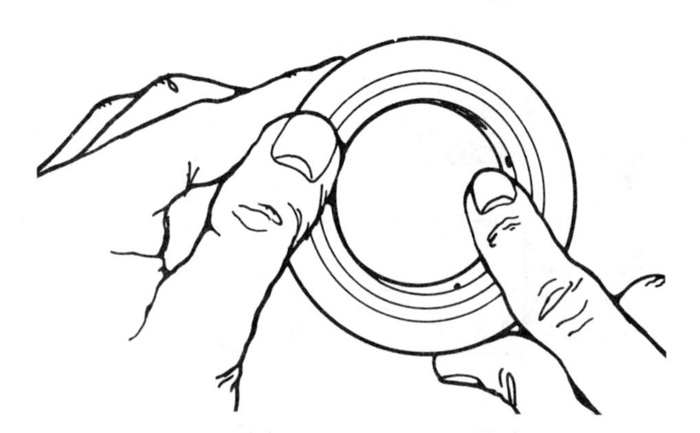

Figure 4-16. Always lubricate the sealing lip of a new seal before installation. Prelubrication will make the seal easier to install and will provide an oil film at the lip during break-in. (Ford)

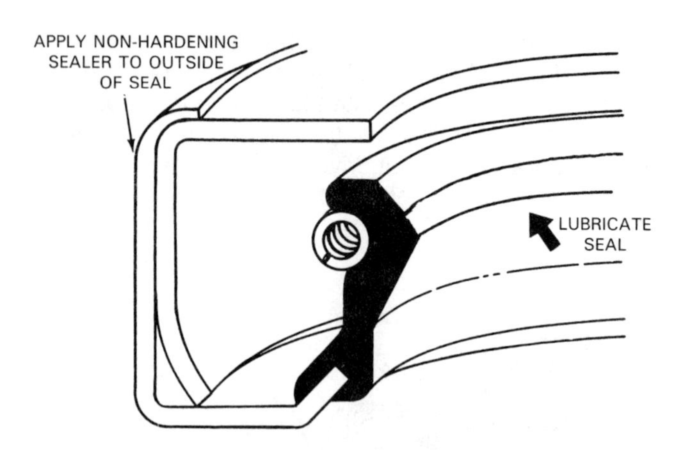

Figure 4-17. If the seal has a metal casing, apply nonhardening sealer to the casing before installation. This step is not necessary if the casing has a flexible coating. (Deere & Co.)

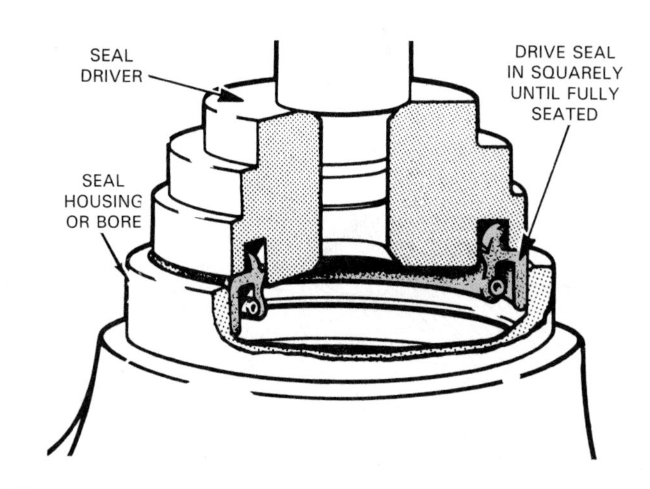

Figure 4-18. An improperly installed seal will be ruined immediately. Watch the seal carefully as it is being installed to ensure that it goes in squarely with the bore. (Ford)

Gaskets

Gaskets are a kind of static seal, used on nonmoving parts. Gaskets used on drive trains range from the simple gasket used to seal a differential inspection cover, to the complex gaskets that seal automatic transmission valve bodies. Gaskets are never used to seal moving parts.

Gaskets used on vehicle drive trains are normally cut from a single piece of gasket material. In contrast, some *engine* gaskets are made of multiple layers of materials. Gaskets are used to prevent the leakage of lubricant between two parts bolted together. There are *external* and *internal* gaskets. ***External gaskets*** keep fluid from leaking out of a component. They also keep out dirt or moisture. ***Internal gaskets*** prevent loss of pressure or crossover of fluids within a component.

Gasket material is made to be flexible. This allows gaskets to seal by conforming to imperfections in mating surfaces. Figure 4-19 is an exaggerated illustration of this. When the parts are tightened, the gasket compresses and deforms, filling in the low spots, scratches, and other imperfections in the mating surfaces. This results in a leakproof seal.

Gasket Materials

The majority of gaskets used in drive trains are made of oil-resistant paper. These are sometimes called *chloroprene gaskets.* This material is often mixed with a substance such as rubber to add flexibility and strength.

Where mating parts will be subject to expansion and contraction, gaskets made from a synthetic rubber called *neoprene* are often used. Natural rubber cannot be used because petroleum-based lubricants, like gear oil and transmission fluid, will make it swell and fail in service.

Some gaskets are made of copper or aluminum. These gaskets have a ridge running around their perimeter. When mating parts are tightened down, the ridge is compressed, providing the seal.

Thick gaskets of cork or felt are usually used in providing seals for assembly covers made of sheet metal. The thickness of the gasket helps to compensate for distortion of the sheet metal cover, Figure 4-20. These thick gaskets are often coated with teflon to reduce sticking when the gasket is removed.

Some gaskets are made from a silicone sealer. These are called ***form-in-place gaskets.*** The silicone normally comes in a tube. It is applied directly to one of the mating part surfaces. The parts are assembled while the silicone gel is still wet to the touch. More will be said about form-in-place gaskets later on in this chapter.

Gasket Problems

Gaskets are sometimes a source of problems. More often, they are blamed for problems really caused by other components. If a gasket *appears* to be leaking, make sure that it is not a cracked part that is leaking or a nearby component leaking onto the gasket area.

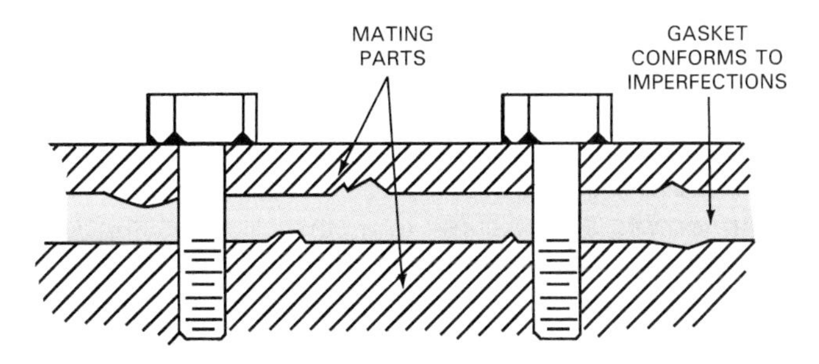

Figure 4-19. Gaskets are made of relatively soft materials. This flexibility enables the gasket to conform to imperfections in the gasket mating surfaces.

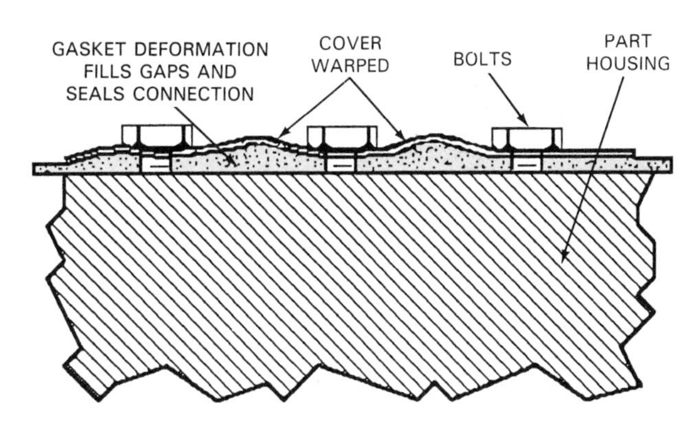

Figure 4-20. Thick gaskets can sometimes compensate for warped sheet metal cover plates or pans.

Some gasket leaks are caused by loose bolts. Tightening the bolts properly may stop the leak. Sometimes, however, pressure from within will tear out a piece of gasket. This is called a ***blown gasket.*** Blown gaskets are usually caused by loose fasteners or warped mating parts that do not tightly hold the gasket. Also, *over* tightening attaching bolts can crush the gasket, resulting in a weak spot that will blow out much more easily.

Gasket Service

Gasket service involves disassembling parts, removing the old gasket, cleaning mating surfaces, and installing a new gasket. Some basic rules for removing and replacing gaskets are:

- *Verify* the cause of the problem before disassembly. Often, what appears to be a gasket problem is a defective part or an improper part installation.
- *Avoid* causing part damage during disassembly or gasket removal. Although the gasket material is designed to compensate for scratches, dents, and nicks, don't push your luck. Aluminum parts are especially easy to damage.
- *Thoroughly* clean the gasket surfaces. If the old gasket must be scraped off, use a dull scraper on the mating surfaces to avoid damaging them. This is especially important for automatic transmission valve bodies, where the slightest scratch can cause an internal leak and pressure loss.
- *Thoroughly* wash the parts with solvent and blow dry with air pressure. *Avoid* using rags to dry internal drive train parts. Rag lint could damage bearings or clog passages and valves.
- Check the new gasket shape. It should conform to the mating part surfaces and the old gasket. Make sure that the gasket contains all of the required holes and that all gasket holes match holes in the mating parts.
- Use sealer *only* if needed, especially where automatic transmissions and transaxles are concerned. If the

manufacturer recommends using sealer, place a thin coat on the gasket.

- Hand-tighten all bolts before using a wrench. Make sure that all bolts will start in their respective holes.
- Tighten the bolts in sequence. If a *tightening sequence* is not given, start from the middle and work toward the edges. Alternate the tightening in a crisscross pattern.
- Tighten all bolts in small steps. After the final tightening sequence, recheck all of the bolts.
- Do *not over*tighten the bolts. This is especially important when installing pans or other sheet metal parts.

Making a gasket

Sometimes, a replacement gasket is not available, and you must make a gasket from a sheet of gasket material. Simple gaskets can be made easily by either of two methods.

One method is to lay out a sheet of gasket material on the component that requires a gasket or lay the component over the gasket material. Use a pencil to trace the outlines of the sealing surface on the gasket material. Use scissors to cut out the gasket material to the desired pattern.

The second method is to lay the sheet of gasket material over the component and use a brass or ball peen hammer to lightly tap the gasket against the sharp edges of the sealing surface. Tap out one or two bolt holes first. Then, thread bolts into the holes to hold the gasket material. Go back to tapping around the sealing surface until you obtain the proper gasket shape. Trim any rough edges with scissors.

Sealants

Sealants, or sealers, are used for various reasons. They can improve the sealing ability of a gasket. They can take the place of a gasket. They can be used to hold a gasket in place during reassembly. Sometimes, sealers are used to improve the sealing ability of dynamic seals and hose connections.

Types of Sealers

There are several types of sealers. They have different properties and are designed for different purposes. Always read the manufacturer's label and a service manual before selecting a sealer. Study use and characteristics of sealer types shown in Figure 4-21.

Hardening sealers

Hardening sealers are used on parts that will remain assembled for long periods of time, possibly the life of the vehicle. They are used on some threaded fittings and fasteners, for example. Sometimes, they are used for filling uneven sealing surfaces. Hardening sealers are usually resistant to heat and chemicals.

Nonhardening sealers and shellac

Nonhardening sealers are used on parts that will be taken apart occasionally, such as transmission pans and differential covers. They are also used on some *threaded fasteners* and fittings and on hose connections. They are resistant to most chemicals and to moderate heat.

Shellac is a gummy and sticky substance that remains pliable. It is frequently used on gaskets of fibrous materials, both for sealing and for holding gaskets in place during assembly. Shellac could be classified as a nonhardening sealer.

Form-in-Place Gaskets

Form-in-place gaskets are sometimes used instead of conventional gaskets. They are applied, in fluid form, directly to parts to be sealed. The gasket is formed when the parts are tightened together and the substance *cures* (dries, or sets up). Form-in-place gaskets are commonly made of either *RTV sealer* or *anaerobic sealer.*

RTV sealer

RTV (room temperature vulcanizing) sealer is silicone sealer. It cures to a material of rubberlike consistency, which forms the gasket. RTV sealer is usually used on thin, flexible flanges.

RTV sealer normally comes in collapsible tubes. The substance is usable for some time after a tube is opened. After the expiration date, the sealer may not cure; in which case, it would not make a good seal. Always check the expiration date on the package before using the sealer. Use a new tube if in doubt.

To apply the RTV sealer, Figure 4-22, lay about a 1/8-in. (3-mm) continuous bead all around the edge of the part. Keep the bead to the inside of the mounting holes–preferably, encircle them. Do not permit any breaks in the bead; a leak will result. Components should be assembled and tightened while the RTV sealer is still wet to the touch. You should have about 10 minutes to get the parts assembled before the sealer dries. Should it be necessary, RTV sealer may be removed with a rag before it cures.

Anaerobic sealer

Anaerobic sealer cures to a plasticlike substance in the absence of air. It will remain fluid as long as it is exposed to the atmosphere. Anaerobic sealer is used with tightly fitting, thick parts, which air cannot penetrate. It is used between two smooth, true surfaces, not on thin, flexible flanges, which may be penetrated by air. Note that anaerobic sealers will not harden in their containers

TYPE	TEMPERATURE RANGE	USE	RESISTANT TO	CHARACTERISTICS
shellac	−65° to 350°F (−54° to 177°C)	general assembly: gaskets of paper, felt, cardboard, rubber, and metal	gasoline, kerosene, grease, water, oil, and antifreeze mixtures	dries slowly sets pliable alcohol soluble
hardening gasket sealant	−65° to 400°F (−54° to 205°C)	permanent assemblies: fittings, threaded connections, and for filling uneven surfaces	water, kerosene, steam, oil, grease, gasoline, alkali, salt solutions, mild acids, and antifreeze mixture	dries quickly sets hard alcohol soluble
nonhardening gasket sealant	−65° to 400°F (−54° to 205°C)	semi-permanent assemblies: cover plates, flanges, threaded assemblies, hose connections, and metal-to-metal assemblies	water, kerosene, steam, oil, grease, gasoline, alkali, salt solutions, mild acids, and antifreeze solutions	dries slowly nonhardening alcohol soluble

Figure 4-21. This table shows some types of sealers and their uses. No one sealer can be used for every sealing job. (Fel-Pro)

because the air space in the containers provides enough air to prevent hardening.

Anaerobic sealer should be applied sparingly. Lay about a 1/16-in. (2-mm) diameter bead on one surface only. It is recommended that the sealer encircle each mounting hole. The parts should be assembled and tightened within 15 minutes of using the sealer.

Note that before any kind of sealer is applied, there are a few preliminary steps that must be taken. All loose material and lubricant must be removed from the part contacting surfaces. To do this, scrape or wire brush the part surfaces at these locations. Check that part surfaces are reasonably flat where the material is to be applied. Use a shop rag and solvent to wipe off any oil and grease on the surfaces. The sealing surfaces must be clean and dry before applying sealer. Finally, before using any kind of sealer, always refer to the manufacturers' instructions to be sure that you use the right kind.

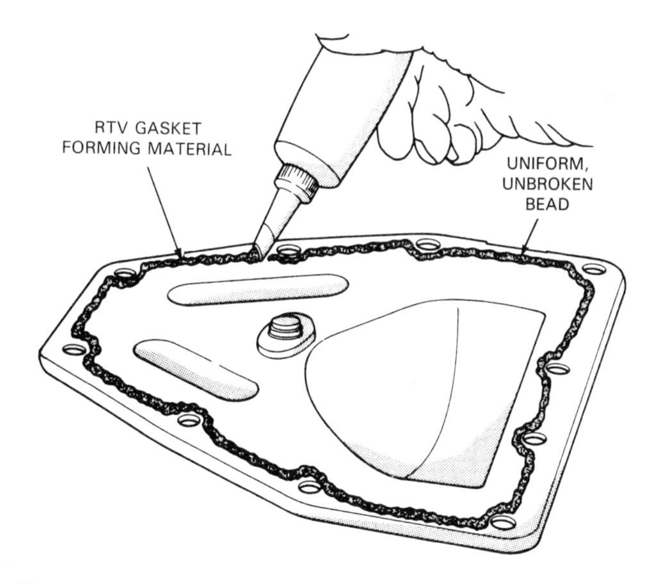

Figure 4-22. RTV gasket forming material can be used to make gaskets on many parts. (Chrysler)

Thread-Locking Compounds and Thread Sealers

Thread-locking compounds are used to prevent threaded fasteners, such as *bolts, screws,* and *nuts,* from loosening. These compounds are a kind of anaerobic sealer. When they are applied to a threaded fastener and the fastener is tightened, the lack of air causes them to harden. Thread-locking compounds make the removal of a fastener very difficult. They should be used only when the vehicle manufacturer recommends them.

Before applying thread-locking compound, threads should be cleaned and sprayed with a special primer. This will reduce the setting time of the compound. Apply the compound, as shown in Figure 4-23. Then, install the fastener.

It will require considerable turning effort to loosen a fastener that was installed with thread-locking compound. The fastener will also be more difficult to remove once the "lock" is broken. This is because the thread-locking compound remains in the threads, which tends to jam the fastener. If a fastener will not loosen, light heat may soften the compound and allow the lock to be broken.

Thread sealers are used to seal any threaded fasteners that extend into the oil-lubricated interior of a drive train component. Oil will seep out through the threads if a sealer

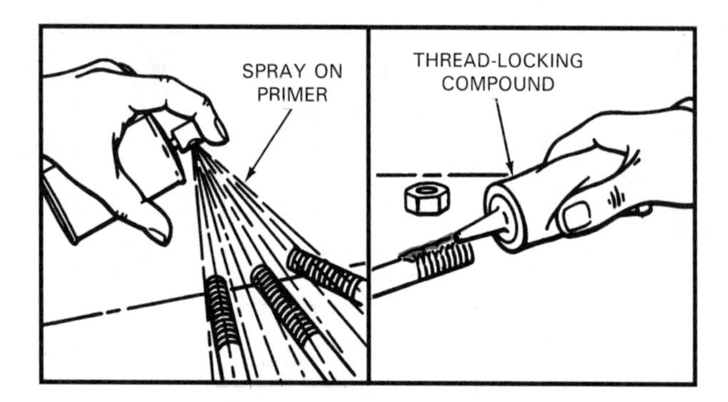

Figure 4-23. Thread-locking compounds are used when a threaded fastener must not loosen. Thread-locking compounds remain liquid in the container and can be easily applied to the threads of the fastener. They harden and form a solid bond when the fastener is tightened. (Deere & Co.)

is not used. Thread sealers should be applied to the fastener threads only after the threads have been thoroughly cleaned.

Adhesives

Adhesives are used for holding parts together. They are *not* used as sealants. Adhesives are sometimes used to hold gaskets in place or small parts together during assembly. Before applying adhesives, make sure that the surfaces to be bonded are clean. Some adhesives are applied in a two-step process; a *setting agent* is applied first, and then, the adhesive is applied. Other adhesives are two-part mixtures applied in a one-step process. These are sometimes called ***epoxies.*** Epoxies are mixed just before application.

Fasteners

Fasteners are devices used to hold, or fasten, parts together. Fasteners can be divided into threaded and nonthreaded varieties. Nuts, bolts, and screws are all types of ***threaded fasteners.*** Snap rings are a type of ***nonthreaded fastener.***

Threaded Fasteners

The most common kinds of threaded fasteners used on drive trains are nuts and bolts. A ***bolt*** is a threaded rod with a *head* on one end. The head is usually six sided (*hex head*), although some special bolts have square or round heads. A ***nut*** has inside threads and, commonly, a six-sided outer shape.

Bolts can be threaded into nuts or parts with threaded holes. A bolt that threads into a part instead of a nut is called a ***cap screw.*** The threads of the bolt and nut or threaded part provide a holding force to keep parts assembled.

Screws resemble bolts, but they are generally smaller than bolts. In addition, they are designed to thread directly into a part or material; they are not designed to accept nuts. Frequently, their heads are designed to accept screwdrivers. ***Machine screws*** are common on drive trains. They are usually smaller than cap screws and are flat across the bottom.

Fastener classification

Fasteners are made in many different sizes, strengths, and thread patterns. Fasteners are classified accordingly.

Screw size terminology. There are a variety of terms used to refer to size of threaded fasteners. The automotive technician must know what these mean. Some of the most commonly used terms relating to fasteners are:

- ***Major diameter.*** The larger diameter pertaining to screw threads. Applies to both internal and external threads.
- ***Minor diameter.*** The smaller diameter pertaining to screw threads. Applies to both internal and external screws.
- ***Bolt size.*** Refers to the major diameter of the bolt threads.
- ***Nut size.*** Refers to the minor diameter of the nut threads.
- ***Head size.*** Size of hex head measured across the *flats.* This is the wrench size.
- ***Bolt length.*** Length measured from the bottom of the bolt head to the end of the threaded section.
- ***Thread length.*** Measure of external thread length.
- ***Thread pitch.*** The distance measured across crests of adjacent threads. *Thread notes* on drawings of metric fasteners give pitch in millimeters. *Conventional* (nonmetric) *fasteners* denote the number of *threads per inch,* which is the reciprocal of pitch. (Conventional threads conform to thread standards in the United States, Canada, and England.)

Thread types. Fasteners used in drive trains use three basic types of fastener threads:

- ***Coarse threads.*** Called *UNC,* which stands for *Unified National Coarse.*
- ***Fine threads.*** Called *UNF,* which stands for *Unified National Fine.*
- ***Metric threads.*** Called *SI,* which stands for *International System of Units* (English translation).

Never attempt to interchange thread types. The threads will be damaged. Metric threads could be mistaken for conventional threads if they are not inspected carefully. A ***thread pitch gauge,*** either conventional or metric, should be used to check any threads that you are unsure of. See Figure 4-24.

In addition to these thread variations, there are *right-* and *left-hand threads* in use on modern vehicles. A bolt or screw with ***right-hand threads*** must be turned clockwise to tighten. Right-hand threads are the more common. A fastener with ***left-hand threads*** must be turned counter-clockwise to tighten. Sometimes, the letter *L* is stamped on fasteners with left-hand threads. Left-hand threads are reserved for special situations.

Bolt grade. ***Bolt grade*** refers to the amount of *pull,* or *tension,* that a fastener can withstand before it stretches or, in some cases, breaks. (Ductile materials will stretch before breaking; brittle materials will break almost right away when a *tensile force* is applied.) Bolt grade, then, is an indication of ***tensile strength,*** which is measured in pounds per square inch. Tensile strengths vary among bolts of the same size and thread pattern. Bolts are made of different metals, some stronger than others.

Bolt head markings are used to identify the tensile strength of a bolt. These markings are called ***grade markings.*** Conventional bolts are marked with lines, or slash marks–the more lines, the stronger the bolt. Grade 8 (6 lines) is the strongest, highest quality. A metric bolt is marked with a numbering system–the larger the number, the stronger the bolt. Bolts range from 4.6 to 10.9. See Figure 4-25.

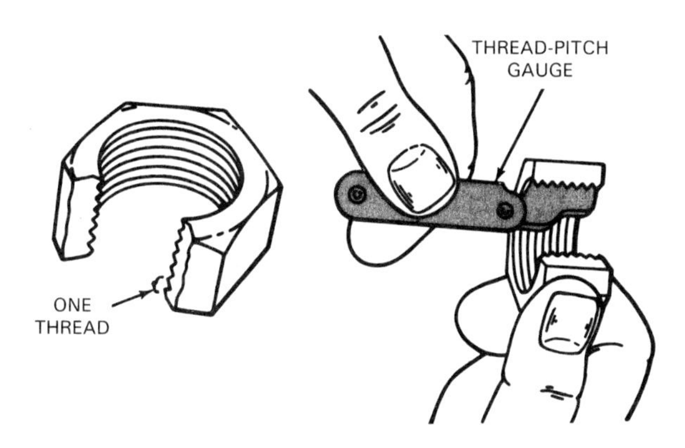

Figure 4-24. A thread-pitch gauge can be used to check thread pitch or number of threads per inch.

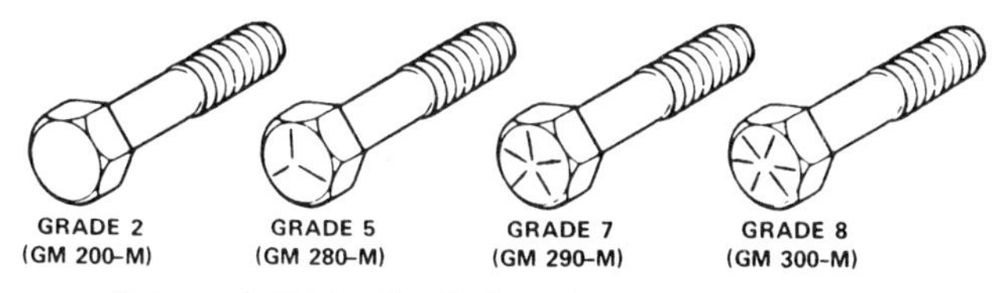

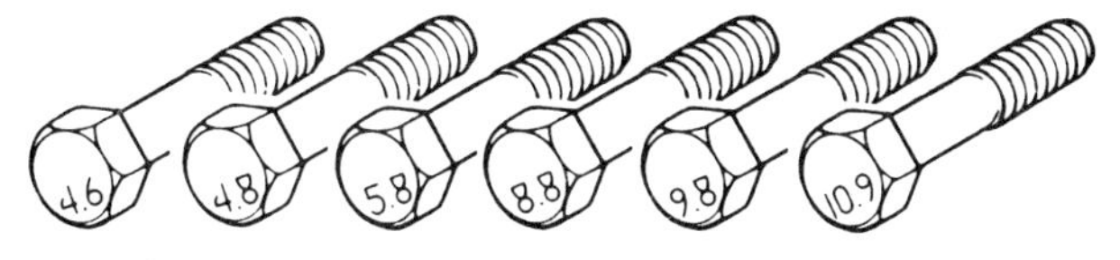

Figure 4-25. It is very important to use a fastener that is strong enough. Always replace a fastener with one that is the same grade or stronger. Bolt grades can be determined by markings on the head of the bolt.

Never replace a higher grade bolt with a lower grade bolt. A weaker bolt can break, causing the failure of a major part. This could lead to a vehicle breakdown and presents a potentially dangerous situation.

Fastener torquing and tightening sequence

Torque specifications are threaded fastener tightening values. These specifications are determined by the vehicle manufacturer. Torque specifications are critical for engine and drive train parts, and other vehicle parts also.

It is very important to tighten all fasteners properly. An overtightened bolt will stretch and possibly break. The threads could also be damaged, or *stripped.* Gaskets could be smashed or broken. An undertightened bolt could work loose and fall out. Part movement could also shear the loose bolt or break a gasket, causing leakage.

Vehicle manufacturers also provides a ***tightening sequence,*** or *pattern,* for threaded fasteners. The sequence assures that parts are fastened evenly. Uneven tightening can cause part warping or breakage and gasket leaks. Tightening is usually done gradually and follows a crisscross pattern, starting in the middle and working outward in steps. This results in an even pressure across the entire mating surface of the parts.

You should follow these rules to properly torque threaded fasteners.

- Pull steadily on the handle of the torque wrench. Jerking the torque wrench will cause excessive *pointer* deflection and invalid readings.
- Clean and oil the fastener threads before tightening.
- *Avoid* using *swivel, extension,* and *crowfoot sockets,* if possible. They will cause invalid readings.
- If using the *flex-bar,* or *beam,* type of torque wrench, look straight at the scale. Viewing from an angle produces a false reading. Do not let the *pointer shaft* contact any other part of the vehicle.
- Obtain and use the manufacturer's recommended tightening sequence and torque specifications.
- Torque the fasteners in steps. Never run one fastener up to full torque before all bolts have been partially torqued.
- Recheck torque–once after all fasteners have been fully torqued and once after component has been run up to normal operating temperatures and cooled off.

Removing broken fasteners

Occasionally, a fastener will break while it is in the part. Various methods of removing broken fasteners are shown in Figure 4-26. Notice that the methods used depend on whether the broken fastener extends from the part or if the break is internal.

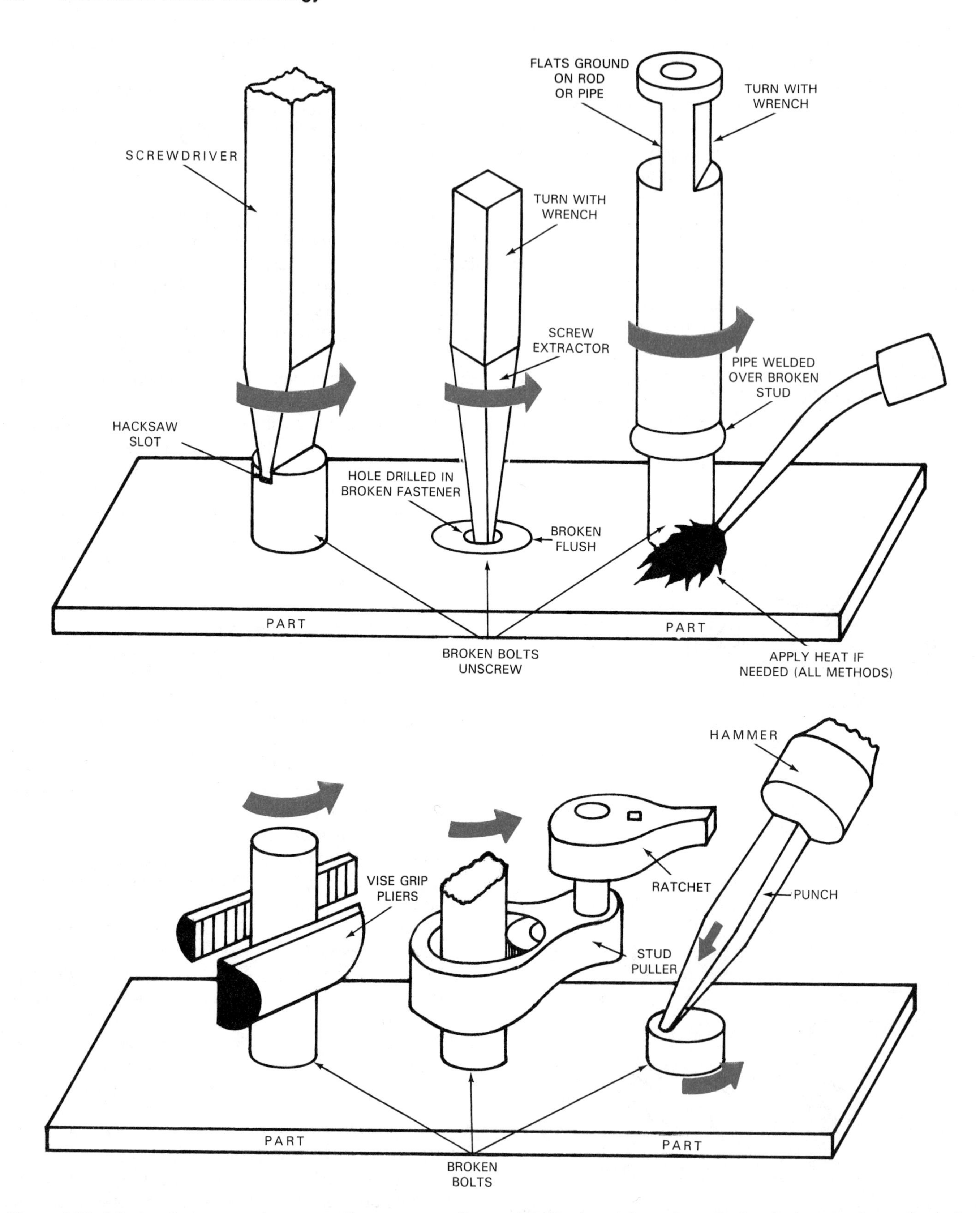

Figure 4-26. A broken fastener can be removed in many ways. Common methods are shown here. Broken fasteners are usually easier to remove if part of the fastener is sticking out of the part.

If a fastener shaft extends from a part, several methods can be used to remove it. A screwdriver slot can be cut into the shaft, and the shaft can be turned out. A pipe or other metal rod can be welded to a broken fastener shaft. The rod can then be turned to unscrew the shaft. Vise grip pliers and stud pullers can be used to tightly grasp the shaft for removal.

If the fastener is broken inside the part, a screw extractor can be used. Screw extractors were briefly described in Chapter 2. Occasionally, the broken fastener can be removed by striking one edge with a chisel. This may get the fastener turning so it can be removed.

Snap Rings

A ***snap ring*** is a split ring used for holding parts on shafts or inside of bores. The snap ring is usually used when clearance is such that it will not permit the use of a threaded fastener. The snap ring fits tightly in or around a groove that is machined to accept it.

The two major kinds of snap rings are *internal snap rings* and *external snap rings.* Figure 4-27 shows the two kinds. Snap rings can be removed and installed with snap ring pliers, as shown. If a snap ring is bent or distorted during removal, it should be replaced.

Drive Train Lubricants

Moving drive train parts require ***lubrication***–a treatment of some sort that coats the parts to reduce friction between them. ***Lubricants*** are materials, usually of a petroleum nature, that provide lubrication. In addition, lubricants may serve cooling, cleaning, sealing, and power transmission functions.

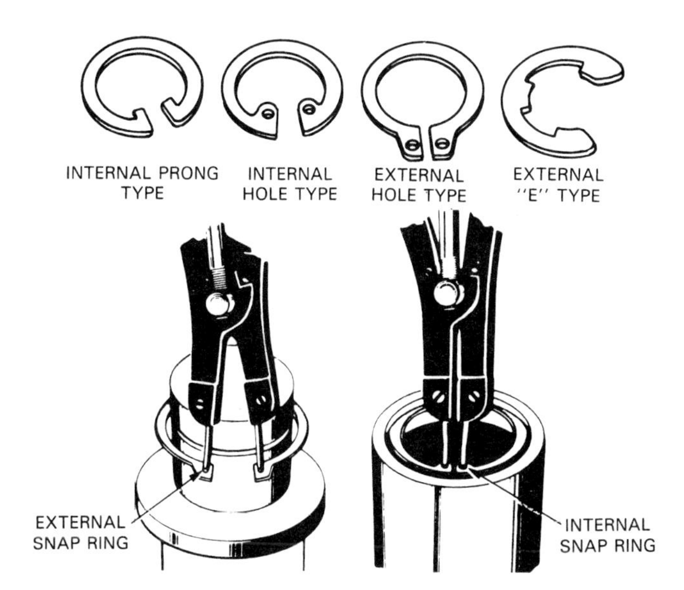

Figure 4-27. There are two kinds of snap rings–internal and external. Snap rings are often used in transmissions and transaxles.

Lubricants are manufactured to meet specific needs of individual drive train components, and different drive train components use different kinds of lubrication. Some components require *lubricating oil,* and others require *grease.* Transmissions, transaxles, and rear axle assemblies use oil. Universal joints and antifriction bearings are lubricated with grease.

Lubricants vary in chemical formula, or composition. They also vary in the *additives* they contain. ***Additives*** are chemicals added to improve some of the *properties,* or characteristics, of the original product. The different chemical compositions and additives yield products that have different *physical* and *chemical properties.* (Physical properties are traits that do not involve chemical change; chemical properties are traits that *do.*)

The most obvious difference between any drive train lubricant is in *thickness,* or *resistance to flow.* This physical property is technically referred to as ***viscosity.*** Viscosity is established through the oil-refining process and through the additives. The viscosity requirement of a drive train lubricant depends upon the application.

Commonly, a lubricant is spoken of in terms of its being a certain *weight;* the weight corresponds with viscosity. A heavy oil is thicker oil; a light oil is a thinner oil. It is given as an ***SAE viscosity grade,*** which classifies a lubricant by its viscosity. A typical SAE grade of *engine* oil, for example, would be SAE 30, called a 30-*weight* oil. The highest grade of engine oil is SAE 50. SAE numbers get larger as viscosity increases, or as the lubricant gets thicker, or heavier.

As mentioned, lubricants also vary in the additives they contain. Additives improve the chemical properties of lubricants. For example, additives may improve a lubricant's resistance to heat. They may reduce oil foaming and, also, *oil oxidation,* the chemical breakdown of oil. Some additives, called *corrosion inhibitors,* help prevent the formation of acids in oils, which can damage metal parts.

Lubricating Oils

Lubricants that exist in a liquid state are called ***lubricating oils,*** or ***lube oils,*** for short. Several kinds of petroleum-based lubricating oils are used in the drive train. These oils can serve different functions. In rear axle assemblies and manual transmissions and transaxles, the major function of the oil is to lubricate drive train parts. In automatic transmissions, the oil must also transmit hydraulic pressure to operate transmission components. In torque converters, the oil transmits the power of the engine. Of course, the oils must be compatible with, or not cause damage to, gaskets, seals, *clutch* and *band facings,* and other such parts. The most common types of lubricating oils used in drive trains are identified in the upcoming paragraphs.

Gear oils

Gear oils are heavy oils that provide lubrication for the gears and bearings in rear axle assemblies and in manual transmissions and transaxles. These oils have high SAE viscosity grades, such as SAE 90 or 140, compared with SAE 30 or 40 for engine oils. This is done to prevent confusion when selecting engine or gear oils.

Note that gear oils are not necessarily always heavier than engine oils, and the lower grade lubricating oils are comparable to some engine oils in viscosity. Engine oils and lubricating oils are graded on different scales. This means that a gear oil and an engine oil having the same viscosity will have quite different SAE viscosity grade designations. For example, an SAE 90 gear oil viscosity can be similar to that of an SAE 40 engine oil. Some vehicle manufacturers even recommend the use of a good quality engine oil in their transaxles, instead of gear oil.

Gear oils were once sold only in single weights. Today's gear oils, however, are available in *multigrade* versions, similar to multigrade engine oils. A gear oil marked SAE 85W-140, for instance, is rated at SAE 85 when cold. The oil contains additives that cause it to maintain enough thickness to rate SAE 140 when it heats up. Multigrade gear oils flow freely between moving parts when cold, but still provide good lubrication when the parts heat up.

Gear oils contain other additives, including friction-reducing and other antiwear agents. One type of additive, in particular, prevents oil from being squeezed out from between helical and hypoid gears as they revolve. Gear oil with this type of additive is called *extreme-pressure,* or *EP, lube.* ***Hypoid gear lube*** is another name for it.

The *American Petroleum Institute (API)* has developed a gear oil rating system using *GL numbers.* GL1 is the lowest grade. GL1 is used only for low-speed, light-duty applications. As the GL numbers increase, the quality of the oil increases. Most gear oils for automotive use are classified GL4, GL5, and GL6.

Special gear oil must be used in rear axle assemblies that use a *limited-slip differential.* One type of limited-slip differential uses clutches, which provide extra traction. To keep the clutches from slipping, a gear oil with special additives must be used. This oil can be used in other differentials and transmissions. Standard gear oils, however, must never be used in limited-slip differentials. Even a small amount will cause the differential clutches to be ruined.

Most manual transmissions, manual transaxles, and rear axle assemblies have holes, capped with fill plugs, on the sides of their cases for adding oil. Before removing the plug to add gear oil, the area around it must be cleaned. Once the plug is removed, the gear oil may be added through the opening. The unit is full when a small amount of oil begins to drip from the opening; at which point, the fill plug is replaced and tightened.

Automatic transmission fluids

Automatic transmissions and automatic transaxles require a special type of lubricating oil. The viscosity of ***automatic transmission fluid (ATF),*** as it is called, is lower than that of manual transmission gear oils. Lighter oil is necessary to operate the automatic transmission hydraulic system and provide efficient power transfer through the torque converter. To provide proper lubrication of the planetary gears and bearings, these fluids contain special EP (extreme-pressure) additives.

Automatic transmission fluids can be divided into different classes, according to the additives that they contain. ATFs contain additives that help them withstand heat, increase their lubricating ability, and reduce foaming. The major difference in the different classes is in the use of additives that affect the coefficient of friction of the fluid. This is a number that describes how much friction the oil allows when it is compressed between two surfaces.

Since automatic transmission *holding members* are always soaked in oil, the coefficient of friction determines how well they can hold. A particular fluid's coefficient of friction is varied according to the kinds of transmissions in which it will be used. Transmission fluid requirements vary according to the kind of holding members and the type of friction material they are made of. The pressures applied on the holding members is another factor.

Dexron III/Mercon. This type of automatic transmission fluid is the latest version in a series of *Dexron* lubricants. This fluid is used on all late-model domestic and imported automatic transmissions. It is recommended as a replacement for all other types of automatic transmission fluid, except those that originally used *Type F,* discussed next.

Type F. This type of automatic transmission fluid was used in Ford vehicles until the late 1970s. It was also used in the automatic transmissions of many imported vehicles, especially those using Borg-Warner automatic transmissions.

Type C-3. This fluid is designed to be used in heavy-duty truck and tractor automatic transmissions and in the hydraulic drive systems of off-road heavy equipment. The C-3 designation is a quality standard developed by Allison Transmissions.

Type A. You may occasionally see a reference to an older kind of fluid, called Type A (Suffix A). This fluid is no longer used. If you are working on an older automatic transmission that calls for Type A, Dexron may be safely substituted.

Transmission fluid is checked and filled through the dipstick tube. The level is always checked with the engine running and the transmission in park or neutral, as required. The transmission fluid should be warm; the dipstick will be too hot to touch comfortably. If the fluid level is low, it should be added in small amounts until the level is correct. Do not overfill an automatic transmission, or the planetary gears will whip the fluid into a foam.

CAUTION! If you are in doubt about the kind of automatic transmission fluid that a particular vehicle uses, do not guess! Different kinds are not compatible with each other, nor are they interchangeable. The proper kind of fluid is printed on the transmission dipstick. Otherwise, check the owner's or service manual for the proper fluid.

Grease

Grease is a lubricant that exists in a solid or semisolid state. Grease provides lubrication over a wide range of temperatures. As it is water resistant, it also works to shield components from moisture and keep metal parts from rusting. Universal joints, front-wheel drive wheel bearings, clutch pilot and throwout bearings, some outer rear axle bearings, and some slip yokes are all examples of components that are lubricated with grease. Note that greases used for these applications are called EP greases. These are designed to provide good lubrication at extreme pressures.

There are a number of different types and grades of grease. The type of grease used depends on the application. In many cases, drive train parts can be lubricated with the same grade of grease that is used for front *suspension fittings.* In other cases, a grease with a special formula that is tailored for a specific application must be used. You should always check the manufacturer's manual for the exact grease to use on a particular part.

Some components are designed to be greased only once, and these will be greased before they are installed on a vehicle. The grease is retained by seals. These parts usually cannot be greased, except when the entire unit is disassembled. Examples include factory-installed universal joints and most grease-lubricated rear axle bearings.

Some parts must be periodically greased. Often, these are provided with a grease fitting for injecting grease. In other cases, such as on front wheel bearings of rear-wheel drive cars, greasing is not as simple. Such parts must be disassembled and cleaned before greasing, according to a specific time or mileage schedule.

Summary

Seals are used to keep lubricant in and contaminants out of components. Seals are used in transmissions, transaxles, and rear axles, etc., to seal rotating shafts and stationary drive train parts. Lip seals are used on rotating or sliding shafts. Ring seals are used to seal stationary parts or hydraulic pistons. There are many kinds of both lip and ring seals. Some seals are protected by dust shields in places where the seal would be exposed to dirt and water.

Boots are used to cover CV joints used in front-wheel drive vehicles. The ends of the boot are fastened by straps or clamps to the drive shafts and joint housings.

Most seal defects are first noticed when the seal begins leaking. A leaking seal should be replaced as soon as possible. There are several methods of removing, inspecting, and replacing seals. When a seal is replaced, all related parts, such as shafts and housings, should be inspected and replaced if defective. In most cases, the replacement seal must exactly match the old one.

Gaskets are used to seal stationary part surfaces. Gaskets are made of relatively soft materials so that they can conform to imperfections in mating parts that they seal. Many different kinds and thicknesses of gasket materials are used in a drive train. Most drive train gaskets are made of single sheets of material.

Many gasket leaks are caused by warped or damaged mating surfaces or loose fasteners. Gaskets are replaced by disassembling the parts, removing the old gasket, cleaning and inspecting the mating surfaces carefully, and installing a new gasket. Check that the new gasket is the right one.

Drive train lubricants can be classified as either oils or greases. Oils are used in major assemblies with many interacting stressed parts. Greases are used in simpler parts where the relative movement is slow or where there are few components.

Gear oils are heavy oils used in manual transmissions, transaxles, and rear axle assemblies. Gear oils were once sold only in single weights. Today's gear oils are available in multigrade versions, similar to multigrade engine oils. A gear oil marked, SAE 85W-140, for instance, has a viscosity rating of SAE 85 when cold and maintains a rating of SAE 140 as it heats up. Multigrade gear oils flow freely between moving parts when cold but still provide good lubrication when the parts warm up to operating temperature.

The API has developed a gear oil rating system using GL numbers. GL1 is the lowest grade. GL1 is used only for low-speed, light-duty applications. As the GL numbers increase, the quality of the oil increases. Most gear oils for automotive use are classified GL4, GL5, and GL6.

Special gear oil must be used in rear axle assemblies that use a limited-slip differential. A gear oil with special additives must be used. Standard gear oils must never be used in limited-slip differentials.

To add gear oil to manual transmissions, manual transaxles, and rear axle assemblies, a plug, usually on the side of the case, must be removed. Once the plug is removed, the gear oil is poured through the opening until a small amount drips from the opening, indicating that the unit is full.

Automatic transmission fluids are thinner than gear oils. Two major kinds of automatic transmission fluids are Dexron II and Type F. Other types are available. The proper fluid should always be used to avoid transmission damage.

Greases are used in some parts of the drive train, such as on universal joints and wheel bearings. In some cases, chassis grease is used. In other instances, special greases should be used. Some parts are greased at the factory. Others must be periodically lubricated through a grease fitting, or by disassembling and packing the part with grease.

Know These Terms

Seal; Lip seal; Dust shield; Ring seal; O-ring; Static seal; Dynamic seal; Seal ring; Boot; Gasket; External gasket; Internal gasket; Form-in-place gasket; Blown gasket; Sealant (Sealer); Hardening sealer; Nonhardening sealer; Shellac; RTV sealer (Room-temperature vulcanizing) sealer; Anaerobic sealer; Thread-locking compound; Thread sealer; Adhesive; Epoxy; Fastener; Threaded fastener; Nonthreaded fastener; Bolt; Nut; Cap screw; Screw; Machine screw; Major diameter; Minor diameter; Bolt size; Nut size; Head size; Bolt length; Thread length; Thread pitch; Coarse thread; Fine thread; Metric thread; Thread-pitch gauge; Right-hand thread; Left-hand thread; Bolt grade; Tensile strength; Grade marking; Torque specification; Tightening sequence; Snap ring; Lubrication; Lubricant; Additive; Viscosity; SAE viscosity rating; Lubricating oil (Lube oil); Gear oil; Hypoid gear lube; Automatic transmission fluid (ATF); Grease.

Review Questions–Chapter 4

Please do not write in this text. Place your answers on a separate sheet of paper.

1. A _____ is a device that keeps a desirable element, such as a lubricant, from getting out of a component.
2. _____ _____ are commonly used to seal rotating-shafts in the drive train.
3. What is the purpose of a dust shield?
4. What are O-rings used for?
5. What, besides the seal itself, should you inspect when replacing a lip seal with many miles of service?
6. What three precautions should you observe during seal removal?
7. Never coat the outside of a lip seal with nonhardening sealer. True or false?
8. Fasteners provide a kind of static seal between non-moving parts. True or false?
9. Cite basic rules you should observe when removing and replacing gaskets.
10. What is RTV sealer?

Certification-Type Questions–Chapter 4

1. Technician A says that seals and gaskets are used to keep lubricants inside of drive train components. Technician B says that seals and gaskets are used to keep dirt and moisture out of drive train components. Who is right?

(A) A only
(B) B only
(C) Both A & B
(D) Neither A nor B

2. All of these statements about seals are true EXCEPT:

(A) they can be used to separate two different fluids.
(B) they are only used to seal rotating or sliding shafts.
(C) boots are a type of seal.
(D) some seals are made of metal.

3. To ensure that a sealing lip makes good contact with a rotating shaft, lip seals often use:

(A) garter springs.
(B) leaf springs.
(C) return springs.
(D) torsion springs.

4. Technician A says that a lip seal with a finger spring is commonly used to seal components lubricated with grease. Technician B says that a finger spring exerts less pressure on the shaft and lip than a garter spring. Who is right?

(A) A only
(B) B only
(C) Both A & B
(D) Neither A nor B

5. Technician A says that lip seals without springs are commonly used to hold back contaminants. Technician B says that lip seals without springs are used in applications where external pressure will force the sealing element against the shaft. Who is right?
- **(A)** A only
- **(B)** B only
- **(C)** Both A & B
- **(D)** Neither A nor B

6. The most common ring seal used in drive trains is the:
- **(A)** square seal.
- **(B)** half-round seal.
- **(C)** triangular seal.
- **(D)** round seal.

7. Technician A says that seal rings are nonpositive seals. Technician B says that seal rings can be made from metal or teflon. Who is right?
- **(A)** A only
- **(B)** B only
- **(C)** Both A & B
- **(D)** Neither A nor B

8. Almost all seal problems show up first as:
- **(A)** noises.
- **(B)** lubricant leaks.
- **(C)** shaft wear.
- **(D)** part overheating.

9. Which of these would be the least likely cause of low-mileage seal failure?
- **(A)** Hardening of the sealing element.
- **(B)** Wrong-sized seal.
- **(C)** Improper seal installation.
- **(D)** Damaged or bent shaft.

10. Technician A says that excessive lateral movement of a shaft may cause a seal to leak. Technician B says that an improperly sized seal will leak immediately. Who is right?
- **(A)** A only
- **(B)** B only
- **(C)** Both A & B
- **(D)** Neither A nor B

11. Technician A says that some lip seals can be removed by carefully prying them from their bore. Technician B says that some static O-ring seals can be removed by using a special puller with a slide hammer. Who is right?
- **(A)** A only
- **(B)** B only
- **(C)** Both A & B
- **(D)** Neither A nor B

12. Before seal installation, a shaft can be polished with:
- **(A)** polishing compound.
- **(B)** steel wool.
- **(C)** crocus cloth.
- **(D)** fine sandpaper.

13. All of these can be used to install a seal in a housing bore EXCEPT:
- **(A)** a hammer.
- **(B)** pliers.
- **(C)** a socket.
- **(D)** a pipe.

14. Gaskets are never used to seal:
- **(A)** aluminum parts.
- **(B)** oil-filled parts.
- **(C)** stationary parts.
- **(D)** moving parts.

15. All of these are used as gasket materials EXCEPT:
- **(A)** oil-resistant paper.
- **(B)** aluminum.
- **(C)** cork.
- **(D)** ceramic.

16. An oil pan gasket is leaking. Technician A says that bolts could have been overtightened, damaging the gasket material. Technician B says that the bolts may be loose. Who is right?
- **(A)** A only
- **(B)** B only
- **(C)** Both A & B
- **(D)** Neither A nor B

17. Technician A says that when servicing an apparent gasket leak, the first step is to clean the gasket surfaces. Technician B says that the first step when servicing an apparent gasket leak is to make sure that the gasket is the cause of the leak. Who is right?

(A) A only
(B) B only
(C) Both A & B
(D) Neither A nor B

18. Which of these can be made relatively easily if an exact replacement is not available?

(A) Gaskets.
(B) Lip seals.
(C) O-rings.
(D) Ring seals.

19. Hardening sealers can be used in all of these applications EXCEPT:

(A) threaded fittings.
(B) uneven surfaces.
(C) permanent assemblies.
(D) hose connections.

20. Technician A says that RTV sealer is commonly used to make form-in-place gaskets. Technician B says that form-in-place gaskets must never be used in place of conventional gaskets. Who is right?

(A) A only
(B) B only
(C) Both A & B
(D) Neither A nor B

21. Anaerobic sealers will harden in the absence of:

(A) moisture.
(B) air.
(C) oil.
(D) gasket compression.

22. Technician A says that epoxies are mixed just before being applied. Technician B says that thread-locking compounds should be used on all threaded fasteners. Who is right?

(A) A only
(B) B only
(C) Both A & B
(D) Neither A nor B

23. All of these are types of threaded fasteners EXCEPT:

(A) nuts.
(B) cap screws.
(C) snap rings.
(D) machine screws.

24. All of these are used to describe threaded fasteners EXCEPT:

(A) major diameter.
(B) thread pitch.
(C) head length.
(D) thread length.

25. Technician A says that most threaded fasteners have left-hand threads. Technician B says that the threads of metric and conventional bolts will interchange even if the head sizes are different. Who is right?

(A) A only
(B) B only
(C) Both A & B
(D) Neither A nor B

26. Technician A says that tightening bolts in the proper sequence is very important when reassembling any drive train part. Technician B says that tightening bolts to the correct torque is very important when reassembling any drive train part. Who is right?

(A) A only
(B) B only
(C) Both A & B
(D) Neither A nor B

27. Technician A says that a bent snap ring should be replaced. Technician B says that when a threaded fastener is broken inside of a part, the part must be replaced. Who is right?

(A) A only
(B) B only
(C) Both A & B
(D) Neither A nor B

28. All of these relate to lubricant viscosity EXCEPT:

(A) weight.
(B) thickness.
(C) resistance to foaming.
(D) resistance to flow.

29. Technician A says that gear oils and engine oils are graded on the same SAE viscosity scale. Technician B says that a special gear oil must be used in limited-slip differentials. Who is right?

(A) A only
(B) B only
(C) Both A & B
(D) Neither A nor B

30. Technician A says that some greasable drive train parts are provided with grease fittings. Technician B says that some drive train parts must be disassembled to regrease them. Who is right?

(A) A only
(B) B only
(C) Both A & B
(D) Neither A nor B

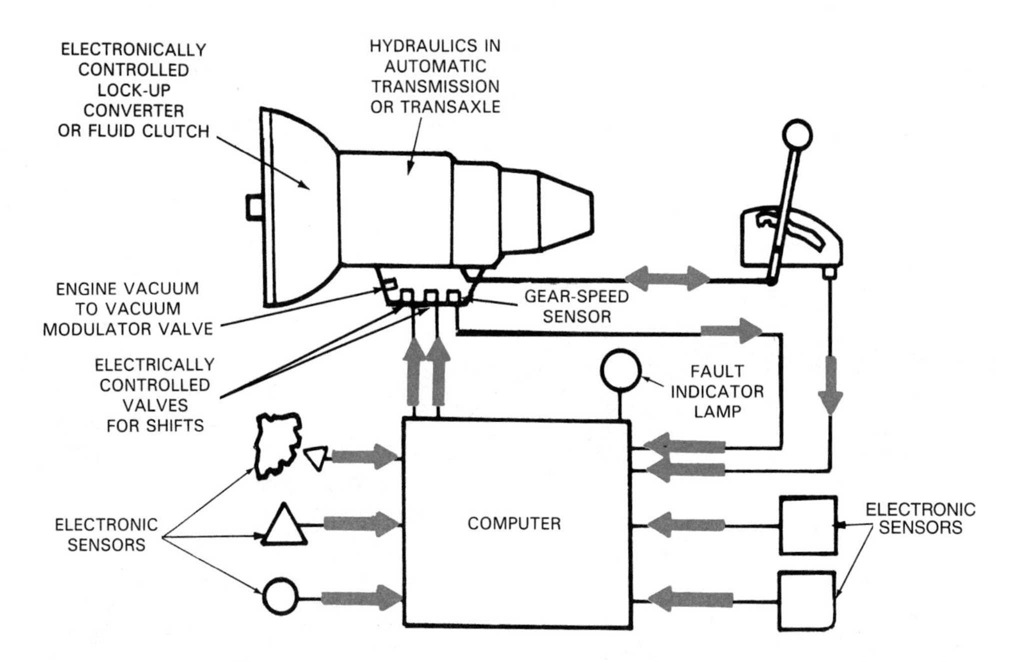

Some late-model vehicles use computers to control automatic transmission shift points. Note flow of data to and from computer and transmission.

Chapter 5

Hydraulics, Pneumatics, and Electronics

After studying this chapter, you will be able to:

- Describe the action of a basic hydraulic system.
- Explain the reason why a hydraulic system can transfer motion and power.
- Discuss different hydraulic components used in a modern drive train.
- Discuss basic pneumatic systems and compare them to hydraulic systems.
- Explain the principles of electricity.
- Identify different types of electrical circuits and components.
- Diagram a vehicle computer system and discuss its various facets.
- Make basic wiring repairs.
- Perform basic electrical tests.
- Relate electrical and electronic components to the operation of a drive train.

Hydraulics, pneumatics, and electronics are widely used in automotive drive trains. Sometimes, such as in certain modern automatic transmissions, all three are used in one component. In these particular units, a *hydraulic system* is used to develop and control hydraulic pressure and flow. Through this system, power is transferred from engine to drive shaft. A *pneumatic system* uses the pressure differential between the outside atmosphere and engine vacuum to control portions of the transmission hydraulic system. An *electronic system* is also involved in controlling the hydraulic system flow. The vehicle's on-board computer sends electric signals to *solenoids* on the transmission. The solenoids control part of the hydraulic system flow, and *sensors* in the drive train feed data back to the computer.

This chapter will identify the basic principles and major components of hydraulic, pneumatic, and electronic systems. This knowledge will enable you to more fully understand the major drive train components discussed in detail in later chapters.

Hydraulics

In general, ***hydraulics*** is the science and technology of liquids at rest and in motion. ***Power hydraulics*** deals specifically with the *transmission of power* through liquids. This branch of hydraulics looks for practical ways to produce and use hydraulic power, which is what a ***hydraulic system*** does.

The major advantages that hydraulic systems have over other kinds of power transmission is that they are simple and provide a compact way to multiply power. Hydraulic power systems operate more quietly than other forms of power transmission. The components last longer because they are continuously lubricated by hydraulic fluid. Further, hydraulic systems are controlled rather easily, and many components are controlled automatically.

Types of Hydraulic Systems

A hydraulic system may be classified as *hydrodynamic* or *hydrostatic.* The type of system it is, described below, depends upon the type of hydraulic power used.

Hydrodynamic systems

Hydrodynamics is the study of *dynamic,* or moving, liquids. In a ***hydrodynamic system,*** power is transmitted when a moving fluid strikes or acts upon a device composed of blades or vanes. The energy in the fluid is transmitted to the blades or vanes, causing them to move. In this kind of system, the moving fluid delivers the force that results in a transfer of power. A water wheel, Figure 5-1, is an example of a hydrodynamic system. *Fluid couplings* and torque converters are examples of hydrodynamic devices used in the drive train. Such hydrodynamic devices are complete in themselves, or are *self-contained*; they are usually replaced as a unit.

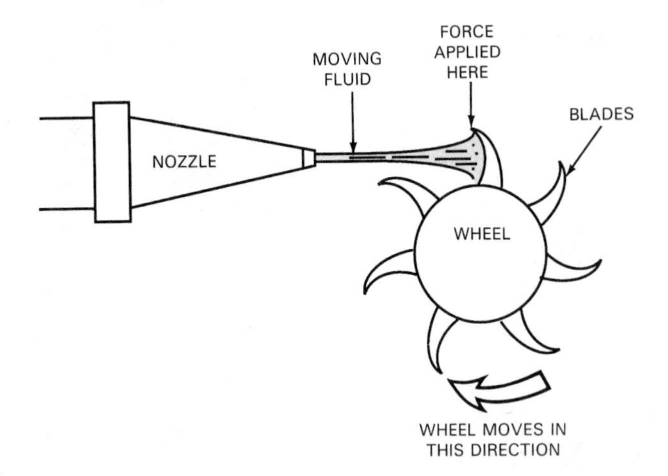

Figure 5-1. Study a simple hydrodynamic system. Moving fluid strikes blades attached to a wheel. The force of the fluid striking the blades causes the wheel to revolve. This principle is used to operate fluid couplings and torque converters.

Hydrostatic systems

Hydrostatics is the study of *static* liquids, or liquids at rest. This branch of hydraulics studies the nature of fluids in regard to pressures exerted on them or by them. In a ***hydrostatic system,*** pressure is applied to liquid confined in the system. Pressure transmitted through the liquid serves as a means for applying force at some other point in the system. Hydrostatic principles are used in hydraulic clutch release mechanisms and in the control systems of automatic transmissions and transaxles. Most of this section will focus on the hydrostatic system and the individual components that go into a typical system. A simple hydrostatic system is shown in Figure 5-2.

Hydraulic Pressure

A common application of hydraulic principles involves transmission of ***hydraulic pressure*** by applying force on an enclosed liquid. The force pressurizes the liquid so that it can perform *work.* The pressure is transmitted undiminished (ideally) to every point in the fluid, throughout the system. The reaction is quite different compared to applying force on a solid object, wherein the force is only transmitted in the same direction as that of the original force.

Hydraulic pressure is applied in many areas of daily life. The systems that deliver water to your home and oil to the critical areas of your engine use hydraulic pressure to perform their jobs. Other vehicle systems–windshield washer and power steering, to name two–work because of hydraulic pressure.

In the vehicle drive train, hydraulic pressure is used to operate automatic transmissions, transaxles, and some transfer cases. The pressure produces a pushing force on some output device–a transmission *servo piston,* for example. In addition, some manual transmission vehicles employ hydraulic pressure–specifically, to operate the *clutch fork* in order to disengage the clutch.

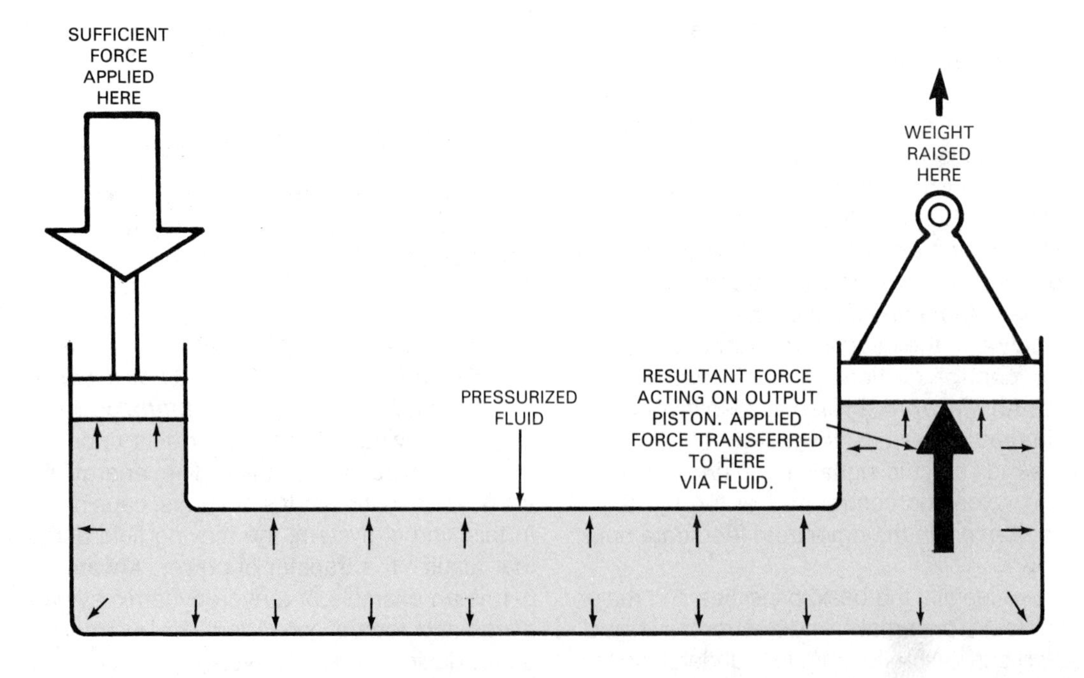

Figure 5-2. A simple hydrostatic system. Force applied to piston on left creates pressure in the closed system. Pressure created in the system can cause the weight on right to be raised. Modifications of this simple system are used in all automotive hydraulic systems.

Pascal's law

To understand hydraulic pressure, you must know that liquids, for all practical purposes, cannot be compressed. They are not like gases, such as air, which can be squeezed down to occupy a smaller space. (*Compressed air* is familiar to anyone who has filled a tire.) Since liquids, such as water or oil, cannot be compressed, any pressure that is placed upon them is immediately translated to all parts of the liquid. As a result, liquids are very useful for transmitting motion.

In Figure 5-3, a cylinder contains a piston on each end. The space between the pistons is filled with fluid. Pushing on either piston will cause movement of the fluid, moving the other piston. The piston that starts the movement is called the ***apply piston.*** The other piston–the piston that is moved as a result of the apply piston–is called the ***output piston.*** This demonstrates that motion may be transmitted by a liquid.

The same principle can be applied to transmit motion from one cylinder to another. In Figure 5-4, two cylinders are connected by a hydraulic line. When the apply piston is moved, the liquid transfers the motion to the output piston. If the cylinders have the same diameter, the distances the pistons travel will be equal.

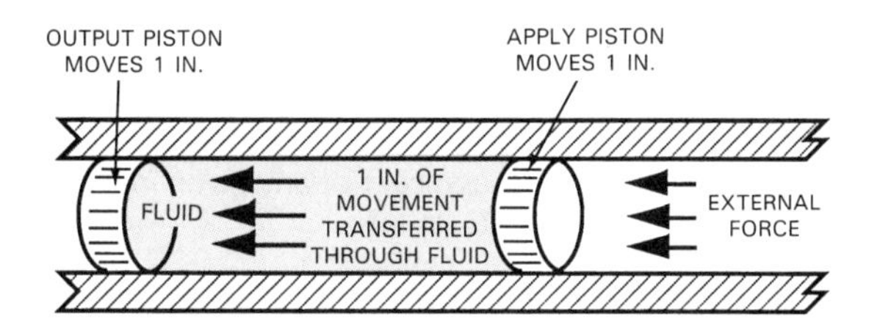

Figure 5-3. Note how motion is transferred through a liquid. Since liquid is incompressible, external force to apply piston will cause a simultaneous movement of the output piston, the distance of which will be the same as that of the apply piston.

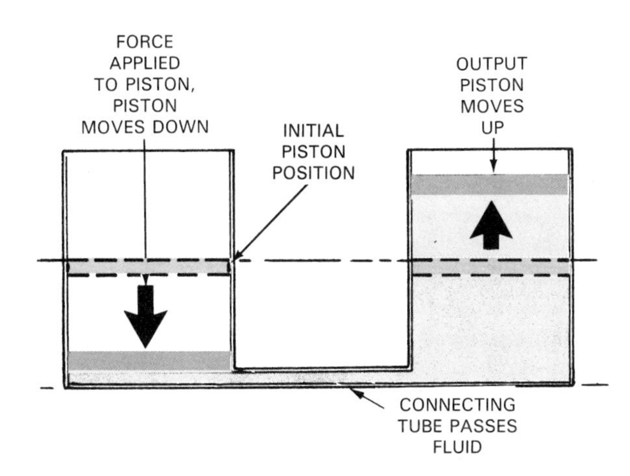

Figure 5-4. Hydraulic pressure can be transmitted through hydraulic lines. Force applied to piston on left created pressure in system, causing output piston to move up.

The underlying principle in these examples just discussed is covered by ***Pascal's law.*** It basically states that a liquid that has been pressurized transmits the same amount of pressure to every part of its container. Shown in Figure 5-5, the pressure is equal at all points and acts equally in all directions, on all areas. This principle was formulated by 17th century physicist *Blaise Pascal.*

Hydraulic pressure calculations

Hydraulic pressure is measured in pounds per square inch (psi) or pounds per square foot (psf) in the English, or customary system. In the metric system, it is measured in kilopascals (kPa). Mathematically, pressure is force per unit area. In other words:

$$P = \frac{F}{A}$$

where, in customary units, P is hydraulic pressure, in psi (or psf), F is force in pounds acting on a given surface, and A is the area of the surface in square inches (or square feet).

Pressure in a simple hydraulic system, then, can be easily determined if you know the value of the applied force and the area of the apply piston. Simply divide the force applied by the area of the apply piston.

In Figure 5-6, a 20-lb. (89-N) force is applied to an apply piston with an area of 2 in.^{2} (12.9 cm^2). Dividing the force by the apply piston area to get pressure gives:

$$P = \frac{F}{A} = \frac{20 \text{ lb.}}{2 \text{ in.}^2} = \frac{10 \text{ lb.}}{\text{in.}^2}$$

The hydraulic pressure in the system is equal to 10 psi (69 kPa).

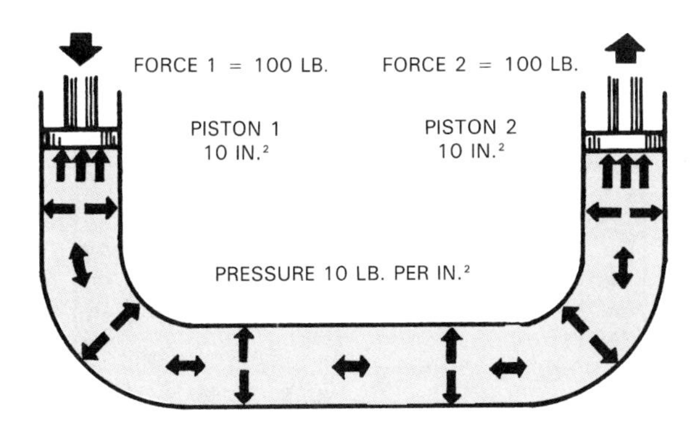

Figure 5-5. An external force applied to a fluid in a closed system creates a pressure that is equal at every spot in the system. This principle is known as Pascal's law. Note that the pressure acts with equal force on equal areas.

To find the force exerted on or by a piston, the formula for pressure is rearranged as follows:

$$F = P \times A$$

Putting in values from Figure 5-6, we can easily determine the value of force at the output piston:

$$F = P \times A = \frac{10 \text{ lb.}}{\text{in.}^2} \times 20 \text{ in.}^2 = 200 \text{ lb.}$$

The force on the output piston is 200 lb. (890 N).

Hydraulic leverage

Compare input versus output forces in both Figures 5-5 and 5-6. You will note, in the latter, that a smaller force develops the same hydraulic pressure (10 psi, or 69 kPa) and this pressure applied to a larger output piston resulted in a greater output force. Note that in obtaining this advantage, output distance (and speed) is sacrificed. For example, in Figure 5-6, the smaller piston must move 10 in. (254 mm) to move the larger piston 1 in. (25.4 mm).

What you find by comparing these two examples illustrates how Pascal's law can be used to develop a hydraulic system that will increase force. This ***hydraulic leverage*** resembles the mechanical leverage that is gained from using a simple lever to lift a heavy object, Figure 5-7. However, where mechanical advantage is obtained by varying placement of a pivot point in a lever system, it is obtained by varying the size of the input and output pistons in a hydraulic system.

The hydraulic jack in Figure 5-8 is a simple example of hydraulic leverage. Due to hydraulic leverage, a person can raise an automobile by hand.

Another application of hydraulic leverage that you may be familiar with is the vehicle brake system. A simple brake

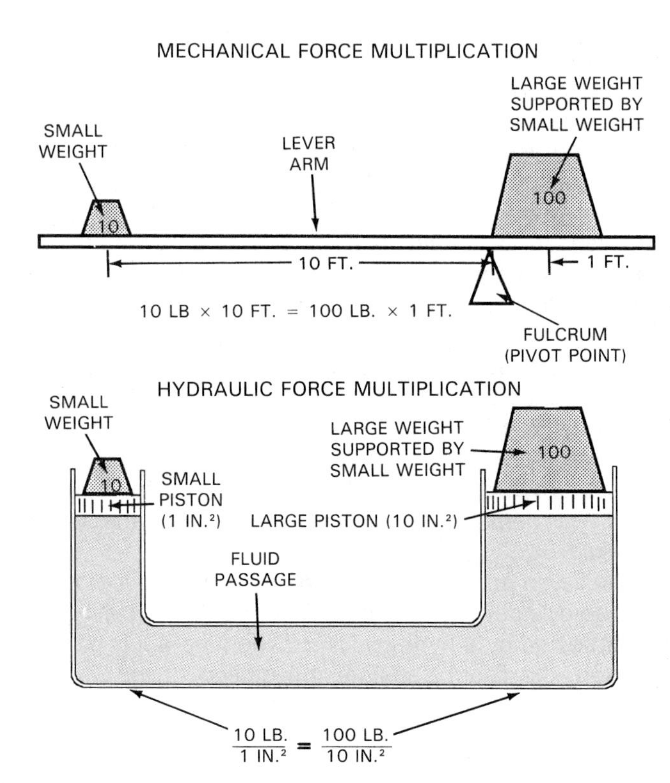

Figure 5-7. The force multiplication in a hydraulic system can be compared to a lever. In the lever system, mechanical advantage is obtained by varying the placement of the pivot point. In a hydraulic system, it is obtained by varying the size of the input and output pistons. In both systems, the increase in force is accompanied by a loss of distance. For any given input, respective outputs travel a shorter distance.

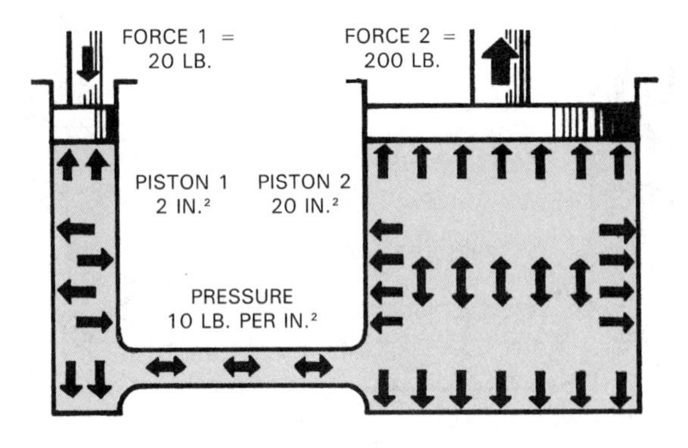

Figure 5-6. Pascal's law, applied to systems with unequal pistons, is an easy way to multiply force. In this example, a 20-lb. force is applied to the 2-in.^{2} piston on left, creating a 10-psi pressure in the system. This pressure acts on the 20-in.^{2} piston on right to produce an output force of 200 psi.

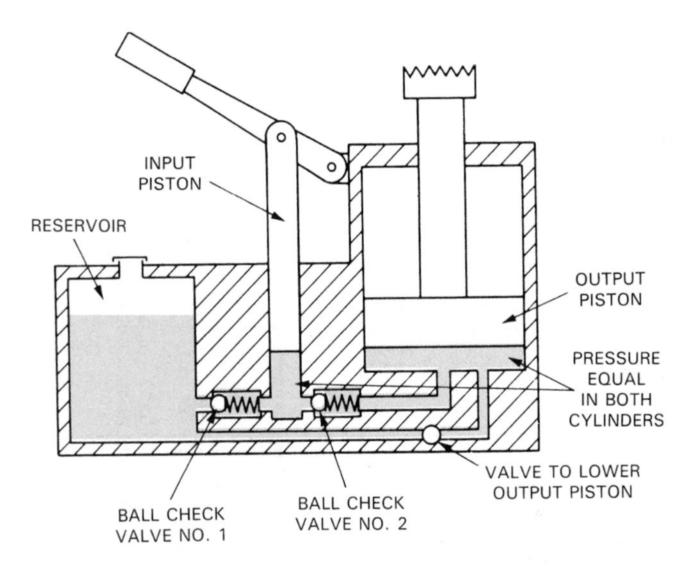

Figure 5-8. A simplified drawing of a hydraulic jack. The piston operated by the jack handle is the input piston. Pressing down on the handle will raise the output piston. Mechanical advantage is obtained because the input piston is smaller than the output piston.

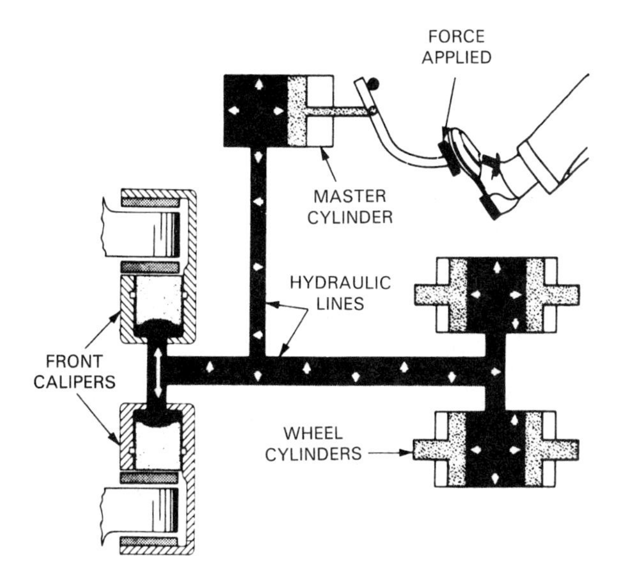

Figure 5-9. A common application of hydraulic principles is found in the vehicle brake system. Pressure created in the master cylinder is transferred to the front calipers and wheel cylinders to apply the brakes. The master cylinder piston is smaller than the brake pistons, and braking force is increased.

system is shown in Figure 5-9. Pressure is developed in the system when the driver steps on the brake pedal, connected to the *master cylinder.* The pressure is transferred to the *front calipers* and rear *wheel cylinders* by hydraulic lines. Force is multiplied due to the difference in size between the master cylinder piston and the brake actuating pistons.

Flow Rate

Fluid movement, or *flow,* is what produces component movement in a hydraulic system. *Hydraulic pumps* (discussed next) actually produce flow, *not* pressure. Pressure develops as a result of the restriction to flow downstream. If there were no restrictions, then pressure at the pump discharge would be very low.

Volume flow rate is typically measured in *gallons per minute (gpm)* or *liters per minute (l/m).* Sometimes, it is measured in *cubic feet per minute (cfm).* Flow rate determines how far and fast an output device, such as a servo piston, can move.

Increasing (or decreasing) volume flow rate impacts hydraulic pressure. The higher the flow rate is in a given hydraulic system, the greater the system pressure is, but also, the greater the pressure drop is through the system. (The increased pressure drop is due to increased frictional losses from increased velocity through the system.) In an automatic transmission, high pressures are needed, and the flow rate is large. A high-*capacity* (gpm) pump is needed to generate required system pressure and compensate for pressure loss.

Hydraulic Pumps

Hydraulic systems rely on ***hydraulic pumps*** to move fluid and pressurize the system. There are many different kinds of hydraulic pumps. They fall into one of two major classes; these are *positive displacement* and *nonpositive displacement.*

Nonpositive displacement pumps are those whose volume does not change during the pumping cycle. The output of this type of pump decreases considerably as system pressure increases. This type of pump is used for lower-pressure, high-volume applications–primarily, for transporting fluids from one location to another. These are hydrodynamic pumps.

Pumps whose volume changes from minimum to maximum to minimum during a pumping cycle (explained later) are ***positive displacement pumps.*** These are hydrostatic pumps. Output from this type of pump is fixed and is not affected by system pressure increases. This type of pump can generate very high pressure. Therefore, it is very useful in fluid power systems. Positive displacement pumps will be the focus of the remainder of this section.

Three kinds of hydraulic pumps are commonly used in automatic transmissions and transaxles. They are called *gear, rotor,* and *sliding-vane pumps.* These pumps have four common features, which are:

- An *inlet port,* where liquid is drawn in from a nonpressurized *fluid reservoir.*
- An *outlet port,* where liquid is discharged into the pressurized system.
- Internal elements–gears, *rotors,* or *vanes,* to draw in and generate hydraulic pressure in the system. The area containing these parts is often called the *pumping chamber.*
- A power source to operate the pump.

In addition to these common features, these pumps all generate system pressure in similar ways. The internal elements rotate in the pumping chamber within the housing. On the inlet side of the chamber, the gear teeth, rotor *lobes,* or sliding vanes move apart as they rotate. On the outlet side, they move together. The side where the elements move apart is connected by an inlet passageway to the system oil reservoir. As the gears or vanes move apart, they create a vacuum. The vacuum draws in hydraulic fluid from the system reservoir. (Actually, fluid is *pushed* into the low-pressure region by atmospheric pressure.) The fluid is carried in spaces between the teeth, lobes, or vanes. As the gears or vanes come back together, the fluid is forced out of the chamber, through an outlet passageway, into the hydraulic system. The flow encounters resistance in the hydraulic system, which causes the buildup of pressure in the system.

Gear pumps

As mentioned, a ***gear pump*** is commonly used in automatic transmissions. The simplest form consists of two

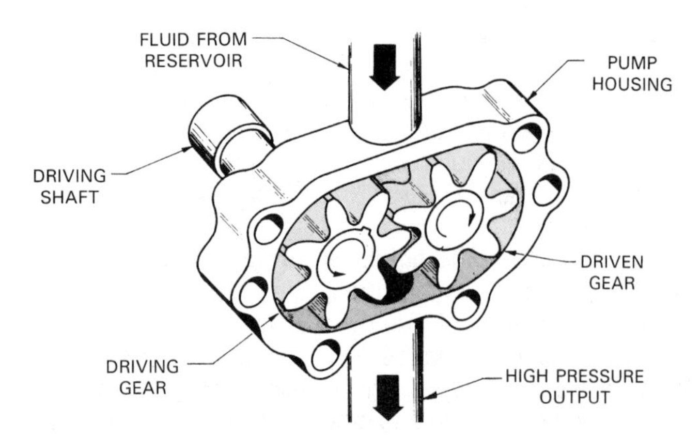

Figure 5-10. The gear pump is a simple form of hydraulic pump. The gears move apart on the inlet side, creating a vacuum. The vacuum draws in fluid from the system reservoir. The fluid is carried around the pumping chamber, between the gear teeth. When the gears come together on the outlet side, the oil is discharged.

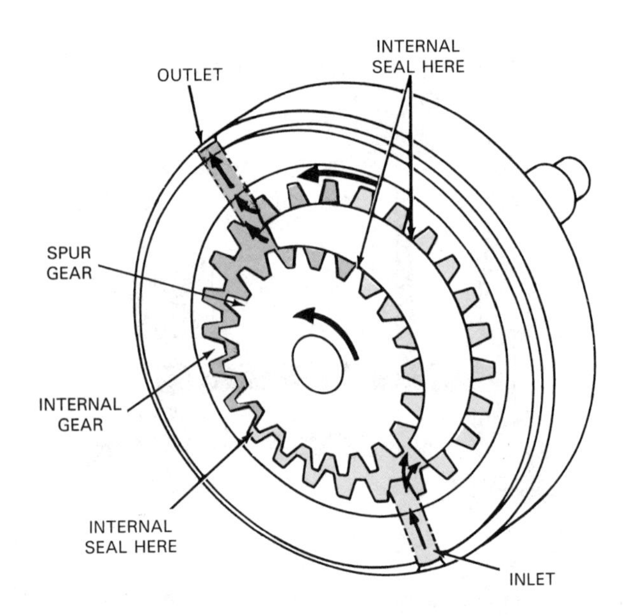

Figure 5-11. The internal gear pump is used on many automatic transmissions. The gears move apart on the inlet side, drawing in fluid. The fluid is carried in spaces between the teeth. When the teeth come together on the outlet side, the fluid is discharged. The crescent section forms a seal between the inlet and outlet sides. (Deere & Co.)

meshed gears inside of a housing, Figure 5-10. The inlet and outlet ports are on opposite sides of the gear meshing point. The *driving gear* is normally driven by the engine, through the torque converter. Since the two gears are meshed, the driving gear causes the other gear to turn. As the gears move apart on the inlet side, they create a vacuum that draws in fluid. The fluid that enters is carried, between the teeth, around the pumping chamber. It is then discharged as the teeth move together on the outlet port side.

One version of the gear pump commonly used in automatic transmissions is the ***internal gear pump,*** or ***crescent pump,*** Figure 5-11. This compact design saves space and reduces system complexity. The internal gear pump consists of a spur gear that meshes with an internal gear inside of the pump housing. The spur gear, driven by the engine through lugs on the rear of the torque converter, drives the internal gear. A crescent-shaped seal extends into the pumping chamber. It seals off the inlet and outlet sides. As the gears rotate, they move apart at the inlet, creating a suction that draws in fluid. The fluid is carried by the teeth to the outlet port and discharged.

Rotor pumps

A ***rotor pump,*** Figure 5-12, is used in some automatic transmissions. It consists of closely fitting *inner* and *outer rotors* and resembles the internal gear pump. The inner rotor is usually driven by the torque converter drive lugs. It causes the outer rotor to turn. As the rotors move apart, fluid is drawn in at the inlet port. The fluid is carried to the outlet port and is discharged when the rotors move together. A close fit between the lobes of the rotors in about the middle of the pumping chamber seals off the inlet and outlet sides.

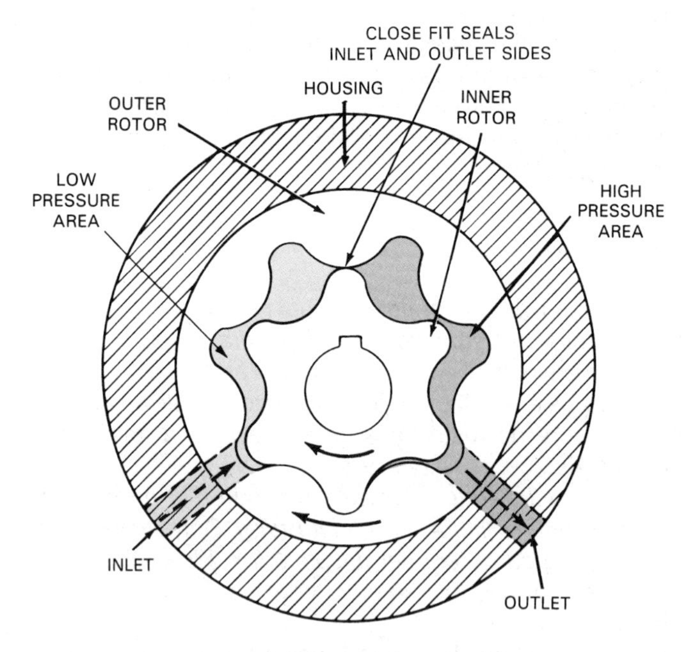

Figure 5-12. The rotor pump is used in some automatic transmissions. The arrangement of the internal and external rotor resembles the crescent pump, only without a crescent seal. Instead, the close fit of the inner and outer rotors forms a seal at the center of the pumping chamber.

Vane pumps

A ***sliding-vane pump,*** or more simply, a ***vane pump,*** is used in many late-model automatic transmissions. It uses less engine power, improving gas mileage. Figure 5-13 shows one type of vane pump called a *balanced vane pump.* It differs in that it has two inlet and two outlet ports. This feature eliminates side loads due to pressure imbalance.

In general, the sliding-vane pump mainly consists of an inner *rotor body,* driven by engine power, and a series of *vanes.* The vanes slide in and out of slots in the rotor body. The pump also has an outer *rotor ring,* with an oval-shaped inner surface.

As the rotor turns, the vanes are thrown outward against the surface of the rotor ring. As the rotation continues, the vanes are forced back into the rotor body by the ring. As the vanes follow the contour, compartments are formed, which continually expand and contract in volume. During this process, fluid from the reservoir is drawn, by suction, through the inlet port into the expanding compartments. It is then carried around and discharged at the outlet port, as the trapped volume of fluid is squeezed out.

One unique feature of the vane pump is that once hydraulic pressure is developed, the pressurized fluid is fed to the vane slots, or the space behind the vanes. The pressurized fluid helps to force the vanes against the housing wall and form a tight seal.

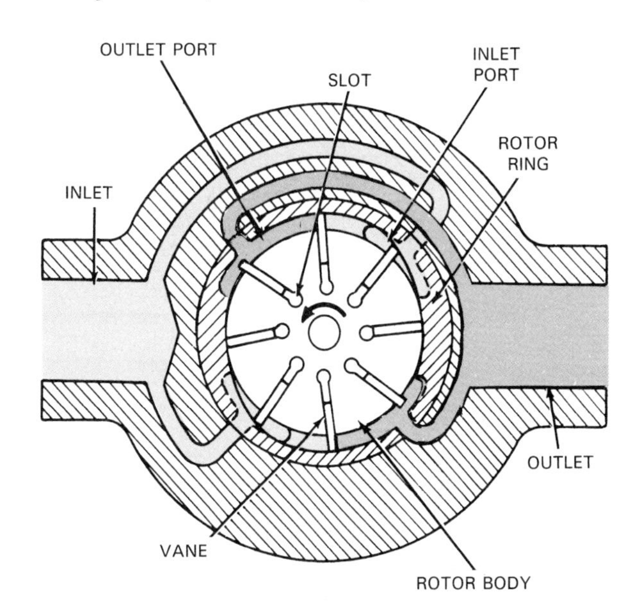

Figure 5-13. The sliding-vane pump is often used in modern transmissions to increase gas mileage. Fluid is drawn in on the inlet side of the pump. The fluid is carried between the vanes and is discharged at the outlet side of the pump. The vane slots are pressurized to keep the vanes in close contact with the rotor ring. (Deere & Co.)

Hydraulic Valves

A pump that would generate enough pressure for a vehicle in idle would create too much pressure at high speeds. To control system pressure, *hydraulic valves* can be used. In addition, these valves can be used to determine where flow will be used in the rest of the hydraulic system. ***Hydraulic valves*** are devices used in a hydraulic circuit to control pressure and, also, direction and rate of fluid flow. In general, these devices help control the entire operation of a hydraulic system. Various types of valves can be used in an automotive drive train.

Check valves

A ***check valve*** permits flow in only one direction. The most basic of check valves is a steel ball, or *check ball,* held within the fluid passageway of a *valve body* (discussed shortly). Flow in one direction through the system and passageway causes the check ball to unseat, permitting flow. Flow in the other direction pushes the ball against the valve seat, blocking flow. A check ball is shown in Figure 5-14.

Check valves are considered to be a type of *directional control valve,* discussed shortly. Directional check valves are used to prevent or restrict fluid flow to parts of the hydraulic system. This is done to more precisely control the action of the hydraulic system.

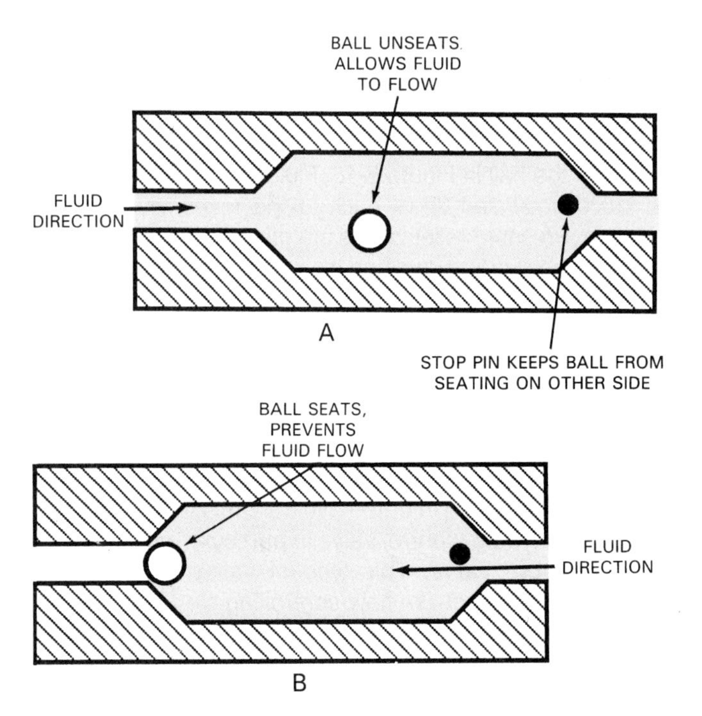

Figure 5-14. The check ball allows fluid to flow in one direction only. A–When fluid is flowing in the direction shown, the ball unseats. This allows the fluid to flow through the valve passageway. B–If fluid attempts to flow in reverse, the ball seats, and fluid flow is blocked. Directional check balls are widely used in automatic transmissions.

Pressure relief valves

The ***pressure relief valve,*** Figure 5-15, is essentially a check valve that opens to exhaust excess pressure. It

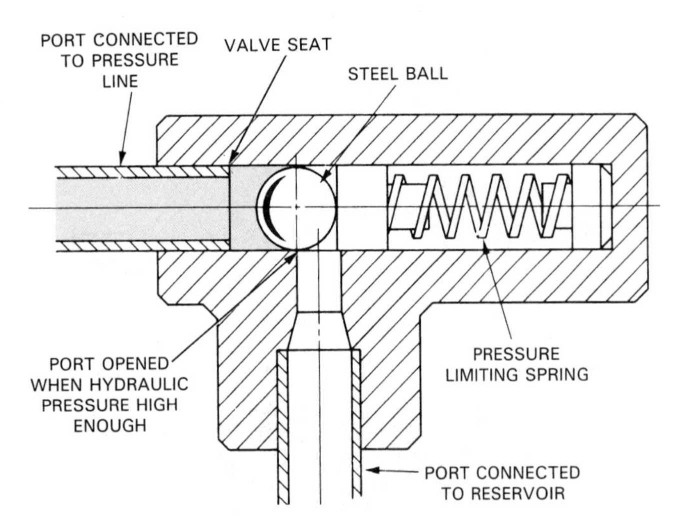

Figure 5-15. The ball in this pressure relief valve is held against the seat by a spring. When fluid pressure becomes greater than spring pressure, the ball is pushed back, allowing fluid to return to the reservoir. Fluid pressure is then reduced to less than spring pressure, and the ball reseats. Some springs are adjustable, making relief pressure adjustable.

keeps pressure from exceeding some preset maximum value. Its purpose is to prevent excess pressure from damaging the hydraulic system. The pressure relief valve is designed to work as a *two-position device*–that is, to be either *open* or *closed.*

The internal valve element is held against its seat by spring pressure. When hydraulic pressure becomes greater than the given spring pressure, the element unseats, as shown by the ball in Figure 5-15. Fluid then passes through the valve body and flows back to the reservoir, reducing pressure. Pressure relief valves are primarily used to relieve excess system pressure when the *pressure regulator valve,* discussed in upcoming paragraphs, is overloaded or malfunctioning.

Control valves

Hydraulic system ***control valves*** are used to regulate the operation of the system and of other hydraulic components. They are used in automatic transmissions and transaxles. One type of control valve in particular is commonly called a ***spool valve.*** This type of valve is so named because of the spool-like flow-controlling element, hereinafter called the *valve.* See Figure 5-16.

Spool valves are installed in an automatic transmission ***valve body,*** which consists of an aluminum or iron casting with internal passageways. Holes are drilled into the valve body to receive the spools. These ***valve bores,*** as they are called, are a very close fit with the valves themselves.

The movement of the valves in the valve body directs pressurized fluid to other parts of the hydraulic system. The valve body is often referred to as the "brain" or "control center" of the hydraulic system. A valve body may contain only a single valve. Most automatic transmission valve bodies contain between 5 and 30 valves. Figure 5-17 is a sectional view of a simple valve body containing a single valve. Note the relationship of the valves to the bores and passageways.

Pressure regulator valves. Some control valves are used to control system pressure; these are called ***pressure regulator valves.*** Pressure regulator valve operation is similar to that of a pressure relief valve. The difference is the pressure *regulator* tries to maintain a

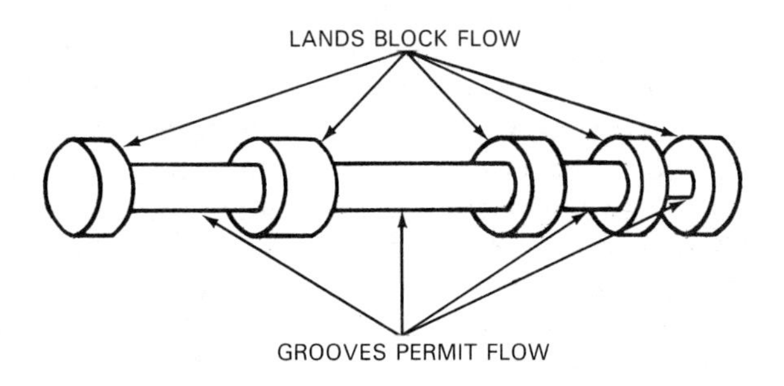

Figure 5-16. Spool valves are so named because they resemble ordinary spools, used to hold wire or thread. The larger-diameter segments of these valves are machined to closely fit the bore of the valve body. These segments of the valve are called lands. The grooves between the lands allow fluid flow.

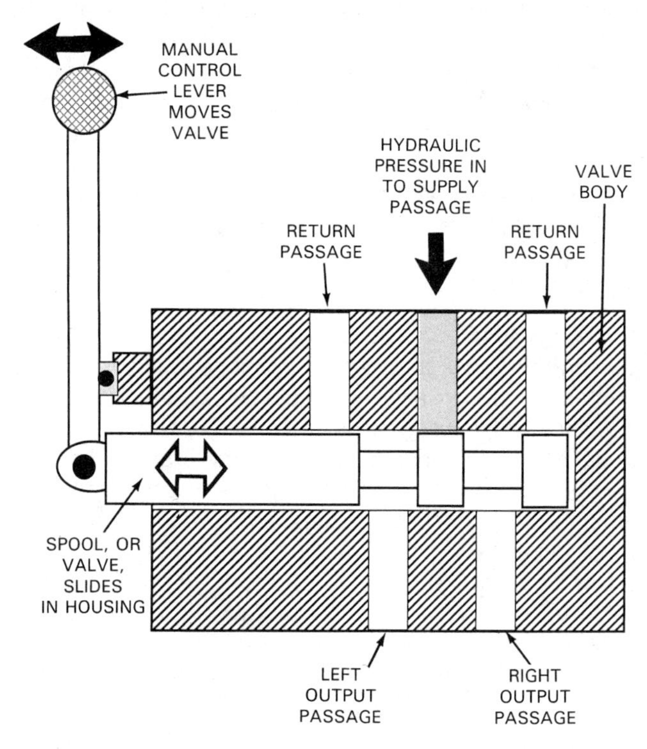

Figure 5-17. Study sectional view of a simple valve body with one valve. Note the relationship of the valve and valve body passageways. Most transmission valve bodies contain many valves. Some valves are manually activated; others are automatic.

constant pressure, and it does this by *modulating,* or moving back and forth. The relief valve, on the other hand, works as a two-position device to keep system pressure from exceeding a specified maximum value.

In operation, the pressure regulator valve, which is a sliding piston, works against a spring. When pressure is low, the valve is positioned so that a port, which is connected to a line returning to the oil reservoir, is closed off. Hydraulic pressure above a certain value causes the valve to begin to move and uncover the port. Some of the fluid is exhausted back to the reservoir, reducing pressure. The valve piston modulates back and forth in the cylinder bore, bleeding off more or less oil in the system to maintain constant output pressure. The valve movement, then, allows a constant pressure to be maintained from a variable pressure source. The pressure setting of the pressure regulator valve is controlled by the valve spring, some of which are adjustable.

Throttle valves. Whereas pressure regulator valves control pressure in response to internal pressure conditions, there are other valves that control system pressure based on some external input. For example, ***throttle valves*** can modify system pressure according to engine load and vehicle speed. This information may be sent to the throttle valve mechanically, by a linkage from the carburetor or throttle body, or pneumatically, by engine manifold vacuum. This input causes modulation of the valve, which compensates for the varying conditions.

Directional control valves. These valves are used to direct the path of flow in a hydraulic circuit. See Figure 5-18. Directional control valves are used in automatic transmissions to control the application of different hydraulic output devices. These valves control transmission shifts, shift quality, and other transmission actions.

The valves themselves may be actuated either by external sources or internally. Externally, the shift lever moved by the driver, engine vacuum operating a vacuum modulator, and electric solenoids are examples. Internal actuation is accomplished by different hydraulic pressures from within the hydraulic system acting on the valve.

Actuators

Hydraulic system output devices are driven by ***actuators.*** These may be pistons actuated by hydraulic pressure, Figure 5-19, or they may be electric solenoids (discussed later). Most hydraulic actuators are held in the *released,* or *normal* (unpressurized), position by one or more springs, called *release,* or *return, springs.* Hydraulic pressure overcomes spring pressure to move the piston. Hydraulic actuators used in automatic transmissions include servo pistons, used to apply friction *bands,* and clutch apply pistons, used to apply friction *clutches.* (These components are discussed in detail in later chapters.) Other hydraulic actuators, such as hydraulic motors, are seldom used on vehicle drive trains.

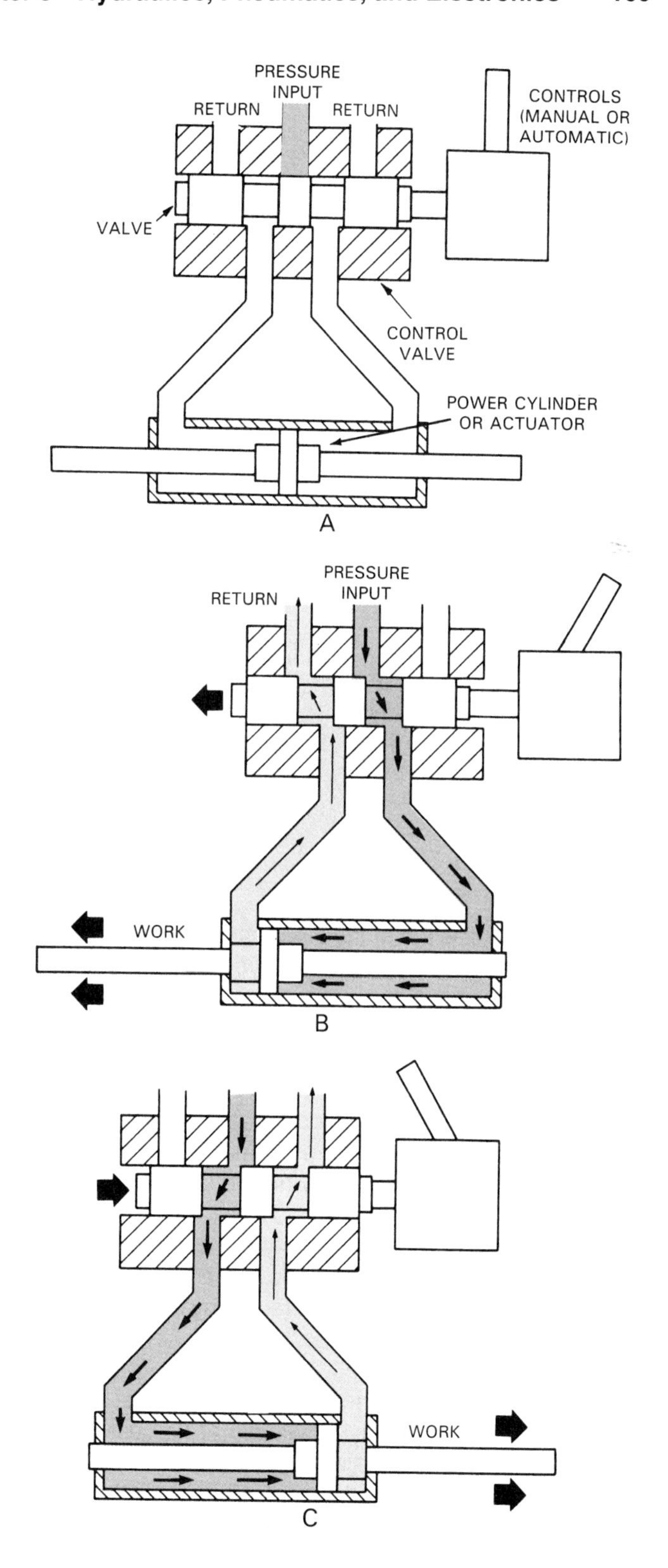

Figure 5-18. Actions of a directional control valve can be summarized by this series of illustrations. A–Valve is in position to prevent flow to the power cylinder. B–In this position, the valve has moved to allow fluid to flow to the right side of the power cylinder. The piston moves left. C–In this position, the valve has moved to allow fluid to flow to the left side of the power cylinder. The piston moves right.

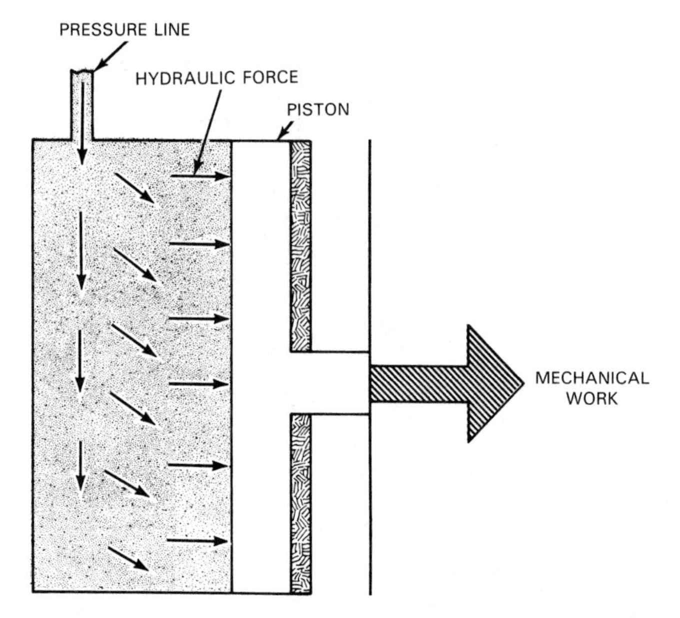

Figure 5-19. Actuators change hydraulic pressure into motion and force. Two types of hydraulic actuators are used in automatic transmissions. The servo piston is used to apply transmission bands. The clutch apply piston is used to apply clutch packs. (Chrysler)

Pneumatics

Pneumatics is the study of the *mechanical properties* (reactions to applied forces) and physical properties of air and other gases. The study is concerned with gases both at rest and in motion. ***Pneumatic systems*** typically use air for a power source.

Since air is compressible, some power must be used to pressurize the air before a compressed-air system can do any work. With a hydraulic system, on the other hand, no power is wasted in pressurizing the system; any pressurizing force applied translates directly into useful work. Therefore, a pneumatic system cannot transfer power as efficiently as a hydraulic system. In addition, pressure and flow in pneumatic systems are affected by air temperature and humidity. These are all disadvantages of a pneumatic system.

One advantage of a pneumatic system is that air, used to transmit power, is always available at no cost. Also, a pneumatic system is an ***open system.*** This means that after the air has been used, it is exhausted to the atmosphere. In this respect, it is unlike a hydraulic system, which is a ***closed system,*** requiring special reservoirs and no-leak designs. In a pneumatic system, air exhausting out of the system does not create a mess, as oil would in a hydraulic system. Further, slight leaks do not affect system performance. The pneumatic system does not need to be as tightly sealed, and components and piping can be lighter and less complex.

Although we seldom think of it, the surrounding air is pressurized by its own weight to about 14.7 psi (101 kPa) at sea level. This is called ***atmospheric pressure.*** The operating principle of any pneumatic system is the difference in pressure between air inside of the system and outside air. This difference in pressure, often called a *pressure differential,* is used to create the force that moves an actuator or spins an air motor.

Types of Pneumatic Systems

There are two types of pneumatic systems: *compressed air* and *vacuum.* If the pressure inside of the system is greater than the outside air, the system is a ***compressed-air system.*** If the pressure inside of the system is a ***vacuum***–that is, if it is less than the outside air, or atmospheric pressure–it is a ***vacuum system.***

Compressed-air system

A common example of a compressed-air system is the air compressor and piping that is found in most repair shops. The compressor is driven by an electric motor. It draws in and compresses air to a higher pressure than atmospheric pressure. This high pressure air is then sent through piping to operate impact wrenches and hammers, clean and dry parts, and inflate tires.

Vacuum system

An example of a device operated by a vacuum system is the ***vacuum modulator valve*** found on some automatic transmissions. The modulator is a chamber, or container, with an internal *diaphragm,* Figure 5-20. The diaphragm is a flexible partition that divides the chamber into two regions, sealing one off from the other. One side of the diaphragm is at atmospheric pressure. The other side is connected to engine vacuum by tubing that leads to the intake manifold.

When the engine is running, downward moving pistons on their intake strokes produce a suction in the intake manifold. This is where the vacuum originates from. The vacuum will vary according to the load that is placed on the engine by changes in throttle opening or weight changes to the vehicle.

Together, atmospheric pressure and engine vacuum act to force the diaphragm in the vacuum modulator toward the vacuum source. However, the modulator is *spring loaded,* and the force of the spring opposes the force caused by the pressure differential acting on the diaphragm. As engine vacuum goes up and down, the diaphragm, acting against the spring, moves back and forth, which causes the transmission valve connected to it to modulate. This, in turn, causes throttle oil pressure to vary as changes in engine load occur, in order to match transmission shift points to engine loading.

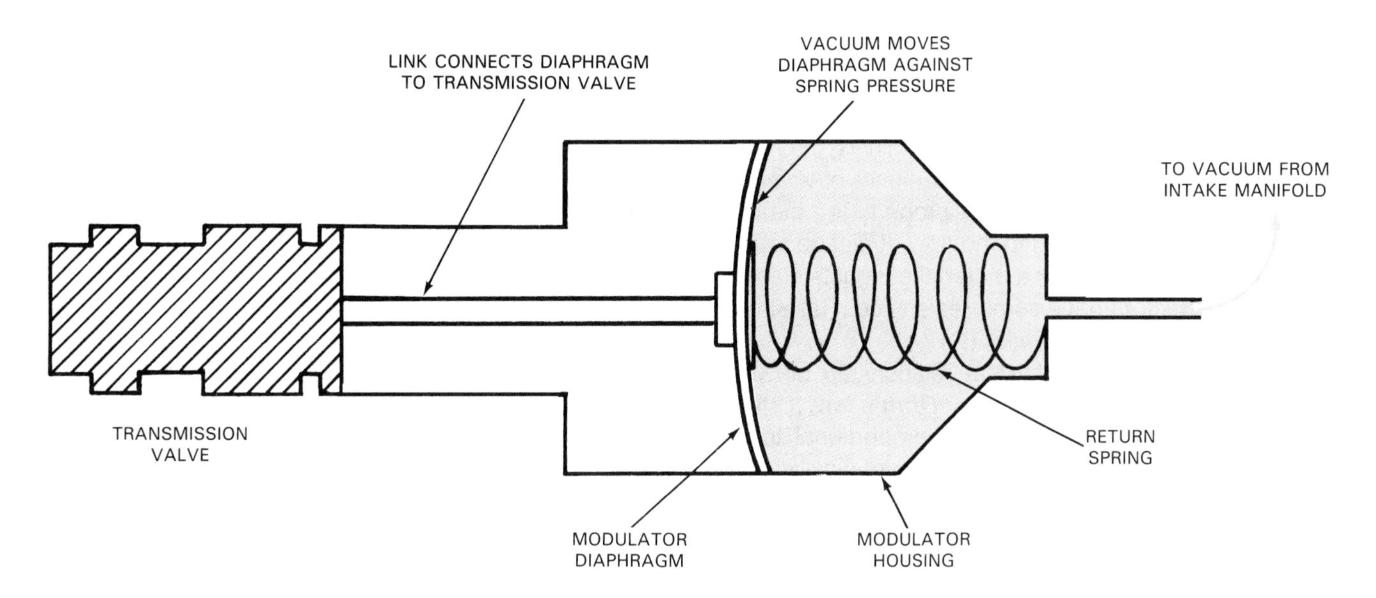

Figure 5-20. The vacuum modulator used on some automatic transmissions is an example of a pneumatic device. Engine vacuum moves the modulator diaphragm against spring pressure. The modulator diaphragm is connected to a valve that modifies transmission fluid pressure. Changes in engine vacuum, caused by load changes on the engine, result in pressure changes in the transmission.

Electronics

Electronics is the study of the behavior of electrons in devices and circuits, including the control and application of electricity. The following is a brief review of electrical theory and practice. If you are not completely familiar with electrical principles, you should study the following paragraphs carefully. Automotive drive trains are becoming more electronic every day!

Fundamental Principles

Atoms are minute particles of *matter* that make up everything in existence. Atoms are composed of even smaller particles. The center, or *nucleus,* of the atom is made up of ***protons,*** which are positively charged, and ***neutrons,*** which have no charge. Revolving around the nucleus are ***electrons,*** which are negatively charged. The electrically neutral atom has the same number of electrons as protons. While protons cannot leave the center of the atoms, electrons can move from one atom to another. ***Electricity*** is the movement of these electrons from one atom to another.

Materials whose atoms can easily give up or receive electrons are good conductors of electricity. Such materials are classified as ***conductors.*** Copper or aluminum easily give up or accept electrons; they are conductors. Materials whose atoms resist giving up or accepting atoms are not good conductors; they are ***insulators.*** Glass and plastic are very reluctant to give up or accept electrons; therefore, they are insulators. Finally, there are ***semiconductors.*** These are materials whose conductive properties lie somewhere between insulators and conductors.

Two factors must be present for electron flow, or ***current.*** First, there must be more electrons in one place than another, or a ***potential difference,*** also known as ***voltage.*** A *voltage source,* which provides energy to an electrical *circuit,* acts as a source of this potential difference. Secondly, there must be a path between the two places. For *sustained* current, there must be a continuous loop, or a *complete* circuit.

In a car or truck, a *battery* or an *alternator* (AC generator) serves as the source of potential difference. The battery or alternator always has two terminals–*positive* and *negative.* (In the alternator, the two terminals alternate between positive and negative.) The negative terminal has more electrons than the positive.

The path between the two terminals can include, for example, the vehicle wiring, the filament of a light bulb, and the internals of a single electrical part or a combination of electrical devices. These devices form the pathway that connects the two terminals, and electrons can flow, from the negative to the positive terminal (as they do in the external circuit).

Basic electrical quantities

The three basic physical quantities of electricity are:

- **Current:** This is the number of electrons flowing past any point in the electrical path in a given time period, or the rate of *charge* flow. It can be considered to be the amount of electricity in the path. Current is measured in units of ***amperes (A),*** which is usually shortened to ***amps.***

- **Voltage:** This is the electrical *force* that makes electrons flow. The greater the difference in the number of electrons, or charge, between two terminals, the higher the voltage is. Voltage is measured in units of ***volts (V).***
- **Resistance:** This is the physical property of a material concerning its opposition to the flow of electrons. It is an indication of how well a material conducts electricity. All conductors have some resistance. Resistance is measured in units of ***ohms*** (Ω).

Note that there is a physical relationship between these three quantities. It is defined by ***Ohm's law.*** It states that current in any circuit is directly proportional to the applied voltage. Ohm's law is stated mathematically by the formula:

$$I = \frac{E}{R}$$

where I is current in amperes, E is voltage in volts, and R is resistance in ohms. Note that this formula shows: For any given voltage, current will increase as resistance decreases. For any given resistance, current will increase with voltage.

Direct and alternating currents

There are two types of current that an electrical system might have. One is ***direct current (DC).*** Direct current is unidirectional. This means that the flow of electrons in a direct-current system is always in the same direction through the system. This is because the terminals of the voltage source in the DC system do not alternate in *polarity,* or from negative to positive. A battery is an example of a DC source. Automotive electrical systems are DC systems.

The other type of current, such as that used in your home, is ***alternating current (AC).*** Alternating current is bidirectional. Electron flow in the AC system *alternates* between one direction and the opposite direction. The flow of electrons changes direction many times every second. This is because the terminals of the voltage source in the AC system alternate in polarity. Since the terminals alternate from positive to negative, current keeps changing direction, as electrons flow to the higher potential, or from negative to positive.

Note that the alternator, used to charge an automotive battery and power the vehicle's electrical devices, produces alternating current. Since automotive electrical systems operate on direct current, the alternating current must be *rectified,* which means changed to direct current. There are electrical devices in the system that perform this function. These devices act like check valves, permitting flow in only one direction through the circuit.

Electrical Circuits

The path that electricity takes is called a ***circuit.*** Any practical electrical circuit has three important elements. These include a voltage source, a complete pathway for current, and a *load,* which offers resistance to limit current. In addition, many circuits have a switch or other means to control the circuit action.

The simple circuit shown in Figure 5-21 consists of a battery, a light bulb, and connecting wiring. The ***battery*** stores chemical energy. This energy is changed into electric current by a chemical reaction in the battery. The reaction creates a potential difference across the terminals of the battery. As a result, electrons flow through the external circuit. The light bulb is made of a special resistance wire that glows when electrons pass through it. Any time that electrons are flowing, the bulb will light. The wiring is made of metals that allow the electrons to pass with very little resistance. This low resistance prevents the electricity from being wasted in places where it is not needed. Most automotive wiring is made of copper, aluminum, or aluminum coated with copper.

Note that on most vehicles, there would be no return wire from the bulb back to the battery; the metal frame serves as the return wire. On modern vehicles, the negative battery terminal is connected by a strap to the frame. The negative terminal is the *ground* terminal. The electrical devices throughout the system are connected directly to the frame. This means that the battery, charging system, and all other electrical devices have their negative terminals connected to a common, negative-ground connection, usually the vehicle frame and body. This type of *one-wire* system in the automobile is termed a ***frame-ground system.*** See Figure 5-22 for an example of how the vehicle frame is used to return electricity to the battery.

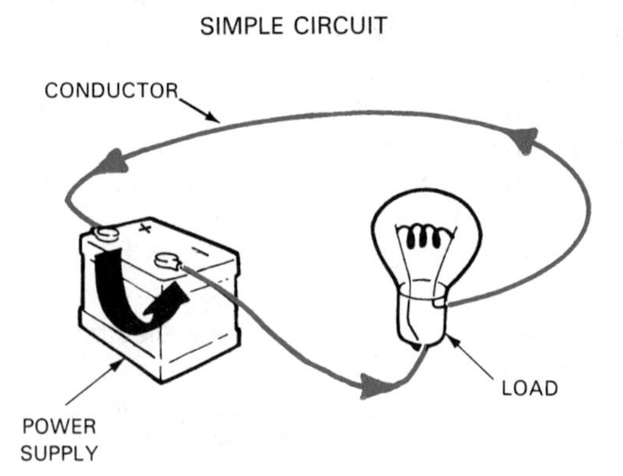

Figure 5-21. This simple circuit consists of a battery, a load resistance (light bulb, in this example), and associated wiring. Electrons flow from the battery, through the wiring and bulb, and back to the battery. This is a complete circuit, in which current can flow and accomplish the job that is was designed to do. (British Leyland)

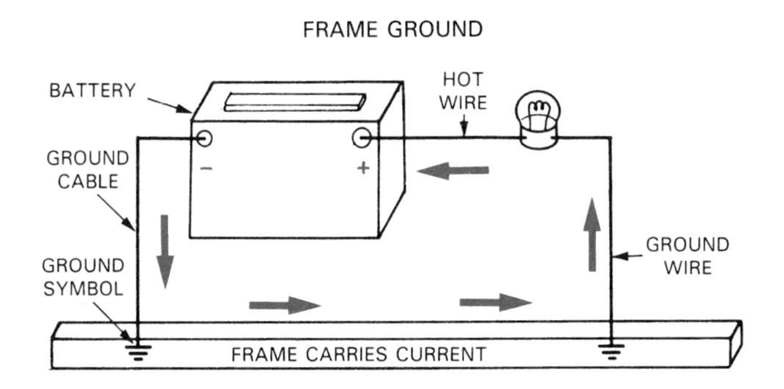

Figure 5-22. Most manufacturers use the vehicle frame to provide the return path, or ground, for the vehicle electrical circuits. This reduces wiring problems. Also, the frame-ground system is easier to trace and service.

Series, parallel, and series-parallel circuits

The simple circuits just described are examples of *series circuits.* These are one of the three basic kinds of circuits. The other two are *parallel circuits* and *series-parallel circuits.* See Figure 5-23.

In the ***series circuit,*** components are connected in a string, or end to end, and current has only one path. The same amount of current flows through every component of the circuit.

In the ***parallel circuit,*** components are connected side to side, and current flows through multiple pathways. Each component provides a path for current. Different amounts of current will flow in different parts of the circuit. This means that total current will be split as it flows into the parallel paths. However, total current leaving the voltage source will be equal to total current returning to the voltage source. If the load resistances have different values, current will be different through each branch. If their values are the same, current will be divided equally through each branch.

Most vehicle circuits are made up of a combination of series circuits and parallel circuits. In these, some components are wired in series, and some are wired in parallel. Such circuits are called ***series-parallel circuits,*** or ***combination circuits.***

Defective circuits

Defective circuits can be divided into two general areas. These include *short circuits* and *open circuits.*

A ***short circuit,*** or ***short,*** is caused when a wire or other electrical component contacts a grounded part of the vehicle. This allows the current to bypass part of the intended electrical circuit to return to the battery. The current bypasses the intended circuit as it presents greater resistance, and the current will take the path of least resistance. Since resistive loads are bypassed, the short circuit usually results in extremely high current. This leads to excessive heat, burnt wires, and possibly, an electrical fire. See Figure 5-24.

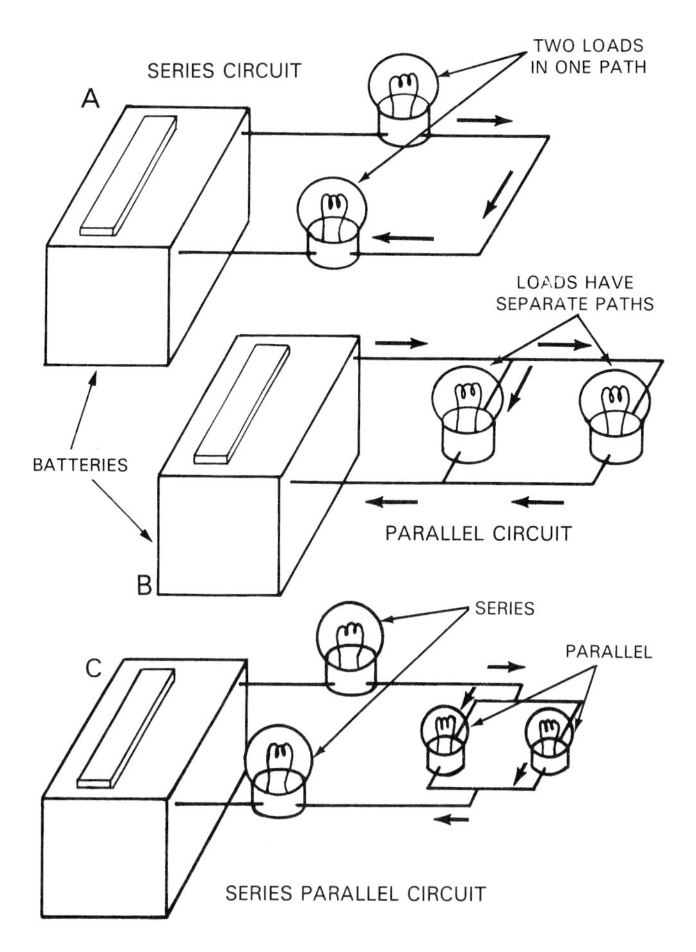

Figure 5-23. There are three basic kinds of electrical circuits. A–The series circuit has only one path for current. The same current flows through all components. B–The parallel circuit has separate current paths for system components. C–The series-parallel circuit is a combination of series and parallel circuits. Current flows in series in some sections of the circuit and in parallel in other sections.

An ***open circuit,*** or ***open,*** occurs when some part of the circuit becomes disconnected, and current cannot flow. See Figure 5-25. Disconnected or broken wires, contacts that are burned or cannot close, loose or dirty connections, and defective electrical components are some causes of open circuits. Note that some short circuits also have a broken connection and may *look* like an open circuit. The difference is that at the break in the short circuit, there is a no-resistance current path to ground.

Electrical Components

The prior discussion has focused on batteries, light bulbs, and wiring as circuit components. Vehicle electrical systems contain many other components. Some of the electrical components that affect the drive train are *fuses* and *circuit breakers, resistors, switches,* and *relays.*

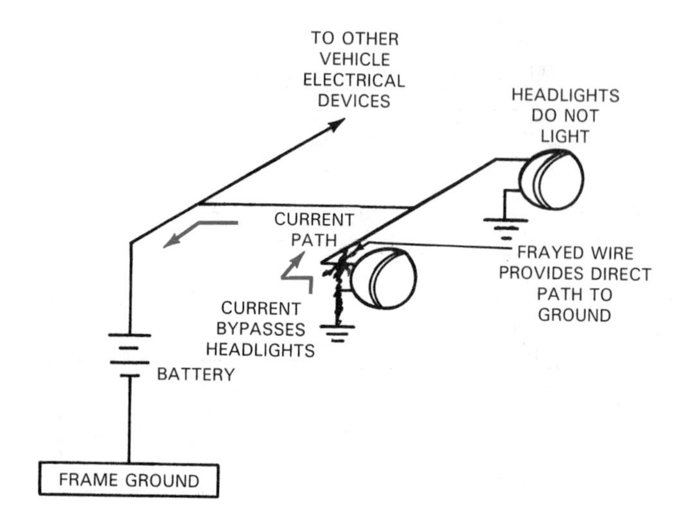

Figure 5-24. A short circuit bypasses part of the intended circuit. This can result in excessive current and overheating. Most shorts are caused by an insulation failure that causes a wire or component to contact a grounded part of the vehicle. Note that in this illustration, neither headlight lights, since there is a short; current bypasses these loads through the frayed wire connected to ground.

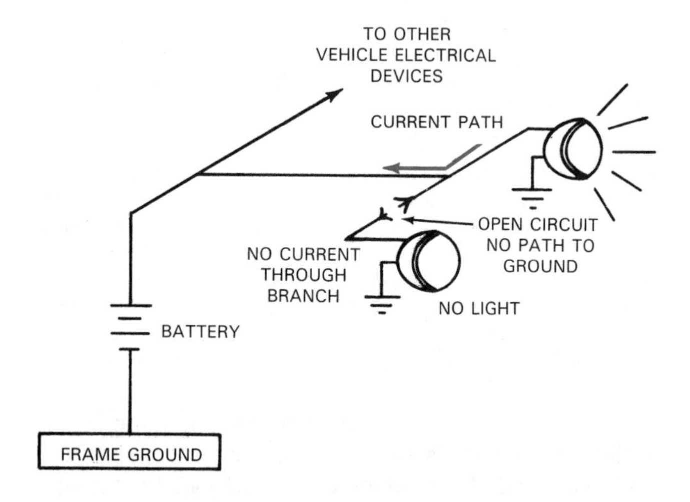

Figure 5-25. An open circuit is a break in the circuit that prevents current. Open circuits can be caused by physical breaks, defective components, or bad connections. Note that in an open series circuit, current is completely cut off. In an open parallel circuit, current is blocked in the branch where the open exists, as shown in this illustration. Current will still flow through the intact part of the circuit.

Fuses and circuit breakers

Fuses and ***circuit breakers*** are used to protect circuits from excessive current, or *overcurrent.* An overcurrent can be caused by a short circuit or an *overload.* An ***overload*** is an overcurrent condition contained within the normal electrical circuit. Overloads occur when a component or components draw more current than the system wiring is designed to handle.

A fuse contains a thin strip of metal that melts at low temperature. The fuse is placed in series with the main circuit, as shown in Figure 5-26. High temperatures produced by overcurrents cause the metal element of the fuse to melt before other components or circuit wiring can be damaged.

A circuit breaker performs the same function as a fuse; however, it is a mechanical device. It is tripped by heat caused by excessive current. It is not ruined when activated, so it does not need to be replaced–simply, reset. All electrical components used with drive trains are protected by a fuse or circuit breaker. These are usually placed in the fuse box.

Resistors

Resistors are placed in a circuit to reduce current to a specific value required for proper operation or for current-limiting protection of other circuit components. Sometimes, they are placed in a circuit to obtain a desired voltage drop. These components reduce current by introducing extra opposition to the flow of electrons. Resistors are made of resistive metals or carbon. They are rated in ohms. Some resistors have fixed values; others are variable.

Switches and relays

Switches are used to connect or disconnect power in a circuit. They provide a convenient way to make or break complete electrical circuits. They are often operated by the vehicle driver. Examples are *neutral safety switches, ignition switches,* and *brake light switches.* Some switches are operated automatically. A *thermal switch,* which is operated by temperature, is an example. A *pressure switch,* which reacts to changes in pressure, is another example.

Relays are a special type of switch. Specifically, they are *electromagnetic* switches. Relays use a small input current to control a larger current in a circuit. The input current passes through the relay coil, creating a magnetic field in the device. The magnetic field causes movement of the relay *armature.* The armature makes or breaks one or more sets of electrical contacts. This action starts, stops, or diverts the larger current, which is wired to the contacts, depending upon how the circuit is wired.

Computers

A ***computer*** is an electronic machine that takes data from an *input device,* processes it through a stored *program,* or set of operating instructions, and sends commands to various *output devices.* You are probably somewhat familiar with computers. On-board computers are increasingly being used to control the drive train. This

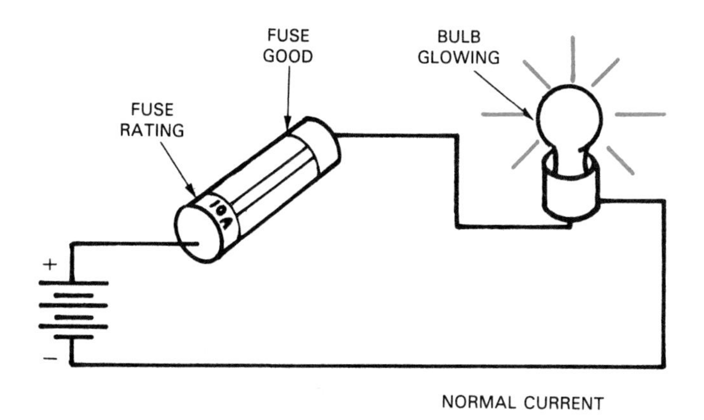

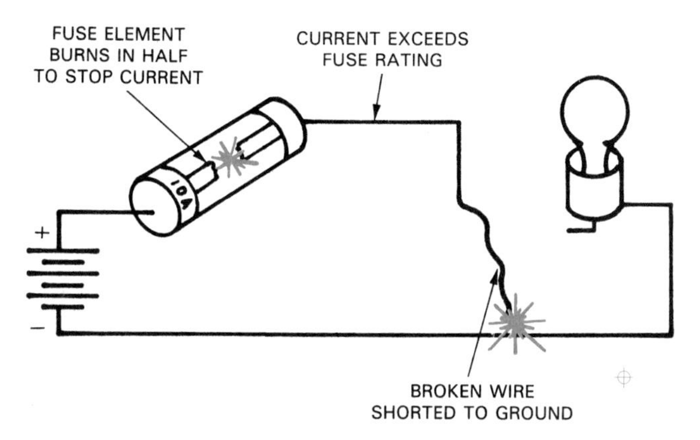

Figure 5-26. Fuses are installed to protect vehicle electrical circuits from excessive current. They are made of a special metal with a low melting point. Excessive current heats the fuse, causing the metal to melt and breaking the circuit.

means that the technician who wishes to service drive trains must be familiar with electronic devices and the basic principles of computer operation.

Transistors and computers

Many electronic components, including computers, make use of ***transistors.*** These devices are key in the operation of computers. In the computer, transistors are largely involved in *switching.* In this capacity, the transistor might be defined as an electronic relay, similar to the electromechanical relay shown in Figure 5-27.

The relay and the transistor perform the same function in Figure 5-27–both are acting as switches. Another example showing where transistors can be used in place of relays is the modern *electronic ignition control module.* It performs the job of old contact points; at the same time, it produces higher voltage with no moving or wearing parts.

Besides having no moving parts, other advantages of the transistor are that it consumes less power and is much smaller than the electromechanical relay. Many transistors can be combined in a small package to perform complex tasks. Tens of thousands may be contained in a single *integrated circuit (IC)* package. ***Microprocessors,*** used in automobile computers, for example, are complete computer processing units held on one IC.

Special- versus general-purpose computers

Special-purpose computers, such as those found on automobiles, operate according to the same basic principles as *general-purpose computers* used in homes and industry. Perhaps the biggest difference between automotive computers and a general-purpose computer, like a *personal computer (PC),* is that the automotive computer is not directly controlled by a person. All computer inputs come from ***sensors.*** These sense vehicle operating conditions. This form of input is only indirectly under driver control. All output devices are operated by the computer, with no direct control by the driver.

Since automotive computer systems are not controlled by the vehicle operator, many of the terms that you may

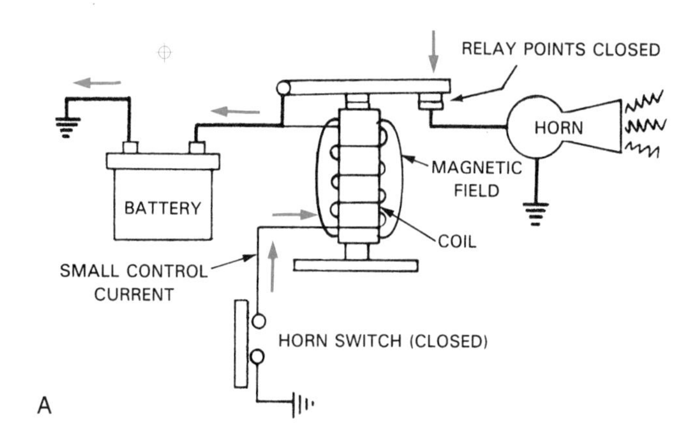

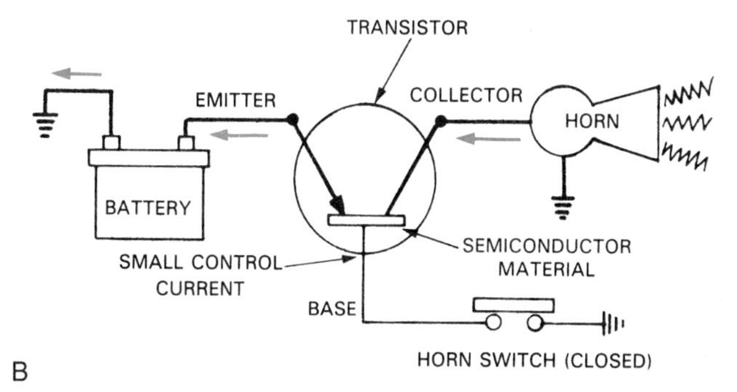

Figure 5-27. Compare the electromechanical relay and the transistor. The relay and the transistor perform the same function in the example–to make the horn operate. Both use a small current to control a larger current. A–When the horn switch is operated, a small current is sent into the relay. A magnetic field is built up in the coil. The field attracts the relay armature, closing the contact points. Battery current flows through the points, causing the horn to operate. B–When the horn switch is operated, a small current enters through the base of the transistor. This changes the semiconductor material in the transistor from an insulator to a conductor. Battery current can then flow through the transistor to the horn.

have heard in connection with the general-purpose computer are not used. PC *hardware* terms, such as *laser printer, disk* and *disk drive,* or *keyboard,* will not commonly apply to the automotive computer. Further, commands such as *copy, delete,* and *edit* will not typically apply. However, some cars use some of the dash buttons to make the computer produce the requested information. When diagnosing problems, the computer can also be triggered to output *trouble codes.*

Automotive computer systems

The basic operation of the automotive computer system can best be understood by studying Figure 5-28. Computer inputs come from sensors. The computer processes these inputs to determine what is happening to the engine. The computer then decides on the commands to send to the output devices.

Analog and digital signals. All late-model automotive computers use *digital* microprocessors, which operate using the ***binary number system.*** This system of numbers is comprised of two digits: 1 (ON) or 0 (OFF). Conveniently, these digits are represented electrically by high or low voltage signals or by the presence or absence of voltage.

Computers readily process ***digital signals.*** These are voltage signals that have distinct values, or that vary in discrete steps. Digital signals may be directly represented by binary digits. A digital input to or output from the computer may be comprised of a single ON/OFF signal or a combination of ON/OFF signals. An example of a sensor providing a digital input signal is a *crankshaft position sensor.* It reports engine piston position to the computer in the form of electrical pulses. An example of a digital output signal required by an output device is the signal sent to operate some fuel injectors. The computer rapidly pulses on and off to precisely control how much fuel is injected into the engine.

Not all inputs and outputs are digital. Some devices transmit or require ***analog signals.*** These are continuously variable signals, which can vary between any range of values. Analog signals will take the form of a varying voltage or current. (Current must be converted into voltage once it gets to the point where it will be used. This is simply done with a resistor.) Analog input signals usually represent physical or chemical variables, such as temperature, oxygen content, mechanical position, etc. A *throttle position sensor* is an example of an input device that provides an analog signal. Few vehicle output devices are controlled with an analog signal. One example would be an instrument computer controlling an analog-type fuel gauge.

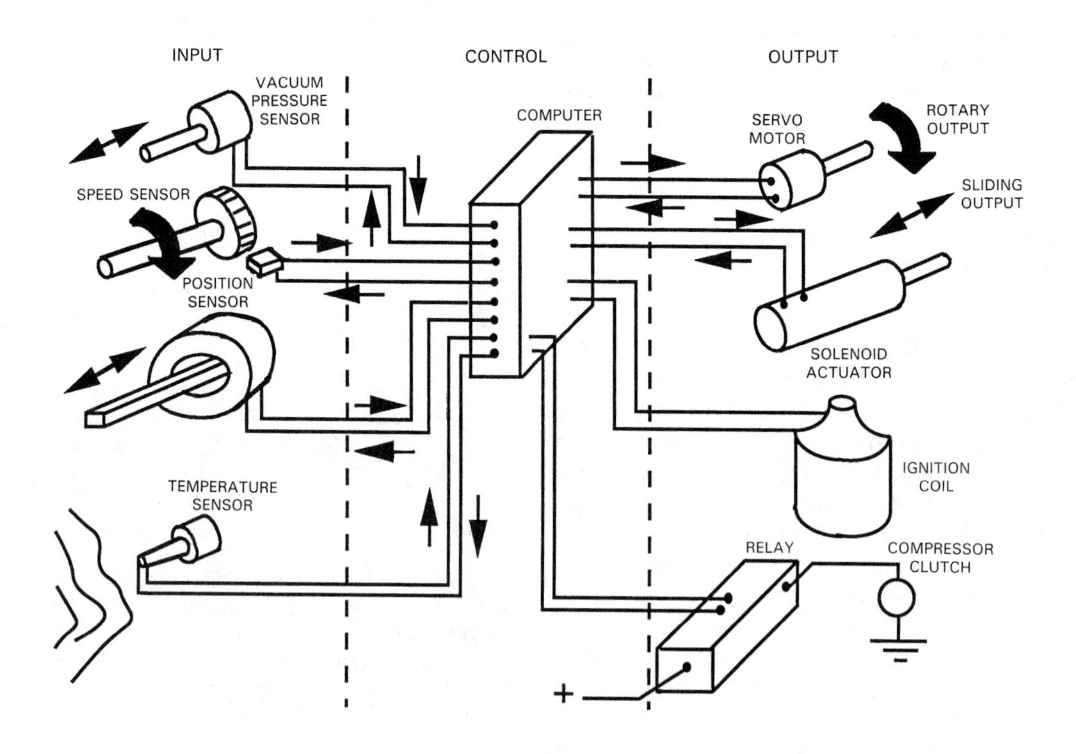

Figure 5-28. Shown is a pictorial representation of an automotive computer. Input sensors (left) send information on engine and other conditions to the computer. Computer (center) processes these input signals and issues output commands. Output commands cause output devices (right) to adjust the engine and drive train components as needed.

Since a computer is only capable of working with binary information, it cannot directly accept or produce signals in analog form. For this reason, computers have *input/output interface circuits* that convert analog information to digital or vice versa. Analog inputs are converted into binary code for use by the computer via an *analog/digital converter.* Digital outputs are converted into analog, as needed, by means of a *digital/analog converter.*

Input devices. ***Input devices*** feed information to a computer. In general, automotive input devices are called sensors. Sensors convert a condition into an electrical signal–either analog or digital, depending on the device. Some sensors read engine conditions. Cooling system temperature, rpm, manifold vacuum, throttle position, and air/fuel ratio are all sensed by different sensors. Other sensors may read outside air temperature and pressure, vehicle speed, the gear that the transmission is in, or whether the air conditioner or headlights are on.

Some inputs are direct electrical connections that do not require a sensor. Examples are electrical inputs from the headlights or air conditioner compressor clutch. Most inputs, however, come from sensors. The sensors are required to convert a physical or chemical variable into an electrical signal. Since they convert information from one form to another, sensors are ***transducers.*** Some sensors produce a signal without requiring electrical excitation; these are called *active transducers.* Other sensors are capable of producing an output signal only when used in connection with an excitation source; these are called *passive transducers.*

Some transducers are nothing more than a switch that is activated upon reaching a preset value of some condition. A low-fluid switch is an example. Many transducers are a type of variable resistor. Changes in the physical condition that they monitor affect the amount of current flowing through them and, therefore, change the electrical signal to the computer. Other transducers generate voltage and act as a variable voltage source.

Output devices. ***Output devices*** are targets for information held in a computer. There are four kinds of output devices used for controlling vehicle operation: motors, solenoids, relays, and coils. Solenoids and motors change electricity into motion, while relays and *coils* control electrical flows.

Solenoids are actuators that convert an electrical signal into back and forth, or linear, motion. These electromagnetic control devices are similar to relays in that they use a magnetic field. However, solenoids use this field to accomplish a *mechanical* task, such as opening a valve or moving a linkage. Relays, on the other hand, use the field to accomplish an electrical task. In both cases, an external source of electricity creates the magnetic field. Fuel injectors are solenoid-operated valves. Solenoids are used to control throttle position. They are involved in control of vehicle emission systems. They operate torque converter *lockup clutches* and are also involved in transmission shifting.

Motors are another type of actuator. These devices are also driven by a magnetic field. They, too, change electric energy into mechanical energy. However, motors change the electrical signal into *rotary* motion. An example is the throttle positioner used on many engines.

A ***coil*** is basically a wire wrapped around an iron, ceramic, or air core. Coils are used in electrical circuits to provide a magnetic field. *Transformers,* used to increase, decrease, or isolate electrical voltage, are comprised of a *primary* and a *secondary* coil. An ignition coil, which is a type of transformer, is a common example.

Relays often serve as computer output devices. They are used to operate high-current devices, such as an air conditioner *compressor clutch.* Relays use a coil to open and close small contact points. With this setup, a small current is used to control a larger current. There is no need for the computer to handle the larger current. In addition, because the computer and output device circuits are isolated, the computer is protected from voltage spikes in the system.

Electrical Tests

As part of drive train service, you may have to perform some electrical tests. These tests require that you be familiar with some basic equipment used for checking electrical systems. The most commonly used test equipment is covered below.

Jumpers

A ***jumper wire,*** or ***jumper,*** is a short piece of insulated wire with an alligator clip on each end. It is commonly used to bypass components or to apply voltage to a component or section of a circuit. It may be used to determine whether current is flowing through a switch, relay, solenoid, electrical connection, wires, etc.

Figure 5-29 shows an example of how a jumper wire is used to check a switch. Prior to jumpering out the switch, the bulb would not light when the switch contacts should have been made, or been closed. The switch is bypassed with the jumper. If attaching the jumper around the switch turns the light on, then the switch is bad. If it does not, the problem is elsewhere.

Test lights

A ***nonpowered test light,*** Figure 5-30, is used to check a circuit for power. The device essentially consists of a probe with a light and a lead with an alligator clip. The device is powered off the circuit. To use the light, the one lead is connected to ground. The probe is then touched to the circuit to check for presence of voltage at that point. If there is voltage, the light will glow. If it does not glow, the circuit is open between the battery and the test point. If there is voltage at the next test point nearer the power source, then the open lies somewhere between the two test points.

A ***self-powered test light*** is used to check for circuit *continuity,* or whether a circuit is complete. This device resembles the nonpowered test light, but it has an internal battery. To use this type of test light, the normal source of power (car battery or feed wire) must be disconnected. The test leads should be connected across the circuit component or components in question. The circuit under test must be isolated, or disconnected from any circuits in parallel. If the light glows when the leads are connected, the component or circuit has continuity. If it does not glow, there is an open somewhere between the two test leads. To prevent damage, do *not* use a self-powered test light to test computer system electronics.

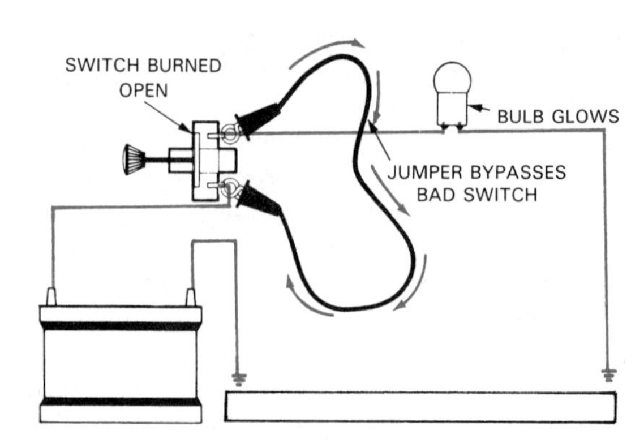

Figure 5-29. A jumper wire can be used to bypass a suspected component. If the circuit begins operating normally, the bypassed component is defective. (Ford)

Multimeters

A ***multimeter*** combines several test devices–*voltmeter, ohmmeter,* and *ammeter*–in one case. Sometimes a multimeter is called a ***VOM*** for *V*olt-*O*hm-*M*illiammeter. Figure 5-31 is a typical modern multimeter. Figure 5-32 depicts the individual function of each of the three meters contained within the multimeter.

The ***voltmeter*** is used to measure the amount of voltage in the circuit. It is normally connected across, or in parallel with, the circuit being checked. The voltage reading can be compared to specifications to determine whether an electrical problem exists.

An ***ammeter*** measures the amount of current flowing in a circuit. Ammeters must be connected in series with the circuit or branch being checked. All current then passes through the ammeter.

A modern ammeter is often an inductive or clip-on design. A clip is slipped over the outside of the wire insulation. The meter senses the magnetic field around the outside of the wire as current flows. This type of ammeter is more convenient to use, since the circuit wiring does not require disconnection.

An ***ohmmeter*** will measure the amount of resistance in a circuit or component. After disconnecting circuit power, the ohmmeter is connected across the wire or component being tested. The circuit under test must also be isolated, or disconnected, from any circuits in parallel. The ohmmeter reading can then be compared to specifications. If too high or low, the part is defective. Remember that the part being tested must be disconnected from the car's battery when using an ohmmeter!

An ohmmeter can be used to test for circuit continuity. However, some ohmmeters may cause damage to some computer system components. Use only instruments recommended by the vehicle manufacturer and always follow recommended procedures.

CAUTION! The ohmmeter has its own power source. *Never* connect an ohmmeter to a live circuit! Doing so can produce a faulty reading and may cause meter damage.

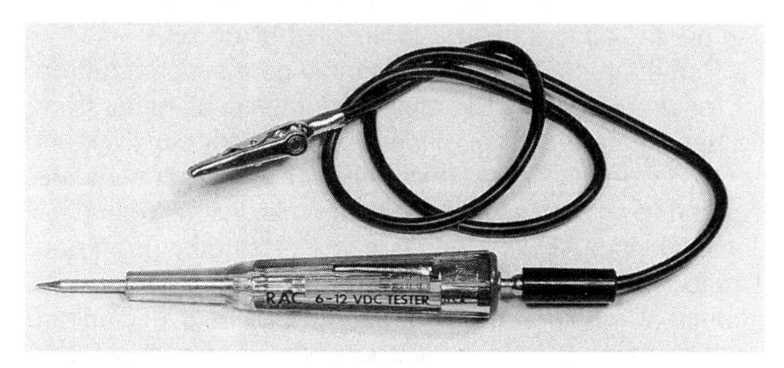

Figure 5-30. Test lights are useful when checking circuits for voltage and continuity. A nonpowered test light, such as the one shown, is used to check for voltage in a circuit. (RAC)

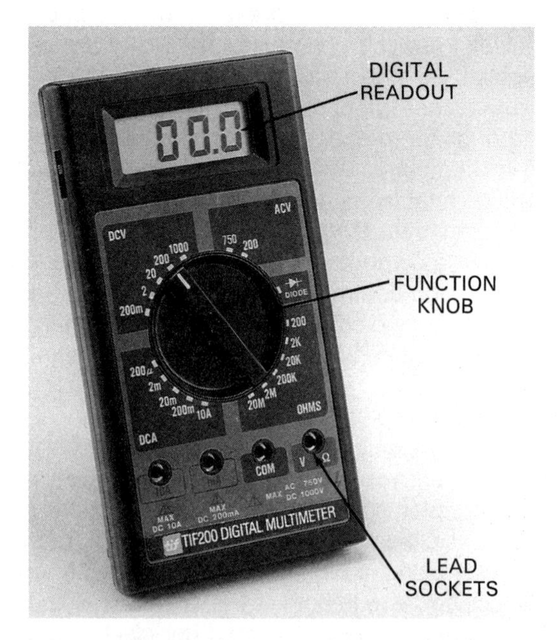

Figure 5-31. The multimeter, often called a VOM, contains several meters. The meter shown is a digital model, called a digital multimeter, and all values are shown as numerical readouts. Other meters are analog types, and values are displayed by a needle indicator and scale. (TIF Instruments)

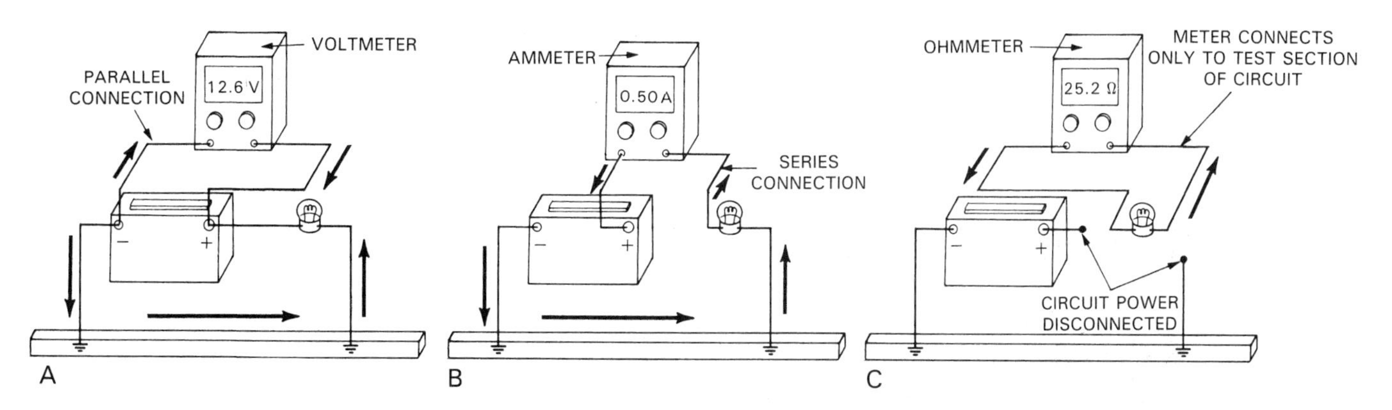

Figure 5-32. The three basic meters used in a multimeter are voltmeter, ammeter, and ohmmeter. A–The voltmeter is connected in parallel to measure voltage, in volts. B–The ammeter connects in series with the circuit to measure current, in amps. C–The ohmmeter is connected across the circuit. It is used with the circuit disconnected from the battery. It measures resistance, in ohms.

Wiring Repairs

Wiring is sometimes damaged during vehicle operation from such things as fires, accidents, and driving over rough ground. Further, wiring is sometimes accidentally cut or crushed while making repairs. Damaged wiring can be repaired easily by *soldering* or by using *crimp connectors.*

CAUTION! Do *not* repair wiring by simply twisting the wires together or installing a loose wire under a bolt. Automotive wiring is subjected to vibration, moisture, oil contamination, and various corrosive elements. Any of these can cause a bad connection if the wires are not properly connected.

Soldering

Soldering is one method used for making electrical connections. This process involves using ***solder,*** which is an alloy of lead and tin. It has a relatively low melting point as compared to other metals. The solder is heated and melted onto the connection to be made.

Soldering is perhaps the best way to connect wires. Once the metal cools, it becomes solidified, forming a joint that is hard to break. In addition, the electrical conductivity of a soldered connection is as good as an unbroken wire.

Solder is applied using a *torch* or a *soldering gun* or *iron.* These tools are used to melt the solder. In addition, ***flux*** should be used when soldering. It is a nonmetallic substance that cleans the metal and helps the solder to flow and bond. Most electrical solder, which comes in wire form, wrapped on a spool, has the flux inside of the wire. Two varieties include *acid-core* and *rosin-core solder.* Rosin-core solder is the only kind that should be used on electrical equipment. Acid-core solder should never be used for electrical repairs because of its corrosive property.

Soldering of wires is easy if the joint to be soldered is properly prepared. The first step is to cut and *strip,* or remove the insulation from, the wire. For soldering purposes, about 1 in. (25 mm) of insulation should be removed. (For other purposes, about 1/4 in. to 1/2 in. [6 mm to 13 mm] of insulation is often all you need to strip from a wire.) The surest way to strip insulation from wires without damaging the wire itself is to use a special stripping tool. Use this tool as shown in Figure 5-33.

Before soldering, twist wires together or twist the wire around the terminal. This forms a good mechanical connection and helps the solder in holding the connection together. Also, make sure that the wires are free from dirt, oil, and grease. If you are using a soldering gun or iron, make sure that the tip is clean and *tinned.* This means it should be coated with a thin layer of solder.

Heat the connection with the torch or gun. Touch the solder to the hot connection and allow it to melt into the connection as you continue to apply heat. This is shown in Figure 5-34. Note that the connection must always be heated first. Never melt the solder and let it drip on the connection! When the soldered joint cools, it should have a smooth surface and be a dull silver color. Pull lightly on the wires or component leads to check your joint. In most instances, you will want to insulate the soldered joint. Use electrical tape or other product designed for this purpose.

Crimping

Crimping is the easiest method of repairing wiring. This process involves using ***crimp connectors.*** These components may be for splicing two wires together or for equipping wire ends with terminals. An assortment of products are available.

Crimp connectors do not require soldering, but they do not make as good an electrical connection either. They are also affected by vibration more, as well as by oil and dirt contamination. Also, crimping should be avoided on low-voltage sensor circuits. All these factors should be kept in mind when deciding how to repair a wire.

It is important to use the proper kind and size of connector. An undersize connector may overheat. An oversize connector may not carry enough current, since the wire will not be properly held. This could also lead to overheating or an open circuit.

To install crimp connectors, a special *crimping tool* is needed. First, however, the wire should be stripped about 1/4 in. (6 mm) from the end. Then, place the connector over the wire end and deform it with the crimping tool. Repeat the process for the other wire end. Make sure that you apply pressure to the right place on the connector, Figure 5-35. As a final step, tug lightly on the connection to ensure that it is tight.

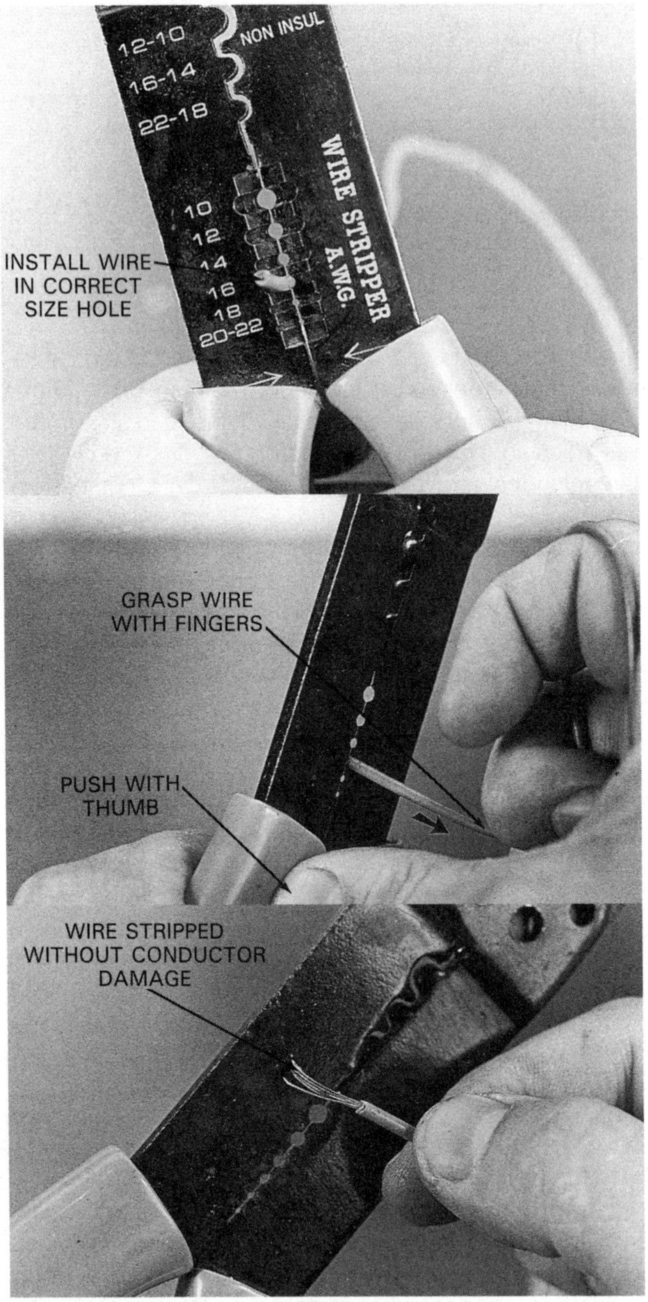

Figure 5-33. To properly strip insulation from a wire, *wire strippers* should be used. First, determine how much insulation should be removed. Then, close the tool around the wire to cut the insulation. Firmly grasp the wire with your fingers and push against the tool with your thumb. The insulation will be pulled from the wire. Note how the insulation is removed without damage to the conductor. Using another method will almost always remove some strands of wire, reducing conductivity of the wire.

Summary

Hydraulics is the science of how power is transmitted through liquids. Hydraulic systems are a simple way to transmit and multiply power. Hydraulics is composed of

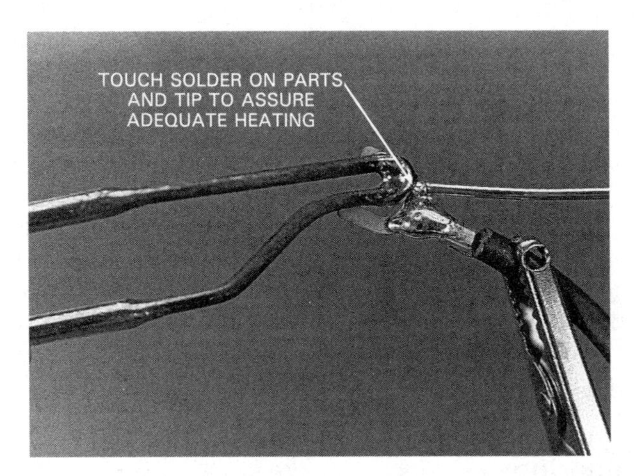

Figure 5-34. Soldering should always be done by heating the work and flowing solder into it. Never drip melted solder on a cold connection; it will not hold!

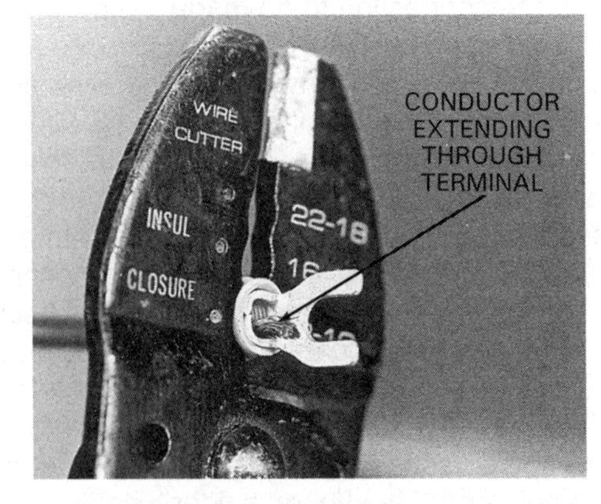

Figure 5-35. To crimp a terminal on a wire, the terminal, wire, and tool must all be in the right position. The wire should extend through the terminal the right amount, and the tool must crimp the terminal in the right place. Always check the connection after installation by lightly tugging on it.

hydrodynamics–the study of moving liquids–and hydrostatics–the study of liquids under pressure.

Liquids, such as hydraulic fluids, cannot be compressed. This principle can be put to use so that hydraulic fluid will transmit pressure and motion.

Pascal's law states that pressure on a fluid in a closed system is equal at every place in the system. Pascal's law can be put to use to multiply force. If the area of the input piston of a hydraulic system is smaller than the area of the output piston, force will be increased at the output piston. This is the principle used in the hydraulic jack and on vehicle brake systems.

Automatic transmissions must develop high pressure and allow for high fluid flow rates. To do this, they require engine driven pumps. The three major kinds of pumps are gear, rotor, and sliding-vane pumps. They all operate by creating a vacuum that draws in fluid and compressing the fluid so that it can be discharged.

Valves are used to control the flow of fluid in the hydraulic system. Check valves are steel balls that control fluid flow. They are used with a spring to control system pressures or in a special chamber to ensure that fluid flows in one direction only.

Spool valves resemble thread spools. They are installed in a valve body to control fluid pressure and flow. Spool valves are operated by external controls or by fluid pressures from other valves.

Actuators change fluid pressure into force. The force and movement that they exert are the end result of hydraulic system operation. Examples of actuators are transmission servos and clutch apply pistons.

Pneumatics is the study of the physical properties of air and other gases. Pneumatic systems can be compressed air or vacuum systems. The automatic transmission vacuum modulator is an example of a drive train pneumatic system.

Atoms are tiny particles, with a center of positive protons surrounded by negative electrons. Electricity is the flow of electrons between atoms. The organization of atoms can produce conductors or insulators, depending on how easily they accept and give up their electrons.

There are three basic kinds of electrical circuits. The series circuit has only one path for current flow and the same current flows through all components. The parallel circuit has separate current paths for system components. The series-parallel circuit is a combination of series and parallel circuits. Most manufacturers use the vehicle frame, instead of separate wires, to provide the ground, or return path, for the vehicle electrical circuits.

Circuits can develop two kinds of problems. Short circuits occur when the current bypasses part of the intended circuit. This causes too much current to flow. Open circuits occur when the circuit is not completed, and current cannot flow.

The technician may encounter several kinds of electrical components when working on drive trains. Fuses are used to prevent electrical overloads. They are made of a soft metal that melts when excessive current begins to flow. Circuit breakers perform the same function mechanically. Resistors reduce current flow in a circuit, and sometimes voltage, to protect other electrical components. Switches open and close circuits. Solenoids and relays are electrically operated magnetic devices. Solenoids perform physical work, and relays send current to other electrical components.

A transistor can perform the same function as a relay. A small current enters through the base of the transistor, and changes the semiconductor material from an insulator to a conductor. A larger current can then flow through the transistor. Transistors are the major component of electronic circuits found on vehicles.

Automotive computers are often used to control transmissions, transaxles, transfer cases, and engines. Automotive computer systems are composed of input sensors, output devices, and the computer itself, often called a microprocessor.

Knowledge of electrical test equipment is essential to the drive train technician. This equipment is needed to check the electrical circuits used on many parts of modern drive trains.

Jumper wires are used to bypass a suspected component. If the circuit then works normally, the bypassed component is defective. Test lights are used to check circuits for voltage and continuity.

Multimeters contain several meters in one case. The different functions are selected by turning a knob or pressing a button on the multimeter face. The three meters used in a multimeter are the voltmeter, ammeter, and ohmmeter.

Broken or damaged wires can be repaired by soldering, or by using crimp connectors. Soldering should be used if possible since a soldered connection will hold and conduct current more efficiently. Crimping is a quick way to repair wiring but should be avoided on low voltage sensor circuits. Wire stripping, soldering, and crimping must all be done carefully to make a good electrical connection.

Know These Terms

Hydraulics; Power hydraulics; Hydraulic system; Hydrodynamics; Hydrodynamic system; Hydrostatic; Hydrostatic system; Hydraulic pressure; Apply piston; Output piston; Pascal's law; Hydraulic leverage; Hydraulic pump; Nonpositive displacement pump; Positive displacement pump; Gear pump; Internal gear pump (Crescent pump); Rotor pump; Sliding-vane pump (Vane pump); Hydraulic valve; Check valve; Pressure relief valve; Control valve; Spool valve; Valve body; Valve bores; Pressure regulator valve; Throttle valve; Actuator; Pneumatics; Pneumatic system; Open system; Closed system; Atmospheric pressure; Compressed-air system; Vacuum; Vacuum system; Vacuum modulator valve; Electronics; Atom; Proton; Neutron; Electron; Electricity; Conductor; Insulator; Semicon-

tron; Electron; Electricity; Conductor; Insulator; Semiconductor; Current; Potential difference (Voltage); Amperes (A) (Amps); Volts (V); Resistance; Ohms (W); Ohm's law; Direct current (DC); Alternating current (AC); Circuit; Battery; Frame-ground system; Series circuit; Parallel circuit; Series-parallel circuit (Combination circuit); Short circuit (Short); Open circuit (Open); Fuse; Circuit breaker; Overload; Resistor; Switch; Relay; Computer; Transistor; Microprocessor; Sensor; Binary number system; Digital signal; Analog signal; Input device; Transducer; Output device; Solenoid; Motor; Coil; Jumper wire (Jumper); Nonpowered test light; Self-powered test light; Multimeter (VOM); Voltmeter; Ammeter; Ohmmeter; Soldering; Solder; Flux; Crimp; Crimp connector.

Review Questions–Chapter 5

Please do not write in this text. Place your answers on a separate sheet of paper.

1. A hydraulic system is an example of an open system. True or false?
2. Explain the basic difference between hydrodynamics and hydrostatics.
3. _____ _____ states that liquid under pressure transmits the same amount of pressure to every part of its container.
4. Describe the basic system that produces hydraulic leverage.
5. List the four common features of hydraulic pumps and state their function.
6. Describe the operation of gear, rotor, and vane pumps.
7. _____ _____ are used to help control the operation of a hydraulic system by controlling pressure and, also, direction and rate of fluid flow.
8. How does a pressure relief valve work?
9. _____ is the study of the physical and mechanical properties of air and other gases.
10. List the three basic physical quantities of electricity and state what each is.
11. Name and briefly discuss two types of defective circuits.
12. How are short circuits and overloads the same and how do they differ?
13. Explain the function of a solenoid.
14. Two devices that control larger load currents with smaller currents are _____ and _____.
15. An electrical signal that varied continuously over a range of values would be:
    a. Digital.
    b. Binary.
    c. Analog.
    d. Direct.
16. _____ convert an elecrical current into rotary motion.
17. How do you use a nonpowered test light?
18. A complete circuit has:
    a. An open switch.
    b. Infinite resistance.
    c. Continuity.
    d. Devices in parallel.
19. Name the three testers commonly incorporated into a multimeter.
20. Describe the procedure for soldering two wires together.

Certification-Type Questions–Chapter 5

1. Technician A says that a torque converter is a hydrodynamic device. Technician B says that hydrostatic principles are used in the control systems of automatic transmissions. Who is right?

(A) A only
(B) B only
(C) Both A & B
(D) Neither A nor B

2. All of these are types of hydraulic devices or systems EXCEPT:

(A) residential water supply.
(B) power steering.
(C) windshield washers.
(D) vacuum modulators.

3. Pascal's law is given by which of these statements?

(A) Pressurized liquid will burst from its container.
(B) Pressurized liquid will distribute no further pressure.
(C) Pressurized liquid transmits equal pressure to all points.
(D) Pressurized liquid transmits unequal pressure to all points.

4. All of these involve the study of pneumatics EXCEPT:
(A) oil pressure.
(B) gases at rest and in motion.
(C) physical properties of gases.
(D) mechanical properties of gases.

5. The two types of pneumatic systems are:
(A) opened and closed systems.
(B) vacuum and compressed-air systems.
(C) hydrostatic and hydrodynamic systems.
(D) pressure regulation and pressure relief systems.

6. Materials containing atoms that can easily give up or receive electrons are:
(A) insulators.
(B) porcelains.
(C) conductors.
(D) semi-insulators.

7. All of these are basic physical quantities of electricity EXCEPT:
(A) current.
(B) voltage.
(C) pressure.
(D) resistance.

8. Ohm's law is given by which of these statements?
(A) Potential difference determines measured voltage.
(B) Pressurized liquid transmits equal pressure to all points.
(C) Electrons flow in the same direction throughout any system.
(D) A circuit's current is directly proportional to applied voltage.

9. An alternator powers a vehicle's electrical devices. Technician A says that the alternator generates power by producing alternating current. Technician B says that automotive electrical systems operate only on direct current. Who is right?
(A) A only
(B) B only
(C) Both A & B
(D) Neither A nor B

10. All of these are basic kinds of circuits EXCEPT:
(A) series circuits.
(B) coaxial circuits.
(C) parallel circuits
(D) series-parallel circuits.

11. Resistors are used in circuits for all of these reasons EXCEPT:
(A) preventing power flow to a specific part.
(B) reducing current to a specific required value for proper operation.
(C) providing current-limiting protection for other circuit components.
(D) obtaining a desired voltage drop.

12. An electromagnetic switch is called a:
(A) relay.
(B) diode.
(C) transistor.
(D) capacitor.

13. Automotive computers operate using the binary number system, which includes the digits:
(A) 0 and 1.
(B) 1 and 2.
(C) 0 thru 5.
(D) 1 thru 10.

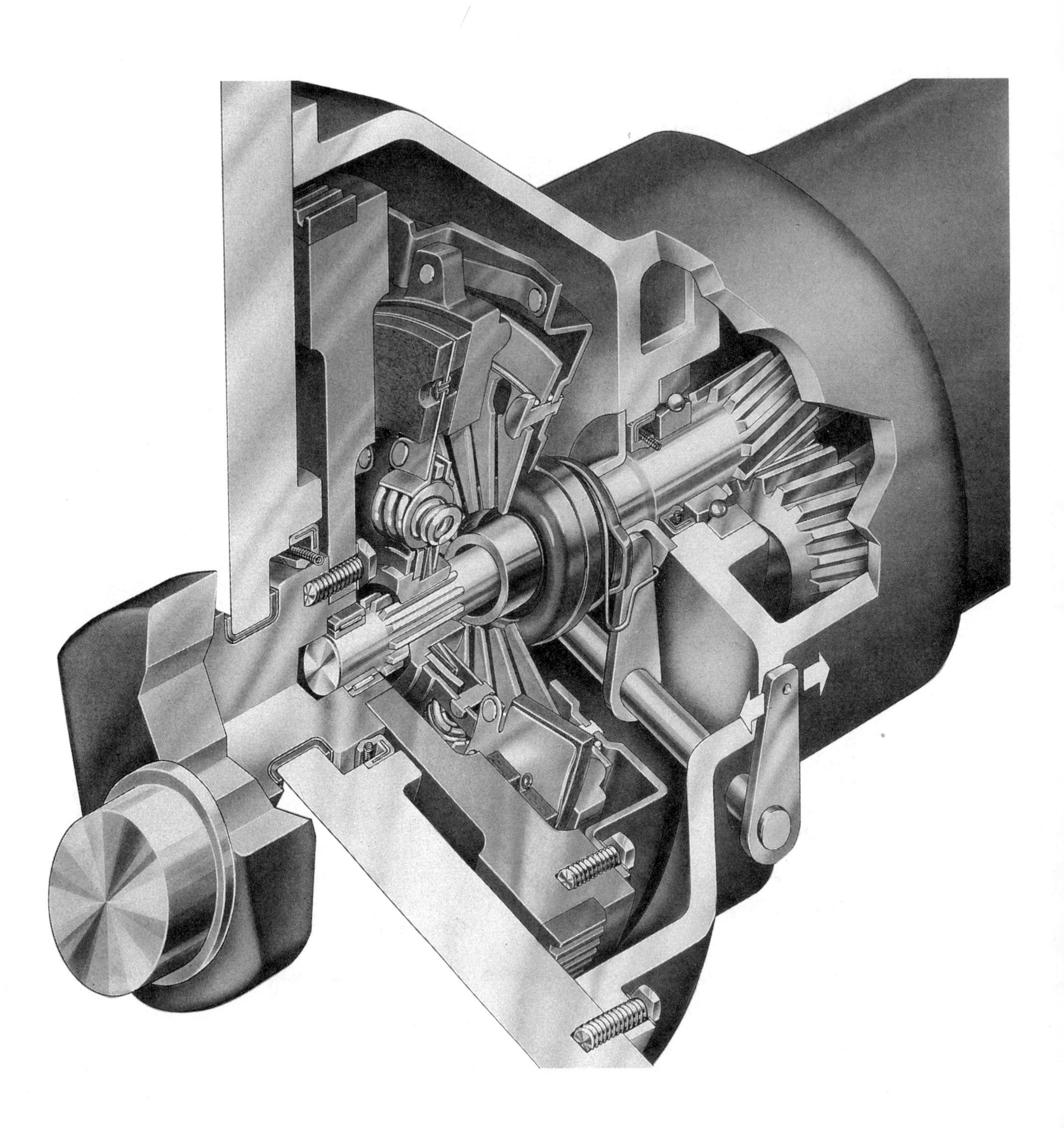

Cutaway shows how clutch looks when assembled. (Luk)

Chapter 6

Clutch Construction and Operation

After studying this chapter, you will be able to:

- Name and describe the basic parts of a clutch.
- Explain how a clutch operates.
- Describe the construction of a flywheel.
- Discuss the function and operation of the transmission input shaft.
- Describe clutch disc construction and operation.
- Name the parts and explain the operation of a pressure plate assembly.
- Describe the construction and operation of the three types of clutch linkage.

This chapter will discuss the parts and operation of the automotive clutch. Each major clutch part is identified and its operation is explained. Understanding the purpose and construction of each part will make it easier for you to adjust or repair a *clutch assembly.* Understanding how all of the clutch parts function together will enable you to properly diagnose clutch problems.

Construction and Operation Overview

A clutch is used on a vehicle having a manual transmission or manual transaxle. (For simplification, throughout this chapter, reference will be made only to manual transmissions. It should be understood that this information generally pertains to both manual transmissions and transaxles.) The purpose of the clutch is to connect or disconnect power flow from the vehicle's engine to the road. A clutch is engaged, or connected, in its normal position. The engaged clutch connects the engine to the rest of the drive train, enabling the vehicle to move under its own power.

A clutch is *disengaged,* or disconnected, manually. This is done by depressing a foot pedal. The pedal is pushed in and then released, momentarily disconnecting the engine from the drive train and road, to permit shifting of transmission gears. When the engine is running and the vehicle is stopped, the clutch pedal is held down to keep the clutch disengaged. Otherwise, the vehicle must be shifted into neutral or the car's engine will stall.

Vehicle manufacturers used several kinds of clutches in the past, but most are no longer used. Most modern cars and light trucks use *single-plate dry clutches.* A typical single-plate dry clutch is shown in Figure 6-1. Its operation involves eight major parts:

- The ***flywheel*** transfers engine power into the clutch. It provides a mounting surface for parts of the clutch. It also serves as an engine vibration damper and a clutch heat absorber.
- The ***manual transmission input shaft*** transfers power flow from the clutch to the transmission.
- The ***clutch disc*** is splined to and transmits power flow to the transmission input shaft. It is held against the flywheel when the clutch is engaged. See Figure 6-2A.
- The ***pressure plate assembly*** puts spring pressure on its *pressure plate* to tightly hold the clutch disc against the flywheel, keeping the clutch engaged.
- The ***throwout bearing*** is pushed along the input shaft into the pressure plate assembly when the clutch pedal is depressed. This causes the force from the pressure plate to be removed from the clutch disc, disengaging the clutch. The disengaged clutch is represented in Figure 6-2B.
- The ***clutch fork*** slides the throwout bearing into the pressure plate assembly.
- The ***clutch linkage*** allows the driver to operate the clutch fork.
- The ***clutch housing*** encloses the entire *clutch assembly* (clutch disc, pressure plate assembly, throwout bearing, and clutch fork). It protects the assembly from damage due to rocks, water, oil, etc.

In operation, the flywheel and pressure plate assembly are bolted together, Figure 6-3, and they revolve as a single unit. They are directly connected to the engine and turn when the engine is running.

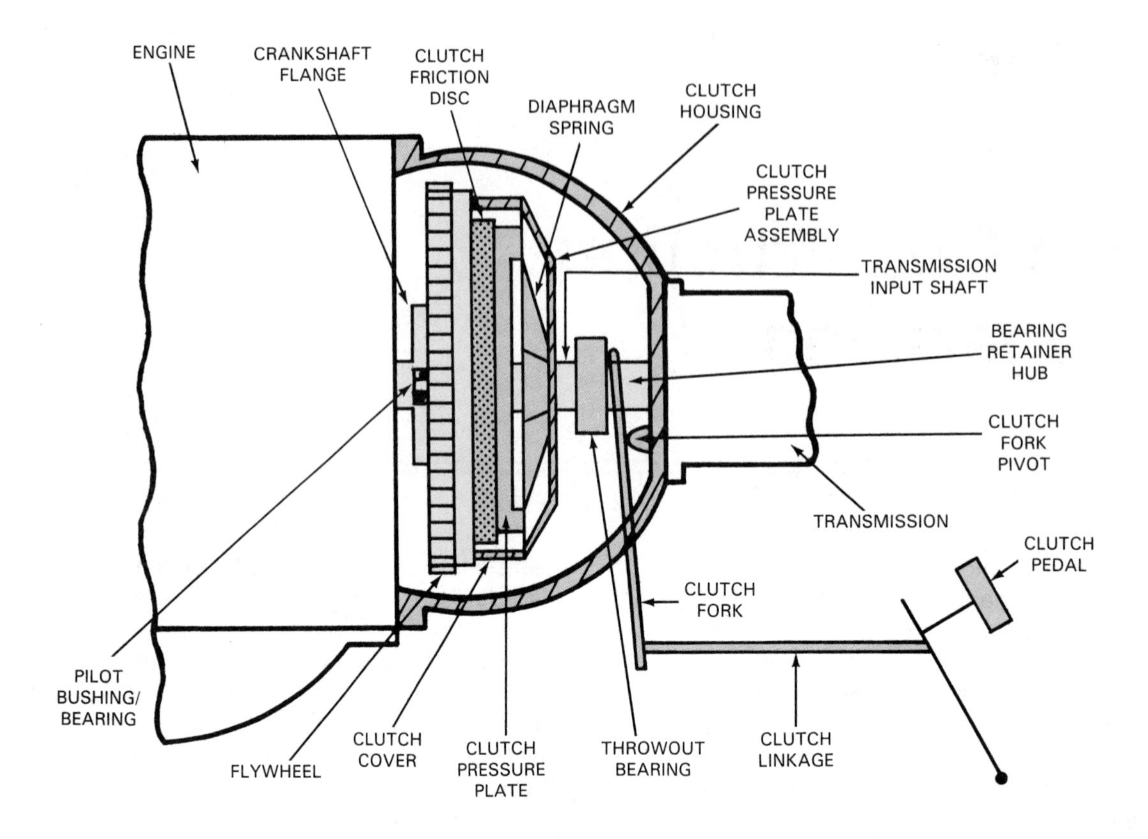

Figure 6-1. Note relationship between various major parts of a typical single-plate dry clutch.

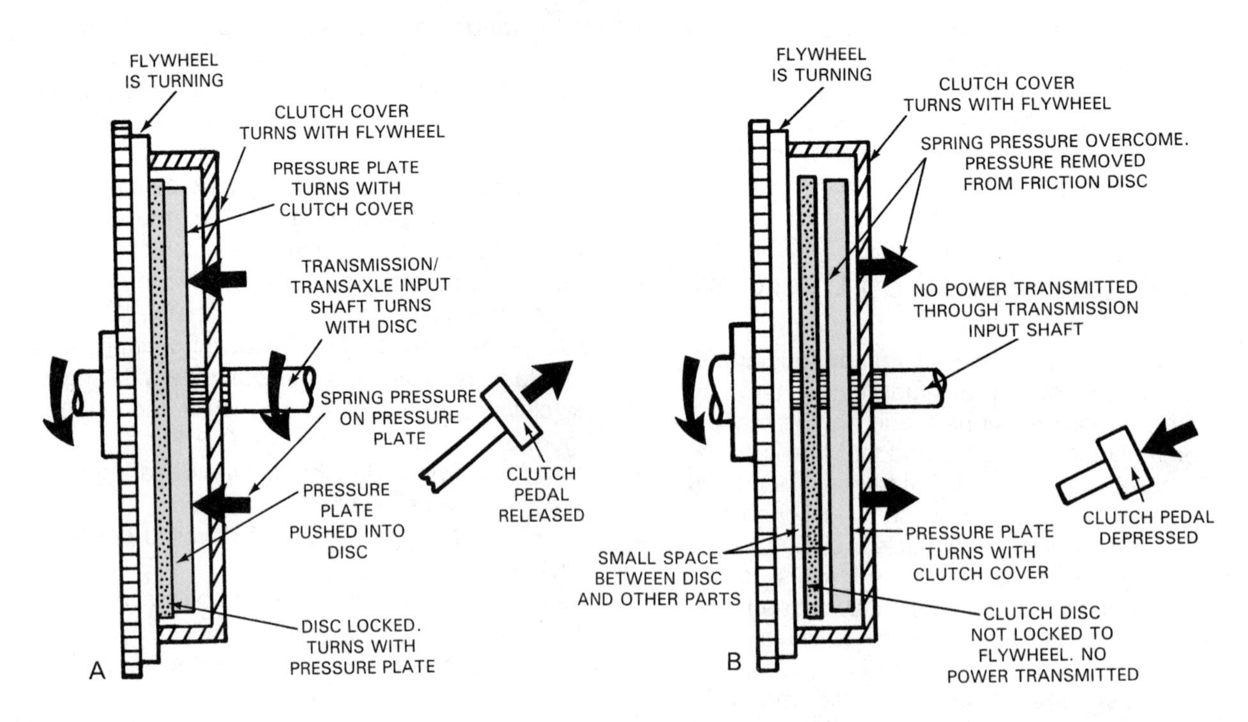

Figure 6-2. Two positions of a clutch. A–Clutch pedal out: clutch engaged. B–Clutch pedal in: clutch disengaged.

When the clutch is engaged, the clutch disc is tightly pressed between the flywheel and pressure plate. The clutch disc, then, turns at the same speed as the flywheel and pressure plate. The clutch disc is splined to the transmission input shaft (Figure 6-3), providing a path from the flywheel, into the transmission, and on to the drive wheels. The clutch disc and input shaft are turning whenever the engine is running and the clutch is engaged.

When the pressure plate moves away from the flywheel, the clutch disc is no longer pressed between the pressure plate and the flywheel. The clutch is disengaged, and power flow is interrupted. With the clutch disengaged and the vehicle at a stop, the clutch disc and transmission input shaft soon slow to a stop, because of drag caused by the transmission bearing and gear loads. The period of time that it takes to stop is called the clutch *spindown* time. Meanwhile, the pressure plate assembly continues to revolve.

Note that the clutch disc and input shaft are always turning *whenever* the drive wheels are turning. While shifting gears, disc and shaft are not being driven by engine *nor* by rotation of drive wheels back up through the drive train. Disc and shaft begin to slow, but they will not come to a rest before the clutch is reengaged.

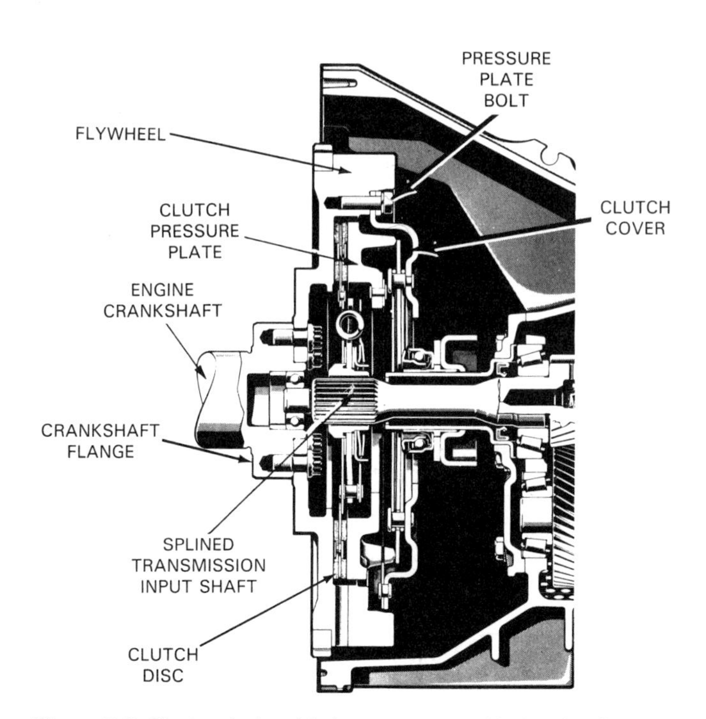

Figure 6-3. Study relationship between parts. Notice that flywheel is bolted to crankshaft flange. Clutch cover of the pressure plate assembly is bolted to flywheel. Crankshaft, flywheel, and pressure plate assembly rotate as a unit. Clutch disc is splined to transmission input shaft. They also rotate as a unit. When clutch is engaged, pressure plate holds clutch disc against flywheel, and all parts rotate as a unit. (Mercedes-Benz)

Detailed Construction and Operation

At this point, you should be familiar with the basic parts and general operation of the clutch. Now, we will take a closer look at each of the eight major parts involved in clutch operation. This will be the focus for the remainder of this chapter.

Flywheel

The flywheel is attached to the rear of the crankshaft, Figure 6-4. It turns whenever the engine is running. A raised ring on the *crankshaft flange* fits into a large hole at the center of the flywheel. Hardened bolts pass through the flywheel and are screwed into threaded holes in the crankshaft flange. The flywheel is usually made of cast iron. A hardened-steel ring gear is usually pressed on or welded to the outside of the flywheel (for engine starting purposes).

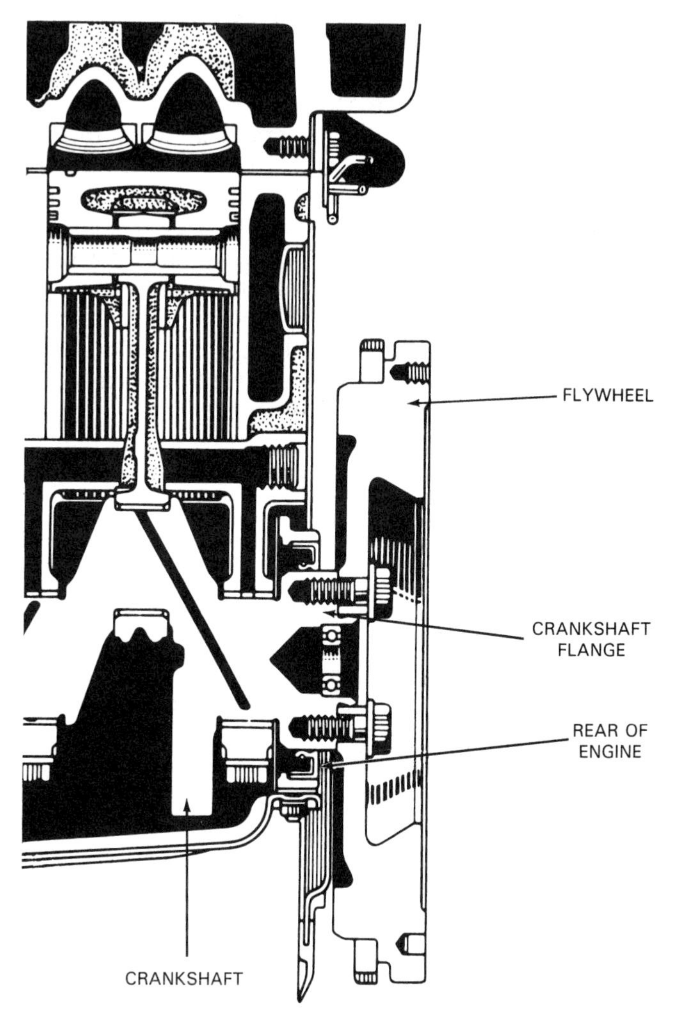

Figure 6-4. This shows the position of the flywheel in relation to the rear of the engine. The flywheel is bolted to the crankshaft flange. (Toyota)

Flywheels are heavy by design. There are two reasons for this. Clutch operation produces heat, and the extra metal can absorb some of this heat and prevent clutch overheating under normal conditions. The weight also makes the flywheel a good *damper,* or *inertia device,* for smoothing out engine vibration.

The surface of the flywheel on the clutch side is machined to a smooth finish. This smooth area is one of the contact faces for the clutch disc. Holes are drilled and threaded into the flywheel face for installing the pressure plate assembly. Some flywheels contain pressed-in *dowels,* which are small metal rods, usually less than 1 in. (25.4 mm) long. The dowels provide for precise alignment of the pressure plate.

Manual Transmission Input Shaft

The manual transmission input shaft is a long shaft that extends from the rear of the engine crankshaft to the inside of the transmission case. The shaft is made from hardened steel. It is supported by bearings at both ends. The input shaft supports the clutch disc and holds it in proper alignment with the other clutch components. Moreover, the input shaft is splined to the central hub of the clutch disc. When the clutch disc rotates, the input shaft must rotate. The input shaft transmits rotation of the clutch disc into the transmission.

The front end of the input shaft is smaller in diameter than other parts of the shaft. This section is sometimes called the *pilot.* The pilot extends about 1 in. (25.4 mm). It fits into the clutch pilot bearing at the rear of the crankshaft.

The ***pilot bearing*** fits into the bore at the center of the crankshaft flange. It can be either a ball or roller bearing or a bronze bushing, referred to more specifically as a ***pilot bushing.*** The bearing provides the front support for the transmission input shaft. Note that it should be lubricated whenever the transmission and clutch are removed from the vehicle.

The shaft diameter increases beyond the pilot, and the next section contains the external splines that mate with the clutch disc splines. The splines allow slight forward and rearward movement of the clutch disc on the input shaft. This sliding action allows the clutch disc to move away from or against the flywheel when it is disengaged or engaged.

After the splined section along the shaft, lies the section for the ***transmission front bearing***–an antifriction bearing that provides the rear support for the input shaft. The shaft extends through the front bearing and into the transmission. The rear of the shaft contains the transmission *main drive gear,* which drives the other gears of the transmission.

The front bearing is held in place by a bearing retainer. The ***front bearing retainer,*** as it is called, also serves as a hub for the clutch throwout bearing and provides a sliding surface for the bearing assembly. The different sections of the input shaft are shown in Figure 6-5. Note that the transmission input shaft is sometimes called the ***clutch shaft.***

Clutch Disc

The clutch disc, also called a ***friction disc,*** is installed between the flywheel and the pressure plate. The job of the friction disc is to transmit power from the flywheel to the transmission input shaft. It must transmit this power with little or no *slippage,* allow for smooth engaging and disengaging, and stand up to high temperatures. The clutch disc

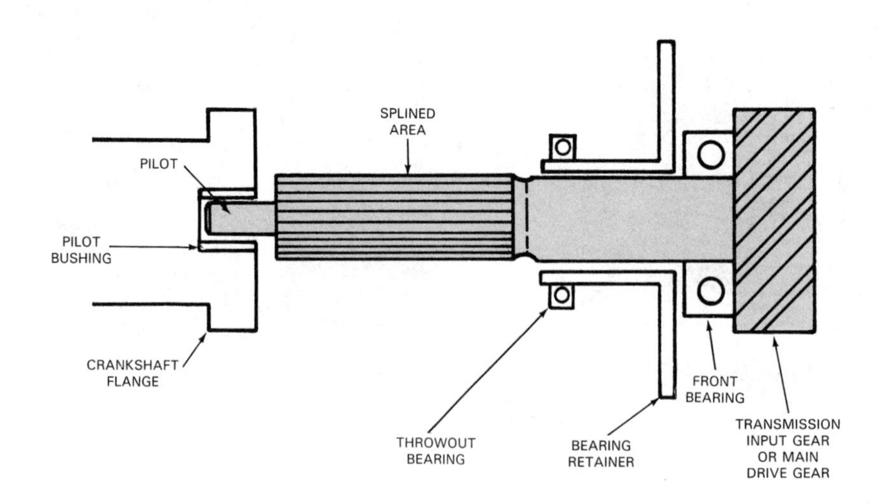

Figure 6-5. Note the different sections along the transmission input shaft.

is made up of an *outer disc assembly* and a central *hub flange.* See Figure 6-6.

Outer disc assembly

The ***outer disc assembly*** is the portion of the clutch disc consisting of a plate of thin steel sandwiched between a friction material, called the clutch disc ***friction facing.*** The facing makes up the part of the disc that actually contacts the flywheel and pressure plate. Friction facings are made of materials that will operate satisfactorily under high temperatures. Materials similar to those found in brake linings are often used. For added strength, the friction materials may be reinforced with wire mesh, or screening.

The friction facing may be held to the steel plate in a couple of different ways. Sometimes, the facing is *bonded* to the steel plate with a high-temperature glue. Often, the facing is *riveted* to the steel plate. The rivets are usually made of brass, which is a softer metal. Using brass rivets reduces damage to the flywheel and pressure plate should the clutch facing wear down to the rivet heads.

The friction facing is often divided into segments. See Figure 6-7. The segments help the clutch to engage smoothly by spreading the friction forces over many small areas instead of one large area. The gaps between the segments provide a suction break between the mating surfaces, which keeps the parts from sticking when the clutch is disengaged. The gaps also provide for more efficient heat removal.

Most friction discs contain ***cushion springs*** under the friction facing. These springs are flexible plates with a slight wave or curve. The plates compress when the clutch is engaged. They absorb some of the shock of engagement.

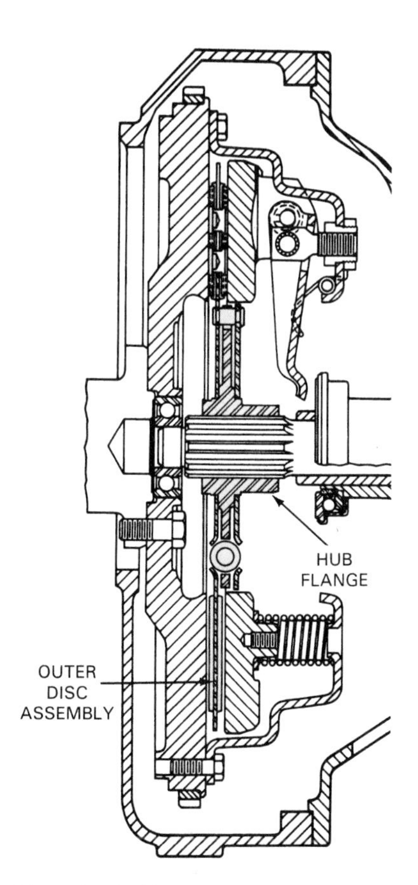

Figure 6-6. The clutch friction disc, made up of an outer disc assembly and a central hub flange, is installed in the clutch between the flywheel and pressure plate. (General Motors)

Hub flange

In the center of the friction disc is a splined hub. The hub splines engage the splines on the transmission input shaft. This ***hub flange,*** as it is called, is usually made of cast iron.

The hub flange and outer disc assembly are riveted together. The rivets are solidly mounted to the hub flange and pass through slotted holes in the outer disc. Coil springs, called ***torsion springs,*** keep the rivets positioned to one end of the disc slots, toward the direction of rotation. Figure 6-7 shows the location of the torsion springs.

Engine power must actually go through the springs to get from the disc to the hub flange. When the clutch is engaged, engine power slightly compresses the springs. As the clutch is being engaged, this will cause the rivets to move in the slotted holes. This spring action cushions the engagement of the clutch. The springs will also smooth out vibrations from the engine.

Note that there are a few clutches intended for heavy-duty use that may not have torsion springs. Also, note in Figure 6-7 that two of the springs appear to be missing.

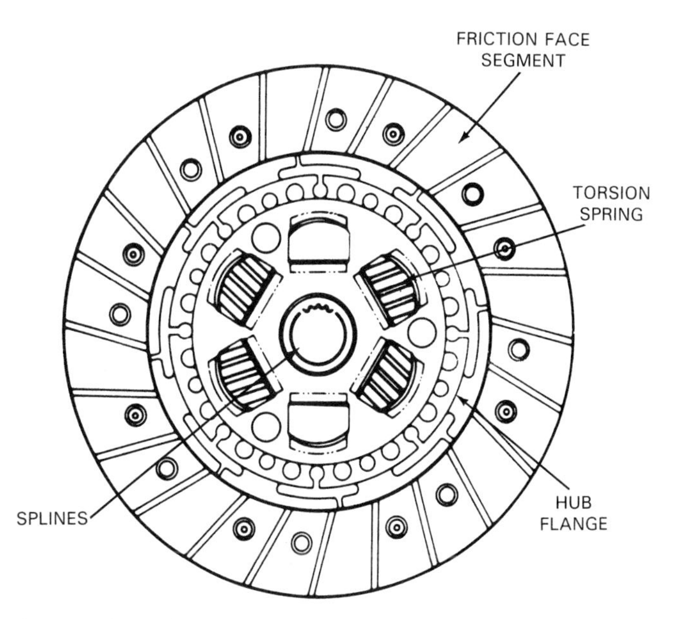

Figure 6-7. Study various parts of the friction disc. Note that there are two kinds of springs to smooth clutch operation. (Chrysler)

This is done deliberately. On many friction discs used with smaller engines, fewer springs are needed.

Pressure Plate Assembly

The pressure plate assembly is made up of a cover, pressure plate, and springs. Refer to Figure 6-8. The ***clutch cover*** is bolted to the flywheel; it turns with the flywheel. In its normal position, the **pressure plate** locks the friction disc between itself and the flywheel, and the entire assembly is driven at flywheel speed. ***Apply springs,*** between the cover and pressure plate, hold the pressure plate against the clutch disc. To unlock the friction disc, the plate is moved away from normal position by the driver. This happens by actuating the ***release levers*** or ***release fingers*** contained within the assembly, thus backing off the pressure plate.

The flywheel and pressure plate assembly are balanced in the factory. The parts have special reference markers. These markers are to be matched when reassembling. This is to prevent an imbalance in the drive train. Sometimes, dowels or an irregular, or *offset,* bolt pattern are used to prevent improper installation.

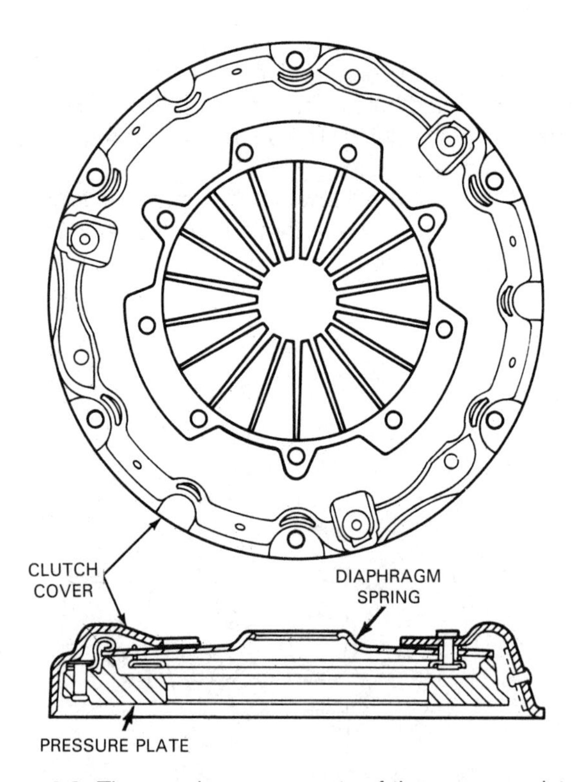

Figure 6-8. Three major components of the pressure plate assembly are shown here. Study the relative positions of the clutch cover, pressure plate, and spring. Note that this pressure plate assembly uses a diaphragm spring. (Chrysler)

Clutch cover

The clutch cover, which covers the pressure plate assembly, is usually made of heavy steel plate, stamped to the proper shape. Holes are located about the cover for rivets or bolts that hold the assembly together or that attach to the flywheel. The cover contains various slots and other openings that function to remove heat. The openings allow air to circulate inside of the assembly.

Pressure plate

The pressure plate is the part of the assembly that actually presses against the clutch disc. The plate is a ring-shaped disc that is made of thick cast iron. The weight of the pressure plate assists the flywheel in heat absorption and vibration damping. The plate is roughly the same size as the outer disc assembly of the clutch disc. The transmission input shaft passes through a center hole in the disc. The size of the hole allows for clearance with the friction disc torsion springs. The side of the pressure plate away from the cover is flat and smooth. This side contacts the friction disc facing area. The other side is machined for installation of the springs and release levers.

Apply springs

The apply springs provide the pressure that holds the pressure plate to the clutch disc. This pressure must be overcome to free the clutch. The springs must be strong enough to keep the friction disc applied but still allow the driver to disengage the clutch without excessive pedal effort. Two types of apply springs include *diaphragm springs* and *coil springs.*

Diaphragm-spring pressure plate assembly. This type of assembly uses a single diaphragm spring. See Figure 6-9. The diaphragm spring is made of spring steel. It is formed in the shape of a dish. This combination produces the spring effect. Radial slits in the diaphragm form numerous release fingers. These act as release levers. They serve to remove the holding power of the spring action. A central hole permits the transmission input shaft to pass through.

The diaphragm spring makes contact with the pressure plate at the outer rim. There are two rings built into the pressure plate assembly. They are located near the outer rim. The rings are called ***pivot rings.*** There is one pivot ring on each side of the diaphragm spring. The rings are held in position by retaining rivets, located about and extending through the clutch cover.

When the pressure plate is in its normal, or engaged, position, the diaphragm spring is in a compressed state. The dish-shaped spring becomes nearly flattened by action against the pressure plate rim and *outer* (cover-side) pivot ring. This happens as force is applied at the pivot ring by tightening the cover bolts. The pressure plate is pushed

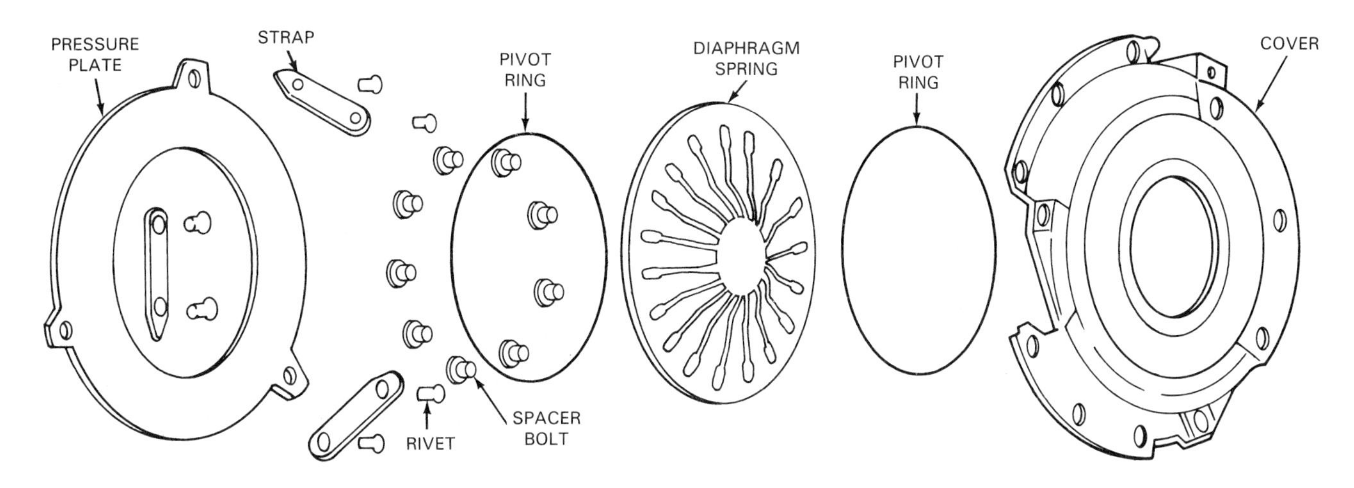

Figure 6-9. Diaphragm-spring pressure plate assembly. (Luk)

into the friction disc because of the resulting spring pressure, Figure 6-10A.

To disengage the clutch, the throwout bearing is pushed *inward,* toward the pressure plate assembly and against the release fingers, Figure 6-10B. The diaphragm pivots on the inner pivot ring. When the center of the diaphragm is pressed inward, the outer rim pivots *outward,* or toward the clutch cover. This action removes the pressure on the clutch disc.

Meanwhile, the pressure plate is drawn back with the outer rim of the diaphragm. Contact between the pressure plate and diaphragm is maintained by special springs (not shown in Figure 6-10) called *holddown springs,* or *retracting springs.* In some designs, the springs are bolted or riveted at one end to the pressure plate, and the other end clips over the outer rim of the diaphragm. In other designs, the springs consist of flexible *drive straps,* made of spring steel. The one end attaches to the pressure plate; the other attaches to the clutch cover. The springs or straps cause the pressure plate to pull away from the clutch disc when the diaphragm is retracted. Whatever the design, the pressure plate and the rest of the assembly continue to rotate with the flywheel.

When the throwout bearing is moved back outward, the diaphragm returns to its original (compressed) position. The pressure plate applies pressure, once again, to the friction disc, and the clutch is reengaged.

Coil-spring pressure plate assembly. This type of assembly uses large coil springs. The springs are pressed between the pressure plate and cover. There may be as few as four springs, but most assemblies have at least ten. The larger the engine and vehicle, the more springs needed. The coil springs are made of hardened spring steel. They can withstand heat and pressure. The relative position of the springs between the cover and pressure plate is shown in Figure 6-11.

In this type of assembly, pressure is removed from the clutch disc by pulling the pressure plate back against spring pressure. Release levers, discussed next, are used for this operation.

Release levers

Release levers are used in a coil-spring pressure plate assembly to overcome spring pressure and pull the pressure plate away from the friction disc. The levers are arranged radially and are hinged to the clutch cover. Most pressure plates use three release levers. The levers pivot on their hinges, just as the diaphragm spring pivots on its pivot rings.

Each release lever is connected at one end to the rear of the pressure plate. The lever is actuated by contact with the throwout bearing at the other end. When the throwout bearing contacts the release lever, the lever moves inward, pivoting on its hinge. This action pulls the pressure plate away from the friction disc, further compressing the coil springs in the process. A typical release lever is shown in Figure 6-12.

Most release levers have additional weight added to their outer ends. The weights are situated such that when the pressure plate assembly is rotating, they tend to be thrown out to a larger diameter. The resulting action causes the weights to exert additional pressure against the pressure plate. This helps the springs to hold the friction disc tightly at high engine speeds. As a result, lighter springs can be used, making it possible to operate the clutch with less effort. Clutches that use this kind of arrangement are called ***semi-centrifugal clutches.***

Throwout Bearing

The throwout bearing, or ***release bearing,*** applies force to the release levers or fingers. In so doing, it disen-

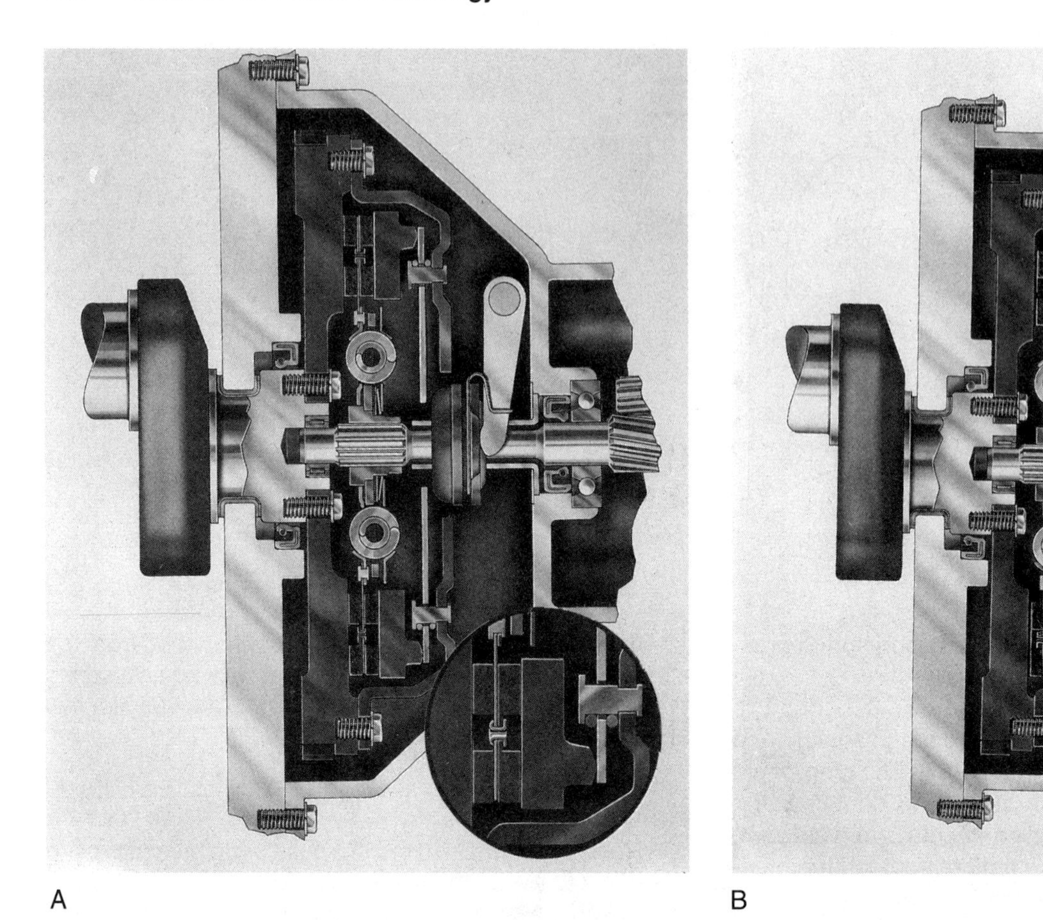

Figure 6-10. Visualize the action with a diaphragm clutch assembly. A–Pressure plate spring tightly holds the friction disc between the flywheel and pressure plate. The clutch is engaged, and power can flow from the engine to the transmission. B–Pressure plate is not pressing friction disc against the flywheel. The clutch is disengaged. Power flow is interrupted. (Luk)

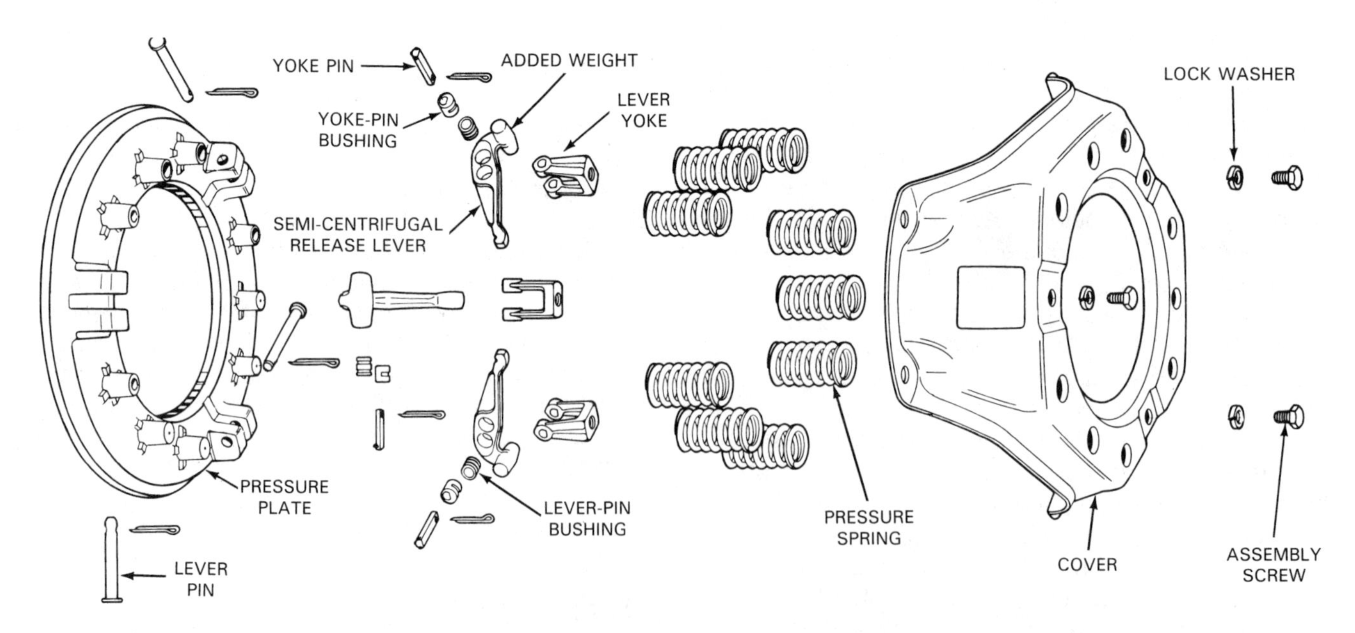

Figure 6-11. Coil-spring pressure plate assembly. (Luk)

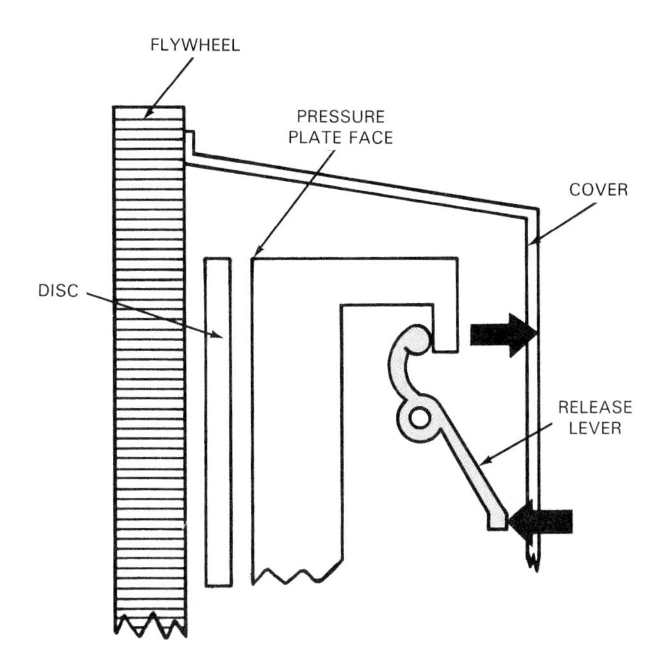

Figure 6-12. Release levers are used on pressure plates that use coil springs. To release the clutch, the lever is pushed in. It pivots on its hinge and pulls the pressure plate away from the friction disc. Although one lever is shown, most pressure plates use three release levers.

gages the clutch. The throwout bearing is designed to operate with a minimum of friction between rotating and stationary points of contact.

The *throwout bearing assembly* consists of the bearing itself and a collar, or sleeve. The bearing is a *thrust*-type of ball bearing, designed to sustain axial loads. It is similar to other thrust bearings in that bearing races are situated axially with respect to each other. The bearing, in some cases, is pressed on a sleeve, or collar. In other cases, it is an integral part of the collar. See Figure 6-13.

The collar slides on the smooth hub that is part of the transmission front bearing retainer. It can rotate on the hub, also. The collar is machined to fit the clutch fork, which operates the throwout bearing assembly, and it incorporates a retainer spring to hold the clutch fork in place and reduce rattles. The entire throwout bearing assembly is greased at the factory, and except for some heavy-duty bearings, it cannot be regreased without removing the transmission.

As the throwout bearing contacts the turning release levers or fingers, the affected bearing race will begin to rotate while the sleeve is retained by the clutch fork. In addition, the sleeve may rotate on the retainer hub, thereby absorbing some of the shock when the throwout bearing first contacts the spinning pressure plate assembly.

When the clutch pedal is released, the clutch *return springs* pull the clutch fork and throwout bearing away from the release levers or fingers. In the engaged position, the throwout bearing does not contact any part of the pressure plate assembly. Figure 6-14 shows the relative positions of the clutch disc, pressure plate assembly, throwout bearing, and clutch fork.

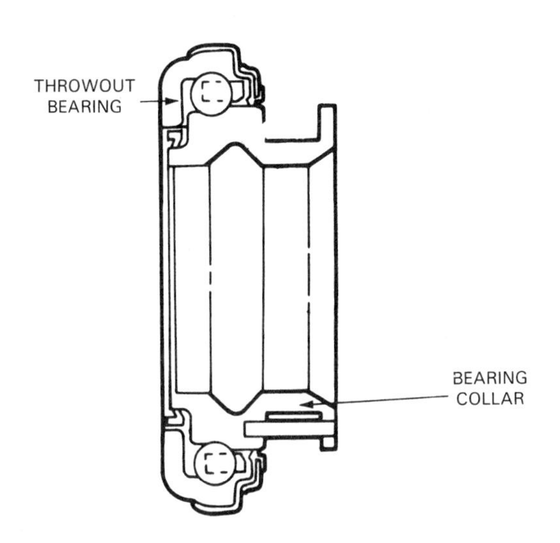

Figure 6-13. Sectional view of a typical throwout bearing assembly in which the bearing is an integral part of the sleeve. When installed, the assembly will slide over the transmission bearing retainer hub. The hub is machined to allow free movement of the throwout bearing. (General Motors)

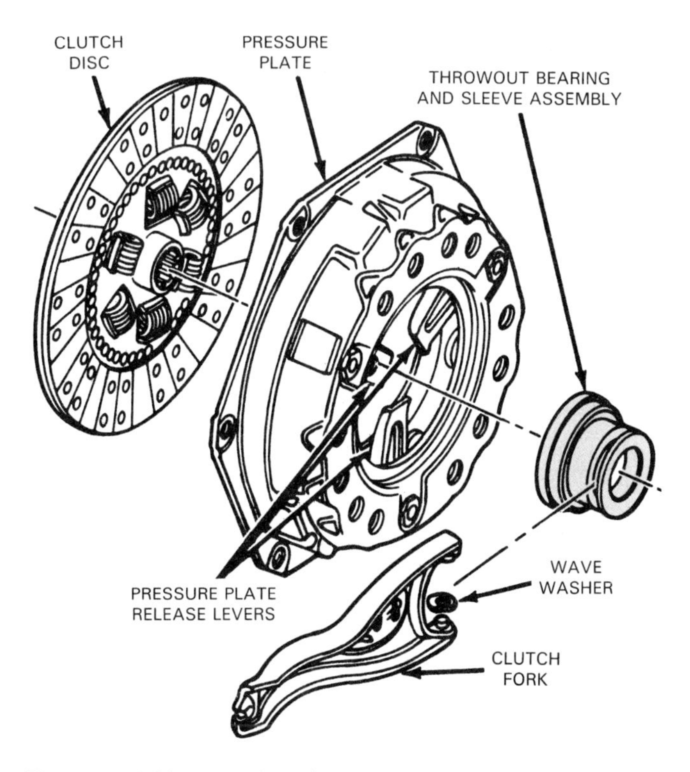

Figure 6-14. Note position of the throwout bearing in relation to other clutch parts. Throwout bearing, which controls operation of the clutch, is operated by the clutch fork. (Chrysler)

Clutch Fork

The function of the clutch fork, or ***release fork,*** is to transfer the force of the clutch pedal and linkage to the throwout bearing. Clutch fork components are composed of hardened steel, pressed into the necessary shape. There are two types of clutch forks.

The lever type of clutch fork pivots on a stationary *ball stud* that is mounted inside of the clutch housing. See Figure 6-15. The fork is held to this ball stud with a *spring clip.* The placement of the pivot point determines the amount of force multiplication relative to the clutch pedal. In addition, the placement will vary depending on the type of clutch release mechanism that is used.

When the outer end of the fork is moved toward the rear of the clutch housing, the inner end will move inward and push the throwout bearing against the clutch release levers or fingers. When the clutch pedal is released, the clutch fork and linkage, aided by return springs, pulls the throwout bearing away from the pressure plate assembly.

The other type of clutch fork is installed on a shaft that extends through the clutch housing. Depending on the installation, the shaft may pass over or under the transmission input shaft and throwout bearing. The fork extends away from its shaft. It is moved inward when the shaft is turned by the clutch linkage. This kind of clutch fork is shown in Figure 6-16.

The clutch is disengaged in the same manner as the lever type. The clutch fork pushes the throwout bearing inward. The throwout bearing frees the pressure plate assembly. To engage the clutch, the clutch linkage pulls the fork and throwout bearing away from the pressure plate assembly.

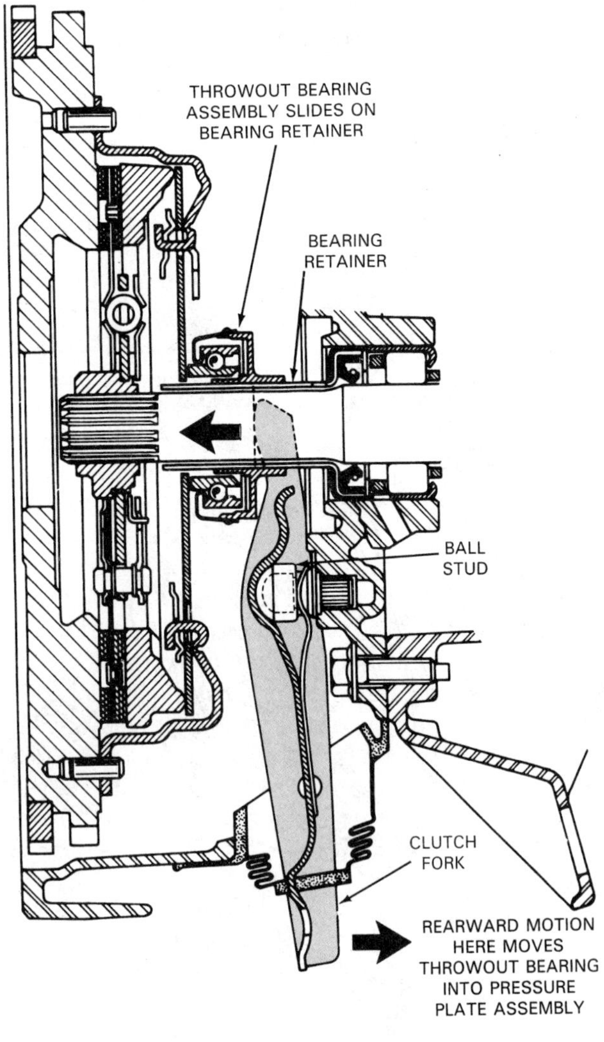

Figure 6-15. A lever-type clutch fork. Note the relationship between clutch fork, ball stud (pivot point), and throwout bearing. (Chrysler)

Clutch Linkage System

The pressure plate apply springs are strong. They must be strong to hold the friction disc firmly and prevent slipping when the clutch is engaged. If not overcome by external pressure, the springs in the pressure plate assembly would hold the clutch in the engaged position at all times. Therefore, some strong force must be found to disengage the clutch. This force is provided through the

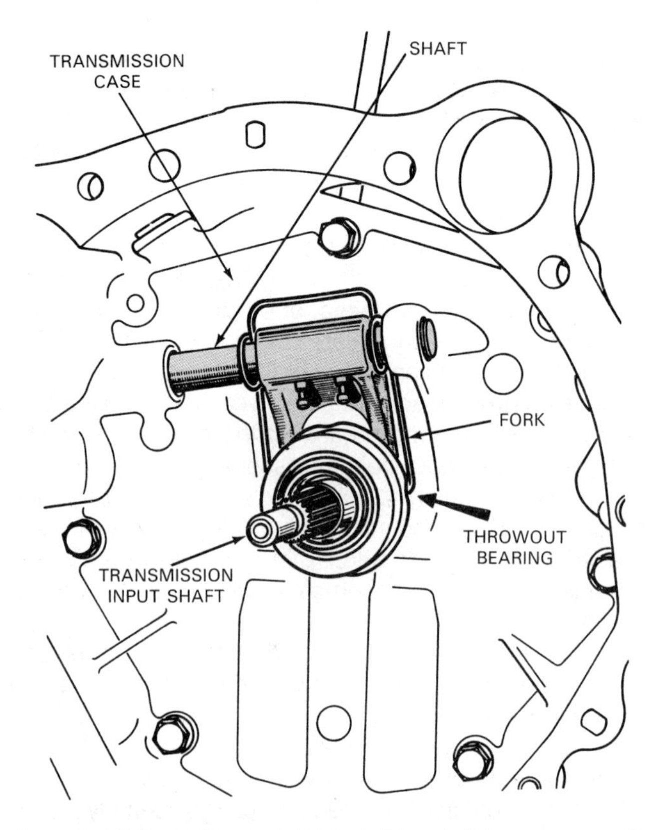

Figure 6-16. A shaft-type clutch fork. The shaft turns, moving the fork and throwout bearing forward. (Renault)

clutch linkage system. The clutch pedal and linkage is considered a part of this system. In addition, some manufacturers consider the clutch fork a part of the clutch linkage system.

Clutch pedal

Vehicle manufacturers decided many years ago that the best way to overcome spring pressure, in order to disengage the clutch, is with a foot-operated ***clutch pedal.*** The clutch fork is operated by the driver's foot via the clutch pedal. This action moves the linkage that moves the clutch fork.

The clutch pedal is a lever with a rubber pad at one end. The other end of the pedal is attached to linkage that connects the pedal to the clutch fork. The pedal is mounted on a bracket under the dashboard. (This bracket may also hold the brake pedal.) The pedal pivots on metal or plastic bushings, which reduces noise and wear. When foot pressure is released, the pedal is returned to its fully released position with the help of the clutch return springs.

The placement of the pivot point determines the amount of force multiplication provided by the lever. The force applied to the clutch is multiplied at the expense of distance. Note that the clutch pedal must be moved about 5 or 6 in. (125 or 150 mm) to move the throwout bearing less than 1 in. (25 mm). The force applied to the throwout bearing is five or six times that exerted on the pedal.

Two terms associated with the clutch pedal are *free play* and *pedal reserve.* ***Free play,*** also called ***free travel,*** is the distance that the pedal will travel when first depressed before the slack in the linkage is taken up and any appreciable resistance is felt by the driver. Once the slack is eliminated, further pedal travel causes the throwout bearing to contact the pressure plate and begin to disengage the clutch. There is usually about 1 in. (25 mm) of free play in the pedal. This distance often relates, by a factor, to the distance between the throwout bearing and release levers or fingers when the clutch pedal is fully released.

Free play should not be confused with ***pedal reserve.*** This is the amount that the pedal can still be depressed beyond the point where the clutch is fully disengaged. The throwout bearing continues further inward, and the pressure plate is moved outward further than is actually needed for full disengagement. When a specification for pedal reserve is given, it usually is about 1 in. (25 mm). Free play and pedal reserve are depicted in Figure 6-17.

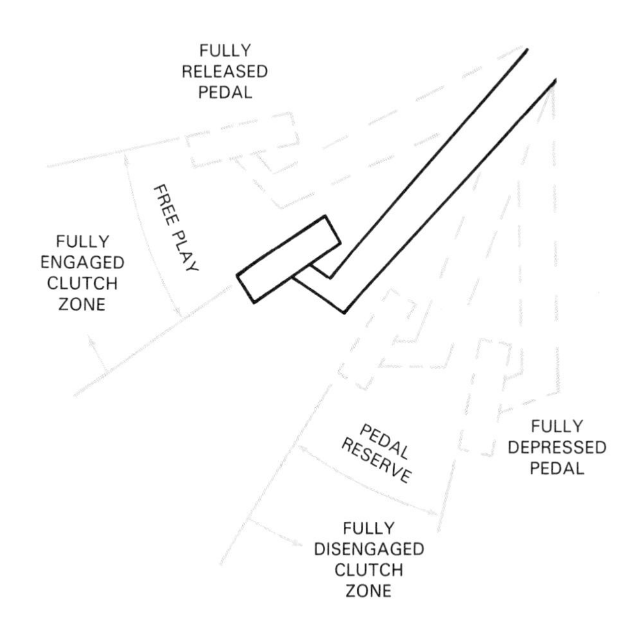

Figure 6-17. Free play is the distance the clutch pedal will move before the throwout bearing contacts the pressure plate. Pedal reserve is the distance that the pedal can be depressed after the clutch is completely disengaged.

Clutch linkage

Clutch linkage transmits the force applied to the clutch pedal to the clutch fork. There are three basic kinds of clutch linkage:

- A mechanical ***rod-and-lever linkage*** depends on a set of rods and levers to transmit movement from pedal to clutch fork.
- A mechanical ***cable linkage*** transmits motion by an enclosed cable between the pedal and the clutch fork.
- A ***hydraulic linkage*** uses the clutch pedal to develop hydraulic pressure. The hydraulic pressure transmits movement to the clutch fork.

All clutch linkages must have some method of adjustment. As the friction disc wears, the pressure plate must move further to compensate. This causes the release levers or diaphragm spring to pivot further, moving inward at the pressure plate and outward at the throwout bearing. As a result, the release levers or fingers will move closer to the throwout bearing, and the clutch free play will decrease. If the clutch wears enough, the free play will disappear completely, and slippage could result. To prevent this, provisions are made for adjusting the linkage so that the throwout bearing can be moved further away from the release levers as parts wear.

Rod-and-lever linkage. This type of clutch linkage, shown in Figure 6-18, has a *pushrod* attached to the clutch pedal. The pushrod moves a lever called the *bellcrank.* The bellcrank operates another pushrod, or *release rod.* The release rod pushes the clutch fork rearward to operate the throwout bearing. The clutch fork pivot point *lies between the throwout bearing and release rod.* Thus, pushing on the clutch fork causes the throwout bearing to move inward. Smooth operation of the entire linkage depends on proper lubrication and alignment of the pivot points.

The release rod is located next to the clutch housing. Its effective length can be changed to adjust the clutch free play.

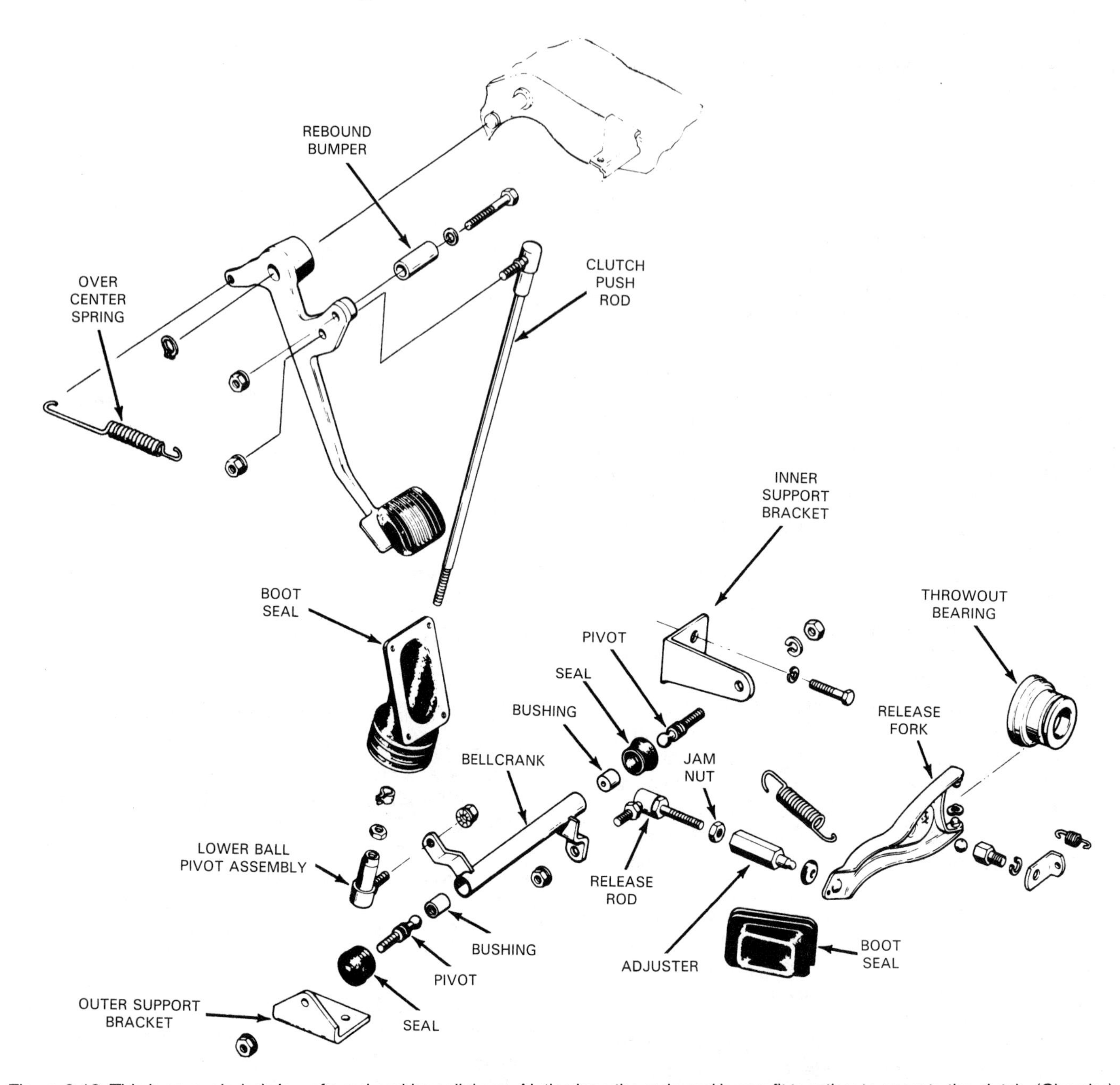

Figure 6-18. This is an exploded view of a rod-and-lever linkage. Notice how the rods and levers fit together to operate the clutch. (Chrysler)

Adjustment is usually made by loosening the release rod adjusting *locknuts* and shortening the release rod. The locknuts are then retightened. Shortening the rod moves the throwout bearing away from the release levers or fingers to increase the free play, which has diminished over time.

Springs will be mounted along the linkage in various locations. ***Return springs*** function to return the clutch fork and linkage to their normal, or released, positions. Other springs may be used for holding certain parts in proper position.

Cable linkage. This type of linkage utilizes a cable. The cable transfers pedal pressure to the clutch fork. To move the clutch fork, the clutch cable is pulled by pressure on the clutch pedal. (This action is opposite from that of a rod-and-lever linkage, wherein the clutch fork is *pushed* by depressing the clutch pedal.) When foot pressure is removed from the clutch pedal, the return springs and pressure plate springs return the pedal to the fully released position. A simple cable release system is shown in Figure 6-19.

With the basic lever-type clutch fork, the cable pulls the clutch fork forward to operate the throwout bearing. The throwout bearing *lies between the cable and pivot point.* Thus, pulling on the clutch fork causes the throwout bearing to move inward. A return spring, which directly opposes the cable, pulls back on the clutch fork when the pedal is released, to disengage the throwout bearing.

Cable linkages have a cable housing, which serves as a guide and protective cover for the cable. The housing holds lubricant to prevent cable sticking. This housing is sometimes called a *sheath.* Cable brackets are located along the length of the sheath to hold the cable in the proper position.

A means for adjusting the cable is located at one end of the cable. The adjustment procedure is similar to that of the rod-and-lever linkage. The effective length of the cable housing is changed to adjust clutch free play. The length is changed by repositioning an adjusting nut. In most cases, free play is increased by *shortening* the effective length of the cable housing, thereby feeding more cable to the clutch fork. This causes the throwout bearing to move further away from the pressure plate assembly.

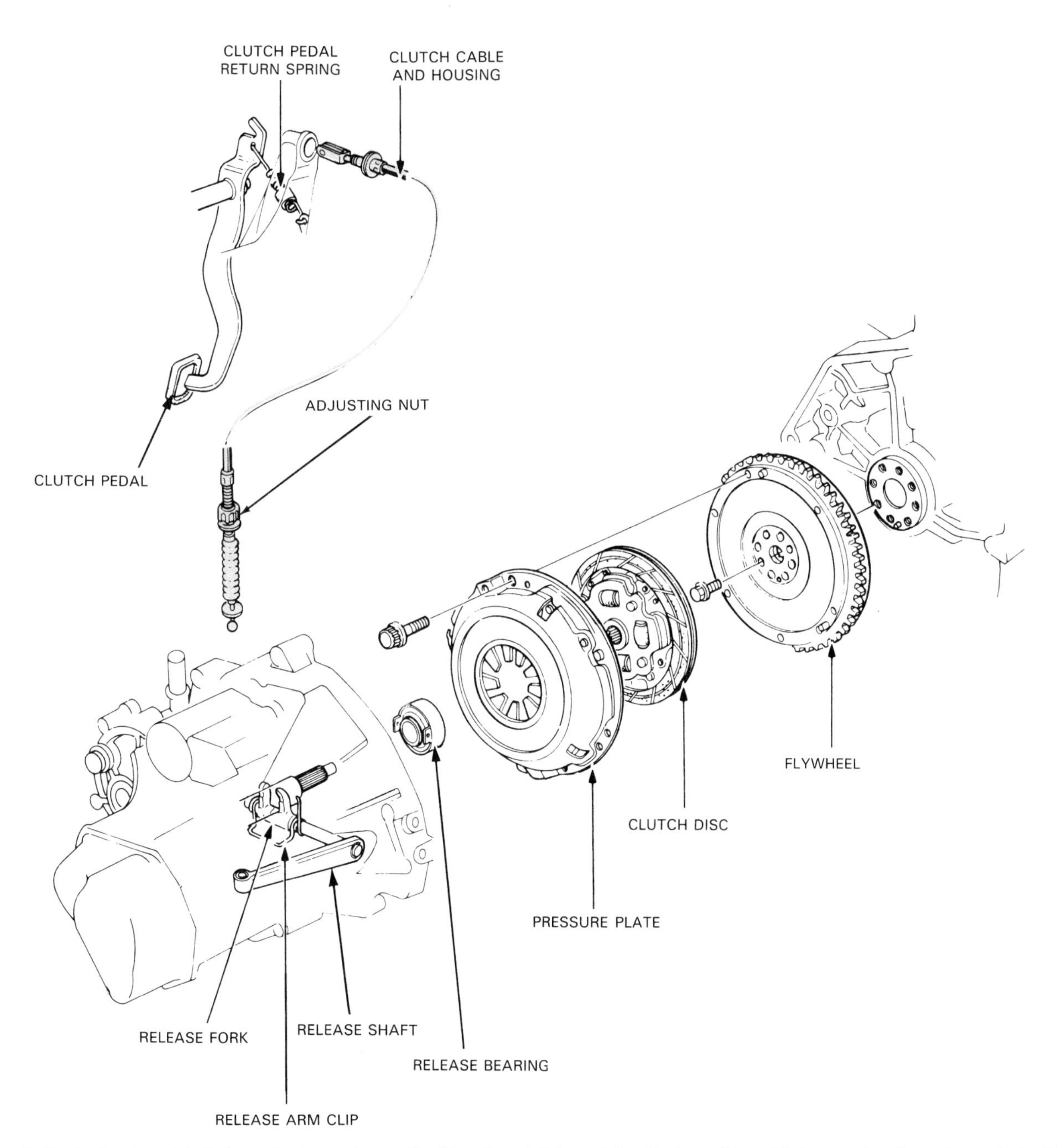

Figure 6-19. A simple cable linkage is shown here. Pushing the clutch pedal pulls the cable, which operates the release fork. (Honda)

Some cable linkages have a provision for self-adjustment to compensate for wear, usually occurring in the cable or clutch disc. The ***self-adjusting cable*** operates through a ***pawl-and-quadrant assembly.*** The mechanism is pictured in the cable linkage shown in Figure 6-20. The *pawl* is a toothed link. It is used to engage teeth that are built into the *quadrant,* which is a pivoting ratchet. It is in the shape of a quarter-circle. It pivots on the pedal bracket. The assembly is mounted under the dashboard.

In a typical assembly, the end of the clutch cable is attached to the far end of the quadrant, and the pawl and quadrant are spring loaded. See Figure 6-21.

When the clutch pedal is depressed, the pawl engages the quadrant teeth. The quadrant is caused to move about its pivot point, and the cable is pulled by this action, thus, disengaging the clutch. The quadrant spring keeps light tension on the cable, and when the clutch pedal is released, the pawl ratchets over the quadrant teeth and returns with the pedal to a stop on the pedal bracket.

As the cable stretches from wear, the self-adjusting clutch picks up the slack. The quadrant automatically takes up the excess because of the quadrant spring. It causes the quadrant to pivot, thus, keeping tension on the cable. Gradually, the quadrant pivots ahead. As it does, the pawl will engage successive teeth on the quadrant.

As the disc facing wears, the inner ends of the release levers or fingers gradually move away from the flywheel, decreasing free play in the process. To compensate, more cable must be fed to the clutch fork to allow it to move further back. The movement of the release levers or fingers is translated to the quadrant. The quadrant adjusts its orientation when the pawl is free of the quadrant, which is when the pedal is in the upmost position. Thus, the quadrant makes adjustments by moving in either direction as cable or clutch wear dictates.

Hydraulic linkage. This type of clutch linkage uses hydraulic pressure to transfer pedal movement to the clutch fork. The system consists of a *master cylinder* and a *slave cylinder* connected by hydraulic lines. In addition, a fluid reservoir is connected to the master cylinder. The cylinders are made of cast iron or aluminum, and the pistons are aluminum. Steel tubing and high-pressure rubber hoses are used for the connecting lines.

The ***master cylinder*** is the controlling cylinder. It develops the system hydraulic pressure. The ***slave cylinder*** is the operating cylinder. It is actuated by the hydraulic pressure developed in the master cylinder. A typical system is illustrated in Figure 6-22.

The master cylinder piston is actuated by a pushrod that is attached to the clutch pedal. Pushing on the pedal

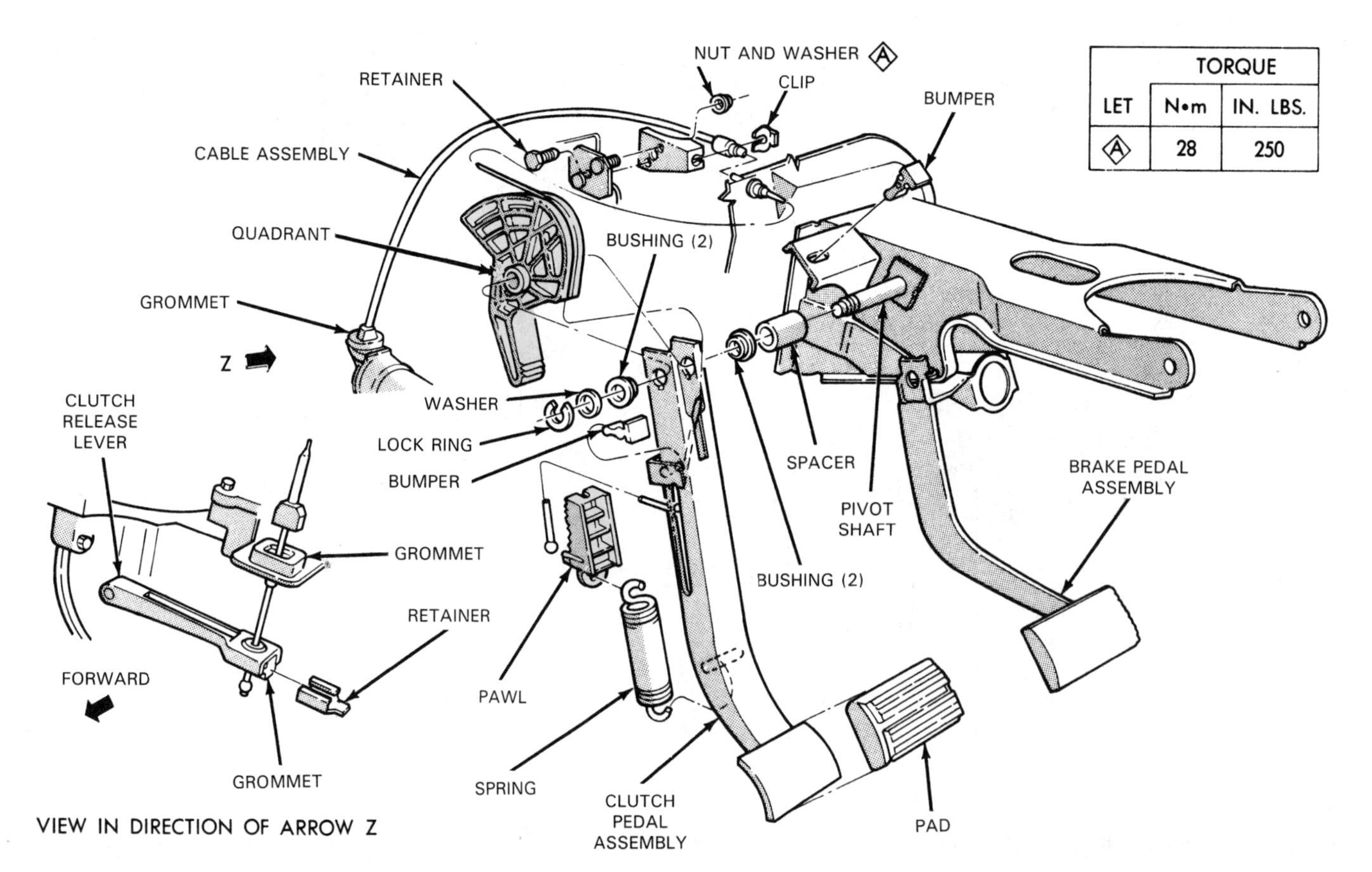

TORQUE		
LET	N•m	IN. LBS.
A	28	250

Figure 6-20. The self-adjusting cable linkage consists of a quadrant and a pawl. The quadrant is mounted under the dash at the clutch pedal bracket. (Chrysler)

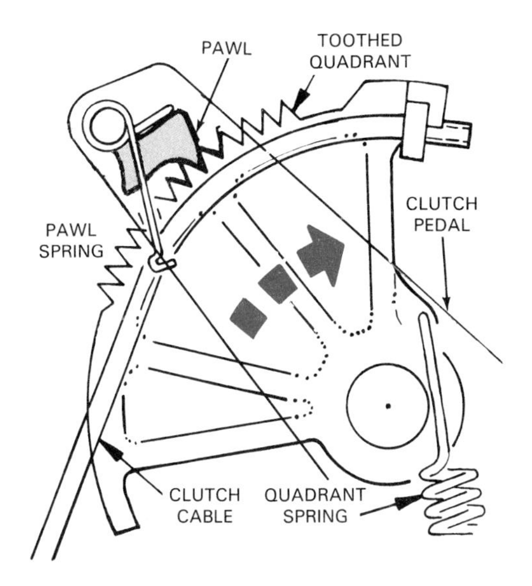

Figure 6-21. Note the action of a self-adjusting cable. The pawl engages teeth on the quadrant. When the clutch pedal is depressed, the pawl moves the quadrant, which pulls the cable. When the clutch is released, the pawl returns to its preset stop point. The cable is always held under tension. This is because any action that would cause it to lose tautness is automatically picked up by the quadrant. The pawl ratchets over the quadrant teeth, and the quadrant assumes its new position. (Ford)

acts on the piston, creating hydraulic pressure. One end of the pushrod is retained within the master cylinder. The rod does not make contact with the piston, however, until the pedal is about 1 in. (25 mm) into its stroke.

The master cylinder is connected to the slave cylinder by hydraulic tubing. The slave cylinder piston is pushed outward by hydraulic pressure. The piston moves the clutch fork by means of a pushrod. See Figure 6-23. The clutch fork is returned, in many cases, by means of a return spring (not shown) that is attached to it.

The piston in the slave cylinder is sometimes larger than the piston in the master cylinder. The pressure developed in the master cylinder acts on a larger piston area in the slave cylinder. This increases the force that the slave cylinder can apply to the clutch fork. In this way, the hydraulic linkage can multiply pedal force.

Two kinds of *piston seals* are used by the master and slave cylinders. These are often referred to as *primary* and *secondary seals,* or *cups.* The ***primary seal*** is used on the fluid end of the piston. Hydraulic pressure will force the primary seal up against the cylinder wall, making for a tight seal. The seal, then, will keep fluid from leaking out so that the system can be pressurized. On the other hand, it will allow leakage when pedal pressure is released and the piston returns to its normal position. Some primary seals

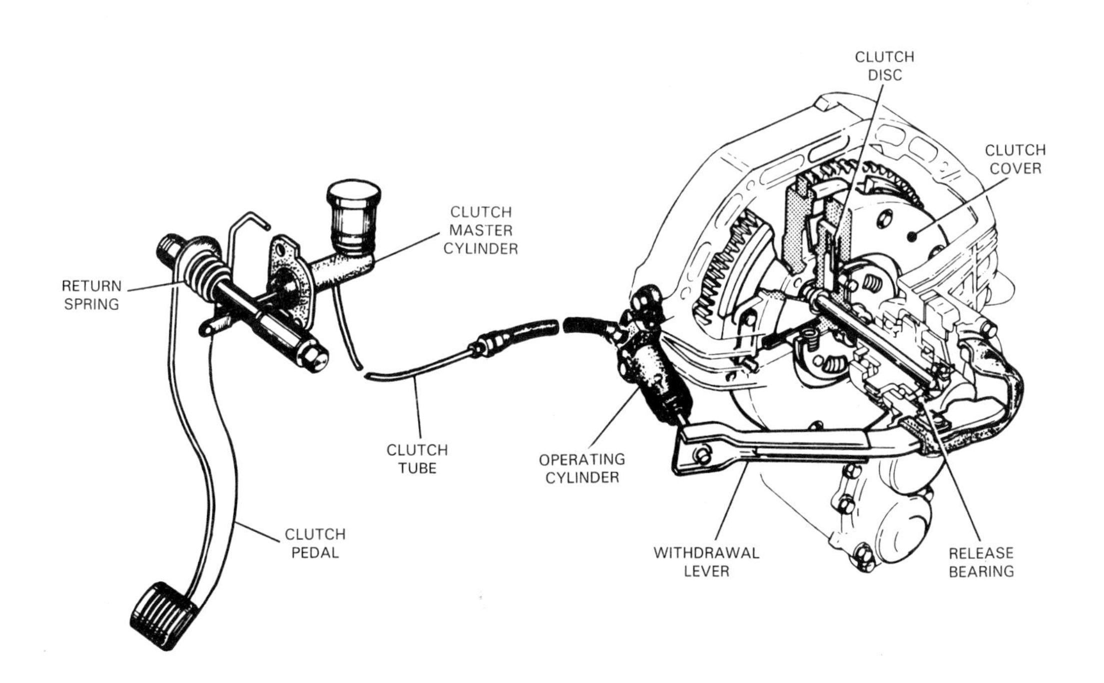

Figure 6-22. Hydraulic linkage is shown controlling clutch of a manual transaxle. Pushing on the clutch pedal develops hydraulic pressure in the master cylinder. This pressure is transmitted through tubing to the slave cylinder. The slave cylinder piston moves out under pressure. This piston movement operates the clutch fork. Free play can be adjusted at the master cylinder or slave cylinder, depending on the particular system. (Subaru)

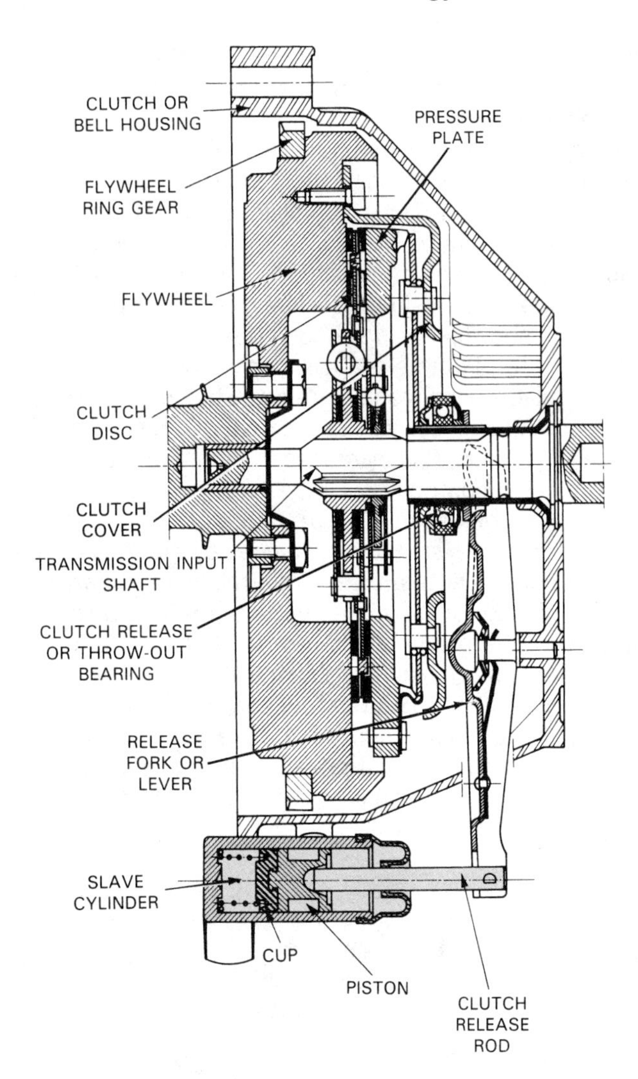

Figure 6-23. Illustration showing slave cylinder used for clutch fork operation. Hydraulic pressure moves piston, which moves pushrod. Pushrod moves the clutch fork. (Peugeot)

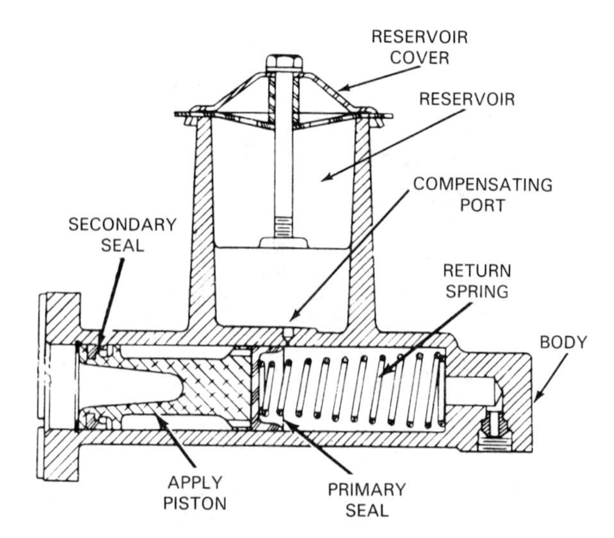

Figure 6-24. Sectional view of clutch master cylinder shows apply piston, return spring, and piston seals. Also shown is fluid reservoir and compensating port.

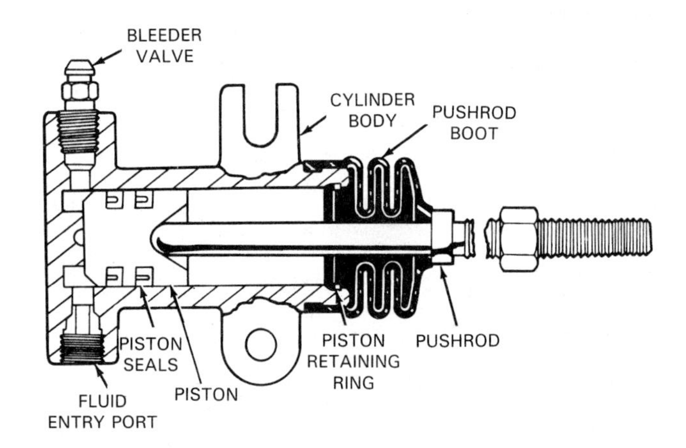

Figure 6-25. A typical slave cylinder contains a piston, pushrod, and seals. Some are equipped with a spring. The outer boot prevents entry of dirt or water. (General Motors)

are lip seals. ***Secondary seals*** are used at the pushrod end of the piston. Their purpose is to keep system fluid from leaking out of the cylinder. Some secondary seals are ring seals.

Since the piston seals and hoses are made of natural rubber, ***DOT 3 brake fluid*** is used as the hydraulic fluid. DOT 3 is the standard established by the Department of Transportation for automotive fluid used in modern brake systems. It is a synthetic fluid. Unlike petroleum-based products, it will not damage rubber components.

The master cylinder contains a *piston return spring.* It also has a *compensating port.* See Figure 6-24. The spring speeds the piston's retraction. The compensating port allows fluid from the reservoir to enter the system.

The slave cylinder, Figure 6-25, is similar in design to the master cylinder; however, it contains no return spring, compensating port, or reservoir. Some slave cylinders do contain a spring. However, it serves to hold the cylinder primary seal in place. It does not function as a return spring.

In operation, when the clutch pedal is pushed and the piston is pushed forward in the cylinder, the compensating port is sealed off from the hydraulic system. As a result, fluid is trapped in the system and prevented from being pushed back into the reservoir.

When the clutch pedal is released, the piston return spring causes the piston to quickly retract. The master cylinder piston retracts more quickly than the slave cylinder piston. As a result, a low-pressure condition develops in the system. Hydraulic fluid will be pushed into the system by atmospheric pressure acting on the surface of the fluid in the reservoir. The fluid enters the system from the spooled region behind the primary seal. This region is supplied by the fluid reservoir via the compensating port.

Due to the pressure differential, fluid leaks from behind the primary seal into the low-pressure zone. By this action, system pressure becomes equalized.

One variation of the hydraulic linkage uses a special design for the slave cylinder. In this variation, the slave cylinder and throwout bearing are combined into a single assembly, as shown in Figure 6-26. The cylinder and apply piston are ring shaped. The entire assembly is installed over the shaft of the front bearing retainer. Hydraulic pressure from the master cylinder causes the ring-shaped piston to push directly on the throwout bearing. When this system is used, there is no clutch fork. The basic layout of this system is shown in Figure 6-27.

Clutch Housing

The clutch housing, or ***bell housing,*** as it is also called, is bolted to the engine. It supports the transmission or transaxle and encloses the clutch. In some designs, the transmission is bolted to the clutch housing. In other designs, the two are combined as a one-piece, or integral, casting. Almost all transaxles are of the latter type. When properly installed, the clutch housing holds the centerlines of the engine crankshaft and transmission input shaft in alignment. The integral design reduces the chance of misalignment between engine and transmission.

Most clutch housings are light aluminum castings; however, some housings used on trucks and older cars are made from cast iron. A sheet metal *clutch access cover* is found on the front of the clutch housing. The cover can be removed for inspection purposes or for clutch repairs.

The clutch housing prevents the entry of dirt, oil, or moisture into the clutch assembly. An opening in the housing, which is used for the clutch fork, is usually sealed by a rubber boot. Most clutch housings have a series of slots, which permits air circulation within the housing. This aids

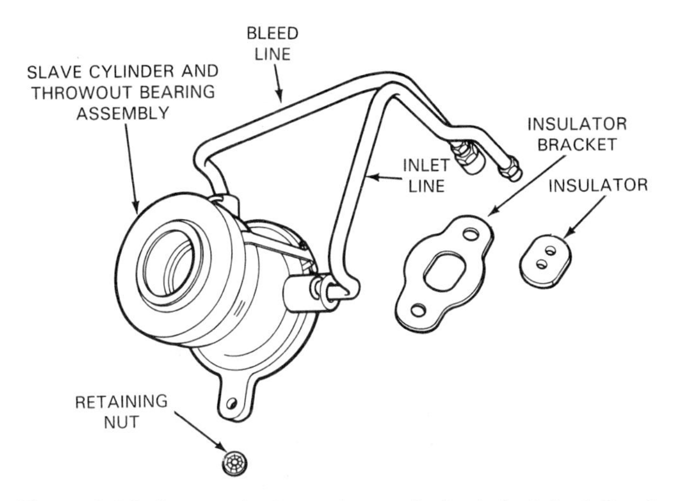

Figure 6-26. An annular-type slave cylinder is installed directly behind the throwout bearing. There is no clutch fork. (Chrysler)

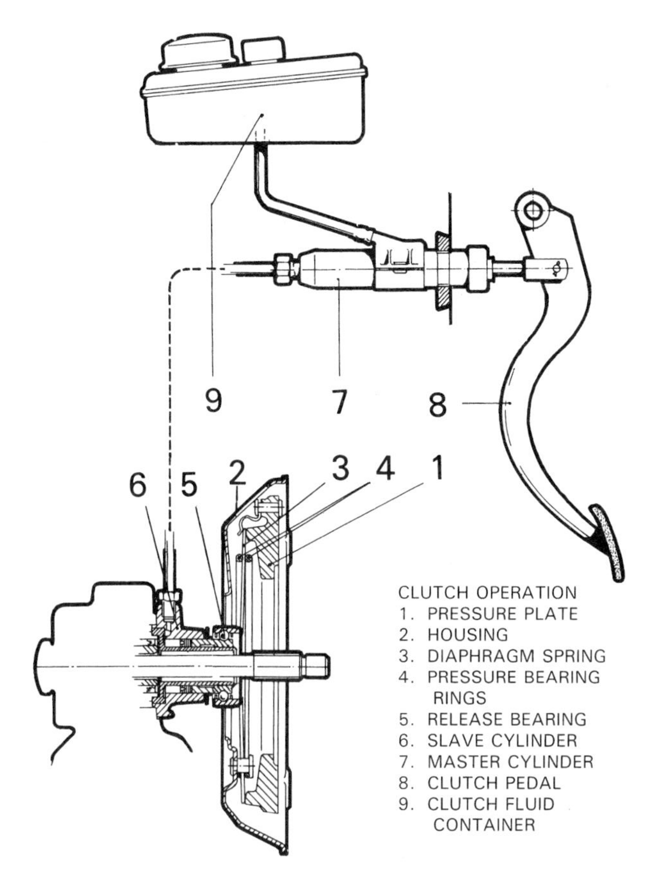

Figure 6-27. Study action of slave cylinder when mounted with throwout bearing. Hydraulic pressure moves the slave cylinder piston outward. Piston pushes directly on throwout bearing, moving it into contact with the pressure plate release levers or fingers. (Saab)

in removal of heat generated by the clutch. Clutch housings have an opening at the bottom so that any oil or water that gets inside can drip out.

On many vehicles, the *starter motor,* or *starter,* is attached to the clutch housing. Inside the clutch housing, the starter *drive pinion* engages the flywheel ring gear to crank the engine. The housing and starter must be carefully aligned to prevent damage to the starter drive pinion and to the flywheel ring gear. In addition, a heavier steel access cover must be used. It adds strength, which is needed to handle the severe forces put on the clutch housing as the starter cranks the engine. This type of access cover is often called a *stiffener plate.*

Summary

The clutch is a device for engaging and disengaging the engine from the road. It is used on every vehicle that has a manual transmission. The clutch allows the engine to operate when the vehicle is not moving. The clutch is also used to disengage the engine and transmission when shifting gears.

The single-plate dry clutch consists of eight major components. The basic action of the clutch assembly consists of pressing a friction disc between a flywheel and pressure plate. When the disc is pressed between the other two parts, all three rotate as a single unit. Power is then transmitted from the engine to the transmission. When the disc is not pressed between the flywheel and pressure plate, power is not transmitted.

The transmission input shaft is mounted on bearings or bushings at each end. The front end of the input shaft extends into the rear of the engine crankshaft. The rear end of the input shaft contains the transmission input gear. The input shaft is splined to the clutch friction disc and transfers engine power into the transmission.

The friction disc is made of a heat-resistant composite material. The disc assembly contains springs that cushion the shock of engagement. The friction disc is splined to the input shaft. The splines allow the disc to move slightly as it is engaged and disengaged.

The pressure plate assembly consists of three major parts. The cover bolts to the flywheel and contains the other parts. The pressure plate provides the contact surface with the friction disc. The springs press the friction disc between flywheel and pressure plate.

There are two kinds of springs used on modern pressure plate assemblies. The diaphragm type is a single dish spring which flexes to apply and release the clutch. Other pressure plate assemblies use multiple coil springs to provide pressure.

The throwout bearing operates the pressure plate release levers or fingers. The throwout bearing slides on a hub built as part of the input shaft bearing retainer. The throwout bearing consists of a ball bearing assembly mounted on a collar. The ball bearing allows the throwout bearing assembly to contact the spinning release levers without damage.

The clutch fork operates the throwout bearing. The clutch fork is a simple lever device that reverses and increases the clutch pedal force. The fork pivots on a ball stud inside of the clutch housing. Some clutch forks are mounted on a shaft. It turns to move the fork.

The clutch linkage allows the driver to operate the clutch mechanism easily. All linkages multiply pedal force. In the process, the distance traveled by the pedal is reduced to a much smaller distance traveled by the throwout bearing.

The rod-and-lever linkage consists of a set of pushrods and levers that transmit clutch pedal pressure to the clutch fork. Pressure delivered at the throwout bearing depends on the placement of the rods and levers. The linkage can usually be adjusted at the clutch fork.

The cable linkage uses a cable in place of pushrods and levers. Pushing the clutch pedal pulls the cable, which pulls the clutch fork. This type of linkage can also be adjusted. Some cable linkages have a self-adjusting feature.

The hydraulic linkage uses hydraulic pressure to transfer pedal pressure to the clutch fork. Pressure developed in the clutch master cylinder is transferred to the slave cylinder, which operates the clutch fork. One type of slave cylinder is built into the throwout bearing assembly, and a clutch fork is not needed.

The clutch itself is enclosed in a clutch housing. This housing protects the clutch parts from dirt, water, and oil.

Know These Terms

Flywheel; Manual transmission input shaft (Clutch shaft); Clutch disc (Friction disc); Pressure plate assembly; Throwout bearing (Release bearing); Clutch fork (Release fork); Clutch linkage; Clutch housing (Bell housing); Pilot bearing; Pilot bushing; Transmission front bearing; Front bearing retainer; Outer disc assembly; Friction facing; Cushion spring; Hub flange; Torsion spring; Clutch cover; Pressure plate; Apply spring; Release lever; Release finger; Diaphragm-spring pressure plate assembly; Coil-spring pressure plate assembly; Pivot ring; Semi-centrifugal clutch; Clutch pedal; Free play (Free travel); Pedal reserve; Rod-and-lever linkage; Cable linkage; Hydraulic linkage; Return spring; Self-adjusting cable; Pawl-and-quadrant assembly; Master cylinder; Slave cylinder; Primary seal (Primary cup); Secondary seal (Secondary cup); DOT 3 brake fluid.

Review Questions–Chapter 6

Please do not write in this text. Place your answers on a separate sheet of paper.

1. What is the function of a clutch?
2. List the eight basic parts of a clutch and tell what role each part serves in clutch operation.
3. The raised ring on the crankshaft flange fits into a hole in the _____.
4. What is the purpose of the pilot bearing?
5. Which part transmits power from the engine flywheel to the transmission input shaft?
   a. Pressure plate.
   b. Throwout bearing
   c. Clutch disc.
   d. Clutch fork.
6. How does the pressure plate assembly work?
7. Briefly describe the two types of clutch forks and their operation.
8. The distance the clutch pedal will travel before resistance is felt by the driver is called _____ _____.
9. In a short paragraph, discuss one of the three types of clutch linkages.
10. How does a self-adjusting clutch work?

Certification-Type Questions–Chapter 6

1. The main purpose of the clutch is to:
- **(A)** absorb heat.
- **(B)** dampen engine vibration.
- **(C)** transfer engine power into the transmission.
- **(D)** connect or disconnect power flow from engine to road.

2. All of these are functions of the flywheel EXCEPT:
- **(A)** transferring engine power to the clutch.
- **(B)** providing a mounting surface for parts of the clutch.
- **(C)** holding the input shaft in proper alignment.
- **(D)** serving as a vibration damper.

3. The clutch linkage is designed to:
- **(A)** slide the throwout bearing into the pressure plate assembly.
- **(B)** hold the clutch disc against the flywheel.
- **(C)** allow the driver to operate the clutch fork.
- **(D)** transfer power flow from the clutch to the transmission.

4. All of these are turning when the engine is running and the clutch is engaged EXCEPT:
- **(A)** the flywheel.
- **(B)** the clutch disc.
- **(C)** the clutch fork.
- **(D)** the input shaft.

5. Technician A says that the flywheel is generally made of cast iron. Technician B says that a hardened-steel ring is usually mounted on the outside of the flywheel. Who is right?
- **(A)** A only
- **(B)** B only
- **(C)** Both A & B
- **(D)** Neither A nor B

6. The manual transmission input shaft:
- **(A)** is made of aluminum.
- **(B)** holds the flywheel in alignment with the clutch components.
- **(C)** extends from the transmission case to the drive shaft.
- **(D)** transmits motion from the clutch disc to the transmission.

7. Technician A says that the clutch disc is located between the flywheel and the pressure plate. Technician B says that the clutch disc transfers power from the engine crankshaft to the flywheel. Who is right?
- **(A)** A only
- **(B)** B only
- **(C)** Both A & B
- **(D)** Neither A nor B

8. Under their friction facings, most friction discs contain:
- **(A)** hub springs.
- **(B)** torsion springs.
- **(C)** cushion springs.
- **(D)** apply springs.

9. For engine power to get from the friction disc to the hub flange, it must actually go through which of these springs?
- **(A)** Apply springs.
- **(B)** Torsion springs.
- **(C)** Cushion springs.
- **(D)** Diaphragm springs.

10. All of these are parts of the pressure plate assembly EXCEPT:
- **(A)** the clutch cover.
- **(B)** apply springs.
- **(C)** the throwout bearing.
- **(D)** a pressure plate.

11. Technician A says that slots in the clutch cover aid heat dissipation. Technician B says that the openings allow air to circulate inside of the assembly. Who is right?
- **(A)** A only
- **(B)** B only
- **(C)** Both A & B
- **(D)** Neither A nor B

12. The pressure plate is a ring-shaped disc that is made of:
- **(A)** cast iron.
- **(B)** aluminum.
- **(C)** high-strength steel.
- **(D)** plastic.

13. Which of these are built into the pressure plate assembly near the outer rim?
- **(A)** Pivot rings.
- **(B)** Clutch rings.
- **(C)** Primary rings.
- **(D)** Pressure rings.

14. In a coil-spring pressure plate assembly, springs are pressed between the plate and the cover. Technician A says that four springs must be used. Technician B says that at least ten are needed. Who is right?
- **(A)** A only
- **(B)** B only
- **(C)** Both A & B
- **(D)** Neither A nor B

15. Release levers are used in:
- **(A)** throwout bearing assemblies.
- **(B)** coil-spring pressure plate assemblies.
- **(C)** outer clutch disc assemblies.
- **(D)** clutch fork assemblies.

16. All of these are part of a throwout bearing assembly EXCEPT:
- **(A)** a collar.
- **(B)** a sleeve.
- **(C)** a thrust bearing.
- **(D)** rod-and-lever linkage.

17. Technician A says that the throwout bearing must be greased occasionally to ensure proper operation. Technician B says that the throwout bearing is generally lubricated at the factory and cannot be greased without removing the transmission. Who is right?
- **(A)** A only
- **(B)** B only
- **(C)** Both A & B
- **(D)** Neither A nor B

18. The function of the release fork is to:
- **(A)** pull the throwout bearing away from the release levers.
- **(B)** connect clutch linkage to throwout bearing.
- **(C)** slide on hub of front bearing retainer.
- **(D)** rotate on retainer hub to absorb spinning pressure shock.

19. Technician A says that the clutch fork components are made of hardened steel. Technician B says that the clutch fork transfers the force of the clutch pedal and linkage to the throwout bearing. Who is right?
- **(A)** A only
- **(B)** B only
- **(C)** Both A & B
- **(D)** Neither A nor B

20. Which of these is used to overcome apply spring pressure and disengage the clutch?
- **(A)** Clutch pedal.
- **(B)** Free play.
- **(C)** Gearshift lever.
- **(D)** Return spring.

21. Technician A says that pedal reserve is the distance that the pedal will travel before the slack in the linkage is taken up and any resistance is felt by the driver. Technician B says that free play is the amount that the pedal can be depressed beyond the point at which the clutch is fully disengaged. Who is right?

(A) A only
(B) B only
(C) Both A & B
(D) Neither A nor B

22. All of these are types of clutch linkage EXCEPT:

(A) pedal linkage.
(B) cable linkage.
(C) hydraulic linkage.
(D) rod-and-lever linkage.

23. Technician A says that all clutch linkages must have some provision for adjustment. Technician B says that the throwout bearing must be moved closer to the release levers as parts wear. Who is right?

(A) A only
(B) B only
(C) Both A & B
(D) Neither A nor B

24. All of these are part of a hydraulic clutch linkage EXCEPT:

(A) master cylinder.
(B) fluid reservoir.
(C) slave cylinder.
(D) pawl and quadrant assembly.

25. Which of these engages the flywheel ring gear to crank the engine inside the clutch housing?

(A) Clutch fork.
(B) Release lever.
(C) Starter motor.
(D) Starter drive pinion.

First things first:
- Make sure you have the right parts.
- Compare disassembled parts with new parts before you install them.

Pay special attention:
1. Always replace the pilot bearing or bushing when in doubt.
2. Check engine and transmission shaft seals. Any leaks are going to ruin the new disc.
3. Examine flywheel and pressure plate contact surfaces for scoring and nicks. Regrind the flywheel friction surface if necessary.
4. Maximum clutch disc axial runout should be 0.5 mm (0.02 in.).
5. Ensure that the clutch disc travels easily along the transmission shaft. Remove residue. Grease spline (small amount of long life grease). Insert shaft horizontally. Do not damage spline. Use only correct alignment tools.
6. Make certain that the side of the disc marked "flywheel side" is toward the flywheel.
7. Check release bearing collar and ensure that it runs easily. If worn down, replace. Grease sliding surface with small amount of long-life grease.
8. Tighten clutch cover mounting bolts no more than two turns at a time, alternating back and forth from one side of the assembly to the other.
9. Check clutch linkage system for wear, proper function, and adequate lubrication. Is the clutch easy to operate?
10. In the case of hydraulic release systems, check the slave cylinder. Bleed the system and change the hydraulic fluid.
11. Check correct centering between the clutch housing and the engine.

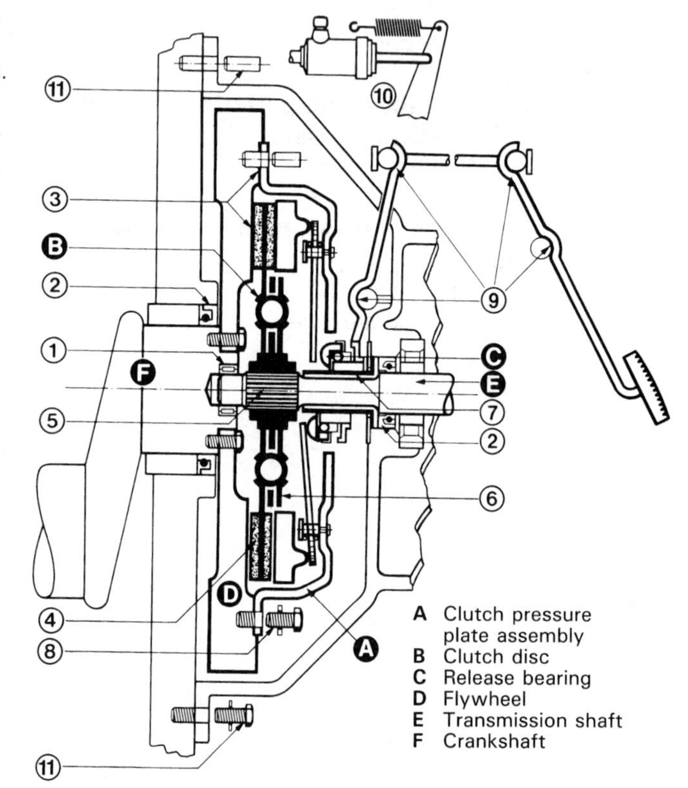

Suggestions for problem-free clutch replacement. Note drawing shows both engaged and disengaged positions of throwout bearing and release levers. (Luk)

Chapter 7

Clutch Problems, Troubleshooting, and Service

After studying this chapter, you will be able to:
- Name and describe the common types of problems that can occur in a clutch assembly.
- Adjust a clutch.
- Remove a clutch assembly.
- Inspect clutch parts for wear and damage.
- Remove and replace a flywheel and a pilot bearing.
- Disassemble, repair, and install various clutch parts.
- Install a clutch assembly.

The purpose of this chapter is to explain how to diagnose and service clutches. In this chapter, clutch problems and their causes will be identified. Procedures performed with the clutch in place, such as *clutch adjustment,* will be explained. Major repair and replacement procedures, which require clutch removal, will also be explained. Figure 7-1 shows typical clutch problems. Study them carefully.

Clutch Problems and Troubleshooting

You cannot properly service a clutch until you know what is wrong with it. In many instances, a clutch problem that seems to be severe is minor and easily remedied. In other cases, the solution involves replacing the clutch assembly or some other major component. Check the easiest, most obvious solutions (such as clutch adjustment) *first,* before removing the clutch. Also, look beyond the clutch for possible solutions. Consider the possibility that the problem is something other than the clutch.

Clutch discs are like brake linings in that they gradually wear out as they operate. Every time that the clutch is engaged or disengaged, some friction material is removed from the clutch disc. The amount of material lost during each action is small, and most clutches will last for 50,000 mi. (80 000 km) or more. It is normal for a clutch to wear out after this mileage is reached.

Often, however, a clutch will not last as long as 50,000 mi. Like any other mechanical part, the clutch can be damaged by careless or abusive operation. Further, it can be damaged by outside factors–factors which cannot be attributed to the vehicle operator. It is important for you, the technician, to find out what caused the premature clutch damage, so that the problem might possibly be prevented in the future.

As stated, clutch problems can occur as a result of careless operation. Driving with a foot resting on the clutch pedal is an example. This careless habit, called ***riding the clutch,*** causes the clutch to slip. As a result, excess heat is generated and premature wearing of clutch friction surfaces occurs. Among the different ways clutches are misused, riding the clutch is the most common. It can put the equivalent of 50,000 mi. of normal wear on a clutch in under 10,000 mi.

Naturally, clutch problems can occur as a result of abuse. An example is increasing engine speed and then suddenly releasing the clutch pedal, allowing the pressure plate apply springs to engage the clutch almost instantly. This is often referred to as ***dumping the clutch,*** or ***popping the clutch.*** It causes instant heat buildup in the clutch disc facings and places tremendous stress on the clutch and the entire drive train.

Clutches usually do not last very long under this type of abuse. Overheated clutch facings, a result of dumping the clutch, can become glazed within seconds. This form of abuse can also cause flywheel, clutch cover, or clutch housing attaching bolts to shear off. Further, it can cause friction disc hub splines to be stripped or input shafts to break.

Placing more load on a clutch than it is designed to handle can also be a source of damage. An example is starting off in a higher, rather than in low, gear. Shifting into a higher gear before the engine has reached the proper rpm will also overload the clutch, as will attempting to change gears at very high engine speeds. Another example is overloading the vehicle to the point that the clutch must be allowed to slip heavily when starting off, just to get the vehicle moving. This is a common problem on vehicles used for towing.

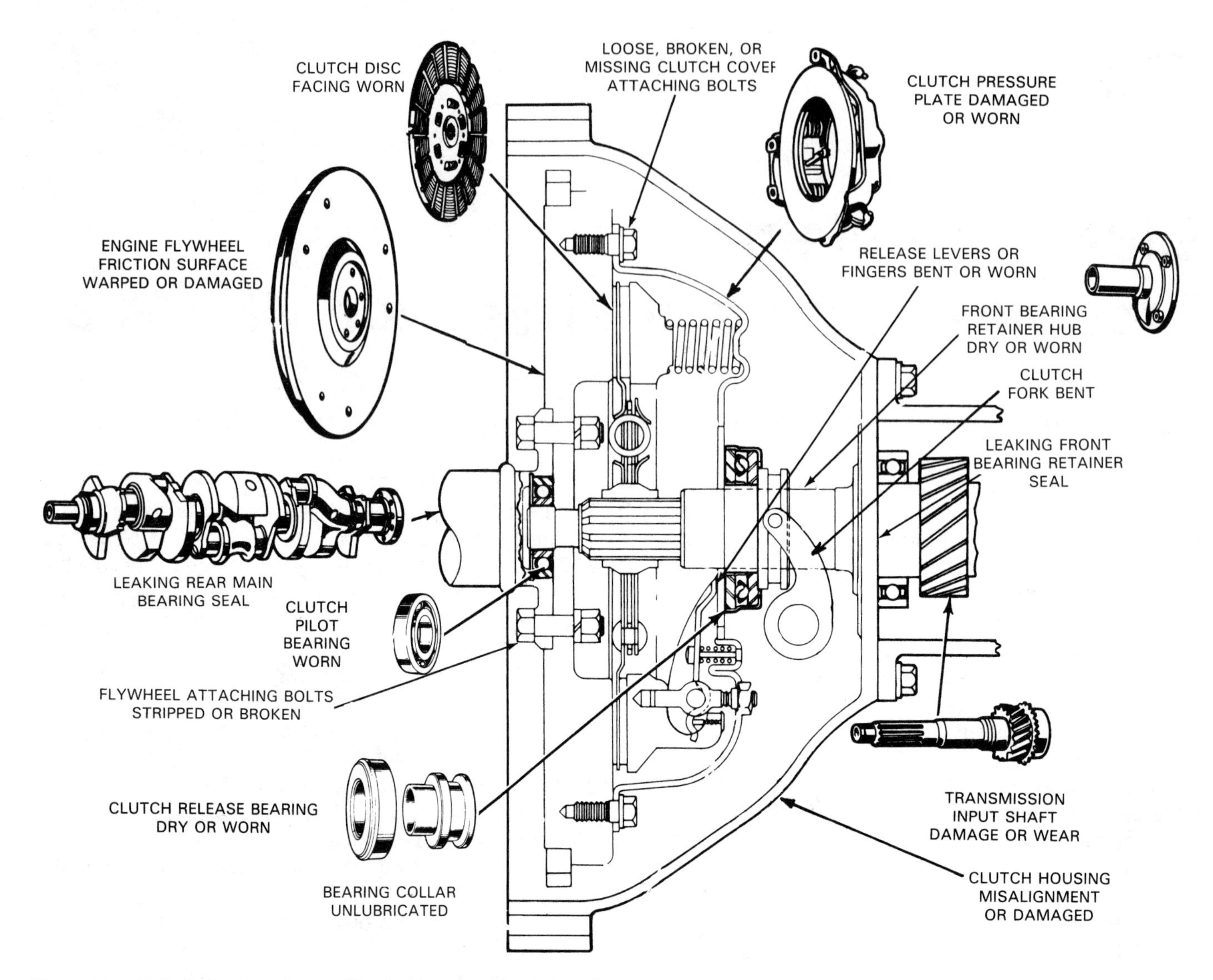

Figure 7-1. Note typical problems with clutch assembly. Some defective parts, such as a leaking engine oil seal, are not part of the clutch assembly but can cause clutch problems. It is very unusual for only one part of a clutch to wear out. You should usually replace any clutch part that you suspect is bad. (Chrysler)

As stated, outside factors can also cause clutch damage. Oil leaks from the engine *rear main bearing seal* or from the transmission *front bearing retainer seal* can reach the friction disc. If the disc gets contaminated with oil, it will *grab* or *chatter* when engaged. If enough oil gets on the disc, it will slip at all times, and the disc surfaces will soon wear out from the resulting friction. Of course, if the disc becomes completely saturated with oil, the surfaces will not wear out, but the disc itself will be useless.

Finally, many clutch problems are due to lack of maintenance or careless service. If the clutch linkage is not adjusted and lubricated on schedule, or if the throwout bearing or clutch hub splines are not properly greased during installation, the moving parts may hang up or operate slowly. The clutch will be subjected to undesirable *slippage,* shortening its life. Slippage and other problems exhibited in clutch operation are presented in detail in the upcoming paragraphs. Figure 7-2 is a table summarizing some common conditions and their causes and corrections.

Clutch Drag

If the clutch cannot be disengaged, or it *fails to release,* you cannot shift gears. In addition, when you stop the vehicle in gear (clutch pedal depressed), as you would normally do, the engine will stall. This condition is known as ***clutch drag.***

Clutch drag varies in degree. Slight drag when the clutch pedal is fully depressed may tend to make the car creep when in gear or cause gear clash when the gears are first engaged. The two most common causes are

CLUTCH TROUBLE DIAGNOSIS

CONDITION	PROBABLE CAUSE	CORRECTION
Fails to Release (Pedal pressed to floor. Shift lever does not move freely in and out of reverse gear.)	a. Improper linkage adjustment. b. Improper pedal travel. c. Loose linkage. d. Faulty pilot bearing. e. Faulty driven disc. f. Fork off ball stud. g. Clutch disc hub binding on clutch gear spline. h. Clutch disc warped or bent. i. Loose pivot rings in cover assembly.	a. Adjust linkage. b. Trim bumper stop and adjust linkage. c. Replace as necessary. d. Replace bearing. e. Replace disc. f. Install fork onto stud. Lightly lubricate fingers at release bearing. Also lube bearing I.D. groove. g. Repair or replace clutch gear and/or disc. h. Replace disc (runout should not exceed 0.020"). i. Replace plate and cover assembly.
Slipping	a. Improve adjustment (no lash). b. Oil soaked driven disc. c. Worn facing or facing torn from disc. d. Warped pressure plate or flywheel. e. Weak diaphragm spring. f. Driven plate not seated in. g. Driven plate overheated.	a. Adjust linkage to spec. b. Install new disc and correct leak at its source. c. Replace disc. d. Replace pressure plate or flywheel. e. Replace pressure plate. (Be sure lash is checked before replacing plate.) f. Make 30 to 40 normal starts. CAUTION: Do Not Overheat. g. Allow to cool–check lash.
Grabbing or Chattering	a. Oil on facing. Burned or glazed facings. b. Worn splines on clutch gear. c. Loose engine mountings. d. Warped pressure plate or flywheel. e. Burned or smeared resin on flywheel or pressure plate.	a. Install new disc and correct leak. b. Replace transmission clutch gear. c. Tighten or replace mountings. d. Replace pressure plate or flywheel. e. Sand off if superficial, replace burned or heat checked parts.
Rattling-Transmission Click	a. Weak retracting springs. b. Release fork loose on ball stud or in bearing groove. c. Oil in driven plate damper. d. Driven plate damper spring failure.	a. Replace pressure plate. b. Check ball stud and retaining. c. Replace driven disc. d. Replace driven disc.
Throwout Bearing Noise with Clutch Fully Engaged	a. Improper adjustment. No lash. b. Release bearing binding on transmission bearing retainer. c. Insufficient tension between clutch fork spring and ball stud. d. Fork improperly installed. e. Weak linkage return spring.	a. Adjust linkage. b. Clean, relubricate, check for burrs, nicks, etc. c. Replace fork. d. Install properly. e. Replace spring.
Noisy	a. Worn release bearing. b. Fork off ball stud (heavy clicking). c. Pilot bearing loose in crankshaft.	a. Replace bearing. b. Install properly and lubricate fork fingers at bearing. c. See Section 6 for bearing fits.
Pedal Stays on Floor When Released	a. Bind in linkage or release bearing. b. Springs weak in pressure plate. c. Springs being over traveled.	a. Lubricate and free up linkage and release bearing. b. Replace pressure plate. c. Adjust linkage to get proper lash, be sure proper pedal stop (bumper) is installed.
Hard Pedal Effort	a. Bind in linkage. b. Driven plate worn.	a. Lubricate and free linkage. b. Replace driven plate.

Figure 7-2. Clutch diagnosis chart. (General Motors)

improper linkage adjustment and lack of lubrication at the moving parts of the linkage.

In some cases, the car creeps during the clutch spindown, wherein the disc continues to spin for several seconds after being disengaged. To determine if this is causing the creeping and not clutch drag, a simple test can be performed (if the transmission is not *fully synchronized*). Depress the clutch pedal, shift into neutral, and wait about 30 seconds. Then, shift into an *unsynchronized gear,* usually reverse. If the gear engages smoothly, the creeping is caused by spindown, which is normal. If the gear clashes, the creeping is caused by clutch drag. (Consult Chapter 8 for explanation on synchronized gears.)

Clutch drag can be caused by a clutch linkage that is loose, disconnected, or inoperable. The clutch fork may be disconnected from the throwout bearing, inside of the clutch housing. A hydraulic linkage may be low on fluid–the reservoir may be empty, or the cylinders or lines may be leaking. There might be air in the system. A disconnected linkage or clutch fork or a dry hydraulic system is characterized by a pedal that requires little, if any, effort to depress. A *spongy* pedal is the result of air in a hydraulic linkage.

Another possible cause of clutch drag is misalignment of the transmission or clutch housing. This can happen as a result of loose bolts or as a result of debris between the mounting faces. Misalignment can cause the linkage or internal clutch parts to stick temporarily. Pedal effort will be normal or higher than normal.

Clutch drag could be caused by a worn or frozen pilot bearing, a sticking or warped clutch disc, or a warped pressure plate or flywheel. A worn disc may become so overheated that it welds itself to the pressure plate. When these defects occur, the clutch linkage is working properly, but the engine and transmission remain mechanically connected.

One of the most common causes of a dragging clutch is too much pedal free travel. With excessive free travel, the pressure plate may back off somewhat, but it will still retain contact with the disc when the pedal is pushed to the floor. Always check clutch linkage adjustment when indications point to a dragging clutch.

Clutch Slippage

Clutch slippage is a condition wherein the full power of the engine does not reach the transmission. The engine speeds up, but the vehicle speed does not increase as it should. The friction disc is not being gripped firmly, and so, it slips between the flywheel and pressure plate as the members rotate.

Specifically, in this section, we are referring to *abnormal* clutch slippage. For smooth shifts, some degree of slippage is desirable. The clutch pedal should not be released too quickly. A controlled release will permit some slippage. As a result, you will not ruin the clutch or other parts of the drive train. In addition, it makes the ride easier on the vehicle occupants.

Clutch slippage begins as a minor problem. At first, it will occur upon initial and hard accelerations. As the problem progresses, slippage will occur on upshifts, downshifts, and upon any kind of acceleration. Left uncorrected, slippage only becomes worse. Eventually, the clutch disc becomes so badly worn that there is not enough friction present to move the vehicle.

A slipping clutch will get very hot. The extreme heat will damage the contact surfaces of the pressure plate and flywheel, causing them to lose their proper finish. This happens because the heat changes the chemical makeup of the metal, reducing strength and heat absorption capability. A slipping clutch should be repaired before the pressure plate and/or flywheel are damaged by excessive heat.

A badly slipping clutch is usually too glazed and worn down to be fixed by adjustment. A clutch disc that is too glazed cannot develop any driving friction between itself and the flywheel and pressure plate. If the clutch disc is completely worn down, the apply springs will not be able to apply enough pressure to the pressure plate to hold it tight against the flywheel. The clutch must be replaced.

A clutch disc that is soaked with oil, such as from a leaking rear main bearing seal, will also slip. In this case, the clutch may not be worn down. If it is soaked with oil, however, it must be replaced.

A clutch will often slip if ever the vehicle is driven through deep water. This is because water enters the clutch housing and soaks the friction disc. The clutch will usually work normally after it dries.

Slippage can also be caused by disc, pressure plate, or flywheel contact surfaces that are warped. If the surface is not perfectly flat and straight, the flywheel or pressure plate will not make good contact with the disc. As a result, it may slip (or chatter). Warped surfaces may be caused by improper manufacturing processes or by excessive heat.

Sometimes, a clutch will slip because the pressure plate apply springs are weak. Weak springs may have been installed in the pressure plate assembly originally, or the assembly may have been intended for a smaller engine and/or vehicle. The springs may have become weak from overheating, or under normal circumstances, the springs may have become weak because of long usage. A clutch pedal that depresses with almost no effort may be a sign of weak clutch apply springs, in which case, there is likely to be slippage.

Clutch slippage can be caused by improper clutch linkage adjustment. In this case, the clutch linkage might be adjusted to where there is no free play, and the throwout bearing is applying pressure to the release levers of fingers. As a result, the clutch is not fully applied, and it will slip.

Note that if the clutch fork return spring is disconnected, the pedal will have no free play. However, this condition will not cause slippage. When the clutch pedal is released after being applied, the apply springs will fully reengage the clutch. The throwout bearing will ride on the release levers or fingers but will not apply pressure on them to cause slippage. Therefore, do not confuse this condition with the lack of free play caused by misadjustment.

Grabbing Clutch

A ***grabbing clutch*** is one that engages with a jerk or shudder, no matter how slowly and carefully it is applied. The effect may be a series of jerks, which is often described as *bucking.* This problem may be severe enough to cause damage to universal joints or other drive train parts.

In the event of this problem, always check the engine and transmission or transaxle mounts before troubleshooting the clutch. Loose mounts can cause a sensation similar to a grabbing clutch. Also, make sure that all clutch housing and transmission bolts are tight.

A grabbing clutch is usually caused by oil deposits disbursed on the friction facing of the clutch disc or by an overly hot or glazed facing. Less common causes are: worn splines on the clutch disc hub flange or transmission input shaft; a warped clutch disc; warped pressure plate or flywheel machined surfaces. Another possible cause is a worn or misaligned clutch fork. This can cause the release bearing to become crooked, or tip on its shaft. The result is an uneven application of pressure on the pressure plate assembly. The bearing collar may begin to wear because of the uneven pressure from the fork.

Clutch Chatter

Clutch chatter is a specific type of vibration that is accompanied by a rapid clunking or rattling noise. It occurs while the clutch is *being* engaged, as opposed to while the clutch is *fully* engaged. It is closely related to clutch grabbing. It differs, however, in that it is *heard* more than *felt,* the actual noise being caused not by the clutch but by the vibration set up in the clutch linkage and drive train. Chatter can be caused by loose or misaligned drive train components. For example, the transmission and clutch housing may be seriously misaligned.

Often, what seems like clutch chatter is really a worn out constant-velocity, or other, universal joint. Other causes of chatter include a misaligned flywheel and/or worn pilot bearing. It is possible that release levers are unevenly adjusted, worn, or both. Sometimes, dust from a worn clutch disc facing may clog the disc segments and cause chatter.

Clutch Vibration

Some clutch defects can result in a vibration that can be felt inside of the vehicle while the clutch is *fully* engaged. This ***clutch vibration*** will vary with engine speed. It differs from chatter, which occurs when the clutch is in the process of being engaged or disengaged. Often, clutch vibration is accompanied by noise; however, the noise is a secondary symptom.

Several checks must be made to pinpoint the source of the vibration. This is to determine if the problem is indeed clutch vibration or vibration from some other source.

Vibration with the vehicle moving may be caused by the engine or any part of the drive train. To narrow down the possibilities, determine when the problem occurs. Does the vibration occur only when the vehicle is moving? If so, the problem is probably not clutch related. Even so, you should check the clutch housing where it attaches to the engine and to the transmission, just to make sure the attaching bolts are not loose. Thoroughly check the drive shaft assembly, drive axles, and engine mounts. Closely inspect conventional or constant-velocity U-joints for any signs of wear. In rare cases, internal problems in the transmission or differential or the transaxle may cause vibration while the vehicle is moving. (Refer to later chapters related to these components.)

If the vibration also occurs with the vehicle stopped (transmission in neutral), the source of the problem is related to the clutch or engine. Make sure that the engine does not have a dead cylinder or other internal problem. Vibration resulting from excessive crankshaft end play can

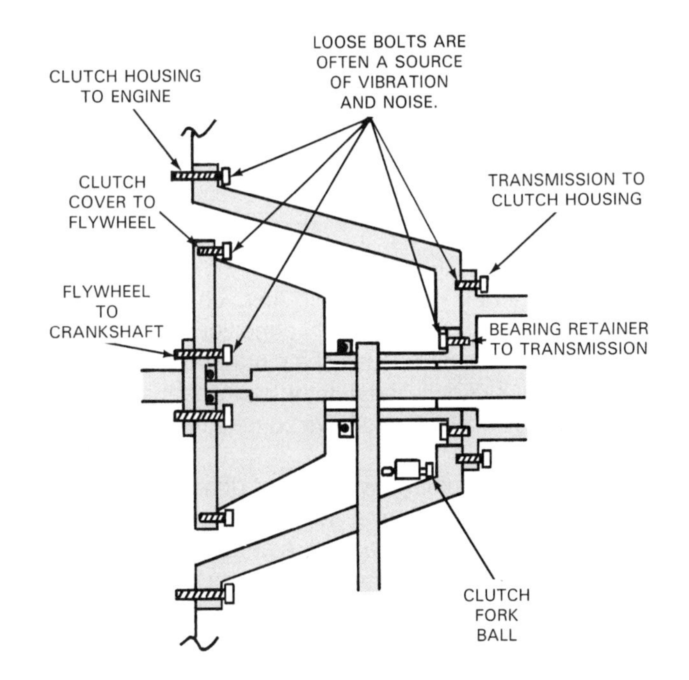

Figure 7-3. Note places where bolts can loosen, causing vibration (and noise). Often, bolts can be tightened to solve the problem. Occasionally, the vibration will damage other parts before the bolts are tightened.

be identified by disengaging the clutch. Pressure from the throwout bearing will push the flywheel and crankshaft forward, taking up the end play and eliminating any corresponding vibration.

If the engine is okay, look for problems indirectly related to the clutch; first, however, check the clutch housing where it attaches to the engine and to the transmission. Check the front bearing retainer where it attaches to the transmission. Ensure that all bolts are tight. Figure 7-3 shows the places where bolts can loosen and cause vibration. Look at the transmission front bearing as a possible cause of vibration.

Inspect the clutch assembly if the foregoing checkpoints have failed to identify the source of vibration. There are two general sources of clutch vibration, as outlined below:

- Some part associated with the clutch assembly can be out of balance. The parts that are heavy enough to be seriously unbalanced are the flywheel and pressure plate. This is more often found after new parts are installed or the flywheel is resurfaced. The chance of a flywheel or pressure plate becoming seriously unbalanced after long use is only slight.
- Some part associated with the clutch assembly is loose or broken. Check attaching bolts of the clutch assembly. The *flywheel attaching bolts* (flywheel-to-crankshaft bolts) may be loose on the crankshaft, or there may be dirt or metal burrs between the crankshaft flange and flywheel. This will cause misalignment of the flywheel. When this happens, the clutch will usually have other symptoms. The clutch may slip or make a knocking noise.

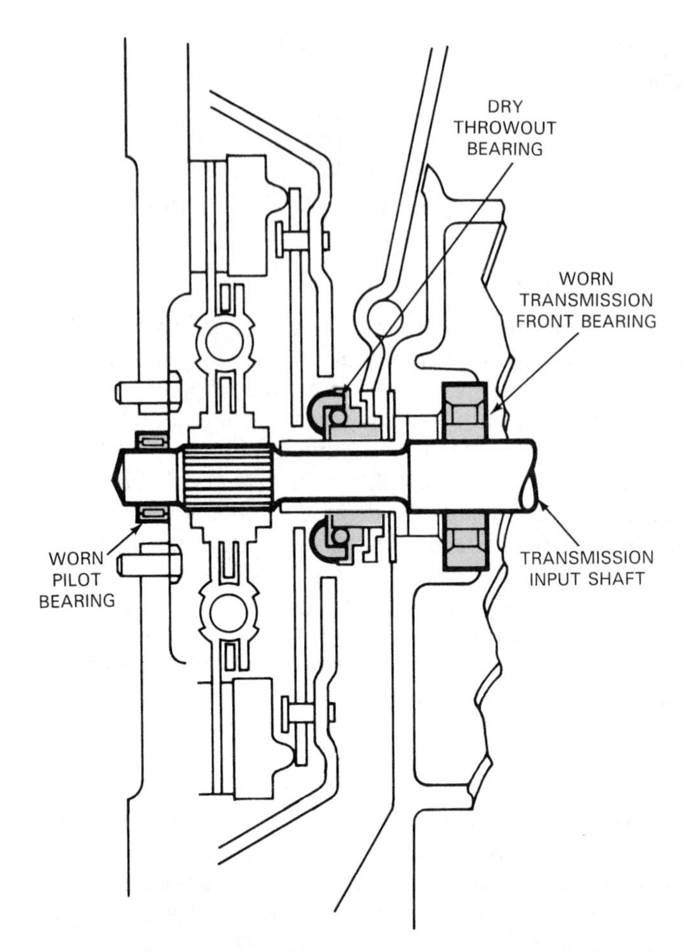

Figure 7-4. Three bearings associated with the clutch are the pilot bearing, the throwout bearing, and the transmission front bearing. Each can be detected as a source of noise by operating the clutch pedal. Wearing or lack of lubrication can cause the bearings to make noise. (Luk)

Abnormal Noise

Often, the clutch is blamed for a noise when some other component is really at fault. Any vehicle component can become noisy under the right conditions. An *abnormal noise* in the clutch can usually be singled out by applying or releasing the clutch pedal. If the clutch is the source of the noise, this action should have some effect on the noise being made.

Rumbling, squealing, whirring, or grinding noises can be caused by a defective transmission front bearing, throwout bearing, or pilot bearing. A defective transmission front bearing will commonly make noise as the clutch drives the input shaft. This happens whenever the clutch is engaged. A defective throwout bearing will make noise as the bearing is applied. A noisy pilot bearing will usually be heard when the clutch is disengaged completely. It will not usually make noise when the clutch is engaged, since the crankshaft and input shaft are turning at the same speed. See Figure 7-4.

The following procedure can be used to isolate defective bearings. Follow these steps in sequence.

1. Depressing the clutch pedal, start the engine and leave it running. With the transmission in neutral, release the clutch pedal, taking your foot completely off the pedal. Any noise that you hear upon engaging the clutch will likely be coming from the transmission front bearing.
2. Push the clutch pedal *only* until all free play is removed. This causes contact with the release levers or fingers, allowing the throwout bearing to rotate without disengaging the clutch. If you hear a noise in this position, typically a whirring or grinding sound, the throwout bearing is probably defective.
3. Push the clutch pedal to the floor. If a noise–typically, a squealing or howling sound–occurs upon disengagement, the pilot bearing is probably defective.

Sometimes, the clutch assembly will have a rattle, especially when the clutch is fully engaged and there is no foot pressure acting on the linkage. A possible source of the rattle is a clutch fork that is loose on its pivot ball. The rattle may be caused by too much clearance between the fork and the groove in the throwout bearing. A loose fit between parts of a rod-and-lever linkage or a missing tension spring can cause the mechanism to rattle. It may be possible to tighten the clutch fork retainer clips or springs to reduce clearances and quiet the linkage.

Rattles can also be caused by weak pressure plate retracting springs, found in the diaphragm-spring pressure plate assembly, or weak or broken clutch disc cushion springs. Rattles stemming from these reasons are usually loudest when the clutch is disengaged.

Loose flywheel attaching bolts will cause movement at the mounting surfaces of the crankshaft and flywheel. This movement causes a deep knocking noise that is often mistaken for an engine *main bearing* or *rod bearing* knock. The noise may be accompanied by vibration. It can usually be reduced by disengaging the clutch. Pressure from the throwout bearing, acting through the release levers or fingers and the clutch cover, will press the flywheel tightly against the crankshaft, quieting the knock.

Hard Pedal Effort

Hard pedal effort is noticed when the clutch pedal is excessively hard to push down. If the pedal is hard to depress, the problem is usually in the linkage. If the driver is strong enough to depress the pedal, it may stay on the floor or rise very slowly when released. Components may wear quickly.

To isolate the cause of the problem, the first step is to disconnect the clutch linkage at the clutch fork and try to move the pedal by hand. If the clutch pedal does not move easily, the problem is somewhere in the linkage.

Where rod-and-lever linkages are concerned, hard pedal effort is due to lack of lubrication or to some part of the linkage that is bent or is loose enough to jam. Sometimes, a rock or other material gets thrown up and jams the linkage.

Clutch pedals of cable linkages usually are hard to depress because moisture has entered the cable sheath. Moisture washes out the cable lubricant and causes corrosion. It is also possible that the cable was misrouted during installation and needs rerouting. If the cable is kinked, as a result of misrouting, it should be replaced.

Clutch pedals of hydraulic linkages are hard to depress because of the master or slave cylinder. The piston sometimes sticks due to internal corrosion. The usual cause, however, is swollen piston seals. This problem is a result of using a petroleum-based oil, such as automatic transmission fluid, instead of DOT 3 brake fluid to refill the reservoir. Petroleum-based oils will cause the seals to swell, jamming the piston in the cylinder. The seals must be replaced.

A less common cause of hard pedal effort in a hydraulic system is a swollen hydraulic hose. If the hose swells so that the fluid passageway becomes fairly constricted, it will take longer to actuate the slave cylinder piston. The swollen hose will slow the disengaging of the clutch. It will also slow reengaging of the clutch and pedal return.

If the clutch linkage worked as it should when it was disconnected–that is, if it moved easily–the problem is in the clutch assembly. The two most common causes are a throwout bearing collar that is sticking on its hub or binding in the clutch fork. This usually occurs after the vehicle has been operated in deep water, wherein water entered the clutch housing and washed off the lubricant.

If the pedal is very hard to work after a new pressure plate assembly is installed, the pressure plate apply springs may be stronger than necessary. This is very common when a heavy-duty or competition (racing) pressure plate is installed. Return springs that are too heavy might be responsible for hard pedal effort. Remember, as you depress the clutch pedal, you are working against return-spring tension.

In-Car Clutch Service

Clutch service commonly requires removal of major components such as the driveline, transmission, and clutch assembly. However, certain clutch problems can be attended to without need of removing these major components. This type of service is categorized as ***in-car service.*** Clutch repairs that fall into this category are clutch adjustment and linkage repairs or replacement. Procedures are given in this section.

Clutch Adjustment

Clutch adjustment involves setting the correct amount of free play in the clutch linkage. Too *much* free play could cause clutch drag, preventing the clutch from ever fully disengaging. In this case, the power flow is not completely cut off from the clutch, and the engine could continue to propel the vehicle. Too *little* free play could cause the clutch to slip. In this case, the throwout bearing rotates continually as it contacts the rotating pressure plate release levers or fingers, and it quickly wears out. With enough pressure on the levers or fingers, the clutch will slip, overheat, and be destroyed.

No matter what type of clutch linkage is used, the free play can be checked by pushing the clutch pedal with your hand. Refer to Figure 7-5 and the next few paragraphs for the proper procedure.

Begin with pedal in fully released position–that is, with the pedal all the way up. Start pushing down on the pedal. The pedal should move easily for about 1 in. (25 mm). In this span, the only pressure that you are working against is that of the return springs. Past this point, the pedal will become harder to push. This is where the throwout bearing contacts the pressure plate release levers or fingers and

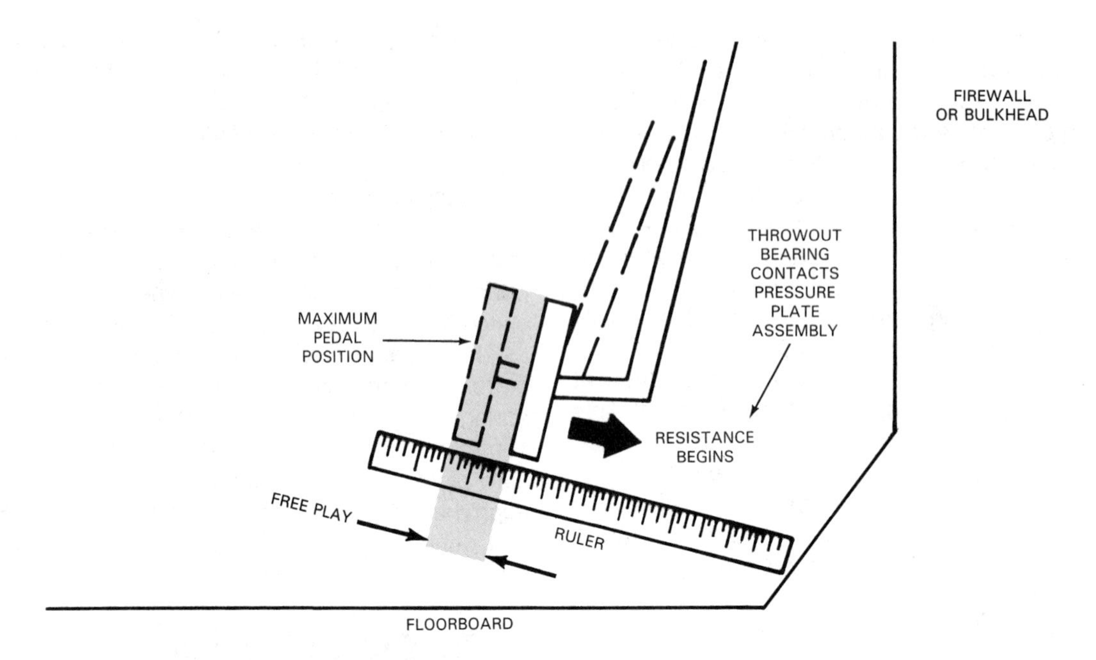

Figure 7-5. Free play can be measured using a ruler. The amount of free play should match the manufacturer's specifications. A general check can be made by observing the pedal movement. Throwout bearing first contacts pressure plate assembly at point where resistance is first felt when pressing on pedal.

you are working against the force of the pressure plate apply springs.

The distance that the pedal moved from the fully released position until it became hard to push is the free play. All vehicle manufacturers specify the proper amount of clutch free play. Free play is adjusted at some place on the clutch linkage. On many vehicles, the clutch has a self-adjusting feature, in which free play is automatically adjusted whenever the clutch pedal is depressed.

Note that a slipping clutch that is badly worn or damaged cannot be repaired by adjustment. If adjustment does not stop clutch slipping, or if enough free play cannot be obtained (indicating worn parts), the clutch assembly must be removed, and worn parts must be replaced.

Rod-and-lever linkage adjustment

A rod-and-lever linkage is a series of links, levers, and rods connecting the clutch pedal to the clutch fork. The means of adjustment is usually provided by a threaded rod that passes through a pivot block. There will be one or two locknuts, Figures 7-6 and 7-7. The adjustment device is located on the linkage at the clutch fork.

To adjust a rod-and-lever linkage, first measure the actual amount of free play at the clutch pedal and compare it against specifications. Use a ruler to make an accurate measurement. If the specifications are not available, 1 in. (25 mm) of free play will usually be close. Before attempting to make the adjustment, check that the linkage is not worn excessively. A loose, sloppy linkage cannot be satisfactorily adjusted.

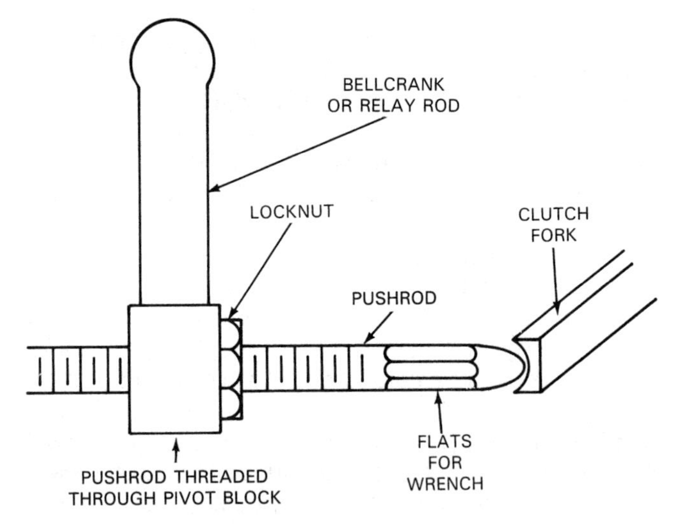

Figure 7-6. This shows a single-locknut adjuster. The locknut is loosened, and the pushrod is turned in or out, to obtain the proper free play.

Next, for the single-locknut adjuster, loosen the locknut that holds the threaded rod to the pivot block. Turn the rod until the proper free play is obtained at the clutch pedal, then, retighten the locknut. For the double-locknut adjuster, loosen the one locknut, depending on which way the adjustment is to be made. Then, turn the other in the same direction, advancing the pushrod, until the proper free play is obtained at the clutch pedal. Retighten the first locknut. Recheck the free play at the clutch pedal. As a final check, road test the car to check clutch operation.

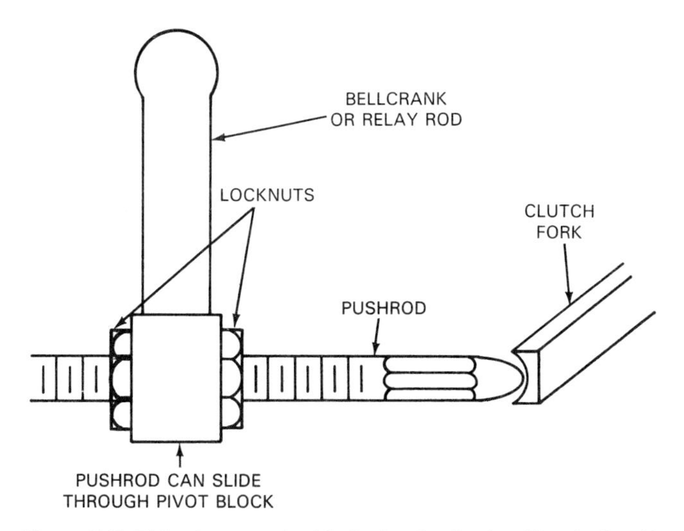

Figure 7-7. This shows a double-locknut adjuster. One locknut is backed away; the other is turned in the same direction to adjust the pushrod length. Note that pushrod slides through hole in pivot block.

Cable linkage adjustment

The cable linkage uses a cable to connect the clutch pedal to the clutch fork. Adjustment is made by a threaded section, which passes through a bracket. This section is located on the end of the cable linkage, at the clutch fork. See Figure 7-8.

Some cables are self-adjusting. If there is no free play on a self-adjusting clutch, the adjustment mechanism is faulty, or the clutch is worn out. Note that in many vehicles with self-adjusting clutch systems, the throwout bearing is always in contact with the release levers or fingers. Such bearings are called ***constant-duty throwout bearings.*** With the clutch pedal fully released, the bearing makes light contact with the release levers or fingers. Even so, there is enough slack in the cable linkage to give some free play at the pedal.

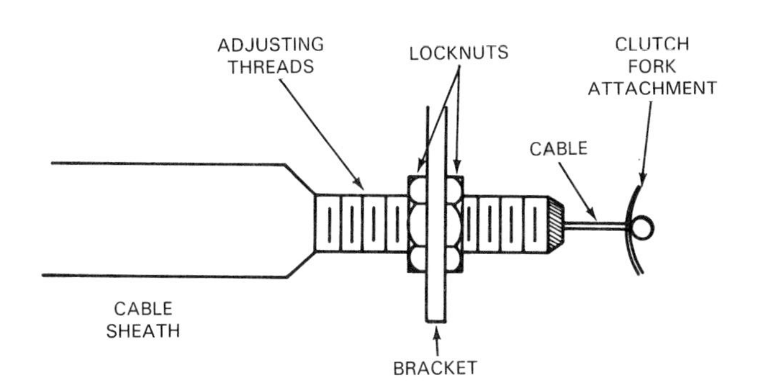

Figure 7-8. Adjusting free play in a cable linkage. The end of the cable is threaded. Loosening the locknuts and moving the cable makes the adjustment.

If the clutch is the manually adjustable type, use a ruler to measure the actual amount of free play at the clutch pedal. Compare the amount against the manufacturer's specifications. If the specs are not available, you can assume that the free play is between 1 in.–1.5 in. (25 mm–38 mm). Also, check that the cable and other linkage works freely and does not bind.

If the free play must be adjusted, loosen the locknut holding the threaded portion of the cable sheath. There are usually two locknuts holding the end of the cable to a bracket. Leave the locknuts loose and alternately turn them until the cable length is correct. Tighten the locknuts and recheck the free play at the clutch pedal. Then, road test the car and check clutch operation.

Hydraulic linkage adjustment

The hydraulic linkage uses hydraulic pressure to transfer pedal movement to the clutch fork. As with the previous adjusters, adjustment is made by a threaded section on the linkage, at the clutch fork. Adjustment to this linkage changes the effective length of the pushrod at the slave cylinder. See Figure 7-9.

Measure the amount of free play at the clutch pedal. Use a ruler to obtain an accurate reading and compare it against specifications. If the specifications are not available, between 1 in.–1.5 in. (25 mm–38 mm) of free play is generally acceptable.

If the free play must be adjusted, first, check the master cylinder reservoir to make sure that the fluid level is sufficient. Add DOT 3 brake fluid if necessary.

CAUTION! *Never* add motor oil, transmission fluid, or any kind of petroleum-based oil to the master cylinder reservoir. Use brake fluid only to prevent major damage to seals!

Loosen the locknut holding the threaded rod to the adjuster. Turn the rod until you obtain the proper free play, as measured at the clutch pedal. Retighten the locknut. Recheck the free play at the clutch pedal. Then, road test the car.

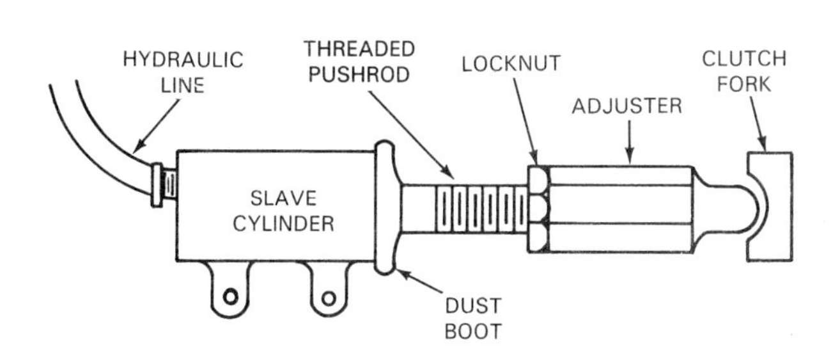

Figure 7-9. Adjusting free play in a hydraulic linkage. Pushrod is often threaded. Shortening or lengthening rod adjusts free play. Pushrod is held by a locknut.

Clutch Linkage Repair and Replacement

Check the clutch linkage for proper operation. One of the most common points of wear in the linkage is the pedal itself. Bushings can wear, or the mounting bracket bolts can loosen. Always check these parts when servicing the clutch linkage or when overhauling a clutch. Figure 7-10 shows typical clutch pedal components.

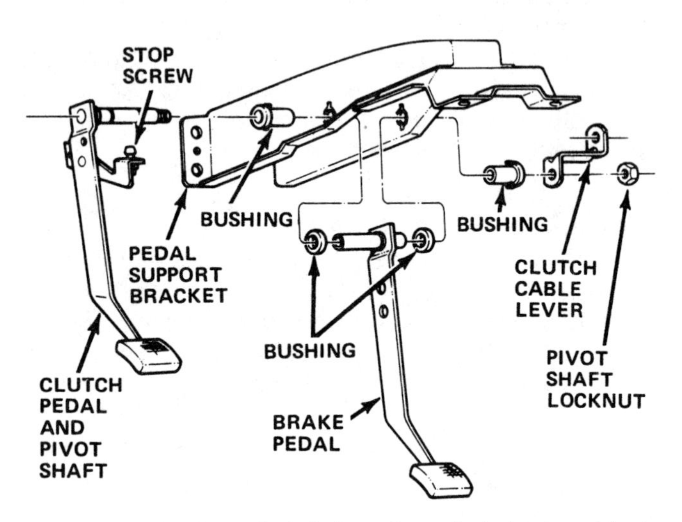

Figure 7-10. Study exploded view of a typical clutch and brake pedal assembly. Pivot points should be carefully checked and lubricated. (Chrysler)

Rod-and-lever linkage

A rod-and-lever linkage usually wears at pivot points. See Figure 7-11. Many pivot points contain metal or plastic bushings. These bushings often wear out.

Sometimes, linkage rods or levers will bend. This may happen if something in the clutch causes excessive resistance in disengagement. Another common cause of problems is a bellcrank that binds or is loose because of worn motor mounts or misalignment of the engine in the vehicle. The linkage must be restored to perfect operating condition before the clutch will operate properly. This may involve lubrication, adjustment, or replacement of linkage parts.

Cable linkage

Cable linkage problems are usually due to a seized or binding cable assembly. Cables can seize due to corrosion or lack of lubrication. A cable may seize or bind if it is caused to kink in its sheath from improper routing. Defective cables should be replaced.

If a cable must be replaced, make sure you route it so there is no possibility of seizing or binding. Use all the original cable brackets and mounting locations if possible. A typical cable arrangement is shown in Figure 7-12.

Hydraulic linkage

Always check the fluid level in the master cylinder reservoir first. If the fluid level is low, check for leaks at the

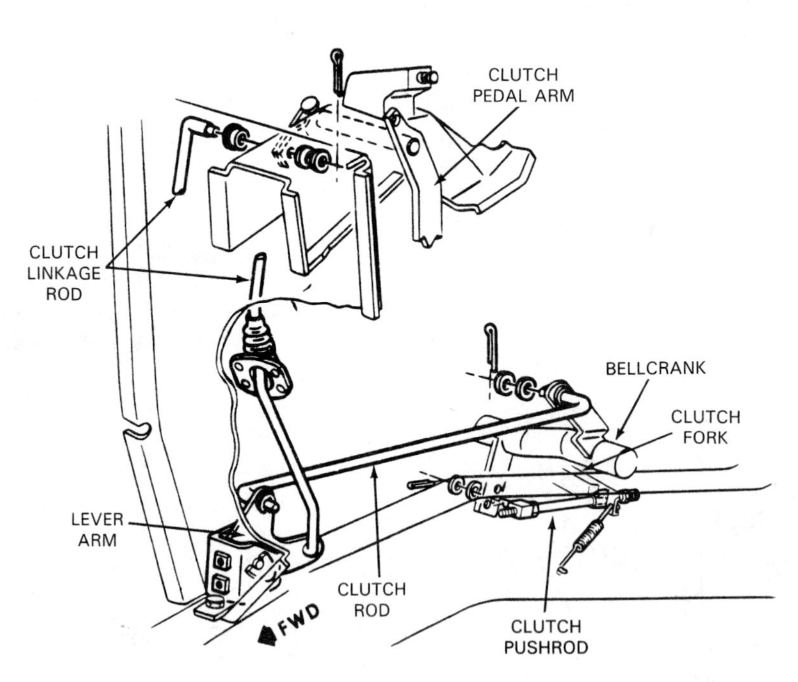

Figure 7-11. Notice pivot points in this exploded view of a rod-and-lever linkage. Most problems in this system are caused by wear, misadjustment, or lack of lubrication at the pivot points. (General Motors)

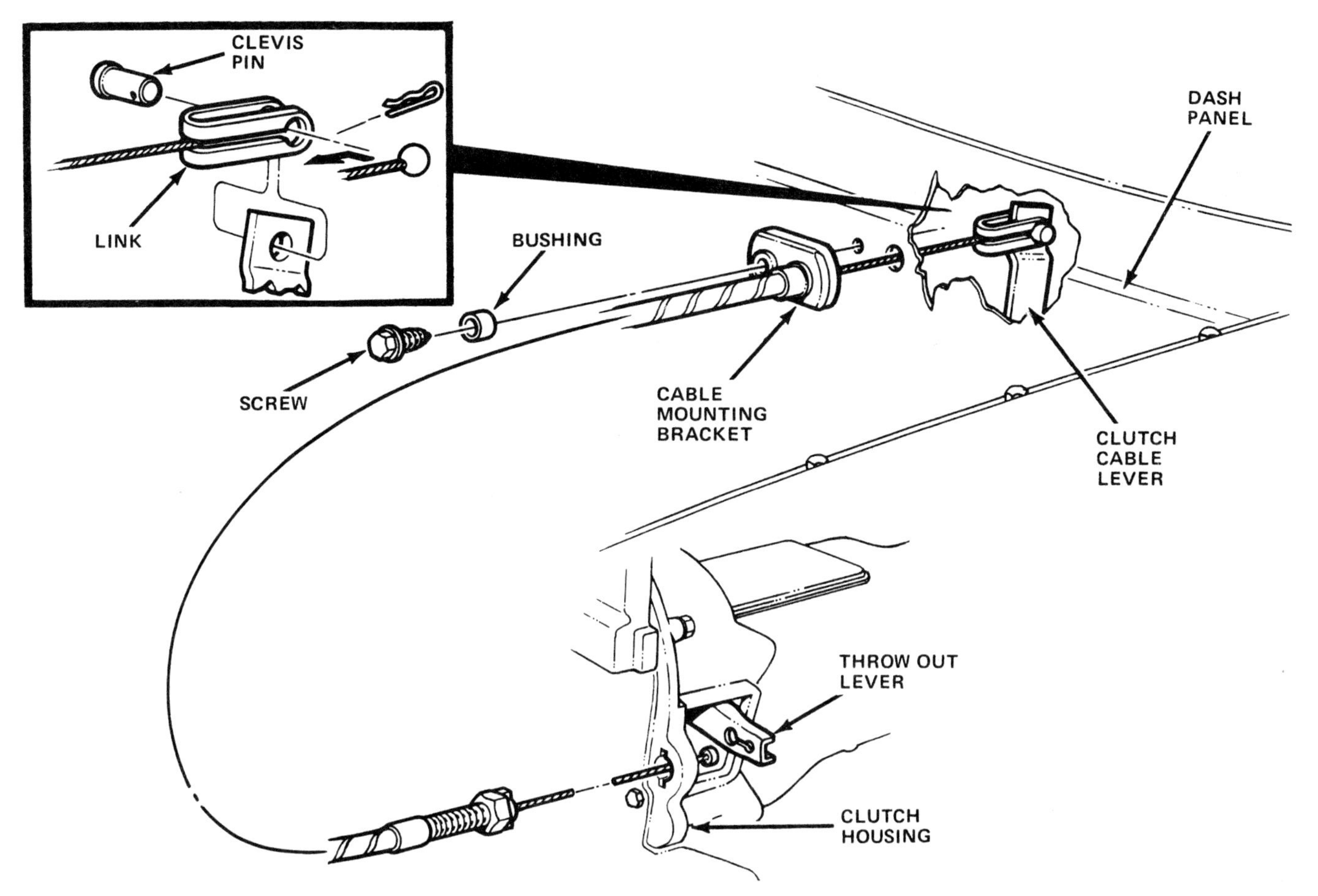

Figure 7-12. This is a typical cable linkage. A replacement cable must be routed so that it does not seize or bind. The original routing is normally the best. (Chrysler)

places indicated in Figure 7-13. Leaks will appear as dark stains on hydraulic system parts. Sometimes the brake fluid, used in the system, is visible at the point of the leak. Leaks can occur in the hydraulic lines or at loose connections. The most common leaks, however, occur at the master or slave cylinder seals.

It is usually easier and cheaper to replace leaking master and slave cylinders. However, in some cases, a replacement cylinder is not available, or the cost is excessive. If a new or remanufactured master or slave cylinder cannot be located, the old cylinder can be rebuilt with the proper seal kit.

Rebuilding the cylinders. Rebuilding a master or slave cylinder is relatively simple. It is similar to rebuilding brake system cylinders. Refer to the exploded views of typical master and slave cylinders in Figure 7-14. Following is a typical rebuilding procedure.

First, remove any rubber boot that covers the free end of the piston. Then, remove the snap ring that holds the piston in the cylinder bore. The piston should slide out of the cylinder. In some cases, the piston will pop out of the cylinder because of a return spring behind it. If the piston is stuck, it can be removed by tapping the cylinder on a wooden block or by carefully applying air pressure to the inlet connection.

WARNING! Air pressure can cause the piston to fly out with great force. Point the open end of the cylinder away from yourself and others and toward rags or some other soft surface before applying pressure.

Once the piston is out, inspect the cylinder for wear or pitting. If the cylinder bore is worn or pitted, the cylinder must be replaced. Remove the old seals from the piston and discard them. Thoroughly clean all parts and allow them to dry.

CAUTION! Do not allow any petroleum-based solvent to remain on the cylinder or piston.

Obtain new seals. Make sure that they are the correct ones. Lubricate them with brake fluid, *never* with petroleum-based lubricants, and install them on the piston. Install the piston in the cylinder. Install the return spring,

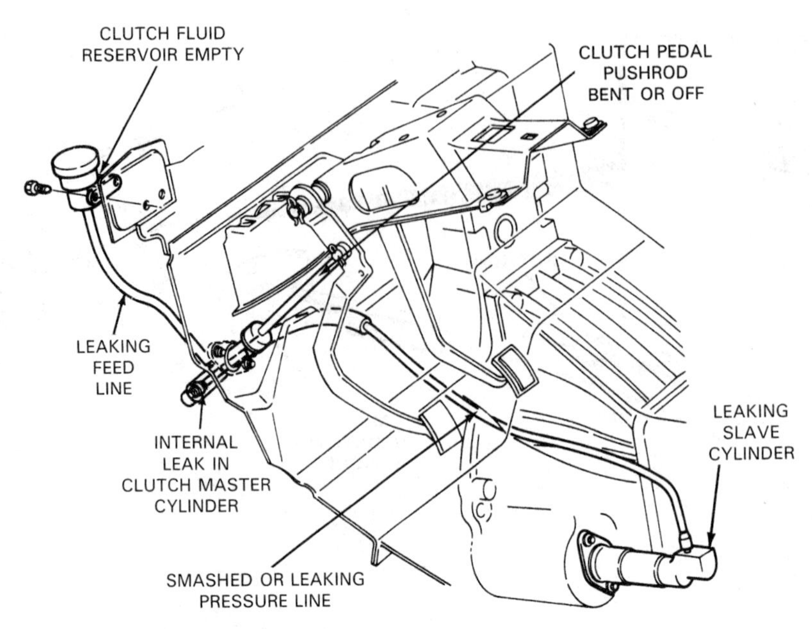

Figure 7-13. Common sources of hydraulic linkage problems are shown. The pushrod for actuating the master cylinder piston may be bent or disconnected, but most problems are caused by low fluid level or by plugged hydraulic lines. Leaks are usually responsible for low fluid levels. Kinks can cause plugging of hydraulic lines. (Ford)

where there is one, also in the cylinder. Be careful not to damage the new seals. Install the retaining snap ring, pushrod, and boot. Then, install the rebuilt cylinder and attach the hydraulic line.

Bleeding the system. There should be no air in hydraulic lines; if there is, it must be removed by bleeding the lines. If the hydraulic system was disassembled for any type of repairs, you must bleed the lines to remove the air. Once the bleeding operation is finished, pressure on the pedal should apply the clutch firmly. There should be no spongy-pedal feel.

The slave cylinder has a ***bleeder valve*** for the purpose of bleeding air. The valve looks like a screw with a small hole running down its length. The hole is closed off when the bleeder valve is tightened down. Loosening the valve a few turns will open the hole and permit bleeding the system of air. Note that some bleeder valves have a cap, as well.

There are two methods to manually bleed air from a hydraulic system. One method is preferred over the other because it minimizes chance of fluid contamination. This preferred method is outlined as follows:

1. Fill the system reservoir with fresh brake fluid. Leave the cover off the reservoir. (Bleeder valve is closed.)
2. If bleeder valve has a cap, remove it at this time. Attach a hose to the end of the bleeder valve. Place the free end of the hose in a clear container.
3. Have an assistant pump several times on the clutch pedal. Then, ask the assistant to hold the pedal down.
4. Open the bleeder valve by loosening it about a quarter or half turn. Make sure that your assistant continues to keep his or her foot on the clutch pedal. The pedal will probably go all the way to the floor.

   Air, or a mixture of air and brake fluid, will come out of the bleeder valve. Keep the outlet of the hose submerged in the fluid as it begins to fill the container. Watch for air bubbles.
5. Close the bleeder valve once the flow stops.
6. Let up on the clutch pedal.
7. Repeat the process until only brake fluid comes out of the bleeder valve. This can be noted by watching the air bubbles emerging from the hose as the system is bled.
8. Top off the system reservoir with fresh brake fluid and put the cover back on. Do not reuse fluid bled from the system as it may be contaminated.

There is an alternative method of bleeding air from a hydraulic system. It can be done without an assistant and without wasting brake fluid; however, expelled fluid, which may be contaminated, is returned to the system. The method is outlined as follows:

1. Open the bleeder valve by loosening it about a quarter or half turn.
2. If bleeder valve has a cap, remove it at this time. Attach a hose to the end of the open bleeder valve. Place the free end of the hose in a clear container of brake fluid.
3. Fill the system reservoir with fresh brake fluid and replace the cap.

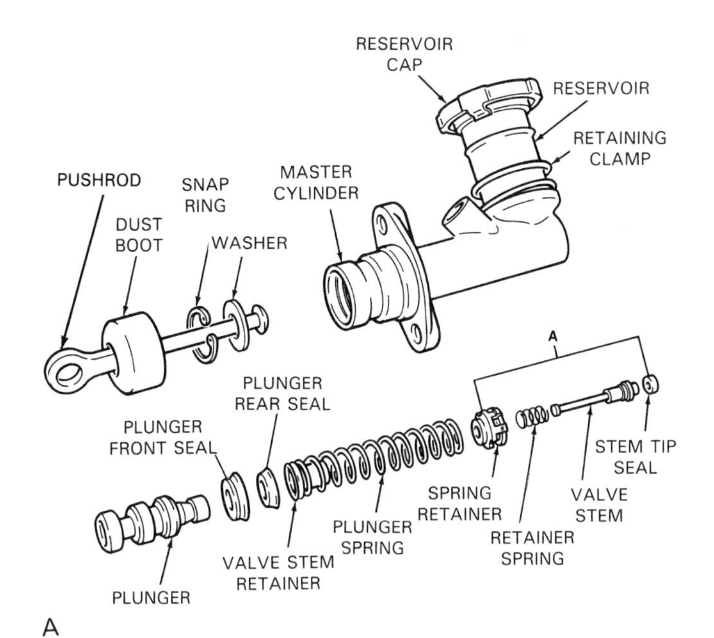

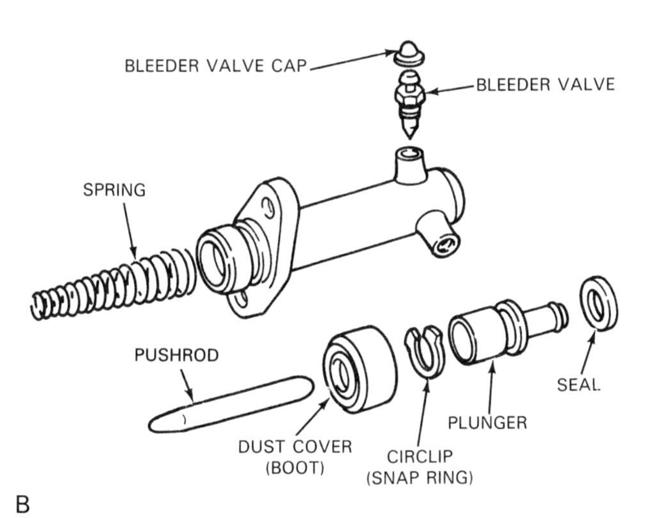

Figure 7-14. Study exploded views of master and slave cylinders. A–A master cylinder is shown. The shape, the included parts, and their layout can vary. The manufacturer's service manual should be consulted for an exact breakdown of parts. B–A slave cylinder is shown. It, too, can vary as to parts, shape, and layout. (Chrysler, General Motors)

4. Pump the clutch pedal until bubbles stop coming out of the hose in the container. You might want to have an assistant watch for the bubbles for you to make the job easier.
5. With clutch pedal released, close the bleeder valve.

Clutch Removal

The exact procedure for clutch removal varies according to the engine and drive train layout. On vehicles with rear-wheel drive and on many vehicles with front-wheel drive, the transmission and clutch are removed from the vehicle without removing the engine. On some front-wheel drive and on most rear-engine vehicles, the engine, clutch, and transaxle are removed from the vehicle as an assembly. When removed from the car, the components are then separated to expose the clutch assembly. On other front-wheel drive vehicles, the engine and clutch are removed, leaving the transaxle in the vehicle. On a few vehicles, the clutch can be removed and replaced without removing any other components.

Following are some *general* procedures detailing how to remove a clutch from a vehicle. You should always refer to the manufacturer's service manual for specific procedures.

Front-Engine, Rear-Wheel Drive

The following procedure details the proper way to remove a clutch from a front-engine, rear-wheel drive vehicle.

1. Disconnect the battery negative cable, Figure 7-15.

WARNING! *Always* disconnect the battery ground cable before working near the clutch. Otherwise, if the starter is accidentally operated, you could be severely injured.

2. Raise the vehicle with an approved hoist or hydraulic jack. If using a hydraulic jack, be sure to install good quality jackstands under the vehicle frame before getting under the vehicle.

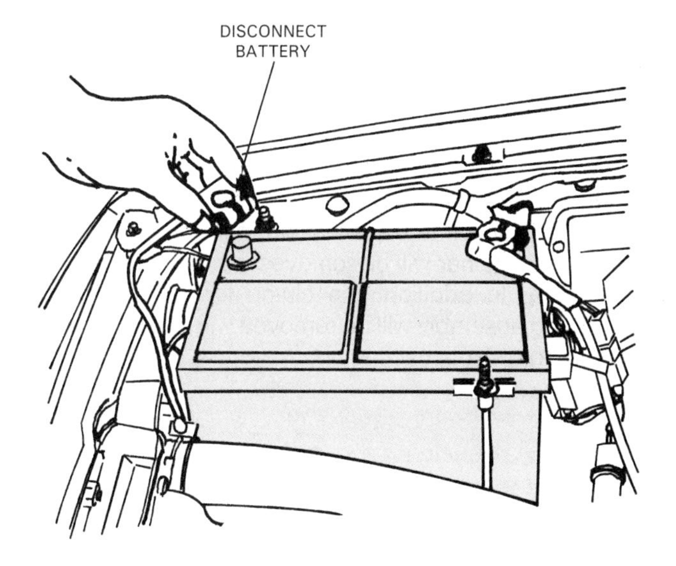

Figure 7-15. The battery should always be disconnected at the negative terminal before beginning any clutch repairs. If the battery is not disconnected, there is always a possibility that the starter will be operated, causing injury. (Subaru)

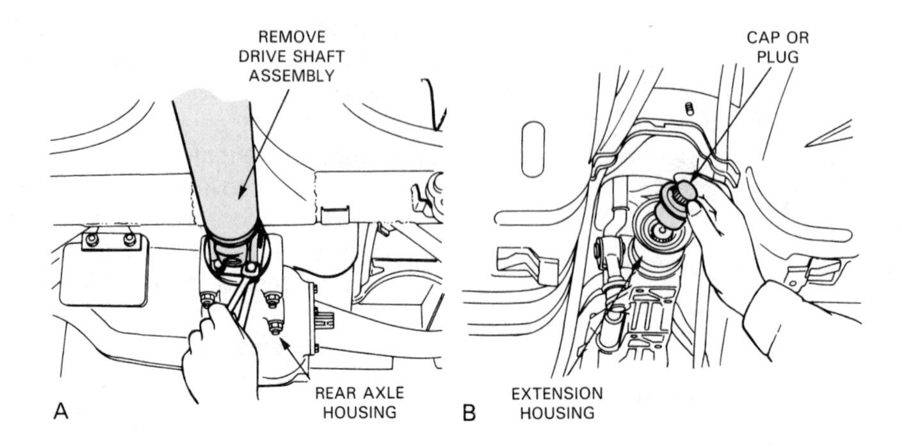

Figure 7-16. First steps to remove clutch. A–Remove drive shaft assembly. B–Cap transmission extension housing to prevent oil loss. Oil leakage can create a safety hazard. (Subaru)

WARNING! *Never* work underneath a vehicle supported only by a hydraulic jack. Always support the vehicle with jackstands.

3. Once the vehicle is properly raised and supported, remove the driveline. (Chapter 15 covers this in detail.) Cap the rear of the transmission to prevent oil dripping. See Figure 7-16.

WARNING! If the vehicle has been operated recently, the engine and exhaust system will be hot. Be careful when working on a hot engine.

4. Disconnect the pushrod or cable and return spring connected to the clutch fork.
5. Remove the transmission. (Chapter 9 covers transmission removal in detail.) On many vehicles, the clutch housing and transmission case are combined in a single casting. In such cases, the clutch housing is removed along with the transmission by removing the unit from the back of the engine.

   Note that the transmission input shaft and front bearing retainer will be removed along with the transmission. In addition, the clutch fork and throwout bearing assembly will be removed with the one-piece, or integral-type, casting.

WARNING! Support the engine when removing the transmission. Usually, the transmission provides some support for the engine. Removing it removes this support.

Also, use a transmission jack, Figure 7-17, to support the transmission. *Never* let the transmission hang unsupported once attaching bolts connecting the transmission to the clutch or engine are removed. This can cause damage to the clutch disc splines or transmission input shaft. Make sure, for this and other reasons, that you support the transmission at all times. If a transmission were to fall, it could cause injury or damage.

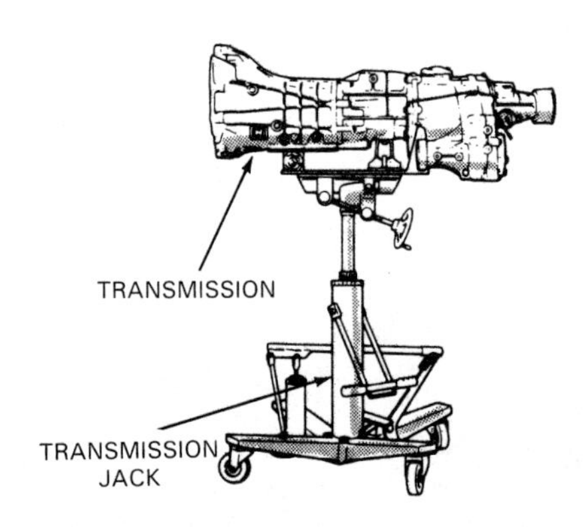

Figure 7-17. Transmissions are heavy. Use a transmission jack to support and lower transmission from vehicle. (Owatonna Tool)

6. Remove as much dust as possible from inside the bell housing with an approved vacuum collection system–designed for use with asbestos fibers, in particular.

WARNING! Some clutch discs contain asbestos–a powerful cancer-causing substance. Avoid breathing dust inside the bell housing or clutch assembly. Do *not* blow dust off with compressed air.

7. Remove the throwout bearing from the clutch fork. If the clutch fork pivots on a ball stud, remove the fork at this time. Most clutch forks are held to the ball stud by spring clips. See Figure 7-18. In many cases, the clutch fork can be removed by pushing (or pulling) the inner end toward the front of the housing while pulling the other end straight out.

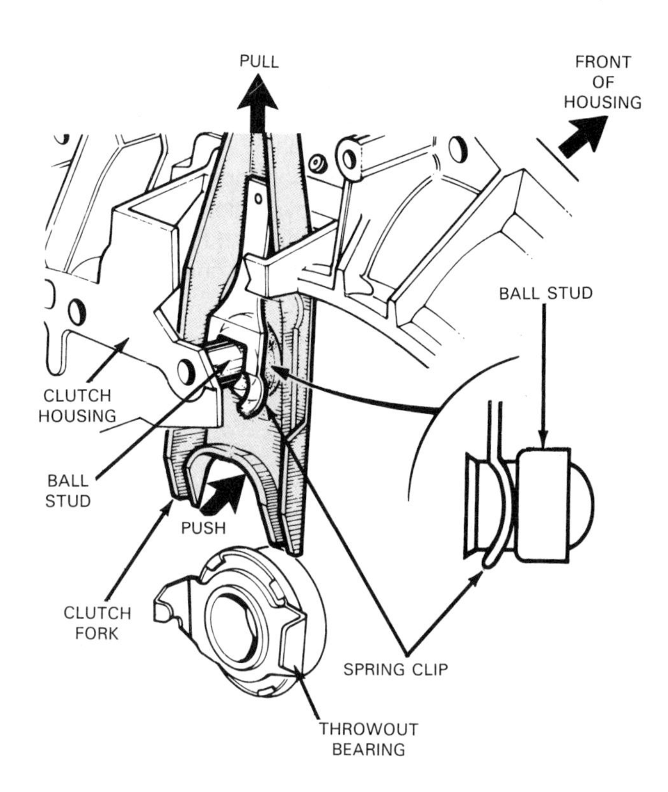

Figure 7-18. Clutch forks that pivot on a ball stud are usually held in place by a spring clip. Remove clutch fork by pushing inner end forward and pulling on outer end. (Chrysler)

8. If still in place, remove clutch housing from the back of the engine. (Integral-type clutch housing was removed with transmission.) To do this, first remove the clutch access cover from the front of the housing. Then, remove the *clutch housing attaching bolts* (clutch housing-to-engine bolts). Vacuum any dust inside the housing.

   Note that housing removal is not necessary in every case. In some vehicles, the clutch assembly may be removed through an opening provided by removal of a clutch access cover.
9. If the pressure plate assembly and flywheel do not have dowels or offset bolt holes for alignment purposes, use a punch to mark the original position of the clutch cover relative to the flywheel. Beforehand, vacuum any dust from the clutch. Typical punch marks are shown in Figure 7-19.
10. Insert a pilot shaft, or clutch alignment tool, through the clutch disc hub and into the pilot bearing, if desired. This will keep the clutch disc from falling out of the pressure plate assembly as the *clutch cover attaching bolts* (clutch cover-to-flywheel bolts) are being removed. Note that an old transmission input shaft makes a good pilot shaft.
11. Loosen the clutch cover attaching bolts sequentially, one turn at a time, until all spring pressure is relieved. If bolts cannot be reached, use a flywheel turner to rotate the flywheel until each bolt is accessible. Once pressure is relieved, carefully remove the bolts and the pressure plate assembly. Then, remove the clutch disc and alignment tool from the flywheel.

WARNING! The pressure plate assembly is fairly heavy. Make sure that you can support its weight before removing the last bolt. If a clutch alignment tool has not been used, be aware that the clutch disc could fall out as the attaching bolts are removed.

Also, if the clutch was slipping, it may be hot for a long time after the vehicle is stopped. Be careful not to burn yourself on a hot pressure plate or flywheel.

In many cases, this will complete clutch removal. You should inspect the pilot bearing and flywheel (outlined in upcoming paragraphs) and determine if they must also be removed.

Front-Engine, Front-Wheel Drive

The following procedure details the proper way to remove the clutch from a front-engine, front-wheel drive vehicle. Note that with a few exceptions, the general procedure, cautions, and warnings duplicate clutch removal for front-engine, rear-wheel drive vehicles.

1. Disconnect the battery negative cable.

WARNING! *Always* disconnect the battery negative cable before working near the clutch. Otherwise, if the starter is accidentally operated, you could be severely injured.

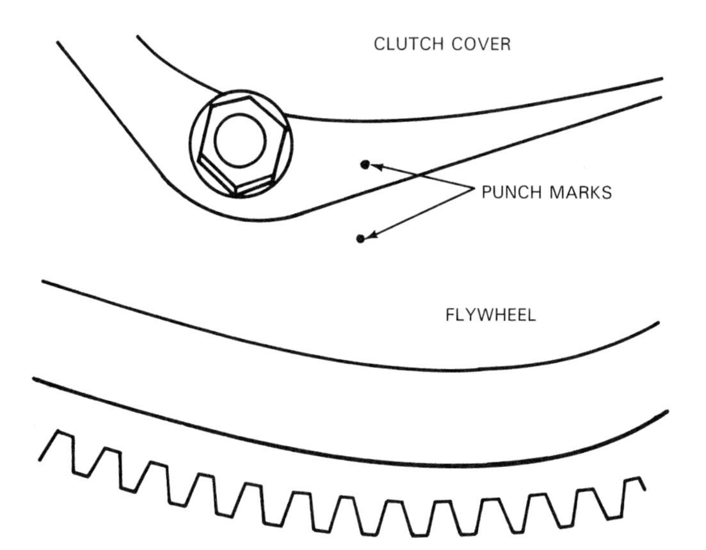

Figure 7-19. To maintain proper balance between reassembled flywheel and pressure plate assembly, make punch marks on both parts before disassembling. This is necessary only if the pressure plate will be reused.

2. Raise the vehicle with an approved hoist or hydraulic jack. If using a hydraulic jack, be sure to install good quality jackstands under the vehicle frame before getting under the vehicle.

WARNING! *Never* work underneath a vehicle supported only by a hydraulic jack. *Always* support the vehicle with jackstands.

3. Once the vehicle is properly raised and supported, remove the drive axles. There are several variations of front drive axle attachment. On some vehicles, the wheels and brake assembly, along with certain steering and suspension parts, must be removed to remove the axles. On other vehicles, the drive axles can be removed without removing these parts. (This is covered in detail in Chapter 23.)

WARNING! If the vehicle has been operated recently, the engine and exhaust system will be hot. Be careful when working on a hot engine.

4. Disconnect the pushrod or cable and return spring connected to the clutch fork.
5. Remove the transaxle. (This is covered in detail in Chapter 19.) On many vehicles, the transaxle and clutch housings are a single unit, Figure 7-20. If this is the case, the complete unit is removed from the engine. On vehicles requiring the engine also be removed for clutch replacement, refer to the manufacturer's service manual.

   Note that the transmission input shaft and front bearing retainer will be removed along with the transaxle. In addition, the clutch fork and throwout bearing assembly will be removed with the one-piece, or integral-type, casting.

WARNING! Support engine with an engine holding fixture when removing the transaxle. This will keep the engine from dropping as you remove the transaxle and will protect the motor mounts.

Also, use a transmission jack to support the transaxle. *Never* let the transaxle hang unsupported once attaching bolts connecting the transaxle and clutch housings are removed. This action can cause damage to the clutch disc splines or transmission input shaft. Make sure, for this and other safety reasons, that you support the transaxle at all times. If a transaxle were to fall, it could cause injury or damage.

6. From here, clutch removal procedure duplicates that for front-engine, rear-wheel drive vehicles. To complete clutch removal, continue with step 6 of that section.

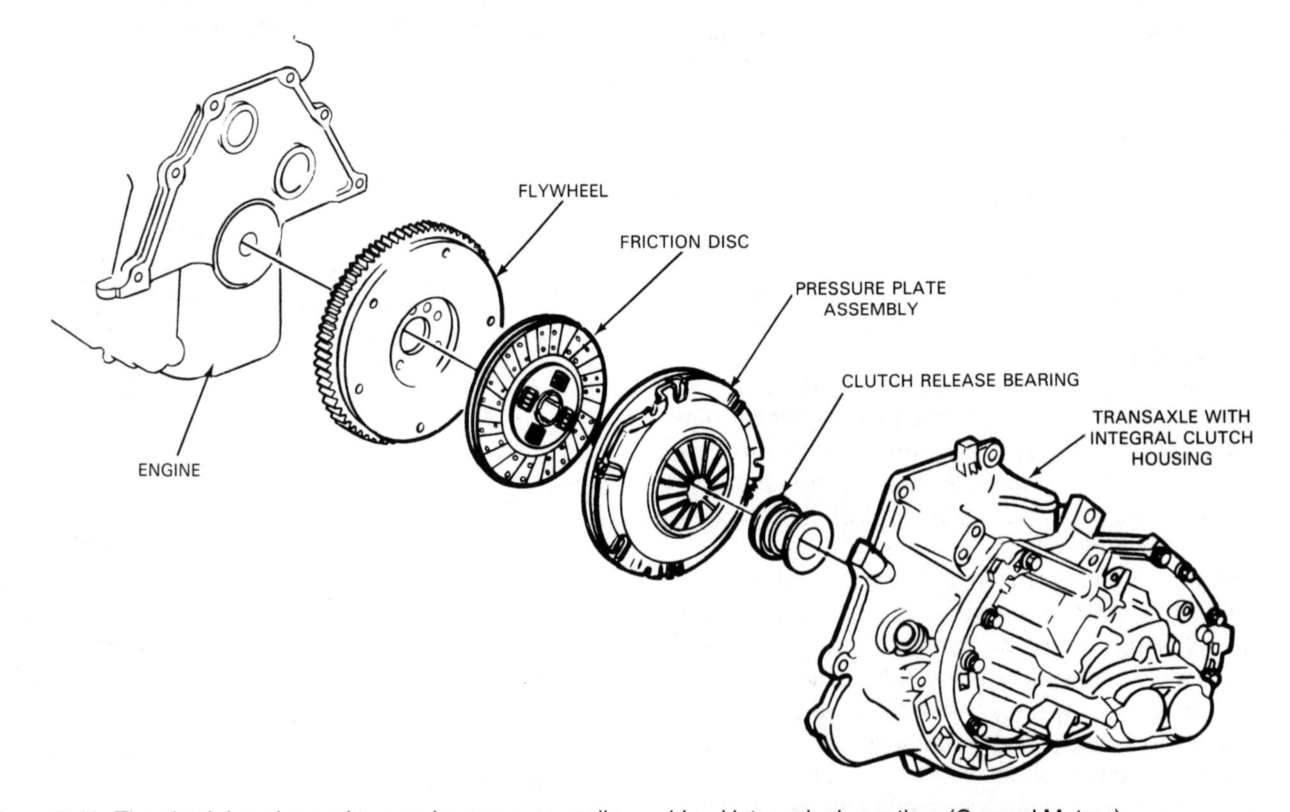

Figure 7-20. The clutch housing and transaxle case are usually combined into a single casting. (General Motors)

Clutch Parts Inspection and Repair

With the clutch removed, each component must be carefully inspected for wear and damage. Be sure all parts are clean before inspection. Clean flywheel face and pressure plate with a nonpetroleum-based cleaner. Do not wash the throwout bearing in any kind of solvent. Do not get grease on the friction disc facings.

Look for any signs of damage or wear, no matter how slight. After the defect is identified, the concerned part(s) may be repaired or replaced. Service procedures for the various related parts of the clutch are the focus of this section.

Pilot Bearing Service

A worn pilot bearing will allow the transmission input shaft and clutch disc to wobble. This can cause clutch vibration, a noisy clutch, and damage to the transmission. Pilot bearings are relatively cheap, and they are easy to change while the clutch is being serviced.

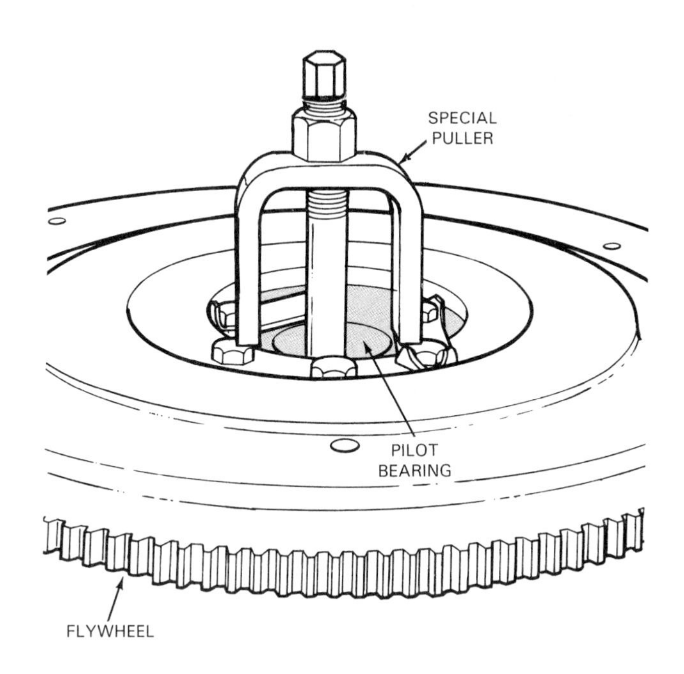

Figure 7-21. The pilot bearing removal tool will make pilot bearing removal easier. (Chrysler)

Bearing inspection

The clutch pilot bearing is not serviced until the clutch is removed. The bearing is often worn out by then. Bearings should be inspected for excessive wear or damage. Some technicians will automatically replace the bearing as a matter of practice. If there is any doubt about a pilot bearing's condition, it should be replaced!

Roller- or ball-type pilot bearings should be inspected for lack of lubrication. As a check of this, turn the bearing and feel for roughness. If it does not turn smoothly, it should be replaced.

In a pilot bushing, wear will show up as looseness between the bushing and the input shaft pilot. A good way to check for looseness is to insert a used input shaft into the bearing and try to wiggle it. If it has too much play, the bushing is worn out, and it should be replaced.

NOTE! To perform this check, you must have the same type of input shaft as is used in the vehicle transmission.

Bearing removal

There are a couple of ways to remove a pilot bearing. One way is to use a *pilot bearing puller tool,* as shown in Figure 7-21. One variation, called a *threaded-tip puller,* is used to remove a pilot bushing; however, it ruins the bearing in the process. The center shaft, or pilot, of this tool has a self-tapping tip. The pilot is threaded into the bore of the bearing. When fully inserted, the legs of the puller are adjusted to seat on the crankshaft flange. As the tool is tightened further, it will pull the pilot bearing from the crankshaft bore. Another variation, called an *expandable finger-tip puller,* is a similar tool used when ball or roller bearings are used.

Another way to remove the old pilot bearing is to pack the recess behind the bearing with heavy grease. Then, install a driver into the pilot bearing; an old input shaft works very nicely. Put on eye protection and strike the exposed end of the input shaft with a hammer. The force from the hammer will travel through the input shaft to the grease. The grease will push the bearing from the crankshaft. See Figure 7-22.

Bearing installation

Before installing the new pilot bearing, slip it over the pilot of the transmission input shaft. This is done to ensure that it is the proper bearing. If it will not slip over the shaft, or if the fit is too loose, this is not the correct bearing.

The next step is to drive the new bearing into the crankshaft bore, Figure 7-23. Use a properly sized driver to avoid damage. Measure the installed depth of the pilot bearing. If this depth is too shallow, the input shaft may contact the bearing. This will damage the bearing, the input shaft, or both. If the bearing is driven in too deeply, it may not make full contact with the input shaft pilot. This is not such a problem with bushings, as they are usually longer. If bushings are driven in too deeply, there is still usually adequate surface contact with the input shaft pilot.

Once the pilot bearing is properly installed, lubricate it with a small quantity of high-temperature grease. Most bearings have a recess behind the bearing. A small amount of grease stored here will find its way to the

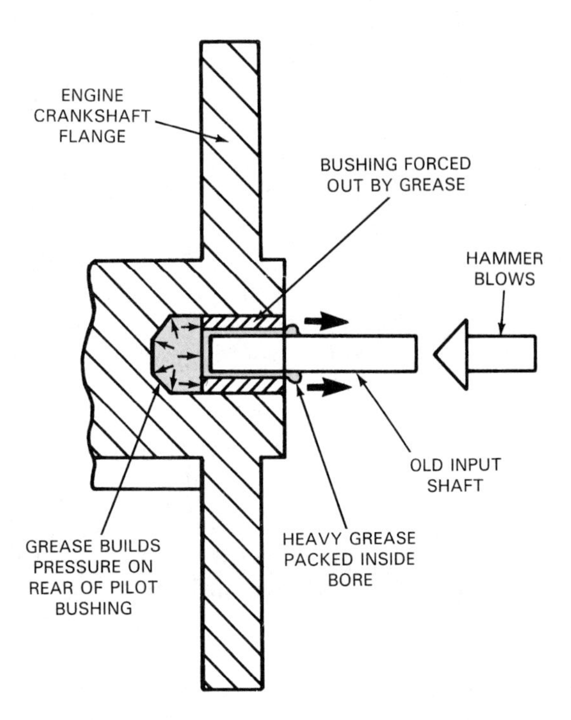

Figure 7-22. A pilot bearing can also be removed by packing grease into the recess behind the bearing. A driver is then placed into the pilot hole and struck. The driver moves into the grease, creating pressure. The pressure drives out the bearing.

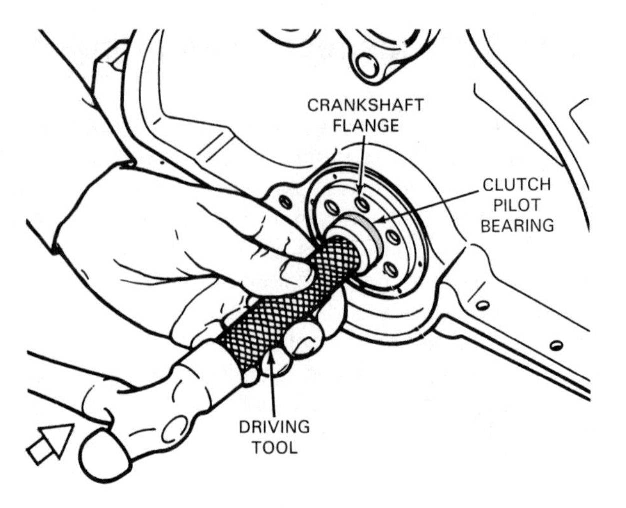

Figure 7-23. A new pilot bearing should be installed with the proper bearing driver. Use a driver that contacts the outer race only when driving a roller- or ball-type pilot bearing. Always be very careful to avoid damage when installing a new bearing. (Ford)

bearing surface, helping to keep it supplied with grease. See Figure 7-24.

CAUTION! Do *not* put too much lubricant on the bearing or in the recess. Excess lubricant will be thrown out of the pilot bearing and can ruin the clutch disc. Note that some bearings are prelubricated and do not require any additional lubrication.

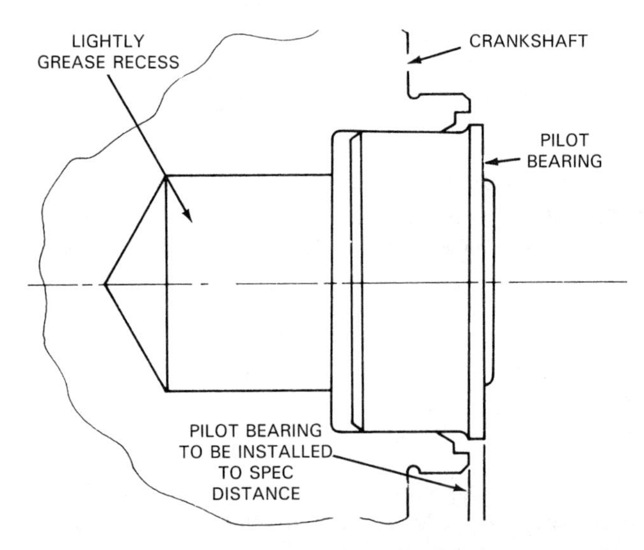

Figure 7-24. Grease the recess behind the pilot bearing but be careful to use only a small amount. Lack of grease will allow the bearing to run dry and be damaged. On the other hand, too much grease will drip out and damage the friction disc. (Ford)

Flywheel Service

The flywheel is large and very heavy, and removing it can be difficult and even dangerous. It should not be removed unless it is in need of resurfacing. It is rare to find a flywheel that is not showing some sign of wear or overheating. Judgment is called for in deciding if it must be resurfaced. In many cases, the flywheel can be reused without machining. If the flywheel is not badly worn, it can be cleaned up by light sanding.

Flywheel inspection

Visually inspect the flywheel for obvious signs of damage. Check the flywheel machined surface for discoloration (caused by overheating), scoring, or extremely shiny surfaces. In some cases, the flywheel may have deep heat cracks. Look for warpage. Inspect the threads in the holes for the clutch cover attaching bolts. If they are stripped, they can be restored. Also, check the flywheel ring gear for damage. If even one ring gear tooth is broken, the ring gear must be replaced. Note that if the ring gear teeth are damaged, you should also check the teeth of the starter drive pinion for damage.

If the flywheel is not obviously damaged, use a straightedge and feeler gauge to check for warping of the contact surfaces. Manufacturers' specifications as to maximum warpage vary. As a general rule, however, if more than a 0.01-in. (0.25-mm) feeler gauge fits under the straightedge, the flywheel must be resurfaced.

A dial indicator can be also used to check for a warped flywheel. In addition, it can be used to check the flywheel *runout.* Push the flywheel toward the engine to remove crankshaft end play. Then, mount the dial indicator on a stationary part of the engine, with the indicator point bearing on a smooth, relatively undamaged part of the machined flywheel surface, as shown in Figure 7-25. Rotate the flywheel and watch the dial to determine how much fluctuation in values there seems to be, which would give an indication of *warpage.* If, rather than a series of fluctuations, the dial shows a steady increase for half a turn and then a steady decrease, flywheel *runout* may be excessive. In either case, the flywheel must be removed and resurfaced or repositioned.

To check warpage, zero the dial indicator, then, slowly turn the flywheel through about a quarter of a revolution. Watch indicator face and note if displacement ever exceeds 0.01 in. (0.25 mm). Without moving indicator, repeat this procedure at least three more times at different positions on the flywheel. If any of the samples produced fluctuations exceeding 0.01 in. (0.25 mm), the flywheel must be resurfaced.

As mentioned, the surface of the flywheel should be checked for ***runout.*** This is a condition of a rotating object, such as a shaft or flywheel, in which the surface is not rotating in a true circle or plane. ***Radial runout*** is a measure of out-of-roundness. It is measured at right angles to the centerline of an object. Excessive radial runout may be caused, for example, by a bent shaft. ***Lateral runout*** is a measure of in-and-out movement, or wobble. It is measured in the direction that is lengthwise to the centerline of the rotating object. In other words, it is measured on the plane surface of the rotating object. Excessive lateral runout can result, for example, if a flywheel is not mounted properly.

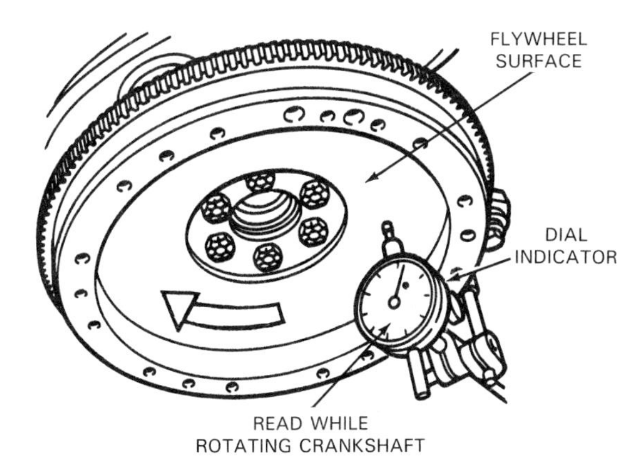

Figure 7-25. Proper mounting of the dial indicator is important. The base must be solidly attached to the rear of the engine. Push on the flywheel to remove crankshaft end play. Position indicator so that plunger just touches the flywheel. The dial is then zeroed. (Ford)

To check for lateral runout, slowly rotate the flywheel while watching the dial indicator. Determine the lowest point on the flywheel, which corresponds to the lowest reading on the dial. Zero the indicator at this point. Move the flywheel until the highest reading is recorded. This measurement is the lateral runout.

Lateral runout should not exceed manufacturer's specifications. If it does, the flywheel may not be mounted properly. Remove it and check for burrs between the flywheel and the face of the crankshaft flange. If no burrs exist, check the runout of the crankshaft flange.

Note that while the dial indicator is mounted, you can also measure crankshaft end play. Push the flywheel forward and rearward and note the displacement on the dial.

Flywheel removal

The easiest way to remove the flywheel attaching bolts is with an impact wrench. If you do not have an impact wrench to remove the flywheel, it must be kept from rotating so the attaching bolts can be loosened. While the bolts are removed, the flywheel can be held in place using a flywheel turner or locked in place with a flywheel holder; alternatively, a block of wood can be placed between the flywheel and a stationary part of the engine.

WARNING! *Always* wear eye protection when using an impact wrench.

Also, a flywheel is very heavy. It can cause injury or severe damage if dropped. Do *not* remove the bolts completely until you have the flywheel secured so that it cannot fall.

Once the flywheel is securely held, the bolts can be loosened, and the flywheel can be removed. Sometimes there is a flat metal spacer between the flywheel and crankshaft flange. The spacer should be saved for reassembly.

After removing the flywheel, check the rear main bearing seal for leaks. If the seal is leaking, oil could reach the new friction disc and ruin it, just as it may have ruined the old disc. A leaking seal should be replaced.

Flywheel repair

Some flywheels may be damaged beyond repair. Others may be repaired and then reinstalled. Some of these may be fixed by replacing the ring gear or by resurfacing. Procedures for these operations are detailed in the next few paragraphs.

Ring gear replacement. In most cases, a flywheel with a damaged ring gear is not repaired but is replaced. At times, however, the ring gear can be removed and replaced to fix the flywheel, Figure 7-26.

Some flywheel ring gears are welded into position. Others are a shrink fit. If it is determined that the ring gear is to be replaced, it can be cut from the flywheel with a

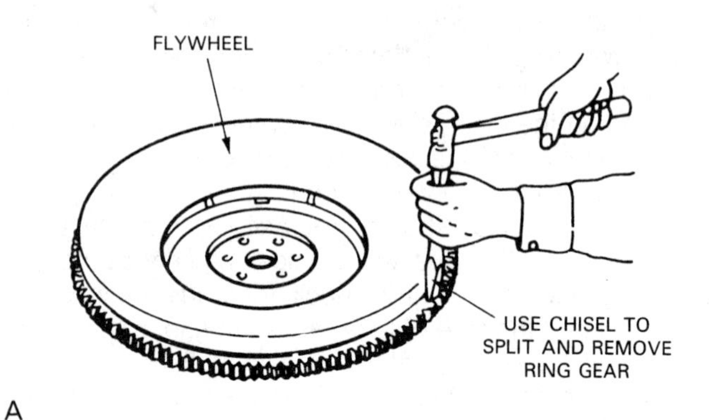

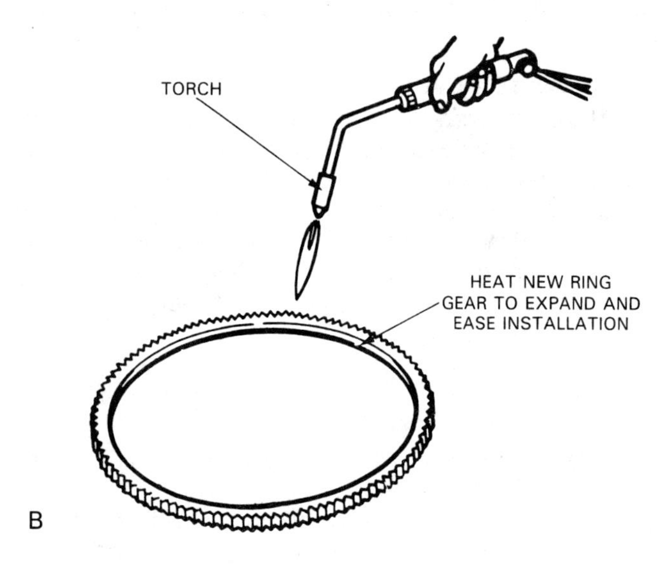

Figure 7-26. Most manual transmission flywheels have a removable ring gear. A–A chisel is being used to remove the ring gear from a flywheel. Be careful not to damage the flywheel. B–You should heat the replacement ring gear with a torch or in an oven. Be careful not to overheat the ring gear, or the metal will lose its strength. C–Place the new ring gear over the flywheel and ensure that it is properly seated. Always wear gloves or use tongs to prevent burning your hands when handling the hot ring gear. (General Motors)

cutting torch or a metal saw. Ring gears that are welded are commonly removed by breaking the welds with a chisel. Those that are a shrink fit can be heated and then removed with a chisel. See Figure 7-26A.

WARNING! Wear eye protection during ring gear removal.

If the new ring gear is to be installed by a shrink fit, you can expand it by heating, Figure 7-26B, or shrink the flywheel by cooling. Usually, the easiest method is to heat the ring gear with a torch. Try to heat the part evenly. Once the ring gear is hot enough (close to but *not* exceeding 450°F [232°C]), it can be placed over the flywheel, Figure 7-26C.

CAUTION! Wear welding gloves when handling the hot ring gear or handle it with tongs.

After the ring gear is placed on the flywheel, it should be checked for proper seating. Make sure that the gear is not warped, since warpage will cause the starter to jam. Once you have determined that the ring gear is properly installed, stake it or weld it in place, as necessary.

Flywheel resurfacing. A cracked, scored, or severely overheated flywheel must be resurfaced. A *severely* cracked flywheel must be replaced. Be sure that the replacement flywheel has its *balance weights* (thickened areas on flywheel) in the proper position. The easiest way to do this is to compare the old and new flywheels. Therefore, never dispose of the old flywheel until reassembly is complete.

Resurfacing a flywheel is a job for a machine shop. Generally speaking, any machine shop that can resurface cylinder heads can resurface a flywheel. The machining process for flywheels is similar to that for cylinder heads.

Flywheels can be resurfaced by turning on a lathe (including a brake lathe), cutting on a milling machine, or grinding on a table with a special grinding wheel. No matter what method is used, the basic process is the same: material is removed until the surface is totally clean and damage-free.

Only a certain amount of metal can be removed, however. If too much is removed, the flywheel will not have sufficient metal to absorb heat. Some flywheels are marked to indicate the maximum amount of metal that can be removed. If it is not marked, the machine shop will usually be able to tell you if there is enough good metal left in the flywheel. As a general rule, about 0.2 in. (5 mm) can be removed from a flywheel that has not been machined before.

NOTE! If you placed alignment marks on the flywheel face, they will be removed during the machining process. Make a duplicate mark at the exact spot on the back of the flywheel. Some flywheels have alignment dowels. These dowels should be removed before the flywheel is sent to the machine shop for resurfacing.

Flywheel installation

Once the flywheel is resurfaced, it should be reinstalled in its original position. First, reinstall any alignment dowels. Solid dowels are installed by driving them in squarely with a soft-faced mallet. A special tool with a shoulder is needed to drive hollow dowels. Drive the dowels in carefully until they are fully seated. If there was a spacer plate between the flywheel and the crankshaft flange, reinstall it before replacing the flywheel.

Align the crankshaft flange and flywheel bolt holes. They are usually spaced so that they will line up at only one location. After they are aligned, start the flywheel attaching bolts. Torque the bolts to the proper tension in a crisscross pattern, Figure 7-27. Use a flywheel holder or a wood block to hold the flywheel during the tightening operation.

CAUTION! Do *not* tighten the bolts with an impact wrench. The bolts must be torqued properly.

The flywheel is held to the crankshaft flange with special hardened bolts. Do *not* use any other kind of bolt. Be careful not to accidentally swap flywheel attaching bolts with any others.

Clutch Disc Service

In most cases, the friction disc is replaced when any kind of clutch teardown is done, as it is relatively inexpensive. If the disc was recently replaced or appears to be in excellent condition, it can be reused. Before deciding to reuse the disc, it should be carefully checked, as outlined below.

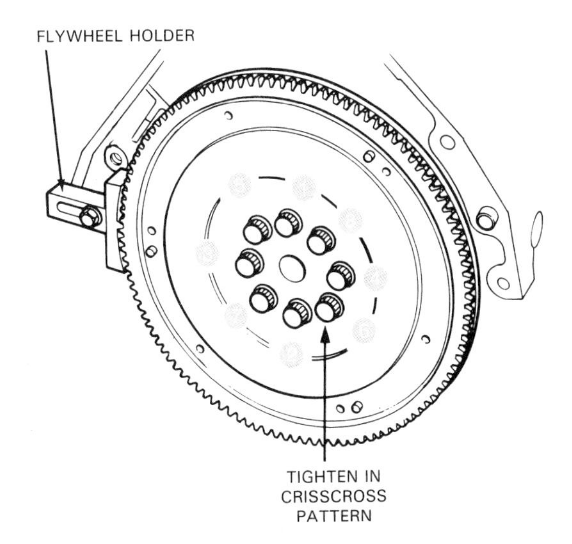

Figure 7-27. Torque the flywheel attaching bolts in a crisscross pattern. Be sure to use the proper bolts and torque them to specification. Do not use an impact wrench to tighten the bolts. (Honda)

WARNING! Asbestos dust is harmful. Avoid breathing any dust from the friction disc, as it can contain large amounts of cancer-causing asbestos.

If you intend to reuse the clutch disc, do not wash it in any kind of cleaning solvent. Solvent will ruin the clutch disc friction facings.

Inspect the clutch disc for thin or cracked friction material. The disc thickness can be checked with a sliding caliper, as shown in Figure 7-28. Usually, however, other indications will be obvious and this step will be unnecessary. Check the disc for loose rivets, oil-soaked friction facings, or broken cushion springs. Make sure that the hub flange splines are not worn and that they slide freely on the splines of the transmission input shaft. If the clutch disc shows any wear or damage, it should be replaced. Figure 7-29 shows various types of clutch disc damage you might find during inspection along with possible reasons damage occurred.

If a replacement disc is to be used, it must be carefully compared to the old one before it is installed. Many clutch problems are caused by installing the wrong replacement disc. It is especially important to check the hub flange size against the old disc. A hub flange that is too large will contact the flywheel and pressure plate, causing clutch disc damage and, also, gear clash. As a final check, slip the replacement disc over the transmission input shaft to make sure that it has the correct splines.

Pressure Plate Assembly Service

The pressure plate assembly is another clutch component that is almost always replaced when the clutch is repaired. You should spend a few moments checking the pressure plate assembly. This will help to determine the cause of the clutch problem that developed and to learn how it can be prevented in the future.

Pressure plate assembly inspection

The pressure plate machined surface develops the same problems as the flywheel surface. Check for wear, scoring, signs of overheating, cracks, and warpage. The pressure plate will usually show more heat damage than the flywheel.

Retaining rivets, for diaphragm-spring pivot rings, should be checked for looseness. Loose rivets indicate long usage or rough clutch treatment.

Elongation of the bolt holes in the clutch cover, where they have become slotted in appearance, is an indication that the clutch cover attaching bolts were loose. The pressure plate assembly may bounce against the flywheel if

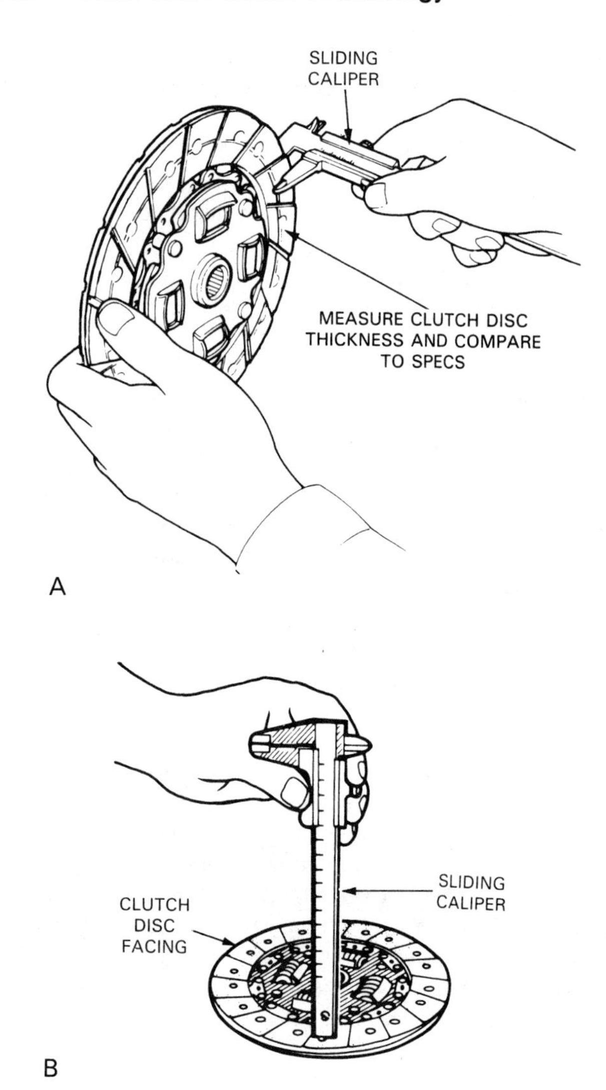

Figure 7-28. Inspecting the clutch disc. A–Clutch disc thickness can be measured with a sliding caliper. If thickness specifications are available, the thickness can be compared to determine whether the disc can be reused. B–Depth gauge on sliding caliper or depth micrometer can be used to determine the distance from the top of the friction disc rivets to the surface of the friction facing. If the rivets are too near the surface, the disc is worn out and should be replaced. (Honda, Chrysler)

these bolts are loose, which would cause the holes to elongate.

Check for weak pressure plate springs and for signs of overheated springs. The release fingers or levers should be checked for wear and misadjustment. If any fingers or levers are lower than others, the pressure plate application will be uneven. The clutch may grab or chatter. Severe wear at the contact point with the throwout bearing indicates that the clutch has seen much use or that there was no free play. Figure 7-30 shows various types of pressure plate assembly damage you might find during inspection along with possible reasons damage occurred. This damage may range from chatter marks on the pressure plate to a scored pressure plate to broken release levers. Being able to readily identify the damage will result in better service times.

Pressure plate assembly repair

Most repair shops choose to completely replace pressure plate assemblies, rather than rebuild them. It is possible, however, with the proper equipment, to take apart a pressure plate assembly and replace just the parts that are worn. This is typically done on an assembly line by large automotive rebuilders. As a general rule, pressure plate assemblies that are put together with nuts and bolts can be rebuilt, while pressure plates held by rivets cannot be rebuilt. If a shop decides to rebuild a pressure plate assembly, it must closely follow the manufacturer's directions. A general procedure for rebuilding a pressure plate assembly is shown in Figure 7-31. Remember to wear eye protection!

Throwout Bearing Assembly Service

The throwout bearing is almost always replaced when the clutch is serviced. It is often a source of problems. If, for some reason, the bearing is to be reused, it should be carefully checked for defects.

CAUTION! If you intend to reuse a throwout bearing, do not wash it in any kind of cleaning solvent. Solvent will dissolve the throwout bearing lubricant, which cannot be replaced.

Throwout bearing assembly inspection

Check the throwout bearing for roughness by attempting to rotate it by hand. If rotation is rough or if it appears to have lost its grease, it should be replaced.

Check the bearing collar for a free (but not loose) fit on the hub of the transmission's front bearing retainer. A loose fit indicates that the collar and/or hub is worn. Check the collar where it contacts the clutch fork. If the collar is worn, the fork is usually worn also. Any worn parts should be replaced. Figure 7-32 shows damage related to the throwout bearing, which you might find during inspection, along with possible reasons damage occurred.

Throwout bearing assembly repair

Some throwout bearings are an integral part of the bearing collar. These are replaced as a unit. Many throwout bearings, however, are a press fit onto the bearing collar. The assembly can be repaired by replacing the bearing or collar–whatever is in need of repair.

Replacement bearings must be pressed onto the collar after the old bearing is removed. To remove the throwout bearing, press it from the collar using a bench vise or press. Adapters may be necessary to properly perform the

A–Burned disc facing–Oil contamination, slipping clutch, insufficent clearance adjustment.

B–Hub splines worn–Improper engine-to-transmission alignment, damaged input shaft, bad pilot bearing.

C–Worn disc facing–Weak pressure plate springs, normal wear, scored or cracked flywheel.

D–Damaged hub splines–Transmission drawn into place with bell housing bolts.

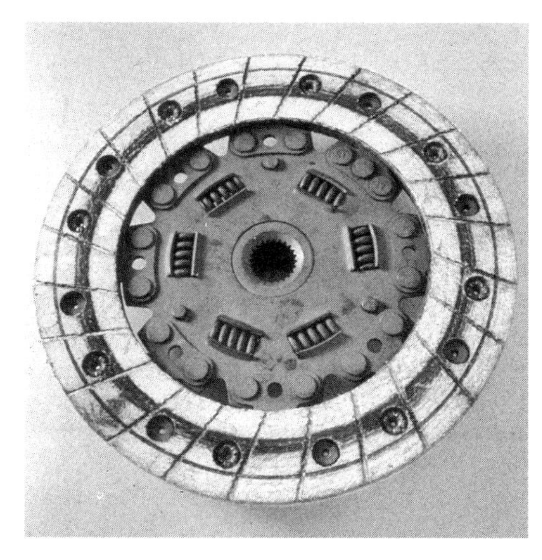

E–Scored disc facing–Flywheel not machined when needed, bad pressure plate reused.

F–Wear on hub–Installed backwards, wrong parts.

Figure 7-29. Study types of clutch disc damage. (Luk)

A–Chatter marks on pressure plate–Oil or grease contamination, binding linkage, loose or soft motor mounts, worn drive line parts.

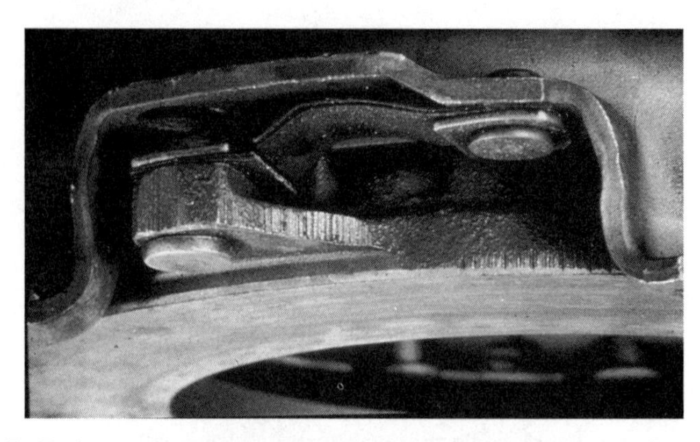

B–Broken pressure plate retracting spring–Normal fatigue, wrong clutch for vehicle, low-quality pressure plate.

C–Pressure plate hot spots or heat checks–Clutch slippage, oil or grease contamination, clutch adjustment too tight, binding linkage, driving habits.

D–Broken release levers–No free play in clutch adjustment, bad release bearing, improper part alignment.

E–Scored pressure plate–Clutch disc worn to rivets, adjustment too tight, clutch slippage, binding linkage.

F–Broken diaphragm spring–Installation error, excessive pedal reserve causing throwout bearing to travel too far during disengagement.

Figure 7-30. Study types of pressure plate assembly damage. (Luk)

pressing operation. Press on the new throwout bearing. It should be fully seated on the shoulder of the bearing collar when complete. Remember! Do not apply driving force through the bearing elements. Refer to Figure 7-33.

After the bearing is in place, rotate it by hand and make sure that it rotates smoothly. It should turn without binding or roughness.

Clutch Fork Service

A bent or worn clutch fork can prevent the clutch from releasing properly. Inspect the clutch fork for such signs of damage. Look for wear at the throwout bearing mount. On lever-type clutch forks, check the pivot-point contact area. Inspect the spring clip that holds the clutch fork to the pivot point. On shaft-type clutch forks, check the pivot shaft for wear.

If inspection reveals a damaged clutch fork, it must be removed (if it is not already) and replaced. The lever-type clutch fork is removed when the clutch is disassembled. The shaft-type clutch fork can be removed by first removing the pivot, or clutch release, shaft. The shaft is usually held by a clip, Figure 7-34A. Once the clip is removed, the shaft will slide out of the housing, and the fork will be removed from the shaft, Figure 7-34B. With the fork assembly apart, you can also check the pivot shaft for wear.

Clutch Housing Service

Clutch housings are not usually a source of problems. It is important, however, that they be inspected for damage. Clean the clutch housing interior and exterior. Then check the housing for cracks and damage to mounting surfaces. Any damage is grounds for replacement.

Also, check the clutch fork ball stud for wear. It should be undamaged and tight. If the stud must be replaced, it can usually be removed from the clutch housing with a large Allen wrench or a differential-plug removal tool.

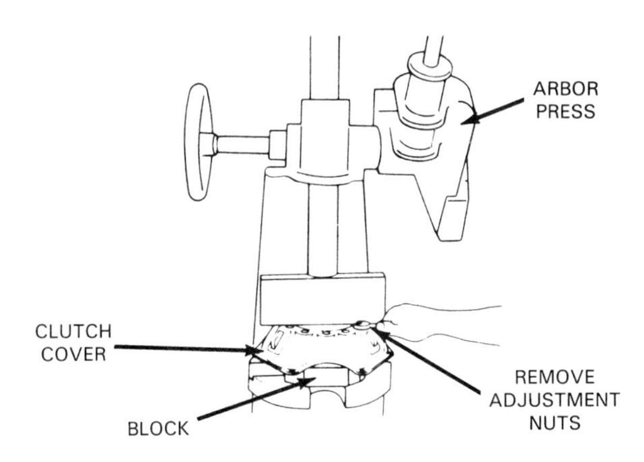

A–Use arbor press to compress pressure plate apply springs so that the assembly can be taken apart. Place block under pressure plate, arranged so that the cover is free to move down. Remove *eyebolt adjustment nuts,* which are used to adjust release lever height, from clutch cover. *Slowly* release arbor press, so cover can be removed without the apply springs flying out.

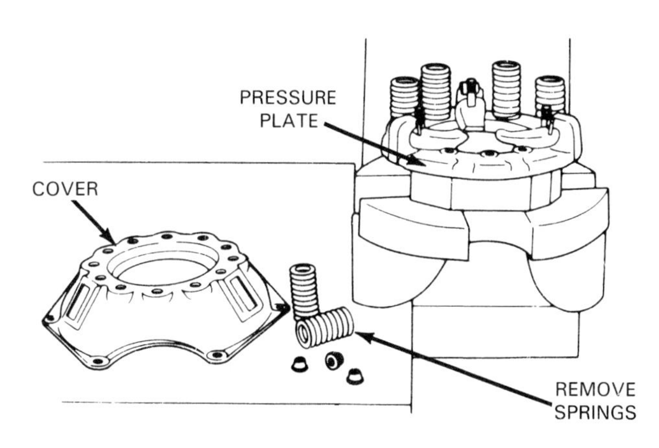

B–Carefully mark positions of all springs and release levers. This will ensure that parts are reinstalled in their original positions. Organize disassembled parts of pressure plate assembly on workbench. Replace all parts that show wear.

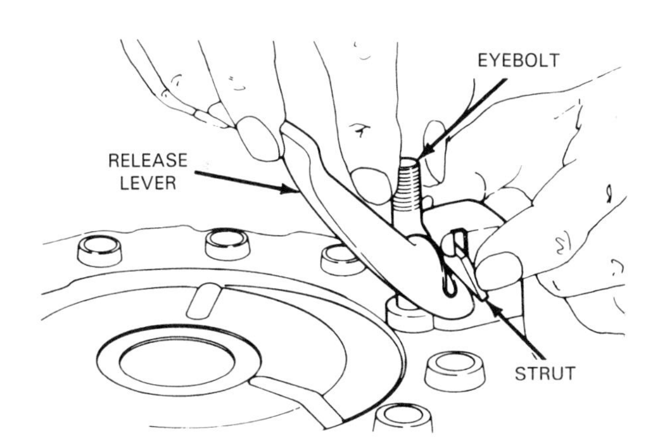

C–Remove release levers. To do this, lift up on lever and remove strut. Keep all parts of each particular lever together.

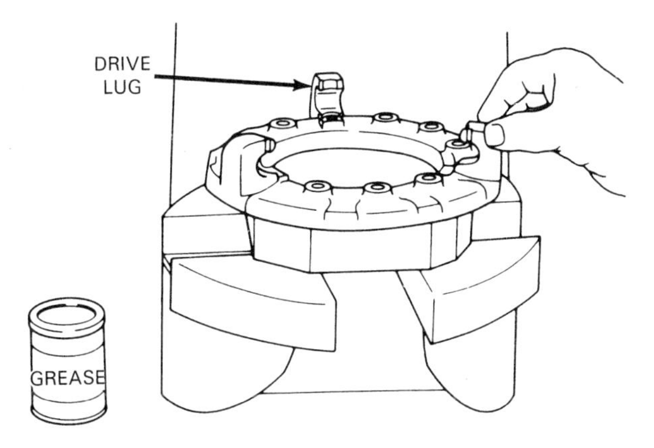

D–Grease pressure plate *drive lugs* with high temperature grease, after inspecting assembly for cracking, wear, and other damage.

Figure 7-31. Procedures A–H illustrate steps in coil-spring pressure plate assembly overhaul. This procedure is general, and the manufacturer's service manual should always be consulted. (General Motors)

(Continued)

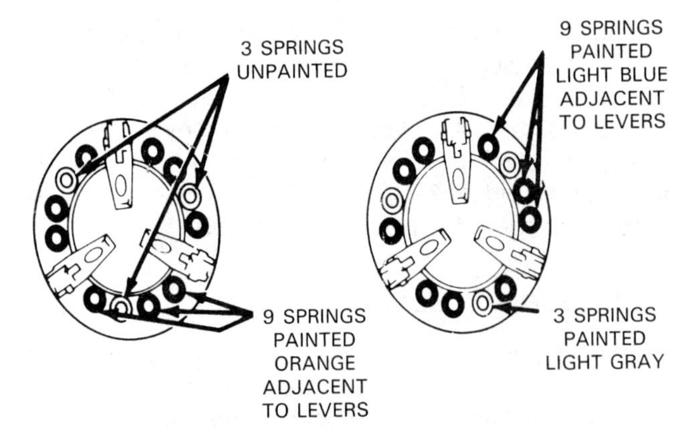

E–To reassemble, install release levers and apply springs. Basically, install release levers by reversing the procedure for removal. If new springs are used, install using sequence shown; otherwise, return springs to their original positions.

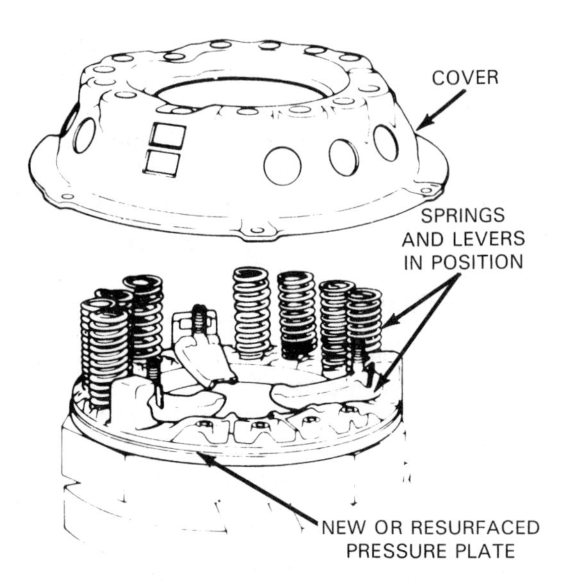

F–With pressure plate assembly positioned on arbor press as per original setup, slowly apply pressure to clutch cover. Drive-lug openings in clutch cover should fit over drive lugs; apply springs should fit into their seats, as shown. Screw eyebolt adjustment nuts onto eyebolts until their tops are flush. Slowly release arbor press and remove assembly.

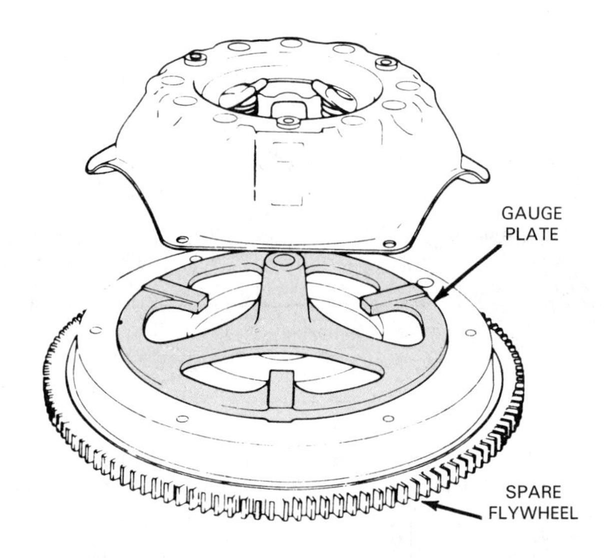

G–To ensure uniform clutch application, release levers must be set to equal heights. Place gauge plate on spare flywheel and place pressure plate assembly over it. The gauge plate occupies the position normally occupied by the clutch disc and simulates installed component positions. Bolt the clutch cover to the flywheel. Note that spacers or shims may be used instead of gauge plate.

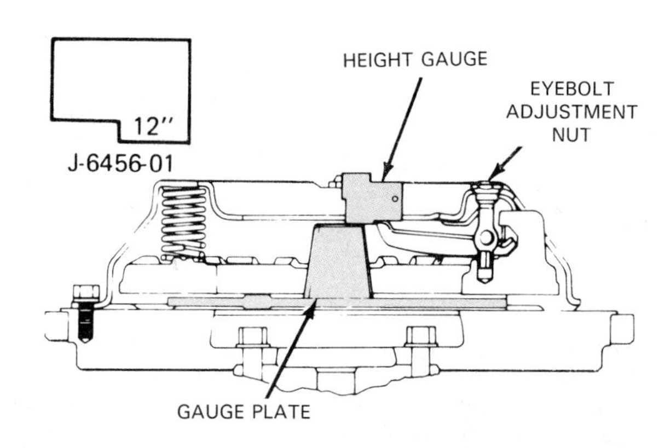

H–Position height gauge on hub of gauge plate. Turn eyebolt adjustment nut until lever is flush with height gauge. Adjust all levers in this manner. Stake all adjustment nuts when finished so they will not loosen in service. In lieu of height gauge, a straightedge can be placed across the cover. Levers can then be set to a height at some specified distance down from the straightedge. Remove clutch cover from flywheel, loosening bolts gradually until spring pressure is relieved.

Figure 7-31. *(Continued)*

Clutch Installation

Following are some general procedures detailing how to install a clutch safely and properly. The exact procedure for clutch installation varies according to the engine and drive train layout. You should always refer to the manufacturer's service manual for specific procedures and techniques to safely and properly install a clutch.

Front-Engine, Rear-Wheel Drive

The following procedure details the proper way to install a clutch in a front-engine, rear-wheel drive vehicle.

A–Dry, worn throwout bearing–Riding clutch, improper clutch adjustment.

B–Worn front bearing retainer hub–Bent fork or forkmount.

Figure 7-32. Study damage related to throwout bearing. (Luk)

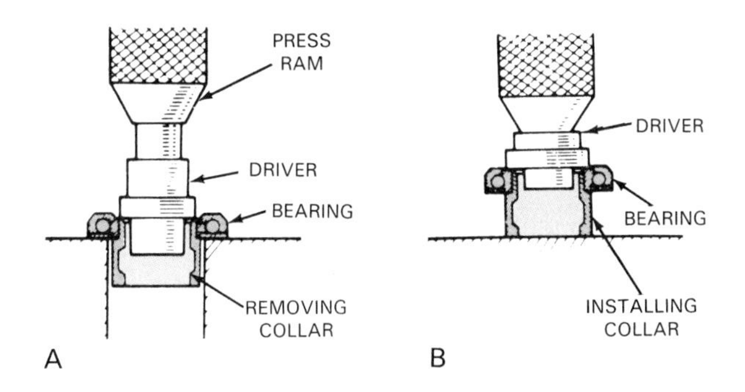

Figure 7-33. Replacing a throwout bearing. A–The throwout bearing can be removed from its collar by using a press. The proper adaptor must be used. In many cases, the bearing can be removed and replaced by using a bench vise. B–This illustrates bearing installation procedure. Be careful not to exert any pressure on the rolling elements or outer race. (Toyota)

1. Pilot bearing and flywheel installations were detailed previously in respective service sections. As a preliminary step, recheck the pilot bearing and flywheel to make sure that they are installed properly.
2. Install the clutch disc and pressure plate assembly onto the flywheel. If one side of the disc is marked *flywheel,* place that side against the flywheel. If not marked, study the disc, flywheel, and assembly to determine which way the long side of the hub should fit into the assembled clutch. It should be fairly obvious. Use your alignment tool, Figure 7-35. It will hold the disc in place and keep the disc hub aligned with the pilot bearing. This way, the input shaft will go in smoothly when you attempt to reinstall it.

CAUTION! When handling clutch parts, make sure your hands are clean. It is especially important that no oil or grease contact the friction disc.

Match holes in the assembly to flywheel dowels, where applicable. Otherwise, align punch marks that you made during removal. Some pressure plate assemblies will be aligned through offset bolt holes. If none of these reference marks are available, turn the pressure plate assembly on the alignment tool, and align bolt holes with the nearest threaded holes in the flywheel.

3. Begin threading in the clutch cover attaching bolts. Install two directly opposite each other, finger-tightening them. The clutch will now support itself. Thread in and finger-tighten the remaining bolts, making sure that all bolts are started properly.

CAUTION! The bolts used to hold the pressure plate assembly to the flywheel are high-strength bolts. Do not use low-tension "hardware-store" replacements, as they will probably break and cause a "clutch explosion."

Finish tightening the clutch cover attaching bolts by torquing them down gradually in a crisscross pattern. Make sure that the pilot shaft does not sag as you tighten the bolts. Putting a slight upward force on the pilot shaft will help to retain the alignment of the clutch disc. Finish tightening the bolts to the proper torque given by the manufacturer. Once the bolts are tightened, make sure that the clutch pilot shaft slides in and out freely. See Figure 7-36.

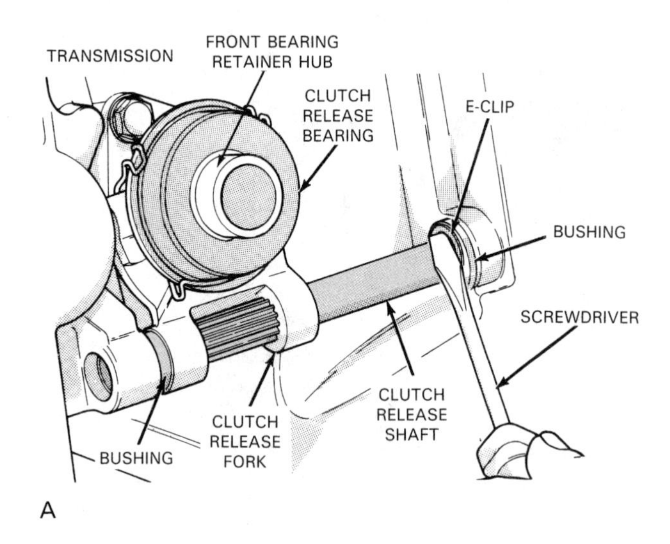

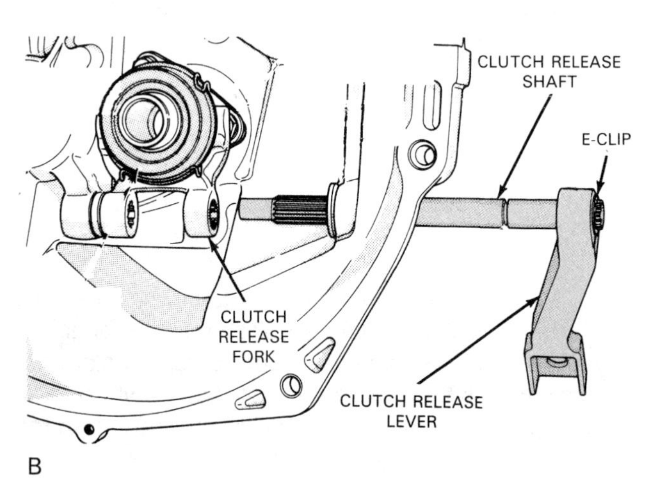

Figure 7-34. This type of clutch fork uses a shaft mounting. It must be disassembled to remove the fork. When the shaft is out of the housing, it should be checked for wear. A–Remove retaining E-clip. B–Remove clutch release shaft. (Chrysler)

4. If so equipped, lubricate the seat of the clutch fork ball stud in the clutch housing. Use high-temperature grease. Install the spring clip in the fork. Force the ball stud into the fork recess so that it is engaged by the clip.

   If the clutch fork is the shaft type, assemble by passing the release shaft through the clutch housing and clutch fork. Lubricate the parts, as necessary, and lock them in place with the retaining E-clip. If lock bolts are used, be sure to torque them properly.
5. Pack the inside of the throwout bearing collar with high-temperature grease. Typically, collars have an outer groove for the clutch fork. The groove should be lightly coated with grease. Refer to Figure 7-37.

   Install the throwout bearing assembly on the fork. Use clips or retaining springs to secure the bearing to the fork. In cases where the clutch and transmission housings are one, it will be necessary to slide the bearing assembly onto the hub of the front bearing retainer first. See Figure 7-38.
6. If the clutch housing was removed, install it now. (A clutch housing that is integral with the transmission case is installed with the transmission.) Before installation, check for dirt or paint that could throw the housing out of alignment. As you tighten the clutch housing attaching bolts, make sure that wires or other vehicle components do not get pinched between the housing and the engine block. Tighten bolts in a crisscross pattern. Torque to manufacturer's specifications.
7. Using a transmission jack, install the transmission as outlined in Chapter 9, sliding the input shaft into engagement with the throwout bearing (where clutch housing is separate), clutch disc splines, and pilot bearing. Use the transmission output shaft to turn the input shaft in order to align the input shaft and clutch disc splines. It may be necessary to slightly shift or wiggle the transmission to get the input shaft through the clutch disc and into the pilot bearing. If the pilot shaft was used properly, the input shaft should enter the clutch assembly without too much difficulty.

CAUTION! If the transmission shows any sign of oil leakage through the front bearing retainer, correct the leak before transmission installation.

   If the clutch housing is part of the transmission, it is bolted to the back of the engine; if separate, the transmission case is bolted to clutch housing. Start the attaching bolts as soon as the input shaft is slid into place and the mating housing surfaces are flush. This will keep the transmission from hanging and damaging the clutch hub splines or the input shaft.

CAUTION! Do *not* use the *bell housing* or *transmission case attaching bolts* to try to draw in a binding transmission. This could bend the clutch disc hub or break the bell housing or transmission case ears. The transmission input shaft could damage the pilot bearing. Serious part damage could result. If the clutch and pilot bearing are installed correctly, the transmission should slide fully into place *by hand.*

8. Reconnect the pushrod or cable and return spring connected to the clutch fork. Then, adjust, repair, or replace, as necessary, the linkage as outlined earlier

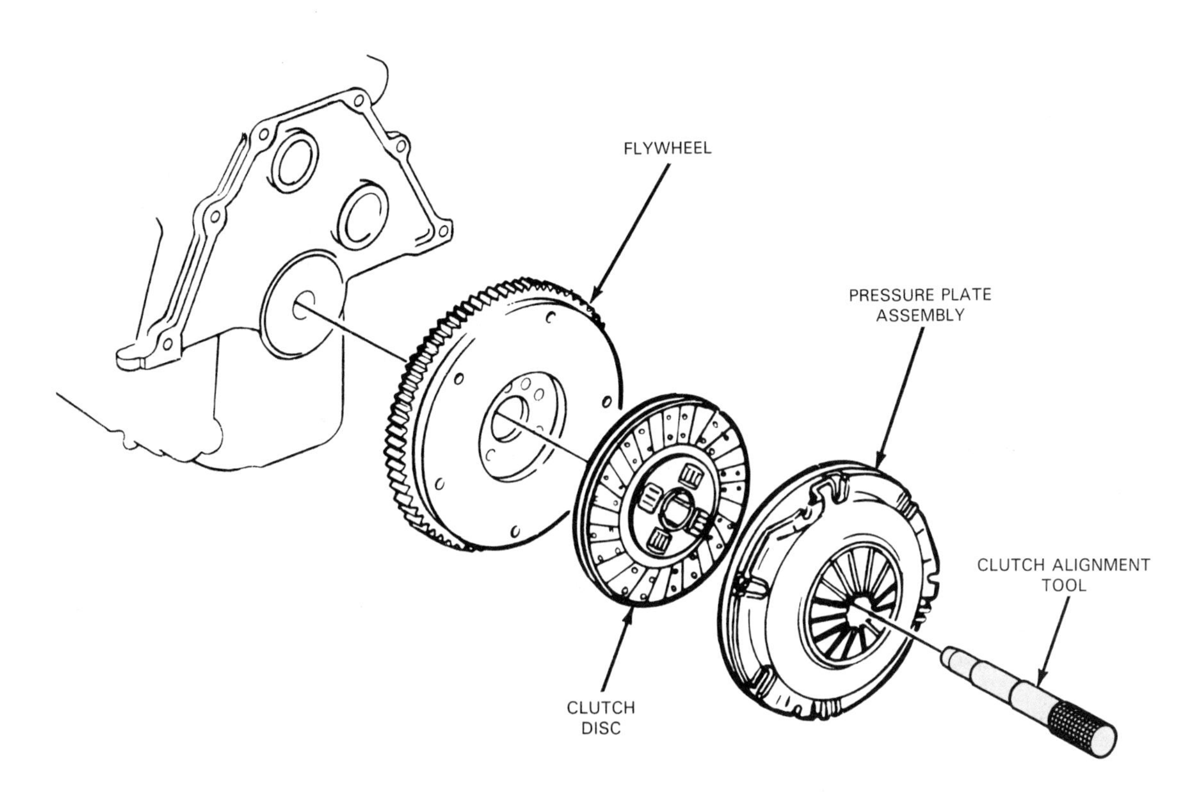

Figure 7-35. Study relative position of clutch parts and clutch pilot shaft. Use of a clutch pilot shaft will make transmission reinstallation much easier. (General Motors)

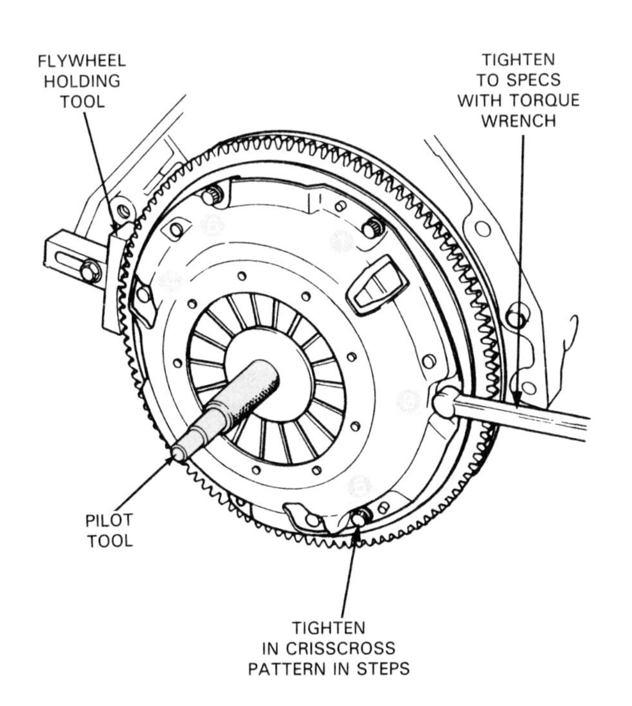

Figure 7-36. The clutch cover attaching bolts should be turned down gradually, in a crisscross pattern. Push up lightly on the pilot shaft during tightening. This will ensure that disc holds its alignment. Bolts should be torqued to specifications. Once the bolts are tight, remove the alignment tool. Try sliding it in and out a few times to see that it does not bind. If it does, the transmission will be difficult to install. (Toyota)

in this chapter. Install any other parts, such as the speedometer cable, shift linkage, or connector wires.

9. Reinstall the driveline as detailed in Chapter 15. After it is installed, ensure that all other drive train connections are made.
10. Check the transmission lubricant and add fluid, if necessary. Reconnect the battery negative cable. Lower the vehicle and road test. During the road test, make sure that there is no slippage, that the clutch engages and disengages smoothly, and that the free play is correct. Operate the clutch at least 25 times to properly seat the clutch mating surfaces. Do not overheat the clutch during this time.

Front-Engine, Front-Wheel Drive

With a few exceptions, the general installation procedure for front-engine, front-wheel drive vehicles duplicates clutch installation for front-engine, rear-wheel drive vehicles. Drive axles are replaced instead of the driveline. Reinstalling the several variations of front drive axles is covered in detail in Chapter 23. Also, in the previous section, substitute references made to the transmission with transaxle. Installing the transaxle is covered in detail in Chapter 19. On vehicles requiring the engine also be removed for clutch replacement, refer to the manufacturer's service manual.

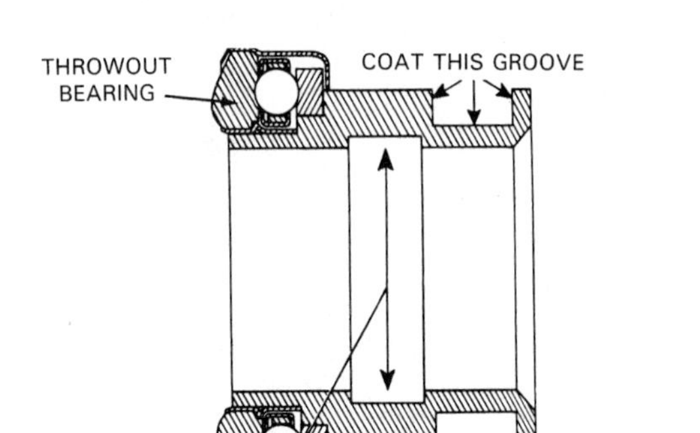

Figure 7-37. If the throwout bearing is not lubricated properly, it will bind on the hub of the transmission's front bearing retainer, or it will wear rapidly. Apply the proper type of grease to the places shown. (Chrysler)

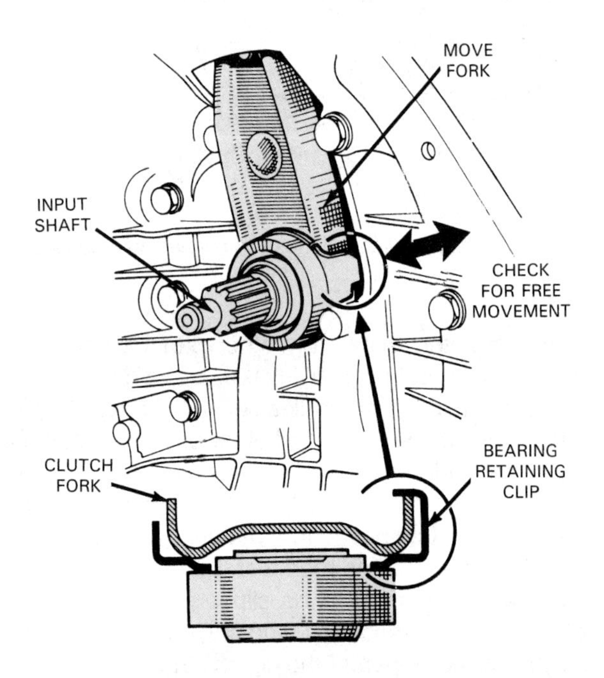

Figure 7-38. In this design, the bell housing is an integral part of the transmission housing. The throwout bearing is placed on the front bearing retainer hub. The clutch fork is installed, and then, the bearing is secured to the clutch fork. Once assembled, the clutch fork and bearing are checked for smooth operation. Note how bearing attaches to clutch fork in this design variation. (Chrysler)

Summary

Clutches can develop problems for many reasons. It is normal for a clutch to wear out after 50,000 mi. (80 000 km). If a clutch wears out sooner, there may be another problem that caused premature clutch failure. The most common cause of early clutch failure is driver abuse.

Clutch problems can be divided into different groups. The most common clutch problem is slippage. Slippage occurs when the clutch is engaged but does not transmit all of the engine power. Slippage can be caused by a worn-out clutch, improper linkage adjustments, or oil on the clutch facings. Less common causes are a warped flywheel, friction disc, or pressure plate, or worn pressure plate apply springs. A slipping clutch should be fixed before excess heat ruins other clutch parts.

Clutch vibration can be caused by loose or unbalanced clutch parts. Loose parts usually also cause noise. Sometimes, what appears to be a clutch vibration is caused by loose engine mounts, loose clutch mounting bolts, or excessive crankshaft end play.

Clutch noises are often caused by defective bearings. The defective bearing can be isolated by operating the clutch pedal. Depending on clutch pedal position, either the pilot bearing, throwout bearing, or transmission front bearing can be at fault. Rattles and knocking noises are almost always caused by loose parts.

Clutch grabbing is usually caused by oil on the clutch facings. Other possible causes are glazed clutch facings or wear and misalignment of the throwout bearing assembly or clutch fork.

Clutch chatter is similar to grabbing, but it is heard more than felt. It can be caused by some of the same defects that cause grabbing. Sometimes chatter is caused by a buildup of clutch dust on the friction facings.

A pedal that is hard to push may be caused by a binding condition in the linkage or a throwout bearing that is sticking on the front bearing retainer hub. A new pressure plate assembly with excessively stiff springs may have been installed.

A clutch that does not release may have a defect in the linkage. Rod-and-lever linkages or cable linkages may be broken or disconnected. A hydraulic linkage may be out of fluid. If the linkage checks out, the problem is inside the clutch assembly. In some cases, the clutch disc may have welded itself to the flywheel or pressure plate.

All types of clutch linkage have some provision for adjusting the free play. Free play is measured at the clutch pedal. The usual free play measurement is between 1 in.–1.5 in. (25 mm–38 mm).

The clutch linkage may need lubrication or repair. Rod-and-lever linkages may require new bushings or realignment of linkage parts. Cable linkages may have problems due to a corroded or kinked cable. The cable must usually be replaced.

Hydraulic linkages can leak, or the hydraulic lines can be kinked or swelled shut. Leaks can be spotted by a visual examination.

Master and slave cylinders can be rebuilt if necessary. After rebuilding, or whenever the hydraulic system has been opened, air must be removed from the system. The process of removing air is called bleeding.

To remove the clutch from a vehicle, other components must be removed. In most cases, the driveline or drive axles must be disconnected from the vehicle. The transmission and clutch housing are then removed to reach the clutch assembly.

To remove the clutch assembly, the clutch cover attaching bolts are removed from the flywheel. Then, the pressure plate assembly and friction disc are removed from the flywheel.

The flywheel and pilot bearing should be checked for defects. A worn or damaged pilot bearing should be replaced. An overheated or warped flywheel can be resurfaced by a machine shop. The flywheel ring gear teeth should be checked for damage. A damaged ring gear can be removed. The replacement ring gear is installed by heating it and placing it over the flywheel.

The friction disc is almost always replaced during a clutch teardown. The disc can be checked to determine the cause of failure. This may help avoid a similar clutch failure in the future.

Always check the new disc to make sure that it is the proper replacement. Many clutch problems after overhaul are caused by using the improper clutch disc.

The pressure plate assembly is usually replaced as part of a clutch overhaul. Some assemblies can be rebuilt. The rebuilding process requires the use of a press.

The throwout bearing should be checked for roughness and lack of grease. Always replace a throwout bearing that shows any sign of defects. The throwout bearing can be pressed from its collar, and a new bearing can be installed. The clutch fork should be checked for wear. Worn forks should be replaced.

Clutch housings do not wear out, but they should be checked for cracks and warping. The housing may contain a ball stud for the clutch fork. The stud should be checked for wear and replaced if necessary.

When reassembling the clutch, always make sure that your hands are free from oil. Oil or grease on the friction disc will cause grabbing or slippage. A pilot shaft should be used to ensure that the disc hub splines and pilot hole line up closely. If proper alignment is not obtained, the transmission will be difficult to install. The clutch cover attaching bolts should always be tightened with the pilot shaft in place.

After the clutch and other drive train parts are reinstalled, the clutch pedal free play should be adjusted to specifications. Road test the vehicle as the final step.

Know These Terms

Riding the clutch; Dumping (popping) the clutch; Clutch drag; Clutch slippage; Grabbing clutch; Clutch chatter; Clutch vibration; In-car service; Clutch adjustment; Constant-duty throwout bearing; Runout; Radial runout; Lateral runout; Bleeder valve.

Review Questions–Chapter 7

Please do not write in this text. Place your answers on a separate sheet of paper.

1. Explain what the phrase *riding the clutch* means. What affects can it have on the clutch over time?
2. Which of the following items concerning the phrase *dumping the clutch* is *not* true?
   a. Pressure plate apply springs engage gradually.
   b. Means increasing engine speed and then suddenly releasing the clutch pedal.
   c. Causes instant heat buildup on clutch disc facings and places tremendous stress on the drive train.
   d. Is a form of clutch abuse.
3. Vibration that occurs when the clutch is being engaged or disengaged is called _____ _____.
4. A deep knocking noise that is similar to an engine main or rod bearing knock is an indication of:
   a. Defective bearings.
   b. A loose clutch fork.
   c. Weak retracting or cushion springs.
   d. Loose flywheel attaching bolts.
5. Clutch adjustment involves:
   a. Resurfacing the flywheel.
   b. Rebuilding the pressure plate assembly.
   c. Setting clutch free play.
   d. All of the above.
6. The very first thing you should do when removing a clutch is remove the driveline or drive axles. True or false?
7. Why should you not let a transmission hang after removing its mounts?
8. Briefly, cite two methods that may be used to remove a pilot bearing from the crankshaft bore.
9. When inspecting a flywheel, what trouble signs should you look for?
10. Briefly summarize typical steps involved in the clutch reassembly process for a front-engine, rear-wheel drive vehicle.
11. A car has a slipping clutch. The car will still move, but the engine races when accelerating. Technician A says to pull the transmission and replace the clutch. Technician B says to adjust the clutch linkage first. Who is correct?
    a. Technician A.
    b. Technician B.
    c. Both Technician A and B.
    d. Neither Technician A nor B.
12. A car shudders when accelerating from a standstill as the clutch pedal is released. Technician A says to check for loose, softened, or broken motor mounts.

Technician B says to check the clutch and flywheel friction surfaces for problems. Who is correct?
a. Technician A.
b. Technician B.
c. Both Technician A and B.
d. Neither Technician A nor B.

13. A driver complains that the clutch is not working properly. The car is equipped with a hydraulic linkage. An inspection shows that the slave cylinder is not responding to pedal action. Fluid level is alright. Technician A says to try bleeding the hydraulic system first. Technician B says to replace the seals in the system first. Who is correct?
a. Technician A.
b. Technician B.
c. Both Technician A and B.
d. Neither Technician A nor B.

14. A clutch makes a grinding sound as the clutch pedal is depressed and the engine is running. Technician A says the problem is probably a worn friction disc facing. Technician B says that the flywheel is probably warped. Who is correct?
a. Technician A.
b. Technician B.
c. Both Technician A and B.
d. Neither Technician A nor B.

Certification-Type Questions–Chapter 7

1. All of these operator actions can cause early clutch failure EXCEPT:
(A) riding the clutch pedal.
(B) disengaging the clutch when decelerating.
(C) dumping the clutch.
(D) overloading the vehicle.

2. Technician A says that light clutch drag can cause gear clashing. Technician B says that clutch disc spindown can be mistaken for clutch drag. Who is right?
(A) A only
(B) B only
(C) Both A & B
(D) Neither A nor B

3. Technician A says that some clutch slippage is necessary for smooth shifts. Technician B says that an oil-soaked clutch disc will slip, even if it is not worn. Who is right?
(A) A only
(B) B only
(C) Both A & B
(D) Neither A nor B

4. All of these can cause clutch slippage EXCEPT:
(A) excessive pressure plate apply spring pressure.
(B) warped flywheel or pressure plate surfaces.
(C) worn friction disc.
(D) misadjusted linkage.

5. Technician A says that clutch grabbing and clutch chatter are the same condition. Technician B says that clutch vibration or chatter could be caused by broken engine or transmission mounts. Who is right?
(A) A only
(B) B only
(C) Both A & B
(D) Neither A nor B

6. If a noise is heard when the clutch pedal is completely released (up position) with the engine running, which of these bearings is probably defective?
(A) Engine rear main bearing.
(B) Clutch pilot bearing.
(C) Throwout bearing.
(D) Transmission front bearing.

7. Technician A says that automatic transmission fluid can be used to refill a hydraulic clutch linkage reservoir. Technician B says that kinked clutch linkage cable can be lubricated to restore normal operation. Who is right?

(A) A only
(B) B only
(C) Both A & B
(D) Neither A nor B

8. All of these are caused by improper clutch adjustment EXCEPT:

(A) clutch drag.
(B) clutch slippage.
(C) grabbing clutch.
(D) throwout bearing wear.

9. Technician A says that a badly slipping clutch usually cannot be fixed by adjusting the linkage. Technician B says that free play at the clutch pedal should be measured before attempting to adjust the clutch linkage. Who is right?

(A) A only
(B) B only
(C) Both A & B
(D) Neither A nor B

10. Opening the slave cylinder bleeder valve while pressing on the clutch pedal will cause the pedal to:

(A) slowly rise.
(B) become hard to push.
(C) go to the floor.
(D) lock up.

11. A pilot shaft is used to align the pilot bearing and the:

(A) pressure plate.
(B) clutch disc.
(C) input shaft.
(D) throwout bearing.

12. Technician A says that one should never work under a vehicle that is supported only by a hydraulic jack. Technician B says that jackstands should always be placed under a vehicle that has been raised on a hoist. Who is right?

(A) A only
(B) B only
(C) Both A & B
(D) Neither A nor B

13. Technician A says that pilot bearings should be carefully checked once the clutch is removed. Technician B says that the flywheel should not be removed unless it is damaged or needs resurfacing. Who is right?

(A) A only
(B) B only
(C) Both A & B
(D) Neither A nor B

14. A flywheel has deep scoring and cracks. Technician A says that any amount of metal can be cut from the flywheel to fix this condition. Technician B says that the flywheel can be lightly sanded and reused. Who is right?

(A) A only
(B) B only
(C) Both A & B
(D) Neither A nor B

15. Technician A says that a clutch disc should be reused unless wear or damage is severe. Technician B says that all replacement clutch parts should be carefully compared to the old parts to help ensure that they will work properly. Who is right?

(A) A only
(B) B only
(C) Both A & B
(D) Neither A nor B

16. Technician A says that a pressure plate must apply evenly to prevent clutch grabbing. Technician B says that the throwout bearing should be replaced if it shows any signs of wear or damage. Who is right?

(A) A only
(B) B only
(C) Both A & B
(D) Neither A nor B

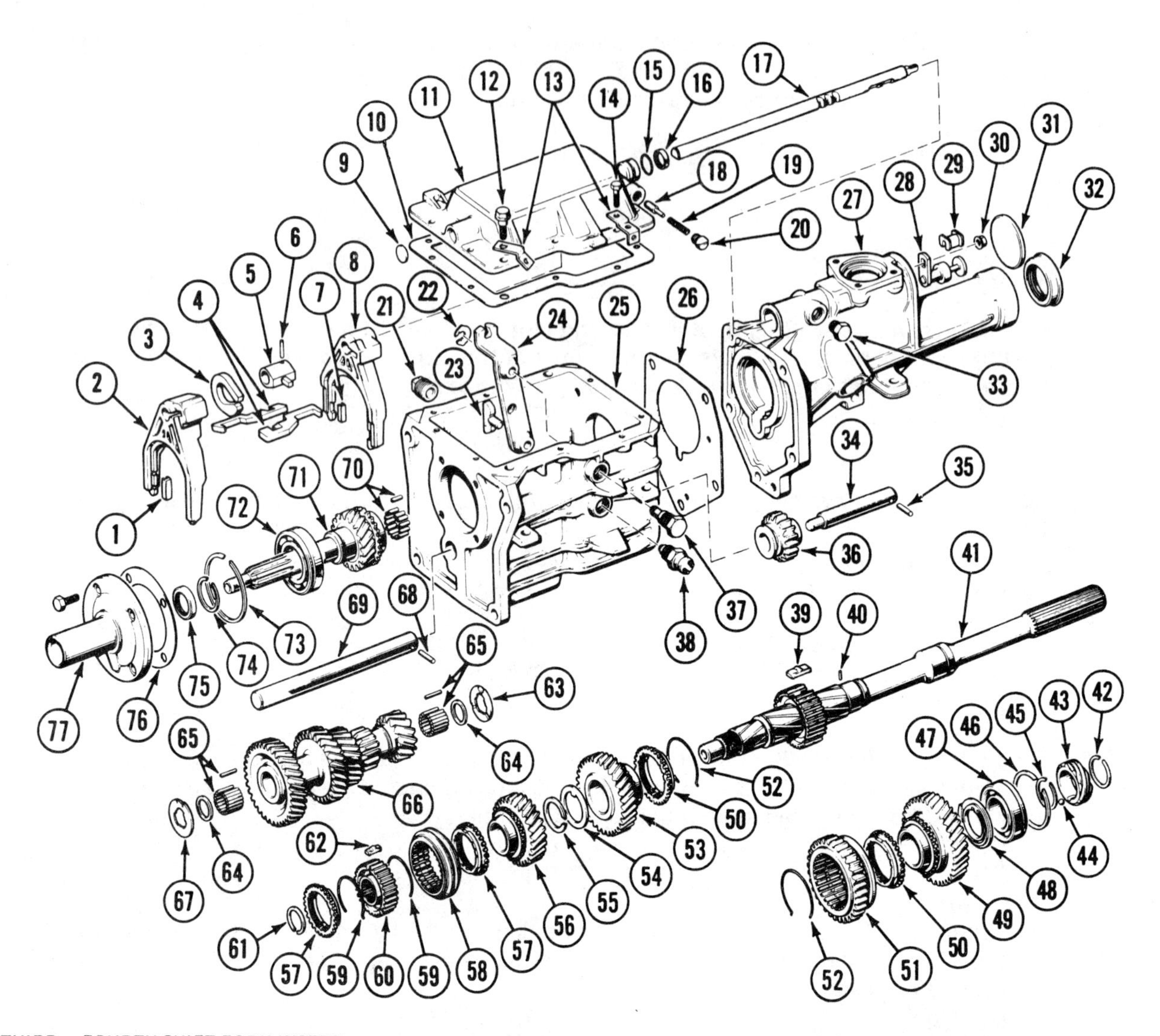

1. THIRD – FOURTH SHIFT FORK INSERT
2. THIRD – FOURTH SHIFT FORK
3. SELECTOR INTERLOCK PLATE
4. SELECTOR ARM PLATE (2)
5. SELECTOR ARM
6. SELECTOR ARM ROLL PIN
7. FIRST – SECOND SHIFT FORK INSERT
8. FIRST – SECOND SHIFT FORK
9. SHIFT RAIL PLUG
10. TRANSMISSION COVER GASKET
11. TRANSMISSION COVER
12. TRANSMISSION COVER DOWEL BOLT (2)
13. CLIP
14. TRANSMISSION COVER BOLT (8)
15. SHIFT RAIL O-RING SEAL
16. SHIFT RAIL OIL SEAL
17. SHIFT RAIL
18. DETENT PLUNGER
19. DETENT SPRING
20. DETENT PLUG
21. FILL PLUG
22. REVERSE LEVER PIVOT BOLT C-CLIP
23. REVERSE LEVER FORK
24. REVERSE LEVER
25. TRANSMISSION CASE
26. EXTENSION HOUSING GASKET
27. EXTENSION HOUSING
28. OFFSET LEVER
29. OFFSET LEVER INSERT
30. OFFSET LEVER RETAINING NUT
31. ACCESS PLUG
32. EXTENSION HOUSING OIL SEAL
33. THREADED PLUG
34. REVERSE IDLER SHAFT
35. REVERSE IDLER SHAFT ROLL PIN
36. REVERSE IDLER GEAR
37. REVERSE LEVER PIVOT BOLT
38. BACKUP LAMP SWITCH
39. FIRST – SECOND SYNCHRONIZER INSERT (3)
40. FIRST GEAR ROLL PIN
41. OUTPUT SHAFT AND HUB ASSEMBLY
42. SPEEDOMETER GEAR SNAP RING
43. SPEEDOMETER GEAR
44. SPEEDOMETER GEAR DRIVE BALL
45. REAR BEARING RETAINING SNAP RING
46. REAR BEARING LOCATING SNAP RING
47. REAR BEARING
48. FIRST GEAR THRUST WASHER
49. FIRST GEAR
50. FIRST – SECOND SYNCHRONIZER BLOCKING RING (2)
51. FIRST – REVERSE SLEEVE AND GEAR
52. FIRST – SECOND SYNCHRONIZER INSERT SPRING (2)
53. SECOND GEAR
54. SECOND GEAR THRUST WASHER (TABBED)
55. SECOND GEAR SNAP RING
56. THIRD GEAR
57. THIRD – FOURTH SYNCHRONIZER BLOCKING RING (2)
58. THIRD – FOURTH SYNCHRONIZER SLEEVE
59. THIRD – FOURTH SYNCHRONIZER INSERT SPRING (2)
60. THIRD – FOURTH SYNCHRONIZER HUB
61. OUTPUT SHAFT SNAP RING
62. THIRD – FOURTH SYNCHRONIZER INSERT (3)
63. COUNTERSHAFT GEAR REAR THRUST WASHER (METAL)
64. COUNTERSHAFT NEEDLE BEARING RETAINER (2)
65. COUNTERSHAFT NEEDLE BEARING (50)
66. COUNTERSHAFT GEAR
67. COUNTERSHAFT GEAR FRONT THRUST WASHER (PLASTIC)
68. COUNTERSHAFT ROLL PIN
69. COUNTERSHAFT
70. CLUTCH SHAFT ROLLER BEARINGS (15)
71. CLUTCH SHAFT
72. FRONT BEARING
73. FRONT BEARING LOCATING SNAP RING
74. FRONT BEARING RETAINING SNAP RING
75. FRONT BEARING CAP OIL SEAL
76. FRONT BEARING CAP GASKET
77. FRONT BEARING CAP

Exploded view of a 4-speed transmission. (Chrysler)

Chapter 8

Manual Transmission Construction and Operation

After studying this chapter, you will be able to:

- Explain the purpose of the manual transmission.
- Describe how a vehicle with a manual transmission is operated.
- Identify the major parts of a manual transmission.
- Discuss the purpose of each major part of a manual transmission.
- Explain the interactions of major manual transmission components.
- Describe how the major parts of a manual transmission are constructed.
- Explain the operation of a synchronizer.
- Diagram power flow through a typical manual transmission.
- Compare manual transmission designs.

This chapter will explain the construction and operation of manual transmissions used on rear-wheel drive vehicles. At one time, such transmissions were the only kind used in passenger cars. Today, *automatic* transmissions and manual and automatic *transaxles* are installed in many vehicles. Although manual transmissions are installed in a smaller percentage of vehicles, you must still be knowledgeable about their construction and operation.

Manual transaxles are similar in design and operation to manual transmissions, and many of the same principles apply. Also, many manual transmission principles are used in automatic transmissions. Understanding this chapter will help you to grasp automatic transmission and manual and automatic transaxle operation, presented in upcoming chapters.

Purpose of the Manual Transmission

The purpose of the manual transmission (or automatic transmission) is to multiply engine torque or reduce engine rpm to match varying forward operating conditions. The transmission uses different gear ratios to achieve these results. (Review Chapter 3 for previous discussion on gear ratio.)

The transmission also provides a way to back up a vehicle. Transmissions make use of idler gears–gears located between drive and driven gears–for this purpose. Idler gears, which do not change gear ratio, cause the transmission output to rotate opposite of the input. In this way, they allow a vehicle to be reversed. (Review Chapter 3 for previous discussion on idler gears.)

Transmissions provide a way to optimize engine power and torque. For one, torque requirements are greater for moving a vehicle from a standstill; however, engines provide less torque at low rpm. Without a transmission, the engine would tend to stall as power was engaged to move the vehicle. High engine rpm would be needed to provide enough torque to move the car from rest, and initial acceleration would be jerky and unacceptable. Secondly, the internal combustion engine develops maximum power over only a very narrow rpm range. This narrow range occurs at relatively high engine speeds–several thousand revolutions per minute. The transmission provides a way to operate the vehicle in this range under many speed and load conditions.

To use available engine power in the most efficient manner, transmissions use several different gear ratios–*first gear, second gear, third gear,* etc. The transmission lower gears multiply (increase) torque while they reduce speed at the *gearbox* output. As vehicle speed increases, less torque multiplication is needed. A higher gear, having less gear reduction, is selected. Output shaft speed increases as a result, allowing engine rpm to be reduced to its maximum power range.

At highway, or *cruising,* speeds, torque multiplication is not a requirement, and high gear (direct drive) is selected. It provides no gear reduction. Transmission output shaft speed again increases. Engine speed may be reduced to operate within its maximum power range again. Some vehicles have an overdrive gear for highway travel.

These have a gear ratio of less than one and allow the engine to run even slower and more efficiently.

Operating the Manual Transmission

In a manual transmission, such as the one shown in Figure 8-1, the gears are selected by the driver. First, the clutch pedal is depressed to disengage the engine from the transmission. Then, the driver moves the hand-operated transmission ***gearshift lever*** to select the proper gear. After the gear is selected, the clutch pedal is carefully released, to engage the clutch, while depressing the throttle pedal. This action allows for a smooth coupling of power between the engine and transmission. The vehicle is then *in gear,* and power is transmitted to the rear wheels.

In the accelerating mode, as the vehicle gains speed, the gearshift lever is moved to the next higher gear. This operation, called ***upshifting,*** brings the vehicle back into its peak operating range. Before moving the lever, the clutch pedal must be depressed, and at the same time, the throttle pedal is let up. After the gearshift lever is moved, the clutch pedal is released while the throttle pedal is pushed down again.

In the decelerating mode, the same basic operation applies; however, the gearshift lever is moved into lower gears. This operation, called ***downshifting,*** brings the vehicle back into the peak operating range. In general, lower gears are used when the load is heavy or when vehicle speed is low. If an increased load, such as a hill, is encountered, the driver may downshift from high gear, for example, down to third or second gear.

When the vehicle is at a stop, the clutch pedal can be held down, or the transmission can be taken out of gear–that is, the gearshift lever can be moved to the *neutral* (no gear) position. Either action interrupts the power flow to the drive wheels.

This summarizes basic external operation of the manual transmission. Once the driver learns to coordinate the movements of the gearshift lever, the clutch pedal, and the throttle pedal, manual transmission operation becomes an unconscious effort.

Construction and Internal Operation

To prepare you for more specific details of transmission construction and operation, refer back to Figure 8-1 and study the parts of the transmission. Learn to identify and locate the fundamental components. These components, which will be presented in detail in this section, include the following:

- The ***manual transmission shafts*** support gears and directly or indirectly transfer rotation from the clutch disc to the driveline.
- The ***transmission gears*** transmit power. They provide a means of changing torque, speed, and direction at the gearbox output.
- The ***synchronizers*** bring certain transmission gears to the same rotational speed as the output shaft before sliding in mesh with them to lock them to the output shaft.
- The ***shift forks*** are pronged units for moving gears or synchronizers on their shaft for gear engagement.

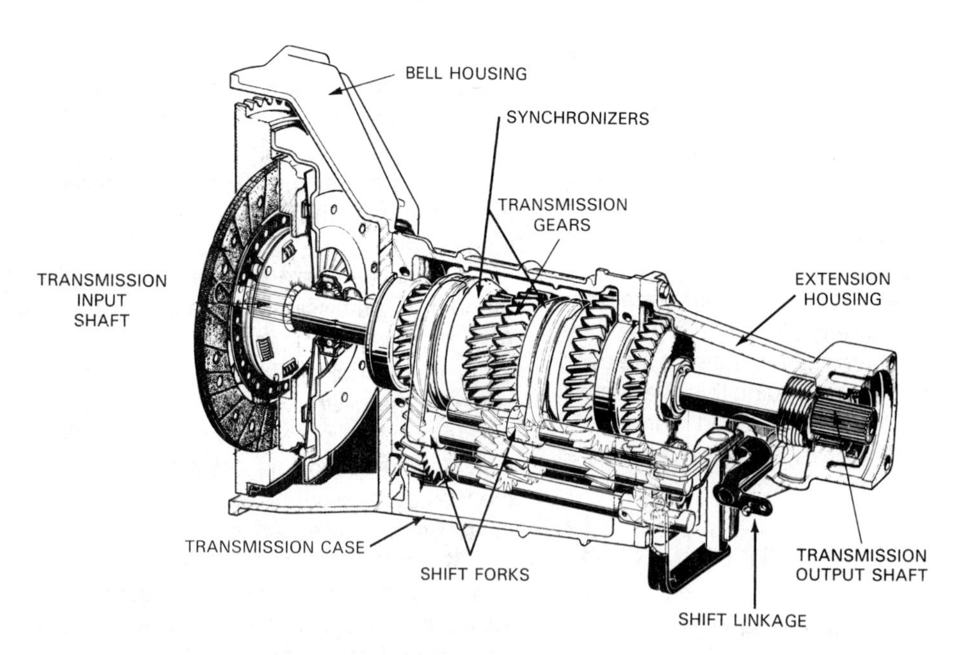

Figure 8-1. Cutaway shows parts of a typical rear-wheel drive manual transmission. Note the position of the transmission in relation to the clutch. The output shaft turns the drive shaft. All rear-wheel drive transmissions are in this relative position–directly behind the engine and directly ahead of the driveline. (Peugeot)

- The ***manual transmission shift linkage*** allows the driver to operate the shift forks.
- The ***manual transmission case*** encloses transmission shafts, gears, synchronizers, shift forks and linkage, and lubricating oil. In some designs, it is combined with the clutch housing.
- The ***extension housing*** encloses the tail end of the transmission output shaft, which extends from the transmission case. It also encloses part of the shift linkage.

In addition to these seven major parts, there are numerous accessory devices mounted on the transmission. Some of these devices, such as the *speedometer assembly,* receive their input from the transmission. These accessories will be discussed in this section.

Manual Transmission Shafts

Transmission shafts are directly or indirectly involved in transmitting motion. A shaft may rotate to perform its function, or it can be a stationary support for a moving gear. Transmission shafts are made from high-strength steel. They are closely machined and polished. Often, they are heat-treated to increase their strength and resistance to wear.

All rotating shafts are supported by bearings attached to the case. These shafts are machined with splines and grooves to mate with other transmission parts, such as gears, bearings, and snap rings.

Nonrotating shafts are machined smooth, so that the gears can rotate on them easily. They are held in place by pins, press fit, or locknuts. In some instances, the shaft is not attached to the case and can rotate in a random manner.

There are many variations of shaft design and placement in manual transmissions. Figure 8-2 shows the four basic shafts found in all manual transmissions; these are the *manual transmission input shaft,* the *countershaft,* the *manual transmission output shaft,* and the *reverse idler shaft.*

Manual transmission input shaft

The manual transmission input shaft, splined to the clutch disc, delivers power into the transmission. Sometimes called the ***clutch shaft,*** this rotating shaft transfers rotation from the clutch disc to the *countershaft gear.* The input shaft, as described in Chapter 6, is supported on the front end by the pilot bearing. Machined on the inner end of the shaft and contained within the transmission case is the transmission main drive gear. The transmission front bearing, which is held by the transmission case, supports the inner end of the input shaft.

The input shaft and front bearing are held in place by the front bearing retainer. Forward shaft movement is prevented by this retainer. Rearward movement of the input shaft is usually prevented by the output shaft. The front bearing is kept from moving by a snap ring, which holds it solidly to the case. In some instances, a snap ring is used for holding the main drive gear to the bearing to prevent lateral shaft movement.

Countershaft

The ***countershaft*** holds several gears of different sizes. It is the central shaft of the countershaft gear assembly. It is located below the input shaft. Normally, the countershaft does *not* turn in the transmission case. Instead, it is locked in place.

Manual transmission output shaft

The ***manual transmission output shaft,*** also called the ***mainshaft,*** is attached to the vehicle driveline. This rotating shaft delivers power to the driveline. This shaft holds the output gears and synchronizers.

The front end of the mainshaft fits into the end of the input shaft, where it rides on needle bearings, sometimes

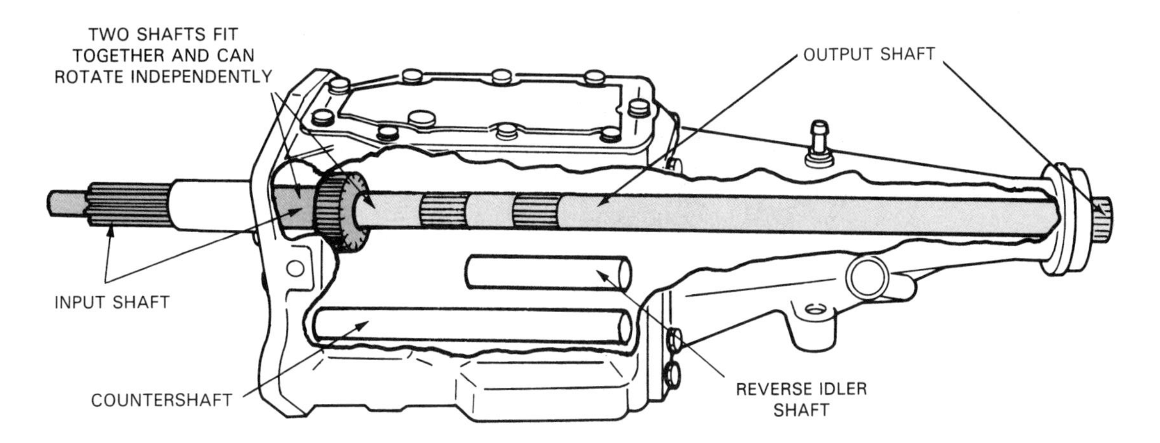

Figure 8-2. This shows the major shafts that are found in manual transmissions. Note that needle bearings are used between input and output shafts since they often rotate at different speeds.

called the *mainshaft pilot needle bearings.* The mainshaft and the input shaft are aligned, so they look like one shaft. The two shafts, however, are free to rotate separately (except in direct drive, where they are locked together and rotate at the same speed). Sometimes, a thrust bearing is used to control end play between the input and output shafts.

The mainshaft extends to the back of the extension housing. The rear end is splined and connects to the driveline. The shaft is supported by one or two antifriction bearings. Typically, the transmission ***rear bearing*** provides the central support for the shaft. The rear bearing is commonly held in place either by the transmission case or the front part of the extension housing; however, in some designs the rear bearing is held in the rear of the extension housing.

Reverse idler shaft

The ***reverse idler shaft*** is a short shaft that supports the reverse idler gear. It normally mounts stationary in the case, about midway between the countershaft and output shaft.

Transmission Gears

Transmission gears transmit power. These are gears of different sizes that are used to obtain different gear ratios–either reducing output speed to multiply torque or increasing output speed to reduce engine rpm. In addition, certain other transmission gears are used to move a vehicle in reverse.

Transmission gears are made of a high-strength steel. They are machined and, then, hardened by a heat-treating process. Two types of gears are commonly used in manual transmissions. Spur gears, which have straight-cut teeth, are used for gears that are only used occasionally. An example is reverse gear. Most other transmission gears are helical gears. Helical gears run more quietly and are stronger because of increased contact of the teeth, which are cut at an angle. Helical gears must be mounted with a minimum of end play, since they tend to slide apart due to thrust forces caused by the angular tooth contact.

Transmission gears are held on transmission shafts. Some transmission gears are an integral part of the shaft; others are splined to the transmission shaft, and others rotate on the shaft. Transmission gears are classified with their associated shaft. The four types include the *main drive gear, countershaft gear, mainshaft gears,* and *reverse idler gear.* See Figure 8-3.

Main drive gear

The transmission input shaft and ***main drive gear*** are a one-piece unit, Figure 8-4. The shaft is splined to the clutch disc. The shaft and main drive gear are turned by the clutch disc whenever the clutch is engaged. When the transmission is in neutral (clutch engaged), the input shaft turns, but the output shaft does not turn. Figure 8-5 shows input shaft, main drive gear, and related parts in exploded view.

Countershaft gear

The ***countershaft gear,*** or ***cluster gear,*** as it is also called, is a series of helical gears machined out of a single piece of steel. An exploded view of a typical *countershaft gear assembly,* which includes the gear, countershaft, and related parts, is shown in Figure 8-6.

The largest gear on the countershaft gear is always the *countershaft driven gear,* which is the forwardmost gear. This gear is always in mesh with the main drive gear, as

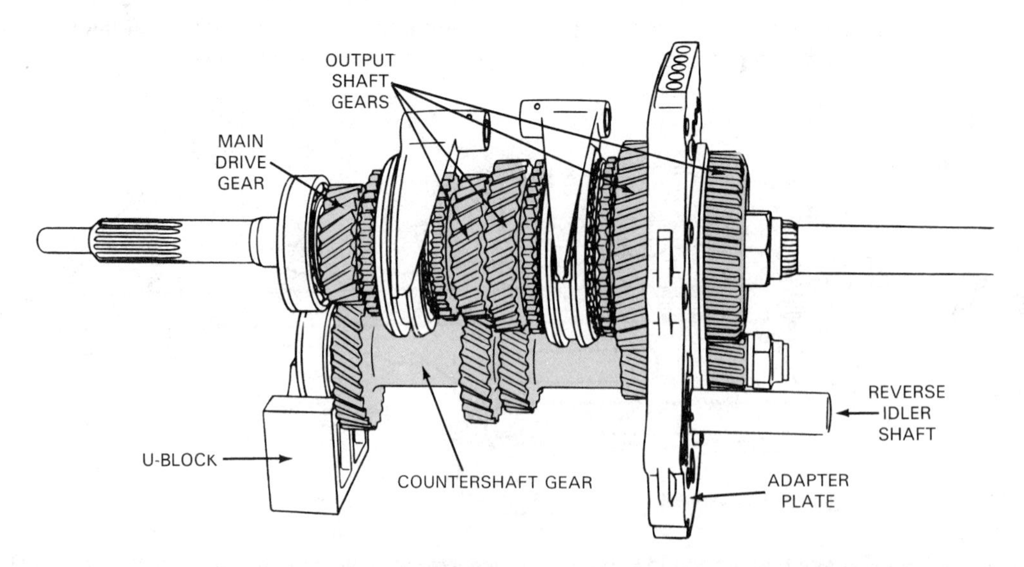

Figure 8-3. Note typical manual transmission gears. Reverse idler gear is not shown. (General Motors)

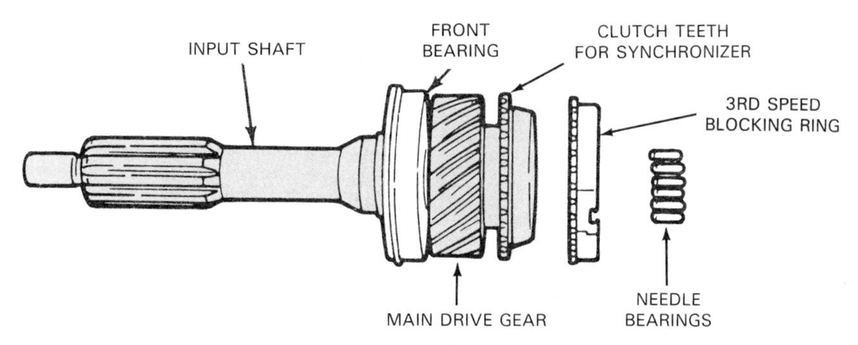

Figure 8-4. Study typical input shaft and its integral drive gear. Most are one-piece, hardened-steel castings. The front bearing is pressed on the input shaft. It can be removed if defective. Needle bearings are placed inside the bore at the end of the input shaft. Output shaft rides on needle bearings. (General Motors)

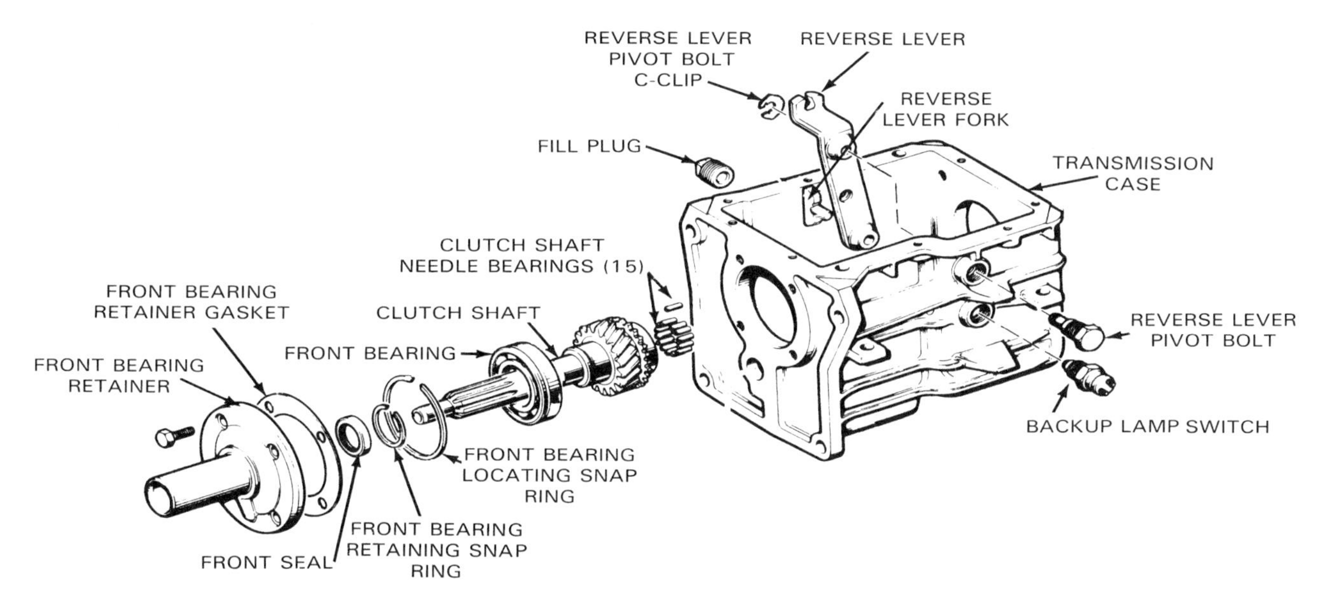

Figure 8-5. Exploded view shows the position of input shaft and related parts in relation to the transmission case. (Chrysler)

shown in Figure 8-7. The entire countershaft gear turns whenever the main drive gear turns.

In addition to the driven gear, there is usually one gear on the gear cluster for each available reduction and overdrive ratio, plus a gear for reverse. There is no gear for direct drive (1:1 gear ratio); the countershaft gear is not used in direct drive. The cluster gear of a *5-speed transmission* with overdrive, for instance, has six gears. In addition to the countershaft driven gear, it has first, second, third, overdrive, and reverse gears.

The size of the individual gears on the cluster increases with each drop in gear ratio. For example, second gear has a smaller ratio than first gear, and it is larger in size than first gear. Third gear has a smaller ratio than second gear; it is larger than second gear.

In most designs, the countershaft gear rotates on a countershaft that is stationary. Needle bearings are used between the cluster gear and shaft to reduce wear. Needle bearings may be loose, or they may be *caged.* Some designs incorporate *caged needle bearings* because they are easier to work with. The bearings do not fall all over the bench during disassembly, and installing the bearings is also much easier.

End play, or back-and-forth movement, of the countershaft gear is controlled by thrust washers placed between the ends of the assembly and the transmission case. In some transmissions, the countershaft gear is supported and end play is controlled through use of tapered roller bearings. Figure 8-8 shows tapered roller bearings used on a countershaft gear.

Some gears are in constant mesh with the countershaft gear; one is the main drive gear, and the other is the reverse idler gear. In modern transmissions, all forward gears of the countershaft are in constant mesh with the mainshaft gears, which themselves are free to rotate. These are sometimes referred to as ***constant-mesh transmissions.*** Reverse gear may be the constant-mesh type, or it may be the type that slides in and out of

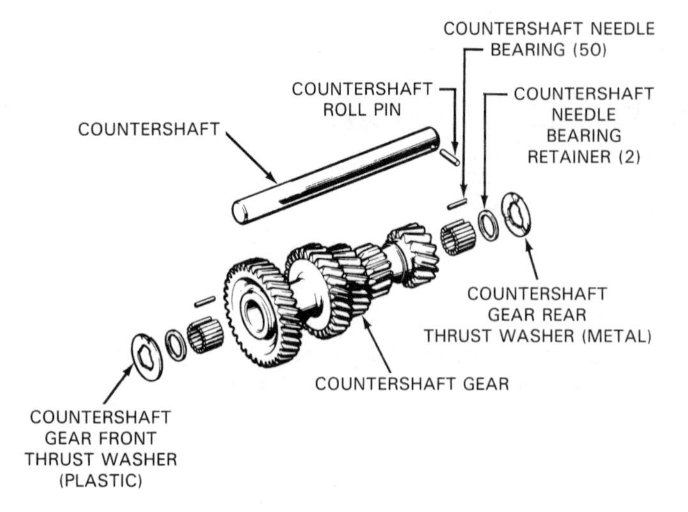

Figure 8-6. Exploded view shows parts of the countershaft gear assembly. This shaft is locked in place by countershaft roll pin. Thrust washers control end play in countershaft gear. Gear rides on needle bearings. (Chrysler)

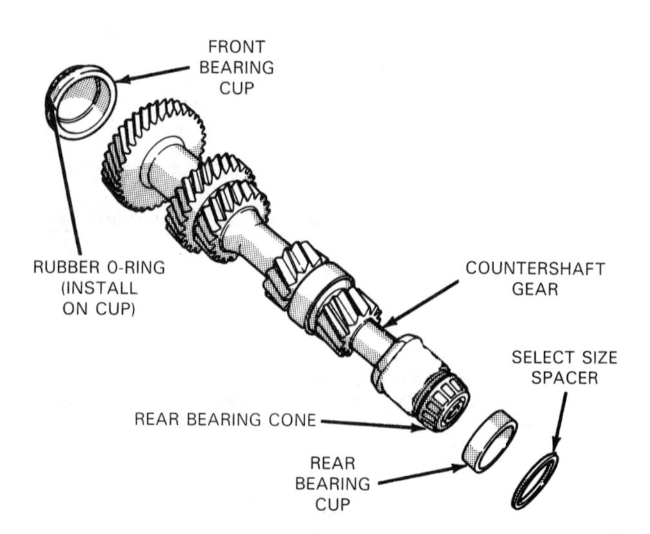

Figure 8-8. The countershaft gear sometimes gets its support from tapered roller bearings on each end. The bearings also control end play. This design does not use a countershaft. (Chrysler)

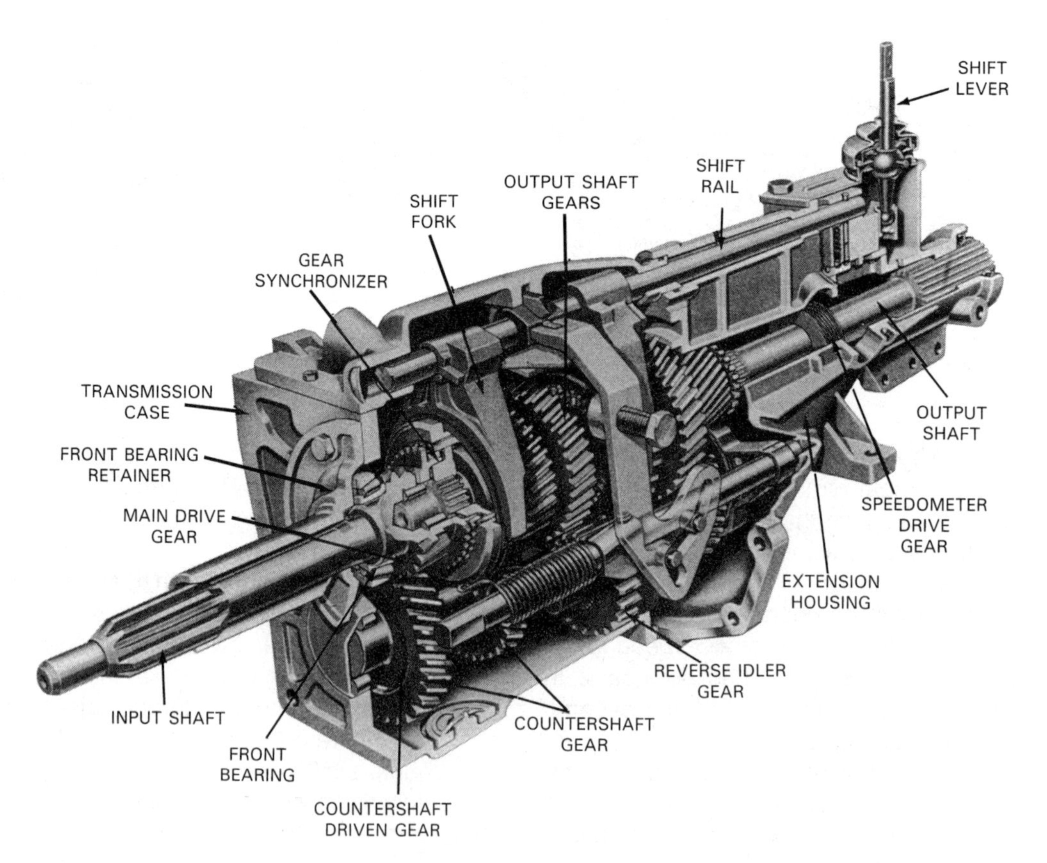

Figure 8-7. Cutaway points out various parts of a modern 5-speed manual transmission with overdrive. Notice how main drive gear fits into the transmission case and how it meshes with the countershaft gear. (Ford)

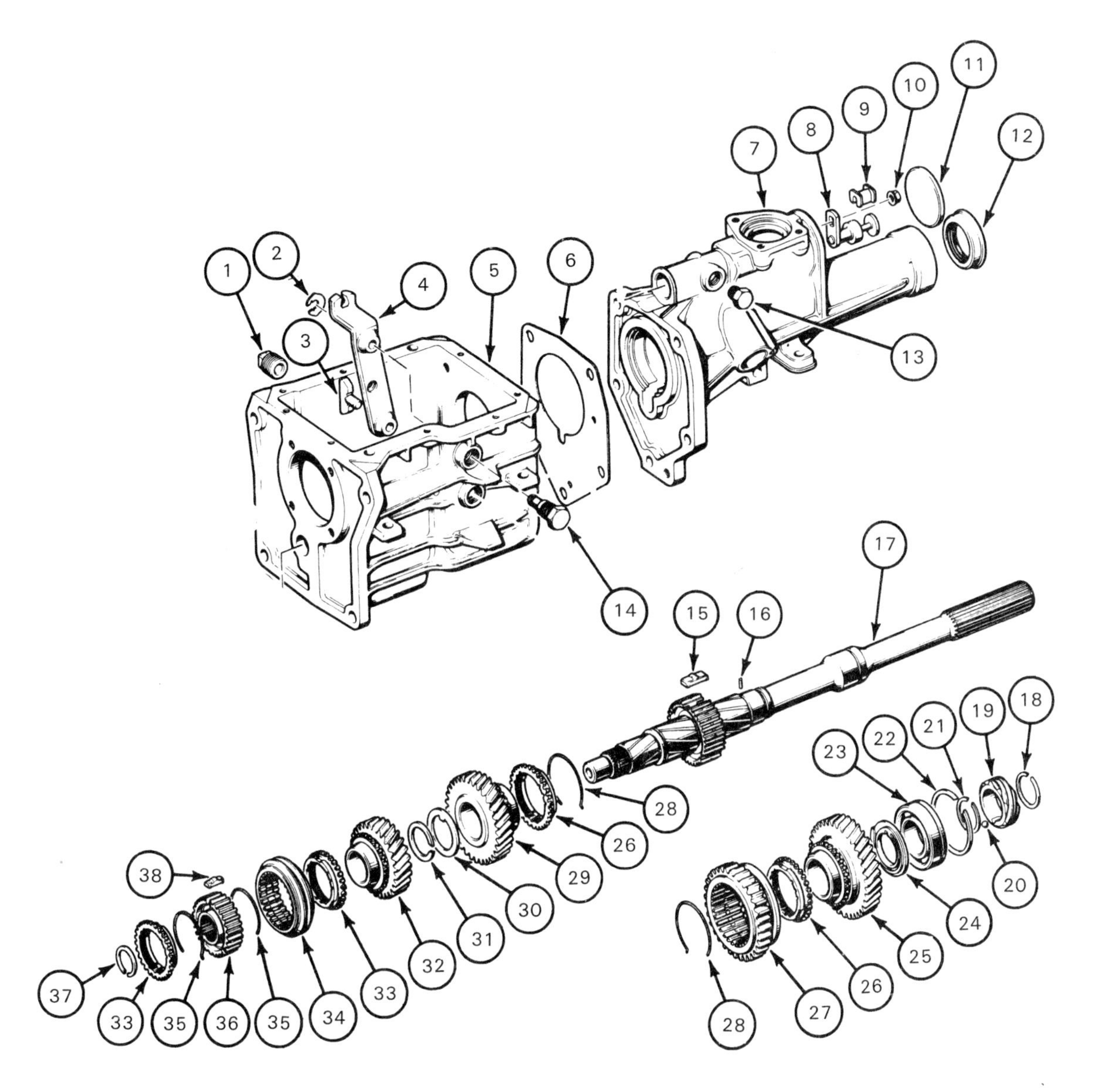

1. FILL PLUG
2. REVERSE LEVER PIVOT BOLT C-CLIP
3. REVERSE LEVER FORK
4. REVERSE LEVER
5. TRANSMISSION CASE
6. EXTENSION HOUSING GASKET
7. EXTENSION HOUSING
8. OFFSET LEVER
9. OFFSET LEVER INSERT
10. OFFSET LEVER RETAINING NUT
11. ACCESS PLUG
12. EXTENSION HOUSING OIL SEAL
13. THREADED PLUG
14. REVERSE LEVER PIVOT BOLT
15. FIRST—SECOND SYNCHRONIZER INSERT (3)
16. FIRST GEAR ROLL PIN
17. OUTPUT SHAFT AND HUB ASSEMBLY
18. SPEEDOMETER GEAR SNAP RING
19. SPEEDOMETER GEAR
20. SPEEDOMETER GEAR DRIVE BALL
21. REAR BEARING RETAINING SNAP RING
22. REAR BEARING LOCATING SNAP RING
23. REAR BEARING
24. FIRST GEAR THRUST WASHER
25. FIRST GEAR
26. FIRST—SECOND SYNCHRONIZER BLOCKING RING (2)
27. FIRST—REVERSE SLEEVE AND GEAR
28. FIRST—SECOND SYNCHRONIZER INSERT SPRING (2)
29. SECOND GEAR
30. SECOND GEAR THRUST WASHER (TABBED)
31. SECOND GEAR SNAP RING
32. THIRD GEAR
33. THIRD—FOURTH SYNCHRONIZER BLOCKING RING (2)
34. THIRD—FOURTH SYNCHRONIZER SLEEVE
35. THIRD—FOURTH SYNCHRONIZER INSERT SPRING (2)
36. THIRD—FOURTH SYNCHRONIZER HUB
37. OUTPUT SHAFT SNAP RING
38. THIRD—FOURTH SYNCHRONIZER INSERT (3)

Figure 8-9. Exploded view shows parts of the mainshaft gear assembly. Output shaft extends back through the extension housing. (Chrysler)

engagement. In some early constant-mesh transmissions, both first and reverse used sliding gears.

Mainshaft gears

The output shaft contains the gears that provide all forward speeds and reverse. These ***mainshaft gears*** mount over the transmission output shaft. See *mainshaft gear assembly* shown in Figure 8-9. The mainshaft gears include one gear for each available reduction and overdrive ratio. In addition, while not the case in this particular design, there is typically a gear for reverse.

Each mainshaft gear (and the main drive gear) is comprised of two sets of teeth. Next to the large helical set of teeth is a set of small, straight-cut teeth. The straight-cut teeth are called *clutch teeth.* Each gear has a tapered, or *cone, surface* next to the clutch teeth.

Typically, each mainshaft gear has a bushing pressed into its inside diameter. The bushing is a close fit with the matching machined area, or *race,* on the shaft. The bushing turns on the race, and friction is reduced by the transmission lubricant.

In modern transmissions, the gears can turn freely (and independently) on the output shaft. Each gear turns at a different speed, depending on the gear ratio between it and its mating countershaft gear. The output shaft itself will not begin to turn until one of the mainshaft gears becomes locked to the shaft. *Synchronizers,* which are splined to the output shaft, are the devices that lock the gears to the shaft. Normally, only one gear will be engaged, or locked to the shaft, at a time.

Reverse idler gear

The reverse idler gear is used to reverse rotation of the output shaft for backing up the vehicle. The gear is turned by the countershaft reverse gear. The gear mounts on the reverse idler shaft, which is normally stationary. Refer to *reverse idler gear assembly* shown in Figure 8-10. Needle bearings are used between the gear and shaft to reduce wear. End play is controlled by thrust washers, which are placed between the ends of the gear assembly and the transmission case. The transmission case often contains a special projection to hold one end of the idler shaft.

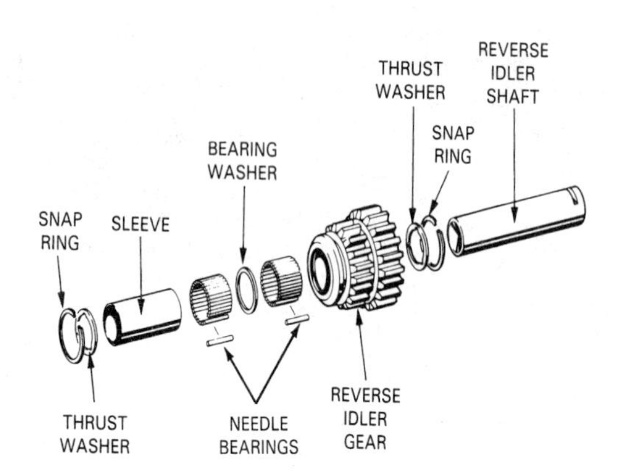

Figure 8-10. Exploded view shows parts of the reverse idler gear assembly. Reverse idler shaft is normally locked in place. Thrust washers control end play. (Chrysler)

Synchronizers

Constant-mesh manual transmissions, used on modern vehicles, make use of synchronizers. These assemblies are required for shifting from one gear to another. They help permit gears to be selected without *clashing,* or *grinding.* They do this by *synchronizing* (matching) the speed of mating mainshaft gears or the main drive gear to their own speed–thus, to output shaft speed–before they engage. (Gear clash makes shifting difficult and causes damage to gears and synchronizers.) Through use of synchronizers, mainshaft gears can be locked to the output shaft, or in the case of direct drive, input and output shafts can be locked together.

Modern transmissions are synchronized in all forward gears. Such transmissions are referred to as ***synchromesh transmissions.*** Reverse gear is not usually synchronized, since the vehicle should be stopped before being shifted into reverse.

When the synchronizer is away from a mainshaft gear, the gear *freewheels,* or spins, on the output shaft. No power is transmitted to the output shaft. When shifting gears, the clutch is disengaged, which means the input

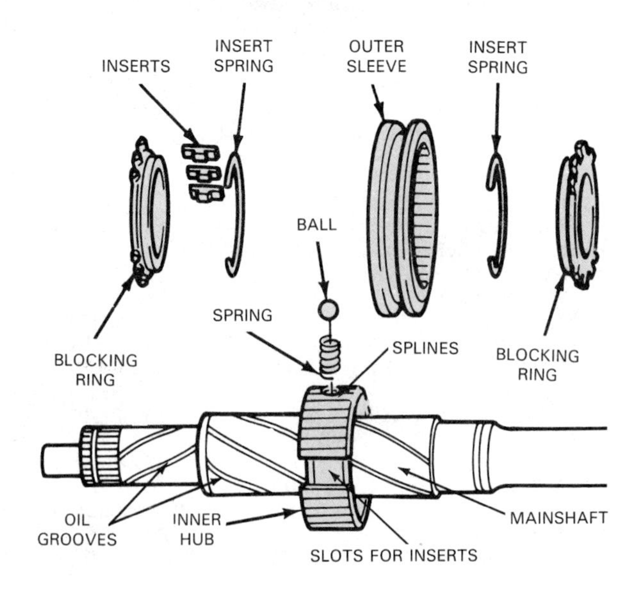

Figure 8-11. Basic synchronizer components. Inner hub is splined to output shaft. Outer sleeve fits over inner hub. Blocking rings allow sleeve to slide into and mesh with mainshaft gear without clashing. Spring-loaded inserts and ball keep sleeve centered on hub until shift fork pushes sleeve on hub. Note that helical lines around mainshaft are oil grooves. (General Motors)

shaft is disengaged from the engine. *Shift forks* slide the synchronizer against a gear, locking the gear to the synchronizer and to the output shaft. The clutch is then reengaged, and power is sent out the transmission to the drive shaft. Synchronizer operation is described in greater detail in upcoming paragraphs.

Synchronizer construction

The most popular type of synchronizer for an automotive manual transmission is the ***block synchronizer.*** This assembly consists primarily of the following components: *inner hub, outer sleeve, inserts, insert springs,* and *blocking rings.* See Figure 8-11.

The ***inner hub*** is splined to the output shaft. It is held in a stationary position between the transmission gears. The hub rotates with the output shaft. The hub also has external splines that mate with the outer sleeve.

The ***outer sleeve*** has internal splines that mate with the external splines of the inner hub. The sleeve can slide back and forth along the splines of the hub. The internal splines of the outer sleeve also mate with the straight-cut teeth of the gears next to which they are mounted.

The ***inserts*** serve, in part, as detents. They fit into slots along the inner hub. The ***insert springs*** fit around the back side of the inner hub, in the recessed area. They push against the inserts, forcing notches on the inserts into engagement with notches in the internal splines of the sleeve. This holds and centers the sleeve on its hub. The inserts slide back and forth with the outer sleeve.

The ***blocking rings*** are toothed brass or copper rings. One is located between each sleeve and hub assembly and mainshaft gear, Figure 8-12; it is not *attached* to either. The blocking ring acts as a clutch to speed up or slow down the mating gears to match the speed of the output shaft.

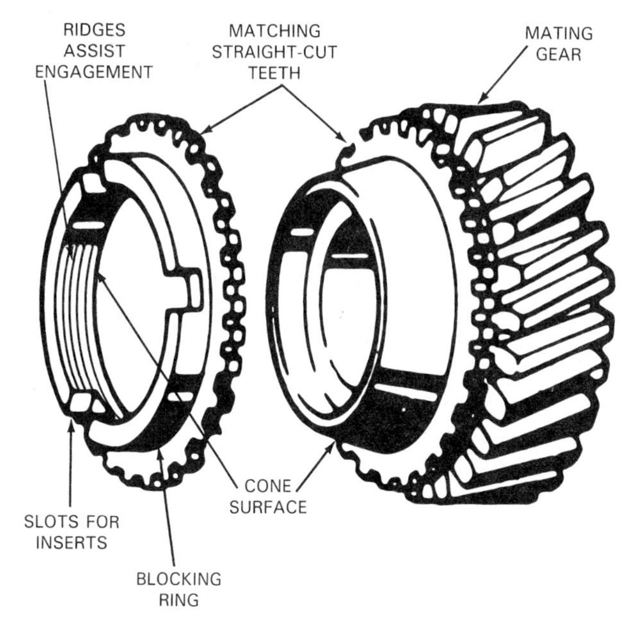

Figure 8-13. A typical blocking ring is a brass ring with external teeth, which match straight-cut teeth of mating gear, and tapered surface, which mates with tapered surface of mating gear. Ridges in tapered surface help synchronizer to solidly engage the cone. (Chrysler)

The teeth on the blocking ring are situated along the outside diameter of the ring. They match the straight-cut teeth of the mating gears, Figure 8-13, and mate with the internal splines of the outer sleeve. The inside diameter of the blocking ring is tapered. This surface fits over the cone surface on the mating gear. The tapered surface of the ring has many small, sharp ridges, which enable the synchronizer to solidly engage the cone surface of the mating gear. Commonly, the inner (assembly) side of the blocking ring fits into the recessed region between the inner hub and the output shaft, where matching slots in the ring engage the inserts of the inner hub.

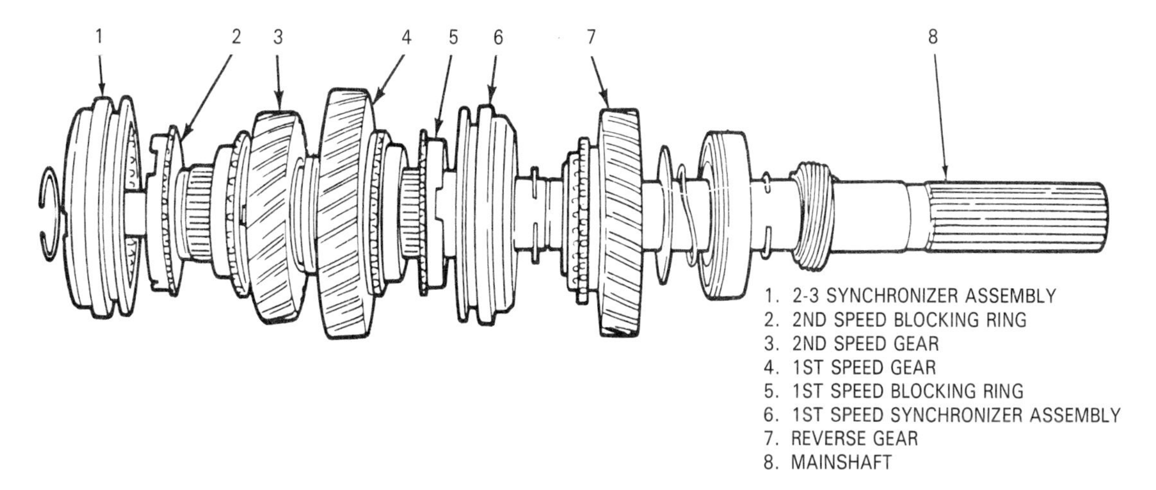

Figure 8-12. This shows location of blocking rings in relation to sleeve and hub assemblies and mainshaft gears. Note that parts would actually be in tighter than shown in this somewhat exploded view. (General Motors)

The blocking ring, when fully engaged, fits between the mating gear and the outer sleeve. However, the sleeve does not move into engagement with the teeth on the gear until the ring and gear are brought to the same speed.

Synchronizer operation

Synchronizer operation is relatively simple. The driver pushes the clutch pedal to disengage the clutch and uses the gearshift lever and linkage to operate the synchronizer outer sleeve. When the outer sleeve is moved toward the gear to be engaged, the internal teeth of the sleeve mesh with the external teeth of the blocking ring. This causes the blocking ring to turn at the same speed as the sleeve, which is turning at output shaft speed. (If the blocking ring is held in place with inserts, it is already turning at sleeve speed.)

As the outer sleeve continues to move, it pushes the blocking ring into engagement with the cone surface of the mating gear. The mating gear is unloaded because no engine power is being delivered while the clutch is disengaged. Friction between the two surfaces brings the mating gear to the same speed as the outer sleeve and, therefore, to the speed of the output shaft. The ring and cone surfaces become solidly engaged. The sleeve continues its forward (or rearward) movement, and it engages the mating gear without clashing. This locks gear, sleeve, hub, and output shaft together. Now, the gear is engaged with the output shaft, and the clutch pedal can be released. Power flows through the gear and on to the rear wheels.

Note that direct drive is provided by engaging the sleeve and blocking ring directly with the cone surface on the main drive gear. Also, note that some transmissions use a single sliding gear for low and reverse. The gear is free to slide back and forth on splines cut into the output shaft. It can be moved to mesh with the low countershaft gear or the reverse idler gear.

Shift Forks

Shift forks, Figure 8-14, are made to slip into grooves cut into the outer sleeves of the synchronizer assemblies. The design makes it possible for the fork to move the sleeve, while allowing the sleeve to rotate.

Some shift forks are made of steel, cast iron, or aluminum. Aluminum shift forks sometimes have nylon inserts at the point where they contact the sleeve. This reduces wear on the contact points of the fork.

Shift fork action is illustrated in Figure 8-15. The shift fork moves in a straight line to push the synchronizer sleeve into engagement with the gear. The sleeve can continue to rotate as it is moved. The shift forks are operated by the transmission shift linkage.

Manual Transmission Shift Linkage

Shift forks are operated by the driver through a series of shafts, levers, links, and sometimes, cables. There are many ways to transfer the force of the gearshift lever to the gears. Some systems are simple, while others are more complex. Shift linkages connect the gearshift lever to the shift forks. They can be divided into two general types of systems–*external* and *internal.*

External shift linkage

External shift linkage is a system that is identified as a series of rods and levers that connect to the outside of the transmission. In addition, some later external systems use cables instead of rods. External shift linkage may be connected to the gearshift lever either on the steering column or on the floor.

At the transmission case, the external shift linkage connects to the inside through shafts that pass through the side of the case. Inside, each shaft connects to a shift lever.

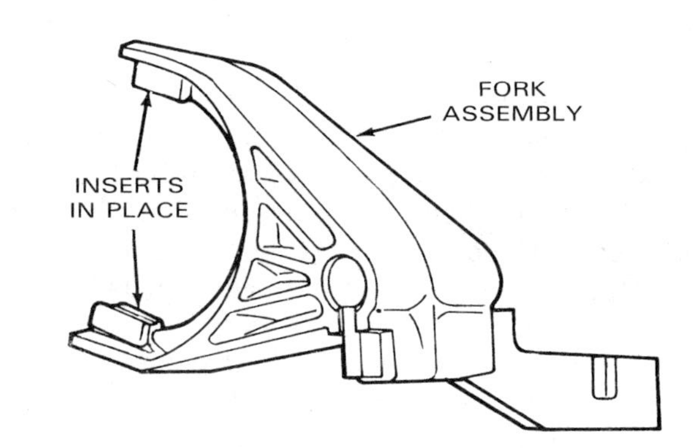

Figure 8-14. Shift forks such as these are used to move the outer sleeves of the synchronizer assemblies. (Ford)

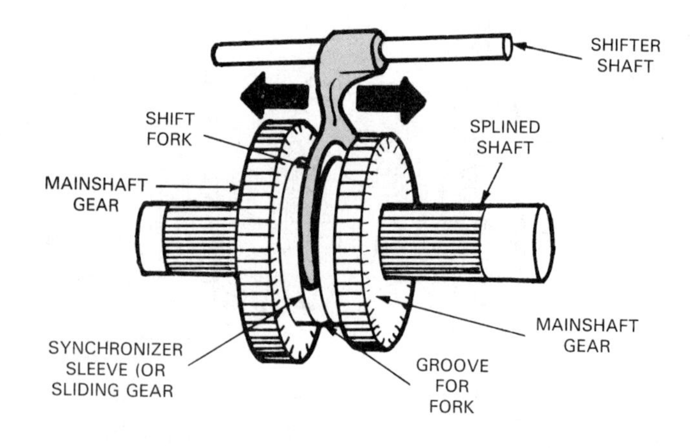

Figure 8-15. Shift forks move in a straight line to move the synchronizer sleeve (or sliding gear) in and out of engagement. The shift fork is operated by the shift linkage.

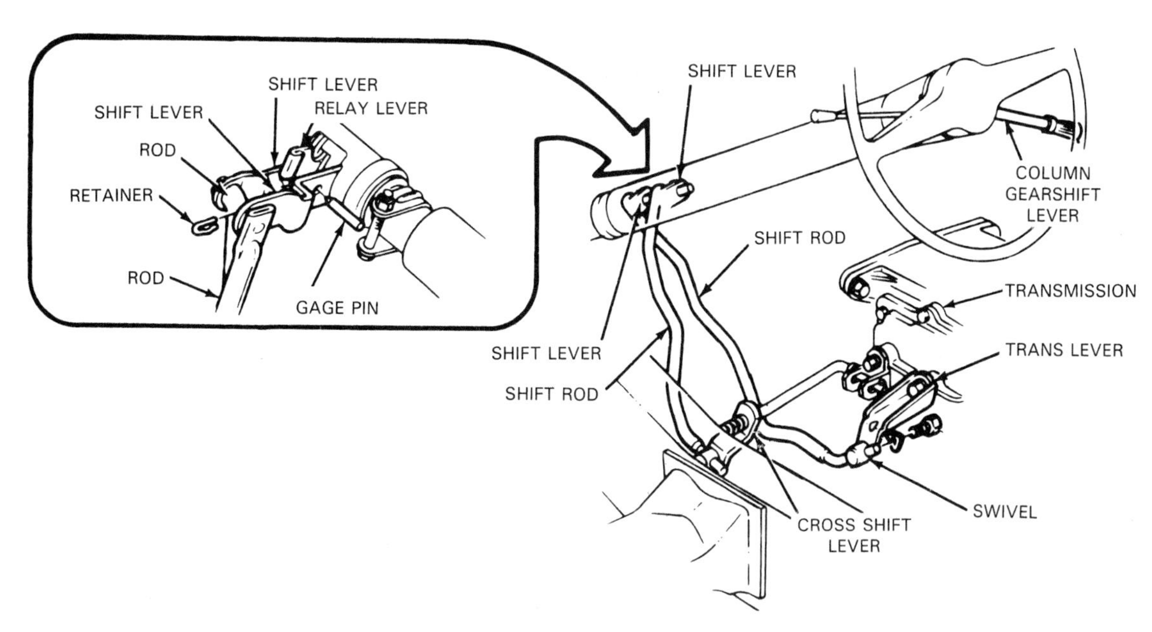

Figure 8-16. The column-shift linkage, with its system of levers and rods, was once commonplace. It has largely been replaced by the simpler floor-shift linkage. Some pickup trucks continue to be equipped with the column-shift linkage. (General Motors)

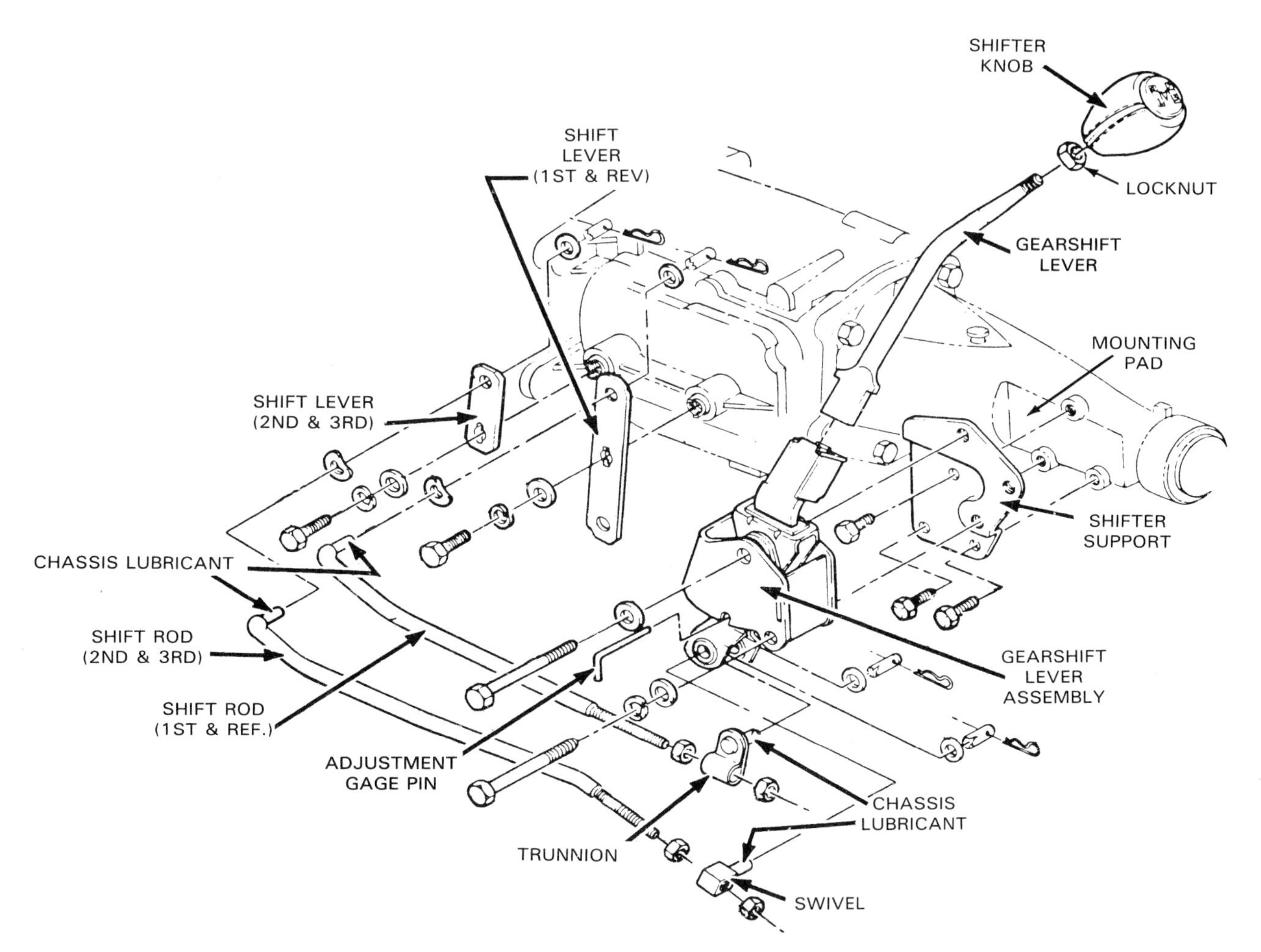

Figure 8-17. The external floor-shift linkage is simpler than the column shift but works according to the same principles. A series of levers and rods transmit motion from the gearshift lever to the transmission. (General Motors)

The shaft rotates to move the lever. The lever moves forward or backward to move the shift fork. In this way, rotation of the shaft is converted to straight-line movement at the shift fork.

The internal shift *lever* usually has some type of spring-loaded detent. The detent is a holding mechanism that keeps the lever and, therefore, the fork and synchronizer sleeve in position when in neutral or in gear. There is also a shift interlock. It ensures that each shift lever is in the neutral position before the other can move the gears. This way, only one gearset can be engaged, or in gear, at a time.

Column-shift linkage. This type of linkage, shown in Figure 8-16, is an example of external shift linkage. It is usually used with a *3-speed transmission.* The gearshift lever is mounted on the steering column and is connected to a *shift tube* inside of the steering column. Moving the gearshift lever causes the shift tube to rotate slightly and move up or down, as necessary. Movement of the shift tube operates one of two *shift levers* at the bottom of the steering column. Each shift lever is connected to a *shift rod,* which transfers movement to the internal transmission shift mechanism. There are usually two shift rods; one rod is used to select low and reverse gears, and the other rod selects second and third gears.

Many column-shift linkages have a *cross shift lever* between the transmission case and the vehicle frame. The cross shift lever redirects shift rod motion and reduces shift linkage looseness. The linkage parts are usually attached with *cotter pins* passing through drilled holes in the shift rods. Washers and bushings are used at linkage pivot points to reduce wear. Linkage adjustments are made by turning threaded *swivels* on the shift rods.

External floor-shift linkage. This type of linkage, shown in Figure 8-17, is usually simpler than column-shift linkage. Floor-shift linkage may be used on 3-speed transmissions; however, it is usually used with transmissions having four or more speeds. The *gearshift lever assembly* mounts to the rear of the transmission, usually on the extension housing. The gearshift lever extends upward

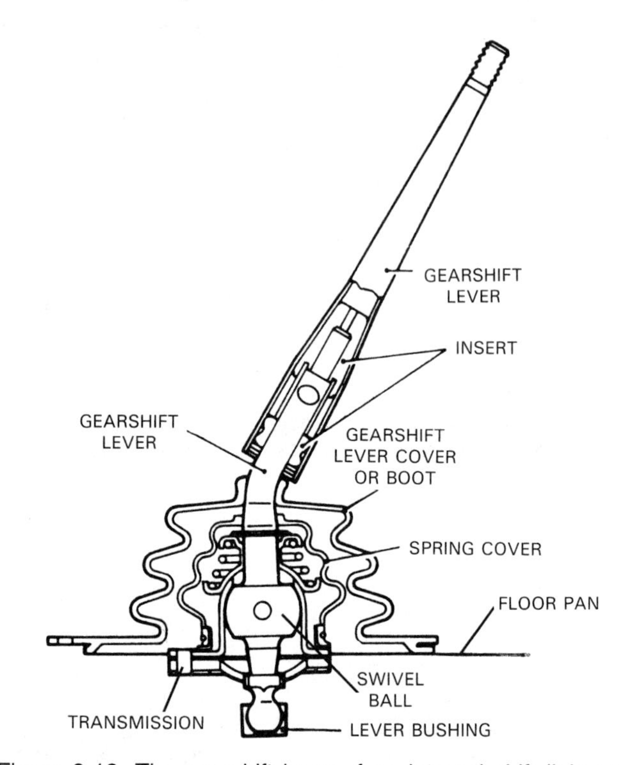

Figure 8-18. The gearshift lever of an internal shift linkage is shown here. Gearshift pivots on internal ball to move shift linkage. Boots at transmission and floor pan prevent loss of lubricant and entry of dirt and water. (Chrysler)

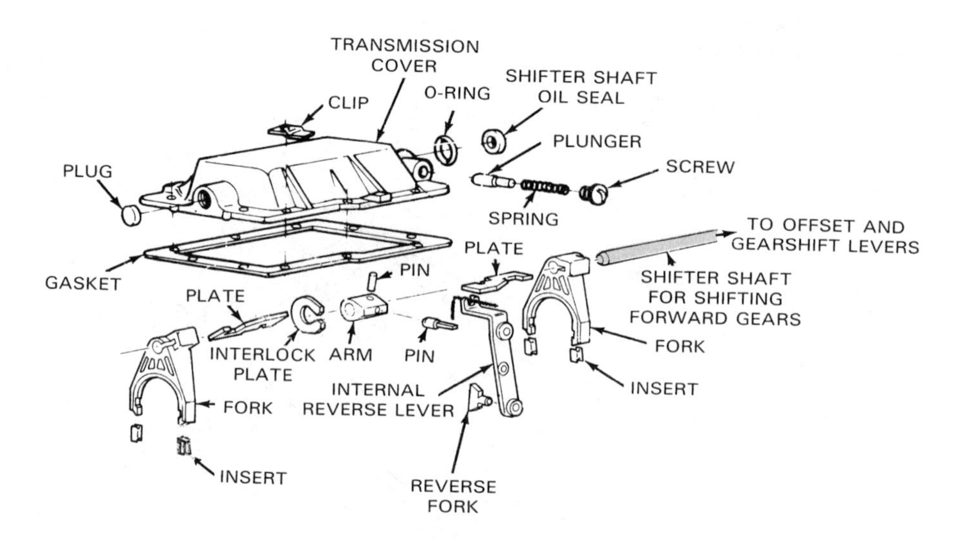

Figure 8-19. A single shifter shaft is used in some manual transmissions. Gearshift lever moves offset lever (not shown) at end of shifter shaft. Offset lever moves shifter shaft, which moves shift forks and, therefore, synchronizer sleeves. Note that shaft extends out of transmission cover, through oil seal, into extension housing. In transmission shown, reverse gear is operated through a separate reverse lever. (Ford)

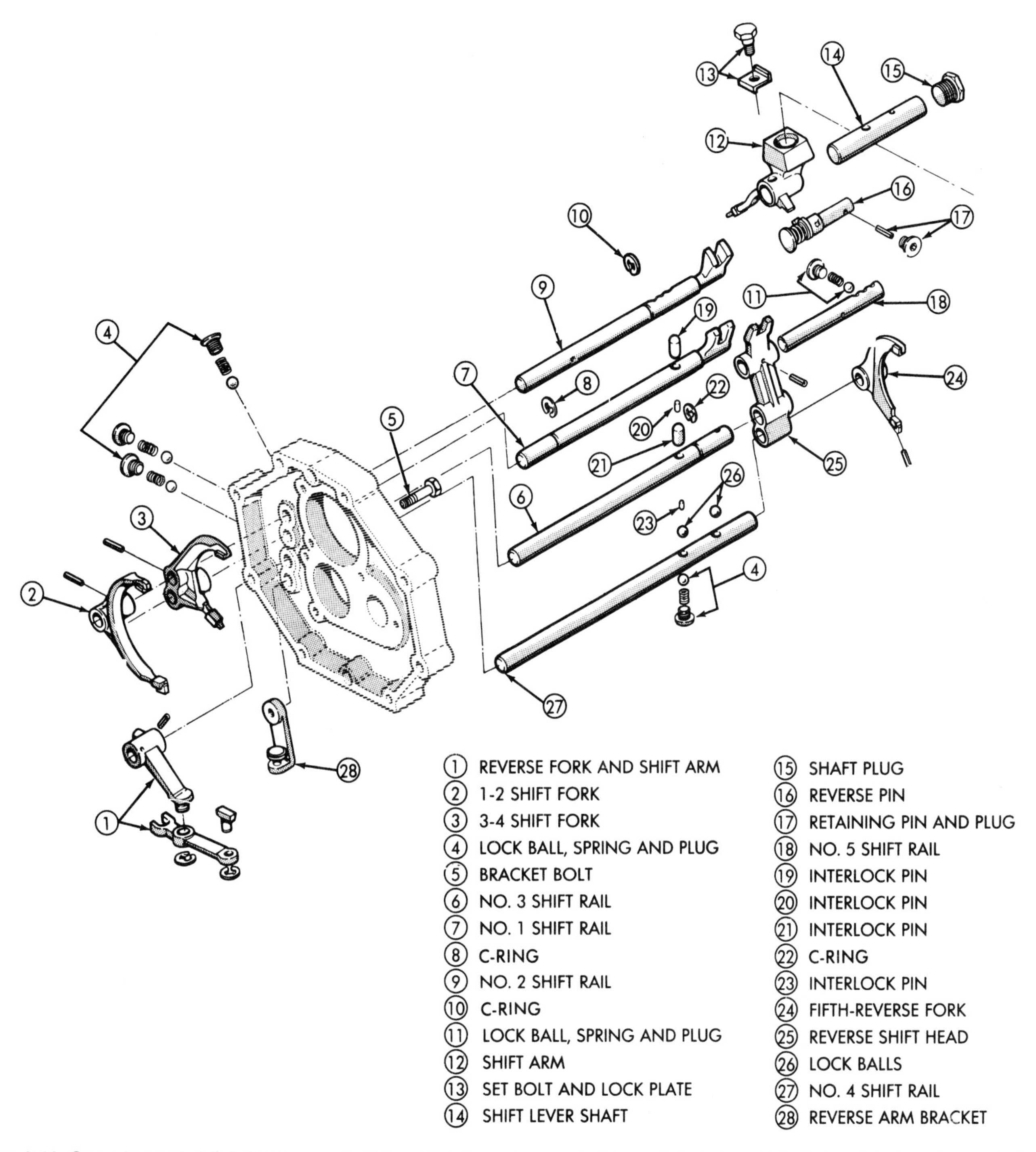

Figure 8-20. Some internal shift linkages use multiple shift rails, as is seen in this exploded view. Note that lock ball, spring, and plug assemblies shown in illustration serve as detents. These *detent ball assemblies* hold the rails into whatever gear is selected. Interlock pins make it impossible to move more than one rail at a time. (Chrysler)

through a hole in the vehicle floor pan. This hole is sealed by a flexible rubber boot, which keeps out dust and water, while allowing gearshift lever movement. The assembly shift mechanism transfers gearshift lever movement to the shift rods. The shift rods operate the shift levers on the transmission case.

As with the column-shift linkage, linkage parts for the external floor-shift linkage are attached with cotter pins through drilled holes in the shift rods. Washers and bushings are used at linkage pivot points to reduce wear, and the linkage can be adjusted by threaded swivels on the shift rods.

Internal shift linkage

Internal shift linkage will be found only on transmissions with floor-mounted gearshift levers. In many instances, the gearshift lever extends down into the extension housing. The transmission and floor pan openings are sealed with rubber boots. Figure 8-18 is an example of such a gearshift lever.

One type of internal shift linkage commonly found employs a single *shifter shaft,* or *shift rail.* See Figure 8-19. The shift forks are positioned along the shaft. When a gear is selected, the lower end of the gearshift lever moves into an *offset lever* at the end of the shifter shaft. Further movement of the gearshift lever causes the shifter shaft to move. The *interlock plate* locks out all gears but the one selected. Thus, the desired shift fork is moved by moving the shifter shaft.

A second type of internal shift linkage employs a *series* of rods, or shift rails, to operate the shift forks. See Figure 8-20. Many transmissions use this system. A shift fork is attached to each shift rail. In addition, the shift rails have notched units attached, directly or indirectly, to their outer ends.

When the driver shifts gears, the bottom of the gearshift lever catches in one of the shift rail notches. As a result, the shift rail and shift fork are caused to move. In this way, transmission gears are changed.

Manual Transmission Case

The transmission case supports the transmission shafts and forms a reservoir for lubricant. It is the main housing, or structure, of the transmission. The front of the case attaches to the clutch housing. In some designs, the clutch housing and case are a single integral casting; in other designs, the case bolts to the back of the clutch housing, Figure 8-21. The transmission case is also the attaching point for the extension housing and external shift linkage. Figure 8-22 illustrates a typical transmission case. There are many variations on this basic design, but the transmission case always serves the same purpose.

Transmission cases can be made of cast iron or aluminum. Many cases contain pressed-in bearings to support rotating shafts. Most manual transmissions have a lubricant *fill plug* and, sometimes, a *drain plug.*

One modern version of the transmission case is shown in Figure 8-23. The case is assembled out of a couple of pieces, including an *adapter plate,* or *center support,* which serves as the back side of the case. It separates the transmission case and extension housing. The transmission shafts and gears are supported at the adapter plate, in addition to front and rear supports. This kind of design makes the transmission easier to service.

The opening in the front of the transmission case is covered by the front bearing retainer, sometimes called the *front bearing cap,* or *front cover.* See Figure 8-24. The front bearing retainer covers the transmission front bearing. The long hub of the retainer serves as the sliding surface for the clutch throwout bearing.

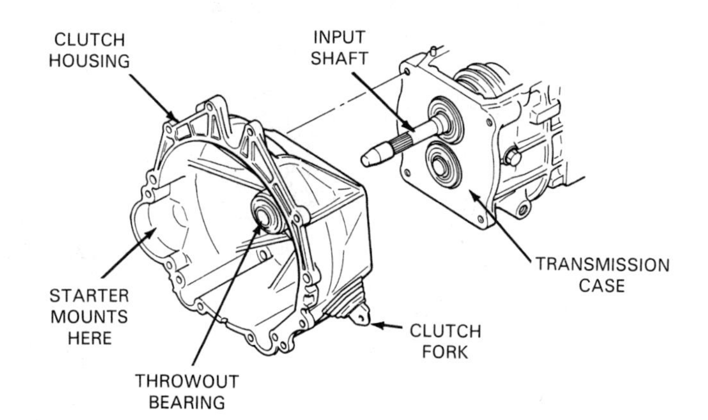

Figure 8-21. The transmission case provides a mounting surface for the clutch housing. Some transmission cases and clutch housings, however, are a combined integral casting. (Chrysler)

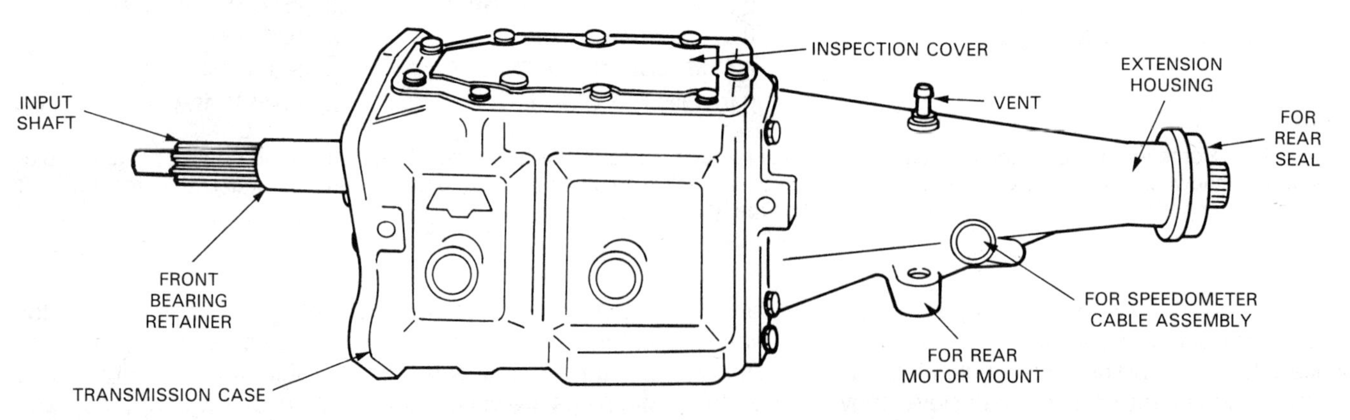

Figure 8-22. The transmission case supports, aligns, and serves as a mounting surface for transmission components. The case also contains the transmission lubricant. (General Motors)

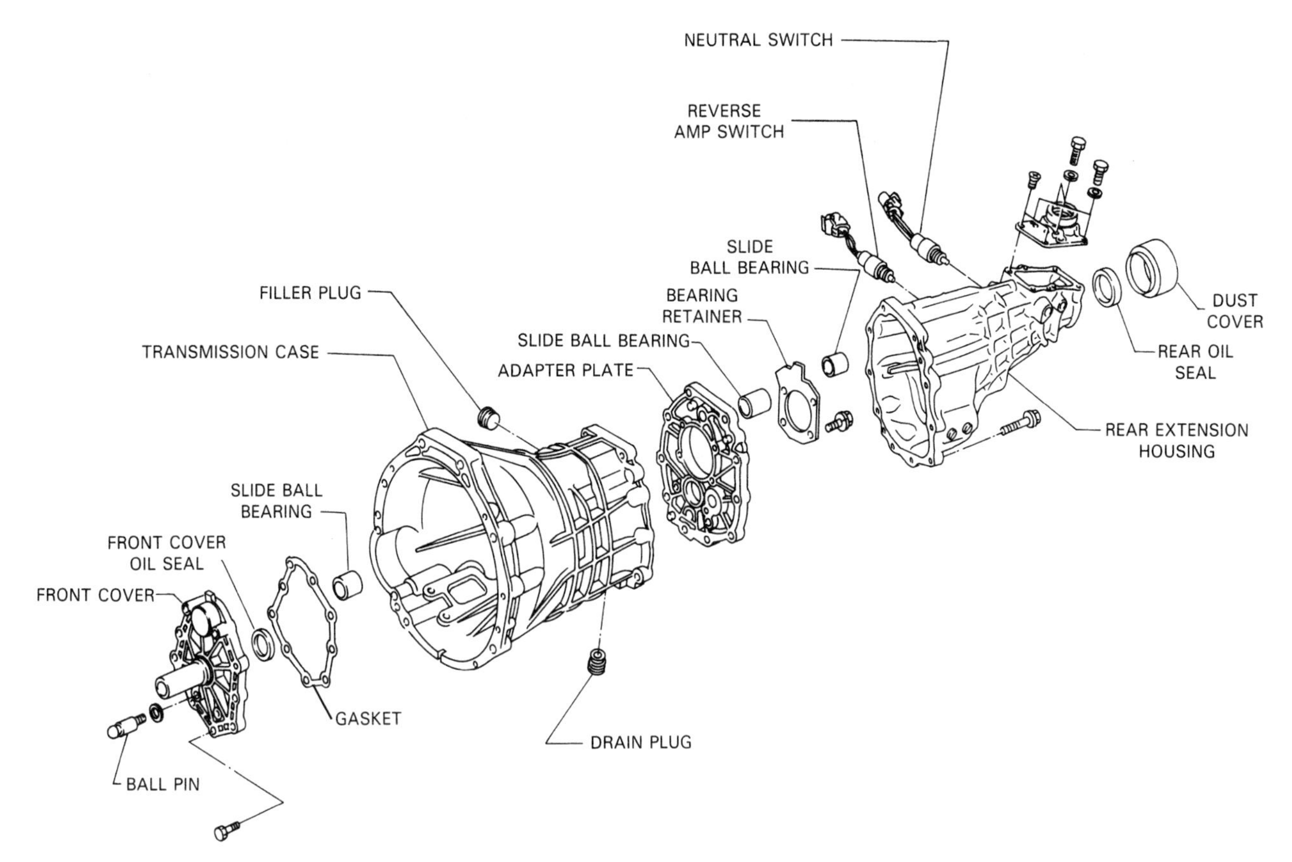

Figure 8-23. This transmission case design makes the transmission easier to service. The gearsets can be more readily exposed. The gear train can then be easily disassembled and repaired. Several manufacturers use variations of this design. (Nissan)

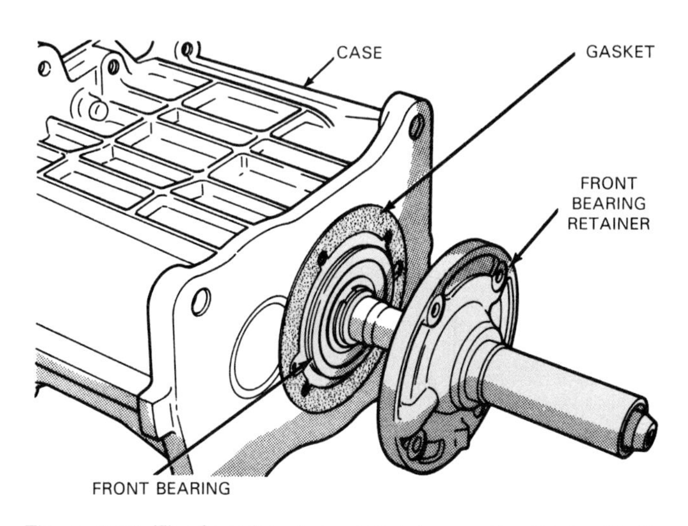

Figure 8-24. The front bearing retainer is usually attached to the transmission case by bolts. It holds the front bearing in place. A gasket and seal prevent lubricant leaks. (Chrysler)

To eliminate lubricant leaks, a gasket is installed between the bearing retainer and case. Also, an oil seal is installed in the retainer, Figure 8-25. The *front bearing retainer seal* prevents oil leakage between the cover and the transmission input shaft.

An *access cover* is commonly built into the top or side of the case. It allows access to the transmission internals for inspection or service. Depending on design, the access cover may simply be a plain, flat plate, or it may be a contoured or panlike shape with built-in shift linkage. (Refer back to Figure 8-19.) On some floor-shift models, the gearshift lever goes directly through the cover. Note that all transmission covers use gaskets.

Extension Housing

The extension housing, made of aluminum or cast iron, provides protection for the end of the transmission output shaft. In some designs, the front of the extension housing holds the rear bearing, which supports the central portion of the output shaft. (In some designs, this bearing is held in the rear of the transmission case. In other designs, a *center bearing* is held in the center support, and the rear bearing is held further back in the extension housing.) The front of the extension housing is bolted directly to the rear of the transmission case. To prevent lubricant leaks, there is usually a gasket between the case and extension housing.

The rear of the extension housing contains a bushing and seal. The transmission *extension bushing* supports and aligns the driveline slip yoke, which in turn, supports the transmission output shaft. The transmission rear seal prevents lubricant leaks and keeps dirt and water out of the transmission. A typical extension housing is illustrated in Figure 8-26.

The *transmission mount,* which supports the transmission and also serves as the *rear engine mount,* is usually installed on the extension housing. The mount is an assembly essentially comprised of a bracket and rubber pad or bushing. A flat surface is machined on the bottom of most extension housings. The surface is drilled to fit the *transmission mount attaching bolts.*

In vehicles with internal shift linkage, the gearshift lever usually extends into the transmission at the extension housing. A boot is used for preventing loss of oil or entry of dirt and water. The boot is usually held in place by a ring or *bezel,* which is attached to the extension housing with bolts or screws. The shift linkage, then, extends from the extension housing into the transmission case.

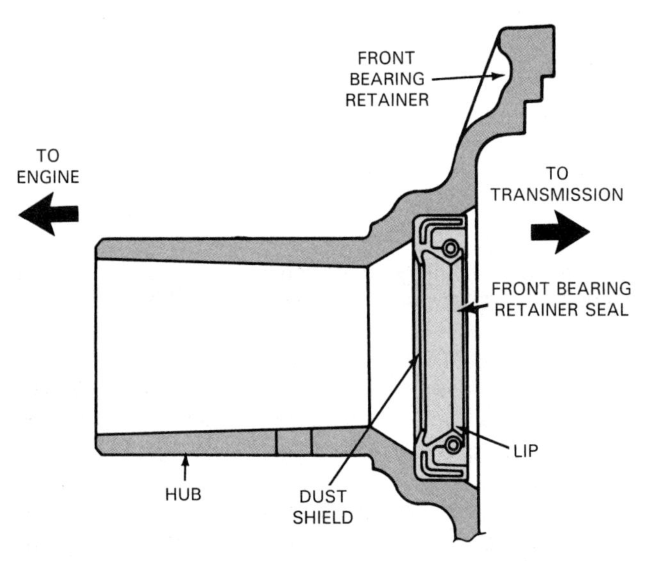

Figure 8-25. This shows a typical front bearing retainer seal. The sealing lip prevents oil loss, and the dust shield keeps out dirt and water. (General Motors)

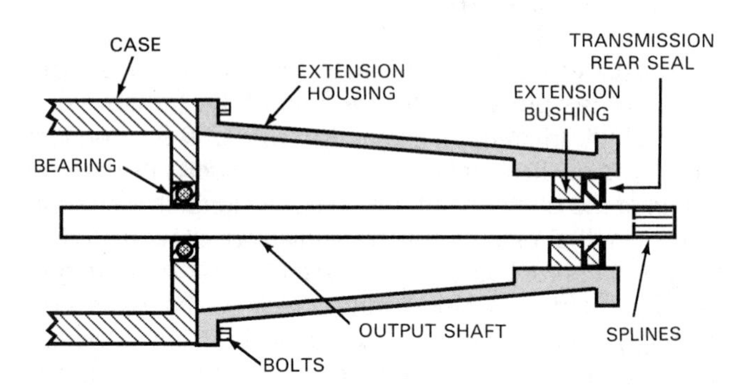

Figure 8-26. The extension housing keeps the output shaft aligned and seals the rear of the transmission. Common extension housing features are extension bushing, rear seal, and housing-to-case gasket.

Accessory Devices

Accessory parts may be bolted to or installed in the transmission case or extension housing. Fill and drain plugs, identification tags, shift levers, and boots are some common examples. Some of the accessories that are often attached to the transmission case or extension housing are shown in Figure 8-27.

Speedometer assembly

The extension housing usually contains the speedometer assembly. It basically comprises the *speedometer adapter* and *pinion gear,* Figure 8-28. The speedometer adapter is installed with bolts or with a single bolt and clamp, similar to a *distributor hold-down clamp.* The speedometer gear mates with a worm gear on the output shaft, and the pinion gear shaft revolves inside of the adapter. The adapter usually has both internal and external seals.

The speedometer assembly drives a cable, which runs through a housing. This *cable assembly* extends from the speedometer adapter in the extension housing to the *speedometer head,* or *speed indicator assembly,* in the dash. Whenever the output shaft turns, the worm gear drives the pinion gear, and the speedometer cable turns. This makes the speedometer head register the road speed of the vehicle.

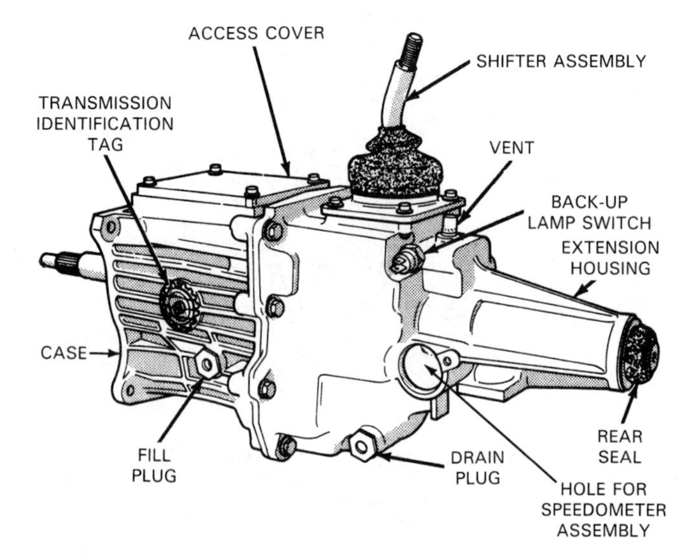

Figure 8-27. Accessory parts may be bolted to or installed in the transmission case or extension housing. Access covers, fill and drain plugs, identification tags, shift levers, and boots are some common examples. In addition, the transmission usually contains a vent somewhere at the top. The vent relieves pressure caused as the transmission lubricant heats up and expands. (Chrysler)

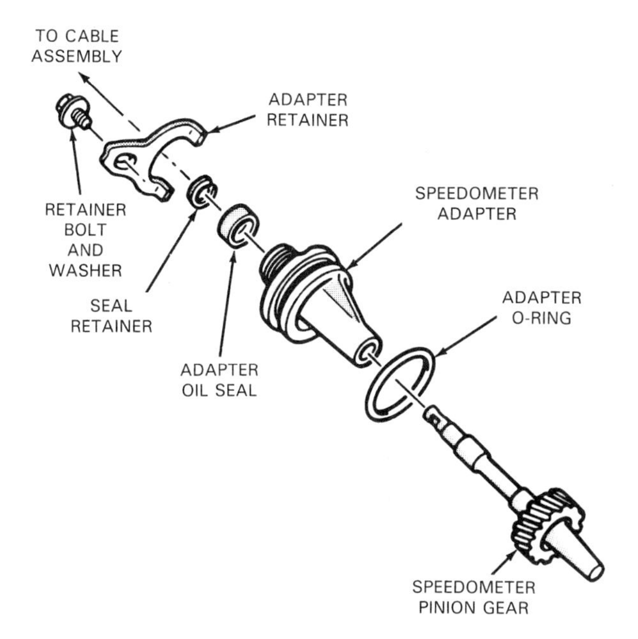

Figure 8-28. The speedometer pinion gear and adapter are usually installed in the extension housing. The adapter in this illustration is held in place by a single bolt and retainer. The speedometer pinion rotates inside of the adapter. (Chrysler)

Switches, sensors, and solenoids

Most late-model manual transmissions will be equipped with one or more switches, sensors, and sometimes, solenoids. Most will be screwed into tapped holes in the transmission case or the extension housing. Solenoids help control transmission shifting. However, other electrical components serve purposes that are unrelated to transmission operation; they tell another system what gear the transmission is in.

For example, one switch is used to turn on the backup lights when the vehicle is shifted into reverse. Another component is used to prevent starter operation if the transmission is in gear. Some sensors tell the on-board computer when the transmission is in a particular gear, so the computer can make adjustments to the engine fuel and ignition systems. Examples of some of these components can be seen in Figures 8-23 and 8-27.

Manual Transmission Gear Oil

Manual transmission gears are lubricated by gear oil contained in the transmission case. The rotation of the gears throws oil onto the gear teeth and other stressed parts. This is called *splash lubrication.*

Manual transmissions are usually filled at the factory with 80- or 90-weight gear oil. In rare instances, heavy 140-weight oil is used. Occasionally, some manual transmissions are filled with automatic transmission fluid by the manufacturer; however, gear oil is recommended for topping off or refills.

Manual Transmission Power Flow

The power flow of manual transmissions can be understood by inspecting Figures 8-29 through 8-33, showing the gear train of a typical manual transmission. Power flow, which is the path that power takes through the drive train, is depicted by arrows passing through the gears that are actually transmitting power.

Reverse Gear

In reverse gear, Figure 8-29, power flows from the main drive gear into the countershaft gear. From the countershaft gear, power flows into the reverse idler gear. The reverse idler gear directs power into the reverse gear on the output shaft. Since the outer sleeve of the rear, or first-and-reverse (1st-rev.), synchronizer has been moved rearward, the output shaft and reverse gear are locked together. Rotation is reversed between the main drive gear and the countershaft gear. It is reversed again between the countershaft gear and the reverse idler gear. The rotation is reversed one last time between the reverse idler gear and the mainshaft reverse gear. Since the rotation has been reversed three times (*odd* number) between input and output shafts, direction of the output shaft is opposite that of engine rotation.

Neutral

In neutral, Figure 8-30, the main drive gear is turning the countershaft gear. All gears that are in constant mesh with the countershaft gear are also turning. This includes first and second gears. On the fully synchronized transmission, it also includes reverse and reverse idler gears. However, since both synchronizer outer sleeves are in the neutral position (centered on inner hub), no power reaches the output shaft, and it remains stationary. The vehicle does not move.

First Gear

In first gear, Figure 8-31, the input shaft's main drive gear turns the countershaft gear, reversing direction. The countershaft gear turns the low gear on the output shaft, reversing direction again, so that the low gear is turning in the same direction as the input shaft. Since the outer sleeve of the rear synchronizer has been moved forward, the low gear is locked to the output shaft. The vehicle is in first gear. The gear ratio is about 3:1, and the car accelerates easily.

Second Gear

In second gear, Figure 8-32, the input shaft's main drive gear turns the countershaft gear, reversing direction.

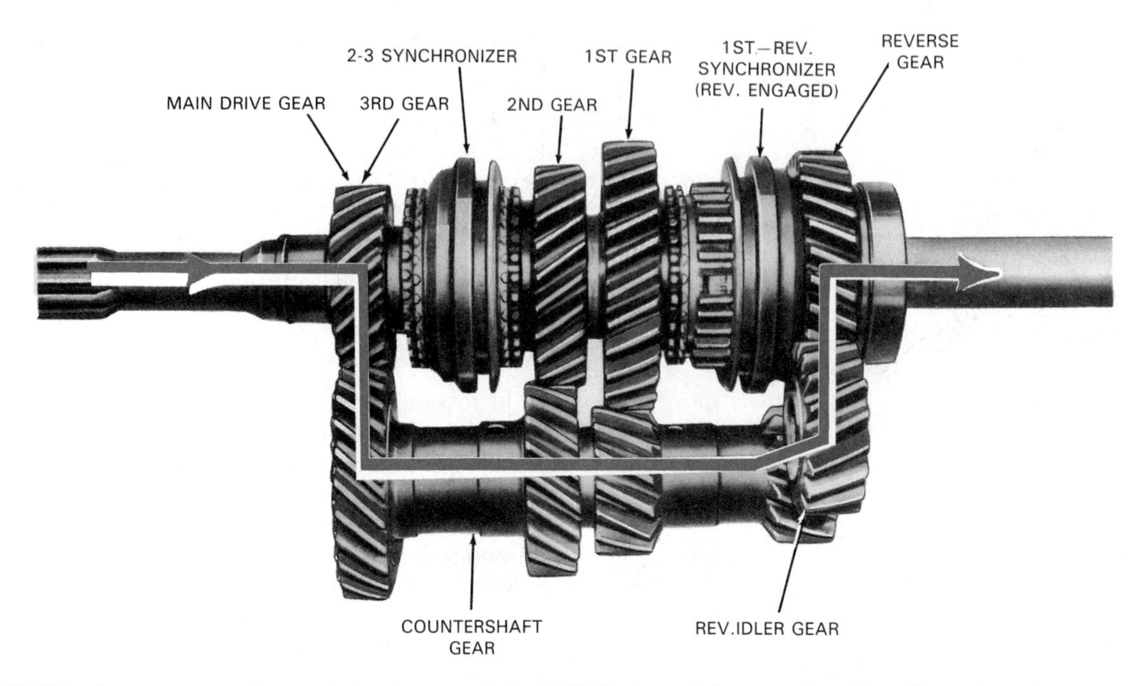

Figure 8-29. This shows a manual transmission in reverse. Front (2-3) sleeve is in neutral position. Rear sleeve has moved rearward to lock reverse gear to output shaft. Power flow is from main drive gear, through countershaft gear to reverse idler gear, and into mainshaft reverse gear. Note that there is no blocking ring on the reverse-gear side of the synchronizer. Thus, reverse is not a synchronized gear. (General Motors)

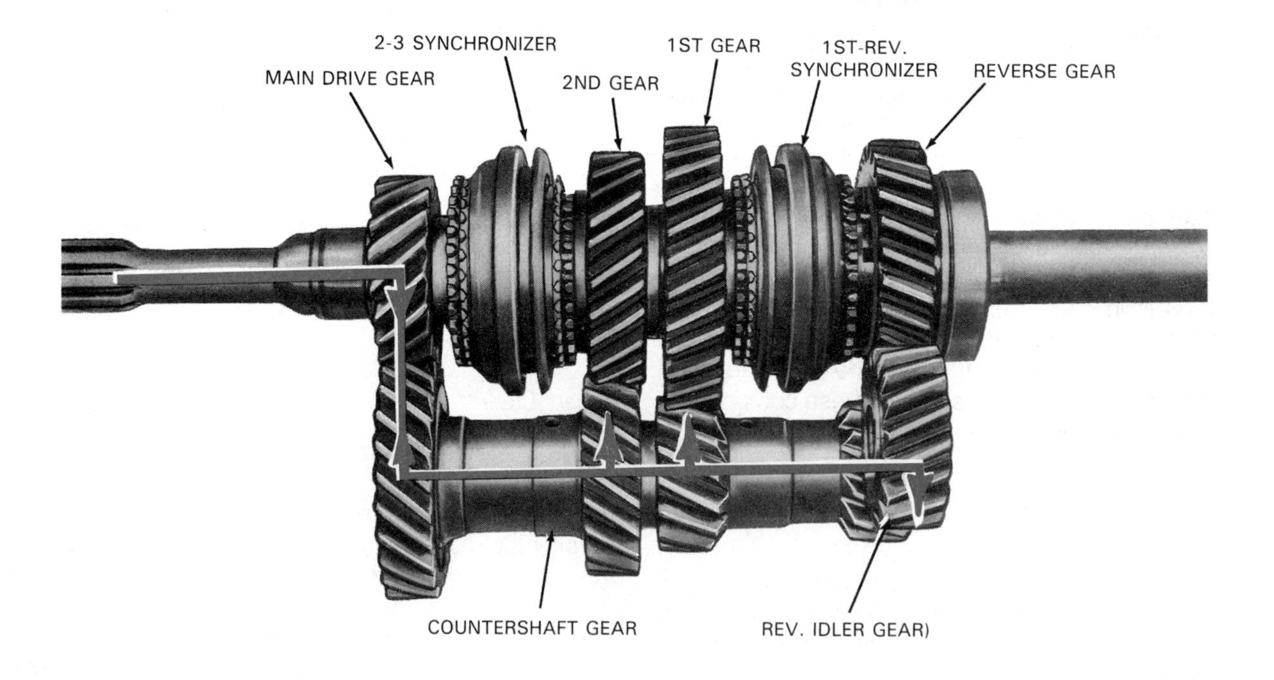

Figure 8-30. In neutral, both sleeves are in neutral position. Main drive gear, countershaft gear, and mainshaft gears are turning but are not connected to mainshaft. Power cannot flow through transmission. (General Motors)

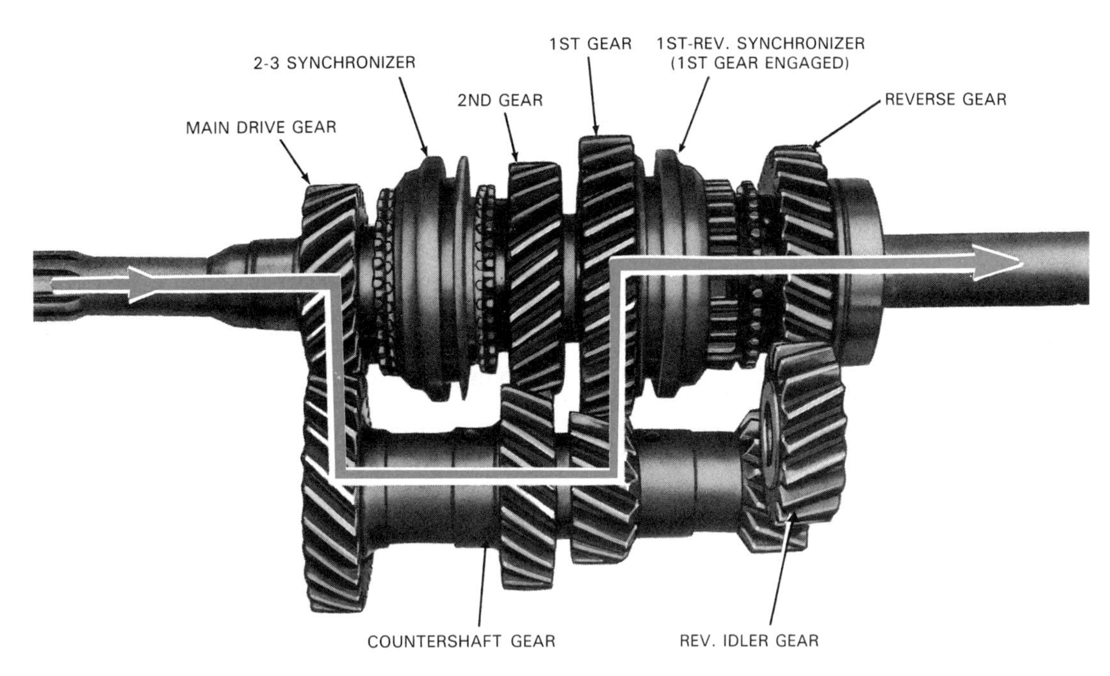

Figure 8-31. In first gear, front sleeve is in neutral position. Rear sleeve has moved forward to lock low gear to output shaft. Power flow is from main drive gear, through countershaft gear, to low gear on output shaft. (General Motors)

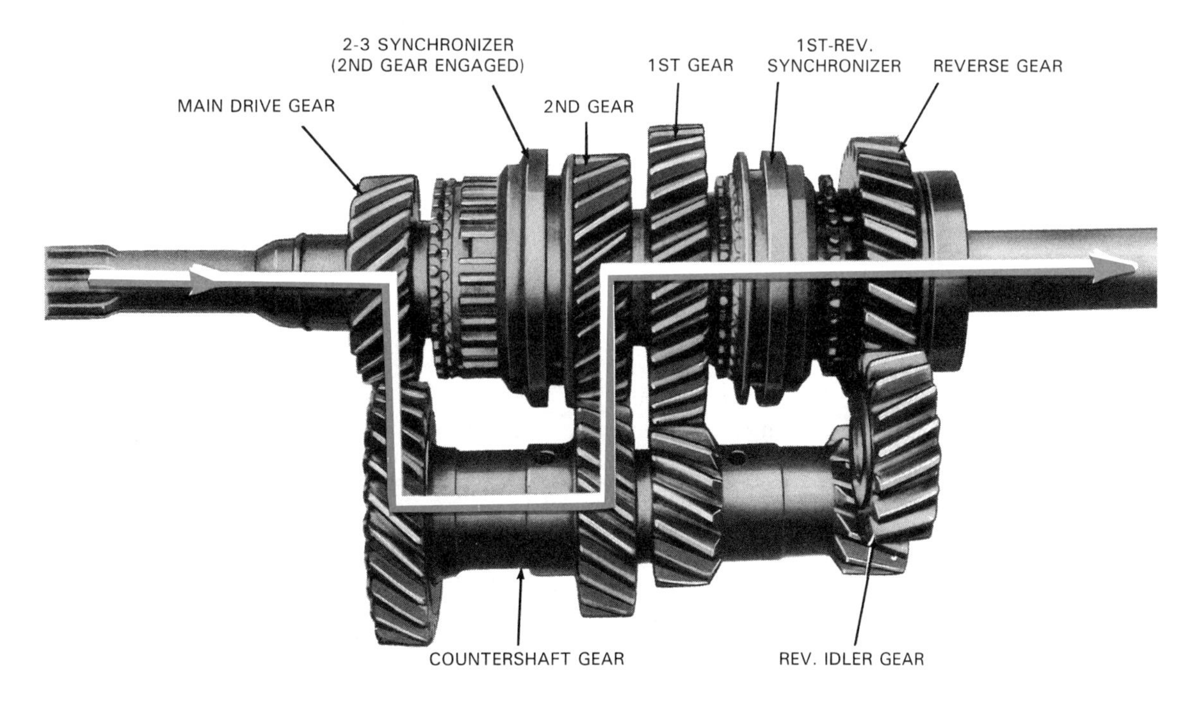

Figure 8-32. In second gear, rear sleeve is in neutral position. Front sleeve has moved rearward to lock second gear to output shaft. Power flow is from main drive gear, through countershaft gear, to second gear on output shaft. (General Motors

The countershaft gear turns the second gear on the output shaft, reversing direction again. This double reversal makes the second gear turn in the same direction as the input shaft. Since the outer sleeve of the synchronizer has been moved rearward, the second gear is locked to the output shaft. The vehicle is in second gear. A typical ratio in second gear is about 2:1.

Third Gear

In third gear, Figure 8-33, the synchronizer outer sleeve moves forward, engaging the main drive gear. This locks the input and output shafts together. This is direct drive, or a 1:1 gear ratio. Notice that the countershaft gear is not used to transmit power in direct drive. This is true of any manual transmission.

Higher Gears

Manual transmissions with more than three forward gears, even those with built-in overdrives, operate according to the same principles as the 3-speed transmission previously presented in this chapter. Notice in the 5-speed transmission shown in Figure 8-34 that all gears, except fourth, use the countershaft gear to transmit power. Fifth gear uses the countershaft gear to obtain overdrive. The fifth gear on the countershaft gear is larger than the fifth gear on the output shaft. This produces an overdrive ratio. A typical overdrive ratio is 0.7:1. The overdrive gear keeps engine rpm down at highway speeds to increase fuel economy and engine service life. Note that this is the typical arrangement with a 5-speed transmission in that fourth gear provides direct drive and fifth gear provides overdrive.

Manual Transmission Designs

Today, there are many manual transmission designs for rear-wheel drive vehicles. Transmissions with more forward speeds provide a better selection of gear ratios. Extra gear ratios are needed for the smaller, low-horsepower, high-efficiency engines of today. Some of the most common transmissions are discussed below.

3-Speed Transmissions

The ***3-speed transmission,*** Figure 8-35, is often found on older cars. These transmissions have three forward gears (gear ratios). In the last few years, 3-speeds have been used strictly in pickup trucks. This kind of transmission often uses external linkage and is usually found in combination with a column shift.

4-Speed Transmissions

The ***4-speed transmission*** has four forward gears. These transmissions were the high-performance models on cars in years past. Today, 4-speeds are used with smaller engines for better economy and performance. The 4-speed transmission is almost always used with a floor

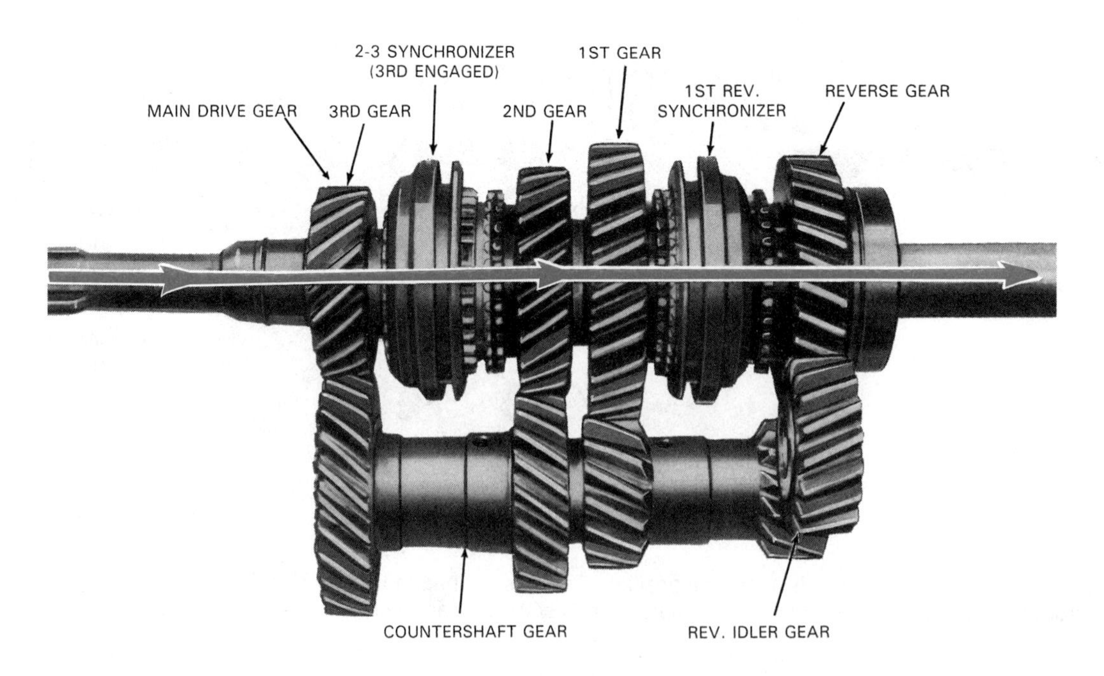

Figure 8-33. In third gear, rear sleeve is in neutral position. Front sleeve has moved forward to lock input shaft to output shaft. Power flow is from input shaft to output shaft. Countershaft gear is not used. (General Motors)

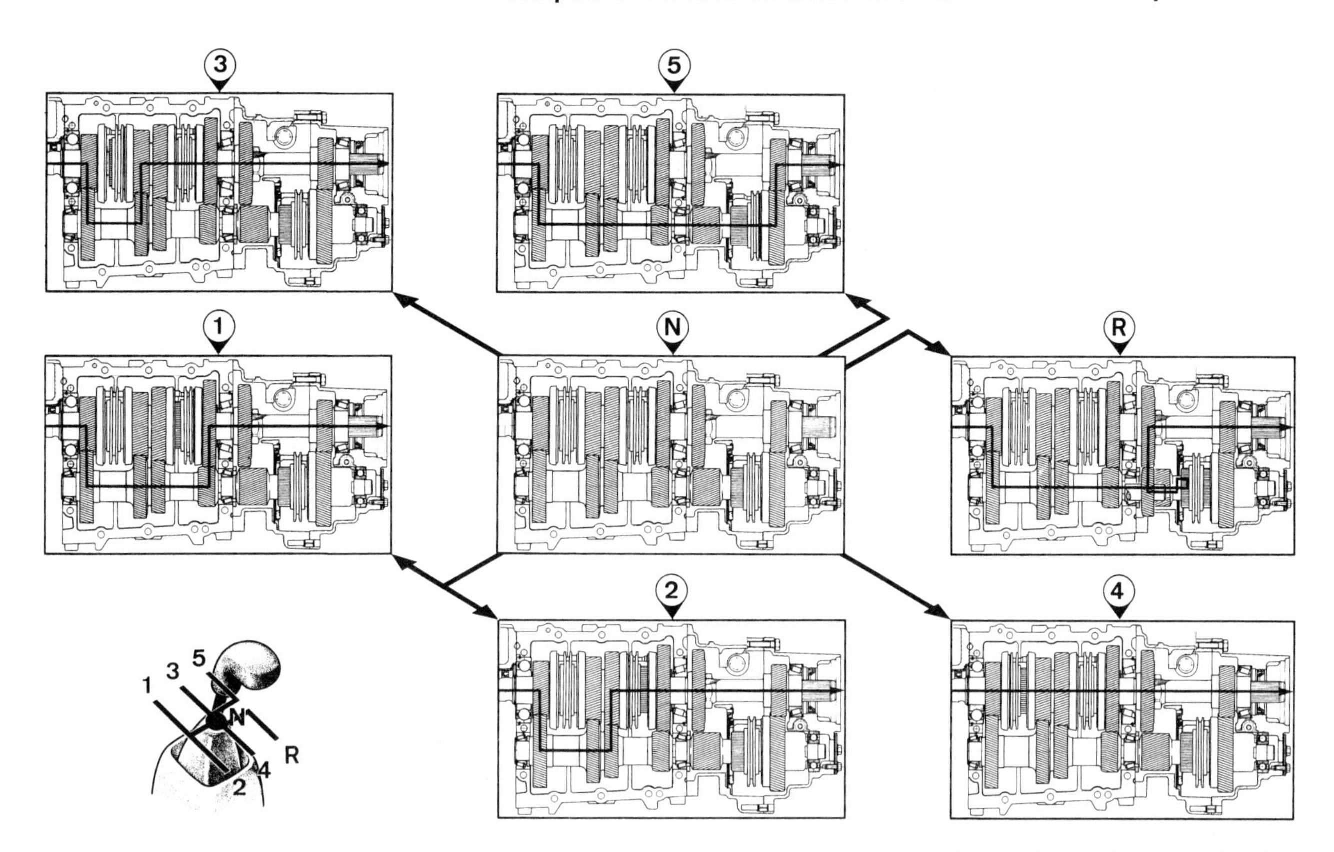

Figure 8-34. This diagram of a 5-speed transmission shows how reduction and overdrive are obtained by running power flow through the countershaft gear. Direct drive is the only forward gear not using the countershaft. Reverse power also flows through the countershaft gear. (Peugeot)

shift. Older 4-speeds usually had external linkage. The linkage on newer models is usually internal. Figure 8-36 shows a common 4-speed transmission.

5-Speed Transmissions

The ***5-speed transmission,*** Figure 8-37, resembles a 4-speed. These transmissions have five forward gears. Fifth gear is usually overdrive. Linkage and case materials are similar to the 4-speed. It is always used with a floor shift. These transmissions are sometimes created from existing 4-speeds by adding extra gears on the countershaft and output shaft. The 5-speed transmissions are becoming increasingly popular because of the improvement in gas mileage with the overdrive gear.

6-Speed Transmissions

The ***6-speed transmission*** might look similar to a 4- or 5-speed transmission; however, it is a completely new design, intended to have six speeds (gear ratios) in a compact case. In some instances, both fifth and sixth gears are overdrives. The linkage is usually internal. A 6-speed is often used on a high-performance engine as a way of obtaining high performance while maintaining acceptable gas mileage.

Summary

Manual transmission gears are selected by the vehicle driver. Operation of a manual transmission is simple. The clutch pedal is depressed, and the proper gear is selected by using the gearshift lever. Then, the clutch pedal is released to allow power to flow to the rear wheels.

There are seven major parts to a manual transmission. These are the transmission shafts and gears, the synchronizers, the shift forks and linkage, and the transmission case and extension housing.

Transmission shafts are used to support gears and to transmit motion. A shaft may rotate, or it can be a stationary support for a moving gear. The major shafts in a manual transmission are the input shaft, countershaft, output shaft, and reverse idler shaft.

Transmission gears may rotate on transmission shafts or may be solidly connected to them. Gears transmit power and provide different gear ratios and a way to reverse the vehicle.

The input shaft and main drive gear are turned by the engine whenever the clutch is engaged. The main drive

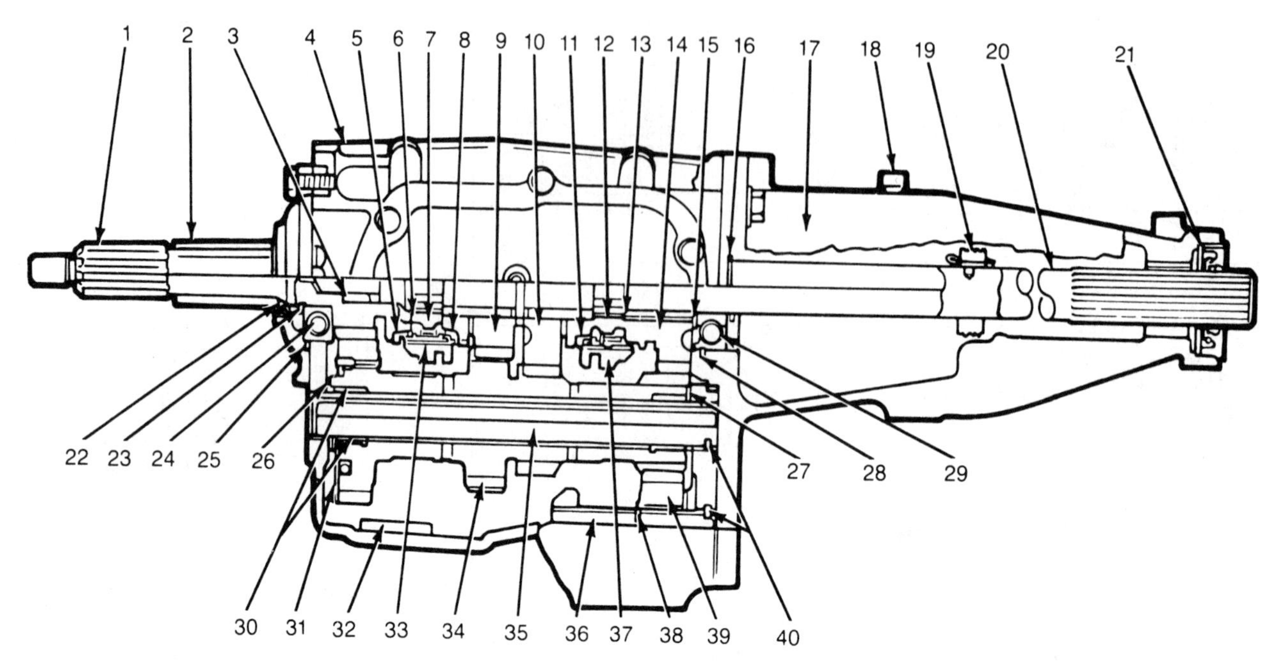

1. CLUTCH GEAR
2. BEARING RETAINER
3. PILOT BEARINGS
4. CASE
5. 3RD SPEED BLOCKER RING
6. 2-3 SYNCH. SNAP RING
7. 2-3 SYNCH. HUB
8. 2ND SPEED BLOCKER RING
9. 2ND SPEED GEAR
10. 1ST SPEED GEAR
11. 1ST SPEED BLOCKER RING
12. 1ST SPEED SYNCH. HUB
13. 1ST SPEED SYNCH. SNAP RING
14. REVERSE GEAR
15. REVERSE GEAR THRUST AND SPRING WASHERS
16. SNAP RING — BEARING TO MAINSHAFT
17. EXTENSION
18. VENT
19. SPEEDOMETER DRIVE GEAR AND CLIP
20. MAINSHAFT
21. REAR OIL SEAL
22. RETAINER OIL SEAL
23. SNAP RING — BEARING TO GEAR
24. CLUTCH GEAR BEARING
25. SNAP RING — BEARING TO CASE
26. THRUST WASHER — FRONT
27. THRUST WASHER — REAR
28. SNAP RING — BEARING TO EXTENSION
29. REAR BEARING
30. COUNTERGEAR ROLLER BEARINGS
31. ANTI-LASH PLATE ASSEMBLY
32. MAGNET
33. 2-3 SYNCH. SLEEVE
34. COUNTERGEAR
35. COUNTER SHAFT
36. REVERSE IDLER SHAFT
37. 1ST SPEED SYNCH. SLEEVE
38. "E " RING
39. REVERSE IDLER GEAR
40. WOODRUFF KEY

Figure 8-35. Vehicles with 3-speed transmissions were once common; however, they have now been eliminated on all but some full-size pickup trucks. (General Motors)

gear is always in mesh with the countershaft gear. This gear turns whenever the main drive gear turns. The output shaft contains gears that provide forward speeds. It also contains reverse gear. The reverse idler gear is in mesh with the countershaft gear. Through the idler gear, rotational direction of the output shaft is made opposite of the input shaft. In this way, the vehicle can be reversed.

Synchronizers are used to select various gears by matching gear speeds to avoid gear clash. They basically consist of a splined outer sleeve and inner hub and blocking rings, which match cone-shaped extensions on mating gears. When the synchronizer is pushed into engagement with the cone, it acts as a clutch, speeding up or slowing down one gear to match the speed of the output shaft.

Most modern transmissions are synchronized in all forward gears. In some transmissions, however, low and reverse gears are selected by sliding the output shaft gears directly into mesh with the countershaft gears.

The synchronizer outer sleeves are operated by the vehicle driver through a series of forks, levers, links, and sometimes cables. Shift forks are made to slip into grooves cut into the outer sleeves. The shift fork moves in a straight line to push the sleeve into engagement with a mating gear. The sleeve can continue to rotate as it is moved.

Shift forks are operated by the driver through a series of shafts, levers, links, and sometimes cables. External shift linkage is a system that is identified as a series of rods and levers that connect to the outside of the transmission. External shift linkage may be connected to the gearshift lever either on the steering column or on the floor.

Internal shift linkage will be found only on transmissions with floor-mounted gearshift levers. Except for the

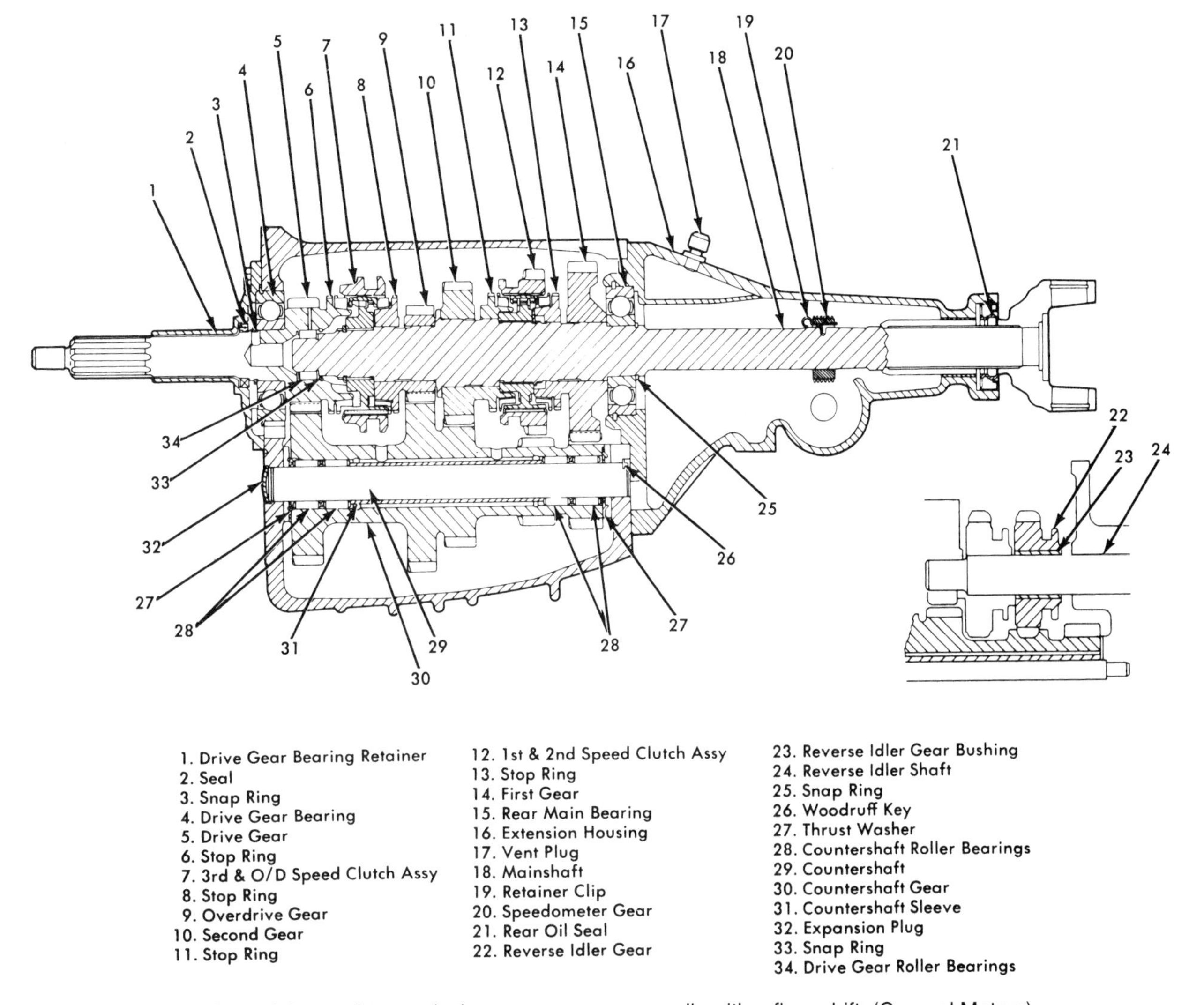

Figure 8-36. Various versions of 4-speed transmissions are common, usually with a floor shift. (General Motors)

gearshift lever, all linkage is internal to the transmission. This type of linkage may employ a single shifter shaft or several shift rails. The shift forks are positioned along the shaft.

The transmission case houses, supports, and aligns the rotating transmission components. It forms a reservoir for lubricant. It is the attaching point for the clutch housing, extension housing, shift linkage, and other components. Many modern transmissions are made with a center support.

The front bearing retainer is installed on the front of the transmission case, around the input shaft. It forms a support for the front bearing. The clutch throwout bearing slides on the nose of the front bearing retainer.

The extension housing supports and protects the end of the transmission output shaft. The front of the extension housing is bolted to the transmission case. The extension housing also contains the speedometer pinion gear and adapter and worm gear. The transmission mount is usually attached to the extension housing.

Most late-model transmissions will be equipped with one or more electrical switches, sensors, and solenoids. Switches and sensors are used to turn on the backup lights or to prevent starting the engine when the transmission is in gear. Sensors are used to provide information to engine control computers. Solenoids help control transmission shifting.

There are many different kinds of rear-wheel drive manual transmissions. Models with only three forward speeds are becoming rare, especially in passenger cars. The 4- and 5-speed models are the most common today. The 6-speed model is increasingly being used to obtain acceptable gas mileage on high-performance vehicles.

Know These Terms

Gearshift lever; Upshifting; Downshifting; Manual transmission shafts; Transmission gear; Synchronizer;

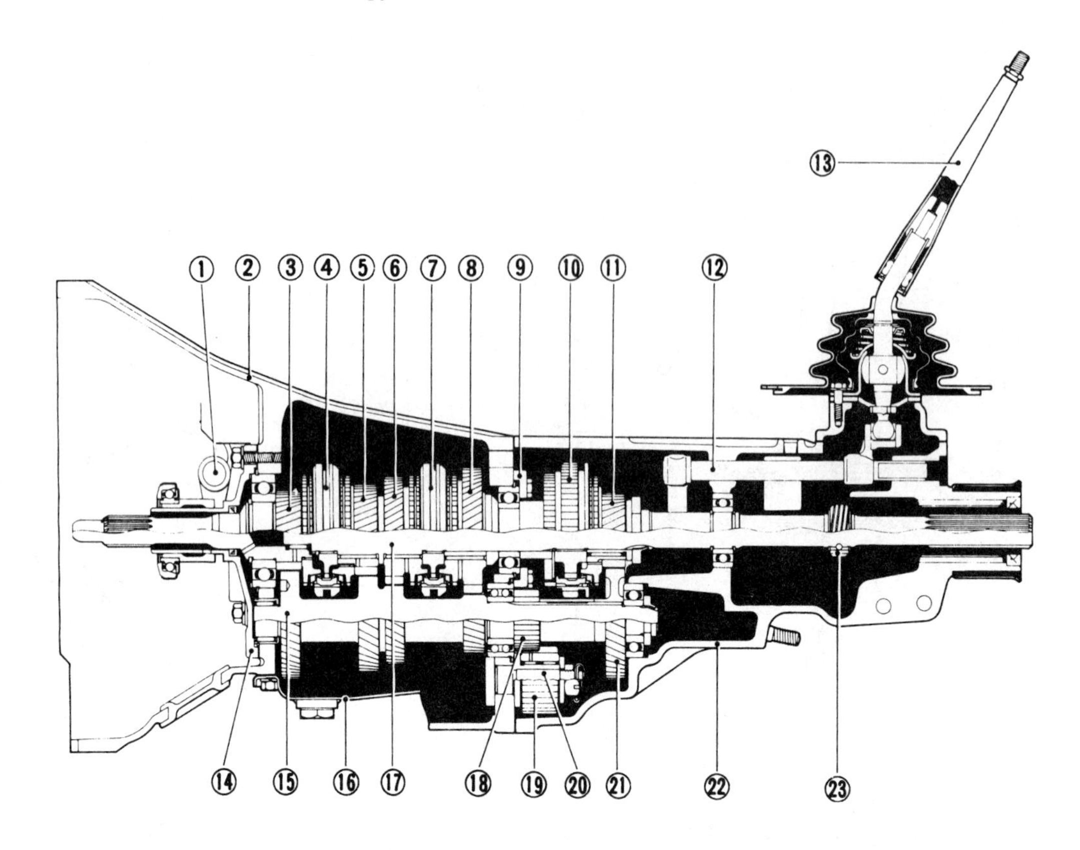

(1) Clutch control shaft
(2) Transmission case
(3) Main drive gear
(4) Synchronizer (3-4 speed)
(5) 3rd speed gear
(6) 2nd speed gear
(7) Synchronizer (1-2 speed)
(8) 1st speed gear
(9) Rear bearing retainer
(10) Synchronizer (reverse and overdrive)
(11) Overdrive gear
(12) Control shaft
(13) Control lever
(14) Front bearing retainer
(15) Countershaft gear
(16) Under cover
(17) Mainshaft
(18) Counter reverse gear
(19) Reverse idler gear
(20) Reverse idler gear shaft
(21) Counter overdrive gear
(22) Extension housing
(23) Speedometer drive gear

Figure 8-37. The 5-speed resembles the 4-speed transmission. Fifth gear is usually an overdrive. In some instances, overdrive has been added to a 4-speed by extending the countershaft gear assembly and transmission case. An extra gear is also added to the output shaft.

Shift fork; Manual transmission shift linkage; Manual transmission case; Extension housing; Clutch shaft; Countershaft; Manual transmission output shaft (Mainshaft); Rear bearing; Reverse idler shaft; Main drive gear; Countershaft gear (Cluster gear); Constant-mesh transmission; Mainshaft gear; Synchromesh transmission; Block synchronizer; Inner hub; Outer sleeve; Insert; Insert spring; Blocking ring; External shift linkage; Column-shift linkage; External floor-shift linkage; Internal shift linkage; 3-speed transmission; 4-speed transmission; 5-speed transmission; 6-speed transmission.

Review Questions–Chapter 8

Please do not write in this text. Place your answers on a separate sheet of paper.

1. What is the purpose of a transmission?
2. The _____ _____ supports the transmission shafts, forms a reservoir for lubricant, and provides an attaching point for other components.
3. The _____ _____ provides protection for the end of the transmission output shaft and is the usual location for the transmission mount.

4. The four types of transmission shafts are the clutch shaft, countershaft, mainshaft, and output shaft. True or False?
5. Why are thrust washers used in transmissions?
6. To lock mainshaft gears to the mainshaft and eliminate gear clash in the process, constant-mesh transmissions use:
   a. Clutch gears.
   b. Sliding gears.
   c. Shift rails.
   d. Synchronizers.
7. In your own words, how do synchronizers operate?
8. Each mainshaft gear is comprised of two sets of teeth. True or false?
9. In an external shift linkage, shifting is accomplished by means of a shift fork attached to a shifter shaft. True or false?
10. What type of lubricant is used in a manual transmission?
11. On a separate sheet of paper, sketch the power flow through a typical 3-speed transmission in each gear position.
12. In your own words, describe the internal action of the 3-speed transmission in each gear position.

Certification-Type Questions–Chapter 8

1. **All of these are found in manual transmissions EXCEPT:**
   (A) input shafts.
   (B) mainshafts.
   (C) countershafts.
   (D) stator shafts.

2. **Rearward movement of the input shaft is usually prevented by:**
   (A) snap rings.
   (B) the output shaft.
   (C) the main drive gear.
   (D) the front bearing retainer.

3. **Which of these gears are used for reverse?**
   (A) Spur gears.
   (B) Helical gears.
   (C) Cluster gears.
   (D) Main drive gears.

4. **Which of these contains the gears that provide all forward speeds and reverse?**
   (A) Input shaft.
   (B) Clutch shaft.
   (C) Output shaft.
   (D) Reverse idler shaft.

5. **Technician A says that each mainshaft gear is comprised of two sets of teeth. Technician B says that the main drive gear has three sets of teeth. Who is right?**
   (A) A only
   (B) B only
   (C) Both A & B
   (D) Neither A nor B

6. **To shift from one gear to another, constant-mesh manual transmissions require:**
   (A) idler gears.
   (B) bearings.
   (C) planetary gears.
   (D) synchronizers.

7. **Technician A says that the outer sleeve of a synchronizer assembly does not rotate until it engages the blocking ring. Technician B says that shift forks move the outer sleeve while permitting sleeve rotation to continue. Who is right?**
   (A) A only
   (B) B only
   (C) Both A & B
   (D) Neither A nor B

8. In an internal shift linkage, which of these locks out all other gears except the one selected?

(A) Locknut.
(B) Blocking ring.
(C) Interlock plate.
(D) Block synchronizer.

9. All of these are used to cover the opening in the front of the transmission case EXCEPT:

(A) the front cover.
(B) the front bearing cap.
(C) the front bearing retainer.
(D) the clutch housing.

10. In reverse gear, rotation between input and output shafts is reversed:

(A) once.
(B) twice.
(C) three times.
(D) five times.

11. On a fully synchronized transmission in neutral, all of these are being driven by the main drive gear EXCEPT:

(A) second gear.
(B) the reverse idler gear.
(C) the countershaft gear.
(D) the mainshaft.

Chapter 9

Manual Transmission Problems, Troubleshooting, and Service

After studying this chapter, you will be able to:

- Name and describe common types of problems that can occur in a manual transmission.
- Adjust external shift linkages.
- Evaluate manual transmission problems to determine if removal is necessary.
- Remove a manual transmission from a vehicle.
- Disassemble a manual transmission.
- Inspect the parts of a manual transmission both visually and by measurement.
- Assemble a manual transmission.
- Install a manual transmission in a vehicle.
- Distinguish between safe and unsafe manual transmission service procedures.

Although manual transmissions are not as common as they once were, many are still installed on modern vehicles. New manual transmissions come in a variety of designs, with as few as three forward speeds or as many as six. The automotive technician must know how to properly diagnose and service these assemblies. If you study the information presented here, you will learn these important skills. In addition, you will have a solid basis for learning how to diagnose and service other gear units, such as front-wheel drive manual transaxles, presented in Chapter 19, and four-wheel drive transfer cases, presented in Chapter 25 of this text; many of the techniques applied in this chapter will also be applied in these later chapters.

This chapter is intended as a general guide to manual transmission diagnosis and service. It covers both in-car service procedures–in particular, servicing the transmission external shift linkage– as well as major overhauls. The information presented here will help you understand the troubleshooting and diagnostic information found in factory service manuals. The removal and installation steps presented in this chapter will be applicable to almost any kind of transmission. Figure 9-1 shows typical transmission problems. Study them carefully.

Manual Transmission Problems and Troubleshooting

The transmission technician should be as good at diagnosing manual transmission problems as he or she is at repairing them. The final outcome of the troubleshooting process should be knowing what the problem is before you begin repairs. If you do not know what is really wrong with a transmission, your chances of repairing it properly are small. The paragraphs below are designed to make troubleshooting transmission problems easier.

Since they consist of mechanical parts, most manual transmissions will wear out, usually because of high mileage. Transmissions may require repair at low mileage because of careless driving habits. For example, attempting to change gears too quickly can chip or break teeth on the gears or synchronizers. Skipping too many gears when downshifting at high vehicle speeds or releasing the clutch pedal before gears are completely engaged can also cause teeth to chip or break. Synchronizer rings, in particular, are made from soft metal–brass, bronze, or copper–and they can be easily damaged by any type of rough shifting.

A lubricant level that is too low is another source of manual transmission problems. Manual transmissions are lubricated by oil that is thrown onto the moving parts as they rotate. This method, called splash lubrication, will work only if the oil in the transmission is at the proper level. If the level is allowed to become too low, the moving parts will be damaged.

Water in a transmission can also cause problems. If the vehicle is driven through deep water, some water may enter through the transmission vent. Water will contaminate the oil and destroy its lubrication value. Gears and

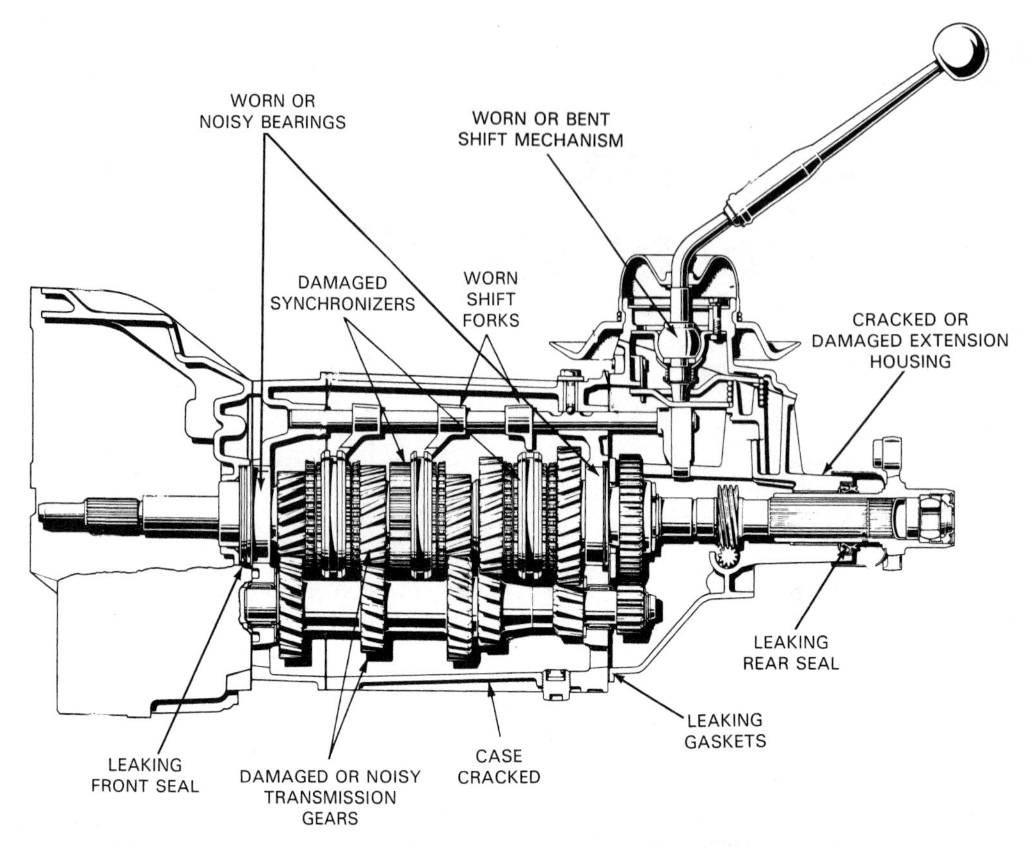

Figure 9-1. Study types of problems found in manual transmissions. (Fiat)

bearings that run in water can become pitted, requiring replacement.

You should start your diagnosis of a manual transmission problem by finding out why the vehicle was brought in for service. Have the owner or driver describe the problem before starting any other diagnostic steps.

Some problems can be diagnosed on a lift, while others require that you take a test drive. When test driving a vehicle, determine in what transmission gear(s) and at what speed the problem occurs. If possible, take the driver of the vehicle with you and have him or her point out the particular problem while it occurs.

Once you have identified the operational problem, check out the easiest possible solution first; you should always proceed from easiest to hardest. For instance, it is always advisable to check the oil level and linkage adjustments before removing and disassembling the transmission. Consider all possibilities, even the most unlikely, and use well-thought-out methods to find solutions; always avoid guesswork, as this is seldom profitable.

Many manual transmission problems can be solved by simple linkage adjustments or by adding oil to the transmission. Every manual transmission that is brought in for service should be checked for clutch- and shift-linkage condition and adjustment. The transmission oil level should be checked, and the proper grade and type of lubricant should be added, if necessary. If oil level is low, you should check for leaks.

Problems exhibited in manual transmission operation are presented in detail in the upcoming paragraphs. Figure 9-2 is a table summarizing some common conditions and their causes and corrections.

Lubricant Leakage

Transmission *lubricant leakage* is an unwanted loss of lubricant from inside of the transmission. A transmission case should be oil-free on the outside. (A slightly rusty case is normal.) Lubricant levels may drop slightly in a manual transmission because some of the lubricant evaporates. However, if the oil level is very low, there is probably a leak somewhere. Most manual transmission oil leaks are first noticed when oil spots show up under the vehicle.

When a vehicle is brought in because of leaking lubricant, you should make sure that the leak is really from the transmission. Engine oil, brake fluid, power steering fluid, or engine coolant could be dripping off the transmission case. Fluids from the front of the vehicle are often carried to the back by airflow under the moving vehicle. If the case has oil on it, but the oil level is normal, something else may be leaking. Finding the exact source of the leak can be difficult. It may be necessary to

MANUAL TRANSMISSION DIAGNOSIS

CONDITION	POSSIBLE CAUSE	CORRECTION
Leaks lubricant	a. Cover loose. b. Cover gasket loose or defective. c. Front bearing retainer loose, broken, or gasket defective. d. Front bearing retainer seal defective. e. Output shaft seal worn. f. Countershaft loose in case. g. Lubricant level too high. h. Shaft expansion plugs loose in case. i. No sealer on bolt threads. j. Damage shifter shaft seal. k. Vent plugged. l. Wrong lubricant. m. Cracked case or extension housing. n. Drain or filler plug loose.	a. Tighten cover. b. Tighten cover and/or replace gasket. c. Tighten retainer, replace gasket or retainer. d. Replace seal. e. Replace seal. f. Replace case. g. Drain to level of filler plug. h. Replace plugs. Use sealer. i. Seal threads. j. Replace seal. k. Open vent. l. Drain and refill with recommended lubricant. m. Replace case or housing. n. Tighten plug.
Shifts hard	a. Excessive clutch pedal free travel. b. Worn or defective clutch. c. Failure to fully depress clutch pedal when shifting. d. Shift cover loose. e. Shift fork, shafts, levers, or detents worn or loose. f. Improper shift linkage adjustment. g. Linkage needs lubrication. h. Linkage binding, bent, or loose. i. Wrong transmission lubricant. j. Insufficient lubricant. k. Excess amount of lubricant. l. Transmission misaligned. m. Front bearing retainer loose or cracked. n. Synchronizer worn, damaged, or improperly assembled.	a. Adjust free travel. b. Replace worn parts. c. Instruct driver. d. Tighten cover. e. Tighten or replace. f. Adjust linkage. g. Lubricate. h. Free, straighten, or tighten as needed. i. Drain and fill with recommended lubricant. j. Add lubricant to filler plug level. k. Drain to level of filler plug. l. Correct alignment. m. Tighten or replace retainer. n. Replace or reassemble synchronizer.
Slips or jumps out of gear	a. Transmission loose or misaligned. b. Clutch housing loose or misaligned. c. Shift linkage improperly adjusted. d. Shift rail detents worn or detent springs weak. e. Synchronizer clutch sleeve teeth worn. f. Loose shifter cover. g. Shift fork, shaft, or levers worn. h. Worn teeth on main drive gear or other gears. i. Worn countershaft gear assembly bearings and/or thrust washers. j. Worn reverse idler gear assembly bushing or bearings. k. Worn output shaft pilot bearing. l. Front bearing retainer loose. m. Other parts striking shift linkage. n. Worn input or output shaft bearings. o. Worn input shaft bushing in flywheel. p. Bent output shaft.	a. Tighten and/or align. b. Tighten and/or align. c. Adjust linkage. d. Replace detents and/or springs. e. Replace synchronizer. f. Tighten cover. g. Replace worn part. h. Replace input shaft or gears. i. Replace countershaft gear bearings, and washers. j. Replace gear, bearings, and shaft. k. Replace rollers, replace shafts if necessary. l. Tighten retainer. m. Make adjustments to provide clearance. n. Replace bearings. o. Replace bushing or bearing. p. Replace shaft.

Figure 9-2. Manual transmission diagnosis chart.

(continued)

CONDITION	POSSIBLE CAUSE	CORRECTION
Gear clash during down-shifting	a. Synchronizer worn, damaged, or improperly assembled. b. Shifting too fast (ramming into lower gear). c. Shifting to a lower gear when road speed is excessive. d. Clutch not releasing properly. e. Excessive output shaft end play. f. (See shifts hard.)	a. Replace or reassemble synchronizer. b. Force into gear with a smooth, slower shift. c. Slow down to appropriate speed before shifting. d. Adjust or repair as needed. e. Adjust end play.
Gear clash shifting from neutral to reverse	a. Insufficient clutch pedal free travel. b. Worn lubricant. c. Engine rpm too high. d. Driver not waiting long enough after depressing clutch. e. Sticking clutch pilot bearing.	a. Adjust free travel. b. Drain and fill with correct lubricant. c. Set to correct idle rpm. d. Instruct driver. e. Replace pilot bearing.
Transmission noisy in neutral	a. Worn or damaged front bearing. b. Worn or damaged gears. c. Lack of lubrication. d. Countershaft gear assembly bearings worn or damaged. e. Output shaft pilot bearing worn or damaged. f. Countershaft gear antilash plate worn or damaged. g. Lubricant contaminated with broken bits of gears, bearings, or other parts.	a. Replace bearing. b. Replace gears. c. Fill to lever of filler plug. d. Replace bearings, countershaft gear, and countershaft. e. Replace all rollers. f. Replace plate or countergear as required. g. Disassemble, clean, and repair transmission.
Transmission noisy in high gear	a. Defective front bearing. b. Defective output shaft bearing. c. Defective synchronizer. d. Defective speedometer gears.	a. Replace bearing. b. Replace output shaft bearing. c. Replace synchronizer. d. Replace speedometer gears.
Transmission noisy in second gear (three speed)	a. Countershaft gear rear bearings worn or damaged. b. Defective synchronizer. c. Constant mesh second gear loose on shaft. d. Constant mesh second gear teeth worn or chipped.	a. Replace countershaft gear assembly bearings. Replace countershaft and countershaft gear if needed. b. Replace synchronizer. c. Replace gear and/or shaft. d. Replace second speed gear.
Transmission noisy in low and reverse	a. Countershaft gear low and reverse gear worn or damaged. b. Low and reverse sliding gear defective. c. Reverse idler gear or synchromesh low gear defective.	a. Replace countershaft gear. b. Replace low and reverse sliding gear. c. Replace reverse idler or synchromesh low gear.
Transmission noisy in reverse	a. Reverse idler bushings worn. b. Reverse idler gear worn or damaged. c. Countershaft gear reverse gear worn or damaged. d. Defective reverse sliding gear (synchromesh low gear).	a. Replace idler gear or bushings. b. Replace reverse idler gear. c. Replace countershaft gear. d. Replace reverse sliding gear.
Transmission noisy in all gears	a. Insufficient lubrication. b. Worn or damaged bearings. c. Worn or damaged gears. d. Wrong lubricant. e. Excessive synchronizer wear. f. Defective speedometer gears. g. Transmission misaliagned. h. Excessive input or output shaft and/or countershaft gear end play. i. Contaminated lubricant.	a. Fill to filler plug. b. Replace bearings. c. Replace gears. d. Drain and fill with recommended lubricant. e. Replace synchronizer. f. Replace gears. g. Correct alignment. h. Adjust end play. i. Disassemble, clean, and repair transmission.

Figure 9-2 *(continued).*

completely clean the case and put the vehicle up on a lift to isolate the exact source.

Once you are sure the transmission is leaking fluid, you must check for the location and cause. A usual cause of leakage is worn or damaged gaskets or seals. Gaskets are found between the case and the extension housing, the front bearing retainer, and the transmission top or side cover. Seals are used at the front bearing retainer, between the retainer and the input shaft. They are also used at the shifter shaft, between the case and extension housing, and at the rear of the extension housing.

Loose bolts are another usual cause of leaks. Other possible causes are loose countershaft or idler shaft sealing plugs, lack of sealer on bolts, a loose drain or fill plug, a hairline crack in the transmission case, or a broken front bearing retainer.

Leaks are sometimes caused by too much oil in the transmission. If the oil level is too high, oil may leak from the transmission vent or from otherwise good seals.

Correcting an oil leak may be simple–that is, if the leak is caused by loose bolts or overfilling. If a seal or gasket is leaking, however, or if other transmission parts are causing the leak, the transmission may require removal to gain access to the parts.

No-Drive Condition

When the transmission will not transmit power from the engine to the drive wheels, a ***no-drive condition*** exists. If only one gear will not transmit power, that gear is probably stripped. If the transmission will not move in any gear, the most likely cause is stripped input shaft and countershaft gears or a broken input or output shaft. Often, the shift linkage is disconnected and will be unable to select a gear. Sometimes, the transmission will be locked in two gears, and the vehicle will be unable to move. This is often caused by a defect in the shift linkage or shift interlock.

If a gear is stripped or a shaft is broken, the engine will idle easily with the clutch engaged and the defective gear selected. If the transmission is locked, engaging the clutch will kill the engine. If the linkage is working properly, the no-drive problem is caused by internal damage, and the transmission must be removed from the vehicle for repairs.

Hard Shifting

Hard shifting is a condition that occurs when the shift lever is difficult to move from gear to gear. In some instances, only one gear is hard to select; in others, all shifting may be difficult.

The usual causes of hard shifting are bent or worn shift linkage and linkage in need of lubrication. Hard shifting may also be caused by binding of the shift rails or forks. Occasionally, hard shifting may be caused by a misaligned transmission case or by a problem in the clutch.

Hard shifting can be diagnosed easily. If the gearshift lever is hard to move even when the engine is stopped, the shift linkage is probably the source of the problem. If the gearshift lever moves easily when the engine is stopped, but shifting becomes hard with the engine running, the problem is probably in the vehicle clutch, or else the case is misaligned.

In transmissions having an external shift linkage, hard shifting may be corrected by adjusting or lubricating shift linkages. In contrast, if *internal* parts of a shift linkage are bent or sticking, the transmission must be removed and disassembled to make repairs. Bad synchronizers, which can be responsible for hard shifting, or clutch problems will also require transmission removal. Misalignment problems may require transmission removal; however, if the case is misaligned because of dirt or other foreign material, it may be possible to clean and realign it without removal.

Gear Disengagement

There are two closely related conditions in which transmission gears and synchronizers disengage on their own, causing a transmission to come out of gear. These conditions are known as *gear jumpout* and *gear slipout.*

With ***gear jumpout,*** the components are fully engaged, but something happens to force them out of engagement, into neutral. In the case of ***gear slipout,*** the gears and synchronizers are never quite fully engaged (but should be). When this happens, gear rotation tends to push the synchronizer outer sleeve and mating gear out of engagement.

Jumpout generally occurs when some outside force overcomes spring pressure in one of the detent holding mechanisms of the shift linkage. Often, there is a problem with the detent mechanism; the springs may be broken, or the detent notches may be worn. The problem could be in the shift rails of an internal shift linkage, for example, or it could be in the gearshift lever assembly. If the problem is a detent, the driver may make a false assumption and relate that the problem occurs in high gear only. The driver may only notice the problem in high gear because this is the only gear in which he or she is not holding on to the gearshift lever.

There are a number of conditions that can cause gear *slipout.* The first step in finding the cause of the problem is to determine which gear(s) are slipping out of engagement.

If the transmission slips out of a reduction gear, begin by checking for problems in the linkage. It could be worn or improperly adjusted. There may be interference between the shift linkage and some other part of the vehicle, or the shift linkage may be binding for some other reason. These actions can limit travel of the shift forks and prevent the synchronizers and mating gears from fully engaging. Linkage interference is often the cause if the problem occurs only when the vehicle is in a gear (high gear included) and accelerated.

If the linkage is all right, check for chipped or broken gear teeth. Defective teeth may be causing the problem (or they may be the result of something else that is causing the problem). Check the splines on the synchronizers. Worn or damaged splines on the outer sleeve or inner hub could be causing slipout. Also, check the pilot bearing and front bearing retainer. A worn or damaged pilot bearing or a front bearing retainer that is broken or loose can cause enough gear movement to push the sleeve and gear apart.

If the transmission slips out of high gear *only,* begin by checking for loose bolts on the transmission, clutch housing, or engine mounts. A related cause is a misaligned transmission case or clutch housing. Other causes include worn linkage or binding shift linkage.

At times, jumpout and slipout problems can be corrected without removing the transmission. Often, problems of external shift linkage can be corrected by lubrication or adjustment. Sometimes, however, linkage parts are too worn to be corrected by adjustment and must be replaced. When the problem is caused by worn or damaged internal transmission parts, the transmission must be removed and disassembled.

Gear Clash

Gear clash is a grinding noise that is made while shifting gears. If a manual transmission develops gear clash, first check for improper clutch adjustment or a faulty clutch condition. A clutch that drags, or does not fully disengage, is the most common cause of gear clash. If the clutch is at fault, the clash should occur in every gear.

If the clutch checks out all right, or if the gear clash does not occur in all gears, check for worn transmission synchronizers, including sleeves, hubs, and blocking rings. Also, check for loose mounting bolts or misalignment, which will sometimes cause gear clash.

Abnormal Noise

An *abnormal noise* in the manual transaxle is noise above normal sound level during operation, when in gear or in neutral. Noises that are most noticeable may sound like whining, rumbling, or roaring. The normal noise level may vary from gear to gear, depending on the gear ratio. Accelerating in low gear from a stop, for example, will cause more noise than accelerating in direct drive at higher speeds.

Transmission noises can be difficult to separate from other vehicle noises. You should always make sure that the problem is in the transmission. Other drive train noises are often blamed on the transmission. Be sure to eliminate the clutch, driveline, and rear axle as possible causes of the problem. If the noise occurs only in one gear, or in neutral with the vehicle stopped (clutch pedal released), you can safely assume that the problem is in the transmission.

If the transmission is noisy in every gear and also in neutral, first check for a low oil level. Another cause of noise in every gear and neutral is a worn or damaged main drive gear or countershaft driven gear. A rumbling noise could be coming from a bad front bearing or bad countershaft needle bearings.

If the transmission is noisy in neutral only, check for a damaged front bearing or damaged needle bearings in the countershaft gear assembly. If the transmission is noisy in high gear only, check for a damaged front or rear bearing or a damaged speedometer gear.

If the transmission is noisy in only one reduction or overdrive gear, check for wear or damage to the gearset of that particular gear. Another possible cause of noise is a set of worn needle bearings in the countershaft gear assembly.

A transmission that is noisy in reverse only may have a worn or damaged idler gear, idler bushings, reverse countershaft gear, or mainshaft reverse gear. If the transmission has a first and reverse sliding gear, wear or damage to that gear may cause noise in reverse and in low.

Some transmissions develop a *clunking* or slapping noise that is heard when the engine is accelerated or decelerated. This clunking noise is caused by excessive backlash, or play, in one or more gears. This can be confirmed by alternately accelerating and backing off in the affected gear. You should check for worn or missing thrust washers causing excessive shaft end play, badly worn gears, worn or misaligned transmission bearings, or loose extension housing bolts. Be sure that you have eliminated the clutch, universal joints, and rear axle assembly as sources of clunking, before removing the transmission. Clunking noises are much more likely to be in other drive train parts than in the transmission.

Correcting most noises will require removing the transmission to gain access to defective gears or bearings. In some instances, such as with loose bolts or misalignment, the problem can be fixed without removing the transmission.

NOTE! An electronic stethoscope or listening device can help to find where noises are coming from. By attaching special clips to the transmission or other assembly, you can hear and pinpoint noises more easily.

In-Car Manual Transmission Service

Correcting some transmission problems does not require removing the transmission. For example, certain transmission leaks can be repaired without removing the transmission, including leaks at the side cover gasket, extension housing-to-case gasket, and rear seal. In some designs, shift forks can be replaced without removing the transmission, provided they are in an accessible cover assembly. Speedometer and cable assemblies can also be replaced without removing the transmission.

External shift linkage is one area of the manual transmission that is vital to proper operation and that can be repaired without transmission removal. As parts wear, the motions of the gearshift lever are not correctly transferred to the transmission. Moving the shift lever may not shift the transmission into the desired gear. The shift linkage requires three basic kinds of service: adjustment, lubrication, and parts replacement. More detail is offered in the next several paragraphs.

Column-Shift Linkage

Column-shift linkage is subject to wear, which causes looseness. In addition, it is subject to binding. Binding can be caused by lack of lubrication, and wear can be accelerated by lack of lubrication. Most column-shift linkage will require adjustment, lubrication, and parts replacement at some time.

To adjust the column-shift linkage, put gearshift lever in neutral. Then, using a hoist, raise the vehicle off the floor. Insert a *shifter alignment tool* through holes in the two levers at the bottom of the steering column. (See *gage pin,* Figure 8-16.) This will lock them in the neutral position.

Most column-shift linkage is adjusted by turning threaded swivels on the shift rods where they connect to the transmission shift levers. Turning the swivel changes the effective length of the shift rod and changes the position of the shift lever at the transmission. The swivels are held stationary by locknuts. To continue adjusting the linkage, loosen the locknuts on the shift levers. Turn each swivel until a slight click can be felt, which means the shift lever is in neutral. (Note that each shift lever has three positions–two gear positions and neutral, with neutral in the middle.) With the shift lever in the neutral position, proper adjustment has been achieved. Tighten the locknuts and remove the alignment tool.

As mentioned, the transmission shift linkage may become loose after a time. Loose linkage can sometimes be corrected by adjusting the shift rod swivels to take up some of the play. Usually, however, loose shift linkage is caused by worn pivot bushings and washers. The bushings and washers must be replaced to restore linkage firmness. After replacement, pivot points on the shift linkage should be lubricated with water-resistant greases that will not attract dust or damage rubber or plastic bushings.

Most problems with the column-shift linkage occur with the shift tube. The tube usually wears where it contacts the gearshift lever or the shift levers at the bottom of the tube. This wear is accelerated by rough shifting. Worn tubes must be replaced. Shift tubes are replaced by removing the steering wheel and the shift levers. The gearshift lever is removed, and the tube is pulled out through the top of the steering column. The steering shaft does not need to be removed.

Floor-Shift Linkage

As covered in Chapter 8, floor-shift linkages can be either internal or external. Many modern floor-shift linkage systems are internal. This type has shift rail(s) installed inside of the transmission; it seldom requires service. The external floor-shift linkage, Figure 9-3, requires the same type of service as the column-shift linkage.

The shift mechanism of the gearshift lever assembly may require adjustment or lubrication. Adjustment is usually accomplished by threaded swivels on the shift rods where they connect to the transmission shift levers. A shifter alignment tool is installed through holes in the shift levers and the shift mechanism during the adjustment process. The shift mechanism should be lubricated with water-resistant grease. The shift linkage rods and levers are relatively trouble-free. Pivot bushings tend to wear out; when they do, they need to be replaced.

Manual Transmission Removal

This section will cover the removal of manual transmissions. The information presented here is general in scope due to the number of different transmissions in use today. It is not intended as a substitute for the factory service manual.

As shown in Figure 9-4, manual transmissions contain many parts. To successfully complete the repair of a manual transmission, careful removal, disassembly, repair, reassembly, and installation are needed. Cleanliness and close inspection of parts are important at all stages of transmission repair.

Before removing the transmission for service, try to determine the likely source of the transmission problem. A transmission that requires service does not always need

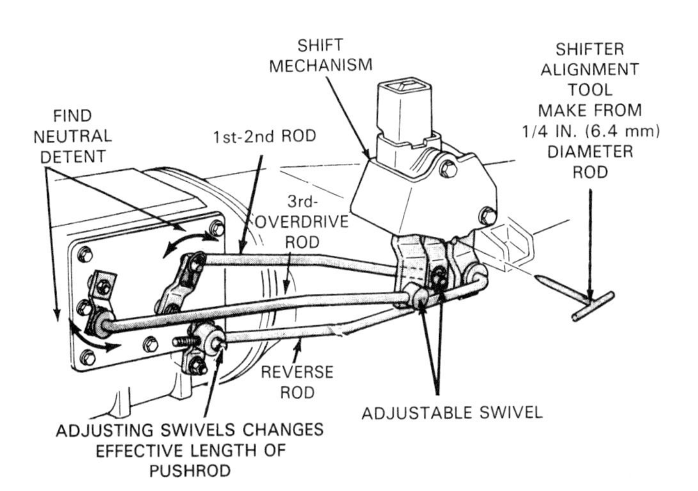

Figure 9-3. To adjust external linkage, put gearshift lever in neutral. Secure shift levers at gearshift lever assembly with shifter alignment tool. Adjust shift rod length so that shift levers at transmission are in position marked by neutral detent. (Chrysler)

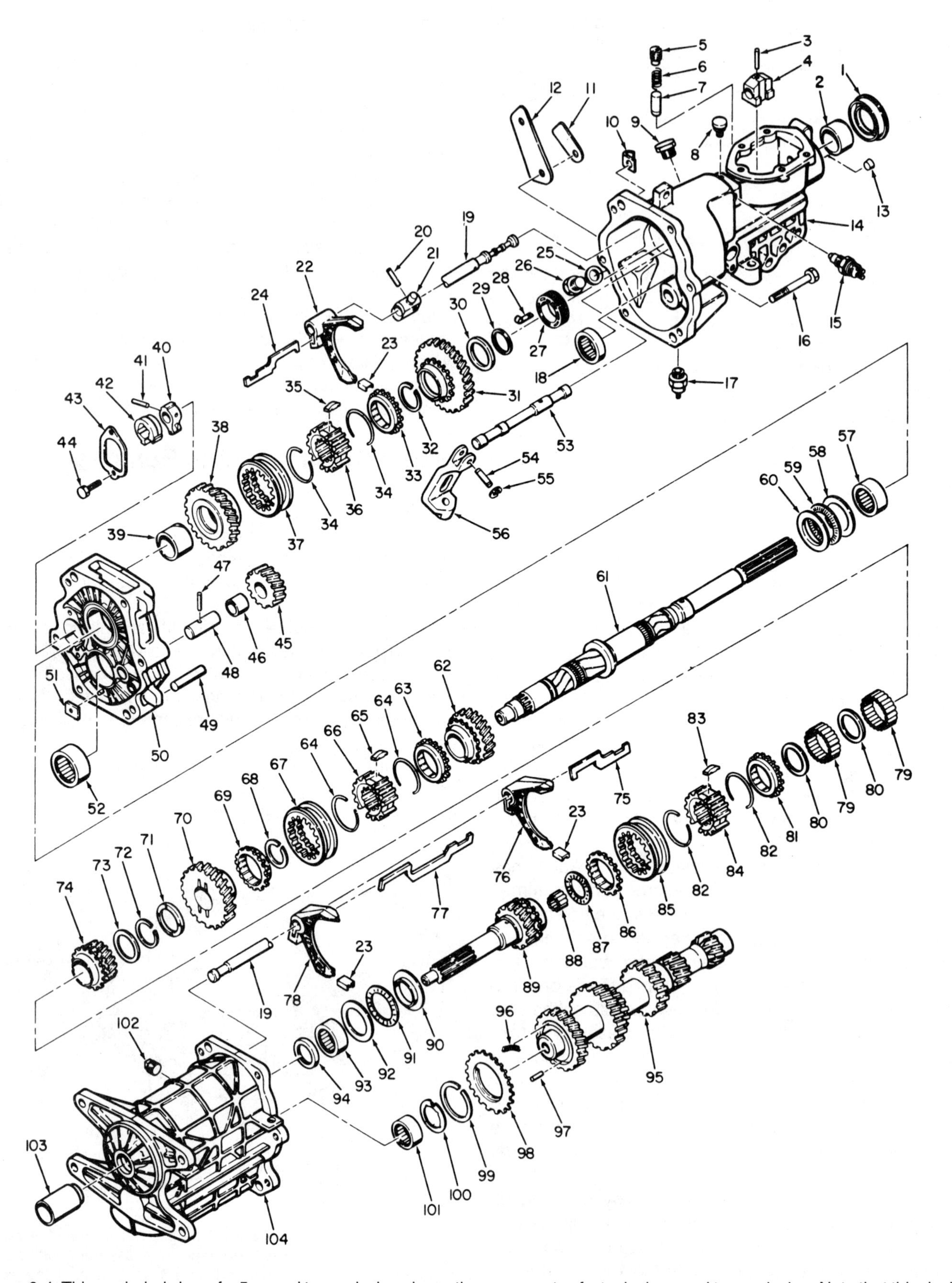

Figure 9-4. This exploded view of a 5-speed transmission shows the many parts of a typical manual transmission. Note that this design has a center support. (General Motors)

(continued)

INDEX NO.		DESCRIPTION	INDEX NO.		DESCRIPTION
1		OIL SEAL	57		NEEDLE BEARING
2		BUSHING	58		THRUST WASHER
3		PIN	59		NEEDLE THRUST BEARING
4		SHIFTER HEAD	60		NEEDLE THRUST RACE
5		THREADED PLUG	61		OUTPUT SHAFT
6		POPPET SPRING	62		3RD SPEED GEAR
7		MESH LOCK PLUNGER	63		BLOCKING RING
8		BREATHER	64		SYNCHRONIZER SPRING
9		SELECTOR LEVER PIVOT	65		SYNCHRONIZER SHIFT PLATE
10		WIRING HARNESS CLIP	66		CLUTCH HUB
11		NAME PLATE	67		CLUTCH SLEEVE
12		BACK-UP LIGHT BRACKET	68		SNAP RING
13		CUP PLUG	69		SYNCHRONIZER BLOCKING RING
14		EXTENSION HOUSING	70		2ND SPEED GEAR
15		SWITCH	71		THRUST WASHER
16		3/8-16 x 3-1/4 HEX HEAD BOLT	72		SNAP RING
17		SWITCH	73		SPACER
18		NEEDLE BEARING	74		5TH SPEED GEAR
19		SHIFT RAIL	75		2ND & 3RD SHIFT LINK
20		SPRING PIN	76		2ND & 3RD SHIFT FORK
21		RAIL SELECTOR	77		4TH & 5TH SHIFT LINK
22		FIRST & REVERSE SHIFT FORK	78		4TH & 5TH SHIFT FORK
23		SHIFT FORK PAD	79		NEEDLE ROLLERS
24		FIRST & REVERSE SHIFT LINK	80		SPACER
25		GASKET	81		SYNCHRONIZER BLOCKING RING
26		9/16-18 PLUG	82		SYNCHRONIZER SPRING
27		SPEEDOMETER GEAR	83		SHIFT PLATE
28		SPEEDOMETER GEAR RETAINER	84		CLUTCH HUB
29		SNAP RING	85		CLUTCH SLEEVE
30		THRUST WASHER	86		SYNCHRONIZER BLOCKING RING
31		1ST SPEED GEAR	87		NEEDLE THRUST BEARING
32		SNAP RING	88		NEEDLE ROLLERS
33		BLOCKING RING	89		INPUT DRIVE GEAR
34		SYNCHRONIZER SPRING	90		NEEDLE THRUST PLATE
35		SHIFT PLATE	91		NEEDLE THRUST BEARING
36		CLUTCH HUB	92		THRUST WASHER
37		CLUTCH SLEEVE	93		NEEDLE BEARING
38		REVERSE GEAR & BUSHING ASSEMBLY	94		OIL SEAL
39		BUSHING	95		CLUSTER GEAR
40		SELECTOR ARM	96		SPRING
41		SPRING PIN	97		SPRING PIN
42		INTERLOCK PAWL	98		GEAR DAMPER
43		SELECTOR ARM RETAINING SCREW	99		SNAP RING
44		1/4-20 x 3/4 HEX HEAD SELF TAPPING SCREW	100		THRUST WASHER
45		REVERSE IDLER GEAR & BUSHING	101		NEEDLE BEARING
46		BUSHING	102		1/2 INCH PIPE PLUG
47		SPRING PIN	103		TRANSMISSION CASE SLEEVE
48		REVERSE IDLER SHAFT	104		TRANSMISSION CASE
49		DOWEL PIN			
50		CENTER SUPPORT			
51		MAGNET			
52		NEEDLE BEARING			
53		SHIFT RAIL			
54		PIN			
55		RETAINING CLIP			
56		SELECTOR LEVER			

Figure 9-4 (*continued.*)

to be removed, as discussed in the preceding section. However, if the source of the problem seems to be internal–defective gears, bearings, or shafts, for example–the transmission will have to come out of the car.

Before performing any kind of transmission service, you must identify the transmission. It is time-consuming and annoying to order parts or obtain service information for the wrong kind of transmission. There are many kinds of manual transmissions, and they are difficult to identify by sight. Often, identical cases will contain different parts. You will usually need a transmission code number to determine exactly what kind of transmission is in the vehicle. Many transmissions have identification tags attached to a cover bolt. If there is no tag, the vehicle identification number on the dash or door jam will often contain a transmission code letter. This tag or code letter will ensure that you get the right parts and information. The service manual will explain the code number.

General Removal Procedures

Following are some *general* procedures detailing how to remove a manual transmission from a vehicle. Usually, a manual transmission of a rear-wheel drive vehicle can be removed from underneath the vehicle without removing the engine. In a few instances, clearances are close, and it is easier to remove the engine and transmission as a unit. This section will assume that the transmission can be removed without removing the engine. Note that you should always refer to the manufacturer's service manual for specific procedures.

1. Disconnect the battery negative cable. This will prevent any accidental electrical shorts.
2. While under the hood, also remove the upper *bell housing attaching bolts* (bell housing-to-engine bolts), if bell housing is part of the transmission case. See Figure 9-5. In some cases, it may be necessary to install an engine holding fixture to support the weight of the engine once the rear engine mount is removed.
3. If the vehicle has an external floor-shift linkage, unscrew the gearshift lever knob and remove the screws holding the rubber boot to the floor. If the vehicle has an internal floor-shift linkage, you may be able to remove the gearshift lever assembly. To do this, remove the boot, remove the bolts attaching the assembly to the extension housing, and pull the lever up and out of the housing. (In either instance, place transmission in neutral first.)
4. Raise the vehicle on a hydraulic lift to get enough clearance to remove the transmission. If you do not have access to a hydraulic lift, always use a good quality hydraulic floor jack and approved jackstands.

WARNING! Never support the vehicle with wooden crates or cement blocks. Be sure the vehicle is secure before working underneath it.

5. Check the engine mounts to ensure that they are not broken, which could be the cause of shifting or clutch problems.

WARNING! Broken engine mounts can allow the engine to fall backwards once the *transmission crossmember,* which provides support for the transmission, has been removed.

6. Drain the transmission oil. Figure 9-6 shows the location of the drain and fill plugs on a typical transmission. It is not absolutely necessary to drain the oil, as a drain plug may not have been provided. However, a full transmission will leak oil when the drive shaft assembly or other parts are removed. Be ready with a drip pan to catch spilled gear oil.
7. Remove the driveline by removing bolts from the rear of the driveline and slipping it out of the extension housing. Mark the driveline at the rear axle assembly so that it can be reinstalled in the same position. (See Chapter 15 for more information on driveline removal.) Once the driveline is out, plug the rear of the extension housing to reduce oil leaking.
8. If the bell housing is part of the transmission case, remove the lower access cover, disconnect the clutch linkage at the clutch fork, and remove the starter (if bolted to the clutch housing).

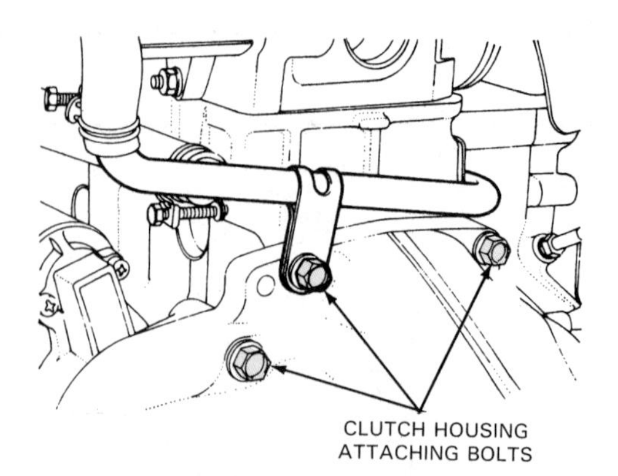

Figure 9-5. As part of transmission removal procedure, remove upper bell housing attaching bolts while the vehicle is still on the ground. (Chrysler)

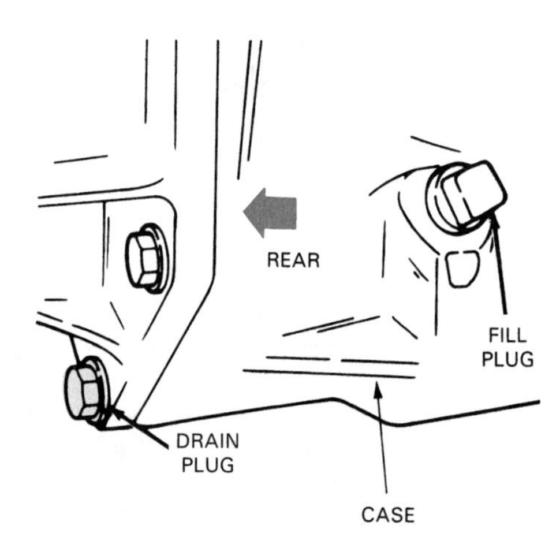

Figure 9-6. As part of transmission removal procedure, drain the oil through the drain plug or through a bolt hole low on the transmission case. (General Motors)

9. If vehicle has external shift linkage, disconnect it at this time. This procedure will vary with the make of transmission. In general, disconnect the shift rods from the shift levers at the transmission case. If an external floor-shift linkage is used, also unbolt the gearshift lever assembly from the case or extension housing. Remove the assembly, pulling the gearshift lever through the hole in the floor pan in the process. Removing the assembly will make for easier handling of the transmission.
10. Remove other miscellaneous items connected to the transmission, such as the speedometer cable assembly. If the transmission has any electrical connections, remove connectors now and place them out of the way. Label them, using masking tape, to simplify reconnection.

11. On many vehicles, especially those with V6 or V8 engines, it may be necessary to remove the exhaust pipes. This will give you more clearance when the transmission is removed. Removing the exhaust pipes will also allow you to lower the back of the engine, as may be required to reach the bell housing attaching bolts.
12. Disconnect any hardware attached to the transmission crossmember, such as the parking brake linkage. Removing the linkage will relieve pressure on the crossmember, caused by the parking brake return spring, and will give more clearance.
13. If an engine holding fixture is not already in place, support the engine with a jack placed under the rear of the engine or clutch housing (where separate from transmission).

WARNING! Never let the engine hang suspended only by the front engine mounts. Failure to support the engine before removing the transmission can result in serious injury.

14. Place a transmission jack (or floor jack) under the transmission. Raise the transmission slightly to remove weight from the crossmember. Then, remove the fasteners holding the crossmember to the vehicle frame and to the transmission, or rear engine, mount. Remove the crossmember.

CAUTION! If you are using a standard floor jack instead of a transmission jack, place a block of wood between the jack pad and the transmission case to reduce movement and possibility of damage to the case. See Figure 9-7.

15. If the bell housing and case are a one-piece design, remove the remaining bell housing attaching bolts, which hold the bell housing to the engine. You may need to tilt the engine and transmission back slightly to get at the bolts.

    If the bell housing and transmission case are separate castings, remove the *transmission case attaching bolts* (transmission case-to-bell housing bolts). Most transmissions will have four bolts, one at each corner.

WARNING! A manual transmission is heavy. Use the transmission jack to support and lower the transmission. If for some reason you do not use a transmission jack, you should have someone help remove the transmission.

Also, do *not* let the transmission hang unsupported once the bell housing or transmission case attaching bolts are removed, as this can damage the clutch disc hub or transmission input shaft. Always leave at least two attaching bolts in place until you have a jack under the transmission or are otherwise prepared to support it.

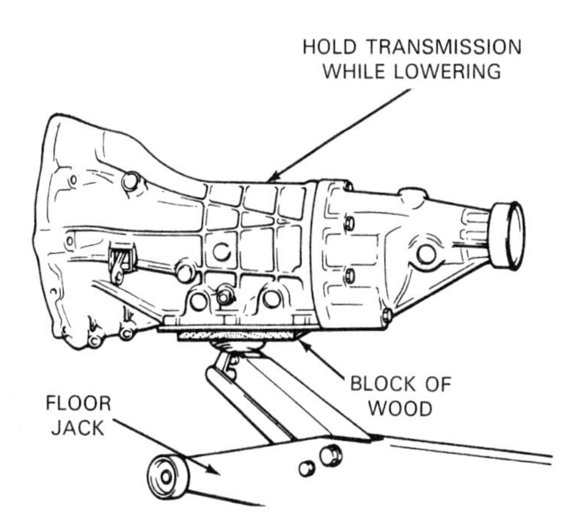

Figure 9-7. Use a transmission jack or a floor jack to support and remove the transmission. When using a floor jack, place a block of wood between the jack pad and the transmission case. Move the transmission to the workbench with the jack lowered. (Chrysler)

16. Slide the transmission straight back until the input shaft clears the clutch assembly or bell housing, as required. In the process, raise or lower the transmission jack, as required, until there is no binding between the splines on the input shaft and the clutch disc hub. When the input shaft is clear of the clutch assembly or bell housing, lower the transmission. Watch clearances and watch for any component that you may have forgotten to disconnect. Transport the transmission to the workbench with the jack in the lowered position.

Manual Transmission Disassembly

Once the transmission is removed from the vehicle, it can be disassembled for cleaning, inspection, and repair. The procedure outlined here is general and does not specifically cover any one transmission. You should always consult the manufacturer's service manual for exact disassembly procedures and specifications.

As you disassemble the transmission, carefully note the position of all parts and lay parts down on the workbench in the same sequence as removed. This will make reassembly easier. If necessary, index mating parts with light punch marks to ensure proper reassembly.

Removing External Components

The first step in disassembling a transmission is to remove the external components. If the gearshift lever assembly has not yet been removed, remove it now, Figure 9-8. If the transmission has an access cover, it should be

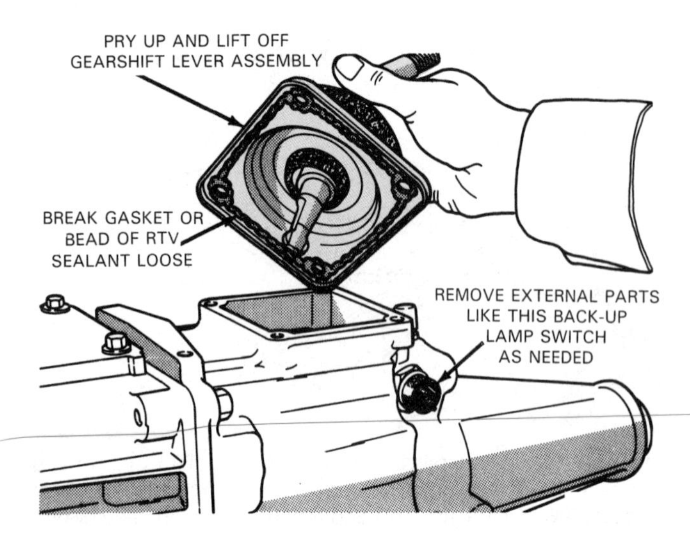

Figure 9-8. This shows the removal of the gearshift lever assembly from the transmission extension housing. Other external parts may also require removal. (Chrysler)

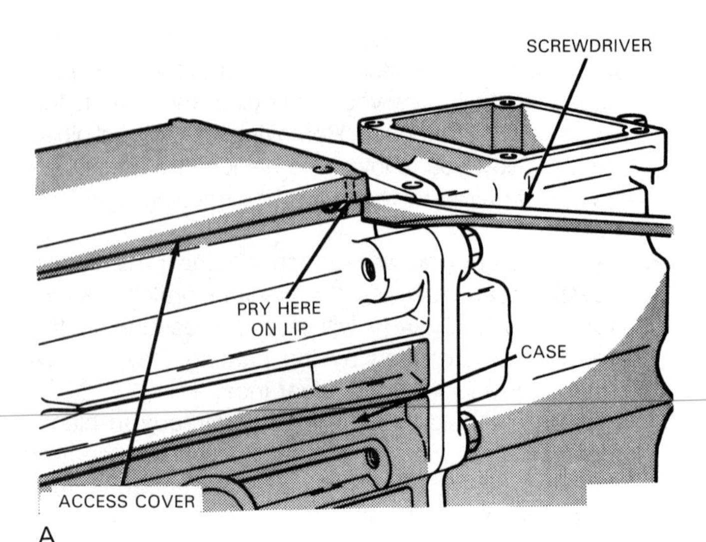

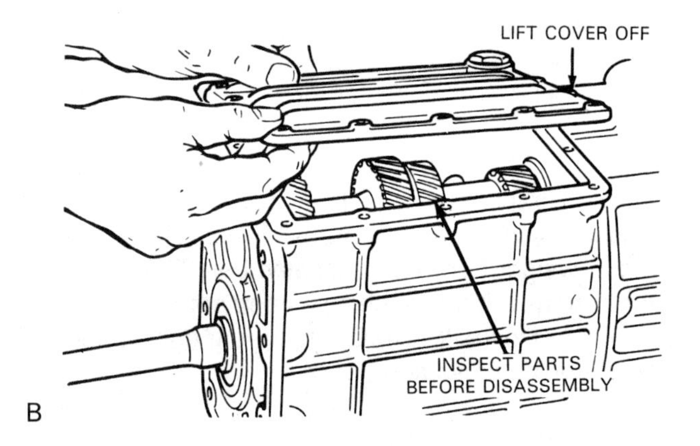

Figure 9-9. Removing the access cover from the transmission case. A–If necessary, the cover can be removed by lightly prying on it with a screwdriver. B–You should inspect the internal parts for obvious problems before removing them. If shift forks are not removed with the access cover, actuate them while turning input shaft. Observe transmission operation through access cover opening. (Chrysler, Ford)

removed, Figure 9-9A. Once the access cover has been removed, the transmission internals can be inspected for obvious damage. See Figure 9-9B.

It is a good practice to observe transmission action with the cover removed. Inspect the operation of the shift forks and the operation of the internal shift linkage. If the shift forks were not removed with the access cover, shift the transmission into each gear by actuating the shift forks. At the same time, rotate the input shaft while inspecting the condition of the gears and also the condition of the synchronizers. Doing this will allow you to observe any operating defects in the gears, the synchronizers, and the internal parts of the shift linkage.

To proceed with the disassembly, mark the case and front bearing retainer, as shown in Figure 9-10. These alignment marks will be used during reassembly to install the retainer in its original position. This step will help ensure proper operation, although it may seem minor.

If the bell housing is a one-piece unit with the case, the throwout bearing and clutch fork will have come out of the vehicle with the transmission, and the bearing will still be on the hub of the front bearing retainer. Remove the bearing from the clutch fork and from the hub. Then, unbolt the retainer and remove it by slipping it off the input shaft. Remove the front bearing retainer seal from its seat within the retainer, Figure 9-11.

After the retainer has been removed, the transmission front bearing should be removed. Some bearings must be removed with a special puller, Figure 9-12. Once the bearing is out, the input shaft can usually be removed, although some input shafts cannot be removed until the mainshaft is out.

Next, remove any external components that would prevent or would otherwise be damaged by extension housing removal, which is the following step. Figure 9-13 shows a shift rail detent ball assembly that would be damaged if it were not removed before the extension housing. The speedometer assembly should also be removed, if it was not taken out during transmission removal. Most speedometer assemblies are held to the case by a single bolt.

After all such components are removed, the extension housing should be removed from the transmission case. To do this, remove the extension housing bolts and break the gasket seal by tapping on the housing with a soft mallet, Figure 9-14. It may be necessary to pry on the housing to break the gasket seal.

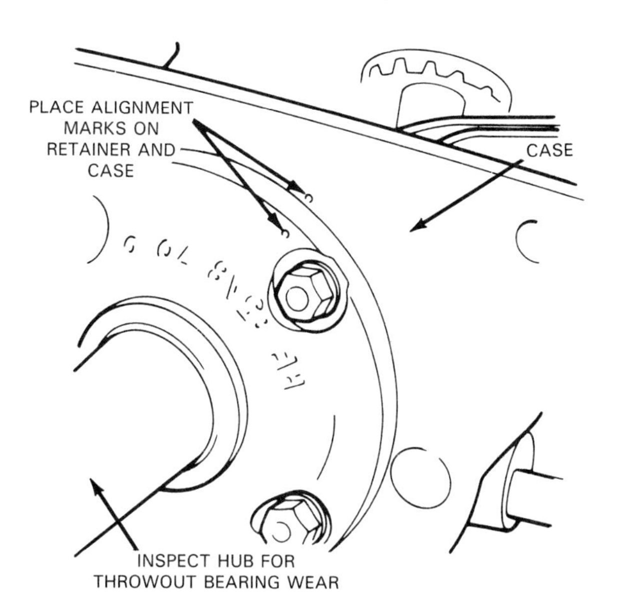

Figure 9-10. Unbolt the front bearing retainer from the transmission case, but always mark the case and bearing retainer first. Slip the retainer over the input shaft. The retainer hub should be inspected for wear caused by the throwout bearing. (General Motors)

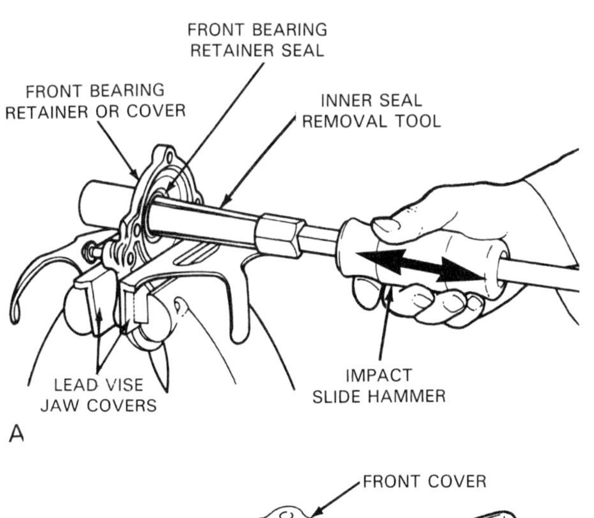

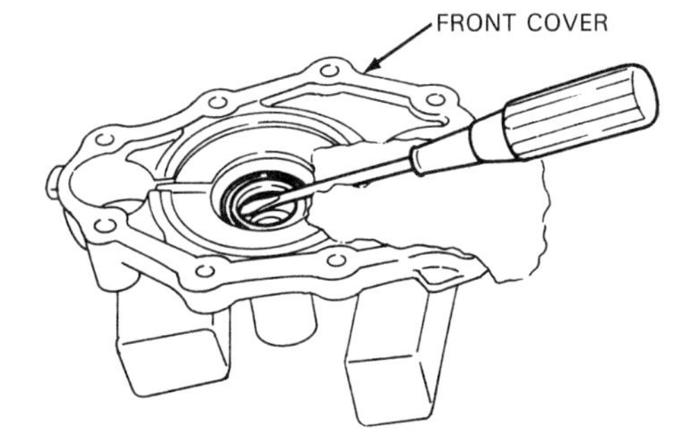

Figure 9-11. Removing the front bearing retainer seal. A–Some seals must be removed with a special puller. B–Some seals can be pried loose with a screwdriver. Ford, Nissan)

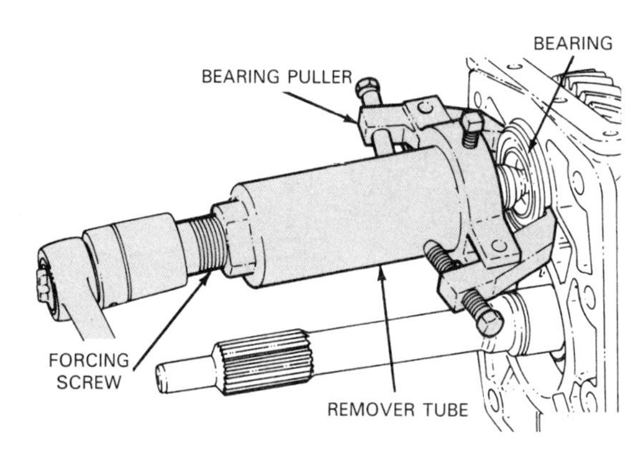

Figure 9-12. Some bearings must be removed with a special puller, as shown here. Other bearings can be removed by removing a large snap ring and then prying the bearing from the case. Note that the bearing puller shown is being used to remove a countershaft bearing. (Ford)

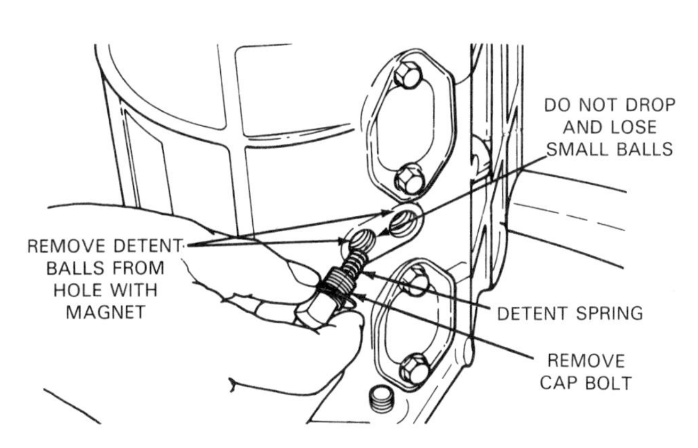

Figure 9-13. Some external parts, such as detent ball assemblies shown here, must be removed before the extension housing is removed from the transmission case. (Ford)

Removing Internal Components

Before removing the internal components, it is a good idea to measure *gear* end play to check for wear. To do this, use a feeler gauge and measure the clearance between each mainshaft gear, Figure 9-15. The standard end play is listed in the specifications. If the end play is out of spec, look for excessive wear on the sides of the gears or thrust washers, once the shaft is disassembled. Replace these components as necessary.

To begin removing internal parts, you must often remove the snap rings that hold gears to their shafts or that hold bearings in their case. Figure 9-16 shows a snap ring being removed with a pair of snap ring pliers.

If the internal parts of the shift linkage and the shift forks were not installed in the access cover, they should

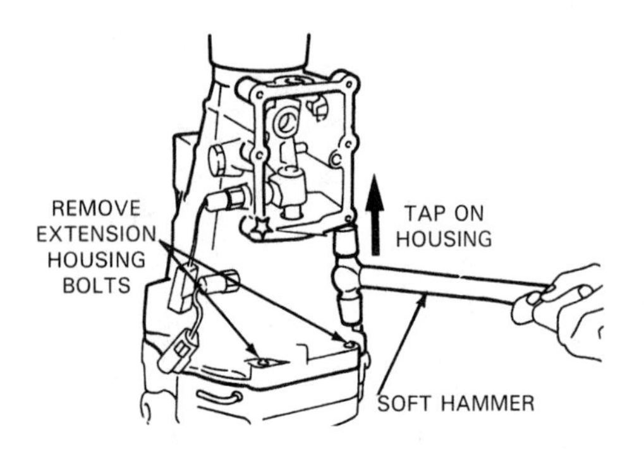

Figure 9-14. Before removing extension housing from transmission case, remove bolts and loosen extension housing by tapping with a soft mallet. (Nissan)

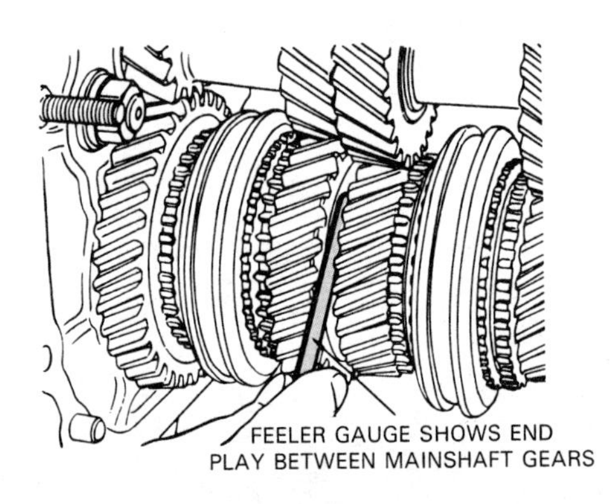

Figure 9-15. Feeler gauges can be used to check the clearance between gears, as shown here. (Subaru)

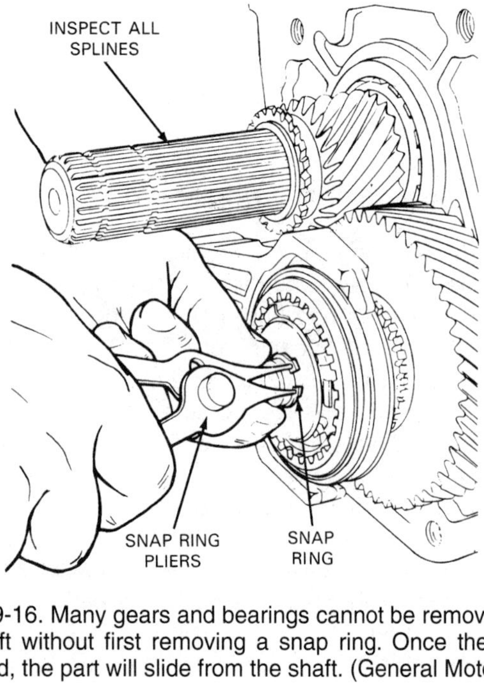

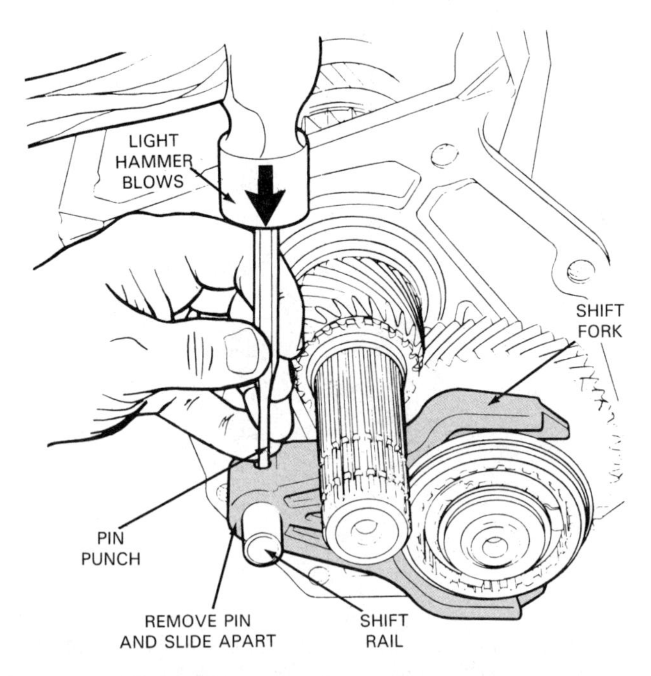

Figure 9-16. Many gears and bearings cannot be removed from the shaft without first removing a snap ring. Once the ring is removed, the part will slide from the shaft. (General Motors)

be removed from the case now. First, mark shift forks and rails with a punch, Figure 9-17. These alignment marks will be used during reassembly to install the parts in their original position. To remove the forks, drive the roll pins from the shift forks and rails. See Figure 9-18.

The next step is to remove the gears inside of the case. If the transmission case is a two-piece type, such as that shown in Figure 9-19, the gears and shafts can be lifted out of the case once the case halves are split. Note that a transmission with this type of case may be referred to as a ***split-case transmission.***

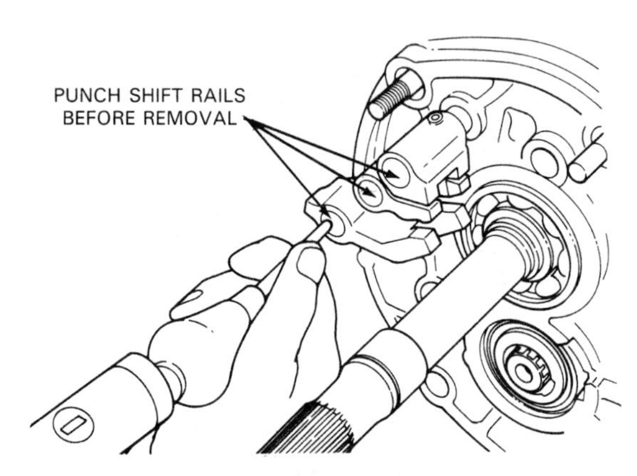

Figure 9-17. Shift rails (and forks) must be marked before removal. If this is not done, they might get installed in the wrong position. (Ford)

Figure 9-18. This shift fork is removed by driving out the roll pin, which holds it to the shift rail. Be careful not to lose the pin. (General Motors)

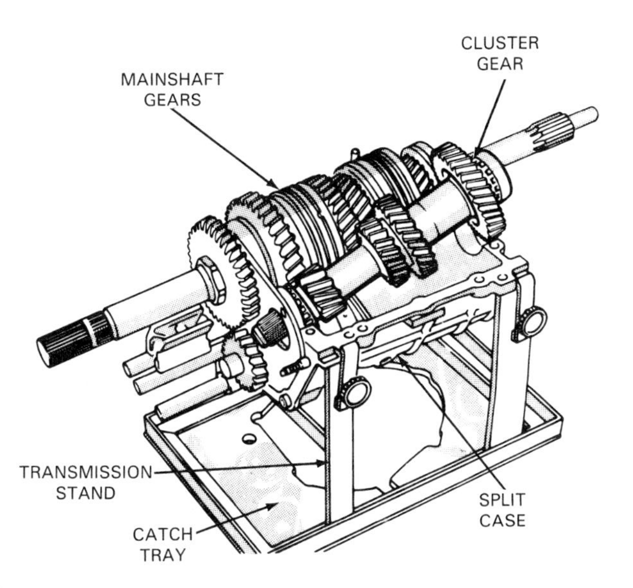

Figure 9-19. A split-case transmission is shown being held on a stand. The top half of the case has been removed. Removing parts from a split-case transmission is easy. The case is opened up, and the parts are simply lifted out. Most transmissions are not this easy to disassemble. (Chrysler)

The gears and shafts of most transmissions with one-piece cases must be removed through a series of steps. Again, the procedure will vary with the type of transmission, and you should consult a service manual for the exact removal sequence.

With some transmission cases, individual gears must be removed from the shaft before the shaft itself can be removed, as shown in Figure 9-20. Once the necessary gears and bearings are removed, the entire mainshaft gear assembly can be removed from the case. This is shown in Figure 9-21. Be aware that loose mainshaft (and countershaft) needle bearings and thrust washers will fall into the transmission case as the shafts are removed. Retrieve all small parts as soon as possible!

The countershaft gear assembly can be removed after the mainshaft is out of the way, by first removing the countershaft. Some countershafts are held in place by special keys, which must be removed before the shaft can be driven from the case. Usually, the shaft is driven from the case using a soft hammer and a *dummy shaft.* Once the countershaft is out, the cluster gear is simply removed from the case.

Note that a ***dummy shaft*** can be made from a round piece of wood. A broomstick or wooden dowel works well. The dummy shaft should be cut to the same length as the cluster gear. The shaft allows you to remove the cluster gear without dislodging the needle bearings from position in their bearing bores.

Also note that with some one-piece transmissions, the countershaft must be removed before the mainshaft gear assembly. In such designs, removing the countershaft first is necessary as it allows the countershaft gear to drop down far enough so that the mainshaft can be removed from the case.

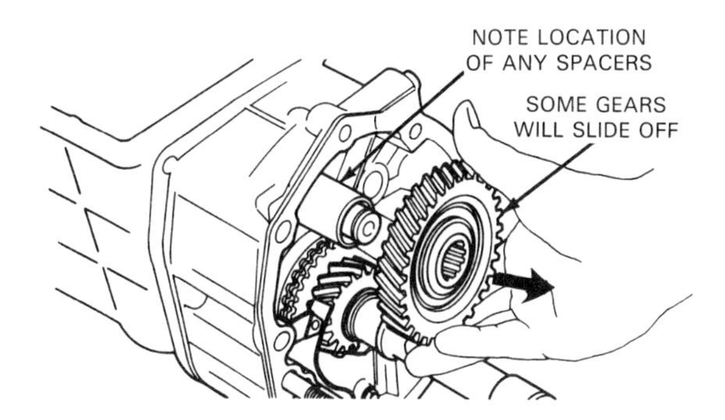

Figure 9-20. In some instances, individual gears must be removed from the shaft before the gear assembly can be removed. (Ford)

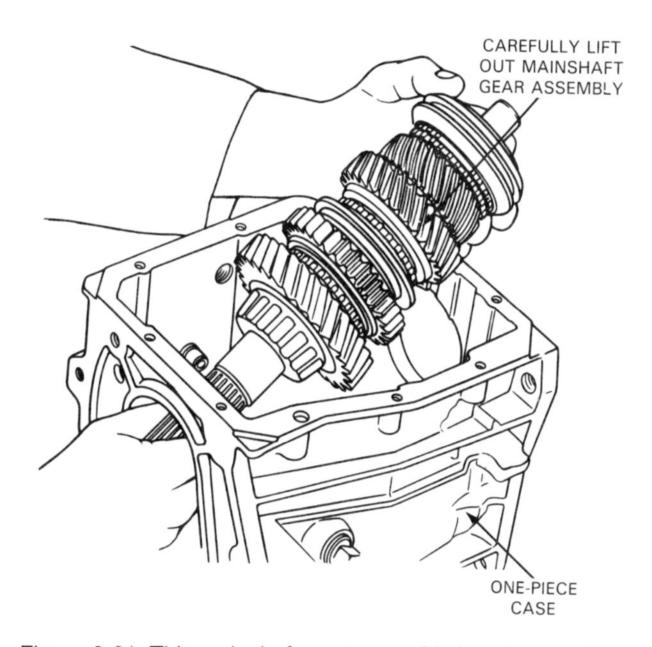

Figure 9-21. This mainshaft gear assembly is pulled out through the opening at the top of this one-piece transmission case. Some mainshafts are removed through the rear. (General Motors)

To remove the reverse idler gear, remove the *reverse idler shaft retainer key* and pull the shaft from the case. Some shafts do not have a key. Remove the needle bearings and thrust washers also.

If the input shaft and main drive gear have not yet been removed, your next step should be to remove any snap rings that hold the front bearing to the transmission case. Then, remove the input shaft and bearing from the case. The bearing may be tight in the case. A light tap on the shaft with a soft hammer should free it.

Disassembling the Components

The next step involved in disassembly is to take apart the components that were removed from the transmission case. After the parts are disassembled, you will need to inspect them. However, it is also important that you examine the parts as they come apart. See Figure 9-22.

To disassemble the mainshaft gear assembly, remove the snap ring from the front end of the output shaft. Then, slide the synchronizers and mainshaft gears off the front end. Remove other snap rings along the output shaft so that all gears and synchronizers may be removed. In a few instances, gears and synchronizers must be pressed from the shaft. Figure 9-23 illustrates a technique that can be used for removing a snap ring. Figure 9-24 shows a mainshaft gear and synchronizer being removed from the output shaft.

Shift linkage, if mounted in an access cover, will require disassembly, as will any bearings that were not removed in previous steps. Seals that are still in place should be removed from their components now. Most seals can be pried from their housings with a screwdriver.

Manual Transmission Parts Inspection

Before ordering parts or beginning reassembly, thoroughly check the transmission components for wear and damage. This includes all moving and stationary parts, since almost any transmission part can be defective in some way.

Begin by thoroughly cleaning the transmission case and all other parts with solvent. When cleaning the case, make sure that you do not lose any parts, such as needle bearings, that may have fallen to the bottom of the case.

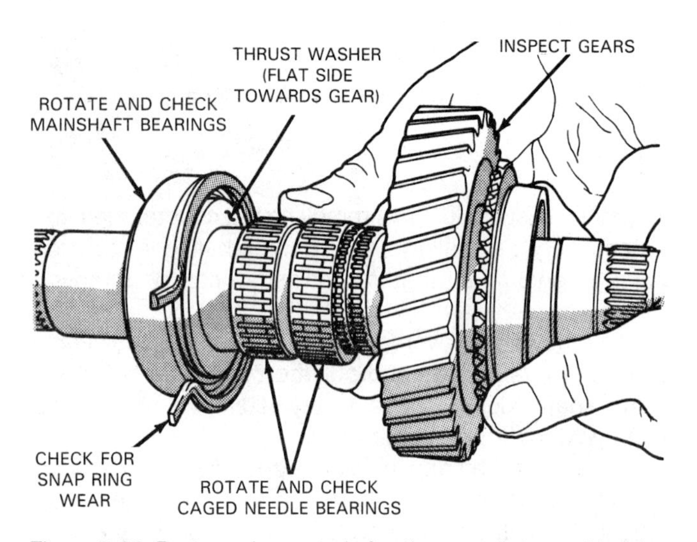

Figure 9-22. Parts are inspected after they are disassembled, but they should also be checked before the assemblies come apart. The parts on this mainshaft are being checked for rotation. (Chrysler)

Clean the bearings thoroughly, but do *not* spin the encased bearings with compressed air. Parts that are cleaned with water should be thoroughly dried and, then, lightly oiled. You should also check related parts that may or may not have been removed with the transmission. The shift linkage, clutch, and driveline universal joints may require servicing.

Visual Inspection

Much of parts inspection is looking for obvious damage by visual examination. Check the synchronizer hubs and

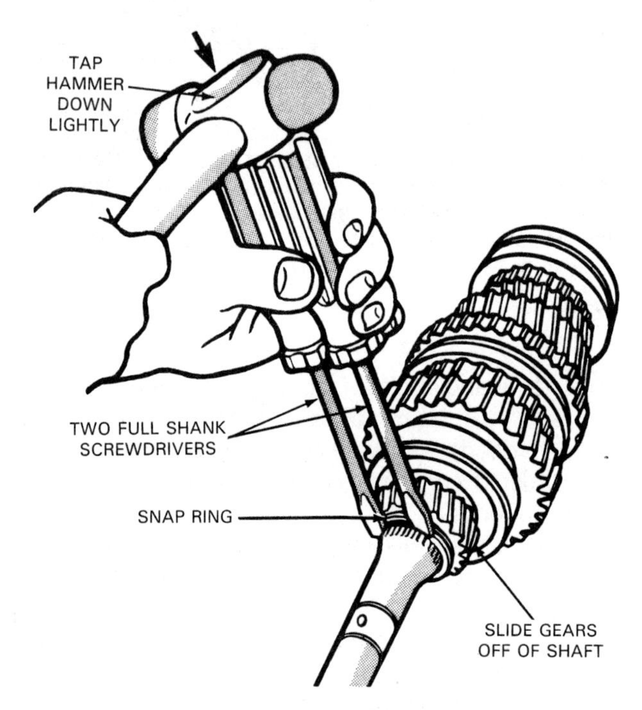

Figure 9-23. This shows fifth gear snap ring being removed from the output shaft. This method of removing a snap ring can be substituted for a special tool. After the snap ring is removed, the gears will often slide from the shaft. Be sure to wear eye protection when removing snap rings! (Chrysler)

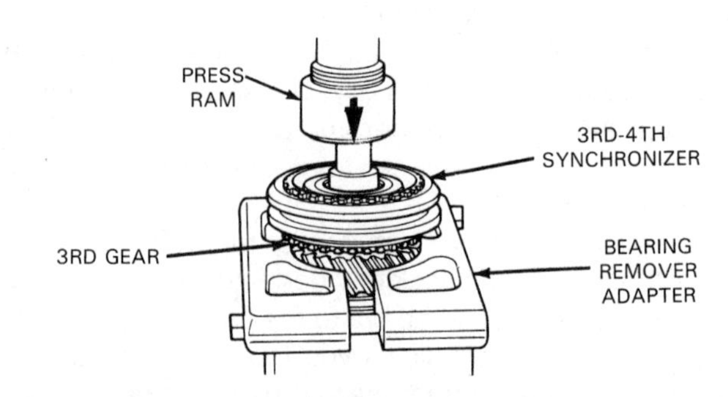

Figure 9-24. Some parts must be disassembled with a press. This shows third gear and the third-fourth synchronizer being pressed from the output shaft. Do not attempt to hammer the parts from the shaft. (Ford)

sleeves for worn or broken teeth, Figure 9-25. Check the grooves in the outer sleeves where the shift forks ride. Also, check the shift forks for wear or distortion. Check the blocking rings for wear on the outer gear teeth and on the tapered ring surfaces that contact the gear cone surface. See Figure 9-26. Worn or damaged parts should be replaced.

Look for signs of gear damage, such as worn, cracked, or chipped gear teeth, Figure 9-27. If any teeth are defective, the gear should be replaced.

Also, inspect the bushings pressed into the gear surfaces. Worn bushings can usually be replaced. However, for some gears, replacement bushings cannot be obtained, and the entire gear must be replaced. Check for wear between each gear and the machined area, or race, it fits over on the shaft. If a gear can be rocked on its shaft, the bushing, the gear, or possibly, the shaft, will need to be replaced.

Thrust washers should be checked for wear and scoring. Also, check snap rings and snap ring grooves for wear or damage. Badly bent snap rings should be replaced. Figure 9-28 shows typical parts of a gear and shaft assembly that should be inspected.

Closely examine all bearings, looking for signs of wear. Ball bearings and roller bearings can be checked by rotating them by hand. If you feel any roughness when turning a clean bearing, the bearing should be replaced. Caged roller bearings can be checked by turning them on their mating bushings. Loose, or uncaged, needle bearings should be checked for flat spots. Bearings with flat spots must be replaced.

The transmission case and all other stationary parts should be checked for cracks, which can cause lubricant leaks. Also, look for any kind of distortion, which can cause misalignment of the moving parts or poor sealing surfaces.

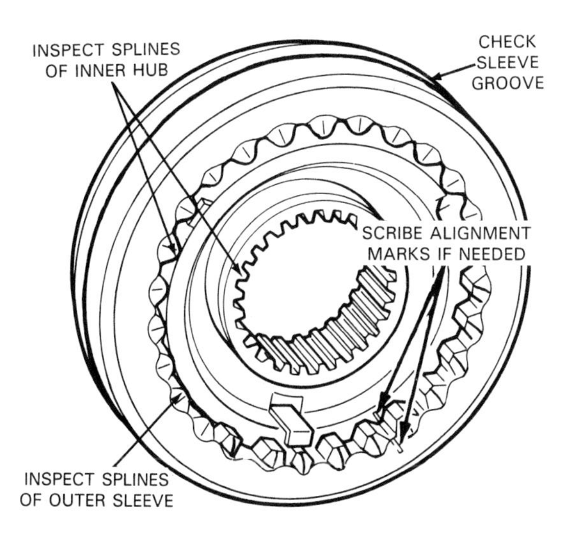

Figure 9-25. The synchronizer hub and sleeve should be carefully checked for wear and damage. Defective synchronizers will cause gear clash, hard shifting, or gear slipout. (General Motors)

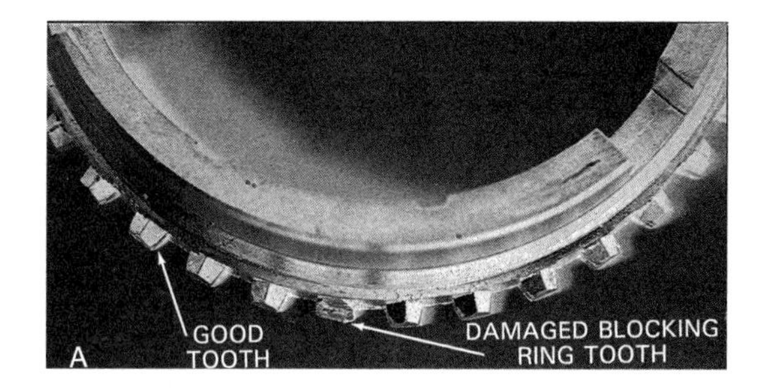

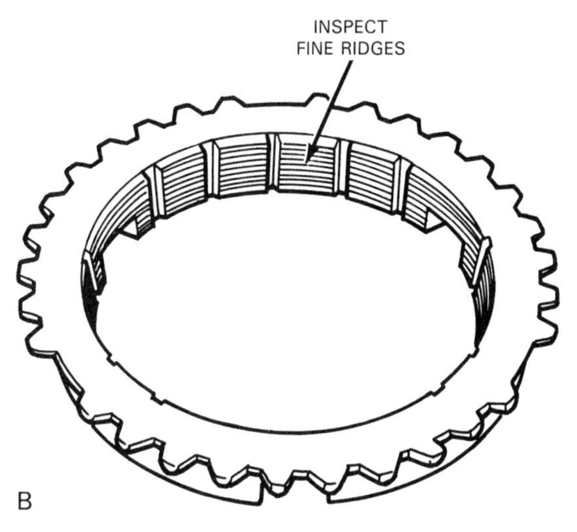

Figure 9-26. Conduct visual checks of synchronizer blocking rings. A–Look for worn or broken teeth. B–Look at fine ridges in tapered surface. Edges should be sharp.

Figure 9-27. All parts should be cleaned and inspected before reassembly. This gear shows signs of wear on both sets of teeth.

Inspect the front bearing retainer hub for wear caused by the throwout bearing. While you are at it, check the clutch fork, ball stud (if applicable), and throwout bearing for wear.

Measurement Inspection

Some transmission parts should be checked with special measuring instruments. Use of feeler gauges, micrometers, calipers, and dial indicators can help to identify defective parts that would otherwise be unnoticed. They can help to ensure a good repair job. Specific measuring instruments mentioned here were covered in detail in Chapter 2.

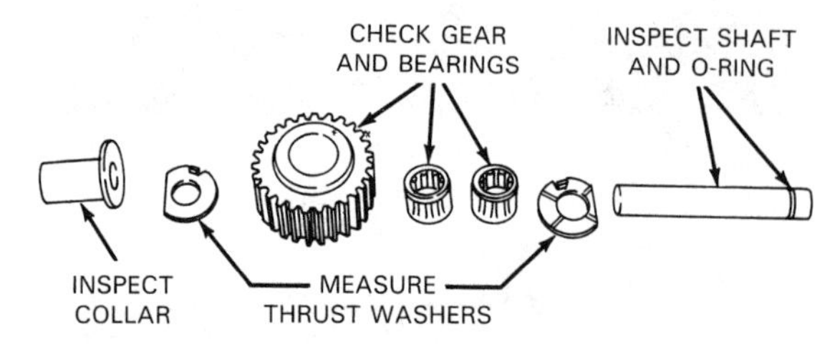

Figure 9-28. These are typical parts that must be inspected during disassembly. Although the idler gear, shown here, is only used in reverse, it can wear out or become damaged. You should inspect all parts of the idler gear assembly and other gear assemblies.

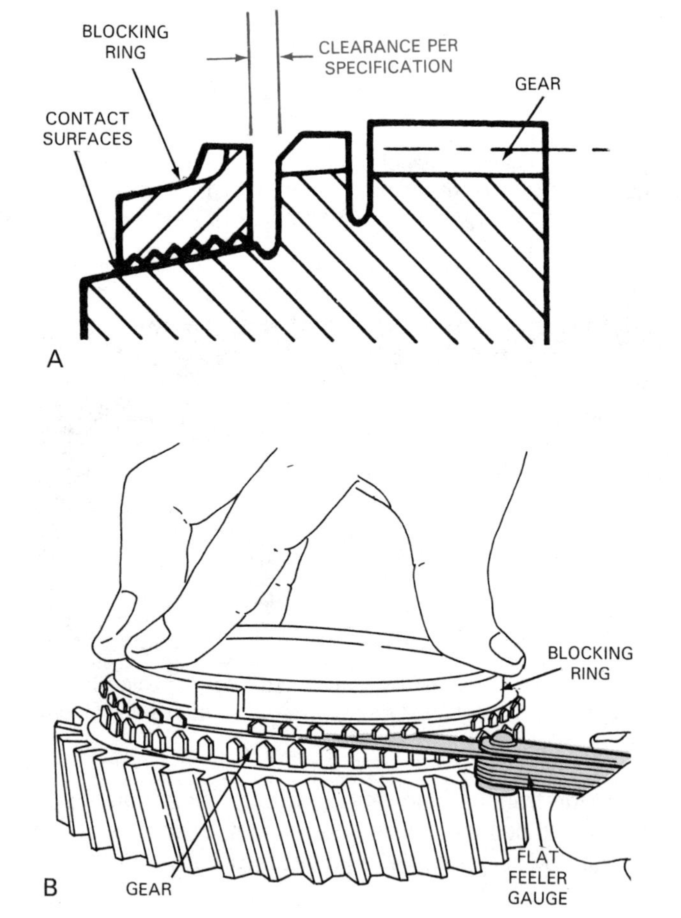

Figure 9-29. Inspecting blocking ring for wear. A–Sectional view shows clearance between blocking ring and mating gear. If the ring moves too far into the mating cone surface, the tapered surface of the blocking ring is worn. B–A feeler gauge can be used to measure clearance between the mating parts. (Ford)

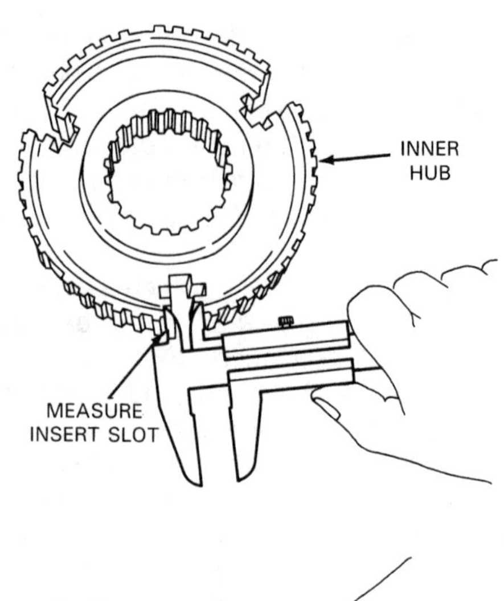

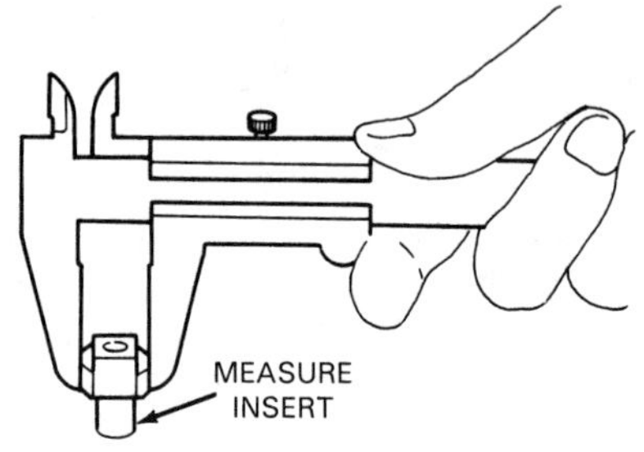

Figure 9-30. Some manufacturers specify a clearance between the inner hub slots and the inserts. A sliding caliper, which can measure inside and outside dimensions, is used in determining this distance.

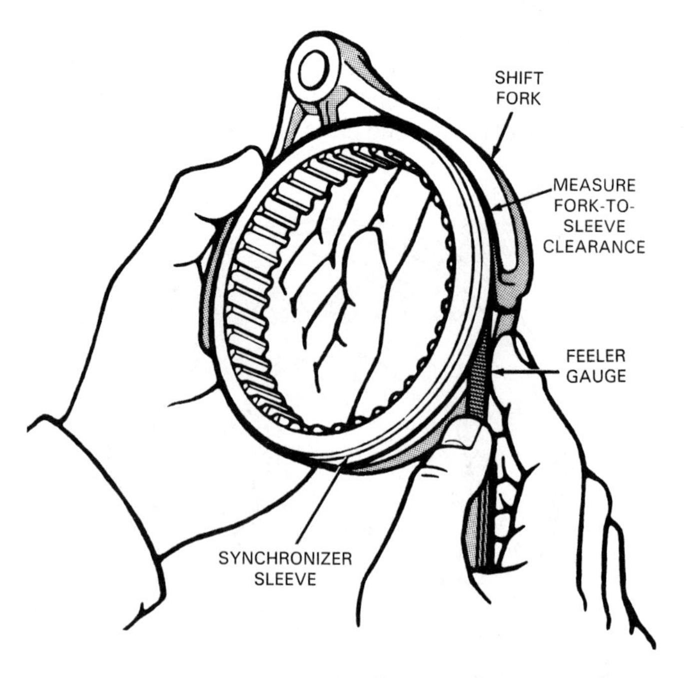

Figure 9-31. Clearance between outer sleeve and shift fork can be checked with a feeler gauge. Note if shift fork has nylon inserts (where fork contacts sleeve), inserts should be replaced when transmission is serviced. (Chrysler)

Blocking rings must maintain a certain clearance between the ring itself and the straight-cut teeth of the mating gear, Figure 9-29A. The ridged surface of the blocking ring is worn if the clearance is too small and the ring has moved too far into the cone surface of the mating gear. To measure the clearance, a flat feeler gauge may be used, as shown in Figure 9-29B.

Some transmissions have a clearance specification for the inserts and insert slots of the synchronizer inner hub, Figure 9-30. The sliding caliper is used to take the inside measurement of the hub slot and the outside measurement of the insert. The difference in the two measurements is the clearance. If clearance is excessive, all parts should be replaced.

Some manual transmission manufacturers provide a clearance measurement between the shift fork and the shift fork groove in the outer sleeve of the synchronizer. This clearance can be measured with a feeler gauge, as shown in Figure 9-31.

Another item to check is the amount of backlash between two gears. This can be checked with a feeler gauge between the teeth or with a dial indicator. Excessive backlash usually means that both gears (or their bushings) are worn and should be replaced.

Other items to check are illustrated in Figure 9-32. A micrometer can be used to check outside diameters of shafts and bushings. This is useful for determining wear on these parts. A sliding caliper can be used to check the thickness of a flange, often done to determine the need for thicker thrust washers. Finally, a dial indicator is useful for checking shaft runout.

Some stationary parts also require measurements. Figure 9-33 shows a depth micrometer being used to check the bearing bore depth of an aluminum housing. This measurement has a direct effect on the amount of end play of the transmission mainshaft. The measurement (and end play) can be controlled by special shims placed at the bottom of the bearing bore.

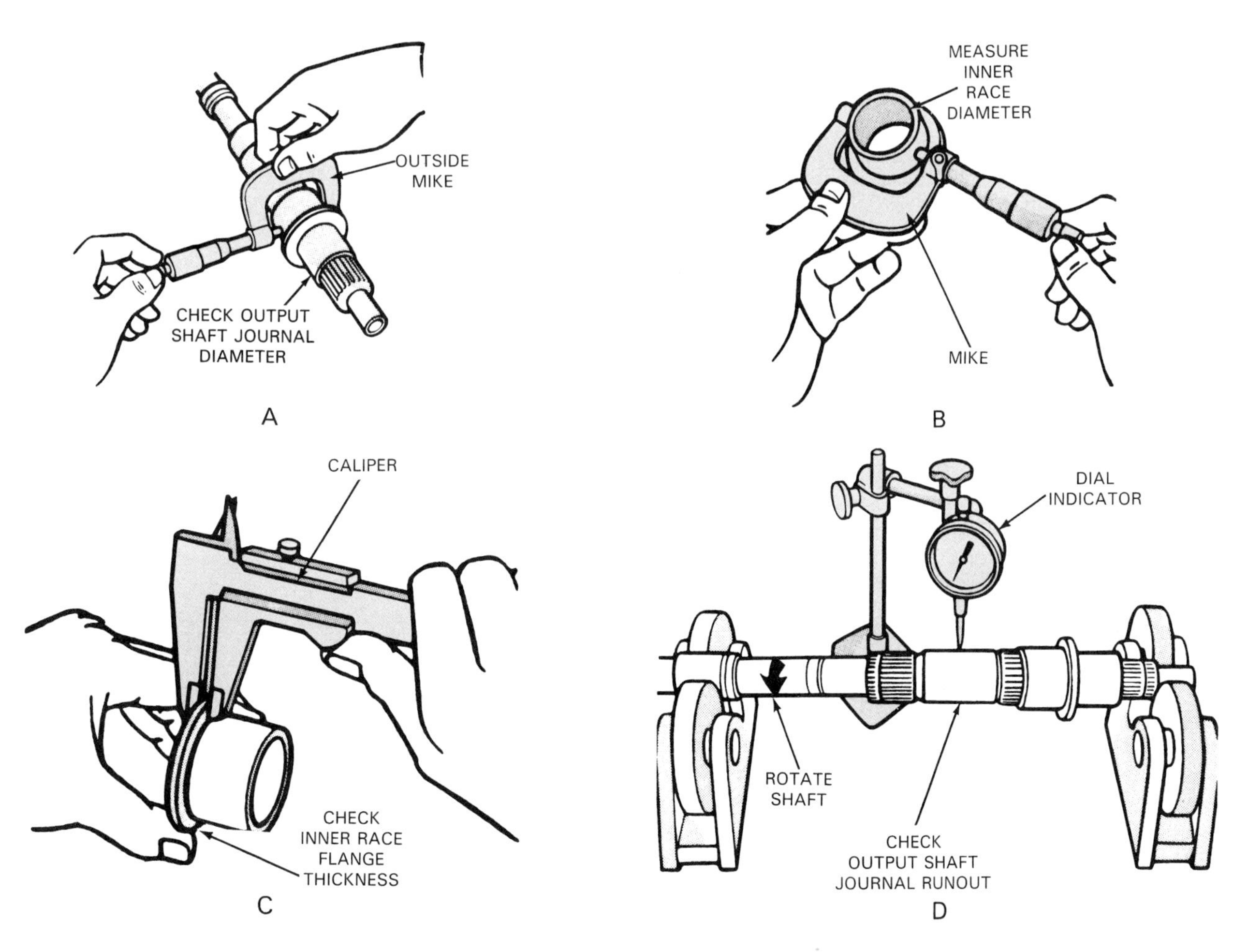

Figure 9-32. Various instruments are used to check different transmission parts. A–Micrometers can be used to check outside diameters of shafts. If the shaft has worn until it is undersized, it must be replaced. B–The micrometer can also be used to check a bushing. A worn bushing should also be replaced. C–A sliding caliper can be used to check the thickness of a flange. The thickness of this flange will affect end play and the fit of the gears and bearings on a shaft. Thrust washers can also be checked with calipers. D–A dial indicator can be used to check shaft runout. If the shaft is bent or has flat spots, it should be replaced. (Chrysler)

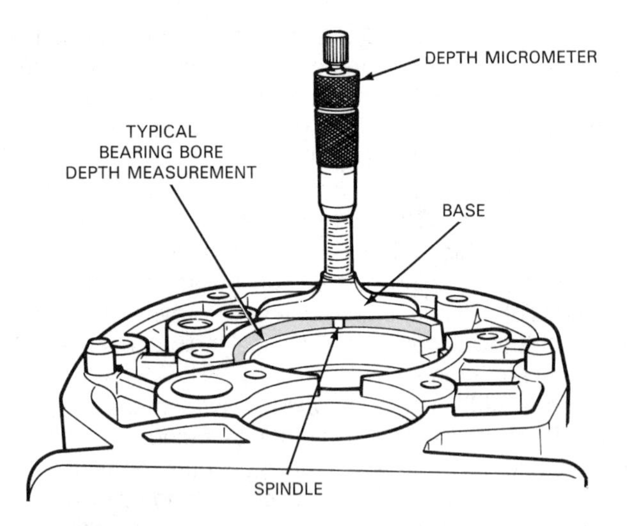

Figure 9-33. The mainshaft bearing bore depth is measured in this transmission case to ensure that the bearing will not be too tight or too loose when the mainshaft is installed. Bore depth is adjusted with shims. (Ford)

Manual Transmission Assembly

This section will give general details of manual transmission assembly. Before beginning the assembly of the transmission, make sure that you have the correct replacement parts by matching them against the old parts. Also, lightly lubricate each transmission part with the proper lubricant.

Assembling the Components

The first step in transmission assembly is to assemble all of the transmission subassemblies. If the transmission uses case-mounted shift forks or rails, in many (but not all) instances, they should be reinstalled now. See Figure 9-34. This step will not apply where the shift linkage is mounted in a top or side access cover.

Assemble the synchronizers, Figure 9-35, and ensure that they operate properly. Then, assemble the gears, synchronizers, and bearings on the mainshaft, Figure 9-36. Sometimes, gears cannot be installed until after the mainshaft is installed in the transmission case, Figure 9-37.

During the assembly process, ensure that the snap rings fit snugly in their grooves. If not, they will allow movement of parts on the shaft or may even pop out of their groove. Figure 9-38 shows a simple check of a snap ring in its groove. For a more exact check, a feeler gauge should be used to measure the clearance between the ring and groove. The measurement is checked against the specifications to see that the maximum groove clearance is not exceeded. If it is, the snap ring is not wide enough. Replacement snap rings are available in varying thicknesses.

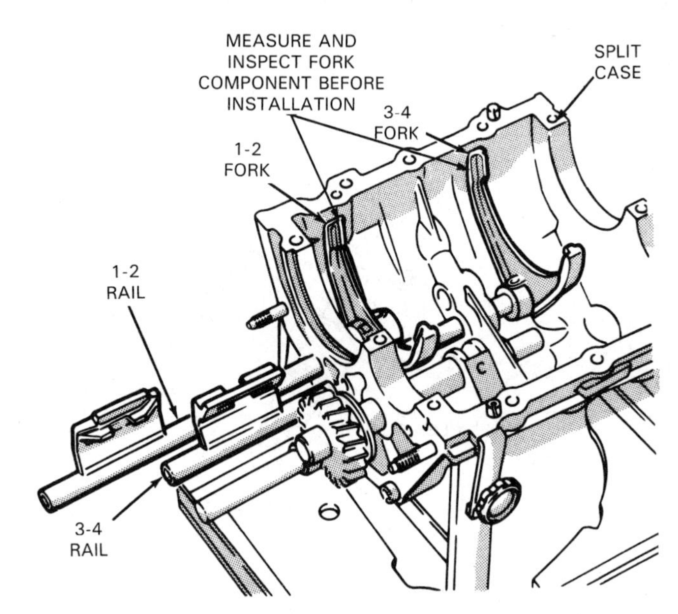

Figure 9-34. To begin reassembly on this transmission, the shift forks and rails are installed. This is not the first step on every transmission. (Chrysler)

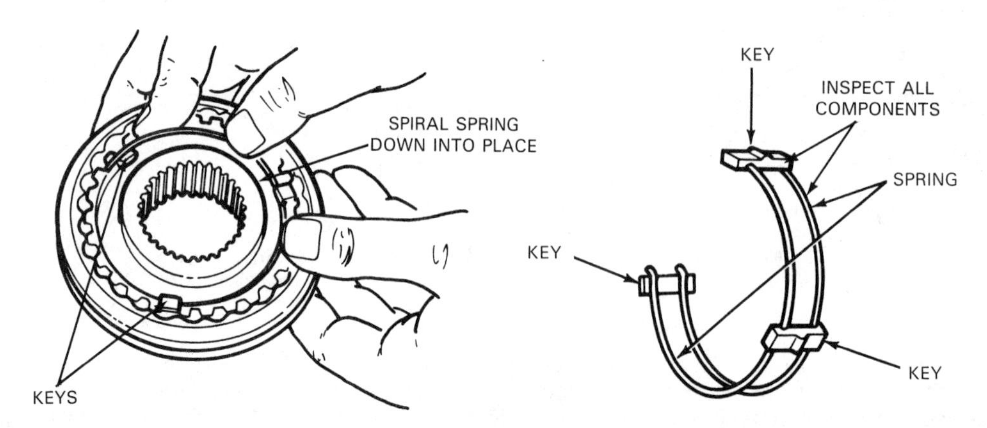

Figure 9-35. The synchronizers should be carefully assembled. The position of the inserts (keys) is critical. (Ford)

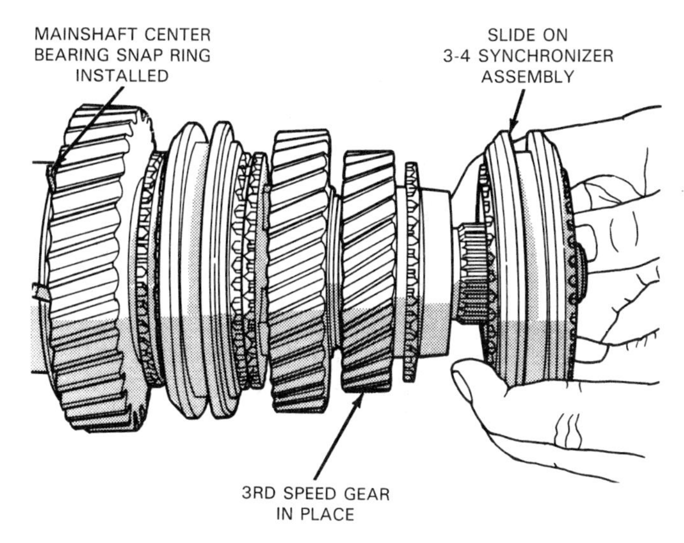

Figure 9-36. This shows the mainshaft gear assembly being put back together. Cleanliness and careful fitting are essential. Never force parts onto the shaft. Note that the number of gears will vary with each type of transmission. (Chrysler)

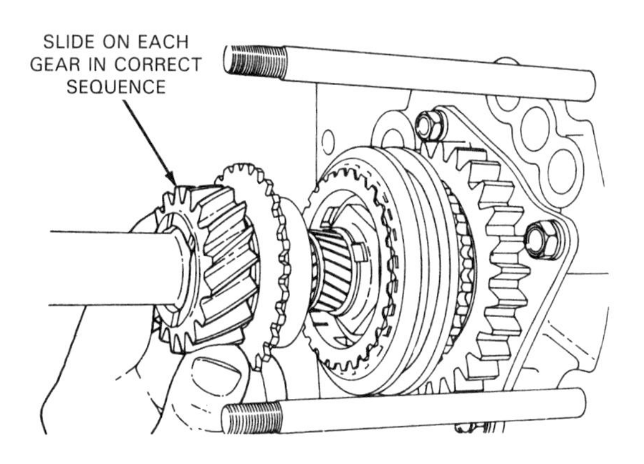

Figure 9-37. The mainshaft assembly shown here is not assembled until the mainshaft is installed in the transmission case. There are many variations to the assembly process, and the factory service manual should always be consulted. (Ford)

After all of the components are installed in the proper order, install the final snap ring on the end of the mainshaft. Then, check that all parts turn freely.

If the mainshaft pilot needle bearings are uncaged, they can be reinstalled in the bore behind the main drive gear, as shown in Figure 9-39. Use heavy grease to hold them in place. Once this is done, set the input shaft aside.

If needle bearings are used in the countershaft gear assembly, position them in the bearing bores at each end of the cluster gear. Use heavy grease and a dummy shaft to hold the needle bearings in place. The dummy shaft will allow you to install the cluster gear in the transmission case without dislodging the needle bearings.

Installing Internal Components

Insert the cluster gear thrust washers onto each end of the transmission housing. Grease can be used to hold the washers in place.

Insert the countershaft gear through the rear opening of the transmission case and set it into position between the end thrust washers. Then, insert the countershaft through the case and push it through the countershaft gear. As the countershaft is pushed in, the dummy shaft will be pushed out.

If the countershaft uses a retainer pin or Woodruff key, set it in place before the shaft is completely installed and push the shaft in until it bottoms out. If the shaft is held with snap rings, install them now. Be careful to install snap rings so they do not pop out.

Note that this procedure involving the countershaft gear assembly does not apply to all designs. Some countershaft gears do not rotate on a countershaft and needle bearings. Instead, the gear rotates on tapered roller bearings, and there is no countershaft. The split-case transmission employs this type of setup. See Figure 9-40.

Once the countershaft gear is secured in position, countershaft gear end play should be checked. The

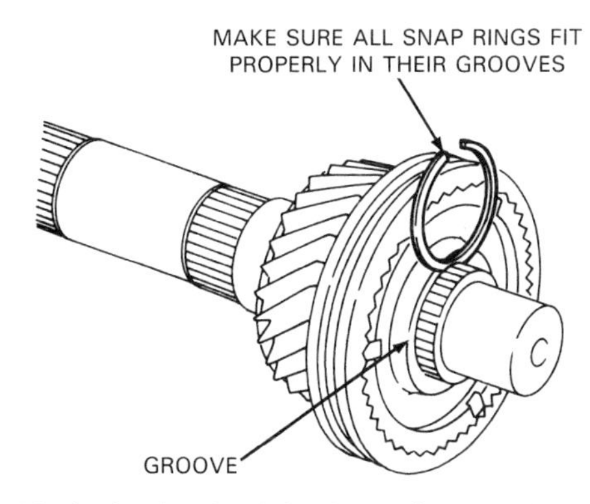

Figure 9-38. A simple check is shown for ensuring that a snap ring fits snugly in its groove. Some snap ring clearances are checked by inserting a feeler gauge of the proper thickness between the ring and groove. (Nissan)

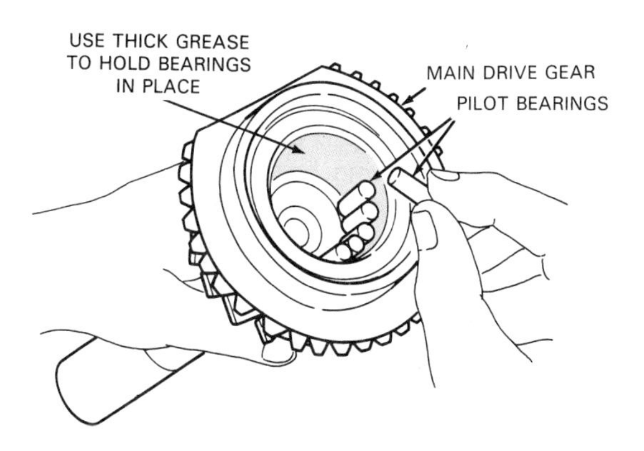

Figure 9-39. The needle bearings used between the main drive gear and the mainshaft are often loose, or uncaged, bearings. Use heavy grease to hold bearings in position while shafts are reinstalled.

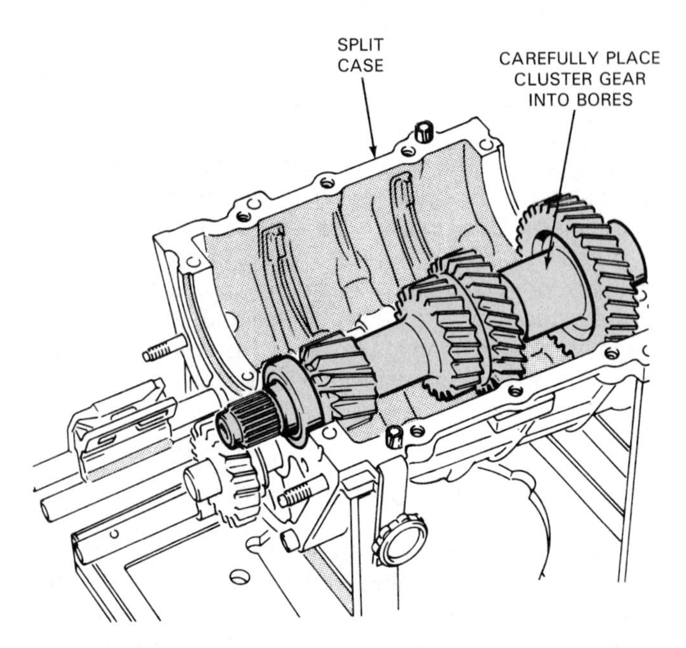

Figure 9-40. Countershaft gears in some designs, such as in this split-case transmission, use tapered roller bearings instead of thrust washers and needle bearings. This design does not employ a countershaft either. The countershaft gear is easily installed in this split-case transmission. Installation is more difficult in transmissions with one-piece cases. (Chrysler)

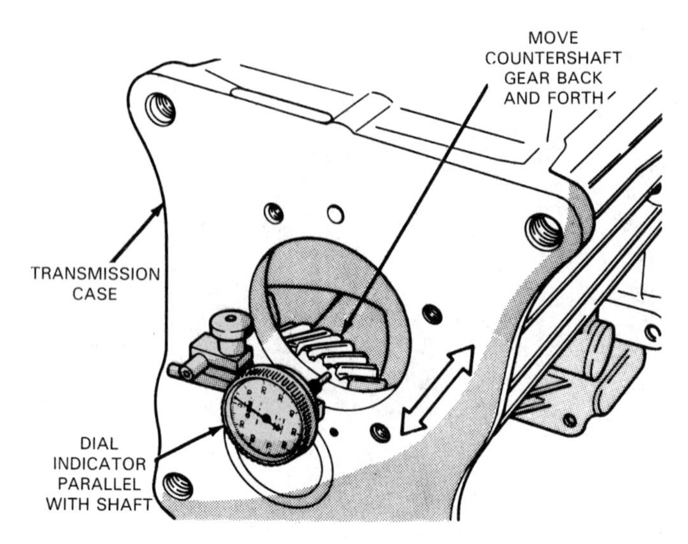

Figure 9-41. Dial indicators are often used to check end play at various points in the reassembly process. End play checks are usually made with the transmission partly assembled. (Chrysler)

measurement is used to select the proper thrust washer or shim to bring end play into specifications. End play is checked with a dial indicator, as shown in Figure 9-41. Install the indicator with the plunger touching the front end of the countershaft gear. Push the gear to the rear and zero the dial indicator. (Make sure that the plunger is still touching the countershaft gear.) Push the gear forward and note the indicator reading. If the end play is greater than specified, you must install thicker thrust washers or shims.

Once the countershaft gear is in place, the reverse idler gear can be installed if it is the type of design that fits into the case. Be sure to install the thrust washers.

With the countershaft gear and reverse idler gear assemblies in place, the mainshaft gear assembly and input shaft can be installed. Place the input shaft in the case. Install the front bearing, being careful not to dislodge the uncaged needle bearings. See Figure 9-42.

Install the mainshaft gear assembly through the rear of the transmission case. Carefully slide the nose of the mainshaft into the pilot bore behind the main drive gear, again, being careful not to dislodge the uncaged needle bearings. Once the mainshaft is in place, install the rear bearing, Figure 9-43. If the rear bearing is held in place by a snap ring, the snap ring should be installed now.

Once internal components are assembled, check their rotation and, once again, measure the gear end play, or clearance. (Refer back to Figure 9-15.) It must be checked to ensure that the gears do not bind when the transmission warms up during use. Proper clearance between the gears installed on a shaft is important to long transmission life. Clearance can be adjusted by changing the thickness of thrust washers or snap rings.

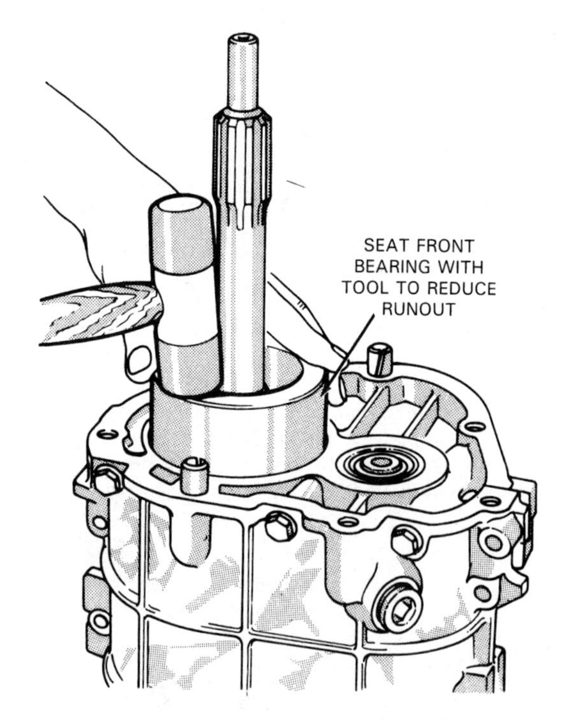

Figure 9-42. The front bearing for the input shaft may have to be pressed into place. Some front bearings may be loosely pressed into place and held by a snap ring. (Chrysler)

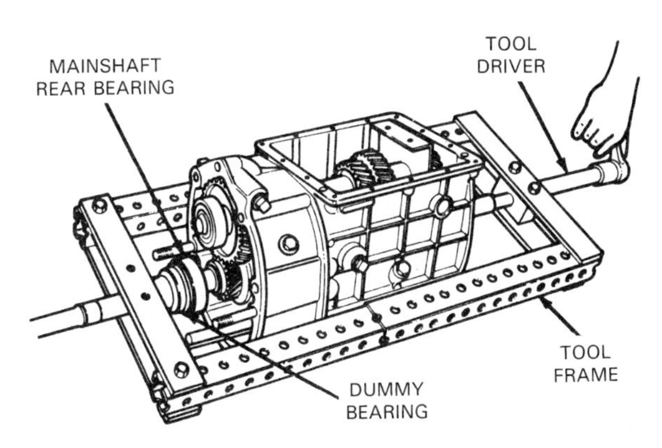

Figure 9-43. Rear bearings may also require pressing into place. This special tool makes the job easier. *Dummy bearing,* shown here, is used to expedite case assembly by aligning shaft as it is put back together. This reduces the chance of damaging the real bearing. Dummy bearing is removed after operation is finished. Some bearings are retained with snap rings. (Ford)

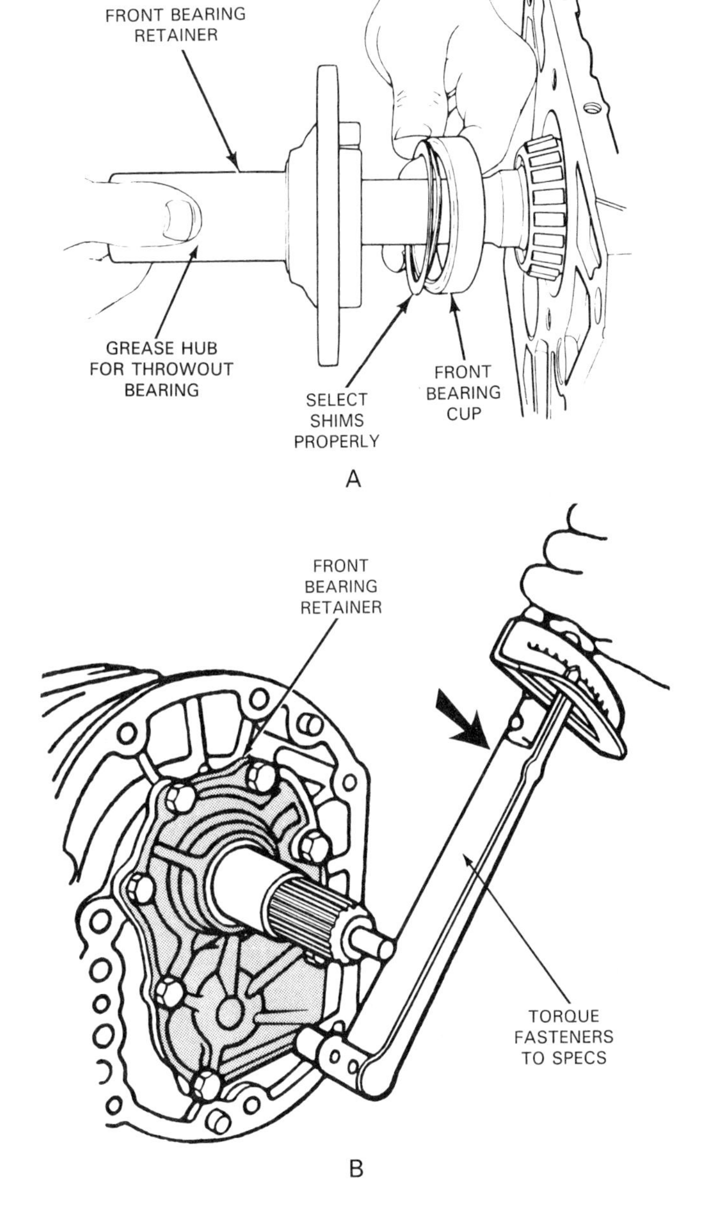

Figure 9-44. The front bearing retainer should be carefully installed. A–When front bearing is a tapered roller bearing, front bearing cup is installed with bearing retainer. Shims are added to obtain necessary preload. They eliminate end play. Retainer hub should be lightly greased before transmission installation. B–The retainer bolts should be tightened with a torque wrench. (General Motors, Chrysler)

Installing External Components

Install the extension housing gaskets. Always use new gaskets (and seals) when reassembling the transmission. If the reverse gear is in the extension housing, it should be installed now. Install the extension housing over the output shaft. Engage the reverse shift fork with the reverse gear, if necessary. Install and tighten the extension housing bolts.

Install the front bearing retainer, Figure 9-44, matching alignment marks made earlier and torquing the fasteners to specifications. Then, check that the transmission input and output shafts turn with no binding. If the shafts do not turn freely, you must find out the reason and correct it.

CAUTION! *Never* install a transmission that is binding internally or that is otherwise incorrectly assembled thinking that the parts will work themselves in and will operate smoothly. The transmission will be severely damaged within the first few minutes of operation!

The shaft end play can be checked once the extension housing and front bearing retainer are tightened in place. This can be done with a dial indicator. Install the dial indicator at the front of the transmission, with the plunger in line with and touching the front end of the input shaft. Push the input shaft to the rear and zero the dial indicator. (Make sure that the pointer is still touching the input shaft.) Then, push the output shaft forward and note the dial indicator reading. If the reading is not within specifications, replace the original thrust washers with thicker or thinner washers, as needed. *Selective thickness thrust washers* are available from dealer parts departments.

Replace the access covers, being sure to align the shift forks with the synchronizer sleeves where necessary. Use new gaskets at all locations. Be certain the transmission drain plug is in place. Reinstall external case components removed as part of transmission disassembly. Included among these components are the speedometer assembly and shift rail detent ball assemblies. Reinstall internal gearshift lever assembly, if applicable. Once this is done, you can check internal shift linkage operation.

If the transmission case and bell housing are an integral casting, lightly grease the throwout bearing and install it on the front bearing retainer hub. Connect the throwout bearing to the clutch fork and, if removed and if applicable, the clutch fork to the ball stud in the bell housing.

Manual Transmission Installation

Transmission installation is the reverse of removal. As with removal, the vehicle should be on a hoist with the engine or bell housing supported. If the procedures are performed properly, the transmission installation should go smoothly. See Figure 9-45. Following are some *general* procedures detailing how to install a manual transmission in a vehicle. Note that you should always refer to the manufacturer's service manual for specific procedures.

1. Check that the throwout bearing is in place on the clutch fork (or input shaft for integral-type casting). Place small amount of grease on the pilot bearing. Throwout bearing surface should also be greased.
2. Raise the transmission with a transmission jack or floor jack, positioning it behind the engine. Carefully align the transmission with the engine. When the transmission is at the proper height and input/output shaft centerlines align with the crankshaft centerline, slide the input shaft into engagement with the throwout bearing, clutch disc hub, and pilot bearing. If the clutch has been disturbed, you may need to align the clutch disc. (Refer to Chapter 7.)

   As the input shaft passes through the clutch disc hub into engagement with the pilot bearing, the transmission case is brought to a tight fit with the bell housing. On models with an integral-type case and housing, the bell housing should snug up tightly to the engine. Alignment dowels should smoothly engage their matching holes.

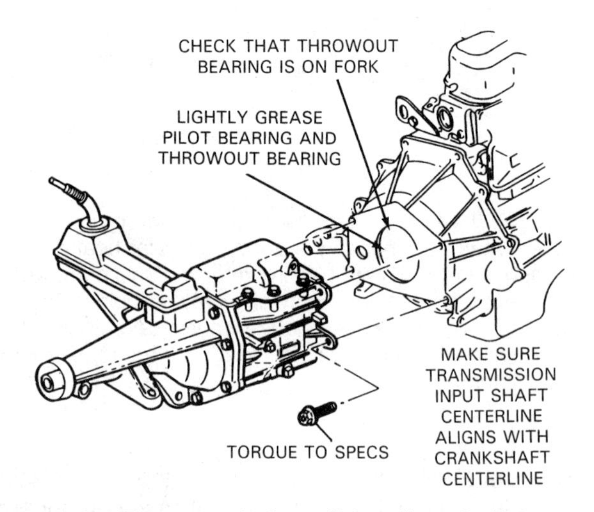

Figure 9-45. The transmission will install easily if the proper procedures are followed. The case of this transmission is separate from the bell housing. General procedures are similar for this type and the one-piece integral casting. (General Motors)

CAUTION! Do *not* try to pull a binding transmission into position by tightening the bell housing or transmission case attaching bolts. This will cause damage to the parts that are binding and can cause the bell housing or transmission case ears to break. If parts are installed properly, the transmission should slide fully into place *by hand.*

3. Install the bell housing or transmission case attaching bolts, as applicable. Ensure that the engine, bell housing, and transmission are properly aligned after all of the bolts are tight. There should be no gaps between any of the mating surfaces. As you tighten, check that no wires, brackets, or hoses are caught between mating surfaces.
4. Raise the transmission slightly with the jack until there is sufficient clearance to install the transmission crossmember. Place the crossmember under the transmission mount and install the attaching bolts. Then, install the attaching bolts that hold the crossmember to the vehicle frame.
5. If external floor-shift gearshift lever assembly was removed, insert the gearshift lever through the hole in the floor pan. Bolt the shift mechanism to the extension housing. Install external floor-shift or column-shift linkage by reconnecting shift rods to shift levers at the transmission case.
6. Reconnect electrical connectors. Install parking brake hardware and exhaust pipes (if removed). Install the speedometer cable assembly. Reconnect the clutch linkage, including return spring, at the clutch fork.
7. Install the driveline, making sure to line up the marks that you made during removal.
8. Fill transmission to the proper level using specified lubricant. Install the fill plug and tighten.
9. If necessary, make preliminary linkage adjustments as specified earlier in this chapter. The clutch linkage may also need adjustment.
10. After all of the drive train parts are reinstalled, lower the vehicle and reconnect any disconnected hardware under the hood, including the battery negative cable. On floor-shift vehicles, install the boot to the floor inside the passenger compartment. Screw the gearshift lever knob back on, if it was removed earlier.
11. After everything has been reconnected, push the clutch pedal a few times to see that the clutch engages and disengages properly. Also, move the gearshift lever to check operation of the shift linkage. Then, road test the vehicle. Make any other adjustments as needed and recheck the transmission oil level.

Summary

To be successfully repaired, manual transmissions require careful diagnosis and following of service procedures. Troubleshooting a transmission is as important as repairing it. The technician should always check out the easiest possible solutions before proceeding to the more complicated possibilities.

Like any mechanical part, the transmission will likely wear out in time. However, if properly operated and cared for, it should last for many thousands of miles. Many transmission problems are caused by careless operation or lack of service. Hard or fast shifting, lack of lubricant, or failure to make periodic linkage adjustments, can cause early transmission failure.

Lubricant leaks are a common transmission problem. Leaks can usually be spotted by visual examination. The technician should make sure that the leak is actually from the transmission before beginning repairs.

Hard shifting can be caused by external or internal transmission problems. Linkage or synchronizer problems can result in a gearshift lever that is hard to move. Slipping or jumping out of gear can also be caused by linkage problems. This problem can also be caused by worn internal parts.

Transmission noises are usually heard as whining, rumbling, or roaring. Transmission noises are usually caused by worn gears or bearings. Often, the noise is coming from another drive train component, such as the rear axle or drive shaft assemblies. Noise in neutral may also be coming from the clutch assembly. Clunking noises are often caused by the driveline universal joints.

Before removing a transmission, make sure that removal is needed. Many transmission problems can be corrected without removing the transmission. Linkage or other adjustments can almost always be made with the transmission installed.

The transmission can usually be removed without removing the engine. To begin removal, disconnect the battery and any other components under the hood, as required. The oil can be drained, and the linkage, driveline, and crossmember removed. The transmission should be carefully lowered from the vehicle and taken to a workbench.

The transmission components are removed from the case in a step-by-step process. Usually the access covers are removed first, followed by the extension housing and the front bearing retainer. The mainshaft gear assembly can then be removed, followed by the input shaft, the countershaft gear assembly, and the reverse idler gear assembly. Carefully observe the position of all parts as they are disassembled. Mark similar parts as necessary. Once the transmission is disassembled, all parts should be thoroughly cleaned.

The transmission internal parts should be carefully checked for wear and damage. Checks can be made by observing parts for obvious damage and by checking for proper clearances with measuring instruments.

The transmission is reassembled in reverse order of disassembly. Check all parts for binding and other operating problems as they are reinstalled in the case. Never attempt to reinstall a transmission with binding shafts. It will be badly damaged.

The transmission should be installed in the reverse order that it was removed. The input shaft should engage smoothly with the clutch disc splines. Do not attempt to force a sticking transmission into position by tightening the transmission case or bell housing attaching bolts. The transmission lubricant should be placed in the transmission after the driveline is installed. The shift and clutch linkage should be installed and adjusted. The components under the hood can be reconnected, and the vehicle road tested.

Know These Terms

No-drive condition; Gear jumpout; Gear slipout; Gear clash; Split-case transmission; Dummy shaft.

Review Questions–Chapter 9

Please do not write in this text. Place your answers on a separate sheet of paper.

1. The following items–*seals, gaskets, plugs, too much oil in case, and damaged parts*–are often the source of what transmission problem?
   a. Gear jumpout.
   b. Noisy transmission.
   c. Gear clash.
   d. Lubricant leakage.
2. A manual transmission problem that occurs when gears and synchronizers do not quite fully engage is known as _____ _____.
3. What should you check when a manual transmission is only noisy in neutral?
4. Which of the following only includes items that can be serviced without removing the transmission?
   a. External shift linkage, access covers, speedometer assembly.
   b. Mainshaft gears, internal shift linkage, speedometer cable.
   c. Front bearing retainer, damaged extension housing, rear seal.
   d. External shift linkage, drain plugs, front bearing retainer seal.
5. In your own words, explain a general procedure for removing a transmission from a vehicle.
6. Simulating transmission operation by shifting through the gears as it sits on the workbench can help to find

problems in the gears, synchronizers, and internal shift linkage. True or False?

7. Name four measuring tools commonly used during a manual transmission rebuild.
8. Why should you always refer to a service manual when rebuilding a manual transmission?
9. A transmission grinds its gears when shifted into third. Technician A says that a synchronizer may be bad. Technician B says that the countershaft gear may have chipped teeth. Who is correct?
   a. Technician A.
   b. Technician B.
   c. Both Technician A and B.
   d. Neither Technician A nor B.
10. A transmission makes excessive noise, but only in second gear. Technician A says to check the lubricant level. Technician B says to check for problems with bearings of the countershaft gear assembly. Who is correct?
    a. Technician A.
    b. Technician B.
    c. Both Technician A and B.
    d. Neither Technician A nor B.

Certification-Type Questions–Chapter 9

1. Technician A says that oil on the outside of the transmission case indicates a lubricant leak. Technician B says that leaks from other vehicle components are often mistaken for transmission leaks. Who is right?

(A) A only
(B) B only
(C) Both A & B
(D) Neither A nor B

2. All of these statements about manual transmissions are true EXCEPT:

(A) many problems can be solved by linkage adjustments.
(B) oil level has no affect on manual transmission operation.
(C) some transmission problems can be diagnosed during a test drive.
(D) water in the oil can cause gear pitting.

3. All of these are causes of a no-drive condition on a manual transmission EXCEPT:

(A) water in the transmission.
(B) stripped gear(s).
(C) broken transmission shaft.
(D) transmission locked in two gears.

4. A manual transmission shifts hard in all gears. Technician A says that worn or misadjusted shift linkage could be the cause. Technician B says that worn or misadjusted clutch linkage could be the cause. Who is right?

(A) A only
(B) B only
(C) Both A & B
(D) Neither A nor B

5. Gear clash in only one gear is usually caused by:

(A) clutch misadjustment.
(B) a worn synchronizer.
(C) a faulty clutch.
(D) low lubricant level.

6. Technician A says that external shift mechanisms should be lubricated with water-resistant grease. Technician B says that internal shift linkage is lubricated by the transmission lubricant. Who is right?

(A) A only
(B) B only
(C) Both A & B
(D) Neither A nor B

7. During transmission removal, which of these could allow the engine to fall backwards when the transmission crossmember is removed?

(A) Broken engine mounts.
(B) Loose transmission case-to-bell housing bolts.
(C) Loose bell housing-to-engine bolts.
(D) Broken transmission mount.

8. Technician A says that some transmission cases and bell housings are one-piece units. Technician B says that transmission cases are often attached to the bell housing by four bolts. Who is right?

(A) A only
(B) B only
(C) Both A & B
(D) Neither A nor B

9. Technician A says that all manual transmissions are internally constructed in the same way. Technician B says that manual transmission external parts should be removed before the internal parts are removed. Who is right?

(A) A only
(B) B only
(C) Both A & B
(D) Neither A nor B

10. Technician A says that a visual inspection is all that is necessary to fully examine manual transmission internal parts. Technician B says that a parts inspection requires that some internal parts be measured using micrometers or feeler gauges. Who is right?

(A) A only
(B) B only
(C) Both A & B
(D) Neither A nor B

11. Transmission end play can be adjusted by using special:

(A) thickness shims.
(B) length bolts.
(C) thickness gaskets.
(D) seal rings.

12. Technician A says that a transmission that is binding internally should be installed and allowed to run in. Technician B says that a transmission that gets hung up while being installed should be pulled into place by tightening the attaching bolts. Who is right?

(A) A only
(B) B only
(C) Both A & B
(D) Neither A nor B

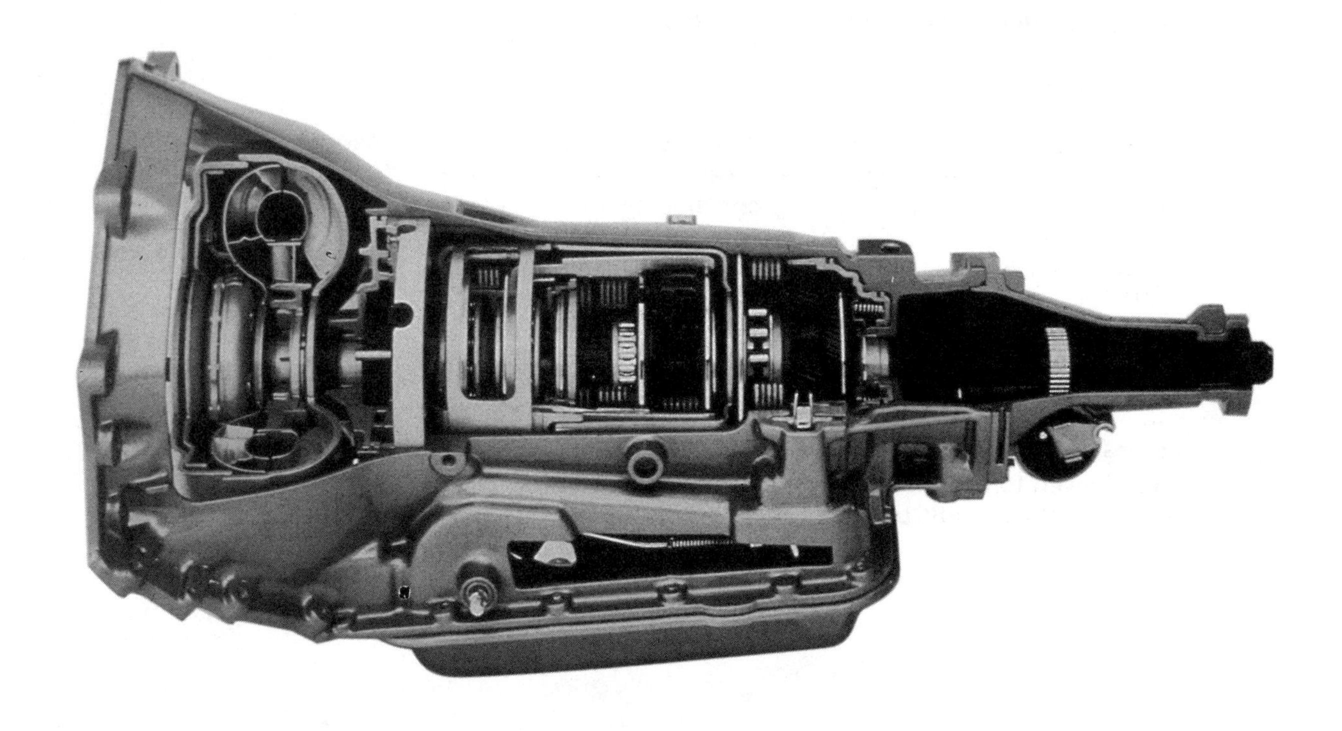

A cutaway of an electronically controlled 4-speed automatic transmission is shown. (General Motors)

Chapter 10

Automatic Transmission Construction and Operation

After studying this chapter, you will be able to:

- Identify the major parts of an automatic transmission.
- Discuss the purpose of each major part of an automatic transmission.
- Explain the operation of a torque converter.
- State various outcomes resulting from driving certain planetary gears and holding others.
- Diagram power flow through a typical automatic transmission.
- Summarize operations of planetary holding members.
- Describe the operation of governors, throttle valves, shift valves, pressure regulator valves, and other hydraulic components.
- Explain how a computer can be used to monitor and control an automatic transmission.

This chapter will explain the construction and operation of automatic transmissions used on rear-wheel drive vehicles. It will explain how automatic transmission components fit together to create a device that eliminates the need of manual operation.

A modern automatic transmission is a hydraulic, a mechanical, and often, an electronic mechanism for providing different gear ratios, reverse, and clutch action. An automatic transmission is complex; however, with a bit of study, it can be understood and serviced.

This chapter will give you background information needed to understand how automatic transmissions operate. By knowing how they are supposed to work, you will be prepared to troubleshoot and service them. This information will also help you understand automatic transaxles, used in front-wheel drive vehicles and presented in a later chapter.

Hopefully, you have thoroughly studied the information in preceding chapters, and you now have a solid grasp of it. An understanding of basic mechanical devices, along with an understanding of hydraulics, electronics, and pneumatics and of manual transmissions, will help you comprehend this chapter.

Construction Overview

Automatic transmissions are made of many separate parts and systems. Refer to Figure 10-1 to see how these parts fit together and their relative position in the transmission. Also, refer to Figure 10-2, which shows an exploded view of transmission components. The parts are numerous. The major automatic transmission parts are:

- The *torque converter* is a hydrodynamic device that uses transmission fluid to transmit and multiply power. The torque converter is mounted directly behind the engine and is turned by the engine crankshaft.
- The ***automatic transmission shafts,*** including the *automatic transmission input* and *output shafts,* transmit power from the torque converter to the driveline. Another transmission shaft provides a stationary support for the torque converter stator.
- The *planetary gearsets* are constant-mesh gears somewhat resembling the solar system. They create different forward gear ratios or provide reverse operation when different members are held stationary and others are driven. Most transmissions use at least two planetary gearsets.
- The ***planetary holding members*** allow gearset operation to occur. The members function by holding a planetary member stationary or applying drive power to it.
- The ***transmission oil pump*** produces the fluid pressure that operates other hydraulic components and lubricates the moving parts of the transmission. It is driven by lugs on the torque converter.
- The ***transmission oil filter*** removes particles from the automatic transmission fluid prior to circulation by the oil pump.
- The ***transmission oil cooler*** is needed to dissipate heat from the transmission fluid, which is generated by the transmission and also picked up from the engine.

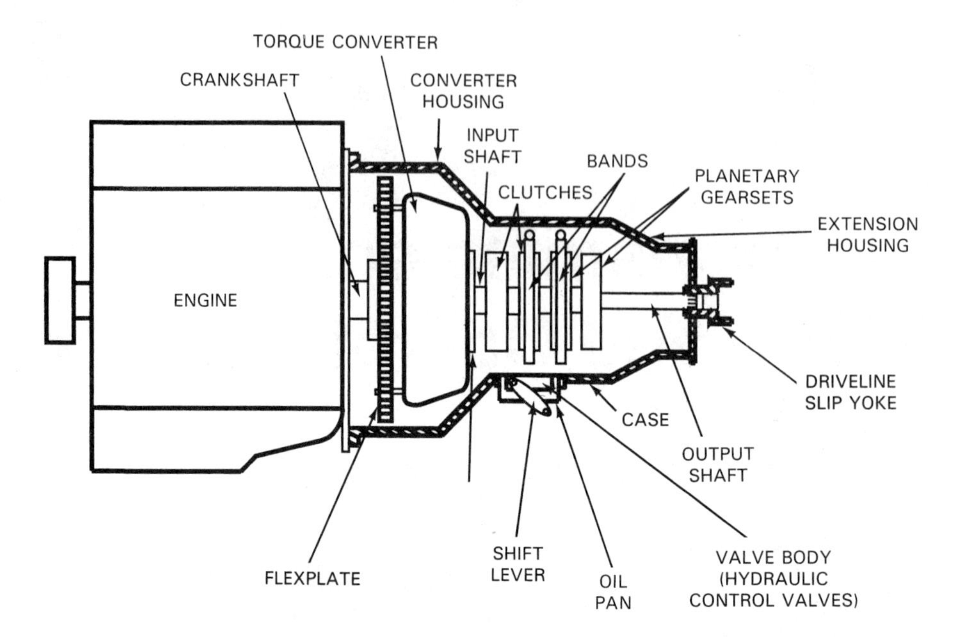

Figure 10-1. Diagram shows most of the major components of an automatic transmission and how they fit together in relation to each other. Compare this diagram with the information in the text.

The transmission oil cooler is often mounted in the engine cooling-system radiator. Separate *external oil coolers,* or *auxiliary oil coolers* as they are sometimes called, may also be installed on the vehicle.

- The *valve body* is a complex casting that serves as the control center for the transmission's hydraulic system. It contains many internal passageways and components, including hydraulic valves.
- *Hydraulic valves* are used in the valve body to control transmission fluid pressure and, also, direction and rate of fluid flow. Specifically, they control application of clutches and bands. In general, hydraulic valves help control the entire operation of the automatic transmission.
- The ***automatic transmission shift linkage*** operates the *manual valve,* thereby forming a direct mechanical connection between the driver and the transmission. Shift *selector levers* can be mounted on the floor or on the steering column.
- The ***converter housing*** is a stationary enclosure that surrounds and protects the torque converter. It is located at the front of the transmission, directly behind the engine. Some converter housings are a single casting with the transmission case.
- The ***automatic transmission case*** encloses the planetary gearsets and holding members and the inner ends of the transmission input and output shafts. Passageways in the case allow hydraulic fluid to travel between hydraulic components. The case bolts to the rear of the converter housing, or they are an integral casting.
- The *extension housing* supports and encloses the tail end of the transmission output shaft. It bolts to the rear of the transmission case.
- The ***transmission oil pan*** is installed on the bottom of the transmission case. It fits over the valve body and serves as a reservoir for transmission fluid.
- *Electronic components,* such as switches and solenoids, are used in the transmission to send signals to other vehicle components or to partially control the transmission hydraulic system.

Torque Converter

The torque converter was preceded in time by a simpler device called a *fluid coupling,* which still finds use in other applications. A ***fluid coupling*** is a hydrodynamic device designed to transmit power through a fluid. The device essentially consists of a *drive member* (impeller) and a *driven member* (turbine), Figure 10-3.

Fluid couplings have some shortcomings. For instance, the two members do not link together well at low speeds. The impeller does a poor job of turning the turbine at low engine speeds and has to build up considerable speed before the turbine will turn. The drive member rotates faster than the driven member and fails to transmit 100 percent of its power. This is a problem of *slippage,* like the slippage of a mechanical clutch. Another shortcoming is a result of the *bounce-back effect* (described later in this section). This effect causes torque reduction through the coupling.

To reduce slippage and to multiply torque, a simple fluid coupling is modified to create a torque converter, as

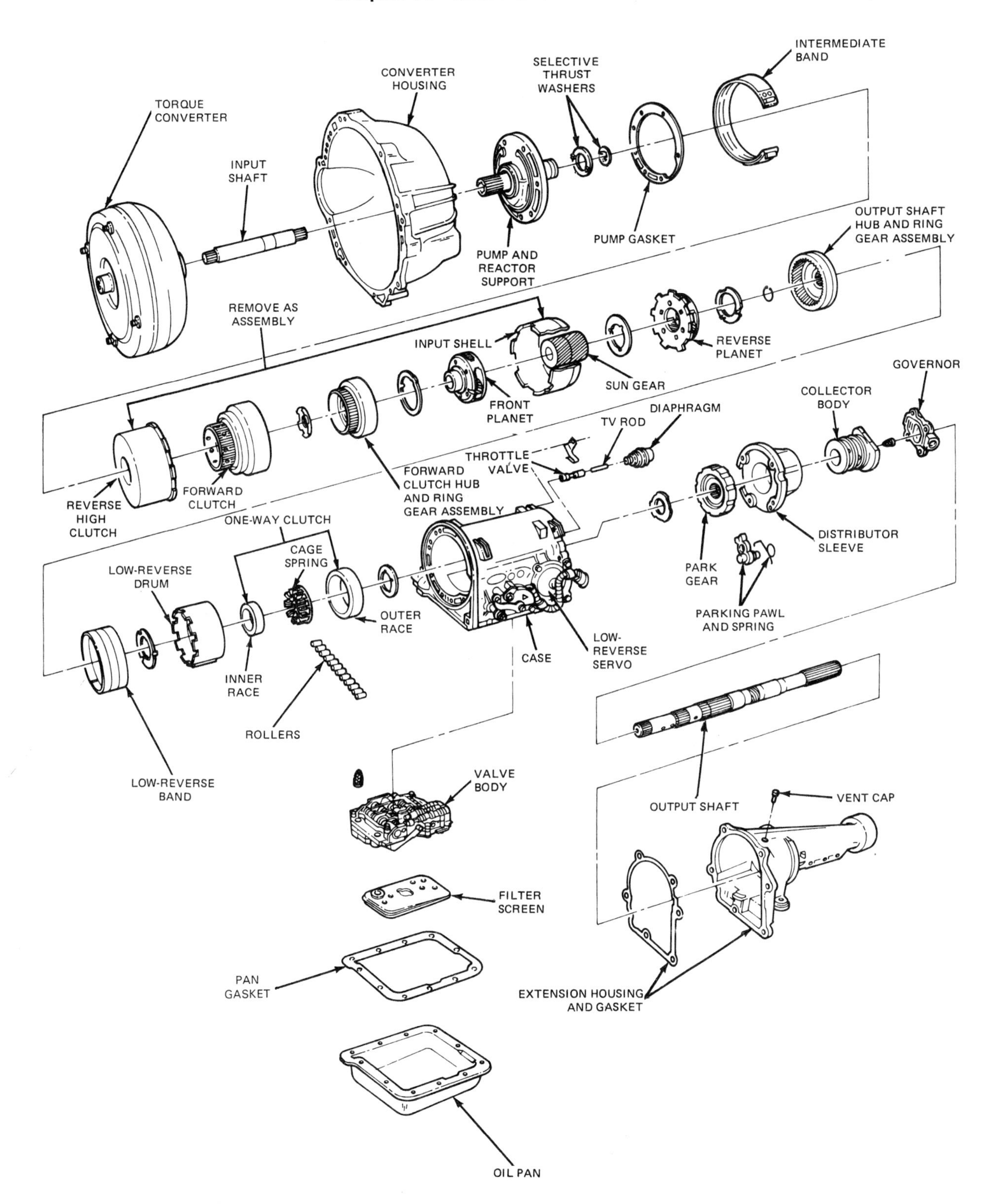

Figure 10-2. This is an exploded view of a typical automatic transmission. (Ford)

used on modern automatic transmissions. The major advantage of fluid couplings and torque converters is that they do not have to be disengaged when the vehicle is stopped and the engine is running.

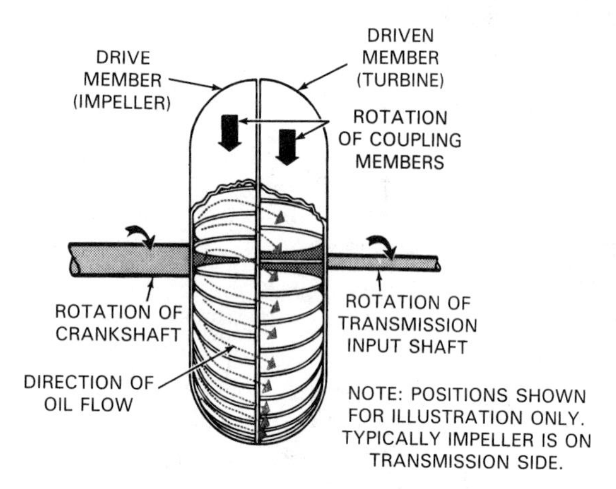

Figure 10-3. Cutaway shows fluid coupling in action. Oil is thrown from the impeller into the turbine. Note member positions are usually opposite of what is shown here.

Torque Converter Construction

A torque converter basically consists of three separate bladed elements inside of a fluid-filled housing. See Figures 10-4 and 10-5. The basic construction of a torque converter includes an impeller, a turbine, a stator, and a converter cover. These and other associated parts are discussed in this section.

Impeller

The impeller, or pump, is composed of a set of curved vanes welded to the inside of a shell that forms the rear half of the converter. The impeller is attached to the converter cover, so it turns whenever the engine turns. It is the torque converter drive member.

Turbine

The turbine is composed of a set of curved vanes welded to a *turbine shell.* The turbine is driven by the impeller through fluid contained within the converter. The turbine comprises the torque converter driven member. The central hub of the turbine contains internal splines that

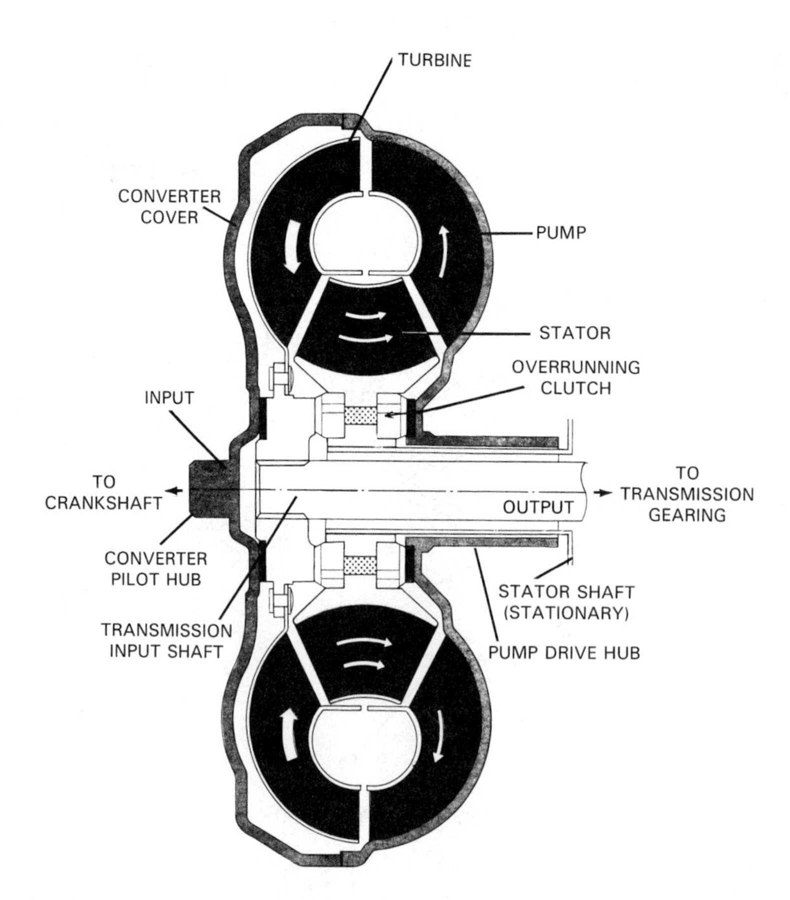

Figure 10-4. Simplified drawing of a torque converter. Crankshaft is fastened (indirectly) to converter cover and impeller. When engine crankshaft turns fast enough, fluid movement rotates turbine element and transmission input shaft. Note stator is mounted on a special type of clutch, which in turn, is mounted on a stationary shaft. (Sachs)

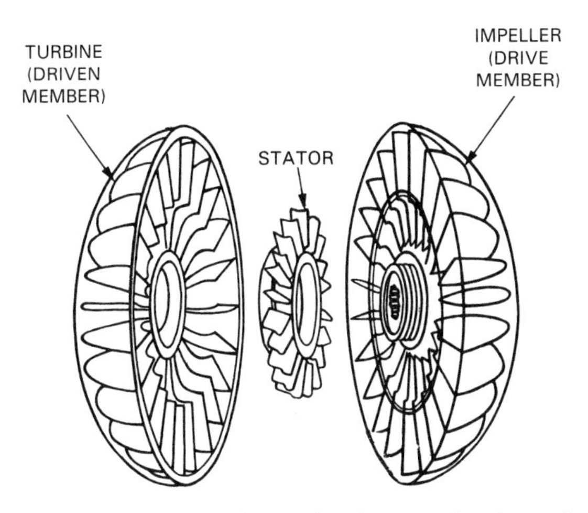

Figure 10-5. Illustration shows the three main elements of a torque converter. The impeller is driven by the engine. The turbine, which drives the transmission input shaft, is driven by the impeller. The stator redirects fluid flow from turbine to multiply torque. Note curvature of vanes is not shown. (Texaco)

fit the external splines of the transmission input shaft, sometimes called the ***turbine shaft.*** The turbine drives the turbine shaft.

Stator

The stator, or guide wheel, fits between the impeller and turbine and is smaller than either. It essentially consists of a set of vanes that are curved in the opposite direction of the turbine vanes. The stator redirects fluid coming from the turbine to the impeller's direction of rotation to help the torque converter multiply power. The stator is what makes a torque converter out of a simple fluid coupling.

The stator is mounted on an *overrunning clutch,* which allows the stator to move under some conditions and keeps it stationary under others. The overrunning clutch is splined to a hollow stationary shaft, called the ***stator shaft,*** connected to the oil pump.

Converter cover

The ***converter cover*** fits over the turbine, attaches to the impeller, and indirectly attaches to the engine crankshaft. Some covers are designed with a ring gear welded to the perimeter. The ring gear engages the starter motor during cranking.

At the center of the converter cover is a round metal extension called the ***converter pilot hub.*** The pilot hub fits into a mating hole at the engine crankshaft flange. The close fit of the pilot hub and crankshaft bore maintains the alignment of the converter and crankshaft centerlines. The converter cover may also have balance weights. In addition, some covers have oil drain plugs for removal of fluid during maintenance.

Flexplate

The converter cover has either tapped holes or threaded studs for connection to a *flexplate,* which is at the rear of the engine. The ***flexplate*** is a lightweight disc with a ring gear, used to engage the starter motor. (The flexplate will not have a ring gear if the converter cover has one.) The flexplate is attached to the engine crankshaft by bolts, referred to as *flexplate attaching bolts* (flexplate-to-crankshaft bolts), in the same manner as a flywheel. The flexplate and converter turn with the engine whenever it is running.

The flexplate is often mistakenly referred to as a flywheel. This is a misnomer, since a flywheel is a heavy disc that smooths engine crankshaft rotation, and the flexplate is a lightweight disc that does nothing for crankshaft rotation. However, the torque converter itself acts as a flywheel in that it is very large and heavy. For this reason, vehicles with automatic transmissions do not require flywheels.

Pump drive hub

Modern torque converters are sealed units, welded together after all internal elements are installed. However, the torque converter has a hollow rear shaft, called the ***pump drive hub,*** that fits over the stator and turbine shafts and into the oil pump at the front of the transmission case. Oil is pumped into the converter through one of the internal shafts, and it exits between the stator shaft and the inside of the hub. Drive lugs on the rear of the hub turn the transmission oil pump. An *oil pump bushing* within the pump housing supports the pump drive hub.

Thrust washers

The torque converter contains several thrust washers between all parts that can move in relation to each other. This means that thrust washers are usually placed between the cover and turbine, the turbine and stator, and the stator and impeller.

Torque Converter Operation

Torque converter operation depends on the movement of transmission fluid. When the engine is running, the converter cover is turning. This causes the impeller to turn, as it is solidly connected to the cover. Since the torque converter is completely filled with transmission fluid, the turning impeller moves the fluid that is caught between the vanes. The impeller causes the fluid to rotate at the same speed and relative direction. Engine power is transferred from the impeller to the rotating fluid.

As the impeller rotates, the fluid is thrown outward, and following the curvature of the impeller shell, the fluid becomes redirected toward the turbine. The fluid leaves the impeller, striking the turbine. The fluid presses forward on the vanes of the turbine, as shown by the small arrows in

Figure 10-6. As a result of this force, the turbine begins to rotate. In this way, the moving fluid transmits motion to the turbine, and engine power is transferred from the fluid to the turbine.

At *idle,* with the vehicle stopped and the transmission in gear, the engine continues to run. If the vehicle had a manual transmission, the clutch would have to be disengaged, or the transmission placed in neutral; otherwise, the engine would die.

With an automatic transmission, the impeller is turned by the engine, causing the fluid to rotate and strike the turbine blades. At low engine speeds, the force of the fluid hitting the turbine is relatively small, and the driver can overcome it by keeping the brakes applied. The turbine does not move.

Note that the energy of the fluid hitting the stationary turbine causes friction instead of motion. This friction changes the energy in the fluid into heat. The heat generated can become intense. For this reason, vehicles with automatic transmissions should not be idled in gear for long periods of time.

During *acceleration,* the force of the fluid leaving the impeller turns the turbine, turning the automatic transmission input shaft. Turning the input shaft, and the resultant operation of the other transmission parts, causes the vehicle to move, but with some slippage.

At *cruising speeds* (maintained speed generally above 35 mph), the impeller and turbine spin at almost the same speed. There is very little slippage, and *torque multiplication* is minimal.

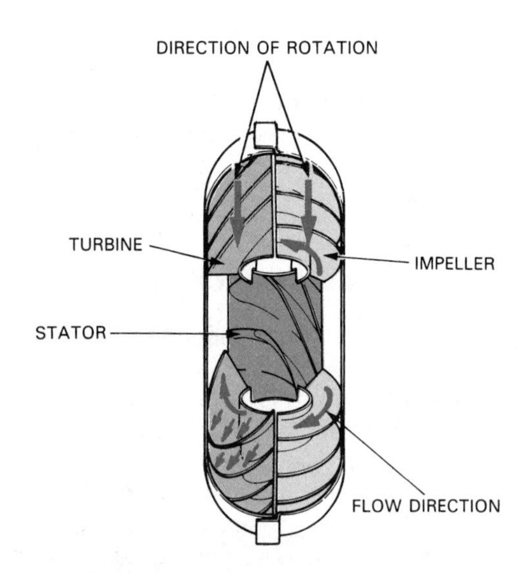

Figure 10-6. Cutaway of torque converter. Heavy arrows show direction of rotation of impeller and turbine. Medium arrows show how fluid circulates between impeller and turbine. Smaller arrows depict the driving force with which the fluid strikes the turbine vanes. Note curvature of impeller, turbine, and stator vanes. (Subaru)

Stator action

As a result of turbine vane curvature, fluid coming from the impeller will deflect, or reverse direction, before it returns to the impeller. The angle of deflection is greatest at conditions of high torque multiplication, decreasing as torque multiplication declines.

When the impeller is moving much faster than the turbine, the fluid still has considerable forward motion as it bounces back to the impeller. Without a stator, return flow would oppose impeller rotation, and it would hit the forward (oncoming) faces of the impeller vanes with a good deal of force. This action, seen in fluid couplings and known as the ***bounce-back effect,*** reduces the effectiveness of the impeller.

To counteract this problem, the stator does the job of reversing the direction of fluid coming off the turbine. Fluid leaving the turbine must pass through the stator before reaching the impeller. The stator blades are curved so that they turn the fluid around. Fluid leaving the stator strikes the impeller in the direction of impeller rotation. Instead of losing engine power in working to reverse the direction of the fluid coming from the turbine, which is what happens without the stator, the impeller is now assisted in rotating. The stator has not only eliminated torque loss due to the bounce-back effect, it has improved efficiency by providing torque multiplication.

Torque multiplication. Just as a small gear can multiply torque by driving a large gear, a torque converter can also multiply torque. Unlike the gearset, however, the torque converter operates through a *continuous range* of "gear" ratios, and torque multiplication varies accordingly.

Torque multiplication refers to the ability of a component, such as torque converter or gearset, to increase output torque above input torque. Specifically, for a torque converter, it refers to the amount of torque at the transmission input shaft above what the engine puts out.

Torque multiplication basically occurs as a result of the redirection of fluid by the stator. Because of the stator, the fluid reenters the impeller in a "helping" direction. The fluid passes through the impeller, and when it reenters the turbine, it actually imparts additional force (torque) to the turbine.

The amount of torque multiplication changes with the difference in rpm between the impeller and turbine–the greater the difference, the greater the multiplication. It follows that the greatest torque multiplication for any given engine speed occurs at initial acceleration. This is the point of maximum slippage, when the turbine shaft is stopped, or *stalled,* and the differences in speed between the two elements are greatest. This point is designated as the ***stall speed,*** given in engine rpm. Maximum possible torque increase is attained at the maximum possible stall speed, the point where engine rpm peaks. Maximum *converter ratios* can be as high as 2.5:1 on some converters; however, most ratios are about 2:1.

The torque multiplication and output (turbine shaft) rpm have an inverse relationship; that is, for a constant engine speed, torque multiplication drops off as turbine speed increases. Thus, as the speed of the turbine is brought up to the speed of the impeller, torque multiplication is continuously reduced, and as the vehicle reaches cruising speed, the converter ratio drops to about 1:1. This says that as the impeller and turbine approach the same speed, torque multiplication drops off to almost zero.

Note that while torque multiplication is greatest at stall speed, power transfer is least efficient. Maximum stall speed is designed into torque converters by the distance established between the impeller and turbine. Decreasing the distance between the two elements reduces maximum stall speed and, therefore, torque multiplication; however, it improves efficiency of power transfer. In contrast, increasing the distance between the two, increases stall speed and maximum torque multiplication, but it reduces efficiency.

Overrunning clutch operation. The stator is installed on the stator shaft over an ***overrunning clutch,*** also called a ***one-way clutch.*** When accelerating, fluid deflects off the turbine blades and reverses direction. Fluid is deflected into the stator, striking the front faces of the vanes. The direction of applied force acts to lock the one-way clutch and, therefore, the stator. When held stationary, the stator can return fluid to the impeller in a *helping* direction, and it can multiply torque.

At cruising speeds, the turbine and impeller begin turning at almost the same speed. During this condition, the fluid enters the turbine at the same speed the turbine is moving, and there is no fluid deflection. The turbine is no longer reversing the fluid that passes through it. When turbine speed approaches about 90 percent of impeller speed, called the ***coupling phase,*** the stator is not needed. In fact, it begins to restrict the flow of fluid from the turbine back to the impeller.

The flow leaving the turbine is now striking the back of the stator blades. As a result, the direction of applied force causes the overrunning clutch and, therefore, the stator, to unlock and freewheel. The stator can then rotate with the impeller and turbine, and it will not restrict the movement of the fluid through the impeller and turbine. The action of the one-way clutch is shown in Figure 10-7.

Lockup Torque Converter

A ***lockup torque converter,*** or ***converter clutch torque converter,*** as it is also called, has an internal friction-clutch mechanism. It is for mechanically locking the turbine to the impeller and converter cover in high gear. It is designed to eliminate the 10 percent slip that takes place between the impeller and turbine at the coupling phase of operation. With this arrangement, the transmission input shaft will travel at the same speed as the engine crankshaft, and efficiency will be 100 percent.

The lockup torque converter offers several advantages. One is improved fuel economy, because engine power is not wasted. Also, engine wear is reduced because the engine can now be operated at a lower rpm to attain the same vehicle speed. Another advantage is that converter operational heat is reduced.

Modern lockup torque converters are equipped with an internal clutch, which serves to lock the input shaft to the converter cover. Figure 10-8 shows a cutaway of a torque converter with a *lockup,* or *converter, clutch.* Typically, the

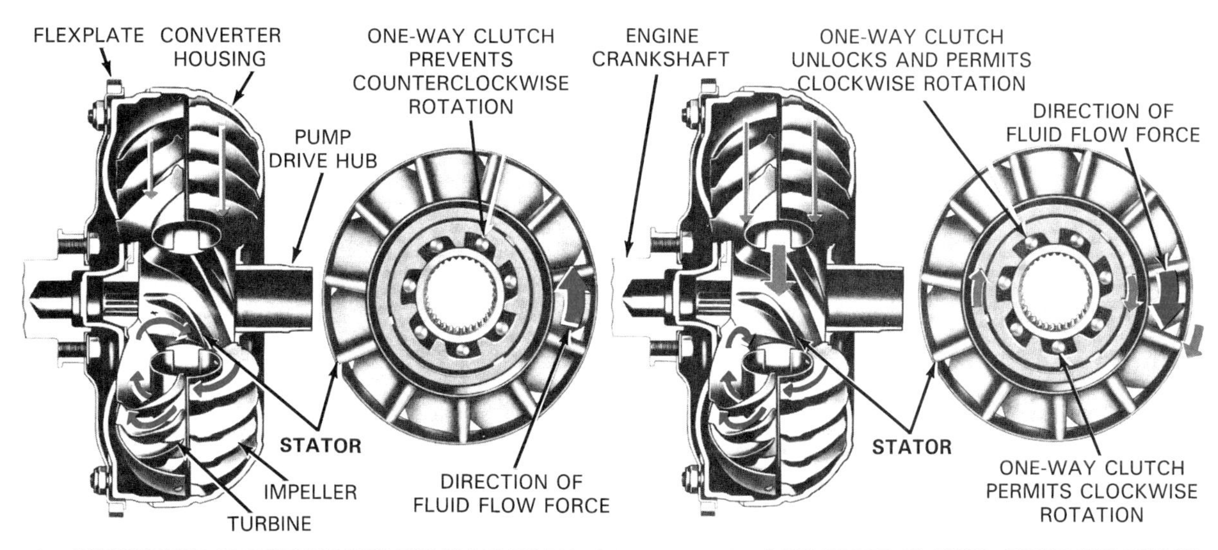

Figure 10-7. Study action of an overrunning clutch mounted between stator and stator shaft. A–In torque multiplication phase, fluid is deflected off of turbine vanes onto front side of stator vanes. This action locks the one-way clutch and holds the stator stationary. B–In coupling phase, deflection (or lack thereof) of fluid is such that fluid strikes back side of stator blades. This causes the stator to spin forward, preventing it from interfering with the fluid flow. (Ford)

clutch assembly consists of a hydraulic *clutch piston,* or *pressure plate,* torsion springs, and a clutch friction material. The piston is splined to the input shaft. The friction material is attached to the converter cover. The torsion springs dampen out engine firing impulses and absorb shock loads that occur during lockup. Figure 10-9 shows an exploded view of a typical hydraulic-type lockup torque converter.

Operation of a typical hydraulic converter clutch is controlled through a *lockup module* and a *switch valve.* Both are housed within the transmission valve body. These mechanisms control the flow of hydraulic fluid to and from the piston.

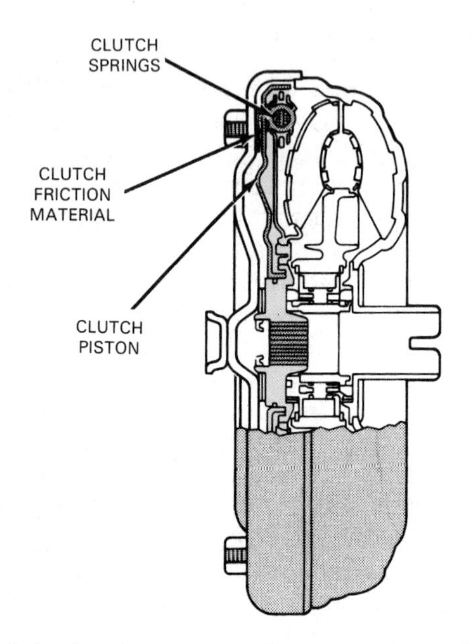

Figure 10-8. Lockup torque converters use an internal clutch to make a direct connection between the engine crankshaft and transmission input shaft. Clutch is splined to the input shaft. When applied by hydraulic pressure, it makes frictional contact with the converter cover. (Chrysler)

In the lower gears, the lockup torque converter operates as a conventional torque converter, Figure 10-10A. When shifted into high gear, or direct drive, transmission fluid is typically channeled through the input shaft into the space between the turbine and clutch piston. Fluid pressure causes the piston to be pushed into the friction material on the converter cover, Figure 10-10B. This action locks the turbine to the impeller. When the transmission shifts out of high gear, fluid pressure and, therefore, the clutch piston are released. The converter returns to conventional operation.

Note that there is another type of lockup torque converter that works by centrifugal force. In this type, shoes made of a friction material are thrown outward against the converter cover, forming a direct mechanical connection between the impeller and turbine.

Combination Torque Converter/Manual Clutch

A number of older-model manual transmission vehicles used a combination torque converter *and* manual clutch in conjunction with a manual transmission. See Figure 10-11. This type of transmission has been referred to as a ***semiautomatic transmission.*** Semiautomatics often had, in addition to the torque converter, some method of automatically selecting manual transmission gears without using the clutch pedal. The converter was filled with fluid and operated in the same manner as torque converters in automatic transmissions. The converter made shifting easier and smoother and allowed some gears to be selected just by moving the gearshift lever–without using the clutch pedal. These transmissions could also be left in

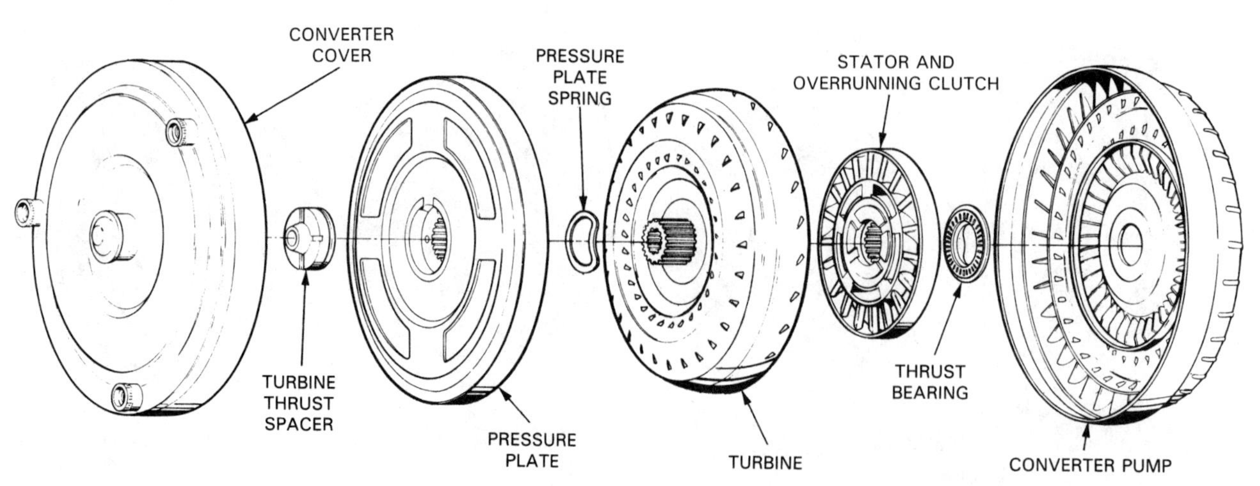

Figure 10-9. This exploded view of a converter clutch torque converter shows relative positions of the pressure plate and its associated spacers and washers. (General Motors)

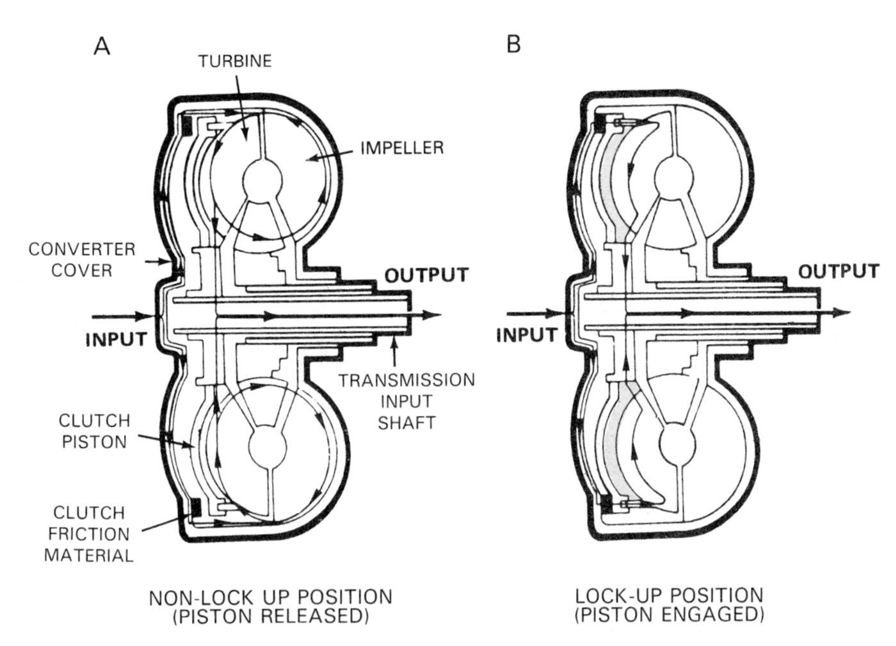

Figure 10-10. Typical lockup torque converter operation. A-In lower gears, no fluid pressure acts on clutch piston. Torque converter operates like conventional unit; impeller drives turbine. B-In high gear, fluid is transferred into piston chamber. Clutch piston is forced into clutch friction material on converter cover. Turbine is mechanically locked to cover and impeller. Crankshaft drives transmission input shaft directly, without slippage. Note that arrows depict power flow. (Chrysler)

gear with the engine running and the vehicle stopped. Although uncommon, semiautomatic transmissions are still used on a few makes of vehicles.

Automatic Transmission Shafts

The two main shafts in the automatic transmission are the *input* and the *output shaft.* These shafts transfer power from the torque converter to the driveline. They are straight shafts made of hardened steel.

The *stator shaft* is a third type of shaft found in the automatic transmission. It is a hollow, stationary shaft that extends from the transmission oil pump. It supports the stator overrunning clutch.

Automatic Transmission Input Shaft

The front end of the ***automatic transmission input shaft,*** or turbine shaft, has external splines that fit into mating internal splines on the turbine of the torque converter, as previously discussed. Power is transferred from the turbine via the input shaft directly (or indirectly) into the front planetary gearset of the transmission. Hydraulic fluid from the valve body to the transmission clutch apply pistons often travels through passageways in the input shaft. The input shaft may contain rings that seal in the fluid going to the clutch apply pistons. The shaft rides on bushings, lubricated by transmission fluid.

Automatic Transmission Output Shaft

The tail end of the ***automatic transmission output shaft*** has external splines that fit into mating splines on the driveline slip yoke. The front end of the output shaft is splined to one or more planetary gearsets. The output shaft centerline aligns with the input shaft. Its front end almost touches the input shaft.

Most output shafts contain the speedometer drive gear. Some speedometer drive gears are machined into the shaft, while others are clipped or pressed onto the shaft. In addition, the *governor,* discussed later in this chapter, may be installed on the output shaft. However, in some designs, the shaft will have a gear, similar to the speedometer drive gear, for driving the governor. The output shaft may also have internal passageways and seal rings for hydraulic purposes.

Stator Shaft

Solidly mounted on the front of the transmission oil pump is the stator shaft, which provides a stationary support for the torque converter stator and overrunning clutch. The overrunning clutch must be solidly supported on a stationary part to work properly. The stator shaft has external splines that fit into mating internal splines on the overrunning clutch, preventing rotation of the inner race of the clutch. The stator shaft is hollow, and the transmission input shaft passes through it. Bushings in the stator shaft support the input shaft.

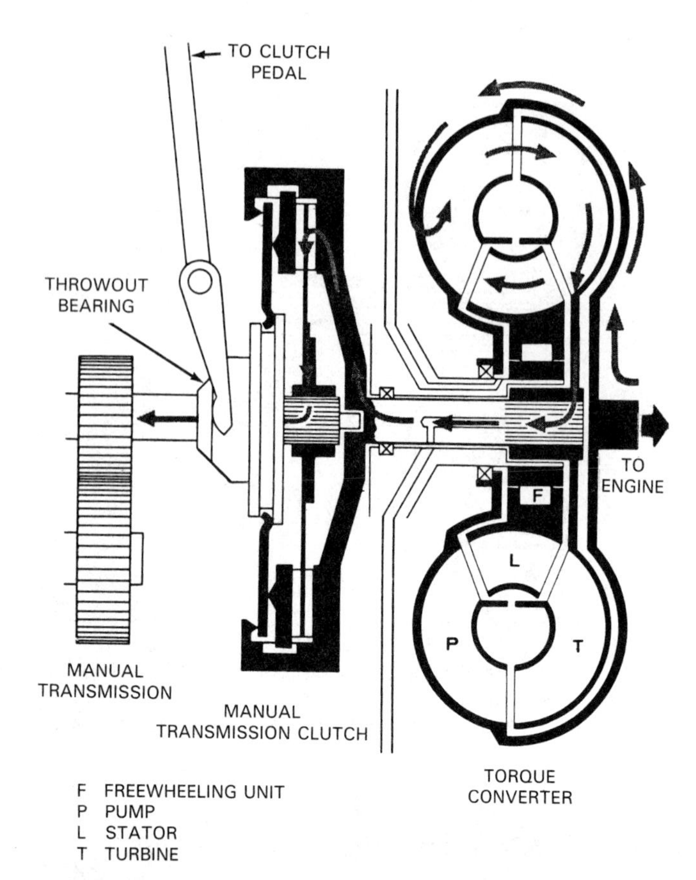

Figure 10-11. Note combination torque converter and clutch. Fluid is means of transferring power, just as in an automatic transmission. Fluid is sealed in converter at the factory. This design is sometimes referred to as a semiautomatic transmission. (Porsche)

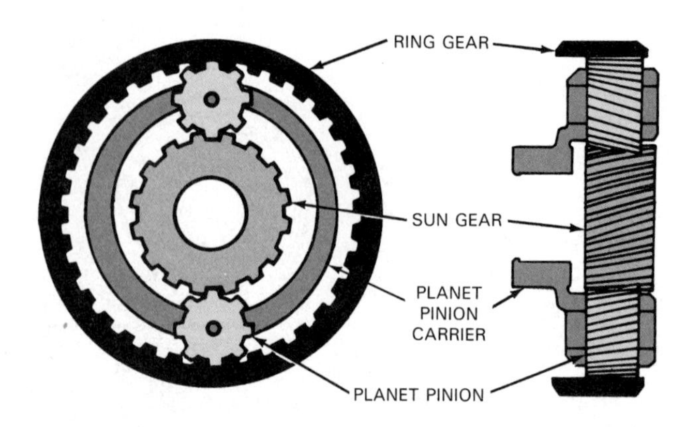

Figure 10-12. A simple planetary gearset consists of a central sun gear surrounded and meshed with planet gears in a planet carrier. A ring gear surrounds and meshes with the planet gears. Different gear ratios can be established, in design, by varying the number of teeth on meshing gears, just as with conventional gearsets. Note that planetary gearsets are compact and tough, and since the gears are constantly engaged, gear clash and broken teeth are not a problem. (Ford)

Planetary Gearsets

The planetary gearset used on modern automatic transmissions consists of a sun gear, several planet pinions, and a planet-pinion carrier. It also consists of a ring gear. These constant-mesh gears are so named because their design somewhat resembles the solar system. Just as Earth or Mars revolves around the Sun, the planet gears revolve around the sun gear. Planetary gearsets were discussed previously in Chapter 3. A simple planetary gearset is shown in Figure 10-12.

Planetary Gearset Power Flow

Power is transferred through planetary gearsets by holding one part of the gearset and driving another. Power exits the third part of the gearset. For example, holding the sun gear and driving the planet carrier causes the ring gear to rotate.

With regard to direction of rotation, it is helpful to remember two points. The first point is that two gears with external teeth in mesh will rotate in opposite directions. The second point is that two gears in mesh, one with internal teeth and one with external, will rotate in the same direction. Understanding these points will help you analyze planetary gearset operation.

Different combinations of gears that are driven and gears that are held stationary produce different gear ratios or actions. Thus, by holding or driving the gears of a planetary gearset, it is possible to obtain:

- *Gear reduction.*
- *Overdrive.*
- *Direct drive.*
- *Reverse gear.*
- *Park* or *neutral.*

Gear reduction

Through planetary gearsets, automatic transmissions can achieve gear reduction. This provides increased torque, accompanied by reduced output speed. One method of obtaining a gear reduction is to turn the sun gear while holding the ring gear stationary. Conversely, another method is to hold the sun gear stationary while driving the ring gear. In either, the planet carrier is the output member. The first method provides the maximum reduction (torque multiplication) that can be achieved in one planetary gearset. The second method produces minimum gear reduction. See Figure 10-13.

Overdrive

Through planetary gearsets, automatic transmissions can achieve overdrive. This provides increased output speed while reducing torque. When the sun or ring gear is held stationary and the carrier is driven, the remaining member (output) will be driven at a faster speed than the planet carrier. This will produce overdrive in a planetary gearset, with maximum speed increase occurring as a result of holding the ring gear stationary. See Figure 10-14.

Direct drive

Through planetary gearsets, automatic transmissions can achieve direct drive. The gearset serves as a solid unit to transfer power. A planetary gearset will act as a solid unit when two of its members are turned at the same speed. The planet gears do not rotate on their axes; they can neither idle nor walk about the sun gear. The entire unit is locked together, forming one rotating unit. See Figure 10-15.

Reverse gear

A planetary gearset is capable of providing output rotation that is opposite of input rotation. This action causes the vehicle to move in reverse direction. The carrier is held stationary. The input shaft drives either the sun gear or the ring gear. A driven sun gear yields torque multiplication at ring gear output. A driven ring gear yields torque reduction at sun gear output. The planet gears act as idler gears. They reverse the direction of rotation between the sun gear and ring gear. See Figure 10-16.

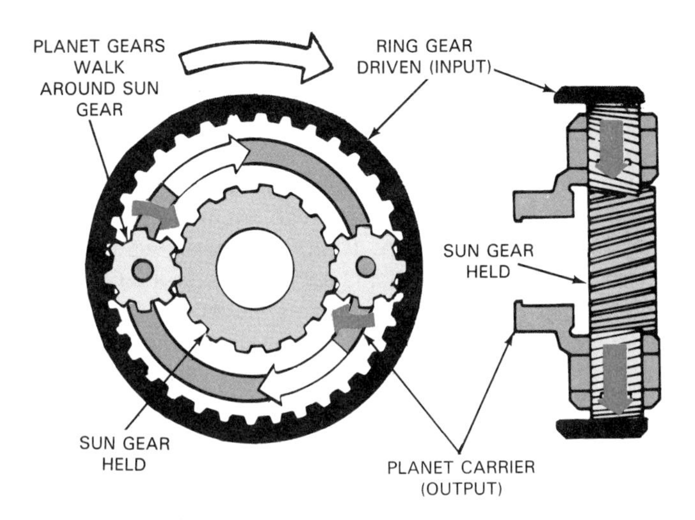

Figure 10-13. Gear reduction is produced with ring gear as input and sun gear held stationary. Planet gears walk around sun gear, taking planet carrier (output) with them. Planet carrier rotates in same direction as ring gear but at a slower speed. Although this arrangement produces good torque, *maximum* torque is produced with the sun gear as input and the ring gear held stationary. (Ford)

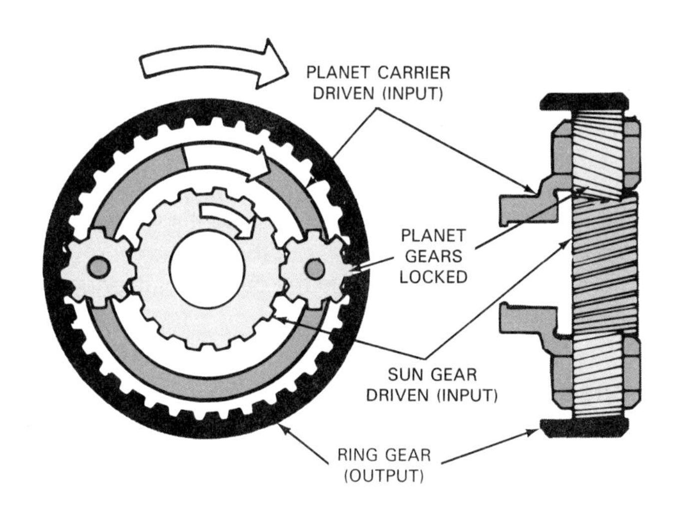

Figure 10-15. In direct drive, driving any two of the gearset elements at the same speed will lock the entire gearset together, making it turn as a unit. (Ford)

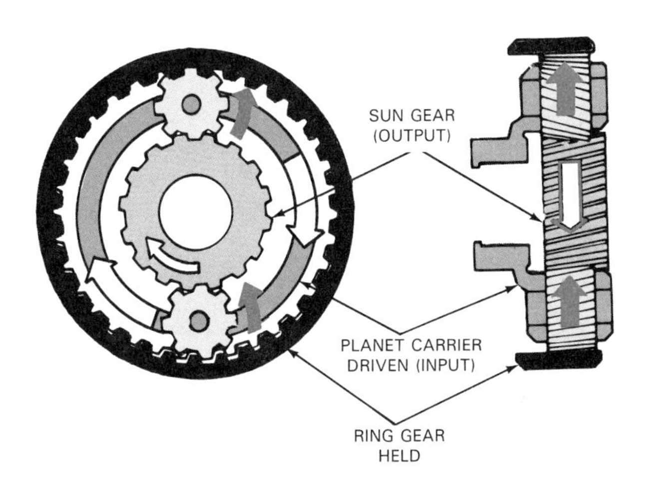

Figure 10-14. Overdrive is produced with the planet carrier as input, ring gear held stationary, and sun gear as output. As planet gears walk around ring gear, they drive sun gear. Sun gear rotates in same direction as planet carrier but at a faster speed. (Ford)

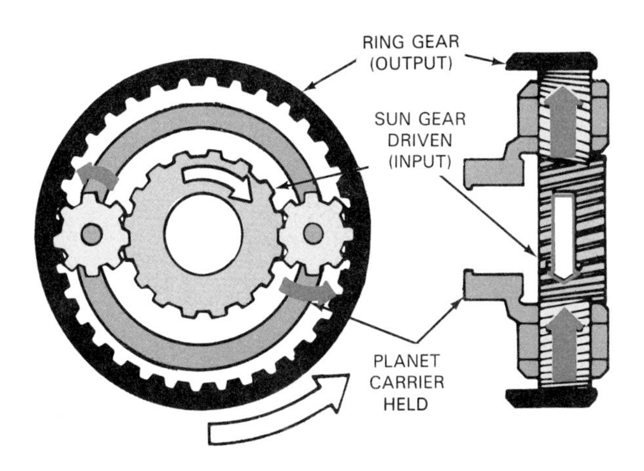

Figure 10-16. Reverse is produced here by holding planet carrier and driving sun gear. Ring gear, as output, turns in reverse direction of input. (Ford)

Park or neutral

A planetary gearset can be allowed to freewheel, stopping power flow. When all of the planetary members are free, the gearset cannot transfer power. This freewheeling condition occurs when an automatic transmission is placed in park or neutral.

Compound Planetary Gearset

A modern 3- or 4-speed automatic transmission uses two or more planetary gearsets linked together, making up what is called a ***compound planetary gearset.*** The arrangement is used because it can provide more forward gear ratios than the basic gearset described previously. There are a couple of variations of compound planetary gearsets in common use, including the *Ravigneaux gear train* and the *Simpson gear train.*

Ravigneaux gear train

The ***Ravigneaux gear train,*** Figure 10-17, is common to automatic transmissions using an overdrive. This design has two sun gears and one planet carrier with two sets of planet gears. Short planet gears engage the *forward sun gear.* Long planet gears mesh with the *reverse sun gear.* A single ring gear, connected to the output shaft, engages the long planet gears.

Simpson gear train

Another compound planetary gearset is the ***Simpson gear train,*** Figure 10-18. The Simpson gear train consists of two separate ring gears, two separate planet carrier sets, and one common sun gear. The long sun gear meshes with both sets of planet gears. The front planet carrier and rear ring gear are splined to the output shaft.

In the gearset shown in Figure 10-18, note the components for handling thrust loads. Since the sun and planet gears are helical, their operation places thrust loads on the moving parts of the transmission. To prevent wear to the moving parts, thrust washers (and plates) are used. The washers absorb some of the thrust loads and also keep moving parts from touching. The average transmission has six to ten thrust washers at various locations. Note the location of thrust washers in the transmission shown in Figure 10-19.

To track *power flow* through a Simpson gear train, refer to Figure 10-20. It shows how power moves from input shaft to output shaft in a typical 3-speed transmission. Figure 10-20A identifies major components. Study each illustration carefully.

Power flow–drive range, first gear. Power flow through the Simpson gear train in first gear, selector lever in *drive,* is shown in Figure 10-20B. Power flows through the front annulus (ring) gear, turning the gear at engine speed and direction of rotation. This movement drives the front planet gears, which are in mesh with the front annulus gear, turning them in the same direction as the annulus gear. The movement of the planet gears drives the planet carrier. The carrier drives the output shaft, as it is splined to the shaft. The planet gears also cause the sun gear to rotate. Note that there is gear reduction through the front gearset.

The sun gear rotates opposite engine rotation, causing the rear planet gears to rotate in the same direction as engine rotation. The rear planet carrier is held stationary. The planet gears cause the rear annulus gear to rotate in the same direction as engine rotation. Torque is then transferred through the rear annulus gear, which is splined to the output shaft. The arrangement produces an additional gear reduction through the rear gearset, and the total

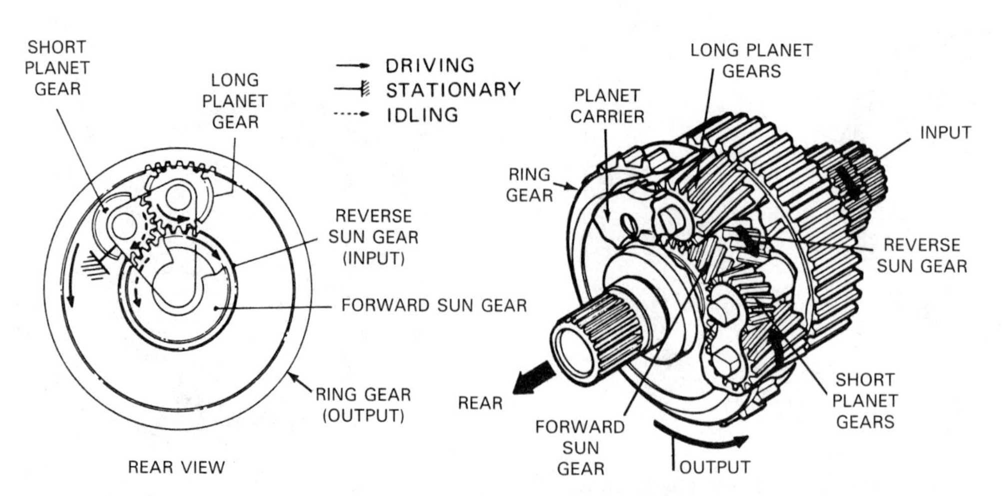

Figure 10-17. A compound planetary gearset, such as the Ravigneaux gear train shown here, consists of two or more planetary gearsets combined into a single unit. A Ravigneaux gear train has two sun gears, two sets of planet gears, sometimes called short and long pinions, and a single ring gear. All forward gears and reverse are derived from this one gearset. Note power flow is depicted for unit in reverse gear. (Subaru)

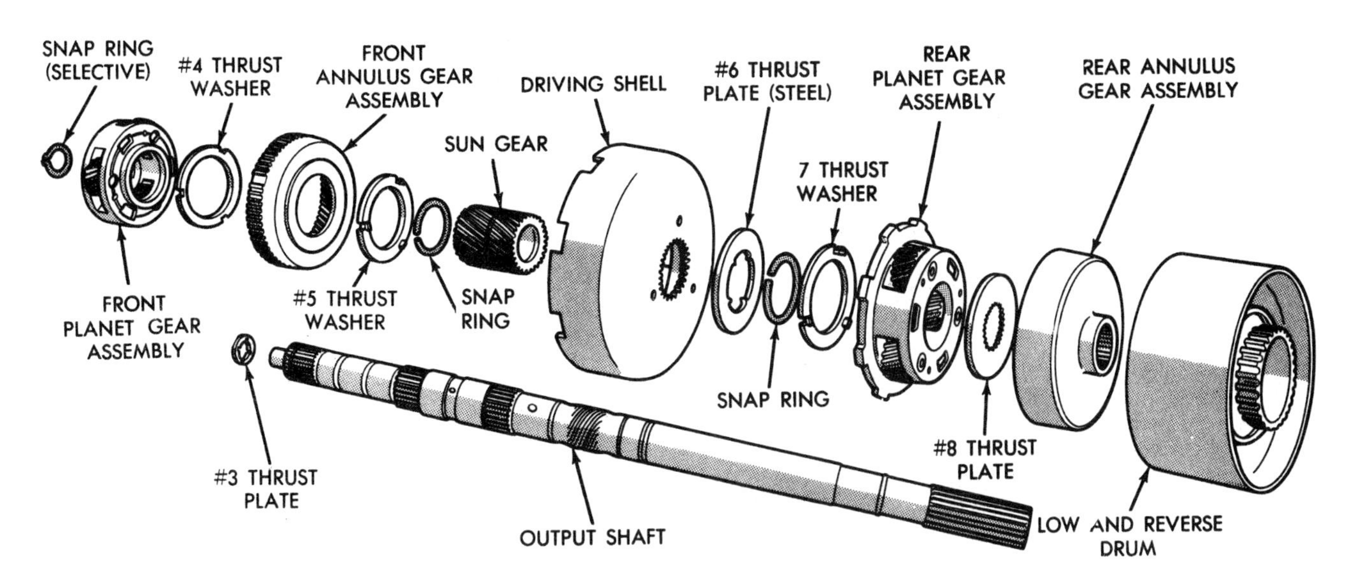

Figure 10-18. The Simpson gear train, shown here in an exploded view, is a type of compound planetary gearset used on modern transmissions. This design consists of a single long sun gear, two sets of planet gears and carriers, and two ring gears. Note the components for handling thrust loads. (Chrysler)

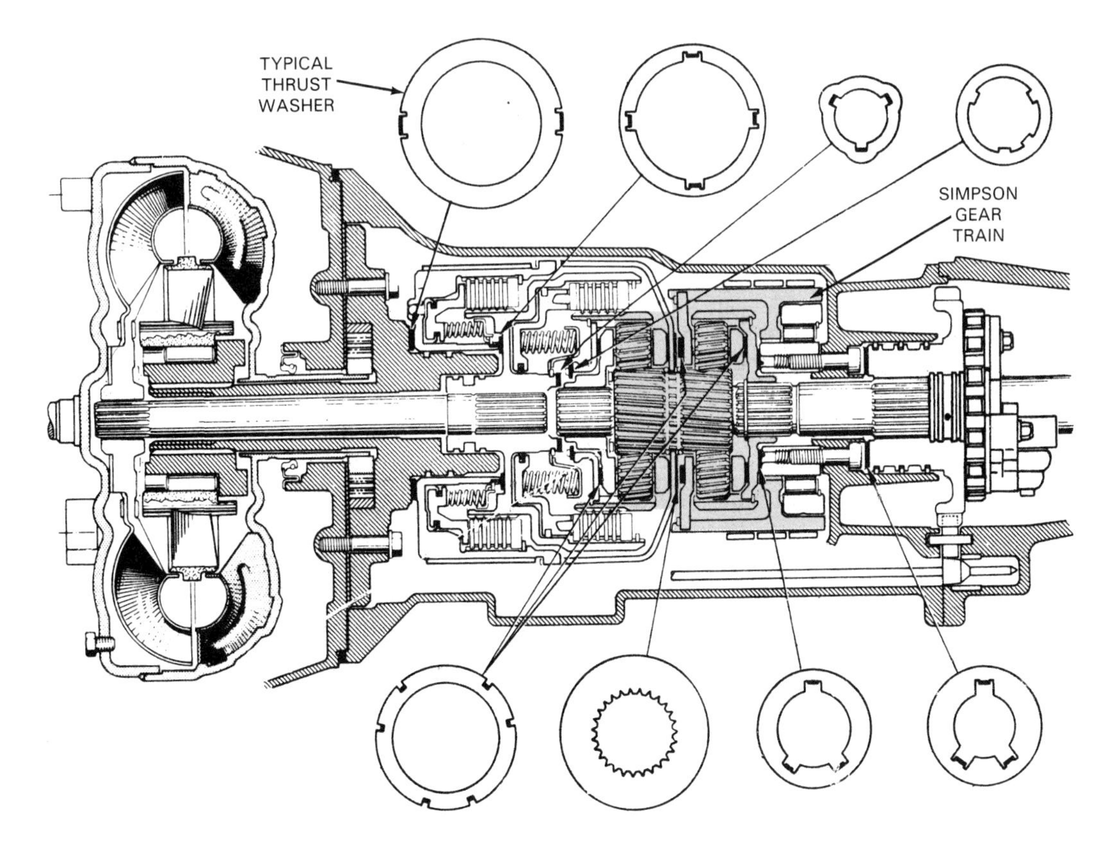

Figure 10-19. Thrust washers, such as the ones shown here, provide wear and thrust surfaces for planetary gears and other moving parts. They keep moving parts from rubbing on each other or on the transmission case. Note Simpson gear train in this transmission. (Ford)

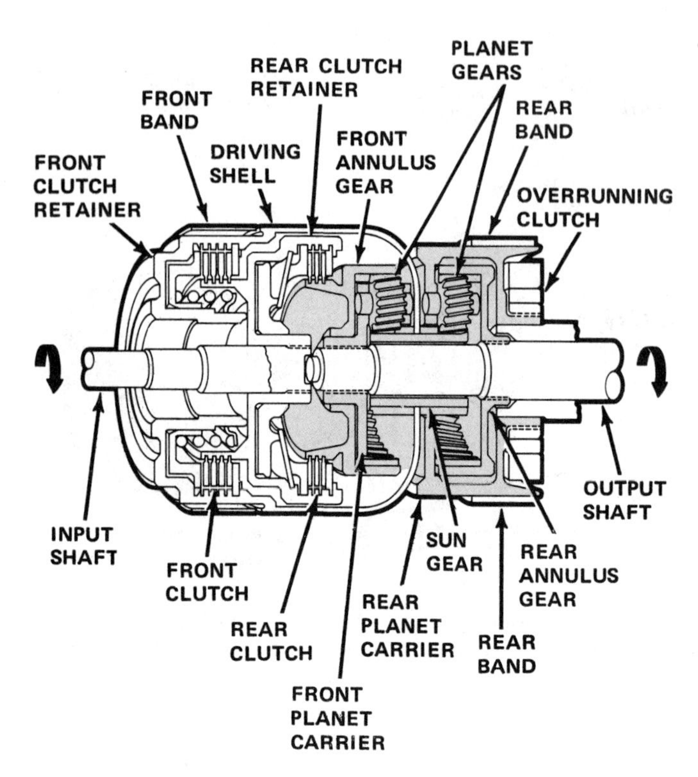

A—Study parts of Simpson gear train. Note gearset is shown here as part of a larger system including clutches and bands.

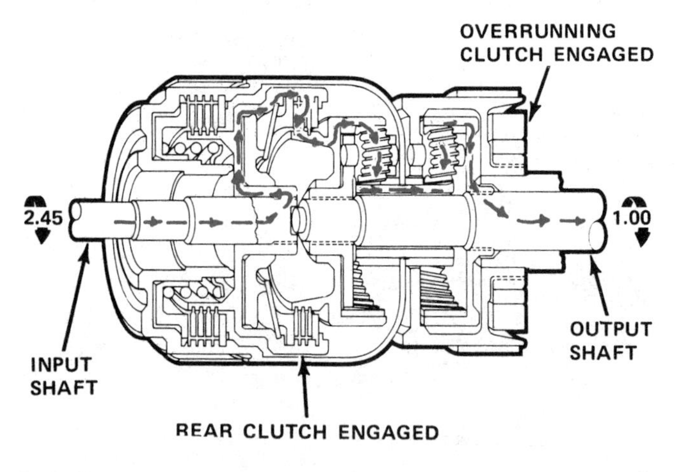

B—In first gear, power flows through front ring gear into planet gears. Planet gears reduce input speed and cause sun gear to rotate opposite of engine direction. Sun gear drives rear planet gears, which further reduce speed. Planet gear rotation is same direction as engine rotation. From planet gears, power enters rear annulus gear, which is splined to output shaft. The double reduction provides a low gear.

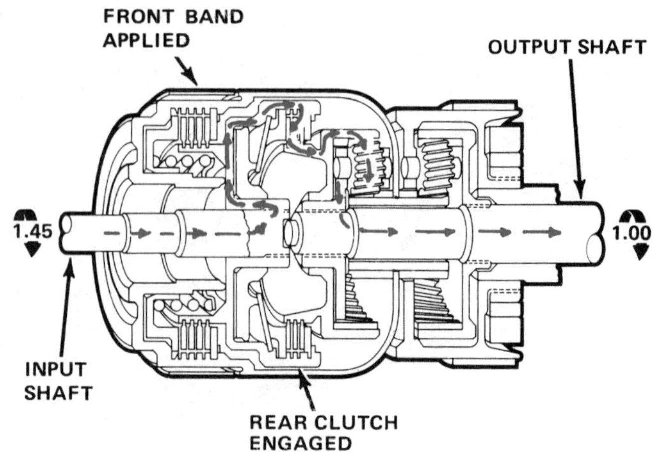

C—In second gear, the sun gear is held stationary. Power from the front annulus gear causes the front planet gears to walk around the sun gear. The planet gears carry the planet carrier and output shaft with them, at a smaller gear reduction ratio than first gear.

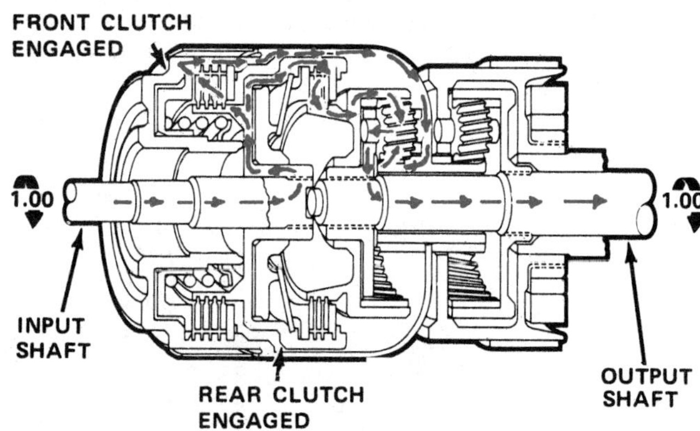

D—In third gear, the sun gear and ring gear are both driven at the same speed, and the gears rotate as a unit. Power is transmitted through the planet gears, to the planet carrier, to the output shaft, splined to the carrier. The rear gearset rotates as a unit with the output shaft but does not transmit torque.

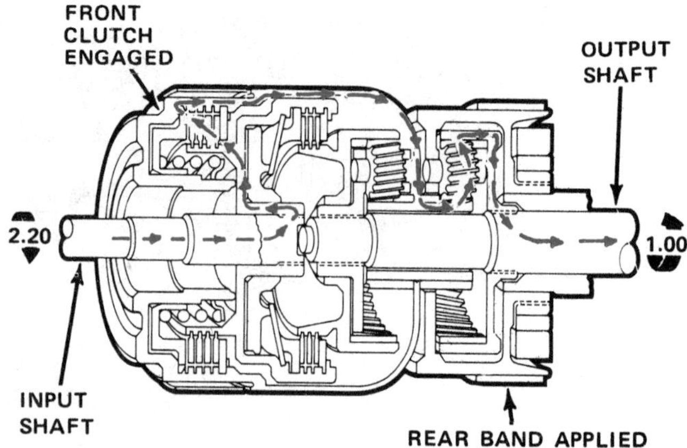

E—In reverse gear, the sun gear is driven by the engine, and the rear planet carrier is held stationary. The rear planet gears are driven by the sun gear and act as reverse idler gears. Power exits the rear annulus gear and output shaft in opposite direction of engine rotation. Front gearset is in idling state and does not transmit power.

Figure 10-20. Power flow through a typical automatic transmission. (Chrysler)

gear reduction through both gearsets is compounded. Since both gearsets are used, a maximum speed reduction and torque multiplication is achieved. Gear ratio in this particular design is 2.45:1, meaning the input shaft must turn 2.45 times to turn the output shaft once.

Power flow–drive range, second gear. Figure 10-20C shows power flow through the Simpson gear train in second gear and selector lever in drive. Power enters the front annulus gear, which drives the front planet gears and planet carrier. The sun gear is held stationary. The front planet carrier, being splined to the output shaft, drives the shaft. Notice that power does not go through the rear gearset. This gives a reduction gear ratio that is not as great as in first gear (1.45:1), but still multiplies power.

Power flow–drive range, direct drive. In third gear, Figure 10-20D, power enters the front annulus gear and sun gear. Both gears are driven at the same speed, locking the planet gears in place. The gearset rotates as a unit. The front planet gears transmit torque to the output shaft through the front carrier, which is splined to the shaft.

The rear annulus gear is splined to the output shaft. It rotates at the same speed as the sun gear. The rear planet gears are locked to the annulus gear and sun gear. The rear gearset rotates as a unit with the output shaft but does not transmit torque. Gear ratio through the compound gearset is 1:1.

Power flow–reverse gear. Power flow in reverse gear is shown in Figure 10-20E. Power enters the sun gear, turning it in direction of engine rotation. The rear carrier is held stationary, and the rear planet gears act as reverse idler gears. Torque is transmitted directly to the rear annulus gear, which is splined to the output shaft. Rotation of the annulus gear and output shaft is opposite engine rotation. The front planetary is in an idling state and does not transmit power.

Planetary Holding Members

Planetary gearsets are controlled by planetary holding members. These are the components of the transmission that apply force to hold or drive other parts. They are what make the gearsets transfer power. The components that accomplish these tasks are *multiple-disc clutches, bands and servos,* and *overrunning clutches.* Figure 10-21 shows a common arrangement of clutches and bands.

Multiple-Disc Clutches

A ***multiple-disc clutch*** is used to hold gearset members in place or to lock, or couple, two elements together for power transfer. The coupled elements may be the input shaft and a planetary member or else two planetary members. Multiple-disc clutches are often called ***clutch packs.*** A typical automatic transmission has two or three clutch packs.

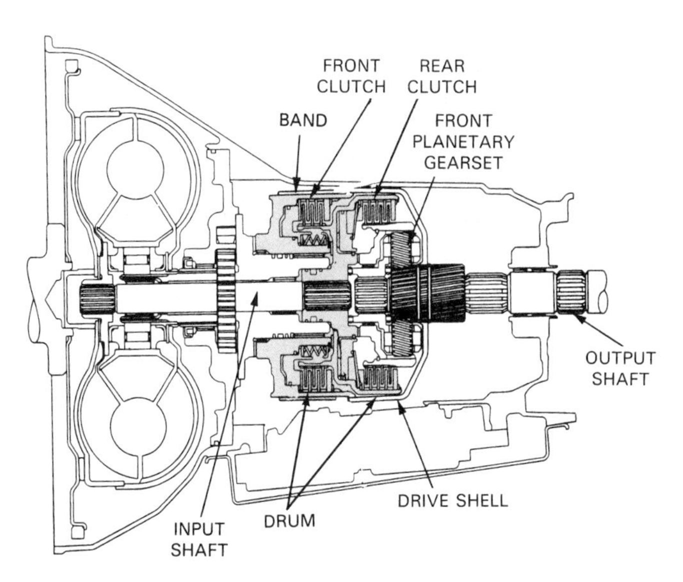

Figure 10-21. This shows positions of multiple-disc clutches and band in a common make of automatic transmission. This arrangement is only one of several used in automatic transmissions. Note position of input shaft, *front clutch, rear clutch,* front planetary gearset, and *drive shell.* Note rear gearset is not shown. (Ford)

As a brief introduction to their operation, the clutch apply piston puts pressure on alternating sets of friction *drive discs* and driven *clutch plates* held within the *clutch drum.* This action locks the clutch elements together and, thereby, transmission elements that are attached to the clutch. Depending on the circumstance, power will be transferred through the clutch, or the clutch will hold a planetary member stationary. A detailed description follows.

Clutch construction

The multiple-disc clutch is made up of several parts. Major components include the clutch drum, drive discs, and clutch plates. Also included are the clutch apply piston, *clutch pressure plate,* and *clutch return spring(s).* In addition, the multiple-disc clutch has an associated *hub* that may or may not be a component of the assembly itself, and there may also be a *drive shell* associated with the clutch pack. Although designs vary, the parts are similar on all makes of transmissions. Refer to Figures 10-22 and 10-23 as you read the next few paragraphs.

Clutch drum. The clutch drum is the housing for all the other clutch components. The outer surface of the clutch drum may be a holding surface for a band (discussed shortly). The inner surface contains large splines, or channels, that mate with teeth on the clutch plates. The clutch drum is also called a *clutch cylinder,* or a *clutch retainer.*

Drive discs. The drive discs, or *friction discs,* are metal plates covered with a friction lining made of a combination of paper and asbestos (or similar material). The lining is similar to the friction material used on a manual clutch, but it is designed to operate in transmission fluid.

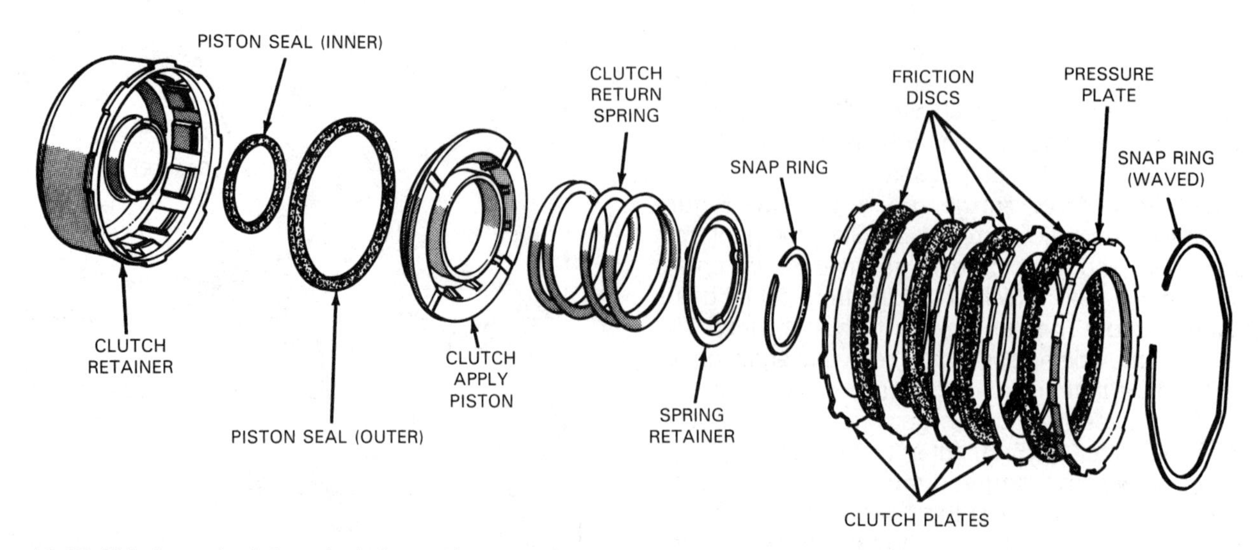

Figure 10-22. This front clutch is typical of a multiple-disc clutch assembly. Hydraulic pressure moves the piston to apply the clutch. The large return spring holds the clutch apply piston against the back of the clutch drum (retainer) when pressure is released. Some clutches use many small coil springs to hold the apply piston. The steel plates have external teeth to engage the drum. The friction discs have internal teeth to engage a splined hub. Note hub for this clutch appears in following illustration. (Chrysler)

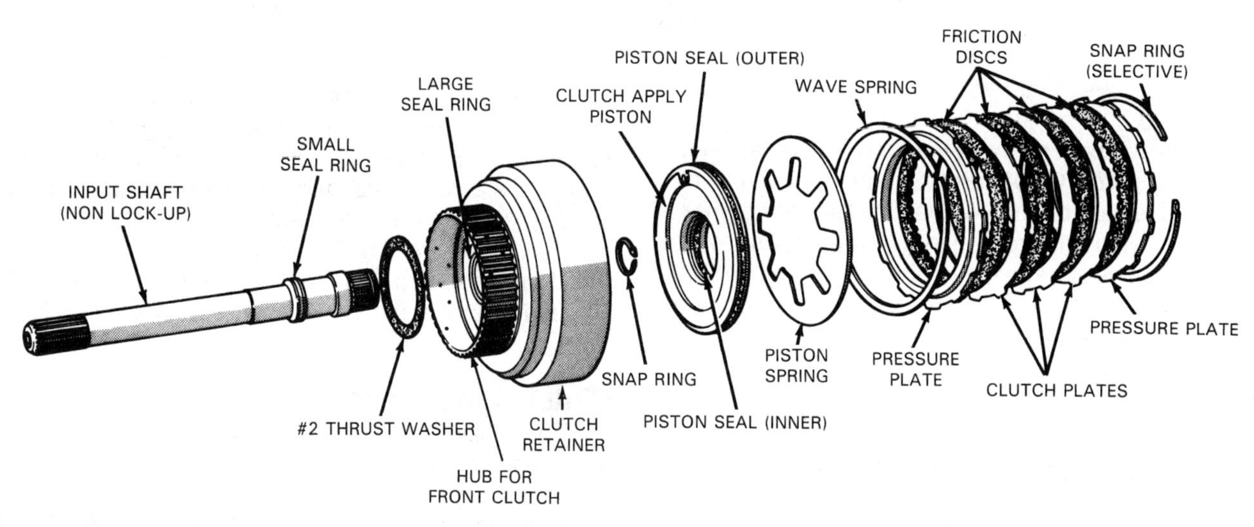

Figure 10-23. This rear clutch is another multiple-disc clutch and a variation of the front clutch. The return spring used to hold the apply piston in place is a large *wave spring,* which takes on a wavy shape when installed. Note rear clutch retainer also serves as hub for front clutch in previous illustration. Hub for rear clutch is served by a member of the front planetary gearset (not shown). (Chrysler)

The discs have internal teeth, or tangs, that engage splines on the clutch hub. Friction discs are sometimes called *composition plates.*

Clutch plates. The driven clutch plates, or *steel plates,* look like the drive discs, but they do not have the friction lining. Also, they have external, rather than internal, teeth. These teeth lock into channels on the inside of the clutch drum. The friction discs and clutch plates form a sandwich of alternating layers. Their purpose is to lock together the drum and hub, and parts connected to each (input shaft or planetary gearset members), when the clutch is engaged.

Clutch apply piston. The clutch apply piston is an aluminum disc that fits into a *piston bore* in the lower portion of the clutch drum. The piston is moved by hydraulic pressure. Movement of the apply piston clamps the friction discs and steel plates together. *Outer* and *inner piston seals* fit around the respective diameters of the piston to prevent fluid loss and consequent pressure loss when the clutch is applied.

Clutch pressure plate. The clutch pressure plate serves as a stop for the set of friction discs and clutch plates when the piston is applied. The pressure plate is held to the drum by a large snap ring.

Clutch return springs. The clutch return spring(s), or *release spring(s),* are used to push the apply piston away from the friction discs and clutch plates after hydraulic pressure is released from the piston. Return springs may be one of several different designs.

Clutch hub. The clutch hub may be comprised of a part within the clutch pack, or it may be part of another component, such as the planetary ring gear. The hub fits into the inside diameter of the set of friction discs and clutch plates. External splines on the hub engage internal teeth of the friction discs. The clutch hub may also be splined to the transmission input shaft or to a part of the gearset.

Drive shell. The ***drive shell,*** or ***input shell,*** is commonly used to transfer power to the planetary sun gear. It is a bell-shaped part made of sheet metal, Figure 10-24. Tangs on the shell fit into notches on the drum of the front clutch, or *direct clutch,* as some refer to it. The shell surrounds the neighboring rear clutch, sometimes called the *forward clutch,* and the front planetary gearset. The shell is splined to the sun gear, and the clutch drum, shell, and sun gear turn together.

In a typical design, the shell drives the sun gear when the front clutch is engaged. When disengaged, the shell and front clutch drum are turned by the sun gear, but they do not transfer power. Also, a band may be tightened around the drum to hold the sun gear stationary. The drive shell is sometimes called the ***sun-gear shell.***

Clutch operation

During clutch engagement, pressurized fluid enters the area behind the clutch apply piston. The fluid forces the piston into the clutch plates and friction discs. The plates and discs become locked together, forming a unit. Hub drives discs, discs drive plates, and plates drive drum; the clutch rotates as a unit to transfer power through the assembly. A simplified illustration is shown in Figure 10-25.

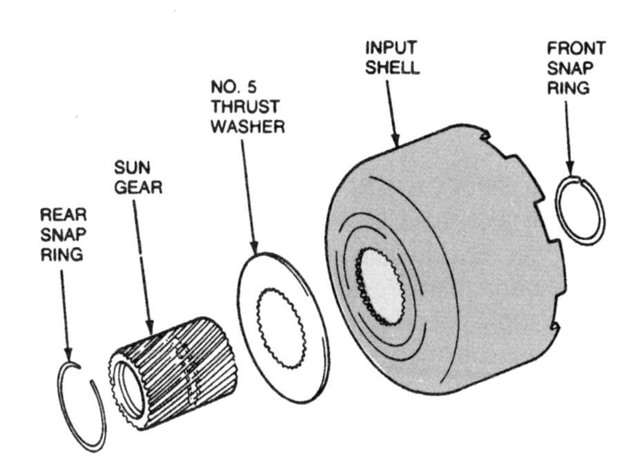

Figure 10-24. The sun-gear shell, or input shell, is used to transfer power from the front clutch drum to the sun gear. It is used on all transmissions using the Simpson gear train. (Ford)

When the hydraulic fluid is exhausted from behind the piston, the return spring pushes the piston away from the steel plates and friction discs. The plates and discs are no longer locked together, and power is no longer transferred through the clutch. The friction discs and steel plates can turn independently. The hub drives the discs, but no power is transferred to the plates and drum. See Figure 10-26.

Bands and Servos

Automatic transmission ***bands*** are flexible-metal friction devices designed for holding planetary gearset members stationary. They are wrapped around clutch drums and can be tightened to stop drum rotation. This action causes desired members of the planetary gearset to be held stationary. Sometimes bands are wrapped directly around a planetary gearset member. For example, the ring gear sometimes has a smooth outer surface for accepting a band. Modern transmissions have one or two bands. (Older transmissions had as many as three.) Bands are applied and released by hydraulic devices called ***servos,*** which essentially consist of a piston and cylinder.

Note that clutches can be attached to the case to hold a planetary gear stationary, replacing a band used for the same purpose. Such a clutch does not connect two rotating members, but one rotating and one stationary. The steel plates are splined to the case, and when engaged, the clutch prevents any movement of the unit it is connected to.

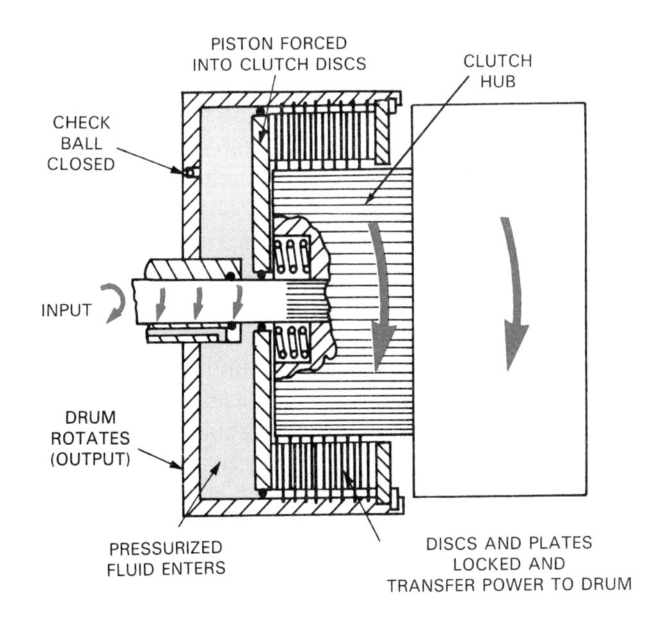

Figure 10-25. Study a multiple-disc clutch being applied. Pressurized fluid enters the space behind the apply piston, pushing it into the clutch plates and friction discs. Input shaft drives the hub. When the plates and discs are jammed together, power flows from the center hub to the drum.

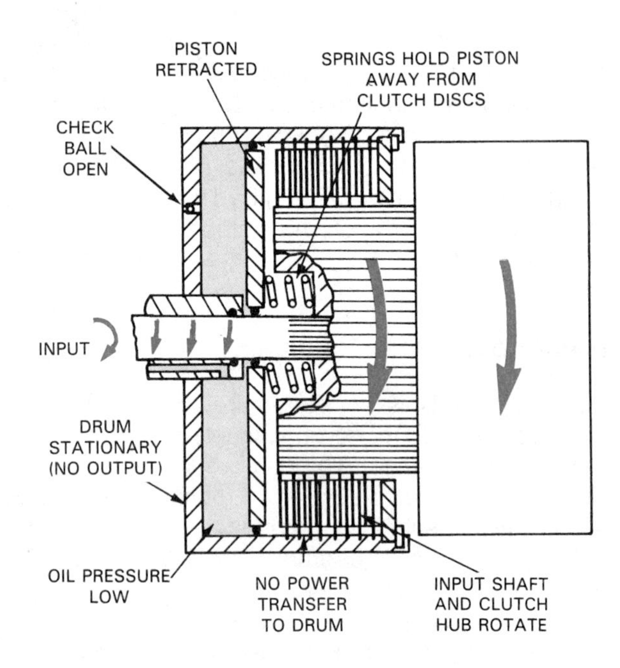

Figure 10-26. In this illustration, the clutch pack is released. When hydraulic fluid is exhausted, the return spring pushes the piston away from the clutch plates and friction discs. Since the plates and discs are no longer engaged together, power does not flow to the drum. Note check ball is open, allowing fluid to escape for quick release of pressure. This is to prevent dragging of plates and discs on release of clutch.

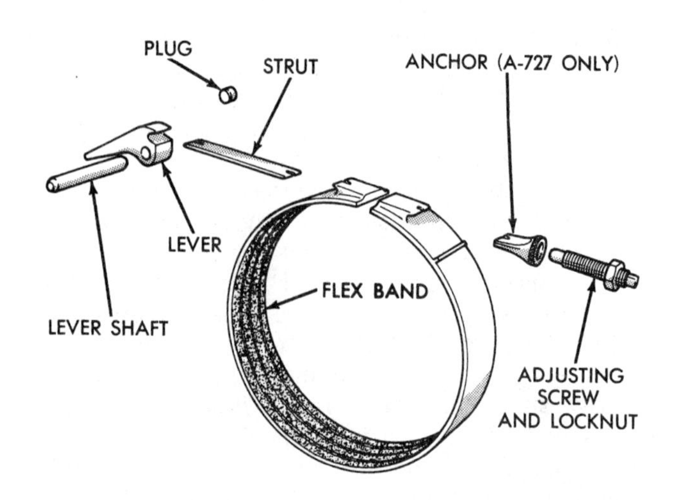

Figure 10-27. Parts of a kickdown band are shown. Note this band is positioned around the front clutch. (Chrysler)

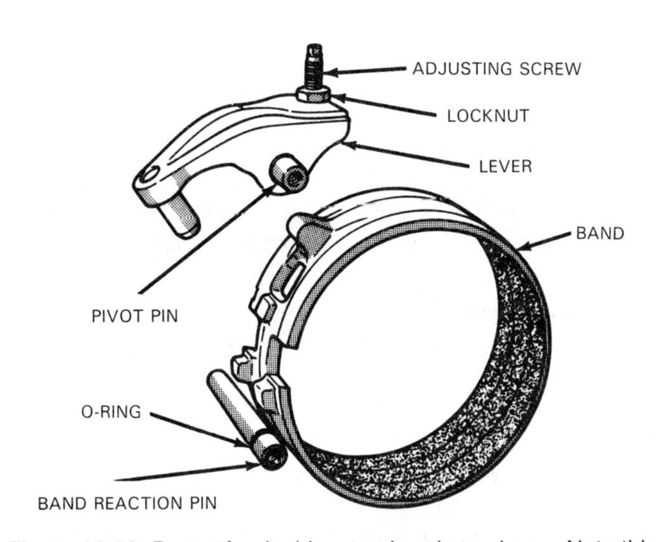

Figure 10-28. Parts of a double-wrap band are shown. Note this type of band gives more uniform clamping action and has greater holding power. It is used only where necessary. (Chrysler)

Band and servo construction

A band is made of flexible-steel strap with a friction lining on its inner surface. The friction material is a combination of paper and asbestos (or similar material). The material used is able to operate in transmission fluid and withstand high temperatures. Bands on modern transmissions attach to the band servo on one end and to the transmission case on the other. The ends of the band are usually called *band ears.* Figure 10-27 shows the *kickdown band* and linkage. Figure 10-28 shows a *double-wrap band* and linkage.

The *servo apply piston* is a metal plunger that operates in a cylinder. The cylinder is machined in the transmission case, or it is a separate assembly mounted inside of the oil pan. Rubber or teflon seals on the outside diameter of the piston prevent fluid leakage and consequent pressure loss. The apply side of the piston usually contains a large return spring. The servo *apply piston pin* actuates the *band lever* attached to the band ear. The servo piston and spring assembly is held in the cylinder by a large snap ring or a bolted cover. Sectional and exploded views of a servo are shown in Figure 10-29.

The band ear across from the servo is attached to a band *anchor pin,* which holds it stationary. The anchor pin may contain a *band adjustment screw.* This screw provides a means of adjusting the *band-to-drum clearance* by moving the band closer to the drum as the friction material wears. See Figure 10-30.

Band and servo operation

Transmission fluid is sent into the apply side of the servo cylinder. Hydraulic pressure moves the apply piston toward the band, compressing the return spring and pushing on the apply piston pin and band ear. Since the other side of the band is anchored to the case, the band tightens around the drum (or planetary member). The friction material locks the band to the drum and stops the drum from turning. Stopping the drum keeps one of the planetary components from revolving. When the oil pressure to the servo is relieved, the return spring pushes the apply piston and piston pin away from the band. The band then releases

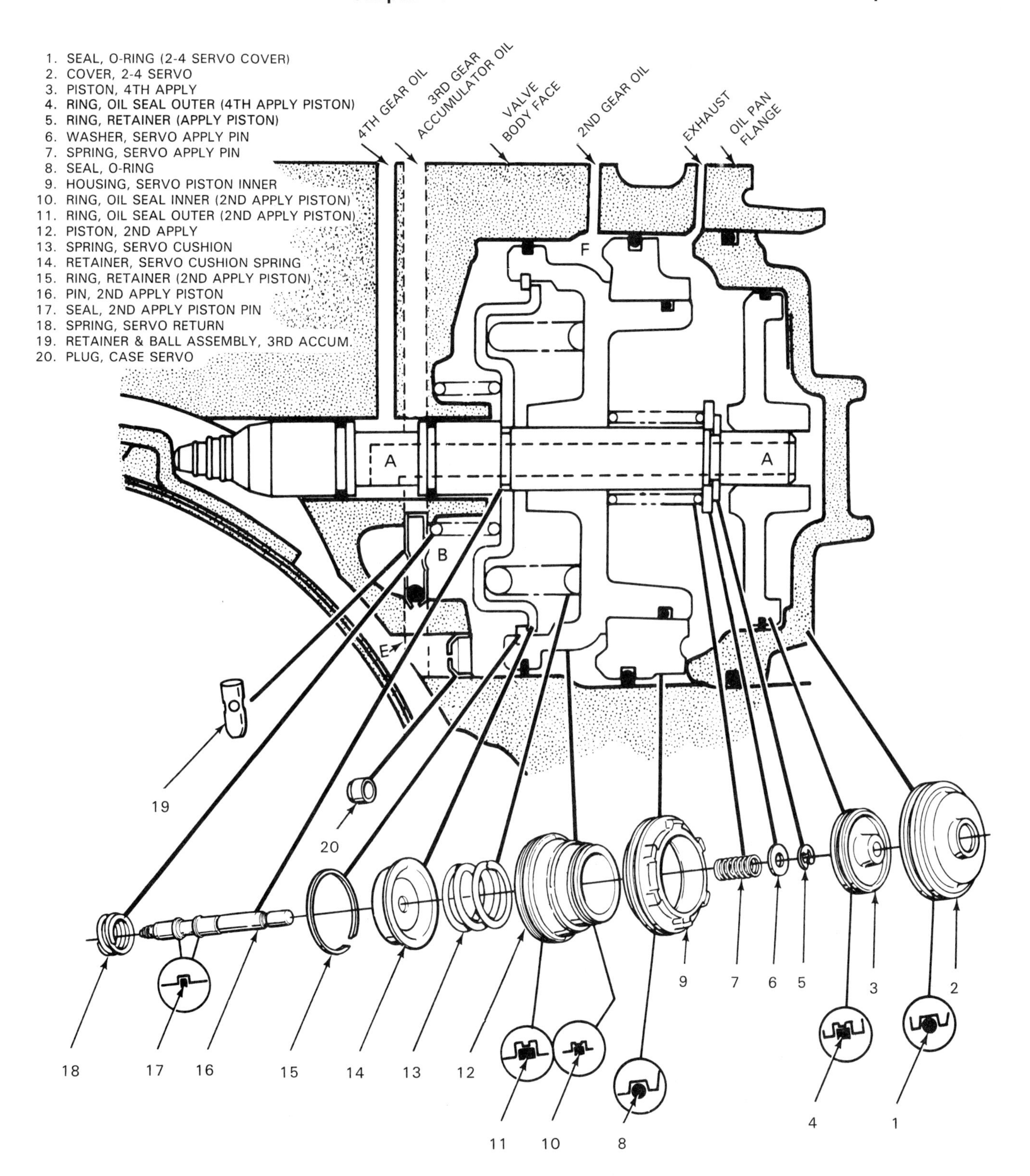

Figure 10-29. Note typical band servo shown in sectional and exploded views. This servo contains extra pistons that act as an accumulator to cushion application of the band. (General Motors)

the drum and/or planetary gearset member. Some servos have an additional *accumulator piston* that is applied before the servo apply piston. The accumulator piston gives an initial application that is fast and soft. It is followed immediately by application of the larger apply piston, which provides greater pressure for firm engagement. The apply piston works against a spring, which serves to delay its application.

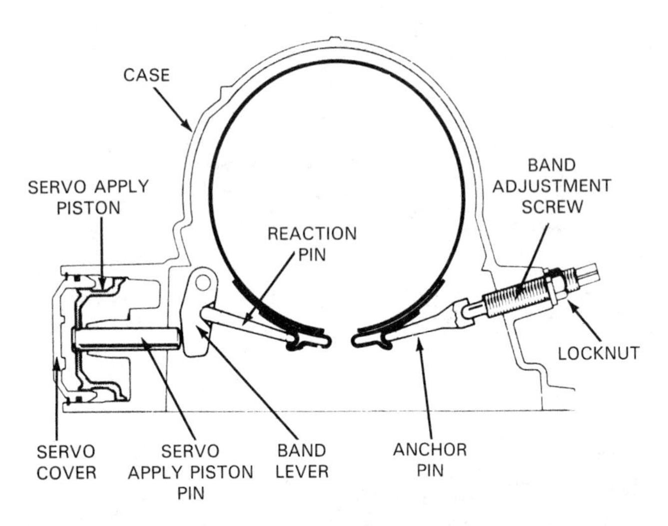

Figure 10-30. This servo is a simple design with one piston. The band, linkage, and adjustment screw are also shown. (Ford)

Overrunning Clutches

Overrunning clutches are mechanical holding members. They can be used to hold a planetary gearset member to improve shift quality and timing. The devices are also used with torque converter stators to improve efficiency, as discussed earlier in the chapter.

There are a couple of different types of overrunning clutches. A typical *roller clutch* consists of inner and outer races, rollers, and springs. The clutch performs its holding function when the rollers become jammed between the inner and outer races. Another type of clutch is known as a *sprag clutch.* This type uses small metal pieces known as *sprags,* instead of rollers, to perform the holding action.

The overrunning clutch works by preventing backward rotation of certain planetary gearset members during shifting. The action of a typical overrunning clutch is illustrated in Figure 10-31. The outer race is usually held to the transmission case, and the inner race is connected to a planetary member. When turned in one direction, the clutch rotates freely, allowing the attached member to rotate freely, too. When turned in the other direction, the clutch locks up and acts as a holding member.

Typically, the one-way clutch is only locked when in drive low gear and manual low gear. Note, however, that the clutch can *overrun* (freewheel) in *drive* low, permitting one planetary member to turn faster than another. This happens when the vehicle goes faster than the engine, as

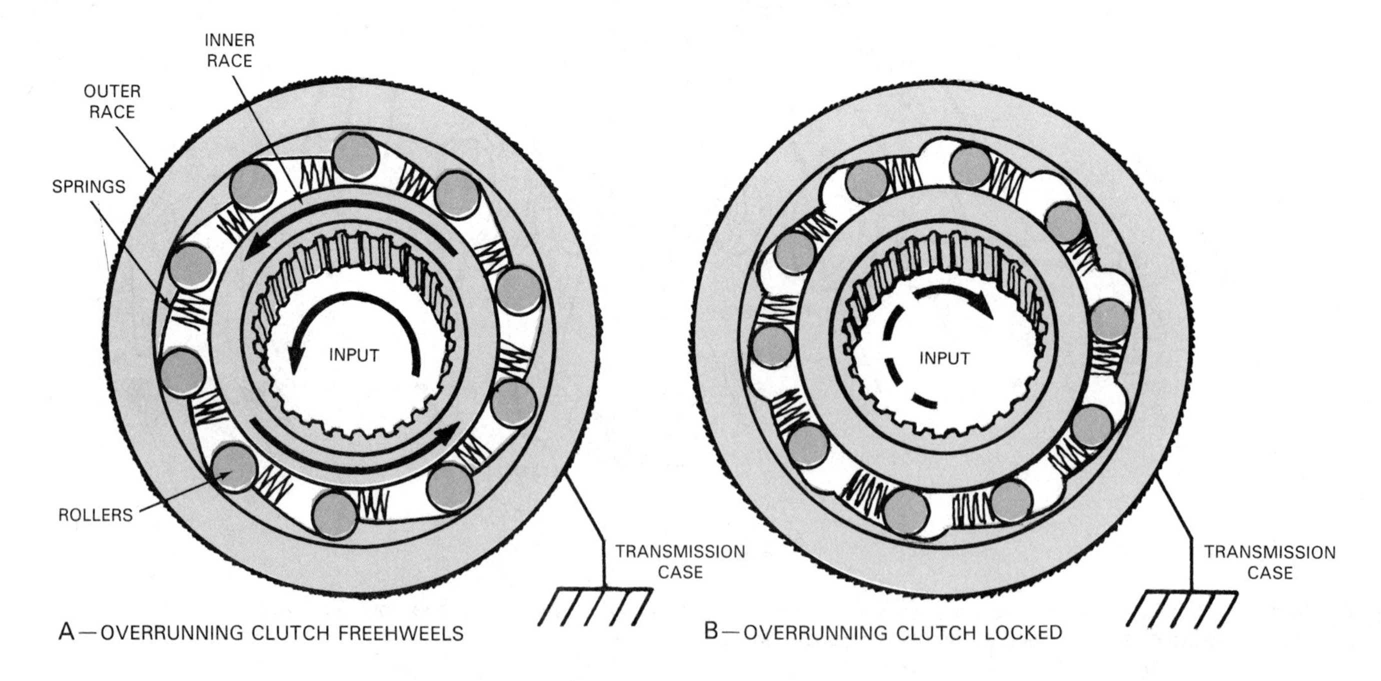

Figure 10-31. The overrunning, or one-way, clutch shown here locks or unlocks according to the direction it is turned. A–When planetary gearset member (input) tries to turn counterclockwise, inner race and attached planetary member turn freely. B–When planetary member tries to turn clockwise, rollers become jammed between inner and outer races. Inner race and attached planetary member are prevented from turning, as outer race is attached to transmission case.

when coasting downhill. In contrast, selecting *manual* low will prevent this overrunning by hydraulically applying an extra holding member. As a result, the engine can be used to slow down the vehicle when going downhill.

In other words, manual low (and manual second) provides ***engine braking.*** Note that this term means using the engine (via engine resistance) to slow down the vehicle, rather than using the brakes. Engine braking does not happen on a coastdown in drive low because the one-way clutch freewheels; the car wheels are, in effect, disconnected from the engine.

Accumulators

While not itself a planetary holding member, an ***accumulator*** is a hydraulic device used in the apply circuit of a band or clutch to cushion initial application. The result is a smoother shift. The accumulator essentially consists of a spring-loaded piston in a cylinder. The piston is pushed to one end of the cylinder by the spring. A hydraulic passageway connects the accumulator to the apply piston of a multiple-disc clutch or band servo.

As the clutch or servo is being applied, fluid in the passageway is received by the accumulator. The accumulator piston begins to move against spring pressure. The expanding cylinder chamber diverts some of the hydraulic fluid from the clutch or servo. As a result, pressure to the clutch or servo does not build up to *line pressure* immediately, and the initial application of the holding members will be soft. After the accumulator spring is fully compressed, full pressure goes to the apply piston, and the clutch or band will be held tightly. Figure 10-32 shows a typical accumulator installed in an *auxiliary valve body.*

Transmission Oil Pump

The transmission oil pump, sometimes called the ***front pump,*** has several basic functions. It generates *system pressure* for the entire hydraulic control system. It keeps the moving parts of the transmission lubricated. It keeps the torque converter filled with transmission fluid for proper operation. (Most of the fluid in an automatic transmission is inside of the converter.) Finally, it circulates transmission fluid to and from the oil cooler for heat transfer.

The transmission oil pump, a positive displacement pump, is driven by the engine through lugs on the back of the torque converter. The relative position of the oil pump in the transmission is shown in Figure 10-33. There are several kinds of transmission oil pumps, all of which consist of elements–gears, rotors, or vanes–that rotate inside of a housing, Figure 10-34. (The different kinds of hydraulic pumps used in automatic transmissions were discussed in Chapter 5.)

The principle of operation is rather simple. As the elements rotate, on one side of the housing they come together, and on the other side they move apart. As they move apart, a low-pressure region is created on the inlet side of the pump. As a result, transmission fluid is drawn from the *sump,* or oil pan, through the transmission filter and pump inlet, into the pumping chamber. The fluid is carried around the chamber to the pump outlet. The fluid

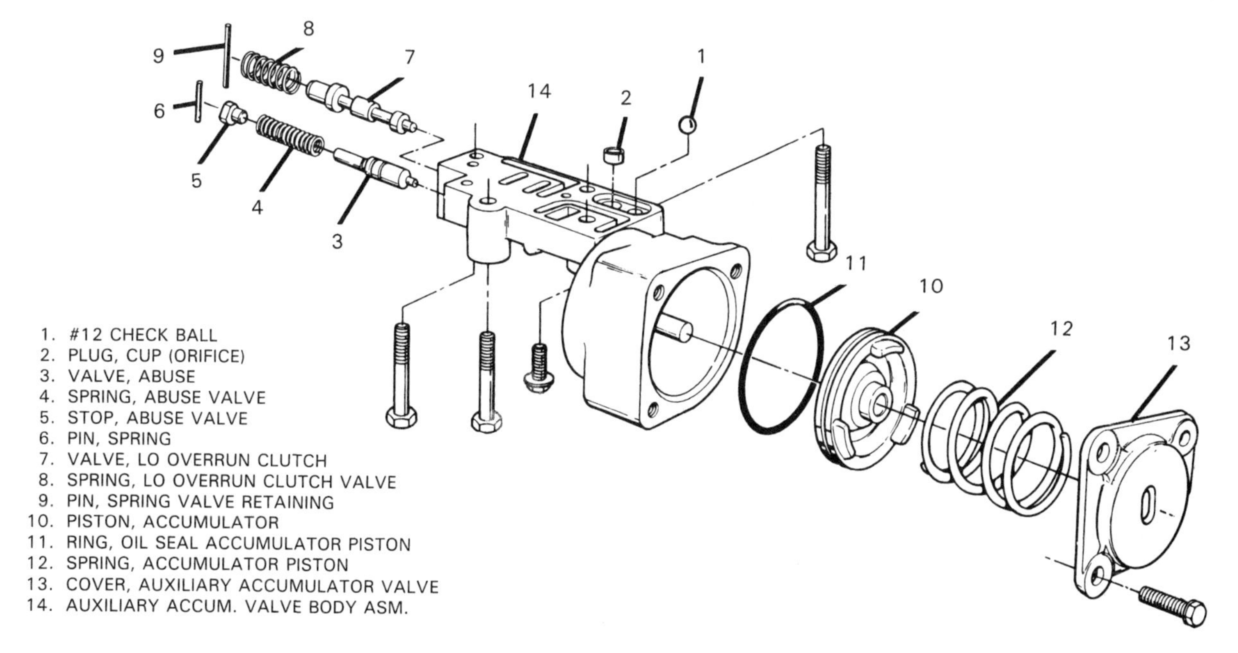

Figure 10-32. Accumulators are used to cushion application of multiple-disc clutches and bands. The accumulator shown here is contained in a separate, or auxiliary, valve body. (General Motors)

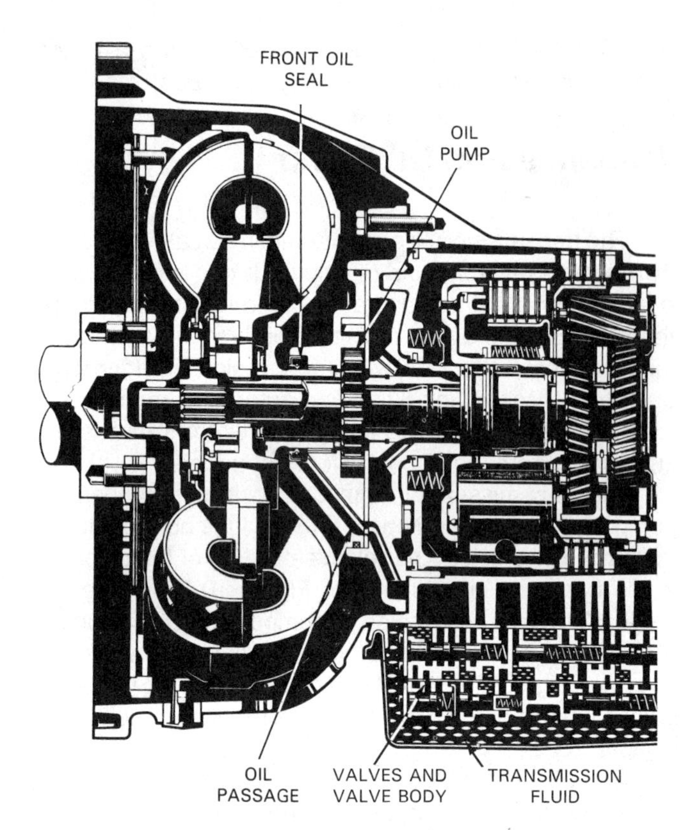

Figure 10-33. The oil pump is often located at the front of the transmission case. It is driven by lugs on the rear of the torque converter. Bolts hold the pump to the case. Some older transmissions also have a rear oil pump, driven by the output shaft. (Mercedes-Benz)

is discharged into the hydraulic system through the outlet as the internal elements come back together.

Note that modern transmissions use only one pump, driven from the engine. However, many older transmissions used two oil pumps. The front oil pump was driven by the engine through the converter. The *rear pump* was driven by the transmission output shaft. The rear pump turned whenever the vehicle was moving. It developed pressure for push starting and for providing lubrication to the transmission if the vehicle was being towed.

Transmission Oil Filter

Transmission oil filters, or ***oil strainers,*** as they are also called, are used to remove particles from the automatic transmission fluid. Fluid being drawn into the oil pump must first pass through the filter. This keeps fluid contaminants from entering the pump or the other hydraulic components. Most filters are made of a disposable filter paper or felt. Some filters are fine mesh screens, which may be cleaned and reused.

The filter is located inside the oil pan. It must be positioned where it will always be submerged in the automatic transmission fluid. Usually, the filter is attached to the bottom of the valve body. It may be held in place by screws, clips, or bolts. Some filters have a built-in pickup tube that goes directly up to the oil pump.

Transmission Oil Cooler

The action of the transmission–the torque converter, in particular–heats the transmission fluid. In addition, the close coupling with the engine serves to add heat to the fluid. This heat must be removed so that the fluid and the transmission do not become overheated. Fluid subjected to overheating will quickly lose its lubricating ability. Further, it will break down into sludge and varnish, which will plug passageways and destroy the transmission.

In addition, excessive heat destroys the nonmetallic materials used in clutch and band friction linings and in seals. As a result, transmission fluid that is overheated can damage these materials or can fail to cool the friction materials properly. Then, not only do the materials become destroyed, but the transmission fluid ends up with particulates that can further act to plug transmission passageways and destroy the transmission.

In general, fluid temperature should be kept below 275°F (135°C). Running at elevated temperatures greatly reduces the life of the transmission fluid. If the transmission is being worked hard, such as in trailer towing, temperatures could exceed 300°F (150°C). Under such conditions, the fluid and filter *must* be changed frequently.

Main Oil Cooler

Fluid in older transmissions was sometimes air-cooled by placing fins on the torque converter. The converter housing was open, so that air could pass through and absorb heat directly from the torque converter. Use of this design has been abandoned.

Fluid in modern automatic transmissions is cooled in the oil cooler. This component consists of a heat exchanger built into a side or bottom tank in the engine cooling-system radiator. Cooler lines connect the transmission to the oil cooler, as shown in Figure 10-35.

The oil cooler is immersed in cooled engine coolant. In operation, hot transmission fluid is pumped from the transmission to the oil cooler. As the fluid passes through the cooler, its heat is transferred to the engine coolant. The cooled transmission fluid then returns to the transmission.

Auxiliary Oil Cooler

When a vehicle is used for trailer towing or other extreme operating condition, auxiliary, or external, coolers may be installed. They consist of direct air-cooled heat exchangers that connect to the existing fluid cooling system. They may be installed to completely replace the oil

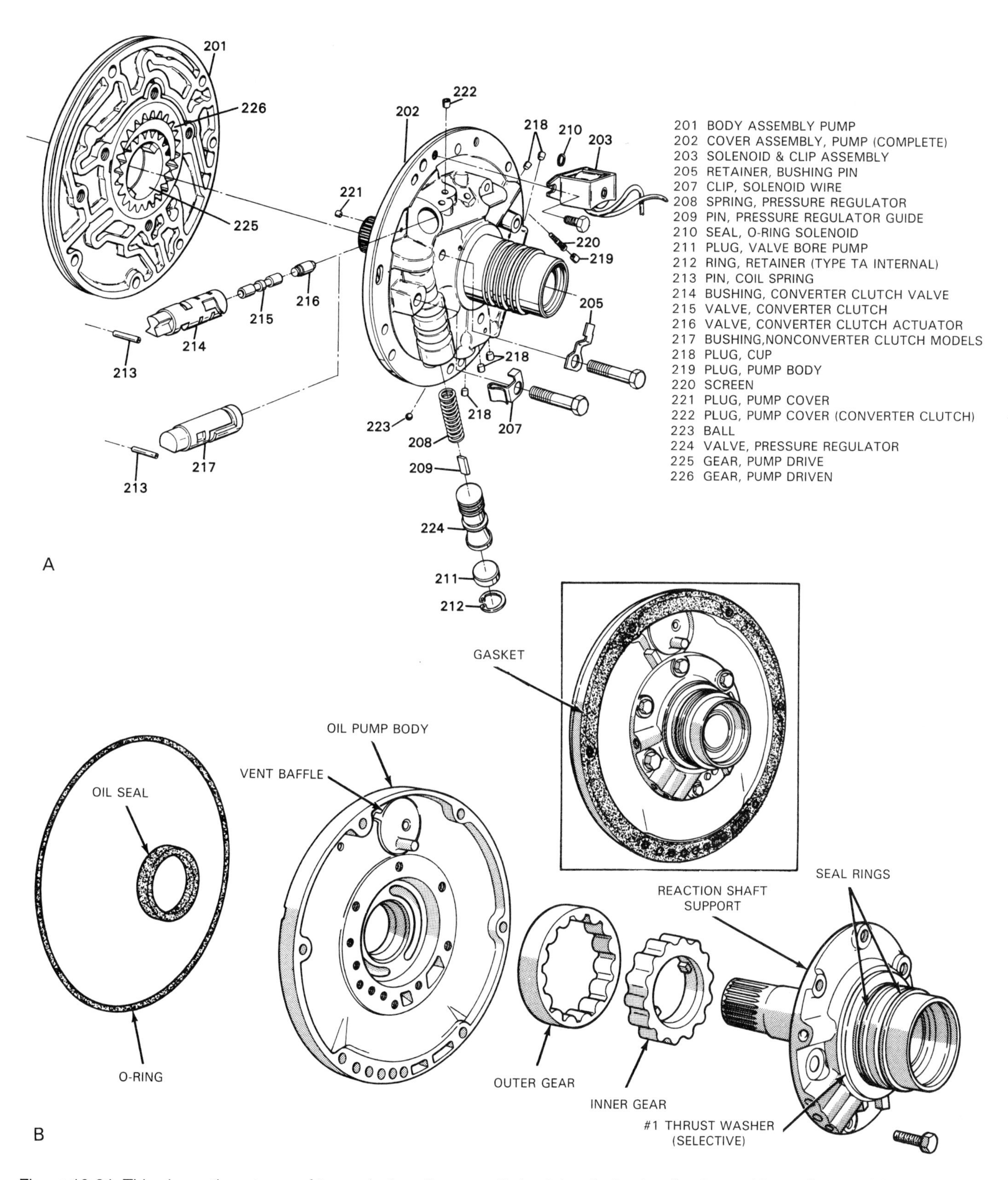

Figure 10-34. This shows three types of transmission oil pumps. Note stator shaft extending forward from oil pump. It is the support for torque converter stator. Shaft extending back from oil pump is the support and fluid feed for some transmission clutches. A–Gear-type oil pump. Pump housing also contains built-in converter clutch and pressure regulator valves. Pressure regulator valve lowers pump pressure by decreasing pump capacity. B–Rotor-type oil pump. Pump cover contains stator and clutch support shafts. (General Motors, Chrysler)

(continued)

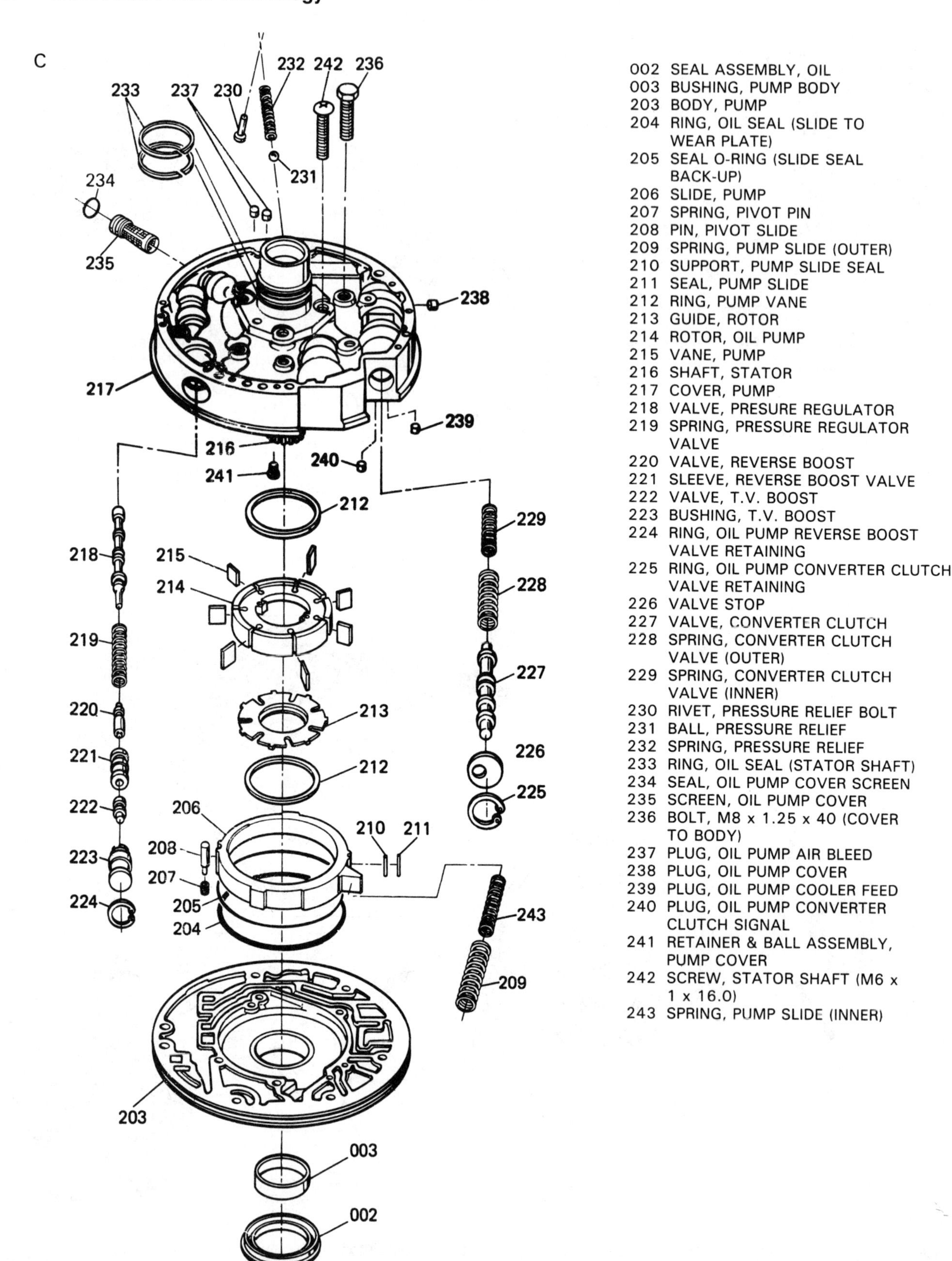

Figure 10-34 *(continued)*. C–Vane-type oil pump. This design also includes built-in converter clutch and pressure regulator valves. (General Motors, Chrysler)

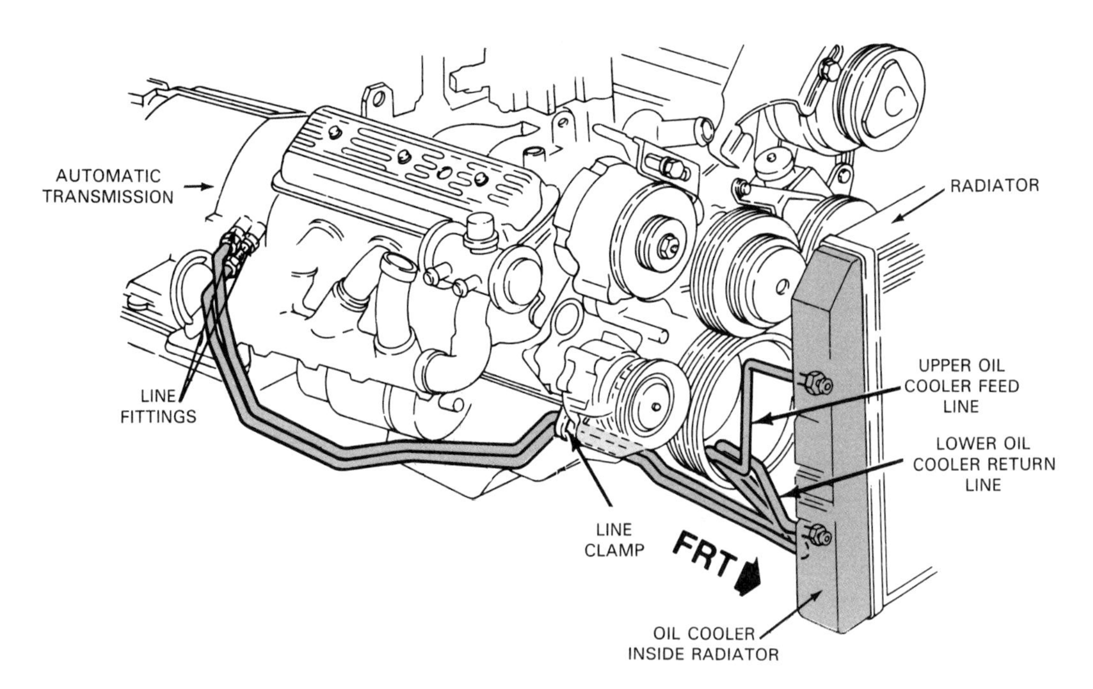

Figure 10-35. Steel tubing connects oil cooler in radiator to transmission hydraulic system. Oil is pumped from torque converter outlet to cooler. Cooled oil is directed through transmission to components requiring continuous lubrication and cooling, such as rear transmission bushings. Excess runs to oil pan, where it is drawn in by the oil pump and distributed elsewhere through the system. (General Motors)

cooler in the radiator, or they may be helper units that further cool fluid that has passed through the radiator oil cooler. They are usually mounted ahead of the radiator.

Valve Body

The valve body is a complex casting made of cast iron or aluminum. It contains many internal passageways and components, including hydraulic valves. The valve body serves as the control center for the transmission's hydraulic system. It is precisely machined, and valve body components are manufactured to exact tolerances. Some are as small as 1 ten-thousandth of an inch (0.0001"), or 25 ten-thousandths of a millimeter (0.0025 mm).

Oil passageways are cast into the valve body during the casting process. After casting, holes are drilled for valves, and one side of the valve body is machined flat. This flat area is intended to closely fit a similar machined area on the transmission case and, sometimes, other transmission parts as well. The close fit will prevent oil leaks that could lead to faulty transmission operation.

The valve body shown in Figure 10-36 is an example of a *one-piece valve body.* This component would bolt directly to the bottom of the transmission case. Other valve bodies are made in two sections. Such *split valve bodies* must be assembled before being installed in the transmission. Auxiliary valve bodies are often bolted to the *main* valve body or to the transmission case. Auxiliary valve bodies are separated from the main valve body because of design considerations, such as clearance restrictions.

Valve bodies contain *spacer plates.* These are sometimes referred to as *separator plates,* or *transfer plates.* (Refer to Figure 10-36.) These are steel plates used to seal off passageways in the valve body and in the transmission case. Spacer plates are also used between the halves of split valve bodies. Holes are drilled in the spacer plate to connect certain passageways of mating components.

The spacer plate holes are sized to give a desired flow rate. In other words, the size of the hole establishes how quickly oil can flow through the passageway. This is to obtain a desired type of shift. Larger holes give quick, hard shifts. Smaller holes restrict oil flow to lengthen shift time and cushion shifts. Check balls are often used to block spacer plate holes, allowing oil to flow in only one direction. Spacer plates are usually installed with gaskets on both sides.

Hydraulic Valves

Hydraulic valves, presented previously in Chapter 5, are the moving parts of the automatic transmission that act as pressure and flow controls, directing fluid to other parts of the transmission. Some hydraulic valves are actuated manually; others are actuated automatically. For instance, some hydraulic valves are operated by pressure inputs

1. OIL STRAINER
2. GASKET
3. VALVE BODY
4. CHECK BALL
5. GASKET
6. TRANSFER PLATE
7. GASKET
9. ACCUMULATOR PISTON
10. SPRING
11. OIL RING
12. ACCUMULATOR
13. RETAINING RING
14. SPRING
15. SERVO APPLY ROD
16. SERVO ADJUSTING SLEEVE
17. SPRING RETAINER
18. SPRING
19. OIL RING
20. SERVO PISTON
21. RETAINING RING
22. ADJUSTING BOLT
23. LOCKNUT
24. RETAINING RING
25. GASKET
28. SERVO COVER
29. REINFORCEMENT PLATE

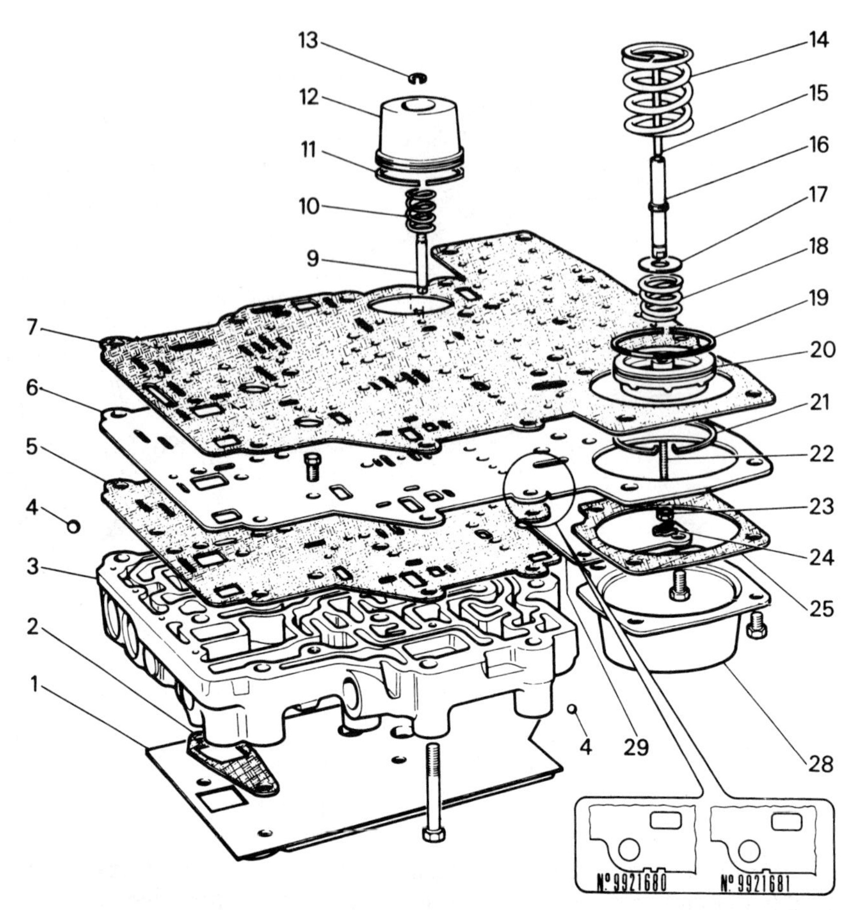

Figure 10-36. Valve bodies contain many of the hydraulic valves of an automatic transmission. The valve body shown here is a one-piece design. Note position of spacer plate and gaskets in the assembly. Also note accumulator and band servo incorporated into this valve body and position of oil filter (strainer). (Fiat)

from other hydraulic valves, which themselves are operated by outside factors, such as vehicle speed. Some valves are located in the valve body, Figure 10-37, while others lie elsewhere in the transmission.

Valves located in the valve body are installed in drilled holes called *valve body bores.* The clearance between the valves and bores is very small, usually less than 2 thousandths of an inch (0.002"), or 50 thousandths of a millimeter (0.050 mm). Valves are held in the valve body bores by snap rings, retaining pins, or retainer plates. Retainer plates are attached to the end of the valve body bores by machine screws or small bolts. In addition, springs, made of heat-resistant steel, are used in the valve body to hold valves in position when they are not in operation.

Springs have other functions as related to hydraulic valves. In some cases, the springs are used to provide constant resistance that hydraulic pressures work against, such as in a pressure regulator valve. In other cases, the springs are used for an initial loading of a valve, to which is added a varying hydraulic *control pressure.* Such valves produce a varying output pressure. In either case, valve springs are carefully sized to achieve a specified pressure. Specific springs are used with each valve, and they cannot be interchanged. For this reason, care should be taken when working on the valve body that springs do not get mixed up.

Types of Valves

There are many types of hydraulic valves used in the transmission. These include *pressure regulator, manual, shift, throttle, governor, detent,* and *check valves.* A working knowledge of these valves will help you understand the internal operation of different transmissions. Following is a brief discussion on each type. (The different valves are explained here on an individual basis. The operation of these valves as integrated into the complete hydraulic system is presented in Chapter 11.)

Pressure regulator valve

The pressure regulator valve controls the pressure of the overall transmission hydraulic system, or system ***line pressure.*** This valve is sometimes located on the pump or

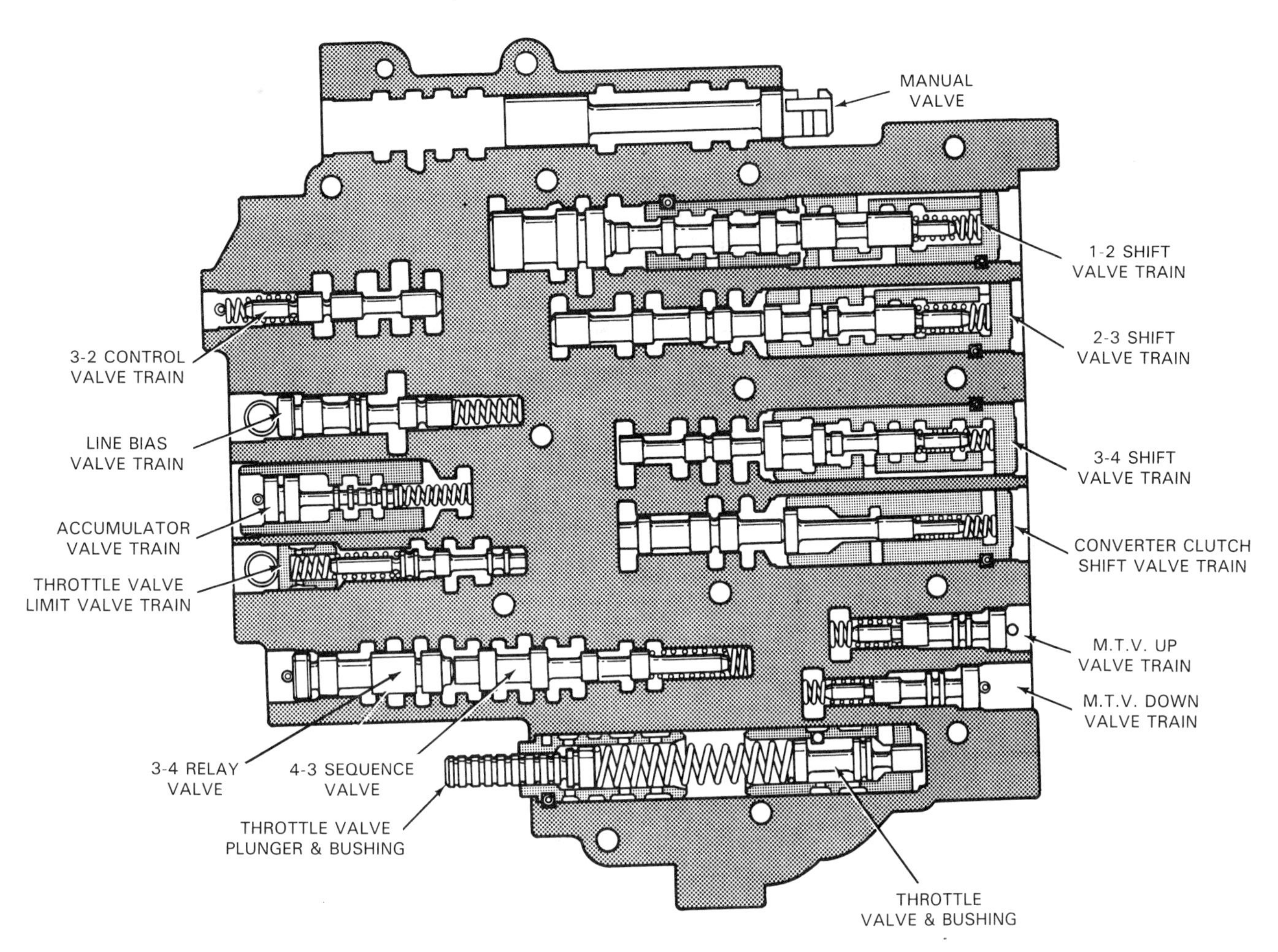

Figure 10-37. Note hydraulic valves in this sectional view of a typical valve body. Notice the difference in valve size and configuration and in placement of the springs. (General Motors)

in a separate valve body, but it is most often found in the main valve body. A few transmissions have a separate regulator valve to control pressure in the torque converter. A typical pressure regulator valve is shown in Figure 10-38.

In operation, pressurized fluid at the oil pump discharge is directed to the pressure regulator valve before going to the rest of the transmission hydraulic system. A heavy spring holds the valve closed against oil pressure from the pump. At idle, the regulator valve is usually closed.

When the engine begins running faster, it turns the oil pump at a faster rate. Fluid pressure rises. When hydraulic pressure exceeds some predetermined value, it forces the valve to move against spring pressure. This opens a passageway back to the oil pan or pump intake. Hydraulic fluid is diverted back through this passageway, and pressure falls back within the limit. This action keeps line pressure constant.

Note that if the pressure were allowed to exceed the set maximum, it would cause hydraulic system malfunctions and damage to parts. Also note that the pressure limit is established by the spring; thus, the stiffer the spring is, the higher the line pressure.

Manual valve

A ***manual valve,*** located in the valve body, is operated by the driver through the shift linkage. It allows the driver to select *park, neutral, reverse,* or different drive ranges. When the shift selector lever is moved, the shift linkage moves the manual valve. As a result, the valve routes hydraulic fluid to the correct components in the transmission, including other hydraulic valves, clutches, and band servos.

Note that when overdrive, drive, or second is selected, the transmission takes over, shifting automatically to meet driving conditions. In low and reverse, the transmission is locked into the selected gear.

Shift valves

Shift valves, located in the valve body, control transmission upshifts and downshifts. Hydraulic pressures

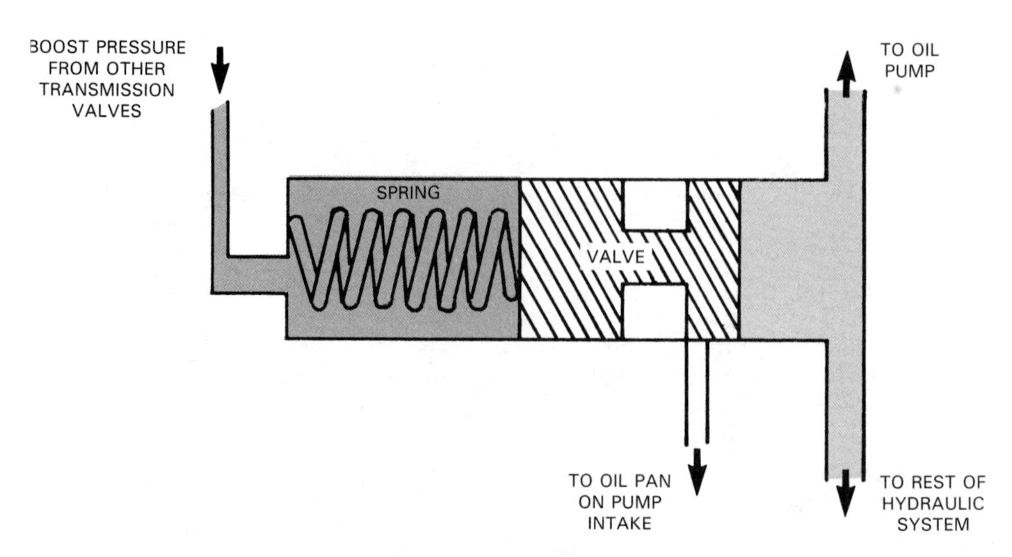

Figure 10-38. The main pressure regulator valve controls hydraulic system pressure by exhausting fluid to the pan or pump intake. Basic pressure settings are determined by a spring that holds the regulator valve closed until a certain pressure is reached. The valve setting is modified by boost pressure from the manual valve and *throttle valve.*

established by the other transmission valves act on the ends of each shift valve. In addition, a spring, located at one end, puts a preload on the valve. Hydraulic pressure acting on one end works against hydraulic pressure *plus* spring pressure acting on the other end. The valve moves back and forth in the bore as pressures on each end change. When pressures on each end are equal, the valve is balanced and does not move.

Line pressure is directed to the center of the shift valve. It is used to apply the planetary holding members. As passageways leading to these devices are uncovered by movement of the shift valve, line pressure will pass through the valve annular groove and be routed to the holding member.

Every transmission will have one less shift valve than it has forward gears. For example, a 3-speed transmission will have two shift valves; a 4-speed transmission will have three.

The sequence of shift valve operation must be carefully controlled so that the transmission upshifts through all available gears. This is done during design by proper sizing of the springs and valve lands. By designing lands of different sizes, or effective *face areas,* hydraulic forces on valves acted on by a common pressure will differ.

Throttle valve

The ***throttle valve*** controls pressure, called ***throttle pressure,*** in the line going to the shift valve(s). Its role, in effect, is to sense how hard the engine is working and to delay upshifting, as necessary. The throttle valve may be located in the valve body, or it may be installed in a separate bore in the transmission case.

Movement of the throttle valve modifies line pressure. The resultant throttle pressure is then transmitted to the shift valve (and a few other valves) at some pressure less than line pressure. Throttle pressure modulates with engine load. As load increases, throttle pressure in the line to the shift valve increases, providing greater opposition to the counter (governor) pressure and delaying the upshift. The throttle valve may be operated by the *throttle valve linkage* or by the *vacuum modulator.*

Throttle valve linkage. The ***throttle valve (T.V.) linkage*** is a mechanical connection between the throttle valve in the transmission and the *throttle lever* on the vehicle throttle body or carburetor, Figure 10-39. The throttle lever, in turn, is connected to the *throttle plate(s),* located in the throttle body or carburetor. Engine power increases with the opening of the throttle plate and decreases with its closing.

The T.V. linkage consists of several levers connected by a series of rods or a cable. Movement of the vehicle accelerator causes the throttle lever (and throttle plate[s]) to move via the *throttle linkage,* which connects the throttle lever to the accelerator. The movement is transferred from the throttle lever to the throttle valve.

The action of the linkage on the throttle valve causes T.V. output pressure to change. Throttle pressure, then, varies as the throttle plate opening is varied, increasing as the throttle plate is opened and decreasing as the throttle plate is closed.

T.V. linkage always has some provision for adjustment. The rod-type linkage is adjusted either at the throttle lever or at the transmission, where the throttle rod enters the case. The cable-type linkage is usually adjusted at the throttle lever.

Vacuum modulator. The ***vacuum modulator,*** discussed in Chapter 5, moves the throttle valve with changes

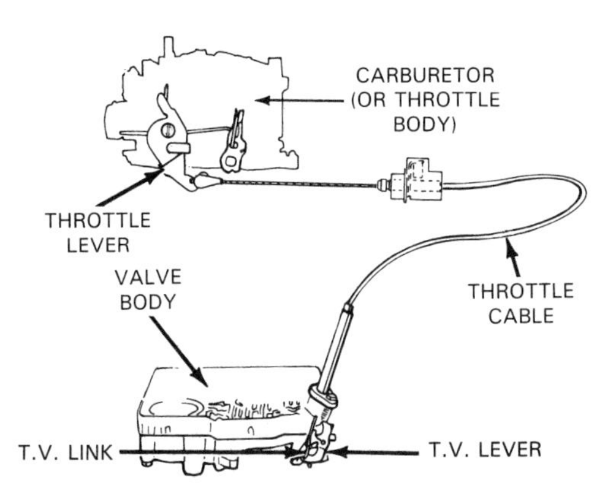

Figure 10-39. The throttle valve on many automatic transmissions is operated by a cable from the engine throttle plate. Changes in throttle plate position change the position of the throttle valve and, therefore, the throttle pressure. Throttle pressure is sent to the shift valves, pressure regulator valve, and sometimes, other valves. (General Motors)

in engine load, causing the throttle pressure to modulate. The basic modulator is a sealed container with an internal vacuum-operated diaphragm. The device is threaded into the transmission case or is held to the case with a bolt or clip, Figure 10-40.

The diaphragm divides the container into two regions, sealing one off from the other. One side of the modulator is connected to engine vacuum by tubing connected to the intake manifold; the other side is at atmospheric pressure.

In operation, the vacuum side of the modulator contains a spring that pushes on the diaphragm, opposing manifold vacuum. Changes in manifold vacuum cause the diaphragm, working against spring pressure, to move back and forth. The other side of the diaphragm is connected to the throttle valve, either directly or through a plunger. The valve is moved by the diaphragm as engine vacuum changes.

Engine vacuum will vary with the load that is placed on the engine by changes in throttle opening and engine load. If the load is high, the manifold vacuum will be low, and throttle pressure will be high. If engine load is low, vacuum will be high, and throttle pressure will be low.

Governor valve

The ***governor valve,*** or ***governor,*** as it is often called, senses output shaft speed to help control shifting. It works together with the throttle valve to determine *shift points,* or the speeds at which the shifts will occur. A typical assembly consists of a driven gear, weights, springs, hydraulic valve, and shaft.

The governor, driven by the transmission output shaft, takes line pressure at the valve and modifies it according to vehicle speed. Greater vehicle speed increases the governor valve output pressure, called ***governor pressure***; lower speed decreases governor pressure. Governor pressure opposes throttle pressure acting on the shift valves. Whichever is the higher pressure–governor or throttle–overrides and controls the shift.

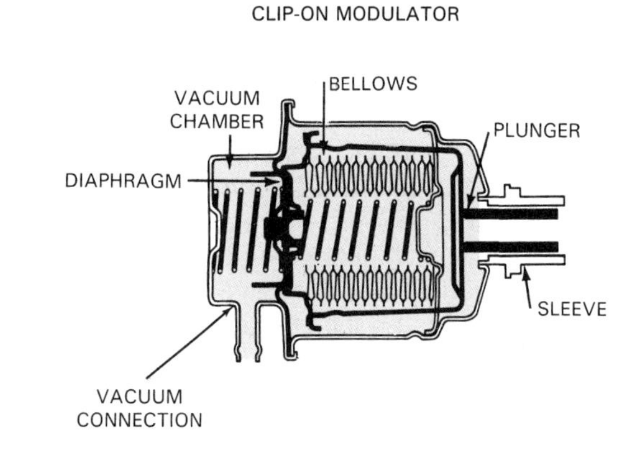

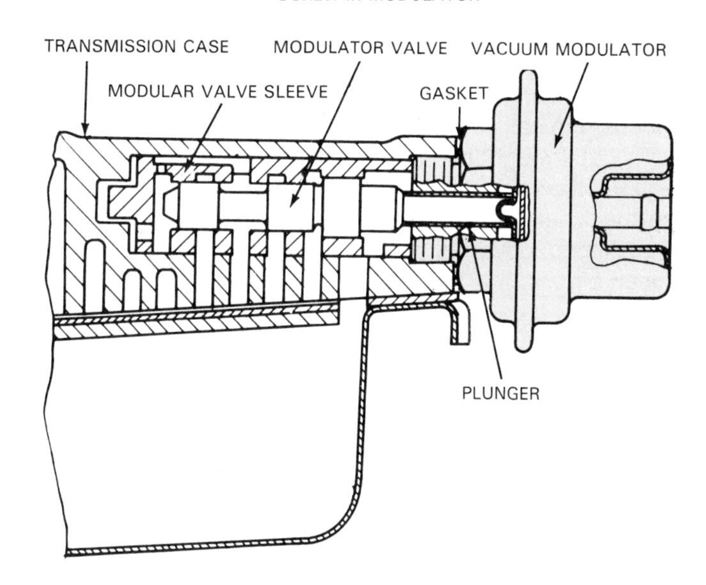

Figure 10-40. Two different kinds of vacuum modulators are shown here. Both use a vacuum-operated diaphragm and spring to affect valve position. Note that internal bellows, which is evacuated, and smaller spring in the one design are used to compensate for power loss that occurs at reduced atmospheric pressures, such as at higher elevations. (General Motors, Fiat)

The governor, as stated, is driven by the transmission output shaft. The assembly may be mounted on the output shaft, or it may be case mounted and driven through gears similar to speedometer drive and driven gears. Either type takes line pressure from the valve body and modifies it according to vehicle speed.

Governors vary in design. Aside from how they mount and are driven, they vary in other aspects of operation. The most common type of governor contains a valve that is moved as a result of weights. As the governor rotates, the weights are thrown outward, and this action causes the valve to move. The weights act against spring tension. Movement of the valve partly opens the governor port to line pressure. Governor pressure builds up rapidly with increasing vehicle speed until about 20 mph (32 km/h). At higher speeds, pressure increases more gradually to prevent it from going too high in the high vehicle-speed ranges. Shaft-mounted and gear-driven versions of this kind of governor are shown in Figure 10-41.

Note that a variation of this valve is without *separate* weights; the weight is an integral part of the valve itself. The motion of the governor, aided by the weight, causes the valve to move outward and open.

Another kind of governor has two openings, or ports, that are served by check balls. Weights are thrown outward as the governor rotates. This action moves the check balls to *close* the openings. See Figure 10-42. Line pressure is fed to this governor through an orifice, which limits the amount of fluid that can flow in the governor circuit. At low speeds, the balls are positioned so that transmission fluid leaks out through the openings. The governor pressure is zero. As the vehicle speeds up, the weights begin to move the check balls inward, allowing governor pressure to build. At maximum governor pressure, the ports are completely closed off.

Detent valve

The ***detent valve,*** also called the ***kickdown valve,*** is used on some transmissions to aid downshifting. This is often needed when a vehicle, already in a higher gear, attempts to climb a hill or accelerate to pass. The valve provides a *forced downshift* by pushing the accelerator all the way to the floor, increasing torque to the drive wheels. The vehicle, in this circumstance, is said to be in ***passing gear.***

The detent valve may be operated through a *kickdown linkage,* connected to the engine throttle, or by an electric solenoid. In operation, the valve increases pressure on the throttle side of the shift valves. The extra pressure forces the shift valves to the downshifted position, putting the transmission in a lower gear.

Detent valves are necessary since changes in throttle pressure are not great enough to cause a downshift at highway speeds. The valve hastens the downshift, placing the vehicle in a lower gear before vehicle speed is slow enough to cause sufficient drop in governor pressure.

Note that some hydraulic circuit designs do not have a separate detent valve. In such designs, kickdown is achieved through the throttle valve. If the throttle valve is moved far enough, a *detent oil* passageway is uncovered, and this oil is then routed to the appropriate shift valves.

Check valves

A common type of check valve in automatic transmissions is the check ball. Check balls are one-way valves; they allow fluid to flow in one direction only. The steel balls are installed in chambers next to holes in the valve body spacer plate.

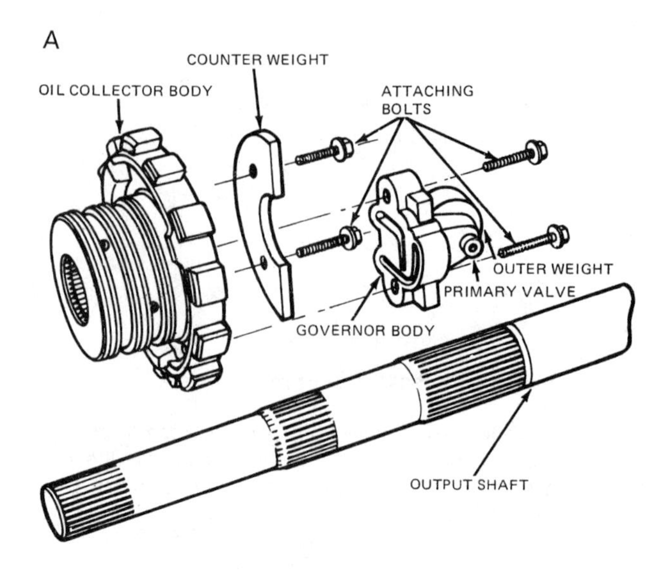

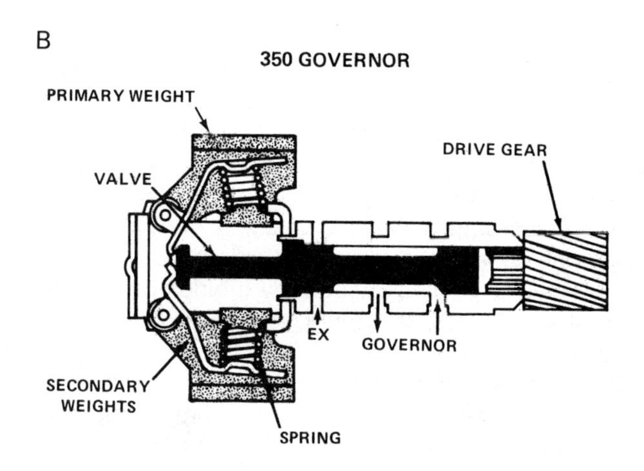

Figure 10-41. This shows two kinds of governors that control governor pressure by permitting restricted flow through the governor valve. A–Output-shaft mounted. B–Case mounted, gear driven. (Ford, General Motors)

In operation, fluid flow in one direction will push the check ball into the spacer plate hole, sealing it. Fluid flow in the opposite direction will push the ball out of the hole allowing fluid to flow.

Note that some transmissions use a *spring-loaded* check ball. These may function to control torque converter pressure, or they may serve in overpressure situations as pressure relief valves.

Automatic Transmission Shift Linkage

The shift linkage is the only direct connection between the driver and the automatic transmission. It makes a

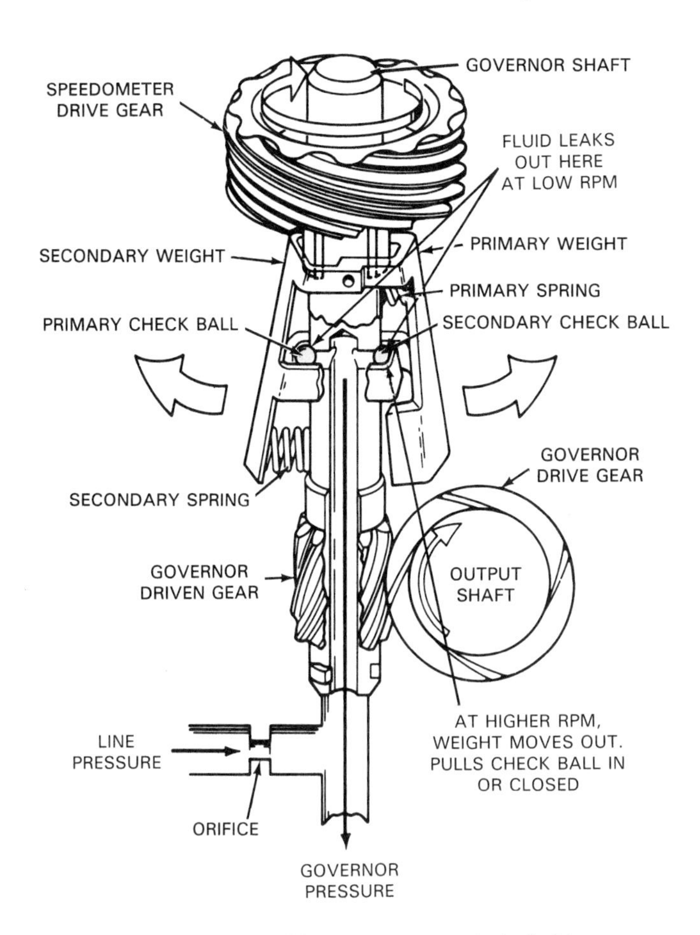

Figure 10-42. *Governor drive gear* on output shaft drives governor. Transmission fluid leaks out past check balls at low rpm. At higher rpm, weights are thrown outward, against spring tension. Check balls are pulled inward, blocking passageway and causing governor pressure to increase. When pressure is great enough to overcome throttle pressure acting on other end of shift valve, upshift occurs. (General Motors)

mechanical connection between the driver and the transmission hydraulic control system. It consists of a series of levers and linkage rods or cables. These transfer motion to the manual valve of the transmission valve body.

Selector Lever

The shift ***selector lever*** in the vehicle passenger compartment can be mounted on the floor or on the steering column. With either arrangement, movement of the selector lever causes the manual valve inside the transmission to move. This action selects the driving range.

Floor-shift selector lever

An assembly incorporating a typical *floor-shift selector lever,* Figure 10-43, is mounted directly on the floor or in a console between the front seats. Levers and rods (or cable) are used to transfer motion from the floor-shift selector lever to the transmission. The selector indicator, or *quadrant,* is usually built into the assembly; however, some are installed in the dash and connected to the transmission by a linkage arrangement.

To move the selector lever between certain positions, a lever *release button* must be depressed. Pressing the release button unlocks the selector lever *latching mechanism.* The latching mechanism typically consists of a *latching spring pin* (or lever) and a detent plate. The pin fits through a slotted hole at the bottom of the selector lever and into the detents in the plate on the mounting base of the shift mechanism. Pressing the release button lowers the pin so that it is no longer in the detent. The selector lever can then be moved. When the button is released, the pin moves into the detent for the new selector position. The latching mechanism is usually designed to allow the selector lever to be moved from a lower gear to a higher gear, and from drive to neutral, without pressing the release button. For other sequences, the button must be depressed.

Column-shift selector lever

A shift linkage incorporating a *column-shift selector lever* uses a series of rods and levers to transfer movement from the steering column to the transmission. The column-shift lever does not have a release button. Pulling the selector lever toward the driver disengages a latching pin from its slot, allowing the lever to be moved to the proper position.

Shift Linkage

Figure 10-44 shows typical shift linkage at a transmission case. Note the *outer manual lever and shaft assembly.* The *outer manual lever* is attached to the shift rod or cable from the shift selector lever. Movement of the outer manual lever is transferred through the shaft passing through the case to move the *inner manual lever.* This lever connects to the manual valve in the valve body. The manual valve opens and closes passageways that direct fluid to the different circuits in the transmission, as needed for the selected operating range.

Note that a hole runs through the shaft of the outer manual lever and shaft assembly in the design of Figure 10-44. Through this hole passes another shaft having levers at each end. The *outer throttle valve lever* connects, by linkage, to the engine throttle, and the *inner throttle valve lever* connects to the throttle valve. Seals at the case and *throttle shaft* prevent transmission fluid leaks.

The inner manual lever is made to hold the manual valve in position. Otherwise, if not held in place, hydraulic pressure and vibration could cause the manual valve to move out of position. A series of detents, sometimes called a *cockscomb,* is part of the inner manual lever. This can also be seen in Figure 10-44. A spring is used to hold a

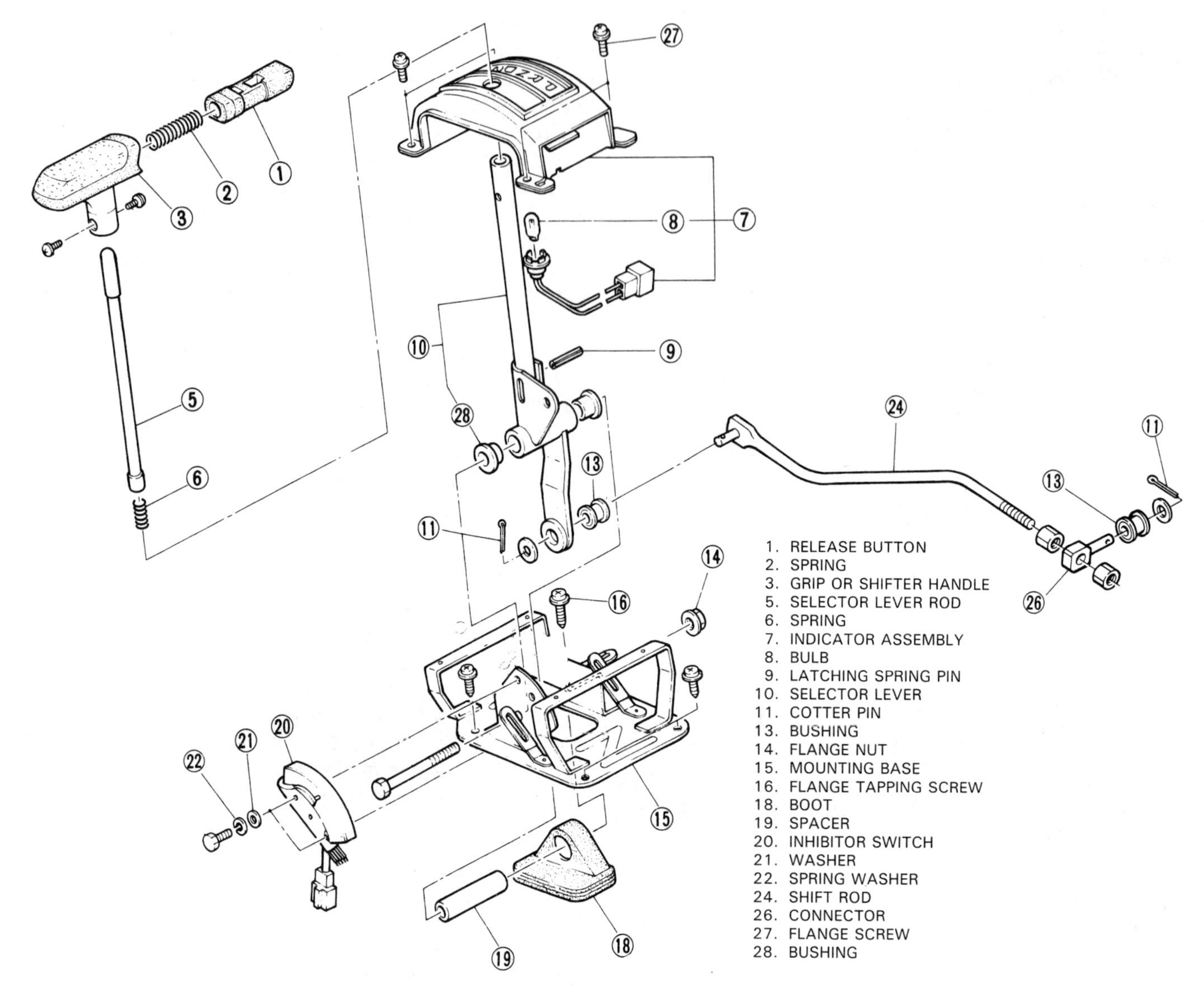

Figure 10-43. A typical floor-shift selector lever is mounted directly on the floor or in a console between the front seats. Shift rod extends to transmission. (Subaru)

roller or ball against the detents. Spring pressure holds the lever in position; however, the spring is not strong enough to prevent movement of the lever when the driver moves the selector lever.

Neutral start switch

The ***neutral start switch,*** which is shown in Figure 10-44, is used mostly with floor-shift selector lever designs. Sometimes called a ***neutral safety switch,*** this switch prevents starting the engine when the selector lever is in any position but park or neutral. The switch may be transmission mounted or console mounted. (The column-shift safety mechanism is usually in the ignition switch. The ignition key will not turn to start the vehicle unless the selector lever is positioned in park or neutral.)

If the engine were to start when the selector lever was in any position but park or neutral, the vehicle would lurch forward, possibly causing an accident or injury. The switch is activated by movement of the shift linkage, and current is allowed only when the selector lever is in neutral or park. Otherwise, the switch contacts are open, and circuit current cannot flow. The vehicle ignition switch is wired to the starter solenoid through the neutral start switch. If the ignition switch is operated when the selector lever is in any drive gear or reverse, the starter motor will not turn.

Note that in some cases, the neutral start switch has an integral backup-light switch. The shift linkage makes contact with a plunger within the switch when the selector lever is in reverse. Moving the plunger causes a second set of contacts within the switch to close, completing the backup-light circuit and causing the lights to go on.

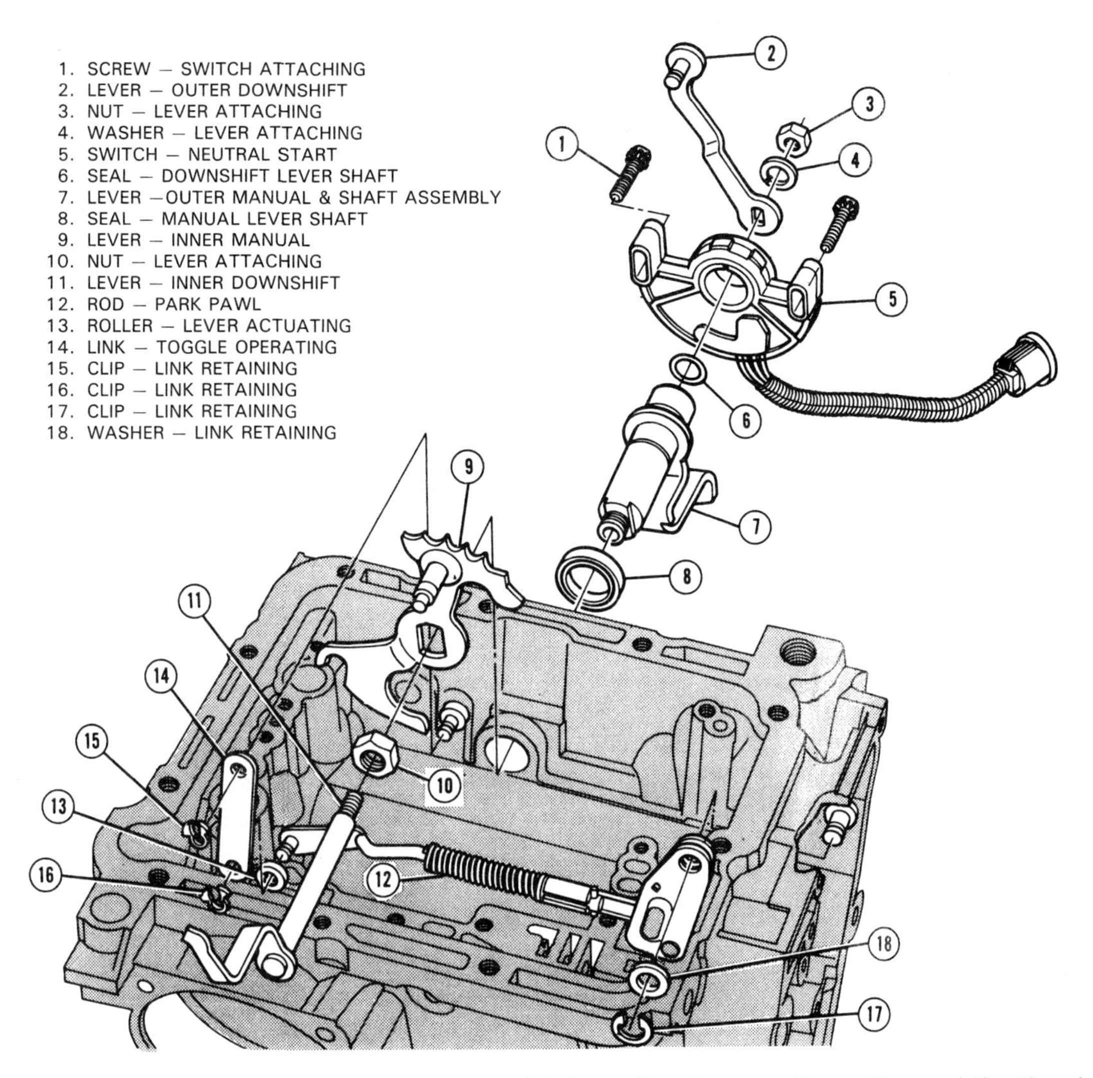

Figure 10-44. This shows outer and inner, case-mounted shift linkage. Also shown are the parking pawl, throttle valve linkage, and neutral start switch. (Ford)

Parking gear

Figure 10-44 also shows part of the linkage that locks up the ***parking gear,*** which is a toothed wheel splined to the output shaft. The linkage extends to the rear of the transmission and locks the output shaft to the case. A lever with a *parking pawl,* Figure 10-45, is attached to the transmission extension housing. The pawl moves in and out of engagement with the parking gear. When the driver shifts to the *park* position, the shift linkage pushes on the parking pawl. The pawl then engages the parking gear, locking the output shaft in place. The linkage also moves the manual valve to neutral.

Converter Housing

The converter housing, Figure 10-46, is a stationary component that covers and protects the transmission torque converter. It is often referred to as the ***bell housing*** because of its bell-like shape. The housing is bolted to the rear of the engine. Many converter housings are combined with the transmission case into a one-piece, or integral, casting. Others are separate and bolt to the front of the transmission case.

Older converter housings were made of cast iron, but most modern housings are cast from aluminum to reduce weight. Most of these have cast ribs at the top for added strength. The front of the housing is machined flat for alignment with the mounting surface on the engine. Holes for the *converter housing attaching bolts* (converter housing-to-engine bolts) and for aligning dowels are often drilled into the front of the housing. Bolts pass through the housing to thread into the rear of the engine block.

Aligning dowels are usually pressed into the rear of the engine block. They fit into the matching holes in the converter housing. The dowels make transmission installation easier and ensure that the centerlines of the transmission

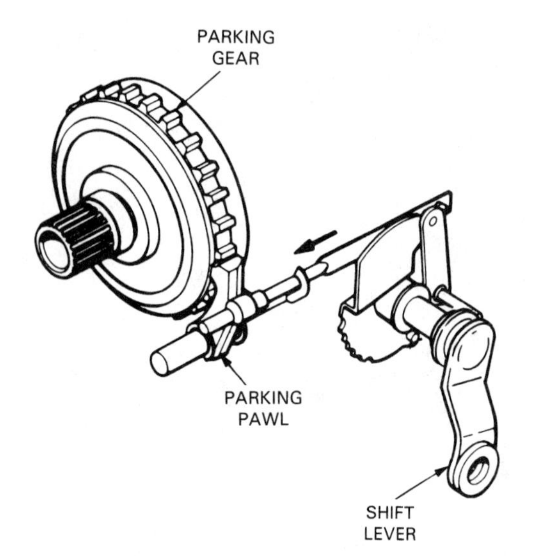

Figure 10-45. The parking pawl is simply a latch that locks into large teeth on the parking gear. Since the pawl is mounted on the case, the output shaft is locked to the case when engaged, and the vehicle will not roll. (Subaru)

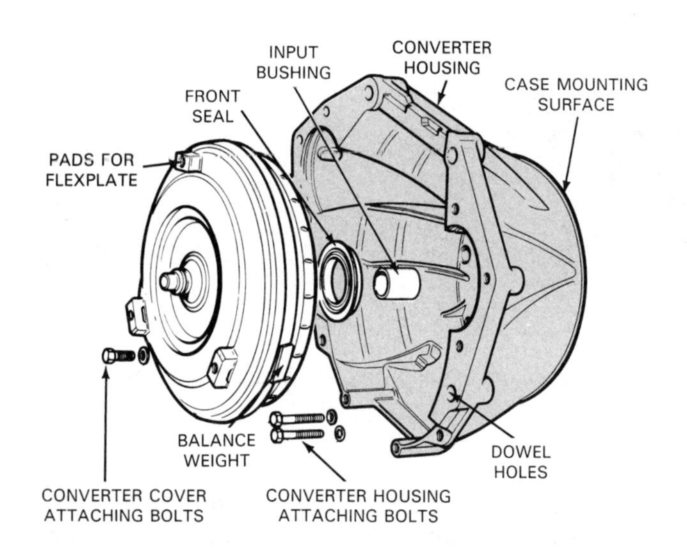

Figure 10-46. The converter housing encloses the torque converter and mounts on the back of the engine. This converter housing is separate from the transmission case. (Fiat)

input shaft and engine crankshaft align. The rear of the converter housing is accurately machined and drilled to mate with the front of the transmission case (for nonintegral casting).

Converter housings may be either fully or partially enclosed. The fully enclosed housing completely surrounds the torque converter. It is usually bolted to the engine through a steel *stiffener plate.* The stiffener plate adds strength to the engine-transmission mounting. The starter motor is often attached to the converter housing through the stiffener plate.

An *inspection cover* is usually installed at the bottom half of the converter housing, directly behind the engine oil pan. The inspection cover can be removed to reach the *converter cover attaching bolts* (converter cover-to-flexplate bolts). When a stiffener plate is used, the inspection cover is usually smaller and is attached directly to the stiffener plate with machine screws.

The partially enclosed housing, or *half housing,* is open at the bottom. This type is shown in Figure 10-47. The bottom of the housing is covered by a *dust cover* made of sheet metal. The dust cover protects the converter from road debris and water. It is removed to reach the converter cover attaching bolts. When a half housing is used, the starter motor is usually mounted on the engine block. The starter drive pinion reaches the flexplate ring gear through a hole in the dust cover. For added strength, this kind of converter housing is usually cast integral with the transmission case.On some vehicles, a *spacer plate* is installed between the engine and converter housing. The spacer plate is installed to maintain the correct distance between the torque converter and the transmission oil pump. On these vehicles, if the spacer plate were not used, the moving parts of the converter and pump would bind, causing severe damage to the transmission.

Automatic Transmission Case

The transmission case is the central part of the transmission. Modern transmission cases are made of aluminum. Many older transmissions had cast iron cases.

The front, rear, and bottom of the transmission case are machined flat and drilled to fit the converter housing, extension housing, and oil pan. The inside of the case is machined to provide a mounting for other transmission parts, such as band anchors, servo pistons, accumulators, clutch plates, center supports, and the oil pump. The bottom of the case is also machined to hold the valve body and, sometimes, servos or accumulators. The rear of the case usually has a pressed-in bearing to support the output shaft. Many transmission accessories, such as shift levers, electrical connectors, *filler tube,* and pressure relief vents, are installed in openings drilled and threaded into the case.

Oil passageways are cast or drilled in most automatic transmission cases, Figure 10-48. They direct flow between other parts of the hydraulic system, such as the oil pump and valve body or the valve body and transmission servo pistons. One passageway directs hot transmission fluid from the torque converter to the transmission oil cooler. Cooled fluid is directed through another passageway to components requiring continuous lubrication and cooling, such as rear transmission bushings. Some passageways also lead out from the case for testing the automatic transmission. These passageways are sealed with pipe plugs.

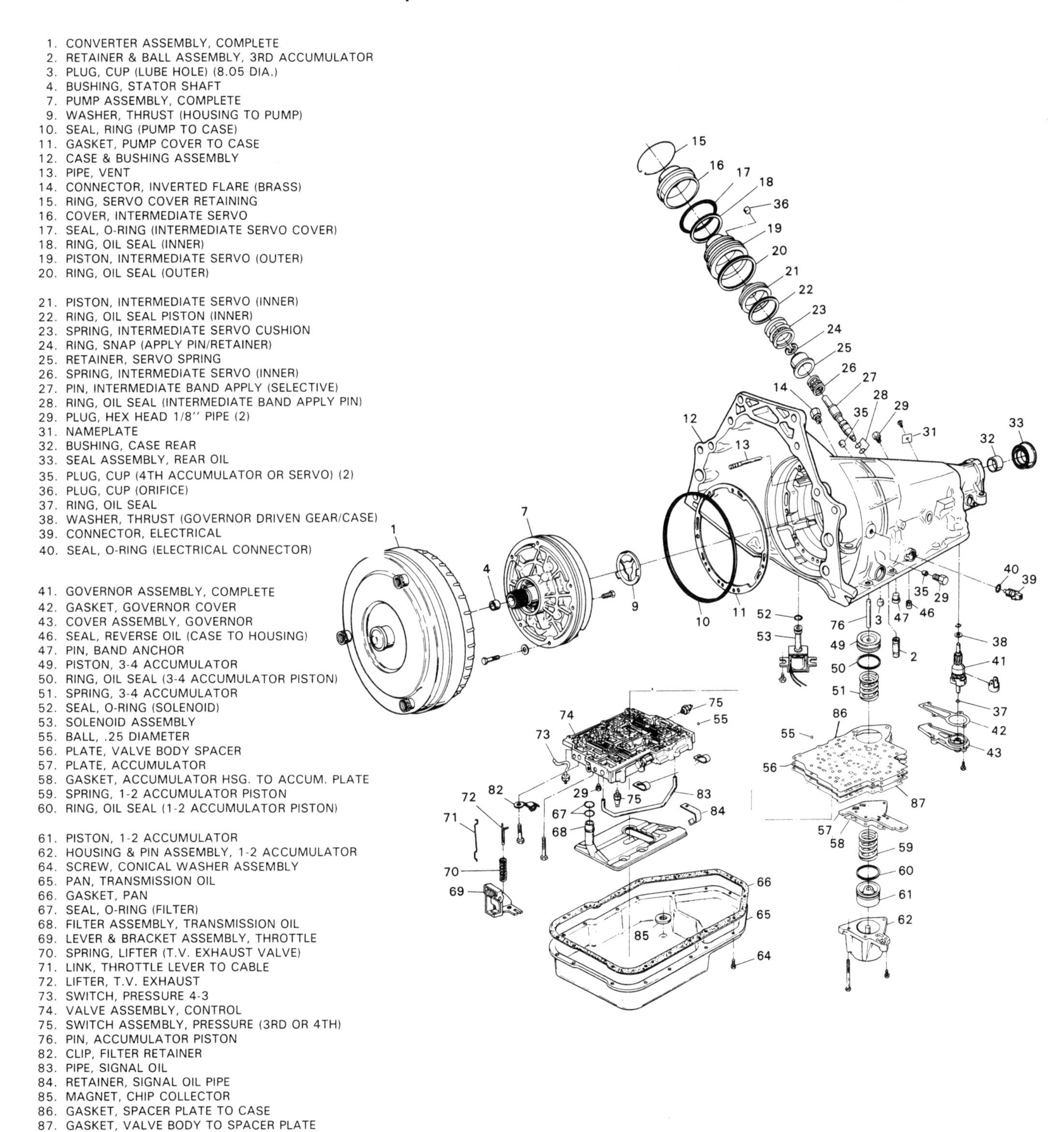

Figure 10-47. This transmission design features a partially enclosed converter housing. (General Motors)

Extension Housing

The extension housing is used to support and protect the tail end of the output shaft. It also contains other internal transmission parts, such as the governor, parking gear and linkage, and speedometer gears. Extension housings are almost always made of aluminum. However, cast iron is sometimes used on heavy-duty applications. Figure 10-49 shows an extension housing.

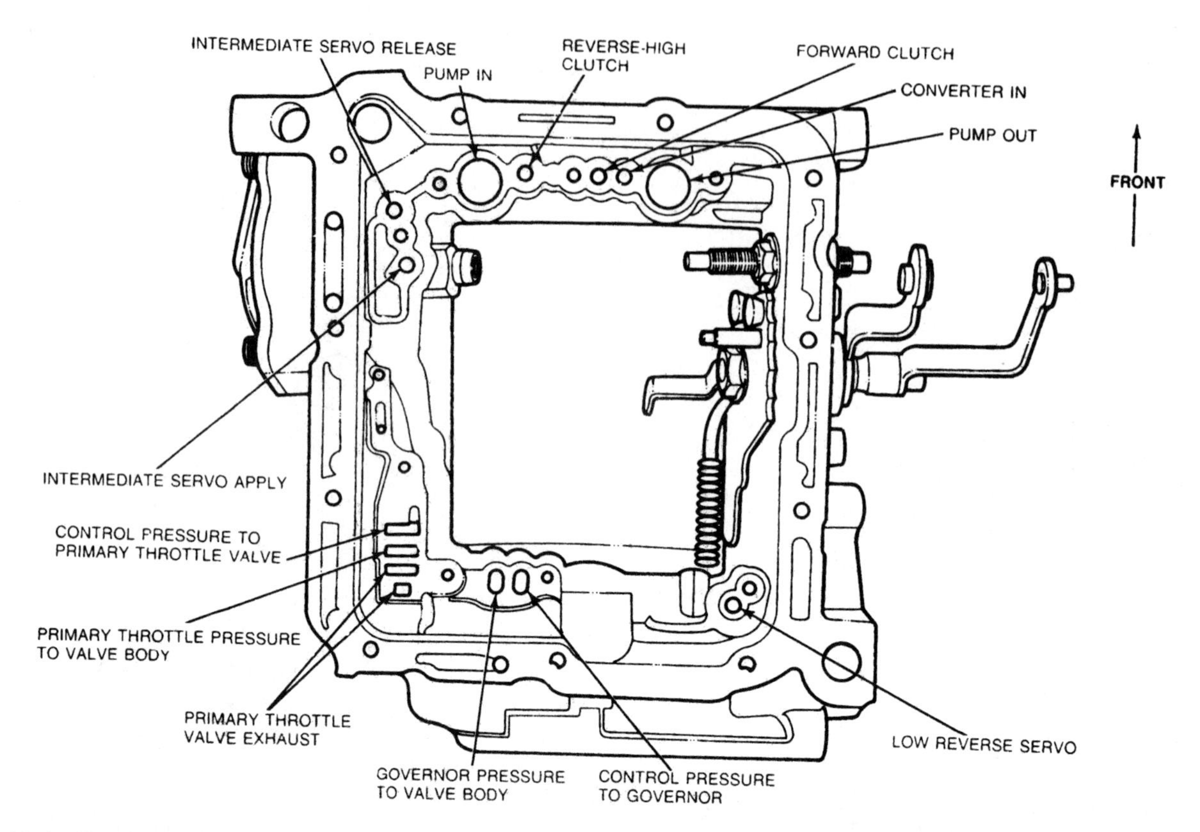

Figure 10-48. Bottom view of a transmission case (pan and valve body removed.) Surfaces are machined flat and fluid passageways of this case are drilled. (Ford)

The rear of the extension housing contains an extension bushing. The bushing is pressed into the housing bore. The bushing supports the driveline slip yoke, which in turn, supports the transmission output shaft. The output shaft turns the slip yoke, which rides on the extension bushing. Some extension housings also use an antifriction bearing to support the output shaft. The extension housing is sealed at the tail end by the transmission rear seal.

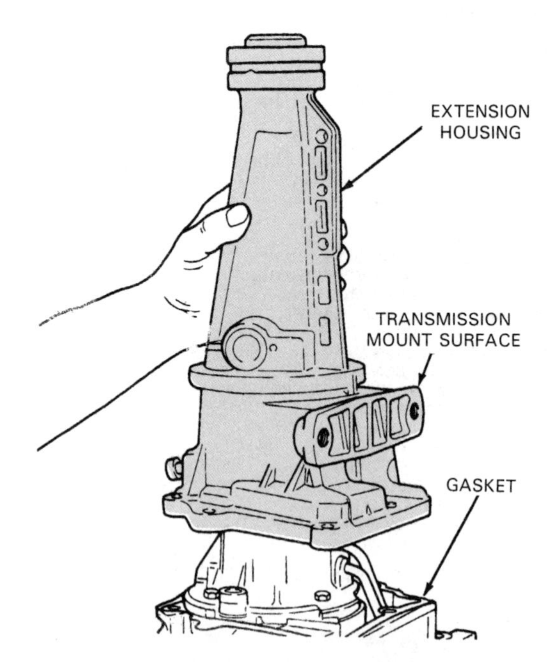

Figure 10-49. The extension housing supports and protects the output shaft. It is bolted to the rear of the transmission housing. (Ford)

Transmission Oil Pan

The oil pan is the transmission fluid reservoir. It is from here that fluid is drawn, to be distributed throughout the transmission. It is also where it ultimately returns. The oil pan is installed at the bottom of the transmission and encloses the valve body, filter, and some other internal parts. The pan holds enough extra fluid to allow for changes in fluid level caused by temperature and road-angle variations and by leaks.

Transmission oil pans are wide and relatively shallow. Their shape allows them to transfer heat from the fluid to air passing under the vehicle. A few have cast aluminum fins to aid cooling even more. Figure 10-50 shows a typical oil pan.

Since it is one of the lowest points on the vehicle, it is subject to damage from gravel and other flying road debris. As a result, oil pans are often made from heavy sheet metal, stamped into the proper shape. They are sometimes made of cast aluminum.

The oil pan is attached to the transmission case by bolts. A gasket is always used between the oil pan and case. Some oil pans contain a drain plug.

The transmission oil filter is installed in the oil pan. Other transmission components are often installed on or in the oil pan, as well. The filler tube is sometimes welded or bolted to the side of the pan. A few used on late-model vehicles have an attached electrical connector that connects a sensor to the computer system. Some pans contain a small magnet to pick up stray metal shavings, which keeps the shavings from entering the hydraulic system.

Electronic Components

Most modern vehicles are controlled, at least in part, by an on-board computer. In addition to controlling the engine and emission systems, the on-board computer receives inputs and sends commands to the automatic transmission. Automotive computer systems, including input and output devices such as sensors and solenoids, were presented previously in Chapter 5.

An example of a computer controlled automatic transmission is shown in Figure 10-51. The system uses many sensors, relays, and solenoids to control transmission operation. The sensor inputs are used by the computer in deciding what commands to send to the transmission. Figure 10-52 shows a couple of electronic input and output devices on a valve body.

Through the electronic hardware, the computer typically monitors engine speed, load, and throttle position, among other variables. It can then provide control for transmission shift points, torque converter lockup, and other functions. This keeps the transmission functioning at maximum efficiency.

A simplified diagram of an electronic control system is shown in Figure 10-53. Input signals from sensors go to the main computer. Output signals from the computer go to solenoids and other output devices. Output device operation affects the operation of the engine and transmission. Engine and transmission operation affects sensor inputs, and the entire process starts over, in an endless circle, or *control loop.*

Switches

Switches are one type of sensor commonly used in the electronic control circuit of the automatic transmission. Switches use electrical contacts to energize or deenergize a circuit or to send ON/OFF signals to the on-board computer. One type of switch is activated by the shift linkage connected to the shift selector lever. This combination switch prevents the starter from operating in gear and turns on the backup lights in reverse gear. Another type is a pressure-operated device that signals a particular transmission gear.

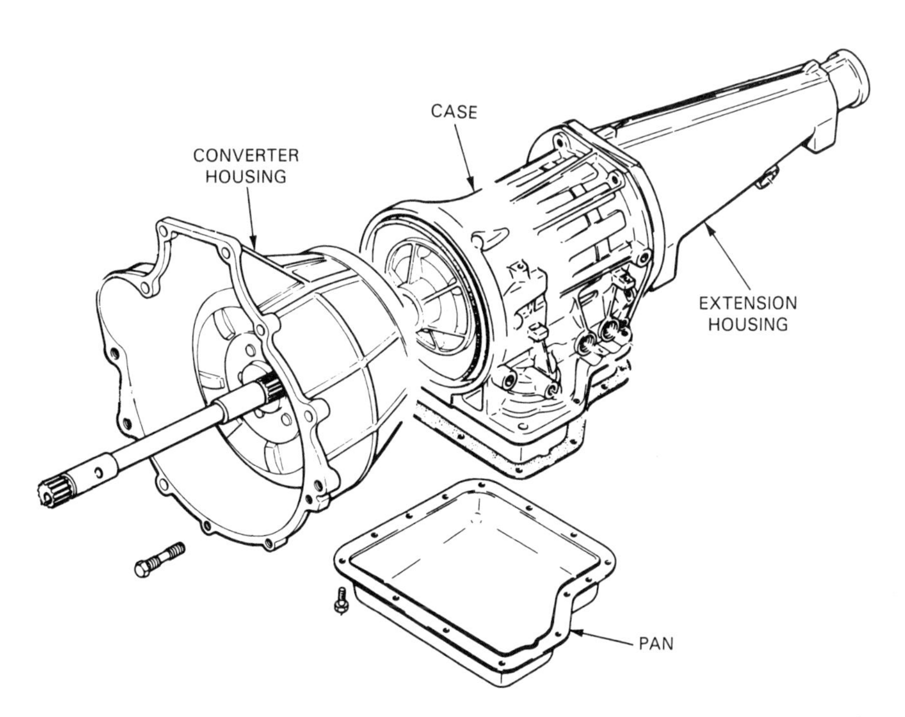

Figure 10-50. This exploded view shows relationship of oil pan to transmission case. The oil pan seals and protects the bottom of the transmission and serves as a fluid reservoir. (General Motors)

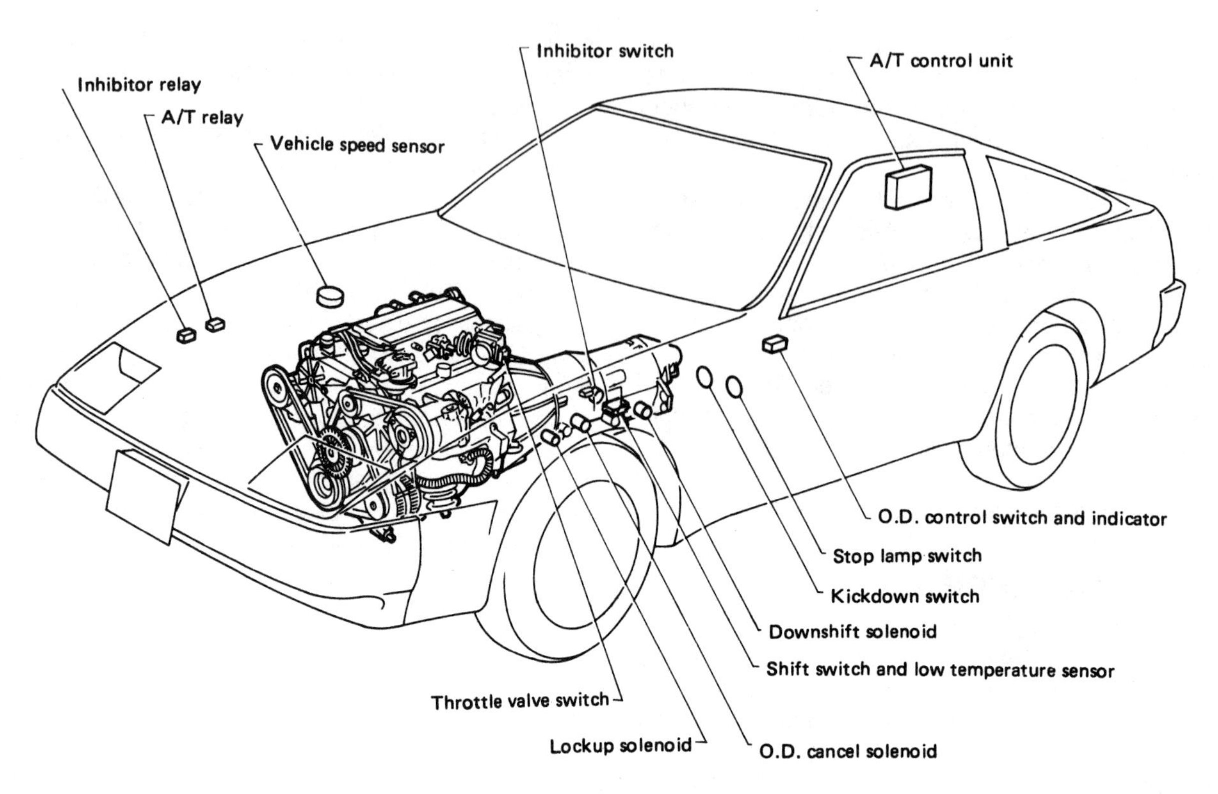

Figure 10-51. Inputs to the computer help it to decide what commands to send to the engine and transmission. Some inputs come from the transmission itself. (Nissan)

The pressure switch, shown on the valve body of Figure 10-52, sends ON/OFF signals to the on-board computer. The computer processes these input signals, along with inputs from many other sensors, and sends output commands to various engine components, emission controls, and back to the transmission.

Solenoids

Computer control of the transmission is accomplished largely through electric solenoids, which control the operation of some transmission hydraulic components. See Figure 10-54. Some solenoids are installed inside the oil pan, against the valve body. These are connected to the outside through a wiring harness having an electrical connector that is mounted on the case. Other solenoids are threaded directly into the transmission case, as shown in Figure 10-55.

A sectional view of a typical solenoid, or ***solenoid valve,*** as it may be called in this instance, is shown in Figure 10-56. The solenoid valve looks and operates much like a small starter solenoid. When energized, the solenoid valve functions by opening a small, self-contained valve or ball. This action causes transmission fluid in the connecting passageway to be exhausted. Loss of pressure in the passageway causes another valve to move, and action from this point is completely hydraulic.

An example of a transmission solenoid is the ***detent solenoid*** used on some automatic transmissions, also called the ***kickdown,*** or ***downshift, solenoid.*** The detent solenoid is energized by closing the *kickdown switch,* which is located near the accelerator. Figure 10-57 is a simplified diagram of a detent solenoid circuit.

Another use of solenoids is the ***lockup solenoid,*** used to engage the torque converter lockup clutch, Figure 10-58. When the solenoid is energized, its integral valve is unseated, causing transmission fluid in the connecting line to be exhausted. Loss of this fluid reduces circuit pressure. This causes the *lockup relay valve* to move in such a way as to apply the clutch. The first lockup solenoids were often operated by vacuum and pressure switches. Today, most lockup solenoids are operated by the on-board computer.

Finally, Figure 10-59 shows a 4-speed transmission valve body that is partially controlled by the computer through solenoids. This design features an ***overdrive solenoid.*** Overdrive system control valves are operated as the overdrive solenoid is energized and deenergized.

Summary

The torque converter is part of an automatic transmission. It transfers power and multiplies torque through a fluid coupling. It is installed at the front of the transmission. The rotating impeller moves the fluid outward and in the same

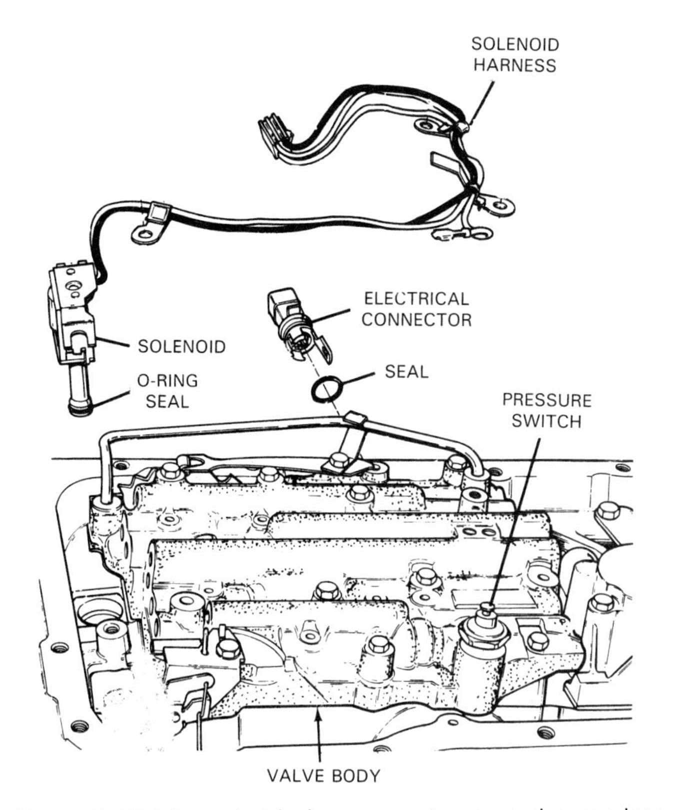

Figure 10-52. Many electrical components are used on modern transmissions. Solenoids, switches, and other electrical parts are often installed on the valve body. Wiring and case connectors make the electrical connections to the rest of the vehicle. (General Motors)

direction as engine rotation. The turbine receives the fluid and reverses its direction by action on its curved blades. As a result, power is transferred from the impeller to the turbine through the fluid. The stator reverses the fluid direction so that it assists the impeller in turning.

On many modern transmissions, a torque converter lockup clutch bypasses the fluid coupling to save energy. The converter clutch is splined to the transmission input shaft. The clutch contacts the inside of the converter cover through a friction surface. When the clutch is applied by hydraulic pressure, it locks the cover and input shaft together.

The input and output shafts send power into and out of the transmission. They connect to the planetary gearsets. The stator shaft is solidly mounted on the oil pump. It provides a stationary support for the torque converter stator and overrunning clutch.

Planetary gears are constant-mesh gears that somewhat resemble the solar system. The gearset consists of a sun gear, planet-pinion gears and carrier, and a ring gear. Different gear ratios are obtained by holding one of the gears and driving another. Power exits through the third member.

Simple planetary gears are combined into compound planetary gearsets for use in automatic transmissions. The Ravigneaux gear train has two sun gears, two sets of planet gears, one planet carrier, and one ring gear. The Simpson gear train has one sun gear, two sets of planet gears, two planet carriers, and two ring gears.

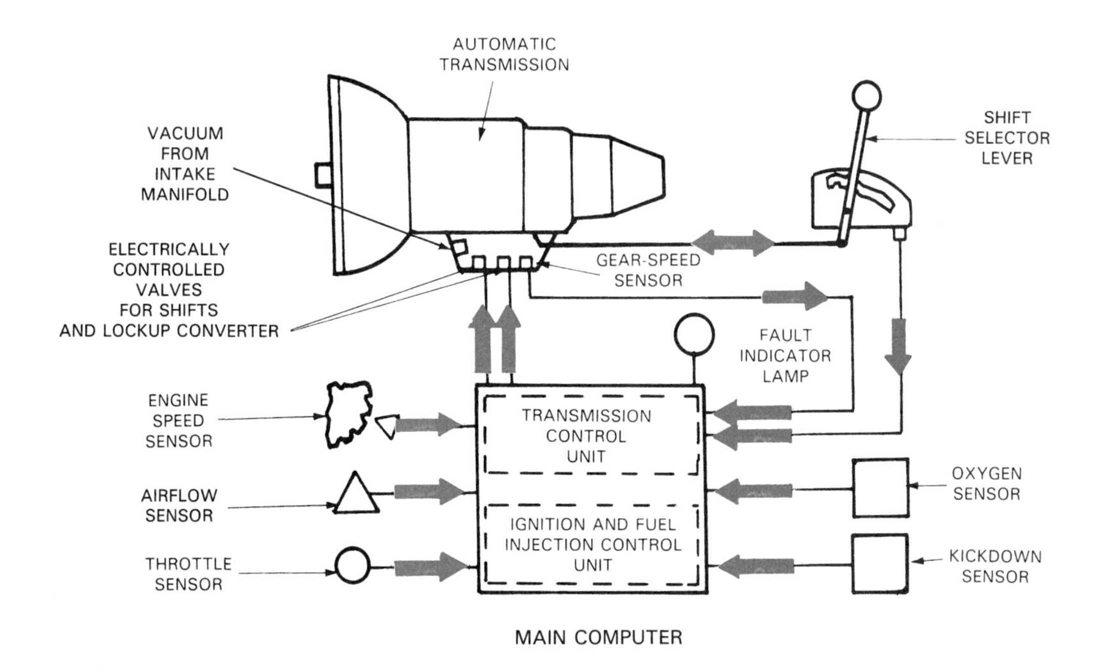

Figure 10-53. This diagram illustrates the flow of data to and from the main computer and the transmission. The computer receives inputs and sends out commands in an endless circle called a control loop. Note that this illustration shows the transmission control unit as part of the main computer. However, many are separate units.

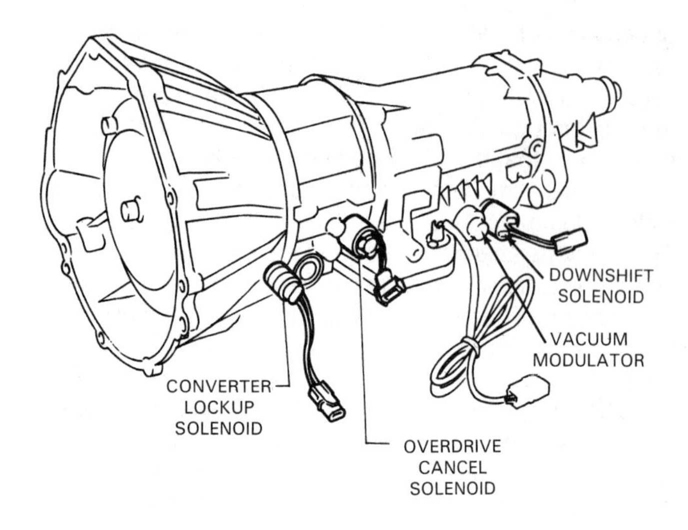

Figure 10-54. A typical modern transmission with solenoid control is shown here. (Nissan)

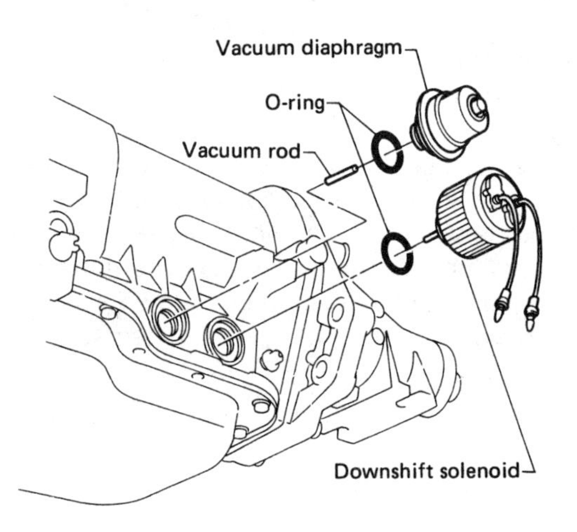

Figure 10-55. Many solenoids, such as the downshift solenoid shown here, are threaded directly into the case. (Nissan)

Planetary gearsets are controlled by holding members called multiple-disc clutches, bands, and overrunning clutches. Clutches and bands are operated by hydraulic pressure. Clutch operating pistons are called apply pistons. Bands are operated by pistons called servos. When engaged, multiple-disc clutches, splined to certain planetary gears, hold parts of the gearset stationary or drive power through parts of the drive train. Bands wrap around drums, holding them stationary when applied. One-way clutches are mechanical devices that either hold or freewheel, depending on what direction power is driven through them.

Accumulators cushion shifts by absorbing some of the fluid being sent to holding members. The resultant pressure in the accumulator compresses a piston, expanding the chamber, causing a gradual pressure buildup in the circuit and cushioning the shift. Transmission oil pumps use gears,

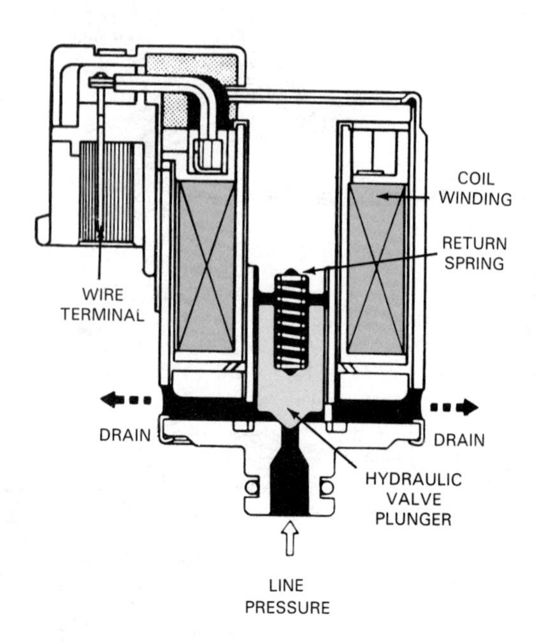

Figure 10-56. A typical solenoid consists of a coil winding, a plunger, and a return spring. The winding, when energized, pulls the plunger against return spring pressure. This causes transmission fluid to be exhausted and activates the particular control circuit. (Chrysler)

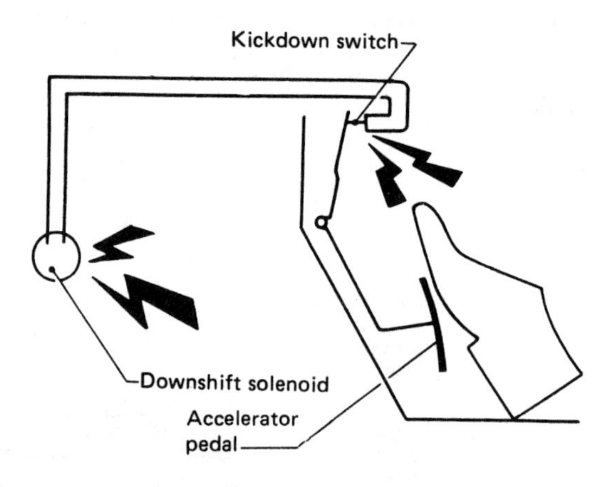

Figure 10-57. Simplified diagram of a typical detent solenoid circuit is shown here. Pressing the accelerator to the floor closes contacts on the kickdown switch and energizes solenoid. Valve opens, and fluid is exhausted, reducing hydraulic circuit pressure. (Nissan)

rotors, or vanes that cause the hydraulic system to become pressurized. The oil pump is usually driven by lugs on the rear of the torque converter. Some older transmissions had a second oil pump that was driven by the transmission output shaft.

Transmission oil filters are used to remove particles from the automatic transmission fluid. Fluid being drawn into the oil pump must first pass through the filter. This keeps fluid contaminants from entering the pump or the other hydraulic components.

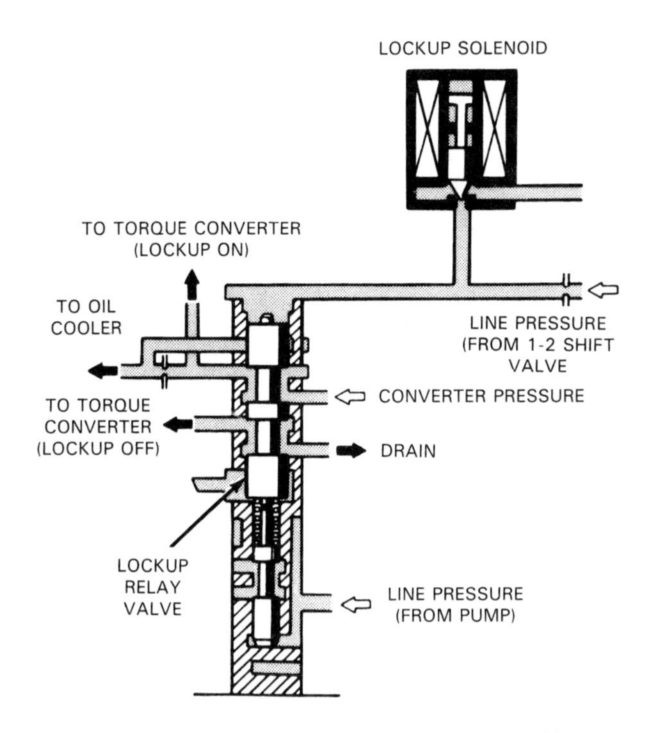

Figure 10-58. The computer controlled lockup solenoid operates the hydraulic portion of the torque converter lockup relay valve on many modern transmissions. (Chrysler)

The transmission oil cooler removes heat from transmission fluid. Fluid that has been overheated does not lubricate properly. Further, it breaks down into sludge and varnish, which plugs passageways. The oil cooler is located in the engine cooling-system radiator.

The valve body contains the manual valve, shift valves, and other hydraulic system valves. Valve bodies usually have several check balls, next to the spacer plate and gaskets.

The pressure regulator valve may be in the valve body, the back side of the oil pump, or in an auxiliary valve body. It regulates line pressure according to spring pressure and inputs from other valves.

The manual valve is operated by the driver through the shift linkage. It allows the driver to select park, neutral, reverse, or different drive ranges. When the shift selector lever is moved, the shift linkage moves the manual valve. As a result, the valve routes hydraulic fluid to the correct components in the transmission.

Shift valves control transmission upshifts and downshifts. Hydraulic pressures from the other transmission valves act on each end of each shift valve.

Throttle valves control the point at which shift valves move to the upshifted position. They can be operated by the T.V. linkage or by the vacuum modulator. Throttle pressure on the shift valves tries to keep them in the downshifted position.

Governor valves produce pressure that varies with vehicle speed. As speed increases, governor pressure increases. Governors can be mounted on the output shaft or on the transmission case. Case-mounted governors are driven by a gear on the output shaft. Governor pressure opposes throttle pressure to upshift the transmission.

The detent valve forces the transmission into a lower gear. This is often needed when a vehicle attempts to climb a hill or accelerate to pass. Detent valves are operated by linkage or by an electric solenoid.

Check balls are one-way valves. They allow fluid to flow in one direction only. The steel balls are installed in chambers next to holes in the valve body spacer plate.

Shift linkage connects the driver to the transmission through levers and rods or cables. The shift linkage also operates the neutral start switch, backup lights, and parking gear mechanism.

The converter housing, transmission case, extension housing, and oil pan are stationary parts of the transmission. They provide mounting surfaces for other transmission components. They also seal in transmission lubricant and protect the internal components from dirt. The converter housing and case are often cast into one unit. The case usually contains drilled or cast oil passageways.

Modern automatic transmissions are often controlled by the on-board computer. The computer takes readings from various sensors and issues commands to the engine and transmission. Most transmission commands are handled by electric solenoids, which affect the transmission hydraulic system.

Know These Terms

Automatic transmission shafts; Planetary holding members; Transmission oil pump (Front pump); Transmission oil filter (Oil strainer); Transmission oil cooler; Automatic transmission shift linkage; Converter housing (Bell housing); Automatic transmission case; Transmission oil pan; Fluid coupling; Turbine shaft (Automatic transmission input shaft); Overrunning clutch; Stator shaft; Converter cover; Converter pilot hub; Flexplate; Pump drive hub; Bounce-back effect; Torque multiplication; Stall speed; Overrunning clutch (One-way clutch); Coupling phase; Lockup torque converter (Converter clutch torque converter); Semiautomatic transmission; Automatic transmission output shaft; Compound planetary gearset; Ravigneaux gear train; Simpson gear train; Multiple-disc clutch (Clutch pack); Drive shell (Input shell, or Sun-gear shell); Band; Servo; Engine braking; Accumulator; Line pressure; Manual valve; Shift valve; Throttle valve; Throttle pressure; Throttle valve (T.V.) linkage; Vacuum modulator; Governor valve (Governor); Governor pressure; Detent valve (Kickdown valve); Passing gear; Selector lever; Neutral start switch (Neutral safety switch); Parking gear; Solenoid valve; Detent solenoid (Kickdown solenoid, or Downshift solenoid); Lockup solenoid; Overdrive solenoid.

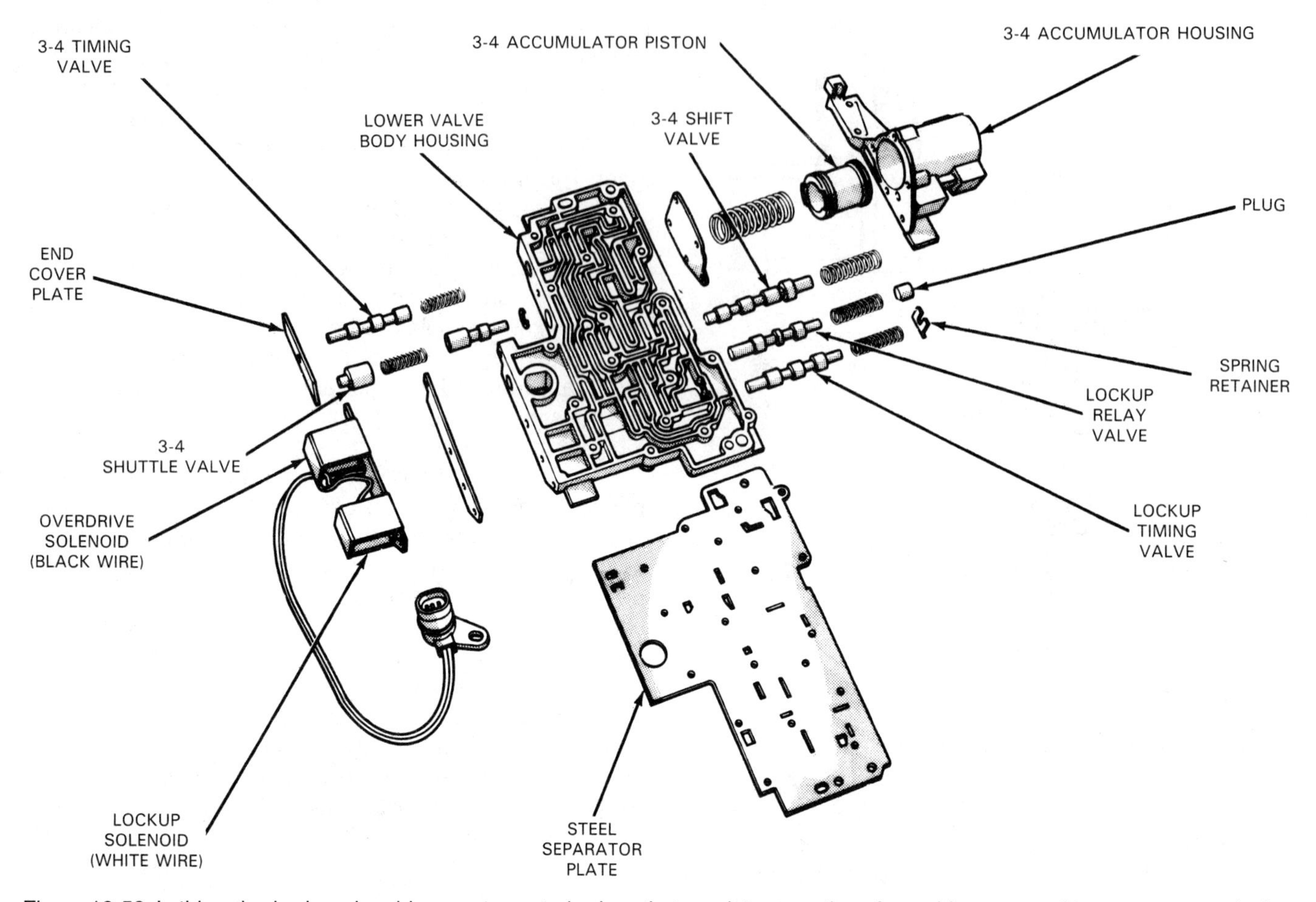

Figure 10-59. In this valve body, solenoids operate control valves that regulate operation of overdrive gear and torque converter lockup clutch. (Chrysler)

Review Questions–Chapter 10

Please do not write in this text. Place your answers on a separate sheet of paper.

1. The three bladed elements of the torque converter are the _____, the _____, and the _____.
2. Briefly summarize the operation of a torque converter.
3. What does a lockup torque converter do and how does one work?
4. List the five actions that can be accomplished by holding or driving the members of a planetary gearset.
5. A _____ _____ _____ combines two planetary gearsets with two separate sun gears or a single common sun gear.
6. Components used to hold gearset members in place or to couple two elements together for power transfer are referred to as:
   a. Torque converters.
   b. Bands.
   c. Servos.
   d. Multiple-disc clutches.
7. The following component is used to operate a band:
   a. Oil pump.
   b. Clutch disc.
   c. Servo.
   d. Clutch apply piston.
8. An overrunning clutch is a mechanical holding member that can be used to hold a planetary gearset member when rotated in one direction. True or false?
9. Cite four functions of an automatic transmission oil pump.
10. The _____ _____ _____ controls overall system pressure by exhausting extra fluid to the oil pan or pump intake.
11. The following valve is used to modify pressure according to vehicle speed:
    a. Manual valve.
    b. Throttle valve.
    c. Governor valve.
    d. Detent valve.
12. The shift linkage transfers motion from the driver to the _____ valve in the valve body.

13. The converter housing turns with the torque converter. True or false?
14. The bottom of the transmission case is machined to fit the valve body. True or false?
15. The _____ _____ supports the tail end of the transmission output shaft.

Certification-Type Questions–Chapter 10

1. **The torque converter is mounted:**
   (A) within the engine.
   (B) directly behind the engine.
   (C) directly in front of the engine.
   (D) on the pump drive hub.

2. **Planetary gearsets do all of these EXCEPT:**
   (A) provide reverse operation.
   (B) resemble the solar system.
   (C) provide torque multiplication.
   (D) drive the torque converter.

3. **Which of these serves as the control center for the transmission's hydraulic system?**
   (A) Drive shell.
   (B) Valve body.
   (C) Turbine shaft.
   (D) Torque converter.

4. **All of these are parts of a torque converter EXCEPT:**
   (A) a one-way clutch.
   (B) a clutch disc.
   (C) a lockup clutch.
   (D) a turbine.

5. **Thrust washers are usually placed between all of these torque converter parts EXCEPT:**
   (A) converter cover and turbine.
   (B) turbine and stator.
   (C) stator and impeller.
   (D) stator and overrunning clutch.

6. **Technician A says that the impeller turns whenever the engine is running. Technician B says that the turbine turns whenever the engine is running. Who is right?**
   (A) A only
   (B) B only
   (C) Both A & B
   (D) Neither A nor B

7. **As a result of turbine vane curvature, fluid coming from the impeller will do which of these before it returns to the impeller?**
   (A) Evaporate.
   (B) Increase flow.
   (C) Decrease flow.
   (D) Reverse direction.

8. **Before reaching the impeller, fluid leaving the turbine must pass through the:**
   (A) stator.
   (B) flexplate.
   (C) drive shell.
   (D) accumulator.

9. **Torque multiplication refers to a component's:**
   (A) decreasing output torque above its input torque.
   (B) increasing output torque above its input torque.
   (C) increasing input torque above its output torque.
   (D) decreasing input torque above its output torque.

10. Which of these is used to mechanically lock the turbine to the impeller and converter cover in high gear?

(A) Multiple-disc clutch.
(B) Lockup clutch.
(C) Overrunning clutch.
(D) Apply springs.

11. All of these shafts are in the automatic transmission EXCEPT:

(A) an input shaft.
(B) an output shaft.
(C) a detent shaft.
(D) a stator shaft.

12. An automatic transmission can achieve overdrive directly through:

(A) a clutch pack.
(B) the converter cover.
(C) the coupling phase.
(D) a planetary gearset.

13. Two or more planetary gearsets can be linked together to form a:

(A) multiple planetary gearset.
(B) compound planetary gearset.
(C) combination planetary gearset.
(D) main planetary gearset.

14. Technician A says that a Ravigneaux gear train is common in transmissions using overdrive. Technician B says that a Ravigneaux gear train contains two sun gears, two planet carriers, and one set of planet gears. Who is right?

(A) A only
(B) B only
(C) Both A & B
(D) Neither A nor B

15. All of these are planetary holding members EXCEPT:

(A) pawl clutches.
(B) bands.
(C) overrunning clutches.
(D) multiple-disc clutches.

16. Multiple-disc clutches are also referred to as:

(A) combination units.
(B) disc packs.
(C) clutch packs.
(D) overrunning clutches.

17. All of these are components in a multiple-disc clutch EXCEPT:

(A) a clutch drum.
(B) a ring gear.
(C) a clutch pressure plate.
(D) drive discs.

18. Automatic transmission bands are wrapped around the:

(A) clutch servo.
(B) pressure plate.
(C) clutch drum.
(D) drive disc.

19. Technician A says that bands are applied and released by servos. Technician B says that servos are hydraulic devices that consist of a cylinder and a piston. Who is right?

(A) A only
(B) B only
(C) Both A & B
(D) Neither A nor B

20. The hydraulic device used in the apply circuit of a band or clutch to cushion initial application is the:

(A) synchronizer.
(B) accumulator.
(C) limiter.
(D) regulator.

21. Technician A says that the transmission oil pump is used to keep the torque converter filled with transmission fluid. Technician B says that the transmission oil pump is used to circulate transmission fluid through the oil cooler. Who is right?

(A) A only
(B) B only
(C) Both A & B
(D) Neither A nor B

22. Technician A says that excessive transmission fluid temperatures will reduce the life of the transmission fluid. Technician B says that temperature will have no effect on the fluid's lubricating ability. Who is right?

(A) A only
(B) B only
(C) Both A & B
(D) Neither A nor B

23. All of these are hydraulic valves used in automatic transmissions EXCEPT:

(A) pressure regulator valves.
(B) manual valves.
(C) shift valves.
(D) recovery valves.

24. Technician A says that the throttle valve may be located in the valve body. Technician B says that the throttle valve may be located in a bore in the transmission case. Who is right?

(A) A only
(B) B only
(C) Both A & B
(D) Neither A nor B

25. The switch that permits starting only when the selector lever is in the park or neutral position is called the:

(A) restrictor switch.
(B) neutral start switch.
(C) ignition limiting switch.
(D) transmission cutoff switch.

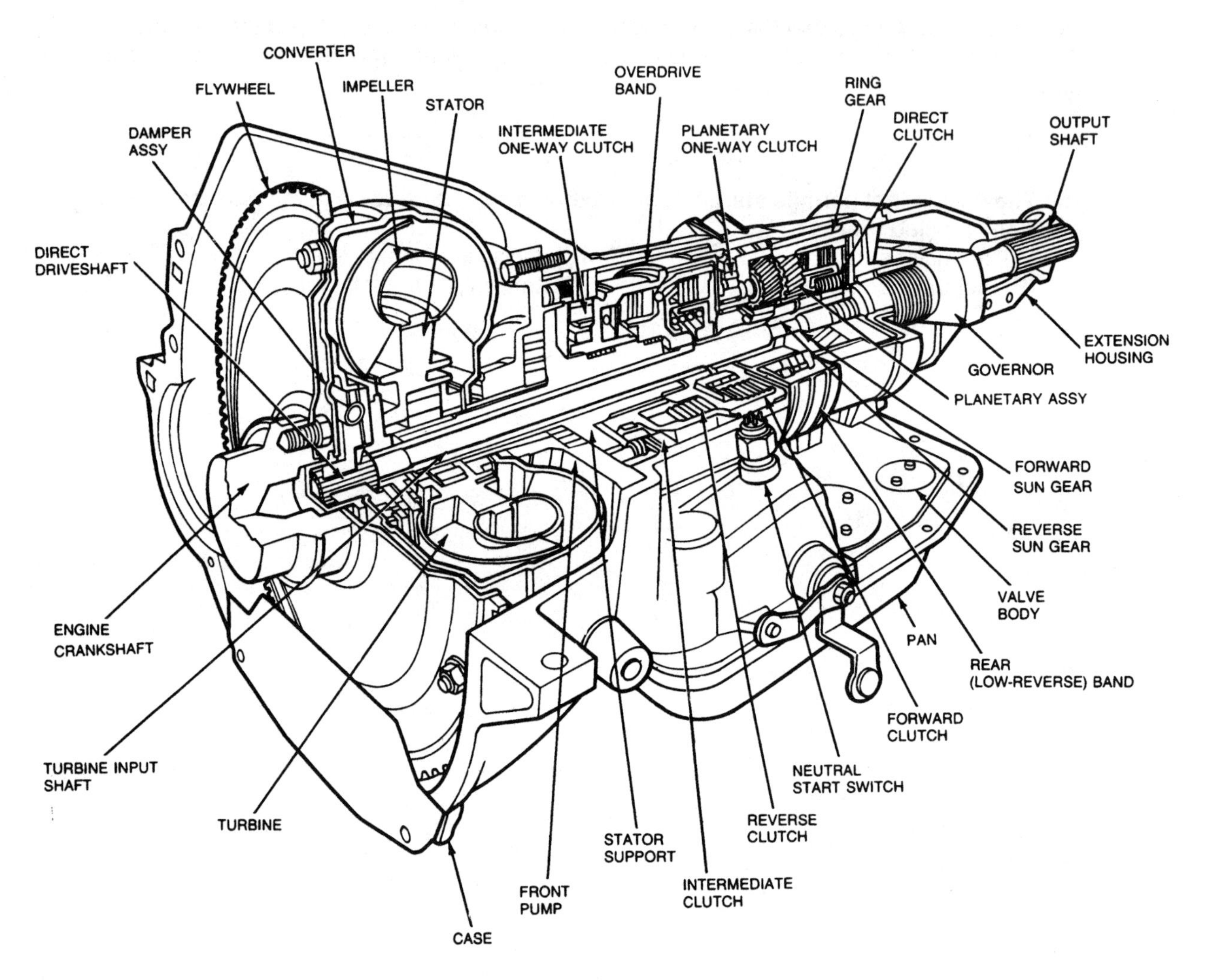

Illustration of a Ford *Automatic Overdrive Transmission* shows the location of the torque converter, front pump, clutches, gear train, and most of the internal parts.

Chapter 11

Automatic Transmission Circuits and Designs

After studying this chapter, you will be able to:

- Describe basic hydraulic circuit operation.
- Trace the path of oil flow through a typical automatic transmission circuit.
- Read and use hydraulic diagrams for automatic transmissions.
- Compare different automatic transmission designs.

This chapter will provide a detailed explanation on the operation of the automatic transmission hydraulic system. It begins with a discussion on basic hydraulic circuit operation. Once this is understood, you will be prepared to follow the operation of a typical modern hydraulic control system. Finally, the chapter will present a few of the many types of automatic transmission designs that exist.

Basic Hydraulic Circuit Operation

The components of a hydraulic circuit must work together to control the automatic transmission's mechanical system and, ultimately, produce movement and affect changes in gear ratios in response to engine and vehicle needs. This, then, is the major function of the complete hydraulic circuit–to apply and release a band or to engage or disengage a multiple-disc clutch in order to accomplish these tasks. It is by this action that the hydraulic circuit regulates the shifting process. In addition to this primary function, the transmission hydraulic circuit has secondary functions, such as routing fluid for cooling and for lubrication purposes.

2-Speed Automatic Transmission

The hydraulic circuit of a simple (2-speed) automatic transmission is shown in Figure 11-1. The complete circuit consists of such items as the oil pump and pan, hydraulic valves, servos and clutches, and connecting passageways. The oil pump produces the system pressure. This pressure is regulated by the pressure regulator valve. From the pressure regulator valve, pressurized fluid, or line pressure, goes to the manual valve in the valve body. The manual valve is controlled by the driver through the shift selector lever and shift linkage.

The transmission having the circuit shown in Figure 11-1 has two planetary holding members that give it low and high gears. The band is applied in low gear, and the clutch pack is used in high gear.

If the driver places the manual valve in drive, and the vehicle is not moving, line pressure goes to the shift valve and also to the throttle and governor valves. The manual valve also directs fluid to the band servo *apply side.* Pressure to the apply side of the servo overcomes the pressure of the release spring and applies the band. This is shown in Figure 11-2.

The output pressure from the throttle valve, or the throttle pressure, will vary, depending on engine load. The throttle valve in Figure 11-2 is operated by a vacuum modulator. Other throttle valves are operated by a throttle valve linkage, or T.V. linkage. With either type, throttle pressure changes with engine load. Heavy engine load causes high throttle pressure; light engine load causes low throttle pressure.

Governor pressure is proportional to road speed. The governor is often driven by the output shaft. It revolves faster as the vehicle speed increases. As the governor speed increases, forces are exerted on the valve, causing it to move and increasing governor pressure. As a result, an increasing pressure signal is sent to the shift valve. When the vehicle slows down, forces exerted on the valve decrease, and it modulates accordingly. As a result, the governor valve puts out less pressure.

Governor pressure and throttle pressure, acting on opposite ends of the shift valve, oppose each other. At low speeds, governor pressure is low; throttle pressure plus spring pressure hold the shift valve in its normal position, to the right in Figure 11-2. With the shift valve in this

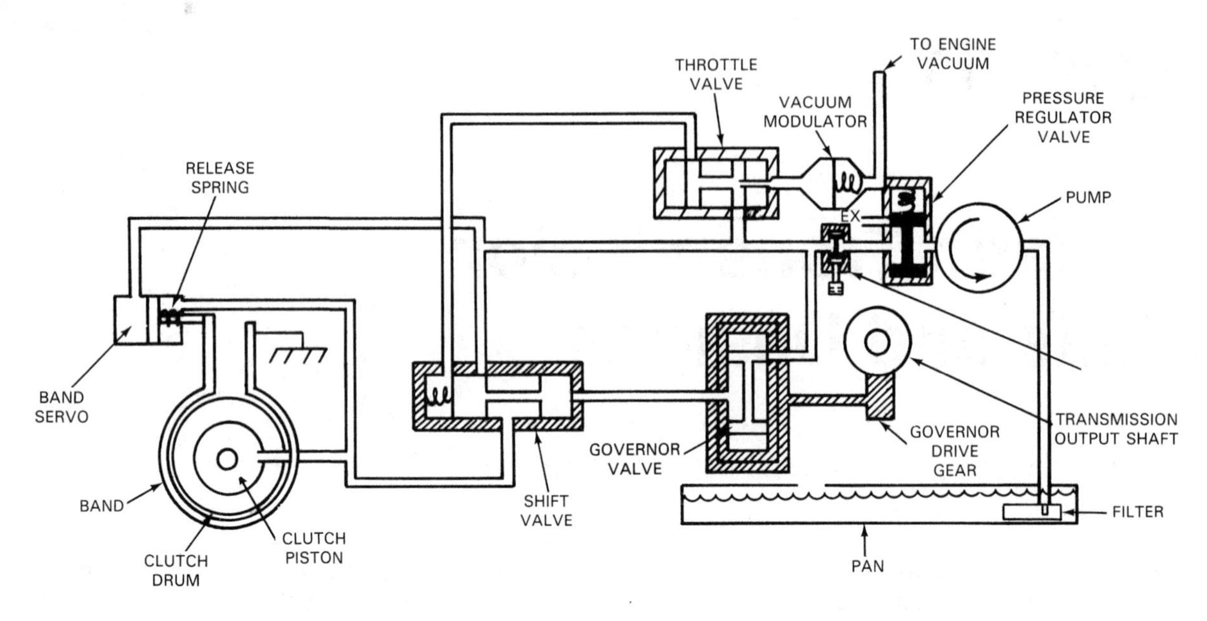

Figure 11-1. This diagram of a simple transmission has one shift valve and two holding members.

position, line pressure from the manual valve is blocked from the clutch pack, and it remains disengaged.

As the vehicle speeds up, governor pressure increases. At a certain vehicle speed, the governor pressure becomes greater than throttle pressure and spring pressure; the shift valve moves, to the left in Figure 11-3. The speed of the vehicle at which the upshift occurs is governed by the throttle pressure. Hard acceleration causes higher throttle pressure, and the governor has to spin faster to put out enough pressure to overcome throttle pressure. This means that the vehicle has to be moving at a higher speed before the upshift can occur. Light acceleration means lower throttle pressure, and a lower vehicle speed will cause enough governor pressure to move the shift valve.

With the shift valve pushed to the left by governor pressure, line pressure from the manual valve can flow

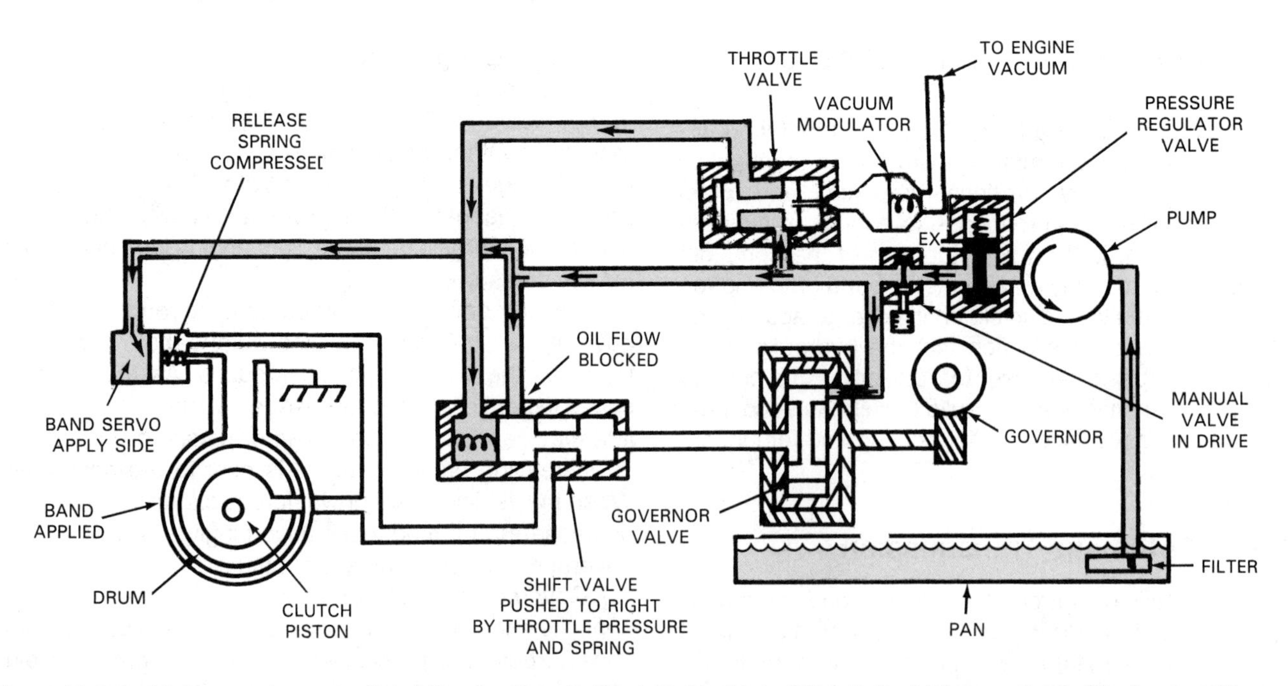

Figure 11-2. In drive range–low gear, the shift valve stays to the right, blocking oil flow to the high gear clutch. The low band is applied through the manual valve. (Note that this is a very simple representation of a transmission hydraulic circuit.)

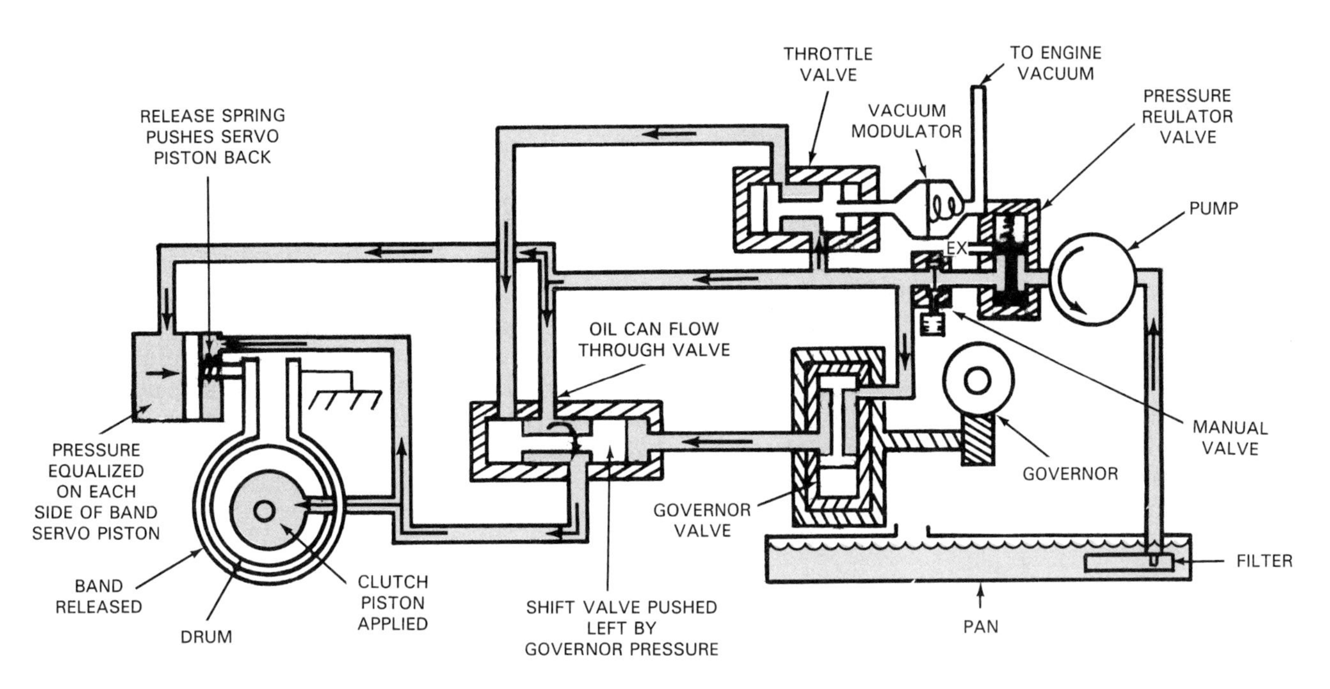

Figure 11-3. In drive range–high gear, the shift valve moves to the left. Oil pressure can flow to the high clutch and the release side of the band servo. This shifts the transmission to high.

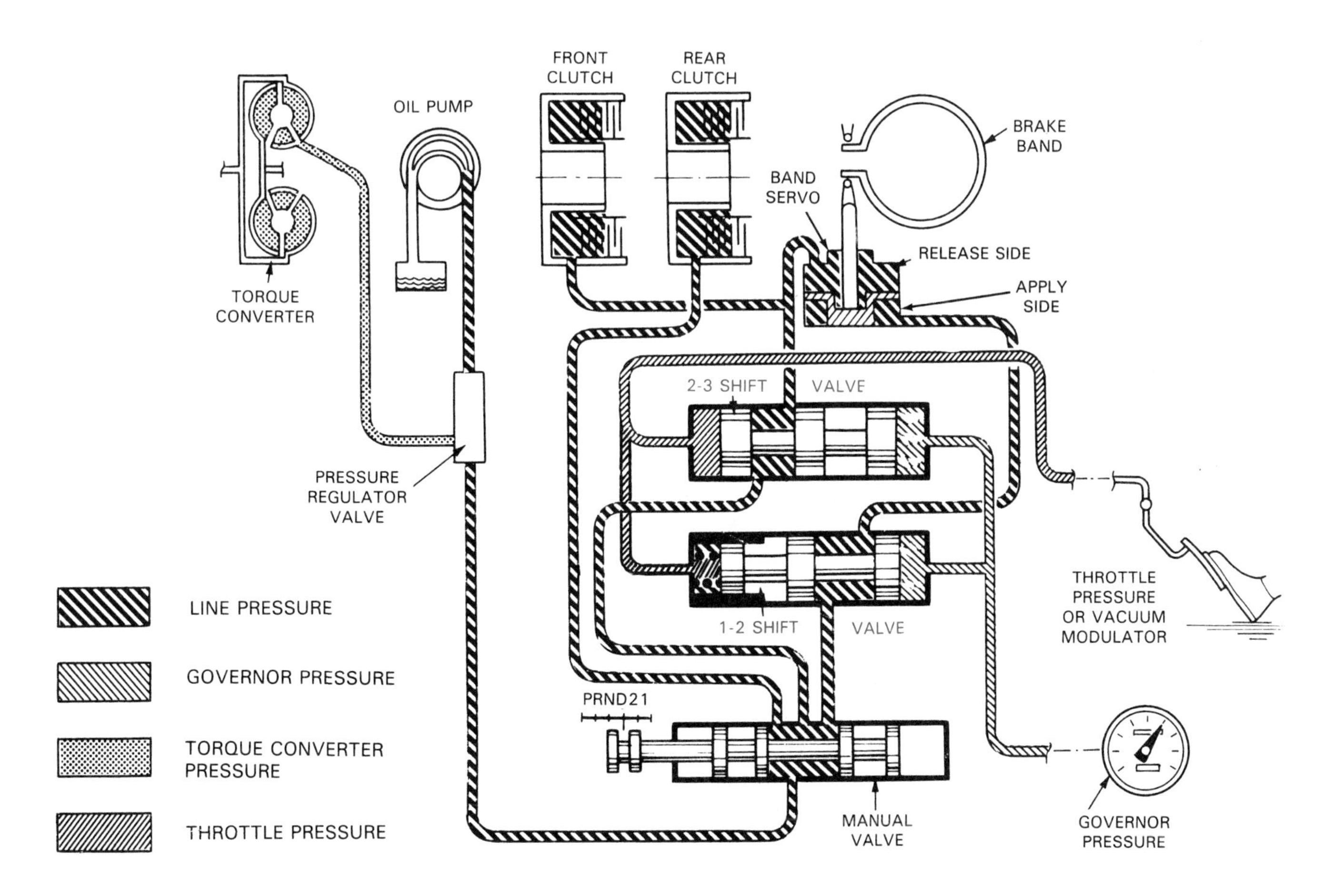

Figure 11-4. A 3-speed transmission has two shift valves. It upshifts in a manner similar to the 2-speed. The shift valves are designed so that the 1-2 shift valve is moved by governor pressure before the 2-3 shift valve. (Nissan)

through the shift valve to cause an upshift. Pressure goes to the clutch piston to apply the clutch and also to the *release side* of the band servo. Oil pressure equalizes on both sides of the servo piston. The release spring pushes the piston away from the band, releasing it. The vehicle is now in high gear.

If the vehicle slows down, governor pressure decreases. At some vehicle speed, depending on throttle position, governor pressure becomes less than the throttle plus spring pressures. The combined pressures will then push the shift valve to the right (in Figure 11-3), causing a downshift. An open throttle (low vacuum pressure) causes high throttle pressure, hastening the downshift. This action causes a forced downshift. A closed throttle (high vacuum pressure) causes low throttle pressure, delaying the downshift. A review of throttle valves, shift valves, and governor valves, presented in Chapter 10, may be helpful if you find that you do not fully understand the operation just described.

3-Speed Automatic Transmission

The hydraulic system of a 3-speed automatic transmission is slightly different from the simple 2-speed unit discussed so far. Figure 11-4 (on preceding page) shows a simple hydraulic control circuit in a 3-speed transmission.

Notice that the ***1-2*** and ***2-3 shift valves*** are constructed so that the throttle pressure side of the 1-2 shift valve is smaller than the throttle pressure side of the 2-3 shift valve. Therefore, less governor pressure is needed to overcome the throttle pressure on the 1-2 shift valve than on the 2-3 shift valve. This arrangement means that the 1-2

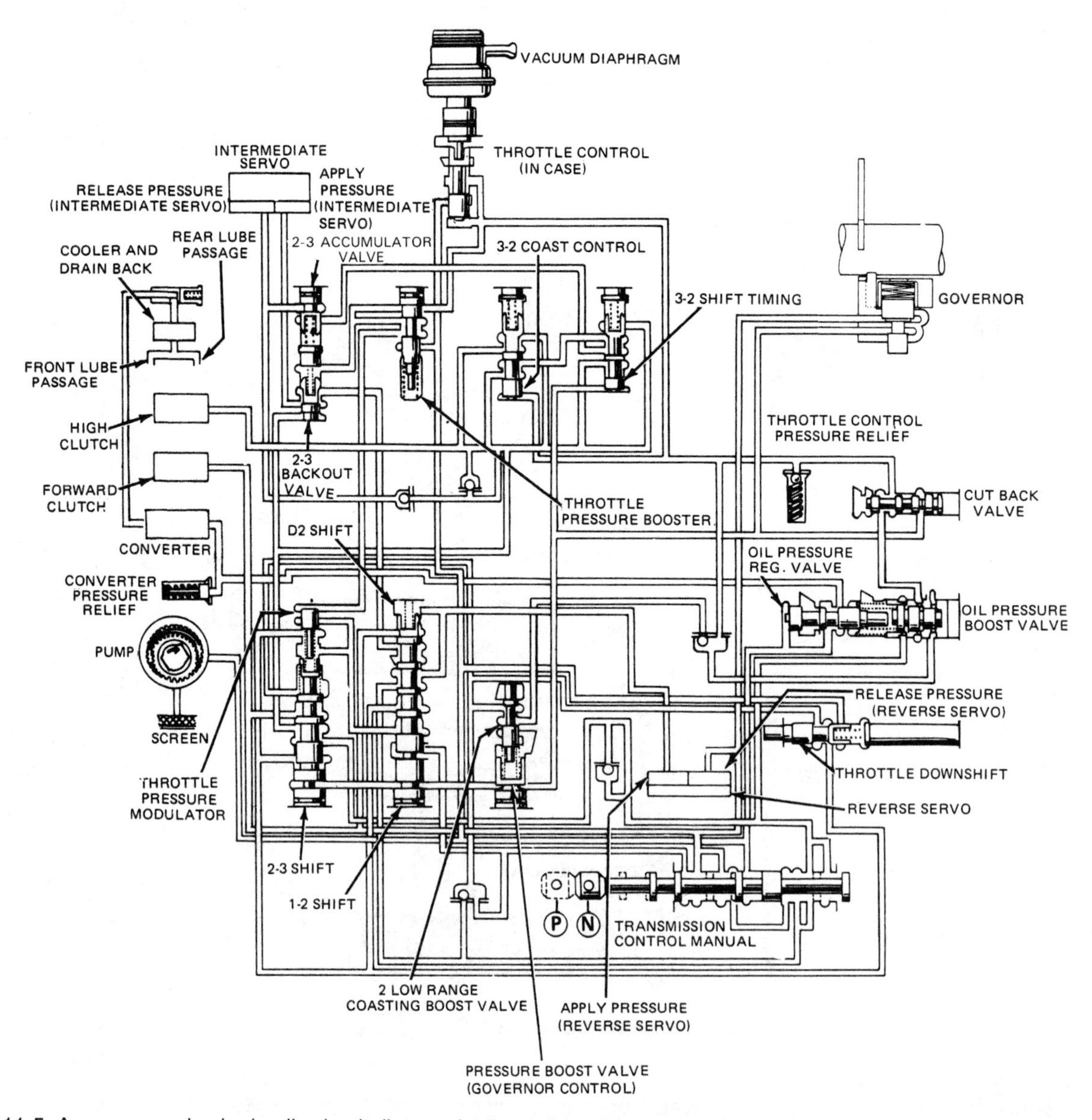

Figure 11-5. A more-complex hydraulic circuit diagram is shown here. The additional valves control shift timing and feel. (Ford)

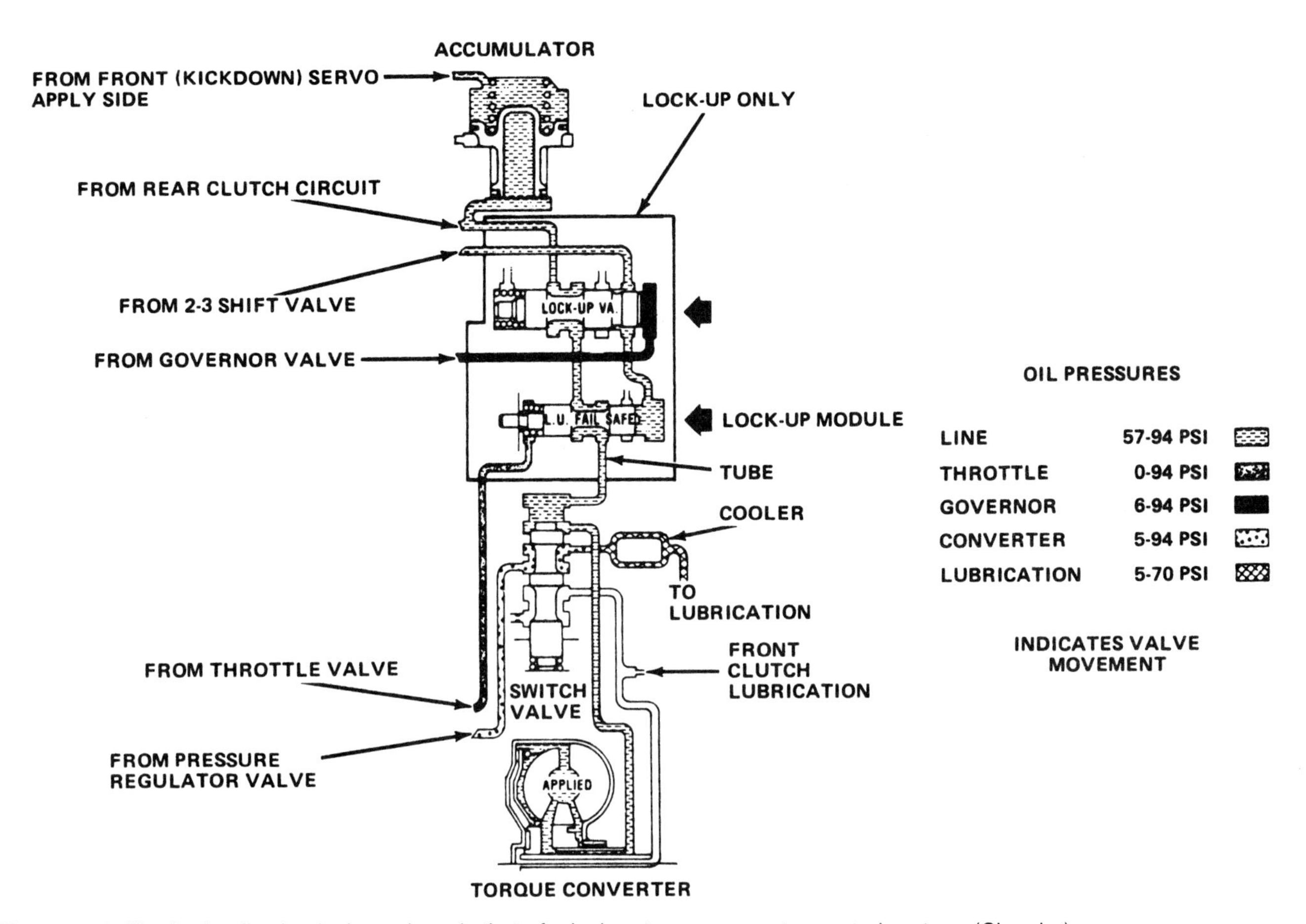

Figure 11-6. The hydraulic circuit shown here is that of a lockup torque converter control system. (Chrysler)

shift valve will always move to the upshifted position before the 2-3 shift valve.

The transmission starts off from rest in drive, and the rear clutch is applied. The vehicle is in first gear. As vehicle speed increases, governor pressure moves the 1-2 shift valve to the upshifted position. Moving the 1-2 shift valve to the upshifted position causes oil to flow to the apply side of the band servo, applying the band and placing the transmission in second gear.

When the vehicle is moving fast enough, governor pressure can overcome throttle pressure and move the 2-3 shift valve to the upshifted position. In the upshifted position, the 2-3 shift valve allows pressure to go to the front clutch and the release side of the band servo. Applying the front clutch and releasing the band puts the transmission in third gear.

When the vehicle slows down, governor pressure drops off, and the 2-3 valve moves, at some point, to the downshifted position. In this position, pressure is blocked from the front clutch and the release side of the band servo. Oil returns to the apply side of the band servo, applying the band and placing the transmission in second gear.

As the vehicle slows down further, governor pressure drops even more. This causes the 1-2 shift valve to move to the downshifted position. Moving the 1-2 shift valve to the downshifted position blocks off oil pressure to the apply side of the band servo, and the transmission goes into first gear.

Figure 11-5 is a more detailed diagram showing a different make of 3-speed automatic transmission. The hydraulic circuit pictured is obviously more complex than the one shown in Figure 11-4. Extra valves are added to more precisely control and cushion automatic and manual shifts.

The *2-3 backout valve* prevents a rough upshift if the driver lifts up on the accelerator just as the shift occurs. The *throttle pressure booster* works with the 2-3 accumulator valve to precisely time and cushion the shift during forced downshifts.

The *cut back valve* uses governor pressure to reduce throttle pressure. This allows the 2-3 shift valve to upshift when the engine speed becomes too high in passing gear. The *3-2 shift timing* valve causes all holding members to briefly disengage during detent downshifts. This allows the engine to speed up for more power in passing gear when the holding members reengage.

Other valves are used to control the torque converter lockup clutch, as shown in Figure 11-6. Oil pressure from the governor, throttle valve, and 2-3 shift valve ensure that the lockup clutch is engaged only at certain speeds, and only when the transmission is in third gear. Other valves are used to regulate converter oil pressure, to boost pressure in reverse and in manual low or manual second gears, or to ensure that the transmission bearings receive enough lubrication.

Modern Hydraulic Control Systems

The hydraulic system of a typical modern automatic transmission is shown in *hydraulic diagrams* of Figures 11-7 through 11-16. It appears to be very complicated, but it can be analyzed and understood now that you know the function of the various components.

Park

Figure 11-7 shows the transmission in park. The engine is running, and the oil pump sends oil to the pressure regulator valve. The pressure regulator valve controls the system pressure. In park, the spring in the pressure regulator valve is the only control of system line pressure.

From the pressure regulator valve, oil travels to the converter clutch *actuator-apply valve.* The oil is directed by the valve to the release side of the converter clutch and on into the torque converter. Oil returning from the converter is routed through the actuator-apply valve to the oil cooler and on to the lubrication system, to keep the transmission bearing surfaces lubricated. From here, the oil returns to the transmission oil pan.

The other oil pressure circuitry in park is simple. Line pressure from the pressure regulator valve goes to the manual valve, and it is stopped at the valve. The parking pawl engages the parking gear to lock the transmission. All clutches and bands are released.

Neutral

In neutral, Figure 11-8, pressurized oil from the pressure regulator valve goes to the converter clutch actuator-apply valve and the manual valve. The converter and lube circuit works the same as when in park. The valve is positioned so that passages leading to the holding members are blocked, keeping clutches and bands in released position.

Oil travels from the manual valve to the throttle valve. Throttle pressure acts on the spring end of the regulator valve. This action serves to boost line pressure according to throttle opening. Throttle pressure acts on the *shift throttle valve,* which serves to limit throttle pressure from exceeding 90 psi (620 kPa). Oil also travels to a few other valves, but it does not have any effect on these valves at this time.

Drive Range–First Gear

When the selector lever is placed in drive (before the vehicle begins moving), the transmission goes into first gear. This is drive range–first gear. See Figure 11-9. Line pressure passes through the manual valve to the forward clutch, applying the clutch. Applying the forward clutch causes the *low roller clutch,* or overrunning clutch, to lock, holding the rear planet-pinion carrier and placing the transmission in first gear. The converter and lube circuit works the same as before.

Oil also flows through the manual valve to the 1-2 shift valve, the 2-3 shift valve, the *1-2 accumulator valve,* the governor, and the throttle valve. Throttle pressure and governor pressure act on opposite ends of the shift valves. The valves are designed so that it takes more governor pressure to move the 2-3 shift valve against throttle pressure than the 1-2 shift valve. Note that this is accomplished in some designs by placing a heavier spring on the throttle side of the 2-3 shift valve. It may also be accomplished by making the face area on the throttle side of the 1-2 shift valve smaller than that of the 2-3 shift valve, as mentioned earlier in the chapter.

Throttle pressure also goes to the pressure regulator valve, where it assists the pressure regulator spring in helping to hold the pressure regulator valve closed. Changes in throttle pressure, therefore, affect the setting of the regulator valve. When throttle pressure increases, the pressure regulator valve raises the line pressure. When throttle pressure decreases, the regulator valve adjusts line pressure back downward. This regulation of line pressure caused by the throttle valve adjusts the firmness of clutch and band application to match the engine power output. Adjusting the shift firmness allows for the smoothest shifts with maximum band and clutch life.

Drive Range–Second Gear

As the vehicle speed increases, the pressure from the governor valve becomes greater than the pressure from the throttle valve (plus spring pressure). The 1-2 shift valve is moved to the upshifted position, placing the transmission in drive range–second gear, Figure 11-10.

When the 1-2 shift valve moves, it sends pressure to the *intermediate servo* to apply the *intermediate band.* However, some of the pressure to the servo is diverted to the *1-2 accumulator piston.* The diversion of this oil causes a gradual buildup of pressure at the servo during band application. This results in a smoother shift. When the accumulator piston bottoms out in its bore, full pressure is applied to the band to hold it firmly.

Applying the intermediate band locks the sun gear to the case, shifting the transmission into second gear. The rear planetary ring gear is splined to the output shaft. It turns with the shaft. The ring gear causes the pinions to walk around the stationary sun gear; the direction of rotation is such that the low roller clutch freewheels so that the planet-pinion carrier rotates without restraint. The converter and lube circuit work the same as before. The forward clutch is still applied.

Drive Range–Third Gear

As governor pressure increases, it overcomes throttle pressure and spring pressure on the 2-3 shift valve and moves the valve to the upshifted position. This places the transmission in drive range–third gear. See Figure 11-11. In this position, *direct clutch oil* passes through the 2-3 shift valve to apply the *direct clutch.* The forward clutch stays applied. Direct clutch oil is fed to the release side of the intermediate servo. This equalizes pressure on both sides of the servo, allowing the servo return spring to decompress and release the band. The action of the servo when it releases absorbs some oil pressure and cushions the direct clutch apply. A separate accumulator is not needed.

In third gear, direct clutch oil also goes to the converter clutch solenoid. When the proper speed is reached in third gear, the solenoid is energized. The energized solenoid closes the exhaust line from the direct clutch. (Normally, the fluid that keeps the direct clutch engaged flows through the clutch and exhausts.) This action sends direct clutch oil to the converter clutch actuator-apply valve, moving it against line pressure. Moving the valve sends oil into the apply side of the converter clutch, causing the clutch piston to engage the converter cover. This causes the clutch and torque converter to lock up. Meanwhile, converter feed oil is directed to the transmission cooler and on to the lubrication system. (Note that diagram of Figure 11-11 depicts converter clutch released.)

When the vehicle drops below the specified engagement speed, the solenoid valve opens to exhaust pressure. The actuator-apply valve is released, opening the line from the apply side of the converter clutch. The pressure on the clutch piston is relieved, and the clutch disengages.

Part Throttle 3-2 Downshift

In Figure 11-12, the throttle pedal has been pushed far enough to move the throttle valve into the *part-throttle, downshifted position.* In this position, the throttle valve can send enough pressure to the 2-3 shift valve to push it into the downshifted position. A special part-throttle passageway sends oil to one of the *annular grooves* of the 2-3 shift valve. The pressure acts upon the associated valve lands so that the net force on the valve opposes governor pressure.

If the vehicle is moving slowly enough, the governor pressure is overcome, and the 2-3 shift valve moves to the downshifted position. This exhausts oil from the direct clutch and from the release side of the intermediate servo. The *direct clutch exhaust valve* controls the clutch release and band reapplication to give a smooth downshift. The transmission goes from third gear back to second since the direct clutch is released and the intermediate band is reapplied. If the converter clutch is applied, loss of direct clutch oil will cause it to release.

Detent Downshift

A detent downshift is provided when the throttle pedal is completely depressed, pushing the throttle valve completely into its bore. This sends *detent oil* to the 1-2 and 2-3 shift valves. See Figure 11-13. Governor pressure will be overcome to force one or both valves into the downshifted position. This will move the transmission from third gear into second or from second gear into first, depending on vehicle speed.

If the vehicle speed exceeds a certain amount, governor pressure will be too great to overcome, and the transmission will remain in third gear. A check ball in the passageway to the direct clutch or intermediate servo unseats to quickly exhaust the apply pressure. Detent pressure through the throttle valve passageway will cause the pressure regulator valve to raise line pressure. Increased line pressure helps the bands and clutches to handle the increased engine output during detent operation.

Intermediate Range

Manual intermediate, or *manual second,* is applied when the driver moves the manual valve to the *intermediate range* position. If the vehicle is in third gear when this range is selected, the transmission will downshift into second gear. Thus, a 3-2 downshift can be accomplished by moving the selector lever into the intermediate range. If the vehicle is stopped when manual second is selected, the transmission will make a normal shift from first to second when the vehicle is accelerated, but it will not shift into third.

Intermediate range is shown in Figure 11-14. When the selector lever is in the intermediate position, *reverse-neutral-drive (RND) oil* will exhaust at the manual valve. RND oil is the feed for direct clutch oil in third gear. With RND oil exhausted, the transmission cannot upshift to third gear regardless of vehicle speed.

Notice that the line pressure is raised by applying boost pressure to the pressure regulator valve. This is done when pressure from the *intermediate boost valve* seats check balls in the intermediate boost and line boost passageways. Seating the check balls applies boost pressure to the spring side of the regulator valve. Additional pressure is needed to hold the intermediate band.

When shift throttle valve pressure is high enough, it will seat one of the check balls against *intermediate boost oil.* Then, pressure from the shift throttle valve will work against the regulator valve, varying line pressure according to throttle opening.

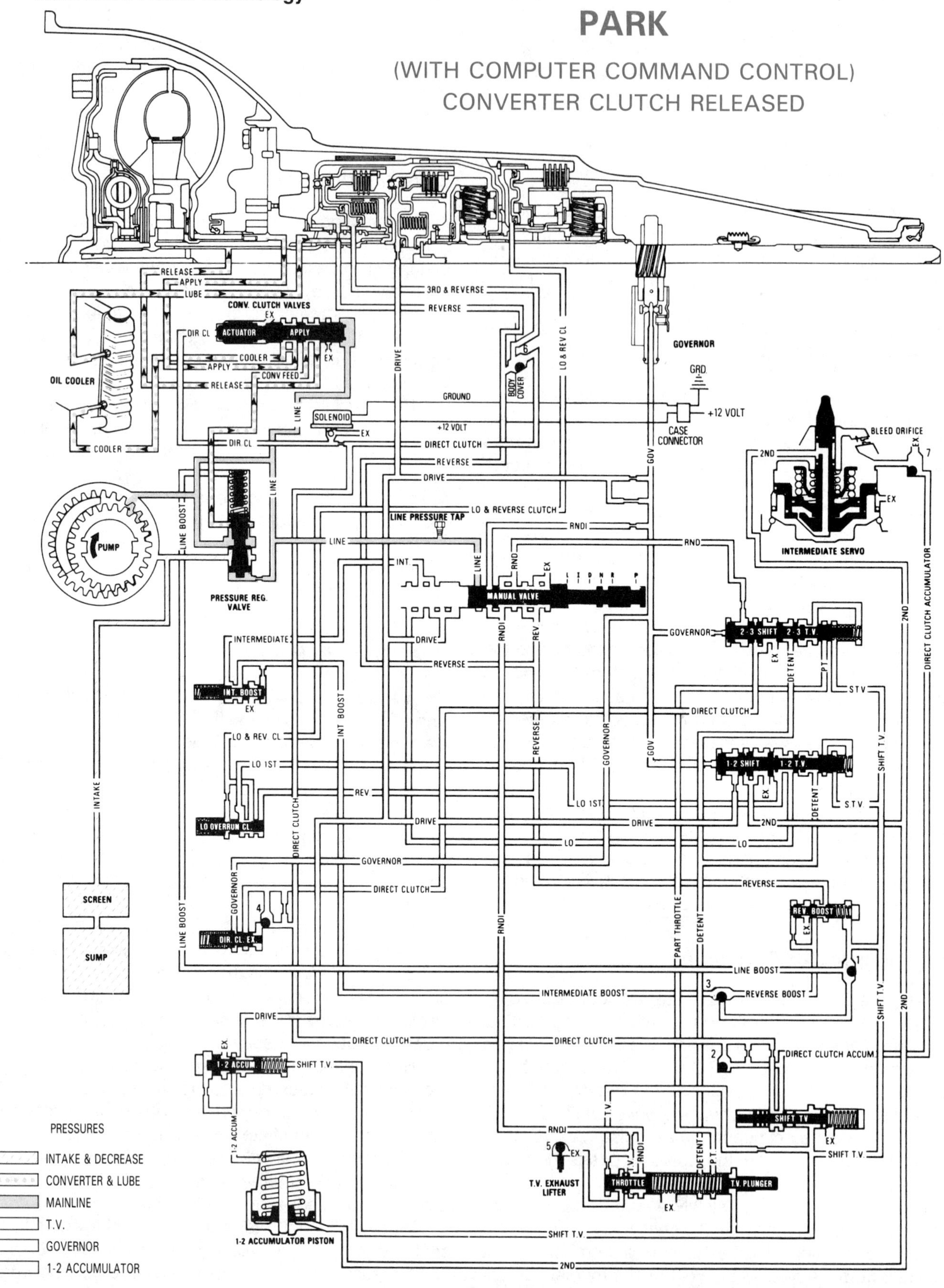

Figure 11-7. This common 3-speed transmission has a hydraulic control system that controls three multiple-disc clutches and a band. The diagram traces path of oil flow when the transmission is in park with the engine running. Converter is filled from the release side. All clutches and the band are released. Parking pawl is engaged with parking gear. (General Motors)

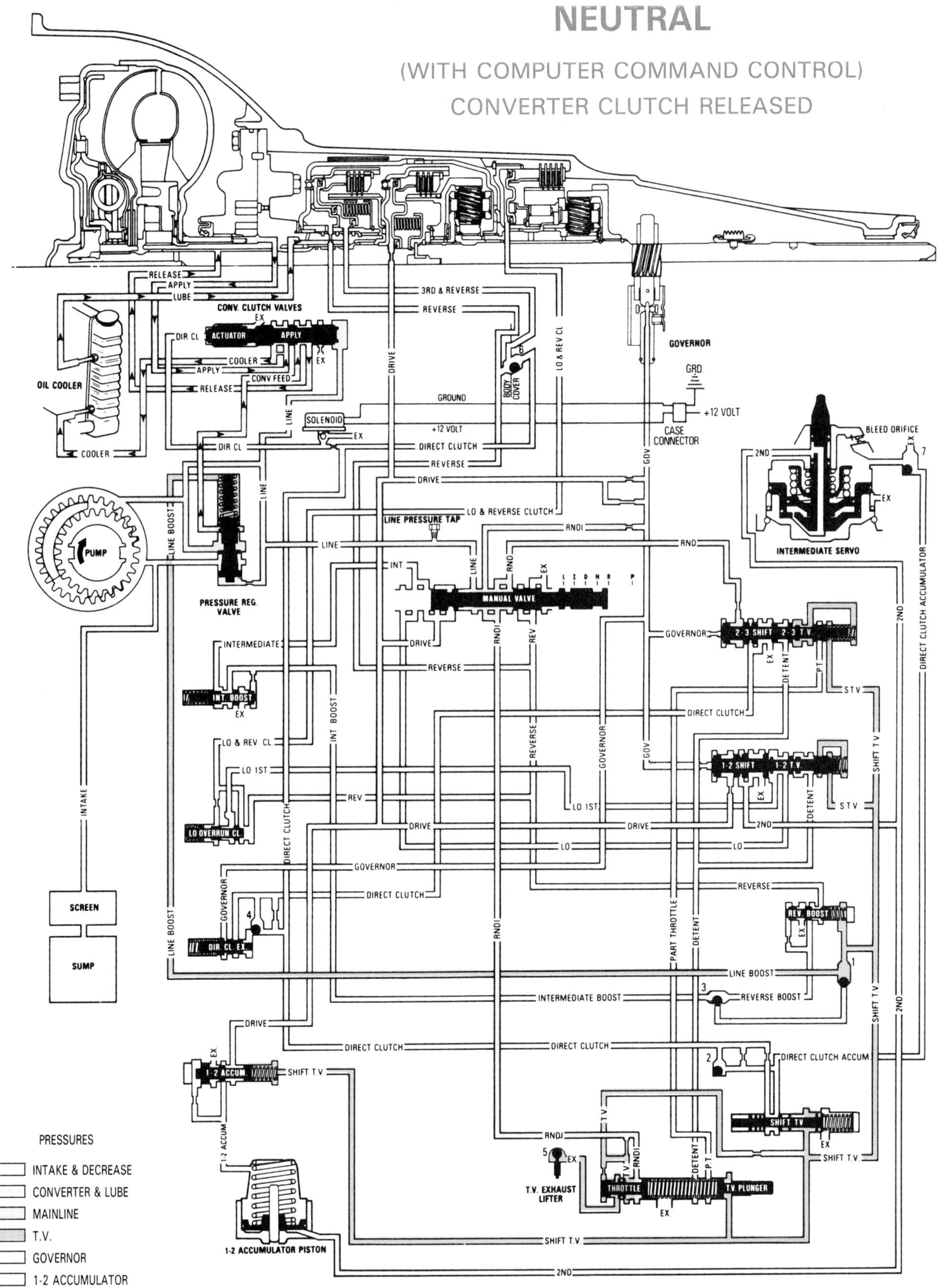

Figure 11-8. In neutral, all clutches and the band are released. The converter is filled. The throttle valve is modifying line pressure by sending a pressure signal to assist the spring of the pressure regulator valve. (General Motors)

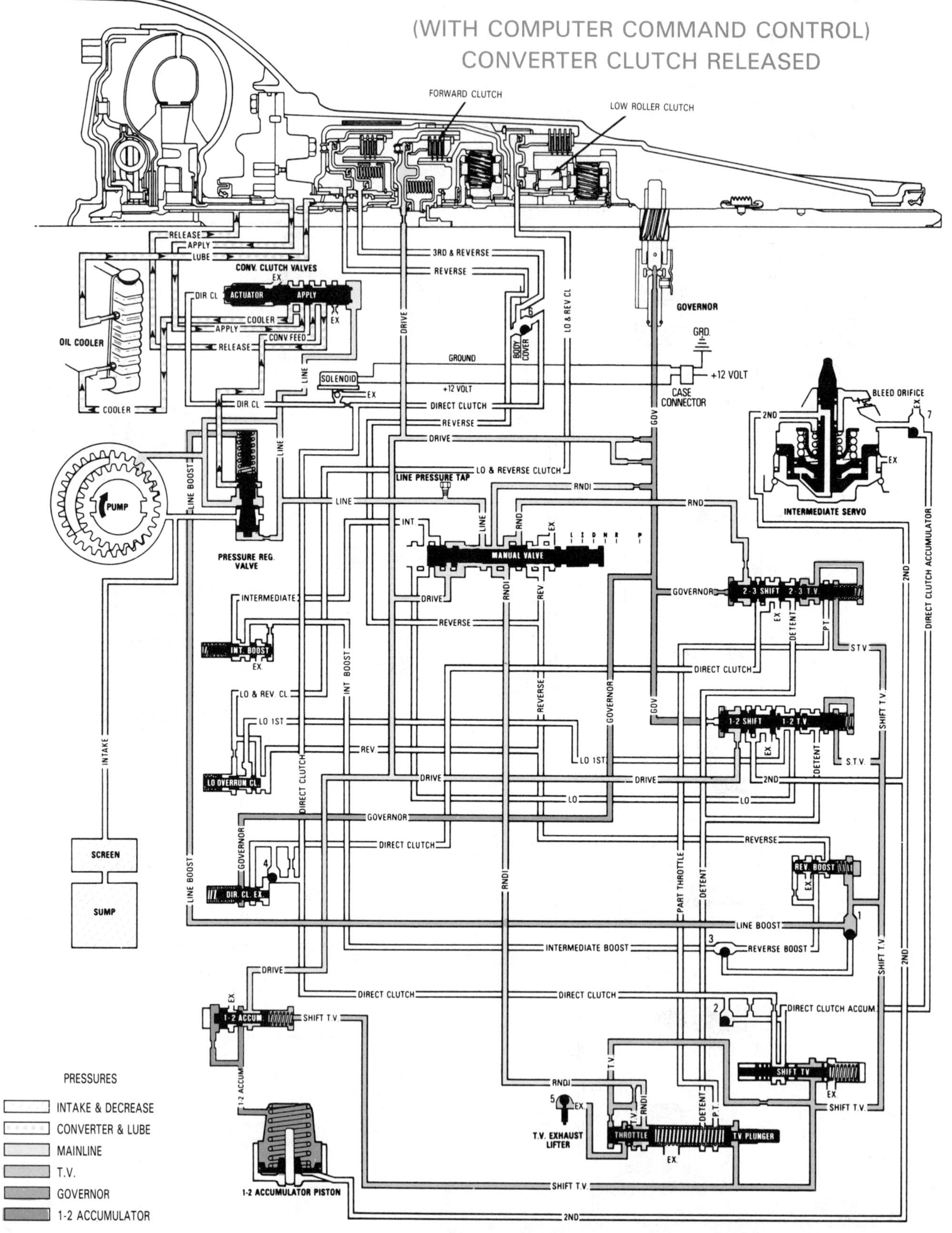

Figure 11-9. In drive with the vehicle not moving, the manual valve applies the forward clutch. The low roller clutch is holding. The converter clutch is released. The governor pressure is not high enough to move the 1-2 shift valve to upshifted position. The vehicle is in first gear. (General Motors)

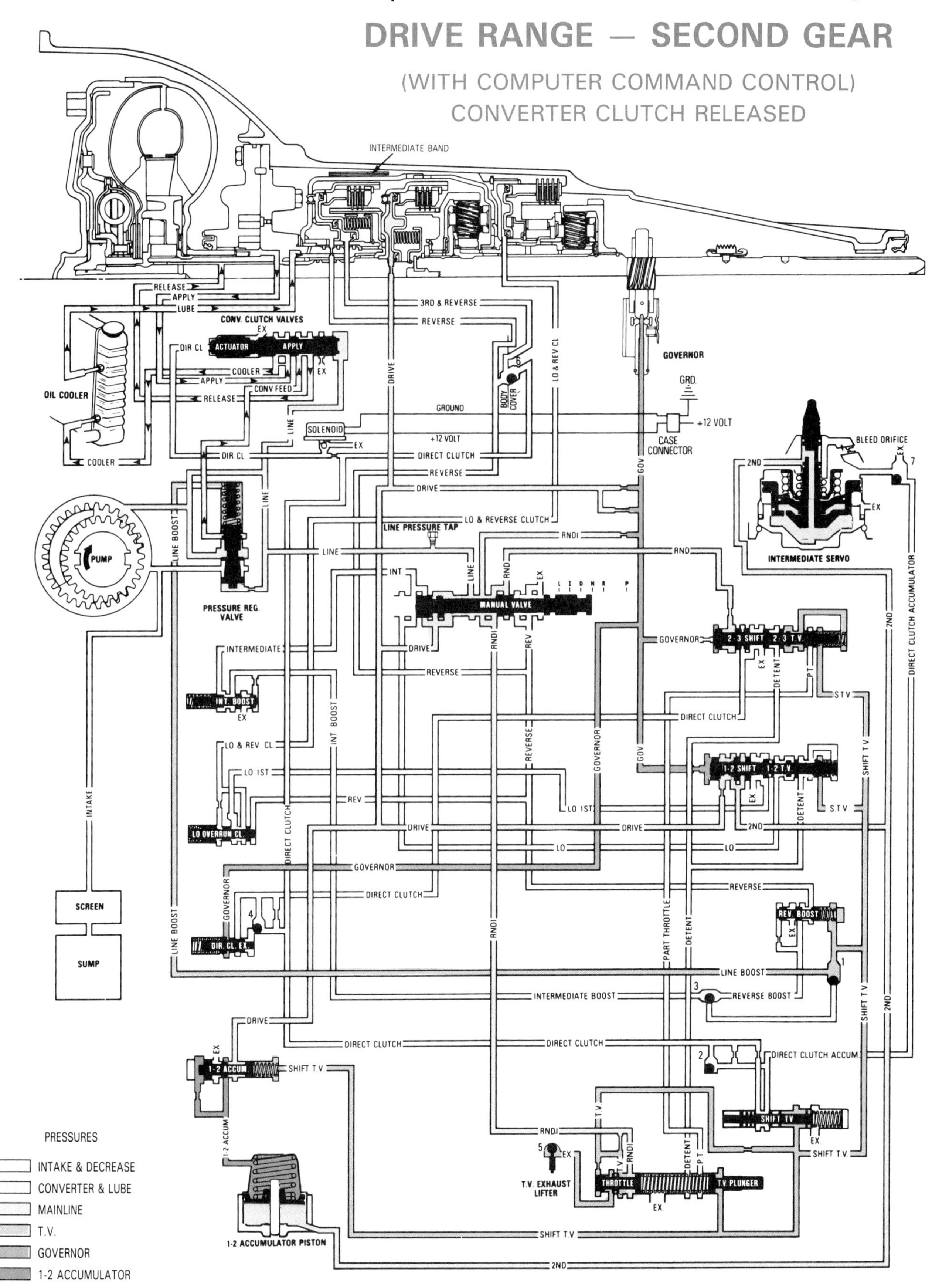

Figure 11-10. In second gear, the 1-2 shift valve has been moved by governor pressure. The valve directs oil to the apply side of the intermediate servo and 1-2 accumulator piston, applying the intermediate band. The forward clutch remains applied. The converter clutch is released and the low roller clutch freewheels. (General Motors)

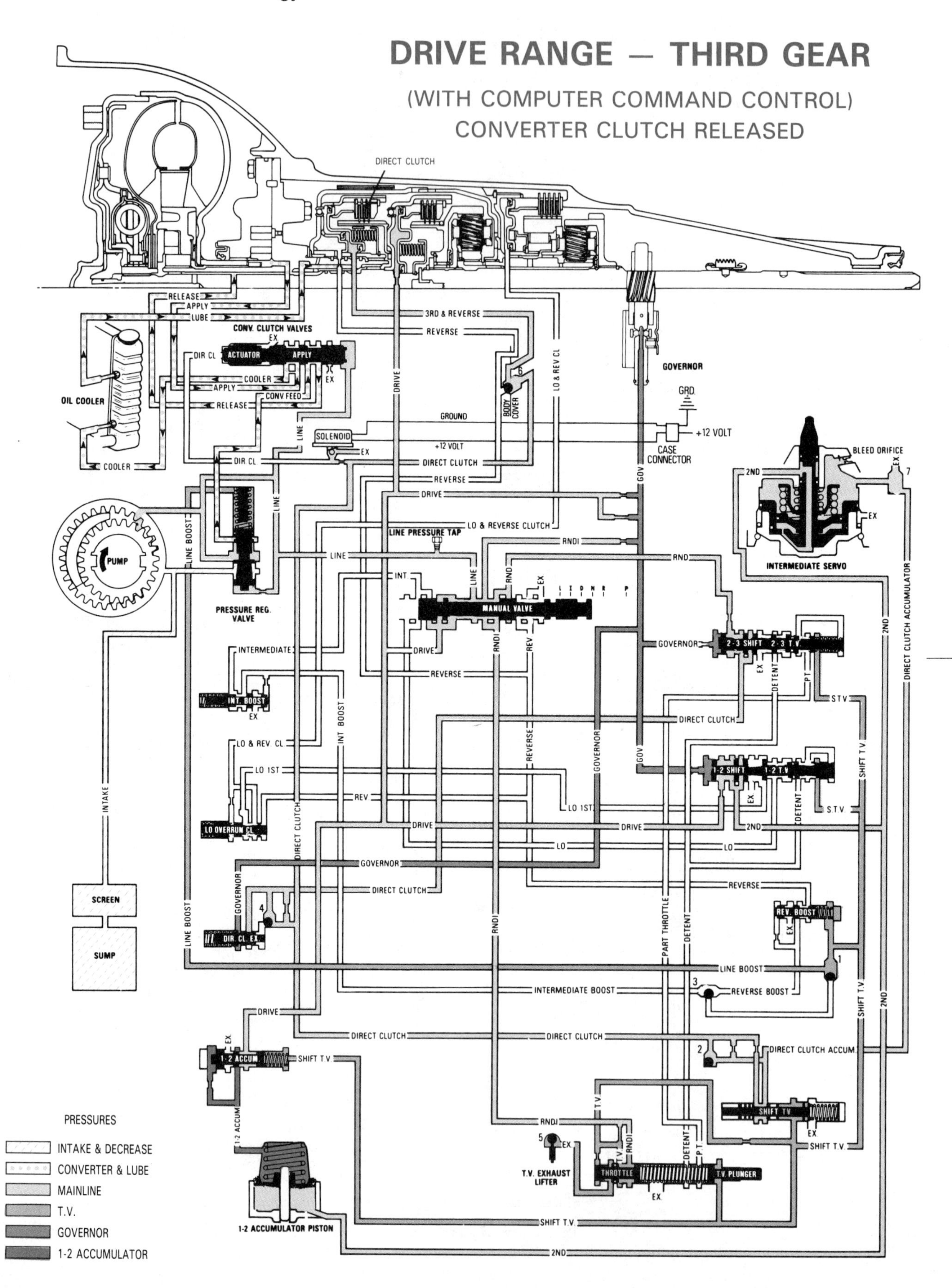

Figure 11-11. In third gear, the 2-3 shift valve sends oil to the direct clutch and to the release side of the intermediate servo. The forward clutch remains applied. Direct clutch oil is exhausted at the converter clutch solenoid. When the proper speed is reached, a computer signal causes the solenoid valve to close the exhaust line. This action diverts the oil to the converter clutch actuator-apply valve, causing the valve to move. Moving the valve sends oil into the apply side of the converter clutch, causing the clutch piston to engage the converter cover. (General Motors)

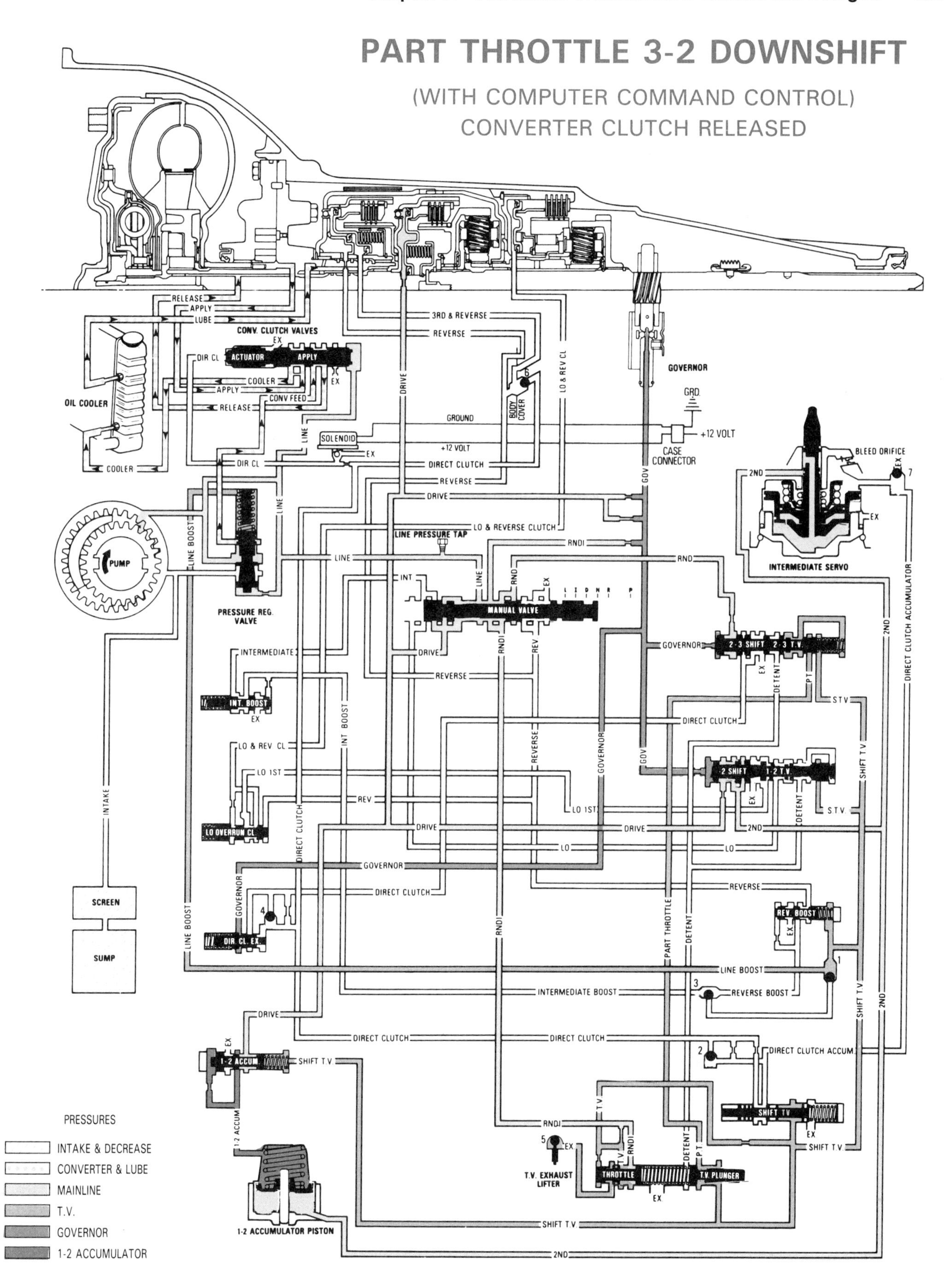

Figure 11-12. If the throttle is partially depressed at low vehicle speeds, a part throttle 3-2 downshift may occur. This places the transmission back in second. (General Motors)

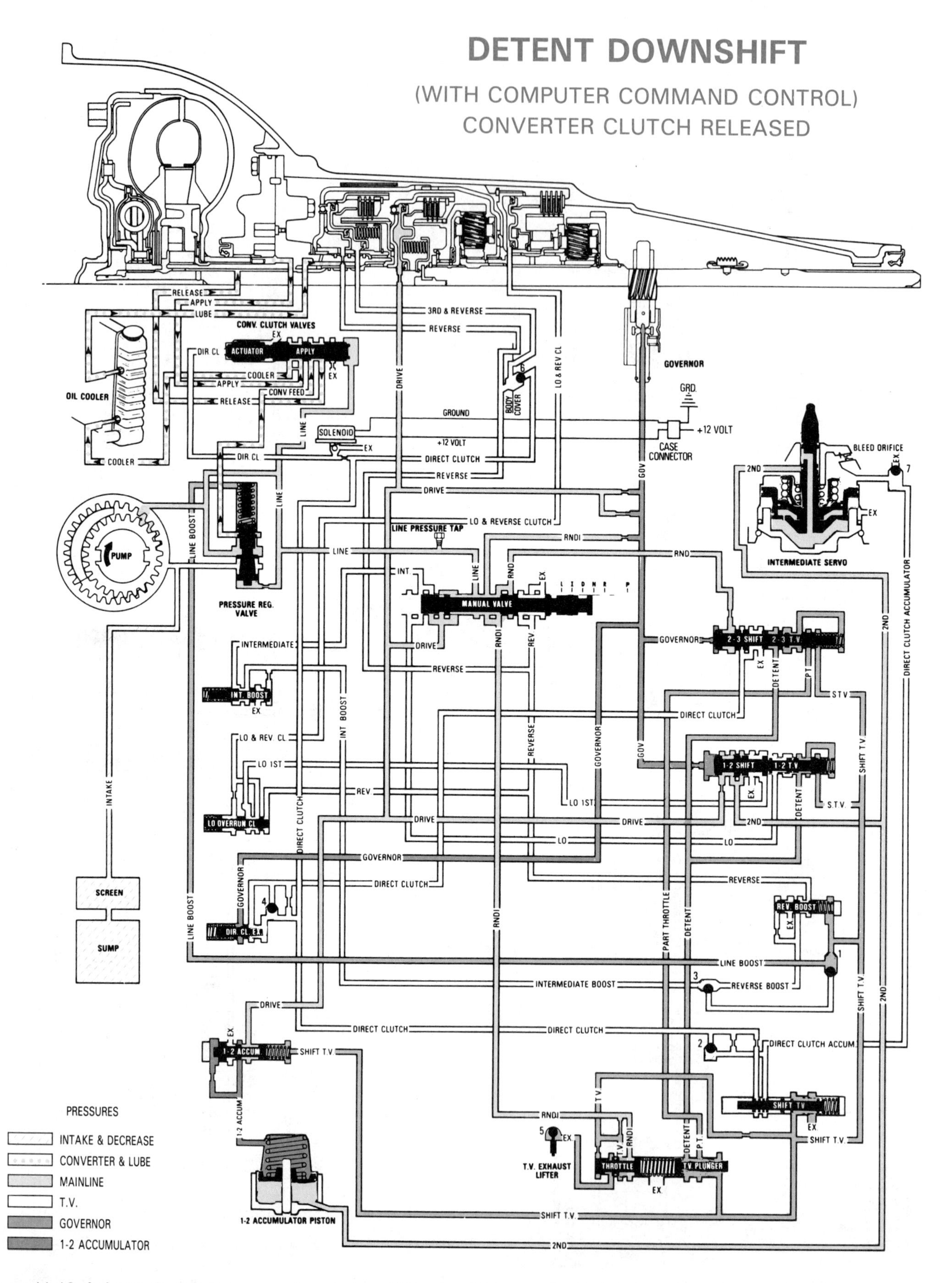

Figure 11-13. A detent downshift causes a forced downshift by sending extra oil to the 1-2 and 2-3 shift valves. The detent valve can be mechanically or electrically operated. Depending on the vehicle speed, the transmission will downshift to either second or first gear. (General Motors)

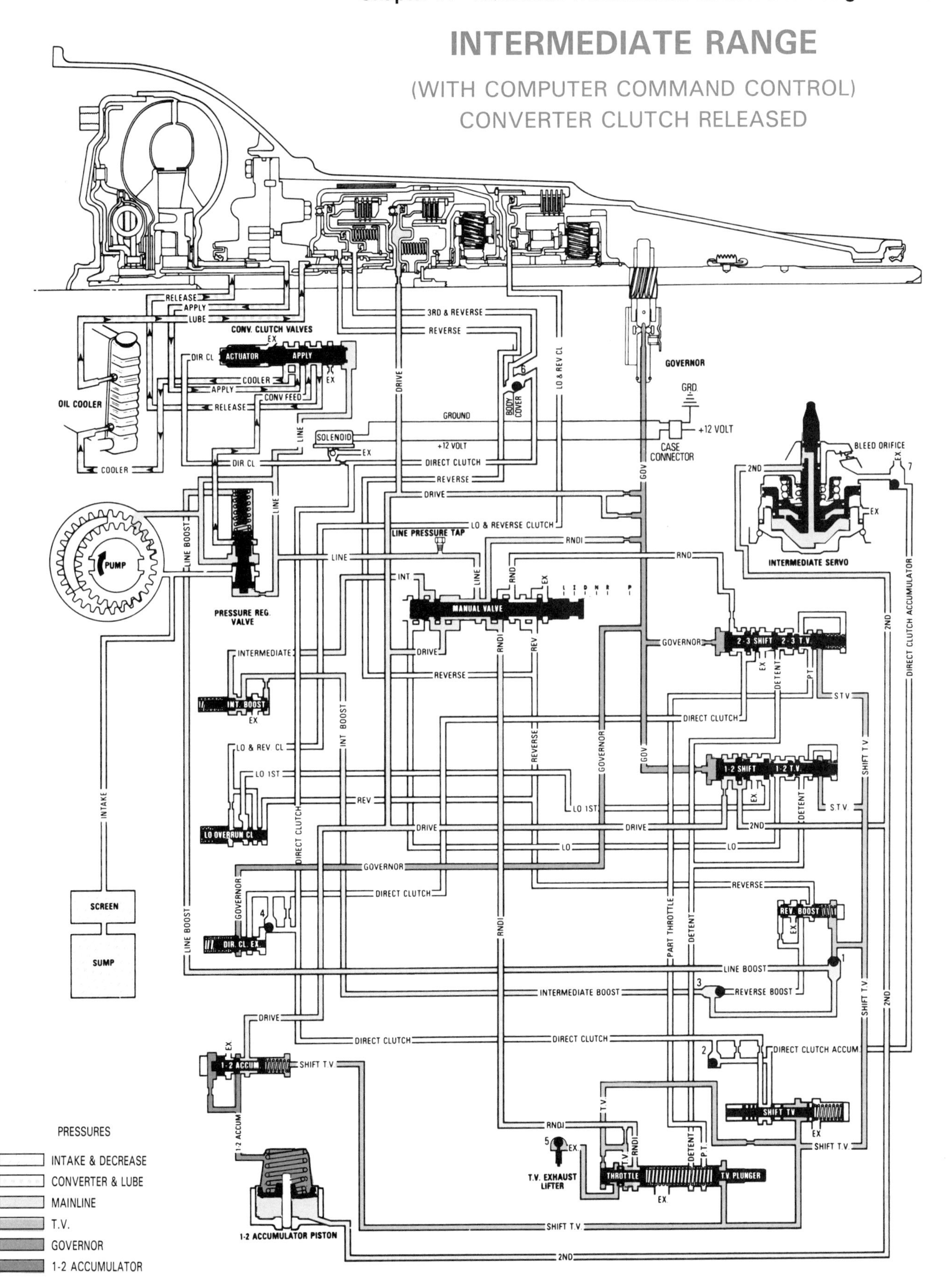

Figure 11-14. Intermediate range is manually selected by the driver. If in third gear, the transmission will downshift to second gear. Accelerating from a stop, the vehicle operates as if it were in drive, but it cannot upshift to third gear. (General Motors)

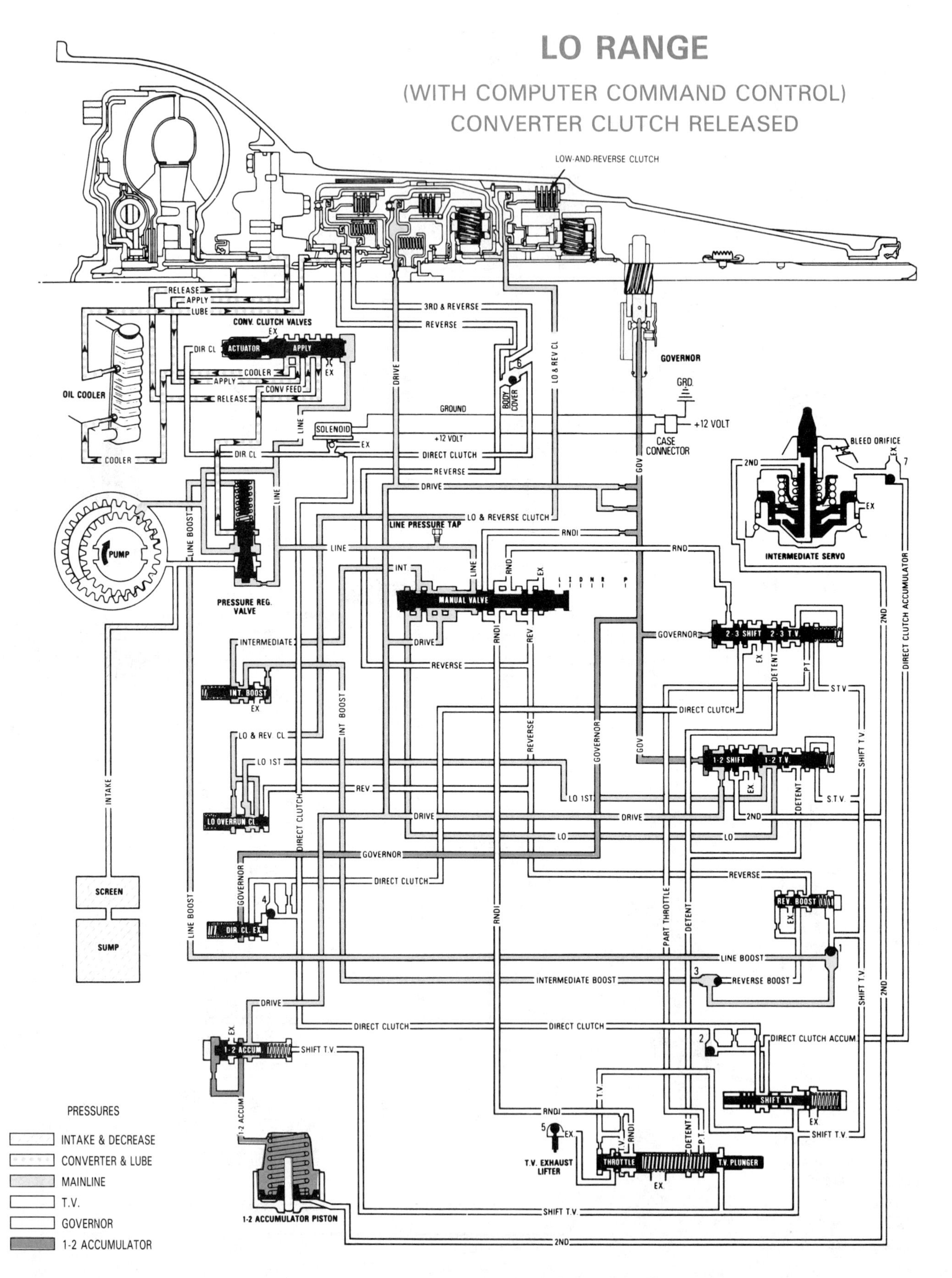

Figure 11-15. Low range is manually selected by the driver. Depending on vehicle speed, the transmission will either downshift directly into first, or it will downshift into second and then into first. Accelerating from a stop, the vehicle will remain in first gear. In low range, the forward clutch is applied, and the low-and-reverse clutch is applied. The low-and-reverse clutch holds the rear planet-pinion carrier just as the low roller clutch does in drive low; however, the low-and-reverse clutch prevents an automatic upshift. (General Motors)

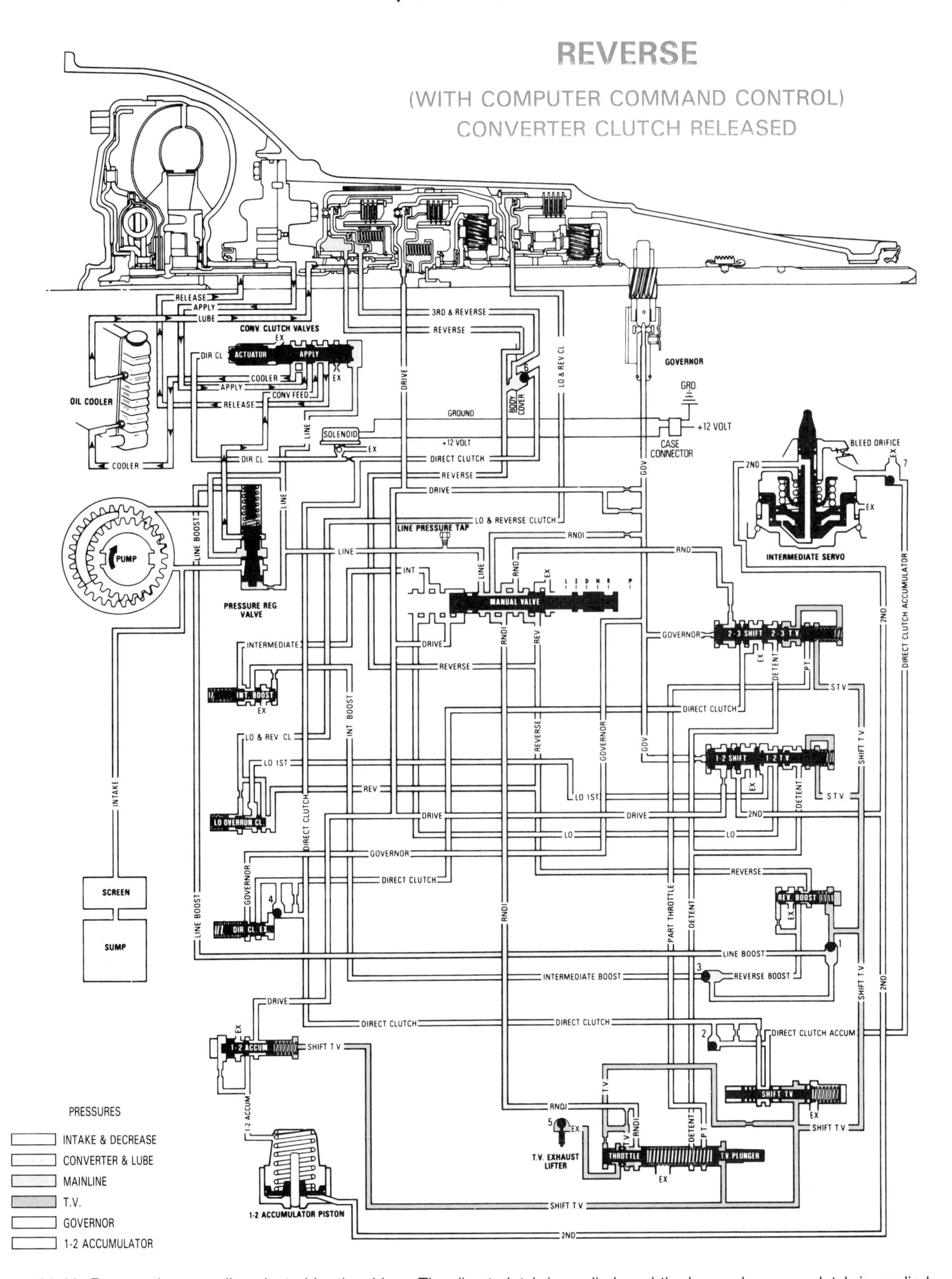

Figure 11-16. Reverse is manually selected by the driver. The direct clutch is applied, and the low-and-reverse clutch is applied. Reverse pressure will be much higher than pressure in the drive ranges. (General Motors)

Low Range

Manual low, or manual first, is applied when the driver moves the manual valve to the low-range position. If the vehicle is in second or third gear when this range is selected, the transmission often shifts into first gear. If vehicle speed is high, governor pressure will prevent a sudden shift to first; the transmission will shift into second until speed drops enough to allow the shift into first. If the vehicle is stopped when manual low is selected, the transmission will remain in low gear. It will not shift into second. Manual low is used for pulling heavy loads and also for maximum engine braking when going downhill.

In manual low, Figure 11-15, the forward clutch is applied. Also, the *low-and-reverse clutch* is applied by the *low overrun clutch valve.* The low roller clutch is applied, but its holding function is being assisted by the low-and-reverse clutch. Line pressure is raised by applying pressure to the spring side of the regulator valve. Additional pressure is needed to hold the low-and-reverse clutch.

The low-and-reverse clutch is applied to lock the rear planet-pinion carrier (and low roller clutch attached to the carrier) in place. It remains applied for engine braking or extra pulling power as long as the selector lever is in low. While it performs the same function as the low roller clutch–holding the rear carrier stationary–it is necessary, since the low roller clutch would hold the carrier only until upshifting occurred. Torque multiplication is the same as in drive low.

Reverse

Reverse, Figure 11-16, is obtained when the manual valve is moved to the reverse position. Oil flows to the direct clutch and to the low-and-reverse clutch, placing the transmission in reverse. Oil pressure is raised by pressure from the manual valve to the back of the regulator valve. Oil also flows to other valves in the transmission but does not have any effect on them.

Automatic Transmission Designs

The following illustrations show some common automatic transmissions. Keep in mind that because of the influx of foreign-made vehicles, there are many more de-

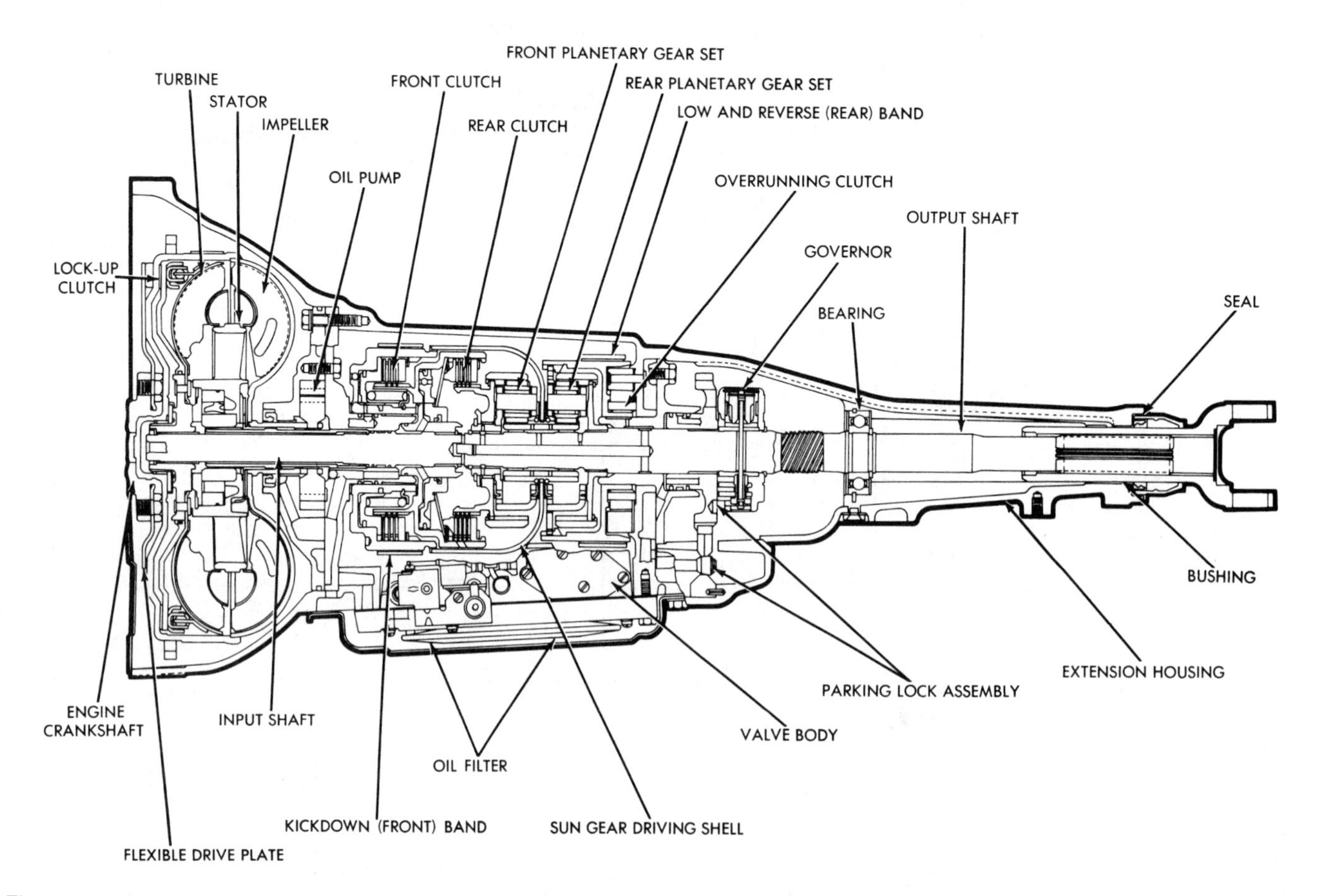

Figure 11-17. The TorqueFlite automatic transmission has been used on many Chrysler products and on many other vehicles. It uses a Simpson gear train and has five holding members. Note that a TorqueFlite A-904-LA transmission is shown here. This particular model has a lockup torque converter. (Chrysler)

sign variations than in the past. This makes it imperative to refer to a service manual for specifics.

Chrysler TorqueFlite

The Chrysler *TorqueFlite* automatic transmission has been used on Chrysler products, including Jeeps and former American Motors cars (*Torque-Command* models), as well as on some imports for years. There are two basic versions of TorqueFlite automatic transmissions used in rear-wheel drive cars. The two versions are similar in construction and in operation. Both are 3-speed units. The *A-904* series is usually used on cars with four- or six-cylinder engines. The *A-727* is more heavily constructed. It is usually used on cars with eight-cylinder engines. The TorqueFlite uses one compound planetary gearset–a Simpson gear train. Planetary holding members include an overrunning clutch, two bands, and two multiple-disc clutches. Later models have a lockup torque converter. A Chrysler TorqueFlite is shown in Figure 11-17. Perhaps you recognize this transmission from its use in previous illustrations.

General Motors Turbo Hydra-Matic 180C

The General Motors *Turbo Hydra-Matic 180C (THM 180C),* Figure 11-18, is a 3-speed automatic transmission with a compound planetary gearset and a converter clutch torque converter. This is one of the few modern 3-speed automatic transmissions that does not use the Simpson gear train. Planetary holding members include a band, an overrunning clutch, and three multiple-disc clutches. This transmission is used in many small General Motors cars and trucks and in some imported vehicles.

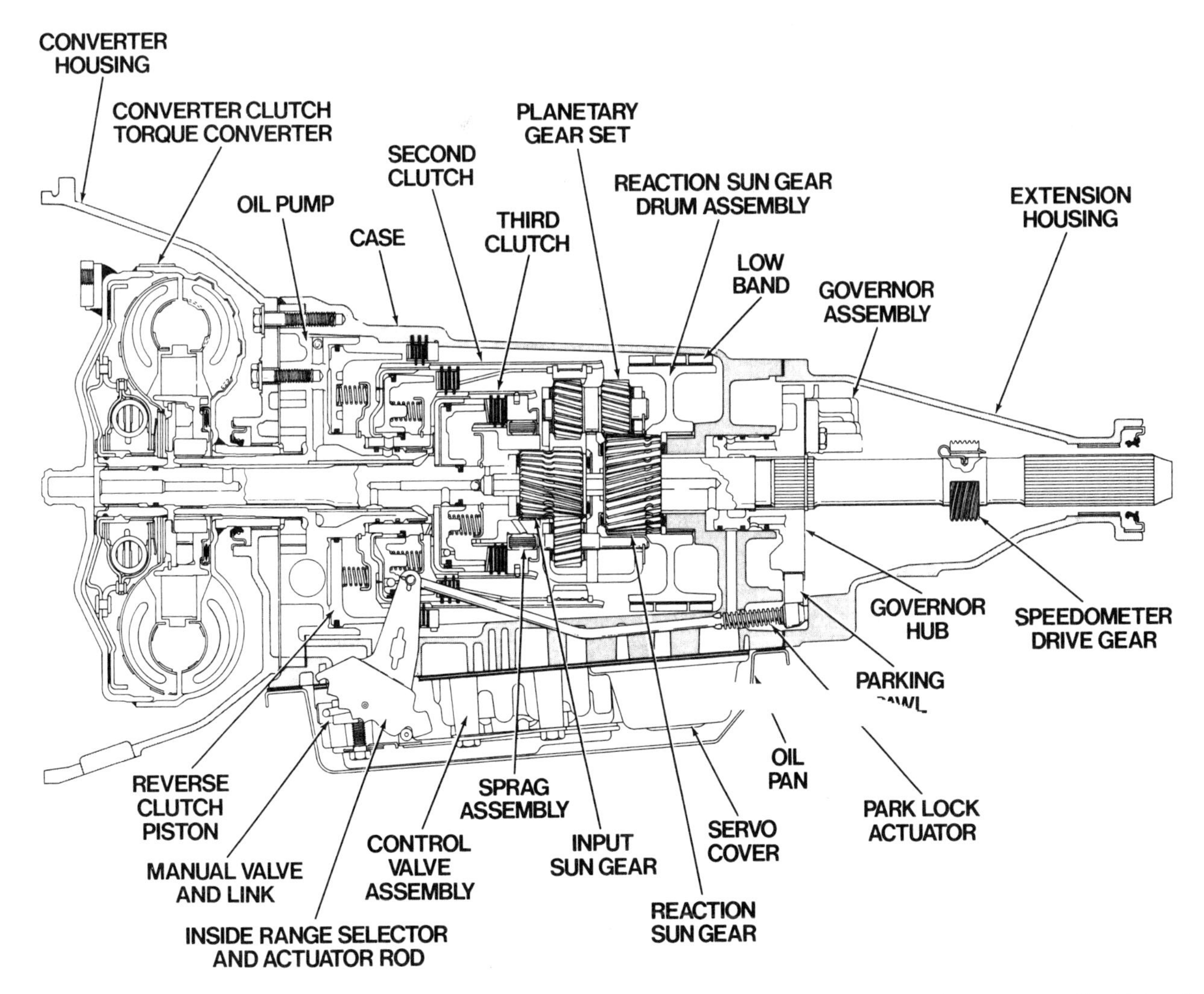

Figure 11-18. The THM 180C is a common 3-speed General Motors automatic transmission. It is used on many small vehicles. It uses a compound planetary gearset and has five holding members. This model has a converter clutch torque converter. (General Motors)

General Motors Turbo Hydra-Matic 200C

The General Motors *Turbo Hydra-Matic 200C (THM 200C)* was featured in Figures 11-7 through 11-16, showing the typical modern hydraulic control system. The THM 200C is a 3-speed transmission with a Simpson gear train. It is used on many General Motors vehicles, including some larger models. Planetary holding members include a band, an overrunning clutch, and three multiple-disc clutches. The THM 200C uses a gear-type oil pump. This model differs from the earlier *THM 200* transmission in that it has a converter clutch.

Note that there is a 4-speed version of the THM 200C–the *THM 200-4R.* It has an extra planetary gearset and a vane-type pump. It has two more multiple-disc clutches and one more overrunning clutch. The fourth speed provides overdrive.

Ford C-5

The Ford *C-5,* Figure 11-19, is a 3-speed automatic transmission with a Simpson gear train and five holding members. These include two bands, two multiple-disc clutches, and an overrunning clutch. The C-5 has a converter clutch torque converter. This particular converter clutch is a centrifugal design, not controlled by the transmission hydraulic system. The C-5 is similar to another design–the Ford *C-4*–except the C-4 does not have a converter clutch.

Ford Automatic Overdrive Transmission

The Ford *Automatic Overdrive (AOD) Transmission,* Figure 11-20, is a 4-speed overdrive transmission. The

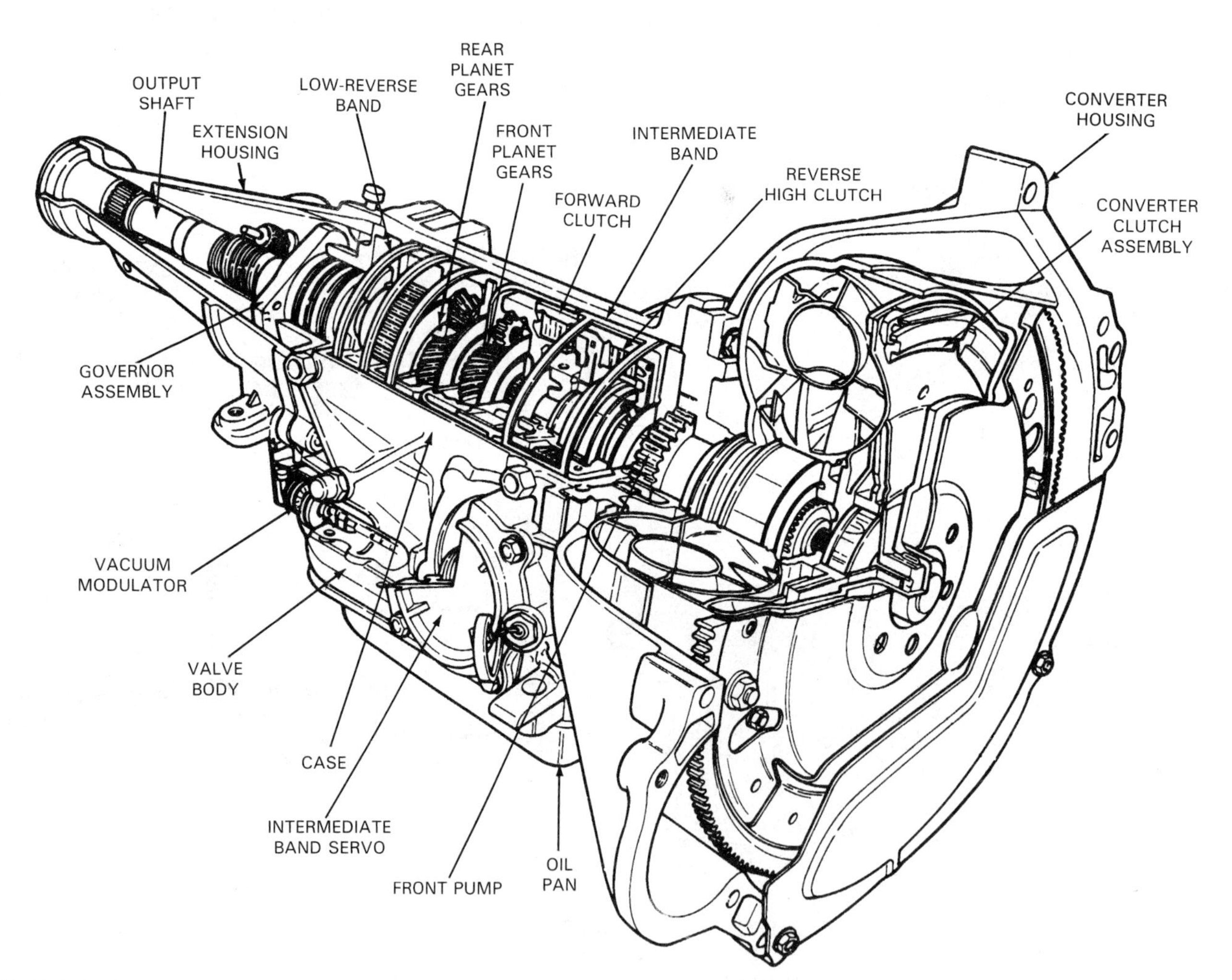

Figure 11-19. The Ford C-5, shown here, has been used for many years. This transmission has a converter clutch torque converter. The converter clutch is not hydraulically applied; it is the centrifugal type. It will engage at a set input shaft speed, regardless of vehicle speed. (Ford)

transmission features a Ravigneaux gear train. Planetary holding members include four multiple-disc clutches, two overrunning clutches, and two bands.

The design of the Ford Automatic Overdrive Transmission is unique in that it has *two* input shafts, which consist of a hollow tube and a solid shaft inside of the tube. The solid input shaft, called the *direct-drive shaft,* is splined on the forward end to the vibration damper on the converter cover. The hollow input shaft, called the *turbine shaft,* is splined on the forward end to the torque converter turbine.

In first and second gear, engine torque is delivered solely through the turbine shaft. In third gear, torque is split between the turbine and direct-drive shafts. Since there is some slippage in the torque converter, less of the engine torque passes through the turbine shaft. As a result, about 60 percent of the power goes through the direct-drive shaft. In overdrive, all power goes through the direct-drive shaft to drive the necessary holding member; the torque converter is completely bypassed, just as in a converter clutch torque converter.

ZF 4-Speed

The European firm *ZF* makes manual and automatic transmissions for many different manufacturers. One model of automatic transmission, Figure 11-21, is a 4-speed with a compound planetary gearset. It uses multiple-disc clutches and overrunning clutches to obtain four forward speeds and one reverse. The model shown also has a converter clutch torque converter.

Warner Transmissions

Borg-Warner (Warner Gear Division) has produced many 3- and 4-speed automatic transmissions from the same basic design. The model shown in Figure 11-22 has four forward speeds and reverse. These are obtained from three planetary gearsets, four multiple-disc clutches, two overrunning clutches, and two bands. The transmission shown also has a lockup torque converter. This transmission, and similar models, have been used extensively in many imported vehicles.

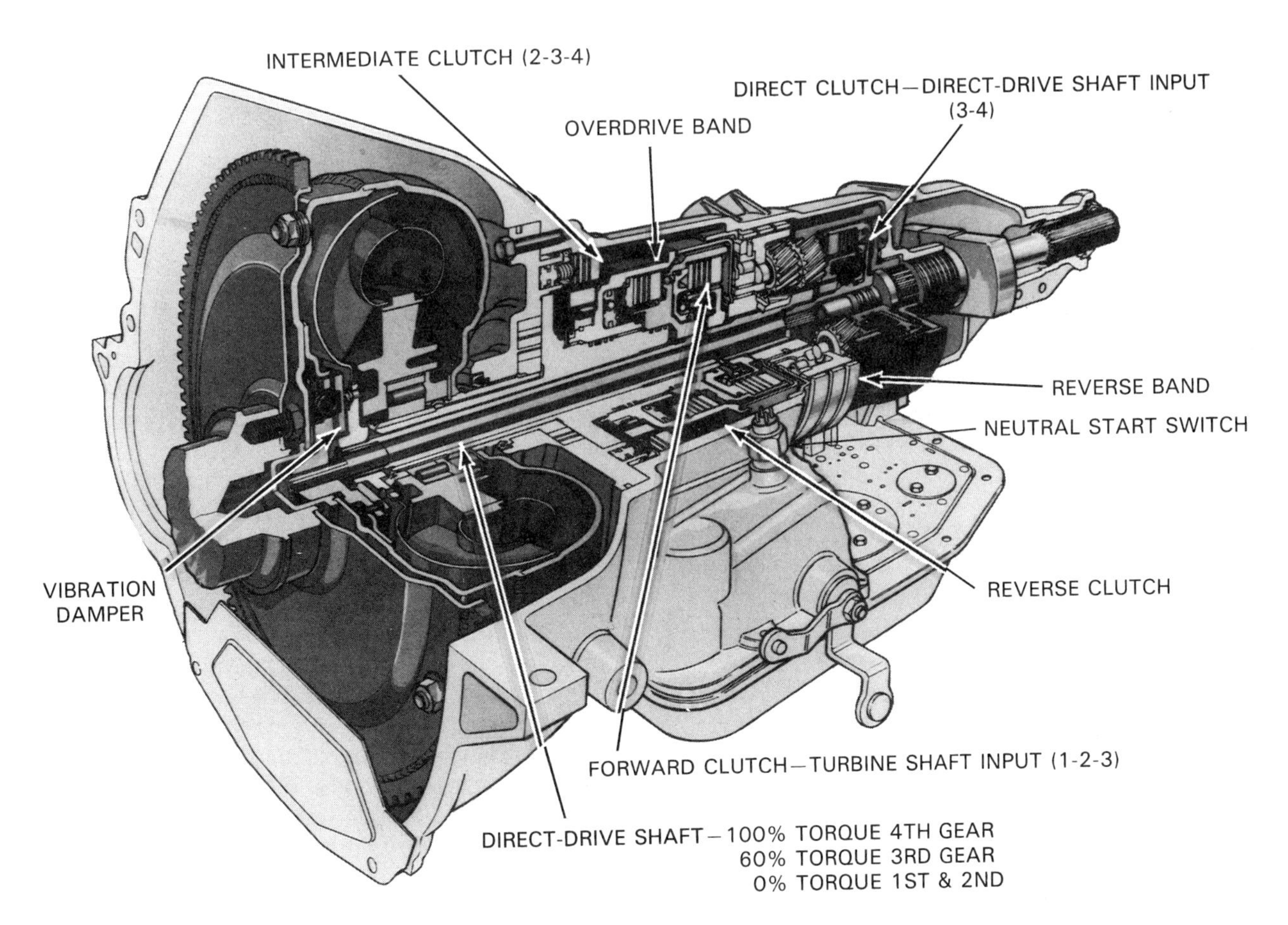

Figure 11-20. The Ford AOD is a 4-speed transmission using two input shafts. Engine power is split between two shafts in third gear. In overdrive, the torque converter is automatically bypassed. (Ford)

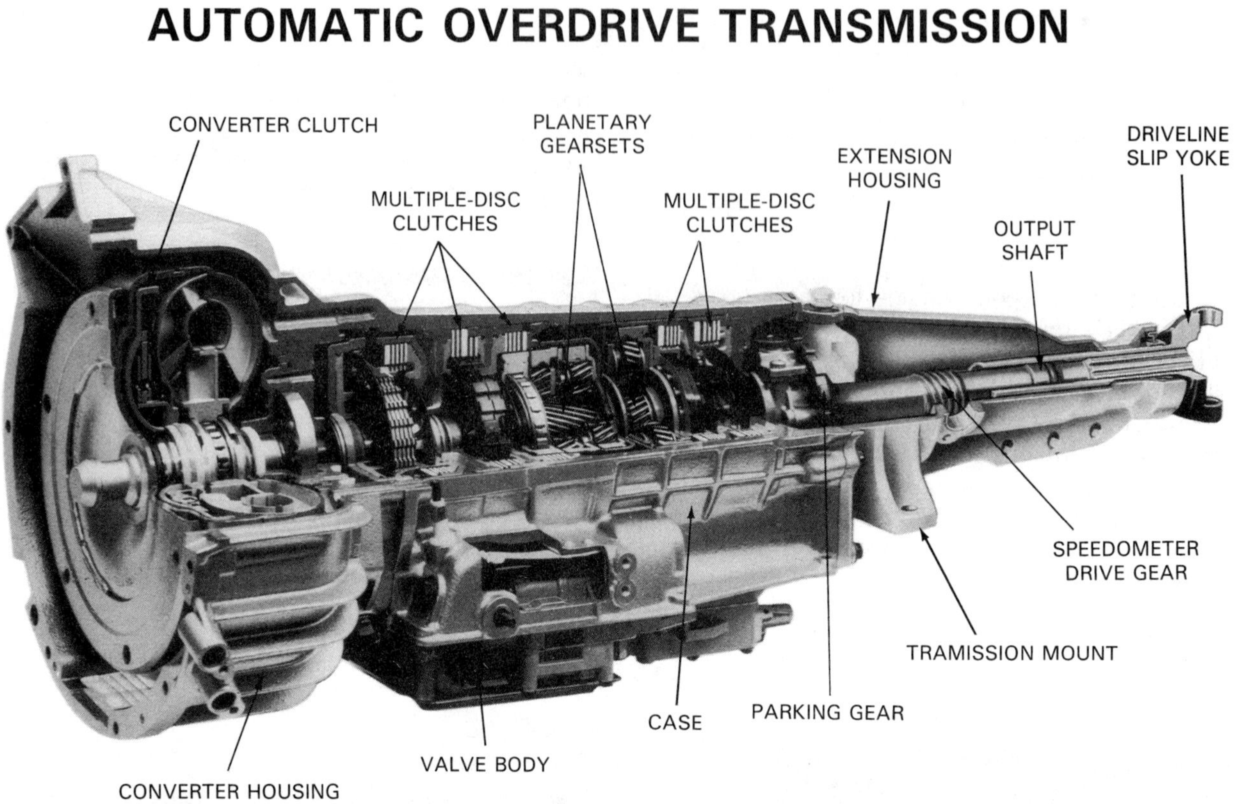

AUTOMATIC OVERDRIVE TRANSMISSION

Figure 11-21. The ZF is a European transmission used in many imported vehicles. (ZF Transmissions)

Toyota 3-Speed and 4-Speed Transmissions

The *A-30* and *A-40* series are produced by Toyota. The A-30 series is comprised of various 3-speed models. The A-40 series was created by adding a separate overdrive unit to the A-30, between the converter and transmission case. Most late-model A-40 series transmissions have lockup torque converters and are controlled by an on-board computer.

Summary

Transmission hydraulic circuits are controlled by hydraulic valves, based on inputs from the selector lever, road speed, and throttle opening. The number of valves and their functions varies with each kind of transmission.

The typical 3-speed transmission has two shift valves, a throttle valve, and a governor valve. Throttle and governor pressures act on opposite ends of the shift valves. When governor pressure exceeds throttle pressure (plus spring pressure), an upshift occurs. The shift valves are designed so that it takes less governor pressure to cause an upshift into second gear than into third gear.

In the part-throttle, 3-2 downshifted position, the throttle valve can send enough pressure to the 2-3 shift valve to push it into the downshifted position. A detent downshift is provided when the throttle pedal is completely depressed, pushing the throttle valve completely into its bore. Governor pressure will be overcome to force one or both shift valves into the downshifted position, moving the transmission from third gear into second or first gear, depending on vehicle speed.

In intermediate range, the transmission is prevented from going higher than second gear. In low range, the transmission will not go above first gear. Moving the selector lever into either of these positions while the vehicle is moving will cause a downshift into the selected range.

Many transmissions are in use today. Most of these are 3- and 4-speed units with lockup torque converters. Most use compound gearsets. Most 3-speed units have Simpson gear trains, while many of the 4-speed units have Ravigneaux gear trains.

Know These Terms

1-2 shift valve; 2-3 shift valve.

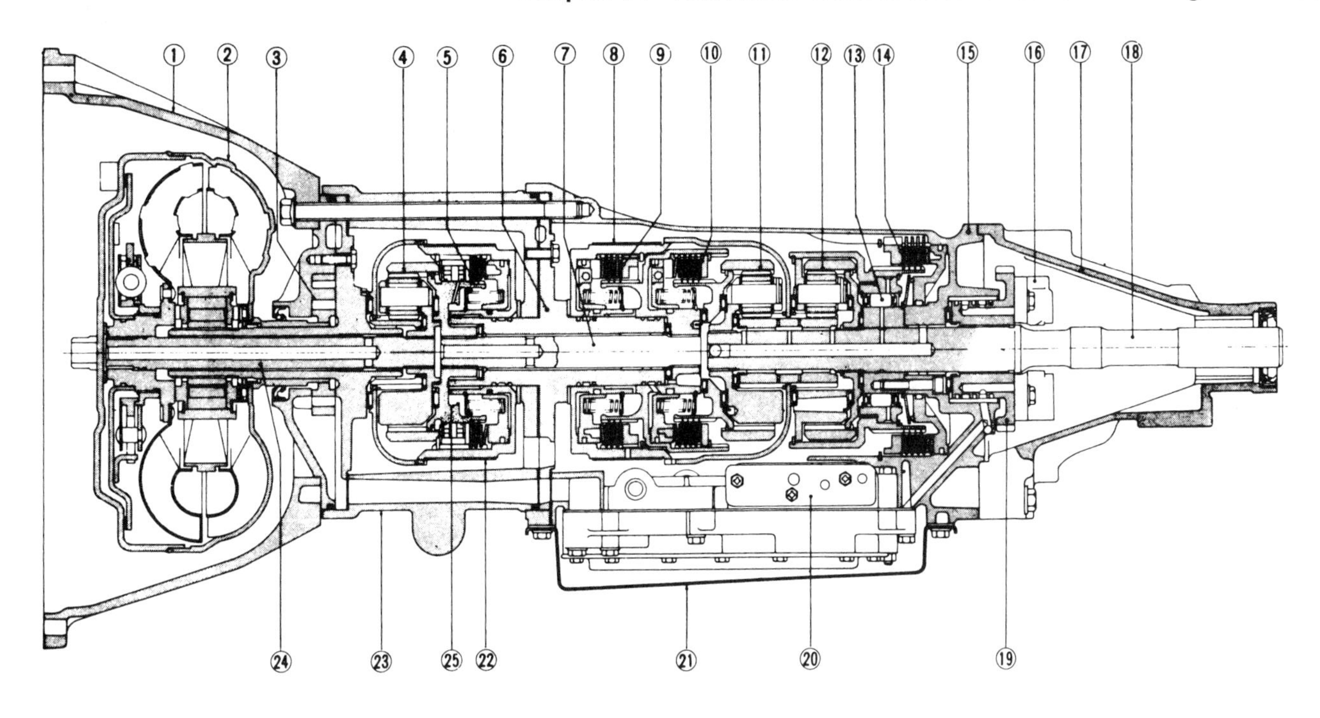

1 Converter housing
2 Torque converter
3 Oil pump assembly
4 O.D. planetary gear
5 Direct clutch
6 Drum support
7 Intermediate shaft
8 2nd band brake
9 Front clutch
10 Rear clutch
11 Front planetary gear
12 Rear planetary gear
13 One-way clutch
14 Low & reverse clutch
15 Transmission case
16 Governor valve assembly
17 Rear extension
18 Output shaft
19 Oil distributor
20 Control valve assembly
21 Oil pan
22 O.D. band brake
23 O.D. case
24 Input shaft
25 O.D. one-way clutch

Figure 11-22. This transmission is one variation of the Borg-Warner series of 3- and 4-speed transmissions. It has a separate section between the torque converter and transmission case for the overdrive unit. (Nissan)

Review Questions–Chapter 11

Please do not write in this text. Place your answers on a separate sheet of paper.

1. The major function of a complete hydraulic circuit is to route fluid for cooling and lubrication purposes. True or false?
2. A 2-speed automatic transmission would have:
   a. One shift valve.
   b. Two shift valves.
   c. Four shift valves.
   d. It depends on the transmission design.
3. The _____ _____ is responsible for system pressure; the _____ _____ _____ regulates system pressure.
4. The speed of a vehicle at which an upshift occurs is governed by:
   a. Line pressure.
   b. Throttle pressure.
   c. Governor pressure.
   d. None of the above.
5. A forced downshift is caused by a closed throttle. True or false?
6. What causes the 1-2 shift valve to move before the 2-3 shift valve?
7. In the THM 200C, line pressure is modulated by boost pressure when the transmission is in park. True or false?
8. When the THM 200C is in first gear, oil flows to the apply side of the converter clutch. True or false?
9. What is the purpose of regulating line pressure with throttle pressure?
10. The THM 200C converter clutch solenoid _____ an exhaust port from the direct clutch when energized.
11. During detent operation, line pressure is:
    a. Increased.
    b. Decreased.
    c. Zero.
    d. None of the above.
12. Why would manual low be used?
13. In manual low, the low roller clutch is used to lock the transmission in low gear until the selector lever is moved. True or false?
14. The Chrysler TorqueFlite uses a Ravigneaux gear train. True or false?

Certification-Type Questions–Chapter 11

1. **All of these are direct functions of the complete hydraulic circuit EXCEPT:**
   (A) to apply or release a band.
   (B) to route fluid for cooling and lubrication.
   (C) to engage or release a multiple-disc clutch.
   (D) to engage or disengage an overrunning clutch.

2. **Which of these controls line pressure?**
   (A) Oil pump.
   (B) Valve body.
   (C) Clutch pack.
   (D) Pressure regulator valve.

3. **Throttle pressure will vary depending on:**
   (A) engine load.
   (B) vehicle speed.
   (C) governor speed.
   (D) pump speed.

4. **A cut back valve uses governor pressure to achieve:**
   (A) reduced throttle pressure.
   (B) increased throttle pressure.
   (C) stabilized throttle pressure.
   (D) a cushioned shift.

5. **When the 1-2 shift valve moves, it sends pressure to the:**
   (A) intermediate servo.
   (B) intermediate band.
   (C) pressure regulator valve.
   (D) governor.

6. **RND oil is the feed for direct clutch oil in:**
   (A) first gear.
   (B) second gear.
   (C) third gear.
   (D) all gears.

7. **All of these are applied in manual low EXCEPT:**
   (A) the forward clutch.
   (B) detent pressure.
   (C) the low roller clutch.
   (D) the low-and-reverse clutch.

8. **All of these are found on a Ford C-4, 3-speed automatic transmission EXCEPT:**
   (A) two bands.
   (B) an overrunning clutch.
   (C) two multiple-disc clutches.
   (D) a converter clutch.

9. **All of these inputs are applied to hydraulic valves in the control of hydraulic system operation EXCEPT:**
   (A) road speed.
   (B) engine torque.
   (C) throttle opening.
   (D) selector lever position.

10. **All of these are found in the typical 3-speed transmission EXCEPT:**
    (A) a throttle valve.
    (B) a governor valve.
    (C) two shift valves.
    (D) an overdrive valve.

11. **Technician A says that shifting into second gear at cruising speed causes the transmission to downshift. Technician B says that third gear cannot be achieved in the intermediate range. Who is right?**
    (A) A only
    (B) B only
    (C) Both A & B
    (D) Neither A nor B

12. When the manual valve is moved to the reverse position, oil flows directly to the:
- **(A)** direct clutch.
- **(B)** overrun clutch.
- **(C)** forward clutch.
- **(D)** low roller clutch.

13. Technician A says that in third gear, engine torque in the Ford AOD transmission is split between two input shafts. Technician B says that in overdrive, the torque converter is completely bypassed. Who is right?
- **(A)** A only
- **(B)** B only
- **(C)** Both A & B
- **(D)** Neither A nor B

14. All of these are part of a transmission hydraulic circuit EXCEPT:
- **(A)** oil pump.
- **(B)** servo.
- **(C)** valve body.
- **(D)** low roller clutch.

15. Technician A says that an automatic transmission will always upshift at some preset road speed. Technician B says that hard acceleration will cause a vehicle to shift at a lower road speed. Who is right?
- **(A)** A only
- **(B)** B only
- **(C)** Both A & B
- **(D)** Neither A nor B

16. Technician A says that throttle pressure acts on a smaller area on the 1-2 shift valve than on the 2-3 shift valve. Technician B says that this design means that the 1-2 shift valve will move to the upshifted position before the 2-3 shift valve. Who is right?
- **(A)** A only
- **(B)** B only
- **(C)** Both A & B
- **(D)** Neither A nor B

A transmission jack, such as the one shown here, is designed for removing, transporting, and installing transmissions when using an overhead lift. (Owatonna Tool)

Chapter 12

Automatic Transmission Problems and Troubleshooting

After studying this chapter, you will be able to:
- Describe typical automatic transmission problems.
- State possible causes of automatic transmission problems.
- Recommend logical troubleshooting procedures.
- Describe typical diagnostic charts.
- Name diagnostic tests used for troubleshooting.
- Explain how a road test should be conducted.
- Explain how to perform a stall test and discuss significance of results.
- Explain how air and hydraulic pressure tests are performed and discuss their usefulness.
- Discuss use of hydraulic diagrams to analyze automatic transmission problems.

Diagnosing automatic transmission problems is as important as fixing them. You should always have at least a general idea of what is wrong before you begin transmission repairs. This chapter will familiarize you with typical transmission problems and troubleshooting techniques. Some of the diagnostic procedures in this chapter can be performed with the transmission installed in the vehicle. Other procedures can be performed only after the transmission is disassembled.

To diagnose transmission problems, you must first have a thorough understanding of automatic transmissions. If you are not familiar with their operation, you should study Chapters 10 and 11 until you feel comfortable with the basics. As depicted in Figure 12-1, the average automatic transmission can have many kinds of problems.

General Troubleshooting Procedure

Automatic transmissions are complex, and it is very important that you proceed in a logical manner while troubleshooting. Malfunctions can be diagnosed by applying logic and following recommended troubleshooting procedures.

You should have the owner or driver describe the problem before taking any other diagnostic steps. Be aware that the driver may be unable to accurately describe the problem. Therefore, if possible, you should *road test* the car with the driver and have him or her point out the particular problem while it occurs. Take note as to what the operating conditions are when the problem occurs.

Note that sometimes the driver will complain about a perceived problem that is in fact a normal condition. For instance, modern transmissions shift more firmly than older models. This is done to improve gas mileage. However, the owner who is used to an older transmission may think that something is wrong.

Before performing the road test, the transmission fluid level and condition should be checked. If the road test indicates a possible control linkage problem, check the various adjustments. This includes the shift linkage. It could also include a throttle valve, or T.V., linkage or a kickdown linkage and the throttle linkage. Adjust linkages as necessary. (Checking fluid condition and linkage adjustment is covered in Chapter 13.)

If the transmission is equipped with a vacuum modulator, check for vacuum, as explained later in this chapter. Also, inspect all electrical connections. If these checks do not turn up any obvious problems, diagnostic checks, such as hydraulic pressure tests, can be made.

Sometimes, the exact problem cannot be determined without disassembling the transmission and making further checks. The oil pan and valve body can be removed, and air pressure can be applied to hydraulic passageways leading to the bands and clutch packs to observe operation. In some cases, it may be necessary to completely disassemble the transmission to make visual confirmation of the problem area.

Refer to *diagnostic charts* prepared for the particular make of transmission as you troubleshoot. They can help you quickly pinpoint the cause of the problem. Experience can also help speed the diagnosis. In many cases, for

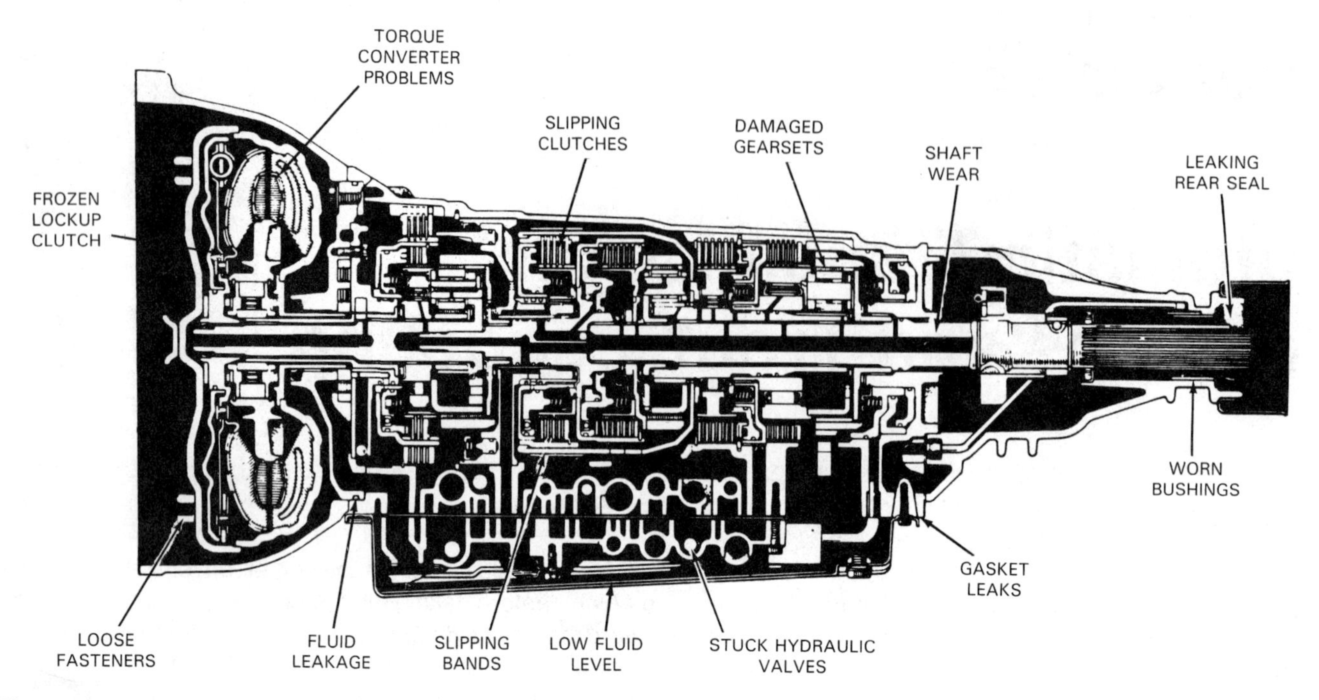

Figure 12-1. Note some of the problems that can occur in an automatic transmission. Although these problems are often complex, they can be easily diagnosed by using logical troubleshooting procedures and by having a thorough knowledge of transmission operating principles. (Chrysler)

instance, a particular kind of transmission has a typical area of failure. In such cases, experience is often the best guide as to the cause and remedy.

As you troubleshoot, make sure that the problem is not caused by the engine or another vehicle system. An engine that needs a tune-up, a plugged catalytic converter or muffler, or a front-end or brake problem may be wrongly blamed on the transmission.

Identifying the System

Often, you will be able to identify a particular system that is causing a problem (or at least you will suspect it is) before you actually pinpoint the cause. You could classify the problem as pertaining to a failure in one of three transmission systems: the *hydraulic system,* the *mechanical system,* or the *electrical system.* Following is a brief discussion of each.

Hydraulic System

Automatic transmission problems often occur due to malfunctions in the hydraulic system, or circuits. Hydraulic system problems are those that develop in the fluid, filter, pump, valves, apply pistons, accumulators, or hydraulic passageways.

The most common hydraulic system problem is a low fluid level. When fluid level is low, the transmission will shift erratically, or it may slip or entirely fail in its function to move the vehicle. A fluid level that is too high causes similar problems in the transmission but for different reasons. Specifics are discussed in Chapter 13.

Other common hydraulic system problems include plugged filters and sticking valves. A plugged filter prevents oil flow. Sticking valves cause pressure or shift problems. Defective transmission oil pumps and torque converters can also cause hydraulic problems. Servo and clutch pistons can be sources of hydraulic system problems, as can accumulators.

Mechanical System

Transmission mechanical problems are those that develop in the mechanical system, both in moving and stationary components. This includes the clutch packs (except for apply piston), bands, overrunning clutches, planetary gears, transmission shafts, and linkage. It also includes the case and related components.

The most common mechanical system malfunction, aside from an out-of-adjustment linkage, is burned and glazed clutch and band linings. Linings in this condition can prevent the holding members from functioning properly, which in turn, can prevent the transmission of power through the unit. As a result, the vehicle will not move. Defective overrunning clutches, broken or stripped planetary gears, and broken shafts are other typical problems. Among stationary mechanical parts, the most common problem is a leaking seal or gasket. A cracked transmission case or extension housing is also a source of problems. A crack in either of these parts will cause an oil leak.

Electrical System

Transmission electrical system problems are usually related to the neutral start switch or detent solenoid. The computer system sensors can also cause problems. Many electrical problems are not directly related to the transmission and can be fixed by replacing fuses or by simple electrical checks. Computer control problems should be diagnosed by checking the entire computer system using suitable testing equipment and troubleshooting guides. Refer to the index for more information on this topic.

Typical Automatic Transmission Problems

Like any other mechanical part, the automatic transmission is subject to certain malfunctions, and each make of transmission has a tendency to experience certain types of problems. Typical operational problems experienced by the different transmissions are outlined below.

Automatic Transmission Slippage

Automatic transmission slippage, discussed briefly in Chapter 10, is a condition that occurs when the transmission does not transmit all of the engine power to the rear wheels. When slippage occurs, the engine will speed up without a corresponding increase in vehicle speed. Sometimes slippage will be constant; other times, the transmission will slip badly when starting off or while shifting and will then suddenly engage completely. Your first diagnostic step (after talking to the driver and checking fluid level) will be to determine if the vehicle slips in all gears or only in one gear. This is determined by making a road test.

If the vehicle slips in every gear, recheck the fluid level. Also, check for a plugged oil filter. A plugged filter will sometimes cause a whining noise that varies with engine speed. In some instances, a transmission with a plugged filter will operate satisfactorily at low speeds but will begin to slip at higher speeds. Changing the filter will often solve the problem.

If the vehicle is equipped with a T.V. linkage, check its adjustment, as described in Chapter 13. Since throttle valve adjustment affects overall transmission pressure, adjusting the linkage to raise the shift points will raise transmission pressures. This increased pressure may eliminate slippage.

If the transmission slips in one gear only, carefully note the gear in which slippage occurs. You can then refer to a *band and clutch application chart* to determine which clutches or bands are applied in this gear. This will tell you where the trouble lies. Note that use of band and clutch application charts is covered later.

Slippage can be caused by too great of a clearance between a band and drum. Adjusting the band, an in-car service procedure, may solve the problem.

Slippage can be caused by burned, glazed, and otherwise worn clutch pack and band linings. Slippage caused by such defective holding members can sometimes be seen by failure of a transmission to retard vehicle speed in a manual reduction gear (engine braking). Adjusting a burned or glazed band will usually not eliminate slippage.

Checking for worn clutch packs or bands usually requires that the transmission be removed from the vehicle and at least partially disassembled so the parts can be visually inspected. Burned clutch packs and bands can often be confirmed by examining the transmission fluid. A burnt odor or sludge in the transmission oil pan is normally caused by failure of the lining material, typically brought on by overheating.

In rare instances, slippage may be caused by a defective torque converter stator or leakage within transmission hydraulic system. These will also be discussed later in this chapter.

No-Drive Condition

When the vehicle has a *no-drive condition,* the transmission fails to transmit power to the rear wheels. The vehicle will either fail to be driven forward in one or all forward gears, or it will fail to move backward, or it will fail to move at all. There are two basic causes for a no-drive condition. One is a binding (locked-up) condition in the drive train or in the brake system. The other is severe slippage within the transmission.

To find out if the no-drive condition is caused by binding, first, raise the drive wheels. Then, race the engine with the transmission in gear and the brakes applied. If the engine strains and continues to strain when the brake pedal is released, the drive train or brakes are binding.

To locate the source of the binding condition, raise the vehicle so the drive wheels are off the ground and remove the drive shaft. If the engine does not strain with the drive shaft removed, the problem is in the rear axle assembly or brake system. If the engine strains after drive shaft removal, the problem is in the transmission: if the parking pawl is not stuck, a clutch pack is welded together, a one-way clutch is jammed, or another internal part is locked.

If binding has been ruled out, the transmission probably will not drive because of severe slippage. If the transmission fluid level is low, add fluid and recheck the transmission operation. If, on the other hand, the fluid level is high, fluid may have drained back from the torque converter. Running the engine for a few minutes may allow the transmission oil pump to refill the converter so the vehicle will move.

If the fluid level is normal, ensure that the shift linkage is connected externally to the outer manual lever at the transmission case and that the inner manual lever is connected to the manual valve at the valve body. If the linkage checks out okay, the problem is inside the transmission. The transmission must be disassembled to check for internal problems.

Shift Problems

Shift problems are those problems that result in improper gear changes. Examples are late or early upshifts, no upshifts or no downshifts, harsh or soft shifts, and no passing gear. Many shift conditions can be corrected easily. Always check the transmission fluid level first.

If the fluid level is okay, adjust the T.V. linkage or check the vacuum modulator, as applicable. Lengthening or shortening the linkage rod or *throttle valve (T.V.) cable* will raise or lower shift points. If the transmission does not upshift, make sure that the kickdown linkage, if there is one, is not jammed in the wide-open throttle position.

If the vehicle is equipped with a vacuum modulator, make sure that the modulator receives full intake manifold vacuum and that the engine is in good condition, since this can affect vacuum development. A simple test is to disconnect the vacuum hose at the modulator with the engine idling. This should cause the engine to speed up and idle roughly. If the engine idle is unchanged, check for a plugged or a leaking vacuum hose or a hose that is not connected to intake manifold vacuum. If the modulator is visibly damaged, or if there is transmission fluid in the vacuum line, the modulator should be replaced.

If servicing the linkage and vacuum modulator does not correct the problem, check the operation of the governor and elements within the valve body. Sticking governors can cause erratic shifting or *no shifts.* Sticking accumulators or valves in the valve body can cause late or early shifts, erratic shifts, hard shifts, or excessively soft shifts. Sludge or metal particles from other transmission parts will often cause these components to stick. In most cases, checking the governor or valve body will require removing the oil pan or extension housing.

Abnormal Noise

An *abnormal noise* coming from the automatic transmission may be a whine, rumble, scrape, or other such noise. Some noises can be heard in every gear, including neutral and park. Other noises occur only in one gear.

If the noise seems to occur in all gears, check the transmission fluid level and oil filter. Noises caused by a plugged filter will be heard in every gear and will become louder and higher pitched as engine speed is increased. This noise may be accompanied by slippage. The noise will stop as soon as the engine is turned off, even if the vehicle is still moving.

Other possible causes of noise occurring in every gear are the torque converter and oil pump. Torque converters can make a whining or scraping noise when the vehicle is starting off or accelerating. It will not be heard when the vehicle is at cruising speed, since all of the converter components are turning at the same speed. Oil pumps will usually run quietly unless they were improperly assembled. Most oil pump noises occur just after the transmission has been overhauled. They will resemble the noise caused by a plugged filter.

If the noise seems to occur in only one gear, the most common cause is the planetary gearsets. Planetary gearset noises can occur only when the gear train is in a gear that causes the planetary gears to turn in relation to each other–reduction, overdrive, neutral, or reverse. Planetary gearsets are quiet in direct drive. Unless they are badly damaged, noisy planetary gearsets will often show no visible signs of problems.

Transmission Overheating

Transmission overheating may occur when transmission fluid temperatures reach in excess of 300°F (150°C). Transmission overheating is not always obvious in itself; it is usually noticed as soft shifts, gear slippage, or delayed gear engagement. When the transmission overheats, the fluid *and* the transmission components are heated to an excess.

As stated in Chapter 10, fluid temperature should generally be kept below 275°F (135°C). Note that occasional operation above 275°F or even 300°F will not permanently damage the fluid or the clutches and bands. However, a constant pattern of fluid temperatures exceeding 275°F leads to broken-down fluid, causing the loss of its lubricating properties. This will show up in the form of plugged valves, burned clutch and band linings, and plugged filters, to name a few.

If you know or suspect a transmission is overheating, you should check the oil cooler. Ensure that fluid is flowing through it. Make sure that it is not clogged or covered with sludge. The engine cooling system should be checked to ensure that it is not operating at excessive temperatures. If the vehicle is being used for towing, the extra load may produce too much heat for the original equipment cooler to handle. In this case, an auxiliary oil cooler should be installed.

Lubricant Leakage

Transmission *lubricant leakage* is an unwanted loss of fluid from inside the transmission. Before attempting to correct the suspected transmission leakage problem, make sure that the leaking fluid is actually coming from the transmission. Keep in mind that automotive fluids are usually carried rearward by the airflow under the vehicle. Thus, finding fluid on the transmission does not necessarily mean the transmission is leaking. Also, while ATF is typically red, fluid color is usually not a reliable indicator for the type of fluid and, therefore, the source of the leak. Engine oil, power steering fluid, and even antifreeze can be mistaken for automatic transmission fluid. After having identified a problem to be in the transmission, you then need to pinpoint the exact location.

If you observe the transmission to be leaking when the engine is running, a usual cause is failure of a seal or gasket that is under pressure when the transmission is operating. Start the diagnostic procedure to determine where on the transmission the leak is occurring by thoroughly cleaning the transmission case. Also, clean the torque converter, pump face, and nearby surfaces with solvent. Removing the transmission inspection cover or dust cover will give you access to these components.

Start the engine and run it for 10 minutes at high idle. Then, examine the case for fresh fluid. Check the rear seal and the *manual lever/shaft seal.* Check other parts on the side of the case, such as the servo, governor, or accumulator covers, oil cooler lines and fittings, and pressure tap plugs. The area around the torque converter may show traces of transmission fluid, streaking outward from the center of the converter where it enters the front pump seal. This area is easily seen from under the vehicle.

If the transmission leaks primarily when the engine is not running, the cause is usually a seal or gasket that is below the level of fluid in the pan. Check the transmission oil pan for a defective gasket or warped sealing surfaces. Also, check the pan for loose bolts or for a possible puncture, caused by road debris.

If the pan is okay, you should clean the transmission case, so you can look for fresh fluid. Then, check seals at the manual lever/shaft assembly, T.V. cable, band adjustment screw, and electrical connectors. Also, check the extension housing gasket, rear seal, and front pump seal.

Most gaskets and seals can be replaced without removing the transmission. An exception is the front pump seal. When changing a rear seal, make sure that the extension bushing is good. Even a new seal cannot adequately protect against leakage where the bushing is defective.

A vacuum modulator can be a source of leakage. Check this component if the transmission fluid level seems to drop without being accompanied by leakage. A ruptured vacuum modulator diaphragm will allow transmission fluid to be drawn into the engine and burned.

Possible Causes of Problems

The automatic transmission has many parts that can fail, as you are probably aware. It is helpful to learn how they become defective, what problems they can cause, and what is needed to repair or replace them. Following is a brief discussion.

Flexplate Malfunction

The flexplate is a steel stamping connecting the crankshaft to the torque converter. It usually lasts the life of the vehicle. Occasionally, it may develop a crack, causing knocking noises and, possibly, vibration that varies with engine speed. The ring gear teeth may also become stripped. The only cure for a defective flexplate is replacement. The transmission must be removed to replace the flexplate.

The bolts holding the flexplate to the crankshaft or to the torque converter cover may become loose. Loose bolts will cause a knocking sound, especially at lower speeds, and will possibly cause vibration. Loose flexplate attaching bolts may also make the starter operate noisily or jam.

Torque Converter Malfunction

The torque converter may slip or make noises. It may also overheat, causing the entire transmission to do the same. If the converter elements begin to break up, metal particles will be carried throughout the transmission. Converter defects are relatively rare. They are usually caused by excessive clearances between the internal elements.

Torque converter noise is a scraping sound that can be heard only when starting off or when accelerating. It tapers off and stops at cruising speeds when all of the internal elements are turning at the same speed.

If the stator overrunning clutch locks up and then fails to release, the transmission will overheat at high speeds. If the clutch fails to lock the stator, the converter will slip at low speeds. If the transmission has heavy sludge deposits, the converter is also full of sludge.

The torque converter is probably not defective if the transmission malfunction occurs in only one gear. Sometimes, transmission fluid will drain back from the converter into the oil pan overnight. This is caused by a sticking valve or oil pump malfunction. It does not mean that the converter is defective.

The only service procedure for a problem inside of a modern welded converter is to replace the converter. The only exception to this is for sludge buildup, which can be removed with a converter flusher. Before replacing the converter, make sure that all other transmission components have been thoroughly checked and eliminated as sources of the problem.

Automatic Transmission Shaft Malfunction

The input and output shafts tend to be relatively trouble-free. Occasionally, a shaft will break, but in normal use, this is rare. Seal rings and seal ring grooves can wear, causing pressure loss and slippage. The transmission must be removed and disassembled to service the shafts. The fit of the shaft in its bushings and the fit of the seal rings should always be checked.

Planetary Gearset Malfunction

Planetary gearset noises can be heard when the gear train is in reduction, overdrive, neutral, or reverse. The gearsets do not make noise in direct drive. Gearset noise may vary as the transmission changes gears. They are usually loudest in low, decreasing as gear reduction decreases. Often, noise is heard in only one gear. The gears may be badly damaged, as shown in Figure 12-2, or they may seem to have no damage at all. Some planetary gearset noise is caused by worn bushings, which cause the gears to go out of alignment. The transmission must be disassembled to replace the planetary gearsets.

Clutch Pack Malfunction

Worn clutch packs almost always cause slippage. In a few cases, the clutch pack may become so hot that it welds itself together, causing the transmission to stick in gear. If the transmission slips in only one gear, the problem is one particular clutch pack. For example, a transmission that drives well in forward but slips in reverse likely has a malfunction of its reverse clutch.

Slightly glazed friction discs may work well when the transmission is cool but slip or chatter once they warm up.

Figure 12-2. Worn or damaged planet gears can result in a noisy transmission. Severely damaged gears will strip off or jam, resulting in a no-drive condition. (BBU, Inc.)

Slippage may also be caused by leaking clutch piston seals, by a missing air bleed or pressure relief check ball, or by tight clutch return spring(s). A weak clutch return spring can cause harsh clutch engagement.

A worn clutch pack can often be confirmed by transmission fluid that has a burnt odor or by sludge in the oil pan. If the sludge is deeper, roughly, than the thickness of a dime, or if it is built up in spots, the transmission should be disassembled and overhauled. It is always necessary to remove the transmission to overhaul clutch packs. Be sure to replace the seals on the clutch apply piston when performing the overhaul.

Band and Servo Malfunctions

Worn bands will slip, especially when they are first applied. A common example is an intermediate (second gear) band that causes slippage as the transmission shifts into second. A band that is severely slipping can rarely be corrected by adjustment. The lining, Figure 12-3, will be too damaged to hold as it should. If the band ear or linkage is broken, the transmission may skip the forward gear or gears that utilize the band. Figure 12-4 shows a band with a broken ear.

Most bands are replaced by removing the transmission and installing a new band. The band servo should also be disassembled, and new piston seals, installed. A few bands can be replaced without removing the transmission from the vehicle. Servos can lose pressure due to leaking seals. Sometimes, the wrong return spring may be installed, causing a soft or harsh shift. Many servos can be disassembled from the outside of the case. Otherwise, they can be accessed by removing the oil pan.

Overrunning Clutch Malfunction

Overrunning clutches will usually work well or not at all. If the one-way planetary holding member fails to lock,

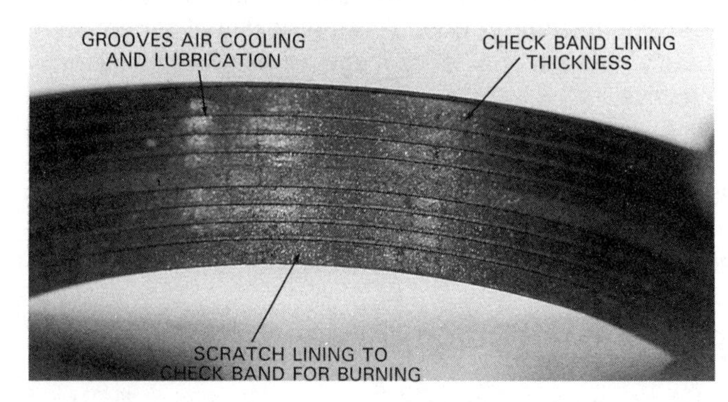

Figure 12-3. The lining of the transmission bands can become worn or glazed. Worn bands can sometimes be adjusted, but a band that is burned and slipping should be replaced. (BBU, Inc.)

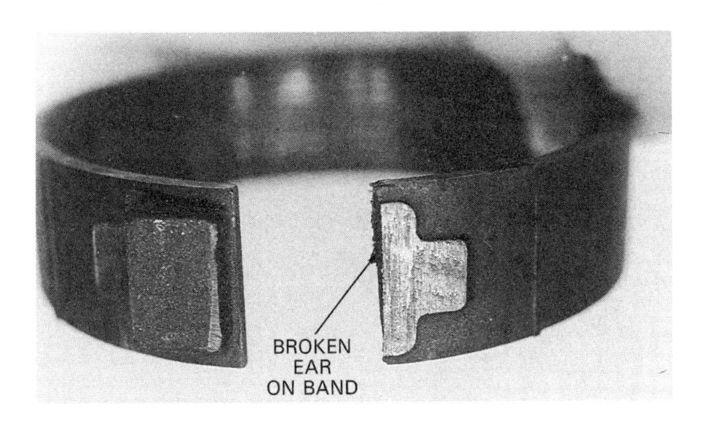

Figure 12-4. A broken band ear results in a no-drive condition in the gear in which the band should be applied. A new band should be installed. (BBU, Inc.)

the automatic transmission will not drive in that gear. The vehicle will not move if the one-way clutch is used for first gear. In addition, the transmission will skip any other gear where it is used. If the one-way clutch fails to unlock, the transmission will jam up except in the gear that normally uses the one-way clutch.

Transmission Oil Pump Malfunction

The automatic transmission front pump can make noises, or it can wear out. Noisy pumps usually make a whining sound. This sound, which is caused by tight pump clearances, will increase as engine speed increases. Noise from a rear pump, if there is one, will vary with road speed.

Sometimes, the transmission oil pump or oil pump bushings will wear and develop excessive clearances. This can cause oil to drain out of the torque converter when the vehicle is not operating. As a sign of this problem, the transmission fluid level will be above the full mark on the dipstick after the vehicle has been standing for several hours. The torque converter will usually refill if the engine is operated for a few minutes.

Another result of excessive pump wear is low oil pressure. Most pumps are designed to generate much more system oil pressure than needed. The pump will be very worn if a reduction in pressure is evident.

A defective rear pump may make it impossible to push start the vehicle. The rear pump, found on many older vehicles, provides pressure to the governor valve. A defective rear pump may also prevent the transmission from upshifting.

Most worn or otherwise defective oil pumps are replaced as a unit. Occasionally, the pump bushing can be replaced in an otherwise good pump. The front pump seal should be replaced as part of any pump service. The transmission must be removed from the vehicle to repair or replace the oil pump.

Transmission Oil Filter Malfunction

The better a filter does its job, the more likely it is to plug up. Dirt and other debris in the fluid gets trapped in the filter. Eventually, it may become plugged. Filters can also plug up because of excess sludge in the transmission, caused by the failure of other parts. A plugged filter will cause a whining noise that varies with engine speed. The transmission may begin to slip at higher engine rpm. As the restriction gets worse, the transmission will begin to slip badly, and eventually, it will not move at all. The transmission oil filter can be replaced by removing the oil pan and removing the screws or clips holding the filter in place. The filter can be checked for debris buildup or obvious damage, Figure 12-5.

Transmission Oil Cooler Malfunction

The oil cooler removes heat from the transmission. Oil cooler failures cause the transmission fluid and the transmission to overheat. Overheating leads to slipping, delayed engagement of the transmission when hot, oil sludging, and eventual holding member failure.

The oil cooler can fail due to internal plugging of the cooler lines, pinched or kinked lines, sludge or debris on the cooler surfaces, or an engine cooling-system problem. The parts of the cooler are largely external. These can develop leaks and rattles. Figure 12-6 shows some typical oil cooler system malfunctions. Cooler malfunctions can usually be solved by flushing out the cooler and radiator. Also, be sure to check engine cooling-system temperature, which is often the source of the cooling problem.

Valve Body Malfunction

Defective valve bodies can cause many hydraulic system problems when valves stick or are improperly

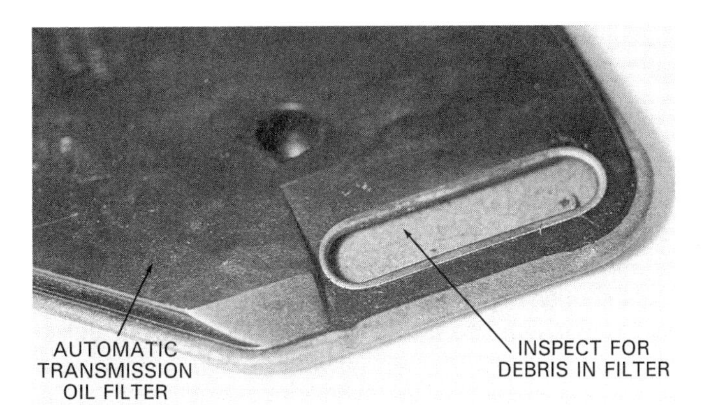

Figure 12-5. A plugged transmission filter can cause many problems, including slippage, noise, and buildup of dirt or metal particles in the transmission and converter. The filter should be replaced at regular intervals and always when the transmission is disassembled for service. (BBU, Inc.)

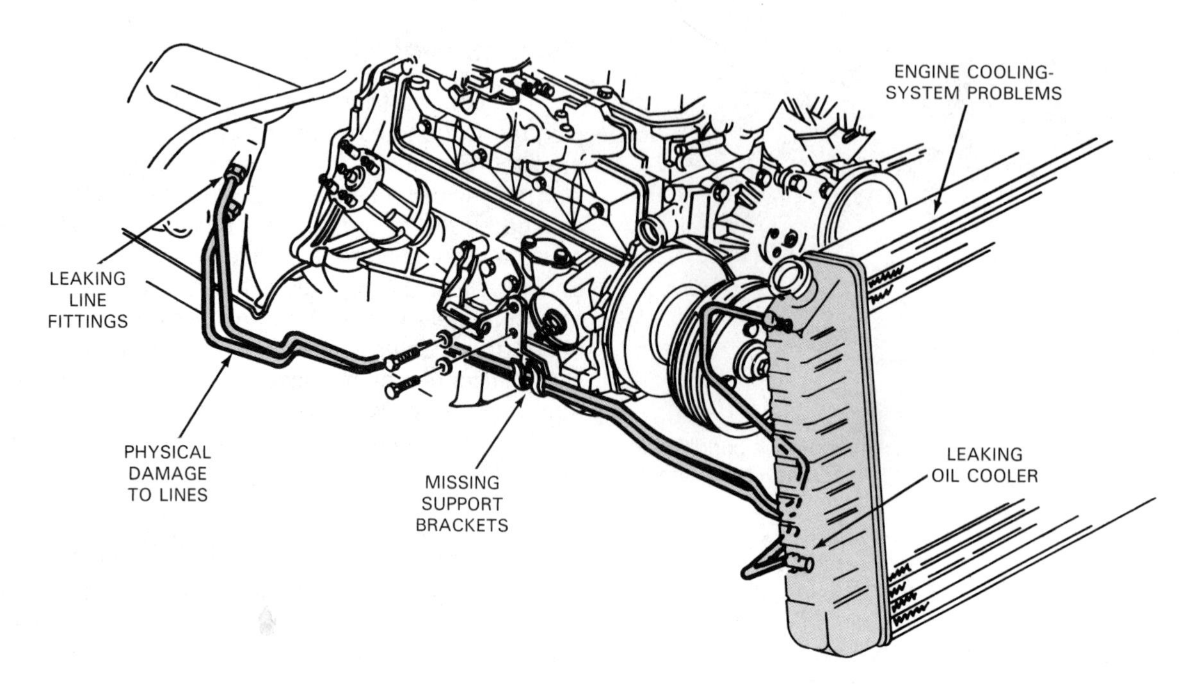

Figure 12-6. The oil cooler can plug up and cause transmission fluid and transmission overheating. The cooler may allow antifreeze to enter the transmission, causing major friction lining damage. The cooler lines can leak or become kinked or crushed. (General Motors)

assembled. Most valve body problems will show up as shift problems, although sticking torque converter or pressure regulator valves can cause pressure problems and slippage. Transmission or torque converter wear may result in the deposit of metal particles in the valve body, causing valves to stick. Note that aluminum valve bodies seem to develop more sticking valves than cast iron valve bodies.

Accumulators in the valve body may leak or stick, affecting shift quality. Check balls may be missing, leading to slippage or erratic shifting.

A valve body can usually be repaired by a thorough cleaning. Most valve bodies can be removed for inspection and cleaning by first removing the oil pan. If any valves cannot be freed by cleaning, the valve body should be replaced.

Vacuum Modulator Malfunction

Many vacuum modulator malfunctions are not in the modulator itself, but in related parts or systems. In many instances, the modulator is not receiving full engine vacuum because of plugged, disconnected, misconnected, or leaking vacuum hoses. See Figure 12-7. A worn out or mistuned engine can also cause the vacuum to be lower than what it should be. In rare instances, the throttle valve may be sticking in its bore.

Of course, the vacuum modulator itself can be damaged. If the modulator is leaking or bent, low or high line and throttle pressures result, which in turn, cause low or high shift points. Low line pressures, specifically, can

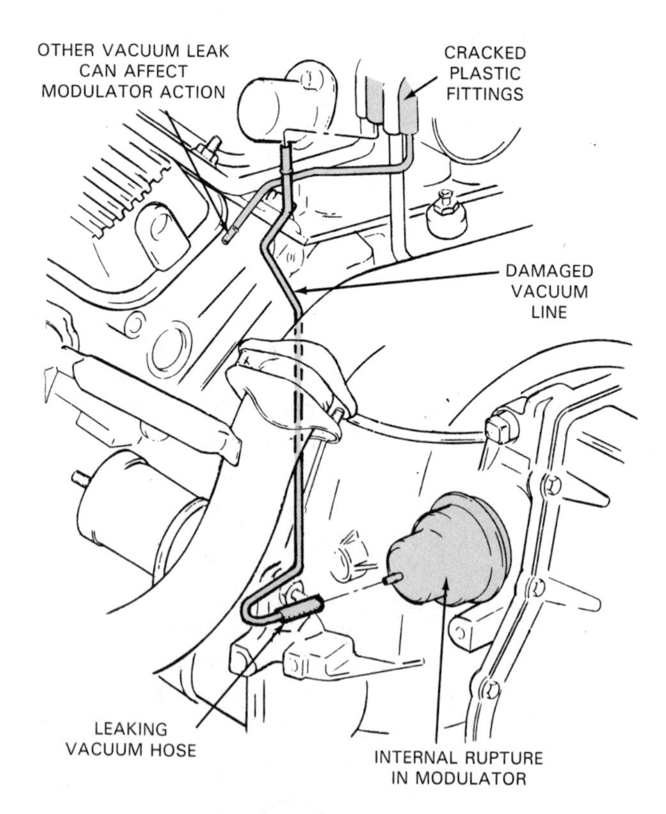

Figure 12-7. Vacuum modulator problems can be caused by a defective modulator or by failure of the modulator to receive full manifold vacuum. Always check the vacuum at the modulator before replacing it. (General Motors)

cause soft shifts, leading to slippage and burning of clutch and band friction linings. A split or hole in the modulator diaphragm, in addition to causing pressure problems, can cause fluid loss, as transmission fluid is drawn into and burned in the intake manifold.

Some automatic transmissions have an adjustable vacuum modulator. In this case, a transmission may be experiencing pressure problems because the modulator is out of adjustment. A simple adjustment may correct the problem.

If the vacuum modulator is bent or leaking, or if adjusting an adjustable type does not fix a pressure problem, the modulator must be replaced. Modulators can be removed from the case without removing any other transmission parts.

Governor Valve Malfunction

The governor valve is often the cause of shift problems. It tends to stick more readily than other valves. The valve, valve bore, and seal rings for the governor can wear, causing pressure loss. Metal particles caused by transmission wear can make the governor stick again–a few miles after it was cleaned.

A few makes of governors incorporate small filters in the governor pressure inlet passage. These filters can plug up and reduce governor pressure.

Governors can usually be restored to service by cleaning. Many governors are mounted on the transmission case and can be easily removed. Other governors are installed on the output shaft. To service these, the extension housing must be removed.

Automatic Transmission Linkage Malfunctions

Many vehicles use a T.V. linkage, consisting of levers connected by a series of rods or a cable, to control shift timing and pressures and to provide a detent for passing gear. Some vehicles have a kickdown linkage. The T.V. linkage or kickdown linkage is connected to the engine throttle, which in turn is connected through the throttle linkage to the accelerator. These linkages are often out of adjustment because of wear or damage. Some linkage is misadjusted at the factory and not noticed by the owner.

The shift selector lever operates the shift linkage, which in turn operates the manual valve. A common problem is the inability to select gears. Most problems can be corrected by adjustment. In some instances, the oil pan must be removed to reconnect the internal linkage to the manual valve.

Most linkages can be adjusted to eliminate problems. Adjustment is covered in Chapter 13. Replacement of linkage can be performed without removing other transmission parts.

Parking Pawl Malfunction

The parking pawl is operated by the driver. A common problem is failure of the parking pawl to hold. Most problems can be corrected by adjustment. In some instances, the oil pan must be removed to reconnect the parking pawl linkage.

Bearing and Thrust Washer Malfunctions

Bearing or thrust washer failures can cause noise and wear of other parts, such as shafts, clutch drums, and gearsets. Defective bushings or thrust washers can allow moving parts to shift enough to cause internal hydraulic pressure loss or external leaks. A worn extension bushing or *output shaft bearing*, which is a large bearing that supports the output shaft, may cause vibration when the vehicle is accelerated.

Bearings and thrust washers are located throughout the transmission. Replacement of some can be simple, while others require the entire transmission be disassembled.

Front Pump Seal and Rear Seal Malfunctions

The front pump and rear seals can develop leaks as they age. The front pump seal is under pressure when the engine is running, and a faulty seal will leak during vehicle operation. See Figure 12-8. A front pump leak may appear as traces (streaks) of fluid directed radially outward from

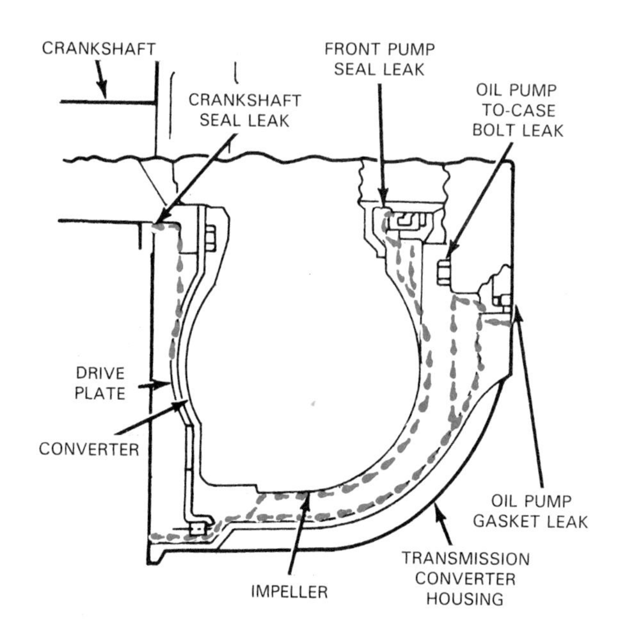

Figure 12-8. The area around the torque converter is often a source of leaks. The front pump seal can leak when the engine is running. Other seals and gaskets can leak when the engine is off–when fluid level is high. Do not confuse transmission leakage with leakage from the rear crankshaft seal. (Chrysler)

the pump seal. The converter will usually be streaked also. The transmission must be removed to service the front pump seal.

The rear seal will leak when the vehicle is moving. Even though there is no pressure behind the rear seal, fluid will be thrown against the rear seal by internal part movement. This splash will make a bad rear seal leak profusely. In addition, the rear seal will sometimes leak when the vehicle is parked facing uphill.

The rear seal can be replaced by removing the drive shaft and prying out the seal. A new seal can then be installed. Always check the bushing located near the rear seal for wear. A worn extension bushing, as it is called, may result in excessive movement of the driveline slip yoke. This will cause a replacement seal to leak as the yoke vibrates in use.

Other Seal and Gasket Malfunctions

There are many case or component seals used on the typical automatic transmission, including manual lever and shaft assembly seals, T.V. cable seals, band adjustment screw seals, and electrical connector seals. The typical transmission also uses a number of gaskets, including oil pan and extension housing gaskets.

Some seals and gaskets are under pressure when the transmission is operating and will leak during vehicle use. Many seals and gaskets will not leak when the engine is running. This is because fluid level is lower when the oil pump is drawing additional fluid from the hydraulic reservoir for the hydraulic system. Leakage will usually occur when the vehicle sits idle for a period of time, such as overnight.

Seals and gaskets must be replaced if they leak. Most can be replaced without removing the transmission from the vehicle.

Automatic Transmission Diagnostic Charts

Unless you are an experienced transmission rebuilder, diagnostic charts can be very useful. ***Diagnostic charts*** will help you to determine the exact cause of a problem. These charts are prepared by the transmission manufacturer. Each chart applies to one specific type of transmission and is designed to match the transmission problem to the specific part malfunction.

There are numerous types of diagnostic charts. An excerpt of a typical automatic transmission diagnostic chart is shown in Figure 12-9. It lists various problems, or conditions, and possible causes in tabular form. Such charts usually take up several pages of a service manual. Another common type takes the form of a flowchart.

A more condensed type of diagnostic chart is shown in Figure 12-10. The *conditions,* appearing down the side of the chart, list many of the problems that are common to this kind of transmission. The *possible causes,* appearing across the top of the chart, list common malfunctions in this transmission. An *X* is placed on the line wherever a certain part could be the cause of the problem. Reading the chart will reveal, for example, that slipping in reverse can be caused by ten different conditions. Projecting up from the *X* to the top of the chart, you can learn what might be causing the problem. The lowest numbered causes are the most likely; the highest numbered causes are the least likely.

As an example, suppose the vehicle had a problem of *no upshift.* Reading to the right, the first *X* encountered is in the fourth column–*hydraulic pressures too low.* This would be the most likely cause of the problem. The cause that would be the least likely lies in column 34–*overrunning clutch inner race damaged.*

Another type of diagnostic chart is the ***band and clutch application chart.*** This chart shows which holding members are applied in a particular gear. This type of chart can be very useful to isolate the defective holding member when the transmission is slipping in a particular gear.

A typical band and clutch application chart is shown in Figure 12-11. This particular chart is used by finding the gear (first, second, etc.) in the two leftmost columns and reading to the right to find the clutches and bands that are used in that particular gear. Note that the same holding members are used in different combinations to obtain the different gears. If, for instance, a transmission takes off well from stop but slips while trying to shift into second, the clutch or band that is applied to obtain second gear is probably at fault.

Sometimes a band or clutch will slip in one gear and not another. For instance, one clutch may slip when applied in high gear but not when used in reverse. The reason for this is that the pressures used in reverse are higher. Even though the friction discs are faulty, the higher reverse pressures apply them tightly, and they do not slip. Lower pressures are used for high gear, allowing the discs to lose their grip and slip. When transmission principles are understood, diagnosing this condition is not too difficult.

Still another type of diagnostic chart is the ***automatic shift diagram,*** Figure 12-12. This chart shows the vehicle's approximate shift points, or points at which the particular transmission should shift. The chart can be compared with actual shift points, and transmission operation, determined accordingly.

Automatic Transmission Diagnostic Testing

The automatic transmission can be tested by many methods, including *road tests, hydraulic pressure tests, stall tests,* and *air pressure tests.* The technician must be guided by experience to determine which of these tests

CONDITION	INSPECT COMPONENT	FOR CAUSE
OIL LEAK	• Oil Pan	– Bolts not correctly torqued. – Improperly installed or damaged pan gasket. – Oil pan gasket mounting face not flat.
	• Filler Pipe	– Multi-lip seal damaged or missing.
	• Filler Pipe Bracket	– Mispositioned.
	• Throttle Valve Cable	– Multi-lip seal missing, damaged, or improperly installed.
	• Rear Seal Assembly	– Damaged or improperly installed.
	• Speedometer Driven Gear	– O-ring damaged.
	• Manual Shaft	– Lip seal damaged or improperly installed.
	• Case	– Line pressure tap plug. – Fourth clutch pressure tap plug. – Porous.
	• Intermediate Servo	– O-rings damaged.
	• Oil Pump Assembly	– Front pump seal leaks: Seal lip cut–check converter hub for nicks, etc.; bushing moved forward and damaged; garter spring missing from seal. – Front pump attaching bolts loose or bolt seal damaged or missing. – Front pump housing O-ring damaged or cut. – Porous casting. – Inspect converter weld area.
	• Vent Pipe	– Transmission overfilled. – Water in oil. – Foreign matter between pump and case or between pump cover and body. – Case porous; front pump cover mounting face shy of stock near breather. – Pump to case gasket mispositioned. – Incorrect dipstick. – Pump shy of stock on mounting faces, porous casting, breather hole plugged in pump cover.

Figure 12-9. The transmission diagnostic chart is a convenient guide to the common problems of a particular transmission. (General Motors)

POSSIBLE CAUSE		HARSH ENGAGEMENT FROM NEUTRAL TO D OR R	DELAYED ENGAGEMENT FROM NEUTRAL TO D OR R	RUNAWAY UPSHIFT	NO UPSHIFT	3-2 KICKDOWN RUNAWAY	NO KICKDOWN OR NORMAL DOWNSHIFT	SHIFTS ERRATIC	SLIPS IN FORWARD DRIVE POSITIONS	SLIPS IN REVERSE ONLY	SLIPS IN ALL POSITIONS	NO DRIVE IN ANY POSITION	NO DRIVE IN FORWARD DRIVE POSITIONS	NO DRIVE IN REVERSE	DRIVES IN NEUTRAL	DRAGS OF LOCKS	GRATING, SCRAPING GROWLING NOISE	BUZZING NOISE	HARD TO FILL, OIL BLOWS OUT FILLER HOLE	TRANSMISSION OVERHEATS	HARSH UPSHIFT	DELAYED UPSHIFT	SLIPS IN REVERSE OR MANUAL LOW
Faulty lockup clutch.	35	X																			X		
Overrunning clutch inner race damaged.	34				X													X					
Overrunning clutch worn, broken or seized.	33								X				X			X	X						
Planetary gear sets broken or seized.	32											X	X	X		X	X						
Rear clutch dragging.	31														X								
Worn or faulty rear clutch.	30	X	X						X				X	X	X								
Insufficient clutch plate clearance.	29														X					X			
Faulty cooling system.	28																			X			
Kickdown band adjustment too tight.	27															X				X			
Hydraulic pressure too high.	26	X																			X		
Breather clogged.	25																		X				
High fluid level.	24																		X				
Worn or faulty front clutch.	23		X	X	X	X		X		X				X								X	
Kickdown servo band or linkage malfunction.	22			X	X	X	X	X														X	
Governor malfunction.	21				X		X	X														X	
Worn or broken reaction shaft support seal rings.	20		X	X	X		X	X		X				X								X	
Governor support seal rings broken or worn.	19				X	X		X														X	
Output shaft bearing and/or bushing damaged.	18																X						
Overrunning clutch not holding.	17								X				X										
Kickdown band out of adjustment.	16					X											X				X	X	
Incorrect throttle linkage adjustment.	15			X	X	X	X	X	X												X	X	
Engine idle speed too low.	14		X																				
Aerated fluid.	13		X	X		X		X	X	X	X							X	X				
Worn or broken input shaft seal rings.	12		X						X		X		X										
Faulty oil pump.	11		X					X	X	X	X	X				X		X		X			X
Oil filter clogged.	10		X	X				X	X		X	X							X				
Incorrect gearshift control linkage adjustment.	9		X		X			X	X	X				X	X					X			
Low fluid level.	8		X	X	X	X		X	X	X	X	X	X					X		X			
Low-reverse servo, band or linkage malfunction.	7		X							X				X									
Valve body malfunction or leakage.	6	X	X	X	X	X	X	X	X	X	X	X	X	X	X			X					
Low-reverse band out of adjustment.	5									X				X		X	X						X
Hydraulic pressures too low.	4		X	X	X	X		X	X	X	X	X	X	X						X	X		
Engine idle speed too high.	3	X																		X			
Stuck lockup valve.	2															X							
Stuck switch valve.	1																			X			

CONDITION

Figure 12-10. The problem and the possible causes are cross-referenced by an *X* or other mark at the appropriate spot on this type of chart. (Chrysler)

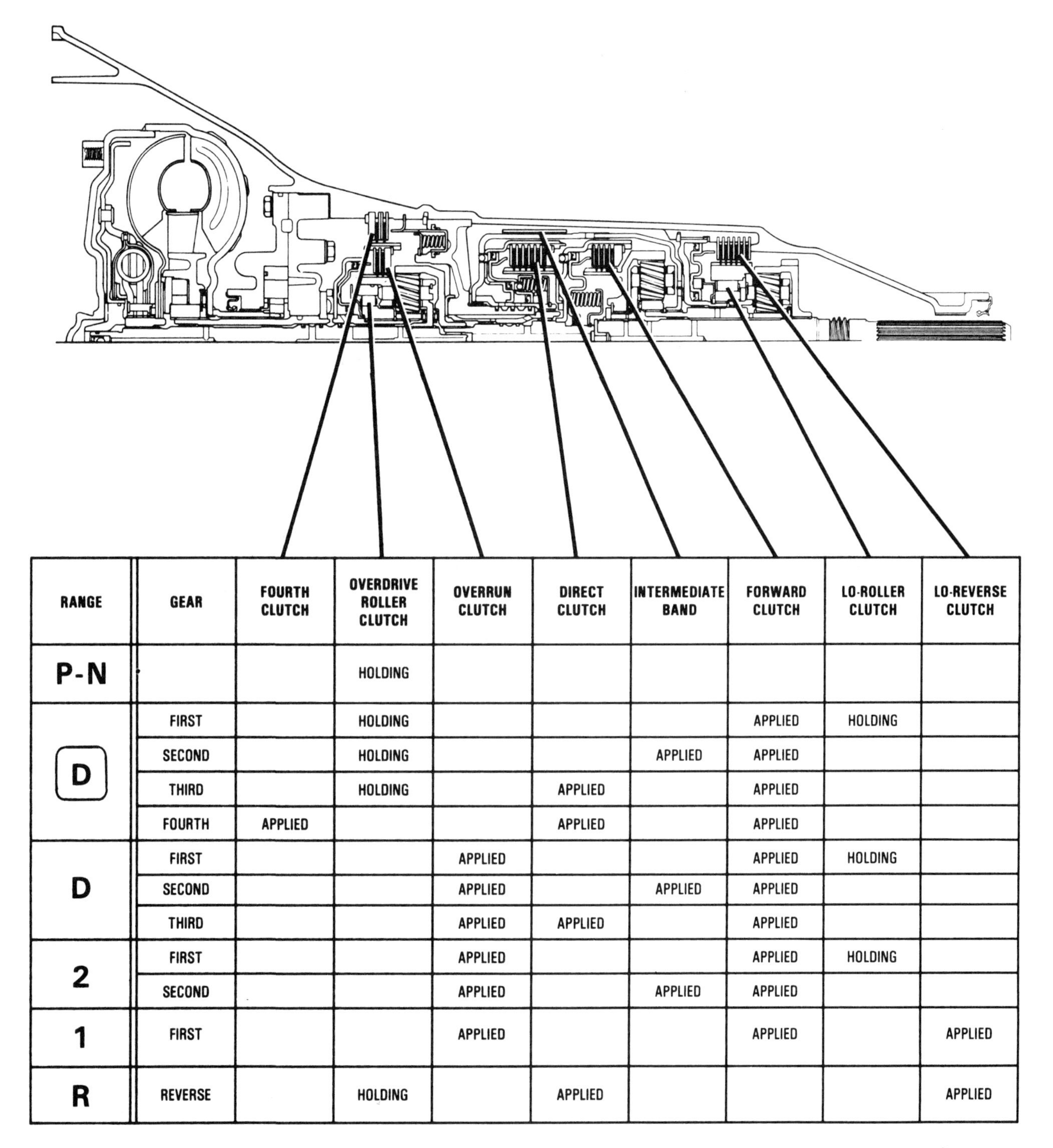

RANGE	GEAR	FOURTH CLUTCH	OVERDRIVE ROLLER CLUTCH	OVERRUN CLUTCH	DIRECT CLUTCH	INTERMEDIATE BAND	FORWARD CLUTCH	LO-ROLLER CLUTCH	LO-REVERSE CLUTCH
P-N			HOLDING						
(D)	FIRST		HOLDING				APPLIED	HOLDING	
	SECOND		HOLDING			APPLIED	APPLIED		
	THIRD		HOLDING		APPLIED		APPLIED		
	FOURTH	APPLIED			APPLIED		APPLIED		
D	FIRST			APPLIED			APPLIED	HOLDING	
	SECOND			APPLIED		APPLIED	APPLIED		
	THIRD			APPLIED	APPLIED		APPLIED		
2	FIRST			APPLIED			APPLIED	HOLDING	
	SECOND			APPLIED		APPLIED	APPLIED		
1	FIRST			APPLIED			APPLIED		APPLIED
R	REVERSE		HOLDING		APPLIED				APPLIED

Figure 12-11. The band and clutch application chart will help the technician find out what holding members are applied in what gear. This will help to isolate problems occurring in a particular gear. (General Motors)

should be performed on a particular transmission. The road test is an exception. It is of critical importance and should *always* be performed if the car is driveable. Procedures for performing these diagnostic tests on automatic transmissions are explained on these next few pages.

Road Test

A ***road test*** involves driving the vehicle to observe transmission operation under actual road conditions. Transmissions should be road tested whenever the transmission is able to drive the vehicle–with the vehicle owner, if possible. Always check the fluid level before beginning the road test and observe all traffic laws during the test.

Start the road test by accelerating normally and comparing shift speeds to the published shift points, such as those shown in Figure 12-12. Decelerate gradually, taking note of downshift speeds. Ensure that the detent and part-throttle downshifts occur at the proper times. During these tests, also take note of the shift feel, missed shifts or slipping, noises, and whatever other problems may be present.

Hydraulic Pressure Tests

A ***hydraulic pressure test*** of an automatic transmission involves connecting one or more pressure gauges to the transmission and observing the hydraulic fluid pressures. Actual pressures are then compared to published values to decide if there is a hydraulic system problem. Hydraulic pressure tests are useful in pinpointing suspected problems or in cases where the problem cannot be determined by any other method.

To perform a hydraulic pressure test, first locate the pressure tap(s) that you will need. Figure 12-13 shows some typical pressure tap locations on one make of transmission. Remove the pressure tap plug(s) and install the pressure gauge. A typical hookup is shown in Figure 12-14. Make sure that the range of the pressure gauge exceeds the maximum pressure that the system can produce. Install a tachometer if the instrument panel does not indicate engine rpm. Start the engine and observe the readings on the pressure gauge, checking pressures in neutral, all drive ranges, and reverse. Make the checks at idle speeds, noting pressures at specified engine rpm. Then, raise the engine speed as specified in the service manual and again note pressure readings.

A hydraulic pressure test procedure is depicted in Figure 12-15. While performing this test on the ground, you should have the parking brake applied. In addition, hold one foot on the brake pedal while depressing the accelerator.

Compare the gauge readings with pressure figures published in the factory service manual. Excessive pressure readings indicate a pressure regulator or pressure relief valve that is stuck closed. Low pressure indicates a worn front pump, a pressure regulator or pressure relief valve that is stuck open, internal leaks, or a plugged filter. If pressure is low in only one gear, the apply piston seals or seal rings are probably leaking.

If the transmission has a T.V. linkage, disconnect linkage rod or cable at the intake manifold and operate linkage by hand, observing the pressure gauge while the engine is running. The transmission should be in drive with the parking brake applied. Line pressure should increase as the cable is moved toward the full throttle position and decrease as the cable is moved to the idle position. This

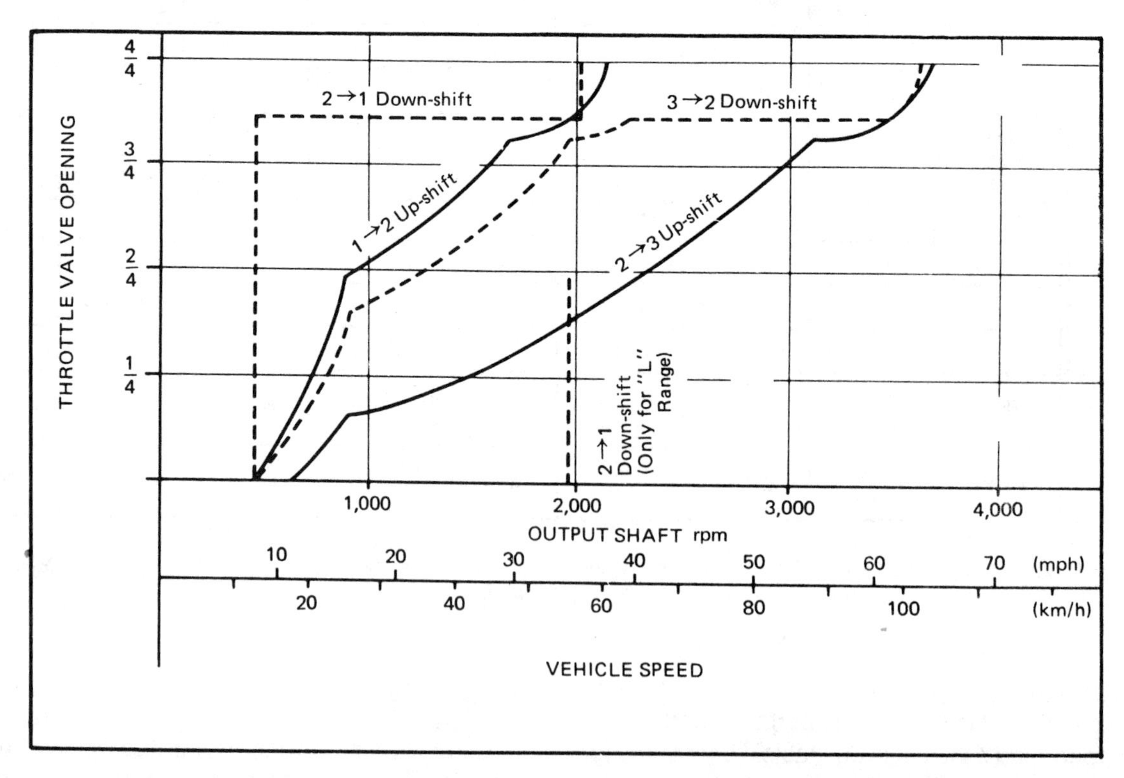

Figure 12-12. The automatic shift diagram shows the approximate points for transmission shifts. These shift points are compared against the transmission's actual shift points. Some manufacturers simply list shift points in their service manuals. (Toyota)

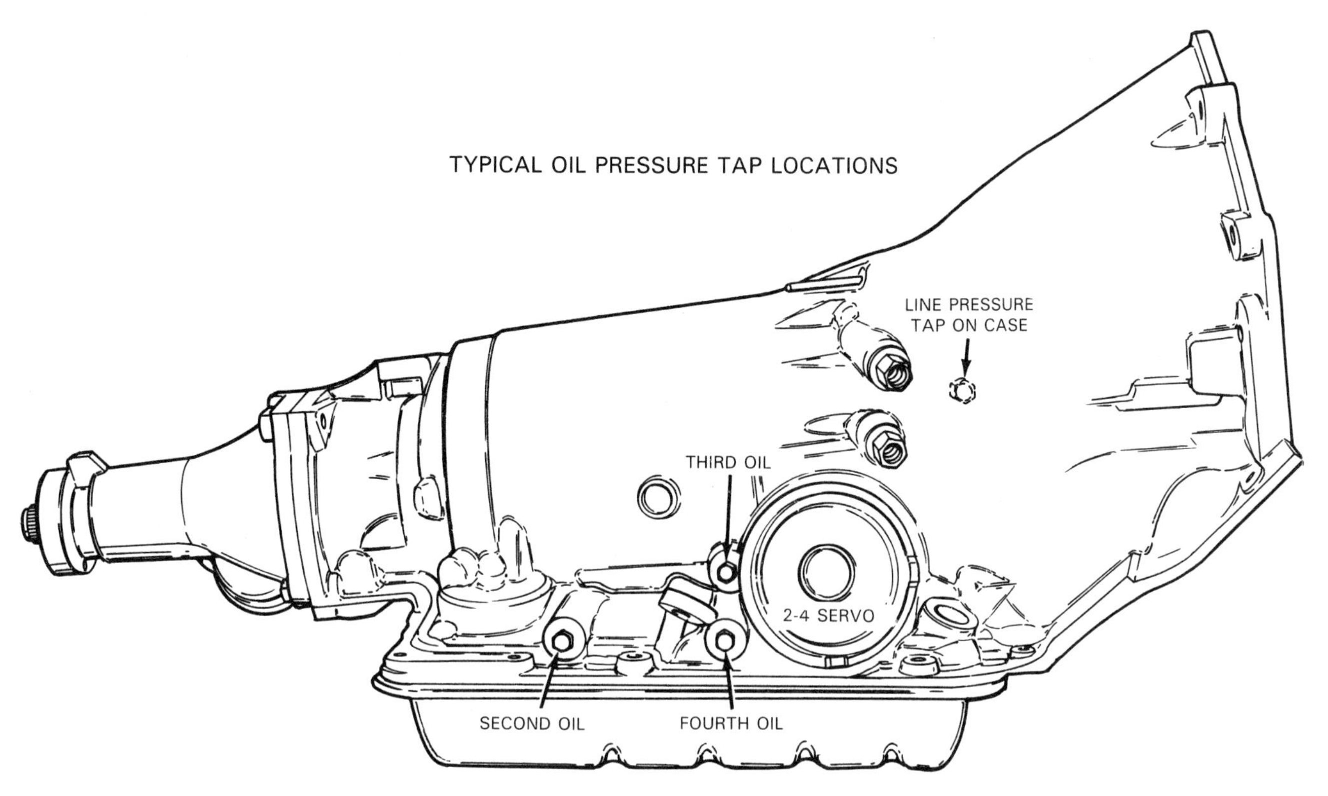

Figure 12-13. Pressure taps for performing control pressure tests are located on various parts of the transmission case. All automatic transmissions will have at least one pressure tap; most have several. (General Motors)

indicates that the throttle and pressure regulator valves are able to respond to the changes in throttle position.

If the transmission has a vacuum modulator, disconnect the vacuum line with the engine running. The transmission should be in drive and the parking brake applied. Pressure should increase, indicating that the vacuum modulator and regulator valve are responding to the loss of manifold vacuum.

Perform a road test with a pressure gauge installed. You must be able to read the gauge from the driver's seat. Therefore, the gauge must have a hose that is long enough to reach the transmission. Observe the gauge needle as each shift occurs. The needle should drop during each shift. A slow pressure rise or low pressure in any one gear is caused by internal leakage from defective piston seals or seal rings in that particular gear. Consult the band and clutch application chart to pinpoint hydraulic leakage to a specific clutch or servo.

The pressure gauge can also be used to check throttle and governor pressure on many vehicles. Consult the service manual for the proper pressure tap locations and pressure specifications.

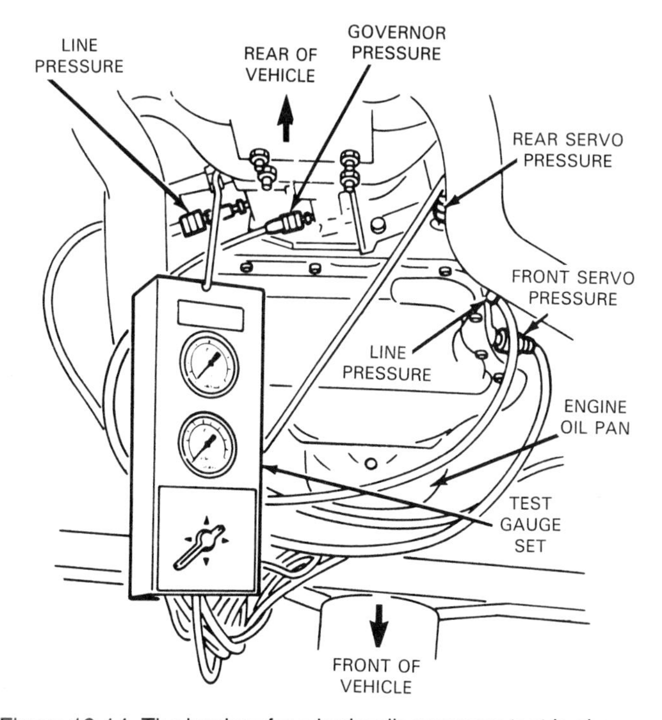

Figure 12-14. The hookup for a hydraulic pressure test is shown here. In this illustration, the test is performed with the vehicle on a hoist. More than one gauge can be installed to test several pressures at the same time. (Chrysler)

Stall Test

A ***stall test*** is a way of loading the engine through the transmission to isolate a problem in the torque converter

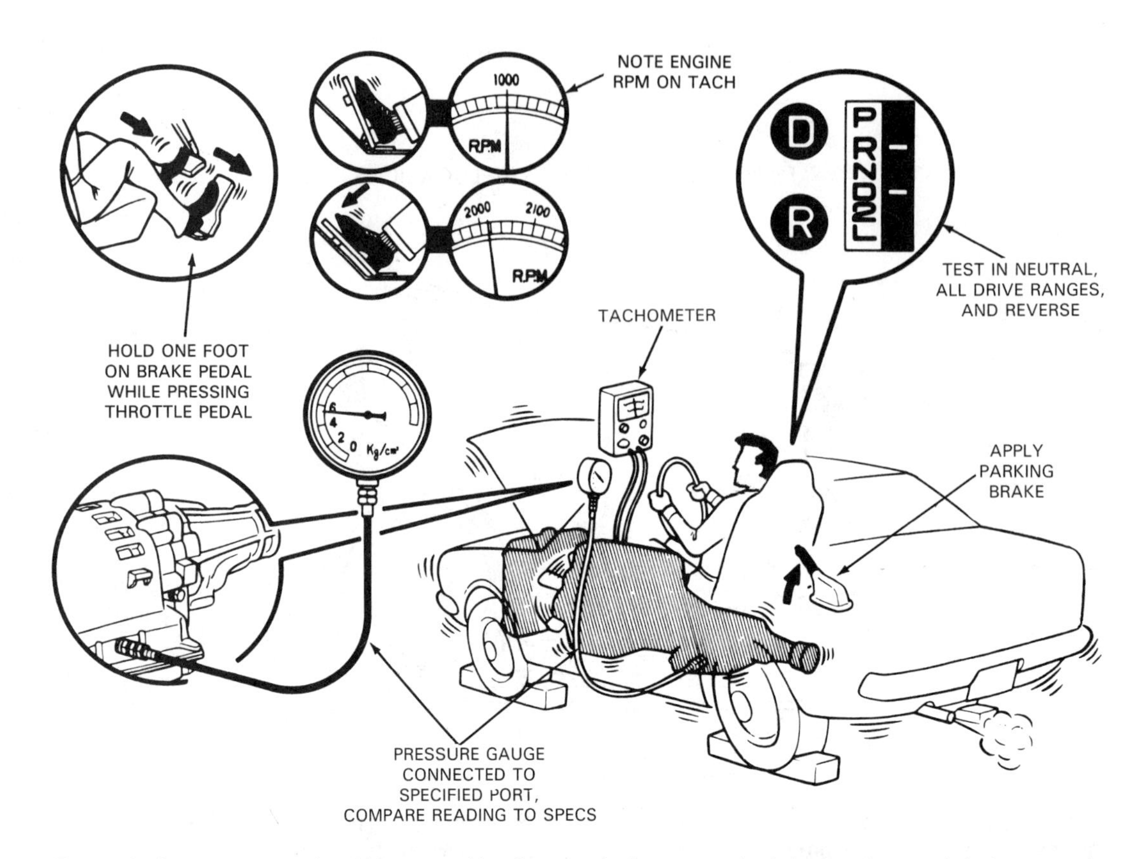

Figure 12-15. Transmission pressures should be tested in all transmission gears, both in the shop and during a road test. The latter allows readings to be obtained while the transmission shifts through all gears at various throttle openings. Pressure readings can then be compared with the factory specifications. (Toyota)

or transmission. Stall tests check the holding ability of the clutch packs and bands and the operation of the stator one-way clutch. The test is sometimes useful to diagnose slipping holding members, low hydraulic pressures, or problems with the torque converter stator.

To conduct a stall test, install a tachometer on the engine, Figure 12-16. Then, start the engine and place the transmission in gear. With the parking brake applied and the brake pedal fully depressed, press the accelerator until the engine speed will not increase. This is maximum stall speed. This speed should not exceed the published figures. If it does, there is a problem in the transmission. Release the accelerator immediately to prevent further damage. A higher than specified stall speed indicates defective holding members or low hydraulic pressures. If the engine will not reach the specified stall speed, the engine requires service, or the torque converter stator is not locking.

Stall tests should be performed in each driving range. Do not run engine at stall speed for longer than 5 seconds because of the rapid heat buildup that results from this test. Allow a cooling period between each test with the transmission in neutral.

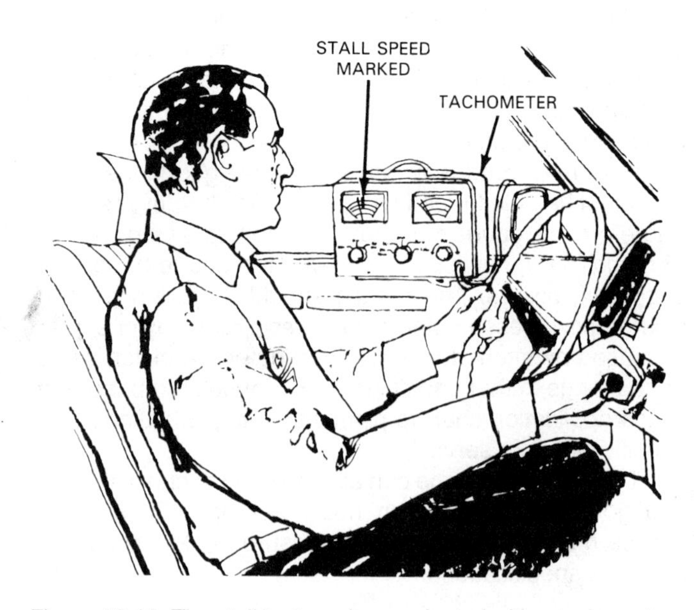

Figure 12-16. The stall test can be conducted with a tachometer. When performing a stall test, always make sure that the parking and service brakes are firmly applied. (Ford)

Some manufacturers do not recommend stall tests because of the stress they put on the engine and transmission, and these manufacturers do not publish stall speed specifications. There is no purpose in conducting a stall test on a vehicle when this is the case.

WARNING! While conducting a stall test, both the parking and service brakes must be fully applied. As an added precaution, block both front wheels. Make sure no one is standing in front of or in back of the car. It could lurch forward or backward as engine reaches maximum rpm, even with the brakes applied and the wheels blocked.

Air Pressure Tests

Air pressure tests involve applying compressed air to the hydraulic pressure ports that supply hydraulic fluid to the holding members, and sometimes, the governor valve and accumulators. See Figure 12-17. In performing air pressure tests, operation of these components is often observed to determine if it is proper or faulty. If faulty, the component may be hung up, or there may be a leak or obstruction within the hydraulic passageway. The clutch and servo apply piston seals, seal rings, or gaskets may not be sealing as they should. An air pressure test may help to determine such problems.

The oil pan and valve body must be removed to perform an air test. Prior to their removal, the transmission fluid must be drained. Once the valve body is removed, compressed air is applied to the proper hydraulic passage, Figure 12-18. Air pressure should be regulated to about 30 psi–40 psi (207 kPa–276 kPa). Air supply must be free of all moisture and dirt.

If compressed air is delivered to a clutch passage, the clutch should make a sharp *thunk* as it applies. The application can sometimes be felt by placing your fingers on the clutch drum as air is applied. Servos will be seen to operate the band as air is applied. A defective clutch or servo apply piston will not apply strongly and may leak air. Some governor valves can also be checked with air pressure. If the governor buzzes when air is applied to the passage, it is free in its bore–meaning, it is not sticking. If the air pressure test indicates that a leak exists somewhere, the transmission must be disassembled to check the apply piston seals and other parts.

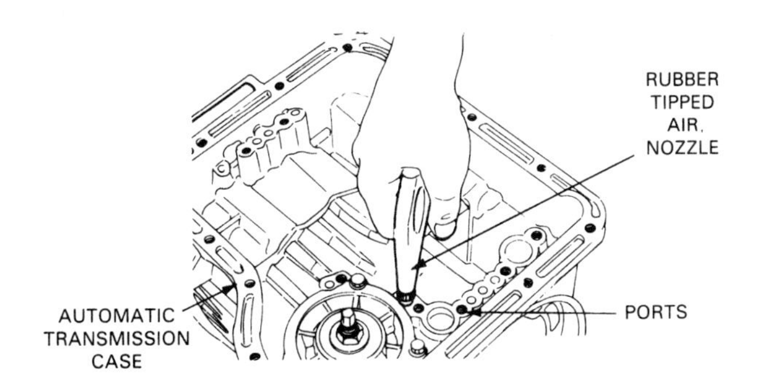

Figure 12-17. An air pressure test is performed to determine if holding members, and sometimes, the governor valve and accumulators, are operable. The test can be used to check for hydraulic system leaks and obstructions. The test is performed by applying air pressure to the ports leading to the clutch pack, band, or other component. The transmission fluid must first be drained and the oil pan and valve body removed. (Nissan)

Using Hydraulic Diagrams to Troubleshoot

Hydraulic diagrams, or oil pressure diagrams, can be used to trace circuits in the hydraulic system. Knowing exactly where the different hydraulic passageways go and what they are for will enable you to understand many seemingly obscure problems. Most service manuals contain at least one hydraulic diagram for the specific make of transmission, and many, as in Chapter 11, contain a diagram for every gear. The hydraulic circuits can be traced to isolate problems and determine the cause of hydraulic malfunctions.

Figure 12-19 shows the hydraulic flow in a common 3-speed transmission. The flow can be traced through the passageways to the manual and shift valves and on to the clutches and bands.

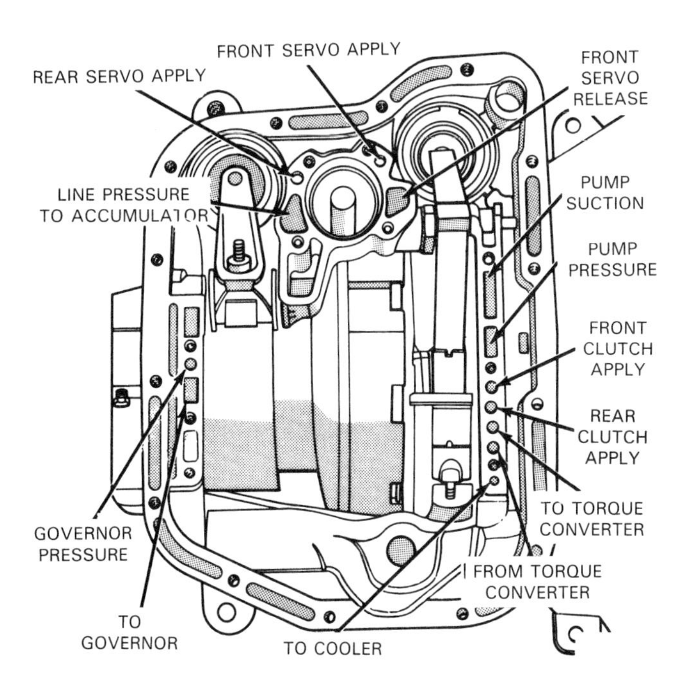

Figure 12-18. Bottom view of a transmission case with oil pan and valve body removed. The factory service manual will usually have a diagram like this, showing where the holding member apply ports are located on the case. Other hydraulic system devices, such as the governor, can also be checked with air pressure. (Chrysler)

Suppose one of the check balls (shown in Figure 12-19 as numbered circles inside of hydraulic lines) were missing. Checking the transmission line pressure would allow you to determine that a missing check ball was causing a transmission malfunction in a certain gear.

If, for instance, the #3 check ball were missing, fluid would be dumped out through the passageway to the manual valve whenever the transmission shifted into third gear. (Refer to Figure 12-19.) This flow would cause a drop in line pressure, as well as prevent full application of the front and rear clutches and full release of the front servo. The transmission would either slip badly in third, or it would not upshift out of second. By connecting a pressure gauge to the transmission and observing a line pressure drop in third gear, you could determine that the #3 check ball was missing.

NOTE! Check balls do not suddenly disappear. A missing check ball would be the result of carelessness during valve body overhaul and would be evident when the vehicle was driven after overhaul.

Notice on this diagram that the fluid used by the 2-3 shift valve to apply the front clutch and release the front band comes from the 1-2 shift valve. Therefore, if the 1-2 shift valve sticks in the upshifted position, the transmission will start in second gear, but the 2-3 shift will be normal. If the 1-2 shift valve sticks in the downshifted position, however, the transmission will stay in first gear, even though the 2-3 shift valve is not sticking. On some transmissions, however, the 2-3 shift valve is fed line pressure, and the transmission would shift from first to third, even though the 1-2 shift valve was stuck. This illustrates the importance of studying the specific hydraulic circuit diagrams for the transmission that you are working on, to avoid spending much time looking for a nonexistent problem. In this case, misunderstanding the hydraulic system could mean wasting time checking the governor or throttle valves.

Figure 12-20 shows the hydraulic flow in a common 4-speed automatic transmission. Extra holding members are required for overdrive. Notice that pressure from the *vacuum throttle valve* is used, in part, to modify the action of the pressure regulator valve. Throttle pressure acts on the regulator valve to assist the spring. Changes in throttle pressure will affect the action of the pressure regulator valve. This helps to explain why problems with the vacuum modulator or throttle valve may cause rough downshifts or slipping.

Valves in this hydraulic system that modify the line pressure and that can affect upshift or downshift quality are the pressure regulator valve, vacuum throttle valve, *pressure modifier valve, throttle back-up valve,* governor valves, and *throttle relief valve.* Shift timing can be affected

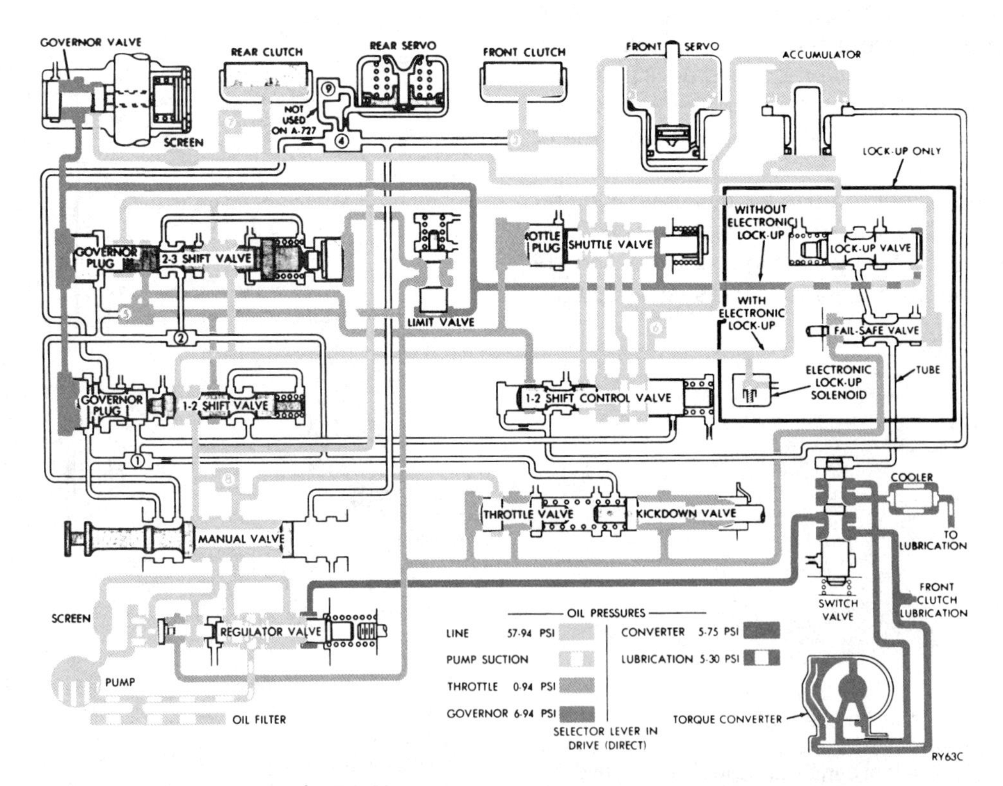

Figure 12-19. Hydraulic diagrams such as this can be used to troubleshoot problems in the hydraulic system. The diagram shows the hydraulic flow of a Chrysler TorqueFlite transmission in third gear (direct drive). (Chrysler)

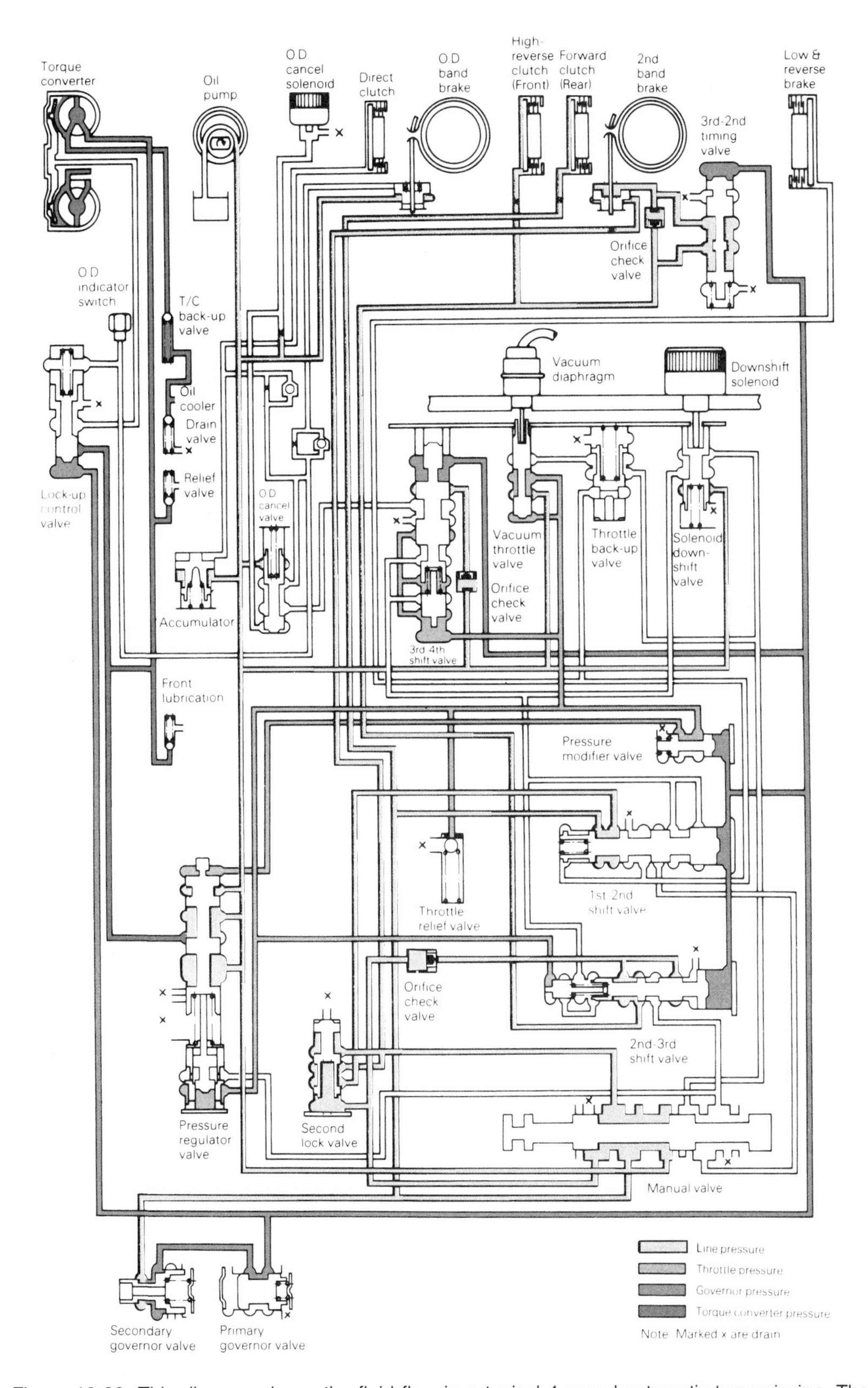

Figure 12-20. This diagram shows the fluid flow in a typical 4-speed automatic transmission. The hydraulic system controls four clutch packs, two bands, and a lockup torque converter. The transmission hydraulic system is acted upon externally by a vacuum modulator and two electric solenoids.

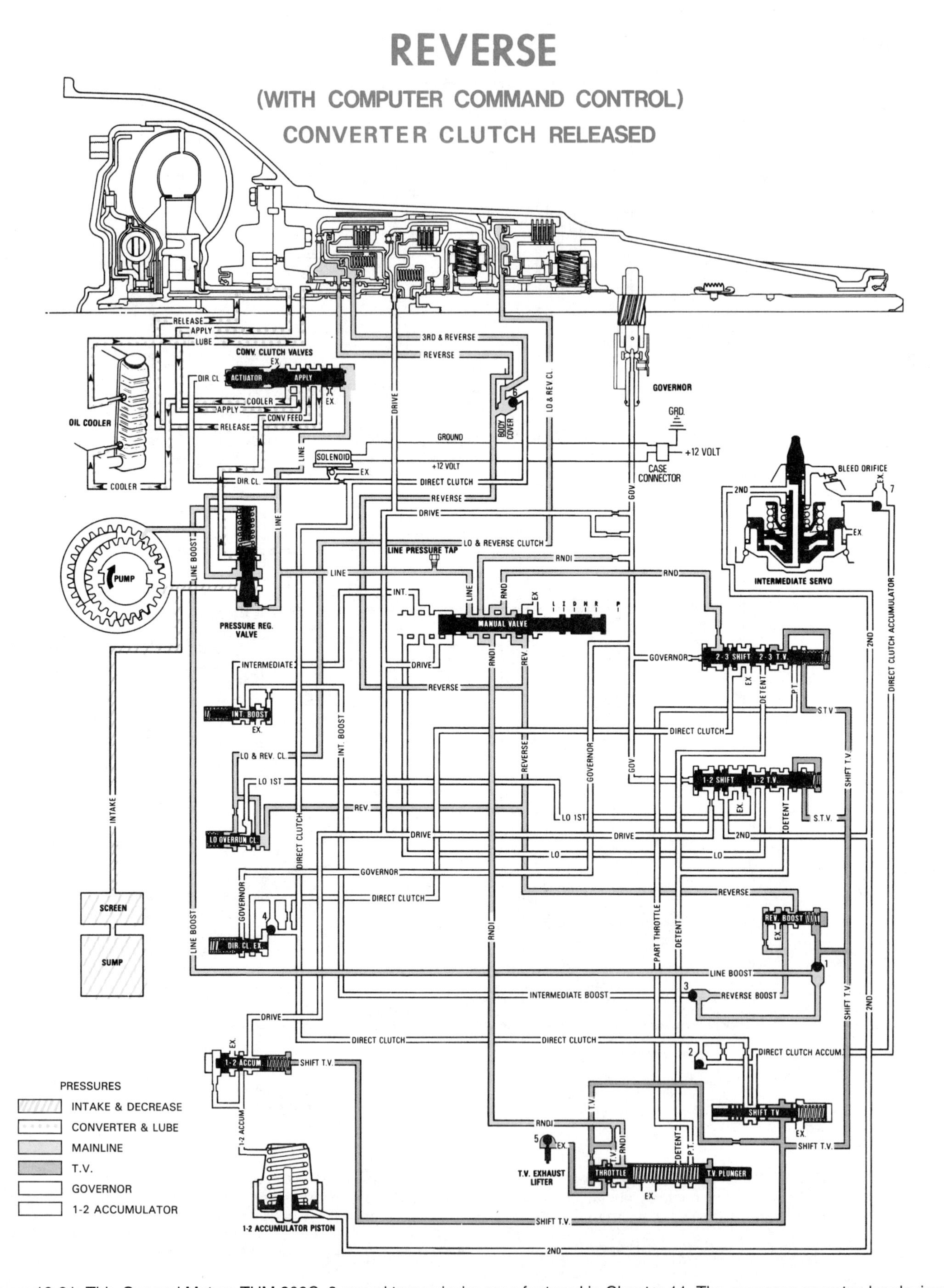

Figure 12-21. This General Motors THM 200C, 3-speed transmission was featured in Chapter 11. The governor operates by closing off the flow of fluid leaking through it as vehicle speed increases. The throttle valve is operated by a T.V. cable. (General Motors)

by the shift valves, governor valves, throttle valves, *3rd-2nd timing valve, solenoid downshift valve,* and throttle back-up valve. Valves that can affect the quality of the shift by modifying pressure during the shift are the pressure modifier valve, any one of the check valves, the accumulator, and the throttle back-up valve. Tracing the fluid flow through these valves can often lead you to a sticking valve that is the source of a slippage problem or a shift timing or quality problem.

Figure 12-21 shows another common 3-speed transmission. The throttle valve is controlled by a T.V. cable. Check ball #5, next to the throttle valve, is designed to seat if the T.V. cable breaks. When the check ball seats, line pressures and shift points will become very high. Knowing this may make it easier to diagnose a high, hard shift on this kind of transmission.

Another feature of this transmission is the torque converter clutch control system. The solenoid is designed to direct pressurized fluid to the converter clutch valve when energized. Therefore, if the converter clutch will not apply, it may be because the solenoid has failed.

Summary

While automatic transmissions are complex, they can be diagnosed and repaired if the technician understands their operation, and uses logical troubleshooting procedures. Before beginning repairs, you must first try to find out what the problem is.

A general troubleshooting procedure is to first have the owner describe the problem. Road test the car with the driver, if possible. Check the transmission fluid level prior to the road test. Check linkage if road test points to possible linkage malfunction and adjust as necessary. Check vacuum at vacuum modulator, as applicable. Inspect electrical connections. Perform such diagnostic tests as hydraulic and air pressure tests, if necessary, and consult diagnostic charts. Make sure that problem is not caused by another vehicle system. Sometimes, transmission disassembly is required if foregoing procedures are inconclusive.

Causes of automatic transmission problems are often identified as belonging to one of three systems. These are the hydraulic system, the mechanical system, and the electrical system. Hydraulic system problems affect the devices that produce and control hydraulic pressure in the transmission. Mechanical system problems affect the shafts, gearsets, holding members, case, etc. Electrical system problems affect components such as solenoids, neutral start switch, and computer sensors.

Typical transmission problems include slippage, no-drive condition, shift problems, noise, overheating, and leakage. There can be many causes of these problems, and the transmission should be thoroughly checked out by a process of elimination to determine the problem.

Typical transmission problems can be attributed to different component malfunctions; the cause of each one of these problems is not unique to any single component. For example, a noisy transmission can be attributed to a broken flexplate, a torque converter malfunction, a damaged planetary gearset, etc.

Diagnostic charts can make troubleshooting easier. One type of diagnostic chart gives common transmission problems with common causes of problems cross-referenced. The band and clutch application chart allows the technician to determine which holding members are applied in which gears. This makes diagnosis of a problem in a particular gear easier. The automatic shift diagram allows the technician to compare transmission shift points against published specifications.

The automatic transmission should be tested before it is disassembled. One such test is the road test, in which the vehicle is driven and transmission action observed. Pressure tests require a pressure gauge. They are conducted to determine whether problems exist in the hydraulic system. Stall tests can be used to determine whether the holding members and torque converter are functioning properly. Stall tests are very hard on the engine and drive train components, and some manufacturers do not recommend them. Air pressure tests are used to determine if the band and clutch apply pistons are operating properly and to spot hydraulic system leaks or obstructions. Air pressure tests are made after the oil pan and valve body are removed from the transmission case.

Automatic transmission hydraulic diagrams can be used to diagnose many problems. They can help the technician spot many unusual problems and locate sources of pressure leaks or obstructions. Manufacturers provide hydraulic diagrams for every type of transmission that they use. Reading a hydraulic diagram takes some practice, but it can be learned with practice.

Know These Terms

Automatic transmission slippage; Diagnostic charts; Band and clutch application chart; Automatic shift diagram; Road test; Hydraulic pressure test; Stall test; Air pressure test.

Review Questions–Chapter 12

Please do not write in this text. Place your answers on a separate sheet of paper.

1. Certain transmissions are prone to certain types of problems. For analyzing problems of these transmissions, the best guide is often:
   a. A band and clutch application chart.
   b. An automatic shift diagram.
   c. A stall test.
   d. Experience.

2. Which of the following would be the most unlikely reason for transmission fluid having a burnt odor or containing sludge?
   a. Clutch pack or band failure.
   b. A faulty oil cooler.
   c. Overheating of fluid.
   d. Out-of-adjustment shift linkage.
3. Which problem is evident when the transmission does not transfer enough power to the drive wheels?
   a. Erratic shifts.
   b. Slippage.
   c. Early upshift.
   d. No passing gear.
4. How would you analyze a noisy transmission?
5. Transmission overheating may occur when the temperature of the fluid reaches roughly ______°F, or ______°C.
6. An automatic transmission slips in second gear only. Technician A says to first do a hydraulic pressure test to check the oil pump and valves. Technician B says to first try to adjust the band. Who is correct?
   a. Technician A.
   b. Technician B.
   c. Both Technician A and B.
   d. Neither Technician A nor B.
7. An automatic transmission makes a whining noise in first gear only. Technician A says to do a hydraulic pressure test. Technician B says the problem is in the gearset. Who is correct?
   a. Technician A.
   b. Technician B.
   c. Both Technician A and B.
   d. Neither Technician A nor B.
8. An automatic transmission does not want to shift into high, engine rpm is excessive, and the transmission stays in second. Technician A says to check the governor. Technician B says to check the torque converter. Who is correct?
   a. Technician A.
   b. Technician B.
   c. Both Technician A and B.
   d. Neither Technician A nor B.
9. When checking for a possible vacuum modulator malfunction, you should check for a ruptured diaphragm and a blocked, leaking, or otherwise damaged vacuum line. True or false?
10. Which one of the following service procedures requires transmission removal?
    a. Adjusting transmission band.
    b. Replacing front pump seal.
    c. Replacing rear seal.
    d. Cleaning valve body.

Certification-Type Questions–Chapter 12

1. Technician A says that automatic transmission fluid level should be checked before road testing the vehicle. Technician B says that a road test may indicate that the vehicle driver is mistaking a normal condition for a problem. Who is right?

(A) A only
(B) B only
(C) Both A & B
(D) Neither A nor B

2. All of these statements about automatic transmission diagnosis are true EXCEPT:

(A) the vehicle driver can accurately describe any problem.
(B) in many cases, the problem can be determined without removing the transmission from the vehicle.
(C) automatic transmissions can have many types of problems.
(D) shift linkage adjustment is a common source of problems.

3. Technician A says that some engine or exhaust system problems can be mistakenly identified as a transmission problem. Technician B says that a no-drive condition may be caused by locked brakes or jammed differential gears, instead of the transmission. Who is right?

(A) A only
(B) B only
(C) Both A & B
(D) Neither A nor B

4. All of these can cause automatic transmission slippage EXCEPT:
- **(A)** a plugged transmission filter.
- **(B)** high pump output.
- **(C)** low fluid level.
- **(D)** worn holding members.

5. Technician A says that burned and glazed clutch and band linings are a common source of automatic transmission problems. Technician B says that many automatic transmission electrical problems require that the transmission be completely disassembled. Who is right?
- **(A)** A only
- **(B)** B only
- **(C)** Both A & B
- **(D)** Neither A nor B

6. Which of these would be the least likely cause of erratic transmission shifts?
- **(A)** Misadjusted throttle valve linkage.
- **(B)** Sticking governor.
- **(C)** Stuck converter stator.
- **(D)** Metal particles in valve body.

7. Technician A says that automatic transmission noise in only one gear is a sign of a plugged oil filter. Technician B says that sticking valves are a common cause of automatic transmission shift problems. Who is right?
- **(A)** A only
- **(B)** B only
- **(C)** Both A & B
- **(D)** Neither A nor B

8. Technician A says that a ruptured vacuum modulator diaphragm will cause transmission fluid level to be low. Technician B says that a ruptured vacuum modulator will cause incorrect transmission shift points. Who is right?
- **(A)** A only
- **(B)** B only
- **(C)** Both A & B
- **(D)** Neither A nor B

9. All of these automatic transmission components can cause noises EXCEPT:
- **(A)** torque converter.
- **(B)** oil pump.
- **(C)** clutch packs.
- **(D)** planetary gearsets.

10. A band and clutch application chart, combined with a road test, can be used to locate a defective:
- **(A)** oil pump.
- **(B)** holding member.
- **(C)** governor.
- **(D)** oil cooler.

11. A vehicle with an automatic transmission slips badly when shifting into second gear. All other gears work properly. Technician A says that the problem could be caused by a leaking second gear servo or improper band adjustment. Technician B says that the problem could be caused by a badly worn second gear band. Who is right?
- **(A)** A only
- **(B)** B only
- **(C)** Both A & B
- **(D)** Neither A nor B

12. A sure sign of an excessively worn pump is:
- **(A)** low oil pressure.
- **(B)** pump noise.
- **(C)** low fluid level.
- **(D)** transmission overheating.

13. A pressure gauge is attached to an automatic transmission pressure tap. The engine is started, and the transmission is shifted through all gears. The transmission has low oil pressure in only one gear. Technician A says that low pressure in only one gear indicates a leaking governor circuit. Technician B says that low pressure in only one gear indicates a worn oil pump. Who is right?
- **(A)** A only
- **(B)** B only
- **(C)** Both A & B
- **(D)** Neither A nor B

14. Technician A says that the transmission oil pan and valve body must be removed to make air pressure tests. Technician B says that air pressure testing requires clean and dry air at 120 psi (800 kPa) minimum. Who is right?

(A) A only
(B) B only
(C) Both A & B
(D) Neither A nor B

15. Technician A says that check balls can dissolve inside of the transmission valve body. Technician B says that hydraulic circuit diagrams can be used to locate sticking valves. Who is right?

(A) A only
(B) B only
(C) Both A & B
(D) Neither A nor B

Chapter 13

Automatic Transmission Service

After studying this chapter, you will be able to:

- Highlight important points regarding transmission fluid service.
- Adjust automatic transmission linkages and bands.
- Remove an automatic transmission from a vehicle.
- Disassemble an automatic transmission.
- Inspect the parts of an automatic transmission.
- Use a service manual to rebuild any type of automatic transmission.
- Install an automatic transmission in a car.

Automatic transmission service is a specialized activity that requires knowledge and careful work habits. This chapter will outline automatic transmission in-car service procedures as well as the major steps and techniques in transmission overhaul. If you have studied the information in Chapter 2 and in Chapters 10 through 12, you will be familiar with the components and tools discussed in this chapter.

Keep in mind that the information contained in this chapter is general. Since there are many kinds of automatic transmissions, you should always consult the appropriate factory service manual for the particular transmission that you are working on.

Quite often, an automatic transmission will not require a complete overhaul. The subassembly repair techniques in this chapter can be used to service many transmission parts without removal or complete teardown of the transmission.

In-Car Automatic Transmission Service

There are a number of service procedures that can be performed without removing the transmission from the vehicle. Obviously, checking fluid level and condition is one; so is changing the transmission fluid and filter. Making adjustments to linkages, the neutral start switch, and the transmission bands are a few other such in-car service procedures.

There are still other service procedures that can be performed while the transmission is still installed. The speedometer pinion gear, rear seal, vacuum modulator, oil cooler lines, extension bushing and output shaft bearing, governor, parking gear and parking lock components, valve body, and accumulator piston can all be serviced without removing the transmission. See Figure 13-1. Many of these procedures are covered in the overhaul section of this chapter, rather than in this section.

Automatic Transmission Fluid Service

Like engine oil, automatic transmission fluid should be checked at specified intervals. Also, some manufacturers will recommend *changing* the fluid at specified intervals. The fluid can become contaminated with metal, dirt, water, and friction material from internal parts. Contaminated fluid can cause rapid part wear and premature transmission failure.

Checking fluid level and condition

The most common hydraulic system problem is a low fluid level. When fluid level is low, air can enter the hydraulic system from the oil pan through the pump intake. When the hydraulic system contains air, hydraulic power will be wasted compressing it. The transmission will shift erratically, or it may slip or fail to move the vehicle. Fluid will become overheated. Proper lubrication cannot be maintained, and parts will be damaged.

If the transmission has been overfilled, the turning planetary gears will churn the transmission fluid into foam. The foaming fluid contains air, causing the same hydraulic system problems as a low fluid level. In addition, foaming can result in fluid escaping from the transmission vent where it may be mistaken for a leak.

TRANSMISSION SERVICE GUIDE

TRANSMISSION COMPONENT	SERVICED WITH TRANSMISSION IN VEHICLE	TRANSMISSION MUST BE REMOVED FROM VEHICLE
Accumulators	X	
Bands		Most Makes
Clutch Pistons		X
Clutch Plates		X
Converter		X
Electrical Switches	X	
Extension Housing	X	
Filter	X	
Governor	X	
Input Shaft		X
Manual Linkage	X	
Modulator	X	
Modulator Valve	X	
Oil Pan	X	
Output Shaft		X
Parking Gear Linkage	X	
Planetary Gears		X
Pump or Pump Seal		X
Seal Rings		X
Servos	X	
Speedometer Gear	X	
Throttle Linkage	X	
Throttle Valve	X	
Valve Body	X	

Figure 13-1. Numerous service procedures can be performed on an automatic transmission without removing the transmission from the vehicle, as shown by this chart.

In addition to the problems mentioned, the introduction of air into the ATF results in severe and rapid oxidizing of the fluid, which seriously affects its properties. One result of fluid oxidation is varnish formation. This problem appears as a sticky, burned-on substance that resembles furniture refinishing varnish.

Figure 13-2 shows typical markings on a dipstick. The fluid level should be checked as specified on the dipstick or in the owner's manual: with the engine running, shift selector lever in park or neutral, and the vehicle on level ground. The transmission fluid should generally be at normal operating temperature. Some dipsticks are marked so that the fluid level can be checked with the fluid hot or cold.

While checking automatic transmission fluid *level,* the *condition* of the fluid should also be inspected. If the fluid smells burnt and is orange or dark brown, the transmission fluid has been overheated. The clutch and band friction linings are probably burned and glazed, impairing transmission operation. Changing the fluid will not solve the problem.

If the transmission fluid is milky looking, it is contaminated with water. Such contamination comes from operating the vehicle in very deep water or from a leak in the oil cooler, allowing water to leak in from the engine cooling-system radiator. A leaking oil cooler will also allow antifreeze to enter the transmission, which will ruin the clutch and band friction linings.

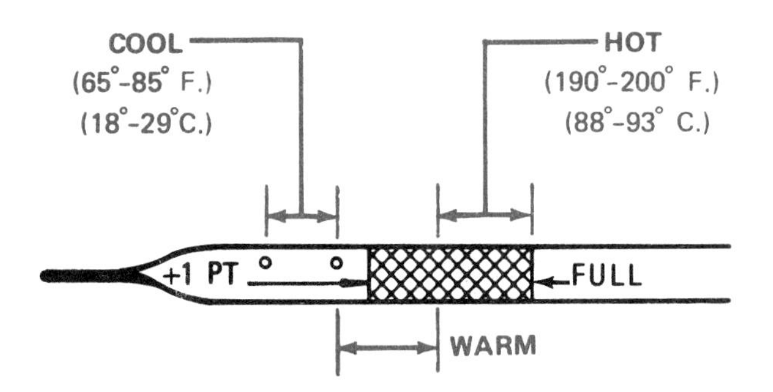

Figure 13-2. Transmission fluid level is important to the proper operation of the transmission. Both hot and cold levels are given on most dipsticks. On the dipstick shown, for example, fluid is at an acceptable level if it falls within the ranges shown at the respective temperatures shown. (General Motors)

Changing fluid and filter

To change the fluid and filter in an automatic transmission, first warm the engine and transmission. Raise the vehicle on a hoist. Position a drain pan under the transmission oil pan.

WARNING! Be careful not to spill the hot transmission fluid. It can cause painful burns!

If a drain plug is not provided, you will need to remove the transmission oil pan attaching screws. Remove all but the front four screws. Pry on the pan to loosen the gasket and allow the pan to drop at the rear. (Transmission fluid may begin to drain at this point.) Loosen the front four screws, allowing the pan to drop further and the fluid to drain. When the fluid stops draining, remove the screws and the transmission oil pan. Examine the fluid before discarding it in an approved receptacle. Take note of any debris.

Remove and replace or clean the transmission filter. Clean the oil pan. Scrape the old gasket off the transmission pan and housing. Position the new pan gasket using an approved sealer. Use sealer sparingly because it could upset the transmission operation.

Start all of the oil pan screws with your fingers. Then, tighten them in a crisscross pattern to their specified torque requirement. Overtightening can split the gasket or distort the oil pan.

Refill the transmission with the correct type and amount of automatic transmission fluid. If required, check the manufacturer's service manual for details. Using a funnel, pour the fluid into the filler tube. Start the engine. With the parking brake on, move the selector lever momentarily to each position. Then, check under the car for leaks. Recheck the fluid level after the transmission is at normal operating temperature.

Automatic Transmission Adjustments

There are several adjustments that can be made with the transmission installed in the vehicle. These relatively simple adjustments will solve many transmission problems. The most common adjustments are made to the following automatic transmission components:

- *Shift linkage.*
- *Throttle, T.V.,* and *kickdown linkages.*
- *Transmission bands.*
- *Neutral start switch.*

Shift linkage adjustment

Shift linkage adjustments should be done whenever the shift selector lever position does not agree with the actual transmission gear selection. The inner manual lever, which holds the manual valve in the proper position, is not adjustable. The only adjustment that can be made to the shift linkage is between the outer manual lever and the shift selector lever. If this adjustment is off, the transmission may not be in the gear position that the driver thinks it is in. Figure 13-3 shows typical shift linkage in both column- and floor-shift versions.

Before attempting any adjustments, inspect the linkage for loose or worn parts. Tighten or replace worn parts before making the adjustment.

Check cables for binding or damage to the sheath. If the cable appears to be damaged from overheating, check the engine ground straps. Disconnected ground straps could cause the vehicle electrical equipment to ground through the cable and melt the sheath.

Exact procedures for adjusting the shift linkage of an automatic transmission vary. Generally, make sure the position of the outer manual lever is synchronized with that of the shift selector lever. For example, if the selector lever is set in park, the outer manual lever must also be set in the park mode. A typical adjustment procedure is given as follows.

If you are sure the shift linkage is in good shape, place the selector lever in park. Then, loosen the locknut holding the *adjuster,* which may be an adjustment swivel, a bracket, or a slot in the linkage rod. Some adjusters are under the vehicle, while others are located under the dash.

Once the locknut is loose, ensure that the shift selector indicator is in park. Next, move the outer manual lever, on the transmission case, all the way to the park, or rear detent, position. With both levers in park, tighten the locknut and recheck shift linkage operation. Note that this procedure is typical of many, except that some manufacturers will specify the selector lever and outer manual lever be placed in a position other than park. See Figure 13-4.

Throttle, T.V., and kickdown linkage adjustments

Linkages should be adjusted whenever the transmission shift points are not correct. The problem could be high

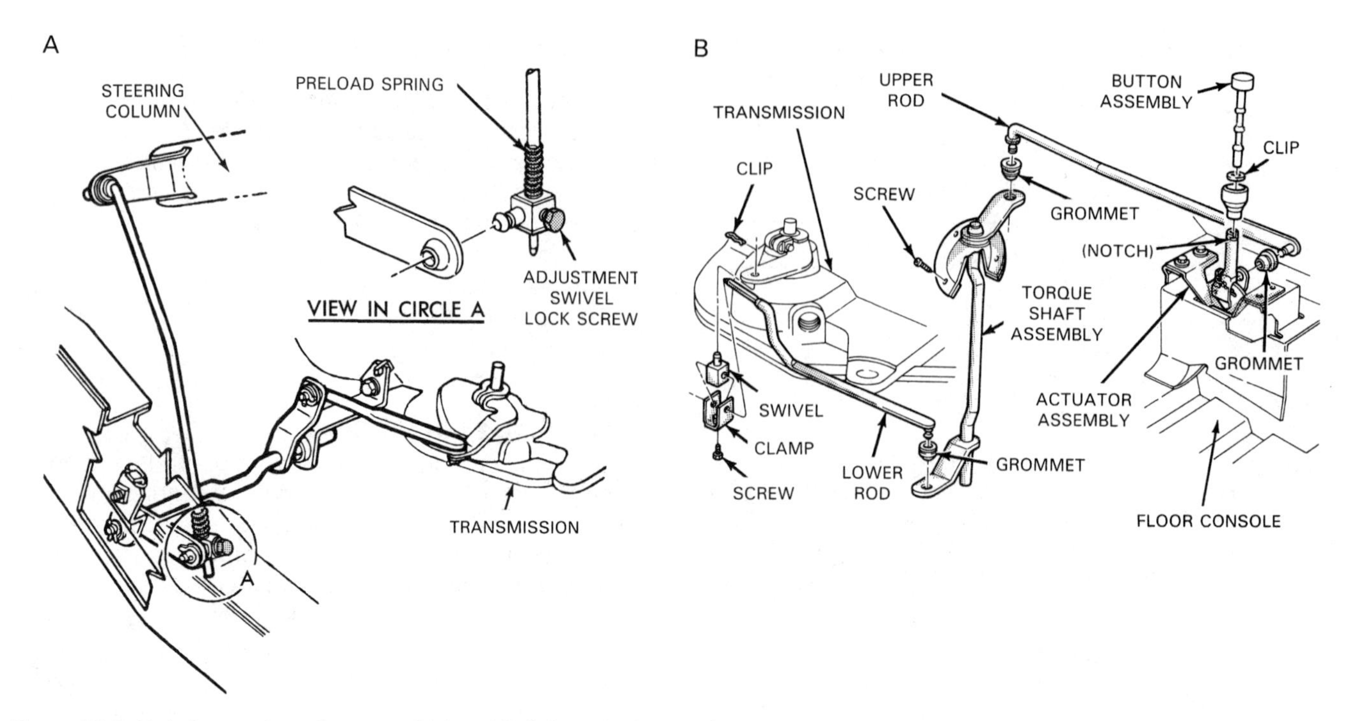

Figure 13-3. Note how automatic transmission shift linkage is designed in a common make of automobile. Other linkage systems are similar. Adjustment of these linkages is made by means of adjustment swivel. A–Linkage with column-shift selector lever. B–Linkage with floor-shift selector lever. (Chrysler)

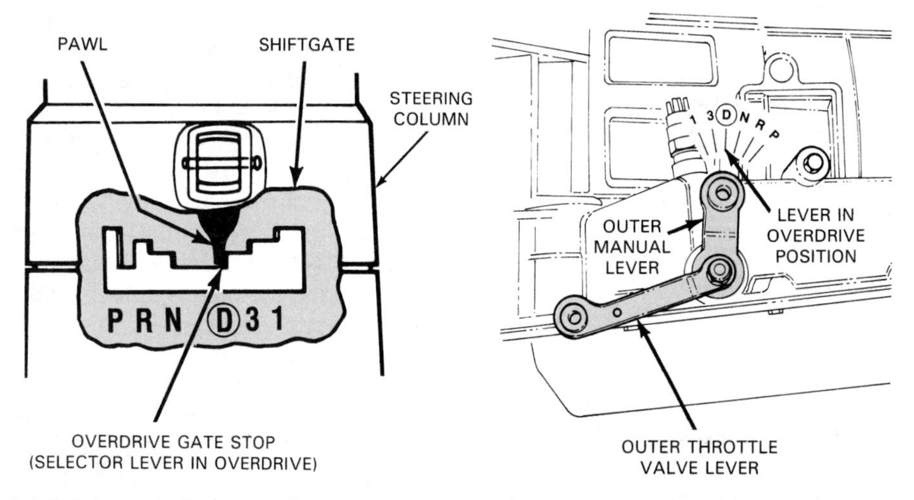

Figure 13-4. To adjust shift linkage, indicator quadrant and outer manual lever are set to the same position–*overdrive* in this particular transmission. The locknut is then tightened, and correct operation is verified in all selected positions. (Ford)

or low shift points. On transmissions without a vacuum modulator, throttle valve (T.V.) linkage, operated by the throttle linkage, controls shift timing and pressures and, sometimes, provides a detent for passing gear. If a transmission has a separate detent valve for passing gear, it may be operated by a kickdown linkage through the throttle linkage. Adjustment procedures are similar for both the T.V. and kickdown linkage.

A typical T.V. linkage assembly is shown in Figure 13-5. Note that some assemblies use threaded swivels to make adjustments. Always check the condition of linkage parts before making any adjustments and replace defective parts.

If the linkage is in good condition, start linkage adjustment by having a helper press the throttle pedal completely to the floor (engine off). Next, check the throttle plate(s) in the throttle body or carburetor. They should be completely open. If they are not, adjust the throttle linkage, which is the linkage between the throttle pedal and throttle plates, until they open fully when the throttle pedal is fully depressed. If adjusting the throttle linkage fails to remedy the problem, the T.V. linkage may be the cause. Disconnect the T.V. linkage. Now, recheck the throttle linkage.

After checking the throttle linkage, loosen the adjustment locknut on the T.V. linkage (or disconnect the swivel retaining clip). Do this while your helper continues to press the throttle pedal completely to the floor. Then, pull the T.V. linkage at the transmission to the wide-open throttle position. (Pull in the direction that the linkage moves when the throttle pedal is depressed.) While holding the linkage in the wide-open throttle position, tighten the locknut (or reinstall the swivel). As a final step, road test the vehicle to check the adjustment.

T.V. cable adjustment is quite similar to that for the rod-type linkage. However, most T.V. cable adjusters are located at the throttle lever of the throttle body or carburetor, and most adjustments can be accomplished by one person. Many adjusters consist of a cable holder that snaps up to release the cable. Many T.V. cables are self-adjusting, whereupon pressing the throttle pedal to the floor completes the adjustment.

Transmission band adjustment

Band adjustment should be done as part of routine maintenance. Although adjusting the band will not cure severe slippage, it is worth trying in cases where the shift is becoming soft. There are many different types of band apply systems and many methods of adjustment.

Begin the band adjustment by loosening the locknut on the band adjustment screw. (If the adjuster is inside of the oil pan, the pan must be removed first.) Next, tighten the band adjustment screw until the band is snug against the drum, counting the number of turns in. This will indicate whether the original adjustment was correct.

NOTE! Hold the locknut stationary as you turn the adjustment screw into the case. Otherwise, it may retighten, resulting in band misadjustment.

Torque the band adjustment screw to the exact torque value specified in the service manual. Afterward, back it off the exact number of turns specified in the service manual. In some cases, it may be possible to back the screw off less than specified to get a tighter band adjustment and a firmer shift. This should not vary more than one quarter to one half turn from factory values. Once the band is adjusted, the band adjustment screw should be kept from turning as the locknut is tightened.

Neutral start switch adjustment

The neutral start switch, for floor-shift and a few column-shift designs, should be adjusted when the vehicle will *not* start in park or neutral or when it *will* start in any *other* gear. With shift linkage properly adjusted, begin the switch adjustment by locating the neutral start switch and loosening its holddown. Then, place the shift selector lever in neutral (parking brake applied) and try to start the engine. If the starter will not operate, rotate the neutral start switch in small steps until the starter operates when the ignition switch is turned. Note that it may be easier to have a helper hold the ignition switch in the start position as you move the neutral start switch.

The switch is sometimes adjusted with an alignment pin, Figure 13-6. The procedure is rather simple. Loosen attaching bolts; place transmission in neutral; then, rotate the switch to align adjustment holes. Insert alignment pin

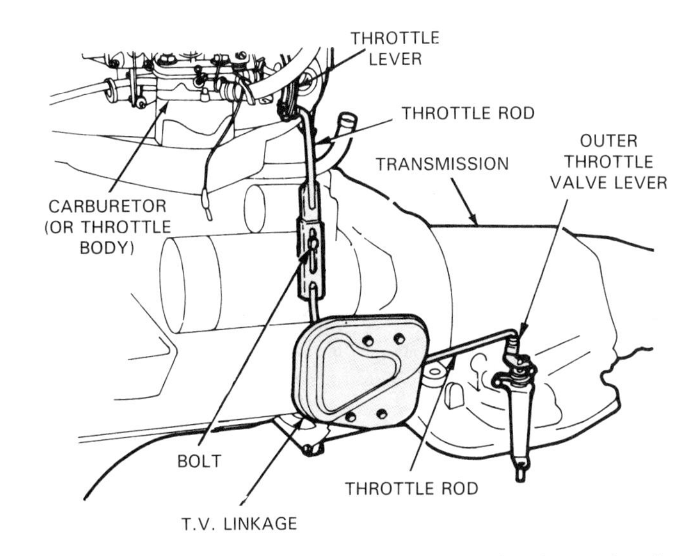

Figure 13-5. This T.V. linkage is adjusted by first loosening the adjuster locknut and then lengthening or shortening the linkage as needed. Always road test the vehicle after any linkage adjustments are made. (Chrysler)

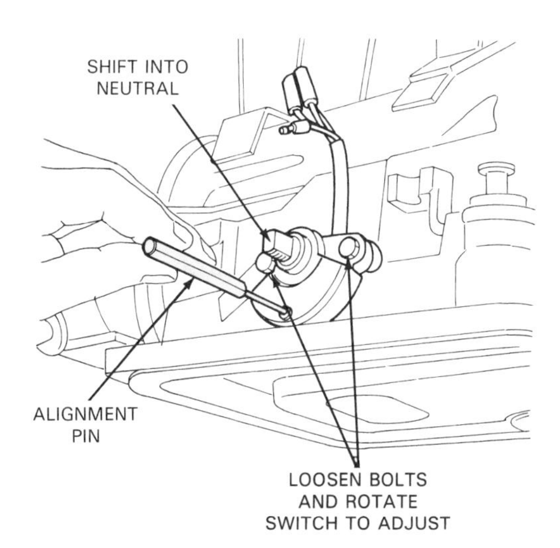

Figure 13-6. An alignment pin makes neutral start switch adjustment easier. Some neutral start switches are adjusted without using an alignment tool. (Nissan)

in holes to the specified depth. Tighten switch holddown and remove alignment pin.

Once the engine starts in neutral, check operation in all other shift quadrant positions. The engine should crank in neutral and park, but in no other positions. Readjust the switch as needed and tighten it down.

Automatic Transmission Overhaul

When your troubleshooting endeavors find major internal problems, the automatic transmission must be removed from the vehicle. The manufacturer's service manual will give accurate instructions on how to remove, disassemble, inspect, rebuild, and install the transmission. Special training is needed to become competent to make major internal repairs on automatic transmissions.

The remainder of this chapter presents *general* procedures explaining how to overhaul a transmission. You should always refer to the manufacturer's service manual for specifics. Note that certain of the procedures included in this section–valve body service, for example–can be performed while the transmission is still on the vehicle. However, they are presented here as part of a complete transmission overhaul.

Automatic Transmission Removal

The first step in overhauling an automatic transmission is to remove the transmission from the vehicle. Removing an automatic transmission is similar to removing a manual transmission.

1. Disconnect the battery negative cable. This will prevent any accidental shorts.
2. While under the hood, also remove the upper converter housing attaching bolts, if possible. See Figure 13-7. In some cases, it may be necessary to install an engine holding fixture at this point, to support the weight of the engine once the rear engine mount is removed.
3. Raise the vehicle on a hydraulic lift to get enough clearance to remove the transmission. If you do not have access to a hydraulic lift, always use a good quality hydraulic floor jack and approved jackstands.

WARNING! Never support the vehicle with wooden crates or cement blocks. Be sure the vehicle is secure before working underneath it.

4. Check the engine mounts to ensure that they are not broken.

WARNING! Broken engine mounts can allow the engine to fall backwards once the transmission crossmember has been removed.

5. Drain the transmission fluid. Draining the fluid *now* will prevent spillage when the driveline and other components are removed. If the oil pan does not have a drain plug, the pan can be loosened, and the fluid allowed to drain out. If you remove the pan to drain the fluid, reinstall it when finished.
6. Remove the driveline by removing bolts from the rear of the driveline and slipping it out of the transmission extension housing. Mark the driveline at the rear axle assembly so that it can be reinstalled in the same position. Refer to Figure 13-8. (See Chapter 15 for more information on driveline removal.)
7. Remove miscellaneous items connected to the transmission, such as the speedometer cable assembly, oil filler tube, and any electrical connectors. Label connectors, using masking tape, to simplify reconnection.

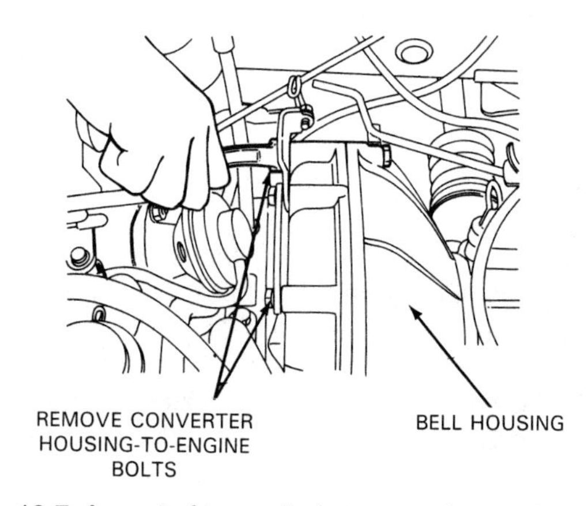

Figure 13-7. As part of transmission removal procedure, remove upper converter housing attaching bolts while the vehicle is still on the ground. (Subaru)

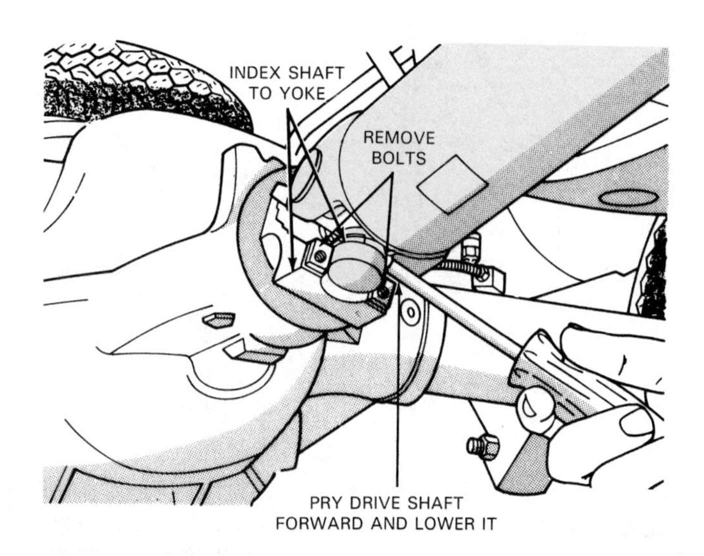

Figure 13-8. To remove driveline, pry the shaft forward and lower the back end. Then, slip it out of the extension housing. (Chrysler)

Disconnect T.V., kickdown, and shift linkages at the transmission. Remove the vacuum line from vacuum modulator, if applicable.

Disconnect the oil cooler lines at the transmission. Use two wrenches to avoid twisting them. The lines will usually leak when they are disconnected, and a pan should be positioned to catch any spilled fluid. Position the lines, linkages, and other parts out of the way to keep them from snagging on the transmission as it is lowered.

8. Remove the torque converter inspection cover or dust cover. Then, remove the converter cover attaching bolts (converter cover-to-flexplate bolts). See Figure 13-9. The flexplate will have to be turned to gain access to all of the attaching bolts. This can be done with a special turning tool, a screwdriver, or by operating the starter with a remote switch, if necessary.
9. If needed, remove the front exhaust pipes to gain clearance for lowering the transmission. Removing the exhaust pipes will also allow you to lower the back of the engine, as may be required to reach the converter housing attaching bolts.
10. Disconnect any hardware attached to the transmission crossmember, such as the parking brake linkage. Removing the linkage will relieve pressure on the crossmember and will give more clearance.
11. If an engine holding fixture is not already in place, support the engine with a jack placed under the rear of the engine.

WARNING! Never let the engine hang suspended only by the front engine mounts. Failure to support the engine before removing the transmission can result in serious injury.

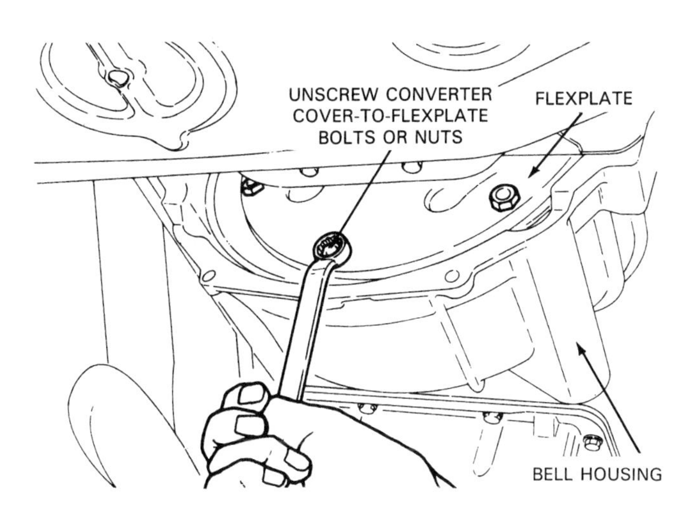

Figure 13-9. Remove converter cover attaching bolts as part of transmission removal. Note inspection shield must be removed to gain access to these bolts. (Chrysler)

12. Place a transmission jack (or floor jack) under the transmission. Raise the transmission slightly to remove weight from the crossmember. Then, remove the fasteners holding the crossmember to the vehicle frame and to the transmission mount, which is located on the extension housing. Remove the crossmember.
13. Remove the starter motor, if it is bolted to the converter housing, and position it out of the way.
14. Remove the remaining converter housing attaching bolts, which hold the transmission to the engine. You may need to tilt the engine and transmission back slightly to get at the bolts.

    Note that brackets, such as those holding the filler tube or wiring harnesses, are sometimes attached to the engine with the converter housing attaching bolts. Take note of their position for reinstallation.

WARNING! Do *not* attempt to remove the transmission without the converter housing. It should not be separated from the transmission until it is safely on the bench, to avoid damage to the torque converter or oil pump.

Also, an automatic transmission is heavy. Use the transmission jack to support and lower the transmission. *Always* have someone help remove the transmission.

15. Slide the transmission straight back and away from the engine. Attach a small C-clamp to the edge of the converter housing or use a *torque converter holding tool* to secure the converter.

WARNING! Failure to secure the torque converter could result in it sliding off and dropping to the floor, possibly causing personal injury or damage to the converter.

16. Lower the transmission. Watch clearances and watch for any component that you may have forgotten to disconnect. Slowly move the transmission to the workbench with the jack in the lowered position. Figure 13-10 highlights some of the activities involved in removing a transmission.

CAUTION! Automatic transmission repairs should *never* be done outside since dust and dirt may blow into the transmission components. Even a tiny bit of dirt can cause a valve to stick and prevent normal operation.

Automatic Transmission Disassembly

Although a typical automatic transmission contains many parts, it can be rebuilt if you follow logical procedures. Transmission disassembly procedures always vary between manufacturers and between different transmissions from the same manufacturer. Always have the proper service manual handy before beginning any disassembly steps.

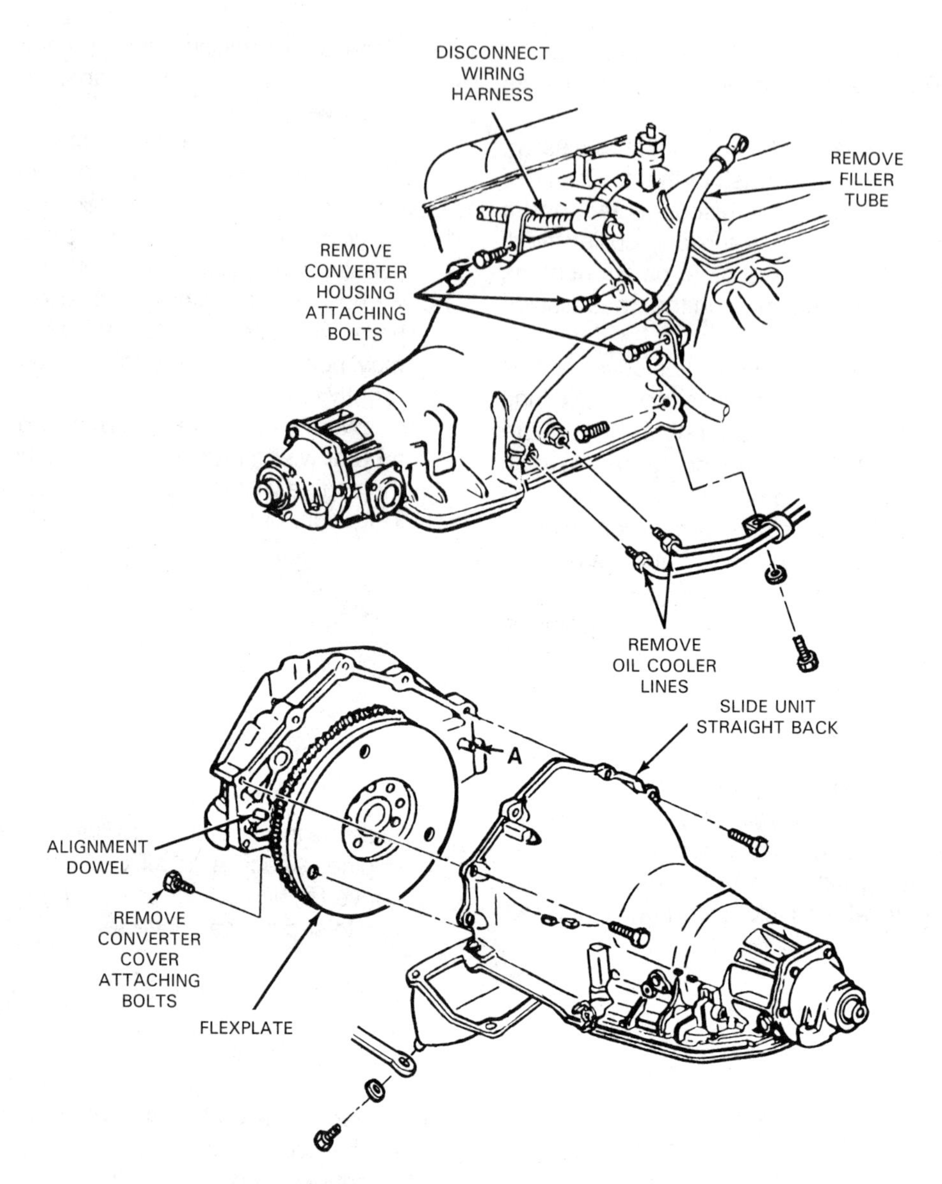

Figure 13-10. Note some of the components that must be removed or disconnected to remove the transmission from the vehicle. (General Motors)

Special service tools are sometimes needed to overhaul an automatic transmission, Figure 13-11. Tools can be ordered from the vehicle dealer or directly from the manufacturer. Sometimes, acceptable substitute tools can be improvised from scrap parts. Refer to Chapter 2 for more information on special service tools.

As you remove the major parts of the transmission, set them aside for further disassembly, as required. Work on one subassembly (pump, clutch pack, servo, for example) at a time. This will help you keep things straight, as an automatic transmission does contain so many parts, Figure 13-12. During disassembly, place all related parts together on the workbench.

To begin the disassembly of the transmission, clean the outside of the transmission thoroughly and mount it in a holding fixture like the one shown in Figure 13-13. If a holding fixture is not available, place the transmission on a clean workbench. The workbench should be constructed so that transmission fluid can drain into a catch basin.

Removing external components

Remove the torque converter holding tool and the torque converter first. Before proceeding with further disassembly, check the transmission end play, as shown in Figure 13-14. The end play should be checked before the transmission is disassembled (and again after it is reassembled). Record the end play reading for later reference.

Remove the transmission oil pan, Figure 13-15. After removing the oil pan, check it for metal particles,

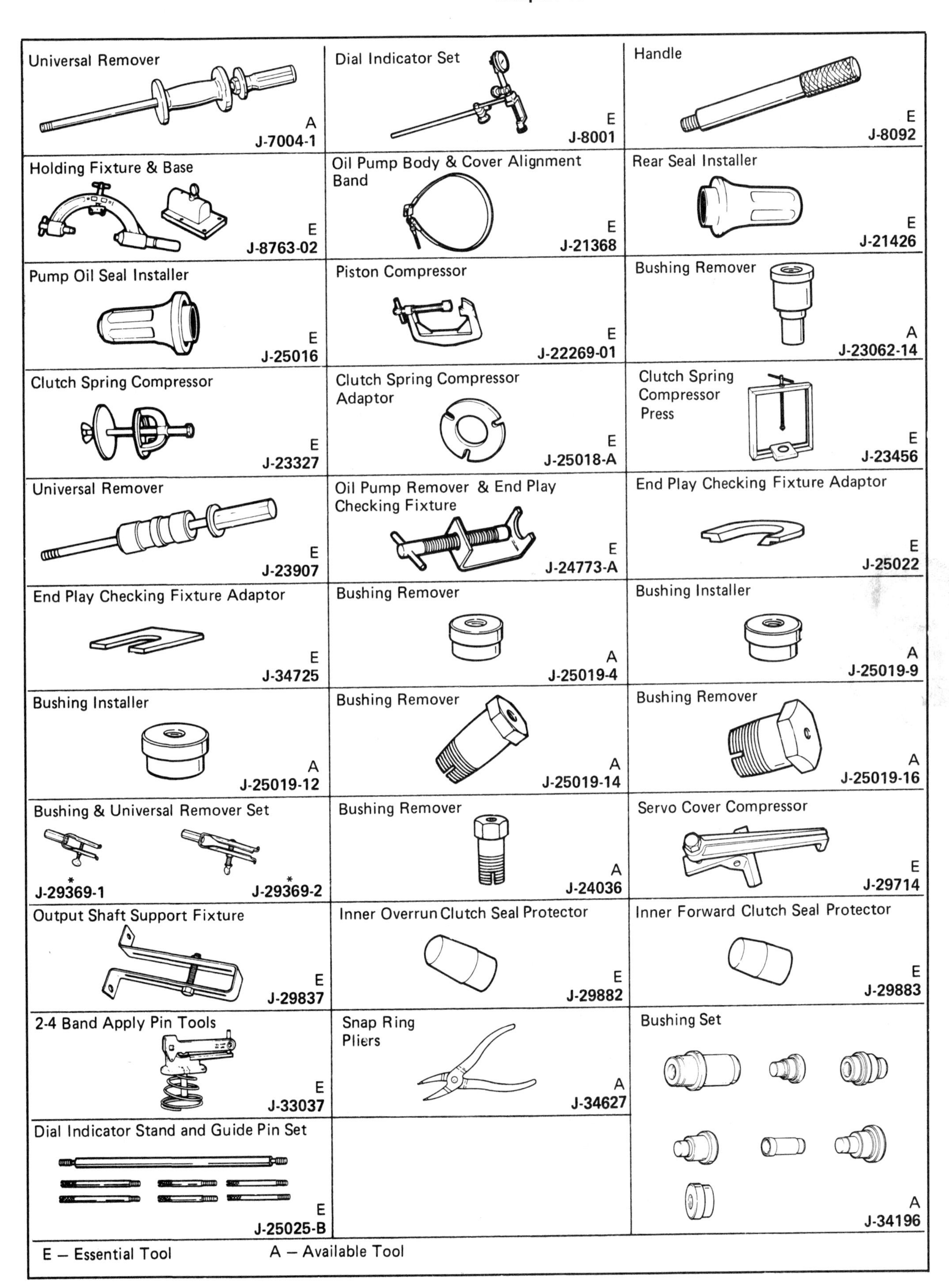

Figure 13-11. Special service tools make transmission overhaul much easier. The tools shown are for one kind of transmission. In many instances, workable tools can be made from scrap parts. (General Motors)

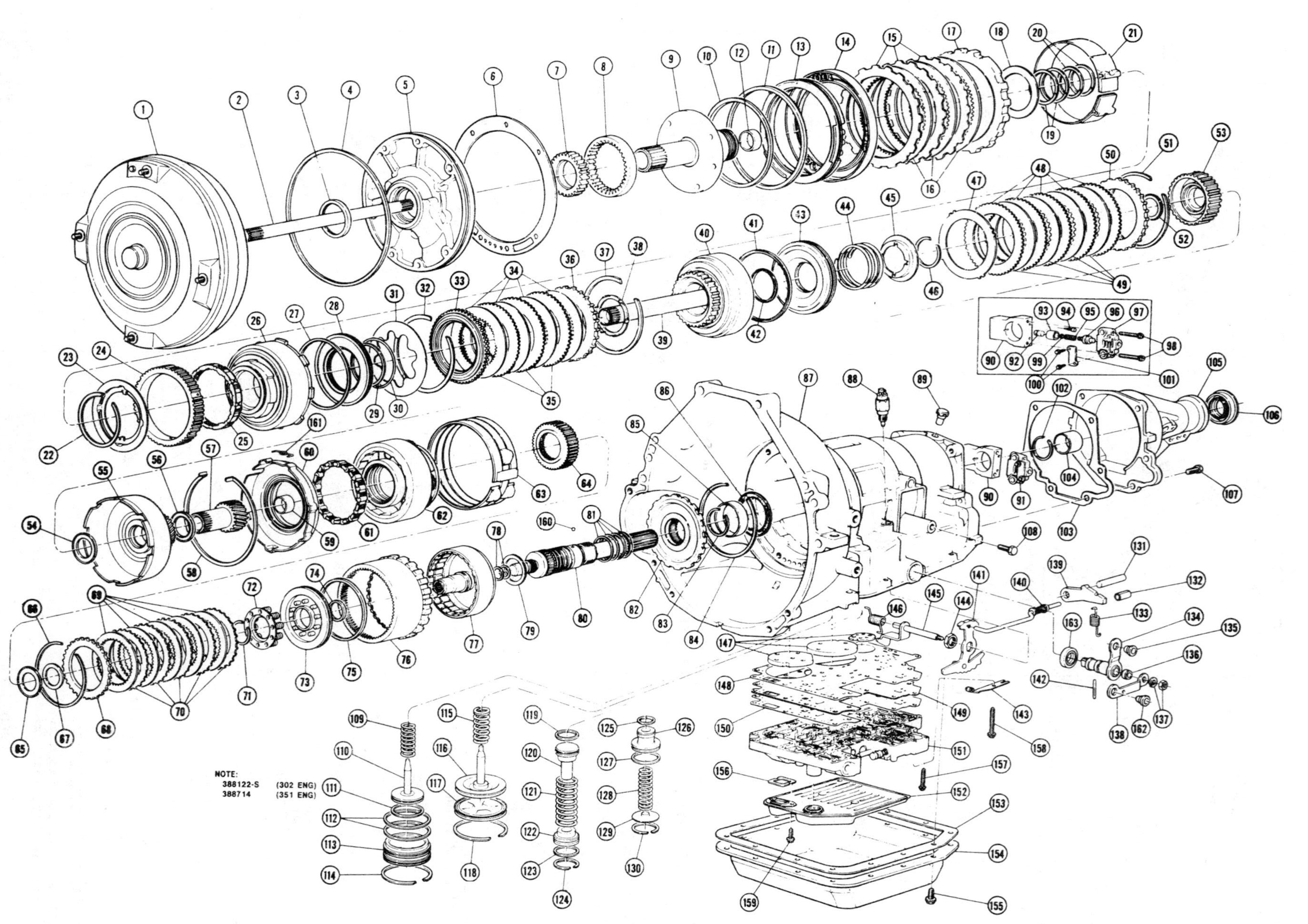

Figure 13-12. Exploded view of a Ford Automatic Overdrive Transmission shows the parts of a typical automatic transmission. Other transmissions are similar. (Ford)

(continued)

Legend:

1. TORQUE CONVERTER
2. DIRECT DRIVE SHAFT
3. FRONT PUMP SEAL
4. FRONT PUMP 'O' RING
5. FRONT PUMP BODY
6. FRONT PUMP GASKET
7. FRONT PUMP DRIVE GEAR
8. FRONT PUMP DRIVEN GEAR
9. STATOR SUPPORT – FRONT PUMP
10. INTERM. CLUTCH PISTON INNER LIP SEAL
11. INTERM. CLUTCH PISTON OUTER LIP SEAL
12. FRONT PUMP BUSHING
13. INTERM. CLUTCH PISTON
14. INTERM. CLUTCH PISTON RETURN SPRINGS & RETAINER
15. INTERM. CLUTCH EXTERNAL SPLINE STEEL PLATES (SEL.)
16. INTERM. CLUTCH INTERNAL SPLINE FRICTION PLATES
17. INTERM. CLUTCH PRESSURE PLATE
18. NO. 1 THRUST WASHER (FRONT PUMP) SELECTIVE
19. STATOR SUPPORT SEAL RINGS (REV. CLUTCH)
20. STATOR SUPPORT SEAL RINGS (FWD. CLUTCH)
21. OVERDRIVE BAND
22. INTERM. OWC RETAINING SNAP
23. INTERM. OWC RETAINING PLATE
24. INTERM. OWC OUTER RACE
25. INTERM. ONE-WAY CLUTCH ASSY.
26. REVERSE CLUTCH DRUM
27. REVERSE CLUTCH PISTON SEAL (OUTER)
28. REVERSE CLUTCH PISTON
29. REVERSE CLUTCH PISTON SEAL (INNER)
30. THRUST RING
31. REVERSE CLUTCH PISTON RETURN SPRING
32. RETAINING SNAP RING
33. REVERSE CLUTCH FRONT PRESSURE PLATE
34. REVERSE CLUTCH INTERNAL SPLINE FRICTION PLATE
35. REVERSE CLUTCH EXTERNAL SPLINE STEEL PLATE
36. REVERSE CLUTCH REAR PRESSURE PLATE
37. REVERSE CLUTCH RETAINING RING (SEL.)
38. NO. 2 THRUST WASHER (REV. CLUTCH)
39. TURBINE SHAFT
40. FORWARD CLUTCH CYLINDER & TURBINE SHAFT
41. FORWARD CLUTCH PISTON SEAL (OUTER)
42. FORWARD CLUTCH PISTON SEAL (INNER)
43. FORWARD CLUTCH PISTON
44. FORWARD CLUTCH PISTON RETURN SPRING
45. RETURN SPRING RETAINER
46. RETAINING SNAP RING
47. WAVED PLATE
48. FORWARD CLUTCH EXTERNAL SPLINE STEEL PLATE
49. FORWARD CLUTCH INTERNAL SPLINE FRICTION PLATE
50. FORWARD CLUTCH PRESSURE PLATE
51. RETAINING SNAP RING (SELECTIVE)
52. NO. 3 NEEDLE BEARING (FWD. CLUTCH)
53. FORWARD CLUTCH HUB
54. NO. 4 NEEDLE BEARING
55. REVERSE SUN GEAR & DRIVE SHELL ASSY.
56. NO. 5 NEEDLE BEARING
57. FORWARD SUN GEAR
58. CENTER SUPPORT RETAINING RING
59. FORWARD SUN GEAR PRECISION BUSHING
60. CENTER SUPPORT PLANETARY
61. PLANETARY OWC CAGE SPRING & ROLLER ASSY.
62. PLANETARY ASSY.
63. REVERSE BAND
64. DIRECT CLUTCH HUB
65. NO. 7 NEEDLE BEARING (DIRECT CLUTCH INNER
66. RETAINING SNAP RING (SELECTIVE)
67. THRUST SPACER
68. DIRECT CLUTCH PRESSURE PLATE
69. DIRECT CLUTCH INTERNAL SPLINE PLATES
70. DIRECT CLUTCH EXTERNAL SPLINE PLATES
71. RETAINING SNAP RING
72. RETURN SPRING & RETAINER
73. DIRECT CLUTCH PISTON
74. DIRECT CLUTCH PISTON SEAL (INNER)
75. DIRECT CLUTCH PISTON SEAL (OUTER)
76. RING GEAR & PARK GEAR
77. DIRECT CYLINDER
78. OUTPUT SHAFT SMALL (2) STEEL SEAL RINGS (DIRECT CLUTCH)
79. NO. 8 NEEDLE BEARING (DIRECT CLUTCH OUTER)
80. OUTPUT SHAFT
81. OUTPUT SHAFT LARGE (4) STEEL SEAL RINGS
82. OUTPUT SHAFT HUB
83. RETAINING SNAP RING (O.P.S. HUB TO O.P.S.)
84. RETAINING SNAP RING (O.P.S. HUB TO RING GEAR)
85. REAR CASE BUSHING
86. NO. 9 NEEDLE BEARING (REAR CASE)
87. CASE ASSY.
88. NEUTRAL START SWITCH
89. VENT CAP
90. GOVERNOR COUNTERWEIGHT
91. BODY ASSY. – GOVERNOR
92. PLUG GOVERNOR
93. SLEEVE GOVERNOR
94. SCREEN ASSY. – GOV. OIL
95. SPRING GOV. VALVE
96. VALVE GOVERNOR
97. BODY GOVERNOR
98. BOLT (GOVERNOR BODY TO COUNTERWEIGHT)
99. CLIP–(GOVERNOR COVER TO GOVERNOR BODY)
100. BOLT (GOVERNOR COVER TO GOVERNOR BODY)
101. COVER – GOVERNOR VALVE BODY
102. RETAINING SNAP RING (GOVERNOR ASSY. TO O.P.S.)
103. EXTENSION HOUSING BRACKET
104. EXTENSION HOUSING BUSHING
105. EXTENSION HOUSING
106. EXTENSION HOUSING SEAL
107. BOLT (EXT. HSG. TO CASE) M6-12.5 X 30 (6 REQ'D.)
108. PIPE PLUG – 1/8-27 DRY SEAL
109. OVERDRIVE SERVO PISTON RETURN SPRING
110. OVERDRIVE SERVO PISTON
111. OVERDRIVE SERVO PISTON SEAL
112. OVERDRIVE SERVO COVER SEAL RINGS
113. OVERDRIVE SERVO COVER
114. RETAINING SNAP RING (O/D SERVO TO CASE)
115. REVERSE SERVO PISTON RETURN SPRING
116. REVERSE SERVO PISTON (SELECTIVE)
117. REVERSE SERVO COVER
118. RETAINING SNAP RING (REV. SERVO TO CASE)
119. 3-4 ACCUMULATOR VALVE SEAL
120. 3-4 ACCUMULATOR VALVE
121. 3-4 ACCUMULATOR VALVE RETURN SPRING
122. 3-4 ACCUMULATOR COVER
123. 3-4 ACCUMULATOR COVER SEAL
124. RETAINING SNAP RING (3-4 ACCUM. TO CASE)
125. 2-3 ACCUMULATOR VALVE SEAL (SMALL)
126. 2-3 ACCUMULATOR VALVE
127. 2-3 ACCUMULATOR VALVE SEAL (LARGE)
128. 2-3 ACCUMULATOR VALVE RETURN SPRING
129. 2-3 ACCUMULATOR COVER
130. RETAINING SNAP RING (2-3 ACCUM. TO CASE)
131. PARK PAWL SHAFT
132. GUIDE CUP
133. PARK PAWL RETURN SPRING
134. MANUAL LEVER
135. GROMMET
136. THROTTLE LEVER OIL SEAL
137. ATTACHING NUT & LOCK WASHER – M8 X 1.25
138. THROTTLE LEVER (OUTER)
139. PARK PAWL
140. PARK PAWL ACTUATING ROD
141. MANUAL LEVER (INNER)
142. ROLL PIN – 1/8 X 0.95 GROOVED
143. DETENT SPRING
144. ATTACHING NUT (MANUAL LVR) – M14 X 1.5 HEX
145. THROTTLE LEVER (INNER)
146. THROTTLE TORSION SPRING
147. VALVE BODY REINFORCEMENT PLATE
148. SEPARATOR PLATE GASKET (UPPER)
149. SEPARATOR PLATE
150. SEPARATOR PLATE GASKET (LOWER)
151. VALVE BODY (MAIN CONTROL)
152. FILTER & GROMMET ASSY. – OIL PAN
153. OIL PAN GASKET
154. OIL PAN
155. BOLT (OIL PAN TO CASE) – M8-1.25 X 15 (17 REQ'D.)
156. OIL FILTER GASKET
157. BOLT (VALVE BODY TO CASE) M6 1.0 X 30 (8 REQ'D.)
158. BOLT (VALVE BODY TO CASE) M6 1.0 X 40 (17 REQ'D.)
159. BOLT (SCREEN TO VALVE BODY) M6 1.0 X 16 (3 REQ'D.)
160. BALL (GOVERNOR DRIVE)
161. SPRING (ANTI CLUNK)
162. GROMMET
163. OIL SEAL ASSY.

Figure 13-12 *(continued).*

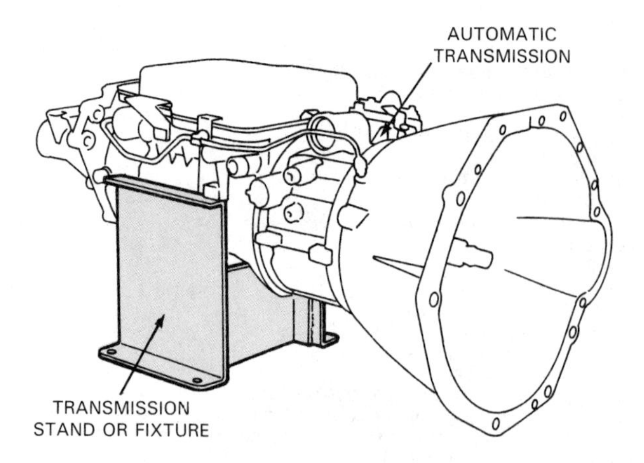

Figure 13-13. It is much easier to overhaul a transmission if it is placed in a special holding fixture. (Nissan)

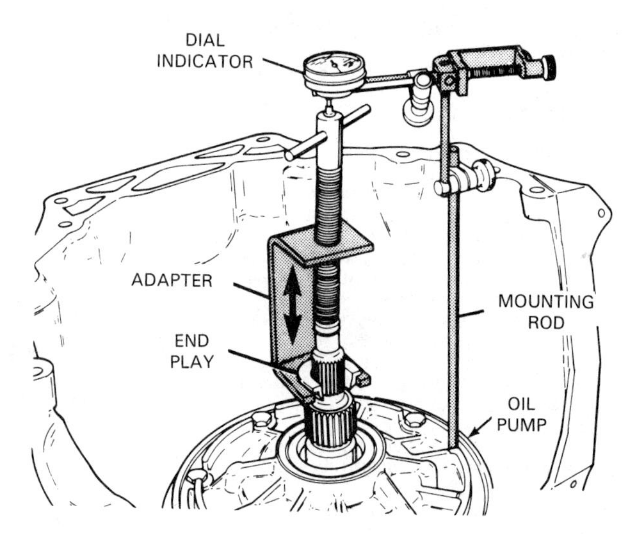

Figure 13-14. Always check end play before disassembling the transmission. The service manual will contain the end play specifications and the procedure for setting up the dial indicators. (General Motors)

varnish buildup, and sludge. Excessive sludge indicates that the holding members are burnt and that fluid had been overheated. Metal particles usually indicate that the torque converter, oil pump, planetary gearsets, or other moving part is wearing badly. If this is the first time the oil pan has been removed for service, a few aluminum particles may show up in the pan, originating from case machining operations at the factory. This is not a cause for concern.

Removing the oil pan gives you access to the filter and valve body. The filter may be pressed into a case passageway and held in place with a clip on the valve body, or it may be bolted directly to the valve body. Remove the filter at this time.

After the filter has been removed, the valve body bolts can be unfastened and the valve body can be

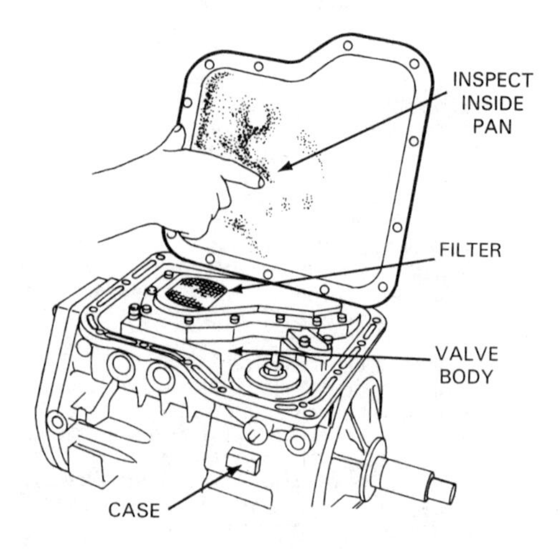

Figure 13-15. Remove the oil pan and check for metal particles, varnish buildup, and sludge. Note that some oil pans have a small magnet to pick up small metal pieces. Be careful not to lose the magnet. (Nissan)

removed from the case. As you lift the valve body from the case, carefully observe the position of all valve body parts, including springs, oil feed tubes, linkage attachments, and check balls, Figure 13-16. At this time, you should also remove the vacuum modulator, pushrod, and throttle valve from the case, if applicable. See Figure 13-17. Place the assembled valve body and related parts and any modulator parts together in a safe spot on the workbench.

Remove the bolts holding the extension housing to the case. The extension housing should pull off over the output shaft after the bolts are removed. A few extension housings contain an output shaft bearing, which supports the output shaft. The extension housing can be removed after the bearing snap ring is expanded, allowing the extension housing to be slid from the bearing. After the extension housing is removed, the governor can be removed. This step applies only to governors mounted on the output shaft, not those mounted in the case.

Remove the front pump seal, as shown in Figure 13-18. It is generally easier to remove it before the oil pump is removed from the transmission. The extra weight of the transmission assembly allows the slide hammer to work more efficiently. The front pump seal should always be changed when the pump is serviced.

The oil pump can be removed next. Manufacturers sometimes recommend removing the input shaft before removing the oil pump. Some transmissions, however, are constructed so that the input shaft is pressed into the front clutch drum. In these designs, the shaft cannot be removed before the pump.

The pump is usually a tight fit in the case. If it is, some pressure must be applied to remove it. Some pumps can be removed by prying them out with a screwdriver or pry bar (once the valve body is removed). Others are removed with slide hammers or special pullers, Figure 13-19.

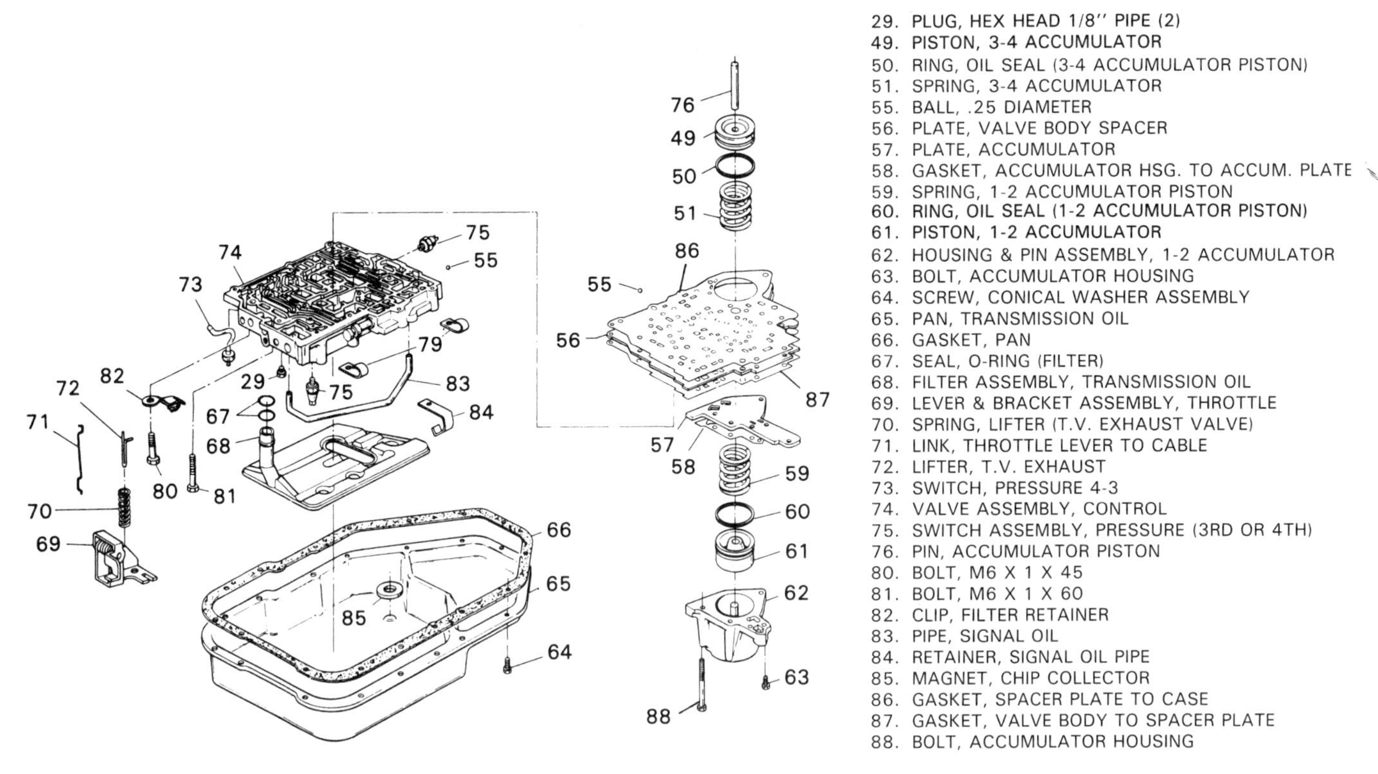

Figure 13-16. The valve body can be removed after the attaching bolts are removed. Many parts, such as spacer plates, accumulator pistons, oil supply tubes, and check balls, are held in place by the valve body. Make sure that you do not lose any of these parts when removing the valve body. (General Motors)

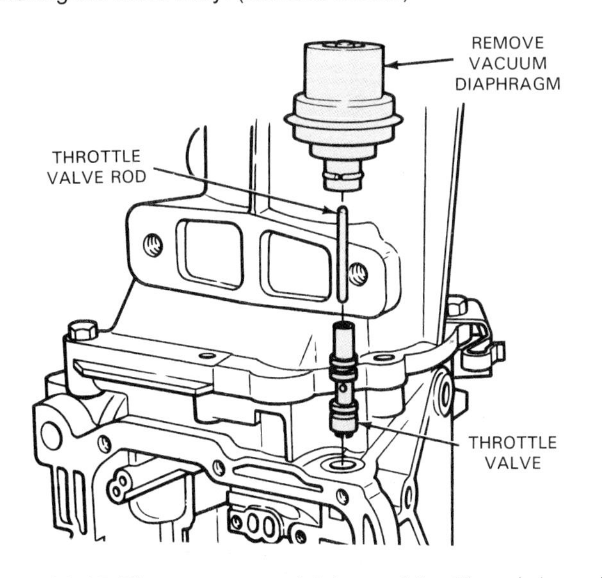

Figure 13-17. The vacuum modulator and throttle valve can be removed now to prevent damage. Do not lose the pushrod, if the modulator uses one. (Ford)

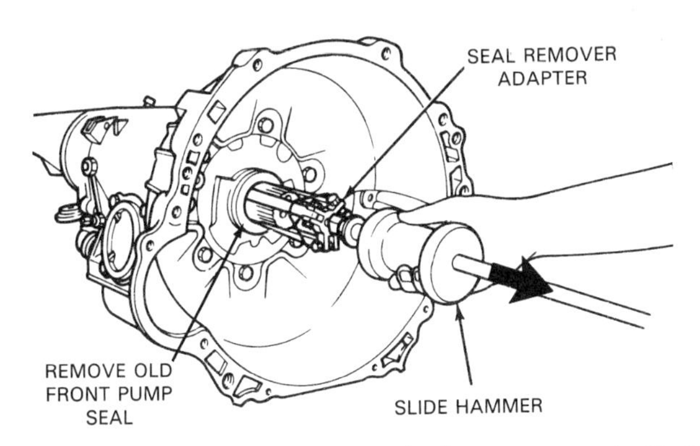

Figure 13-18. It is easier to remove the front pump seal with the transmission assembled. The weight of the transmission allows the slide hammer to exert more force on the seal. (Ford)

When removing the oil pump, always check for thrust washers that may fall out of place. A few oil pumps have a clutch apply piston installed in their inner face. In this case, a set of friction discs and clutch plates is installed directly behind the pump. These parts should be removed and placed on the pump for later service.

Removing internal components

With the oil pump out of the way, the internal parts, such as the input shaft, bands, clutch packs, planetary gearsets, and output shaft, can now be removed. Removal steps are similar for all transmissions. The major factor affecting disassembly is whether or not the transmission uses a center support.

On transmissions without center supports, removing internals is usually a simple matter once the front pump is removed. A typical procedure might involve loosening the front band adjuster, sliding the band out of the case, and

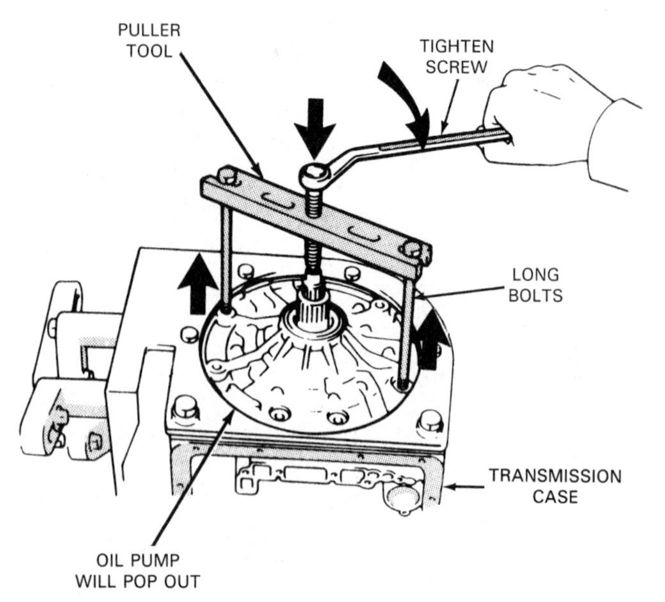

Figure 13-19. Some pumps can be removed by pushing them out from the rear. On many transmissions, however, the pump cannot be reached from the rear, and these must be removed with a slide hammer or special puller. (Ford)

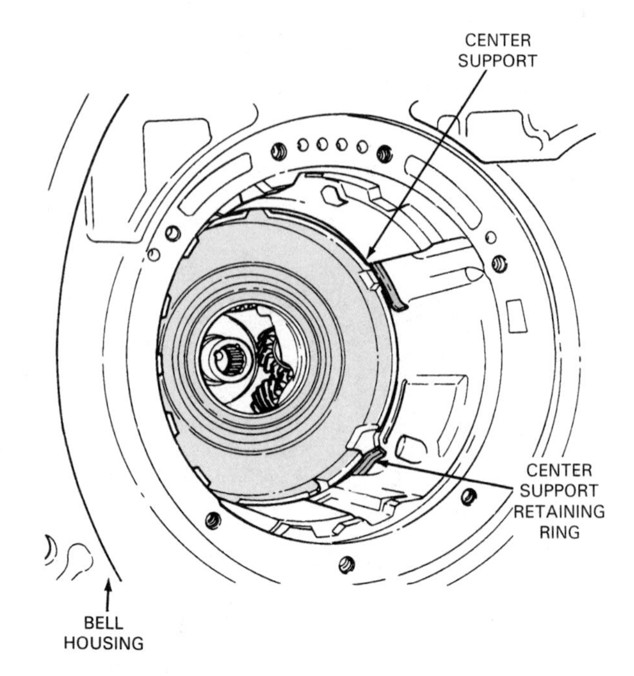

Figure 13-20. Many modern transmissions have a center support. The support shown here is held in place with a retaining ring, which must be removed before removing the rear planetary gearset and holding members and the output shaft. (Ford)

then, removing the forwardmost clutch packs, front planetary gearset, and input shell as an assembly. Be sure to save all thrust washers as they come out with these parts.

Once the front components are removed, any rear planetary gearsets and holding members and the output shaft can be removed. Rear gearsets are removed through the front of the case. The output shaft will usually come out the rear of the case.

Note that some clutch packs are held in place by a snap ring installed on the sun gear, inside of the input shell. The ring must be removed before the clutch pack and other internal parts can be removed. In addition, some rear planetary gearsets are held in place by a snap ring on the front of the output shaft. This snap ring must be removed in order to completely disassemble these components. Make sure that the shaft will not fall out of the case after the snap ring is removed. Also, do not distort or stretch the snap rings when removing them. Use snap ring pliers or other suitable tool.

As mentioned, some automatic transmissions have a center support. A center support in an automatic transmission may be retained by one or two case bolts under the valve body and, sometimes, by bolts on the outside of the case. Other center supports are held in place by a retaining ring in the case, Figure 13-20. The center support keeps the rear clutches and gearsets in place. Always make sure the center support bolts or snap rings have been removed before attempting to remove the center support and mechanical components.

CAUTION! Attempting to drive the rear clutch, planetary gearset, and shaft assemblies from the case without first removing the center support fasteners will severely damage these mechanical components, the center support, or the case.

Once the center support is removed, the rear planetary gearsets, bands, clutches, and output shaft can be removed. The gearsets, holding members, and output shaft will usually be removed through the front of the case. It may be necessary to remove the parking gear to remove the output shaft.

Once the mechanical components mentioned are removed, any internal case-mounted accessories can be removed. Examples of these parts include some servos, accumulators, clutch apply pistons (usually used at rear of case), and governors. (If these parts are not the type mounted inside the case–for example, if they mount inside the oil pan–they will be removed with the other external components.)

Note that many servos and accumulators are under strong spring pressure. Always consult the service manual before removing any retaining rings or bolts. It may be easier to leave the servos and accumulators installed in the case until it is time to inspect them and replace their seals. If the transmission is filled with sludge and debris, however, the parts should be removed at this time for a thorough cleaning.

Automatic Transmission Parts Cleaning

All automatic transmission parts must be cleaned thoroughly. This is very important because small dirt particles or varnish formation can cause a passageway to plug, a valve to stick, or a seal to leak.

Carefully scrape all old gasket material from the transmission case, oil pan, and other parts. Pay attention to the gaskets on hydraulic control assemblies such as the front pump, valve body, and spacer plates and remove all gasket material.

Clean all metal transmission parts, including the case, with a safe solvent. It may be necessary to soak parts overnight. Do *not* clean clutch friction linings or seals with solvent. To avoid having a confusing pile of parts, always remember to work on one subassembly at a time while cleaning.

WARNING! *Never* clean parts with gasoline or other flammable solvents. A fire could result!

Dry the cleaned parts with compressed air (if possible) or allow them to air dry. Do *not* use rags or shop towels to dry the parts because they will leave lint on the parts. It is better to leave slight deposits of solvent on the parts than dry them with rags or shop towels. The lint left behind can cause valves to stick. Once cleaned, cover disassembled parts with a clean lint-free cloth if they are to sit out overnight.

Drain the torque converter. If there is a drain plug, remove it and set the converter upright with the drain at the lowest point. On converters without a drain plug, lay the center opening face down to drain. This will allow most of the old transmission fluid to drain out.

If the transmission parts were very dirty or heavily coated with varnish, then the torque converter will be full of dirt and varnish. A torque converter that is filled with dirt or metal particles can be cleaned with a converter flusher. (Converter flushers were introduced in Chapter 2.) A heavily varnished converter, however, is difficult to clean and is often replaced.

If there is evidence of fluid contamination, the transmission oil cooler and cooling lines must be flushed out. One way to do this is to use an oil cooler and line flusher (introduced in Chapter 2) and follow up with a blast of compressed air directed into one of the cooling lines. Have a drain pan available to catch solvent coming out of the other line. Do *not* apply full air pressure because this could rupture the oil cooler.

Automatic Transmission Parts Inspection and Repair

Once the transmission is disassembled and cleaned, the parts can be inspected, replacement parts can be ordered, and subassemblies can be rebuilt. In this section, you will find that parts inspection and repair are treated as an integral procedure of each subassembly. In actual practice, the transmission parts should be inspected during and immediately after the transmission is taken apart and cleaned. In this way, you can order replacement parts right away, eliminating lost time while waiting on parts. You should inspect transmission parts one subassembly at a time. Failure to do so can lead to mass confusion and could turn your transmission overhaul into a nightmare!

During inspection, it is not absolutely necessary to check the ***soft parts***–parts such as gaskets, lip seals, O-rings, seal rings, and friction discs. These are normally replaced during an automatic transmission overhaul. Checking these parts, however, will often help you to determine what caused the transmission to fail. Automatic transmission ***overhaul kits*** are designed to contain all of the necessary soft parts for a rebuild.

Transmission ***hard parts,*** as they are called, must be inspected. These are parts that are not normally replaced as part of an overhaul. Examples include torque converters, shafts, gearsets, bands, and clutch drums. Replacement parts for these must generally be ordered separately. (In some instances, clutch plates, bands, filters, and vacuum modulators are included in an overhaul kit. Always check the contents of an overhaul kit before ordering additional parts.)

Automatic transmission shaft service

Automatic transmission shafts, including input, output, and stator shafts, require little service. Check seal ring grooves and bushing surfaces. Inspect shaft splines for wear or damage. Any shaft with worn or damaged splines should be replaced. If you suspect that a shaft is bent or warped, it can be checked with *V-blocks* and a dial indicator, as shown in Figure 13-21.

Planetary gearset service

Check the teeth of the planetary gears for wear, nicks, or chipping. Slightly nicked gears can cause noises and vibration, and eventually, they will fail completely. Splines found on the gearset members should also be checked for wear and damage. Check the clearance between the planet gears and the planet carrier to determine the amount of end play, Figure 13-22. Planetary gearsets often contain bushings or needle bearings that wear out. These should be carefully checked, as well. Any parts that are worn or damaged should be replaced.

Clutch pack service

One of the most important parts of automatic transmission overhaul is proper checking and rebuilding of the clutch packs. These operations are discussed in the next few paragraphs.

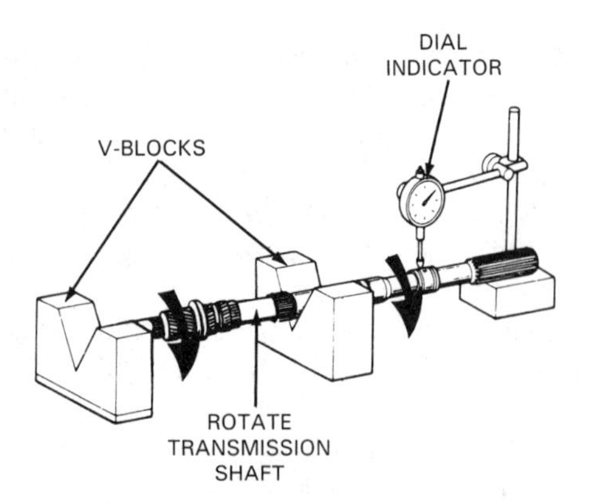

Figure 13-21. Input and output shafts can be checked for bends or warpage using V-blocks and a dial indicator. Never attempt to straighten a bent or warped shaft. (Toyota)

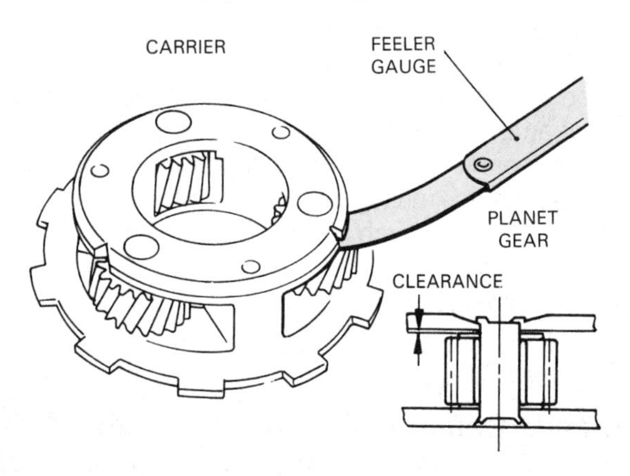

Figure 13-22. The planet gears and planet carrier can be checked for wear with a feeler gauge, as shown here. Disassembly is not required to make this check. (Nissan)

Clutch pack inspection. To check the steel plates, take a screwdriver and remove the snap ring that holds them and the friction discs in the clutch drum; then, remove the plates and the discs. Keep all of the steel plates together, noting if there are any *reaction plates* (special thicker steel plates) at the top and bottom of the stack. Do not interchange one set of clutch pack parts with another.

Inspect the steel plates, Figure 13-23. Look for discoloration, caused by overheating and often a sign of a badly slipping clutch pack. Look for worn or scored plates, a sign of a prolonged slippage condition. Also, check the tangs for wear or damage. If there is any doubt about the condition of the plates, they should be replaced. Steel plates are sometimes included in a transmission overhaul kit.

Friction discs are normally replaced as part of a transmission overhaul. These parts are usually included in a transmission overhaul kit. Worn or damaged discs may be charred, glazed, or heavily pitted. The friction lining may scrape off easily with your fingernail. In some instances, all of the friction material will be missing from the friction discs.

Check the condition of the channels on the inside surface of the clutch drum and the splines on the clutch hub. Worn or damaged parts should be replaced.

Clutch pack rebuilding. Rebuilding a clutch pack requires the use of a special spring compressor, Figure 13-24. The tool pushes down on the spring retainer, compressing the clutch return spring(s) and allowing the retaining snap ring to be removed. With the snap ring removed, the rest of the clutch pack can be disassembled, Figure 13-25. The procedure is detailed in the next few paragraphs.

To finish clutch pack disassembly, install the spring compressor and compress the return spring. Remove the

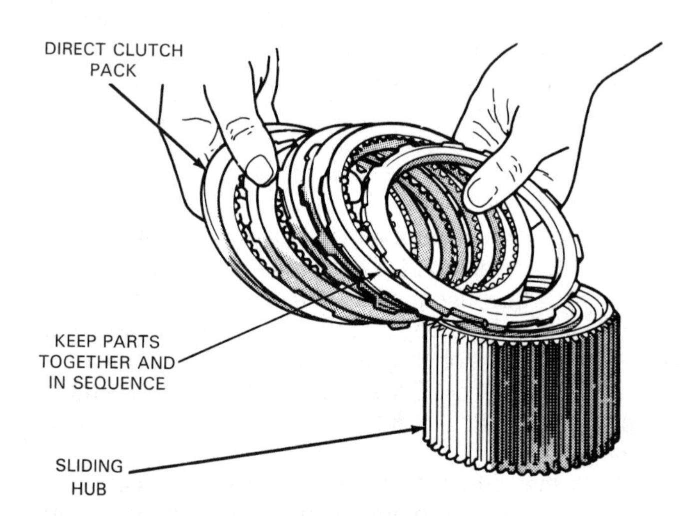

Figure 13-23. Inspect the clutch plates for wear, burning, scoring, and other damage. Also, check the drum, bushings, hub splines, and thrust washers. Replace parts that show any signs of such wear or damage. (Chrysler)

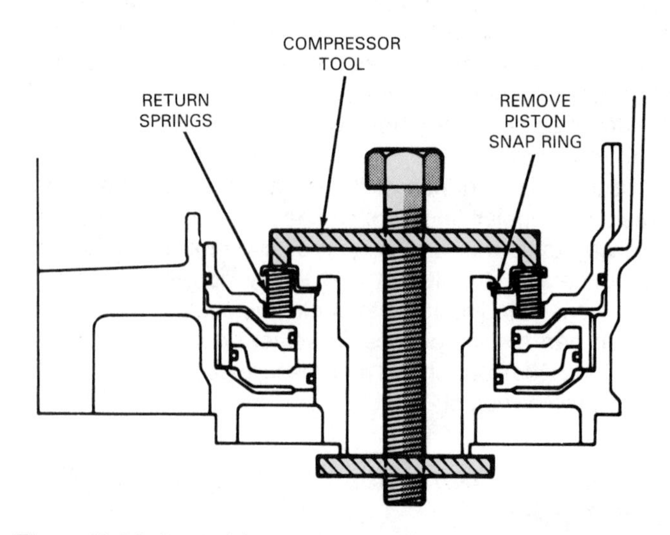

Figure 13-24. A special compressor tool must be used to remove the clutch apply piston in order to get at the seals of the clutch assembly. Universal tools are available that fit all types of assemblies. (Chrysler)

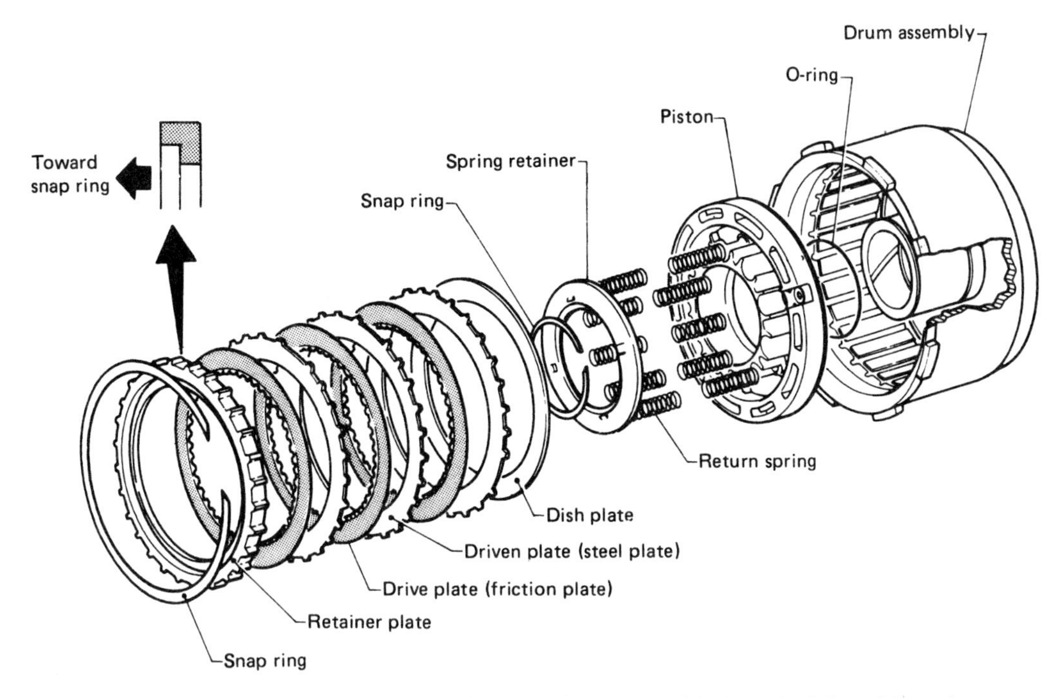

Figure 13-25. Exploded view of a clutch pack shows parts that must be removed during rebuilding. (Nissan)

retaining snap ring with snap ring pliers or a screwdriver. Release the spring compressor and remove the spring retainer and return spring.

Remove the clutch apply piston from the drum. If the piston sticks in the drum, it can sometimes be removed by slamming the drum downward on a wooden workbench or a block of wood. This will often jar the piston loose. Compressed air can also be used to blow the piston from the drum.

WARNING! When using compressed air to remove a piston, point the open end of the drum away from yourself and others. Direct the piston into rags or other soft surface so that it does not fly out and become damaged or cause injury. Also, use a regulated amount of air pressure.

With the piston out of the drum, remove the piston seals. Hard or cracked seals indicate that the clutch has been overheated. If there is also a seal around the interior (hub) of the clutch drum (inner piston seal), replace it. Then, thoroughly clean the piston and piston bore.

Obtain new piston seals. Check them against the old ones to ensure that you are installing the proper seals on the piston. It is also a good idea to check the fit of the new seals by placing them in the piston bore. They should fit snugly. A bulge, or buckle, in an outer piston seal means the seal diameter is too large. A gap left between an outer piston seal and piston bore means the diameter is too small.

Before installing the new piston seals, lubricate them with transmission fluid. Petroleum jelly can also be used, but other types of oil or grease can damage the seals. Carefully install the new piston seals on the clutch drum hub and clutch piston, as applicable. The new seals should fit snugly, but not *too* snugly. Lip seals must be installed on a piston so that the sealing lip will be directed toward the hydraulic pressure, or the back of the drum.

Install the clutch piston in its bore. The type of piston seal determines the installation method. If O-ring seals are used, the piston can be pressed into the bore, after being thoroughly lubricated.

A piston using a lip seal must be worked into place with a feeler gauge or a special seal installation tool. A satisfactory tool can be made from a stiff piece of wire and a small length of copper tubing. See Figure 13-26. Once the piston is in place, check its installation by trying to turn it. If you cannot turn the piston, the seal is not properly situated.

Once you are sure the piston is properly installed, put the piston return spring and spring retainer back in position on the piston. Then, use the spring compressor tool to move the return spring and retainer past the groove for the retaining snap ring. Install the snap ring, and make sure that it is snug in its groove. See Figure 13-27.

If the clutch pack uses a reaction plate next to the apply piston, install it now. Then, install the friction discs and steel plates, alternating them and using new parts as required. If the clutch pack uses a reaction plate at the top, install it before installing the snap ring.

CAUTION! A dry friction disc will burn severely the first time the clutch is applied. To prevent this, soak new friction discs in clean transmission fluid before installing them. Allow them to soak for about 15 minutes or until they stop bubbling. Always use the correct type of automatic transmission fluid. The vehicle owner's manual should list the proper ATF. The fluid type may also be stamped on the transmission dipstick.

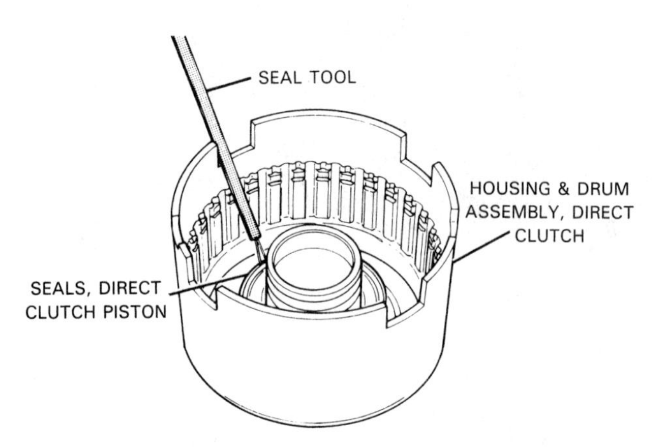

Figure 13-26. This special tool can be fabricated from copper tubing and wire to make lip seal installation easier. This tool will be much less likely to cut the new seal than a feeler gauge. (General Motors)

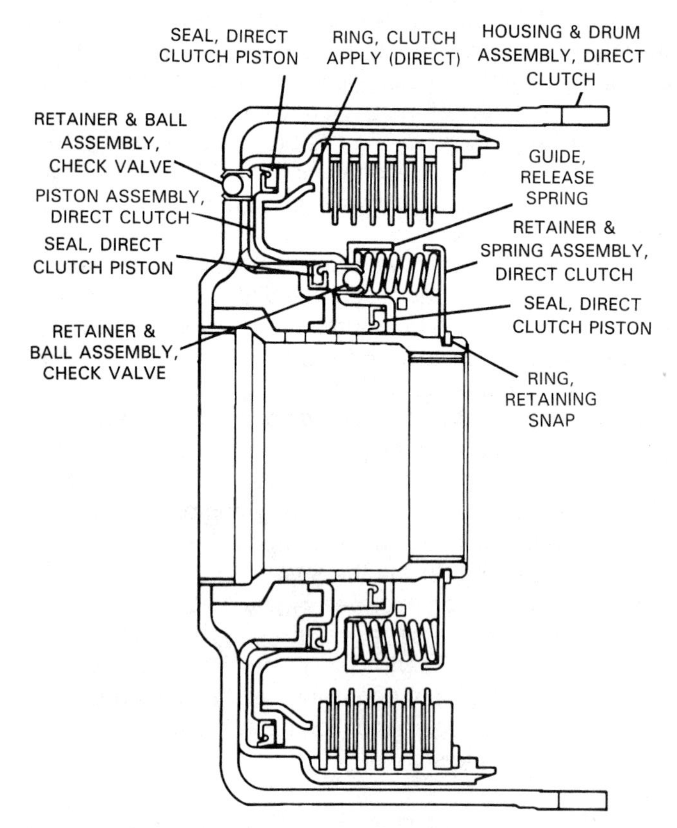

Figure 13-27. Place the retaining snap ring in its groove to secure the return spring and clutch piston assemblies. (General Motors)

With the clutch pack now assembled, measure the ***clutch clearance,*** or the distance between the pressure plate and outer snap ring. Clutch clearance is checked with a feeler gauge, Figure 13-28. The exact placement of the feeler gauge will vary between makes. Most vehicle manufacturers have clearance adjustment kits, consisting of snap rings or pressure plates of varying thicknesses. Substituting a different thickness snap ring or pressure plate, as needed, will adjust the clearance.

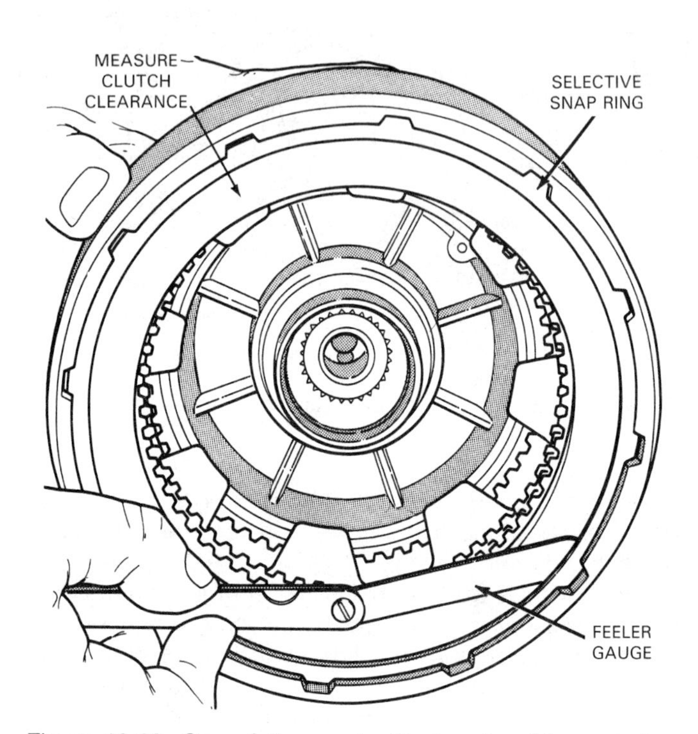

Figure 13-28. One of the most critical parts of transmission overhaul is checking the clutch pack clearance. Incorrect clearances must be fixed by replacing the pressure plate or snap ring with one of the proper thickness. (Chrysler)

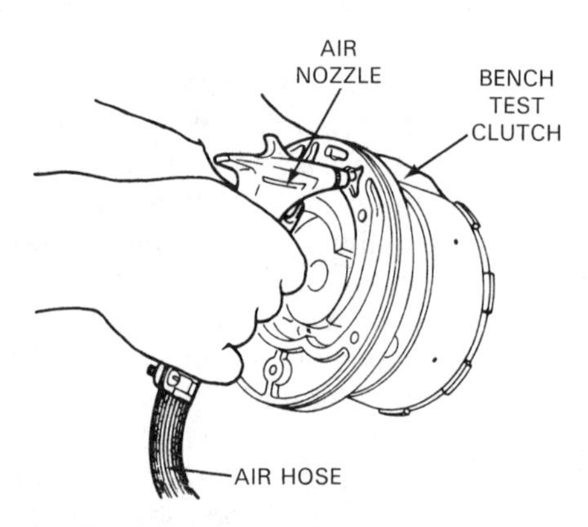

Figure 13-29. Clutch pack operation should be checked with air pressure before transmission assembly. This will ensure that the clutch pack operates properly when it is installed. The clutch plates and friction discs should lock together when air is applied to the proper passage. (Subaru)

Clutch clearance is an indicator of how tightly the plates and discs are packed. It is important because it affects transmission operation. If the clearance is too large, the clutch will not be applied tightly and may slip. If the clearance is too small, the shift may be rough, or the clutch may be lightly applied when it should be released. Proper clearance helps to assure good shifts and long clutch life.

Clutch pack operation can often be checked with compressed air, Figure 13-29. Air is directed to the fluid inlet port of the clutch, applying the piston. Piston action can be observed to verify that the clutch will operate when installed in the transmission.

CAUTION! *Never* apply air pressure to the clutch piston unless the clutch plates and friction discs are installed. Air pressure may push the piston far enough to jam it in its bore.

Repeat the clutch pack rebuilding sequence for every clutch pack in the transmission.

Band service

Check band friction linings for overall wear. Look for burn marks, glazing, and nonuniform wearing. Check to see that grooves are still visible and check for flaking. If any material can be scraped off, the band should be replaced. Look for cracks or embedded metal particles. Also, check band for cracked ends or distortion. See Figure 13-30.

The surface that the band rides on should also be inspected. If it shows any signs of scoring or wear, it should be turned down on a lathe. If turning will remove more than a few thousandths of an inch (hundredths of a millimeter) of surface, the clutch drum or gearset member that the band rides on should be replaced. A surface that is shiny must be sanded with emery cloth to remove the shine. Sand *around* the surface perimeter if a paper-lined band will be used. Sand up and down if a fabric-lined band will be used. A new band should always be used when the band's mating surface has been sanded.

CAUTION! Soak new bands in clean ATF for about 15 minutes or until they stop bubbling. Always use the correct fluid.

Servo service

Servos seldom require more service than the replacement of the seals and, if used, the cover gaskets. The servo should be disassembled and checked, however, Figure 13-31. Many servos are held to the transmission case with a snap ring. A special service tool is required to compress the servo return spring so that the snap ring can be removed.

To rebuild a servo, begin by removing the servo cover. To do this, remove the servo cover retaining bolts or depress the servo cover and remove the snap ring. Do not try to remove the snap ring until the cover has been pushed in and is out of the way. A *servo-cover depressor* may be used for this purpose.

With the snap ring out of the way, remove the servo cover and pull the servo apply piston and return spring from the servo cylinder. Some servos have more than one piston and more than one spring. Note the position of the parts as you remove them. Next, remove the seals from the servo piston(s), noting the direction that sealing lips on lip seals, if used, face.

Note that a few pistons will be tight in their cylinder. These may be removed by *carefully* applying air pressure to the *servo release port.* See Figure 13-32.

Check the servo piston and piston pin for wear. Normally, servo parts do not wear very much. If a spring height measurement is given in the service manual, check it with a sliding caliper, as shown in Figure 13-33. If the spring is not within specification, replace it.

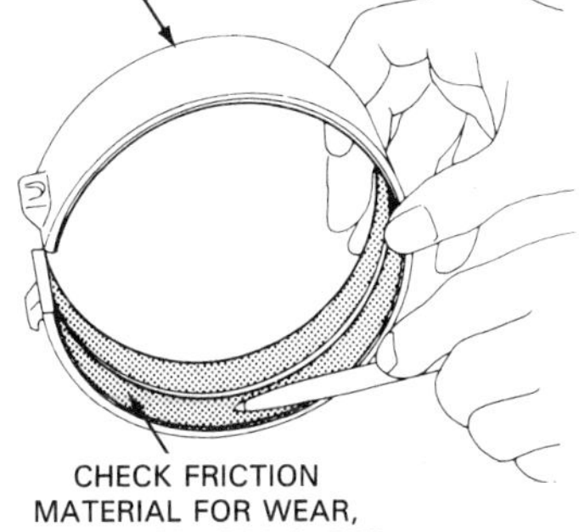

BANDS ARE FREQUENTLY REPLACED

CHECK FRICTION MATERIAL FOR WEAR, BURNING, CRACKING, ETC.

Figure 13-30. Check bands for wear, burning, cracked linings, etc. A band that shows any sign of wear should be replaced. Also, check the drum where the band rides.

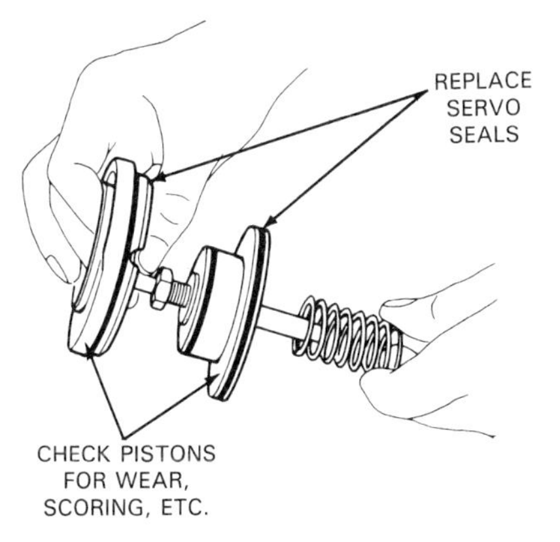

Figure 13-31. Most servos can be serviced by replacing the seals. The servo should be checked for any signs of wear. Return servo pistons and springs to their original positions.

After cleaning all servo parts, select, lubricate, and install the new seals on the servo apply piston and, if there is one, on the accumulator piston. There may be a seal on the servo apply piston pin. This seal should also be replaced. Then install the servo piston(s) in the cylinder. Make sure that all servo pistons and springs are reinstalled in their original locations. Select the proper servo cover O-rings or gasket and reinstall the servo cover. An exploded view of a typical servo assembly is shown in Figure 13-34.

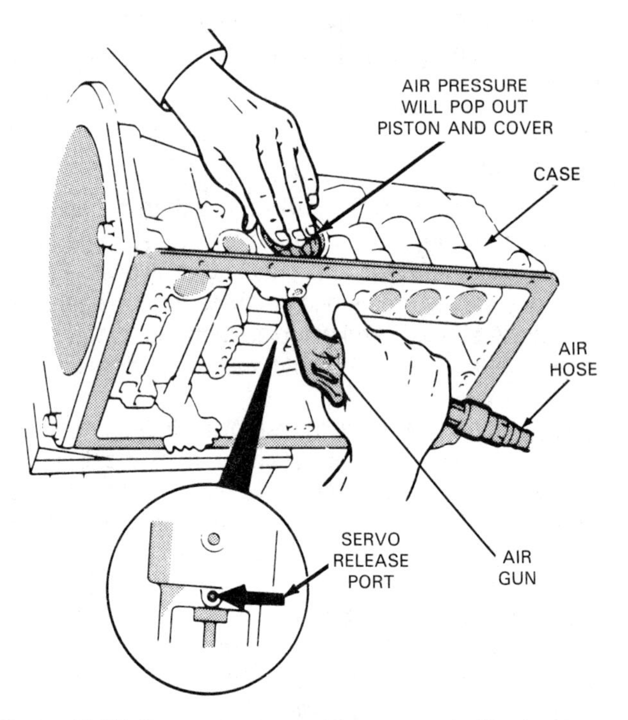

Figure 13-32. Some servos must be removed with air pressure, as shown in this illustration. Always make sure that the servo does not fly out of the case when air pressure is applied. (Chrysler)

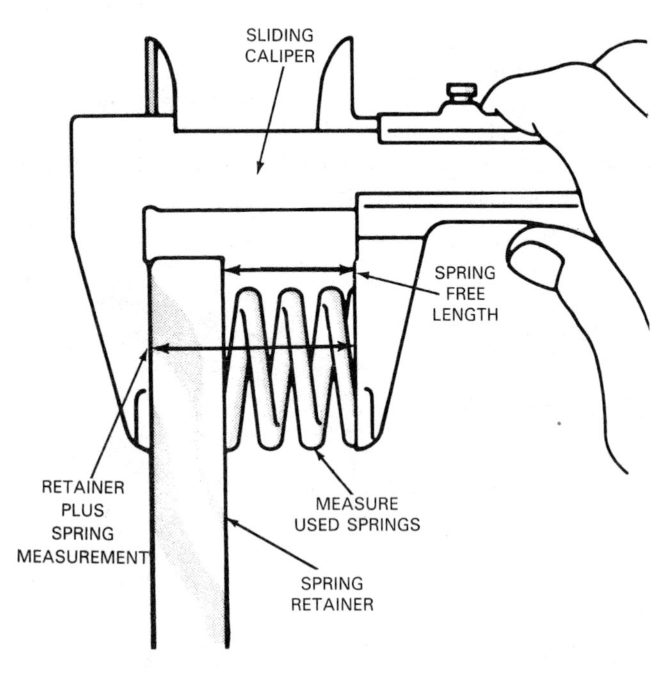

Figure 13-33. Servo return spring length can be measured with a sliding caliper. Undersized springs should be replaced. (Chrysler)

Overrunning clutch service

All overrunning clutches in the transmission should be checked to ensure that they turn freely in the direction that they should and that they lock up in the other direction. See Figure 13-35. Any overrunning clutch that turns in both directions must be replaced. Also, any clutch showing signs of wear should be replaced. Always be careful when replacing an overrunning clutch to ensure that it turns in the proper direction when installed.

Accumulator service

Rebuilding an accumulator is similar to rebuilding a servo. Remove the accumulator cover, if applicable, and pull the accumulator piston and return spring from the accumulator cylinder. Inspect all parts. Look for nicks, burrs, or scoring. Replace the accumulator piston seals and reinstall the piston and return spring. Reinstall the cover, if used.

Transmission oil pump service

The front pump must be disassembled to check it properly. Most pumps are held together by bolts. Some pumps also contain the pressure regulator valve and other valves.

Once the front pump is apart, check it for wear where the gears, rotors, or vanes ride against the stationary pump body. Badly worn, scored, or pitted pumps can usually be detected visually. Scratching the suspect area with your fingernail will usually confirm the presence of wear or scoring. In addition, the condition of the pump can be assessed with the help of feeler gauges, as shown in Figure 13-36.

Since the front pump is designed to provide much more pressure than needed, the technician is often tempted to reuse a slightly worn pump. If you are reusing an old pump, make sure the wear is slight and that the parts are reinstalled in the same position as they were originally. The pump internal elements usually have factory marks to ensure proper installation.

Always replace the front pump seal, no matter how good it appears. Also, replace the *pump bushing,* if needed. See Figure 13-37. The pump bushing is critical, and the proper removal and installation tools should always be used. The bushing should be staked in place to prevent it from moving.

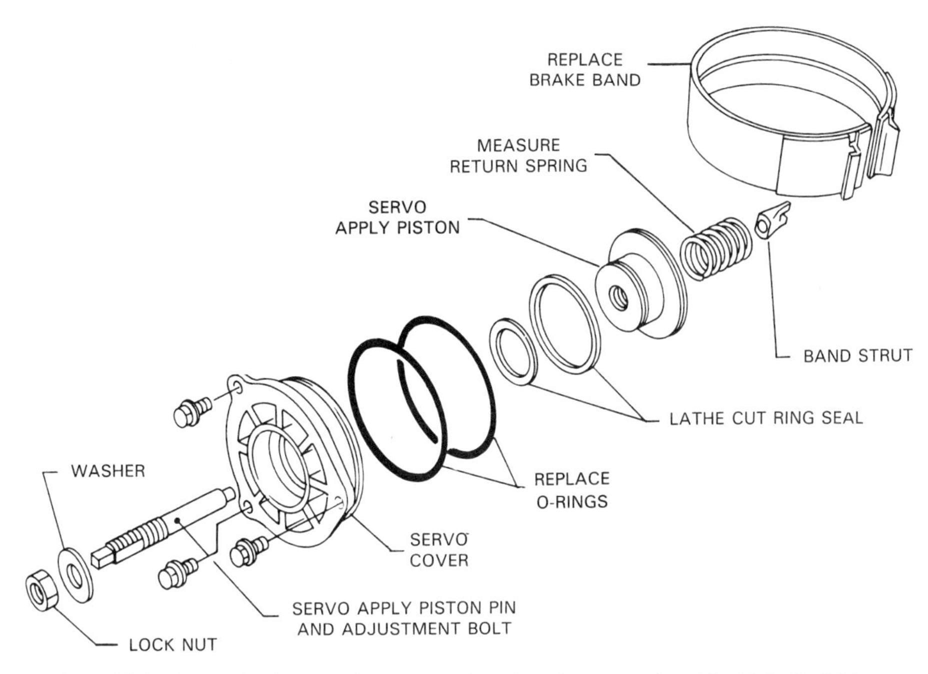

Figure 13-34. This shows the layout of parts on a band and servo assembly. Note that this servo cover is attached with bolts rather than a snap ring.

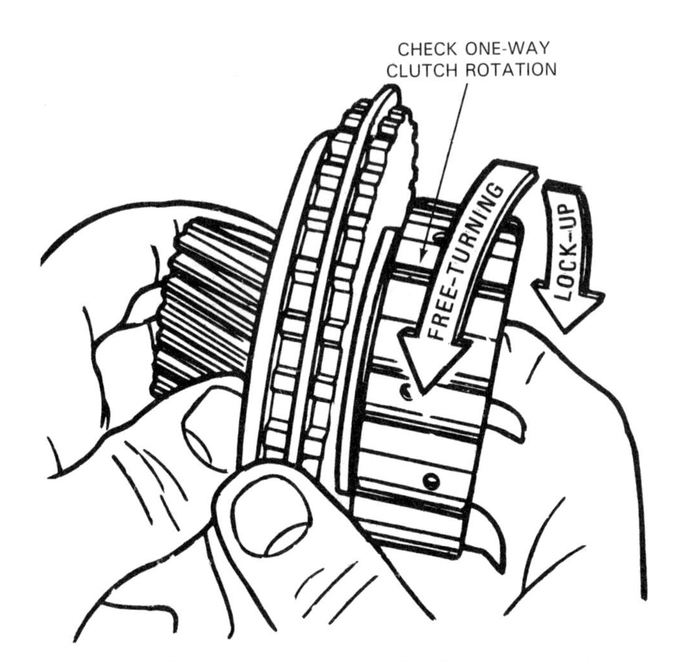

Figure 13-35. Check every one-way clutch for wear and proper action. One-way clutches should turn in one direction only, as specified. (General Motors)

Valve body service

The valve body should be thoroughly cleaned in cleaning solvent to remove dirt and metal particles. The casting should be checked for porosity, cracks, and damaged machined surfaces. Orifices and passageways must be clean and free from obstructions. Valves should be checked for burrs, nicks, and scores. Springs should be checked for distorted or otherwise damaged coils.

Check all valves for free movement in their bores. Any valve that does not move freely in its bore is sticking and can cause problems. Valves can be manually moved in their bores against spring pressure. If the spring does not return the valve to its original position, the valve is sticking.

To repair a sticking valve, the valve must be removed from the valve body. Most valves are held in the valve body by small retainer plates, shown in Figure 13-38. To avoid mixing up the valves, remove one retainer plate at a time. Clean and reassemble the valves under that retainer plate before going on to the next one.

Sticking valves can sometimes be loosened by lightly polishing the valve lands with crocus cloth. Some manufacturers recommend using a microfine lapping compound instead of crocus cloth. Always polish in a rotary motion, or around the surface. Be careful not to polish the sharp edges, as these help to keep the valve from jamming on small pieces of dirt. Cleaning the valve *bore* may also restore a valve to proper operation. If proper operation cannot be restored, the valve or even the valve body assembly must be replaced.

It is not necessary to completely disassemble the valve body if you just want to check the valves. However, if the transmission is very dirty or sludged up, the entire valve body must be disassembled. Always consult the service manual for a valve diagram, such as in Figure 13-39.

Figure 13-40 shows relative positioning of valves and springs. This illustrates how they should be installed in the valve body. When taking the valve body apart, watch for

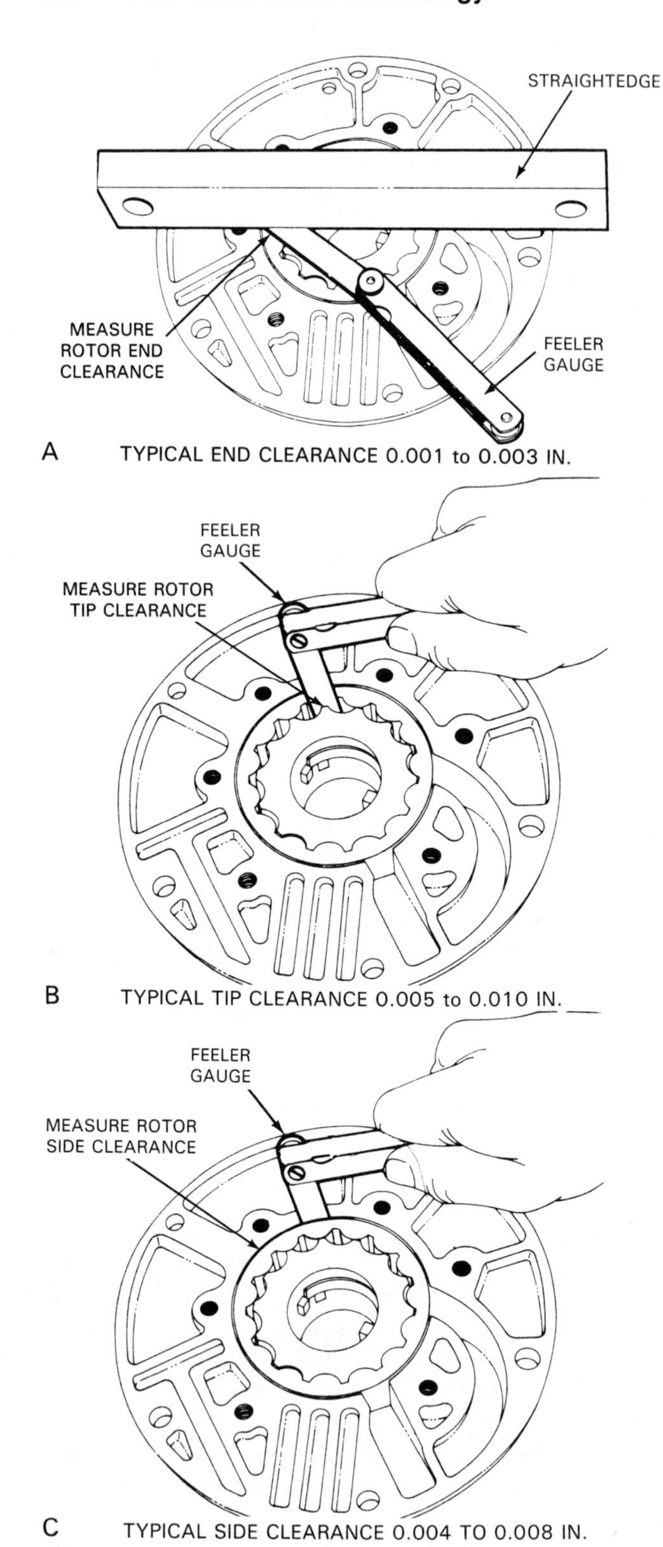

Figure 13-36. These three checks will determine whether the pump can be reused. These checks apply to gear and rotor pumps. Always refer to the service manual for wear specifications. A–Checking clearance between rotor or gear faces and straightedge laid across pump body (end clearance). B–Checking clearance between tips of inner and outer rotor lobes or gear teeth (tip clearance). C–Checking clearance between outer rotor or gear and pump body (side clearance). (Chrysler)

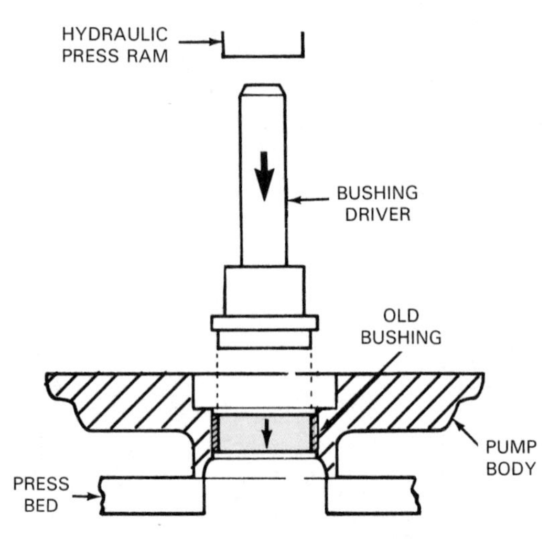

Figure 13-37. Worn pump bushings can be replaced with new bushings. The old bushing is pressed out, as shown here. Always check the service manual for exact specifications and use the proper bushing removal and installation tools. (Ford)

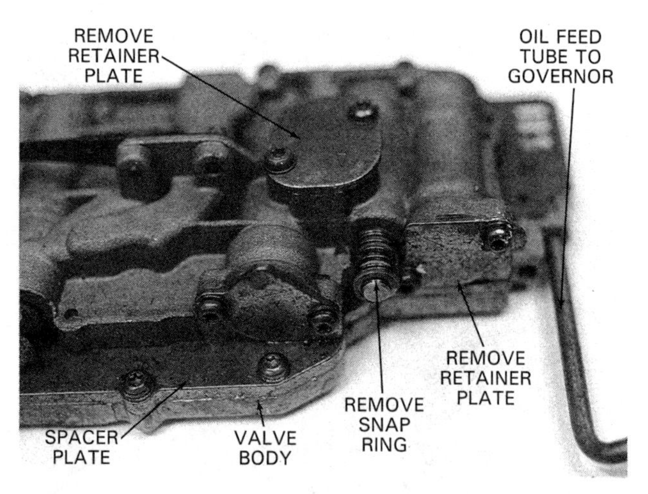

Figure 13-38. Overhauling the valve body should be done carefully. Remove one retainer plate and valve at a time, if possible. Then, reinstall the valve exactly as it was originally. If a valve is suspected of sticking, it can be checked for free movement in its bore. Split valve bodies should be opened to replace the spacer plate gaskets and for more thorough cleaning. (BBU, Inc.)

check balls, springs, and other small parts that must be reinstalled.

Check the spacer plates for warping, baked-on gasket material, deep scratches, or pounded out check ball orifices. Damaged spacer plates should be replaced.

Also, carefully check the inner manual lever, which holds the manual valve in position. Check the teeth on the cockscomb to make sure they are not worn, Figure 13-41. Check the spring-loaded roller or ball to make sure that it puts pressure on the teeth and that it holds the lever in position.

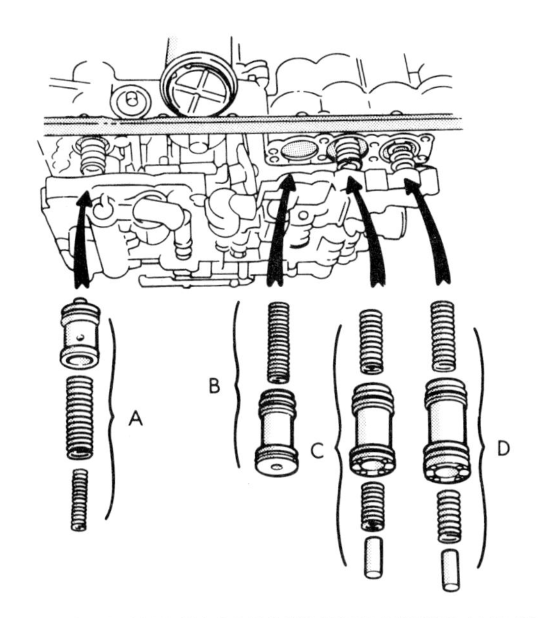

A. OVERDRIVE CLUTCH ACCUMULATOR PISTON AND SPRINGS
B. OVERDRIVE BRAKE ACCUMULATOR PISTON AND SPRINGS
C. SECOND CLUTCH ACCUMULATOR PISTON, SPRINGS, AND SPACER
D. SECOND CLUTCH ACCUMULATOR PISTON, SPRINGS, AND SPACER

NOTE: PISTON HEIGHT AND DIAMETER ARE OUTLINED IN THE SPECIFICATIONS SECTION.

Figure 13-39. If you must remove all of the valves to clean a very dirty valve body, always have a service manual handy. This is so you can replace the valves and springs in the bores exactly as they came out. (Chrysler)

Vacuum modulator service

If the transmission has a vacuum modulator, check for a dented neck and for transmission fluid on the vacuum hose side. If the neck is bent, or if there is fluid on the vacuum side, the modulator must be replaced. As a further check, use a hand-powered vacuum pump to place a vacuum on the modulator, Figure 13-42. If the modulator will not hold vacuum, replace it.

The replacement modulator may look different from the original model. This is common with replacement modulators. It does not indicate that the part is wrong.

Governor service

The major trouble in governors is sticking weights and valves. To inspect the governor, check the weights and valve and make sure they move freely. Look for scored surfaces and make sure mating surfaces are smooth and flat. The valve can be cleaned and polished in the same way as the valves in the valve body. If a sticking weight or valve continues to stick after cleaning and polishing, or if you have any doubts about the condition of the governor, replace it.

Some governors contain small screens to filter out dirt or metal particles. These screens should be thoroughly cleaned. If a screen is torn or otherwise damaged, it should be replaced.

Seal ring service

Seal rings are critical to proper transmission operation since they seal in hydraulic fluid going from the valve body to the bands and clutch packs. Seal rings can be made of metal or teflon. They should always be replaced during an overhaul!

After removing the old seal rings, check the ring grooves. They must not be worn. Look for nicks or ridges at the groove wall, which could prevent the new ring from seating on the wall as it should, allowing leaks. Grooves can usually be checked with the new ring and a feeler gauge, Figure 13-43.

Check the bore surfaces that the seal rings ride against, as well. These mating bore surfaces must be smooth, with no ridges or signs of wear, and they must be round. Place the replacement ring squarely in the bore to see that it conforms to the bore and that the bore is round. The ring should touch the bore all the way around.

Parts having worn or damaged seal ring grooves or bores should be replaced. Always install new seal rings carefully, using plenty of transmission fluid as lubricant.

Bearing and thrust washer service

All transmission bearings and thrust washers should be checked for wear or scoring. General procedures for checking bushings and washers were discussed in Chapter 2. Figure 13-44A shows the procedure for checking bushing wear in a sun gear. Figures 13-44B and 13-44C show the procedure used to replace a bushing. Note the use of special tools to remove the old bushing and install the new one.

There are many thrust washers in a modern automatic transmission. Some of these thrust washers are made in selective thicknesses to control end play. All thrust washers should be checked with a micrometer to ensure the proper end play. Use thicker or thinner washers as needed. For instance, if the end play was 0.030 in. (0.762 mm) too large, a 0.030 in. thicker thrust washer should be obtained. If end play was insufficient, a thinner selective thickness washer would be needed.

Miscellaneous case parts service

Many small parts are installed in the case, including electrical switches and other sensors, electrical case connectors, small screens in the pump output passages, seals, and check balls. These parts are as important as the major transmission parts. They must be cleaned and carefully checked for wear and damage. Electrical parts should be checked according to the manufacturer's service manual, using the proper electrical test equipment. Seals, such as

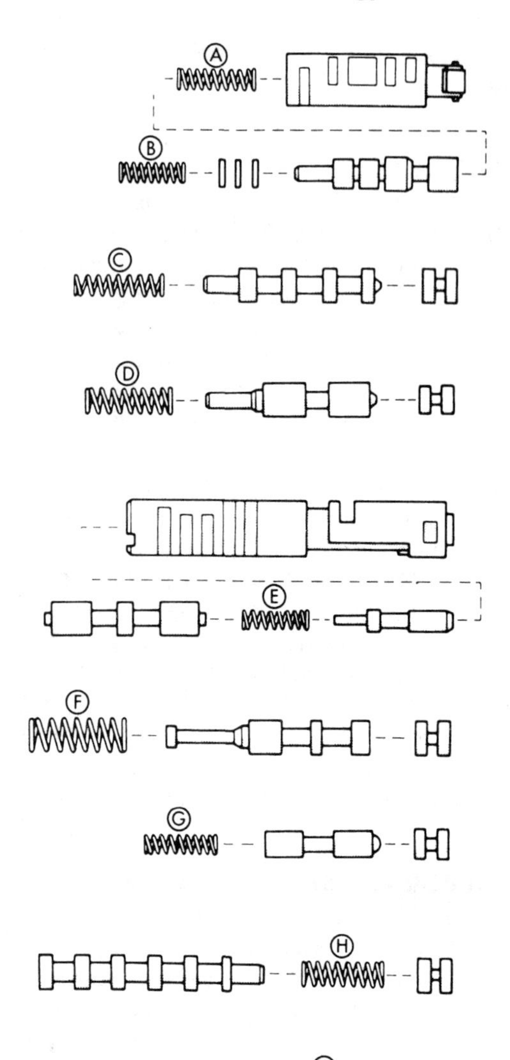

Spring	Free Length
(A) Downshift Plug	27.3 mm (1.074 in.)
(B) Throttle Valve	20.6 mm (0.811 in.)
(C) 3-4 Shift Valve	30.8 mm (1.212 in.)
(D) Second Coast Modulator Valve	25.3 mm (0.996 in.)
(E) Lockup Relay Valve	21.4 mm (0.843 in.)
(F) Second Regulator Valve	30.9 mm (1.217 in.)
(G) Cut-Back Valve	21.8 mm (0.858 in.)
(H) 2-3 Shift Valve	30.8 mm (1.212 in.)
(J) Low Coast Modulator Valve	27.8 mm (1.094 in.)

Figure 13-40. Exact replacement of the springs is critical to proper transmission operation. Spring lengths or number of coils can be compared to the service manual chart. Some springs are color coded. (Chrysler)

those used on the T.V. cable or manual lever and shaft assembly, should always be changed.

Automatic Transmission Assembly

Once all automatic transmission subassemblies have been rebuilt, they can be installed in the transmission case. Always refer to the manufacturer's service manual for the exact assembly procedures. It will give procedures and specifications for the exact type of transmission. Exploded views, such as shown in Figure 13-45, are very useful during assembly.

CAUTION! Remember to soak the friction discs, bands, and all rubber parts in fresh transmission fluid before installation.

Begin assembly with the rear of the transmission case. If the case contains a low-and-reverse clutch, and if not done prior to this, disassemble and clean it. Install new seals on the clutch apply piston. Install the piston and the friction discs and steel plates. Make sure the clutch clearance is correct and install the retaining snap ring. If the transmission uses a rear band, install the band in the case but do not install the band apply linkage. See Figure 13-46.

Install the rear planetary gearsets and output shaft. Be careful to align all parts before installing any snap rings or bolts. After the snap rings or bolts are in place, make sure the output shaft will turn.

If a transmission center support is used, install it now. Install the center support snap ring and/or bolts and make sure the output shaft will still turn.

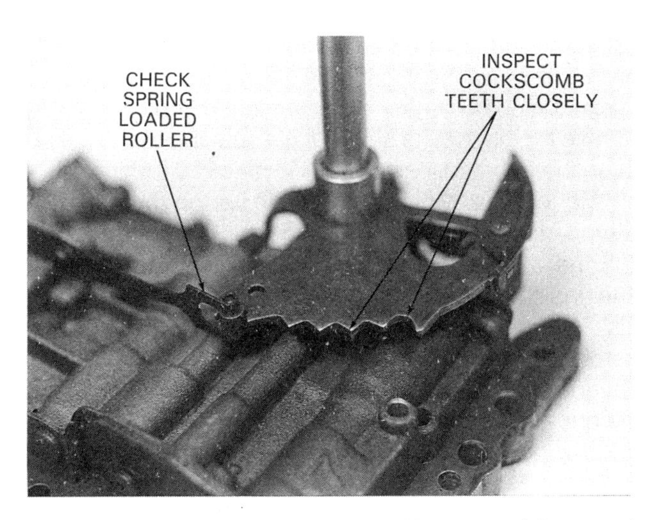

Figure 13-41. Check the inner manual lever carefully to make sure that parts are not worn and that it operates properly to select shift positions. (BBU, Inc.)

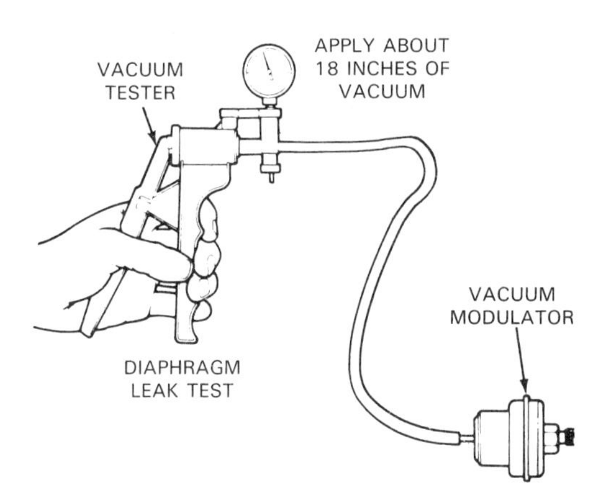

Figure 13-42. A hand-powered vacuum pump can be connected to a vacuum modulator to check the diaphragm. (Ford)

Many transmissions require that the clutches be assembled before they are installed in the case. Figure 13-47 shows direct and forward clutches fitting together as an assembly. The height of the assembly is being checked to ensure that it meets specification. With some transmissions, the clutches and front gearsets will be assembled prior to installation. See Figure 13-48. Install any such assembly at this point. Also, if the transmission has a front band, install it now.

If the transmission uses a set of clutch plates and friction discs directly behind the front pump, install them now, making sure to alternate discs and plates. Next, place the proper gasket on the case and install the front pump. As you slowly tighten down the pump fasteners, check that the input and output shafts can be turned by hand. If either shaft begins to bind, you must find the cause and correct it before proceeding.

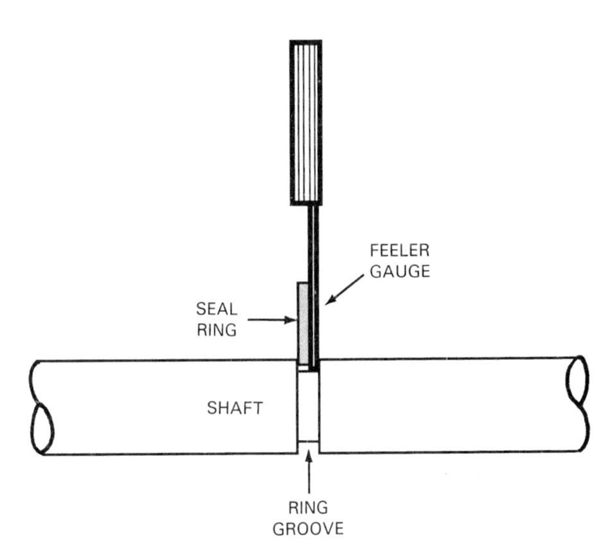

Figure 13-43. Seal rings and grooves are vital to proper operation of transmission hydraulic system. They should be carefully checked to ensure that they can do a good job of sealing. Using a feeler gauge as shown here, the maximum end clearance of a new seal ring in its groove should be about 0.003 in.–0.006 in. (0.07 mm–0.14 mm).

CAUTION! *Never* install a transmission on a vehicle if the shafts are binding. The transmission will be badly damaged.

Check the end play using a dial indicator, as shown in Figure 13-49. If the end play is not within specifications, the original selective fit thrust washer must be replaced.

If not done already, install the servos in the transmission case. Then, install the band apply linkage and adjust the bands to specifications. Some bands require special tools to perform the adjustment. Install the governor in the case or on the output shaft, as applicable, and install the extension housing. Install a new seal on the extension housing, Figure 13-50.

Before installing the valve body, perform an air pressure test on the transmission to make sure the clutch apply pistons and servos are working. Air pressure testing was discussed in Chapter 12. Figure 13-51 shows air pressure being used to check piston stroke on one kind of transmission.

After air pressure tests are complete, install the valve body-to-case spacer plate, the spacer plate gaskets, and any check balls in the transmission case. Carefully match the holes in the spacer plate and the new spacer plate gaskets to ensure proper fluid flow.

Install the valve body and a new oil filter. Then, install the transmission oil pan using a new gasket. Turn the transmission right side up in preparation to install the torque converter.

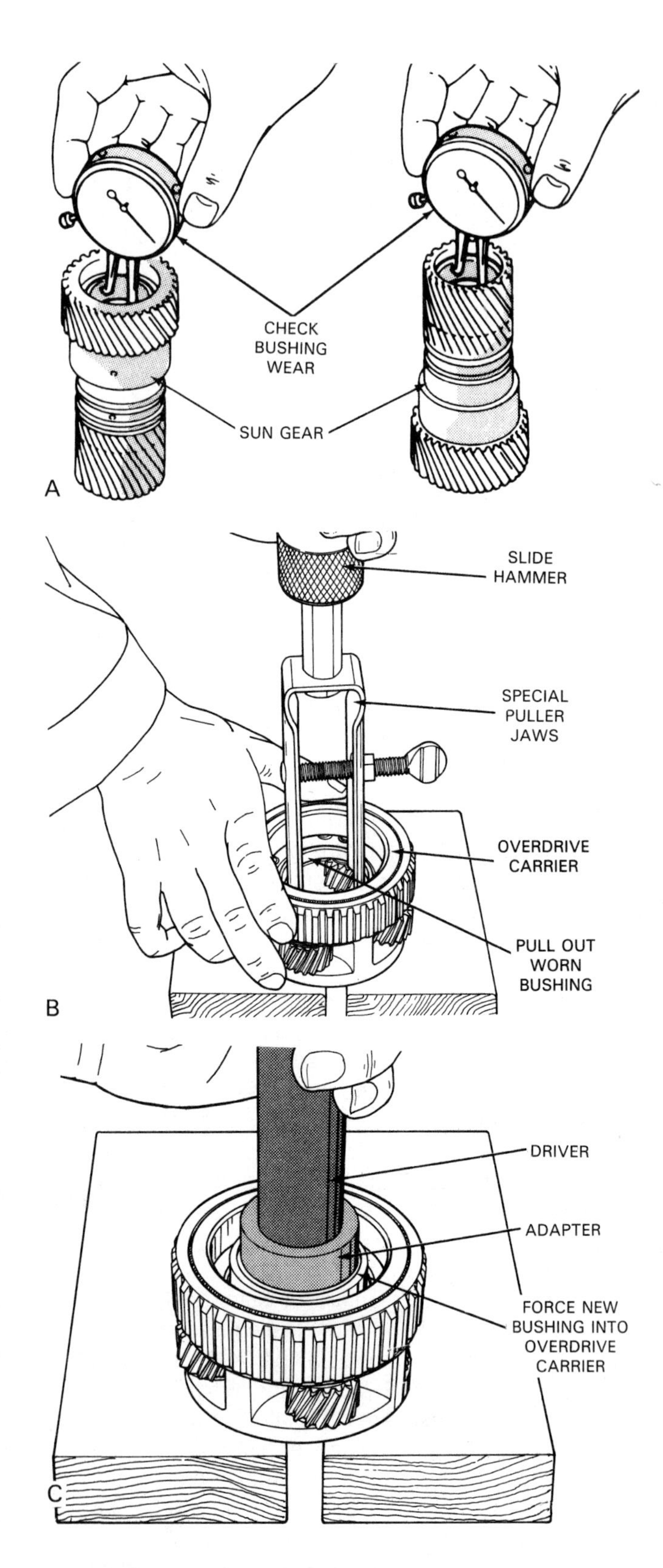

Figure 13-44. Check all bushings for wear and replace those that are worn. A–Checking a sun gear bushing. B–Removing a planet carrier bushing. C–Installing a new planet carrier bushing. (Chrysler, General Motors)

The converter can be checked prior to installation for proper end play as shown in Figure 13-52. The tool shown has an expanding pilot that can be tightened firmly into the splined hole for the turbine shaft. Make sure the tool is fully seated on the hub and lock the tool in place by turning the *threaded inner post.* Mount a dial indicator on the tool. A typical procedure calls for positioning the plunger so that it contacts the converter shell. With the indicator in place, zero the indicator. Then, pull up on the checking tool handles. The reading obtained is total end play of turbine and stator. End play must be within specifications; if not, the converter must be replaced.

The one-way clutch can also be checked. This can be done by placing a snap ring pliers down through the pump drive hub and expanding them into the internal splines on the one-way clutch. (These splines mate with external splines on the stator shaft.) Applying torque to the pliers should turn the stator easily in one direction and not so easily in the other.

If the converter passes the aforementioned tests, it is almost ready for installation. First, check the torque converter for wear at the hub and install about 2 1/8 qt. (2 L) of fresh transmission fluid. See Figure 13-53.

Next, place the torque converter over the input shaft and push it into the converter housing. The internal splines of the converter will move into contact with the splines of the stator shaft and input shaft. It may be necessary to wiggle the converter while pushing on it to get it to engage with the transmission shafts. The converter may also need to be rotated to line up the lugs on the pump drive hub with the slots on the front pump. The lugs should fully engage the front pump. To verify that the torque converter is properly installed, some manufacturers recommend checking it with a measuring rule and a straightedge, Figure 13-54. The reading is then compared against factory specifications.

CAUTION! Under no circumstances should you install the converter on the flexplate while it is disassembled from from the rest of the transmission. The splines of the transmission shafts and converter can be damaged.

Automatic Transmission Installation

Transmission installation is basically the reverse of removal. As with removal, the vehicle should be on a hoist with the engine supported. Following are some *general* procedures detailing how to install an automatic transmission in a vehicle. Note that you should always refer to the manufacturer's service manual for specific procedures.

1. Secure the torque converter with a torque converter holding tool or a C-clamp. Lubricate the converter pilot hub and crankshaft bore with multipurpose grease.

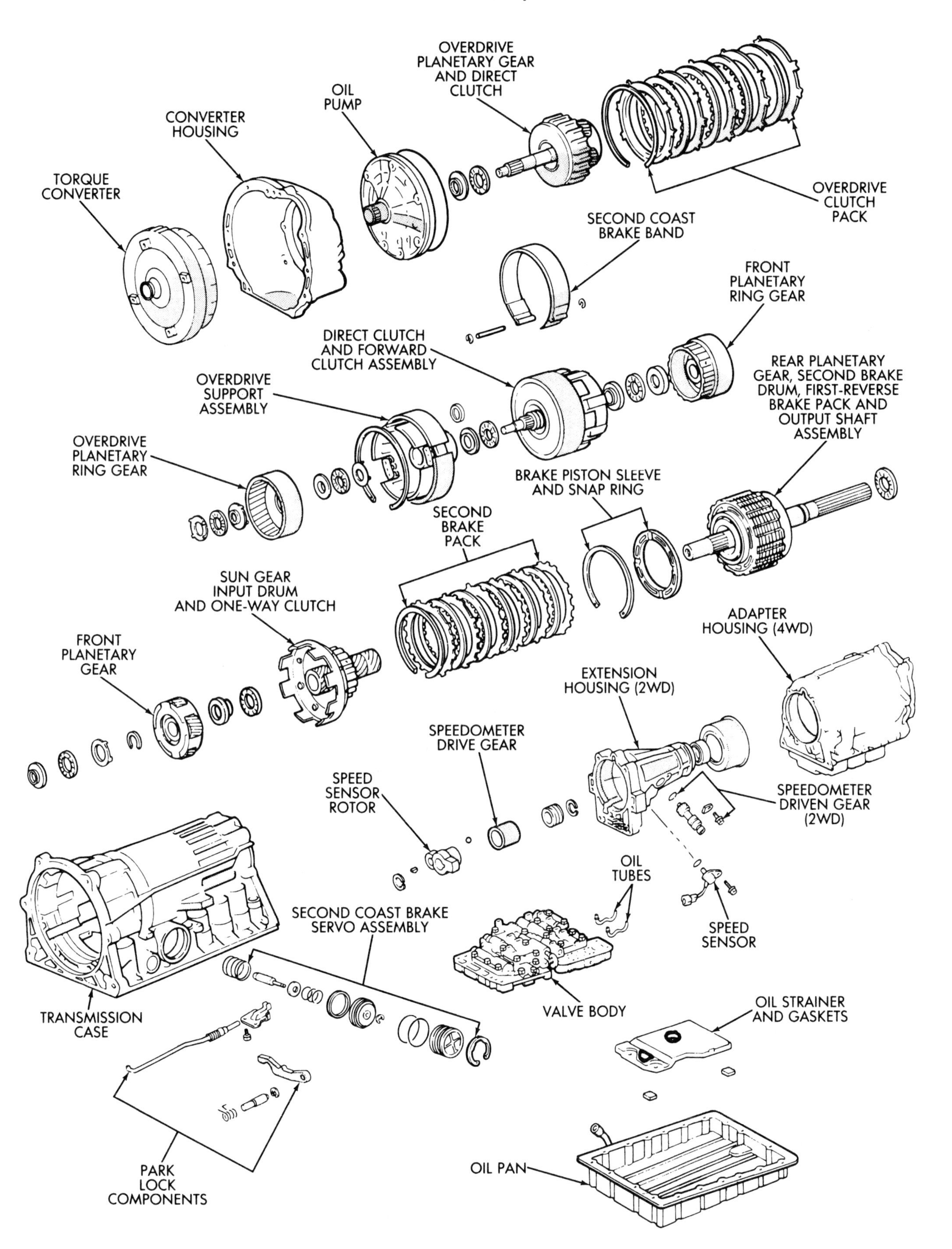

Figure 13-45. Note some of the parts in a typical automatic transmission. All transmissions should be carefully assembled, or they will not operate properly. (Chrysler)

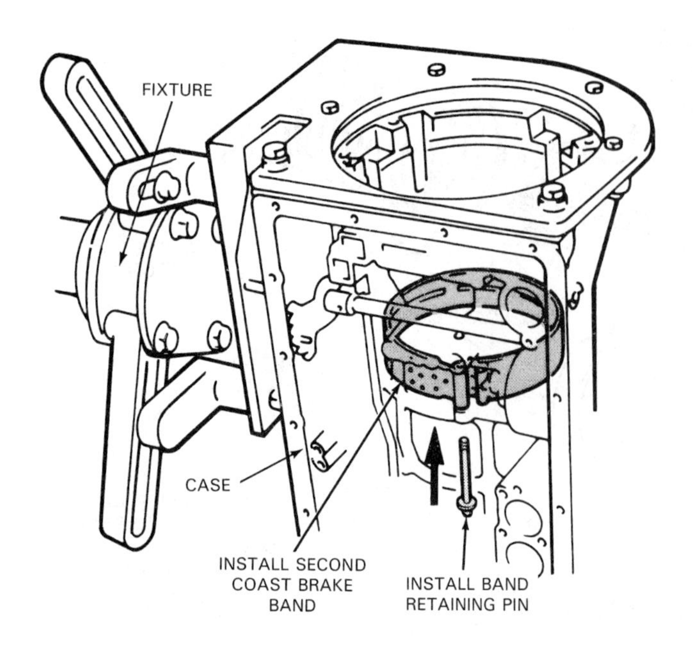

Figure 13-46. This shows the location of a band in the case on one type of transmission. In most instances, the band should be installed before installing the drum or gearset member that it will ride on. (Chrysler)

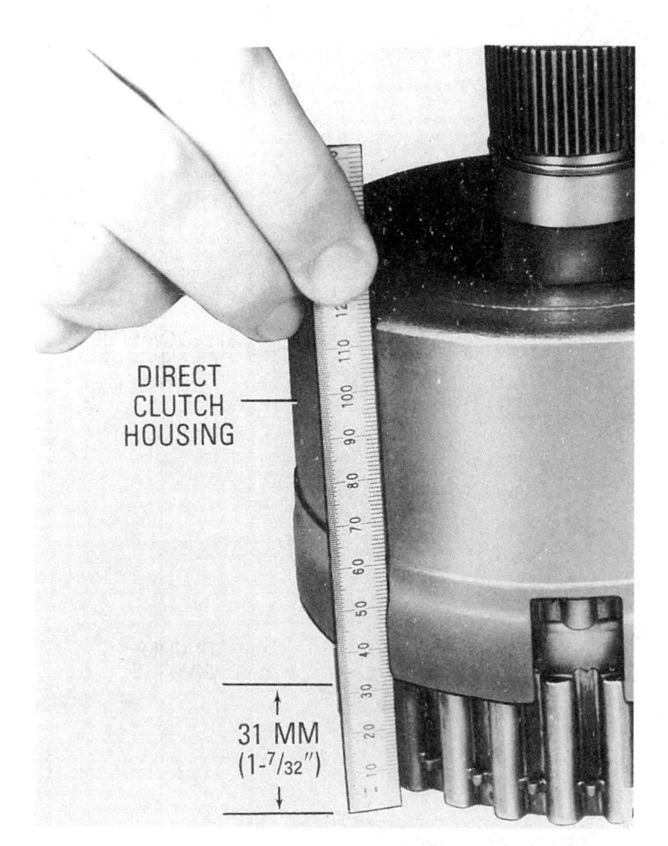

Figure 13-47. It is often necessary to assemble the forwardmost clutches and to check the height of the assembly. The height check ensures that all parts are properly installed. (General Motors)

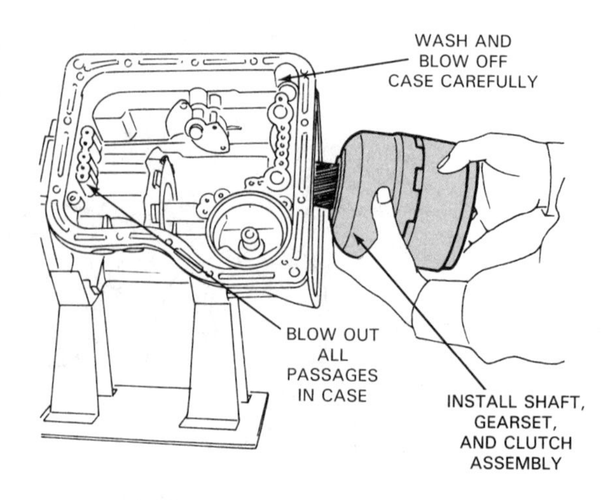

Figure 13-48. A clutch and gearset assembly is shown being installed into the transmission case. Check that parts are properly seated before proceeding with measurements. (Nissan)

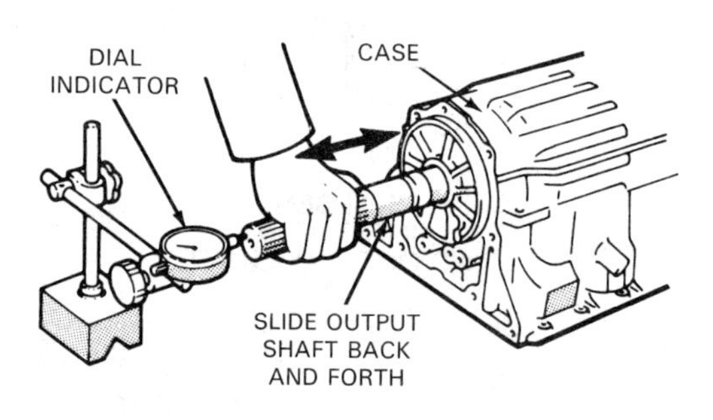

Figure 13-49. After assembling all of the drive train parts, recheck end play to ensure that you have installed the proper selective thickness thrust washer. Also, make sure that both shafts turn easily. (Chrysler)

2. Raise the transmission with a transmission jack or floor jack to the level of the engine. Then, carefully work the transmission forward until the converter pilot hub enters the crankshaft bore and the converter housing snugs up to the rear of the engine. (Remove the converter holding tool when necessary.) The engine often has alignment dowels to aid in installation. If the torque converter is held by studs and nuts, align the studs with the proper holes in the flexplate.
3. Once the converter housing is flush with the rear of the engine, install at least two of the converter housing attaching bolts.

After the bolts are installed, make sure the torque converter is not jammed. If it is, it is not properly aligned or engaged with the front pump. The converter should turn freely, and there should be a slight clearance between it and the flexplate.

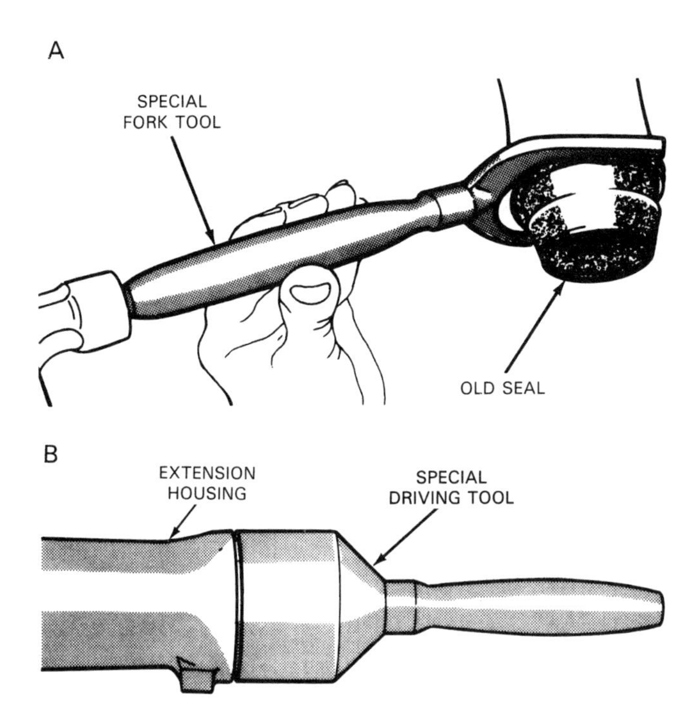

Figure 13-50. After installing the extension housing, install a new rear seal. Worn extension bushings should also be replaced. A–Removing the old seal with a special service tool. B–Installing a new seal with the proper driver. (Chrysler)

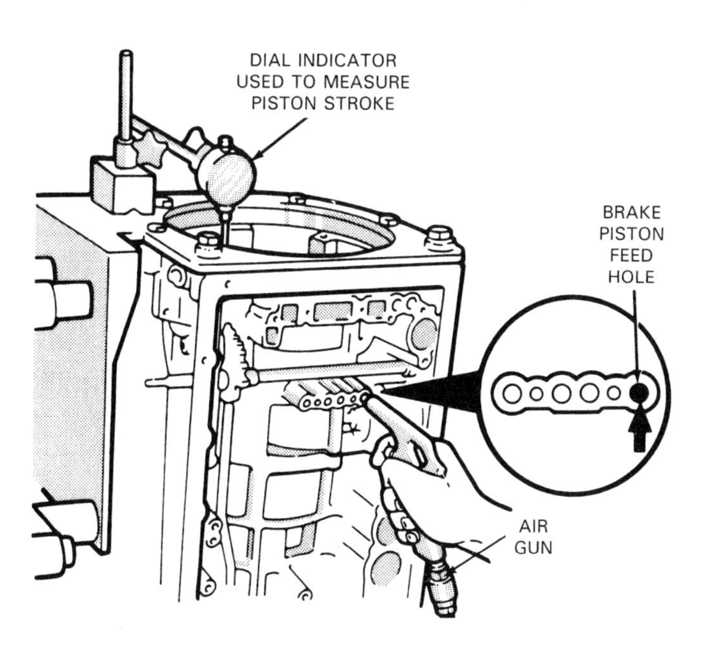

Figure 13-51. Always perform an air pressure test on the transmission before installing the valve body. Band movement or a sharp *thunk* from the clutch packs indicates that the air pressure is operating the holding members properly and that all seals and passages should channel fluid properly. (Chrysler)

If the converter installation is proper, install the remainder of the converter housing-to-engine bolts. Torque all bolts to specifications.

4. Install and tighten the two lower converter cover attaching bolts (or nuts) to specifications. Rotate the torque converter and install the remaining bolts.

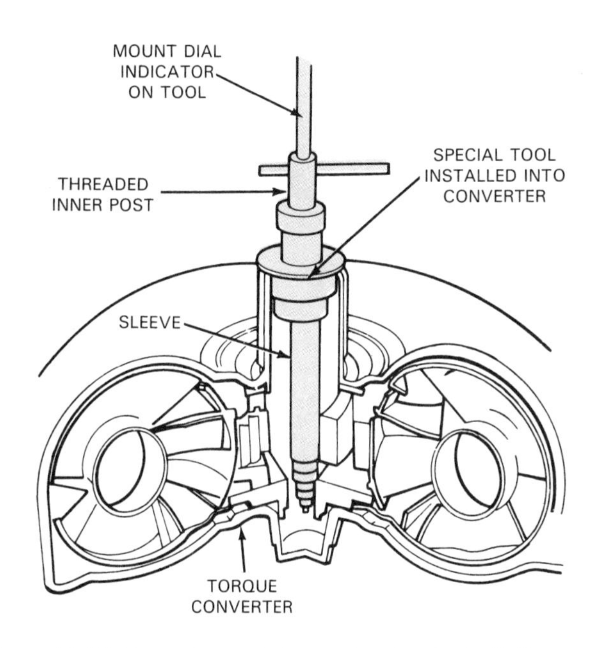

Figure 13-52. Converter end play can be checked with the special service tool shown here. A rough check can sometimes be made by inserting a pair of snap ring pliers into the splined turbine hub and pulling. Observing the plier movement will show how much end play there is. Operation of the stator one-way clutch can also be checked. (Ford)

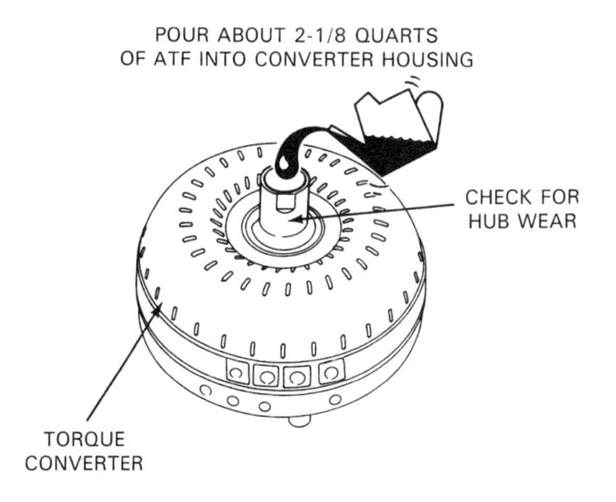

Figure 13-53. Once the torque converter has been cleaned, check it out before installing it. The pump drive hub can be checked for wear, and the internal operation of the converter can be checked. Add measured amount of transmission fluid to torque converter. (Nissan)

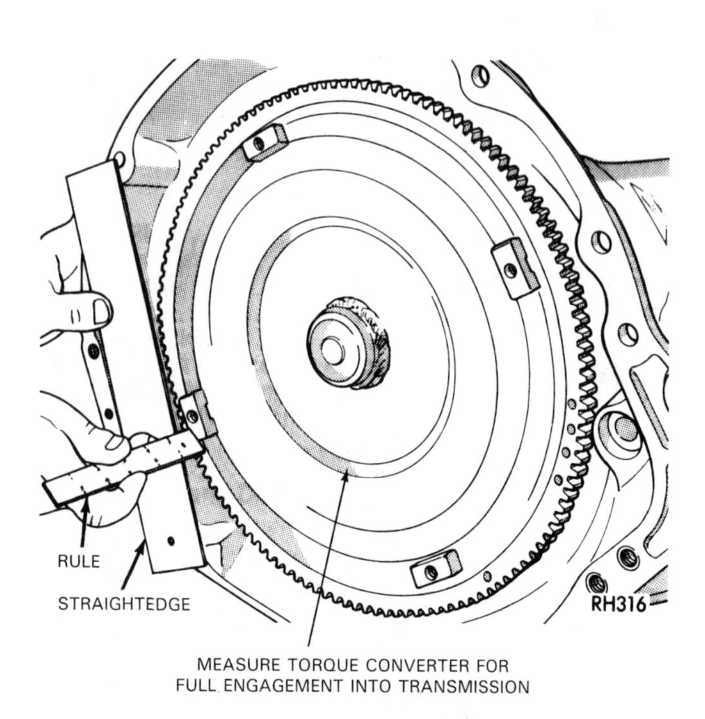

Figure 13-54. After installing the torque converter, ensure that it is completely seated on the input shaft and stator shaft. Also, the lugs on the pump drive hub should be fully engaged in the front pump. Measuring the installed depth of the torque converter with a rule and straightedge, as shown, helps verify that installation is correct. (Chrysler)

5. Raise the rear of the transmission and position the crossmember under the extension housing. Then, install the attaching bolts for the crossmember to the extension housing and the vehicle frame.
6. Install miscellaneous items, such as the parking brake linkage, oil cooler lines, modulator vacuum line, electrical connectors, speedometer cable assembly, and T.V., kickdown, and shift linkages. Install the exhaust pipes, if they were removed.
7. Lubricate the driveline slip yoke; then, install the driveline, making sure to line up the marks that were made before removal.
8. Lower the vehicle but leave the wheels off the ground so that the drive wheels are free to turn. Reinstall the battery negative cable. Check the vacuum lines at the engine intake manifold. Also, check the various linkages to ensure that they are properly adjusted.
9. Fill the transmission with the proper ATF. It is advisable to overfill a dry transmission by about 2 quarts (or 2 liters), *as measured on the dipstick,* before starting the engine.
10. Start the engine and immediately recheck the transmission fluid. Add fluid until the proper level is reached. A sucking noise from the oil filler tube is a good indication that more fluid is needed. Be careful not to overfill the transmission.
11. Place the shift selector lever in reverse and observe whether the wheels turn in reverse direction. Then, place the transmission in drive and allow it to upshift and downshift several times by operating the throttle pedal. This will check the transmission for proper shifting and will allow transmission fluid to lubricate the moving parts. Move the selector lever through all quadrant positions.
12. Lower the vehicle and recheck the fluid level. Perform a road test. Check operation in all gears and shift quadrant positions, noting any noises, presence of slippage, and improper shift points. Look for signs of leakage and recheck the fluid level.

    If it seems that the transmission is operating improperly, it can be diagnosed using the techniques covered in Chapter 12. If the shift points are incorrect, always check the T.V. and the throttle linkage or the vacuum modulator, as applicable. Also, check vacuum connections and engine condition on vehicles with vacuum modulators.

Summary

Although automatic transmissions are complicated, they can be fixed if you apply your knowledge of transmissions and work carefully. In many cases, the transmission does not have to be removed from the vehicle to perform repair operations.

Simple adjustments can often solve transmission problems. Linkages can be adjusted while the transmission is installed. Many bands can be adjusted from the outside of the case. Some transmission bands can be adjusted after the oil pan is removed. Vehicles should always be road tested after adjustments are made.

The first step to overhauling an automatic transmission is to remove the transmission from the vehicle. Automatic transmissions and manual transmissions are removed in a similar manner. The automatic transmission has more connections, and it is much heavier. Therefore, caution should be used when removing one. An approved transmission jack should be used if available.

Once the transmission is out of the vehicle, remove the torque converter and place the transmission case in a holding fixture or on a clean workbench.

Before beginning the disassembly procedure, make sure that you have the proper tools and a service manual that covers the transmission model to be overhauled. Clean the outside of the transmission thoroughly and check shaft end play. Then remove the oil pan, filter, and valve body. Inspect and save all small parts. After the valve body is off, the front pump and transmission internal parts can be removed. Removal of the parts will vary according to the transmission's internal construction and whether or not it has a center support.

Carefully note the condition of all parts as they are removed from the case. Then remove any other accessory parts such as servos, accumulators, and modulators from the case.

After disassembly, thoroughly clean all internal parts and make further inspections for wear or damage. The converter and oil cooler lines should also be cleaned. Determine what replacement parts will be needed and order them. Most parts that are usually replaced are provided in overhaul kits. Hard parts, such as pumps and gears, will have to be ordered separately. In some overhaul kits, certain hard parts, such as bands and vacuum modulators, are included.

The clutch packs are among the most critical parts of the transmission. They should be carefully inspected and serviced. Disassemble the clutch packs and inspect the plates for wear. Friction discs are usually replaced as part of an overhaul. Steel plates are often replaced.

Check bands for wear, cracking, or burning. Any defective bands should be replaced. Band servos should be disassembled, cleaned, and inspected. After installation of new seals, the servos are ready to be reinstalled. Accumulators are rebuilt in the same manner as servos.

The one-way clutches should be checked to ensure that they turn in one direction only. Planetary gearsets should be checked for obvious defects, such as chipped teeth. Also, check all associated bushings, bearings, and washers at all locations in the transmission.

Check the transmission shafts. Inspect input and output shafts for wear, bending, or warpage. Also check all of the case parts, such as screens, valves, and seals, for wear and damage. Seals should always be replaced.

To reassemble the transmission, start by soaking all new friction discs and bands in transmission fluid. Then, install the output shaft, rear planetary gearsets, clutches, bands, and other components at the rear of the transmission case. Remember to reinstall all snap rings in their original position. Then, install the center support, if the transmission is equipped with one.

Install the front planetary gearsets and clutches as complete assemblies, as required. Measure the assembled height to ensure proper assembly. Install any bands or case-mounted clutches. Then, install the oil pump. Check the transmission end play and make sure that both shafts turn.

Install the governor on the output shaft or in the case, as applicable. Then, install the extension housing and other output shaft parts.

Install the servos and adjust the bands. Also install any case-mounted accumulators, electrical switches, and other case parts. Then air pressure test the transmission. Install parts such as the valve body, spacer plates, check balls, filter, etc., that are installed inside of the oil pan. Install the oil pan.

Check the converter for wear or internal damage. Then, install it on the transmission. Make sure that the converter is installed properly on the transmission shafts and that the pump drive hub is fully engaged in the front pump.

Installing the transmission is the reverse of removal. Make sure that the converter is not jammed after the converter housing is tightened to the engine. Make all linkage connections and adjust them as needed. Lower the vehicle, leaving the rear wheels free to turn.

Add fluid to the transmission and start the engine. Add additional fluid until the dipstick reading shows a normal level. Then, operate the transmission through all gears with the wheels off the ground. Observe transmission shifting, listen for noises, and look for leaks. If the transmission seems to operate properly, take the vehicle for a road test and observe operation. Make all final adjustments and recheck linkage and vacuum connections.

Know These Terms

Soft parts; Overhaul kits; Hard parts; Clutch clearance.

Review Questions–Chapter 13

Please do not write in this text. Place your answers on a separate sheet of paper.

1. Replacing a valve body may be done as an in-car service procedure. True or false?
2. Problems resulting from an overfilled transmission are entirely different from those resulting from an underfilled transmission. True or false?
3. When removing a transmission, what can be done to gain access to the converter cover attaching bolts?
4. To prevent injury, make sure that none of the _____ _____ are broken before removing the crossmember.
5. When rebuilding an automatic transmission, you should work on one subassembly at a time, so that parts do not get mixed up. True or false?
6. What can be learned from inspection of the oil pan and its contents?
7. It is usually easier to remove the front pump seal after removing the pump. True or false?
8. Many servos and accumulators are under strong spring pressure, and caution should be observed during disassembly. True or false?
9. Describe what a transmission component having varnish formation would look like.
10. Gaskets, seals, and friction discs are examples of _____ parts.
11. Torque converters, transmission shafts, and planetary gearsets are examples of _____ parts.

12. When polishing a valve, you should try to round the edges because sharp edges tend to cause the valve to get hung up in the bore. True or false?
13. Seal rings should always be replaced during a transmission rebuild. True or false?
14. What visual signs indicate that a clutch pack has been slipping?
15. If clutch clearance is too large, the clutch pack may not apply tightly, and it may slip. True or false?

Certification-Type Questions–Chapter 13

1. Technician A says that a malfunctioning automatic transmission always requires a complete overhaul. Technician B says that any dirt or metal particles left in the transmission after overhaul will eventually reach the filter and will not cause any problems. Who is right?

(A) A only
(B) B only
(C) Both A & B
(D) Neither A nor B

2. The most common automatic transmission hydraulic system problem is:

(A) low pump output.
(B) internal leaks.
(C) sticking valves.
(D) low fluid level.

3. To drain a transmission pan without a drain plug, the technician should remove all pan bolts EXCEPT:

(A) one at each corner.
(B) the rear four.
(C) the front four.
(D) two on each side.

4. Technician A says that some shift cables can be damaged if the engine ground straps are disconnected. Technician B says that the throttle linkage on some vehicles with modulators is used only for passing gear. Who is right?

(A) A only
(B) B only
(C) Both A & B
(D) Neither A nor B

5. Technician A says that the band adjustment screw must be tightened against the drum and, then, backed off a certain number of turns to properly adjust the band. Technician B says that the band adjustment screw must be held as the locknut is tightened. Who is right?

(A) A only
(B) B only
(C) Both A & B
(D) Neither A nor B

6. The automatic transmission must be removed from the vehicle to service all of these components EXCEPT:

(A) the oil pump.
(B) the valve body.
(C) clutch packs.
(D) thrust washers.

7. Technician A says that the torque converter can be left installed on the engine when the transmission is removed. Technician B says that the torque converter can be installed on the engine flywheel before the transmission is reinstalled. Who is right?

(A) A only
(B) B only
(C) Both A & B
(D) Neither A nor B

8. Technician A says that end play should be checked before the transmission is disassembled. Technician B says that a few aluminum particles in the oil pan do not indicate a severe transmission problem. Who is right?

(A) A only
(B) B only
(C) Both A & B
(D) Neither A nor B

9. All of these tools can be used to remove the front pump from the transmission case EXCEPT:

(A) a special puller.
(B) a slide hammer.
(C) a hammer and punch.
(D) a pry bar.

10. Technician A says that center supports may be held in place by one or several bolts or by a retaining ring. Technician B says that to aid disassembly, you should remove the rear clutches, gearsets, and shaft assemblies while the center support fasteners are in place. Who is right?

(A) A only
(B) B only
(C) Both A & B
(D) Neither A nor B

11. All of these automatic transmission parts are normally replaced as part of an overhaul EXCEPT:

(A) gaskets.
(B) seal rings.
(C) friction discs.
(D) torque converters.

12. Technician A says that transmission hard parts are usually replaced during an overhaul. Technician B says that a transmission overhaul kit usually contains all of the needed soft parts. Who is right?

(A) A only
(B) B only
(C) Both A & B
(D) Neither A nor B

13. Technician A says that new friction discs should be soaked in clean transmission fluid before being installed. Technician B says that compressed air is used to check clutch pack clearance. Who is right?

(A) A only
(B) B only
(C) Both A & B
(D) Neither A nor B

14. The usual reason for servicing band servos is to replace:

(A) springs.
(B) pistons.
(C) seals.
(D) snap rings.

15. Technician A says that a one-way clutch that turns in both directions can be reused. Technician B says that a one-way clutch showing any signs of wear should be replaced. Who is right?

(A) A only
(B) B only
(C) Both A & B
(D) Neither A nor B

16. Technician A says that all valves should be checked for free movement in the valve body. Technician B says that valves should be removed and reinstalled one at a time to reduce the possibility of mixing up valves or springs. Who is right?

(A) A only
(B) B only
(C) Both A & B
(D) Neither A nor B

17. Transmission end play may be corrected by replacing the:

(A) case gaskets.
(B) needle bearings.
(C) thrust washers.
(D) snap rings.

18. Technician A says that an air pressure test should be performed before the valve body is installed. Technician B says that the transmission end play should be checked before the valve body is installed. Who is right?

(A) A only
(B) B only
(C) Both A & B
(D) Neither A nor B

19. Operation of the stator one-way clutch:

(A) can be checked by applying air pressure to the torque converter assembly.
(B) can be checked by turning the clutch hub with snap ring pliers.
(C) can be checked by shaking the torque converter while listening for rattles.
(D) cannot be checked.

20. A newly overhauled transmission is being operated with the drive wheels off of the ground. A sucking noise is coming from the oil filler tube. Technician A says that the fluid level is low. Technician B says that the oil filter is plugged. Who is right?

(A) A only
(B) B only
(C) Both A & B
(D) Neither A nor B

Chapter 14

Driveline Construction and Operation

After studying this chapter, you will be able to:

- Explain the purpose of a drive shaft assembly.
- List the major parts of a typical driveline.
- Explain the function and operation of a slip joint.
- Explain the function and operation of a universal joint.
- Name and describe different types of universal joints.
- Discuss Hotchkiss drive and torque-tube drive.
- Describe the function of a center support bearing in a two-piece drive shaft.
- Briefly explain the concept of driveline angles.

On rear-wheel drive vehicles, the drive shaft assembly, or driveline, forms the connection between the transmission and the rear axle assembly. The driveline is a critical part of the drive train. This is because the power transfer occurs while the rear axle assembly moves up and down as the vehicle moves over the road surface. The driveline must transfer this power smoothly, and it must be strong enough to absorb the shock of rapid power and speed changes.

This chapter will discuss the fundamental principles of drivelines. The principles discussed here will be of use when servicing the driveline.

Drive Shaft Assembly

The driveline, Figure 14-1, consists of a slip yoke, front universal joint, drive shaft, and rear universal joint. The slip yoke connects to the transmission output shaft, and the rear universal joint connects to the differential through the differential pinion yoke or flange. An exploded view of a typical drive shaft assembly is shown in Figure 14-2. Notice

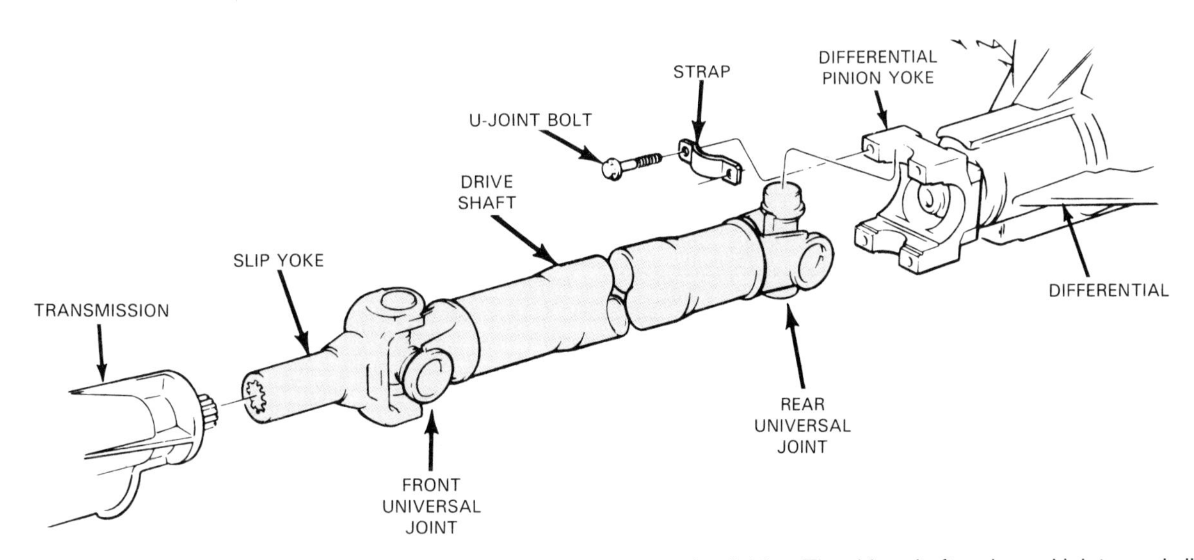

Figure 14-1. This is a typical modern driveline used on a vehicle with rear-wheel drive. The drive shaft, universal joints, and slip yoke are all part of the driveline. The transmission output shaft and differential pinion yoke are also shown. These are the front and rear attaching points for the drive shaft. Note that the shaft is hollow to reduce weight and lessen vibration problems. (General Motors)

that this drive shaft contains balance weights to reduce vibration.

Drive shafts come in many sizes. The length of the drive shaft varies between vehicles. Sometimes, vehicles of the same model will have different length drive shafts because various-size transmissions are used. Drive shaft length is measured from the center of each yoke hole on the drive shaft ends. See Figure 14-3.

On a modern vehicle, both front and rear axles are designed to move to keep the vehicle chassis level with the road surface. This is done to provide a smooth ride. On a rear-wheel drive vehicle, the entire rear axle housing and both rear wheels move up and down in relation to the chassis. The rear axle assembly is connected to the chassis by a suspension system that allows considerable movement, and the assembly moves a great distance over bumps and potholes.

The driveline transfers power from the transmission to the rear axle assembly. It must transfer this power as the rear axle assembly moves up and down, which changes the distance between the transmission and the assembly. The effective length of driveline must change to compensate for movement of the assembly; since the transmission is solidly attached to the frame of the car, the driveline must be able to flex or bend as the rear axle assembly moves. A simple drive shaft, by itself, would break. To provide flexibility, flexible joints are used.

The power flow through the driveline comes into the slip yoke from the transmission. From the slip yoke, the power flows into the front universal joint and, then, into the drive shaft. Power flows from the drive shaft into the rear universal joint. From here, it flows into the rear axle assembly.

Drive Shaft

The drive shaft, or propeller shaft, is the central part of the drive shaft assembly. The universal joints and other parts are attached to it. The drive shaft consists of a steel tube with yokes welded to or pressed on each end. The shaft has no moving parts. Sometimes, a two-piece drive shaft is used (explained in greater detail toward the end of the chapter).

The drive shaft yokes are the mounting points for the universal joints. The shaft is balanced when it is manufac-

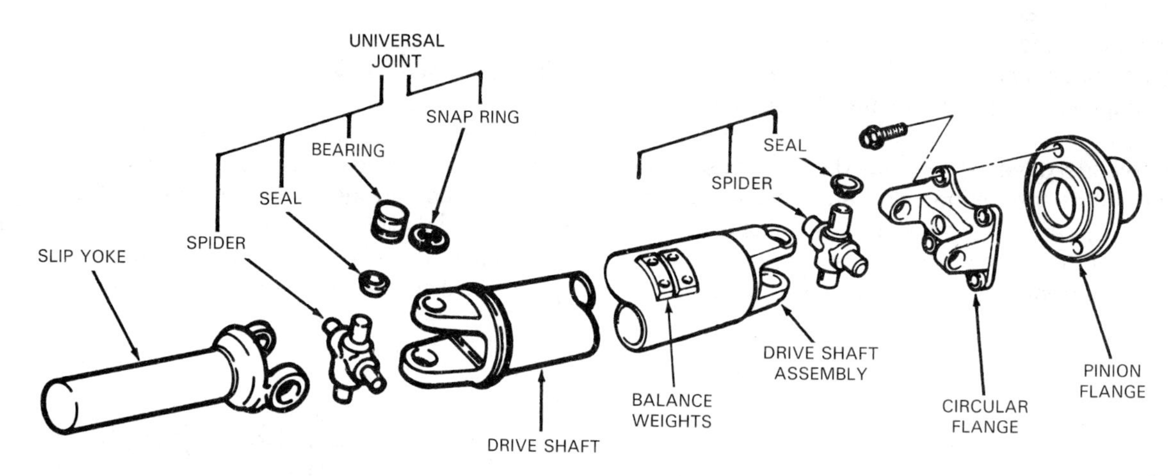

Figure 14-2. Exploded view shows parts of the driveline and the differential pinion yoke. (Ford)

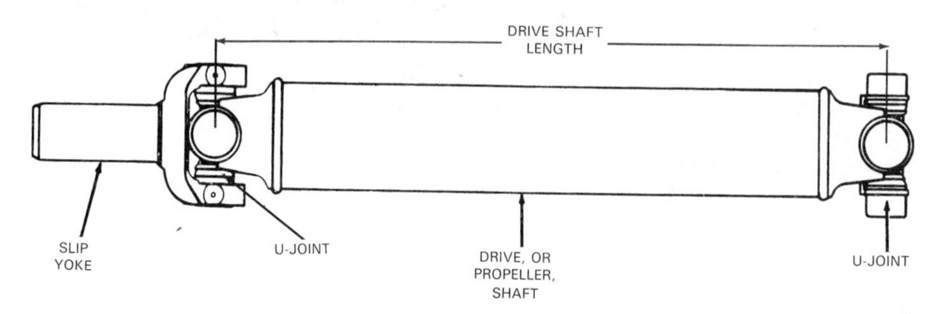

Figure 14-3. Drive shaft length is measured from the center of the front U-joint to the center of the rear U-joint. Drive shaft must be the proper length, or it will not work as it should. A short drive shaft could fall out when the vehicle hits a pothole and the rear axle assembly drops. An overly long shaft will bind when the rear axle assembly rises as the vehicle goes over a bump. (General Motors)

tured. In many cases, balance weights are welded to the shaft. Shafts used on some trucks and a few large cars may have heavy, vibration-damping rings attached to them.

Slip Yoke

As the rear axle assembly moves up and down as the vehicle goes over bumps, the effective distance between the transmission and rear axle assembly will vary. To allow the drive shaft to adjust to these changes in length, a slip yoke is used where the driveline enters the rear of the transmission.

The slip yoke is a steel tube, with a yoke at one end. The yoke attaches to the front universal joint. The internal part of the tube is splined to match external splines on the transmission output shaft. The outside diameter of the tube is machined smooth, Figure 14-4. It forms a close fit with the transmission extension bushing and rear seal. Figure 14-5 shows a weight, called *high inertia damper,* that is incorporated into some slip yokes to absorb vibration.

The relationship of the slip yoke to the transmission output shaft and extension housing is shown in Figure 14-6. In operation, the slip yoke slides on the transmission output shaft whenever the rear axle assembly moves and changes the effective distance between itself and the transmission. The action of the slip yoke causes the effective length of the driveline to change as necessary.

Universal Joints

Most rear-wheel drive vehicles use a swivel connection, or coupling, called a universal joint, or U-joint. Universal joints allow the driveline to transmit power smoothly and constantly even though the rear wheels and axle are moving up and down over bumps and dips. Manufacturers use several kinds of universal joints.

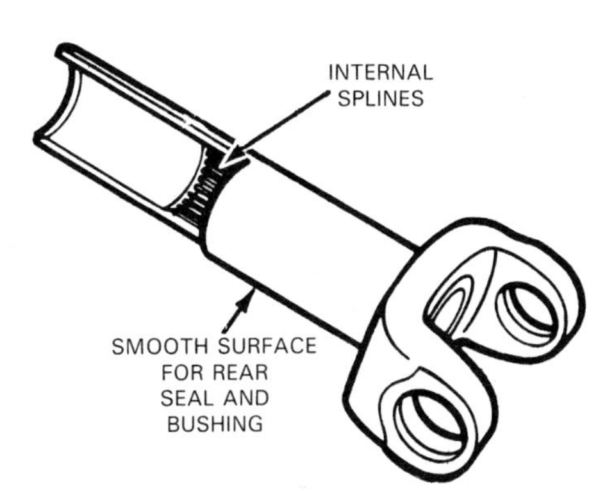

Figure 14-4. The slip yoke is designed to compensate for changes required in driveline length. It is made of hardened steel. One end of the yoke has internal splines that form a close fit with splines on the transmission output shaft. The outer surface is smooth to closely mate with transmission extension bushing. The other end has a yoke, which accepts the front universal joint. (Ford)

Cross-and-roller U-joints

Most rear-wheel drive vehicles use the ***cross-and-roller U-joint,*** which consists of a center *cross,* or *spider,* and four *cups* (or *caps*) containing needle bearings. The cups are machined to fit into the drive shaft yokes, the slip yoke, and the differential pinion yoke, Figure 14-7. The needle bearings allow the yokes to swivel on the cross ends with minimum friction. This kind of joint is sometimes called a ***Cardan U-joint,*** or ***conventional U-joint.***

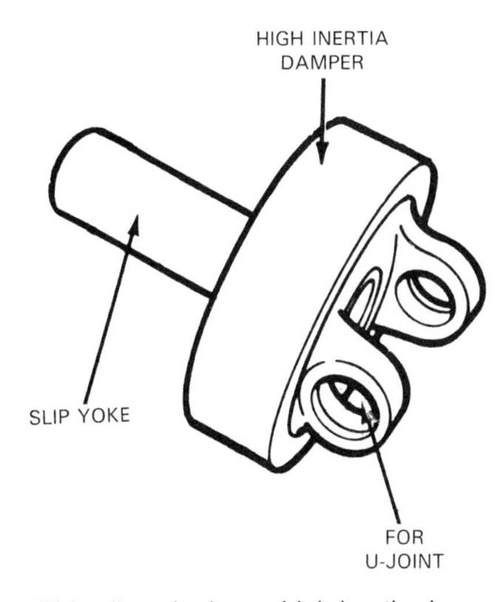

Figure 14-5. This slip yoke has a high inertia damper, made up of a heavy weight, to absorb vibration. (Ford)

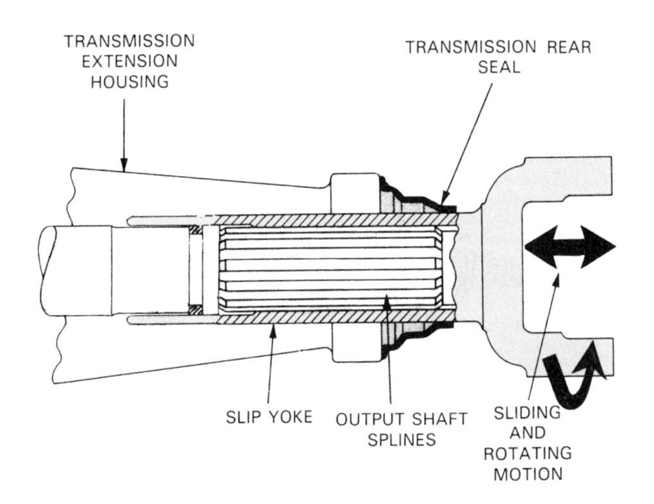

Figure 14-6. Study broken-out section to see how slip yoke installs into rear of transmission. Note that splines are splash lubricated with oil from transmission. Rear seal prevents leakage of transmission fluid past bushing (not shown) and slip yoke. (Ford)

Figure 14-8 is an exploded view of a typical cross-and-roller U-joint. The cross ends are called *trunnions.* Since the needle bearings ride on the trunnions, the trunnion surfaces are machined to a smooth finish.

A hole in each trunnion leads to the center of the cross, Figure 14-9. A grease fitting often connects to the center of the cross. Grease, injected through the fitting, flows through the holes in the trunnions to each needle bearing. Some universal joints are sealed at the factory and cannot be greased unless they are disassembled.

The cups may be held in the yokes by retainer snap rings or clips, U-bolts, bolted straps, or bolted covers. Some vehicles use injected plastic to hold them in place. Some of the different retaining methods are shown in Figure 14-10.

The ends of the cups contain seals that keep the grease in place and prevent the entrance of water and dirt. Figure 14-11 is a sectional view of a typical cup seal.

The action of the universal joint during operation of the vehicle is a two-way motion. The universal joint rotates with

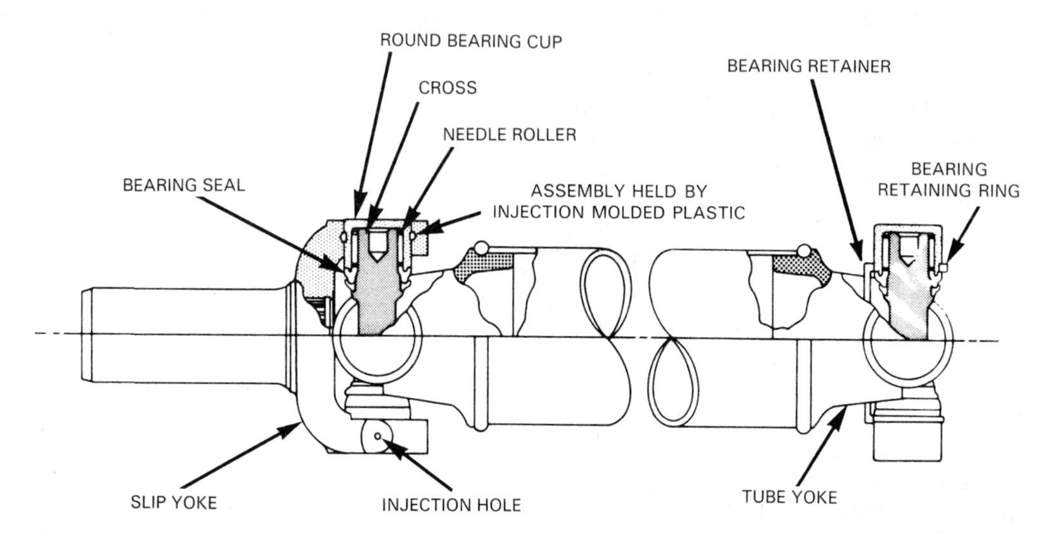

Figure 14-7. Study broken-out sections of front and rear universal joints. Notice how bearing cup and rollers relate to cross and yokes. In this particular installation, the front cups are held in the yokes with injected plastic. This is a common retaining method on many vehicles. To remove the U-joint, the plastic is heated and the cup is pressed out of the yoke. (General Motors)

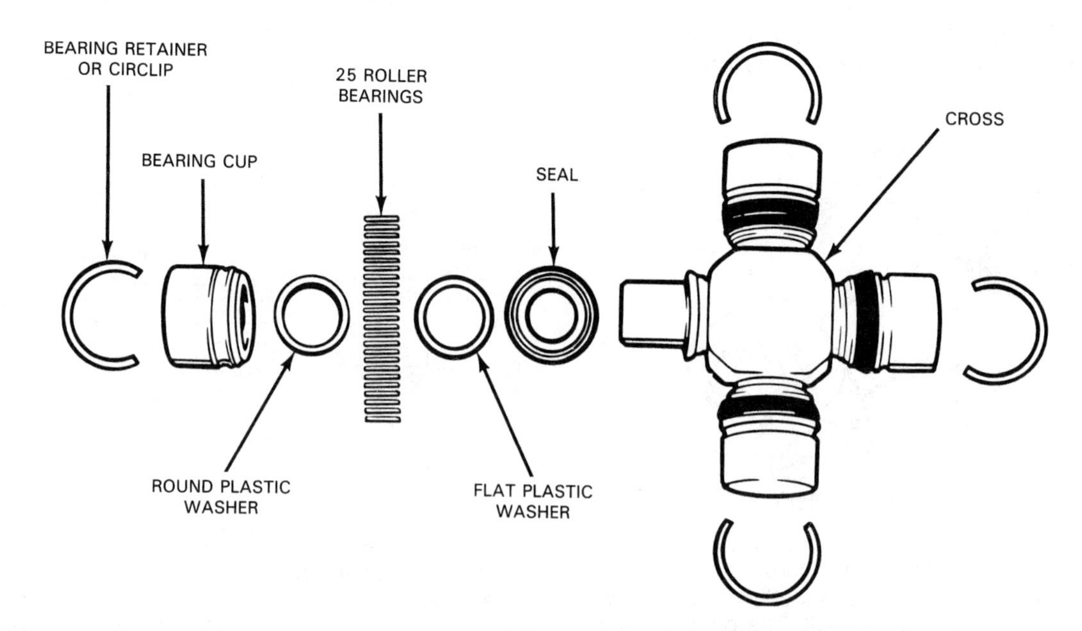

Figure 14-8. Note exploded view of a cross-and-roller type universal joint. Bearing cups are retained in their yokes by clips in this installation. Needle bearings and spacers are lubricated by heavy grease. The seal keeps lubricant in and keeps water and dirt out. (General Motors)

its driving shaft. It also swivels to compensate for changes in *driveline angles.* (Driveline angles are discussed later on in this chapter.)

The conventional U-joint is almost always driving at an angle. As a result of the angle, the speed of the drive shaft will rise and fall for any given transmission output shaft rpm. The drive shaft speed will rise and fall two times for every revolution, as shown in Figure 14-12.

The fluctuation in drive shaft speed is created at the front U-joint. It is cancelled at the differential pinion yoke by the rear U-joint if: (a) the U-joint at the output shaft and drive shaft and the U-joint at the drive shaft and differential pinion yoke form equal angles; (b) the U-joints are aligned *in phase.* When in phase, the ears on the front drive shaft yoke align with those on the rear drive shaft yoke, and those of the slip yoke align with the pinion yoke. Under these circumstances, the reaction at the rear U-joint will be opposite to that at the front U-joint. As a result, the drive shaft speed fluctuation produced by the front U-joint is cancelled by the rear U-joint. The end effect is a fairly constant rotation and a smooth transfer of power delivered to the differential drive pinion gear.

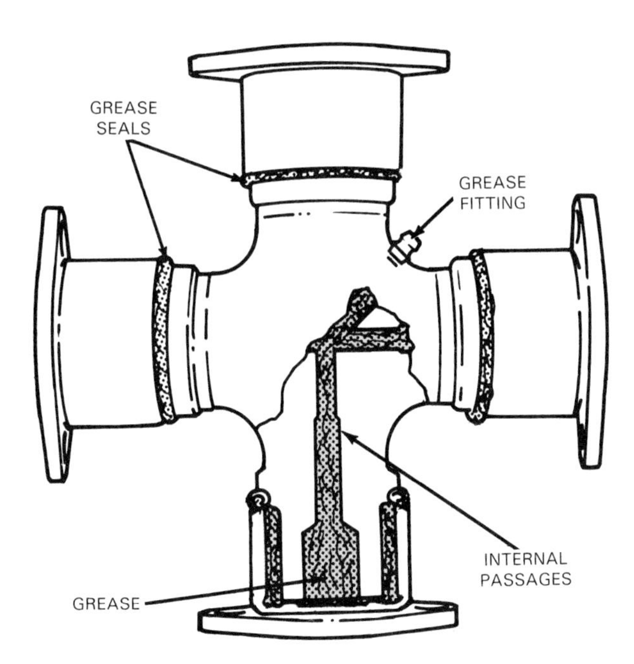

Figure 14-9. Internal passageways in this cross-and-roller universal joint act as a grease reservoir. They allow all of the cups to be greased from a single grease fitting. Grease travels from fitting to each cup and, then, to needle bearings. (Chrysler)

Constant-velocity U-joints

One disadvantage of the conventional U-joint is the uneven transfer of power through angles between drive and driven shafts, resulting in vibration, among other problems. Aligning the drive shaft yokes reduces the problem on most rear-wheel drive cars and trucks. However, some larger rear-wheel drive vehicles use a more complex arrangement to further reduce vibration.

Constant-velocity U-joints, or CV joints, provide an extremely smooth power transfer. However, they are more complicated than the conventional U-joint. There are several different types of CV joints.

Double Cardan U-joints. A CV joint used on some rear-wheel drive automobiles is the ***double Cardan U-joint,*** Figure 14-13. This joint is constructed with a cross-and-roller U-joint at each end of a special coupling yoke, called a *center yoke.* A central ball-and-socket connection acts as a centering device. The name of this CV joint comes from the two Cardan U-joints that are used. A driveline may use one double Cardan U-joint and one conventional U-joint, or it may use two double Cardan U-joints.

Figure 14-14 illustrates the action of the double Cardan U-joint. The construction of this joint allows a fluctuating output speed to be compensated for within the center yoke. This CV joint design splits the driveline angle equally between the two shafts. This is accomplished through the ball-and-socket centering device. The equal driveline angles offset each other, and speed changes at the output of the first cross-and-roller U-joint are counteracted by speed changes at the second.

Ball-and-trunnion U-joints. Another type of universal joint designed for constant velocity is the ***ball-and-trunnion U-joint.*** This design is not used on modern vehicles,

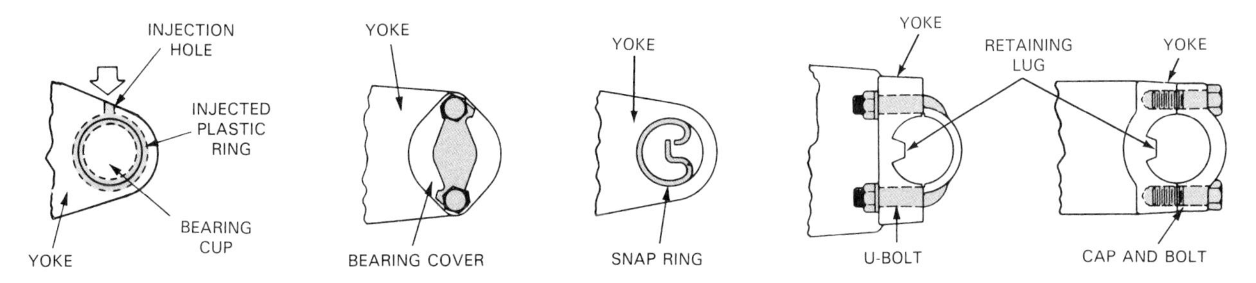

Figure 14-10. Different ways that a bearing cup can be retained are shown here. Injected plastic and clips or snap rings are the most common cup retainers used on automobiles. The retaining lugs shown (far right) are used to align the bearing cups in relation to the pinion flange. Note that bolted cups are found on truck universal joints. (Dana)

but it was popular on some older vehicles. This joint, in addition to allowing the driveline to bend, also allowed fore-and-aft (forward and backward) movement of the propeller shaft. Therefore, a separate slip yoke was not necessary when this type of joint was used.

Rzeppa joints. The Rzeppa joint is a type of CV joint that uses a series of ball bearings that are held by a slotted cross assembly. The balls are held in slotted channels and will rotate in any direction. The Rzeppa joint can transmit power through changing angles by allowing the balls to move up and down on the cross as torque is transmitted through them. This type of joint is often used on front-wheel drive vehicles. It is sometimes seen on drive axles of rear-wheel drive vehicles with independent rear suspension. These will be covered in greater detail in a later chapter.

Tripod joints. The ***tripod joint*** is a type of CV joint that is often used as the inner bearing on front-wheel drive vehicles. It resembles a ball-and-trunnion U-joint, but it uses a *three-pointed cross pin.* (This will also be explained in a later chapter in greater detail.) This joint is sometimes used on drive axles of rear-wheel drive vehicles with independent rear suspension.

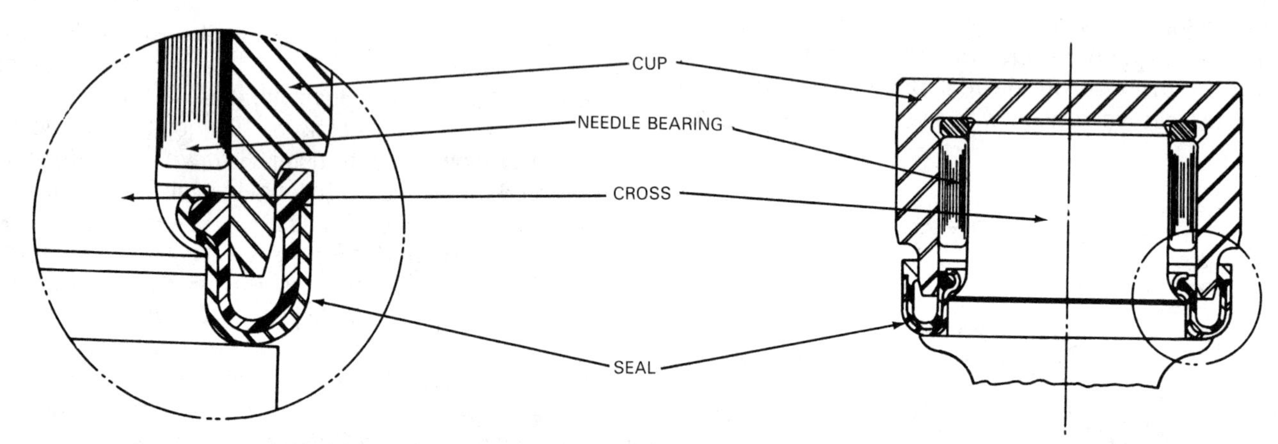

Figure 14-11. Construction of bearing cup seal keeps grease in and keeps dirt and water out. Seal can withstand slight internal pressure, but it will be blown out and ruined by overgreasing. (Dana)

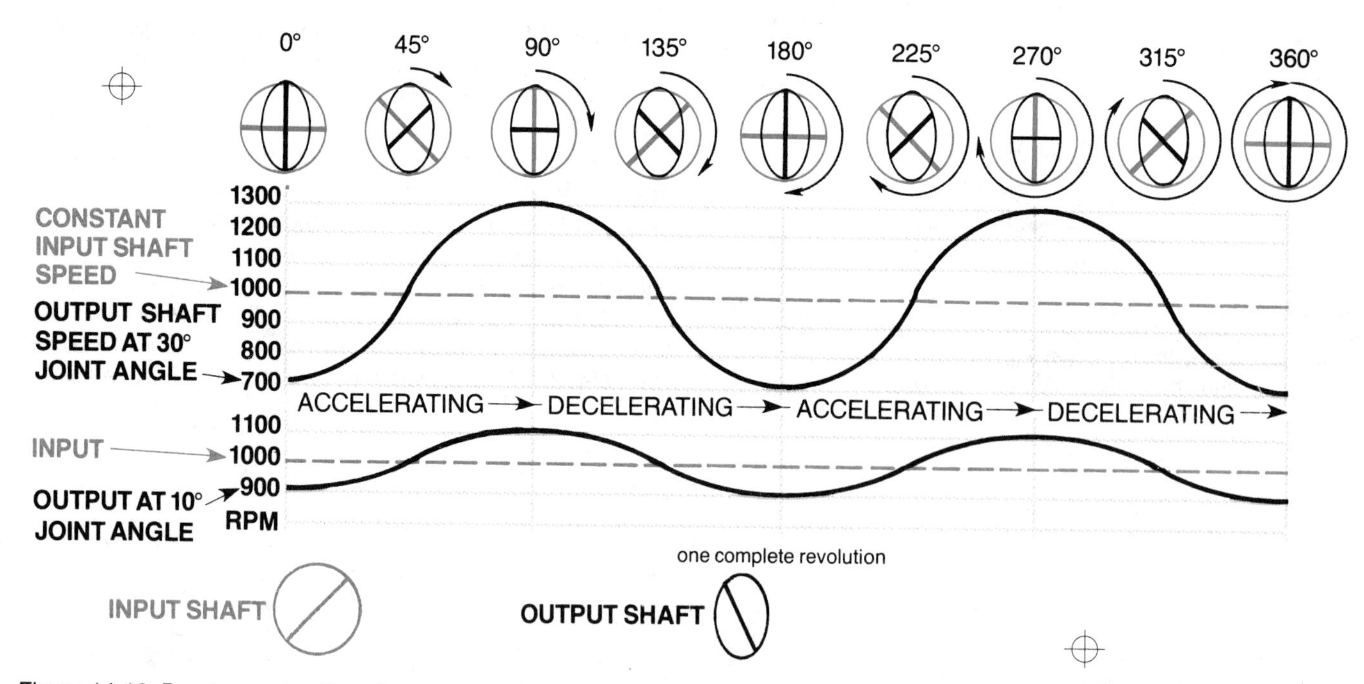

Figure 14-12. Due to construction of cross-and-roller universal joint, drive shaft (output shaft in this illustration) speeds up and slows down twice during every revolution. Range of speed fluctuation increases as joint angle increases. This may be seen by comparing output shaft speeds for 10° and 30° joint angles given in this illustration. Note that if joint angles are excessive, vibration may be felt in the passenger compartment. (Perfect Circle)

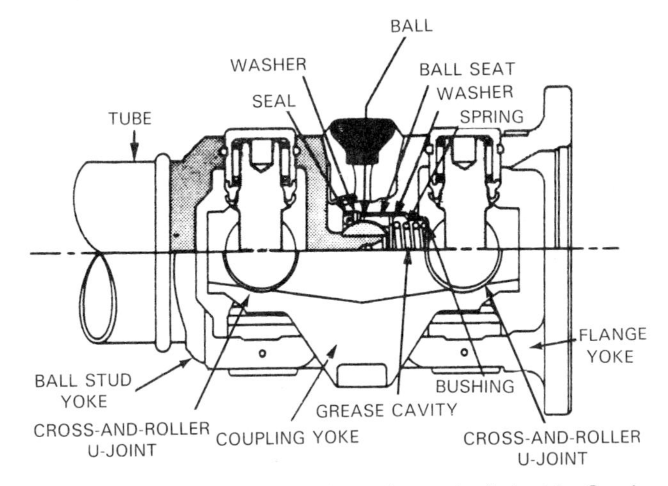

Figure 14-13. This detailed view of a typical double Cardan U-joint shows that it is simply two Cardan U-joints connected to each other through a center yoke. The universal joints and the ball-and-socket connection require periodic lubrication. (General Motors)

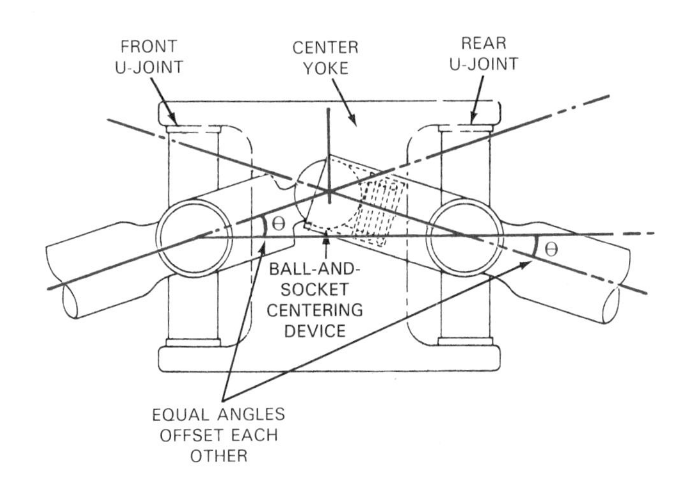

Figure 14-14. The construction of the double Cardan universal joint cancels out speed fluctuation in the driveline. Speed fluctuation caused by front universal joint is cancelled out by the rear universal joint because driveline angles (q in illustration) are equal. Centering ball keeps both shafts aligned, and all speed fluctuations occur in the center yoke. (General Motors)

Differential Pinion Yoke/Flange

The ***differential pinion yoke*** connects to the drive pinion gear in the rear axle assembly. The yoke has internal splines that fit external splines on the *drive pinion gear shaft.* The yoke is held to the shaft by a large locknut.

On some rear axle assemblies, the yoke is a two-piece assembly, consisting of two mating flanged sections. This particular design is referred to as a *differential* ***pinion flange,*** or a ***companion flange.*** The one flange holds the rear universal joint, and the other connects to the pinion gear. The arrangement attaches the rear universal joint to the pinion. The flange is purposely heavy to absorb shock and act as a vibration damper. The differential pinion yoke/flange is covered in more detail in Chapter 16.

Classes of Drivelines

Drivelines can be divided into two general classes. These are the *Hotchkiss drive* and the *torque-tube drive.* The Hotchkiss version is the most common. The torque-tube design was used on some makes of automobiles for many years. It is uncommon at present.

Hotchkiss Drive

Hotchkiss drive uses an *open shaft* (as opposed to an *enclosed* shaft) with two or more U-joints, Figure 14-15. Thus far, this type has been the focus of the chapter–a single shaft, a slip yoke and U-joint at the transmission, and a U-joint at the differential. The driveline can flex at both ends of the drive shaft–at the front and the rear universal joints. Hotchkiss drive is lighter, cheaper to produce, and easier to service. In late model vehicles, it is used almost exclusively.

A variation of Hotchkiss drive is shown in Figure 14-16. A constant-velocity U-joint is used. The slip yoke is at the rear of the shaft, instead of at the front where it usually is. This kind of slip yoke usually has one spline that is larger than the others. This is to ensure that the parts get reassembled in their original alignment.

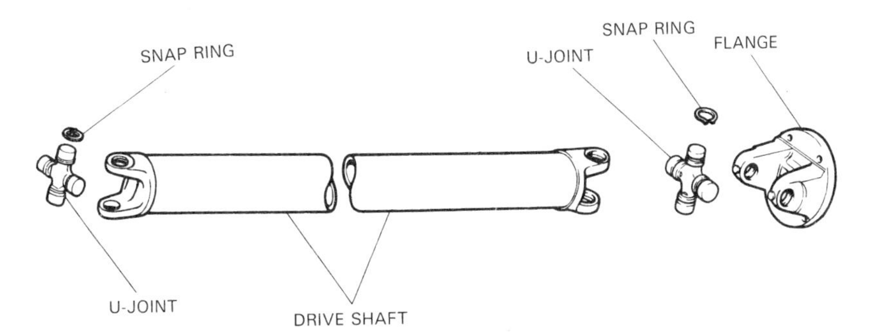

Figure 14-15. Hotchkiss drive uses this type of drive shaft. It is sometimes called an open drive shaft since it is not enclosed in a tube or housing. Most Hotchkiss designs have two universal joints, one at each end of the shaft. (Fiat)

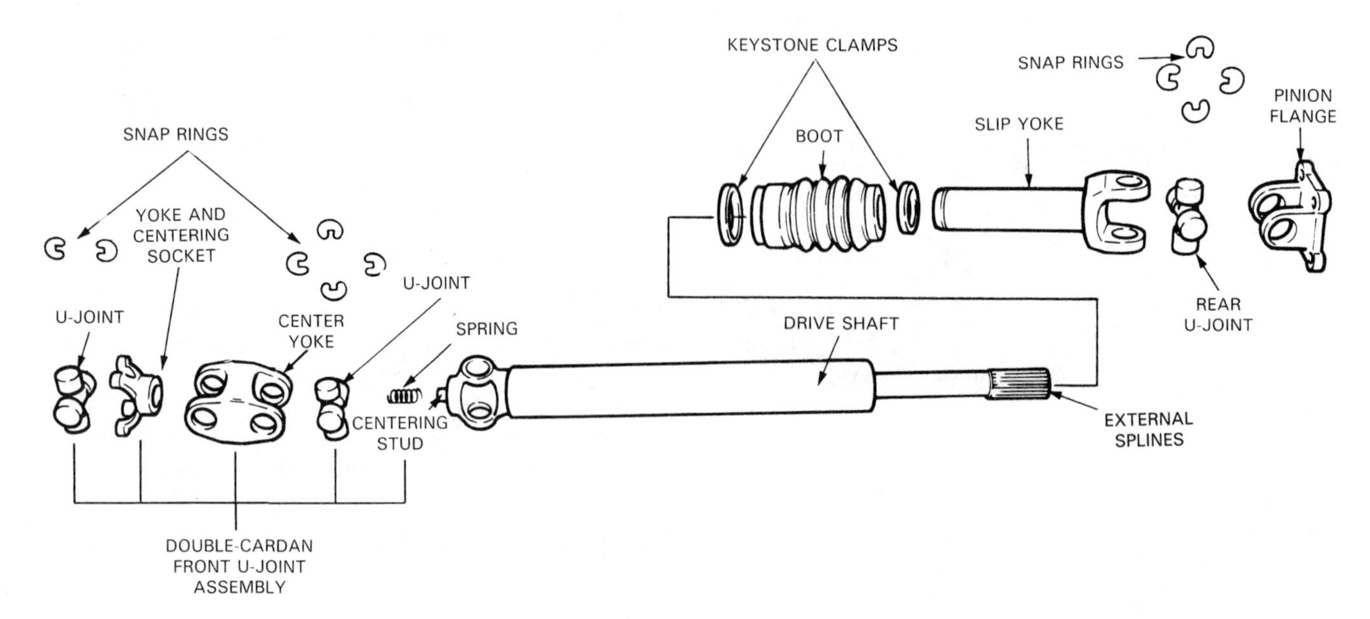

Figure 14-16. Note variation of Hotchkiss drive. It uses a CV joint at the front to reduce vibration. An unusual feature of this drive shaft is that the slip yoke is part of the drive shaft. (Ford)

Torque-Tube Drive

Torque-tube drive has a drive shaft that is solid steel and enclosed in a tube, designed to control *rear-end torque.* The tube is solidly connected to the rear axle assembly with bolts. In many designs, it is connected to the transmission through a ball joint, about which it pivots.

The steel drive shaft connects to the transmission output shaft through a universal joint. The other end of the shaft connects to the drive pinion gear through a coupling or splined connection. There is a universal joint at only one end. Often, it is a ball-and-trunnion U-joint.

The driveline can flex only at the front end of the drive shaft, at the U-joint between the transmission and drive shaft. The rear axle assembly moves up and down in an arc as it pivots about the U-joint, and the distance between the transmission and rear axle assembly is always the same. Since the drive shaft length does not have to change, a slip yoke is not needed.

Torque-tube drive was used on many vehicles. It was discontinued because the tube added extra weight to the vehicle, and because it was necessary to remove the tube and rear axle assembly to gain access to the universal joint or drive shaft.

Two-Piece Drive Shafts

A ***two-piece drive shaft*** is a drive shaft that is divided into two sections, with a universal joint between the front and rear sections. A type of two-piece drive shaft is shown in Figure 14-17. The two-piece drive shaft uses three universal joints. These universal joints are identical to those used with a one-piece drive shaft. With three universal joints, driveline angles are reduced at each universal joint compared to what they would be if only two were used. This reduces vibration.

Two-piece drive shafts were once widely used as a means of reducing driveline vibration. They were sometimes used when the design of the vehicle frame made it impossible to use a single drive shaft. Today, most two-piece drive shafts are found on large trucks and on some long-bed pickups.

The two-piece drive shaft is a variation of the Hotchkiss drive. The two shafts are steel tubes with yokes on all four ends. They resemble shorter versions of the single drive shaft. See Figure 14-18.

Two-piece drive shafts require a *center support bearing* to hold the rear of the front shaft and the front of the rear drive shaft, Figure 14-19 (and Figure 14-17). The inner race of the bearing is pressed on or solidly bolted to the front drive shaft. The bearing is usually greased and sealed at the factory. The outer race of the bearing is held by a thick rubber ring that is fastened to the vehicle frame or underbody by a bracket. The rubber ring acts as a cushion to absorb vibration.

The two-piece drive shaft must have a slip yoke. Since in many designs, the front shaft is kept from moving back and forth by the center support bearing, the slip yoke is usually installed behind the support bearing. Some two-piece drive shafts have two slip yokes.

Driveline Angles

All rotating parts have an ***axial centerline.*** This is the central axis around which the rotating parts revolve, or the

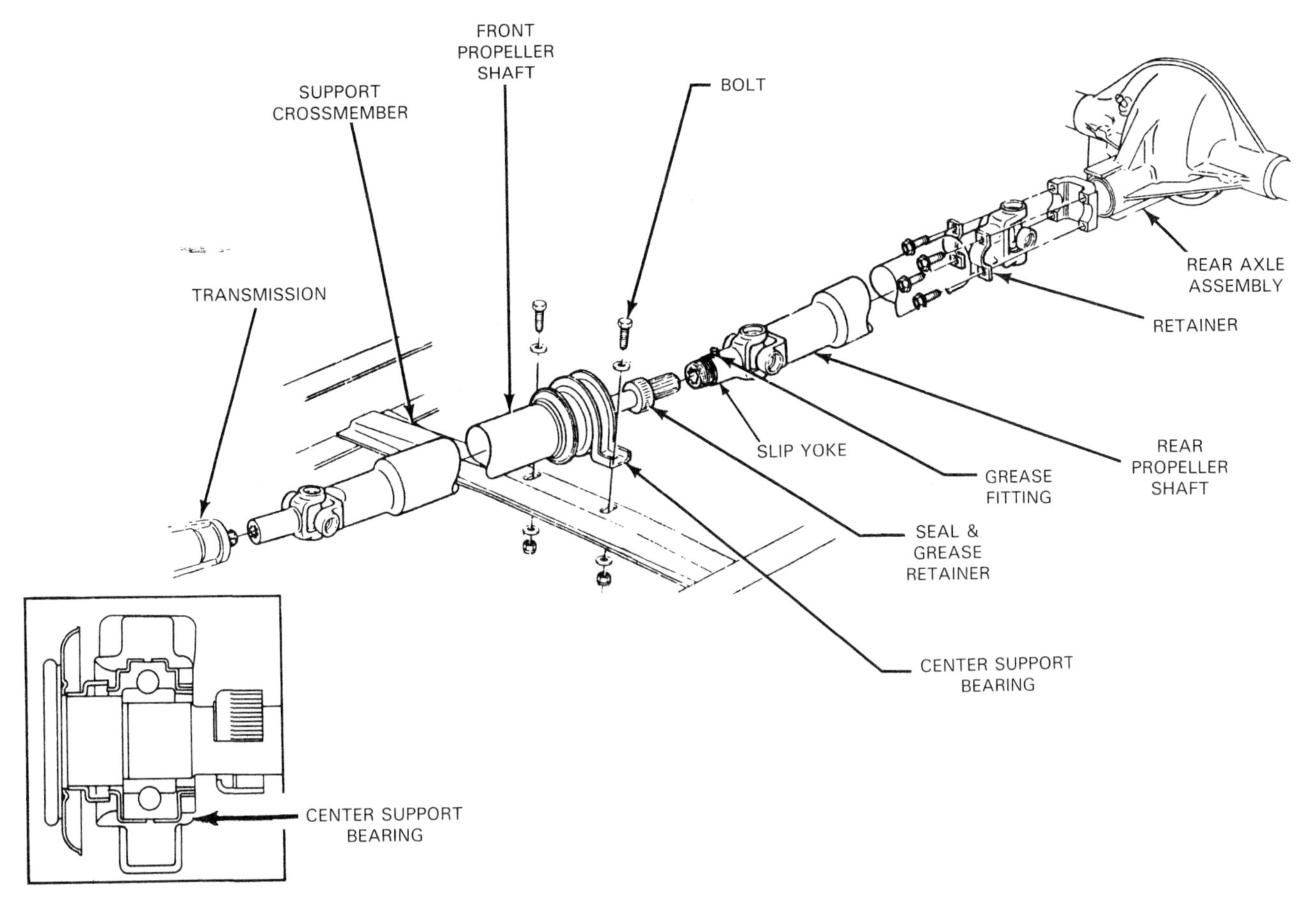

Figure 14-17. Two-piece drive shaft, shown here, consists of two drive shafts and three universal joints. Center support bearing keeps drive shafts aligned. This system was used on many cars in past to reduce vibration or because frame design prevented use of single drive shaft. (General Motors)

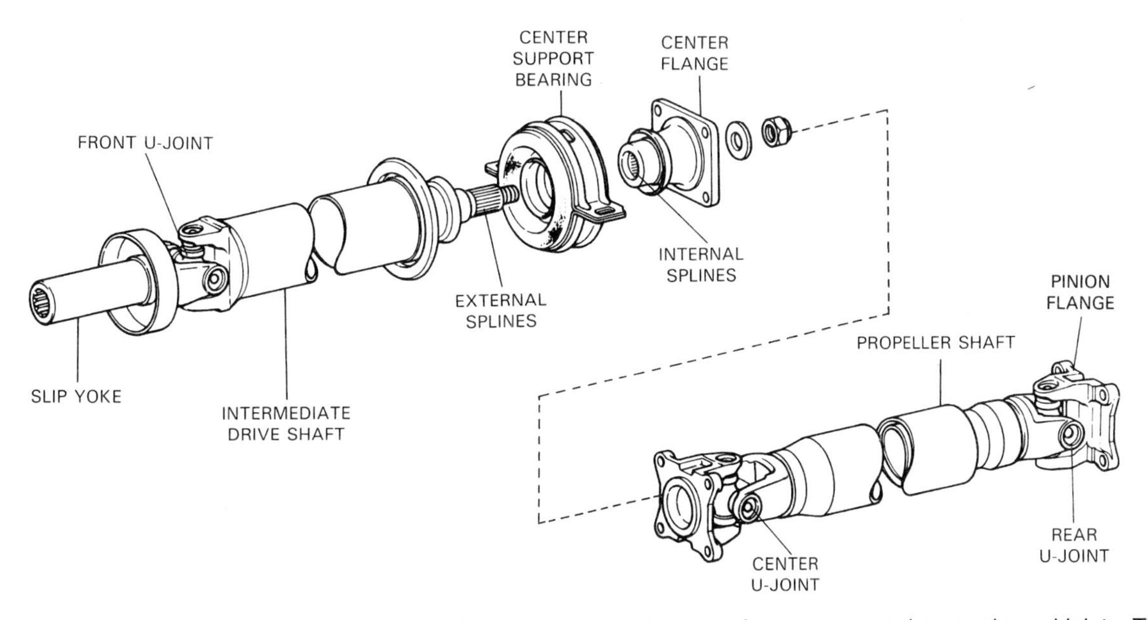

Figure 14-18. Note relative positions of parts of two-piece drive shaft. Also, note front, center, and rear universal joints. The design pictured has two slip yokes–one up front and one at rear of front shaft, behind center support bearing. The bearing is solidly mounted and will not allow any axial movement of the front shaft. The rear slip yoke permits back-and-forth movement of the rear shaft. The front slip yoke is used to simplify shaft installation.

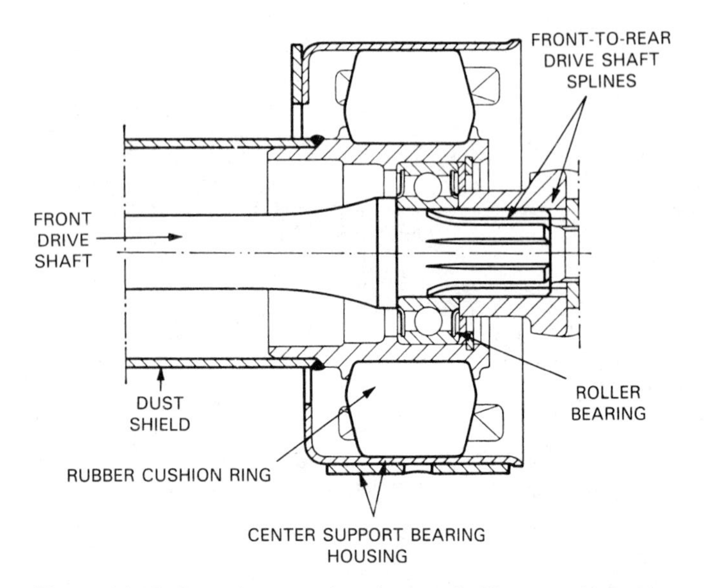

Figure 14-19. Center support bearing used with a two-piece drive shaft is usually a permanently greased ball bearing. Some heavy-duty bearings can be greased through a fitting. Bearing is installed near the rear of the front drive shaft. Most center support bearings are pressed on the shaft, but some may be held by bolts or other kind of fastener. Some bearings simply slip onto the shaft. (Fiat)

rotational axis. Sometimes, several devices with rotating parts are coupled together. An example of this is the vehicle transmission, drive shaft, and rear axle assembly.

When rotating parts are coupled together, care must be taken to ensure that the centerlines are closely matched. ***Perfect alignment,*** Figure 14-20A, would mean that all of the centerlines lined up exactly, forming one long centerline. Under actual conditions, perfect alignment is seldom achieved. Any deviation from a perfect alignment means that there is an angle between the centerlines, as shown in Figure 14-20B. These angles are measured in degrees, sometimes shown by a ° symbol.

Figure 14-20B shows the ***driveline angles,*** or ***operating angles,*** on a front-engine, rear-wheel drive vehicle. These are the angles between the centerlines of the transmission output shaft and drive shaft and between the drive shaft and the drive pinion gear shaft. The driveline angles are also referred to as the ***U-joint angles*** because the inclination of the driveline changes at the universal joints. The angles are measured at the front and rear U-joints and compared to specifications.

Note that the vehicle manufacturer's specification for U-joint angle is usually small–about 1° or 2°. If the angle is greater than specified, the speed fluctuations of the driven

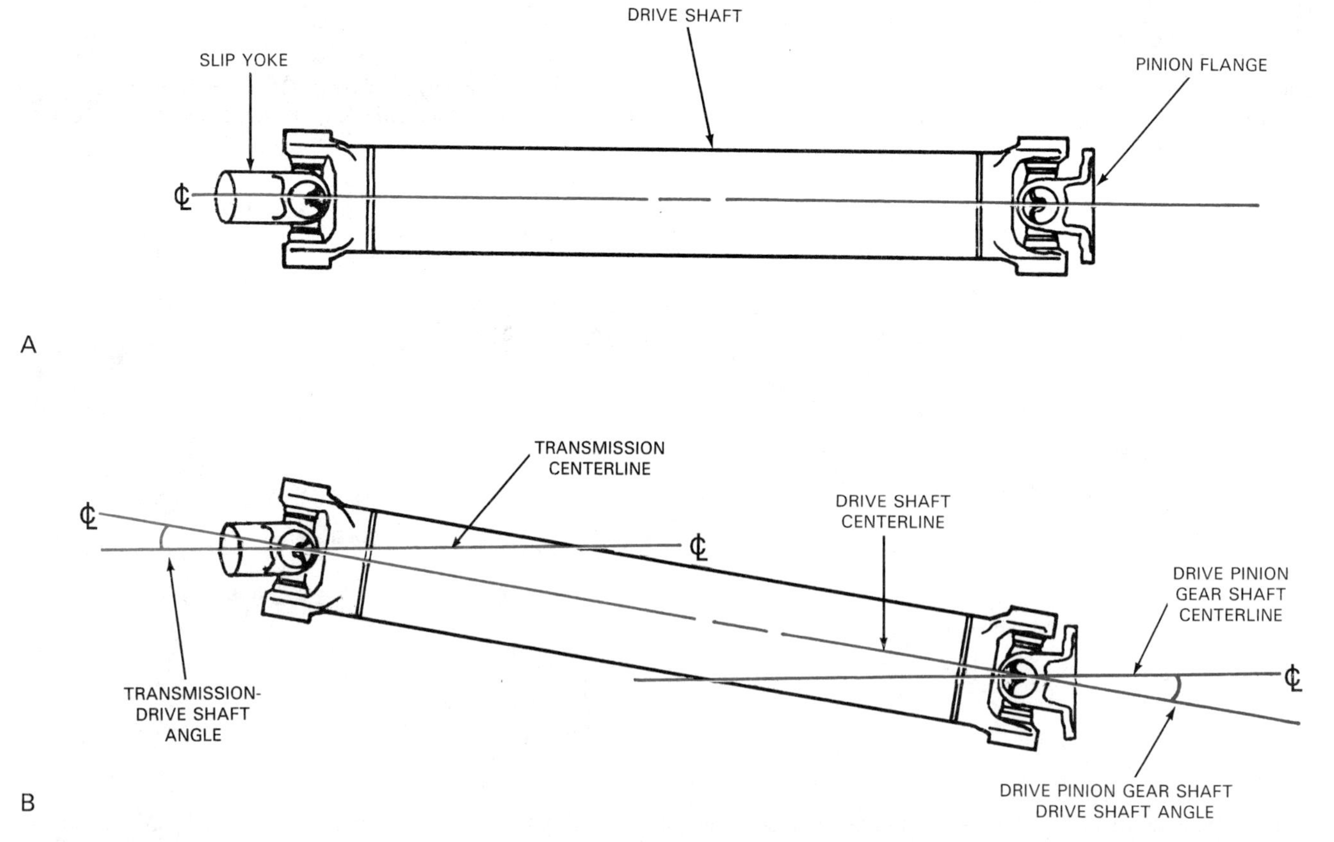

Figure 14-20. Driveline angles. A–The ideal driveline angle would be 0°, or a perfect alignment. B–As actually designed, however, the driveline will have some angles, due to relative placement of the transmission and rear axle assembly.

shaft will increase in range. The drive shaft may have an objectionable vibration. If a U-joint angle becomes too large, the joint itself can break.

Driveline angles are determined by carefully measuring the inclinations of the transmission output shaft, drive shaft, and drive pinion gear shaft. The driveline angles are then determined from these measurements. Chapter 15 contains detailed procedures for checking driveline angles.

Note that since the rear axle assembly moves in relation to the transmission, the driveline angle will vary with any movement of the vehicle suspension. Factors that affect the suspension and, therefore, the driveline angle include: added weight in the vehicle; road conditions (bumps, dips, and potholes); vehicle acceleration or deceleration; suspension system wear or alterations. For this reason, it is doubly important that the original driveline angles be kept as close as possible to the manufacturer's specifications.

Summary

The driveline must transfer power from the transmission to the rear axle assembly. At the same time, it must make changes in angle and length to compensate for changes in position of the rear axle assembly.

The driveline is composed of the drive shaft, universal joints, slip yoke, and balance weights. The shaft is a one-piece steel tube. It is hollow to reduce weight. The other driveline parts are attached to the shaft.

The drive shaft must be rugged to absorb shocks. It must be carefully balanced to reduce vibration.

The slip yoke is a splined shaft that is designed to slide on the transmission output shaft. This sliding action allows it to change length as required.

The universal joints allow the driveline angles to change as the rear axle assembly moves over the road surface. The most common universal joint is the cross-and-roller or Cardan type. It consists of a cross and four bearing cups. The cups attach to the drive shaft yokes by various attachment methods. Needle bearings between the cross trunnions and cups reduce friction. There may be a grease fitting to allow lubrication of the needle bearings.

The cross-and-roller U-joint can cause drive shaft vibration. Angles between two shafts result in the driven shaft speeding up twice and slowing down twice during each revolution. The cross-and-roller U-joint must be carefully installed to reduce vibration.

To further reduce vibration, some other kinds of universal joints are used. The double Cardan U-joint is composed of two cross-and-roller U-joints connected by a center yoke. Vibration forces are cancelled within this yoke.

The differential pinion yoke/flange attaches the rear universal joint to the rear axle assembly. It may be bolted directly to the drive pinion gear shaft, or it may be a separate part bolted to the pinion flange.

The two major classes of drivelines are the Hotchkiss and the torque tube. The Hotchkiss drive is the most common. It is often called an open driveline. The Hotchkiss design has at least two universal joints.

The two-piece drive shaft is a type of Hotchkiss design. It has two separate shafts and three universal joints. The slip yoke is usually between the front and rear shafts. Two-piece shafts have a center support bearing at the rear of the front shaft. This ball bearing is factory greased and sealed. It is attached to the vehicle through a flexible mounting to reduce the transfer of drive shaft vibration.

Proper driveline angles are critical to vibration reduction and long component life. These angles must be as close to zero as possible.

Know These Terms

Cross-and-roller U-joint (Cardan U-joint, or Conventional U-joint); Double Cardan U-joint; Ball-and-trunnion U-joint; Tripod joint; Differential pinion yoke; Pinion flange (Companion flange); Hotchkiss drive; Torque-tube drive; Two-piece drive shaft; Axial centerline; Rotational axis; Perfect alignment; Driveline angle (Operating angle, or U-joint angle).

Review Questions–Chapter 14

Please do not write in this text. Place your answers on a separate sheet of paper.

1. Drive shaft _____ is measured from center to center of each yoke hole.
2. What is a slip yoke?
3. Most rear-wheel drive vehicles use:
   a. Cross-and-roller U-joints.
   b. Cardan U-joints.
   c. Conventional U-joints.
   d. All of the above.
4. The cross ends of a Cardan U-joint are called _____.
5. This type of universal joint uses several steel balls to allow rotation:
   a. Ball-and-trunnion.
   b. Cardan.
   c. Rzeppa.
   d. Pivotal.
6. Torque-tube drive is the most common. True or false?
7. Hotchkiss drive uses an _____ shaft with two or more _____ _____.
8. Where is a center support bearing used, and why?
9. What are U-joint angles?
10. Give four factors that affect driveline angles.

Certification-Type Questions–Chapter 14

1. On rear-wheel drive vehicles, the transmission and the rear axle assembly are connected by the:
- **(A)** driveline.
- **(B)** output shaft.
- **(C)** pinion flange.
- **(D)** drive pinion gear.

2. All of these are parts of the driveline EXCEPT:
- **(A)** a slip yoke.
- **(B)** a drive shaft.
- **(C)** a drive pinion gear.
- **(D)** U-joints.

3. The driveline transfers power from the transmission to the:
- **(A)** axial centerline.
- **(B)** torque-tube drive.
- **(C)** front axle assembly.
- **(D)** rear axle assembly.

4. Which of these is used where the driveline enters the rear of the transmission, allowing the driveline to change its effective length?
- **(A)** U-joint.
- **(B)** Slip yoke.
- **(C)** Companion flange.
- **(D)** Differential pinion yoke.

5. All of these are drive train components consisting of a center spider and four cups containing needle bearings EXCEPT:
- **(A)** Cardan U-joints.
- **(B)** conventional U-joints.
- **(C)** cross-and-roller U-joints.
- **(D)** double Cardan U-joints.

6. Fluctuation in drive shaft speed is created at the:
- **(A)** rear U-joint.
- **(B)** front U-joint.
- **(C)** slip yoke.
- **(D)** axial centerline.

7. A double Cardan U-joint design splits the driveline angle equally between two shafts. Technician A says that this is accomplished through the joint's ball-and-socket centering device. Technician B says that the joint's center yoke is responsible. Who is right?
- **(A)** A only
- **(B)** B only
- **(C)** Both A & B
- **(D)** Neither A nor B

8. All of these universal joints are designed for constant velocity EXCEPT:
- **(A)** a tripod joint.
- **(B)** a Rzeppa joint.
- **(C)** a ball-and-trunnion joint.
- **(D)** a cross-and-roller joint.

9. In the rear axle assembly, the differential pinion yoke connects the:
- **(A)** slip yoke.
- **(B)** pinion flange.
- **(C)** drive pinion gear.
- **(D)** two-piece drive shaft.

10. When comparing drivelines, Hotchkiss drive can be described as all of these EXCEPT:
- **(A)** easier to service.
- **(B)** using an enclosed shaft.
- **(C)** lighter in weight.
- **(D)** cheaper to produce.

11. Since the driveline length of a torque-tube drive does not have to change, which of these is not needed?
- **(A)** Slip yoke.
- **(B)** Pinion yoke.
- **(C)** Front U-joint.
- **(D)** Pinion flange.

12. A two-piece drive shaft uses:

(A) one U-joint.
(B) two U-joints.
(C) three U-joints.
(D) no U-joints.

13. Technician A says that the slip yoke has external splines that mate with the transmission output shaft. Technician B says that the slip yoke fits closely with the transmission extension bushing and rear seal. Who is right?

(A) A only
(B) B only
(C) Both A & B
(D) Neither A nor B

14. Technician A says that all cross-and-roller U-joints have grease fittings. Technician B says that some universal joints are sealed at the factory and cannot be greased unless they are disassembled. Who is right?

(A) A only
(B) B only
(C) Both A & B
(D) Neither A nor B

15. All of these are used to hold bearing cups in place EXCEPT:

(A) retainer snap rings.
(B) bolted straps.
(C) anaerobic sealer.
(D) injected plastic.

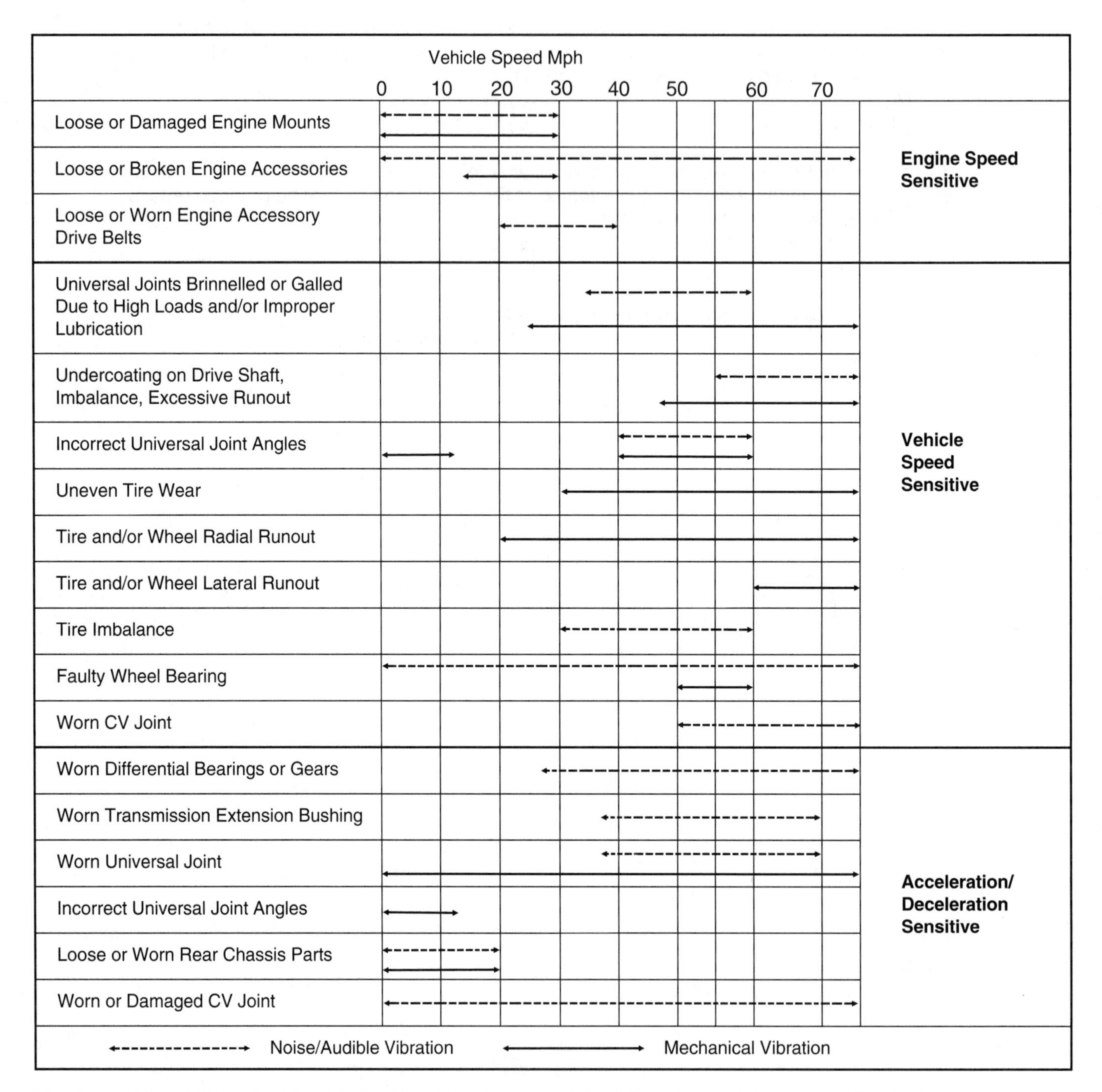

Vibration problems? This list, while not exhaustive, identifies some of the defects that could be causing vibration in your vehicle.

Chapter 15

Driveline Problems, Troubleshooting, and Service

After studying this chapter, you will be able to:

- Identify driveline-related problems.
- Inspect a drive shaft assembly both visually and with instruments.
- Locate the source of typical driveline problems.
- Perform in-car driveline service procedures.
- Remove a driveline from a vehicle.
- Service universal joints.
- Install a driveline in a car.

Although some of the newest drive shafts of rear-wheel drive vehicles are composites of graphite and aluminum, most are hardened-steel tubes. These shafts are rugged and do not wear out. Drive shafts can, however, become bent or unbalanced. This can happen from such things as an accident, driving over very rough ground, or anything else that would cause a heavy blow to the shaft.

Most driveline problems are caused by the universal joints. U-joints installed at the factory normally do not have grease fittings; these cannot be greased without being taken apart. Eventually, the original grease washes or leaks out, and the U-joints begin to wear and become noisy. Worn-out U-joints are a common problem.

At first, a loose or noisy U-joint will not cause other problems. However, if the vehicle is driven for long periods with a defective U-joint, the joint will wear until it finally comes apart. When this happens, the driveline will come undone at the transmission or the rear end (rear axle assembly). The vehicle will then be stranded.

When a rotating drive shaft becomes disconnected, it can damage itself and the underside of the vehicle. The drive shaft will usually get bent or dented and have to be replaced. Fuel and brake lines, electrical, and rear end parts can also be damaged or destroyed. For these reasons, you should replace a defective U-joint as soon as you know about it.

This chapter will enable you to diagnose, inspect, and service U-joints–and drivelines, in general. Figure 15-1 shows the various places where driveline problems occur. Study them carefully.

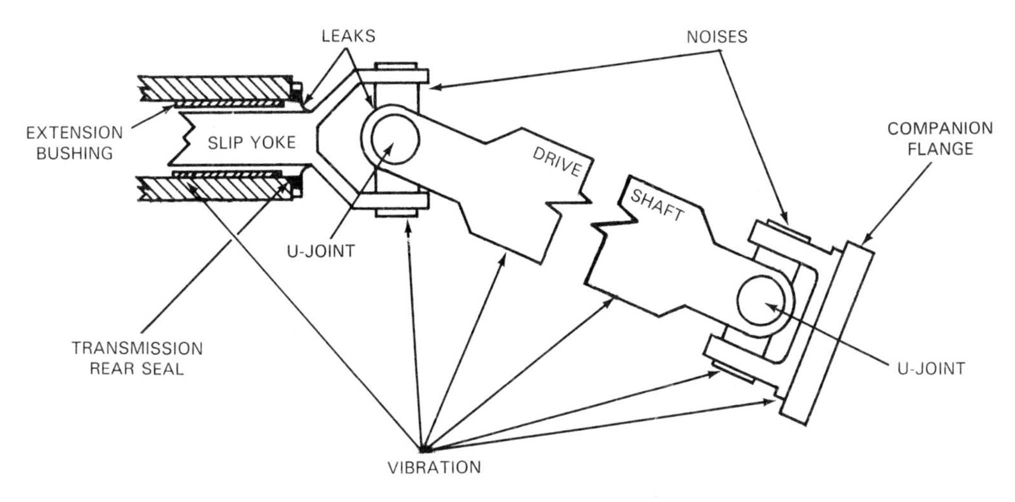

Figure 15-1. Note typical problems of a drive shaft assembly.

Driveline Problems and Troubleshooting

Diagnosing driveline problems can be very challenging. The source of a problem can be hard to pinpoint, and it can be very expensive–not to mention, embarrassing–to make repairs or replace parts and not solve the problem.

Driveline malfunctions can result from misaligned parts, improper machining, and external damage to the drive shaft. They are sometimes difficult to troubleshoot, since the problems that they cause are similar to problems caused by other assemblies. To arrive at a proper diagnosis, check every part of the driveline–and assume nothing. Also, check other possible causes of the problem. Examples include a bent wheel or a cupped tire; they can cause vibrations similar to those caused by a bent drive shaft.

Always try to obtain as much information as possible before beginning any repairs. Find out exactly why the vehicle was brought in for service. Whenever possible, test drive the vehicle with the owner to make sure that you agree about the nature of the problem. Also, *driveline troubleshooting charts* are available. Be sure to consult these for possible solutions.

Check out the easiest possible solutions first, before proceeding to more difficult solutions. For instance, if a grease fitting is provided, a squeaking U-joint can sometimes be silenced by greasing the joint. If the grease does not help, then the U-joint should be replaced.

There are two major types of driveline problems. These problems, presented in upcoming paragraphs, are:

- Abnormal noise–caused by worn U-joints or bearings.
- Driveline vibration–caused by worn or damaged parts or shaft imbalance.

NOTE! In addition, driveline problems can lead to leaks of transmission oil or U-joint bearing grease.

Abnormal Noise

Abnormal noises in the driveline include squeaks, clunks, whines, and similar sounds. They can usually be traced to the defective part without much difficulty.

Squeaks are usually caused by a dry U-joint. They are usually heard when driving in reverse. Squeaking is often an early indication that a U-joint is going bad.

A clunking noise when reversing direction or at low speeds under light acceleration can be caused by too much *backlash* somewhere in the drive train. Note that in Chapter 3, ***backlash*** was defined in relation to gears. In a broader sense, backlash is a condition of clearance or movement between two closely mating mechanical parts. It is the amount of play between these parts.

Before checking the driveline, always make sure that the excessive backlash is not inside the differential. If the differential is okay, you should check the driveline, concentrating on the U-joints, slip yoke, and center support bearing (if so equipped). Also, check the transmission extension bushing.

Excessive backlash often shows up in U-joints. Worn U-joints can sometimes be spotted by grease leaking from them or by excessive play between parts when wiggling or rotating the drive shaft. If the drive shaft has a center support bearing, make sure that the attaching bolts have not worked loose. The extension bushing can be checked for wear by trying to wiggle the slip yoke while it is in position. Too much play in any one of these parts could be causing a clunking noise.

Clunks, in rare cases, can be caused by worn splines on the slip yoke or on the transmission output shaft. The slip yoke and output shaft can be checked by removing the driveline and visually inspecting the splines for wear. Check all other possibilities first, however.

Rubbing, whining, or whirring noises can be caused by the drive shaft rubbing on a stationary part of the vehicle (the exhaust system, for example) or by a defective center support bearing. One clue that a drive shaft is rubbing is a change in sound when the vehicle is driven over bumps.

Driveline Vibration

Vibration is the rapid back-and-forth or side-to-side movement of a rotating part. *Driveline vibrations* can originate from any of its associated parts.

In terms of the drive shaft, there are two types of vibration–*axial* and *radial.* ***Axial vibration,*** Figure 15-2, is a lengthwise, or back-and-forth, movement of the shaft along its rotational axis. Back-and-forth movement is strictly limited by the design tolerances built into the drive train. It does not cause problems unless the U-joints are very worn or loose. ***Radial vibration*** is a side-to-side movement of a rotating part along its rotational axis, like that shown in Figure 15-3. It is the most common cause of vibration in an automotive drive shaft.

Drive shaft vibrations occur whenever the vehicle is moving, and the vibrational pulses occur at regular intervals. The vibrations can be measured in terms of *frequency,* or *cycles per second,* and *amplitude.* Note that the frequency varies in direct proportion to drive shaft speed.

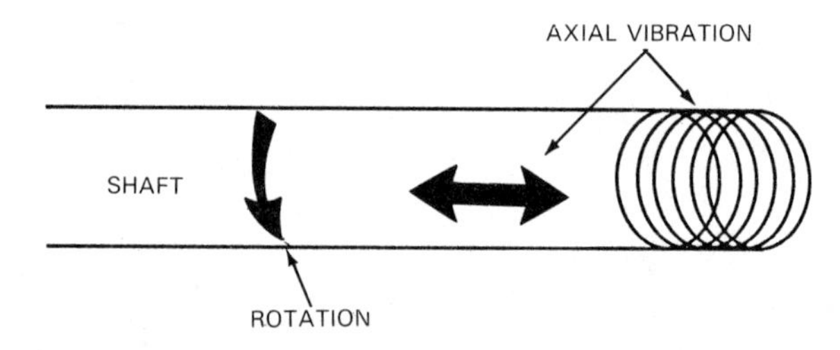

Figure 15-2. Axial vibration is vibration back and forth in relation to shaft rotation. It is not usually a problem on drive shafts unless U-joints are severely worn.

Also, the greater the amplitude, or deviation from *exact center,* the stronger the vibration is.

Figure 15-4 graphically represents the vibration of the shaft shown to its left. The graph depicts one *cycle* of vibration. Points B and C represent the maximum distance of shaft deviation from exact center. It begins with the shaft centered at point A. As the shaft turns, it goes through point B, reverses direction, and passes through point A to point C. At C, it reverses again and returns to point A. The number of cycles in a given time period (frequency) increases with drive shaft speed.

Driveline vibrations can be felt in the steering wheel, brake pedal, or in the vehicle body. As a first step, make sure that the vibration is not caused by problems in the tires, wheels, or brake system.

NOTE! The drive shaft turns at higher speeds than the wheels, and as a result, drive shaft vibration will occur at a higher frequency than wheel- and tire-produced vibration.

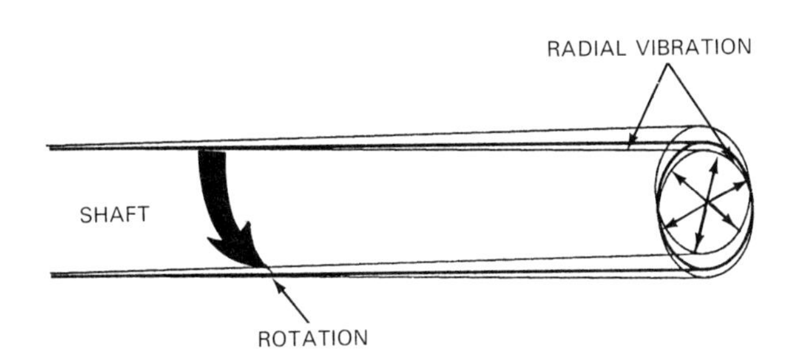

Figure 15-3. Radial vibration is the more common drive shaft vibration problem. It is a side-to-side movement in relation to shaft rotation. It can often be felt in the passenger compartment. It can be caused by worn U-joints or by a bent or otherwise damaged drive shaft. Misalignment or anything else that contributes to an unbalanced condition can cause vibration.

Sometimes, a vibration can be traced to an engine problem, such as a dead cylinder or flywheel imbalance. One way to check this is to shift the transmission into neutral at cruising speeds and allow the engine to idle. If the vibration decreases, the problem is in the engine. If the vibration stays the same, the problem is elsewhere.

Once you are sure that the source of the problem is in the driveline, first check the U-joints for excessive backlash or other signs of wear. Worn U-joints, especially the front joint, will cause a vibration when the vehicle is accelerated between 30 mph and 40 mph (48 km/h and 64 km/h).

Check for dirt or undercoating stuck to the drive shaft. Foreign material on the outside of the drive shaft can be easily spotted and removed. Check for missing balance weights. Also, check to see if the shaft is bent. Any of these conditions can cause drive shaft imbalance, resulting in vibration.

Other possible causes of vibration are a loose fit between driveline yokes and U-joint bearing cups or a broken lug on one of the yokes. Vibration can also be caused by a defective center support bearing on vehicles with two-piece drive shafts. A loose or worn center support bearing will almost always be noisy.

The slip yoke or transmission extension bushing can also wear out, causing a vibration that is at its worst when the vehicle is accelerated. A worn slip yoke or bushing will almost always cause a leak at the transmission rear seal.

Driveline Service Precautions

You should always observe safety rules and certain other precautions when working around drivelines. To

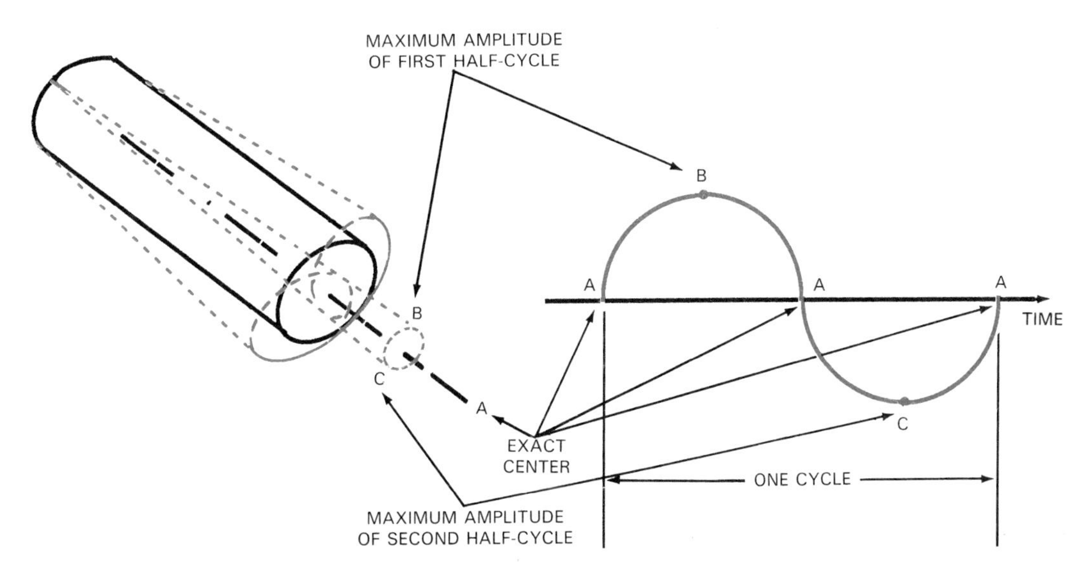

Figure 15-4. One cycle of vibration is represented by this waveform. Frequency is the number of cycles occurring in a certain period of time. Amplitude is the strength of the vibration.

review information that was highlighted in Chapter 2, some general precautions for driveline service are:

- *Always* support the vehicle in a safe manner. If a hoist is not available, support the ends of the rear axle housing on jackstands after raising the vehicle with a suitable jack. *Never* depend on *any* jack to keep the vehicle raised while working.
- *Stay away* from rotating driveline parts and rear wheels when the engine is running.
- Do *not* wear loose clothing or jewelry that could be caught by the rotating drive shaft assembly. Long hair should be tied up or otherwise secured.
- Wear eye protection whenever you are under the vehicle! (This is very important if you must hammer on any part of the shaft.)
- Make sure that the engine won't be accidentally started when you are working on or closely inspecting the drive shaft.
- The drive shaft is heavy and awkward. Do *not* allow it to drop when attachments to the rear axle assembly are removed. This could damage the shaft or transmission extension housing. It could also hurt you!
- Make sure that you do *not* damage the drive shaft while making other repairs. You can damage a drive shaft by dropping it, striking it against solid objects, or by clamping it too tightly in a vise.
- It is easy to deform a drive shaft yoke when removing a U-joint. *Always* carefully align and support all bearing removal tools before pressing or hammering. Make sure that the drive shaft is securely held before beginning repairs.
- Always reinstall the bearing cup retainer clips. A retainer clip that will not seat is a symptom of another problem, such as a misplaced bearing needle or a deformed yoke. A bearing cup without a retainer clip will be thrown out of the yoke at normal speeds.

Driveline Inspection

Having gathered preliminary information from the vehicle owner, taken a test drive, and consulted a troubleshooting chart, the next step is to inspect the installed driveline. This section will explain the right way to inspect the driveline. It involves performing a *visual inspection* and, sometimes, a *measurement inspection.*

Visual Inspection

Much of driveline inspection is looking for obvious damage by visual examination. Looking underneath the vehicle may be all that is required to pinpoint a problem. After the source of the problem is identified, the defective part or parts may be repaired or replaced.

Drive shaft inspection

To check the drive shaft for worn parts, raise the vehicle on a hoist or place it on jackstands to gain access to the drive shaft. Place the transmission in park (automatic transmissions) or in gear (manual transmissions). Do not apply the parking brake. Visually inspect the drive shaft for undesirable conditions, such as undercoating, dirt, dents, and missing balance weights. Also, look for cracked welds and for contact with other parts.

U-joint inspection

Look for signs of wear when checking the U-joints. To check for U-joint wear, attempt to rotate the drive shaft while watching the U-joint yokes. No movement should occur between the cross and yoke, either at the front yoke or the rear yoke (although the entire assembly may rotate slightly). Attempt to wiggle the shaft up and down and from side to side. Try sliding the shaft back and forth. This is shown in Figure 15-5. Again, no movement should occur at either U-joint. If a cross moves inside its yoke, the U-joint must be replaced.

If the U-joint passes the above inspection, visually check that all snap rings, U-bolts, or other attachments are installed tightly. Also, make sure that none of the U-joint bearing cups are loose in their yokes. Check CV joints using this same procedure.

Slip yoke and extension bushing inspection

The slip yoke and transmission extension bushing can be checked by a procedure similar to that used to check the front and rear U-joints. Try to move the slip yoke up and down and from side to side while watching the transmission extension housing. Any movement of the slip yoke should be slight. Typically, it should be no more than 1/16 in. (1.5 mm).

If slip yoke movement is greater than this, or if the transmission rear seal is leaking, remove the drive shaft. You can then inspect the yoke and bushing. Look for wear or scoring. The bushing will usually show much more wear than the yoke.

Center support bearing inspection

On drivelines with a center support, the bearing is checked by trying to move the drive shaft yoke up and down and from side to side at the center support. There should be almost no movement between the drive shaft and bearing.

If the center support passes this test, put the transmission in neutral and turn the drive shaft. If you feel any roughness as you turn the shaft or hear any noises from the center support bearing, the bearing should be replaced. The bearing can also be checked with a stethoscope to detect noise with greater accuracy.

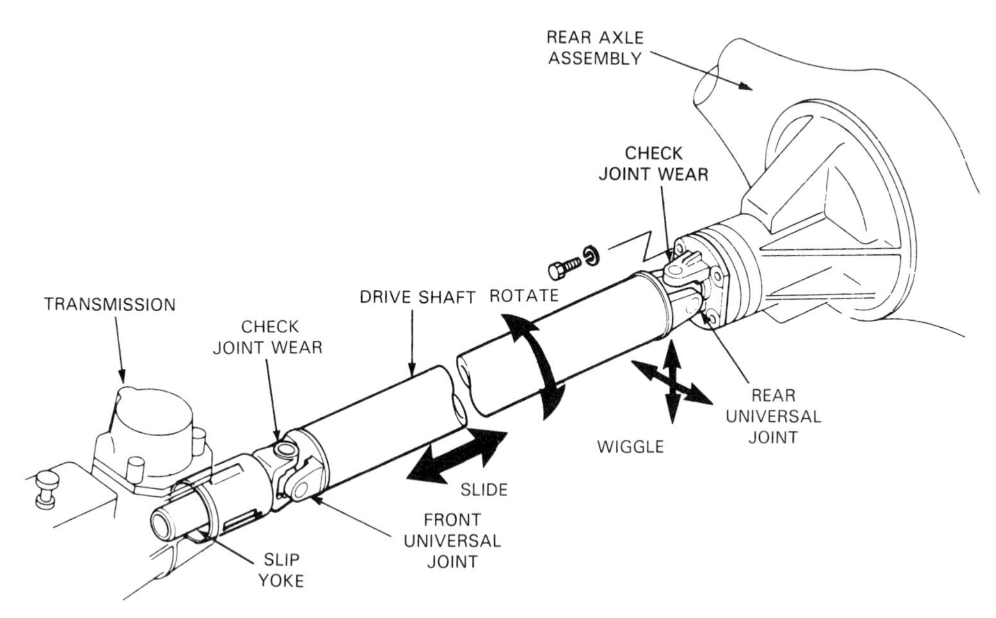

Figure 15-5. Note how to check a U-joint for wear. You should try to manually rotate, slide, and wiggle the drive shaft while watching the U-joint for movement. If the U-joint shows any movement, it must be replaced. (Mazda)

WARNING! Use extreme caution when using a stethoscope to check for drive train noises. Be careful not to touch spinning parts.

Measurement Inspection

Sometimes it is necessary to use special measuring instruments to locate the source of a driveline problem. Instruments most commonly used are the dial indicator and the inclinometer. These instruments, introduced in Chapter 2, are used to measure drive shaft radial runout and U-joint angle.

Drive shaft runout

The drive shaft should theoretically turn in a perfect circle, with every point along the shaft surface turning in a perfect circle about the rotational axis. In real life, the shaft will vary slightly from perfect rotation. This variation, or how much a part moves off-center as it rotates, is called runout, as discussed previously in Chapter 7. If the runout is too large, the drive shaft will be unbalanced, and it will vibrate.

Most drive shafts will never require runout checking. However, sooner or later, most automotive technicians will need to perform this task when servicing a driveline. There are two reasons to check drive shaft runout:

- If the vehicle has a high-speed vibration problem and all other causes have been eliminated.
- If the drive shaft has been struck going over rough ground or the vehicle has been in a collision.

Check the drive shaft for straightness and excessive runout by first placing the transmission in neutral. Then, raise the vehicle on a hoist or with a jack so that the rear wheels are free to turn. Support the ends of the rear axle housing with jackstands, as necessary, to keep the transmission and rear axle assembly in normal alignment. The drive shaft must *not* be at a sharp angle during runout measurement.

NOTE! If the vehicle is raised so the rear axle assembly is hanging unsupported, the test results will not be accurate. Also, if the wheels cannot rotate freely, the runout check cannot be made.

After raising the vehicle, select a spot on the front of the drive shaft to attach a dial indicator. Manually rotate the drive shaft to make sure that the spot selected does not have any weights, flat spots, weld splatter, or any other variations that would cause an incorrect reading. Mount the indicator so that its plunger contacts the drive shaft. The indicator can be mounted on a tall floor stand under the car or clamped to the underside of the vehicle.

Rotate the drive shaft slowly by hand. Sometimes this can be done by turning a rear wheel. Observe and record the maximum variation of the dial indicator. Reposition the dial indicator at the center and rear of the drive shaft and take readings in the same way as before. A two-piece drive shaft would need more readings. See Figure 15-6.

Maximum drive shaft runout, at any point, should typically be no more than 0.040 in. (1 mm). If the runout is more than specifications, remove the driveline from the differential pinion yoke, rotate it a half turn, and reinstall it. Then, recheck the runout.

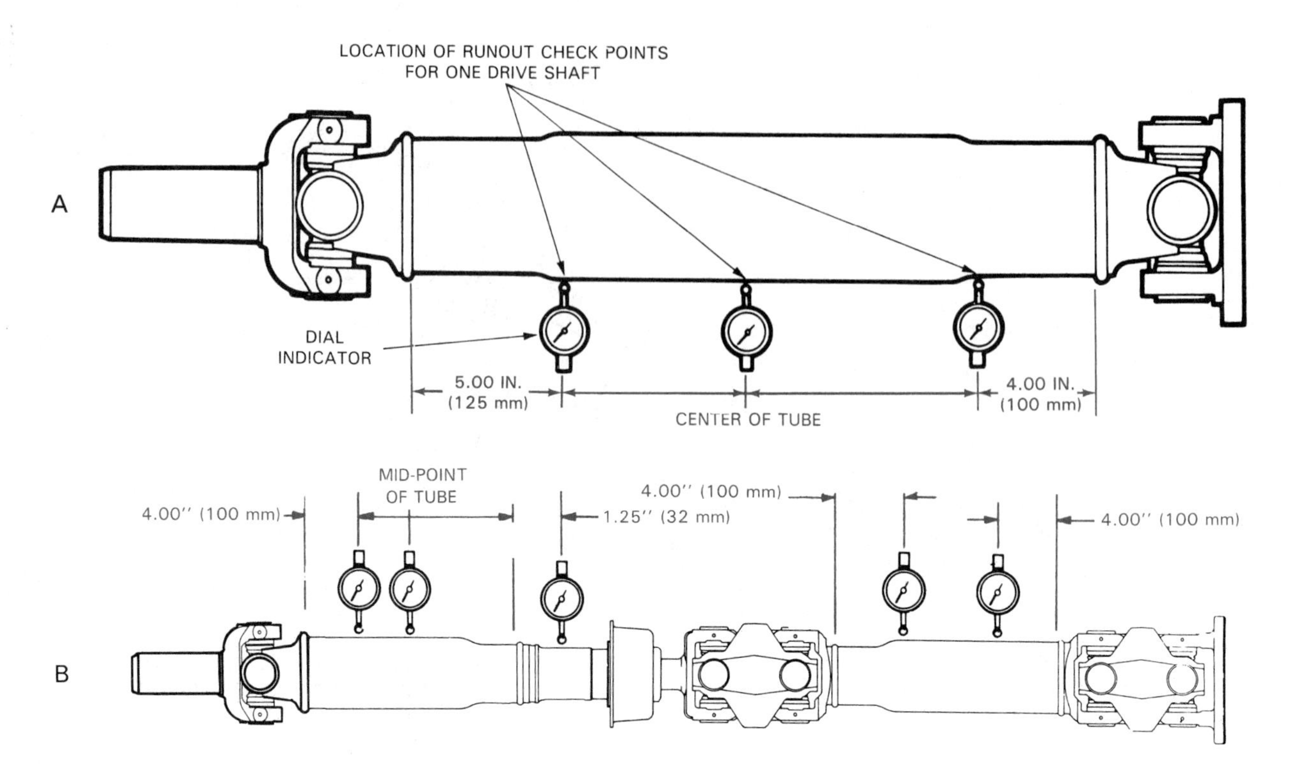

Figure 15-6. This shows usual locations where dial indicator is placed to check drive shaft radial runout. Any variation from factory specifications means that corrections must be made to drive shaft. A–Measuring runout on one-piece drive shaft. B–Measuring runout on two-piece drive shaft. (General Motors)

If the runout is still too great, replace the drive shaft or send it to a drive shaft specialty shop for repairs. In some rare cases, the differential pinion flange is warped or improperly machined, causing the excessive runout problem. If the drive shaft specialty shop says that the drive shaft is okay, the pinion flange should be checked. Do not remove the flange from the drive pinion gear for this check.

Driveline angle

Original driveline angles can change over time, and they do. Angle changes can be caused by sagging springs at the front or rear and wear and settling of front or rear suspension bushings. Vehicle modifications can also affect original driveline angles. Installation of air shocks and shortening or lengthening the driveline in order to use a different transmission are two examples. Further, collision damage can alter the original driveline angles of a vehicle.

When driveline angles fall out of spec or change so that the two U-joints are not of the same approximate angle, driveline vibration is likely to occur. If the inclination of the drive shaft is too great, the U-joints may break on hard acceleration.

Driveline, or U-joint, angles are set at the factory by the design of the drive train and chassis. They can be changed in the field by adjusting the rear axle assembly or by adding or removing *shims* between the transmission and the transmission mount.

Initial preparation. Before checking driveline angles, you must make sure that the rear axle housing is at its *curb-height* specification (normal position, or ride height) relative to the vehicle body. The wheels do not have to turn freely, but you must have some way to rotate the drive shaft by 90°.

The ideal situation is for the vehicle to be over a pit or on a *drive-on platform hoist.* You can use another safe way to raise the vehicle, as long as the rear axle housing is in the proper relative position and the vehicle is at its normal weight. The gas tank should be full, and there should be nothing heavy in the car. Before beginning, bounce the vehicle up and down to take any *windup* out of the springs.

If your vehicle specifications call for checking the distance between the top of the rear axle housing *axle tubes* and the bottom of the chassis, do so now, before checking the U-joint angles. If the distance is too much, you may have to add some weight to the vehicle to obtain the proper readings. If the distance is too short, the springs are sagging and should be replaced.

Clean off any dirt, undercoating, and rust from all of the U-joint bearing cups. This will help ensure an accurate measurement. Then, take the angle readings.

Using an inclinometer. Checking driveline angles requires a measuring tool called an ***inclinometer.*** See Figure 15-7. This tool measures inclinations of drive train parts, relative to a horizontal reference plane. The readings are then used to calculate the difference between inclinations of two mating parts. This difference is the U-joint angle. See Figure 15-8.

You can begin by placing the inclinometer on the bearing cup at the drive shaft rear yoke. Make sure that the bearing cup is facing downward, or pointing straight up and down. Next, assuming a gauge with a *level bubble,* center the bubble in the sight glass and record the measurement.

The next step is to rotate the drive shaft 90° and place the gauge on the bearing cup at the differential pinion yoke/flange. Center the bubble in the sight glass and record the measurement. *Subtract* the smaller reading from the larger reading if the slopes of the two connected components run in the same direction. If the slopes run in *opposite* directions, *add* the readings. The answer is the rear U-joint angle.

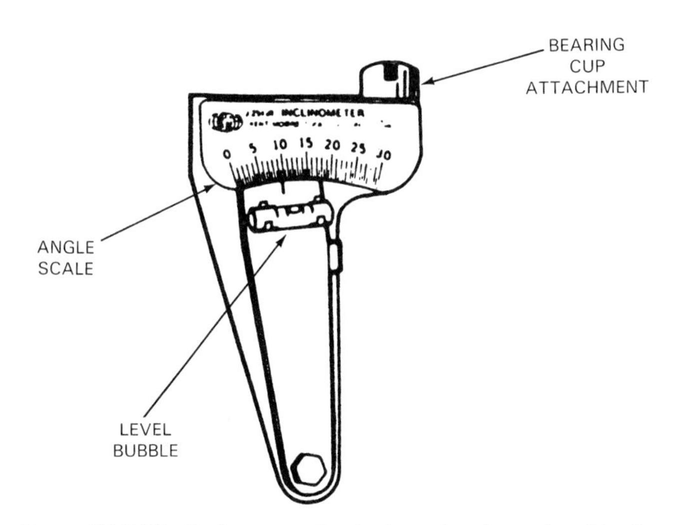

Figure 15-7. The inclinometer is a tool used to determine driveline angles. The inclinometer shown is a typical model recommended by many vehicle manufacturers. (General Motors)

After obtaining the rear U-joint angle, measure the front U-joint. Start by placing the inclinometer on the bearing cup at the drive shaft front yoke. Make sure that the bearing cup is straight up and down. Center the bubble in the sight glass and record the measurement.

Then, rotate the drive shaft 90° and place the gauge on the bearing cup at the slip yoke. Center the bubble in the sight glass and record the measurement. Subtract the smaller figure from the larger figure, or add the two figures, as required. This will give you the front U-joint angle.

Note that with certain inclinometers, the U-joint angle may be read directly from the inclinometer. Refer to Figure 15-9.

In-Car Driveline Service

If a vehicle is experiencing driveline problems, certain adjustments can be made with the driveline installed that might correct the problem. It may be that a driveline angle is too severe or that the drive shaft is out of balance. Either problem can sometimes be fixed by adjusting the angle or balancing the shaft. Of course, it could be that the driveline components simply need to be greased. These procedures are detailed in this section.

Adjusting Driveline Angles

After measuring both front and rear U-joint angles, you can compare the two against specifications. If the U-joint angles are less than the maximum allowable angles, no adjustment is necessary. However, if the U-joint angles are greater than the maximum allowable, they must be adjusted.

One way to adjust driveline angles is to add or remove shims from the transmission mount. A ***shim*** is a thin sheet of metal that is placed between two surfaces to keep them apart. Generally, adding shims at the trans-

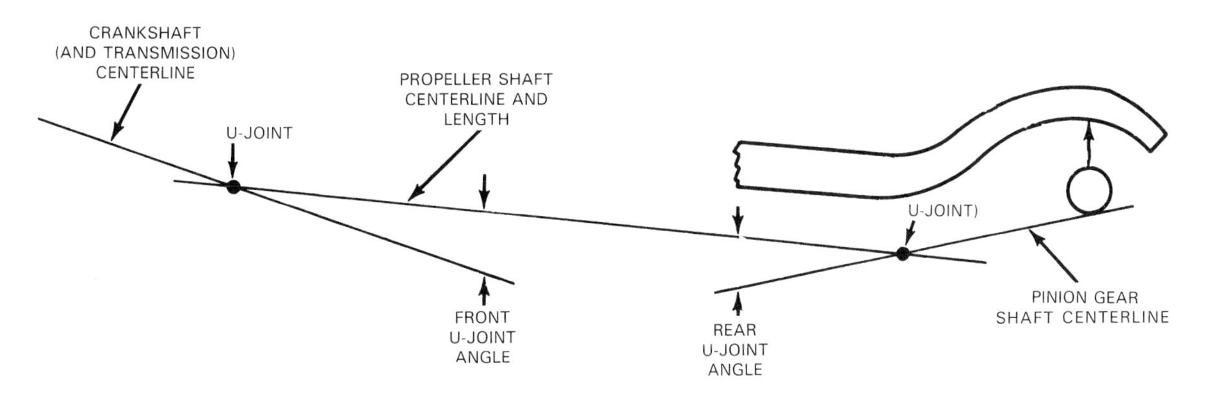

Figure 15-8. U-joint angles of a typical drive train are shown here. Inclinations of transmission output shaft, propeller shaft, and drive pinion gear shaft, obtained through use of an inclinometer, are needed to determine U-joint angles. U-joint angles are the angles between centerlines of these components. They may be determined by adding or subtracting inclination readings, as required. (Chrysler)

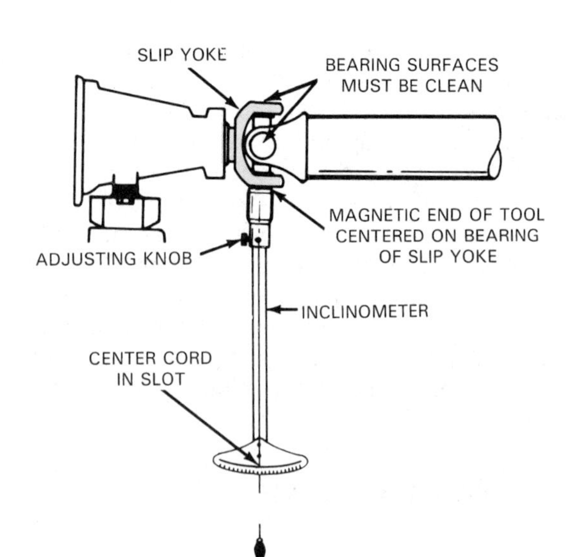

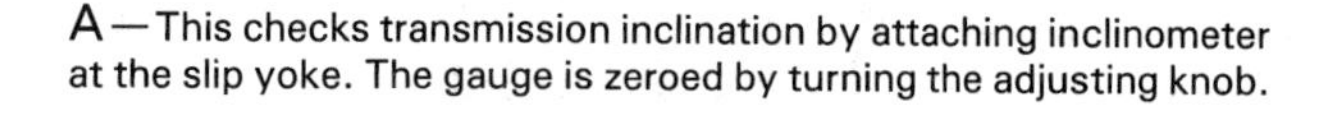
A—This checks transmission inclination by attaching inclinometer at the slip yoke. The gauge is zeroed by turning the adjusting knob.

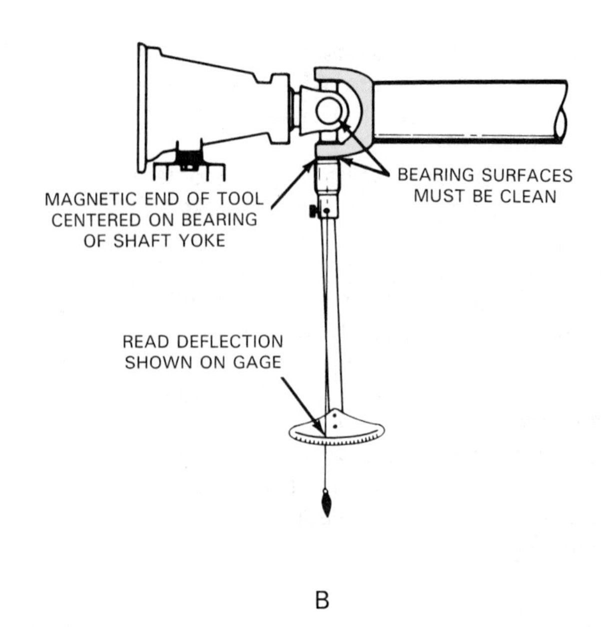

B—This checks drive shaft inclination by attaching inclinometer at the drive shaft front yoke. With the gauge having been zeroed, the front U-joint angle is read directly off the scale.

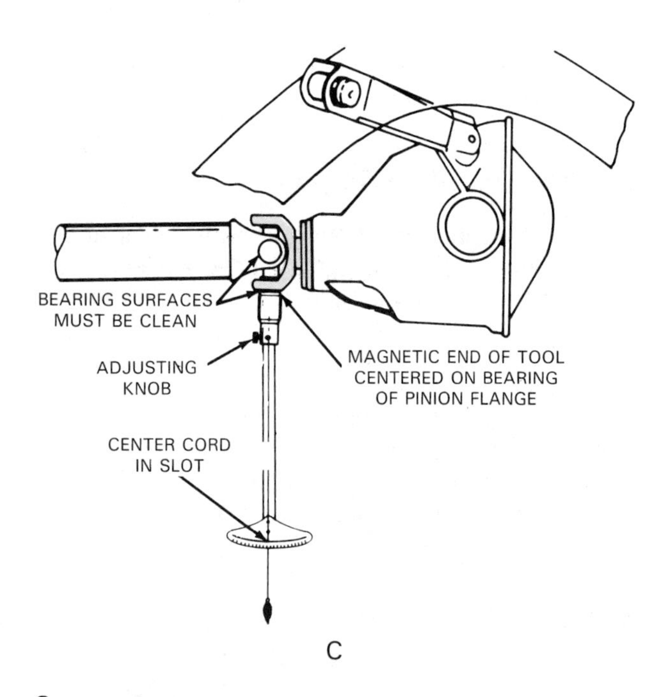

C—This checks inclination of the drive pinion gear shaft by attaching inclinometer at the pinion flange. The adjusting knob is turned to zero the gauge.

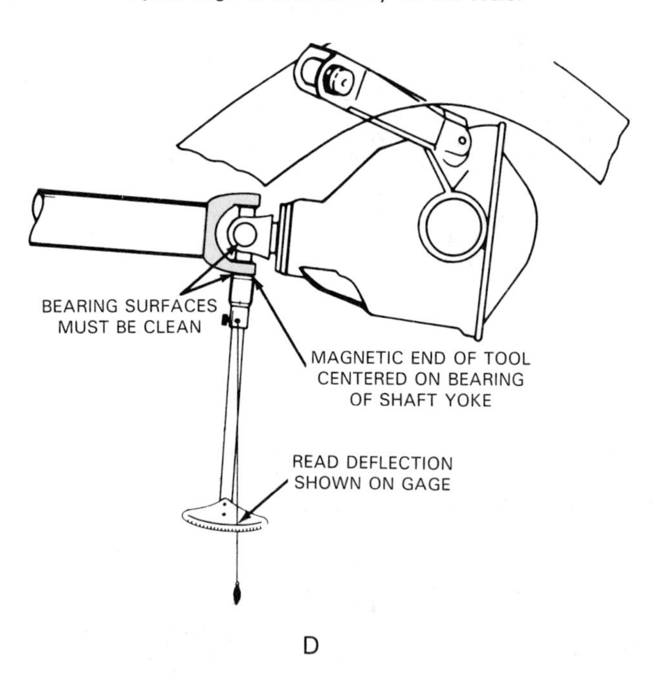

D—This checks drive shaft inclination by attaching inclinometer at the drive shaft rear yoke. With the gauge having been zeroed, the rear U-joint angle is read directly off the scale.

Figure 15-9. This shows the places where drive train inclinations are measured. This type of inclinometer has an adjusting knob for zeroing the gauge, permitting direct reading of U-joint angles. Other inclinometers do not have this feature, and the U-joint angles must be calculated from the readings. The gauge manufacturer and service manual should always be consulted to check U-joint angles. (General Motors)

mission mount will reduce the angle of the front U-joint, while it will increase the angle of the rear U-joint. Removing shims will have the reverse effect. The front U-joint angle will change about twice as much as the rear U-joint angle. This makes transmission mount shimming an effective adjustment when the front U-joint requires the most correction.

Sometimes, the *control arms* attaching the rear axle assembly to the vehicle chassis can be loosened and the rear axle assembly moved up or down to adjust the rear U-joint. This movement is limited, however, and may have to be done in combination with transmission mount shimming to get the proper angle. If the vehicle has rear leaf springs, the rear U-joint angle is sometimes adjusted by placing wedges between the top of the spring leaves and the bottom of the rear axle housing.

A final adjustment, when all else fails, is to change the front engine mounts. The original mounts gradually compress and may cause driveline angles to fall out of spec. New mounts will usually restore the original engine and transmission height and correct angle changes. Shimming the front engine mounts is impractical and often causes mount or fastener failure.

Balancing a Drive Shaft

Balance in a rotating part is achieved when all forces cancel each other out. A *balanced drive shaft* has weight distributed equally around its centerline. The only way a drive shaft can typically lose its original factory balance is if a balance weight gets thrown off, the shaft gets badly dented, or undercoating gets sprayed on the shaft. Drive shafts seldom go out of balance; when they do, they can usually be rebalanced by either mechanical or electronic means using standard screw-type hose clamps.

In preparation, the vehicle should be raised on a hoist with the rear axle housing supported and the wheels free to rotate. See Figure 15-10. After raising the vehicle, check the U-joints for wear and check the drive shaft for extreme bends or dents.

If the U-joints and the drive shaft do not show any damage, remove both rear wheels. This way balance or imbalance of wheels will not affect rotation of drive shaft. Reinstall the wheel lug nuts with the flat sides inward. This will hold the brake drums or discs in place.

Next, mark and number the drive shaft at four spots, 90° apart. Make the marks at the rear of the shaft, ahead of any balance weight, as shown in Figure 15-11. Then, start the engine and shift the transmission into drive. Run the vehicle at approximately 55 mph (89 km/h), noting the amount of vibration. Turn off the engine.

WARNING! Use extreme caution when working around a spinning drive shaft. Do not let any part of your body come in contact with the driveline. Watch that your hair does not get caught up in a U-joint.

Install two hose clamps near the rear of the propeller shaft, Figure 15-12. Align screws of both clamps at one of the four marks that you made on the shaft. Tighten the clamps securely. Make sure that the clamps will not hit any stationary part of the vehicle when the drive shaft is rotating. Again, start the engine and shift the transmission into drive. Run the vehicle at approximately 55 mph and note the amount of vibration.

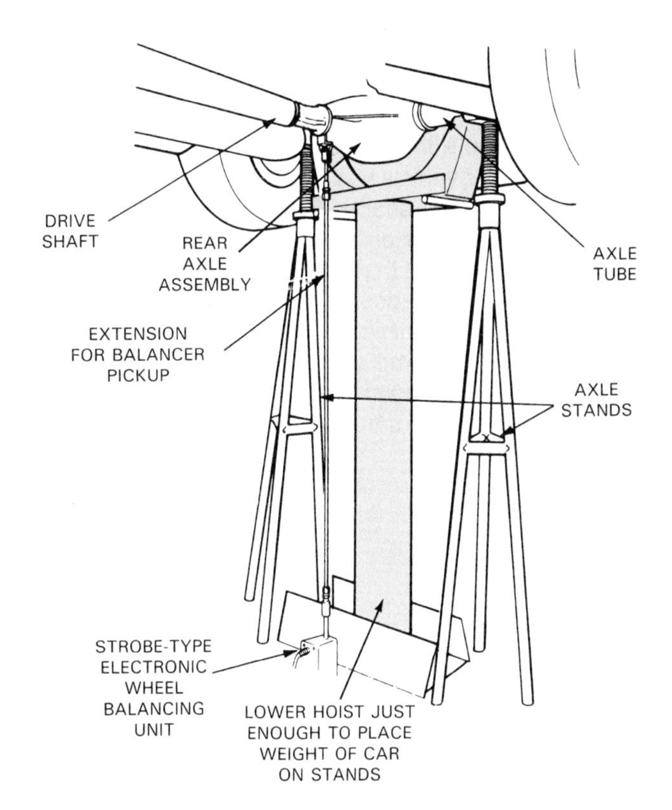

Figure 15-10. Before balancing the drive shaft, raise the vehicle on a hoist. Rear axle housing must be supported, and rear wheels free to rotate. For electronic balancing, support axle tubes with axle stands, lowering *twin-post hoist* (shown here) so that it just clears the axle. This way, sensitivity of operation is not destroyed. (General Motors)

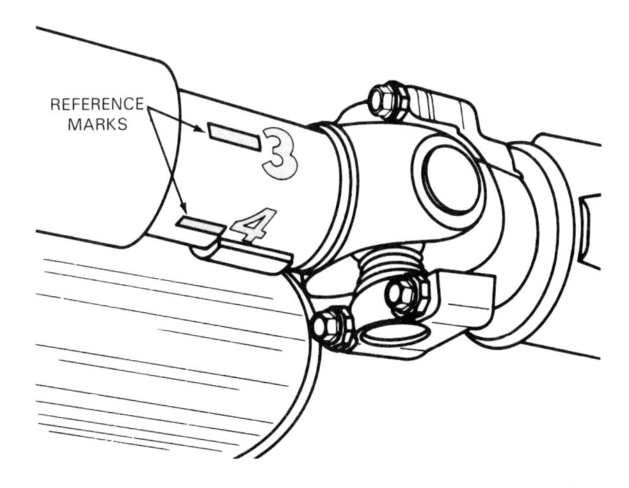

Figure 15-11. This shows the way to make preliminary reference marks on a drive shaft for balance checking. The vehicle is then run at a specified speed with rear wheels off the ground, and the amount of vibration is noted. (General Motors)

CAUTION! Do not operate the vehicle on the hoist for long periods of time to avoid overheating the transmission or engine.

Press on the brake pedal to stop drive shaft rotation. Then, shift into park, stop the engine, and rotate the clamp to the next mark on the drive shaft. Again, run the engine to 55 mph and note the amount of vibration. Repeat this procedure until the vehicle has been tested with the clamps at all four spots on the shaft.

Decide which point produced the least vibration and reposition the clamps at that point. Next, rotate the clamps away from each other about 45°–one clamp rotated each way from the point of minimum imbalance. This is shown in Figure 15-13. Run the vehicle again and note if the vibration has been reduced. Continue to move the clamps away from each other in smaller amounts until the vibration is at its minimum.

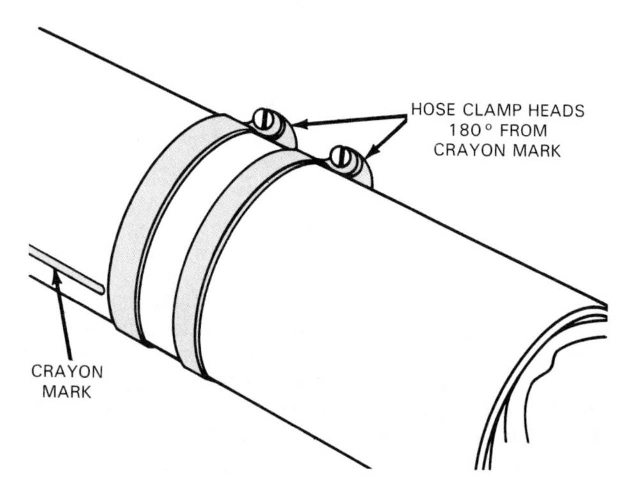

Figure 15-12. Hose clamps are installed on drive shaft for purposes of checking balance. Vehicle is then rechecked for greater or lesser vibration at same speed. Wheels are still off ground. (General Motors)

In some cases, it may be necessary to use just one clamp at the point of minimum imbalance. In other cases, to obtain a good balance, it may be necessary to use three clamps, with the third clamp between the two outside clamps. If three hose clamps do not solve the problem, replace the drive shaft or have it rebuilt (new shaft installed between yokes).

Reinstall the wheels and road test the vehicle. Often, a slight vibration felt with the vehicle on a hoist may not be detectable with the vehicle on the ground.

Greasing the Driveline

Modern drivelines seldom need any type of maintenance except for periodic greasing. Driveline parts that have grease fittings should be lubricated periodically. This applies to some U-joints, slip yokes, and center support bearings. Although these driveline parts do not move great distances, they are under heavy pressures during vehicle operation and require good lubrication.

Since the driveline parts are underneath the vehicle, they are exposed to dirt and water from the road. These elements cause rapid wear and rusting. Grease and good sealing prevents the entry of dirt and water, lengthening the service life of driveline parts.

A good rule is to grease all driveline fittings whenever the front end is greased. Remember that most original equipment U-joints do not have grease fittings.

U-joint greasing is a simple procedure, requiring the same grease gun as used for front-end lubrication. U-joint grease fittings, often called *Alemite,* or *Zerk, fittings,* are identical to those used on front-end parts. The greasing procedure is simple:

1. Wipe off the grease fitting with a rag.

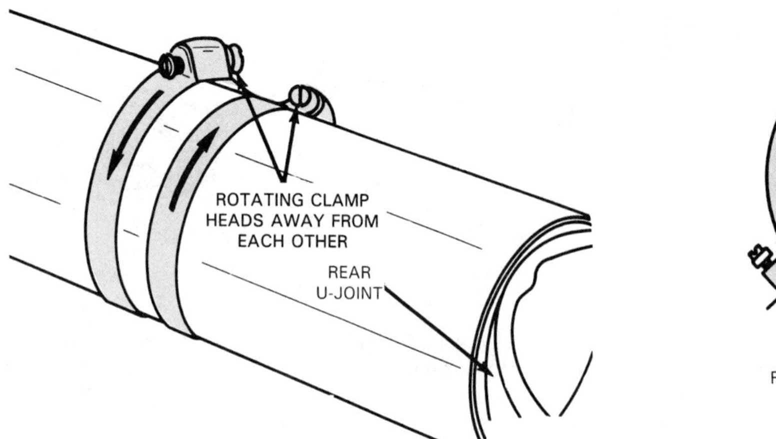

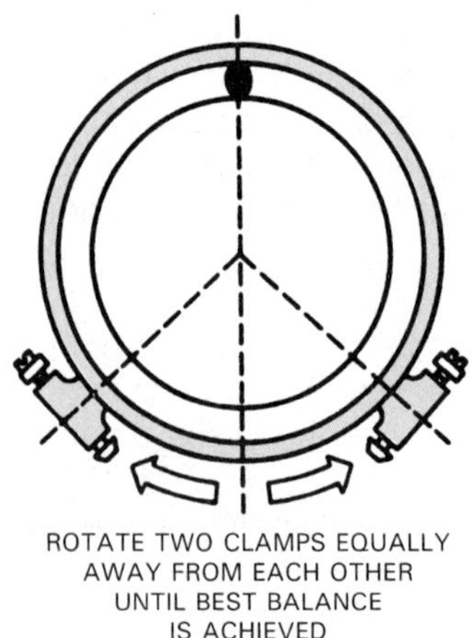

Figure 15-13. This shows the next step in drive shaft balancing. Hose clamps are rotated and the vehicle is rechecked to obtain the location of least vibration. If vibration cannot be reduced beyond a reasonable amount, the drive shaft tube must usually be replaced. (General Motors)

2. Install the grease-gun nozzle over the grease fitting.
3. Pump in grease until the bearing cup seals swell slightly.
4. Remove the grease-gun nozzle and wipe off excess grease.

Some fittings require special adapters. In other cases, the grease is injected through a hole by a special grease nozzle. The greasing procedure is the same as for standard grease fittings. If the fittings have dust caps or seals, these should be put back after greasing is completed.

Do not *over* grease the U-joint. Pumping in grease until it flows out of the seals can destroy the seals. Water and dirt can then quickly enter the U-joint and destroy it.

Slip yokes and center support bearings that have grease fittings are greased in the same manner as U-joints. The same precautions against overgreasing should be observed.

Driveline Removal

Inspection of the installed driveline may have been enough to pinpoint the cause of a particular problem a vehicle is experiencing. On the other hand, your inspection efforts may have proven inconclusive; in which case, further inspection is required. In either event, the driveline will have to be removed. The driveline must also be removed before the transmission can be removed or before making repairs to the rear axle assembly.

It is easy to remove a driveline. The following removal steps apply to Hotchkiss-type drivelines. Consult the proper service manual to remove a torque-tube driveline.

1. Place the vehicle transmission in neutral.
2. Raise the vehicle in a safe manner.
3. Place match marks at the rear of the drive shaft and differential pinion yoke or flanges. This will guarantee that the driveline will be reinstalled in its original position. Balance will not be affected if installed in the same position.
4. If the driveline has a center support bearing, remove the bracket and save the attaching hardware. In rare cases, the center support is inside of the frame.
5. Remove the fasteners holding the rear of the driveline to the rear axle assembly. This may mean unbolting and separating a companion flange, as in Figure 15-14, or it may mean removing two U-bolts secured with nuts or two straps secured with cap screws. If a CV joint is used, it is probably attached to a companion flange with four cap screws.

NOTE! Do not let loose bearing cups fall off their trunnions once U-bolts or straps are removed. The cups can be held in place with masking tape or a large rubber band if they are very loose. This will keep the needle bearings from spilling out.

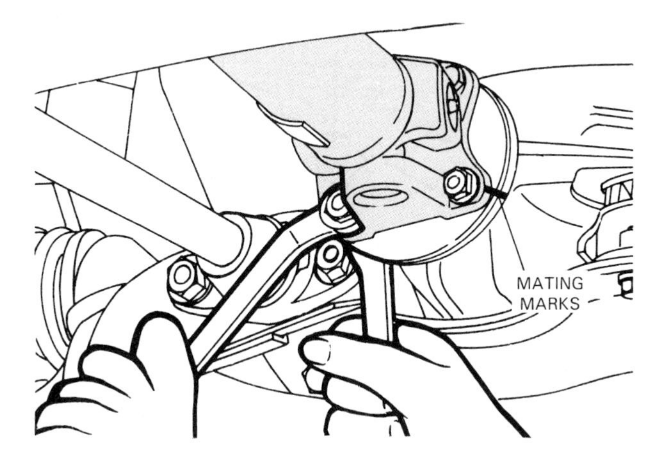

Figure 15-14. This shows the driveline being detached at the rear axle assembly. Note how mating flanges are marked before rear of driveline is unbolted. (Chrysler)

6. Lower the rear of the drive shaft assembly until it is below the pinion yoke or flange. For drivelines that were fastened with U-bolts or straps, slide the drive shaft assembly forward until the rear U-joint clears the differential pinion yoke. The slip yoke will slide about 1 in. (25 mm) into the transmission. If the U-joint sticks to the pinion yoke, it can be lightly pried loose with a large screwdriver. Do not forget to hold the drive shaft while doing this.

CAUTION! Do *not* let the full weight of the driveline hang on the slip yoke. Support the assembly to prevent damage to the extension housing and bushing and the front U-joint.

7. Pull the drive shaft assembly out of the transmission. Some oil may drip out of the rear of the transmission when the assembly is removed, so have a container ready. Drips can be prevented by placing a plastic plug or an old slip yoke in the rear of the transmission or by slightly lowering the front of the vehicle.

Cross-and-Roller U-Joint Service

Once the driveline is removed, you can disassemble it. Prior to disassembly, you can *bench check* the U-joints, and after disassembly, you can further inspect them, if necessary, to pinpoint the cause of a particular problem. If a U-joint is defective, you must install a replacement.

U-Joint Disassembly

Prior to disassembly, quickly bench check the U-joints. Checking a U-joint on the bench is similar to checking it on the vehicle, Figure 15-15. The major advantage to bench checking is that the joint can be more closely inspected. It

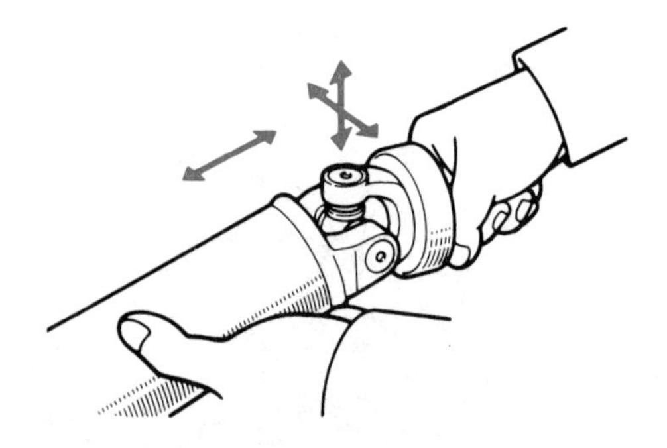

Figure 15-15. Note how to bench check a U-joint for wear. Pull on and wiggle parts attached to U-joints, as shown, while watching for movement. Also, rotate the parts on their trunnions. (Toyota)

allows for bending the joints at angles greater than possible when the drive shaft is on the vehicle. The joints can be closely inspected for excessive play. Move the yokes and U-joints to the limits of their travel and note any roughness, noise, or binding. After you have assessed the joint action, you are ready to remove the U-joints from their installation.

Before trying to remove the U-joints, place the drive shaft assembly on a stand or bench so that the drive shaft is horizontal. The drive shaft yoke can be placed in a vise to keep it from slipping off the bench. Be careful not to clamp it too tightly!

Start by putting on your safety glasses and removing the retainer snap rings or clips for the bearing cups. The U-joints may have internal metal clips that fit over a shoulder in the bearing cup. Alternately, they may have clips that fit into a groove on the outside of the yoke. Use screwdrivers, snap ring pliers, or needle-nose pliers to remove them.

A few original equipment U-joints use injected plastic to hold the cups into the yokes. The plastic can be melted with a torch just before removing the cups. Replacement U-joints for injected plastic retainers are held with internal metal clips.

Bearing cups of large trucks and some cars may be held with U-bolts, straps, or covers. These can be removed by unscrewing the fasteners.

Note that if you are removing a front U-joint or any U-joint between two shafts, you *must* scribe match marks on the neighboring parts. For instance, mark the slip yoke and the front part of the drive shaft. This is so they can be reassembled in the same position relative to each other.

Next, obtain two sockets from a socket wrench set. One should have an inside diameter that is slightly larger than the bearing cup. This socket will act as a receiver for the bearing cup. The other should have an outside diameter that is slightly less than that of the bearing cup. This socket will act as a bearing cup driver. Note that special bearing removal sleeves or that two short segments of pipe can be used instead of the sockets.

Position the sockets in a press of some sort to drive the bearing cups out of their yokes. The driving can be done with several different types of tools, including:

- A vise, as in Figure 15-16.
- A C-clamp or a C-frame driver, as in Figure 15-17.
- An arbor press, as in Figure 15-18.

Alternately, a hammer and driver can be used to drive bearing cups from their yokes, as in Figure 15-19. It is helpful to place a socket on the bench to receive the bearing cup, placing one of the yoke lugs against the socket. Place the driver over the opposite bearing cup. Carefully strike the driver with a hammer until the cross moves down and forces the bottom bearing cup out of the yoke.

WARNING! Do *not* use a socket as a driver when hammering. The socket may shatter when struck with a hammer.

Repeat the removal process on the opposite bearing cup. If the cup does not come completely out, you can remove the cross from its yoke and, then, use a brass drift to force out the bearing cup. Repeat this operation on the other yokes as needed.

U-Joint Inspection

Normally, a U-joint is replaced anytime it is disassembled. If the joint is relatively new, however, you can inspect, clean, lubricate, and reassemble it.

Figure 15-16. A vise and two sockets can be used to remove bearing cups and U-joints. Place socket as shown and slowly tighten the vise. Small socket presses bearing cup on opposite side out of yoke and into large socket. If cup fails to come out completely, clamp it in a vise just until it is snug and tap the yoke from the underside with a plastic mallet. Remove other bearing cup by repeating entire procedure. Then, remove cross. (Chrysler)

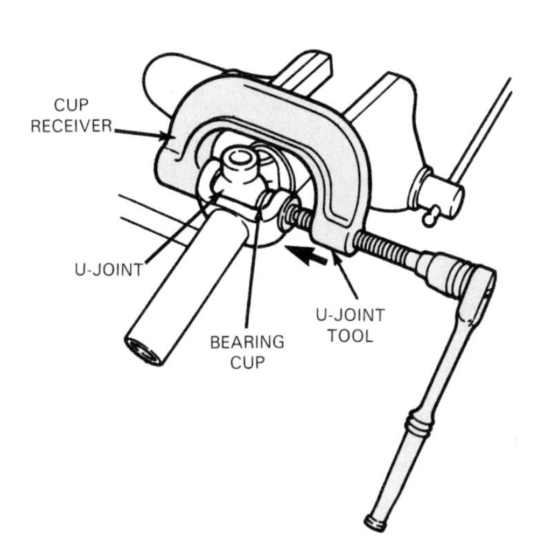

Figure 15-17. This special service tool, resembling a C-clamp, is used to remove bearing cup and cross. Turning screw on U-joint tool drives bearing cup on opposite side out of yoke and into cup receiver. This special tool is self-contained, eliminating the need for sockets. Note that a C-clamp and two sockets could be used in its place. (Ford)

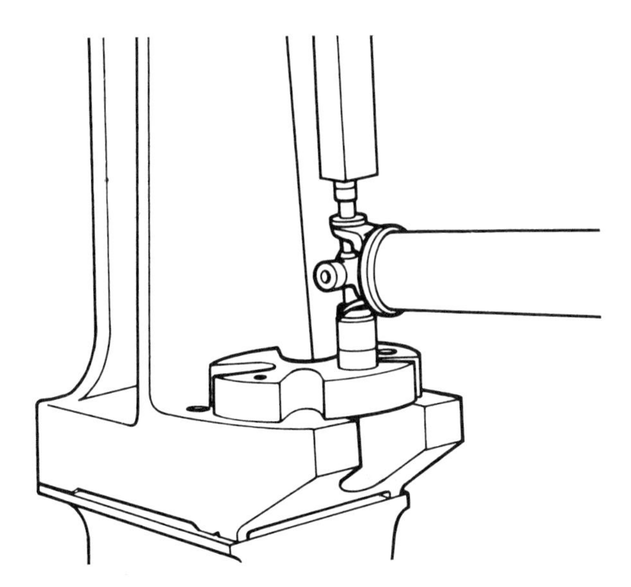

Figure 15-18. This shows an arbor press being used to remove a bearing cup from a yoke. (General Motors)

Inspect the U-joint for wear or rusting and clean the roller bearings and other parts in solvent. The U-joint trunnions should be perfectly smooth and round. There should be no signs of cracking, spalling, or pitting. If there is any sign of such wear or damage anywhere on the U-joint cross, it should be replaced. If rust powder is found in any U-joint bearing cup, replace the U-joint. It makes no sense to go to the trouble to remove and disassemble a U-joint and, then, reinstall a defective joint; U-joints are not that expensive.

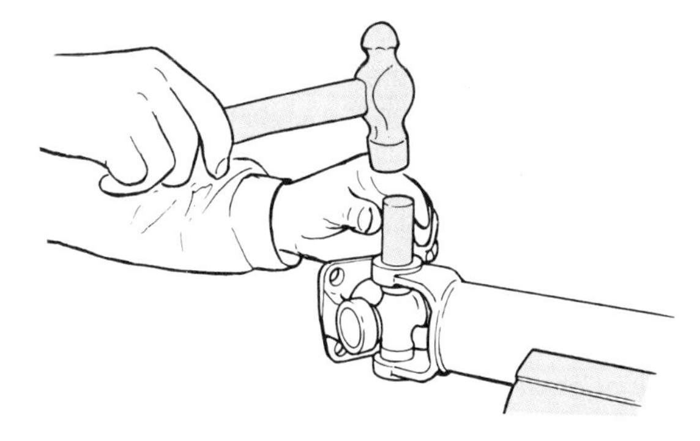

Figure 15-19. In this illustration, bearing cups and U-joint cross are removed with a hammer, a driver, and a vise (used to hold the drive shaft). This procedure should always be done carefully since the chance of damage or personal injury is great. Always wear eye protection when using a hammer on a U-joint and never take a hammer to a socket. (General Motors)

While the driveline is disassembled, check the other driveline parts for wear, rusting, or damage. Make sure that the yokes do not have any burrs caused by removing the bearing cups. Burrs should be removed with a file or sandpaper.

It is easier to check for a bent drive shaft when the drive shaft is in the vehicle. It can, however, be done with the shaft removed by placing it between two V-blocks and checking with a dial indicator. Check in the same places as on an installed drive shaft. If the drive shaft appears to be bent, you must send it to a drive shaft specialty shop for tube replacement.

U-Joint Installation

New U-joints come pregreased. They do not require additional greasing until some time after they are installed. You will need to use grease, however, to retain bearing needles if they have been removed from a bearing cup. If you are sure a used U-joint can be reused, add some grease to each bearing cup after cleaning them and before installation.

If a U-joint disassembled for inspection is found to be usable, begin bearing cup installation by lining the cup with the U-joint needle bearings. To prevent the needles from falling out, always handle the bearing cups gently and use grease to hold the needles in place. Install new seals on each cup and make a final check just before installing the cups onto their trunnions.

Note that the major difficulty when installing a U-joint is having bearing needles fall out of position. They fall to the bottom of the cup and prevent the bearings from fully seating. If this happens, the retainer snap rings or clips will not fit into their grooves, and you will have to disassemble the U-joint and start again.

To install a bearing cup onto its trunnion, push the cup about a quarter of its length into either lug of the front yoke.

See Figure 15-20. Push carefully by hand and check for proper alignment. Make sure that the bearing needles are still in place. Then, position the U-joint cross in the front yoke and into the bearing cup. Hammer or press the cup onto the trunnion, until the bearing cup is flush with the outside of the yoke and fully seated on the cross trunnion. See Figures 15-21 and 15-22.

CAUTION! If the bearing cup fails to press into place with normal pressure, take apart the joint and check the needle bearings. A needle may have fallen over in the cup. If left uncorrected, the U-joints and drive shaft could be ruined.

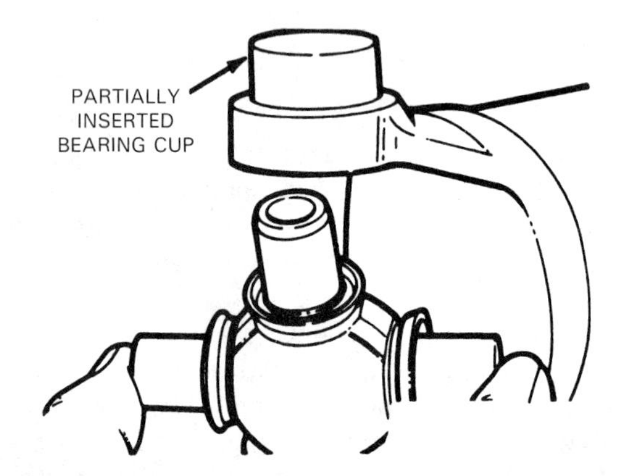

Figure 15-20. Once the bearing cup is partially inserted, it should be checked to ensure that the needle bearings have not fallen out of place. Then, the cross should be installed and the bearing pressed into place on the trunnion. (General Motors)

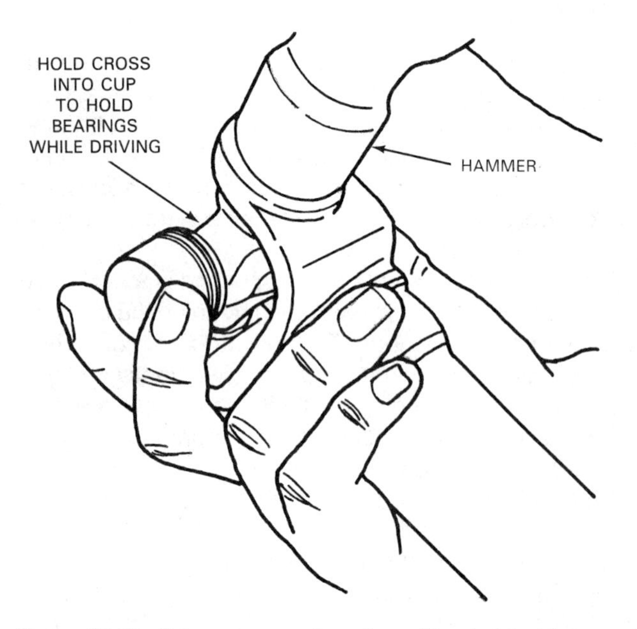

Figure 15-21. Drive cup over trunnion without disturbing needle bearings. Make sure cup installs fully. If you have to use excess force, needle has probably dropped down. Disassemble and start over! (Chrysler)

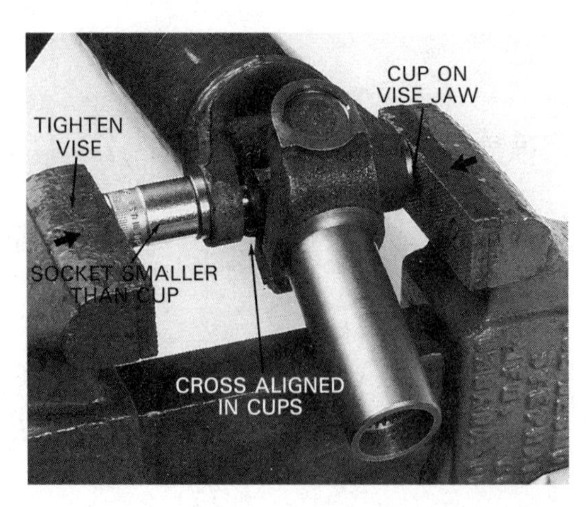

Figure 15-22. A vise and one small socket may be used to press bearing cups into yoke. Press in until you can install one snap ring. Then, press other cap in and install other snap ring. (Chrysler)

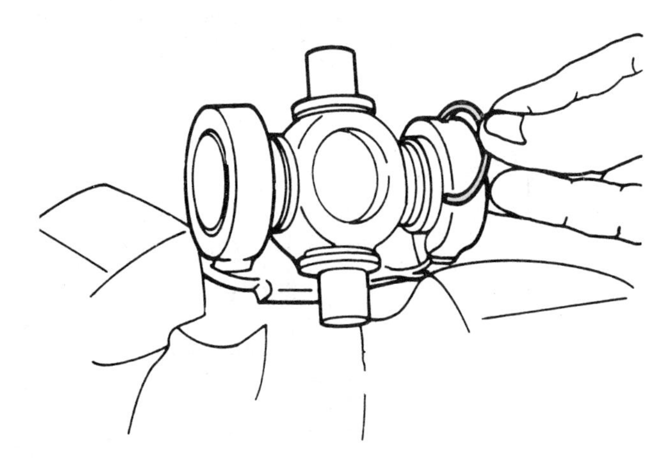

Figure 15-23. Fit retainer clips or snap rings back into their grooves. (General Motors)

Install another bearing cup into the yoke lug opposite to the one just worked on, making sure that the bearing needles do not jam on the trunnion. Hammer or press the cup onto its trunnion.

Install the retainer snap rings or clips now, to ensure that the bearing cups are *fully seated*. See Figure 15-23. Afterwards, make sure that the U-joint can move freely. If the U-joint is jammed, it can usually be freed up by lightly tapping it with a hammer.

Install the rest of the bearings in the same manner, making sure to align the marks that you made during disassembly. Check rotation, as before, to make sure the U-joint moves freely. Bearing cups of rear U-joints held by straps or U-bolts may be placed on their trunnions and secured with tape until the drive shaft assembly is reinstalled. If a new U-joint has a grease fitting, add a small amount of grease to displace any air inside of the cross assembly.

CAUTION! Do *not* overgrease as this will destroy the U-joint seals.

Double Cardan U-Joint Service

A double Cardan U-joint is serviced using the same general procedures just discussed. However, you must mark more parts to assure alignment upon reassembly.

If you are rebuilding a double Cardan U-joint, lubricate the ball and socket with chassis grease. Once this is done, the CV joint is reassembled by engaging the ball and socket, aligning any reference marks. On many of these CV joints, you will need to install the U-joint bearing cups while holding the spring-loaded ball-and-socket assembly together. This is the only way to make sure that the yokes and bearings are properly aligned.

Center Support Bearing Service

Servicing the center support bearing usually involves replacing it. To remove the center support bearing, unbolt its supporting bracket from the vehicle chassis. The drive shaft must then be removed from the vehicle in the usual manner.

Once the drive shaft is out of the vehicle, remove the rear yoke from the front drive shaft by removing the attaching bolts. Then, remove the center support bearing assembly from the shaft. This can usually be accomplished by pressing the center support bearing at the center race. Save all of the washers and spacers.

If the center support bearing is a separate part of the assembly, remove it from the assembly and obtain a new bearing. Install the new bearing by pressing it onto the front shaft until it bottoms. Some bearings are a slip fit and do not require pressing.

Be sure to reinstall any washers or spacers that were originally removed. After the bearing is fully seated, reinstall the rear yoke on the shaft.

Driveline Installation

Before installing the driveline, check all of your repairs. Also, check the condition of the extension bushing and rear seal. Lightly oil the outside of the slip yoke with the proper type of transmission fluid. Double check your alignment marks on the drive shaft and slip yoke (if previously disassembled).

Having made these preliminary checks, proceed with driveline installation as follows:

1. Slide the slip yoke onto the transmission output shaft. Be careful not to damage the rear seal.
2. If the vehicle has a center support bearing, align it with the proper spot on the vehicle chassis and loosely install the fasteners and other retaining hardware.
3. Position the rear U-joint on the differential pinion yoke or position the companion flanges, as applicable. Make sure that you line up the marks that you made during removal. Tighten fasteners, installing U-bolts or straps, as applicable. Ensure that bearing cups align properly in the pinion yoke. Be careful not to overtighten the fasteners and break them.
4. If the vehicle has a center support bearing, tighten the fasteners for the retaining bracket.
5. Lower the vehicle and road test.

Summary

Drive shafts are durable and do not ordinarily need much service. However, other driveline parts, such as U-joints and center support bearings, often wear out and require replacement. A defective U-joint should be replaced as soon as it is discovered. If it is not replaced, the drive shaft can fall off, causing severe damage and even an accident.

The two major types of driveline problems are noises and vibration. Noises can usually be traced to the offending part without much trouble. Drive shaft vibrations are often difficult to isolate and may be confused with other sources of vibration. The drive shaft can vibrate because of worn or loose U-joints or center support bearing. Vibration can also be caused by a bent or out-of-balance drive shaft, excessive U-joint angles, or a worn out slip yoke or extension bushing.

Precautions are necessary to prevent damage to the drive shaft or personal injury. Always clamp the drive shaft securely but without denting it. Wear eye protection. If any shafts are being separated, mark them so they can be reinstalled in the same relative position.

U-joints can be checked by attempting to rotate the drive shaft back and forth while watching the yokes. Look for excessive movement or play between the cross and bearing cups. Then, the two yokes are moved up and down and from side to side. Again, any excessive movement means that the U-joint is bad.

Once the U-joints have been inspected, check for problems with the other driveline parts, such as the slip yoke and center support bearing. Also, check the extension bushing.

After performing a visual inspection, you can perform measurement inspections, including checking drive shaft runout and driveline angle. The runout check is used to determine whether or not the drive shaft is straight. Driveline angle inspection provides a way to check angles through the U-joints.

Adjusting driveline angles, balancing a drive shaft, and greasing a driveline are all in-car service procedures. These procedures may correct a vehicle's driveline problem.

Drive shafts require very little maintenance. All parts with grease fittings should be lubricated when the front end is lubricated. Greasing is a simple operation, using the same type of grease gun as used on the front-end parts. Some U-joints, including CV joints, must be lubricated by using a special adapter on the grease gun.

The driveline must be removed so that it can be repaired. It also must be removed to repair the transmission, clutch, or rear axle assembly. Removing the drive shaft is a simple procedure. However, care must be taken to ensure that the heavy drive shaft assembly does not fall, causing damage or personal injury.

The U-joints are removed by pressing or hammering them out of the drive shaft yokes. This operation requires a socket (or other device) with an inside diameter larger than the bearing cup and another tool with an outside diameter smaller than the bearing cup.

U-joints should be carefully disassembled and inspected if you intend to reuse them. Replace U-joints that show any sign of wear, rusting, or other damage.

When installing new U-joints, take care that the needle bearings stay in place. Carefully reassemble the U-joints and make sure that all bearing cups are fully seated and shafts are reinstalled according to their alignment marks. All of the retainer clips or snap rings should be seated.

The center support bearing should be replaced when it gets noisy or loose. Most center support bearings must be pressed on and off, after the center U-joint connector is removed.

Drive shaft installation is the reverse of removal. You must be careful not to damage the transmission rear seal and to tighten all of the attaching hardware securely.

Know These Terms

Backlash; Axial vibration; Radial vibration; Inclinometer; Shim.

Review Questions–Chapter 15

Please do not write in this text. Place your answer on a separate sheet of paper.

1. Most driveline problems are caused by the _____ _____.
2. Most drive shaft vibrations are:
   a. Axial.
   b. Radial.
   c. Terminal.
   d. Bilateral.
3. If a U-joint shows any sign of looseness, it should be:
   a. Greased.
   b. Cleaned.
   c. Repacked.
   d. Replaced.
4. The _____ _____ is used to check drive shaft runout.
5. The _____ is used to check driveline angles.
6. Screw-type hose clamps are used to assist with:
   a. Balancing the drive shaft.
   b. Adjusting driveline angles.
   c. Universal joint service.
   d. Measuring shaft runout.
7. If you pump grease into a U-joint until it flows out of the seals:
   a. It is well greased.
   b. It will fail immediately.
   c. It might ruin the seals.
   d. It might ruin the needles.
8. If the bearing cup retainers will not seat fully:
   a. Disassemble the U-joint and locate the cause.
   b. Press the bearing cups together even harder.
   c. Strike the U-joint yokes with a large hammer.
   d. Leave it alone and do not be concerned.
9. A drive shaft assembly squeaks when backing up. Technician A says to first measure the driveline angles. Technician B says to first try greasing the U-joints. Who is correct?
   a. Technician A.
   b. Technician B.
   c. Both Technician A and B.
   d. Neither Technician A nor B.
10. Bearing cups need to be removed from a drive shaft yoke. Technician A says that injected plastic may be holding the bearing cups to their yoke. Technician B says that internal retainer clips may be holding the bearing cups to their yoke. Who is correct?
    a. Technician A.
    b. Technician B.
    c. Both Technician A and B.
    d. Neither Technician A nor B.

Certification-Type Questions–Chapter 15

1. **All of these statements about driveline problems are true EXCEPT:**
   (A) all driveline problems are caused by vibration.
   (B) U-joints are a common source of problems.
   (C) a bent wheel can be confused with a driveline problem.
   (D) driveline malfunctions can result from misaligned parts.

2. **A rear-wheel drive vehicle vibrates during acceleration, and the problem appears to be in the drive shaft assembly. Technician A says that the first thing to check is the driveline angles. Technician B says that the first thing to check is the condition of the U-joints. Who is right?**
   (A) A only
   (B) B only
   (C) Both A & B
   (D) Neither A nor B

3. **Technician A says that a worn U-joint can cause a clunking sound when reversing direction. Technician B says that a loose U-joint can sometimes be repaired by greasing the fitting. Who is right?**
   (A) A only
   (B) B only
   (C) Both A & B
   (D) Neither A nor B

4. **Which of these is the most likely cause of a squeaking noise coming from the drive shaft assembly?**
   (A) Bent shaft.
   (B) Dry U-joints.
   (C) Worn U-joints.
   (D) Worn slip yoke.

5. **Technician A says that dirt or undercoating on the drive shaft will not affect the smoothness of driveline operation. Technician B says that the slip yoke will usually wear much more than the transmission extension bushing. Who is right?**
   (A) A only
   (B) B only
   (C) Both A & B
   (D) Neither A nor B

6. **All of these statements about checking drive shaft runout are true EXCEPT:**
   (A) the drive shaft should theoretically turn in a perfect circle.
   (B) the drive shaft should be at a sharp angle during runout checking.
   (C) most drive shafts should never require a runout check.
   (D) all other drive shaft problems should be eliminated before the runout is checked.

7. **Technician A says driveline angles can be changed by normal spring wear or by a collision. Technician B says that excessive driveline angles can cause broken U-joints. Who is right?**
   (A) A only
   (B) B only
   (C) Both A & B
   (D) Neither A nor B

8. **Original equipment U-joints do not normally have:**
   (A) needle bearings.
   (B) retainers.
   (C) grease fittings.
   (D) a center cross.

9. Technician A says that to ensure that a drive shaft will not move while removing U-joints, the shaft should be tightly clamped in a vise until it bends slightly. Technician B says that before removing a U-joint between two shafts, the shafts should be marked so that they are reinstalled in the same position. Who is right?

(A) A only
(B) B only
(C) Both A & B
(D) Neither A nor B

10. All of these can be used as adapters to remove U-joint bearing cups EXCEPT:

(A) large sockets.
(B) special removal sleeves.
(C) brass drifts.
(D) short pipe segments.

11. A new U-joint is being installed in a drive shaft assembly. The bearing cup cannot be pressed far enough into the yoke to install the retainer snap ring.Technician A says that the problem could be caused by a needle bearing that has fallen to the bottom of the cup. Technician B says that the problem could be caused by a snap ring that is too thick. Who is right?

(A) A only
(B) B only
(C) Both A & B
(D) Neither A nor B

Chapter 16

Rear Axle Assembly Construction and Operation

After studying this chapter, you will be able to:

- Explain the purpose of a rear axle assembly.
- Identify the major parts of a rear axle assembly.
- Describe the differential drive gears and related parts.
- Calculate rear axle ratio.
- Compare differential and rear axle assembly design variations.
- Describe the operation of a standard differential and of the various types of locking differentials.

The rear axle assembly is used on rear-wheel drive vehicles. This assembly, consisting of the differential assembly, rear drive axles, and housing, is the final leg of the drive train. It is often called the final drive, or rear end. The rear axle assembly is often mistakenly called the differential. The differential is only part of the rear axle assembly.

The basic design of rear axle assemblies has been adopted by all manufacturers for many years. There are several variations, but all operate according to the same basic principles. The major difference between rear axle assemblies depends on whether the vehicle has *solid-axle rear suspension* or *independent rear suspension.* ***Solid-axle rear suspension*** incorporates rigid and nonflexing drive axles and axle tubes; both sides move as one solid unit in response to outside influences. ***Independent rear suspension*** incorporates jointed drive axles (no axle tubes); they allow for flexibility and independent axle movement.

This chapter is designed to identify and explain the construction and operation of various rear axle assemblies. The material in this chapter will provide a basis for understanding how to properly troubleshoot and repair rear axle assemblies.

Construction and Operation Overview

The rear axle assembly, previewed in Chapter 1, includes the differential assembly, rear drive axles, and rear axle housing. Rear axle assemblies are subjected to heavy loads from the engine and road. They are ruggedly constructed and seldom fail. The most common rear end failures are axle bearing failures. A typical rear axle assembly is shown in Figure 16-1.

In a rear axle assembly, engine power enters the drive pinion gear from the driveline and differential pinion yoke/flange. The pinion gear, in mesh with the ring gear, causes the ring gear to turn. The interaction of the ring and pinion turns the power flow at a 90° angle and also reduces its speed. Power from the ring gear flows through the differential case, spider gears, and side gears to the drive axles. The drive axles transfer power from the differential assembly to the rear wheels.

The bearings and rear axle housing are key components of the rear axle assembly. They are designed to support and align the differential assembly and the drive axles. Notice that the bearings and axle housing are large, heavy-duty parts. This is to ensure that they will stand up under hard usage.

Seals and gaskets are also very important to the operation of the rear axle assembly. Seals are used at the differential pinion yoke/flange and at the outer drive axles. Gaskets are used at housing interfaces to provide a tight seal from the outside.

Figure 16-2 is an exploded view of a common type of rear axle assembly. Notice the relationship of the internal parts to the housing and to each other. Note that the rear axle housing and drive axle designs will be different when the vehicle has independent rear suspension. Also, when the rear axle assembly is equipped with a limited-slip differential, it will contain more parts. These features will be discussed later in this chapter.

Differential Assembly

The differential assembly in a rear-wheel drive vehicle has three functions. The first, and most obvious, is to redirect the power flow to drive the rear wheels. The power

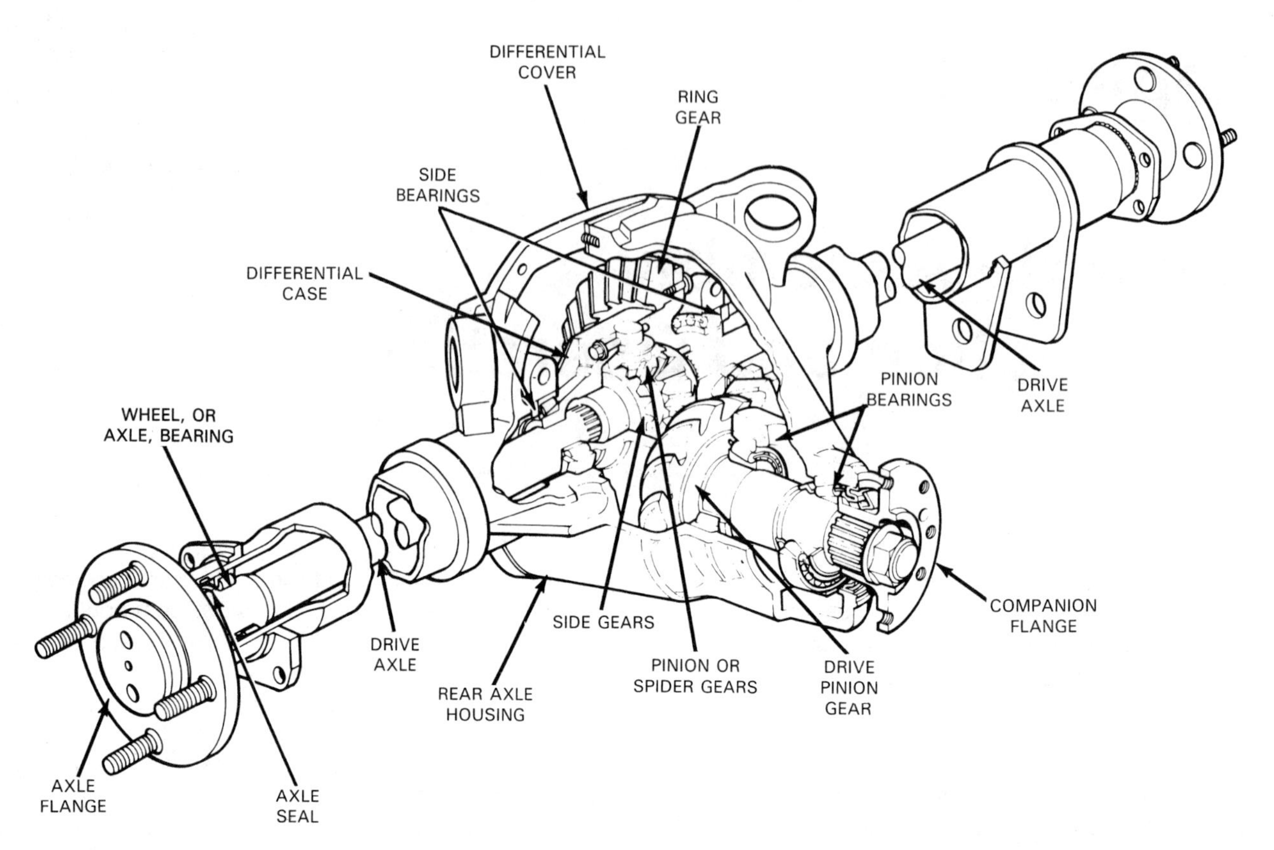

Figure 16-1. Most rear axle assemblies contain the same parts as shown in this cutaway. Note that some drive axles differ from this basic design. (Ford)

flow must make a 90° turn between the driveline and the rear wheels. This is accomplished in the differential assembly by the drive pinion and ring gears.

The second function of the differential assembly is to multiply engine power, reducing speed at the output in the process. If there were no gear reduction (1:1 gear ratio), the vehicle would accelerate very slowly. In some cases, the engine would be unable to move the vehicle. At the very least, gas mileage would be harmed, since the engine would not reach its most efficient rpm range. For this reason, the ring and pinion, by design, provides a reduced speed at its output. The reduction is between 2:1 and about 5:1, depending on the engine size, vehicle weight, and intended use of the vehicle.

The third function of the differential assembly is to allow the vehicle to make turns. If the assembly did not make allowances for the different speeds of the rear wheels during turns, one tire would have to lose traction with the ground to turn corners. The differential assembly allows the vehicle to make smooth turns.

The differential assembly is comprised of numerous parts, including the *differential drive gears, pinion bearings,* differential case, spider and side gears, and *side bearings.* See Figure 16-3. These parts and their function are described in this section in detail.

Differential Drive Gears

The ***differential drive gears*** are comprised of the ring and pinion, Figure 16-4. These hypoid gears redirect power flow by 90° and multiply engine power. The number of teeth in the ring gear compared to the number of teeth in the drive pinion gear sets the rear axle ratio. For instance, if the ring gear has 40 teeth, and the pinion gear has 10 teeth, the ratio is 40:10, or 4:1. The ring gear always has more teeth than the pinion gear. Rear axle ratios can always be determined by dividing the number of teeth on the ring gear by the number of teeth on the pinion gear.

Drive pinion gear

The ***drive pinion gear,*** Figure 16-5, is a hardened-steel gear with an integral shaft. It is machined to mesh with the ring gear, and it rotates the ring gear. The end opposite the gear has external splines that fit internal splines of the differential pinion yoke/flange. The gear is supported by two tapered roller bearings, called ***pinion bearings.***

By design, the axial centerline of the drive pinion gear lies below that of the ring gear. With this design, the pinion gear is placed lower in the rear axle housing. This is done to lower the drive shaft and, therefore, the drive shaft hump

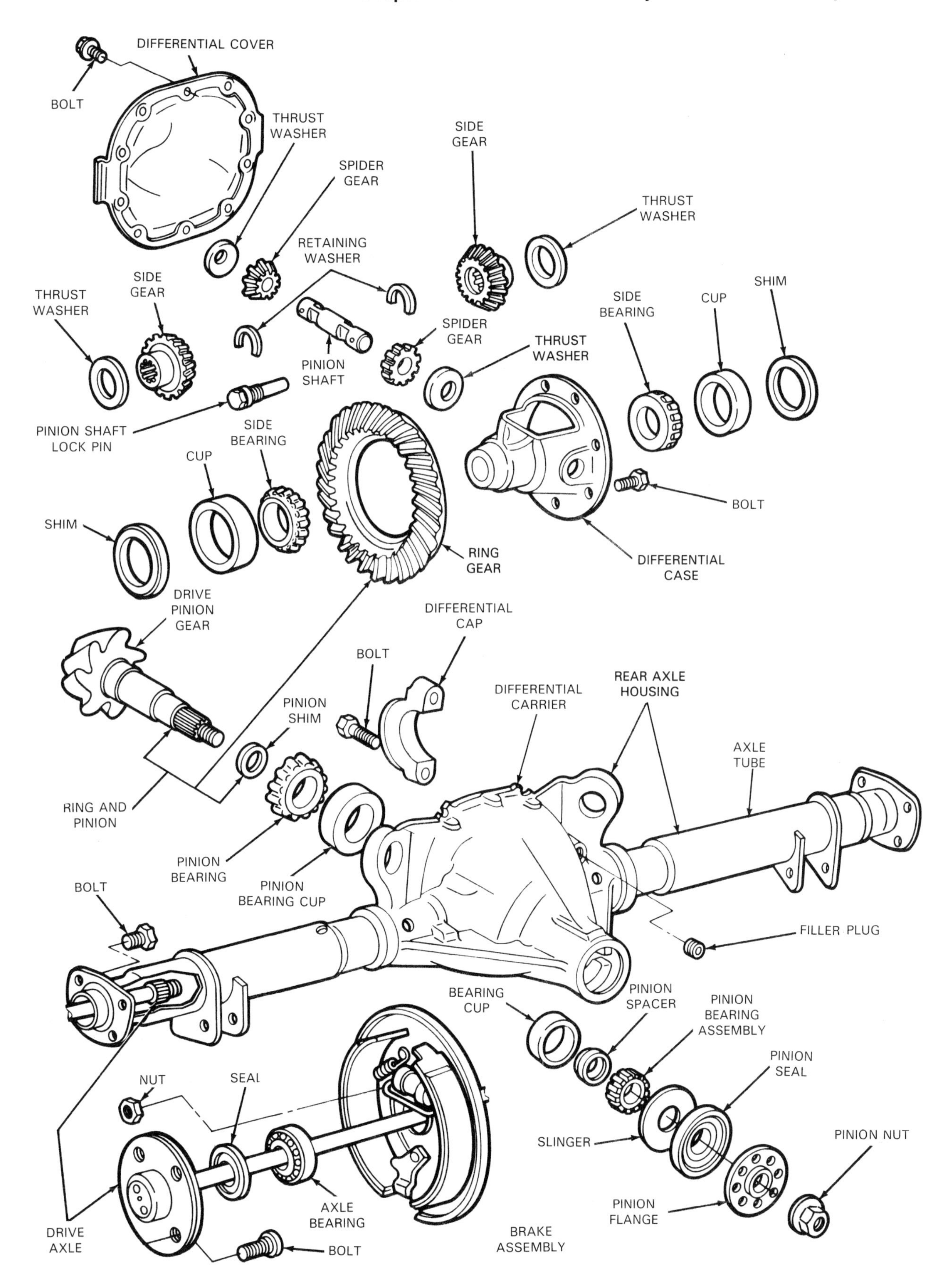

Figure 16-2. This is an exploded view of rear axle assembly shown in Figure 16-1. (Ford)

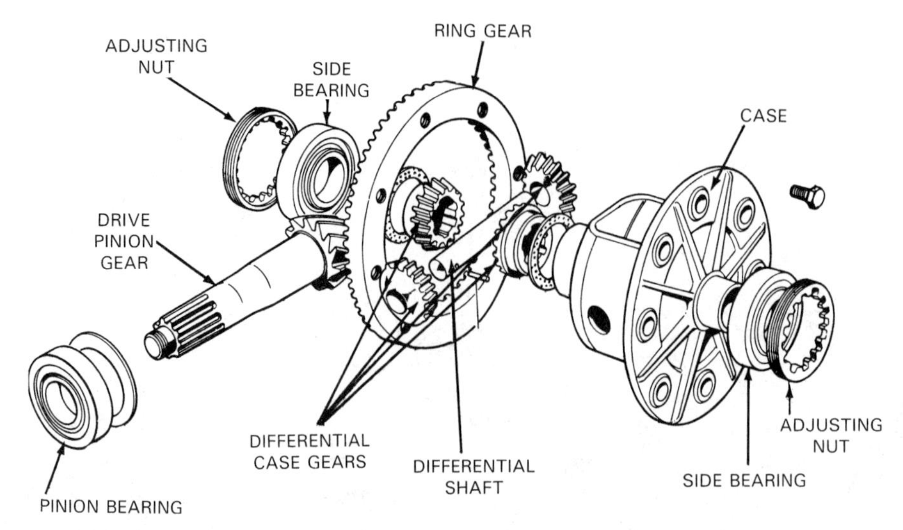

Figure 16-3. This shows relative positions of parts of differential assembly. Interaction of various parts of differential may be more easily understood by studying this illustration. (Subaru)

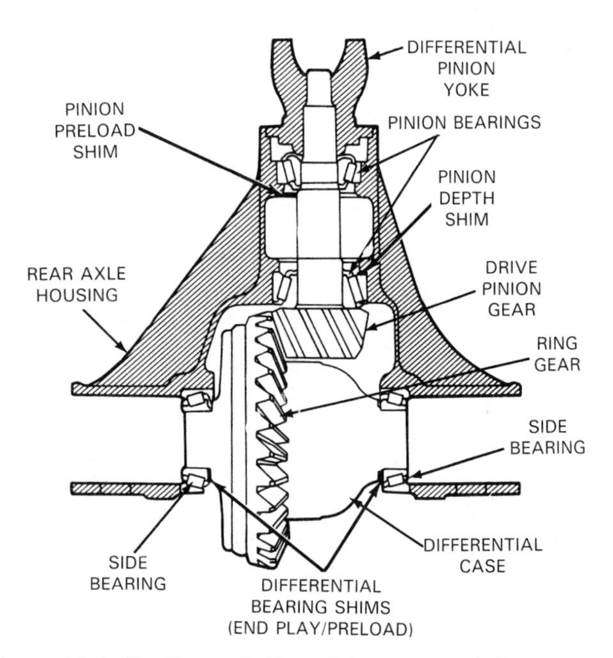

Figure 16-4. Positions of drive pinion gear and ring gear are always about the same. Two *pinion bearings* and two side bearings are always tapered roller bearings that must be carefully adjusted. Some pinion gears have a third bearing–a *pinion pilot bearing*–for support. Bearings and adjusting devices (shims or adjusting nuts) are usually located as shown. (Chrysler)

in the vehicle passenger compartment. The spiral design of the gear teeth allows the gears to mesh with a sliding motion, creating a smoother power transfer. As a result of the sliding action, the gears must have a good supply of the proper lubricant. Gears of this type of gearset are called hypoid gears.

The *rear pinion bearing* is pressed on to the drive pinion gear shaft at the gear end. The *front pinion bearing* is often a slip fit on the smaller end of the shaft. The outer races, or bearing cups, of both bearings are pressed into the rear axle housing.

Either a solid spacer or a *collapsible spacer* (*crush washer*) is used to set the *pinion bearing preload.* The ***collapsible spacer*** is designed to be slightly compressed when the drive pinion gear is installed in the rear axle housing. The spacer maintains a mild pressure between the front and rear pinion bearings, making it possible to accurately adjust the bearing preloads.

The differential pinion yoke/flange has internal splines that fit the external splines on the drive pinion gear shaft. See Figure 16-6. The rear of the yoke/flange, where it fits into the rear axle housing, is machined smooth. This is the sealing surface for the *pinion seal.* The yoke/flange is held to the pinion gear shaft by a large nut and washer that threads onto the shaft. This nut is a type known as a ***jam nut.*** The top threads of the nut are deformed to tightly grip the threads on the drive pinion gear shaft. This is an interference fit. Tightening the nut also adjusts the pinion bearing preload.

The pinion yoke is machined to accept the bearing cups of the rear universal joint. The cups are either pressed in and held with snap rings, or they are attached to the yoke with U-bolts or bolted-on straps.

Pinion flanges are simply a two-piece yoke, joined by mating flanges. The outer section has the yoke; the inner has the external splines for the pinion gear shaft. These companion flanges, as they are also called, would be separated at the flanged section to remove the drive shaft assembly, Figure 16-7.

The position of the pinion gear relative to the ring gear must be set exactly. Otherwise, the gears will be noisy and will wear out quickly. The position of the pinion in the housing must be carefully adjusted so that it contacts the

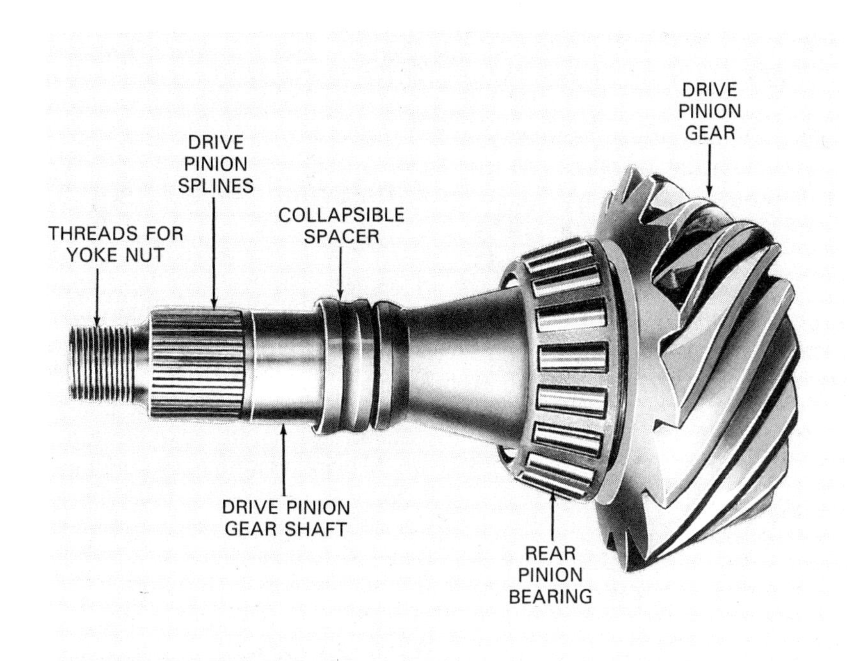

Figure 16-5. A typical drive pinion gear is shown here. Rear pinion bearing is pressed on drive pinion gear shaft. Collapsible spacer is used to aid in pinion bearing installation. Threads and splines at front of pinion are used for installing differential pinion yoke. (General Motors)

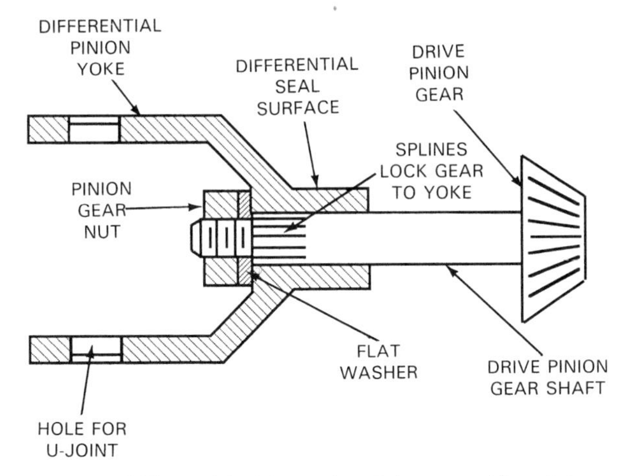

Figure 16-6. Differential pinion yoke slides over drive pinion gear shaft and is secured by pinion gear nut. Tightening nut also preloads pinion bearings. Outer surface of pinion gear shaft seals against front oil seal.

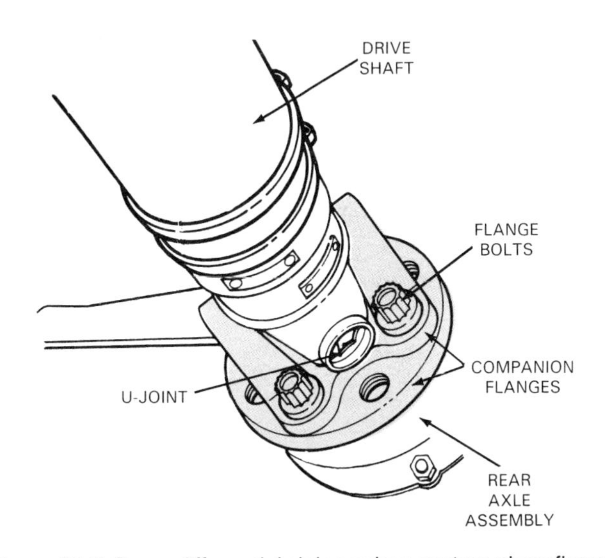

Figure 16-7. Some differential pinion yokes are two-piece flanged assemblies, as shown here. This type of design is referred to as a differential pinion flange, or companion flange. (Ford)

ring gear at exactly the right tooth depth. To make this adjustment to the *ring and pinion clearance,* a ***pinion shim*** is installed in the housing, behind the rear bearing cup. The thickness of this shim determines the depth of the pinion in the housing. This shim is installed at the factory when the rear end is assembled. It must be checked for proper thickness whenever the pinion gear is removed.

Figure 16-8 shows the position of the pinion shim on most rear axle assemblies. This figure also shows the relative position of the collapsible spacer in the pinion assembly.

Ring gear

The ***ring gear,*** Figure 16-9, transfers power from the drive pinion gear to the differential case. Both the ring gear and the case are machined to fit together tightly when the bolts are installed. The bolts holding the gear to the case pass through the holes in the case. They are threaded into tapped holes in the back of the ring gear.

Since the ring and pinion gear teeth must mesh accurately to transmit motion without noise or damage, the position of the ring gear is important. Automotive technicians should be familiar with gear terminology that will be encountered while adjusting the differential assembly to obtain correct gear positions. The *convex side,* or *drive side,* and the *concave side,* or *coast side,* of the ring gear are pointed out in Figure 16-9A. These terms will be used when differential gears are adjusted. The tooth parts that must be carefully adjusted are identified in Figure 16-9B. The terms *heel* and *toe* will be used extensively for ring and pinion adjustment.

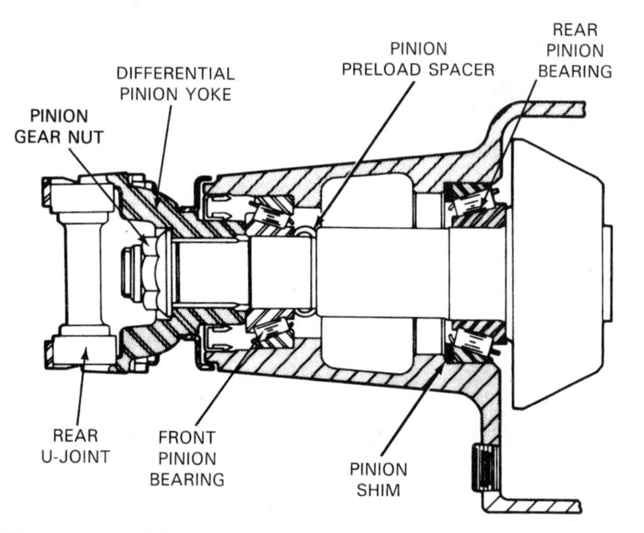

Figure 16-8. Pinion shim and preload spacer locations are shown here. Proper pinion adjustment is critical. Adjusting nut, preload spacer, and depth shim are all critical to proper pinion adjustment. (Chrysler)

Differential Case Assembly

When a vehicle makes a turn, the *outer wheel* travels a greater distance than the *inner wheel*; the arc (or radius) of the turn is greater at the outer wheel. If the rear drive axles were simply connected together, both wheels would have to travel an arc of the same length during a turn. Since this is impossible, one of the tires would have to lose traction, or slip, to make the turn. If the tire did not slip, it would skip over the road surface–a condition called ***wheel hop.***

The purpose of the differential case assembly is to allow the vehicle to make turns without wheel hop. It does this with an arrangement of gears that allows the rear wheels to turn at different speeds. Two basic kinds of differential case assemblies used to accomplish this task are the *standard differential* and the *locking differential,* of which there are several types.

Standard differential

The ***standard differential,*** also called a ***single-pull differential,*** is composed of meshing spider and side gears enclosed in a differential case. See Figure 16-10.

The standard ***differential case*** is usually a one-piece unit. The ring gear is bolted to the case. The case is usually made of cast iron. Occasionally, it is made of aluminum. Side bearings are usually pressed onto the case.

The ***spider gears,*** made of hardened steel, are held in place by a steel shaft which passes through the differ-

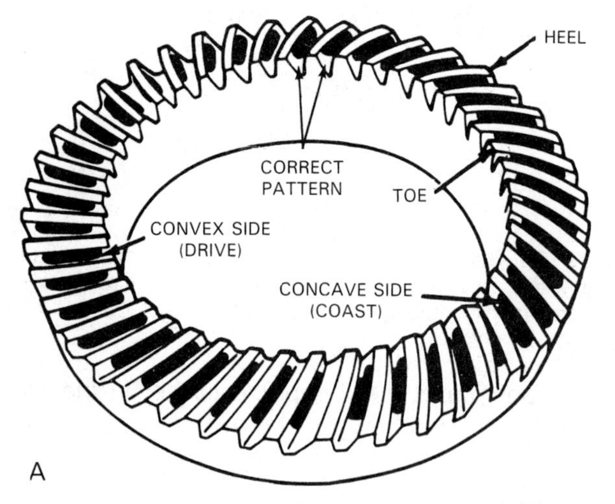

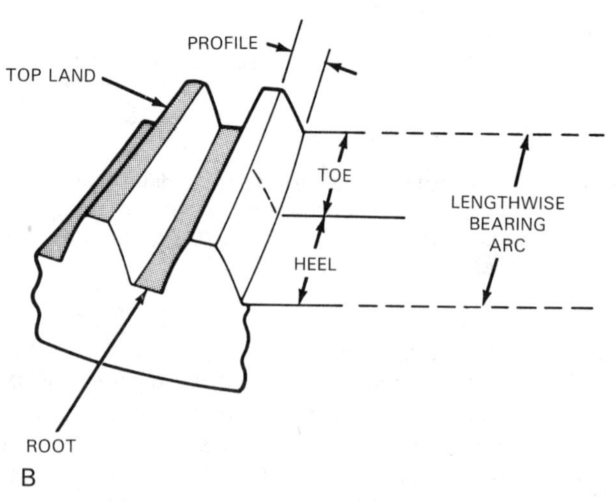

Figure 16-9. When installed, ring gear is bolted to differential case, and it meshes with drive pinion gear. (General Motors, Chrysler) A–The ring gear has convex and concave sides. Convex side is the drive side. It contacts pinion gear when vehicle is accelerating. Concave side is the coast side. It contacts pinion gear when vehicle is decelerating. B–Gear terminology will be important when differential assembly is serviced. Proper heal and toe contact is critical to quiet operation and long life.

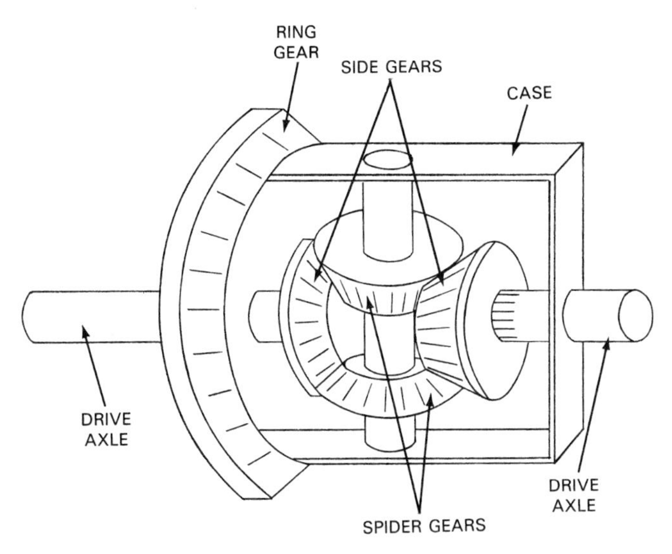

Figure 16-10. Basic components of differential case assembly are shown here. Ring gear is bolted to case, and spider gears and side gears are mounted inside. On most differential assemblies, side bearings are pressed onto case. All differentials contain same general parts. Function of parts will be identified in the text.

ential case and the center of the spider gears. This *pinion shaft,* as it is called, is attached to the case with a bolt. Spider gears are also called ***pinion gears.***

Spider gears mesh with ***side gears,*** also made of hardened steel. When the ring gear and differential case turn, the spider and side gears also turn. Power flow is through the case, into the spider gears, and on into the side gears. The side gears are splined to the drive axles. They transfer power to the drive axles and rear wheels. Side gears are also called ***axle end gears.***

Some heavy-duty differentials contain four spider gears and two pinion shafts. In this design, there is a center hole in one of the shafts. The other shaft passes through it. The side gears are splined to the drive axle. On some differentials, the side gears contain *C-locks,* which hold the axles in place. See Figure 16-11.

The spider and side gears are bevel gears. Power transfer through the bevel gears causes them to be forced away from each other. This causes high thrust forces on the backs of the gears, where they contact the differential case. Hardened-steel washers are usually installed between the back of the gears and the case. These washers provide a sliding surface and reduce wear. See Figure 16-12.

Figure 16-13 shows the operating states of the differential while driving straight ahead and while driving around a corner. In Figure 16-13A, the vehicle is moving straight ahead, and both wheels are traveling at the same speed. The spider and side gears rotate with the case but do not move in relation to it. The entire case assembly rotates as a unit.

When the vehicle makes a turn, the axles and the side gears begin turning at different speeds. The outer wheel–the left wheel, in the case of a right turn–turns faster than the inner wheel, and the left side gear turns faster than the right side gear. See Figure 16-13B. As a result of the different axle speeds, the spider gears begin to rotate. The left side gear, which is moving faster than the right side gear, drives the spider gears, causing them to rotate on, or walk around, the right side gear.

Note that the differential case speed on turns is the average of the side gear speeds. This is because one side gear is rotating faster than the case and the other side gear is rotating slower than the case. In Figure 16-14, when the vehicle makes a turn, the action of the differential allows the outer wheel to turn at 110 percent of case speed, while the inner wheel turns at 90 percent of differential case speed. These percentages will vary with the radius of the turn.

Locking differential

The standard differential works well in most situations. However, on very slippery surfaces, such as icy or muddy roads, lack of traction can cause the rear wheels to slip. This is because the standard differential will drive the wheel with the least traction.

If one drive wheel is on dry pavement, and the other is on ice or mud, the ring gear and differential case will drive the spider gears. However, the spider gears will not drive both side gears. When the spider gears are driven by the differential case, they will walk around the side gear related to the wheel on dry pavement. As a result, the spider gears drive the slipping wheel, and the vehicle will not move. The standard differential sends almost all engine power to the slipping wheel.

To overcome this problem, locking differentials are used. ***Locking differentials*** overcome traction problems by sending some power to both wheels, while allowing the vehicle to make normal turns. There are several different types of locking differentials, including limited-slip, *ratchet,* and *Torsen differentials.*

Limited-slip differential. There are two basic types of ***limited-slip differential.*** The more prevalent of the two is the *clutch-plate differential,* which uses multiple-disc clutches. The other type is the *cone differential,* which uses a cone clutch. These differentials have various brand names, including *Positive Traction, Sure-Grip, Anti-Spin* and *Traction-Lok.* Due to their complexity and higher cost, only about 2 percent of rear-wheel drive automobiles and about 10 percent of trucks have limited-slip differentials.

An example of a common ***clutch-plate differential*** is shown in Figure 16-15. The most obvious difference between this limited-slip differential and a standard differential is the clutch packs placed between the side gears and the differential case.

The clutch friction discs are made of steel covered with a friction material. The clutch plates are made of steel. The discs and plates are alternately splined to the side gear, Figure 16-16, and *dogged* (meaning tabs fit into grooves)

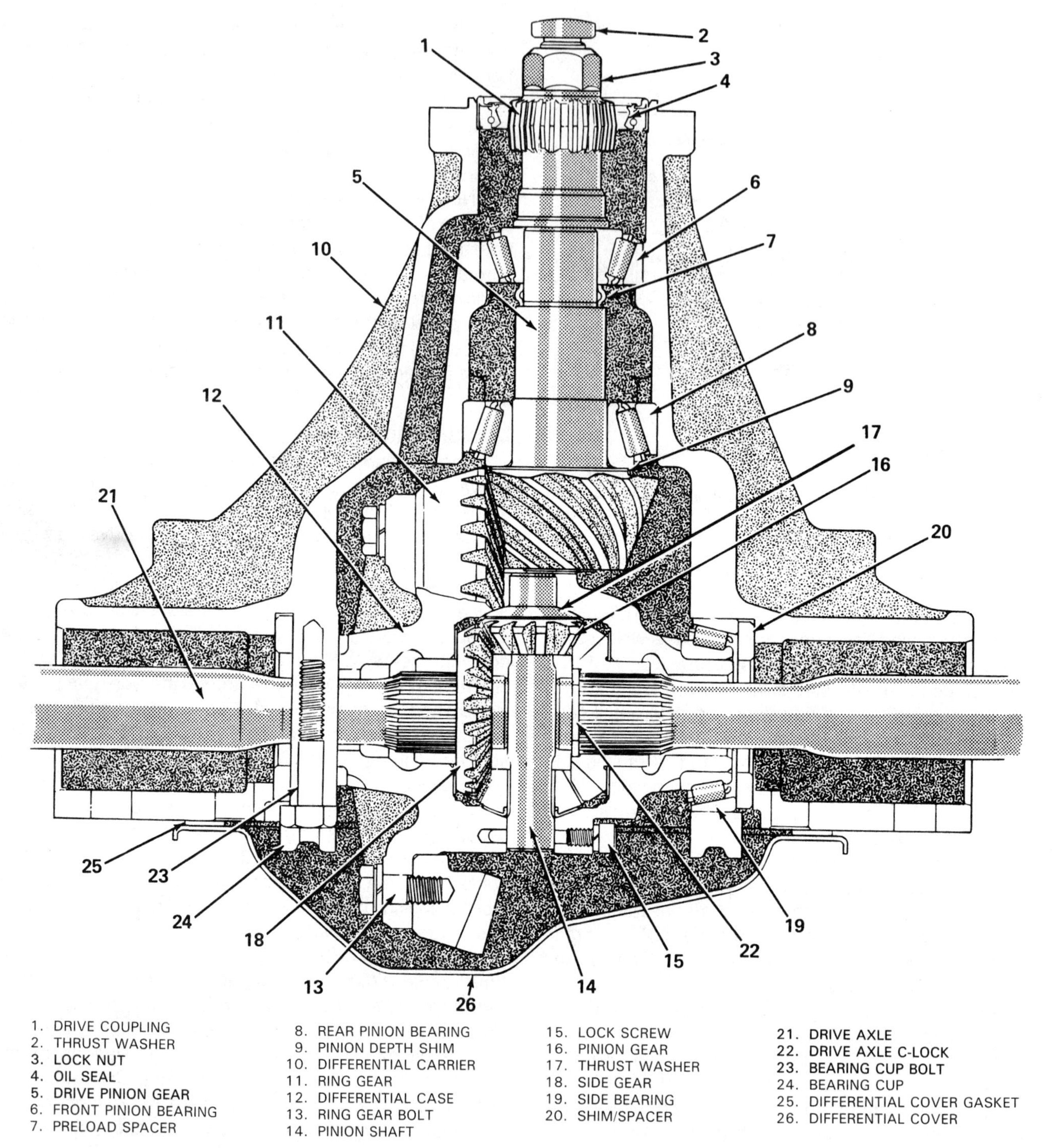

Figure 16-11. This shows a section view of a typical differential contained in rear axle assembly. Differential case is installed in rear axle housing. Ring gear, which is attached to differential case, meshes with drive pinion gear. Relative positions of parts are similar on all differentials. Note C-locks on differential side gears. They retain drive axles in housing. (General Motors)

to the differential case. Grooves in the discs or plates are for better grabbing power.

Figure 16-17 shows the moving parts of a limited-slip differential. The spider gears, side gears, and other parts are very similar to that of the standard differential. The differential case of the limited-slip differential is often made in two parts to allow for clutch pack removal, as shown in Figure 16-18.

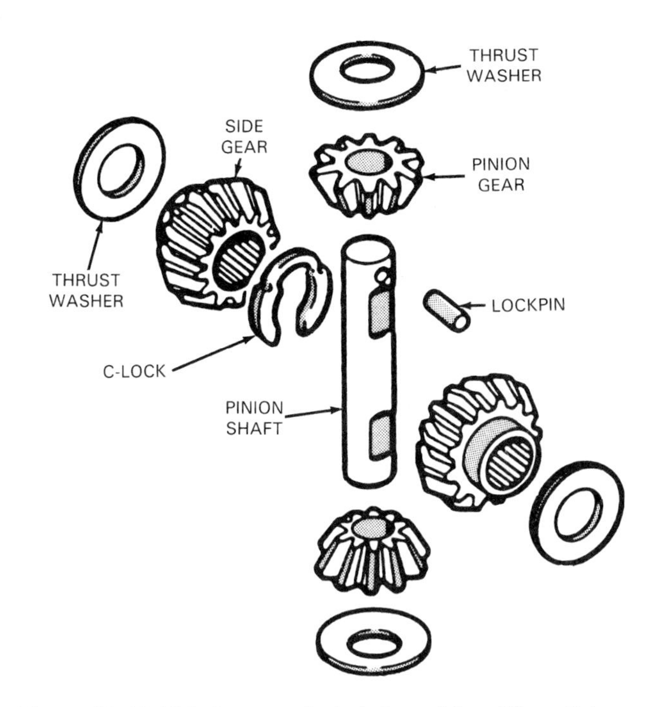

Figure 16-12. This is an exploded view of the differential gears. Note relationship of side and spider gears to each other. Also note thrust washers that separate gears from case and C-lock that holds axle shaft in place. Pinion shaft is held in place by pin that passes through both shaft and differential case. (Chrysler)

The discs and plates are applied by the preload springs and by the mechanical pressure of the spider gears on the side gears. Since the spider and side gears are bevel gears, their teeth try to come out of engagement when the differential is transmitting engine torque. This creates a pushing action on the side gears, forcing them outward against the differential case. The outward pressure of the side gears presses the friction discs and steel plates together between the side gears and the case. Whenever the discs and plates are pressed together, the splined and dogged connections ensure that the side gear and differential case are locked together.

The operation of a limited-slip differential is shown in Figure 16-19. When the vehicle is moving straight ahead, Figure 16-19A, the limited-slip differential operates in the same manner as a standard differential. The rear wheels and the differential case are turning at the same speed. The clutch packs are applied, but are not needed.

In Figure 16-19B, the vehicle is losing traction at one wheel. Since this wheel is slipping, the spider gears are not pressing tightly on the side gear of the slipping wheel. Since the side gear is not being pressed toward the case, the clutch pack of the slipping wheel is not pressed tightly together.

Since there is a normal tendency for the side gears to move away from the spider gears under load, the other side

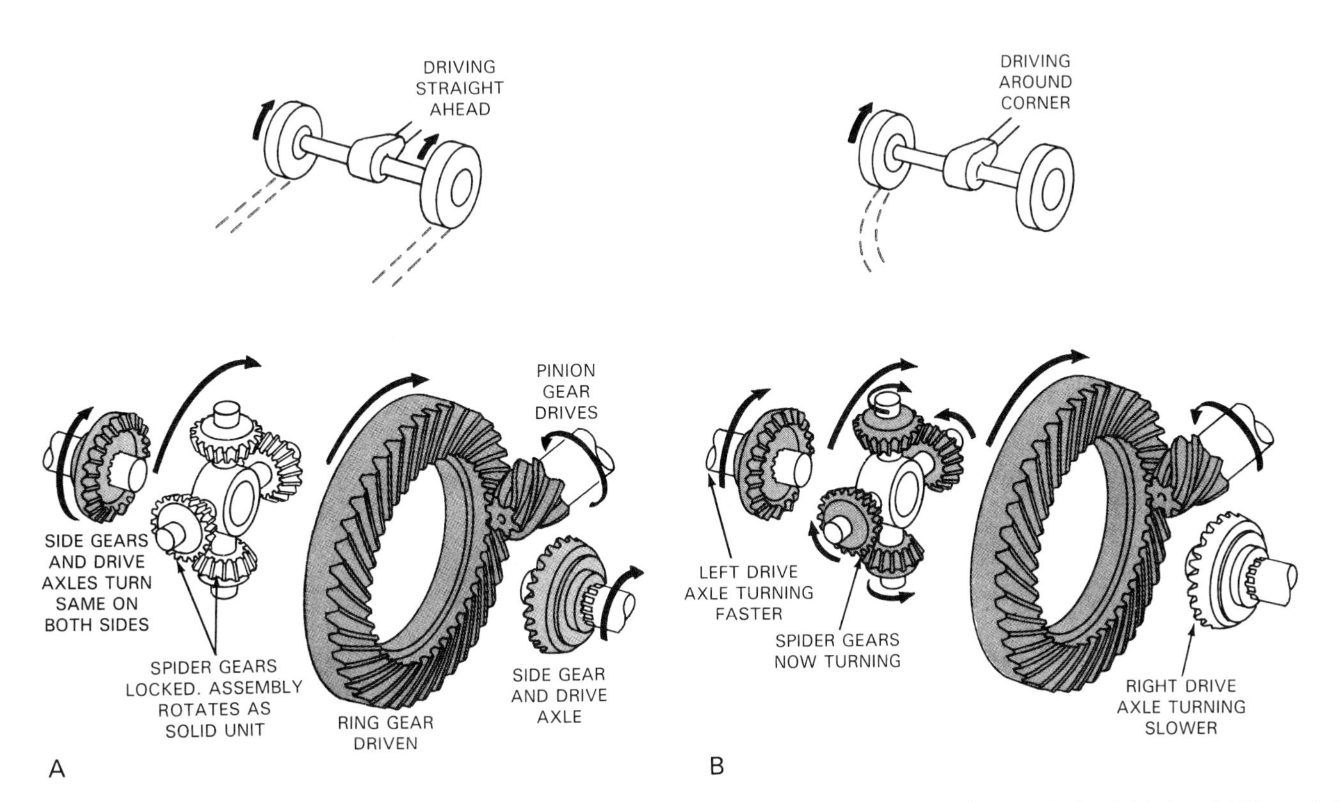

Figure 16-13. Differential action is shown here. Note use of four spider gears, rather than the regular two. A–Straight ahead: Differential case gears turn as a unit. Both drive axles and differential case are turning at same speed. B–Right turn: Left axle is moving faster than right axle. Left side gear drives pinion gears. Pinion gears turn and walk around right side gear. Note that the differential works the same way for a left turn, except the action of left and right sides, just described, is reversed. (Deere & Co.)

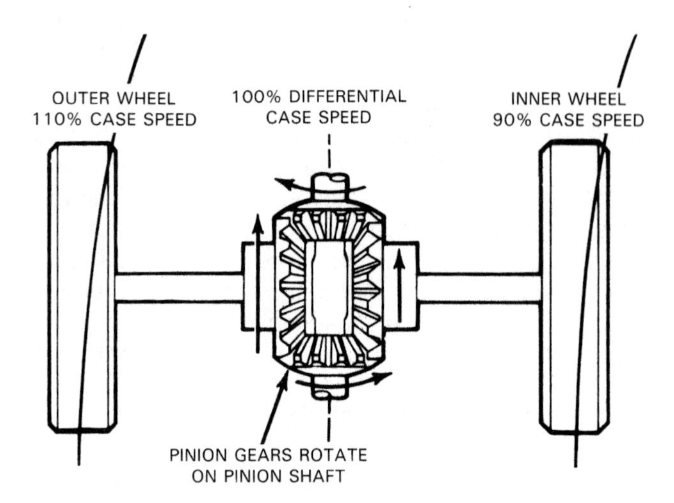

Figure 16-14. Speed of the differential case on turns is average of side gear speeds. (Chrysler)

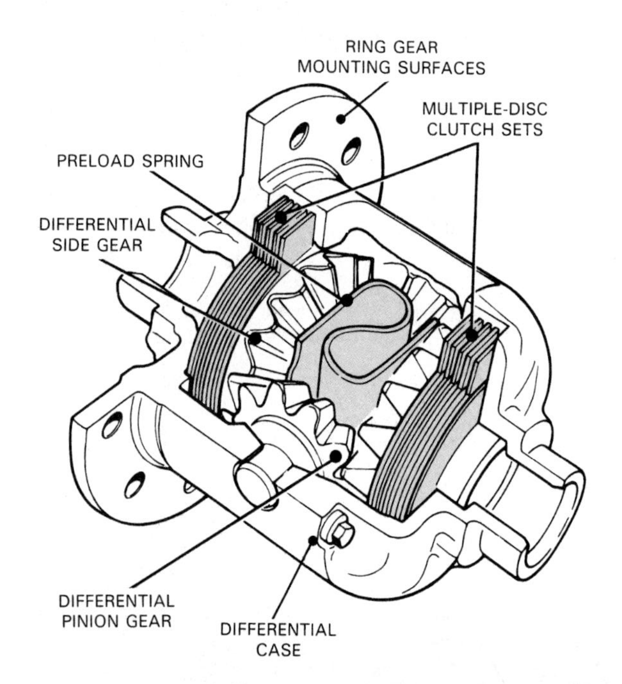

Figure 16-15. Study assembled view of clutch-plate differential. Clutch packs are sandwiched between side gears and differential case. Preload spring applies initial force to clutch packs but still allows enough slippage in clutch pack for normal differential operation. (Ford)

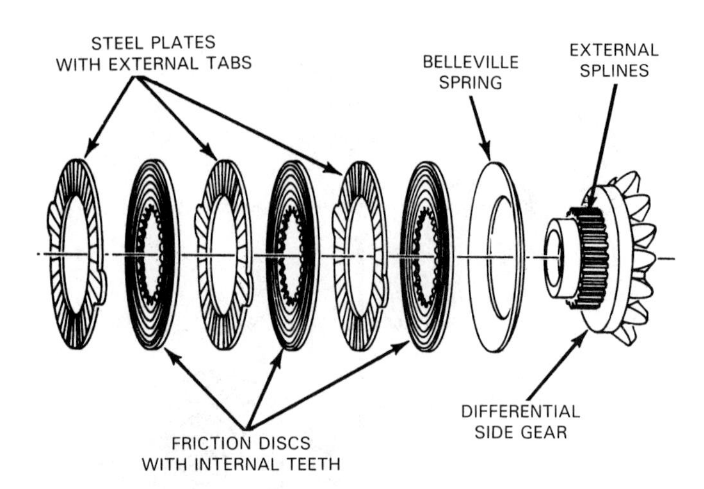

Figure 16-16. Exploded view shows clutch pack of limited-slip differential. Each clutch pack has same number of clutch discs and plates. Note internal teeth on friction discs and external tabs on steel plates. Grooves in discs and plates reduce chance of slippage. (Chrysler)

gear moves outward and away from the spider gears. It is under load because its related wheel has traction. The pressure on this side gear causes the related clutch pack to be pressed tightly together. The side gear is locked to the case by the clutch pack, and power is delivered to the wheel with traction.

The clutch pack is designed to slip when some preset torque value is reached. When the vehicle is making a turn, a high torque, caused by the outer wheel rotating faster than the case, causes the clutch pack to slip. This allows the differential to operate in the same manner as a standard differential when making turns. The discs and plates slide against each other–discs turning with side gears, plates turning with case–allowing different rotating speeds between case and side gears and, therefore, between rear wheels.

Figure 16-20 shows a ***cone differential,*** which is another version of the limited-slip differential. In place of clutch packs, friction-lined cones are used. The operation is similar to that of the clutch-plate differential. Preload spring and side gear pressures force the cone into a dished depression in the differential case. Friction tries to lock the cone and, therefore, the side gear to the case, sending power to the wheel with the most traction. Figure 16-21 is an exploded view of the cone differential.

Note that both clutch plate and cone differentials require special limited-slip gear oil. Using ordinary gear oil in limited-slip differentials will cause the discs and plates or cones to slip and vibrate during turns.

Ratchet differential. The ***ratchet differential,*** nicknamed a *Detroit locker,* uses a series of cams and ramps to direct power to the wheel with the most traction. Its operation depends on relative wheel speed, rather than on wheel traction. The ratchet differential transfers power through a set of teeth that can be engaged and disengaged. This kind of engaging teeth system is sometimes called a *dog clutch.* The series of cams and ramps disengage the teeth of the dog clutch on the side of the wheel with the least traction. An example of the ratchet differential is shown in Figure 16-22.

For straight-ahead driving, Figure 16-22A, both sets of teeth are engaged, and the differential case and wheels turn at the same speed. During turns or when one wheel loses traction, Figure 16-22B and 16-22C, the speed

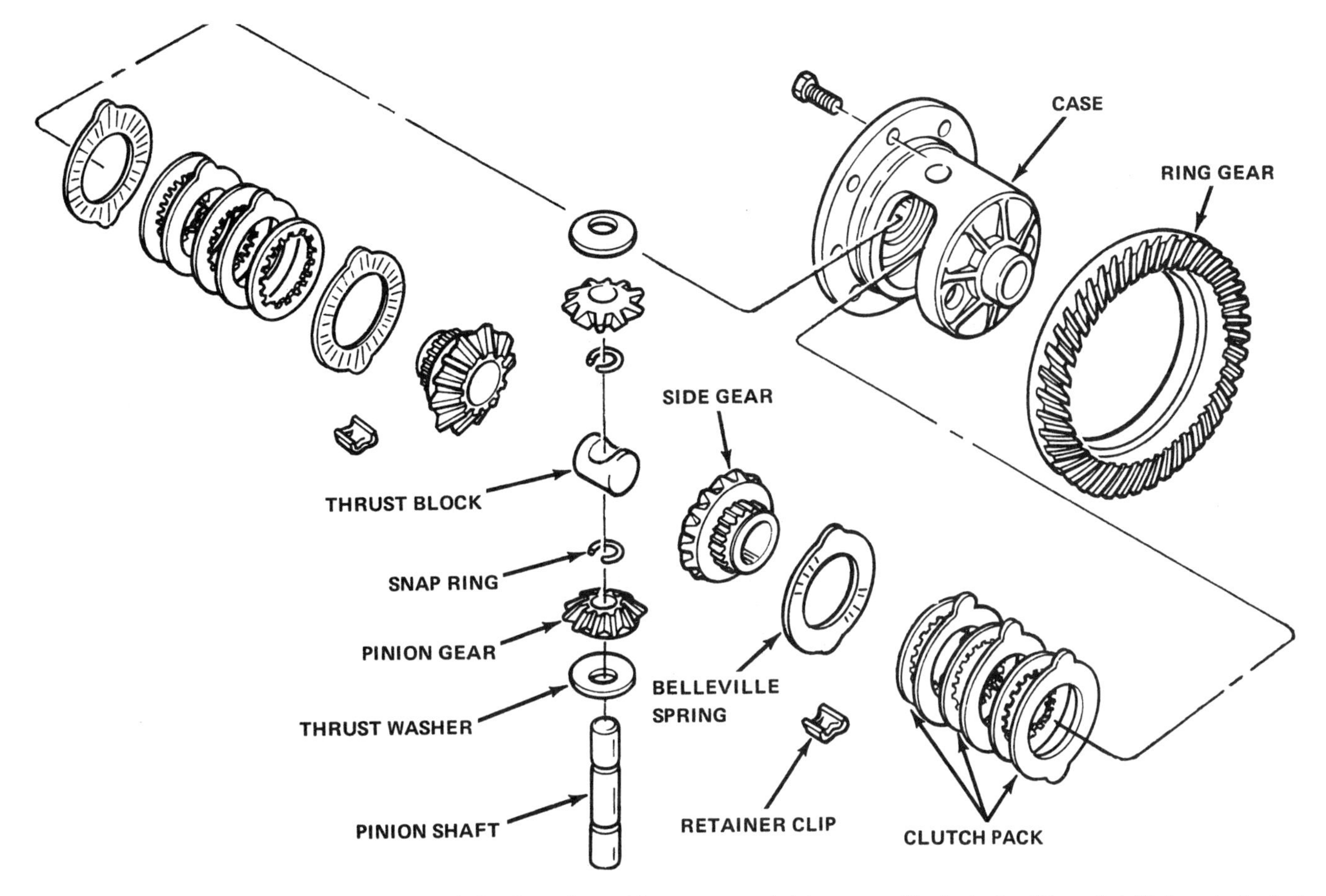

Figure 16-17. This shows relative positions of clutch packs, spider gears, and side gears of limited-slip differential. Notice similarity to standard differential. (Chrysler)

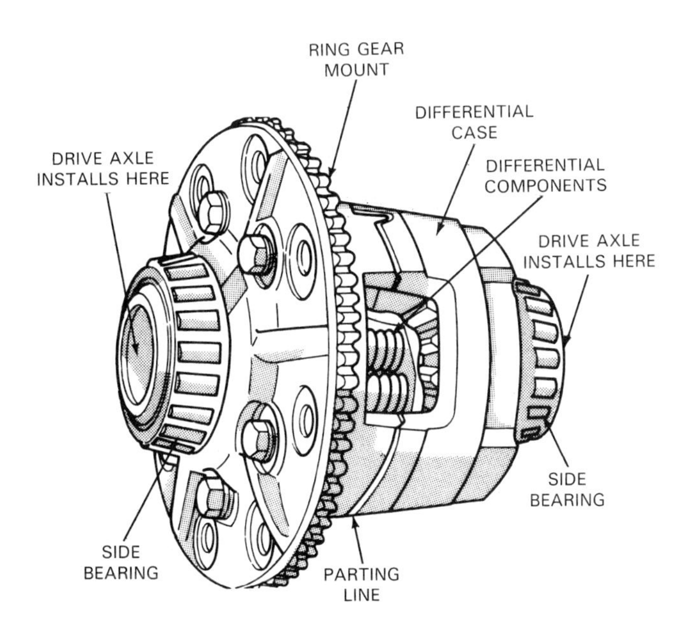

Figure 16-18. The differential case of a limited-slip differential is often made in two parts. Note parting line on case of this limited-slip differential, used to separate the two halves. (Chrysler)

difference between the wheels causes the internal cam and ramp to disengage the teeth on the side of the faster moving wheel. All power is then sent through the other wheel.

Since the faster moving wheel is always the one that is slipping, power always goes to the wheel with traction. On turns, the loss of power to the outer wheel is not noticeable. This design is durable and does not require special gear oil, but it is often rough and noisy in operation. It is usually used in off-road and racing vehicles.

Torsen® differential. The ***Torsen differential*** is a locking differential using complex *worm gearsets.* The gearsets include *worms* (drive gears) and *worm wheels* (driven gears). The Torsen differential has been available since the 1960s as a high-performance replacement unit for standard differentials. It is now being offered as original equipment on some European cars. The basic mechanical principle of this differential is that while the worm can drive the worm wheel, the worm wheel cannot drive the worm.

As shown in Figure 16-23, the Torsen differential has two central worms. For purposes of clarity, these will be referred to as *axle gears.* One axle gear is attached to each axle shaft. Worm wheels ride on and are driven by the axle

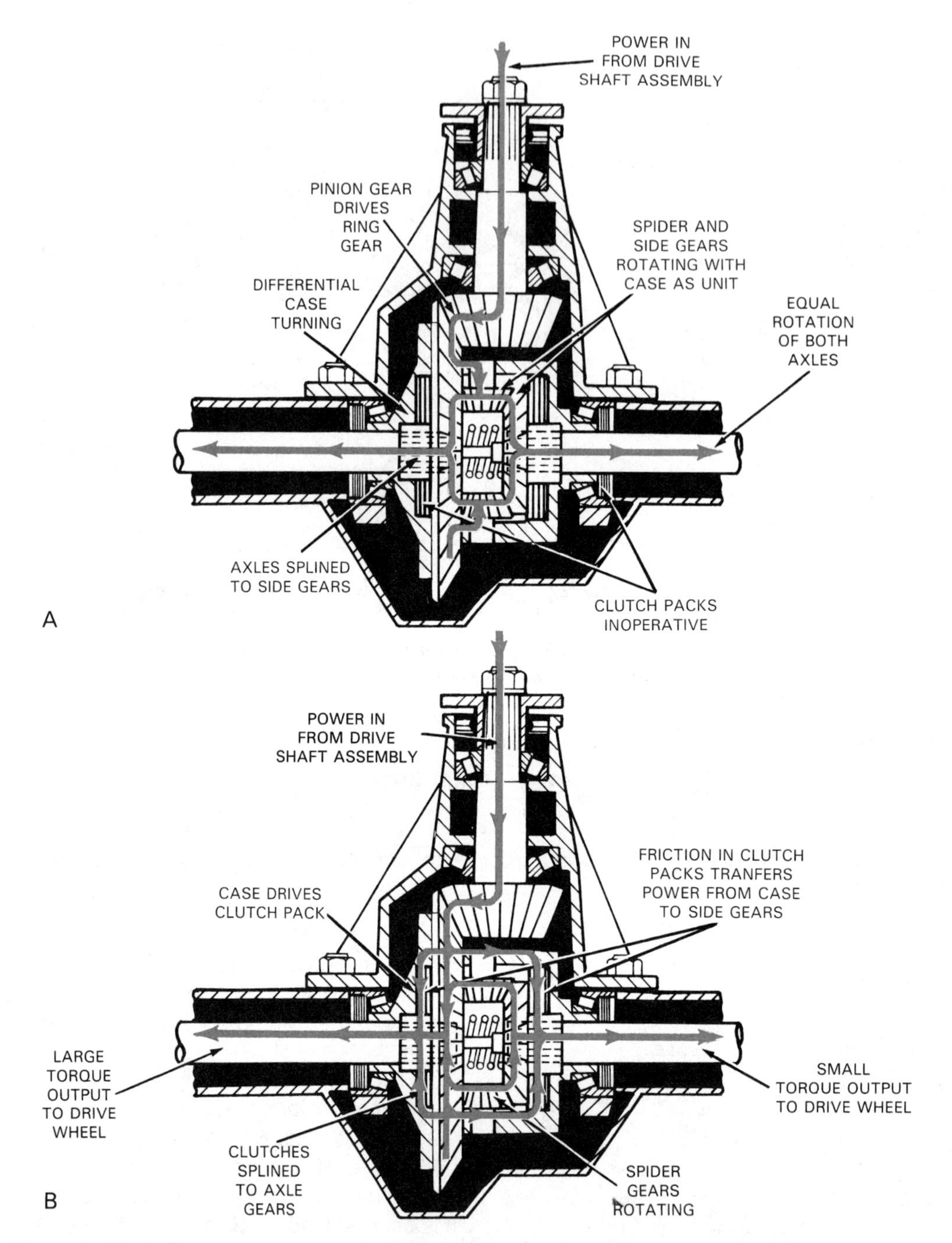

Figure 16-19. Study action of limited-slip differential. A–Traction on both wheels. Differential parts are locked together and rotate as a unit. Clutch packs are not operating. B–One wheel slipping. Pressure on side gear of wheel with traction causes discs and plates of related clutch pack to grab, sending most of engine power to that wheel.

gears. The worm wheels are held in place by the differential case. Spur gears machined on the ends of the worm wheels mesh and form the only connection between the two axle shafts. Engine power drives the differential case, and the worm wheels, held by the case, turn with it. The worm wheels cannot turn the axle gears, so they lock themselves to the gears. In this way, power is transmitted; the axle gears and axles are locked to the case, and they rotate with it.

During straight-ahead operation, the differential assembly operates like a standard differential; all internal gears turn as a unit. When the vehicle makes a turn, or when one drive wheel is slipping, the relative speed of the drive wheels and, therefore, of the axles, change. This speed change is transmitted from the faster axle to the slower one by the action of the meshing spur gears.

The axle gear on the faster axle can drive the respective worm wheels. This driving force is transferred from the spur gears on the faster turning worm wheels to the spur gears on the slower turning worm wheels. Engine power is transferred from the faster to the slower worm wheels by the interaction of the gears. The worm wheel on the slower side still cannot drive the slower axle gear, but it can transfer the increased power from the faster wheel as pressure. This pressure increases the amount of power sent to the slower axle gear and axle. It does not turn the axle gear, but it does allow it to turn with more force.

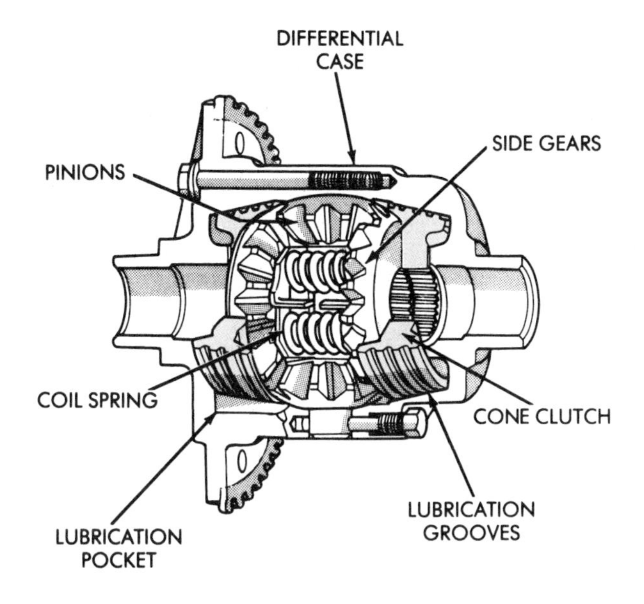

Figure 16-20. Study construction of cone differential. Operation of this limited-slip differential is similar to clutch-plate differential. Pressure on side gear of wheel with traction causes cone to be pressed into dished area of differential case, locking case to drive axle on that side. (Chrysler)

Rear Axle Housing: Solid-Axle Rear Suspension

The rear axle housing is the container and support for the other parts of the rear axle assembly. It also forms a reservoir for the rear end lubricant. The housing accommodates suspension system attachment. Further, most rear axle housings are used to support the stationary parts of the rear brake assemblies.

The rear axle housing associated with solid-axle rear suspension consists of a central housing, or ***differential carrier,*** and ***axle tubes,*** which enclose the drive axles and extend to the rear wheels. (Vehicles with independent rear suspension will not have axle tubes.) Rear axle housings will have a vent to relieve pressure buildup. They will also have oil drain and fill plugs. See Figure 16-24.

Most rear axle housings are made of steel. Steel axle tubes are pressed and welded into the housing or are cast integral with the housing. The axle tubes usually have an integral flange at the outer end. The flange provides a mounting surface for the *brake backing plate* and an *axle retainer plate.*

Since the rear axle housing is a solid structure, it moves up and down with the wheels as they move over bumps and holes. To control this movement, the rear axle housing is attached to the vehicle body through an arrange-

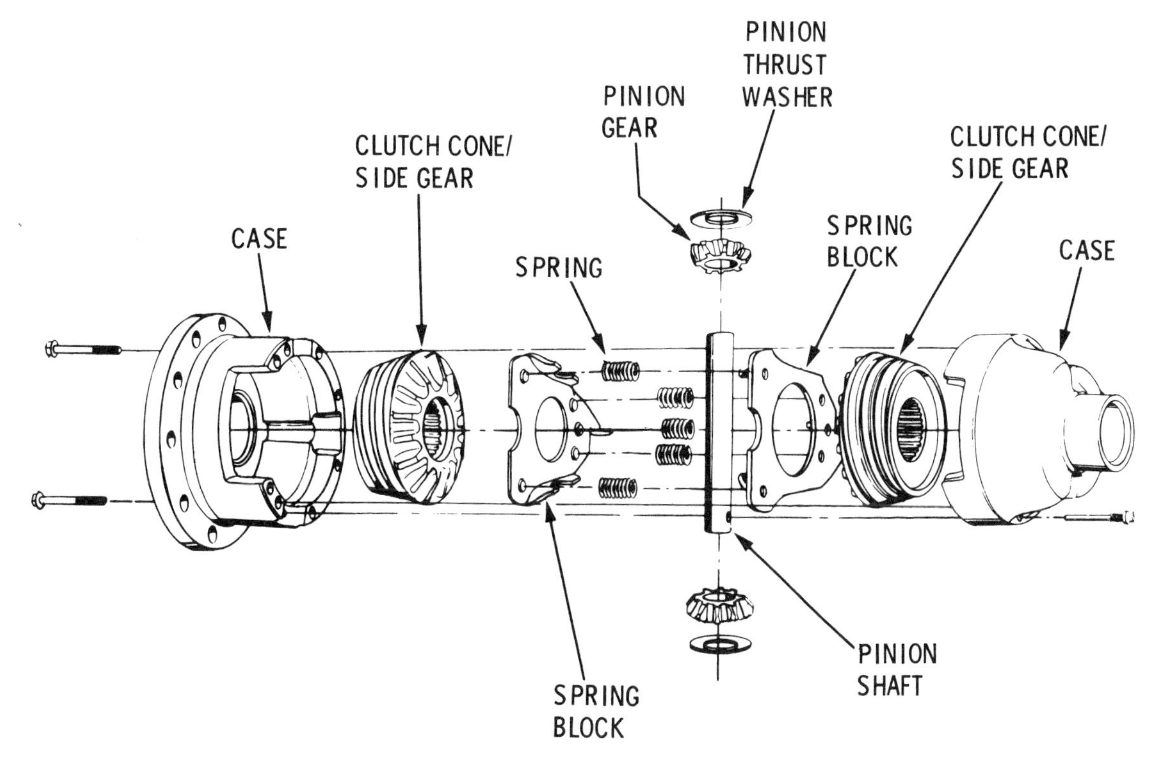

Figure 16-21. Exploded view of cone differential shows relationship of parts. Grooves in cones help to solidly engage case. (Chrysler)

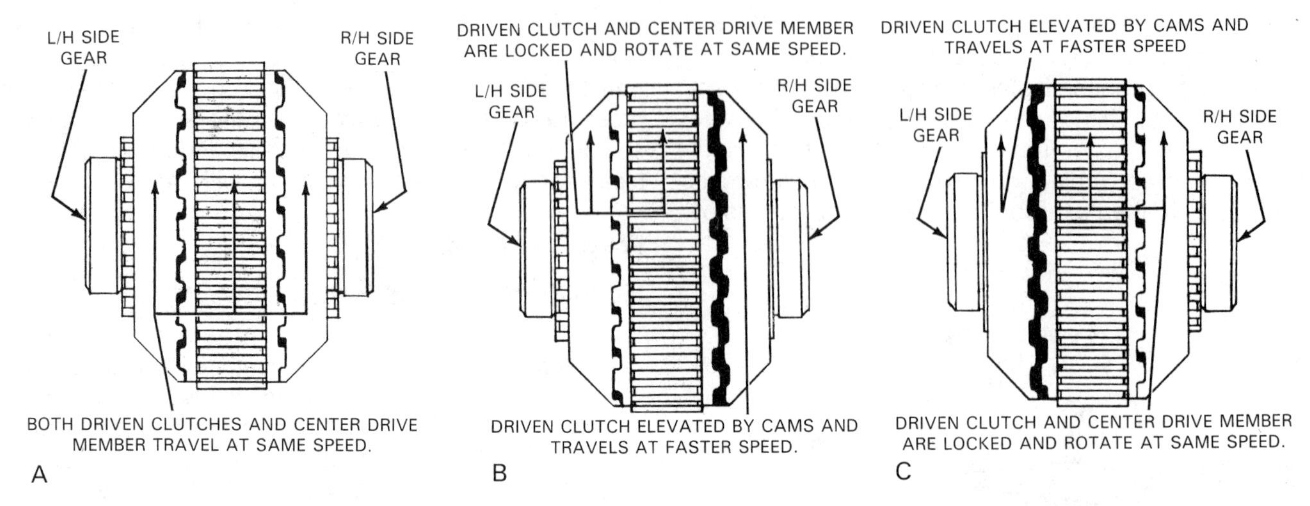

Figure 16-22. Ratchet differential uses matching sets of teeth on each side of differential case. Teeth are engaged and disengaged to transfer power. (Ford) A–Differential is straight-ahead operation. Teeth are engaged on both sides of case, and power is transferred equally to each wheel. B–When vehicle makes left turn, greater speed of right wheel causes internal cam on right side of case to take right-side teeth out of engagement. All power goes through left axle and wheel. C–When vehicle makes right turn, greater speed of left wheel causes left-side cam to take left-side teeth out of engagement. All power goes through right axle and wheel.

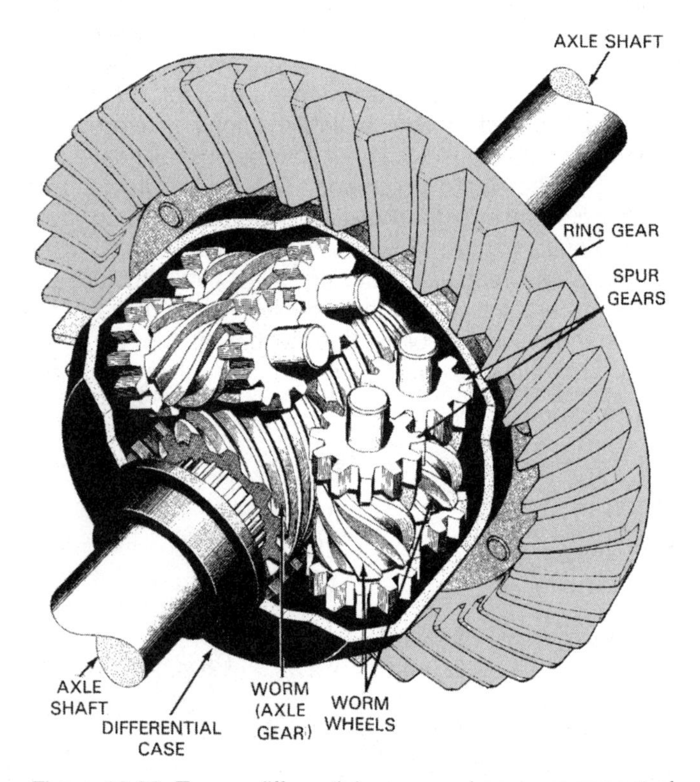

Figure 16-23. Torsen differential uses a unique arrangement of gears to transfer power. This differential has been available as high-performance aftermarket replacement for about 25 years. It is now being offered as original equipment on some European vehicles. Operation of this differential is complex. (Torsen)

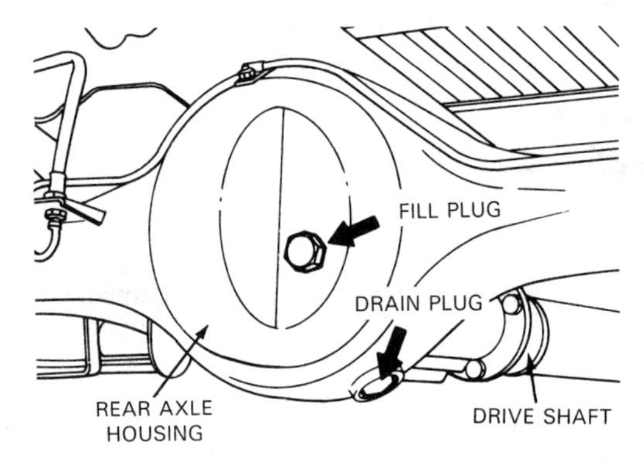

Figure 16-24. Rear axle assemblies will almost always have fill plugs, but not necessarily drain plugs. On differential without drain plug, inspection cover or carrier must usually be removed to drain oil. Oil can sometimes be drained by removing lowest inspection cover attaching bolt. (Chrysler)

ment of springs, shock absorbers, and control arms, Figure 16-25. These parts align the rear axle assembly to the vehicle while isolating most of the axle movement. Leaf springs support the axle and hold it in alignment, eliminating the need for control arms. When coil springs are used, separate control arms must be used to maintain rear axle alignment.

Two kinds of rear axle housings are used on vehicles without independent rear suspensions–*removable carrier* and *integral carrier.* Both types will be discussed in this section. (Rear axle housings used with independent rear suspension will be discussed in the section that follows.)

Removable Carrier

The ***removable carrier,*** Figure 16-26, has a separate housing for the differential assembly. It can be unbolted and removed from the rest of the rear axle housing after

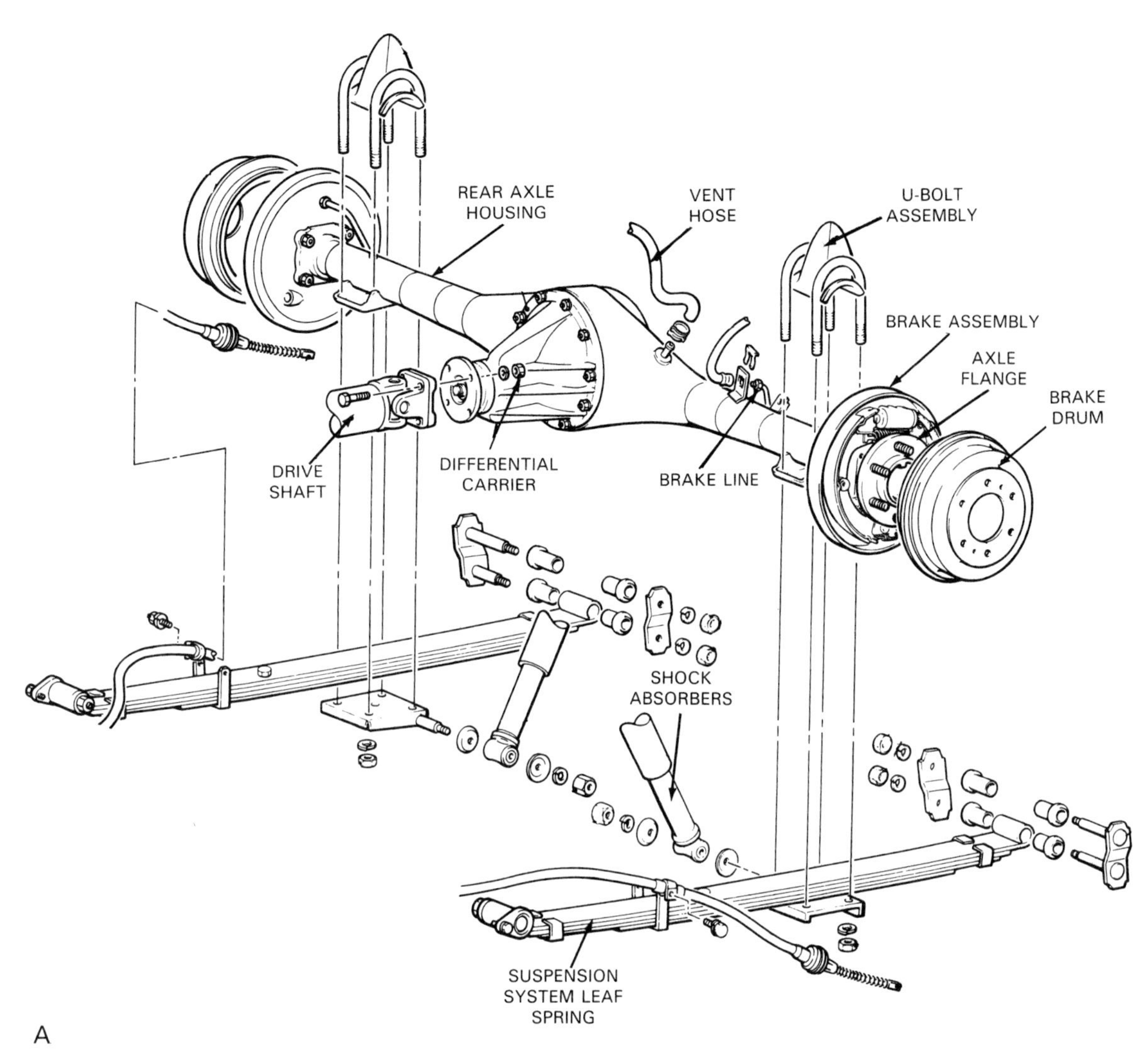

Figure 16-25. Two methods of suspending rear axle assembly of vehicle with solid-axle rear suspension are shown. Both methods involve use of conventional shock absorbers. A–Leaf springs support and align rear axle assembly. This method simplifies assembly, but makes removal of rear end difficult. (Chrysler)

(continued)

the drive axles are removed. All of the internal differential parts, then, will be removed with it. Differential assemblies housed in this kind of carrier are, in general, easier to service, since repairs can be done on the bench instead of on the vehicle.

Figure 16-27 shows a typical removable carrier. The carrier mounting flange is where the carrier attaches to the rest of the rear axle housing. Usually, threaded studs are installed in the housing. The studs pass through holes in the mounting flange when the carrier is installed. The carrier is then tightened in place by installing and tightening nuts over the threaded studs. This attaching method makes it easier to align and reinstall the carrier. A gasket is always used between the carrier and axle housing.

The differential pinion bearings are installed in the carrier in the pinion bearing bores. When tapered roller bearings are used, the bearing cups are tightly pressed into the bores. Some removable carriers have an extra support bearing at the end of the pinion gear, called a ***pinion pilot bearing.***

Figure 16-27 also shows the attaching points for the differential ***side bearings,*** also called ***case bearings.*** The side bearings are held in place by bolted, U-shaped caps. Most differential side bearing mounts have a provision for adjusting the *side bearing preload.* This adjustment is usually made with a threaded *end cap,* or *adjusting nut.* The end cap is tightened against the bearing cup until the proper preload is attained.

Drain and fill plugs may be mounted on the differential carrier or on the rear axle housing, depending on the particular manufacturer. The ribs on the front of the carrier strengthen it without adding a great deal of weight to the assembly.

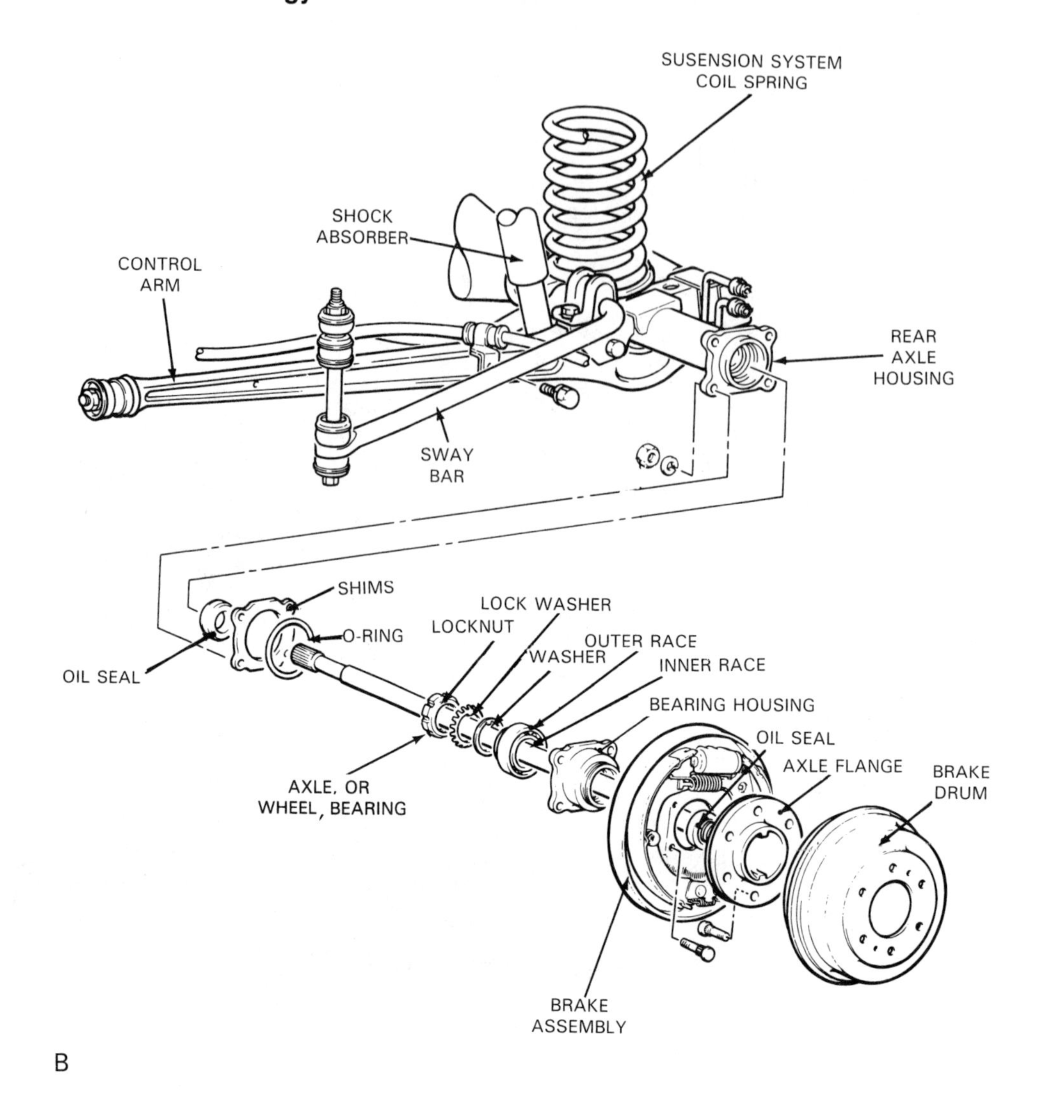

Figure 16-25 *(continued).* B–Coil springs require use of control arms and sway bars to maintain alignment. This method makes for easier rear end removal. (Chrysler)

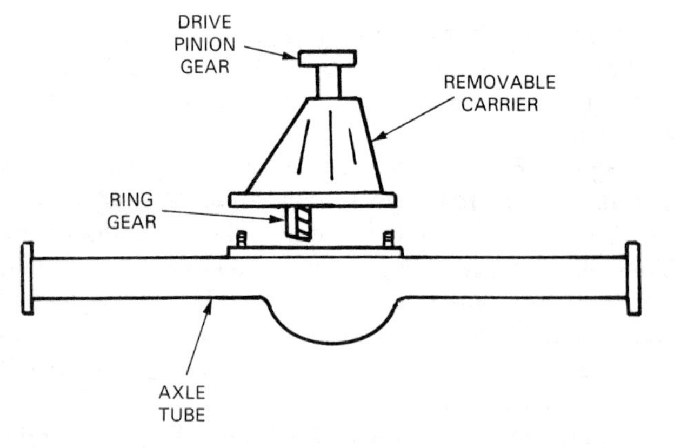

Figure 16-26. Removable-carrier type of rear axle housing is a two-piece assembly. Carrier contains differential parts. It can be unbolted and, after drive axles are removed, serviced on bench.

Integral Carrier

The ***integral carrier,*** as the name implies, is an integral part of the rear axle housing. See Figure 16-28. This type of rear axle housing has a sheet metal or cast metal inspection cover, sometimes called the *differential cover.* The inspection cover can be removed to service the rear end components. Service operations must be performed under the vehicle, since the carrier cannot be separated from the rest of the rear axle housing.

Figure 16-29 shows a typical integral carrier. Notice that almost all of the rear end components are installed inside of the rear axle housing. Most of these can be removed through the opening that is kept closed off by the differential cover. The cover is sealed with a gasket of some sort.

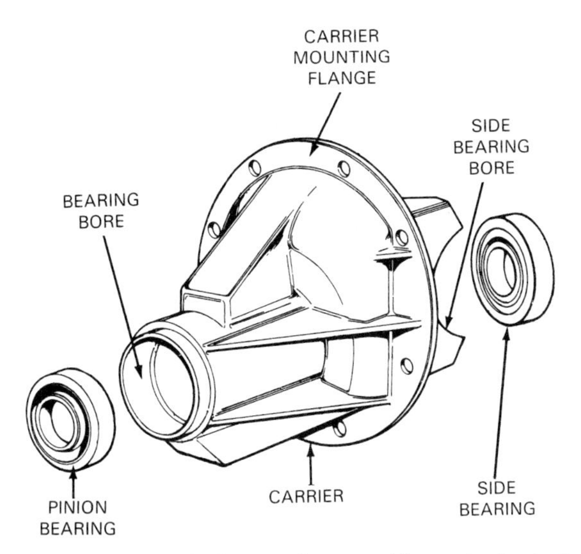

Figure 16-27. Mounting flange of removable carrier is designed to seat against axle portion of rear axle housing. Carrier is attached with studs and nuts. Gasket is always installed between carrier and axle housing. (Subaru)

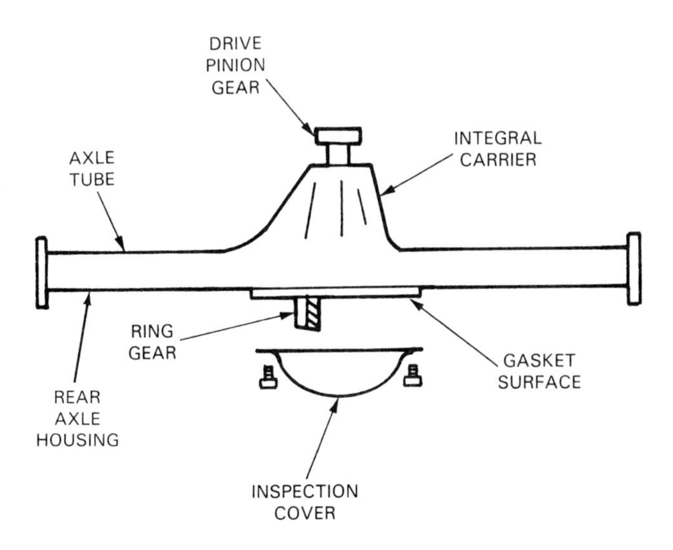

Figure 16-28. Rear axle housing with integral carrier is essentially a one-piece unit. All moving parts are inside of housing. Differential parts are reached for service by removing inspection cover at rear. This type of axle is usually serviced under vehicle.

The pinion front and rear bearing cups are pressed into the carrier portion of the rear axle housing. Integral carriers do not normally have a pinion pilot bearing.

Differential side bearings are installed in the integral carrier in the same manner as on a removable carrier. The side bearing preload adjustment is sometimes made with shims. These shims are placed between the bearing cup and the rear axle housing. In other instances, the preload adjustment is made with a threaded end cap, as on the removable carrier.

Rear Axle Housing: Independent Rear Suspension

On vehicles having independent rear suspensions, a modified rear axle housing is used. Figure 16-30 shows such a housing. Notice that the housing has no axle tubes. The drive axles resemble drive shaft assemblies to some degree, complete with conventional or constant-velocity universal joints. This design allows each wheel to react independently to the road surface, improving ride quality and handling.

The housing has oil seals to prevent oil loss where the axles enter the housing. The internal construction of the housing is identical to the previously discussed carriers. To reduce vibration and strengthen the drive train, a torque tube is sometimes used to attach the front of the housing to one of the vehicle crossmembers. (Torque tubes were explained in Chapter 14.)

A modified rear axle housing is also used on some front-wheel drive vehicles. An example of this kind of housing is shown in Figure 16-31. This kind of housing is found on front-wheel drive vehicles where the engine is installed conventionally (longitudinally). Usually, the axle housing is attached directly to the transmission case, and the drive pinion gear attaches directly to the transmission output shaft. No drive shaft is used. The internal construction of the housing is very similar to that of rear-wheel drive vehicles.

Rear Drive Axles

The rear drive axles transfer power from the differential assembly to the rear wheels. There are two major kinds of drive axle designs. One is the *solid drive axle,* shown in Figure 16-32. The other is the *independently suspended drive axle,* shown in Figure 16-33.

Solid Drive Axle

A ***solid drive axle,*** or ***live axle,*** as it is sometimes called, is a hardened-steel shaft. See Figure 16-34. Each rear axle assembly has two. Splines on the *inboard* (inner) end of each axle mate with internal splines on the differential side gear to which it is connected. An ***axle flange*** at the *outboard* (outer) end of each axle acts as a *wheel hub.* It provides the mounting surface for the brake drum or rotor and the wheel. The brake assembly and wheel are installed directly on the flange *wheel studs.*

Each shaft is supported on the outboard end by an ***axle bearing,*** also called a ***wheel bearing.*** The axle bearing can be pressed on the shaft or installed in the axle tube.

Axle bearings that are pressed on the shaft are usually packed with grease. An *axle seal* is pressed into the housing behind, or on the inboard side of, the bearing. The

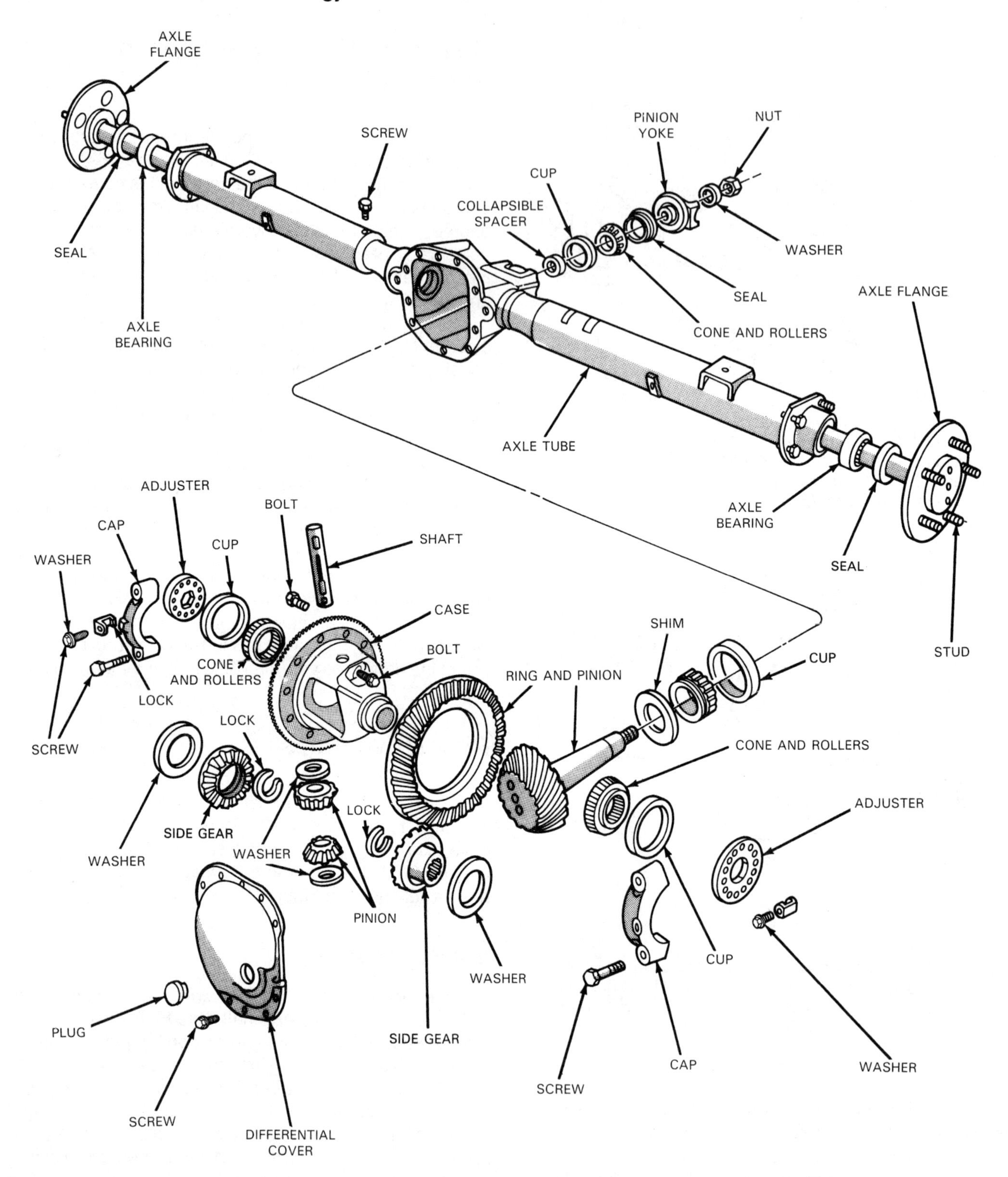

Figure 16-29. This is an exploded view of integral carrier rear axle assembly. Notice that most moving parts fit inside of rear axle housing. Differential cover often contains fill plug. (Chrysler)

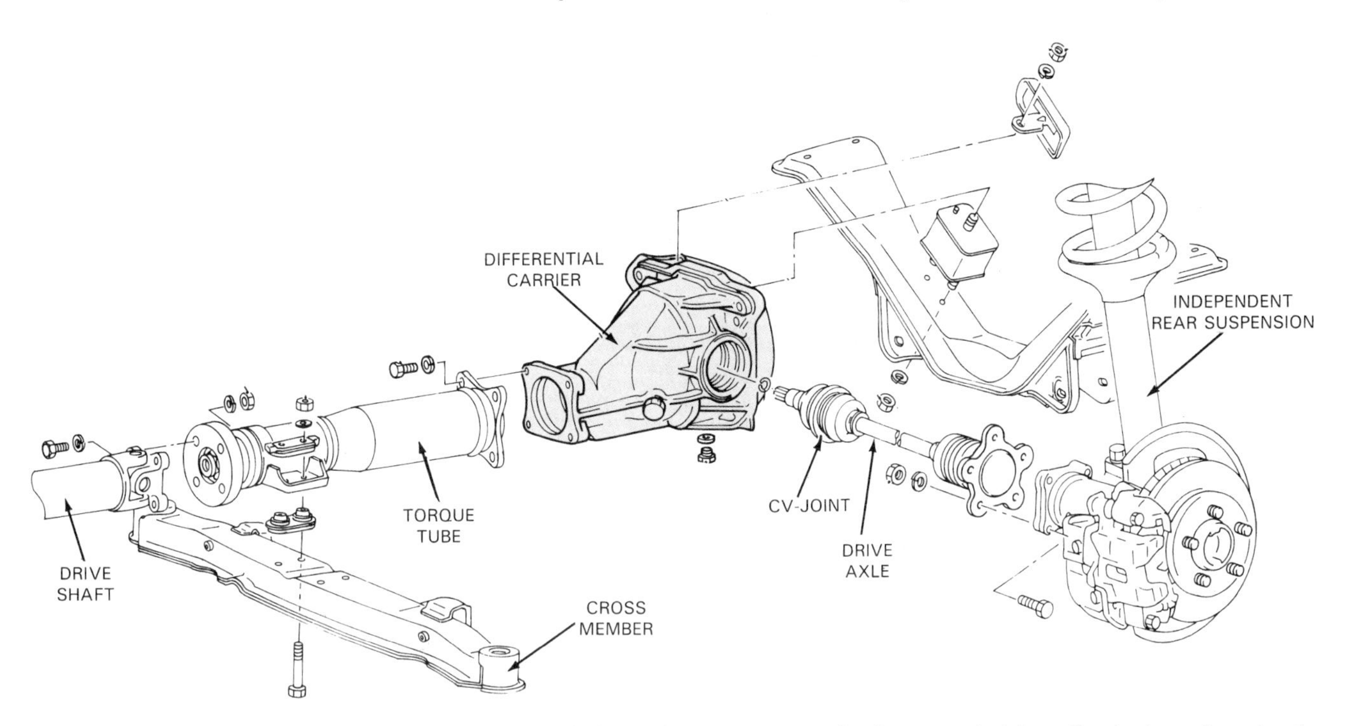

Figure 16-30. Rear axle housing used on vehicles with independent rear suspension has no axle tubes. Carrier has oil seals where drive axles enter. A torque tube is often used at front of housing to increase rigidity and reduce vibration. (Chrysler)

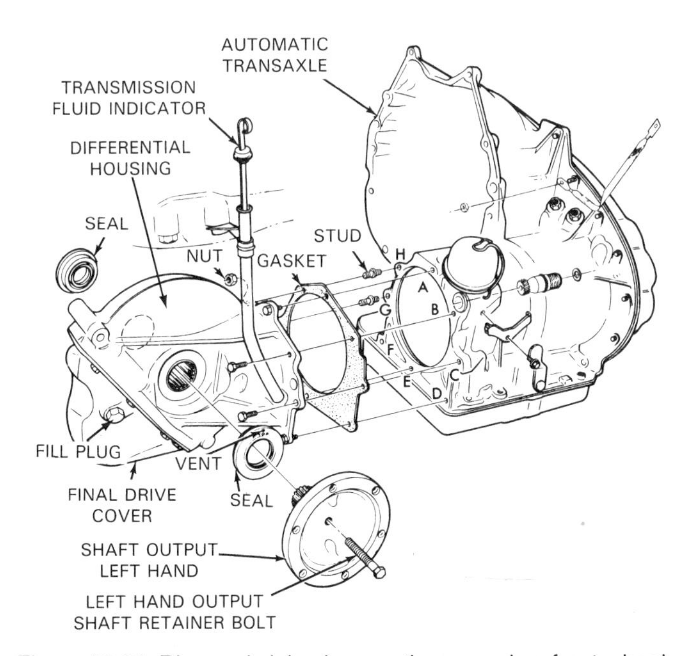

Figure 16-31. Ring and pinion is sometimes used on front-wheel drive vehicles. With conventional engine mounting on front-wheel drive vehicle, conventional rear axle assembly is used. Assembly, then, operates in same manner as rear-wheel drive unit. (General Motors)

lip of the seal seats against a machined area of the shaft. This seal keeps rear end lubricant from reaching the bearing. An outer seal prevents water and dirt from leaking through the outer ends of the rear axle housing and entering the bearing.

Axle bearings that are installed in the housing are lubricated by rear end lubricant (gear oil). When the vehicle makes a turn, lubricant is thrown outward from the carrier, reaching the axle bearing. An axle seal is installed in front of, or on the outboard side of, the bearing to keep lubricant from leaking out from the outer ends of the rear axle housing.

Axles having bearings that are pressed on will have a bearing retainer, or an ***axle collar,*** to keep the bearing in place. In addition, some will have a spacer to keep the bearing at the proper distance from the end of the axle. The ***axle retainer plate*** holds the axle and axle bearing into the axle tube.

Semi-floating axles

Solid drive axles can be *semi-floating* or *full-floating.* Most automobiles and light trucks have semi-floating axles. In the ***semi-floating axle,*** the weight of the vehicle passes through the axle bearing to the drive axle and on to the wheel and tire. Figure 16-35 shows three versions of the semi-floating axle.

Figure 16-35A shows a semi-floating axle using a ball bearing. This is a pregreased bearing. There is an axle seal behind the bearing. The bearing is pressed onto the shaft. The bearing and axle are held in the housing by an axle retainer plate, mounted on the outer end of the rear axle housing. The retainer plate and bearing control end play during turns.

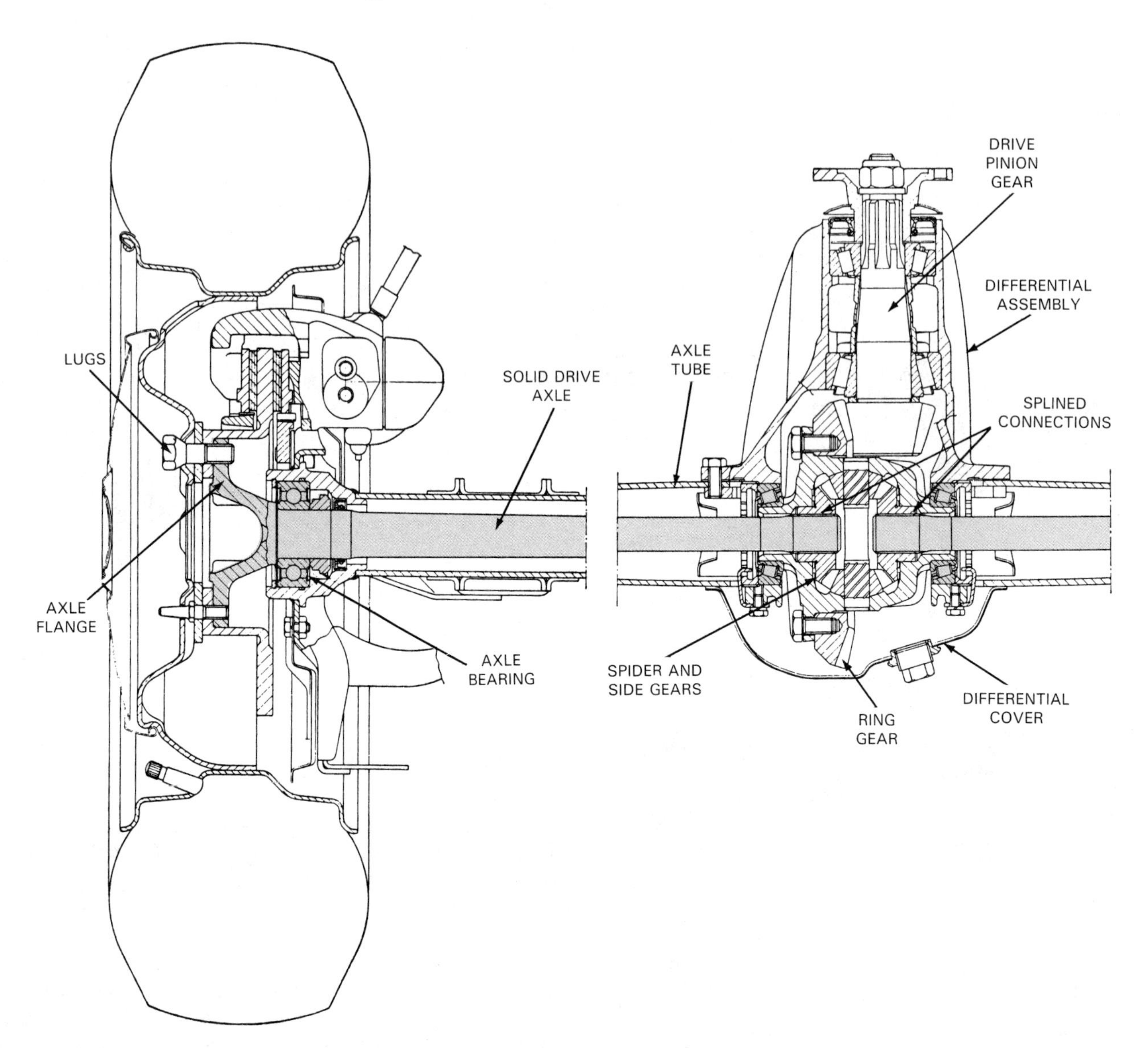

Figure 16-32. Axle shaft used on solid drive axle, or live axle, is single piece of steel, supported on both ends. Outer support is provided by axle bearing, and inner support is provided by differential side bearings. Note that the differential has been rotated 90° for purposes of illustration. (Fiat)

Figure 16-35B shows a roller bearing version of the semi-floating axle. This bearing is lubricated by rear end lubricant. The axle seal is installed in front of the bearing. When this kind of bearing is used, the axle is held in the housing by a clip on the inboard end of the shaft, at the differential assembly. This kind of axle is sometimes called a C-lock axle, from the shape of the locking clip. End play on turns is controlled by the fit of the axle shaft between the C-lock and the other parts of the differential assembly.

Figure 16-35C shows a semi-floating axle using a tapered roller bearing. When this type of bearing is used, there is usually some provision for adjusting the bearing preload to control end play. This is generally done by using ***axle shims*** or by turning an adjusting nut. Tapered roller bearings may be packed with grease or lubricated from the rear axle housing, depending on the particular manufacturer's design.

In Figure 16-35C, notice the use of the tapered axle. This is one of two methods used to secure a wheel hub to its axle. The tapered end wedges into a tapered hole in the wheel hub, and the key keeps the axle from rotating in the hub. The other method, mentioned earlier, has the wheel hub (axle flange, in this case) solidly mounted to the axle.

The design of the semi-floating axle causes weight loads to be placed on the axle. These loads will shift as the axle rotates, placing flexing stresses on the shaft. On automobiles and light trucks, the loading is not serious, and the axles will usually last the life of the vehicle.

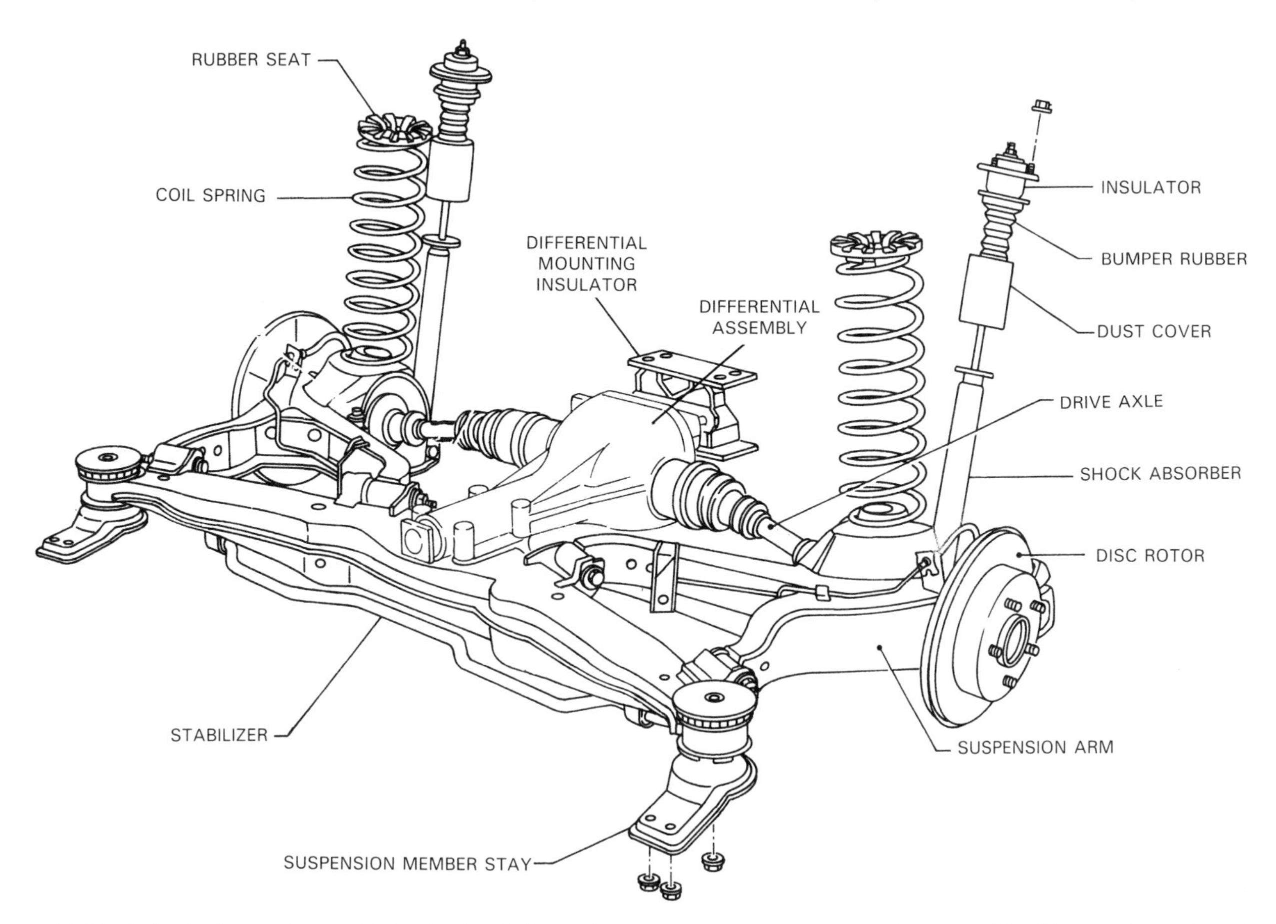

Figure 16-33. Axle shafts used on vehicles having independent rear suspensions somewhat resemble drive shafts. A flexible joint, such as a CV joint, is used on each end of each shaft.

Full-floating axles

If the rear end will be subjected to heavy loads, such as the rear end of a large truck might be, a ***full-floating axle*** is used. Figure 16-36 shows an example of a full-floating axle. With this design, the axle drives the wheel but does not carry any of the vehicle weight. The weight passes through the bearings on the wheel hub. The wheel hub absorbs the stresses. This design reduces the stresses on the shaft, prolonging its life. Full-floating axles are not used on light duty vehicles because of their extra cost and complexity.

Independently Suspended Drive Axle

Independently suspended drive axles, used on vehicles with independent rear suspension, resemble miniature drivelines. The axle consists of a central shaft with flexible joints and *stub axles* on each end. The flexible joints–either conventional universal joints or CV joints–allow each wheel to move independently of the vehicle body and of each other.

A typical independently suspended drive axle arrangement is shown in Figure 16-37. Although they look different, these axles transfer power in much the same manner as solid drive axles.

Figure 16-38 is an example of how an independently suspended drive axle and wheel hub are assembled. The hub is firmly attached to the suspension control arm. The inner portion of the hub rotates inside of a bearing and acts as a mounting flange for the wheel and brake assembly. The stub axle is splined to the hub and drives it. The universal joint allows free movement of the suspension control arm. Some splined axles can slide to compensate for changes in axle length when the rear suspension moves up and down.

Summary

All rear axle assemblies have the same basic design and operate by the same principles. Rear end variations depend on whether the vehicle has a solid-axle or independent rear suspension, removable or integral carrier, semi-floating or full-floating axles, and standard or limited-slip differential.

The major parts of the rear axle assembly are the differential assembly, rear axle housing, drive axles,

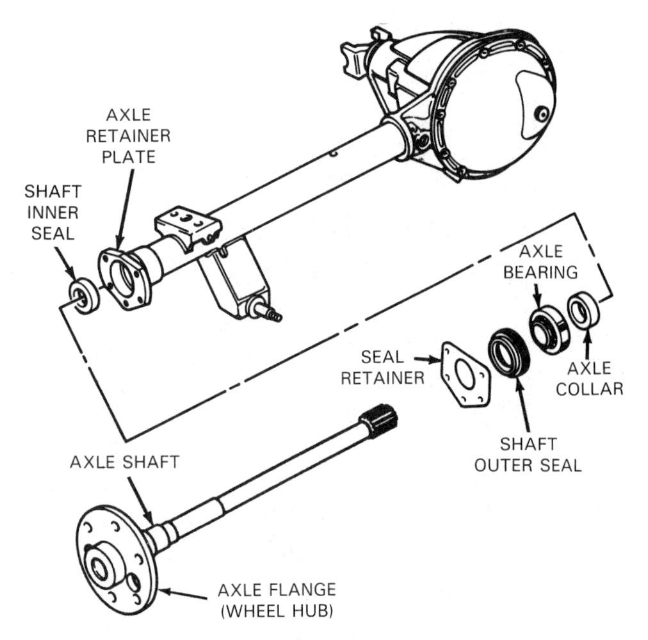

Figure 16-34. Drive axle and related components are shown here. External splines on inboard end of axle mate with matching internal splines in differential side gear. Axle flange is mounting surface for brake drum or rotor and wheel. Bearing is kept in place by axle collar. Axle retainer plate keeps axle and bearing retained in axle tube. (Chrysler)

bearings, and seals. Engine power enters the drive pinion gear through differential pinion yoke and driveline. The drive pinion gear turns the ring gear. The interaction of the ring and pinion turns the power at a 90° angle and reduces its speed. The ring gear is bolted to the differential case. Power flows from the ring gear into the differential case, which transfers it to the spider gears. The spider gears transfer the power to side gears, which transfer it to the drive axles and rear wheels.

The differential assembly has three purposes. It redirects the driveline rotation in a 90° angle, reduces rotating speed to increase power, and allows the vehicle to make turns without wheel hop or axle breakage.

The relative positions of the ring and pinion gears must be set exactly, or the gears will be noisy and wear out prematurely. The position of the ring and pinion in the case and in relation to each other must be carefully adjusted.

The differential case assembly allows the vehicle to make turns without wheel hop. It has an arrangement of gears that allow the rear wheels to turn at different speeds. There are two kinds of differential case assemblies, standard and locking.

The standard differential is composed of meshing spider and side gears, enclosed in a differential case. The ring gear is bolted to the case. Power flow is through the case, into the spider gears, and on to the side gears. The side gears are splined to the drive axles. They transfer power to the drive axles and rear wheels.

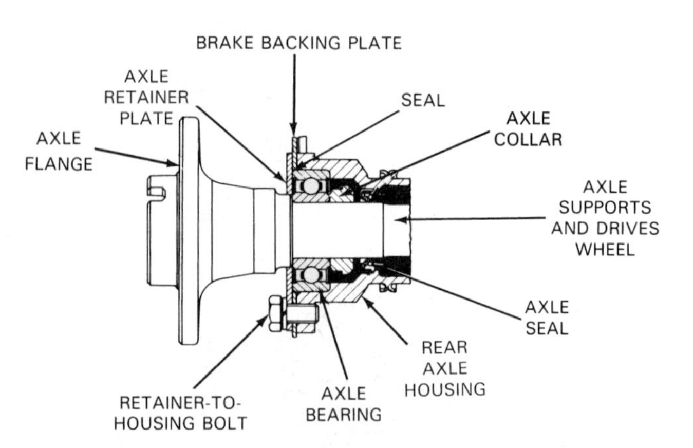

A—Ball bearing semi-floating axle. Ball bearing is retained on shaft by pressed on axle collar. Bearing and axle are held in housing by bolted retainer plate.

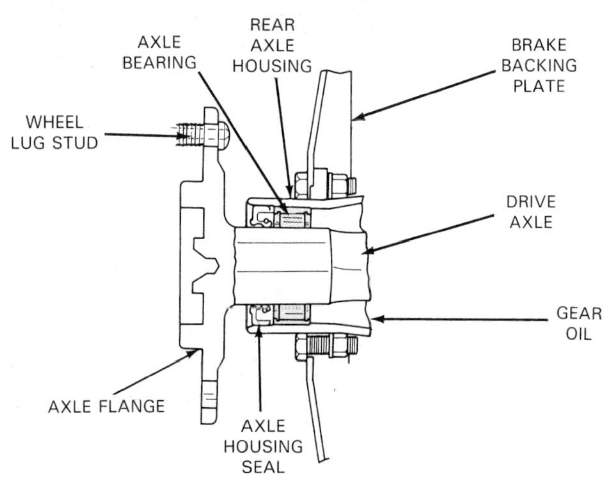

B—Roller bearing semi-floating axle. The major difference between this design and that of ball bearing is shaft locking method. This axle is retained by C-lock at inside of shaft. C-lock attaches axle to differential. Bearing plays no part in keeping shaft in place.

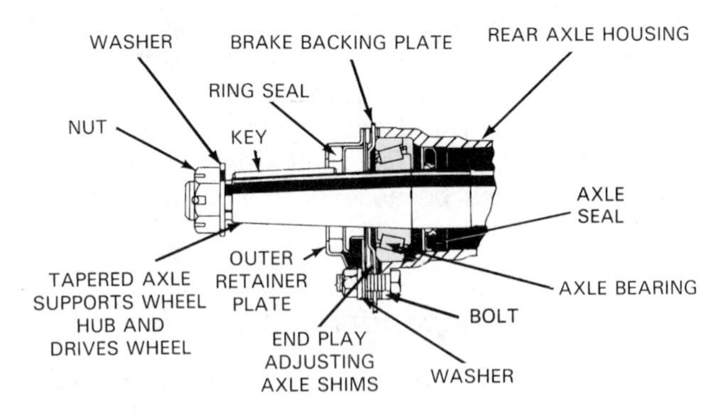

C—Tapered roller bearing semi-floating axle. Bearing preload is adjust by shims or adjusting nut. Axle is retained in same manner as ball bearing, except without axle collar. Shoulder on axle keeps axle from sliding past bearing.

Figure 16-35. Semi-floating axle is most common shaft and bearing design used on cars and light trucks. Bearing passes vehicle weight through axle shaft and out to wheel. Axle drives and supports vehicle. (Fiat, General Motors, Deere & Co.)

When driving on slippery surfaces, the rear wheels of a vehicle with a standard differential will often slip. This is because the differential will always drive the wheel with the least traction. To overcome this problem, various kinds of locking differentials are used. They increase traction by sending power to the wheel with the most traction.

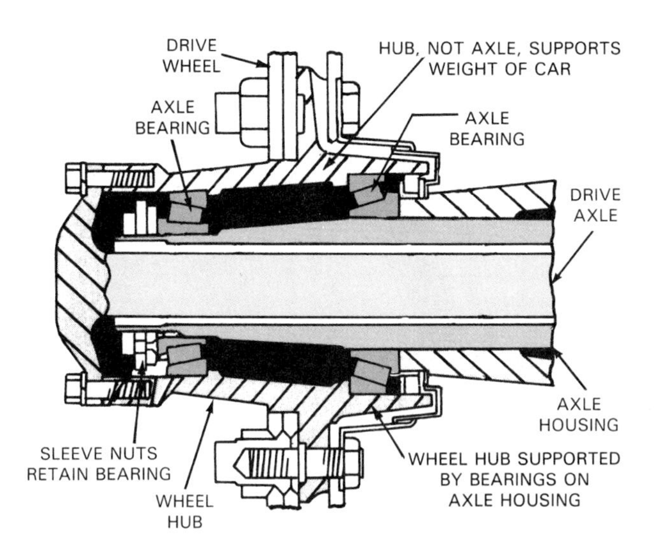

Figure 16-36. Full-floating axle is used on trucks and other vehicles that carry heavy loads. Bearings on hub transmit vehicle weight from rear axle housing to wheel hub and wheel without loading axle. The only job of axle is to propel vehicle. (Deere & Co.)

The most common locking differential is the limited-slip differential. One type uses clutch packs placed between the side gears and the differential case. Friction discs are splined to side gears; steel plates are dogged to case. The clutch packs are pressed together by the pressure of the spider gears on the side gears.

When the vehicle is moving straight ahead, the limited-slip differential operates like a standard differential. The rear axle parts are all turning at the same speed, and the clutch packs are not used. When a wheel starts slipping, the difference in pressure on the side gears causes the clutches to apply. The difference in traction between the inner and outer wheels is not a factor during normal turns, and the friction discs and steel plates slip over each other.

Another version of the limited-slip differential uses cones instead of clutch packs. Operation is similar to the clutch-plate differential.

The ratchet differential has a series of internal cams and ramps that direct power to the wheel with the most traction. Its operation depends on relative wheel speeds, rather than wheel traction. The ratchet differential transfers power through a set of teeth which can be engaged and disengaged.

The Torsen differential uses an arrangement of worms and worm wheels to transfer power. On turns or when one

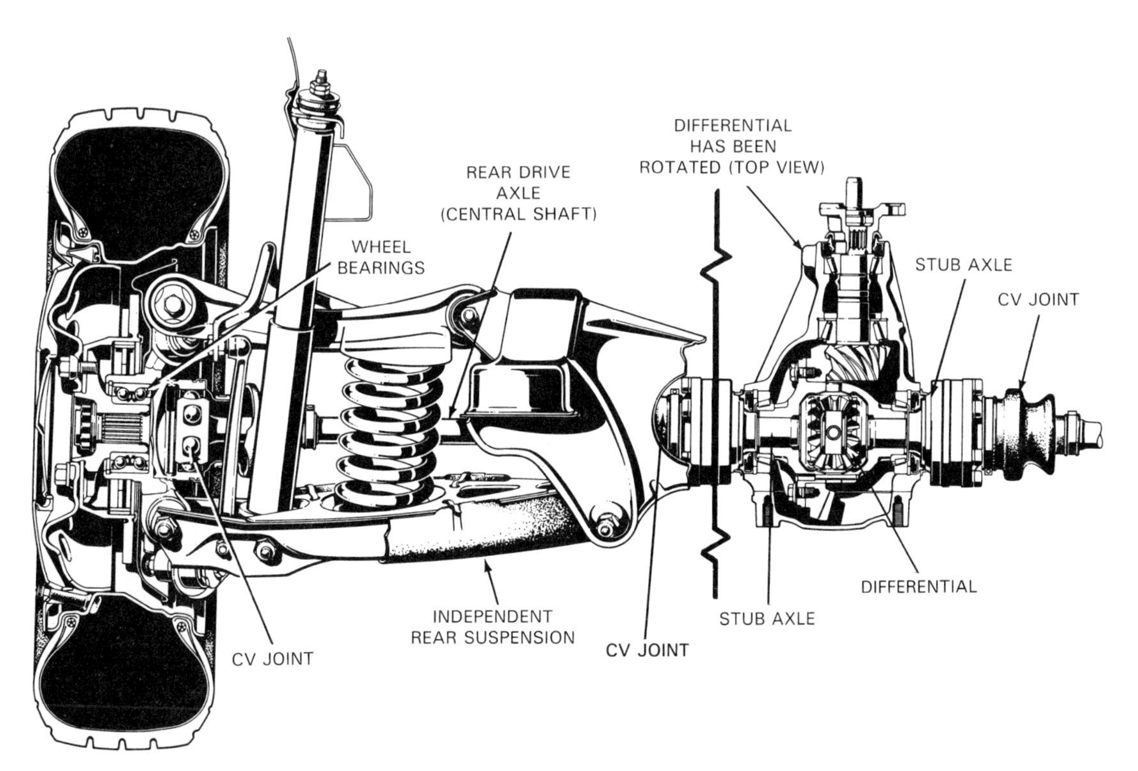

Figure 16-37. Drive axle of a vehicle with independent rear suspension consists of three shafts and two U-joints. Central shaft is connected through U-joints to a short shaft, or stub axle, on either side. Stub axles are splined to wheel hub and side gears. Note that the differential has been rotated 90° for purposes of illustration.

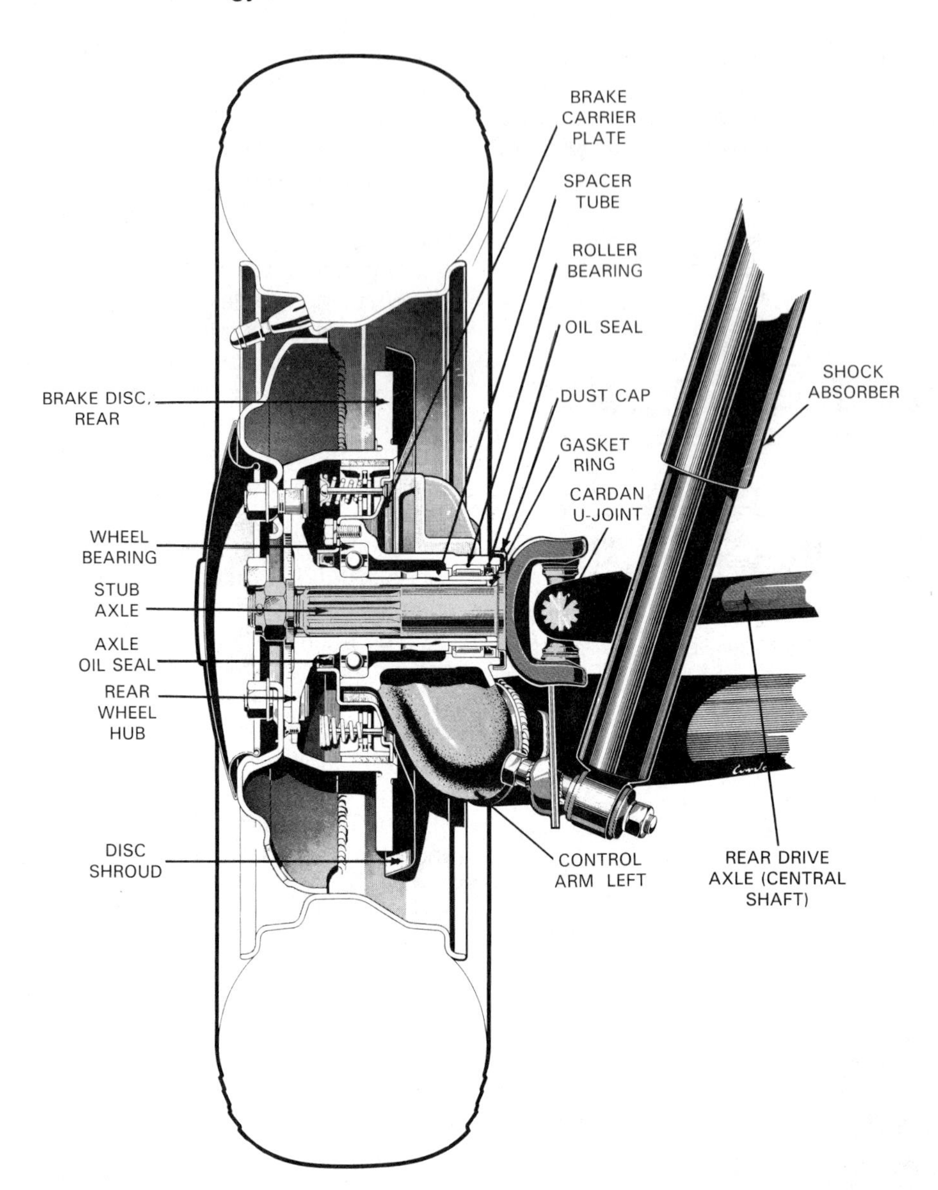

Figure 16-38. This shows how stub axle and universal joint of independently suspended drive axle are installed to hub and wheel of a vehicle with independent rear suspension. (Porsche)

wheel is slipping, the axle gear and worm wheel arrangement transfers power from the faster wheel to the slower wheel.

The rear axle housing encloses and supports the other parts of the rear axle assembly and forms a reservoir for the rear end lubricant. The rear brake assemblies are usually attached to the rear axle housing. The rear axle housing is attached to the vehicle body by the suspension system.

Two kinds of rear axle housings are used on vehicles without independent rear suspensions. The removable carrier type has all of its moving parts, except the axles, in a carrier which can be unbolted from the rear axle housing. The integral carrier type is a one-piece unit. It is serviced by removing a sheet metal inspection cover, located at the rear of the housing.

A modified rear axle housing is used on vehicles having independent rear suspension. The internal construction of the housing is identical to those used on live axles. A version of this housing is used on some front-wheel drive vehicles that have the engine mounted longitudinally.

The rear drive axles transfer power from the side gears to the rear wheels. Drive axles can be solid or independently suspended. Solid axles are splined with, and supported by, the side gears at the inboard end. The outboard end is supported by axle bearings. The axle bearing can be pressed onto the shaft or installed in the housing. Bearings that are pressed onto the shaft are usually packed with grease. Bearings that are installed in the housing are lubricated by rear end lubricant. Seals are used to keep lubricant from leaking out of the rear axle housing.

Solid axles can be semi-floating or full-floating. In the semi-floating axle, the weight of the vehicle goes through the axle bearing to the shaft, and out to the wheel. In the full-floating axle, the axle drives the wheel but does not carry any of the vehicle weight. Most passenger cars have semi-floating axles.

Independently suspended drive axles resemble drivelines. They consist of a central shaft with flexible joints and stub axles on each end. The flexible joints can be conventional U-joints or CV joints. They allow each wheel to move independently of the vehicle body and of each other.

Know These Terms

Solid-axle rear suspension; Independent rear suspension; Differential drive gears; Drive pinion gear; Pinion bearing; Collapsible spacer; Jam nut; Pinion shim; Ring gear; Wheel hop; Standard differential (Single-pull differential); Differential case; Spider gears (Pinion gears); Side gears (Axle end gears); Locking differential; Limited-slip differential; Clutch-plate differential; Cone differential; Ratchet differential; Torsen differential; Differential carrier; Axle tube; Removable carrier; Pinion pilot bearing; Side bearings (Case bearings); Integral carrier; Solid drive axle (Live axle); Axle flange; Axle bearing (Wheel bearing); Axle collar; Axle retainer plate; Semi-floating axle; Axle shim; Full-floating axle; Independently suspended drive axle;

Review Questions–Chapter 16

Please do not write in this text. Place your answers on a separate sheet of paper.

1. Which of the following items does not belong with the others?
   a. Rear axle assembly.
   b. Final drive.
   c. Differential.
   d. Rear end.
2. What is independent rear suspension?
3. Which rear end components change the direction of power flow by 90°?
4. How can rear axle ratio be determined?
5. If the drive pinion gear has 10 teeth and the ring gear has 35 teeth, what is the rear axle ratio?
6. Describe the construction of a drive pinion gear.
7. Which of the following parts is used to set pinion bearing preload?
   a. Jam nut.
   b. Crush washer.
   c. Lock washer.
   d. Castle nut.
8. The convex side of ring gear teeth is the _____ side, and the concave side of ring gear teeth is the _____ side.
9. Explain the function of the rear axle housing.
10. Describe the two major kinds of carriers.
11. In the full-floating axle, the weight of the vehicle passes through the axle bearing to the drive axle and on to the wheel and tire. True or false?
12. The differential case assembly contains two _____ gears and two or four _____ gears.
13. The limited-slip differential will drive the wheel with the least traction. True or false?
14. In terms of their construction, what is the major difference between a standard differential and a limited-slip differential?
15. The Torsen differential is a locking differential that uses:
    a. A multiple-disc clutch.
    b. A cone clutch.
    c. A dog clutch.
    d. A worm gearset.

Certification-Type Questions–Chapter 16

1. Technician A says that every rear axle assembly has a housing, a differential assembly, and rear drive axles. Technician B says that every rear axle housing has axle tubes. Who is right?

(A) A only
(B) B only
(C) Both A & B
(D) Neither A nor B

2. The most common rear axle assembly failures are:

(A) axle bearing failures.
(B) pinion yokes failures.
(C) cracked spider gears.
(D) stripped ring and pinion gears.

3. All of these are primary functions of the differential assembly EXCEPT:
- **(A)** multiplying engine power.
- **(B)** allowing the vehicle to make turns.
- **(C)** supporting and aligning the drive axles.
- **(D)** redirecting power flow to the rear wheels.

4. Rear axle ratio can be found by dividing the number of teeth on the ring gear by the number of teeth on the:
- **(A)** side gear.
- **(B)** spider gear.
- **(C)** drive pinion gear.
- **(D)** axle end gear.

5. All of these are used to set the pinion bearing preload EXCEPT:
- **(A)** a solid spacer.
- **(B)** a crush washer.
- **(C)** a collapsible spacer.
- **(D)** the rear pinion bearing.

6. The ring gear transfers power directly from the drive pinion gear to the:
- **(A)** axle flange.
- **(B)** differential case.
- **(C)** differential carrier.
- **(D)** differential pinion yoke.

7. A rear-wheel drive vehicle cannot be driven because one of its drive wheels is parked on ice. Technician A says that the ring gear and differential case will drive the spider gears. Technician B says that the differential spider gears will walk around the side gear related to the wheel on dry pavement. Who is right?
- **(A)** A only
- **(B)** B only
- **(C)** Both A & B
- **(D)** Neither A nor B

8. Locking differentials overcome traction problems by sending power to:
- **(A)** the wheel with traction.
- **(B)** both wheels.
- **(C)** the slipping wheel.
- **(D)** the wheel bearings.

9. All of these are locking differentials EXCEPT:
- **(A)** Torsen differentials.
- **(B)** ratchet differentials.
- **(C)** limited-slip differentials.
- **(D)** MacPherson differentials.

10. All of these functions are served by the rear axle housing EXCEPT:
- **(A)** determining the depth of the drive pinion gear in the carrier.
- **(B)** forming a reservoir for rear end lubricant.
- **(C)** accommodating suspension system attachment.
- **(D)** supporting stationary parts of rear brake assemblies.

11. All of these types of drive axles are found on rear-wheel drive vehicles EXCEPT:
- **(A)** full-floating axles.
- **(B)** Rzeppa axles.
- **(C)** semi-floating axles.
- **(D)** independently suspended axles.

12. Major differences among rear-wheel drive vehicles with solid-axle rear suspension include all of these EXCEPT:
- **(A)** conventional versus constant-velocity U-joints.
- **(B)** removable versus integral carrier.
- **(C)** semi-floating versus full-floating axles.
- **(D)** standard versus limited-slip differential.

Chapter 17

Rear Axle Assembly Problems, Troubleshooting, and Service

After studying this chapter, you will be able to:

- Identify and describe typical rear end problems.
- Explain how to check and verify rear end troubles.
- Remove, service, and install drive axles.
- Remove a removable carrier from a rear axle assembly.
- Remove an entire rear axle assembly from a vehicle.
- Explain when rear end service is necessary.
- Properly disassemble a rear axle assembly.
- Check and adjust parts of a rear axle assembly.
- Install a removable carrier in a rear axle assembly.
- Assemble and install a rear axle assembly in a vehicle.

Rear-wheel drive vehicles and a few front-wheel drive automobiles have an axle assembly using hypoid ring and pinion gears with a differential case assembly. Although the hypoid type of axle assembly is relatively trouble-free, it can develop problems and require servicing.

This chapter will acquaint you with the methods of diagnosing and repairing rear axle assemblies. Since major rear end repairs are relatively rare, it is sometimes difficult to gain hands-on experience with the internal parts of one. This chapter presents steps in removing, overhauling, adjusting, and installing an entire rear axle assembly. This information will be needed when you encounter your first rear axle assembly. Also included in this chapter is information on common rear end service tasks, such as axle bearing and seal service.

The information presented here applies to rear axles with solid-axle rear suspension and independent rear suspension. It can also be used to service the drive axle assemblies found on vehicles with front-wheel drive where the engine is conventionally mounted. Refer to Figure 17-1 to see where rear end problems can occur.

Rear End Problems and Troubleshooting

Major problems in rear axle assemblies are relatively rare. Most internal parts will last for a long time, but once in a while, they will fail. The usual causes of failure are high mileage and/or abuse. Some parts, such as the axle bearings and seals, will require repairs at some time in the life of the vehicle.

When major problems do occur, it will be obvious that the assembly needs repair work. Sometimes, drive axles will break because of high shock loads, and no power may be sent to the road. Axles held by pressed-on retainers (axle collars) can slide out of the housing due to retainer failure. Ring and pinion gears can be stripped because of high mileage or because of overloading or other abuse. Sometimes, rear axle assemblies will overheat because of improper lubricant or severe overloading.

To begin diagnosis, gather information about the problem. Find out why the vehicle was brought in for service. Test drive the vehicle–with the owner, if possible. Road test the car on a smooth surface. Use your understanding of operating principles to determine if the problem is actually in the rear end or if the malfunction is in another system. For example, worn U-joints or transmission gears can produce symptoms similar to those caused by faulty rear end components.

Unless the rear axle assembly has an obvious problem, such as a cracked housing or severe oil leak, always check the fluid level first, after the road test. Rear end lubricant service is discussed following this section. If adding lubricant does not improve the operation of the rear axle assembly, refer to the manufacturer's service manual for troubleshooting information.

The most commonly experienced rear axle problems include noises, vibrations, and leaks. These problems are presented in detail in the upcoming paragraphs.

Abnormal Noise

An *abnormal noise* in the rear axle assembly will be a clunk, whine, roar, or rumble. Before disassembling the rear axle assembly, make sure that the problem is not in

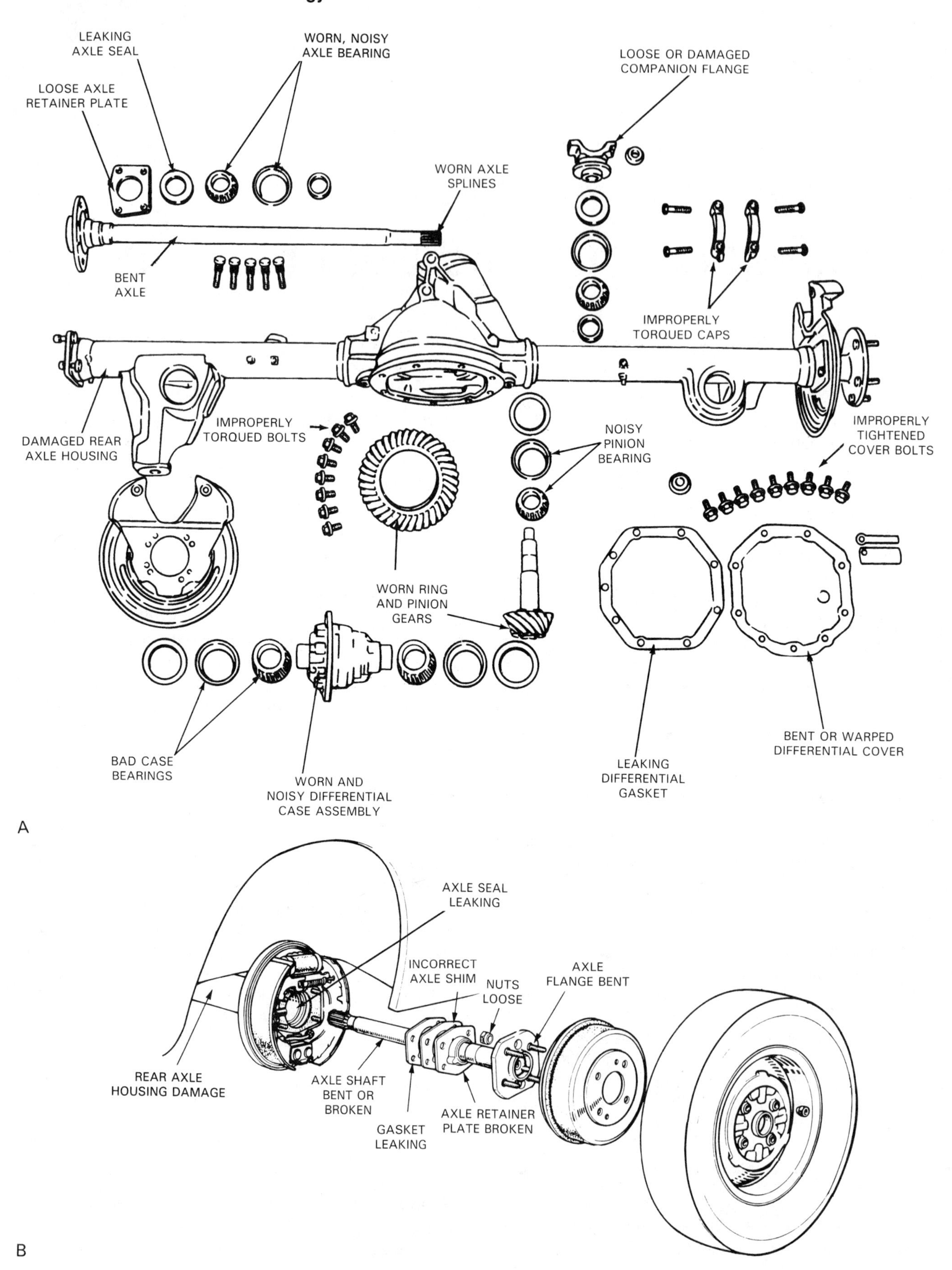

Figure 17-1. The rear axle assembly contains many parts, all of which can become defective. A–This is an exploded view of a complete rear axle assembly. Note where parts may wear out or break. B–Axle bearings and seals often wear out or leak lubricant. (General Motors, Toyota)

the clutch, transmission, or driveline. Other sources of noise are the front wheel bearings, tires, and brakes. As you diagnose the problem, these must be considered as possible causes.

A defective rear axle assembly may make a clunking noise. This will be a low-pitched, sometimes metallic-sounding, thumping noise. It will occur when the vehicle is first accelerated or when the driver takes his or her foot off the accelerator. Sometimes, the noise will occur during turns or when changing lanes.

A clunking noise will be caused by excessive play between moving parts in the rear axle assembly. Excessive play can be caused by improper ring and pinion clearance, side gear and spider gear clearance, or side gear and drive axle clearance. It can also be caused by wear in the differential case, wear of the pinion shaft, or worn or improperly adjusted bearings.

Improper ring and pinion clearance attributed to worn ring and pinion gears is the most common cause of a clunking noise. However, when the noise occurs upon making turns, it is usually caused by worn spider or side gears or by excessive ***axle end play,*** Fig 17-2. Clunking noises in the rear axle assembly require the disassembly or removal of the assembly for parts replacement and adjustment.

Whining noises–high-pitched sounds that occur when the vehicle is moving–usually vary with road speed. If a whining noise occurs when driving straight ahead, the problem can be worn or misadjusted ring and pinion gears or worn bearings. Worn bearings usually make a lower-pitched sound than worn or misadjusted gears. Whining noises coming from the drive pinion gear are usually higher pitched, since it is rotating at a higher speed than the other rear end parts.

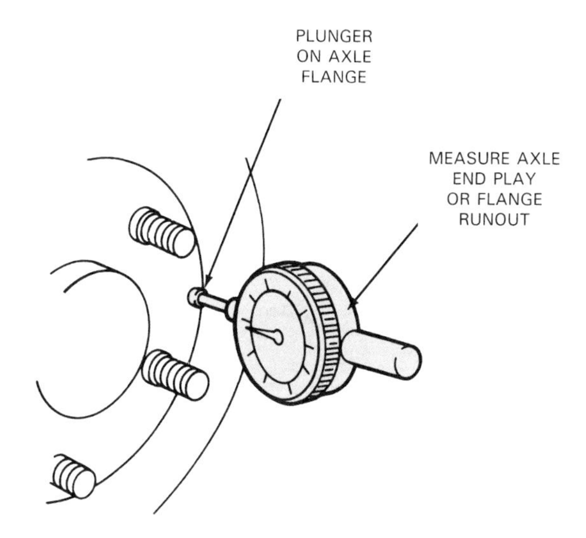

Figure 17-2. Use a dial indicator to measure axle end play. Mount indicator so that plunger is parallel with the axle centerline. Push axle in as far as it will go, then zero the indicator. Pull the axle out and note reading on indicator face. Compare reading to factory specs. Note that this setup can also be used to measure axle flange lateral runout. (Ford)

If the whining noise occurs when accelerating, but not when coasting, the rear pinion bearing may be defective. If the whine occurs when coasting, but not when accelerating, the problem could be the front pinion bearing. If it seems to change pitch when the vehicle is swerved from side to side, the case or axle bearings could be the cause. If the noise occurs only when the vehicle is making a turn, the spider and side gears are probably the source of the problem. Whining noises usually mean that the rear end must be disassembled so that parts can be changed.

Rumbling or roaring noises almost always indicate worn or damaged bearings. The axle bearings at the ends of the axle usually develop these problems. Swerving the vehicle from side to side may cause the bearings to change tone, confirming the problem. If the axle bearing is worn bad enough, it will cause the axle seal to leak. The axle bearings can be changed easily, but replacing the differential bearings requires disassembling the rear end.

Use a stethoscope to find out where the rear end noise is coming from. This will help you determine the cause of the noise. For example, if the sound is the loudest on the outer end of the rear axle housing, it will tell you that the axle bearing is probably bad.

Rear End Vibration

Rear end vibrations are high-frequency vibrations that seem to be coming from the rear of the vehicle when it is moving. Vibrations are almost always caused by bent axles or by bent axle flanges, usually caused by severe impact, such as hitting a curb or being hit by another vehicle. To diagnose rear end vibration, the wheels must first be eliminated as the cause of the vibration, usually by switching wheels from the front to rear or by using a dial indicator.

Bent axle flanges can be checked visually or with a dial indicator without removing the axle from the vehicle. The dial indicator will measure *axle flange lateral runout.* Use the basic setup shown in Figure 17-2, but rather than pull out on the axle flange, rotate it and note deviation as you do. Runout should generally not exceed 0.030 in. (0.76 mm). If it does, the axle flange is bent, and the axle should be replaced.

Bent axles are harder to detect than bent flanges. An axle that is hard to remove from the housing may be bent, but this is not a positive indication. The axle must usually be checked with a dial indicator. It is placed between V-blocks and rotated. As the axle is rotated, radial runout is measured with the dial indicator. If the axle has more than approximately 0.030 in. (0.76 mm) of runout, it should be replaced.

WARNING! Do not try to straighten rear axles. They are hardened and can snap violently!

Lubricant Leakage

Rear end *lubricant leakage* is an unwanted loss of fluid from inside the rear axle housing. Lubricant can leak at the pinion seal, the carrier or the inspection cover gaskets, the axle seals, or the oil fill plug. Further, it can leak from cracks in the housing or axle tubes.

Oil will be seen on the ground under the leak, and it will have stained the area around the leak, making the leak easier to spot. A leak at the rear axle housing vent may mean that the assembly has been overfilled with rear end lubricant. It could also be an indication of water infiltration. Water could have entered through the vent if the vehicle was driven through deep water, causing the fluid level to rise to the point of leakage.

Test for presence of water by dripping some of the oil onto a hot exhaust manifold or by heating with a match or lighter. Pure oil will smoke; a mixture of oil and water will sizzle.

Some leaks can be repaired easily by replacing the faulty seal or gasket. The pinion seal cannot be replaced without removing the pinion yoke/flange, which means that the pinion bearing preload must be reset. If not weldable, cracked housings require that the complete rear axle assembly be disassembled and a new carrier be installed.

If the rear axle housing has been overfilled, oil can be drained. If the housing is contaminated with water, it must be thoroughly flushed out and refilled with the proper type of oil.

Limited-Slip Differential Problems and Troubleshooting

The limited-slip differential can develop the same problems as the standard differential, plus some special problems with the limited-slip clutches. Since they are not operating when driving straight ahead, problems related to the limited-slip feature usually occur when the vehicle is accelerating on slippery surfaces or turning a corner. Many of the problems are caused by overloading or by adding the improper oil to the rear axle assembly. Limited-slip differentials require lubricants that contain special additives.

The most common problem experienced by vehicles with limited-slip differentials is chattering or vibration on turns. Sometimes, the rear end will make a low-pitched moaning noise when the vehicle is turned. These problems are caused by worn clutches or by use of the wrong oil; either way, the rear end must be disassembled, and the clutches must be replaced. Further, if the wrong oil was used, the differential must also be completely flushed out and refilled with the proper oil.

If the vehicle fails to drive properly on slippery surfaces, or does not perform any better than a standard differential, the clutches are probably worn. If they are, they should be replaced.

Rear End Lubricant Service

Many automobile manufacturers recommend that rear end lubricant be checked or replaced at specific intervals. In addition, checking and filling the fluid is important when diagnosing problems in the rear axle assembly. Oil fill plugs are located on the inspection cover or on the front side of the differential carrier.

The fluid level, with the rear end at its normal position, should just wet the bottom of the fill plug threads. See Figure 17-3A. Some manufacturers recommend allowing the fluid level to drop as much as 2 in. (51 mm) below the plug before adding fluid. If the oil level is too high, the rear axle housing may have been overfilled, or the vehicle may have been driven through deep water, allowing water to enter through the vent.

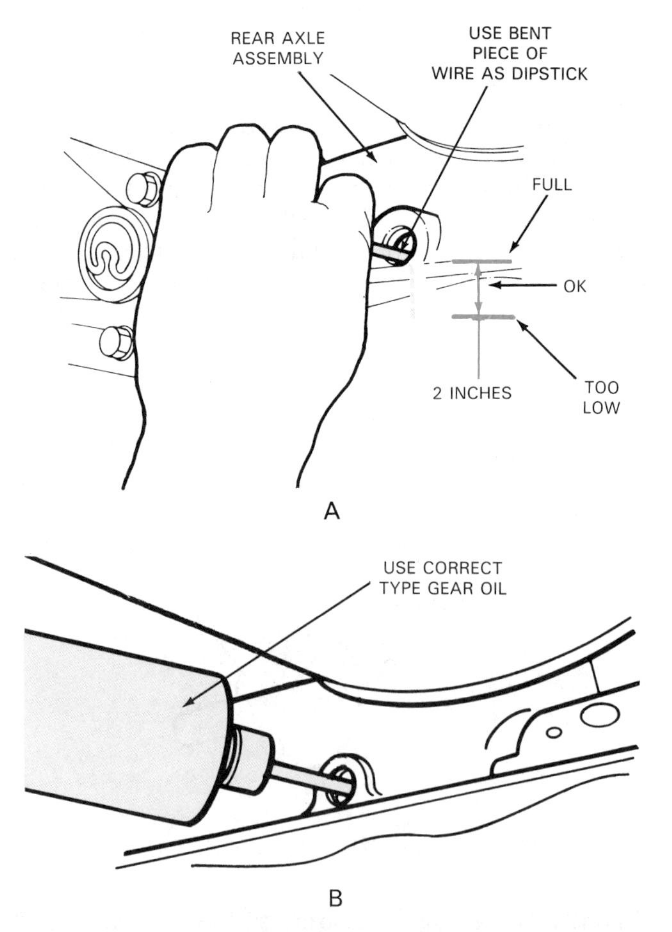

Figure 17-3. Checking and filling lubricant is important when diagnosing rear end problems. A–Fill plugs are usually located on side of carrier or on differential cover. Some rear axle housings will also have drain plugs. B–Always use correct type of oil in rear axle assembly. Squeeze bottles make oil filling easier. (Ford)

Fluid can be added as shown in Figure 17-3B. Always make sure that you are adding the proper kind of lubricant. Most modern rear axle assemblies use 80- to 90-weight gear oil. Many trucks and off-road vehicles use oils as high as 140-weight.

Limited-slip gear oil can usually be used in a standard rear axle assembly. Check the manufacturer's service manual to determine the proper lubricant. Do not use engine oil or ATF in a rear axle assembly.

If the rear axle assembly has a limited-slip differential, special limited-slip oil must be used. Never use ordinary gear oil in a limited-slip differential.

Rear Drive Axle Service

Since the ends of the rear drive axles are exposed to the outside, they are subject to more damage than other components. The axle ends are subjected to flying road debris, snow, ice, and careless handling. Axle flanges can be bent by road hazards, wheel studs can be broken or stripped by overtightening, and the axles themselves may require new bearings and seals. Auto accidents are also a common cause of damage.

There are many kinds of rear ends installed in modern vehicles. The axle attachment methods and the bearing and seal types are different. Axle lengths and diameters also vary.

It is important that the technician determine exactly what kind of rear axle assembly he or she is dealing with. Most assemblies have an ID tag installed on one of the carrier bolts or on one of the cover bolts. The information on the tag will tell the person at the parts counter what kind of rear axle assembly is installed in the vehicle. If the tag is missing, the information may be *stamped* on the carrier or contained in the VIN on the dash. Sometimes, it is necessary to bring in the old parts so that they can be compared to the replacement parts.

Drive Axle Removal

Some solid drive axles are held in the rear axle housing by a retainer plate bolted onto the housing. Other solid drive axles are locked to the side gear by a small clip, or C-lock. Removal methods are different for each type.

Independently suspended drive axles, in most cases, resemble a drive shaft assembly in their installation. These are held in place by one of several methods. Consult the proper service manual for exact removal procedures of independently suspended drive axles.

Retainer-type axle

The following steps detail the proper way to remove a retainer-type drive axle.

1. Disconnect the battery negative cable.
2. Raise the vehicle on an approved hoist or with a hydraulic jack. Secure the vehicle on jackstands if using a jack.
3. Remove the rear wheel.
4. Remove the brake drum, Figure 17-4. Mark it so that it can be reinstalled in the same position.
5. Locate the nuts that hold the retainer plate to the rear axle housing by turning the axle flange by hand. The nuts can be removed by inserting a socket and extension through access holes in the axle flange. Loosen nuts with a socket wrench. Note that some retainer plate nuts are removed from the back side of the retainer plate.

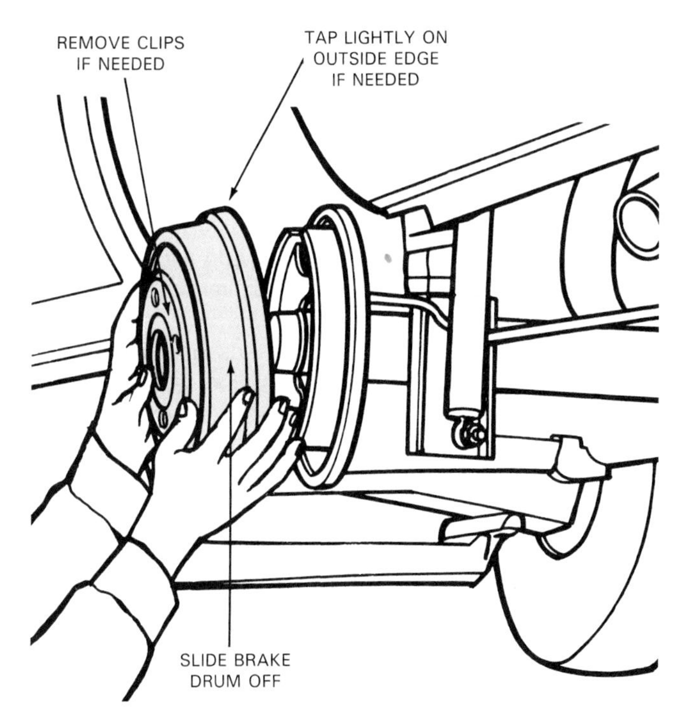

Figure 17-4. After removing the wheel, remove the brake drum, as shown here. (Ford)

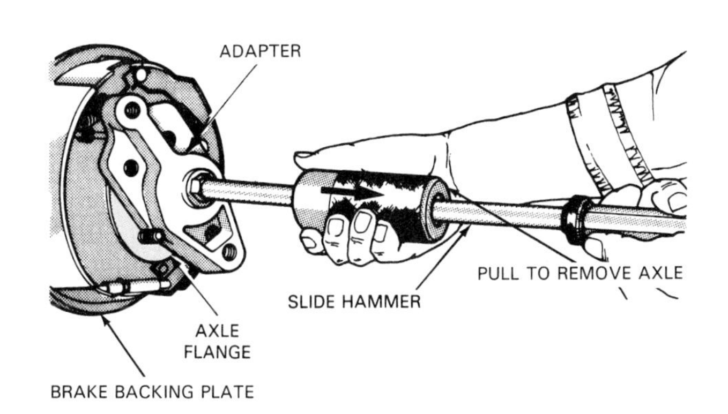

Figure 17-5. Remove the axle with a slide hammer, if necessary. (Chrysler)

6. Once the nuts are loose, the flange, axle, retainer plate, and bearing can be pulled out of the housing as a unit. If the axle and bearing stick in the housing, a slide hammer can be used to loosen the axle. See Figure 17-5.

CAUTION! When pulling out the axle, make sure that the retainer plate does not hang up on the brake backing plate, causing the backing plate to pull away from the rear axle housing. This could damage a brake line.

C-lock axle

The following steps detail the proper way to remove a C-lock drive axle.

1. Disconnect the battery negative cable.
2. Raise the vehicle on an approved hoist or with a hydraulic jack. Secure the vehicle on jackstands if using a jack.
3. Remove the rear wheel.
4. Remove the brake drum, marking it so that it can be reinstalled in the same position.
5. Place an oil pan beneath the differential carrier. Loosen cover bolts or drain plug and allow the lubricant to drain. Remove the differential cover after the lubricant has drained.
6. Remove the pinion shaft, which holds the spider gears, by unscrewing the lock bolt. Push the shaft out of the differential case. See Figure 17-6. You may need to rotate the case so that the shaft clears the rear axle housing. The spider gears can be left in place.
7. Push the axle in toward the vehicle. This will allow the C-lock to be removed from the axle, which is the next step. See Figure 17-7. In many cases, the C-lock will fall out of position once the shaft is pushed inward.
8. Return pinion shaft to its location in the differential case and tighten the lock bolt. This will hold the differential gears in place.
9. Pull the axle from the housing, Figure 17-8. If the axle sticks, use a slide hammer to loosen it.

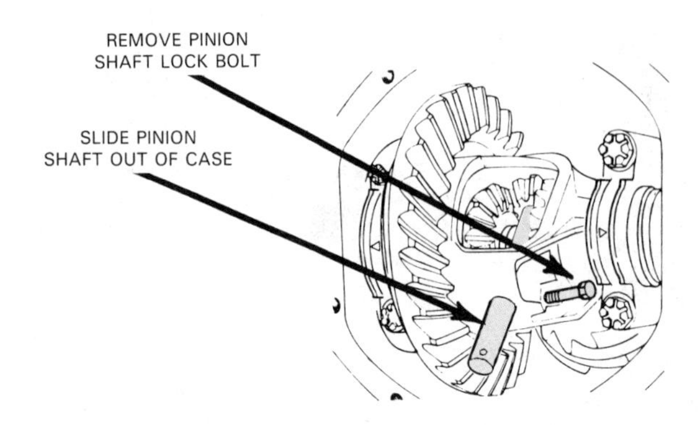

Figure 17-6. To remove a C-lock axle, the pinion shaft in the differential case must be removed. (Ford)

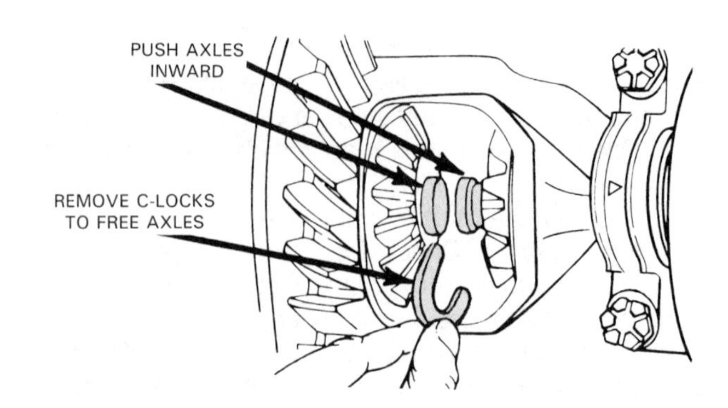

Figure 17-7. Push axles inward and remove C-locks. (Ford)

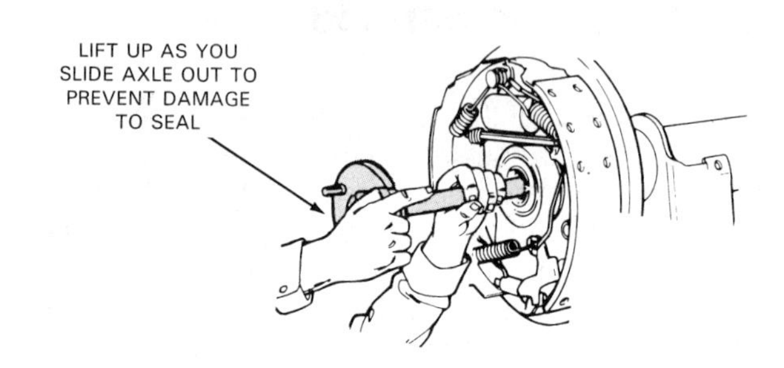

Figure 17-8. Remove axle from axle tube. Do not let axle hang in tube, as seal or bearing may be damaged. Do not let axle splines damage seal as you remove axle. (Ford)

Axle Bearing Service

Axle bearings are replaced in different ways, according to the kind of axle. To replace the axle bearings on a retainer-type axle, the axle collar must be removed by breaking it off with a chisel. See Figure 17-9. Never remove the collar by cutting or expanding it with a torch. Heat will be conducted to the axle and will ruin it.

WARNING! Wear eye protection when breaking an axle bearing or collar off of its axle.

Once the collar is off, the bearing can be pressed off as shown in Figure 17-10. Inspect the shaft for any damage to the bearing area or to the splines. Decide if it can be reused.

Inspect the new bearing and collar to make sure that they match the old parts. The retainer plate can then be reinstalled, and the new bearing and collar can then be pressed onto the shaft. The collar should require approximately 2 tons (4000 psi, or 27.58 MPa) of pressure for proper installation. If it goes on with less pressure, the shaft should be replaced, as it may be worn.

The C-lock axle bearing is installed in the rear axle housing. It can be removed by first removing the seal in the housing and then prying out the bearing. If the bearing is stuck in the housing, you may need a slide hammer with external jaws to remove it.

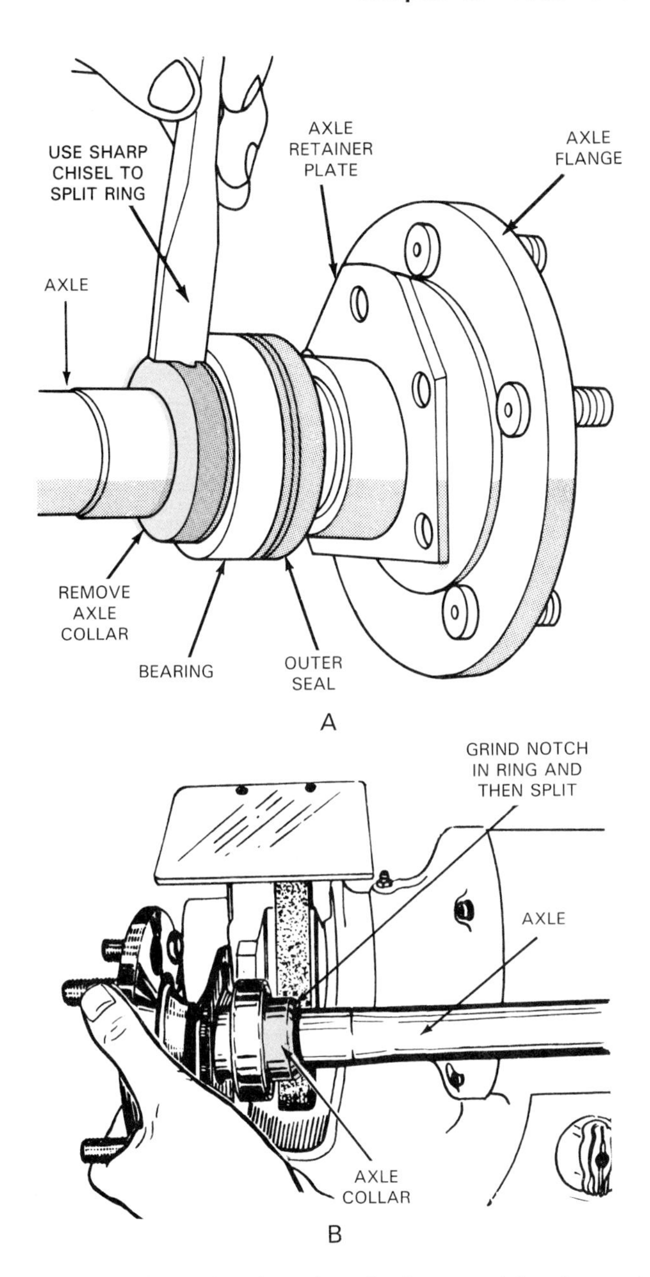

Figure 17-9. Removing the axle collar from a retainer-type axle is shown. A–Remove collar with a chisel. B–Collar will be easier to remove if a spot is first ground down on a grinder. (Chrysler)

Before installing the new bearing, check the axle, Figure 17-11. If the axle is worn where it contacts the bearing, the axle should be replaced. The replacement bearing should be lubricated and lightly tapped into place. Ensure that the bearing seats flush against the mounting shoulder in the rear axle housing.

Wheel Stud Service

The wheel studs are pressed into the axle flange. These threaded studs can be easily replaced if they become broken or stripped. First, remove the wheel and brake drum, marking their original position to maintain balance. Then, drive the defective stud from the flange with a hammer and punch. Do not forget to remove the stud if it falls behind the brake backing plate.

To install the new stud, insert the threaded portion through the hole from the backside of the flange. Push the stud in as far as it will go. Then, install a washer and one of the wheel lug nuts on the stud, as shown in Figure 17-12. Tighten the nut. The larger, unthreaded portion of the stud will then be drawn into the flange hole. After the stud is fully seated, reinstall the brake drum and wheel.

This replacement procedure can also be used with the axle removed from the vehicle. However, if the axle is removed from the vehicle, it may be easier to install the new wheel stud with a hydraulic press. Place the stud in the flange hole and, using the proper adapters, press the stud securely into place.

Axle Seal Service

Axle seals should be replaced whenever other axle service, such as bearing replacement, is performed. Seals should also be replaced when they start leaking. The most obvious sign of a leaking axle seal is oil dripping from the brake backing plate. Usually, the inside of the tire will also be streaked with oil.

It is important to check the brake linings when replacing a leaking seal. Oil leaking from the seal will be absorbed by the linings, destroying their frictional properties and making the brakes unsafe. Slightly oily linings can be sanded and returned to service, but soaked linings should be replaced.

The axle seal can be removed by first removing the axle, as discussed earlier. The seal can then be removed by prying with a screwdriver or other tool, or removing it with a slide hammer equipped with external jaws. Do not gouge the inside of the rear axle housing.

Before installation, check the new seal for proper fit in the axle tube bore and on the axle. Some seals are precoated with sealant on their outside diameter. If the seal is not precoated, a light coat of nonhardening sealer should be placed around the outside diameter. The seal can then be driven into place with a seal driver.

Figure 17-13 shows a typical seal replacement procedure. If you do not have the proper driver, be extra careful that the seal is installed without doing damage to it. Do not dent or bend the seal.

The seal should be installed so that it is square with the axle tube bore. There is usually a shoulder for the seal to bottom against. Before installing the axle, lightly oil the lip of the seal with the proper kind of lubricant. When installing the axle, make sure that you do not drag it across the seal.

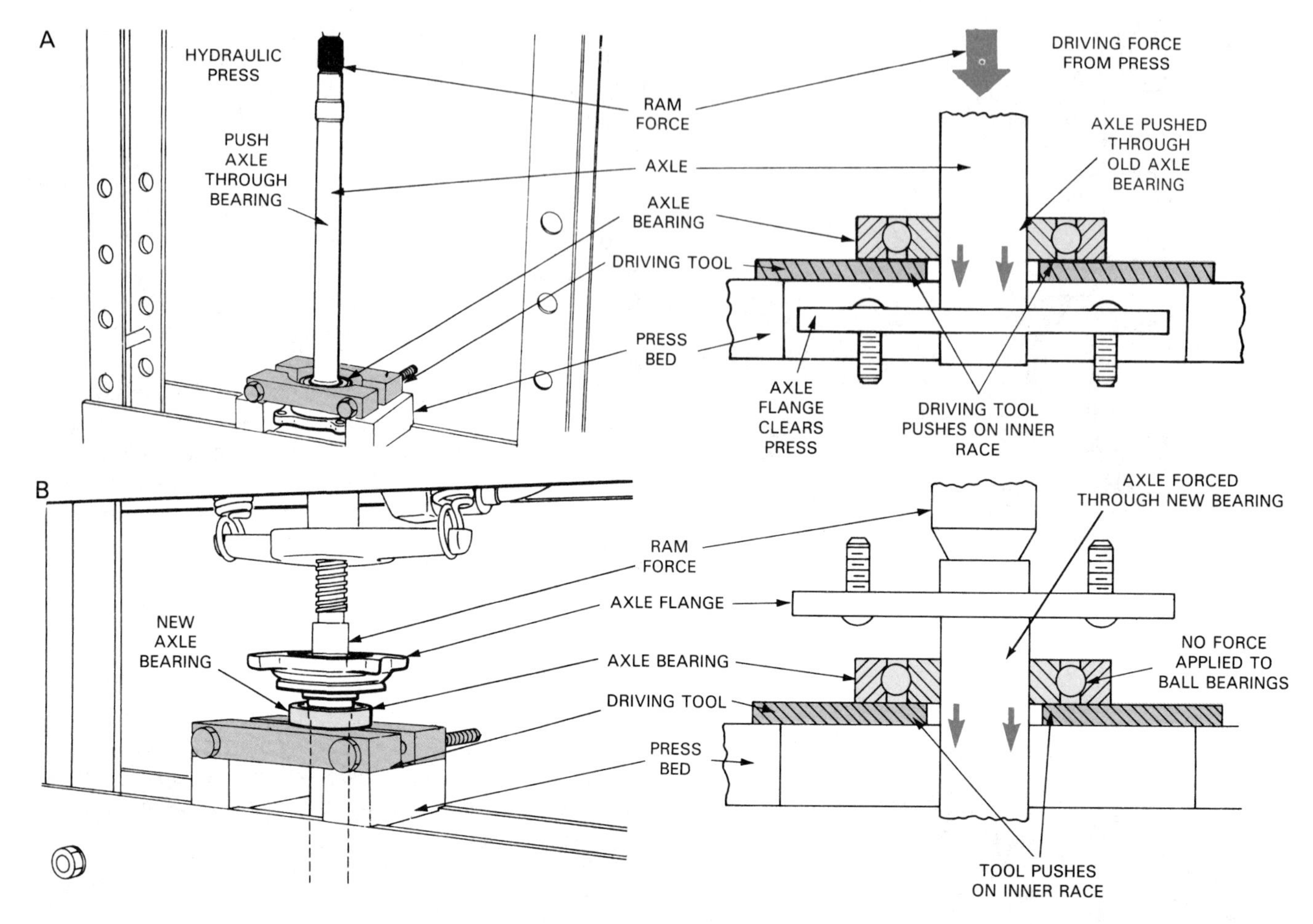

Figure 17-10. This shows steps in axle bearing removal and installation, using a hydraulic press. A–Pressing off old bearing. Note bearing should be supported by proper bearing adapter so press can push axle through bearing. B–Pressing new bearing onto shaft. Press until bearing bottoms on axle shoulder. Again, note use of proper bearing adapter. (Chrysler)

Drive Axle Installation

Following are some general procedures detailing how to install solid drive axles properly. Exact procedures vary, but basically it is the reverse of removal. You should always refer to the manufacturer's service manual for specific installation procedures–for solid *or* independently suspended drive axles.

Retainer-type axle

The following steps detail the proper way to install a retainer-type drive axle.

1. Raise the vehicle on an approved hoist or with a hydraulic jack. Secure the vehicle on jackstands if using a jack.
2. Remove brake backing plate momentarily and check *axle tube flange* to see that surface is clean and free of old gasket material. Install bolts through flange. Then, place a new gasket over flange.
3. Position brake backing plate and brake assembly over gasket and flange. Position outer gasket over brake assembly.
4. Lubricate splines on drive axle with proper lubricant.
5. Install drive axle into axle tube, being careful not to damage axle seal. Engage axle splines with side gear splines, pushing in axle as far as it will go.
6. Position axle retainer plate over axle tube flange bolts. Install nuts over bolts and torque to manufacturer's specifications.
7. Check axle end play and adjust if necessary. (Refer to Figure 17-2.)
8. Install brake drum and rear wheel.
9. Lower the vehicle. Check the brakes. Test drive the vehicle to ensure proper operation.

C-lock axle

The following steps detail the proper way to install a C-lock drive axle.

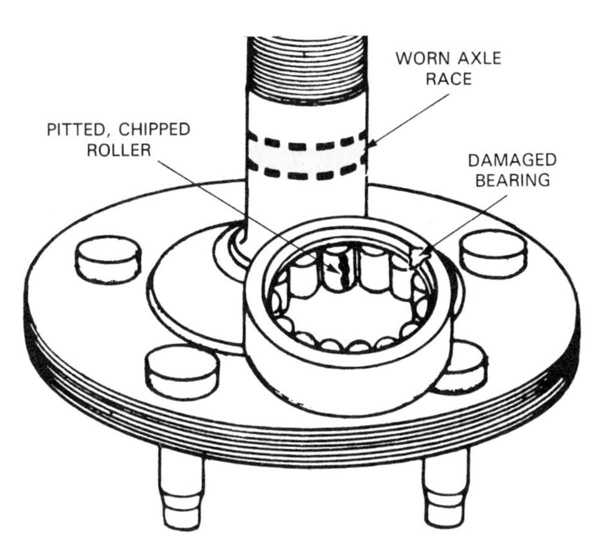

Figure 17-11. Carefully inspect C-lock axle race, or surface where axle bearing rollers contact. If axle is worn or damaged where it rides on bearing, axle should be replaced. Note axle bearing, shown here, is a common source of trouble. The trouble usually arises from an inadequate supply of lubrication. (Ford)

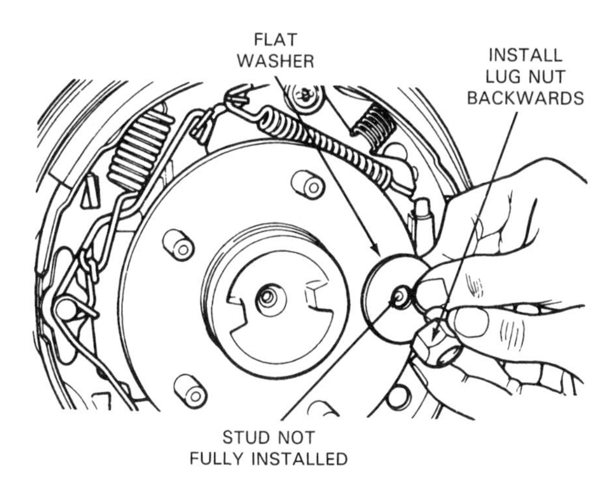

Figure 17-12. A new wheel stud can be installed on wheel flange by tightening it into place with a lug nut. A washer should be installed temporarily to distribute pulling force and ensure that stud is completely installed. (General Motors)

1. Raise the vehicle on an approved hoist or with a hydraulic jack. Secure the vehicle on jackstands if using a jack.
2. Remove the lock bolt or pin and the pinion shaft from the differential case assembly.
3. Lubricate splines on drive axle with proper lubricant.
4. Install drive axle into axle tube, being careful not to damage axle seal. Engage axle splines with side gear splines, pushing in axle until the end extends past the side gear.
5. Insert the C-locks in the grooves in the axle. Pull the axle outward to seat the C-locks.

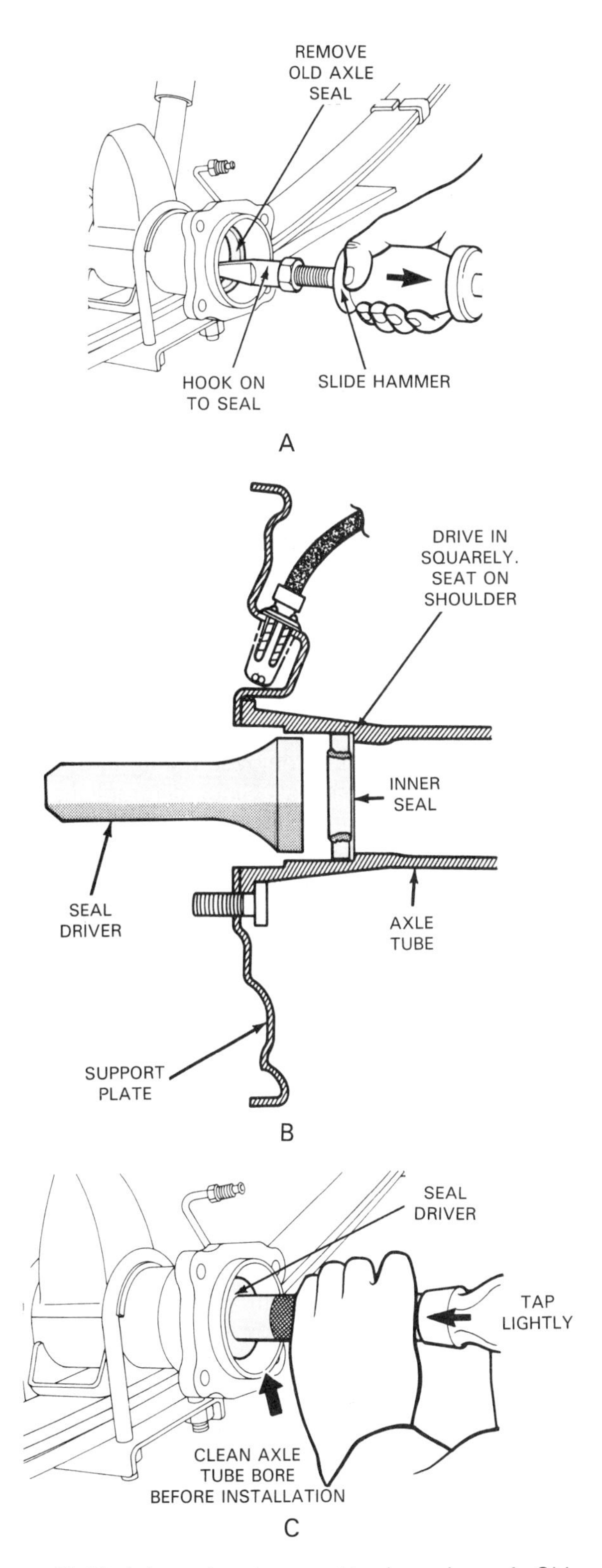

Figure 17-13. Axle seal replacement is shown here. A–Old seal is removed with a special puller. Sometimes, seal can be removed with a prying tool, but care must be taken not to damage housing. B–New seal (shown here as sectional view) should be installed with a seal driver that matches seal diameter. C–Gently tap new seal into place. (Chrysler)

6. Install the pinion shaft and lock bolt or pin. Torque bolt to proper specifications.
7. Check axle end play and adjust if necessary. (Refer to Figure 17-2.)
8. Scrape off any old gasket material from the differential cover and its mating surface on the rear axle housing. Install a new gasket and use a nonhardening gasket sealer. Note that some manufacturers recommend using form-in-place gaskets. The bead of this type of gasket should be laid evenly and should completely encircle all bolt holes.
9. Tighten the differential cover bolts in a crisscross pattern. Torque the bolts to manufacturer's specification.
10. Add the proper type of rear end lubricant. Replace the fill plug and torque to specifications.
11. Install brake drum and rear wheel.
12. Lower the vehicle. Check the brakes. Test drive the vehicle to ensure proper operation.

Differential Assembly Service

This section covers the removal, repair, and reinstallation of differential assemblies. Since differential assemblies are not frequently repaired, it is difficult to build up a body of experience. This section will help you to learn the steps involved in repairing a differential assembly so that you will be ready to repair one when the time comes.

Differential Assembly Inspection

Before removing and disassembling the differential, check for problems while it can still operate. This is not possible in cases of severe damage, such as with stripped gears or broken shafts, but it will be helpful in other cases to confirm a diagnosis.

To check the differential assembly, first raise the vehicle so that both rear wheels are off the ground and free to rotate. Ensure that the vehicle is solidly supported and cannot work its way from the lift or jackstands. Start the engine and place the transmission in gear. Then, listen for noises. A stethoscope can be used to pinpoint the exact location of the noise.

WARNING! Be extremely careful when working around the turning wheels and driveline. Cover long hair and do not wear loose clothing. Never touch spinning parts with your stethoscope!

As an additional check, pull on parking brake cable on one side while the engine is still running. This will stop one wheel while allowing the other wheel to turn. Do the same on the other side. This procedure may isolate the problem to one side or the other.

An additional check can be made on limited-slip differentials. The wheels do not have to be removed. The vehicle should be raised so that only one rear wheel is off the ground. Do *not* start the engine and make sure that the transmission is in *neutral.* Attach a torque wrench to the raised wheel using an adapter, as shown in Figure 17-14. Try to turn the axle with the torque wrench.

The axle should *break away,* or begin to move, when the torque reading approaches approximately 30 ft.-lb. (41 N-m). This value is determined by the pressure on the clutch pack due to the preload spring(s). The clutch pack begins to slip when applied torque exceeds this value. The torque reading can be higher than 30 ft.-lb., but it should not be much lower. Improper rear end lubricant could be responsible for a low reading. If the reading is lower, the clutches will need to be replaced. Note that once the axle starts turning, the torque reading will be reduced. This is normal, as long as the torque remains constant through a full revolution of the axle.

WARNING! Do not start the engine and place the transmission in gear with one wheel on the ground. The limited-slip differential will attempt to drive the wheel on the ground, possibly causing the vehicle to be driven off of the lift or jackstands.

Finally, if the differential is the type that has an integral carrier, remove the inspection cover and closely inspect the rear axle lubricant for metal particles or for presence of water. This will give you a clue to possible internal problems.

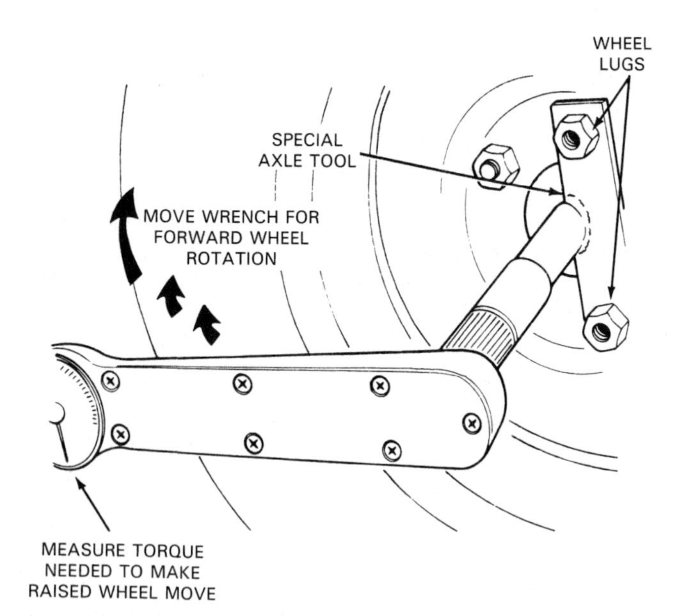

Figure 17-14. A limited-slip differential can be checked with a torque wrench. Wrench is attached to raised wheel with a special adapter. Other wheel should be on the ground. If torque is lower than specified, clutches are worn out, or the wrong lubricant has been used. (Ford)

Then, the gears can be thoroughly cleaned with solvent, Figure 17-15.

After the cleaning process, inspect the ring and pinion for wear and for excessive play between the gears. Check the pattern of wear. Typical ***ring gear contact patterns*** are shown in Figure 17-16. Most of the problem areas will be caused by incorrect adjustment. Even if the contact pattern is correct, the teeth may be excessively worn or pitted. Also, check for wear on the spider and side gears and for excessive play between them or for other obvious problems.

Differential Assembly Removal

The exact procedure for differential assembly removal varies according to the type of differential. Differential assemblies having removable carriers are easier to remove from the vehicle, since the entire rear axle assembly does not require removal. If the differential carrier is the integral type, it is normally easier to leave the unit installed in the vehicle for service. However, it is usually harder to work on the internal components while under the vehicle.

Removal procedures for a differential assembly from a vehicle with independent rear suspension are similar to those for solid-axle rear suspension; however, there are still differences between them. Figure 17-17 shows some of the components that must be disconnected to remove a rear axle assembly using independent rear suspension.

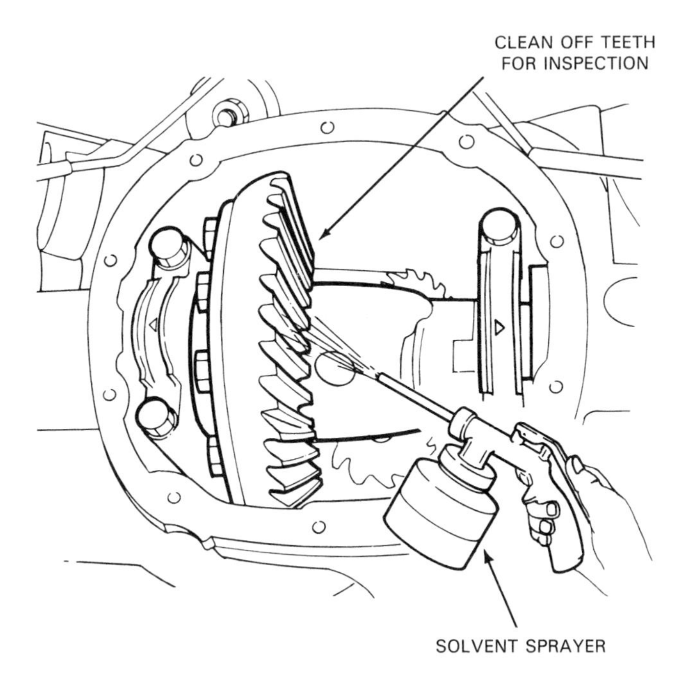

Figure 17-15. To inspect rear end parts, they must be thoroughly cleaned. Use a petroleum-based solvent and air dry. Never clean any parts with water. (Ford)

Removable carrier

The following procedure details the proper way to remove a differential assembly from a rear axle assembly having a removable carrier.

1. Disconnect the battery negative cable.
2. Raise the vehicle on an approved hoist or with a hydraulic jack. Make sure that it is properly supported.
3. Remove the rear wheels.
4. Remove the brake drums, marking them so they can be reinstalled in the same position.
5. Remove the nuts that hold the axle retainer plates to the rear axle housing. (Most drive axles of rear axle assemblies with removable carriers are secured with retainer plates.)
6. Pull the axles approximately 1 ft. (30 cm) out of the rear axle housing (or remove them completely).
7. Mark the drive shaft assembly for proper reinstallation and remove it.
8. Remove any hardware, such as brake line brackets, attached to the carrier.
9. If the carrier has a drain plug, remove it and allow the oil to drain.
10. Support the carrier on a jackstand. Then, loosen the nuts holding the carrier to the rest of the rear axle housing and pry the carrier away from the housing. Allow the oil to finish draining.
11. Remove the nuts and, with at least one other person to assist you, slide the carrier away from the housing and lower it to the floor. See Figure 17-18.

WARNING! A cast iron removable carrier is surprisingly heavy for its size. Be ready to support its weight during removal.

Integral carrier

For convenient access to a differential assembly having an integral carrier, it is sometimes better to remove the entire rear axle assembly. Figure 17-19 shows some of the hardware that must be removed to accomplish this. The following procedure details the proper way to remove the rear axle assembly of a vehicle with either solid-axle rear suspension or independent rear suspension.

1. Remove the battery negative cable.
2. Raise the vehicle on an approved hoist or with a hydraulic jack. Make sure that it is properly supported.
3. Remove the rear wheels.
4. Remove the brake drums if there is a danger of them falling off the axle flanges. If not, they can be left installed. Mark brake drums, if they are removed, so they can be properly installed.
5. Mark the drive shaft assembly for proper reinstallation and remove it. See Figure 17-20.
6. Raise the rear axle housing slightly and support it with a floor jack or a transmission jack, as required. This

CHECKING AND ADJUSTING GEAR TOOTH CONTACT PATTERN

DRIVE SIDE | **COAST SIDE**

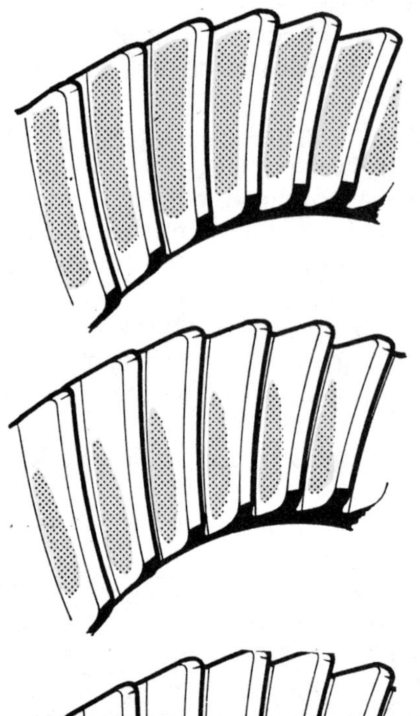

CORRECT MESH

The contact pattern is uniformly distributed over both tooth faces, drive and coast.

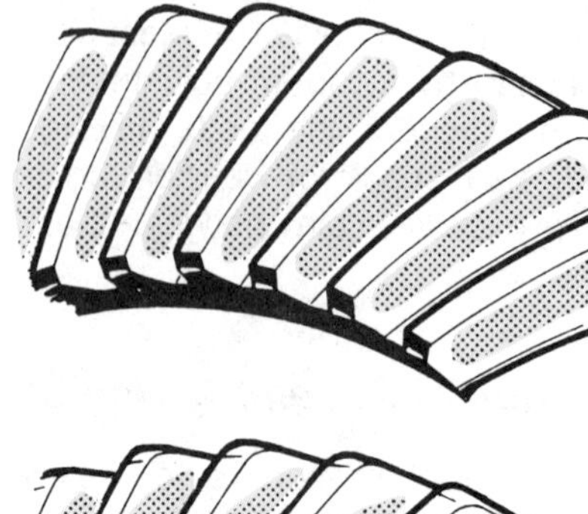

INCORRECT MESH

Divide side: contact on top of tooth and towards center.

Coast side: contact on heel of tooth and towards center.

Move ring gear away from pinion, using thinner side bearing shim.

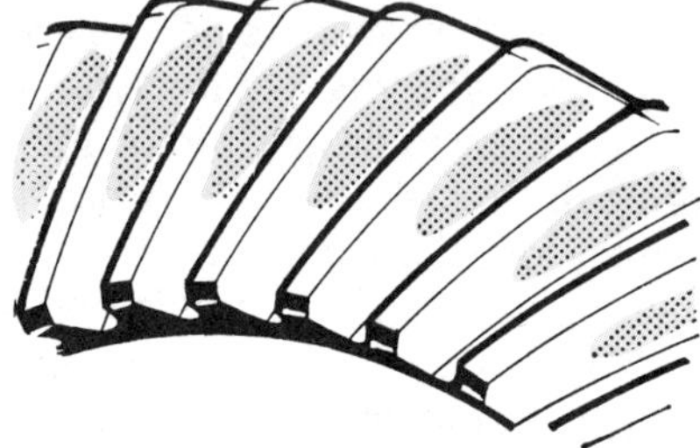

INCORRECT MESH

Divide side: Toe contact, localized at root.

Coast side: Heel contact, localized at root.

Move pinion away from ring gear, using thinner drive pinion gear shim.

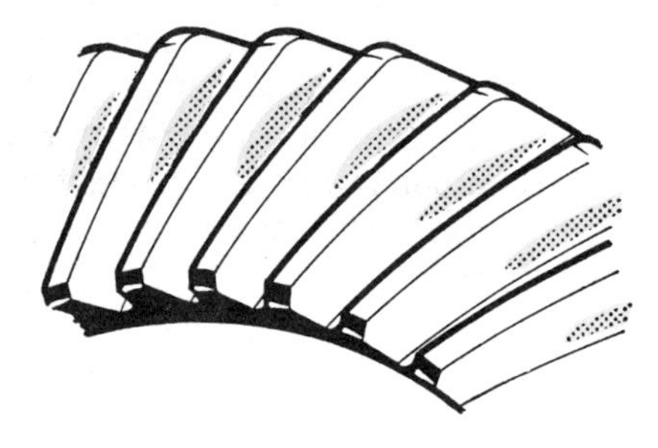

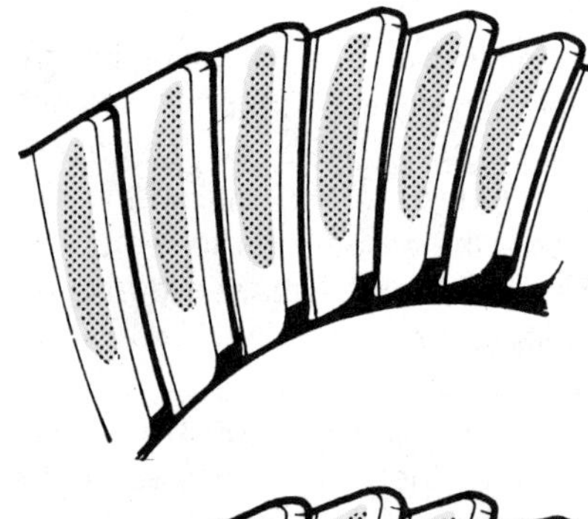

INCORRECT MESH

Drive side: Heel contact, towards center of tooth.

Coast side: Toe contact, towards center of tooth.

Move ring gear closer to pinion, using thicker side bearing shim.

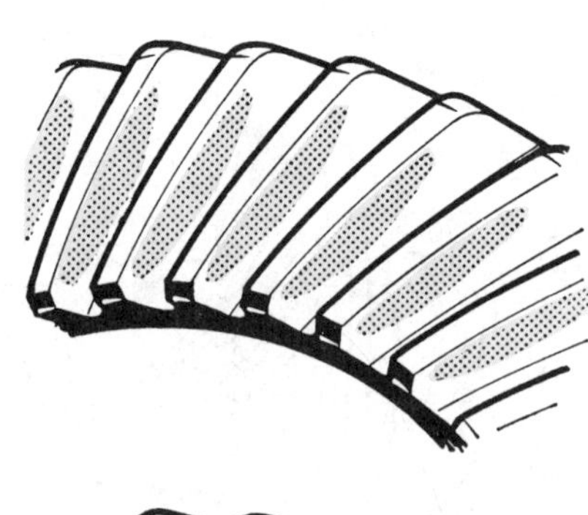

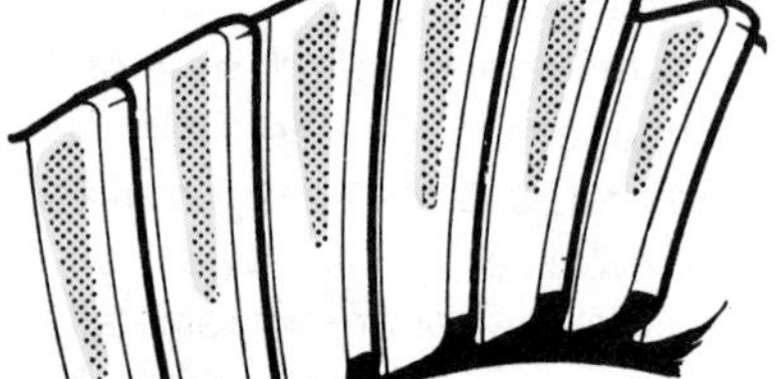

INCORRECT MESH

Drive side: Heel contact, localized on crest of tooth.

Coast side: Toe contact, localized on crest of tooth.

Move pinion closer to ring gear, using thicker drive pinion gear shim.

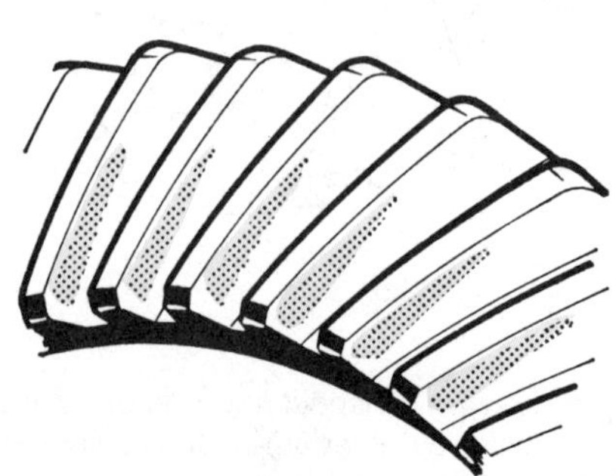

Figure 17-16. This chart shows typical ring gear contact patterns, which show patterns of tooth wear. The chart can be used to determine what must be done to repair the rear axle assembly. Worn gears cannot be fixed by readjustment. (Peugeot)

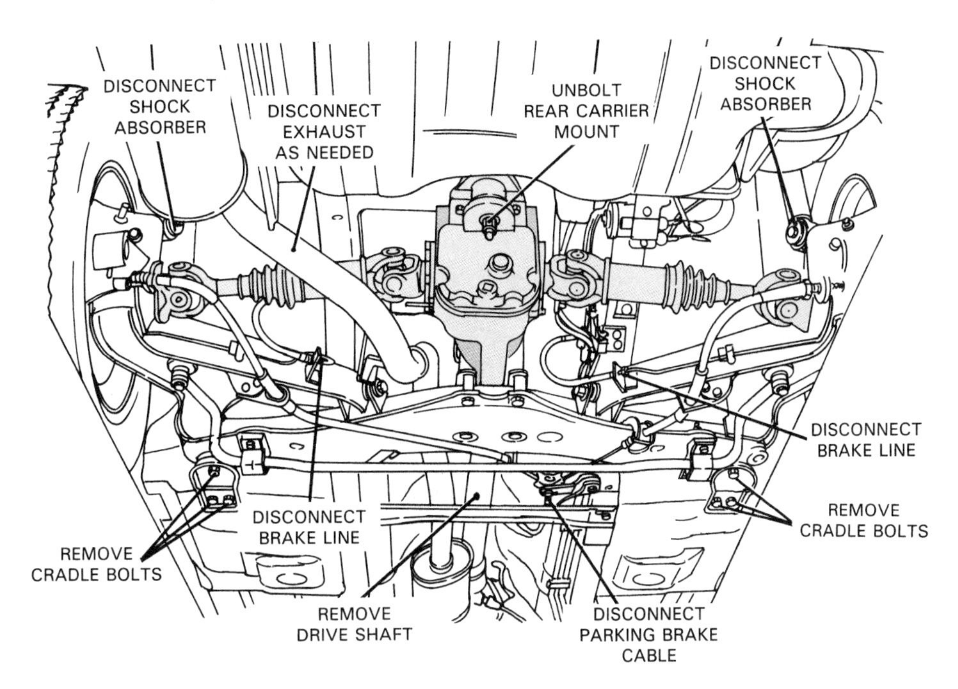

Figure 17-17. To remove a rear axle assembly, other vehicle attachments and hardware must be disconnected. This shows typical components that must be disconnected to remove a rear axle assembly used with independent rear suspension. (Subaru)

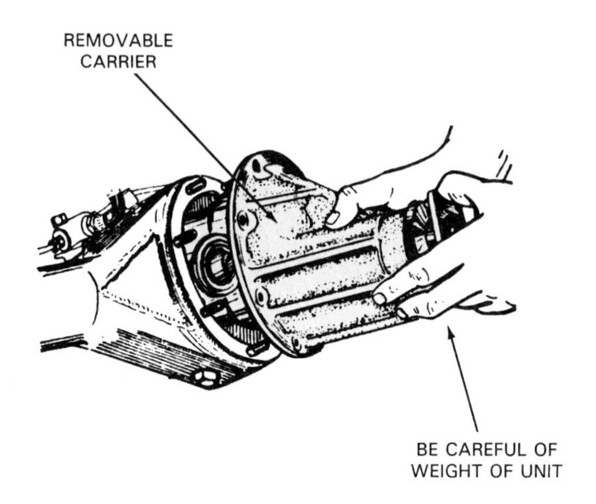

Figure 17-18. Slide the removable carrier back and lower it. The carrier is very heavy. Have a helper assist you and be ready to support its weight. (Chrysler)

will remove the tension from the springs and allow the attaching hardware to be removed. It will also keep the rear axle assembly from dropping once attaching hardware is removed.

WARNING! All tension should be removed from the rear suspension springs before spring attaching bolts are removed. These springs are very powerful, and they can cause personal injury or serious equipment damage.

7. Remove shock absorbers and spring attachments. Leaf springs are usually attached to the housing by U-bolts. Leaf springs will fall free once U-bolts are removed. Coil springs are usually held in place by the weight of the vehicle and the travel limits of the shock absorbers. Coil springs will fall free once the shocks are removed and the rear axle assembly drops down far enough. Watch so springs do not fall on you or anyone else.
8. Remove any stabilizer bars or other stiffening bars and any strut arms. (Strut arms, sometimes called track bars, are usually used with coil spring systems and are located between the vehicle body and housing. Refer back to Figure 17-19.)
9. Disconnect brake lines and parking brake cables. Cap the brake lines to retain as much of the brake fluid as possible.
10. Remove the exhaust pipes if they pass below or close to the rear axle housing.
11. If the rear axle assembly is part of an independent rear suspension, remove the drive axles. See Figure 17-21. Also, most independent rear suspension systems use a cradle to hold the assembly in place. It should be unbolted, along with the rear differential carrier mount, after all other attachments are removed. In a few instances, the entire assembly–differential carrier, drive axles, and cradle–are removed as a unit. (Refer back to Figure 17-17.)

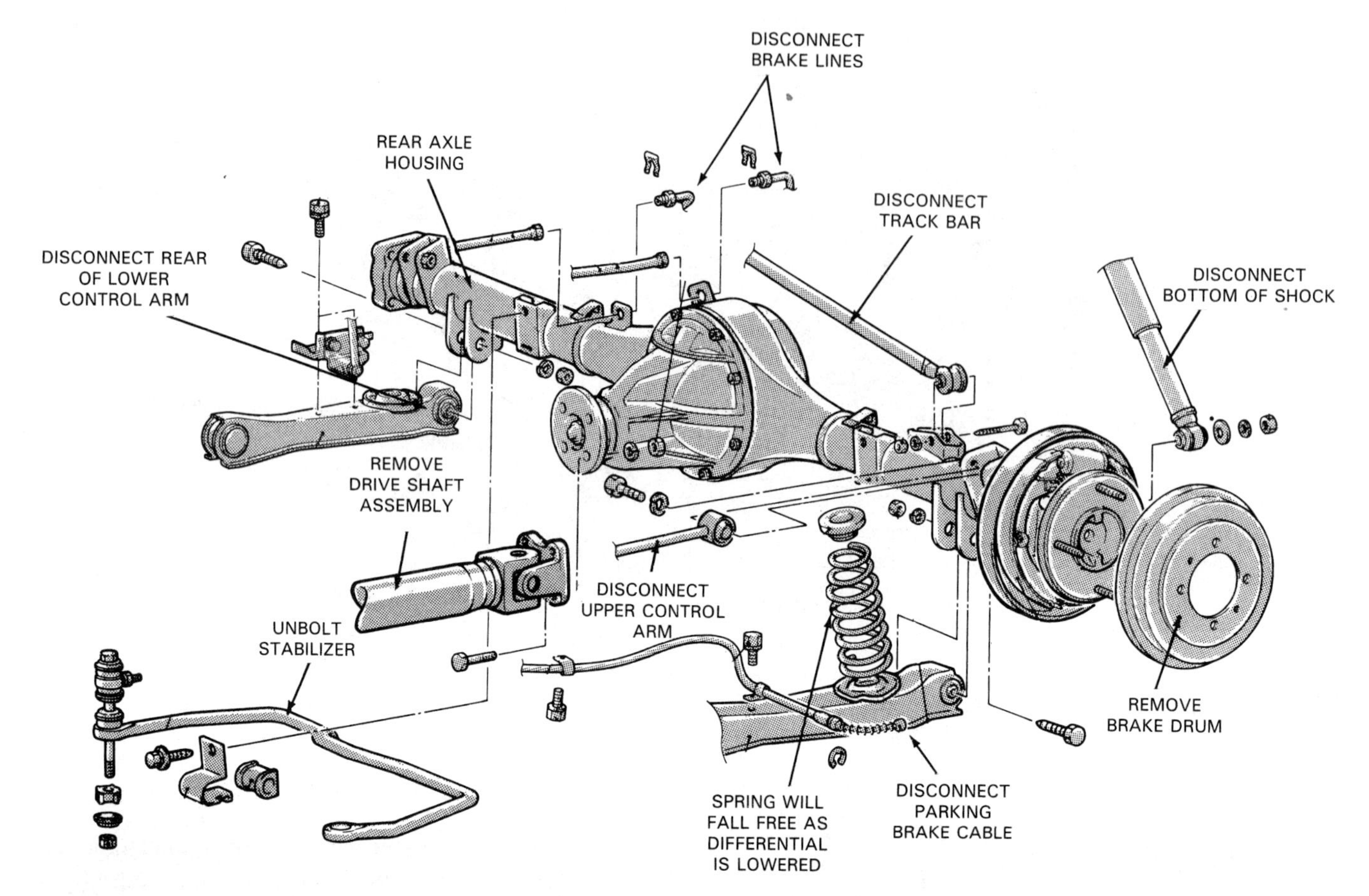

Figure 17-19. A typical solid drive axle is shown here. Note parts that must be disconnected from the rear axle housing before it can be removed. Note that this assembly has a removable carrier, and it may not be necessary to remove the entire housing. (Chrysler)

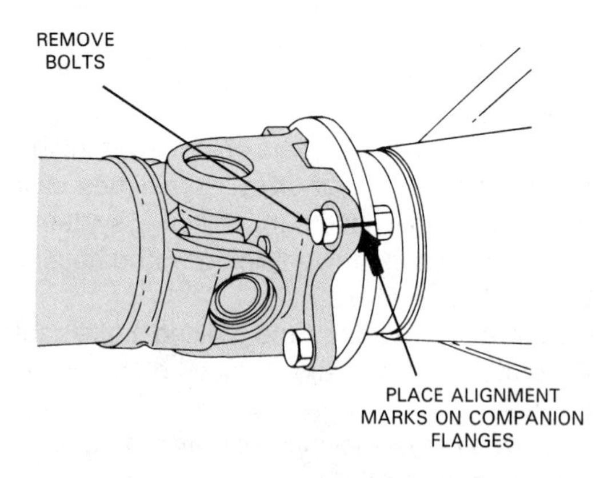

Figure 17-20. The driveline must be removed before removing the rear axle assembly. Place match marks at the rear of the drive shaft and differential pinion yoke or flanges to help assure proper reinstallation. (Chrysler)

12. Once all attachments have been removed, carefully lower the rear axle assembly from the vehicle, Figure 17-22. The oil does not have to be drained until the vehicle is on the bench.

WARNING! Remember that the rear axle assembly is heavy and awkward. You should have another person help you hold on to the assembly as it is lowered with the jack. Be careful not to let the assembly fall during removal. Severe damage or injury could result.

Differential Disassembly

Before beginning the disassembly, move the entire assembly to a bench or other place where it can be conveniently serviced. If possible, the assembly should be installed in a fixture, Figure 17-23, to make it easier to work on. (Disregard this step for differentials with integral carrier that is to be serviced on the vehicle.)

Remove drive axles if they have not been removed as of yet. Follow procedure detailed earlier in the chapter for retainer-type axle or C-lock axle removal, as applicable; however, in the latter case, do not put pinion shaft back after removing C-locks.

Once C-lock axles are out of the way, the spider gears and side gears can be removed. Turning one of the side gears with the pinion shaft removed will cause the spider gears to rotate out of the differential case. Both spider and

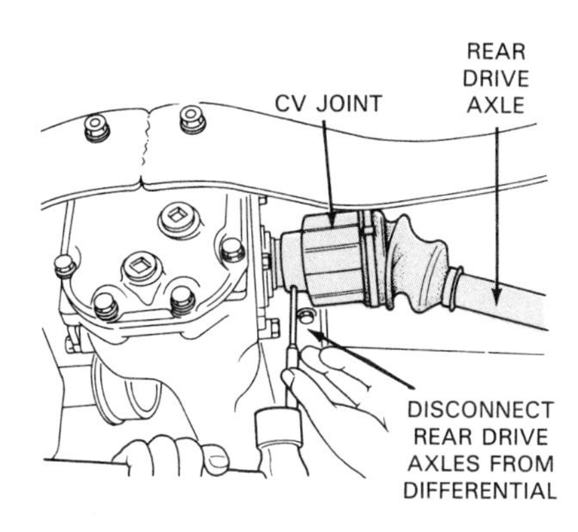

Figure 17-21. Remove independently suspended drive axles before removing rear axle housing. Removal methods vary. In design shown, a roll pin is first removed from inner stub axle with a punch and hammer. The next step would be to mark and remove axle flange bolts and to remove flange from outer wheel hub assembly. Then, the inner stub axle would be pried from differential carrier with a large screwdriver or other tool. (Chrysler, Subaru)

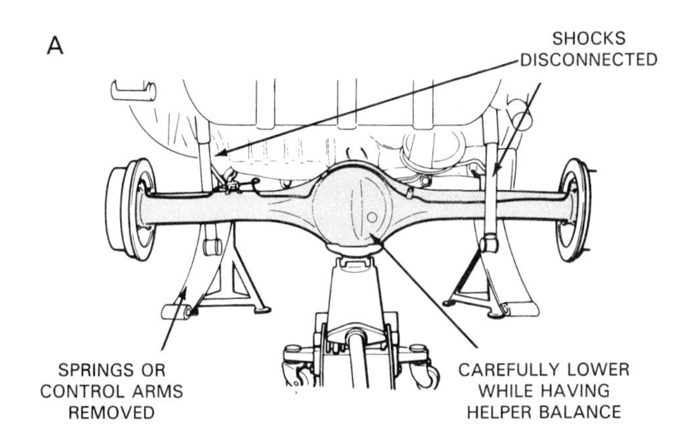

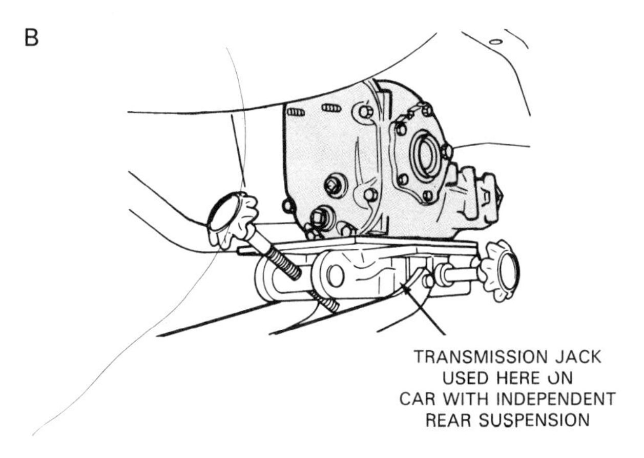

Figure 17-22. Carefully lower the rear axle assembly from under the vehicle. A–A rear axle assembly with a solid rear axle is shown being lowered from the vehicle with a floor jack. B–The rear axle housing from a vehicle with independent rear suspension is shown being lowered with a transmission jack. (Subaru)

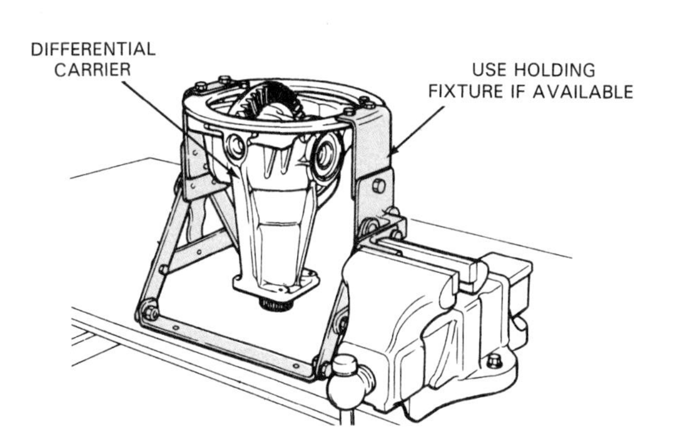

Figure 17-23. If a holding fixture is available, differential should be bolted to fixture for ease of servicing. If not available, assembly should be securely placed on a workbench. (Chrysler)

side gears will have thrust washers between them and the case. These washers should be kept with the gears to avoid losing them and to simplify reassembly. For retainer-type axles, the removal of spider and side gears does not have to be done until after the differential case is removed from the differential carrier.

The procedure for removing the differential case from the carrier is shown in Figure 17-24. Mark the side bearing caps, if needed. Unbolt bearing caps on shim-type differential case (integral carrier) and remove the case. The shim-type case will have to be pried out of the carrier with a crowbar or other long metal bar. If threaded end caps are used instead of shims, remove them before removing the bearing caps and differential case. The bearing cups will be positioned on the side bearings and will be removed with the differential case. After the case is removed, remove any shims from the carrier.

WARNING! If you are working under the vehicle, do not allow the differential case to fall on the floor when it comes loose.

Once the differential case is removed from the carrier, the drive pinion gear can be removed. The first step in drive pinion gear removal is to remove the large nut holding the pinion yoke/flange. The flange must be held to loosen and remove the nut. Once the nut is removed, the flange will slide out from the splines on the drive pinion gear shaft. If the flange is stuck, it can be loosened by lightly tapping the pinion gear with a brass or plastic hammer. The pinion gear can then be removed from the rear of the carrier. In some instances, the pinion gear is a snug fit with the front pinion bearing, and it must be lightly tapped through the bearing.

Pry the pinion seal out of the differential carrier with a large screwdriver. When the seal is removed, the front pinion bearing and any adjusting shims can be pulled out from the front of the carrier. Remove the rear pinion bearing from the pinion gear shaft using a press and adapters.

Usually, the side bearings will be replaced when the rear axle assembly is overhauled. If you intend to reuse the bearings, do not remove them from the differential case and take care not to switch the bearing cups. A bearing showing any signs of wear or damage should be replaced. Figure 17-25 illustrates a method of removing the bearings using a gear puller.

WARNING! *Always* wear eye protection when attempting to remove bearings, since metal chips are likely to fly around. Do not remove the bearings by cutting them with a torch, since the differential case is likely to be overheated and ruined.

The ring gear should not be removed from the differential case unless it is definitely defective or unless the drive pinion gear must be replaced. If the ring gear must be replaced, remove the bolts holding it to the case. After it is unbolted, it can be lightly tapped loose with a brass hammer or plastic mallet, Figure 17-26.

NOTE! The ring and pinion are a matched set. They should always be replaced as a set. Mixing new and used gears will result in a noisy gearset that will fail quickly.

If spider and side gears have not yet been removed from the differential case, remove them now. First, as previously stated, remove the pinion shaft lock bolt or pin and drive or slide the pinion shaft from the case, Figure

A
ARROW CAST INTO BEARING CAP
PUNCH MARK IF NEEDED

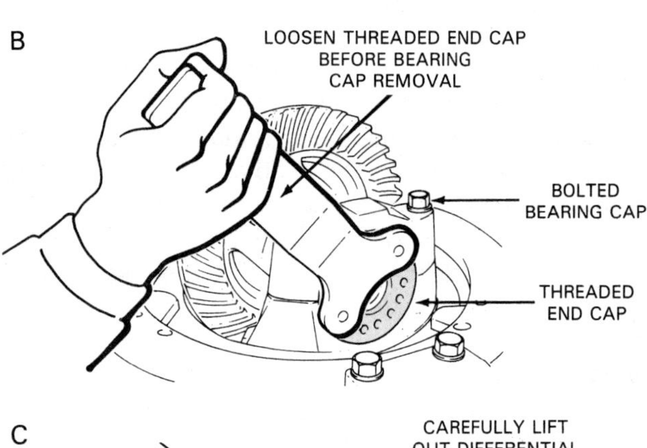

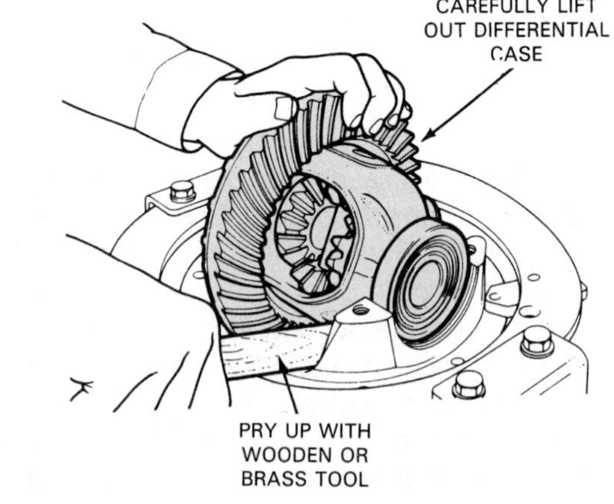

Figure 17-24. Steps to remove differential case are shown. A–Mark caps and carrier with a punch, if needed. Some bearing caps are already marked. If case is the shim type, bearing caps can then be unbolted and case can be removed. B–If threaded end caps are used for adjustment, they should be removed before bearing cap removal. C–Once bolted bearing caps are removed, case can be lifted from differential carrier. (Ford, Chrysler)

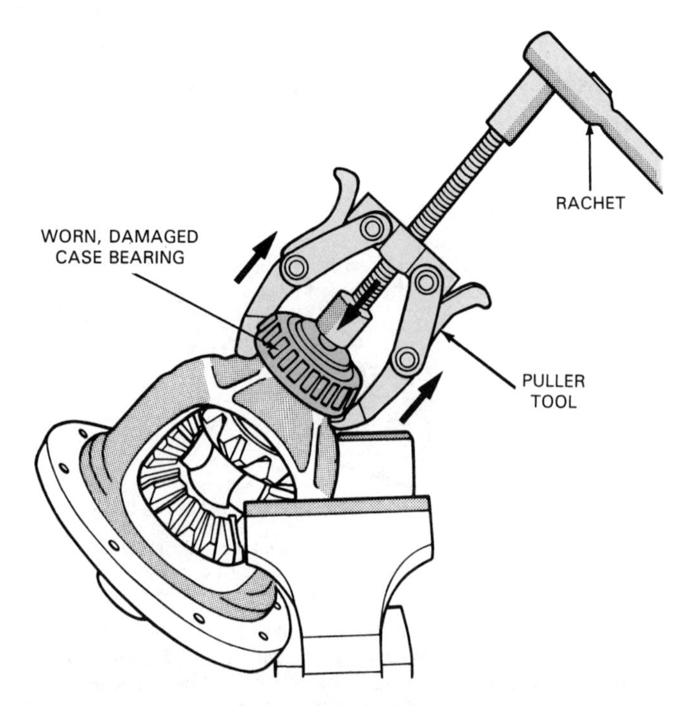

Figure 17-25. A puller tool such as this can be used to remove case bearings from differential case. Bearings should be removed only if they are to be replaced. (Chrysler)

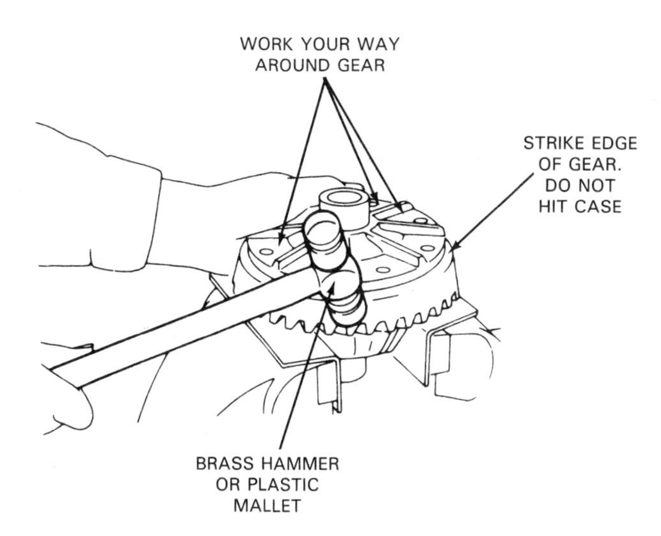

Figure 17-26. The ring gear should be removed from case only if it is to be replaced. After bolts are removed, ring gear can be removed by tapping it lightly with a hammer. (Nissan)

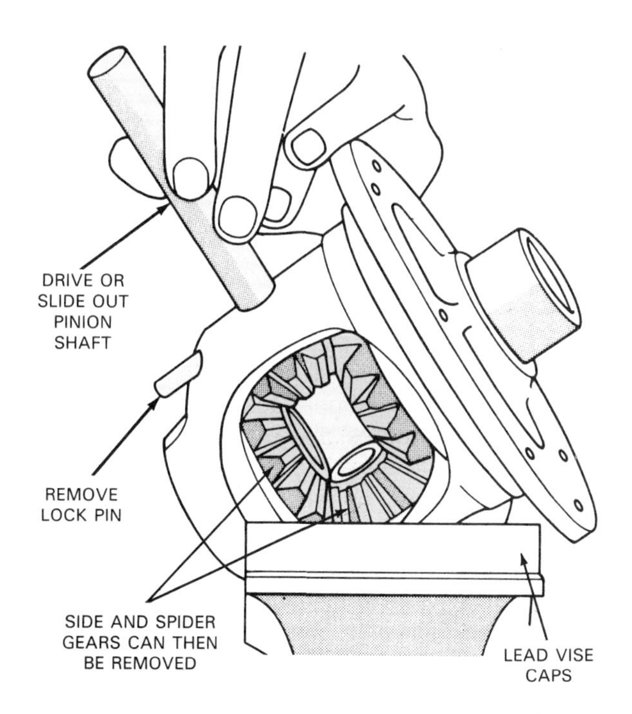

Figure 17-27. Note removal of lock pin and pinion shaft to gain access to spider and side gears in differential case. (Chrysler)

17-27. Turn one of the side gears until the spider gears rotate out of the case. Then, remove the side gears and thrust washers.

Limited-slip differential disassembly

Most limited-slip differential cases are made in two pieces. The clutches can be removed by removing the bolts holding the case together. After the case is apart, the components can be lifted out. Carefully note the position of all limited-slip components as you disassemble the case.

Figure 17-28 illustrates the parts of a limited-slip differential case assembly. Standard differentials are similar, but they do not contain the limited-slip clutches. Every part should be inspected for wear and damage.

Differential Assembly Parts Inspection

When checking the rear end components, pay particular attention to the bearings and bearing outer races, since they are a common source of rear end noise. Also, check all of the gears for wear and scoring. Check for pitting on gear and bearing surfaces, often caused by water in the oil. Make sure that the side gear and axle splines are not worn. Check the limited-slip differential clutches for wear, usually noticed as shiny contact surfaces or extreme smoothness. Once the differential has been checked, obtain new gaskets and seals. Obtain other new parts as needed.

Differential Reassembly

The first step in differential reassembly is replacing the pinion outer races, or bearing cups, if needed. The old bearing cups can be driven out of the differential carrier with a hammer and drift, as shown in Figure 17-29A. Be careful not to gouge the housing surfaces when removing the bearing cups.

Figure 17-29B shows the correct method of installing new bearing cups. If a special tool is not available, the cups must be very carefully driven into place. The bearings are closely machined, and the slightest mark on the bearing cup will ruin it.

After the bearing cups are in place, install the front pinion bearing and pinion seal in the carrier. The bearing should be lightly lubricated before installation. The pinion seal should be installed with a special seal installation tool, as shown in Figure 17-30. Do not forget to install any bearing spacers, or shims, that may have been present between the bearing and the carrier.

Press the rear pinion bearing on the drive pinion gear shaft. Use a new bearing and new drive pinion gear if needed. If a new pinion gear is needed, you must also replace the ring gear. Figure 17-31 shows the ring and pinion markings of a typical gearset. Always check the markings of replacement gearsets to ensure that the proper ring and pinion are being installed. The number of teeth on each gear should match the number of teeth on each of the old gears, unless you are intentionally modifying the rear axle ratio.

There is usually a steel shim between the pinion gear and the rear pinion bearing. This shim adjusts the pinion gear depth in the differential carrier. Pinion gear depth has a critical effect on the contact pattern between the pinion and ring gears. The shim is matched to the carrier. You should reinstall the same one, unless the carrier has been

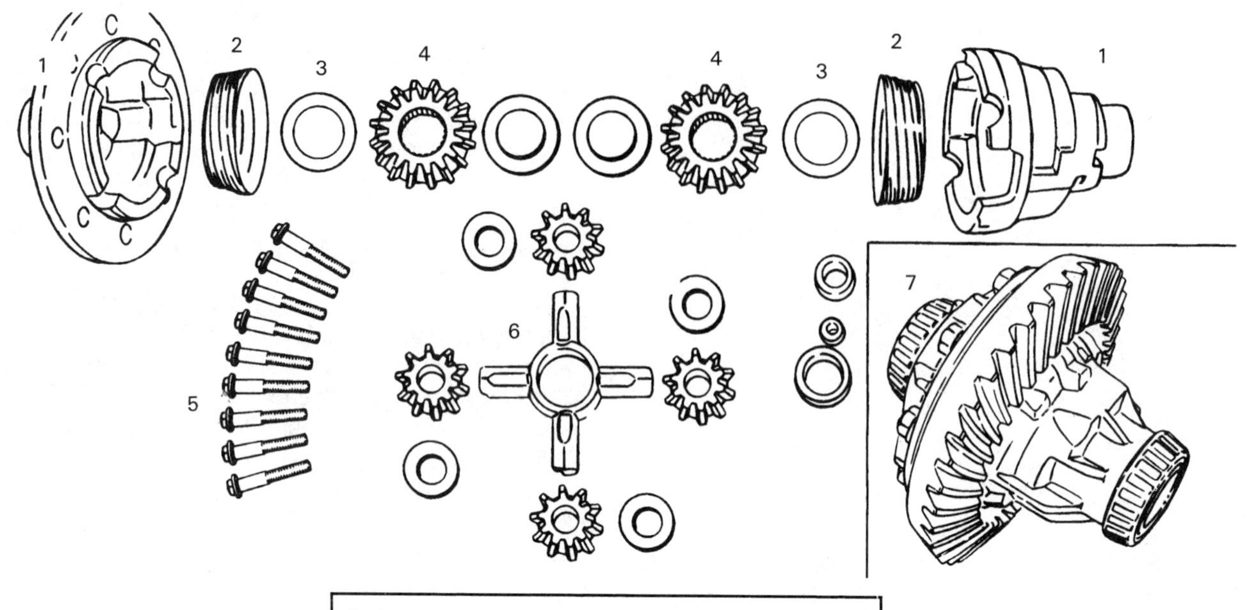

1. Inspect case for wear and cracks or other damage.
2. Check cone friction surfaces for wear.
3. Measure thrust washer wear.
4. Inspect side gear teeth and splines.
5. Make sure none of the bolt threads are damaged.
6. Check spider gears and other parts for wear.
7. Inspect before, after, and during repairs.

Figure 17-28. Exploded view of a limited-slip differential shows parts that must be cleaned and inspected and, if necessary, replaced. Standard differential case assemblies are similar. (General Motors)

changed. If a new carrier is used, a special procedure must be followed to select the proper replacement shim.

To install the rear pinion bearing, use a hydraulic press. The bearing must be completely bottomed on the pinion gear and shim. Lightly lubricate the bearing before installation.

To install the drive pinion gear in the differential carrier, place a new collapsible spacer or solid spacer, as required, ahead of, or on the forward side of, the rear pinion bearing. A collapsible spacer places the proper preload on the pinion bearings, and a new one should always be used. A solid spacer can be reused.

Install the drive pinion gear in the differential carrier by sliding the pinion gear shaft through the front pinion bearing. Then, install the pinion yoke and a new nut. Tighten the nut until the pinion gear has no back-and-forth play. See Figure 17-32A.

As the pinion nut is tightened, it will slightly compress a collapsible spacer or contact a solid spacer and preload the bearings. See Figure 17-32B. The final bearing preload can normally be measured with an inch-pound torque wrench, Figure 17-32C. Typical preload readings are 4 in.-lb.–11 in.-lb. (0.5 N-m–1.3 N-m). Always check the service manual for the exact reading. You must be very careful to reach the exact preload figure without going beyond it. If you overtighten the pinion nut, the pinion gear will have to be removed, and a new collapsible spacer will have to be installed.

With the drive pinion gear assembly in place, you are ready to assemble and install the differential case assembly. Before permanently installing the spider and side gears, check the clearances, as recommended by the manufacturer. Typically, this will involve checking the clearance between the splines of the side gears and the axles. This is determined by measuring the side gear-to-axle play with a dial indicator, Figure 17-33A. It will also involve checking the spider gear-to-side gear backlash. This is measured by checking the play between the two gears, Figure 17-33B. The thrust washers should be in place when doing this test.

After checking clearances, the spider and side gears can be installed permanently. Lubricate the differential case components with rear end lubricant, including the thrust washers, gears, and pinion shaft. Then, reinstall the parts, starting with the side gears. After the side gears are in place, the spider gears can be walked into position as shown in Figure 17-34. Once these are in place, install the pinion shaft and lock bolt or pin.

NOTE! The internal parts of a differential case assembly used on C-lock axles cannot be reinstalled in the case until the case is reinstalled in the differential carrier.

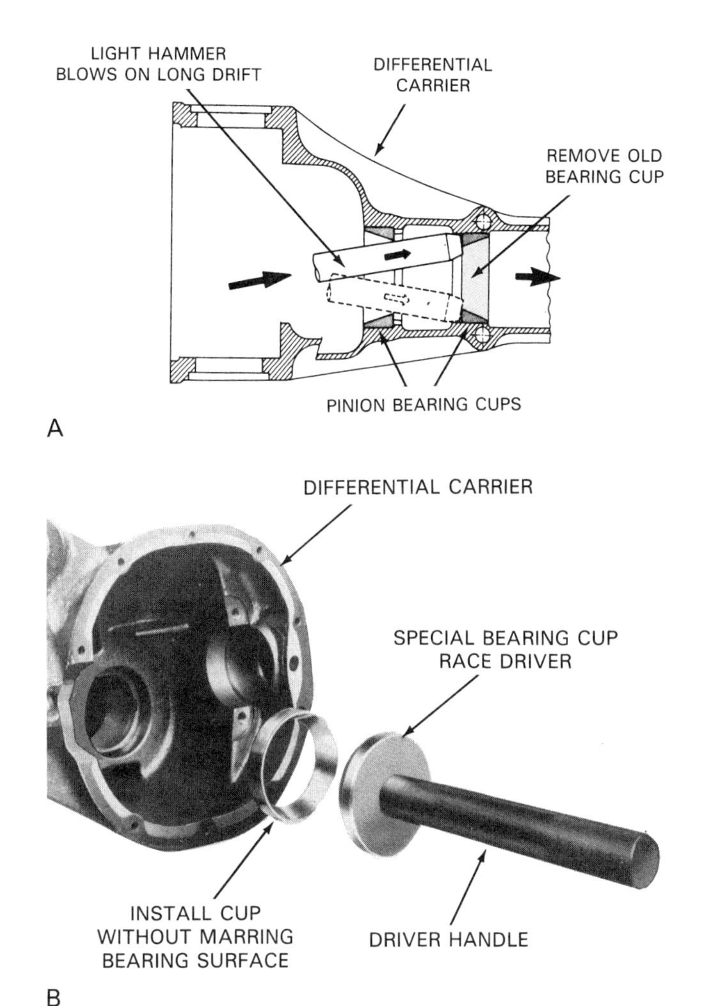

Figure 17-29. Study bearing cup replacement. A–Bearing cups can be removed with a hammer and a long punch or drift. Do not gouge differential carrier with tools during removal. B–A special tool should be used to install replacement cups. Do not damage cup surfaces. (Subaru, General Motors)

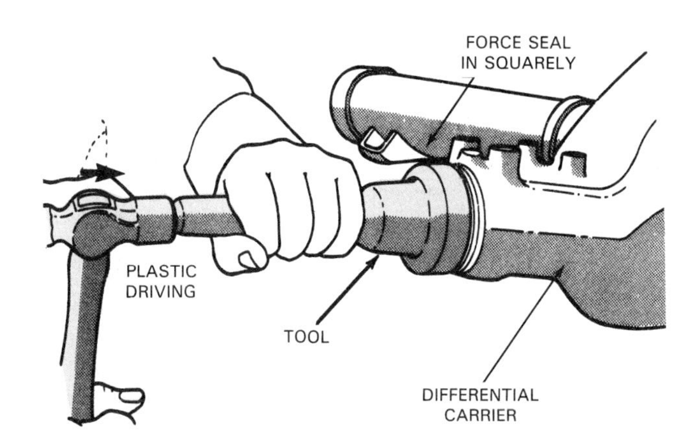

Figure 17-30. A new pinion seal should be installed with a special seal driver and a hammer. Seal should be completely seated against housing. (Chrysler)

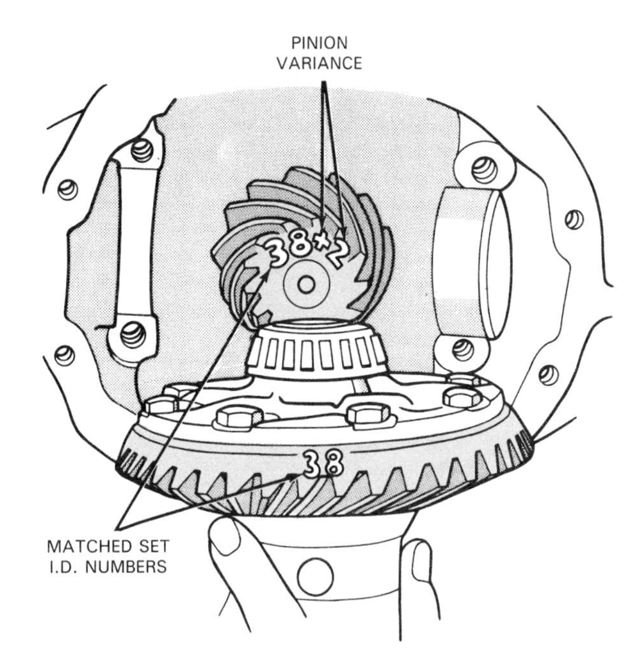

Figure 17-31. Ring and pinion gears must always be installed in matched sets. Markings on replacement sets should be checked to ensure that they have been matched at the factory. (Chrysler)

If the differential is the limited-slip type, reassembly is relatively easy. First, lubricate all parts, including clutch friction discs and steel plates, Figure 17-35. Assemble the parts in one half of the differential case. Then, reassemble the two halves of the case. See Figure 17-36. The clutches can be bench checked with a torque wrench, Figure 17-37. The readings should be the same as they were for the differential while installed in the vehicle (30 ft.-lb., or 41 N-m minimum).

If a new ring gear is being used, install it now on the differential case. The ring gear may be keyed or the bolt pattern may be arranged so that there is only one way to install the ring gear on the case. The gear should be placed on the case and lightly tapped into place with a soft hammer or a block of wood. Be very careful not to chip or nick the gear teeth.

Install and tighten the attaching bolts in a crisscross pattern. This is done so that one side is not tightened before the other side is bottomed out. Torque the bolts to the specified torque values.

Note that if the differential is the type used with independent rear suspension, as shown in Figure 17-38, you will need to select the proper number of shims to preload the side bearings. The original number of shims will usually work. See the manufacturer's service manual for more information on adjusting.

If new side bearings are to be installed, install them now. Position each bearing cone on hub of differential case. Use an arbor press in conjunction with a bearing driver to press the bearings into position. See Figure 17-39.

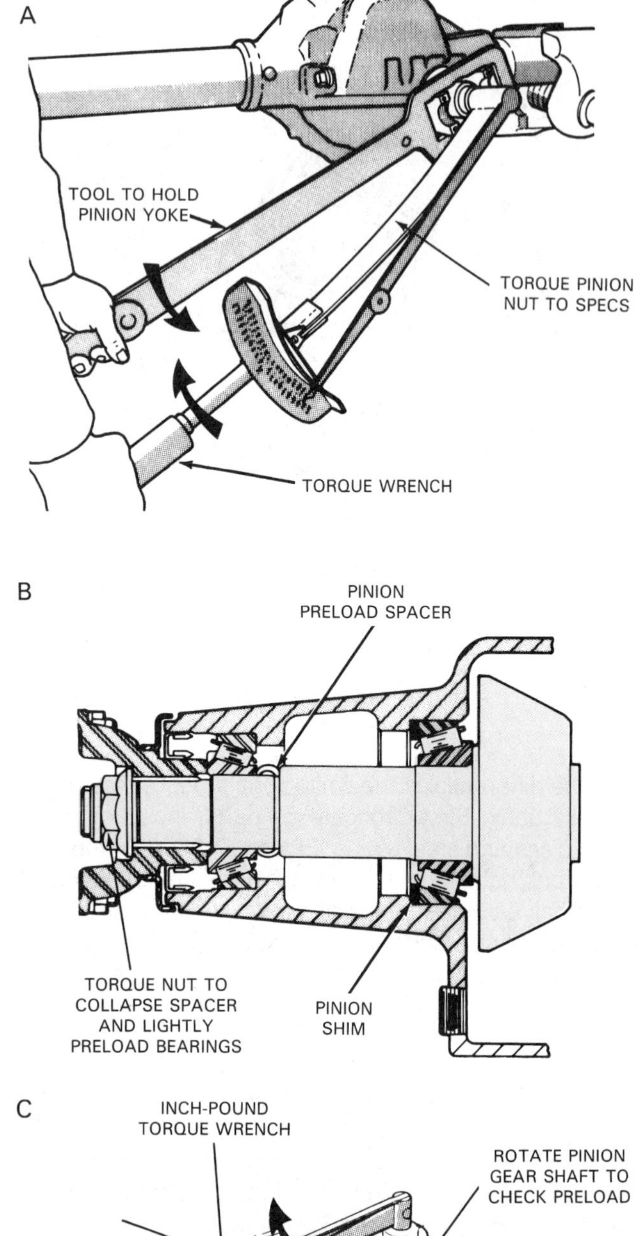

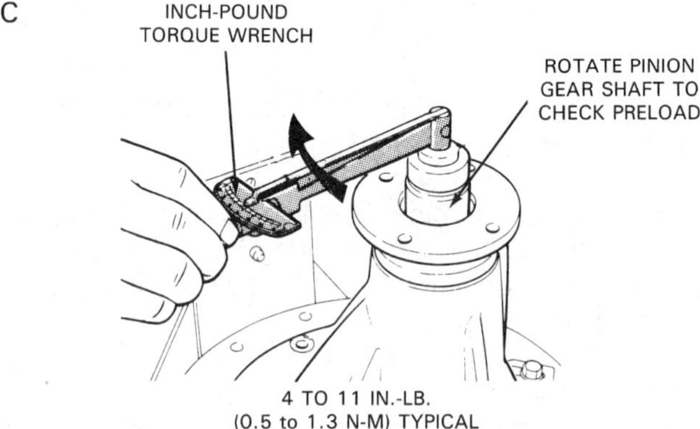

Figure 17-32. Note important details concerning installation of the drive pinion gear. A–Tighten large nut on pinion gear shaft to collapse space and preload bearings. B–Pinion gear assembly will look like this when properly assembled. Collapsible spacer preloads pinion bearings. C–Proper preload will cause a slight drag on pinion gear shaft when it is rotated. The drag is measured in inch-pounds or newton-meters. (Chrysler)

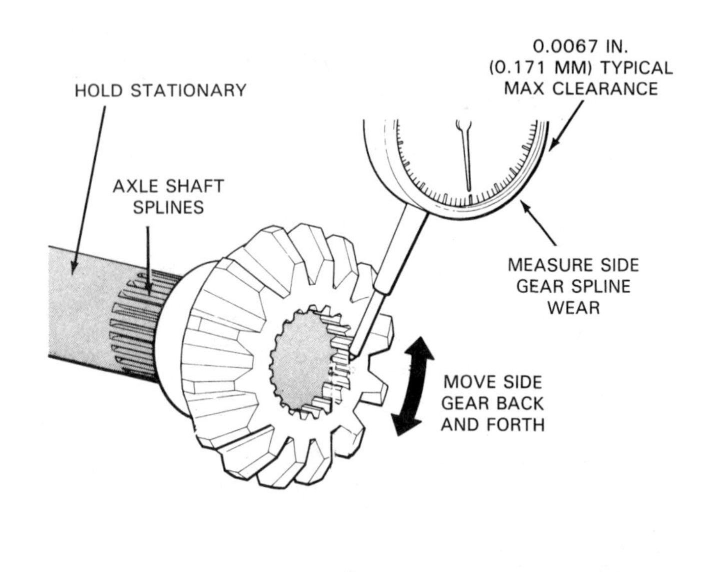

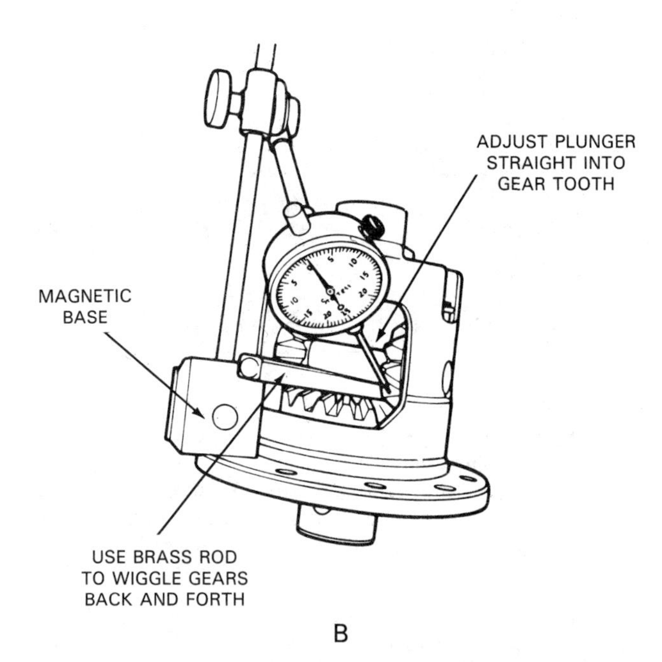

Figure 17-33. Side gears and spider gears must be checked for wear. Consult service manual for proper clearance specifications. Not all manufacturers provide specifications for these measurements. A–Side gear-to-axle clearance can be measured with a dial indicator to check for looseness between the two. B–Temporarily assemble thrust washers and differential gears to measure spider gear-to-side gear backlash. (Chrysler, General Motors)

Lightly lubricate the side bearings. The differential case can now be lowered into the differential carrier, Figure 17-40. The bearing cups and adjusting shims, if used, should be installed with the case. Since the adjusting shims are tightly fitted into the case to preload the bearings, they may need to be tapped into place with a soft hammer. Differential carrier spreading tools are

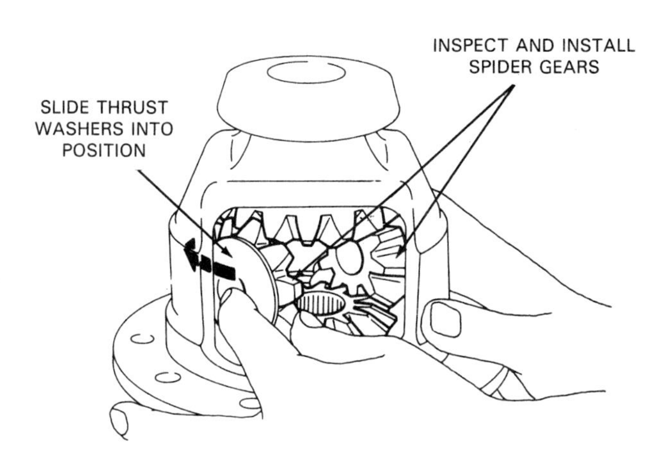

Figure 17-34. Spider gears and thrust washers must be installed into differential case as a unit. Spider gears should be placed on side gears and, then, rotated into case. Do not allow thrust washers to fall out of position. (Nissan)

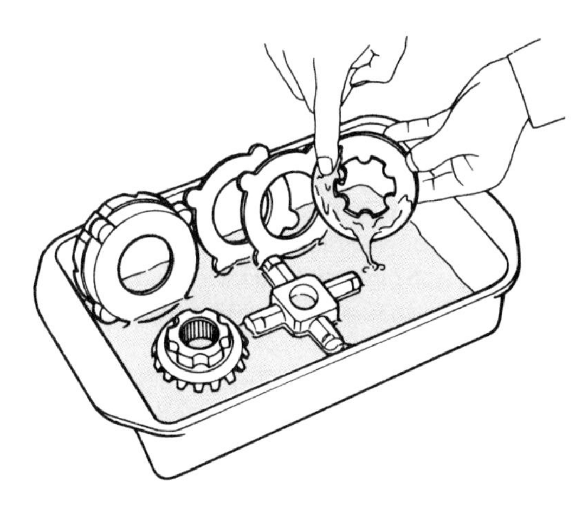

Figure 17-35. Lubricate differential case assembly parts, including limited-slip clutches as shown here. Submerging in a pan of lubricant is an ideal way to lubricate parts. Be sure to use recommended limited-slip differential gear oil, where applicable. (Nissan)

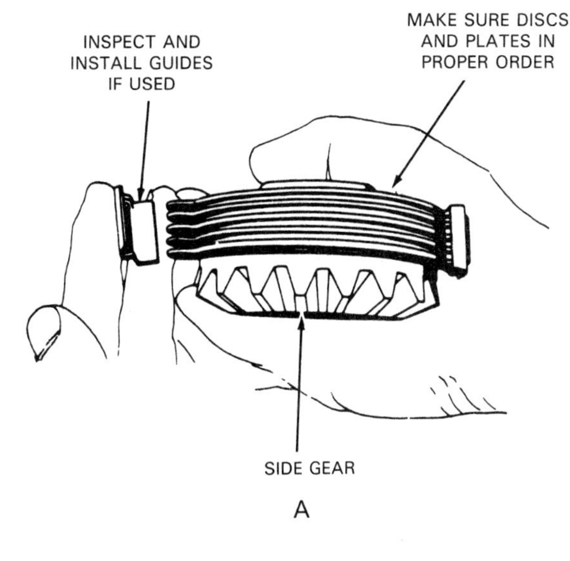

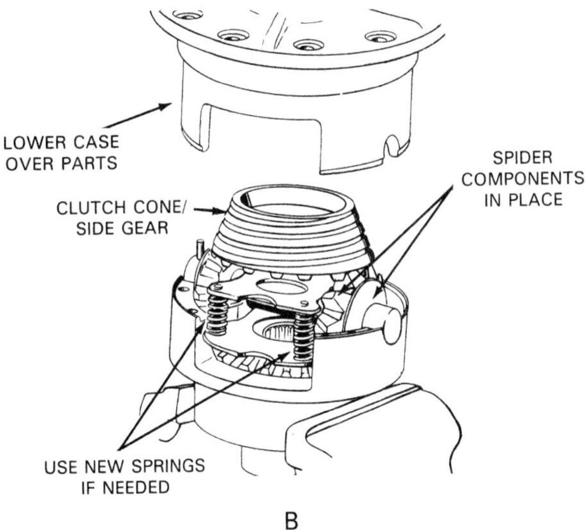

Figure 17-36. This shows the differential case assembly of a limited-slip differential. A–Multiple-disc clutches should be stacked in the case, alternating friction discs and steel plates. B–Cone clutches must be engaged with driving lugs on side gears. Place other half of case over assembled parts. (General Motors)

available to slightly expand the carrier for easier shim installation.

It may be necessary to add or subtract shims to obtain the proper bearing preload. This procedure is discussed in the next section. After the bearing cups and shims are in place, install the bearing caps and torque the retaining bolts. After the bolts are tight, rotate the differential case to ensure that the bearings are not binding.

Differential Assembly Adjustments

There are several adjustments that must be made when assembling a differential. Bearing preloads, *ring gear backlash, ring gear runout,* and ring gear contact patterns are extremely critical. These must be checked and, if necessary, adjusted.

Side bearing preload

Side bearing preload is the amount of force pushing the side bearings of the differential case together. Side bearing preload is critical. If it is too low, bearings are too loose. Differential case movement and ring and pinion gear noise can result. If preload is too high, the side bearings are too tight, and bearing overheating and failure can result.

When end caps are used, the caps are usually tightened until all of the play is out of the bearings. Then, each cap is tightened a certain amount to preload the bearings. Lock plates are installed to prevent the end caps from backing off, Figure 17-41.

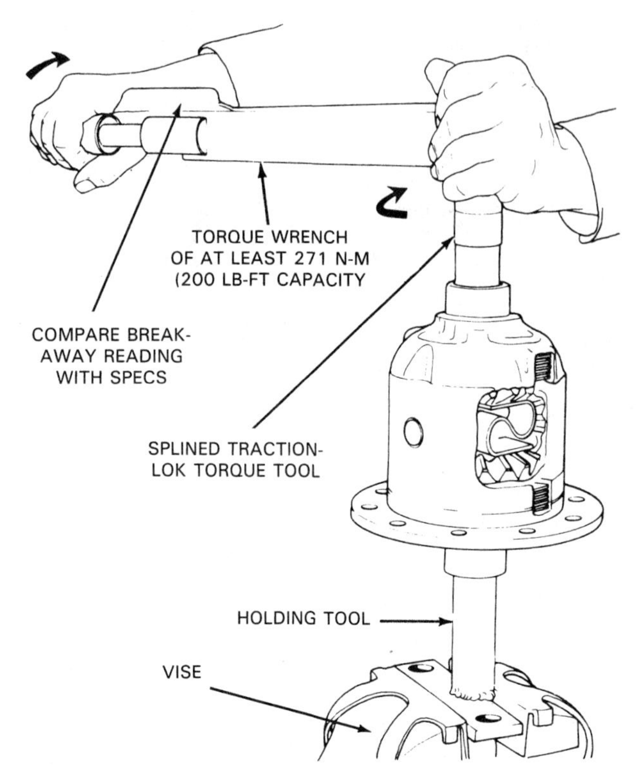

Figure 17-37. After reassembly, a limited-slip differential can be checked with a torque wrench. The procedure is the same as that performed with rear end assembled. It may be necessary to fabricate a special tool to hold and turn side gears. (Ford)

When shims are used, you may need to use a feeler gauge to check the clearances between the side bearings and differential case. One method of accomplishing this task is shown in Figure 17-42. The method shown will let you calculate the correct shim thickness to preload the side bearings properly. Refer to a shop manual for more specific procedures.

Ring gear backlash

After the differential case is installed, use a dial indicator to check ***ring gear backlash***–the clearance, or play, between the ring and pinion gears. To check ring gear backlash, place the plunger of a dial indicator on a ring gear tooth, as shown in Figure 17-43. Lock the pinion gear in place and rock the ring gear back and forth. The movement registered by the dial indicator is the backlash. A typical ring gear backlash setting is about 0.005 in.–0.012 in. (0.13 mm–0.31 mm).

Ring gear backlash can be changed. For a differential case assembly with adjusting nuts (end caps), this is accomplished by turning the nuts, as shown in Figure 17-44A. For an assembly adjusted with shims, proper-sized replacement shims must be used to adjust the backlash, as shown in Figure 17-44B. Adjustments must be made to each side of the case so that they offset each other, to maintain the proper bearing preload. Ring gear backlash may need to be readjusted after the gear tooth pattern is checked, as will be explained shortly.

Ring gear runout

Ring gear runout–the amount of wobble when the ring gear is rotated–should also be checked after the differential case is installed. Check ring gear runout by placing the dial indicator on the back side of the ring gear, as shown in Figure 17-45. Rotate the ring gear. The dial indicator movement is the runout. Runout should not exceed about 0.005 in. (0.13 mm).

Ring gear runout can usually be corrected only by replacing the ring gear and differential case. However, once in a while, runout can be corrected by changing the position of the ring gear on the differential case. Also, if a new ring gear was installed on the case, some dirt or metal shavings may have been caught between the gear and the case. Removing and cleaning both surfaces may eliminate excessive runout.

Ring gear contact pattern

Once the ring gear backlash and runout are checked, check the ring gear contact pattern. The tooth pattern between the ring and pinion is critical to quiet operation and long gear life. Refer to Figure 17-46 for the contact pattern checking procedure.

Begin the procedure by coating the ring gear teeth with white grease. ***Machine blue,*** also called ***Prussian blue,*** can be used in its place. Machine blue is a deep blue dye mixed in a greaselike substance. Use a small stiff bristle brush to apply a light even coat of the white grease or machine blue.

Use the drive pinion gear to rotate the ring gear through one complete revolution in the normal forward direction and one revolution in reverse. Apply a load to the back side of the ring gear. This will create a contact pattern on both sides of the ring gear teeth. Inspect the pattern on the entire ring gear. If the pattern is normal–that is, if the contact area is in the middle of both sides–no further adjustments are necessary. Otherwise, the positions of the ring gear and drive pinion gear must be adjusted.

If incorrect contact is caused by too much or too little backlash, adjust side bearing shims or end caps, as previously described. Be sure to maintain the proper bearing preload. If the drive pinion gear shim is incorrect, the unit will require disassembly to remove the rear pinion bearing and change the pinion shim. Always recheck the contact pattern after making adjustments.

When finished, wipe off the excess grease or machine blue. It is not necessary to remove all of it, since it will mix with the rear end lubricant. The differential assembly is now ready for reinstallation.

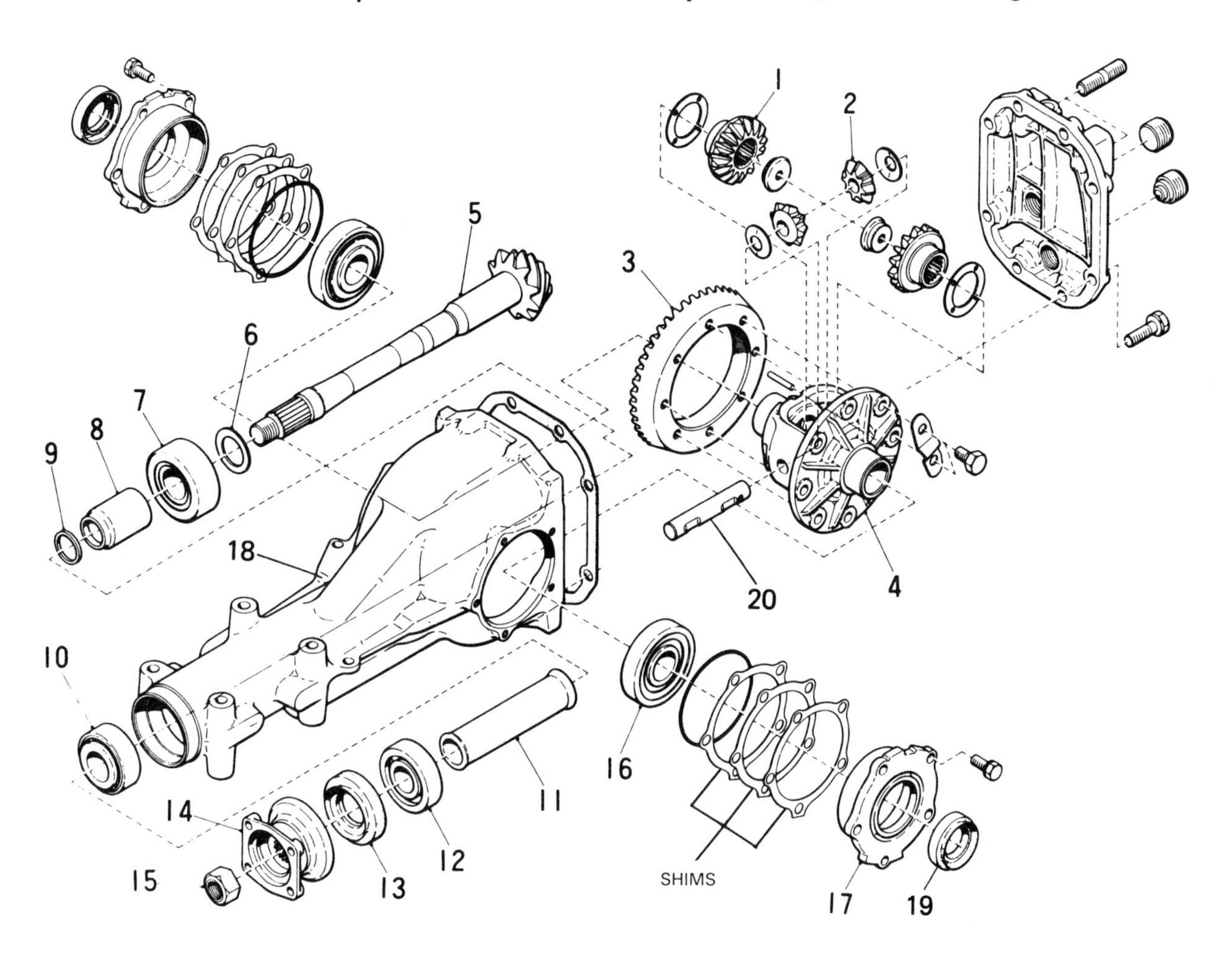

1. SIDE GEAR
2. PINION MATE GEAR
3. DRIVE GEAR
4. DIFFERENTIAL CASE
5. DRIVE PINION
6. PINION HEIGHT ADJUSTING WASHER
7. REAR BEARING
8. PRELOAD ADJUSTING SPACER
9. PRELOAD ADJUSTING WASHER
10. FRONT BEARING
11. SPACER
12. PILOT BEARING
13. OIL SEAL
14. COMPANION FLANGE
15. PINION NUT
16. SIDE BEARING
17. SIDE BEARING RETAINER
18. DIFFERENTIAL CARRIER
19. SIDE OIL SEAL
20. DIFFERENTIAL PINION SHAFT

Figure 17-38. Exploded view of a rear axle assembly used with an independent rear suspension shows shims that are used to adjust side bearing preloads. Other parts resemble those used on standard rear axle assemblies. (Subaru)

Differential Assembly Installation

Following are some general procedures detailing how to install a differential assembly safely and properly. The exact procedure for installation varies according to the type of differential, but basically it is the reverse of removal. You should always refer to the manufacturer's service manual for specific procedures.

Removable carrier

The following procedure details the proper way to install a differential assembly from a rear axle assembly having a removable carrier.

1. Place a new gasket over the studs on the rear axle housing, where the differential carrier mates with the rest of the housing. Make sure that no old gasket is stuck on either of the mating surfaces.
2. With the help of an assistant, raise the carrier into position and slide it over the studs. Place a jackstand under the carrier to support it.
3. Install the nuts and tighten them in an even pattern to prevent binding. Torque the nuts to the proper specifications.
4. Install the drive axles in the rear axle housing. (Refer to drive axle installation section, occurring earlier in the chapter.)

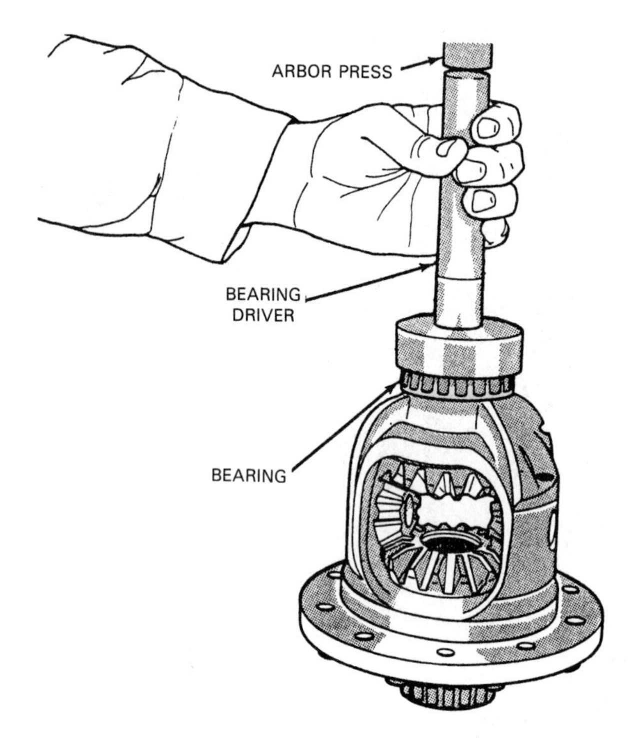

Figure 17-39. This shows installation of side bearing using an arbor press. Note bearing cone taper is away from differential case. Also, never exert pressure against the bearing cage. Bearing damage may result. (Chrysler)

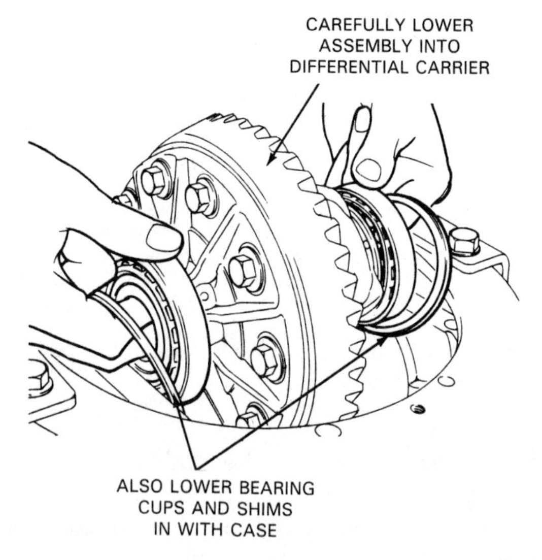

Figure 17-40. Carefully lower differential case assembly, bearing cups, and shims into differential carrier as a unit. It may be necessary to use a special carrier spreading tool to allow shims to slip into place. (Chrysler)

5. Install brake drums and wheels.
6. Install the drive shaft assembly, aligning match marks made during disassembly.
7. Install any hardware that was attached to the carrier.

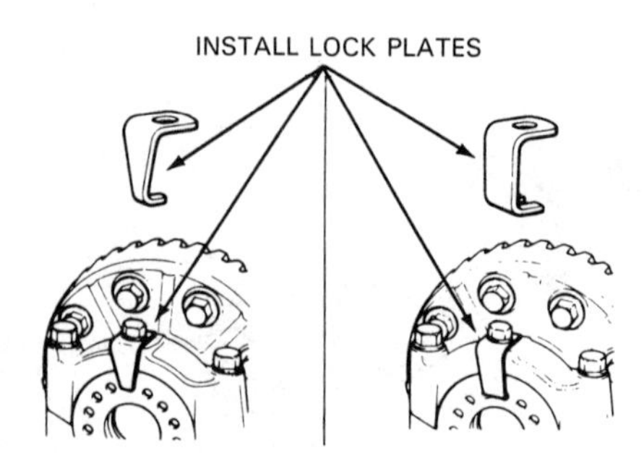

Figure 17-41. Install lock plates to keep end caps from backing off and coming out of adjustment. (Chrysler)

8. Lower the vehicle. Check the brakes. Test drive the vehicle to ensure proper operation.

Integral carrier

Integral carriers do not generally need to be reinstalled, unless the entire rear axle assembly was removed to service the assembly on a workbench. The following procedure details the proper way to install a rear axle assembly of a vehicle with either solid-axle rear suspension or independent rear suspension.

1. Install solid drive axles in rear axle housing. (Refer to drive axle installation section, occurring earlier in the chapter.)
2. Check the differential cover for damage or warping (if not already installed). Scrape off any old gasket material from the cover or its mating surface on the rear axle housing. Install a new gasket and use a nonhardening gasket sealer. Note that some manufacturers recommend using form-in-place gaskets. The bead of this type of gasket should be laid evenly and should completely encircle all bolt holes.
3. Tighten the differential cover bolts in a crisscross pattern. Torque the bolts to manufacturer's specification.
4. Place the rear axle assembly under the vehicle and raise it into position. Have a helper assist you and use a floor jack or transmission jack to lift the assembly. Support rear axle housing with a jackstand. Then, install all rear axle housing attaching hardware.
5. Install independently suspended drive axles.
6. Reconnect miscellaneous items, such as brake lines, parking brake cables, exhaust pipes, and suspension components.
7. Install the drive shaft assembly, aligning match marks made during disassembly.
8. Install brake drums and wheels.

WARNING! If the brake lines were removed, bleed the brakes before road testing the vehicle.

9. Add the proper type of rear end lubricant.

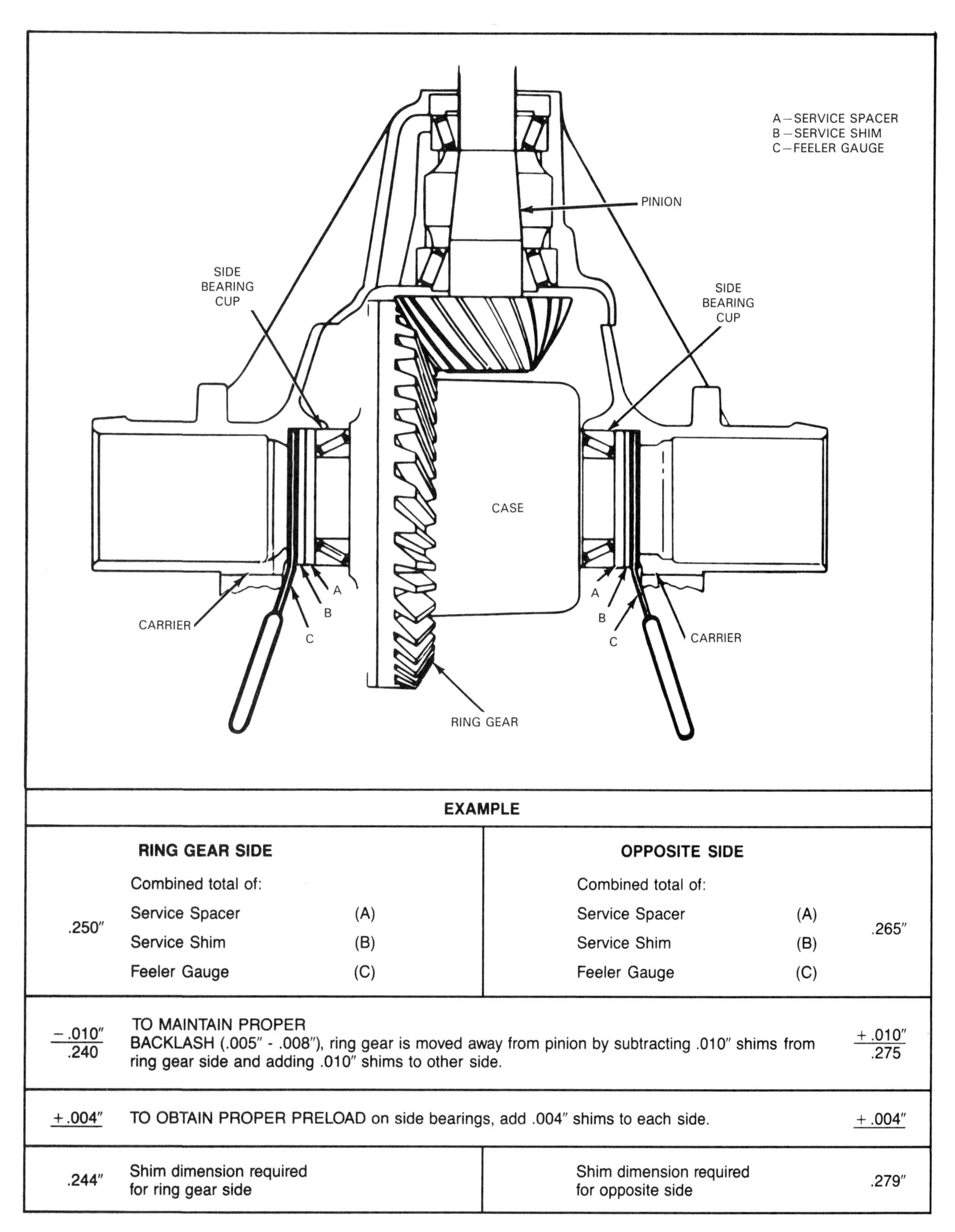

EXAMPLE

	RING GEAR SIDE		OPPOSITE SIDE		
	Combined total of:		Combined total of:		
.250″	Service Spacer	(A)	Service Spacer	(A)	.265″
	Service Shim	(B)	Service Shim	(B)	
	Feeler Gauge	(C)	Feeler Gauge	(C)	
−.010″ / .240	TO MAINTAIN PROPER BACKLASH (.005″ - .008″), ring gear is moved away from pinion by subtracting .010″ shims from ring gear side and adding .010″ shims to other side.				+.010″ / .275
+.004″	TO OBTAIN PROPER PRELOAD on side bearings, add .004″ shims to each side.				+.004″
.244″	Shim dimension required for ring gear side		Shim dimension required for opposite side		.279″

Figure 17-42. This shows procedure for adjusting bearing preload and, also, ring gear backlash on a shim-type differential case. To maintain proper preload, always keep total shim thickness the same. For example, if you must subtract 0.01 in. from one side, add 0.01 in. to the other side. (General Motors)

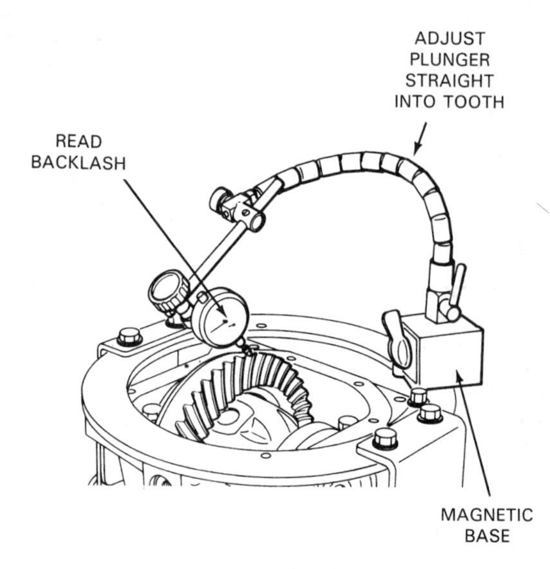

Figure 17-43. Ring gear backlash can be checked with a dial indicator. Check backlash with drive pinion gear locked in place. Backlash should be within the manufacturer's range. (Chrysler)

10. Lower the vehicle. Check the brakes. Test drive the vehicle to ensure proper operation.

Note that if properly assembled, most rear axle assemblies will at least allow the vehicle to be driven. The main concern will be noise. To check for noises, find a quiet road with smooth pavement. If the rear end is quiet at about 35 mph–45 mph, it is working properly, and everything is in order.

To check limited-slip clutches, make several slow turns in a quiet area, such as an empty parking lot. If the vehicle turns smoothly, without shuddering or noises, the clutches are working properly. Sometimes, the limited-slip clutches will make noise or work erratically for the first few miles, until the clutch surfaces have time to wear in.

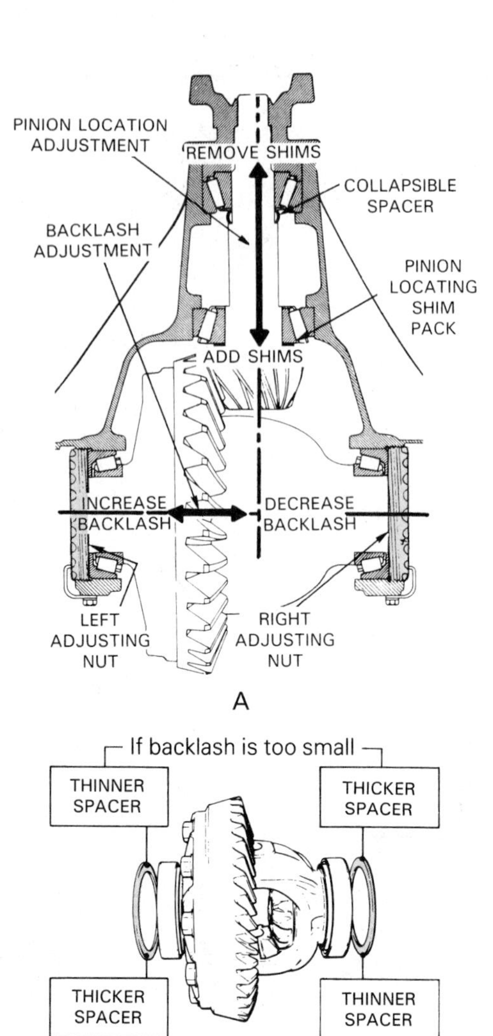

Figure 17-44. Ring gear backlash can be changed by turning adjusting nuts or adding thicker or thinner shims. A–Turn left adjusting nut (in illustration) inward to decrease backlash. Turn nut outward to increase backlash. Be sure to maintain proper bearing preload. B–Add thicker shim to left side and thinner shim to right side to decrease backlash. Do the opposite to increase backlash. (Ford, Chrysler)

Summary

The technician should always begin rear end troubleshooting by checking the oil level. Add oil as needed. If this does not solve the problem, proceed to check other possible causes. Most rear end problems fall into three main areas: noises, vibrations, or leaks.

Rear end noises include clunking, whining, and rumbling. Noises can be caused by worn or misadjusted gears, defective bearings, or excessive clearance between gears. Make sure that the noise is actually coming from the rear end, since other components can make similar sounding noises.

Vibrations are usually caused by bent axles or axle flanges. The source of the vibrations should be carefully determined, since vibration can be caused by many other parts of the vehicle.

Oil leaks can be caused by seal or gasket failure, overfilling or water contamination, loose bolts, or cracked housings. They can usually be spotted by oil drips and stains from the defective part.

Problems in limited-slip differentials usually show up on turns. Defective clutches or the wrong oil will cause the rear axle assembly to shudder or make noises when turning corners. Limited-slip differential repair usually involves disassembling the rear end.

The axles are removed by two different methods, depending on whether retainer-type or C-lock axles are used. C-lock axles require partial disassembly of the differential, while the retainer type do not.

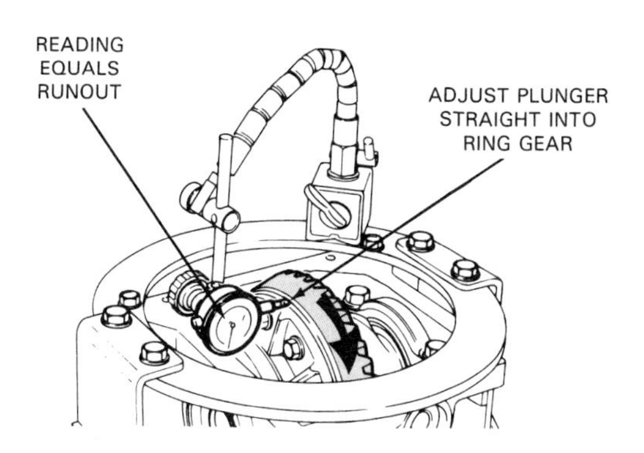

Figure 17-45. Measure ring gear runout by placing a dial indicator against the back side of the ring gear and rotating the ring gear. (Chrysler)

Axle bearing removal procedures are also different. Bearings for C-lock axles must be removed from the housing, while those for retainer-type axles must be removed from the axle. Axle studs can be replaced on or off the vehicle.

Axle seals should be replaced when they begin leaking or when performing any other drive axle service procedure. If a seal is leaking badly, the brake linings should also be checked. The old seal can be pried out with a special tool or screwdriver. The new seal should be installed with a special driver so that it is flush against the housing.

Before removing the rear axle assembly, it should be checked to ensure that an internal problem exists. The vehicle should be operated on a lift to determine the exact problem. Limited-slip differential clutches can be checked with a torque wrench.

The methods of removing the rear axle assembly vary according to type. In many cases, the integral carrier design can be serviced from under the vehicle. In other cases, it is easier to work on the rear axle assembly on a bench.

Rear end disassembly starts in the differential case area. The side and spider gears can be removed if rear axle assembly has C-lock axles. Then, the side bearing caps are removed, and the case is lifted or pried from the case. Once the case is removed, the drive pinion gear can be removed by removing the nut holding the pinion flange to the pinion gear. The pinion gear and bearings can then be removed from the case.

All parts should be inspected for wear and damage. New gaskets and seals should be obtained. Bearing cups should be installed very carefully. The rear pinion bearing is changed, keeping the original spacer. The drive pinion gear can then be installed, and the flange nut can be tightened.

The side and spider gears and limited-slip components, if applicable, can be reinstalled in the differential case. The case is then installed in the differential carrier, the bearing caps are tightened, and the backlash and runout are checked. Adjustment varies according to whether the side bearings are preloaded with end caps or shims. Then, the ring gear contact pattern is checked and adjusted as needed. The ring and pinion gears may both require adjustment.

After the rear end is reassembled and adjusted, it can be reinstalled in the vehicle. Installation is the reverse of removal. Do not forget to add the proper kind of lubricant to the differential and to bleed the brakes if the brake lines were disconnected.

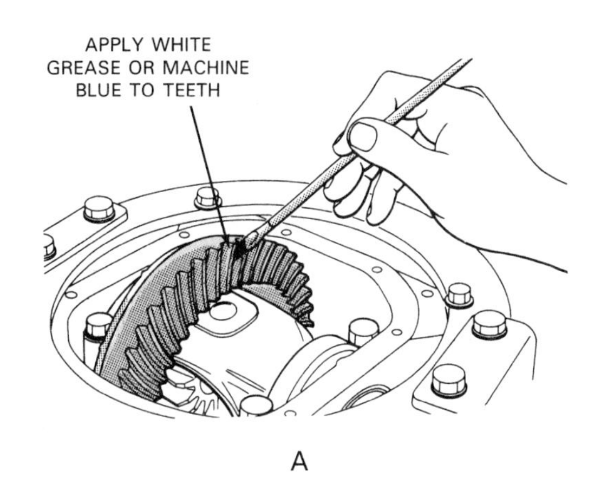

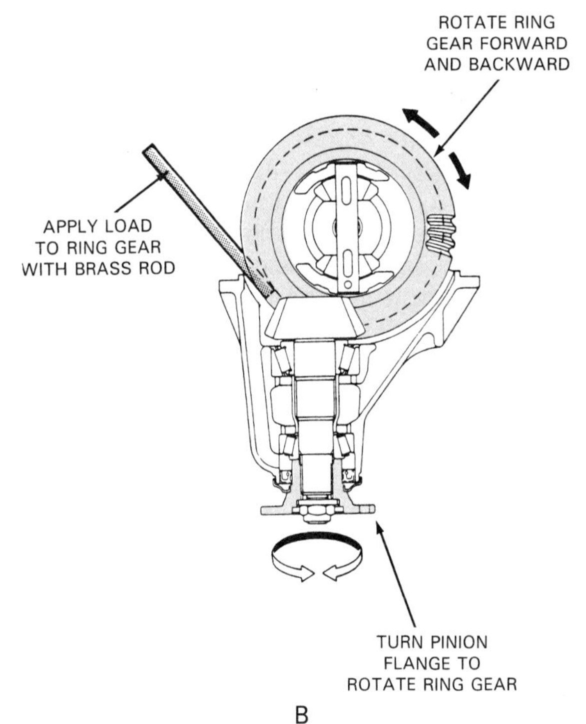

Figure 17-46. This shows procedure for checking ring gear contact pattern. A–Apply white grease or machine blue to teeth. B–Rotate ring gear by turning pinion yoke/flange. Apply load to back side of the ring gear at same time. Turn ring gear through one complete revolution forward and one complete revolution backward. Contact pattern can then be checked. (Chrysler)

The vehicle should be road tested for proper operation. Most problems after overhaul will be in the area of noises. Some limited-slip differential components may require some run-in to operate properly.

Know These Terms

Axle end play; Ring gear contact pattern; Side bearing preload; Ring gear backlash; Ring gear runout; Machine blue (Prussion blue).

Review Questions–Chapter 17

Please do not write in this text. Place your answers on a separate sheet of paper.

1. What should you check first with most rear end problems?
2. Whining noises usually occur when the vehicle is first accelerated. True or false?
3. Never try to straighten a bent rear axle. True or false?
4. An unusual noise seems to be coming from somewhere in the rear axle assembly. Technician A says to pull on each parking brake cable while the engine is running to isolate the problem. Technician B says to listen for the noise with a stethoscope. Who is correct?
   a. Technician A.
   b. Technician B.
   c. Both Technician A and B.
   d. Neither Technician A nor B.
5. Oil is found to be leaking from the rear axle housing vent. Technician A says that it looks like the rear axle assembly has been overfilled. Technician B says that the vehicle may have been driven through high water. Who is correct?
   a. Technician A.
   b. Technician B.
   c. Both Technician A and B.
   d. Neither Technician A nor B.
6. A vehicle with a limited-slip differential makes a low-pitched moaning noise when the vehicle is turned. Technician A says that the wrong lubricant may have been used. Technician B says that the clutches may be worn. Who is correct?
   a. Technician A.
   b. Technician B.
   c. Both Technician A and B.
   d. Neither Technician A nor B.
7. A leaking axle seal can make a car's brakes unsafe. True or false?
8. In order to remove a retainer-type axle, the differential cover must be removed. True or false?
9. A C-lock axle bearing is installed in the axle tube. True or false?
10. A drive pinion gear is found to be defective. Technician A says that a new ring gear must be installed, along with a new pinion gear. Technician B says to install a new pinion gear and use the old ring gear. Who is correct?
    a. Technician A.
    b. Technician B.
    c. Both Technician A and B.
    d. Neither Technician A nor B.
11. Where collapsible spacers are used, it does not matter if preload specifications are exceeded when torquing the pinion nut. True or false?
12. Side bearing preload is adjusted by means of:
    a. Shims.
    b. Adjustable end caps.
    c. Collapsible spacers.
    d. Both a and b.
13. Ring gear _____ is the clearance, or play, between ring and pinion gears.
14. Ring gear _____ is the amount of wobble in the ring gear as it rotates.
15. To adjust a differential assembly where tooth contact is localized on the crests of the drive side and coast side tooth faces:
    a. Move the ring gear away from the drive pinion gear using a thinner side bearing shim.
    b. Move the ring gear closer to the drive pinion gear using a thicker side bearing shim.
    c. Move the drive pinion gear closer to the ring gear using a thicker pinion gear shim.
    d. Move the drive pinion gear away from the ring gear using a thinner pinion gear shim.

Certification-Type Questions–Chapter 17

1. All of these abnormal noises are typical of those coming from faulty rear axle assemblies EXCEPT:

(A) whining.
(B) roaring.
(C) clunking.
(D) squeaking.

2. All of these statements about rear drive axles are true EXCEPT:

(A) bent axles are a common cause of vibration.
(B) vibrations in rear drive axles are always accompanied by noise.
(C) bent axles can be checked with a dial indicator.
(D) if an axle flange is bent, the axle should always be replaced.

3. A rear-wheel drive vehicle has a severe lubricant leak from the center of the rear axle assembly. Technician A says that the cause could be a leaking pinion seal. Technician B says that the cause could be loose bolts on the differential cover. Who is right?

(A) A only
(B) B only
(C) Both A & B
(D) Neither A nor B

4. Technician A says that automatic transmission fluid can be used to refill a limited-slip differential. Technician B says that standard 80- to 90-weight gear oil can be used to refill a limited-slip differential. Who is right?

(A) A only
(B) B only
(C) Both A & B
(D) Neither A nor B

5. Technician A says that the drive axle can be loosened by using a slide hammer after the retainers are removed. Technician B says that the differential must be partially disassembled to remove a retainer-type drive axle. Who is right?

(A) A only
(B) B only
(C) Both A & B
(D) Neither A nor B

6. Which of these parts need not be removed to remove a C-lock from a drive axle?

(A) Pinion shaft lock bolt.
(B) Pinion shaft.
(C) Spider gears.
(D) Differential cover.

7. All of these statements about servicing axle bearings for retainer-type axles are true EXCEPT:

(A) they should be removed by first cutting the axle collar with a torch.
(B) they should be removed by first cutting the axle collar with a chisel.
(C) once the axle collar is removed, the bearing is removed by pressing it from the shaft.
(D) if the new bearing presses on with less than 4000 psi (27.58 MPa) of pressure, the shaft should be replaced.

8. Technician A says that the axle must be removed to replace a wheel stud. Technician B says that a press can be used to replace a wheel stud if the axle is removed from the vehicle. Who is right?

(A) A only
(B) B only
(C) Both A & B
(D) Neither A nor B

9. Using which of these tools with a hammer is the best way to install an axle seal?

(A) Correct-sized pipe.
(B) Wooden block.
(C) Seal driver.
(D) Correct-sized socket.

10. Technician A says that a limited-slip differential can be checked for proper operation with a torque wrench. Technician B says that improper lubricant can cause limited-slip differential slippage. Who is right?

(A) A only
(B) B only
(C) Both A & B
(D) Neither A nor B

11. Which of these should be removed last when removing a rear axle assembly?

(A) Rear springs.
(B) Rear shock absorbers.
(C) Drive shaft.
(D) Rear wheels.

12. If the ring gear backlash is improperly adjusted, which of these could occur?

(A) Side bearing overheating.
(B) Pinion bearing wear.
(C) Lubricant leaks.
(D) Rear end whining or clunking.

13. All of these statements about ring gear contact pattern are true EXCEPT:

(A) adjust the side bearing shims or end caps if incorrect contact pattern is caused by too much or too little end play.
(B) changing the pinion shim may be necessary to establish the correct contact pattern.
(C) using a heavier-weight lubricant will correct many improper contact patterns.
(D) proper ring gear contact pattern is critical to quiet operation and long gear life.

14. Technician A says that when installing a gasket for the differential cover, an anaerobic sealer should be used. Technician B says that form-in-place gaskets should never be used for the differential cover. Who is right?

(A) A only
(B) B only
(C) Both A & B
(D) Neither A nor B

Chapter 18

Manual Transaxle Construction and Operation

After studying this chapter, you will be able to:

- Identify the major parts of a manual transaxle.
- Describe the construction of a typical manual transaxle.
- Explain the operation of a typical manual transaxle.
- Trace the power flow through manual transaxles used with transverse and longitudinal engines.
- Compare manual transaxle design variations.

Although rear-wheel drive vehicles are still used today, front-wheel drive vehicles with transaxles are by far more common. The drive train of the front-wheel drive vehicle is much lighter than that of the rear-wheel drive vehicle, since it eliminates the drive shaft assembly and rear axle housing. Placing all of the drive train parts in the front of the vehicle also allows the manufacturer to eliminate the transmission hump found on the floor of rear-wheel drive vehicles. This increases passenger room.

Manual transaxle operating principles are similar to those of a manual transmission and differential assembly. The major difference in a transaxle is that all parts are usually together in one housing. The manual transaxle has the same kinds of gears, synchronizers, and shift linkage as a manual transmission. It contains the same type of differential case assembly, performing the same job as that found on a rear axle assembly. The greatest differences occur in the shape of the transaxle case and the placement of the parts within the case. An obvious external difference is two *CV axles* connecting to the transaxle, as opposed to the driveline that connects to the rear-wheel drive manual transmission.

This chapter will familiarize you with the parts and operating principles of manual transaxles. Many parts resemble those used on a rear-wheel drive vehicle, although they are arranged differently. This chapter will indicate what parts are different from those of a rear-wheel drive vehicle. It will also explain how power flows through the two major types of transaxles, as well as the differences in parts and layout between the two types. This chapter also briefly describes transaxles used on modern rear-wheel drive vehicles.

Purpose of the Manual Transaxle

A manual transaxle (or automatic transaxle) is a drive unit that combines a transmission and a differential assembly in one case. See Figures 18-1 and 18-2. Transaxles can be used on rear-engine, rear-wheel drive vehicles and on front-wheel drive vehicles; most are found on vehicles with front-wheel drive. The transaxle is attached to the *rear* (output end) of the engine. The transaxle and engine assembly is attached to the vehicle frame and body through flexible engine mounts. The flexible mounts reduce vibration transfer from the engine and transaxle to the passenger compartment.

The basic job of the transaxle is to transfer power from the engine to the drive axles. The transaxle must also be able to provide the proper gear ratio for varying operating conditions and allow the drive wheels to make turns. Power is received from the engine through a clutch assembly.

The transmission portion of the transaxle is used to select the proper gear ratio for any driving situation and to move the vehicle in reverse. The differential portion of the transaxle allows the vehicle to make turns while power is being delivered. The differential assembly drives two CV axles that turn the front drive wheels.

Operating the Manual Transaxle

From the driver's point of view, operation of the manual transaxle is identical to that of the rear-wheel drive manual transmission. Gears are selected by the driver, through a gearshift lever and manual clutch. The driver pushes the clutch pedal to disengage the clutch and moves the gearshift lever to the desired gear position. Moving the gearshift lever operates the linkage that engages gears inside of the

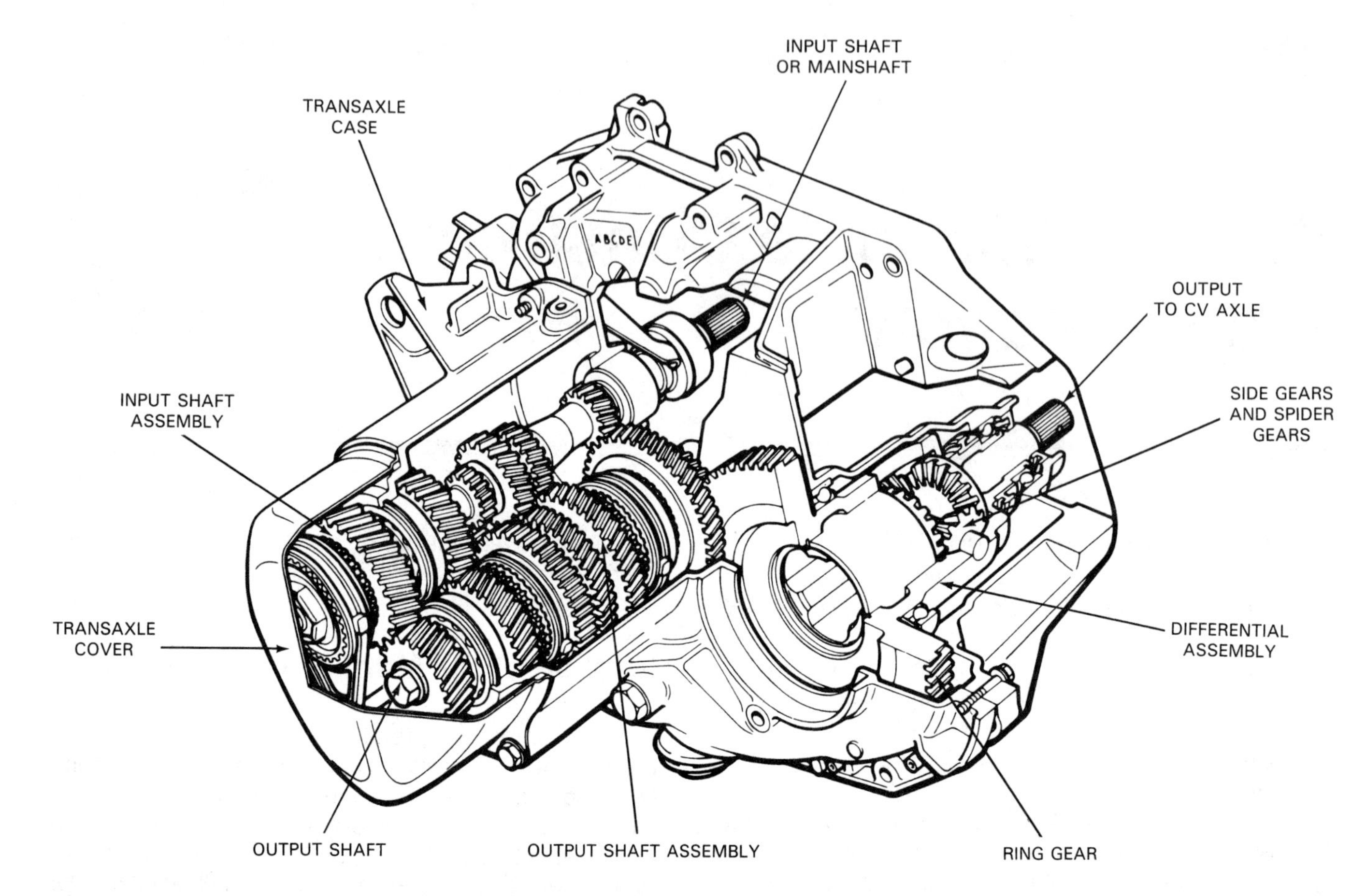

Figure 18-1. Cutaway shows major transaxle parts, including case and covers, input and output shafts and gears, and differential assembly. All transaxles, while varying in details, have these basic parts. Note that this particular transaxle is used with a transverse engine. (Chrysler)

transmission. The driver then releases the clutch pedal to engage the clutch. With the clutch engaged, the gears can transfer power through the transaxle to the output shaft. The vehicle can then move in the gear position that was selected.

Construction and Internal Operation

There are two basic kinds of front-wheel drive transaxle construction. Which type is used depends on engine placement in the vehicle. Engine installation can be transverse (crosswise) or longitudinal (lengthwise). (Refer back to Figure 18-1 to see transaxle used with a transverse engine.) Major manual transaxle components will be presented in detail in this chapter. They include the following:

- The ***manual transaxle shafts*** support gears and directly or indirectly transfer rotation from the clutch disc to the transaxle differential assembly.
- The *transaxle transmission gears* transmit power. They provide a means of changing vehicle torque, speed, and direction.
- The *synchronizers* bring certain transmission gears to the same rotational speed as their associated shaft before sliding in mesh with the gears and locking them to the shaft. Synchronizers are used to prevent gear clash. They operate in the same way as synchronizers on rear-wheel drive manual transmissions.
- The *shift forks* are pronged units for moving transmission gears or synchronizers on their shaft for gear engagement.
- The *manual transaxle shift linkage* connects the gearshift lever to the shift forks. Linkage consists of shift levers and rods or cables.
- The ***manual transaxle differential assembly*** receives power from the transaxle transmission. The differential assembly contains differential drive gears, spider gears, and side gears. It allows the vehicle to make turns, allowing each drive wheel to rotate at a different speed. Construction and operation is very similar to that of a rear-wheel drive vehicle.
- The ***manual transaxle case*** encloses transaxle shafts, gears, synchronizers, shift forks and linkage, and lubricating oil. It also encloses the differential assembly.

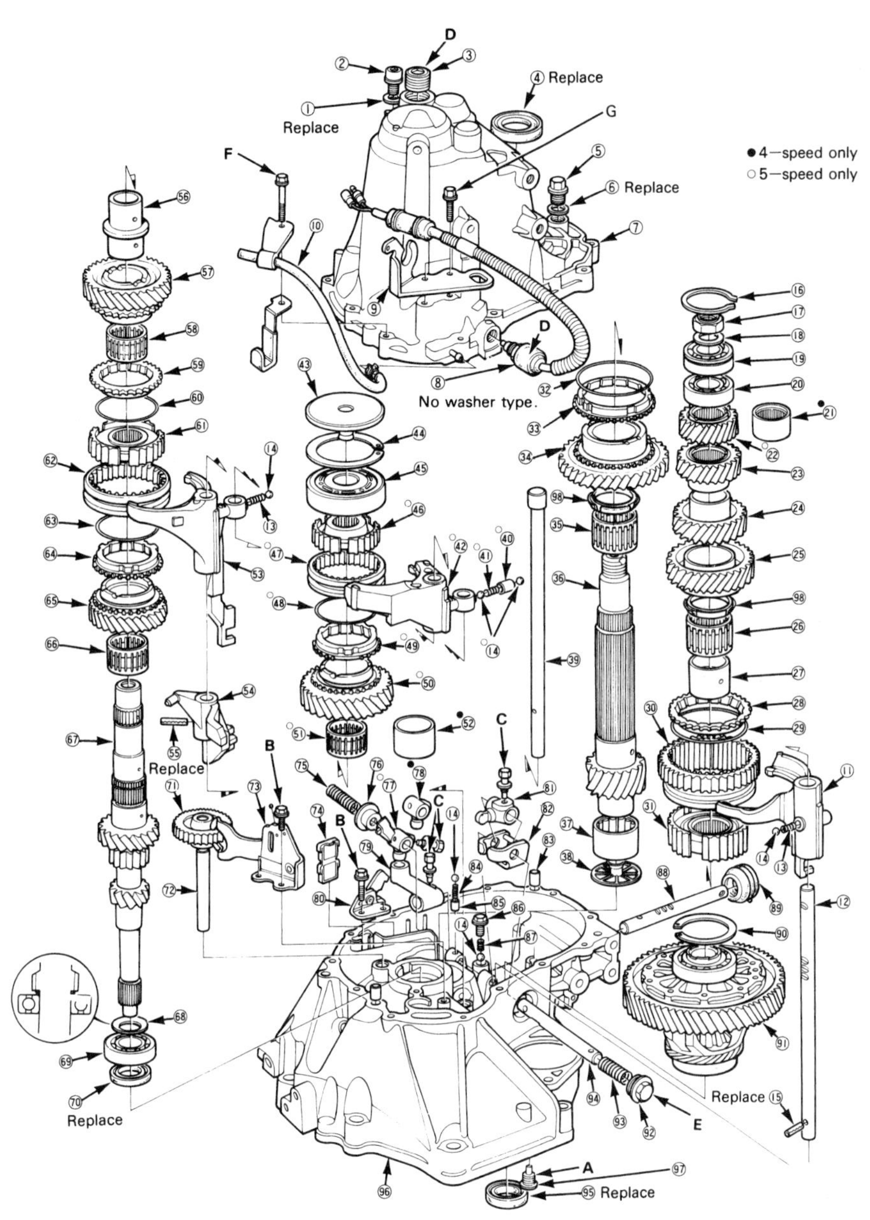

1. WASHER
2. OIL DRAIN PLUG
3. 32 mm SEALING BOLT
4. OIL SEAL
5. OIL FILLER BOLT
6. SEALING WASHER
7. TRANSMISSION HOUSING
8. BACK-UP LIGHT SWITCH
9. CLUTCH CABLE BRACKET
10. BREATHER TUBE
11. 1st/2nd SHIFT FORK
12. 1st/2nd SHIFT FORK SHAFT
13. SHIFT FORK SPRING
14. STEEL BALL
15. SPRING PIN
16. SNAP RING
17. COUNTERSHAFT LOCKNUT
18. WASHER
19. BALL BEARING
20. NEEDLE BEARING
21. COLLAR (4-speed only)
22. COUNTERSHAFT 5th GEAR
23. COUNTERSHAFT 4th GEAR
24. COUNTERSHAFT 3rd GEAR
25. COUNTERSHAFT 2nd GEAR
26. NEEDLE BEARING
27. SPACER COLLAR
28. SYNCHRO RING
29. SYNCHRO SPRING
30. REVERSE GEAR
31. SYNCHRO HUB
32. SYNCHRO SPRING
33. SYNCHRO RING
34. COUNTERSHAFT 1st GEAR
35. NEEDLE BEARING
36. COUNTERSHAFT
37. NEEDLE BEARING
38. OIL BARRIER
39. 5th/REVERSE SHIFT FORK SHAFT
40. ROLLER (5-speed only)
41. 5th DETENT SPRING (5-speed only)
42. 5th SHIFT FORK (5-speed only)
43. OIL GUIDE PLATE
44. THRUST SHIM
45. BALL BEARING
46. SYNCHRO HUB (5-speed only)
47. SYNCHRO SLEEVE (5-speed only)
48. SYNCHRO SPRING (5-speed only)
49. SYNCHRO RING (5-speed only)
50. 5th GEAR (5-speed only)
51. NEEDLE BEARING (5-speed only)
52. COLLAR (4-speed only)
53. 3rd/4th SHIFT FORK
54. SHIFT PIECE
55. SPRING PIN
56. SPACER COLLAR
57. 4th GEAR
58. NEEDLE BEARING
59. SYNCHRO RING
60. SYNCHRO SPRING
61. SYNCHRO HUB
62. SYNCHRO SLEEVE
63. SYNCHRO SPRING
64. SYNCHRO RING
65. 3rd GEAR
66. NEEDLE BEARING
67. MAINSHAFT
68. SPRING WASHER
69. BALL BEARING
70. OIL SEAL
71. REVERSE IDLER GEAR
72. REVERSE IDLER SHAFT
73. REVERSE SHIFT HOLDER
74. MAGNET
75. REVERSE SELECT SPRING
76. REVERSE RETURN SELECT
77. SHIFT ARM C (5-speed only)
78. SHIFT ARM C (4-speed only)
79. SHIFT ARM A
80. REVERSE LOCK CAM
81. SHIFT ARM B
82. INTERLOCK
83. DOWEL PIN
84. SPRING
85. SPRING COLLAR
86. SPRING BOLT
87. SPRING
88. SHIFT ROD
89. BOOT
90. SHIM
91. DIFFERENTIAL ASSEMBLY
92. 28 mm PLUG
93. 1st/2nd SELECT SPRING
94. SHIFT ARM SHAFT
95. OIL SEAL
96. CLUTCH HOUSING
97. INTERLOCK GUIDE BOLT
98. FRICTION DAMPER

Figure 18-2. Exploded view of transaxle used with transverse engine shows internal parts in detail. (Honda)

Manual Transaxle Shafts

Manual transaxles contain several shafts, including input, output, and reverse idler shafts. These shafts are made of hardened steel. They must fit in the case with very close tolerances. In some designs, the shafts rotate on ball bearings, straight roller bearings, or needle bearings. Thrust washers, placed between gears on these shafts and the transaxle case, control end play. In other designs, shafts rotate on tapered roller bearings to control end play. Manual transaxle shafts are arranged differently in different transaxles, but their basic function is the same.

Manual transaxle input shaft

The ***manual transaxle input shaft*** receives power from the engine through the clutch. On transaxles used with transverse engines, Figure 18-3, the input shaft is a single shaft, splined to the clutch friction disc, and extending from it into the transaxle transmission. On transaxles used with longitudinal engines, the input shaft may be connected to the clutch through an additional shaft and a drive chain or drive gears. This is shown in Figure 18-4. The input shaft is turned by the engine whenever the clutch is engaged.

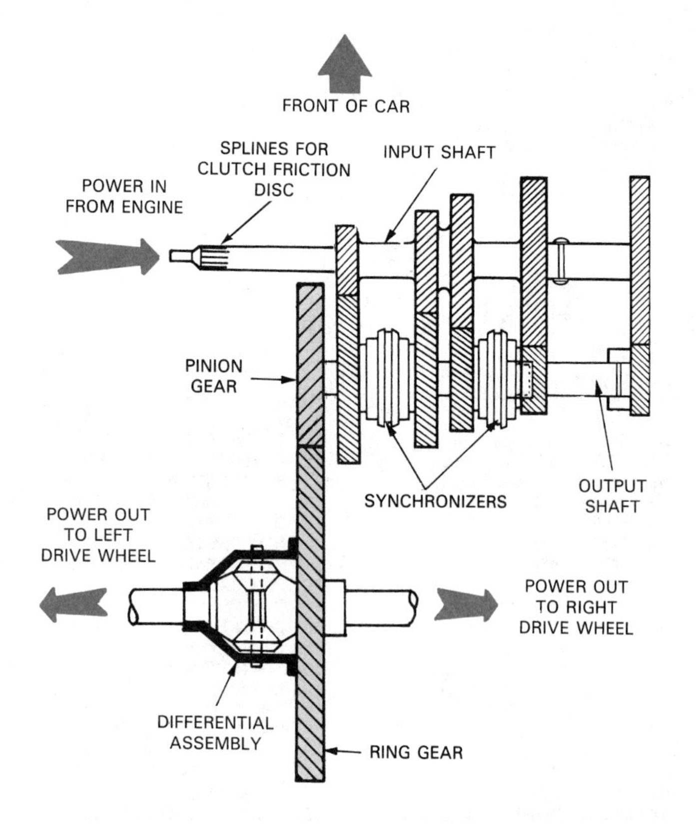

Figure 18-3. With transverse engine, crankshaft and input shaft centerlines are in alignment. Also, crankshaft and drive axles are oriented in same direction so that diverting power flow by 90° in the differential is not necessary.

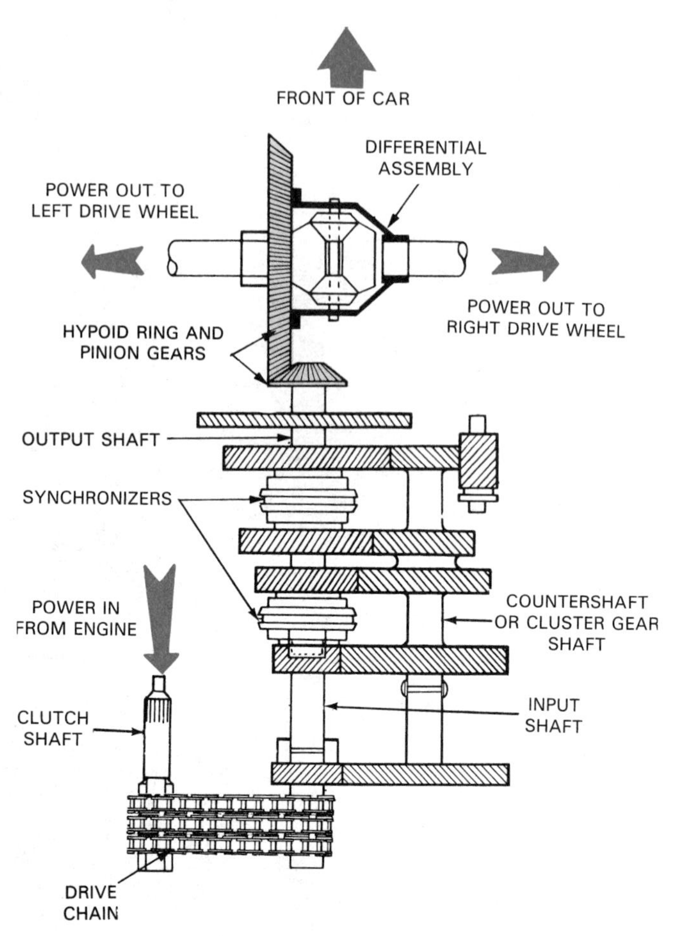

Figure 18-4. With longitudinal engine, crankshaft and input shaft have different centerlines. Drive chain is used to transfer power from clutch. Hypoid ring and pinion must be used to divert power flow at differential assembly.

The transaxle input shaft holds gears and synchronizers, which are mounted on the shaft in the same manner as on a rear-wheel drive manual transmission. Excessive shaft movement is prevented by front and rear bearings and by thrust washers. The input shaft is sometimes referred to as the *mainshaft* or the *input cluster gear.*

Manual transaxle output shaft

The ***manual transaxle output shaft*** is a shaft containing gears that mate with the input shaft. This rotating shaft, which holds output gears and synchronizers, delivers power to the transaxle differential. The output shaft is sometimes referred to as the *mainshaft,* the *countershaft,* or the *pinion shaft.*

Note that some transaxle designs have an output shaft and a separate countershaft. This design is usually seen in transaxles used with longitudinal engines. Such output shafts use a hypoid drive pinion gear to engage the ring gear in the differential portion of the transaxle.

Reverse idler shaft

Some transaxles have a reverse idler shaft. This is a short shaft that supports the reverse idler gear, an additional gear used to reverse output shaft direction so the vehicle can back up. The reverse idler shaft normally mounts stationary in the case.

Transaxle Transmission Gears

The gears of the transmission portion of the manual transaxle are similar to the gears of a rear-wheel drive manual transmission. Spur gears or, more often, helical gears are mounted on the transaxle transmission shafts. Some gears are cut directly onto the shaft, making them an integral part of the shaft; others are splined to the transmission shaft. Other gears rotate on the shaft, riding on a bushing or needle bearings placed between the gear and the shaft.

The gears of the transaxle transmission are used to transmit power and to provide different gear ratios, as well as reverse. Transaxle gears, like rear-wheel drive transmission gears, are made of machined high-strength steel. The gears are heat-treated for extra strength. Figure 18-5 shows typical manual transaxle transmission gears.

Some gears are in constant mesh, while others are not. Which gears are permanently in mesh varies with the transaxle manufacturer. The number of gears on the input and output shafts also varies according to the manufacturer's design.

The size of the gear is an indication of the ratio of the gearset: the smaller the gear on the input shaft and the larger the gear on the output shaft, the greater the reduction. On the input shaft, for instance, the gear for low is smaller than the gear for second. This can be seen in Figures 18-6 and 18-7.

Manual transaxle gears are usually lubricated by gear oil. A few transaxles use automatic transmission fluid as the lubricant. The rotation of the gears distributes the oil to the moving parts of the transaxle. In addition, oil passages in the shafts, Figure 18-8, permit lubrication of gear bushings and needle bearings.

A few transaxles are equipped with special washers that pump oil to the gears and bearings. These washers are made of spring steel. As the driver shifts gears, the direction of the thrust loading on the input and output shafts changes, causing these washers to flex. As they flex, the washers create a pumping action that distributes oil to critical areas of the transaxle.

Synchronizers

Before freewheeling gears can be locked to the turning shaft on which they ride, they must be turning at almost the same speed as the shaft. This is done in the transaxle transmission with gear synchronizers, Figure 18-9. The synchronizers are placed along the shaft between these gears, and they are splined to the shaft. They match the speed of mating gears to their own speed (output shaft speed) in the process of locking a gear to its shaft. If the gear and synchronizer speeds were not synchronized in this way, they would grind against each other before meshing, causing gear clash. As a result, shifting would be difficult and noisy. Further, the synchronizers or gears could be damaged.

Manual transaxle transmissions are synchronized in all forward gears. Reverse gear is unsynchronized. Con-

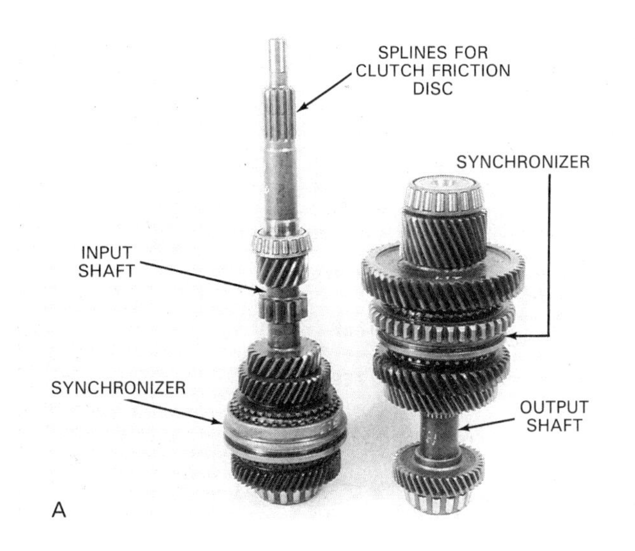

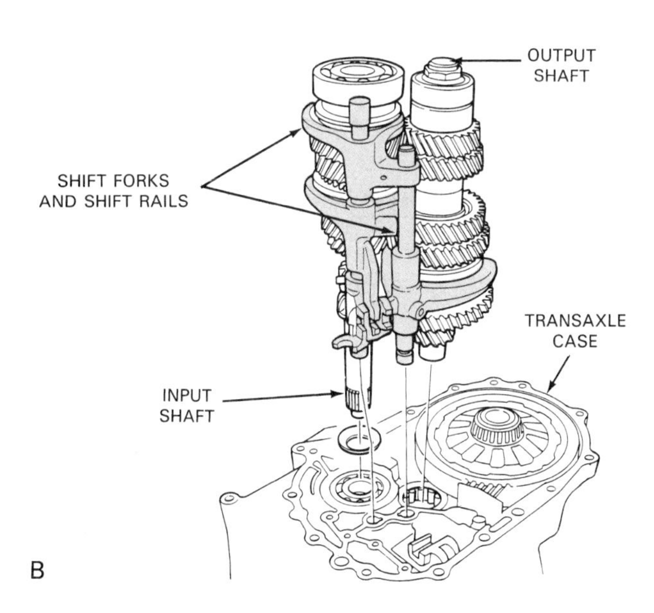

Figure 18-5. Note gears of transaxle transmission input and output shafts shown here. A–Note gears, bearings, and synchronizers on these input and output shafts. B–In addition to gears, note shift rails and shift forks on these input and output shaft assemblies. (General Motors, Honda)

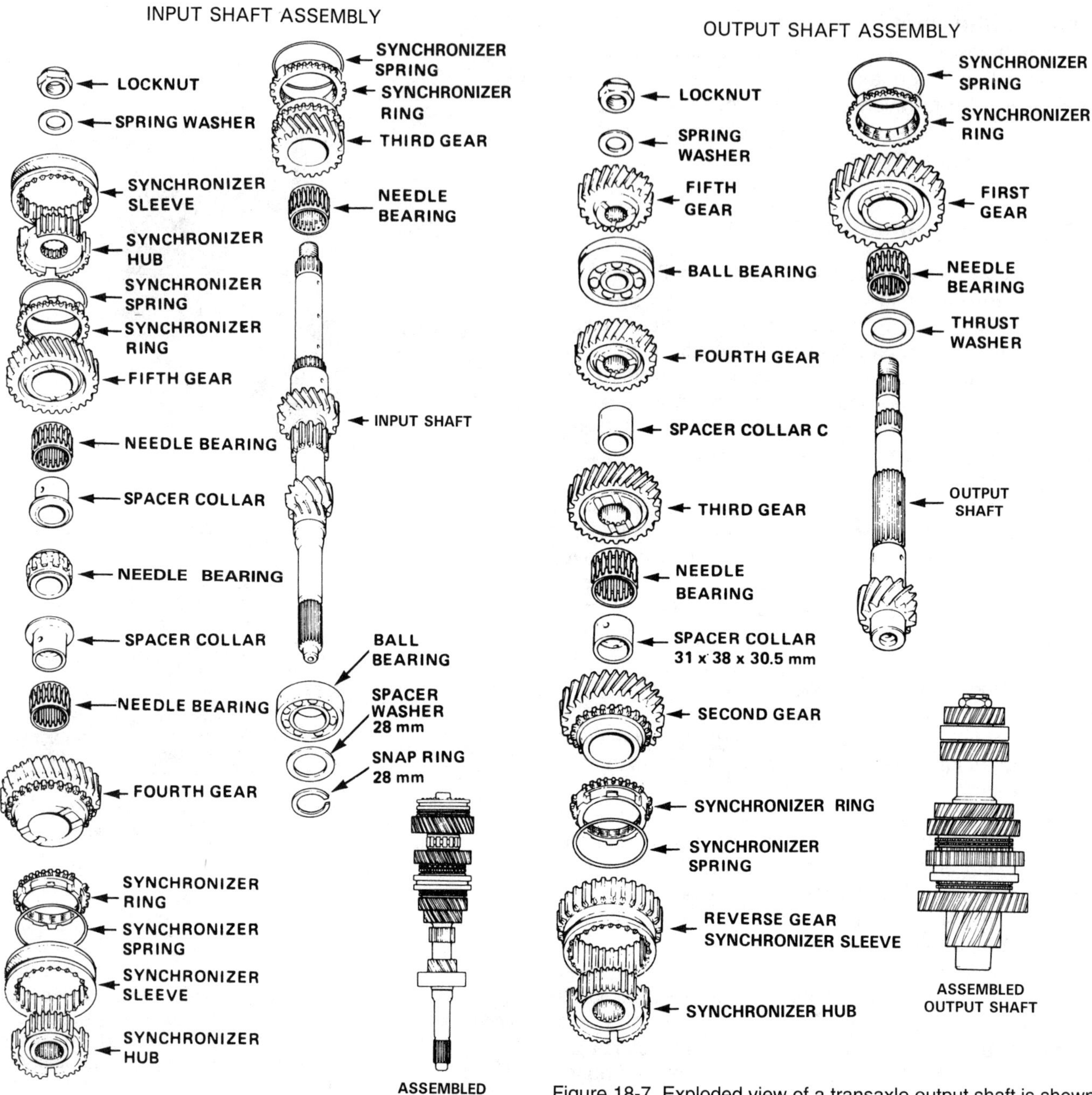

Figure 18-6. Exploded view of a manual transaxle input shaft is shown here. (Honda)

Figure 18-7. Exploded view of a transaxle output shaft is shown here. (Honda)

struction and operation of transaxle synchronizers is essentially the same as that of the manual transmission. Refer to Chapter 8 for a detailed discussion. One difference between synchronizers of transaxles and those of manual transmissions is that transaxle synchronizers may be located on both input and output shafts. You will recall that in the manual transmission, they were found only on the output shaft.

Shift Forks

Shift forks, like the one shown in Figure 18-10, are made to slip into a groove cut into the outer sleeve of a synchronizer assembly. They are usually made from steel or cast iron. Some shift forks have nylon inserts at the point where they contact the outer sleeve, to reduce wear and noise.

Shift fork action is identical to that of shift forks in a rear-wheel drive manual transmission. The shift fork moves in a straight line to push the synchronizer outer

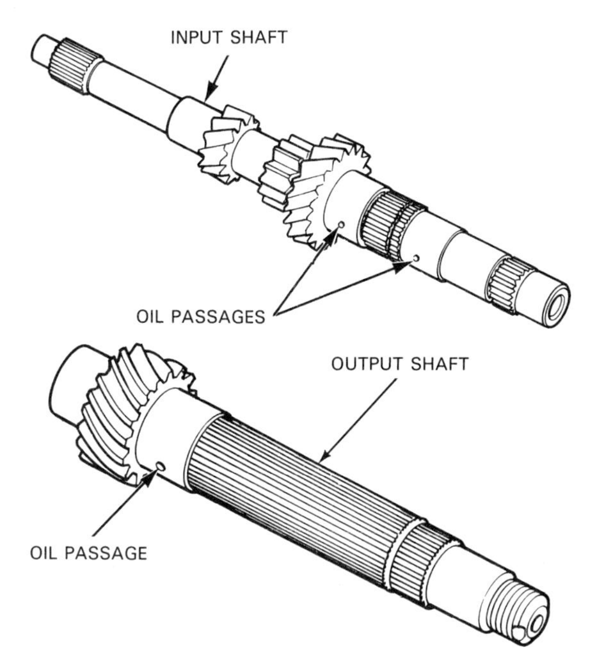

Figure 18-8. Note oil passages, used for lubricating bearings, in these input and output shafts. (Honda)

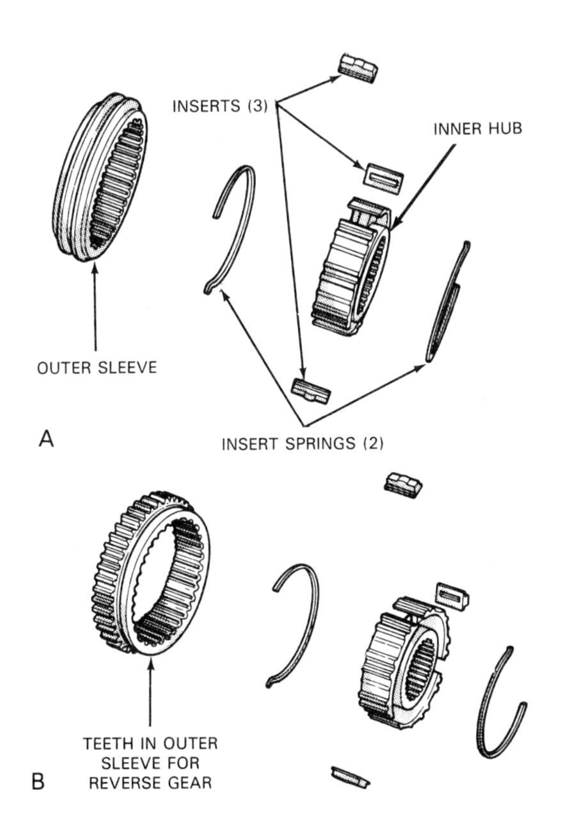

Figure 18-9. Synchronizers (shown here) act as clutches, bringing mating gears to the same speed before they are fully engaged. Note blocking rings, considered part of the synchronizer assembly, are not shown here. A–Note likeness of synchronizer to that found in rear-wheel drive manual transmission. B–The gear cut into the outer sleeve of this synchronizer serves as reverse gear. (Chrysler)

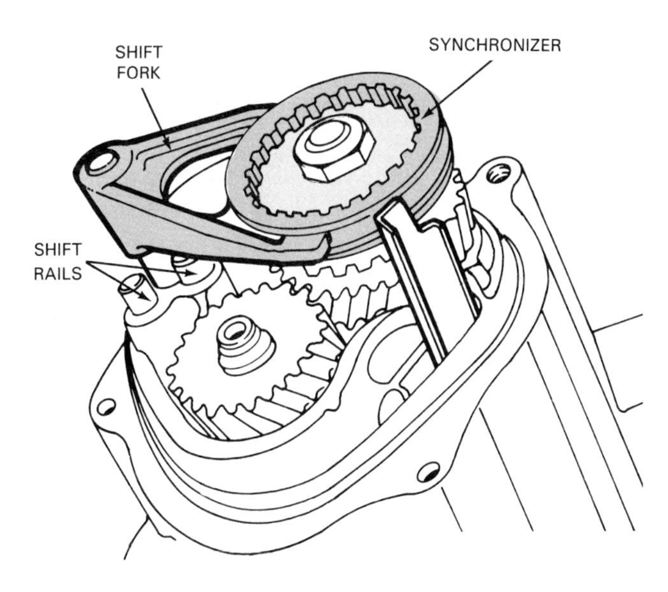

Figure 18-10. A shift fork, such as the one shown here, fits into the synchronizer outer sleeve. The fork can slide the sleeve forward or backward as the sleeve rotates with the shaft. (Chrysler)

sleeve and blocking ring into engagement with the mating gear. The design permits the synchronizer assembly to continue to rotate as the outer sleeve is moved. The shift fork, in conjunction with the synchronizer, selects one gear when it is pushed forward and another gear when it is pushed backward. The shift forks are operated by the transaxle transmission shift linkage.

Manual Transaxle Shift Linkage

The transaxle shift forks are operated by the vehicle driver through a series of shafts, levers, links, and sometimes, cables. There are many ways to transfer the force applied to the gearshift lever to the transaxle transmission gears. Most transaxle shift linkages are very simple both in construction and in operation. The gearshift lever and the components that make up a manual transaxle shift linkage are discussed in this section.

Gearshift lever

Vehicles with front-wheel drive manual transaxles use a floor-mounted gearshift lever, Figure 18-11. The gearshift lever may have a knob with a button, which is depressed to obtain reverse. The lever may alternately have a sliding ring, which is moved upward to obtain reverse. Either mechanism prevents shifting into reverse while moving forward. Note that the lever extends down to external shift linkage through a hole in the vehicle floor pan. This hole is sealed by a flexible rubber boot. Sometimes, two boots are used.

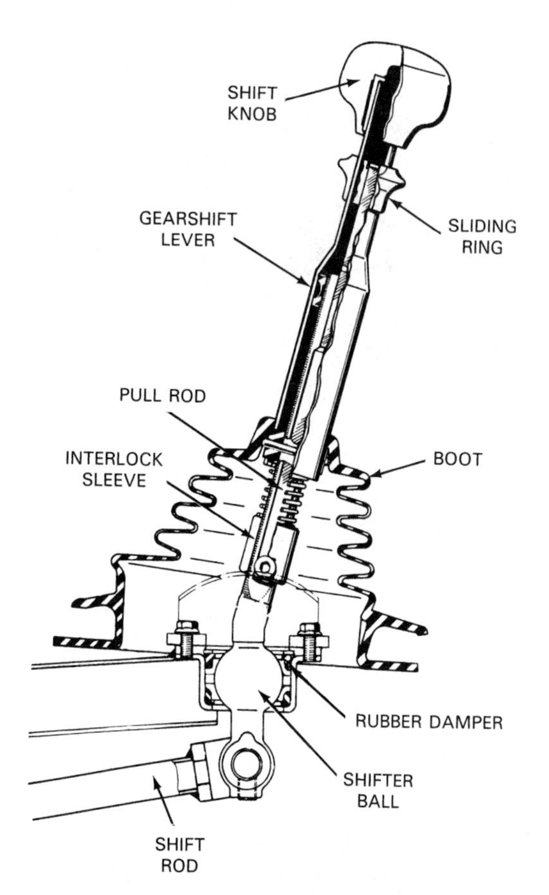

Figure 18-11. An example of a typical gearshift lever is shown here. Gearshift levers are always floor mounted on manual transaxle vehicles. The boot keeps dirt and water out of the passenger compartment. The sliding ring is pulled up to move the lever to the reverse position, but not all gearshift levers have such a feature. (Volvo)

External shift linkage

The gearshift lever operates a shift mechanism under the dust boot. Since the transaxle is placed far ahead of the passenger compartment, external linkage must be used to connect the shift mechanism and transmission internals.

External shift linkage consists of a series of levers and rods or cables that connect the gearshift lever with the transaxle internal shift linkage. A typical rod-and-lever linkage system is shown in Figure 18-12. A typical cable linkage system is shown in Figure 18-13. The system shown has two selector cables. Other systems may use only one cable, or as many as three. The shift linkage system shown in Figure 18-14 uses only a single rod to connect the gearshift lever assembly to shift levers and rods at the transaxle. A few linkage systems use a combination of rods and cables to change gears.

Rod-and-lever linkage parts are attached with cotter pins through drilled holes in the shift rods, in the same manner as they are on a rear-wheel drive manual transmission. Washers and bushings are used to reduce wear. The linkage can be adjusted by threaded swivels on the shift rods. Cable linkages have adjustment mechanisms that resemble those used to adjust a clutch cable linkage. Some shift linkage systems have provisions for lubrication, but most do not.

Internal shift linkage

The shift fork is operated inside of the transaxle by a series of levers, and more often, a shifter shaft or shift rails that slide in holes drilled in the case. See Figure 18-15. The shift rail system is used on many transaxles and transmissions. It consists of a series of rods (shift rails), which operate the shift forks. The rails move in a straight line. The shift forks are attached directly to the shift rails.

An exploded view of a shift rail system is shown in Figure 18-16. This system has four shift forks, which are used to select the transaxle's six forward gears and one reverse gear. Notches seen along the shift rails are for detents. These spring-loaded devices keep the levers and, therefore, the fork and synchronizer sleeve in position when they are not being moved by the driver.

Note that one kind of detent is inside a separate cover, installed in the case with a snap ring, Figure 18-17. This design serves the same purpose as other detent systems, but it can be serviced from outside the case.

Manual Transaxle Differential Assembly

The transaxle differential assembly is the part of the transaxle that allows the front wheels to turn at different speeds when the vehicle is turning a corner. The action of the differential assembly is the same as that of a rear axle assembly. When the vehicle makes a turn, the differential assembly allows the outer wheel to travel further than the inner wheel. This is necessary because the arc (or radius) of the turn is always greater at the outer wheel. If the axles were simply connected together, both wheels would be forced to rotate the same amount during a turn. To do this, one of the wheels would have to lose traction to make the turn. The transaxle differential assembly allows the vehicle to make turns without wheel hop. For further discussion on the general construction and operation of differential assemblies, refer back to Chapter 16.

Differential drive gears are used in all transaxle differential assemblies. This gearset typically consists of helical ring and pinion gears. However, some transaxle designs have a hypoid ring and pinion, to divert power flow by 90°. The drive pinion gear is on the transaxle output shaft. It meshes with the ring gear on the differential case.

The transaxle differential assembly is also composed of meshing spider and side gears, enclosed in a differential case. Figure 18-18 is an exploded view of a typical transaxle differential assembly. The differential case is a one-piece unit, made of cast iron or aluminum. The ring gear is

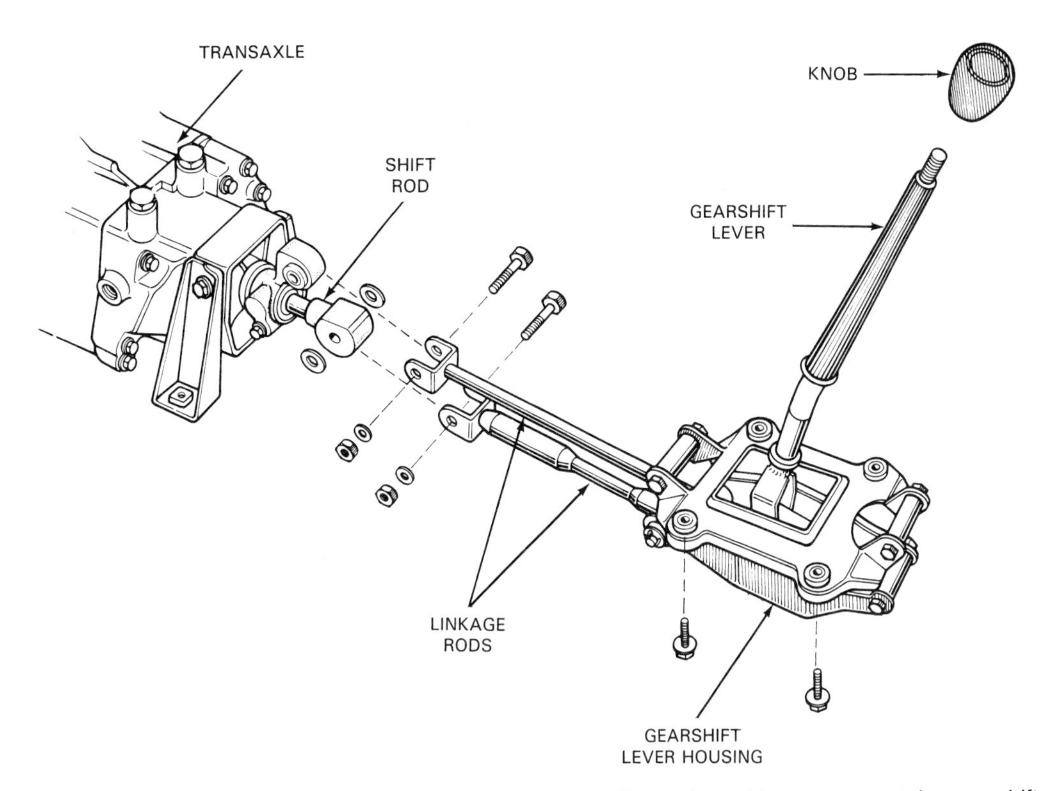

Figure 18-12. This is a rod-and-lever type of shift linkage. The rods and levers connect the gearshift lever with the transaxle internal shift linkage. (Toyota)

bolted to the case. It meshes with the drive pinion gear, located on the output shaft of the transaxle transmission. The number of teeth on the ring and pinion gears determines the final drive ratio of the transaxle.

The differential assembly shown in Figure 18-19 illustrates nonidentical side gears, found on many transaxle differential assemblies. Since the CV axles are of different lengths or may have an intermediate shaft, different side gear designs may provide the most efficient way to transfer power. This is the main visual difference between many transaxle differentials and differential assembies used on rear-wheel drive vehicles.

The differential assembly is often lubricated with the same oil that is used to lubricate the transmission gears. Some differentials, especially the hypoid types used with longitudinal engines, may have a separate lubrication sump. This type of differential usually uses a heavier grade of gear oil than the transmission does.

Manual Transaxle Case

The case of a manual transaxle protects and aligns the gears and other moving parts. The transaxle case also contains transaxle lubricant and keeps dirt and water out of the gears and bearings. The inboard ends of the CV axles and the stationary parts of the transmission are mounted on the case. Most of the bearings that support the gear and axle shafts are mounted in the case.

There may be one or more sheet metal or cast aluminum access covers on the transaxle case, which can be removed for servicing the transaxle. In some instances, these covers can be removed to service certain internal parts without removing the transaxle from the vehicle. Many transaxle cases are designed in two pieces. When the transaxle requires service, the case can be split to gain access to the internal parts.

Most manual transaxle cases of front-wheel drive vehicles are made of aluminum. The case is machined to precisely fit the rear, or output end, of the engine, Figure 18-20, and to accept the internal and external transaxle parts. A spacer plate may be installed between the transaxle and engine to strengthen the assembly. This is often done when the transaxle is the mounting surface for the engine starter.

The clutch housing is usually cast integral with the transaxle case. Further, the housing for the differential portion of the transaxle is usually an integral part of the case. Some differentials, however, especially those on transaxles with conventionally mounted engines, have separate housings, allowing for easier service of the internal parts. Many separate differential housings are made of cast iron for added strength.

Figure 18-21 shows three other views of the manual transaxle featured in Figure 18-20. Notice that the housing for the differential assembly is integral with the case, just as the clutch housing is. This particular case has two sheet

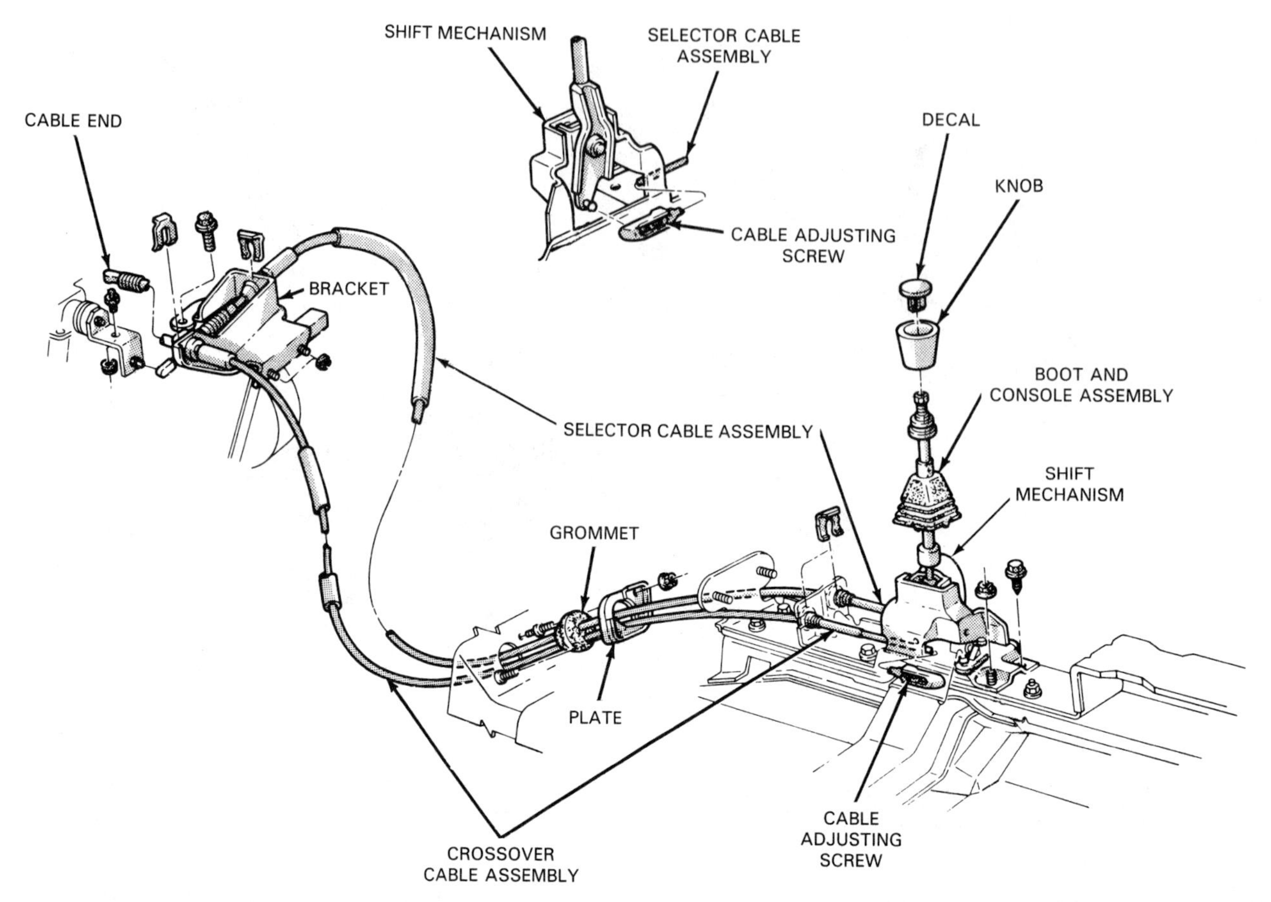

Figure 18-13. Study cable type of shift linkage. The system shown has two sheathed cables that connect the gearshift lever to the transaxle internal shift linkage. (Chrysler)

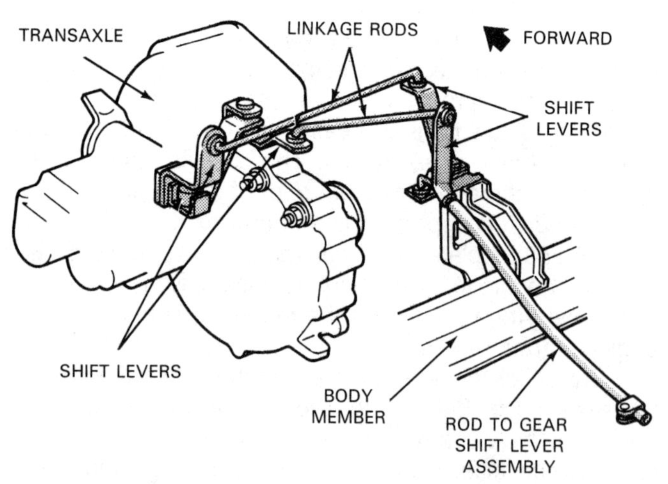

Figure 18-14. A single rod extends from the shift mechanism at the gearshift lever assembly to linkage at the transaxle. The rod can move back and forth, and it can also rotate. This movement is transferred to the levers and rods mounted at the transaxle. (Chrysler)

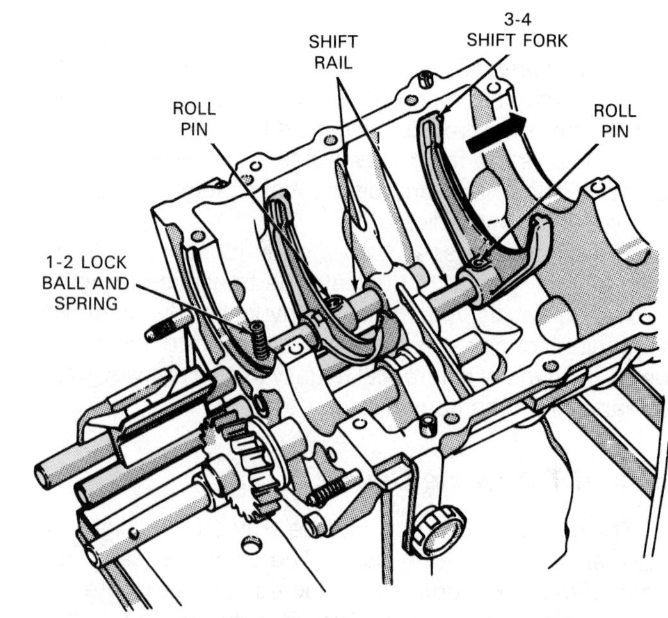

Figure 18-15. Some transaxle internal shift linkages use multiple shift rails, as is seen in this view. Notice relative positions of shift rails and shift forks inside a transaxle case. (Chrysler)

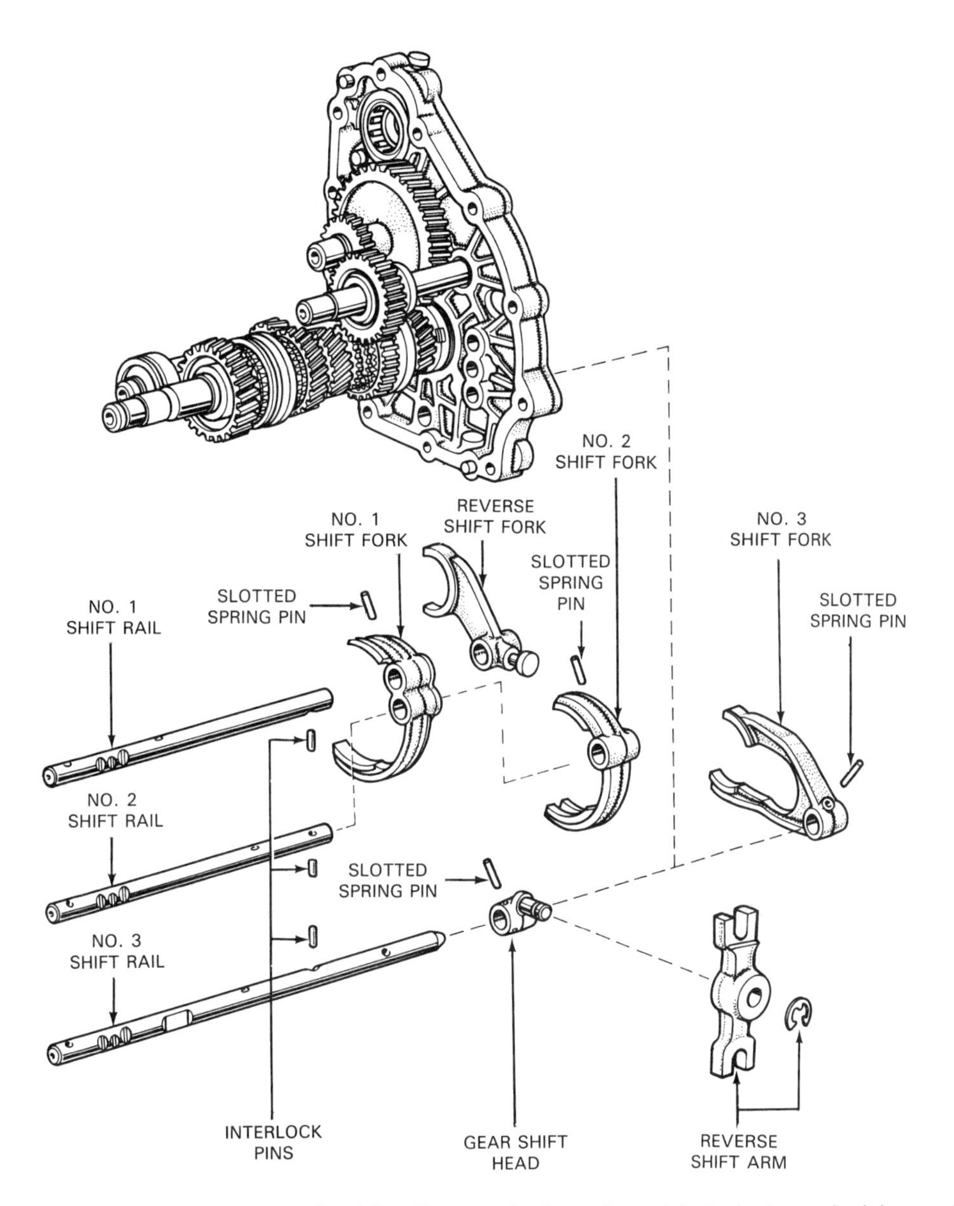

Figure 18-16. An exploded view of a shift rail system is shown here. Interlock pins make it impossible to move more than one rail at a time; this ensures that only one gear can be engaged during a shift. (Toyota)

metal covers, which can be removed to service the transaxle internal parts. Among other items, the case contains a lubricant fill plug and vent.

Note that the openings in the transaxle case for the input shaft and CV axles are equipped with lip seals. These are used for preventing lubricant leakage. Stationary parts, such as the transaxle case halves or covers, are sealed with gaskets. Openings in the case, such as for speedometer, linkage, and switches, may be sealed simply when these components are threaded into the case. In some instances, O-rings may be used on such components to help provide a tight seal.

Manual Transaxle Power Flow

Power flow through a transaxle mounted on a transverse engine runs parallel with the crankshaft; there is no change in angle once leaving the engine. Engine power enters the input shaft of the transaxle transmission. Power is engaged and disengaged through the clutch assembly. The input shaft drives the output shaft through the shaft gearsets. The drive pinion gear, located on the end of the output shaft, drives the ring gear on the differential case. From the ring gear, power enters the spider and side gears and exits through the CV axles to the front wheels.

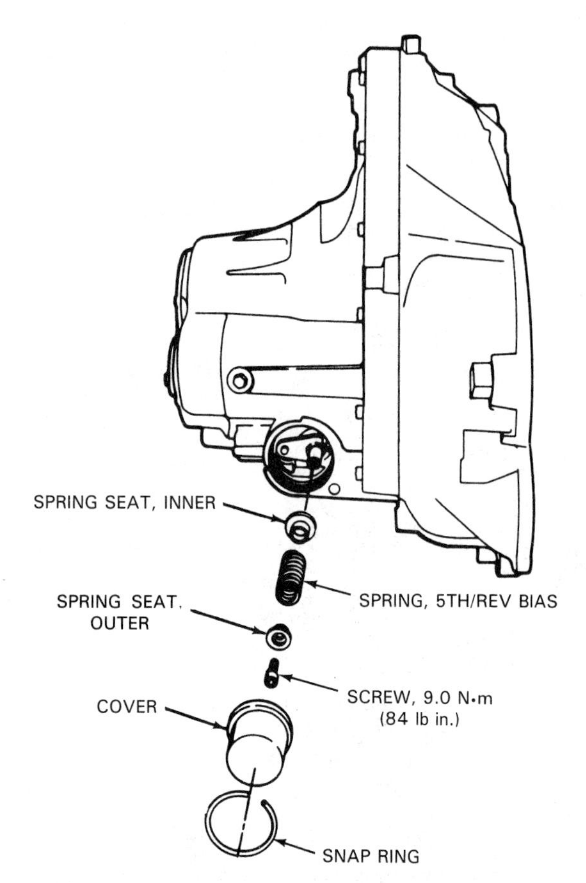

Figure 18-17. Detents fit into notches on shift rails to keep shift forks and shift rails in place once desired gear is selected. The spring puts pressure on the shaft to keep it in place. (General Motors)

On a transaxle attached to a longitudinal engine, power flow must make a right-angle turn at the differential assembly, as it does in a rear-wheel drive vehicle. This type of transaxle essentially has two input shafts. One shaft delivers power from the clutch to a drive chain. The drive chain then drives the input shaft of the transaxle. The input shaft runs parallel to the clutch shaft. (Refer back to Figure 18-4.) Note that some transaxles of this type use two large gears instead of a drive chain to transfer power to the transaxle input shaft.

Power flow, with this arrangement, then enters a countershaft, through the shaft gearsets. The countershaft drives a third transaxle shaft–the output shaft. The output shaft drives the differential pinion gear, which is solidly attached to the end of the output shaft. The pinion gear drives the differential ring gear and case assembly. The ring and pinion is the hypoid type, like the ones used on front-engine, rear-wheel drive vehicles. The design causes the power flow to make a 90° turn. From the differential side gears, power exits through the CV axles to the front wheels. Note that this type of transaxle more closely resembles the rear-wheel drive manual transmission.

Detailed Look at Power Flow

A detailed look at power flow through a typical transaxle used with a transverse engine is illustrated in Figures 18-22 through 18-27. This transaxle is a model with a basic two-shaft (input and output) arrangement. It is used with a transverse engine.

Figure 18-22 shows the gears in neutral. The clutch is engaged and is turning the input shaft, but power is not being transmitted to the output shaft. The gears on the output shaft are being turned by those on the input shaft, but both synchronizers are centered. The synchronizers must be moved off of center to engage the gears riding on the output shaft with the shaft itself.

In Figure 18-23, the transaxle is in first gear. The rear synchronizer remains centered. The front synchronizer has been moved toward the front of the transaxle, engaging first gear on the output shaft with the shaft itself. Power flows through the input shaft to the first gear on the input shaft. From here, it flows to the first gear on the output shaft, through the shaft to the differential drive pinion gear, and on to the ring gear. The combination of the number of teeth on the particular input and output shaft gears produces a 3:1 ratio in first gear. From the ring gear, power flows through the differential case assembly and out to the front wheels through the CV axles.

Figure 18-24 shows the transaxle in second gear. The rear synchronizer is centered. The front synchronizer has been moved toward the rear of the transaxle, connecting second gear on the output shaft with the shaft itself. Power flows through the input shaft, to the output shaft, through the differential drive gears and case assembly, and out through the CV axles. A gear ratio of 2.5:1 is achieved in second gear.

In Figure 18-25, the transaxle is in third gear. The front synchronizer is centered. The rear synchronizer has been moved toward the front of the transaxle, connecting the third gear output shaft gear with the output shaft. Power flows through the input shaft third gear to the output shaft third gear to drive the output shaft. From the differential pinion gear, power flows to the ring gear and differential case assembly. A gear ratio of 2:1 is achieved.

Fourth gear is shown in Figure 18-26. The front synchronizer remains centered, while the rear synchronizer is pushed back to connect fourth gear on the output shaft with the shaft itself. Since the particular input and output shaft gears have the same number of teeth, a gear ratio of 1:1, or direct drive, is achieved. Power flows through the gears, to the pinion and ring gears, and out of the differential assembly to the CV axles and front wheels.

Reverse is shown in Figure 18-27. Both synchronizers remain centered as the reverse idler gear slides into position to connect the input and output shafts. The idler gear causes the output shaft to rotate in the opposite direction, causing the differential side gears and CV axles to turn backwards.

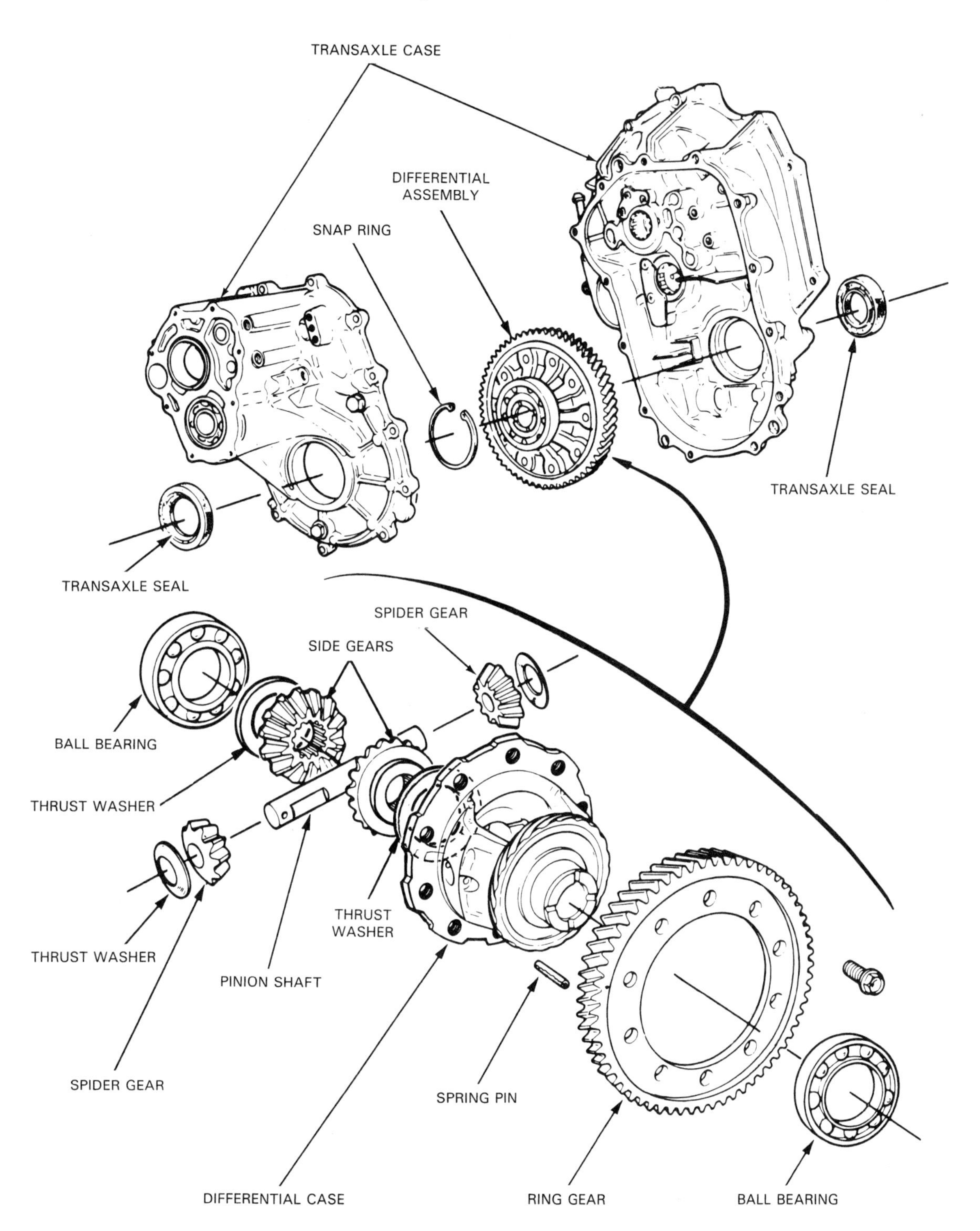

Figure 18-18. This exploded view of a transaxle differential shows the relative positions of the various components. Note likeness of this differential to that of a rear-wheel drive vehicle. This transaxle is unusual in that it uses ball bearings for side bearings. (Honda)

Manual Transaxle Designs

Although manual transaxles are similar in their basic design, many minor differences exist between models. Some of these differences may be seen by examination of Figures 18-28 through 18-30.

Figure 18-28 shows a typical General Motors manual transaxle for use with transverse engines. This particular transaxle has five forward speeds, all synchronized.

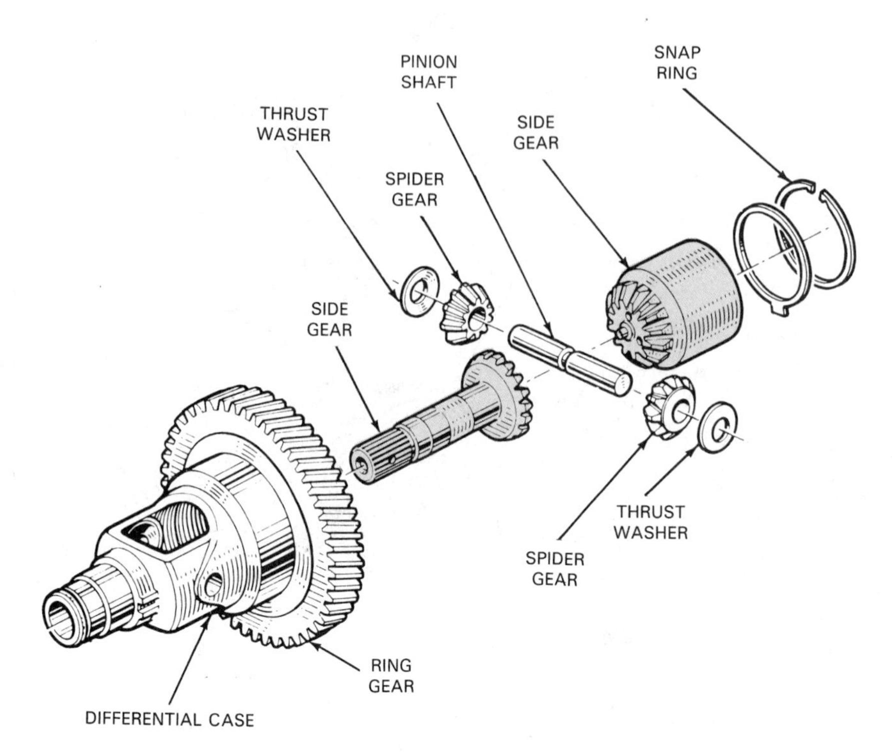

Figure 18-19. This exploded view of a differential assembly shows the differences that can exist between side gears. The side gears have the familiar beveled-cut teeth, designed to mate with the beveled spider gear teeth. (Chrysler)

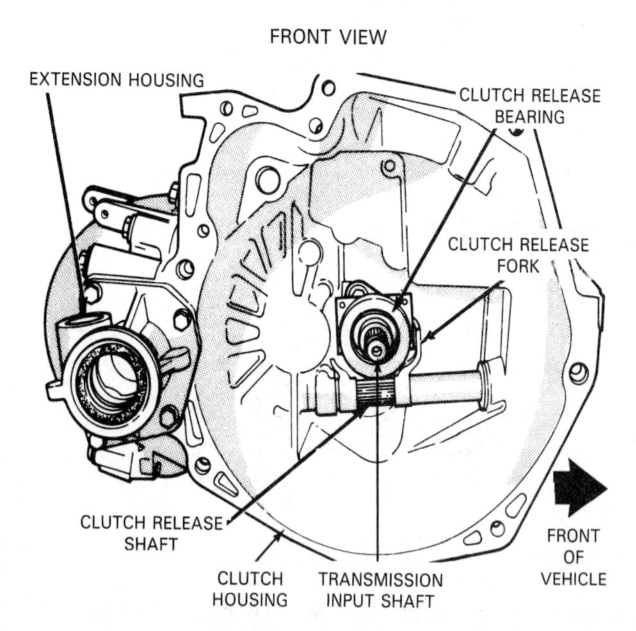

Figure 18-20. This is the *front* view of a popular make of manual transaxle used with a transverse engine. The clutch housing is integral with the transaxle case. It is machined to precisely fit the rear of the engine. Note clutch release (throwout) bearing, clutch release shaft and fork, and input shaft. Extension housing contains opening for one of CV axles. Note transaxle orientation with respect to front of vehicle. (Chrysler)

Figure 18-29 is a 4-speed manual transaxle used on many Chrysler front-wheel drive vehicles with transverse engines. It is fully synchronized in all forward gears. Note the selector shaft extending through the top of the case. It operates the shift rails. Also note the unusual location of the clutch release bearing. The clutch pushrod extends through the center of the input shaft to actuate the clutch.

The Ford transaxle shown in Figure 18-30 has an overdrive fifth gear. The input shaft contains a cluster gear, which is integral with the shaft. Synchronizers are not used on the input shaft of this design, whereas they were on those of the previous two transaxle designs.

Rear-Wheel Drive Transaxle

The most well-known rear-wheel drive transaxles were the VW Beetle and the Chevrolet Corvair. At one time, these vehicles were common, but both are now out of production. Transaxles are still occasionally used on modern rear-wheel drive vehicles–some Porsche models, for example. The internal design and construction of rear-wheel drive transaxles is similar to that of front-wheel drive vehicles. The major classes of rear-wheel drive transaxles are those installed with a rear engine, and those used on a front-engine vehicle.

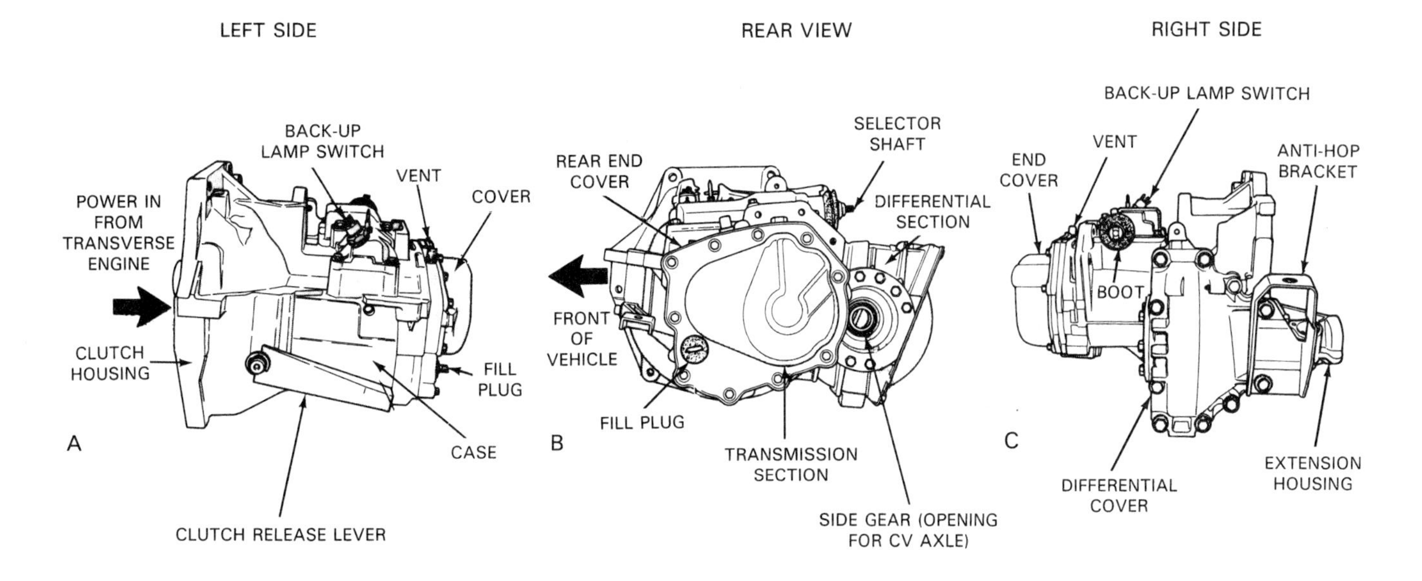

A—Left side view shows where power from transverse engine enters clutch housing. Note vent, fill plug, and other features shown on this view.

B—Rear view shows transmission and differential sections of transaxle. Note rear end cover and transaxle output opening for CV axle. Also note transaxle orientation with respect to front of vehicle.

C—Right side view shows differential cover, end cover, and one of the mounting brackets.

Figure 18-21. These views of the previously illustrated transaxle show various other features of the case. (Chrysler)

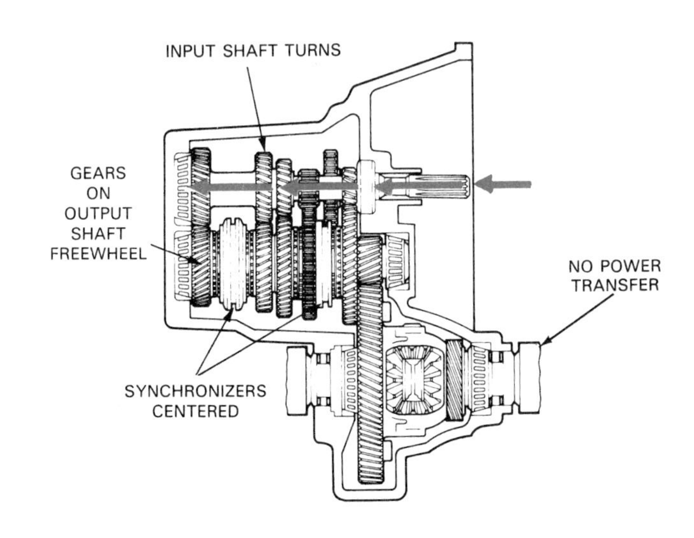

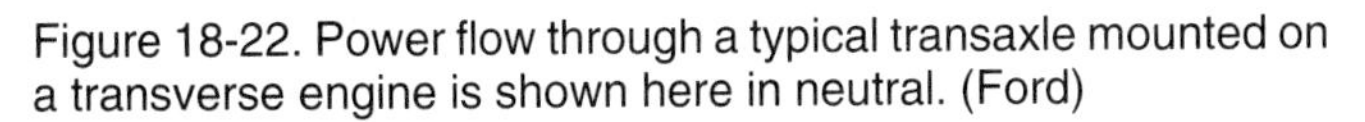
Figure 18-22. Power flow through a typical transaxle mounted on a transverse engine is shown here in neutral. (Ford)

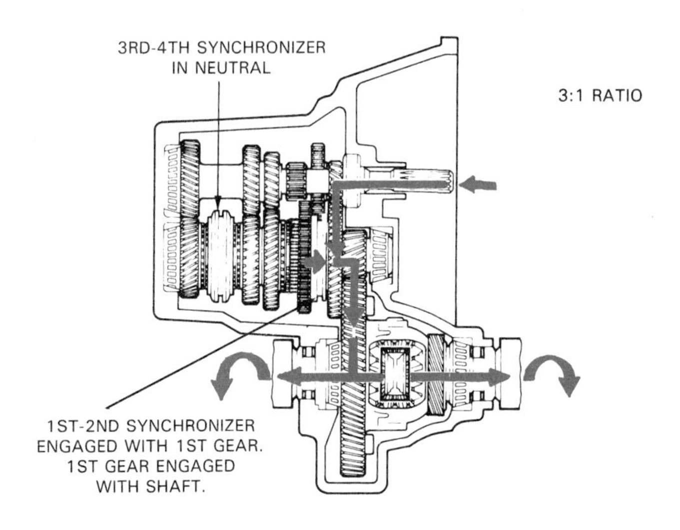

Figure 18-23. Power flow is depicted with the same transaxle in first gear. (Ford)

Figure 18-31 shows a typical rear-wheel drive manual transaxle used with a rear-mounted engine. With this construction, the engine is placed at the extreme rear of the vehicle, and power flows forward to reach the transaxle. After passing through the transaxle, the power flow exits to the rear drive wheels.

Figure 18-32 shows a typical transaxle used with a front-engine, rear-wheel drive vehicle. The clutch, as well as the transaxle transmission and differential assembly, are mounted in a single housing. The transaxle receives its power from a drive shaft that is connected to the front-mounted engine.

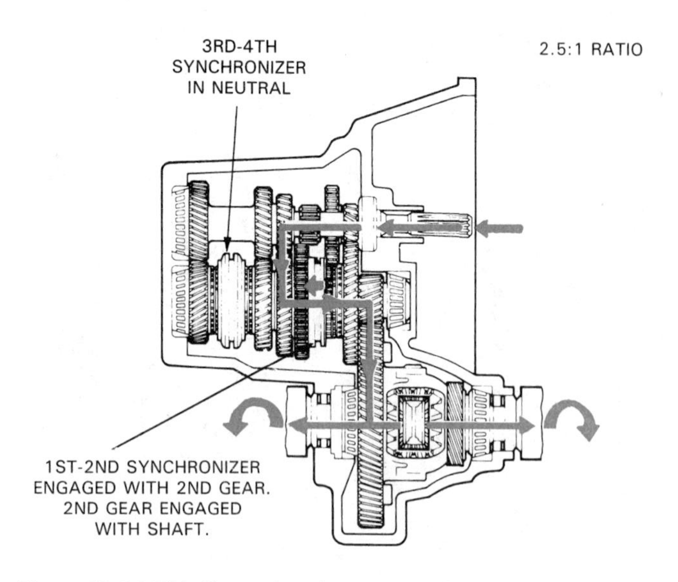

Figure 18-24. This illustration shows power flow through the same transaxle in second gear. (Ford)

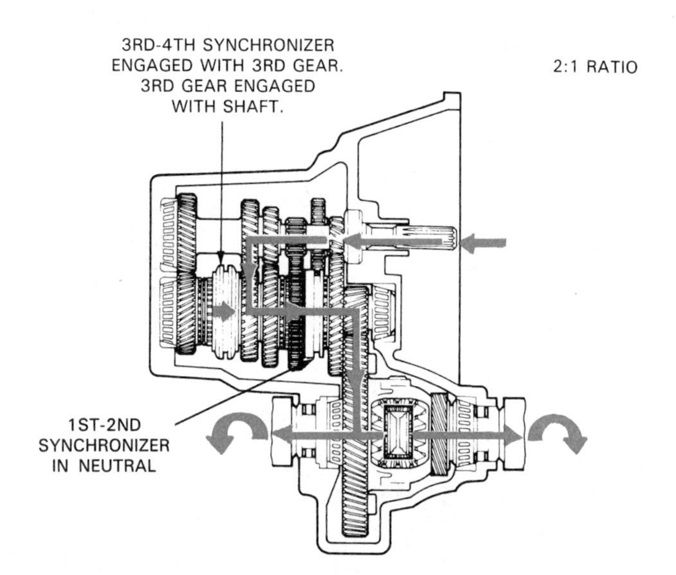

Figure 18-25. Power flow is shown here with the transaxle in third gear. (Ford)

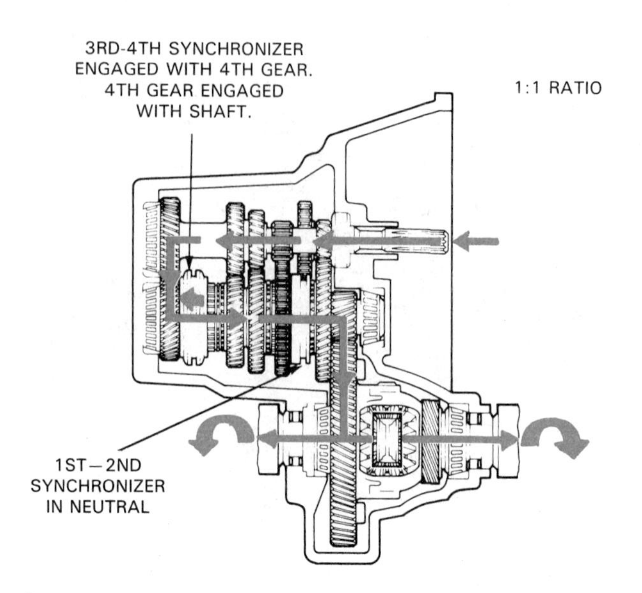

Figure 18-26. This shows power flow through the transaxle in fourth gear. (Ford)

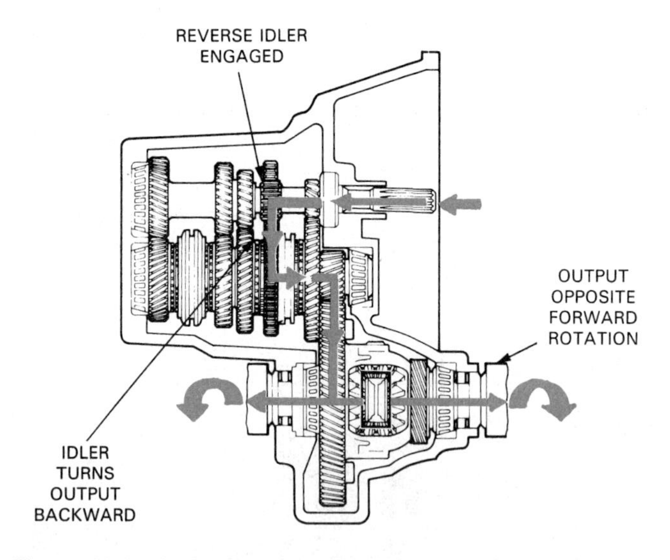

Figure 18-27. In reverse, the reverse idler gear is engaged. It causes the output shaft and differential assembly to turn in the reverse direction. (Ford)

Summary

Transaxle construction and operation is similar to that of manual transmissions and differential assemblies used in rear-wheel drive train vehicles. The transaxle has the transmission and differential contained in one case. Transaxle design and power flow varies, according to how the engine is placed in the vehicle. Transverse engines use a transaxle that allows power to flow in a straight line, without changing angles. The transaxle used in a vehicle with a longitudinal engine more closely resembles a rear-wheel drive manual transmission. There is usually a drive chain or gears to connect the input shaft and clutch. This transaxle uses hypoid ring and pinion gears.

Many of the manual transaxle transmission parts are similar to a manual transmission. Major transmission shafts of the typical transaxle include an input shaft and a countershaft that serves as the output shaft. The differential pinion gear is on the end of the output shaft. Synchronizers and shift forks operate just as they do on rear-wheel drive transmissions.

All transaxles have a ring and pinion gearset. On transaxles used with transverse engines, the ring and pinion are straight-cut or helical gears. If the transaxle is

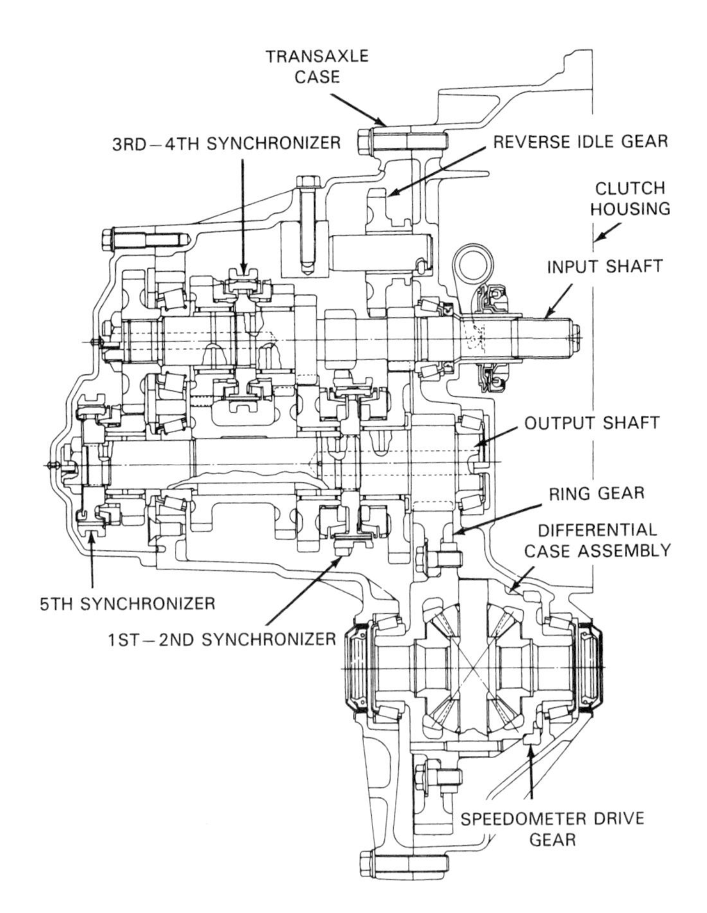

Figure 18-28. This 5-speed transaxle is used with transverse engines. (General Motors)

used with a longitudinal engine, the ring and pinion are a hypoid set. The difference in the number of teeth in the ring and pinion determines the final drive ratio.

The shift linkage consists of a series of links, levers, rods, and sometimes, cables, which allow the driver to select gears. Gearshift levers used for transaxles are floor mounted. Moving this lever allows the external shift linkage to move shift rods and forks inside of the transaxle case. These rods and forks move the synchronizers.

The main stationary parts of the transaxle are the case and covers. The case is made of cast aluminum. The case may be a one-piece unit, or it may be designed to be split apart for service. The covers are made of sheet metal or cast aluminum. They cover various operating parts of the transaxle and are designed to be removed for transaxle service.

Know These Terms

Manual transaxle shafts; Transaxle differential assembly; Manual transaxle case; Manual transaxle input shaft; Manual transaxle output shaft.

Review Questions–Chapter 18

Please do not write in this text. Place your answers on a separate sheet of paper.

1. List and explain the eight major parts of a manual transaxle.
2. A _____ engine has the engine sitting sideways in the engine compartment.
3. In your own words, describe the power flow through a typical manual transaxle mounted on a transverse engine.
4. Which one is made to slip into a groove that runs radially around the outer sleeve of the synchronizer?
   a. Shift linkage.
   b. Shift fork.
   c. Synchronizer inner hub.
   d. Synchronizer blocking ring.
   e. None of the above.
5. Explain a shift rail system.
6. A _____ is a spring-loaded holding mechanism.
7. A synchronizer blocking ring has external teeth on its outside surface. True or false?
8. The only place synchronizers are *not* found in a typical transaxle are:
   a. First gear.
   b. Direct drive
   c. Overdrive gear.
   d. Reverse gear.
9. The shaft arrangement in a transaxle attached to a _____ mounted engine is similar to that found in a manual transmission.
10. How are differential gears lubricated?

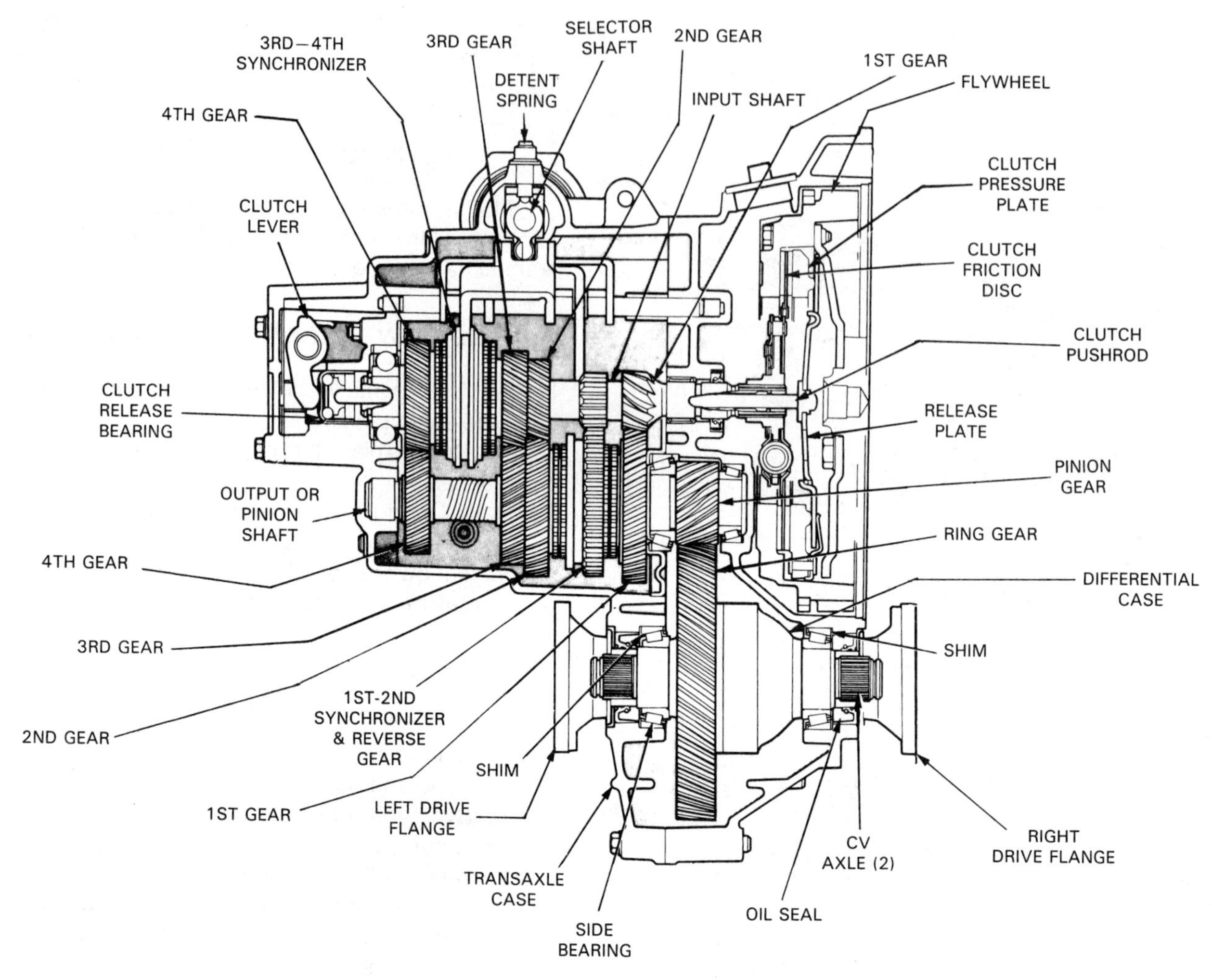

Figure 18-29. A 4-speed transaxle used with transverse engines is shown here. This transaxle has a unique feature in the position of the clutch release bearing. (Chrysler)

Certification-Type Questions–Chapter 18

1. The drive train of a front-wheel drive vehicle, as compared to that of a rear-wheel drive vehicle, eliminates the need for all of these EXCEPT:

(A) rear axle housing.
(B) transmission hump.
(C) drive shaft assembly.
(D) differential unit.

2. The basic construction of a transaxle in any front-wheel drive vehicle depends on:

(A) presence or absence of a differential assembly.
(B) gear ratio.
(C) engine placement in the vehicle.
(D) how power flows through the transaxle.

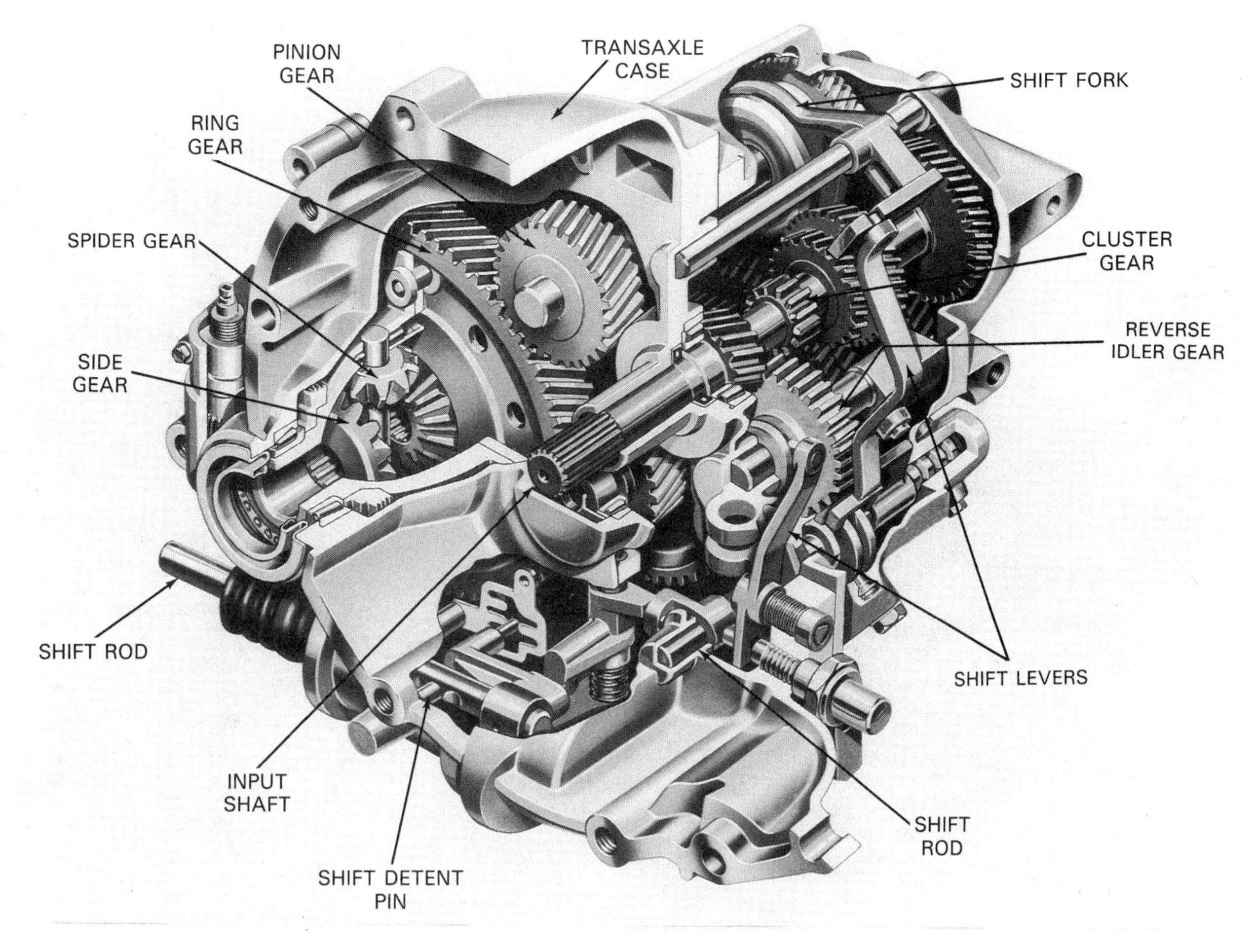

Figure 18-30. Cutaway shows inside of a 5-speed manual transaxle with overdrive. (Ford)

3. Manual transaxle shafts include all of these EXCEPT:
(A) stator shafts.
(B) input shafts.
(C) output shafts.
(D) reverse idler shafts.

4. Gears of the transaxle transmission are used for all of these EXCEPT:
(A) distributing oil.
(B) transmitting power.
(C) traction control.
(D) backing up a vehicle.

5. Before freewheeling gears can be locked to the turning shaft on which they ride, they must turn at almost the same speed as the shaft. Technician A says that the combined power of input and output shafts achieves this. Technician B says that this is accomplished through the synchronizers. Who is right?
(A) A only
(B) B only
(C) Both A & B
(D) Neither A nor B

6. All of these are functions of the manual transaxle EXCEPT:
(A) diverting power flow from the transverse engine by 90°.
(B) selecting the proper gear ratio or reverse.
(C) allowing the vehicle to make turns.
(D) transferring engine power to the drive axles.

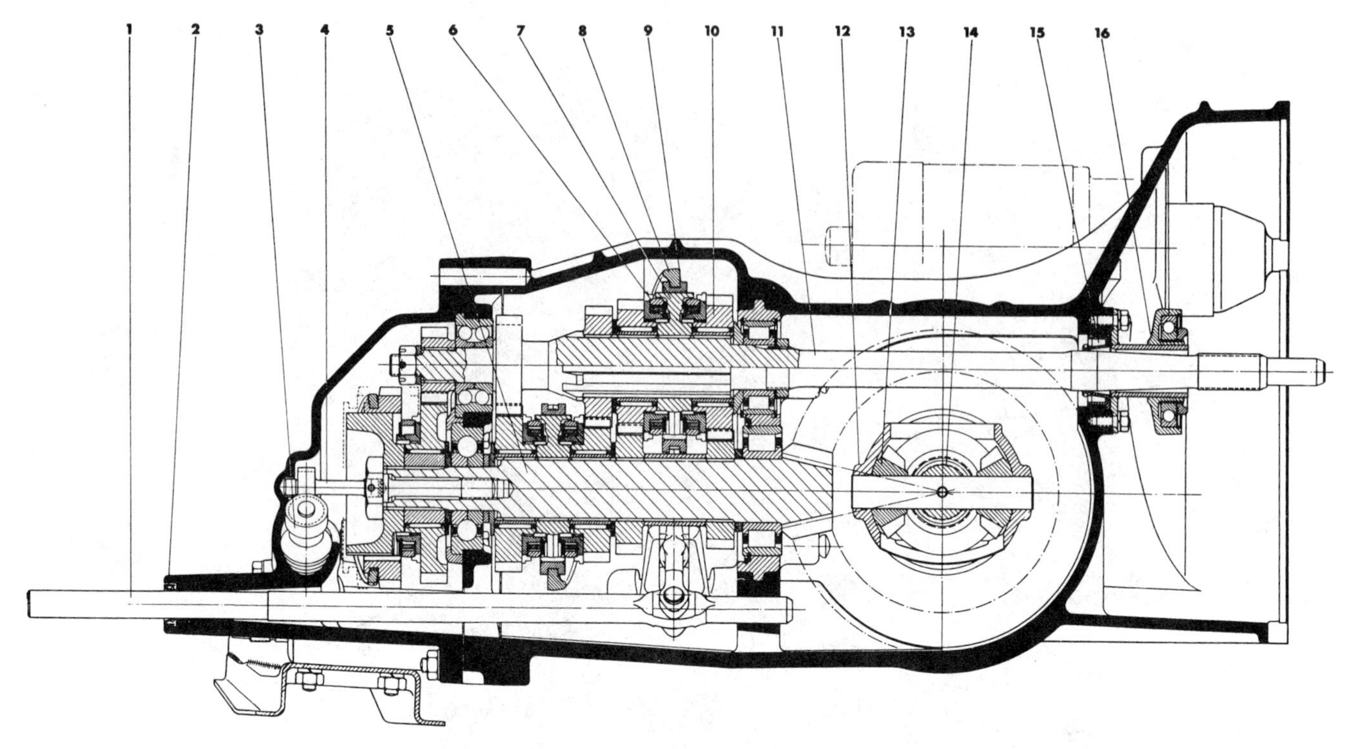

CROSS-SECTION VIEW OF TRANSMISSION AND DIFFERENTIAL

1. SHIFT ROD
2. SHIFT ROD OIL SEAL
3. TACHOMETER DRIVE
4. GEAR SHAFT
5. PINION SHAFT
6. SYNCHRONIZER BLOCKING RING
7. SPIDER
8. SHIFT FORK
9. SYNCHRONIZER OUTER SLEEVE
10. GEAR 1, 5TH SPEED
11. INPUT SHAFT
12. DIFFERENTIAL CASE
13. BEVEL SPIDER GEAR
14. SIDE GEAR SHAFT
15. INPUT SHAFT OIL SEAL
16. CLUTCH THROWOUT BEARING

Figure 18-31. Study rear-engine transaxle. On this design, the engine is installed at the rear of the vehicle, and power comes forward to enter the transaxle. (Porsche)

7. A shift fork is made to slip into a groove cut into the outer sleeve of the:

(A) drive chain.
(B) input shaft.
(C) synchronizer.
(D) gear bushing.

8. Components involved in shifting a manual transaxle include all of these EXCEPT:

(A) the spacer plate.
(B) synchronizers.
(C) the gearshift lever.
(D) shift rails.

9. Differential case gears for front-wheel drive vehicles with transverse engines are typically driven by:

(A) chain drives.
(B) helical ring and pinion gears.
(C) hypoid ring and pinion gears.
(D) worm gears.

10. All of these functions are served by the manual transaxle case EXCEPT:

(A) protecting gears.
(B) aligning moving parts.
(C) enclosing the clutch.
(D) holding transaxle lubricant.

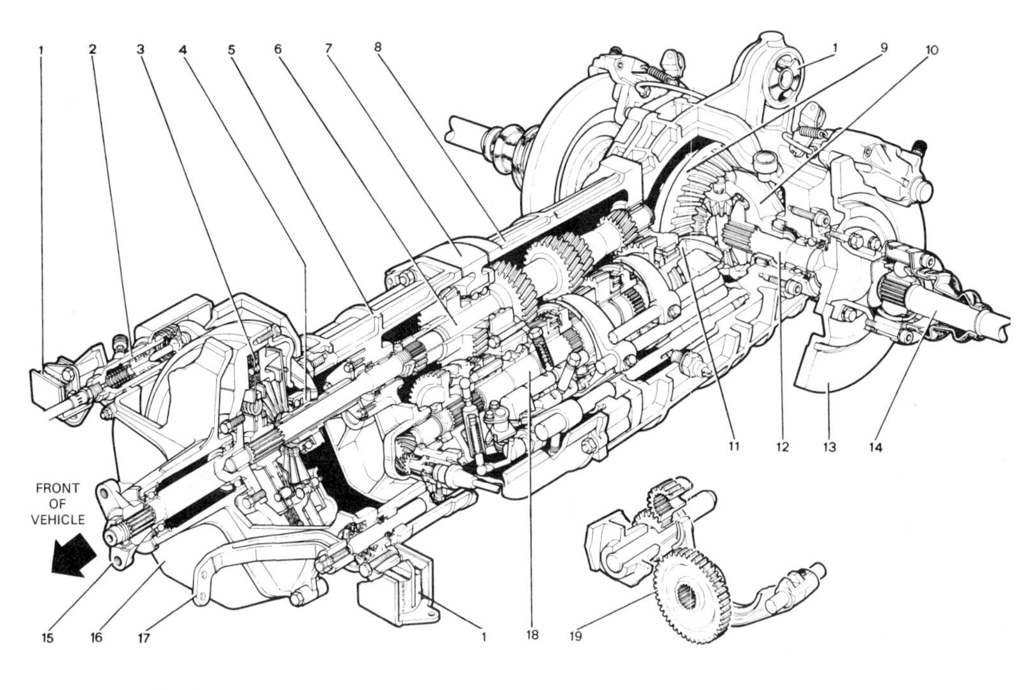

1. ANTI-VIBRATION MOUNTINGS
2. CLUTCH SLAVE CYLINDER
3. CLUTCH PLATE
4. RELEASE BEARING
5. CLUTCH–GEARBOX HOUSING
6. MAINSHAFT
7. INTERMEDIATE FLANGE
8. GEARBOX–DIFFERENTIAL HOUSING
9. CROWN OR RING WHEEL
10. DIFFERENTIAL CASE
11. DRIVE PINION GEAR
12. DIFFERENTIAL OUTPUT SHAFTS
13. BRAKE DISC
14. DRIVE SHAFT
15. UNIVERSAL JOINT YOKE CONNECTING CLUTCH SHAFT TO PROPELLER SHAFT
16. CLUTCH HOUSING
17. REAR GEAR CHANGE LEVER
18. PINION SHAFT
19. REVERSE GEAR COMPONENTS

Figure 18-32. This is a rear-mounted transaxle used on a front-engine vehicle. Power is delivered to the transaxle through a drive shaft ahead of the clutch assembly. (Alfa Romeo)

11. Power flow to the manual transaxle is engaged and disengaged by means of the:

(A) CV axles.
(B) clutch assembly.
(C) drive shaft assembly.
(D) differential assembly.

12. A manual transaxle used with a longitudinal engine will have which of these at the end of the output shaft?

(A) Drive chain.
(B) Reverse gear synchronizer.
(C) Hypoid ring and pinion gearset.
(D) CV axles.

This all-wheel drive vehicle offers superior traction under all conditions. It is equipped with a 5-speed manual transmission. (Porsche)

Chapter 19

Manual Transaxle Problems, Troubleshooting, and Service

After studying this chapter, you will be able to:

- Name and describe common types of problems that can occur in a manual transaxle.
- Remove a manual transaxle from a vehicle.
- Disassemble a manual transaxle.
- Inspect the parts of a manual transaxle.
- Assemble a manual transaxle.
- Install a manual transaxle in a vehicle.

Most cars now come equipped with front-wheel drive, and manual transaxles are common. Manual transaxles come in varying designs–with four, five, or six forward gears. The modern automotive technician should know how to diagnose and repair the different kinds of manual transaxles.

This chapter is a general guide to manual transaxle diagnosis and repair. It is intended to build upon the information you gained in previous chapters. The information in this chapter will help you understand transaxle diagnostic information contained in factory service manuals. If you study the information presented here, you will know the general techniques of servicing all types of manual transaxles. Figure 19-1 shows typical transaxle problems. Study them carefully.

Manual Transaxle Problems and Troubleshooting

If you do not know what is wrong with a transaxle, how can you fix it? You must become skilled at finding out what the problem is before beginning repairs. This section is designed to make transaxle troubleshooting easier.

Since they consist of mechanical parts, most manual transaxles will wear out eventually. Often, this is because of high vehicle mileage. Sometimes, transaxles require repair at low mileage because of careless driving habits. For example, changing gears too quickly, downshifting at high speeds, or releasing the clutch pedal before gears are completely engaged can damage the transaxle. These actions can chip or break teeth on the transaxle transmission gears or damage the synchronizers.

Low fluid level or water contamination can also cause harm to the transaxle. If the transaxle lubricant level becomes too low, moving parts will wear prematurely. If the vehicle is driven through deep water, the water can enter the transaxle through the vent. Water will contaminate the oil. As a result, the gears and bearings will be destroyed through pitting and poor lubrication.

You should always begin diagnosing manual transaxle problems by finding out why the vehicle was brought in for service. Get the customer to describe the transaxle problem. When test driving a vehicle, determine in what gear and at what speed the problem occurs. If possible, take the customer with you and have him or her point out the particular problem while it occurs.

During a road test, try to isolate transaxle problems from CV axle, engine, tire, and other vehicle problems. For example, rumbling or roaring noises that increase and decrease with road speed are usually caused by bad wheel bearings or defective tires. Vibrations that cycle according to vehicle speed are usually caused by the CV axles. Slippage is caused more often by the clutch than the transaxle. Problems that seem to occur in one gear only are usually caused by the transaxle itself.

WARNING! When road testing a vehicle, *always* obey all traffic regulations. Also, try to choose uncongested roads or empty parking lots for testing. Do not become so engrossed in diagnosing the transaxle that you end up having an accident.

Once you have identified the problem, check out the easiest possible solution first; you should always proceed from easiest to hardest. Consider all possible causes, even the most unlikely.

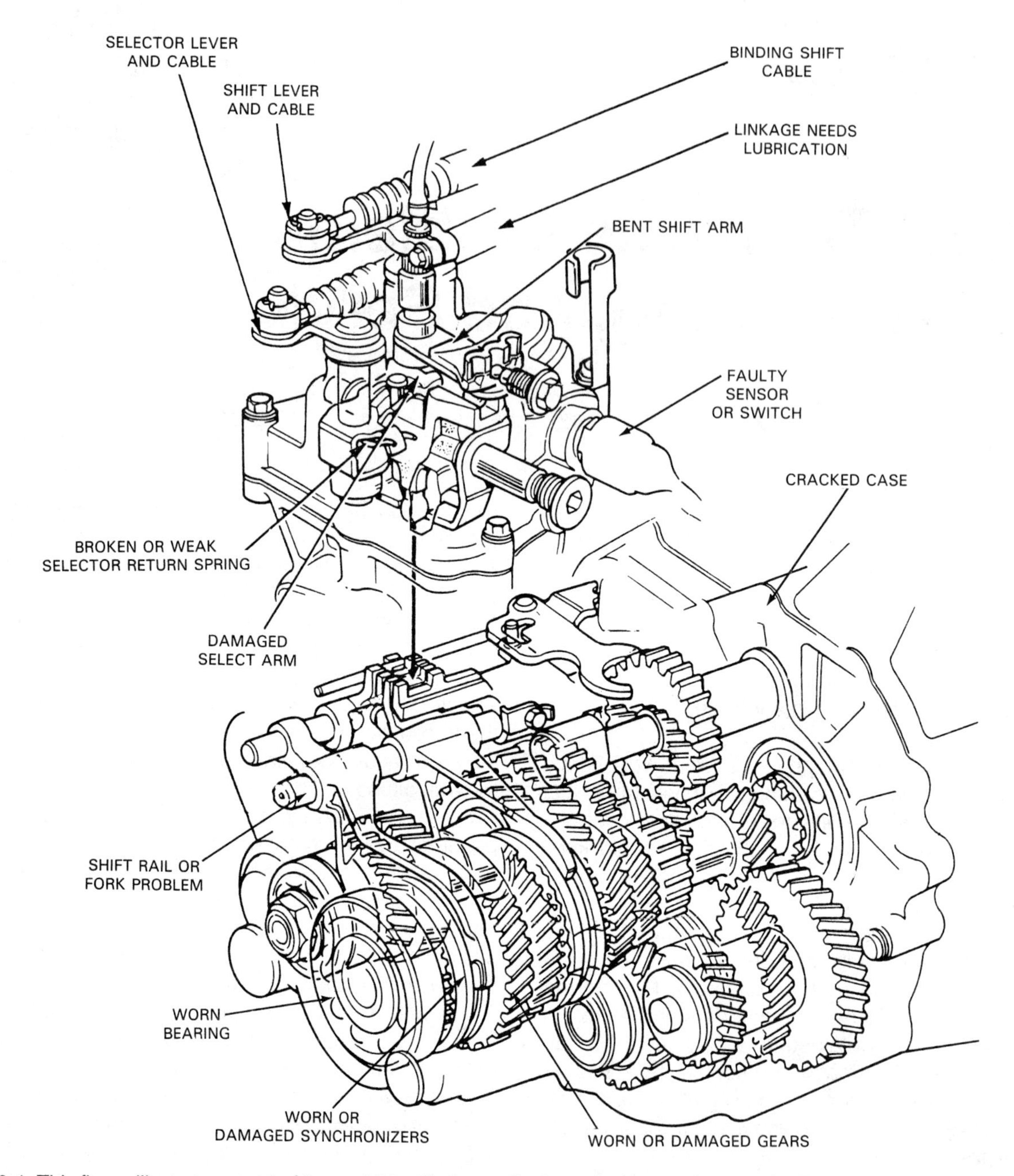

Figure 19-1. This figure illustrates some of the problems that can affect a typical manual transaxle. These problems can be diagnosed using logical troubleshooting techniques. Note selector cable, seen here, transmits right and left movements of gearshift lever to shift forks. Shift cable transmits fore and aft movements of gearshift lever. (Honda)

Many manual transaxle problems can be solved by simple linkage adjustments or by adding transaxle lubricant. Every manual transaxle that is brought in for service should have its clutch and shift linkage adjustments and its fluid level checked. A manual transaxle may have a dipstick or just a fill hole for checking the level. If needed, the proper grade and type of oil should be added, Figure 19-2. If the oil level is low, you should check for leaks.

NOTE! If the transaxle has a separate reservoir for differential lubricant, the fluid level in *it* should also be checked.

Common problems encountered in the manual transaxle are similar to those seen in the manual transmission and rear axle assembly. Following are some of these problems and their possible solutions.

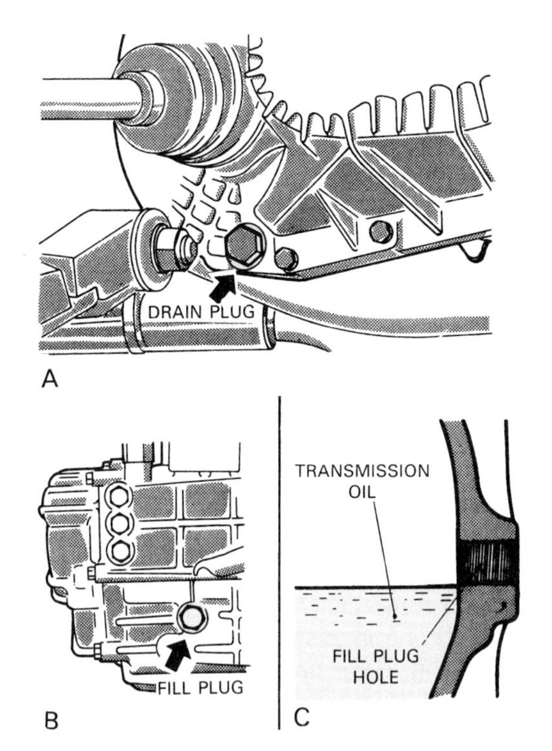

Figure 19-2. Transaxle lubricant levels must be maintained to avoid severe damage. Some transaxles must have lubricant changed periodically. Consult the manufacturer's service manual for instructions on how to check the level and to ensure that you use the right kind of lubricant. A–Typical drain plug location. B–Typical fill plug location. C–Lubricant level should just wet the bottom of the fill plug threads. (Chrysler)

Lubricant Leakage

Lubricant leakage is an unwanted loss of lubricant from the transaxle. Transaxle lubricant levels may drop slightly over long periods of time because of evaporation. If the lubricant level is very low, the transaxle is probably leaking.

Always make sure that the leak is actually coming from the transaxle. Engine oil, brake fluid, power steering fluid, or engine coolant could be dripping down onto the transaxle. If the transaxle case is wet, but its oil level is normal, something else may be leaking.

Once you are sure the leak is coming from the transaxle, check to determine the location and cause. Transaxle leaks are most commonly found at the case and end cover gaskets; sometimes the case or cover has been gouged or otherwise damaged. Other causes of leaks are bad CV axle or shift rail seals, lack of sealer on bolts, loose drain or fill plugs, and cracks in the transaxle case. Leaks are often caused by a high lubricant level in the transaxle. High fluid level can cause oil to leak from good seals or from the transaxle vent.

If a seal or gasket is leaking, or if defects in other transaxle parts are causing the leak, the transaxle may have to be removed to gain access to the parts. The CV axles can usually be removed to replace the axle seals without removing the transaxle.

No-Drive Condition

A *no-drive condition* exists when engine power cannot be transmitted through the transaxle. The most common cause of a no-drive condition is a broken shaft or stripped gears on the input shaft. Sometimes, the transmission will be locked in two gears and will not be able to move. This is often a linkage problem. The linkage may be broken or disconnected and, therefore, may be unable to engage the gears, or there may be a problem with the linkage interlock.

If a shaft is broken or a gear is stripped, the engine will run easily with the clutch engaged in the particular gear. If the transmission is locked, engaging the clutch will kill the engine. Most no-drive problems will require the removal and disassembly of the transaxle.

Hard Shifting

Hard shifting occurs when the gearshift lever is difficult to move from gear to gear. Sometimes, the problem occurs in only one gear. In most cases, however, all gears will be difficult to select. The usual causes of hard shifting are bent or worn external shift linkage or linkage in need of lubrication or adjustment. Hard shifting may also be caused *inside* the transaxle by binding shift rails or forks. Some hard shifting problems are caused by clutch problems.

Sometimes, hard shifting results when the transaxle case and engine are misaligned or when the engine or transaxle mountings are misaligned. Most front-wheel drive vehicles have at least one mount from the top of the engine to the body. This mount very commonly wears out, becomes loose, or even breaks, causing hard shifting.

Hard shifting can be diagnosed easily. If the gearshift lever is hard to move even when the engine is stopped, the linkage is probably the source of the problem. If the gearshift lever moves easily when the engine is stopped, but becomes hard to move when the engine is running, the problem may be in the clutch or the transaxle synchronizers.

In many cases, hard shifting can be reduced or corrected by adjusting or lubricating the shift linkage. The shift linkage is a common source of trouble on transaxles, since the linkage must be long to connect the gearshift lever with the transaxle case. If internal parts of the shift linkage are bent or sticking, the transaxle must be removed and disassembled to make repairs.

Synchronizer or clutch problems will require transaxle removal. If the transaxle and/or engine are misaligned, it may be possible to realign them without removing the transaxle.

Gear Disengagement

Gear disengagement is a condition in which transaxle transmission gears and synchronizers disengage on their own, causing a transaxle to come out of gear. The problem may be more specifically one of gear jumpout or gear slipout. The two conditions are closely related. With gear jumpout, the components are fully engaged, but something happens to force them out of engagement, into neutral. In the case of gear slipout, the gears and synchronizers are never quite fully engaged (but should be). When this happens, gear rotation tends to push the synchronizer outer sleeve and mating gear out of engagement.

Jumpout generally occurs when some outside force overcomes spring pressure in one of the detent holding mechanisms of the shift linkage. Often, there is a problem with the detent mechanism: the springs may be broken, or the detent notches may be worn. The problem could be in the shift rails of an internal shift linkage, for example, or it could be in the gearshift lever assembly. If the problem is a detent, the driver may make a false assumption and relate that the problem occurs in high gear only. The driver may only notice the problem in high gear because this is the only gear in which he or she is not holding on to the gearshift lever.

There are a number of conditions that can cause gear slipout. The first step in finding the cause of the problem is to determine which gear(s) are slipping out of engagement.

If the transmission slips out of a reduction gear, begin by checking for problems in the linkage. It could be worn or improperly adjusted. There may be interference between the shift linkage and some other part of the vehicle, or the shift linkage may be binding for some other reason, such as from a tight shift lever seal on the transaxle. These actions can limit travel of the shift forks and prevent the synchronizers and mating gears from fully engaging. Linkage interference is often the cause if the problem occurs only when the vehicle is in a gear (high gear included) and accelerated.

If the linkage is alright, check for chipped or broken gear teeth. Defective teeth may be causing the problem (or they may be the result of something else that is causing the problem). Check the splines on the synchronizers. Worn or damaged splines on the outer sleeve or inner hub could be causing slipout. Also check the pilot and input shaft bearings and bearing retainer. A worn or damaged pilot bearing or an input shaft bearing or retainer that is broken or loose can cause enough gear movement to push the sleeve and gear apart.

If the transaxle slips out of high gear *only,* begin by checking for loose transaxle case-to-engine bolts, which hold the transaxle case to the engine. Also, check the engine mounts, especially the top mount. Often, a worn top engine mount will not be obvious, and it will require removal to make sure that it is not defective. Another cause is a misaligned transaxle case, or sometimes, the entire engine and transaxle assembly are misaligned. Other causes include worn or binding shift linkage.

At times, jumpout and slipout problems can be corrected without removing the transmission. Often, problems of external shift linkage can be corrected by lubrication or adjustment. Sometimes, however, linkage parts are too worn to be corrected by adjustment and must be replaced. When the problem is caused by worn or damaged internal transmission parts, the transmission must be removed and disassembled.

Gear Clash

Gear clash is a grinding noise that is heard when a gear is first engaged. For this problem, always check the clutch adjustment and condition before working on the transaxle. A clutch that does not fully disengage is the most common cause of gear clash, especially if the clash occurs in every gear. If the clutch checks out alright, or if the clashing occurs only in one gear, check for worn transaxle synchronizers. In a few cases, loose mounting bolts or misalignment will cause gear clash.

Abnormal Noise

An *abnormal noise* in the manual transaxle is a noise above normal sound level during vehicle operation. Noises that are most noticeable may sound like whining or rumbling and may change from gear to gear, varying directly with the reduction ratio. Low gear, with its higher gear ratio, for instance, will usually be more noisy than high gear.

Front-wheel drive vehicles normally have higher noise levels than rear-wheel drive vehicles, since the moving parts are mostly concentrated at the front. Be sure to eliminate the engine, clutch, CV axle, and wheel bearings as possible causes of the problem. If the noise occurs in one gear only or in neutral with the vehicle stopped, you can safely assume that the problem is in one of the gears or synchronizers in the transaxle.

If the transaxle is noisy in every gear, including neutral, you should first check for a low oil level. Other defects that could cause noise in every gear (and sometimes neutral) are worn or damaged transaxle transmission gears. Rumbling noises are often caused by worn or damaged bearings on the input or output shafts or in the differential assembly. Noise in every gear when moving may be caused by a defective speedometer gear.

A clunking or slapping noise that is heard when the engine is accelerated or decelerated may be caused by excessive backlash in the moving parts, most commonly in the differential unit. This can be due to worn or missing thrust washers, badly worn gears, worn or misaligned bearings, or loose bolts. Be sure to check the clutch, engine mounts, and CV axles before removing the transaxle. Clunking noises are much more likely to be coming from the CV axles than the transaxle.

Correcting most noises will require removing the transaxle to gain access to defective gears or bearings. In some cases, such as with loose bolts or misalignment, the problem can be fixed without removing the transaxle.

WARNING! Always keep safety uppermost in your mind when checking for transaxle problems. Be cautious around rotating parts when testing the transaxle with the vehicle on a lift. The CV axles and tires rotate at high speeds and can cause severe injury.

Always support the lower control arms so the front wheels are at or near their normal position in relation to the vehicle body. Operation of the CV axles at excessive angles by allowing them to hang will cause severe CV joint damage and possible injury to anyone near the vehicle. When testing a front-wheel drive vehicle on the lift, never exceed 30 mph (48 km/h).

In-Car Manual Transaxle Service

Do not remove the transaxle until you are sure that it must be removed. In many cases, the transaxle can be serviced without removing it from the vehicle. For instance, the shift linkage, axle seals, and speedometer gears can be serviced without need of removing the transaxle.

External shift linkage is one area of the manual transaxle that is vital to proper operation and that can be repaired without transaxle removal. Procedures for adjusting transaxle shift linkage are similar to procedures for adjusting linkage on rear-wheel drive manual transmissions. Front-wheel drive vehicles always use a floor-mounted gearshift lever, and there are no column-shift mechanisms to deal with. The floor-mounted shift linkage is usually simple and direct. The linkage, however, is usually not contained inside of the transaxle in the manner of many rear-wheel drive transmissions, since the transaxle is located well ahead of the passenger compartment.

Many transaxles use cables to connect the shift mechanism of the gearshift lever assembly to the transaxle gears. Most cable adjustment mechanisms are located at the gearshift lever assembly. Adjustments can be made by removing the console or shift boot to reach the adjustment mechanism. If the vehicle uses a shift rod mechanism, the adjustment must usually be made under the vehicle or at the transaxle under the hood.

Manual Transaxle Removal

If the transaxle has internal problems, such as defective gears, bearings, or shafts, it will generally have to come out of the car. Note that in some instances, however, certain internal parts can be serviced on the vehicle by removal of a sheet metal or aluminum access cover.

Most manual transaxles can be removed from underneath the vehicle without removing the engine. A few vehicles are designed so that the transmission and engine must be removed as a unit. On others, underhood clearances are so close that it is easier to remove them this way. In this case, you must have a good engine hoist and know how to use it.

Following are some *general* procedures detailing how to remove a manual transmission from a vehicle. The removal process presented in this chapter assumes that the transaxle can be removed without removing the engine. Keep in mind that the removal process is much different from that used on a rear-wheel drive transmission. Note that you should always refer to the manufacturer's service manual for specific procedures.

1. Disconnect the battery negative cable to prevent shorting wires or accidentally operating the starter.
2. While under the hood, also remove the upper transaxle case-to-engine bolts, clutch and shift linkages, electrical connectors, and ground straps. In many cases, it will be necessary to install an engine holding fixture to remove the engine weight from the lower engine and transaxle mounts. A typical engine holding fixture is shown in Figure 19-3.
3. Raise the vehicle to get enough clearance to remove the transaxle. If you do not have access to a hydraulic lift, always use a good hydraulic floor jack and approved jackstands.

WARNING! Never support the vehicle with wooden crates or cement blocks. Be sure the vehicle is secure before working underneath it. Also, do not allow the CV axles to hang.

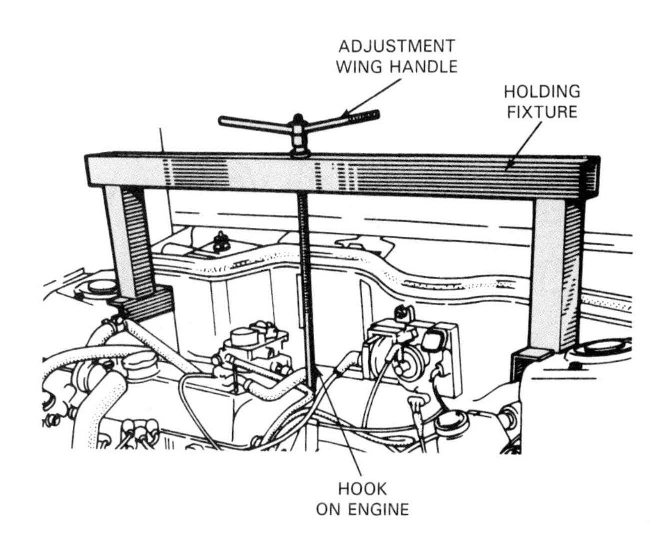

Figure 19-3. This engine holding fixture can be used to remove engine weight from the engine and transaxle mounts. (Chrysler)

4. Check the engine mounts to ensure that they are not broken. Broken mounts could be the cause of shifting or clutch problems.
5. Remove the splash shields and any other hardware that will be in the way during CV axle or transaxle removal. Then, remove or disconnect one of the CV axles. In some cases, it is possible to remove the CV axle from the transaxle without removing it completely from the vehicle, as in Figure 19-4. Be sure to securely wire the axle in place to prevent damage. CV axle removal is covered in detail in Chapter 23.

CAUTION! Be sure to remove the CV axles in the proper manner, or the axles or other vehicle parts can be damaged.

6. Once one of the CV axles is removed, allow the transmission oil to drip into a drip pan. Then, remove the other CV axle from the transaxle. Plug the CV axle openings to reduce oil leakage.
7. Remove the speedometer cable assembly.
8. Remove the dust cover in front of the clutch and flywheel. Remove the starter, if it is bolted to the transaxle clutch housing.
9. Remove the exhaust pipes if necessary to gain clearance for transaxle removal. Also, on some vehicles, it may be necessary to unbolt and swing out the front frame extension or engine cradle on one side to remove the transaxle. On other vehicles, some suspension parts must be removed to gain enough clearance to lower the transaxle. Consult the specific service manual to be sure that you perform this procedure properly.
10. If an engine holding fixture is not already in place, support the engine with a jack placed under the rear of the engine.
11. Loosen the remaining transaxle case-to-engine bolts.

WARNING! Always leave at least two bolts holding the transaxle to the engine until you have a transmission jack under the transaxle. Although most transaxle cases are aluminum, they are heavy. You should always use a transmission jack to support and lower the transaxle.

12. Place the transmission jack under the transaxle and raise it slightly to remove transaxle weight from the lower transaxle mounts. See Figure 19-5. Then, remove the lower transaxle mounts and any other parts holding the transaxle to the vehicle body.
13. Remove the last bolts holding the transaxle to the engine. Raise or lower the transmission jack until there is no binding between the splines on the transaxle input shaft and the clutch disc.

CAUTION! Do not let the transaxle weight rest on the clutch assembly. This will damage the clutch disc hub and possibly the transaxle input shaft.

14. Slide the transaxle away from the engine, until the input shaft clears the clutch disc. Then, lower the transaxle, watching clearances, which are usually close on front-wheel drive vehicles. Also, watch for any component that you may have forgotten to disconnect, especially any electrical connectors, ground wires, or other wiring connected to the transaxle.

    As soon as the transaxle is clear of the vehicle, lower it to the floor and transport it to the workbench

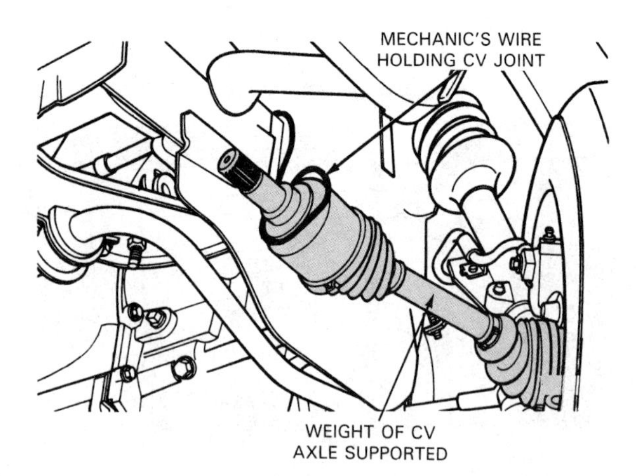

Figure 19-4. The CV axles are usually removed from the vehicle to remove the transaxle. A few can be removed from the transaxle without being disconnected at the wheels, as shown here. The CV axle should be tied to the vehicle body to prevent excessive CV joint angles and damage. (Ford)

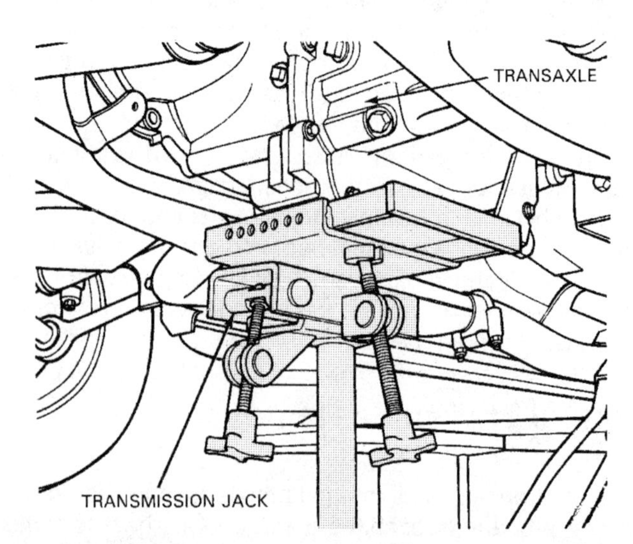

Figure 19-5. A transmission jack must be used to remove the transaxle. Always secure the transaxle to the jack before removing the final transaxle case-to-engine bolts. (Ford)

in the lowered position. If the transaxle is very dirty, you may want to remove some of the outside dirt before beginning disassembly.

Manual Transaxle Disassembly

Once the transaxle is removed, it can be disassembled for cleaning, inspection, and repair. The procedure outlined here is general and does not specifically cover any one transaxle. You should always consult the manufacturer's service manual for exact disassembly procedures and specifications. In addition, the tools needed for servicing vary with each transaxle. These tools are covered in detail in the service manual. A typical set of special service tools is shown in Figure 19-6.

If possible, mount the transaxle in a special holding fixture before beginning disassembly. If a holding fixture is not available, make sure that the transaxle is securely mounted on the workbench. One way to secure the transaxle is to tighten one of the projecting parts of the case in a vise, letting the rest of the transaxle weight rest on the workbench.

Removing External Components

Start the disassembly by removing the external parts of the transaxle, such as the speedometer driven gear, Figure 19-7. If the clutch throwout bearing and clutch fork were removed from the vehicle with the transaxle, remove the throwout bearing from its hub and from the clutch fork; then, remove the clutch fork from its installation. Remove any sheet metal or cast aluminum covers.

Before taking the case apart, shift the transaxle through its gears. This will make it easier to spot any defects in the shift linkage. Then, take the case apart. Many transaxle cases are made in two halves or as a center support with end covers. Remove the through bolts and pry the case halves apart. Sometimes, the case halves can be separated by lightly tapping them with a soft-faced hammer, Figure 19-8.

Note that some cases contain ball bearings or tapered roller bearings to support the shafts. The shafts may be held to the bearings with large nuts that resemble rear axle pinion nuts. These nuts must be removed before the case can be removed. Some retaining nuts have left-hand threads. Make sure that the bearings are not damaged during disassembly.

NOTE! As you disassemble the transaxle, carefully note the position of all parts. If necessary, mark mating parts to ensure proper reassembly.

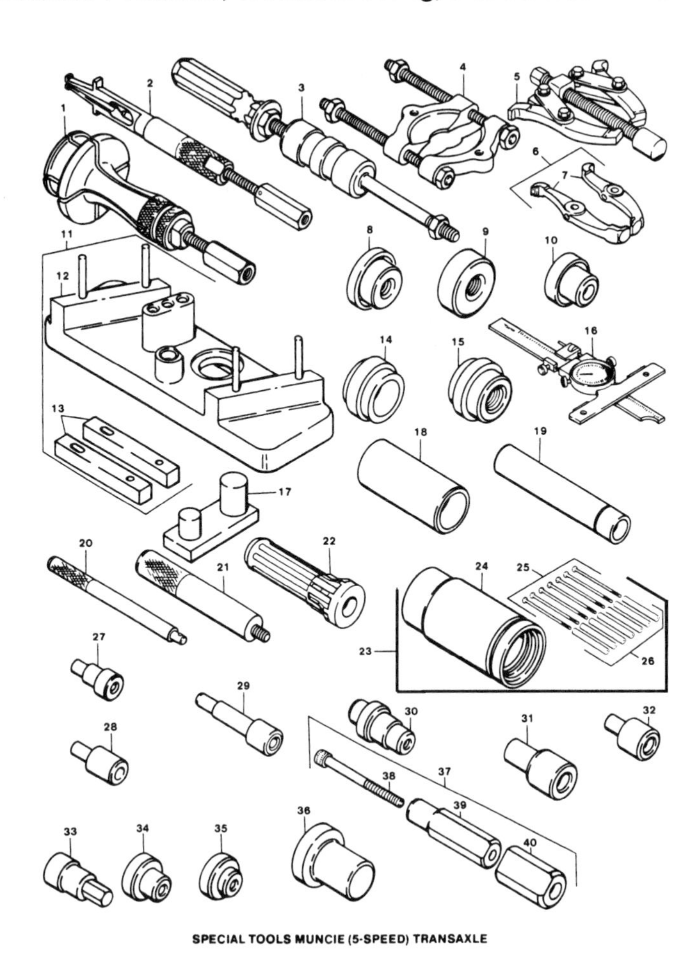

SPECIAL TOOLS MUNCIE (5-SPEED) TRANSAXLE

TOOL NAME	ITEM NUMBER
UNIVERSAL DRIVER HANDLE	21
SIDE BEARING PULLER ADAPTER, DIFFERENTIAL	34
BEARING REMOVER	5
BEARING REMOVER LEG (TWO)	7
BEARING REMOVER LEG SET	6
INPUT/OUTPUT SHAFT GEARS REMOVER	4
DIFFERENTIAL INNER BEARING INSTALLER	10
DIFFERENTIAL AND OUTPUT SHAFT BEARING CUP INSTALLER	15
SIDE BEARING PULLER ADAPTER, DIFFERENTIAL	35
SLIDE HAMMER AND ADAPTER SET	3
METRIC DIAL DEPTH GAGE	16
SHIM SELECTION SET	23
DIFFERENTIAL SHIMMING GAGE	24
SHIM SELECTION SET SPACERS (SEVEN)	26
BOLTS, M8 x 1.25-6G / LENGTH - 160 mm (SEVEN)	25
SEAL AND RACE, DIFFERENTIAL, INSTALLER	8
SHIFT SHAFT SEAL INSTALLER	22
INPUT BEARING ASSEMBLY REMOVER/INSTALLER/SHAFT SUPPORT BEARING INSTALLER	36
SHIFT SHAFT BEARING REMOVER	27
SHIFT RAIL BUSHING REMOVER/INSTALLER	37
ADAPTER TO DRIVE HANDLE	40
INSTALL/REMOVER ADAPTER	39
CAP SCREW, 1/4-20 x 2 1/2	38
REVERSE SHIFT RAIL BUSHING INSTALLER	28
BEARING RETAINER BOLT HEX SOCKET	33
CLUTCH SHAFT INNER REVERSE SHIFT RAIL BUSHING REMOVER	2
CLUTCH SHAFT INNER BUSHING INSTALLER	31
SLIDING SLEEVE BUSHING REMOVER/INSTALLER	29
CLUTCH SHAFT UPPER BUSHING REMOVER/INSTALLER	19
OUTPUT SHAFT BEARING CUP REMOVER	1
SHIFT DETENT LEVER BUSHING REMOVER/INSTALLER	30
DIFFERENTIAL BEARING CUP REMOVER	9
INPUT/OUTPUT SHAFT REMOVER/INSTALLER ASSEMBLY SET	11
PALLET	12
DISASSEMBLY ADAPTER (TWO)	13
INPUT SHAFT REMOVER/INSTALLER PRESS TUBE	18
OUTPUT SHAFT REMOVER/INSTALLER ADAPTER PRESS TUBE REDUCER	14
INPUT/OUTPUT SHAFT BEARING REMOVER	17
SHIFT SHAFT BEARING INSTALLER	32
UNIVERSAL DRIVER HANDLE	20

Figure 19-6. The special service tools shown here are used to repair a common make of manual transaxle. The type of tools needed for the job will vary with the type of transaxle. (General Motors)

Once the case has been taken apart, the transaxle internals can be inspected for obvious damage. Figure 19-9 shows a common transaxle gearset as it appears after the upper half of the case has been removed.

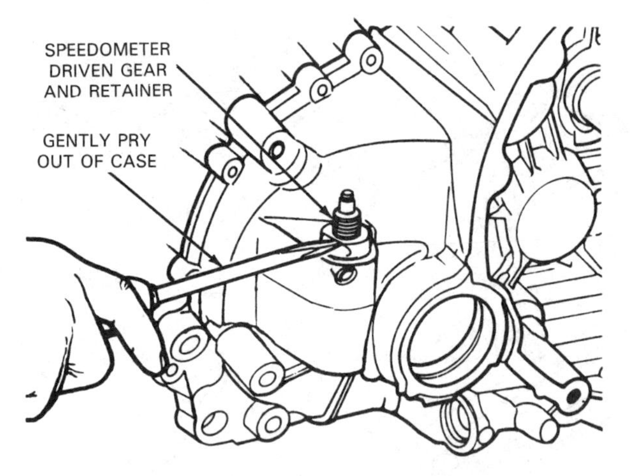

Figure 19-7. Remove all external parts before opening up the transaxle case. Many parts are connected to the transaxle shafts and gears, and they will be damaged if not removed beforehand. (Ford)

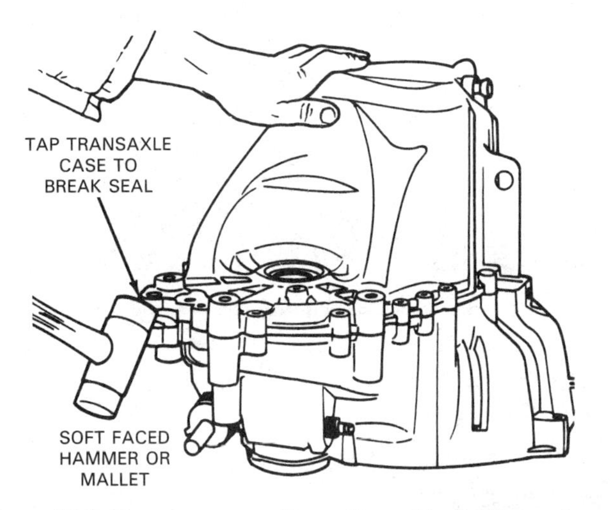

Figure 19-8. Many cases must be split apart by lightly tapping on them after the retaining bolts are removed. Some cases are two-piece units, while others consist of a central support housing with two outer ends. (Ford)

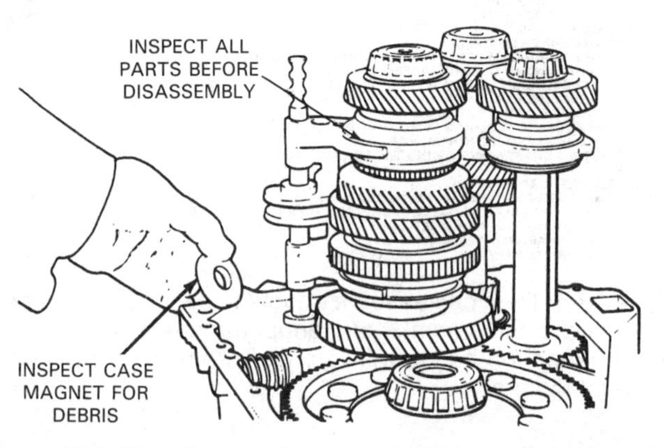

Figure 19-9. After the case is removed, the internal parts can be inspected and removed. All internal parts should be carefully checked for wear and damage before disassembly. (Ford)

Removing Internal Components

Transaxles are often easier to disassemble than manual transmissions, since most moving parts are revealed once the case halves are disassembled. To begin removing internal parts, you must often remove the snap rings that hold gears to their shafts. The gears can then be removed. In some instances, certain gears are held in place by special fasteners. Figure 19-10 shows the methods necessary to remove gears pressed to a shaft or shafts held in the case by a retaining bolt.

The differential case assembly can also be removed when the transmission internal parts are removed. In some designs, it it removed first; in others, it is removed following the transmission internal parts. Figures 19-11 through

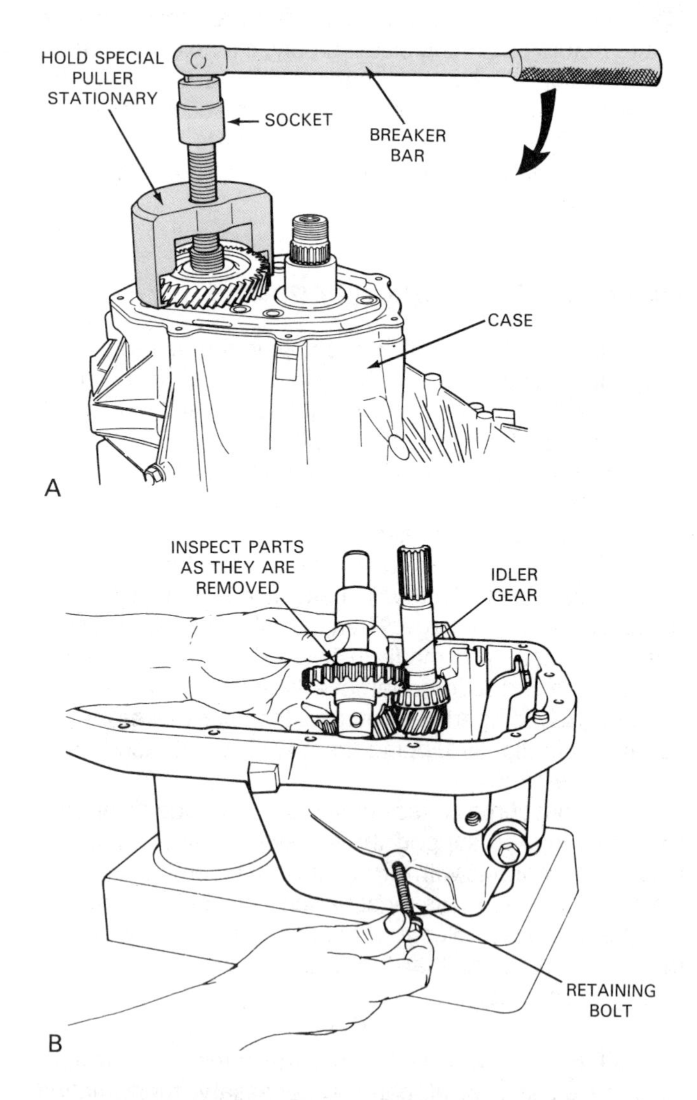

Figure 19-10. Transaxle shafts and gears are installed by many methods–some simple, others more unusual. Removal methods vary accordingly. A–Pressed-on gears may be removed with a special puller. B–Removing a special retainer bolt will release a shaft and gear in this transaxle. (General Motors)

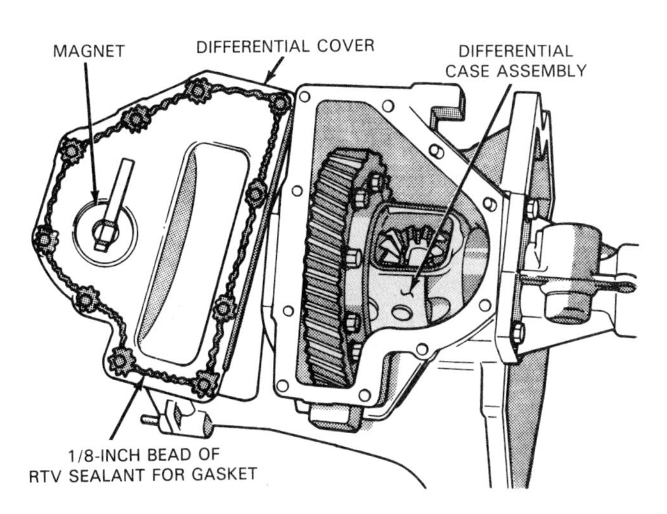

Figure 19-11. The transaxle differential case assembly is often installed under a special cover, as shown here. It must be removed to gain access to this differential unit. The cover contains a magnet to pick up metal particles. (Chrysler)

19-13 show removal of a differential case assembly from a common make of transaxle.

Most differentials are simple units, using helical gears for the ring and pinion. If the transaxle is used with a longitudinal engine, it probably has a hypoid ring and pinion. This type of differential assembly is overhauled in the same manner as one on a rear-wheel drive vehicle. The differential used with the longitudinal engine is in its own housing, separate from the transmission, and it does not require disassembly unless it, too, is defective. Many hypoid-type differentials used with front-wheel drive vehicles develop case cracks, so they must be checked closely.

Disassembling the Components

The next step in transaxle overhaul is to disassemble the individual components. Disassembly techniques are similar to those used to overhaul manual transmissions.

Some transaxle bearings are pressed onto shafts and must be removed by using a press and the proper adapters, Figure 19-14. In general, do not remove the bearings unless they are going to be replaced. Many tapered roller bearings used on transaxles have outer races, or bearing cups, placed in the transaxle case. These should not be removed unless the bearing is replaced. Always mark the bearing cups if they are removed. This will ensure that they are reinstalled in their original positions, Figure 19-15.

Gears may need to be pressed off their shafts, while other gears are integral with their shaft. Most gears are held onto their shaft by snap rings. Always pay particular attention to the synchronizer blocking rings during gear removal, to prevent damage to the soft brass. The gears and synchronizers should be laid on the workbench in order of

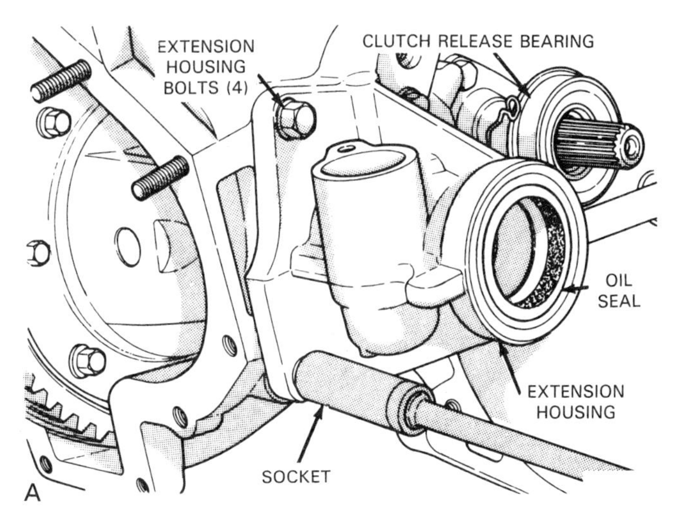

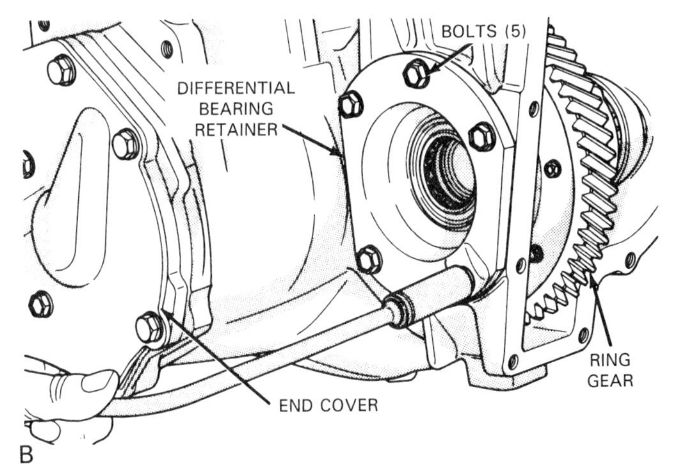

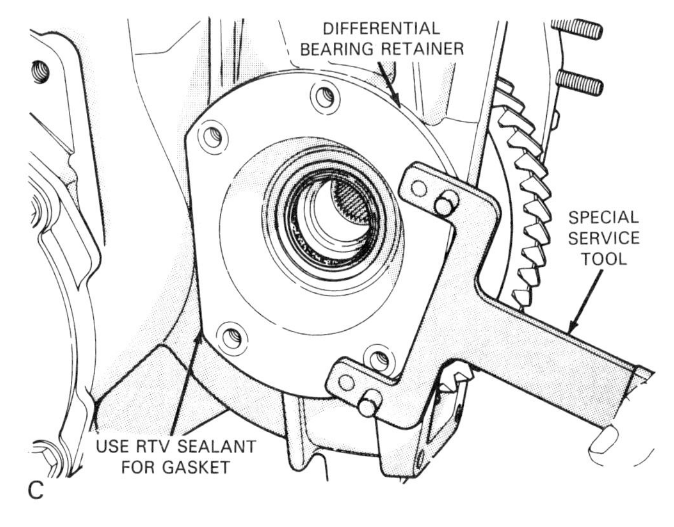

Figure 19-12. The differential assembly can be removed by following the factory service procedure. A–Removal of a CV axle extension housing is shown here. B–After the extension housing is removed, the differential bearing retainer bolts can be accessed. C–With bolts removed, the bearing retainer is then rotated to remove it. (Chrysler)

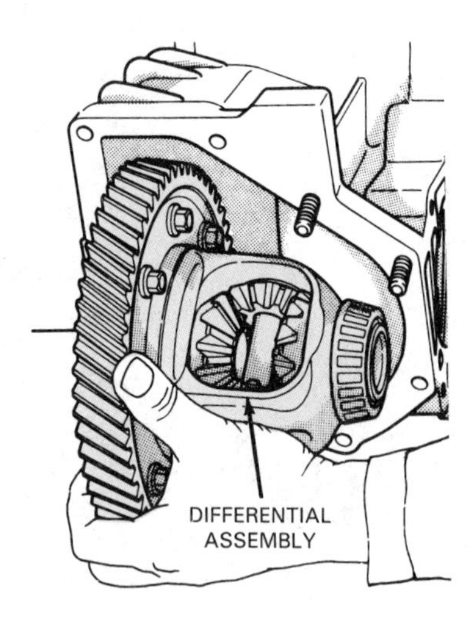

Figure 19-13. After the differential bearing retainer is removed, the differential case assembly is lifted out through the differential cover opening. (Chrysler)

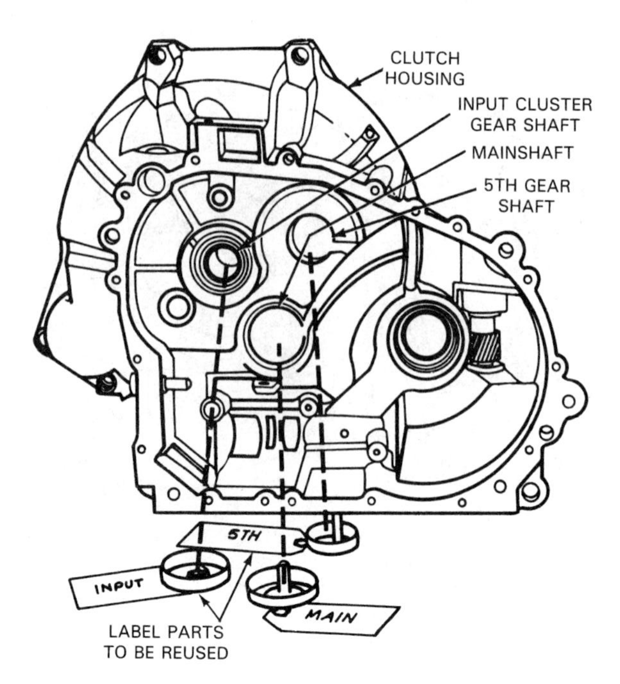

Figure 19-15. Various bearing cups placed in the transaxle case are shown here. Do not mix up any bearing cups that are the same size, since bearings and cups have worn in to fit each other. (Ford)

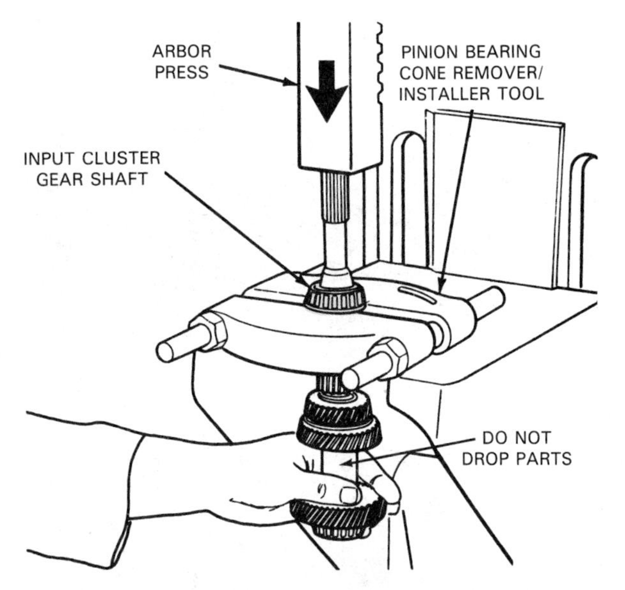

Figure 19-14. Some transaxle bearings are pressed onto the shaft that they support. The bearings must be removed with a press, as shown here. (Ford)

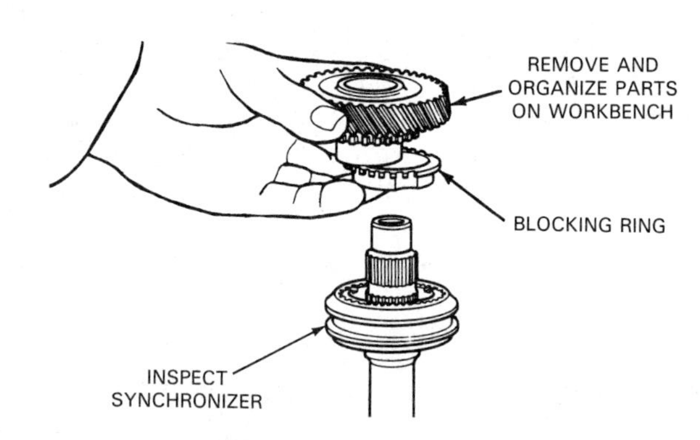

Figure 19-16. Carefully remove the synchronizers and gears from their shafts, and keep all parts in order on the workbench. Check parts for wear and damage as you disassemble them. (Ford)

disassembly. See Figure 19-16. Careful disassembly of parts will make reassembly easier. Inspect all parts as you disassemble them. This will enable you to determine which parts will require replacement.

Other parts that require disassembly are the internal shift linkage and, also, any bearings that were not removed in previous steps. Damaged shift forks or shift rails may require further disassembly, Figure 19-17.

The differential case assembly should also be disassembled for inspection. If necessary, the differential side bearings can be removed by using the proper puller, Figure 19-18. Since the removal process usually damages the bearings, they should not be removed unless they are going to be replaced.

Transaxle seals should be removed from their housings once all other parts are removed. Most seals can be pried from their housings with a screwdriver. Be careful when removing seals not to nick or dent any gasket sealing surfaces or seal housings.

Manual Transaxle Parts Inspection

Almost any transaxle part can wear out or become defective in some way. You should check every part for possible problems. Begin by thoroughly cleaning the

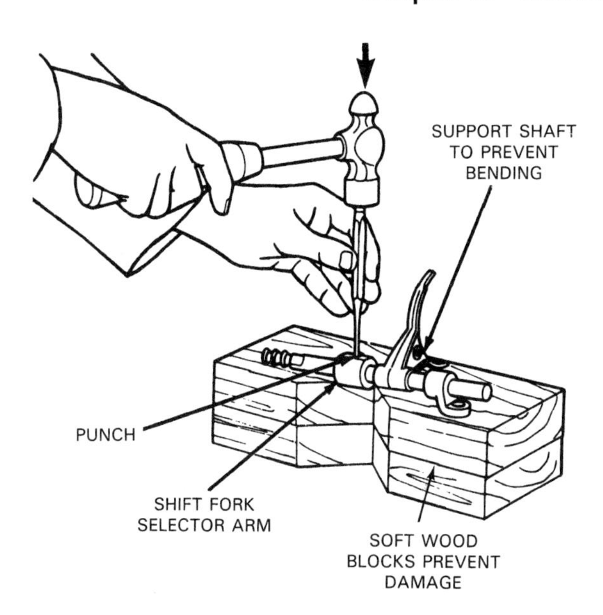

Figure 19-17. Most shift forks are attached to the shift rails with retaining pins. It is necessary to tap out the pin to disassemble the parts. (Ford)

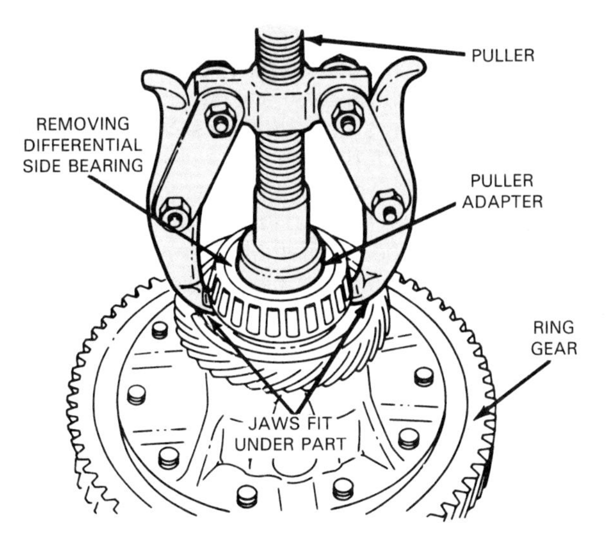

Figure 19-18. The differential side bearings are pressed into place, just as on a rear-wheel drive differential. The bearings must be removed with a puller, as shown here. Do not remove the bearings unless they are being replaced. (General Motors)

transaxle case halves and all internal parts with fresh cleaning solvent. When cleaning the cases, make sure that no small parts, such as needle bearings, are lost. If you clean any parts with water, dry them thoroughly with compressed air and coat them with oil.

While inspecting the transaxle, you should also check related parts that may still be attached to the vehicle. Examples are the external shift linkage, clutch disc and pressure plate assembly, flywheel, crankshaft pilot bearing, and CV axles.

Visual Inspection

Visual inspection can be performed to uncover any obvious damage. Since transaxles have smaller gears and shafts than most manual transmissions found on rear-wheel drive vehicles, they are more likely to become worn or damaged. You should check shaft splines and snap ring grooves carefully, Figure 19-19. Any signs of wear on these areas mean that the shaft should be replaced.

Check the synchronizers for wear or broken teeth and check the grooves in the outer sleeves to look for wear caused by contact with the shift fork. Check the gears for worn, cracked, or chipped gear teeth, Figure 19-20. If any teeth on a gear are defective, the entire gear should be replaced. Also, inspect the transaxle drive chain for wear.

Inspect all ball and needle bearings for wear or looseness. The transaxle bushings and thrust washers should also be checked for wear and scoring. Do not forget to check the condition and fit of all bearing surfaces, where they contact and turn against their supporting shaft. See Figure 19-21.

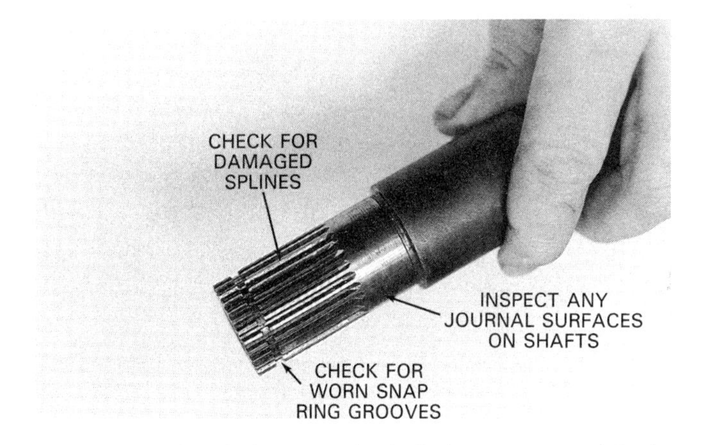

Figure 19-19. Check the transaxle shafts for wear, especially the splines and snap ring grooves. (Chrysler)

Figure 19-20. Check gears for damage at the teeth. This gear has both drive and synchronizer teeth. Both sets of teeth should be carefully checked. (Chrysler)

The differential will require inspection when the transaxle is overhauled, unless contained in a separate housing. The differential gears should be checked for wear or damage, and the differential case should also be checked for problems. If necessary, the ring gear can be removed from the differential case, Figure 19-22. Any damaged or worn differential parts should be replaced. Replace seals or gaskets used for separating differential lubricant from transaxle lubricant.

The transaxle case and all other stationary parts should be checked at this time for cracks and any distortion, which may cause leaks, poor sealing surfaces, or misalignment of the moving parts. Also, check the clutch throwout bearing and any clutch linkage that is attached to the transaxle case.

Figure 19-21. Bushing surfaces should be checked for wear. Replace any gear with worn out bushing surfaces, even if the teeth are in good condition. (Chrysler)

Measurement Inspection

Many transaxles have parts that should be checked with special measuring instruments. Feeler gauges, micrometers, sliding calipers, and dial indicators are often necessary to identify defective parts that would otherwise be unnoticed. Use of these instruments will ensure a good repair job. Instruments of measurement are covered in Chapter 2. Measurement checks performed are similar, in many cases, to those for overhauling manual transmissions and differential assemblies of rear-wheel drive vehicles. Consult Chapters 9 and 17 for typical examples.

Manual Transaxle Assembly

This section will give general details of manual transaxle reassembly. Before beginning reassembly, make sure that you have the correct replacement parts by comparing them to the old parts. Always use new gaskets and seals and lubricate all of the moving parts before reassembling them.

Install the differential case assembly–before or after assembling the transaxle transmission, as required. To put the transmission back together, reassemble all of the transmission gears, synchronizers, and shafts. Once assembled, ensure that synchronizers work properly. Reassemble the shift forks and shift rails and install them on the input and output shaft gear assemblies; then, place the entire assembly in position in the case, Figure 19-23.

Note that some gear assembly parts may need to be installed individually into the case, as shown in Figure 19-24. Some shift forks, for example, are installed after the gears are in place. In this instance, be sure to properly align them before tightening the set screw, as in Figure 19-25.

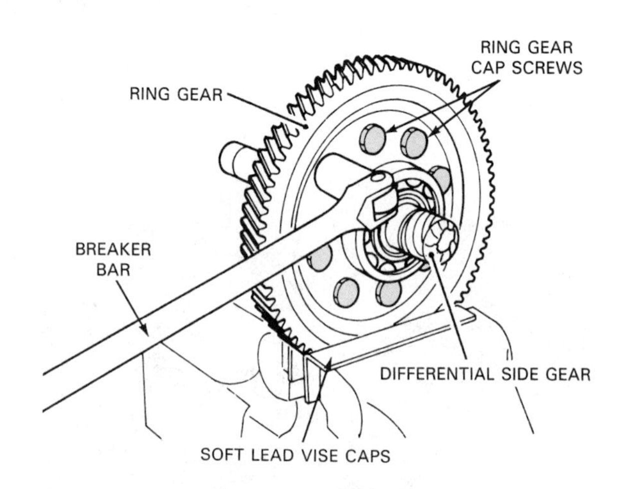

Figure 19-22. The differential case and ring gear can be taken apart by removing the cap screws. This should not be done unless one of the components must be replaced.

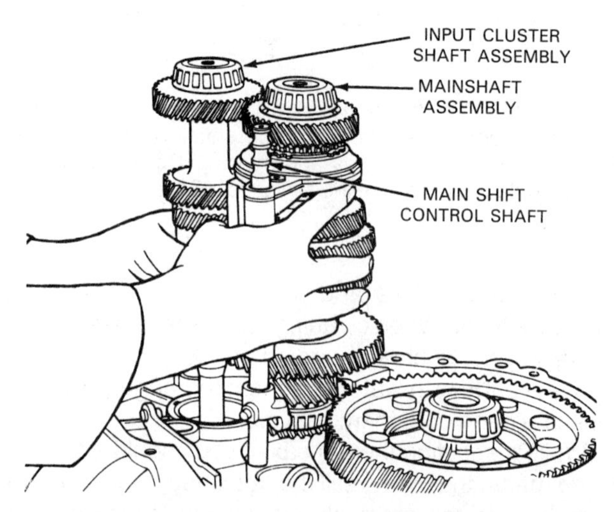

Figure 19-23. Shift forks and rails and input and output shaft gear assemblies are lowered into position as a unit. Be sure to align all parts before installation. (Ford)

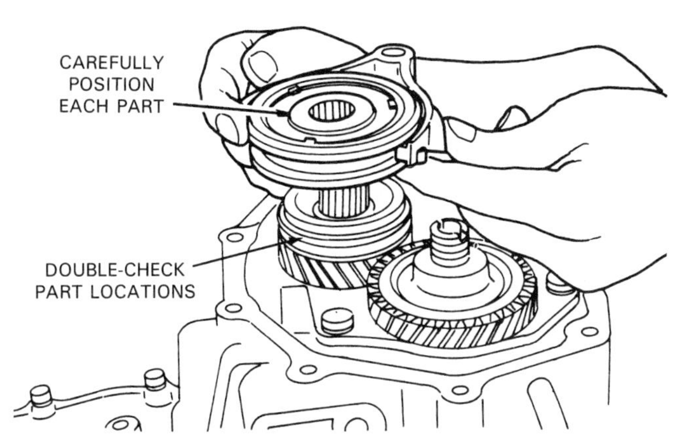

Figure 19-24. Some transaxle shaft components must be reassembled with the shaft installed in the case. You should always recheck part positions before completing reassembly.

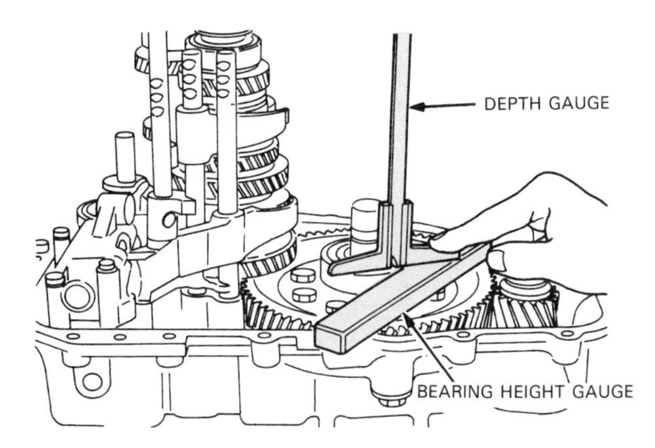

Figure 19-26. End play and other adjustments are critical on manual transaxles. Always take measurements according to the directions given in the service manual and correct conditions that are not to specification.

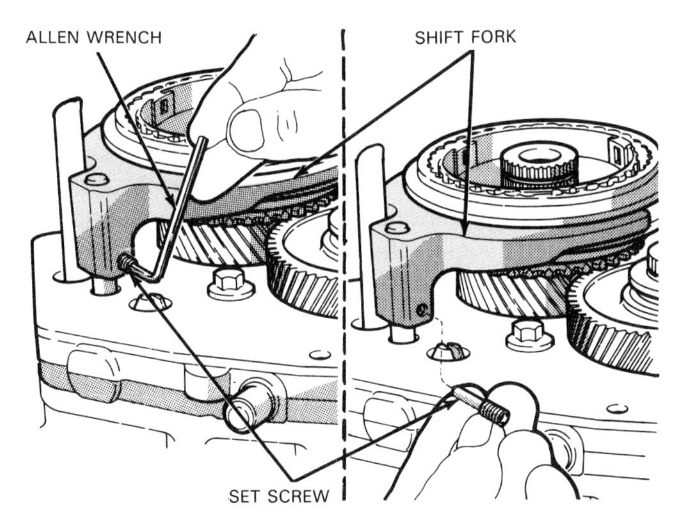

Figure 19-25. The shift forks must be carefully reinstalled to prevent shifting difficulties. It is possible to inadvertently install a shift fork upside down. Consult the service manual if you have any doubts.

Reinstall all washers in the position and direction shown in the factory service manual. Some washers used with manual transaxles function as oil pumps, delivering oil to the shafts as they flex during shifts. To function properly, these must be installed as specified.

In many transaxles, shaft end play must be checked before the case halves are reassembled. One method of checking end play makes use of a dial indicator, as on a rear-wheel drive transmission. Figure 19-26 shows another method of checking end play that involves the use of special measuring tools. Adjusting the end play can be done with adjusting nuts or by installing thrust washers of different thicknesses.

Once all internal parts have been installed, the case halves may be reassembled. Always use the proper kind of sealer or gasket. Do not substitute gasket material for RTV sealer or replace a gasket with a thicker one. Doing so will change the end play adjustments on the transaxle shafts, and it will affect the preload on the tapered roller bearings. Figure 19-27 shows assembly of two case halves. After tightening the case halves, make sure that the transaxle shafts turn properly. Check shaft end play once the transaxle is reassembled.

If the transaxle uses a separate housing for the internal shift linkage, install it now. Figure 19-28 shows a typical housing. Be careful to align the shift levers and forks before tightening the bolts. Then, check that the linkage operates properly.

Install the input shaft, if it was not already installed, and install its bearing housing assembly. Also, install and tighten any outer housings, such as the extension housings sometimes used to hold the seals at the CV axles, Figure 19-29.

Make sure that the transaxle input and output shafts turn without binding. If the shafts do not turn freely, you must find the reason and correct it.

CAUTION! Do not install a transaxle with binding shafts without correcting the problem. The transaxle will be damaged within the first few minutes of operation.

Reinstall any external case components, including electrical connectors and linkage. Install the clutch throwout bearing and clutch fork, if they were removed from their installation.

Manual Transaxle Installation

The transaxle installation procedure is the reverse of removal. If the installation is performed properly, the transaxle should be easily reattached to the engine and vehicle. Following are some *general* procedures detailing

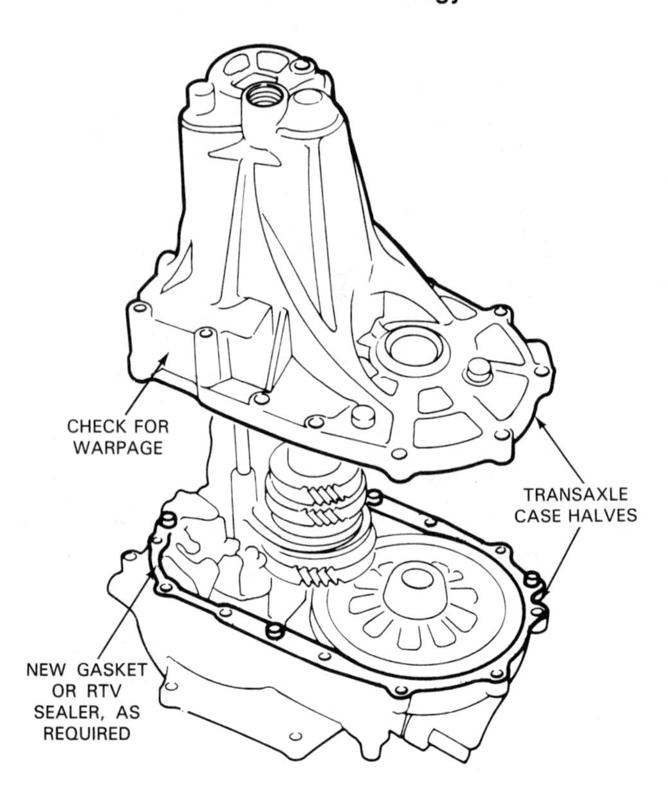

Figure 19-27. The case can be reassembled after all internal parts have been replaced. Use the right kind of gasket or sealer on the case sealing surfaces to prevent leaks and incorrect bearing preloads. Remove all old gasket material from sealing surfaces before installing new material. (Honda)

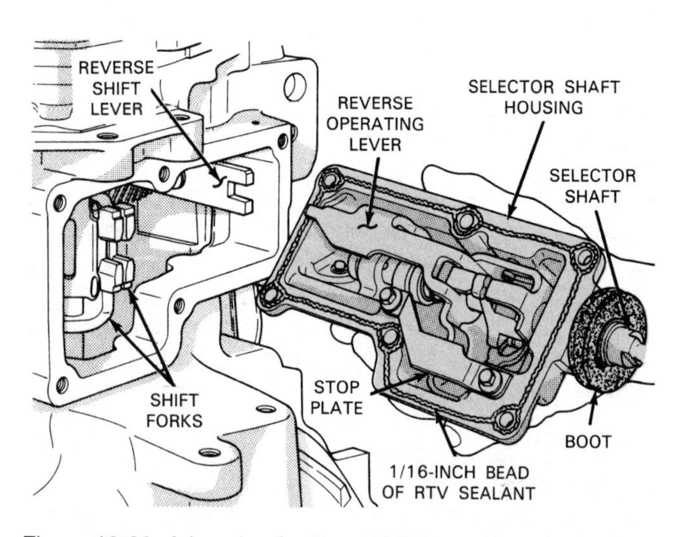

Figure 19-28. A housing for the shift linkage, if used, should be installed carefully, so that the parts line up correctly. If the parts are installed incorrectly, the transaxle gears could be damaged. (Chrysler)

how to install a manual transaxle in a vehicle. Note that you should always refer to the manufacturer's service manual for specific procedures.

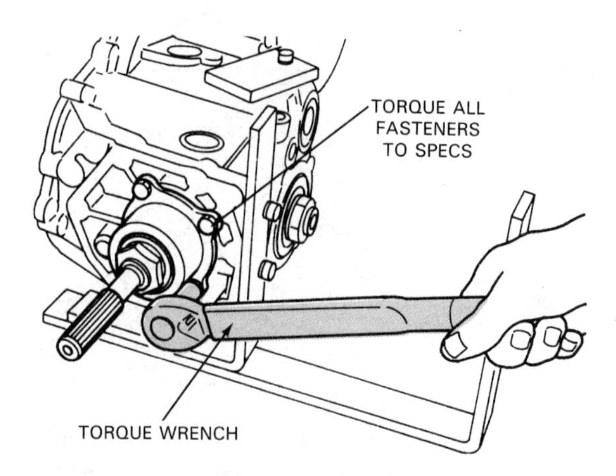

Figure 19-29. Install all extension housings with the proper gaskets or sealer and torque them to specifications.

1. Begin by raising the transaxle with the transmission jack. When the transaxle is at the proper height, slide it into engagement with the throwout bearing, clutch disc hub, and crankshaft pilot bushing. If the clutch has been disturbed, you may need a clutch pilot shaft to align the clutch disc. (This was covered in Chapter 7.) If necessary, turn the engine crankshaft to help the input shaft engage the clutch. It may be necessary to remove an access cover to reach the front crankshaft pulley.

   As the input shaft passes through the clutch disc hub into engagement with the pilot bushing, the transaxle clutch housing should fit closely onto the engine. Any alignment dowels should smoothly engage their matching holes.

CAUTION! Do not try to pull a binding transaxle into engagement with the engine by tightening the bolts. This will cause damage to the clutch disc and input shaft and could possibly break the aluminum transaxle case.

2. Install the transaxle case-to-engine bolts. Make sure that the engine and transaxle are properly aligned when all of the bolts are tight. There should be no gaps between any of the mating surfaces. As you tighten, check that no wires, brackets, or hoses are caught between the engine and transaxle.
3. Install and tighten the engine mounts and other fasteners holding the transaxle to the vehicle frame.
4. Install the starter, if it was removed, and the dust cover. Install the speedometer cable assembly and install the exhaust pipes, if they were removed.
5. Install the CV axles, making sure to match up the alignment marks made during removal. If the CV axles are held in the transaxle with retaining clips, replace the clips with new ones to ensure that the axle does not pop out when the vehicle suspension moves. Do

not stretch the clips with the thought of trying to ensure that they hold, since this will make the CV axle impossible to remove the next time that it needs service.

6. Install any other frame and body parts that were removed for clearance.
7. Lower the vehicle and reconnect the transaxle shift and clutch linkages, electrical connectors, and any other components that can be reached from under the hood. Make sure that you reinstall any ground wires, since they are critical to the proper operation of the vehicle electrical system. Remove the engine holding fixture, if used. Install the battery negative cable last.
8. Make preliminary shift linkage adjustments, if necessary. The clutch linkage may also require adjustment.
9. Fill the transaxle with the proper type of lubricant. If the unit has a separate reservoir for differential lubricant, refill it also.
10. After everything has been reconnected, operate the clutch pedal to ensure that the clutch engages and disengages properly. Also, move the gearshift lever to check operation of the shift linkage. Then, start the vehicle and road test. Make any linkage adjustments as needed and recheck the transaxle fluid level(s).

Summary

Manual transaxle repair is similar to rear-wheel drive manual transmission repair. Since transaxles have mechanical parts, most will eventually wear out. In many cases, the transaxle will develop problems ahead of time because of abuse or neglect. Most manual transaxle problems are easy to find if logical troubleshooting procedures are followed. The technician must carefully analyze symptoms to find the exact problem.

Safety should always come first when working on the vehicle. Support the front suspension when driving the vehicle on a lift, so that the CV axles are not damaged. When road testing, always obey traffic laws and try to use uncongested areas for testing.

Some transaxle parts can be replaced without removing the transaxle from the vehicle. Before beginning transaxle removal, disconnect the battery negative cable. When removing the CV axles, be careful that they are not damaged.

Once the transaxle is removed from the vehicle, it should be installed securely on a clean workbench. The external parts should be removed before the case halves or end covers are removed. Once the case halves are removed, the gears and shift linkages can be disassembled. The differential assembly should also be disassembled for inspection.

Inspect all parts for wear and damage. Check the shafts, gears, and bearings. Any defective parts should be replaced. Reassemble the transaxle carefully, following all manufacturers' instructions. If the transaxle gears bind or do not shift easily after reassembly, find and correct the problem. Do not expect binding parts to wear in!

Reinstall the transaxle carefully. Make sure that the CV axles are carefully installed to prevent damage. Refill the unit with fluid and adjust the clutch and shift linkages before test driving the vehicle.

Review Questions–Chapter 19

Please do not write in this text. Place your answers on a separate sheet of paper.

1. You should always begin diagnosing manual transaxle problems by:
   a. Checking the level of the lubricant.
   b. Adjusting the shift linkage.
   c. Test driving the vehicle.
   d. Finding out why the vehicle was brought in for service.
2. What are two relatively minor procedures that will solve many manual transaxle problems?
3. A vehicle with a manual transaxle is experiencing hard shifting. Technician A says to check for bent or worn shift linkage. Technician B says to check for linkage in need of lubrication or adjustment. Who is correct?
   a. Technician A.
   b. Technician B.
   c. Both Technician A and B.
   d. Neither Technician A nor B.
4. A driver complains of a manual transaxle coming out of gear. Technician A says a detent mechanism may be faulty. Technician B says to check the shift linkage adjustment. Who is correct?
   a. Technician A.
   b. Technician B.
   c. Both Technician A and B.
   d. Neither Technician A nor B.
5. A car with a manual transaxle is experiencing gear clash. Technician A says to check the synchronizers first. Technician B says to check the clutch adjustment first. Who is correct?
   a. Technician A.
   b. Technician B.
   c. Both Technician A and B.
   d. Neither Technician A nor B.
6. If a manual transaxle is noisy in every gear, including neutral, first check bearing preloads. True or false?
7. Before assuming a transaxle is leaking lubricant, what should you consider?
8. When making a road test, you should try to isolate transaxle problems from CV axle, engine, tire, and other problems. True or false?
9. Most manual transaxles must be removed with the engine, as a unit. True or false?

10. In your own words, explain a general procedure for removing a transaxle.
11. Why must you not let the weight of a manual transaxle rest on the clutch assembly?
12. In general, transaxles have _____ gears and shafts than most rear-wheel drive transmissions.
13. A transaxle case is to be reassembled. Technician A says that you must not replace RTV sealer with a thick gasket on the transaxle case. Technician B says that you can use a thicker gasket than recommended to help ensure that the case does not leak. Who is correct?
    a. Technician A.
    b. Technician B.
    c. Both Technician A and B.
    d. Neither Technician A nor B.

Certification-Type Questions–Chapter 19

1. **All of these statements about transaxle problems are true EXCEPT:**
   (A) downshifting at high speeds can damage the synchronizers.
   (B) high fluid level will reduce parts wear.
   (C) water in the lubricating oil will cause gear pitting.
   (D) wear of moving transaxle parts can be attributed to high mileage.

2. **A front-wheel drive vehicle is noisy in first gear only. Technician A says that the first thing to check is the clutch adjustment. Technician B says that the first thing to check is the condition of the vehicle tires. Who is right?**
   (A) A only
   (B) B only
   (C) Both A & B
   (D) Neither A nor B

3. **Which of these would be the least likely cause of transaxle gear clash?**
   (A) Misadjusted clutch.
   (B) Worn shift linkage.
   (C) Low lubricant level.
   (D) Damaged synchronizers.

4. **A vehicle with a manual transaxle has a no-drive condition. The engine continues to run when the clutch is released (up position). Technician A says that the no-drive condition could be caused by locked transaxle gears. Technician B says that the no-drive condition could be caused by stripped transaxle gears. Who is right?**
   (A) A only
   (B) B only
   (C) Both A & B
   (D) Neither A nor B

5. **All of these service operations can be performed while the transaxle is installed in the vehicle EXCEPT:**
   (A) clutch replacement.
   (B) clutch adjustment.
   (C) axle seal replacement.
   (D) shift linkage replacement.

6. **Technician A says that some internal transaxle parts can be serviced by removing an access cover. Technician B says that it is usually easier to remove the engine when the transaxle must be removed. Who is right?**
   (A) A only
   (B) B only
   (C) Both A & B
   (D) Neither A nor B

7. **Technician A says that engine and transaxle misalignment can cause hard shifting. Technician B says that broken engine or transaxle mounts can cause gear slipout. Who is right?**
   (A) A only
   (B) B only
   (C) Both A & B
   (D) Neither A nor B

8. **Transaxle gears may be held to their shaft by all of the following EXCEPT:**
   (A) snap rings.
   (B) integral construction.
   (C) a press fit.
   (D) case-hardened bolts.

9. **Synchronizer blocking rings should be handled carefully, since they are made of:**
   (A) plastic.
   (B) brass.
   (C) aluminum.
   (D) cast iron.

10. **Technician A says that the CV axles should be allowed to hang from the the transaxle so that they may be easily removed. Technician B says that the engine should be supported by a special fixture before the engine and transaxle mounts are loosened. Who is right?**
    (A) A only
    (B) B only
    (C) Both A & B
    (D) Neither A nor B

11. **If the transaxle differential assembly is separate from the transmission assembly, it should be disassembled:**
    (A) whenever the transaxle is removed from the vehicle.
    (B) when the transaxle is disassembled.
    (C) only if the technician suspects that it is defective.
    (D) only if it is leaking lubricant.

12. **A reassembled transaxle has improper end play. Technician A says that the end play is measured by the use of dial indicators and special measuring tools. Technician B says that the end play is adjusted by selective thickness shims or by turning adjusting nuts. Who is right?**
    (A) A only
    (B) B only
    (C) Both A & B
    (D) Neither A nor B

13. **All of these statements about manual transaxle shift linkages are true EXCEPT:**
    (A) some foreign vehicles have column-mounted gearshift levers.
    (B) shift linkage may be cable operated.
    (C) linkage adjustment mechanisms are often found in the gearshift lever assembly.
    (D) shift linkage rods or cables are longer than those on rear-wheel drive transmissions.

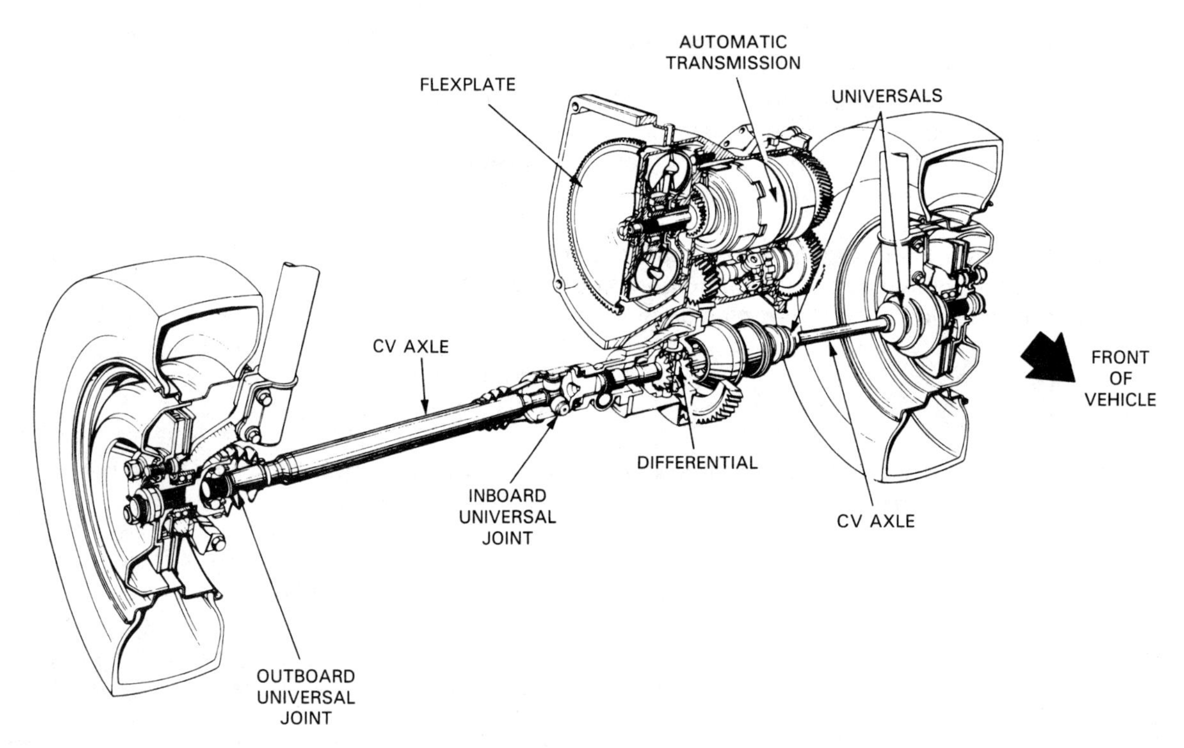

This illustration shows the position of the automatic transaxle in the front-wheel drive vehicle. Note the positions of the engine flexplate, transmission section, and differential section. Also note the transaxle unit in relation to the CV axles and the front wheels. This is used with a transverse engine. The transaxles used with longitudinal engines are slightly different. (Chrysler)

Chapter 20

Automatic Transaxle Construction and Operation

After studying this chapter, you will be able to:

- Identify the major parts of an automatic transaxle.
- Discuss the purpose of each major part of an automatic transaxle.
- Explain overall operation of an automatic transaxle.
- Trace power flow through automatic transaxles used with transverse and longitudinal engines.
- Compare automatic transaxle design variations.
- Describe the operation of a continuously variable transmission.

Automatic transaxles are commonly used on many domestic and imported vehicles. Automatic transaxle operating principles are similar to those of an automatic transmission and differential assembly of a rear-wheel drive vehicle. The major difference in an automatic transaxle is that all parts are usually together in one housing.

This chapter will familiarize you with the parts and operating principles of automatic transaxles. Although they are arranged differently, many parts are similar to those of an automatic transmission and differential assembly of a rear-wheel drive vehicle. This chapter will indicate what parts are different. It will also explain how power flows through the two major types of transaxles, as well as the differences in parts and layout between the two types.

Before beginning this chapter, make sure that you thoroughly understand the information in previous chapters. Pay particular attention to the information in Chapters 10 and 11; understanding these chapters will make grasping the material in this chapter much easier.

Construction and Basic Operation

Automatic transaxles are made of many separate parts and systems. Study Figures 20-1 and 20-2 to see how automatic transaxle parts fit together and their relative position in the transaxle. These components, which will be presented in detail in this chapter, include:

- The *torque converter* uses transmission fluid to transfer and multiply power. The torque converter is mounted directly behind the engine and is turned by the engine crankshaft.
- The ***automatic transaxle shafts*** serve various functions. Some transmit power from the torque converter to the transaxle gear train and differential assembly. One shaft provides a stationary support for the torque converter stator. Another shaft is used to drive the transaxle oil pump.
- A *chain drive* is used on some transaxles. It transfers power from the turbine shaft, which is splined to the central hub of the torque converter (as in an automatic transmission), to the transmission input shaft.
- The *planetary gearsets* are constant-mesh gears somewhat resembling the solar system. They create different forward gear ratios or provide reverse operation when different members are held stationary and others are driven. Most transaxles use at least two planetary gearsets.
- The *planetary holding members* allow gearset operation to occur. The members function by holding a planetary member stationary or applying drive power to it.
- The ***transaxle oil pump*** is a positive displacement pump that produces the fluid pressure that operates other hydraulic components and lubricates the moving parts of the transaxle. It is driven by lugs on the torque converter or by splines on the end of an *oil pump shaft.*
- The *transaxle oil filter* removes particles from the automatic transmission fluid prior to circulation by the transaxle oil pump.
- The *transaxle oil cooler* is needed to dissipate heat from the transmission fluid, which is generated by the transaxle and also picked up from the engine. The transaxle oil cooler is often mounted in the engine cooling-system radiator. External oil coolers may also be installed on the vehicle.

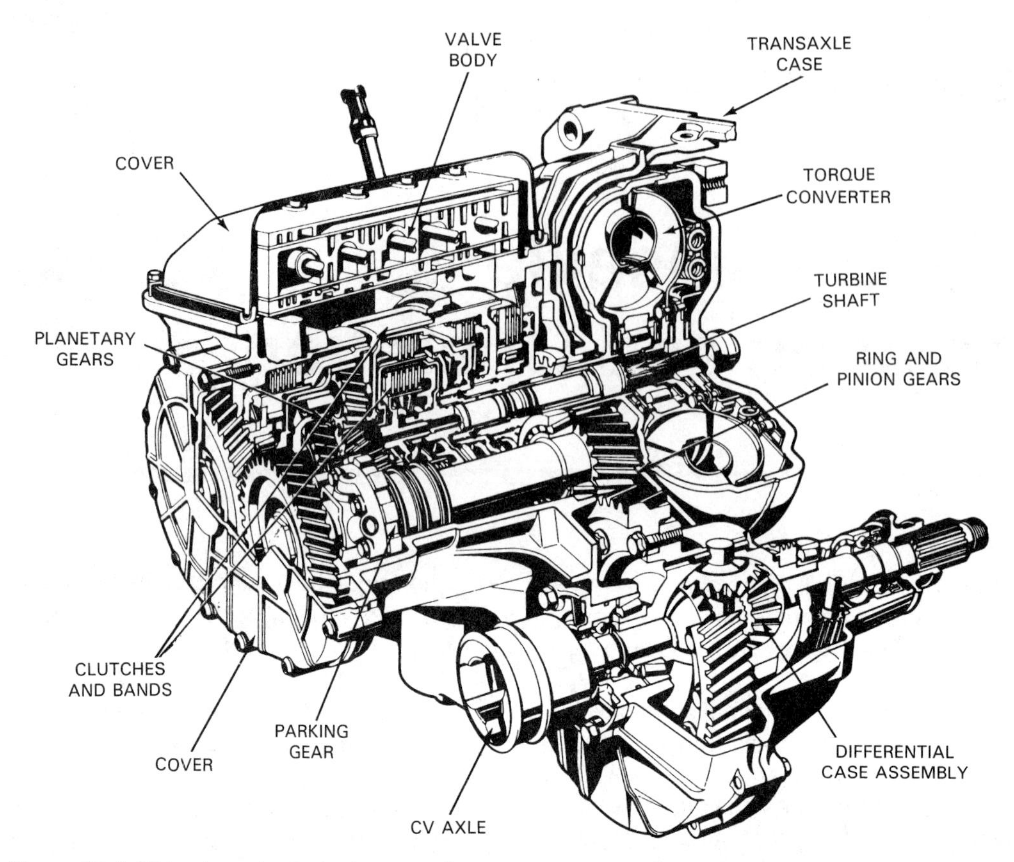

Figure 20-1. This view of a typical automatic transaxle shows the major components of transaxles. Many of the components are identical in operation to those of automatic transmissions on rear-wheel drive vehicles. (Saab)

- The *valve body* is a precision casting that serves as control center for the transaxle's hydraulic system. It contains the manual valve, shift valves, and other valves. It may contain the pressure regulator valve.
- *Hydraulic valves* are used in the valve body to control pressure and, also, direction and rate of fluid flow. Specifically, they control application of clutches and bands. In general, these devices help control the entire operation of the automatic transaxle.
- The *automatic transaxle shift linkage* operates the manual valve, thereby forming a direct mechanical connection between the driver and the transaxle. It also controls operation of the parking pawl.
- The ***automatic transaxle differential assembly*** receives power from the transaxle transmission. The differential assembly contains spider gears and side gears. It may contain differential drive gears.
- The ***automatic transaxle case*** encloses the transaxle transmission and differential. Passageways in the case allow hydraulic fluid to travel between hydraulic components.
- The ***transaxle oil pan*** is installed on the bottom of the transaxle case. It serves as a reservoir for transmission fluid.
- *Electronic components,* such as switches and solenoids, are used in the transaxle to send signals to other vehicle components or to partially control the transaxle hydraulic system.

Torque Converter

The torque converter, Figure 20-3, is a power-transfer device using automatic transmission fluid as the transmission medium. As in the automatic transmission of a rear-wheel drive vehicle, the converter is located behind the engine flexplate. Some converters have a pump drive hub, with lugs to drive the transaxle oil pump; others do not. The torque converter drives the turbine shaft, called the transaxle input shaft in some designs.

Most of the torque converters used in modern transaxles are lockup torque converters. These have an internal clutch to make a direct mechanical connection between the turbine shaft and engine at certain speeds and gears. A simple explanation of lockup torque converter operation is given in Figure 20-4.

When the pressure between the clutch piston and the converter cover is exhausted, pressure behind the clutch piston pushes it into engagement with the cover. This forms

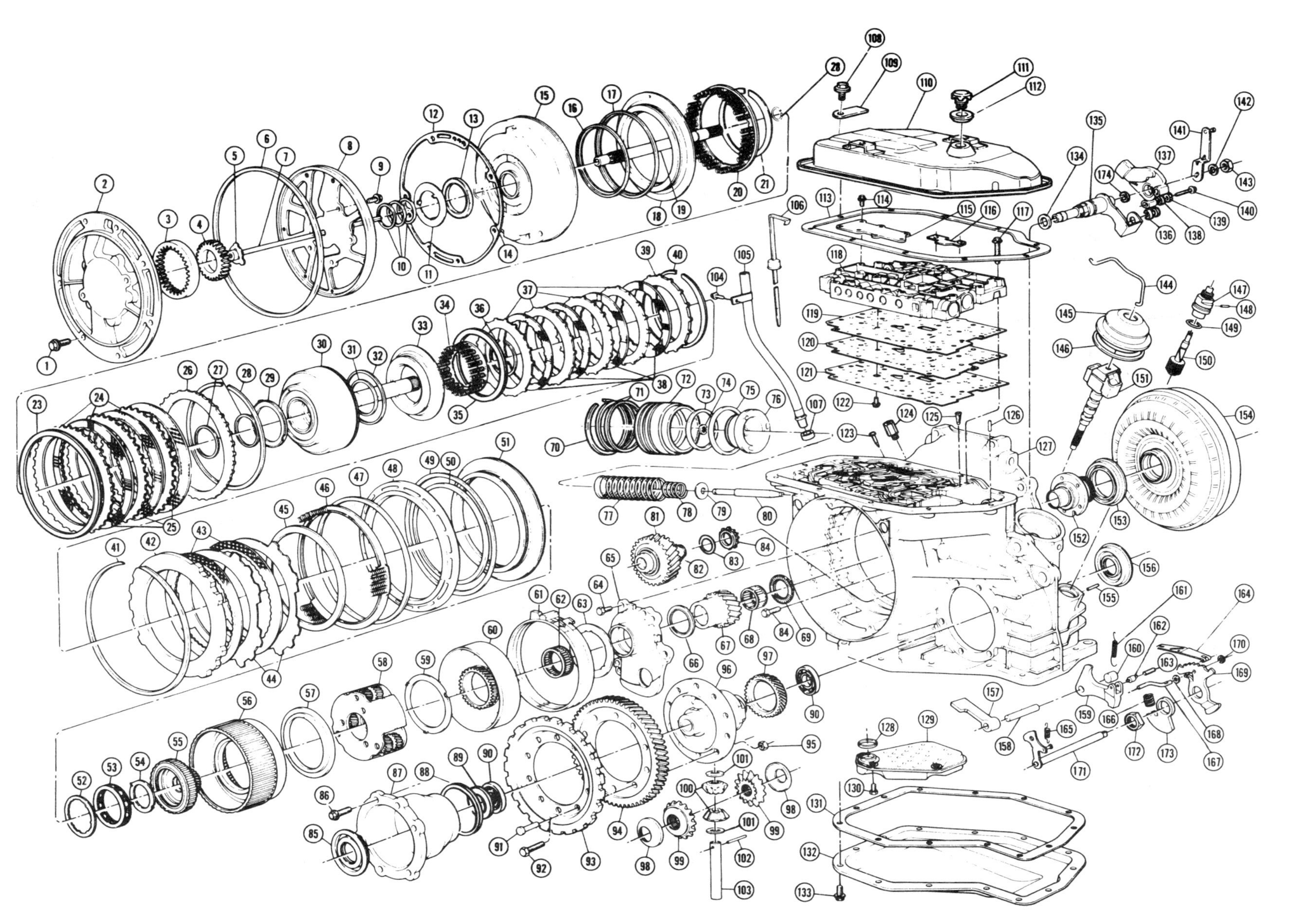

Figure 20-2. This exploded view shows the relationship of various automatic transaxle parts. Studying an exploded view is a good way to become familiar with the parts of an automatic transaxle and to see how they relate to each other. Note that a complete parts list is found on the following page. (Ford)

(continued)

1. BOLT & WSHR. ASSY. (7A103 TO 7005) M6-1 X 40 (7 REQ'D.)
2. BODY & SLEEVE ASSY. — OIL PUMP
3. GEAR — OIL PUMP DRIVEN
4. GEAR ASSY. — OIL PUMP DRIVE
5. INSERT — OIL PUMP DRIVE GEAR
6. SEAL — OIL PUMP
7. SHAFT — OIL PUMP DRIVE
8. SUPPORT & BSHG. ASSY. — OIL PUMP
9. BOLT (7A108 TO 7F370) M6-1 X 16MM LG. (5 REQ'D.)
10. SEAL — INTERM. CLUTCH — INNER (TEFLON)
11. OIL PUMP THRUST WASHER (SELECTIVE)
12. GASKET — OIL PUMP
13. BRG. ASSY. — INTERM. CLUTCH DRUM THRUST
14. RING — 17.0 RETAINING RD. WIRE EXTERNAL
15. CYLINDER — INTERM. CLUTCH
16. SEAL — INTERM. CLUTCH PISTON — INNER
17. SEAL — INTERM. CLUTCH PISTON — OUTER
18. PISTON — INTERM. CLUTCH
19. SHAFT — INTERM. CLUTCH
20. RET. & SPRING ASSY. — INTERM. CLUTCH
21. RING — 111.76MM RETAINING EXTERNAL
22.
23. SPRING — REV. CLUTCH CUSHION
24. PLATE — INTERM. CL. EXT. SPLINE
25. PLATE ASSY. — INTERM. CL. INT. SPLINE
26. PLATE — INTERM. CLUTCH PRESSURE
27. SEAL — INTERM. CLUTCH — OUTER (TEFLON)
28. RING — RETAINING INT. (SELECTIVE)
29. BRG. ASSY. — DIRECT & INTERM. CLUTCH
30. CYL. SHAFT & RACE ASSY. — DIRECT CLUTCH
31. SEAL — DIRECT CL. PISTON — INNER
32. SEAL — DIRECT CL. PISTON OIL — OUTER
33. PISTON — DIRECT CLUTCH
34. RET. & SPRING ASSY. — DIRECT CLUTCH
35. SPRING — DIRECT CLUTCH CUSHION
36. RING — 59.5MM RETAINING, EXTERNAL
37. PLATE — DIRECT CLUTCH EXT. SPLINE (AS REQ'D.)
38. PLATE ASSY. — DIRECT CL. INT. SPLINE (AS REQ'D.)
39. PLATE — DIRECT CLUTCH PRESSURE
40. RING — RETAINING INT. (SELECTIVE)
41. RING — RETAINING INT.
42. PLATE — REVERSE CLUTCH PRESSURE
43. PLATE ASSY. — REV. CL. INT. SPLINE
44. PLATE — REV. CLUTCH EXT. SPLINE
45. SPRING — REV. CLUTCH CUSHION
46. SPRING & RET. ASSY. — REV. CLUTCH
47. SEAL — 196.0MM
48. PISTON — REVERSE CLUTCH
49. SEAL — REV. CL. PISTON — OUTER
50. SEAL — REV. CL. PISTON — INNER
51. CYLINDER — REVERSE CLUTCH
52. BEARING — ONE-WAY CLUTCH
53. SPRING & ROLLER ASSY. — ONE-WAY CLUTCH
54. WASHER — DIRECT CL. CYL. THRUST
55. GEAR ASSY. — 1ST-3RD REVERSE SPEED
56. GEAR ASSY. — INTER. & REV. CL. RG.
57. RACE & BRG. ASSY. — PLANT THRUST REAR
58. PLANE ASSEMBLY
59. WASHER — PLANETARY THRUST — FRONT
60. DRUM & SUN GEAR ASSY. — LOW INTERM.
61. BAND ASSY. — LOW INTERM.
62. BEARING ASSY. — TRANSFER
63. WASHER — INTERM. SUN GR. THRUST
64. BOLT — M8-1.25 X 25.0 HEX FLANGE HD.(5 REQ'D.)
65. HOUSING — FINAL DRIVE GEAR
66. BRG. ASSY. — FINAL DRIVE GEAR THRUST — REAR
67. GEAR — FINAL DRIVE INPUT
68. BRG. ASSY. — FINAL DRIVE INPUT GEAR
69. BRG. ASSY. — FINAL DRIVE GEAR THRUST — FRONT
70. RING — 103.5MM RET. FLAT INTERNAL
71. SEAL — LOW & INTERM. BAND SERVO PISTON COVER
72. COVER — LOW/INTERM. BAND SERVO
73. SEAL — LOW/INTERM. SERVO PISTON — SMALL
74. RING — 15.8MM RETAINING EXTERNAL
75. SEAL — LOW/INTERM. SERVO PISTON — LARGE
76. PISTON — LOW & INTERM. SERVO
77. SPRING — LOW/INTERM. SERVO PISTON
78. SPRING — SERVO PISTON CUSHION
79. WASHER — 9.7MM X 30 X 2.5 FLAT STEEL
80. ROD — LOW/INTERM. SERVO PISTON NOT AVAILABLE
81. GEAR & BRG. ASSY. — IDLER GEAR
82. SHAFT — IDLER GEAR
83. SEAL — 22.8 X 1.6 O-RING
84. NUT — M25 X 1-12 POINT
85. SEAL ASSY. — TRANSAXLE — DIFF.
86. BOLT — M8-1.25 X 30 HEX FLANGE HD. (6 REQ'D.)
87. RETAINER — DIFF. BEARING
88. GASKET
89. SHIM — DIFF. BEARING (SELECTIVE)
90. BALL BEARING — DIFF.
91. RIVET — M10 X 38 SOLID FLAT HD. (REF. ONLY — PRODUCTION)
92. BOLT — M10 X 1.5 X 40 HEX HD. (10 REQ'D) SERVICE ONLY
93. GEAR — OUTPUT SHAFT PARK
94. GEAR — FINAL DRIVE OUTPUT
95. NUT — M10 X 1.5 HEX PLT. — (10 REQ'D.) SERVICE ONLY
96. DIFF. ASSY. — TRANSAXLE
97. GEAR — SPEEDO DRIVE
98. WASHER — TRANSAXLE DIFF. SIDE GR. THRUST
99. GEAR — TRANSAXLE DIFF. SIDE
100. PINION — TRANSAXLE DIFF.
101. WASHER — TRANSAXLE DIFF. PINION THRUST
102. PIN — 4.75MM X 38.1MM
103. SHAFT — TRANSAXLE DIFF. PINION
104. BOLT — M6-1 X 12 HEX FLANGE HD.
105. TUBE ASSY. — OIL FILTER
106. INDICATOR ASSY. — OIL LEVEL
107. GROMMET (SEAL FILLER TUBE TO CASE)
108. BOLT — M6-1 X 14MM LG (10 REQ'D.)
109. IDENTIFICATION TAG
110. COVER ASSY. — MAIN CONTROL
111. VENT ASSY. — MAIN CONTROL COVER
112. GROMMET — MAIN CONTROL COVER
113. GASKET — MAIN CONTROL COVER
114. BOLT — M6-1.0 X 45 HEX FLANGE HD. (7 REQ'D.)
115. PLATE — MAIN OIL PRESS. REG. EXH.
116. PLATE — TRANS.
117. BOLT — M6-1.0 X 40 HEX FLANGE HD. (20 REQ'D.)
118. CONTROL ASSY. — MAIN
119. GASKET — MAIN CONTROL (BET. 7A092 & 7A008)
120. PLATE — CONTROL VALVE BODY SEP.
121. GASKET — MAIN CONTROL (BET. 7A008 & 7006)
122. BOLT — M6-1 X 12 HEX FLANGE HD. (2 REQ'D.)
123. PIN — TIMING (2.3L ONLY)
124. COMM. ASSY. PUSH-IN
125. SCREEN ASSY. — GOV. OIL
126. PIN — 3.2MM X 25.65 DOWEL HRDN.
127. CASE & HSG. ASSY.
128. GASKET — OIL FILTER
129. FILTER ASSY. — OIL
130. BOLT — M6-1 X 14MM LG (3 REQ'D.)
131. GASKET — OIL PAN
132. PAN — OIL
133. BOLT — M8-1.25 X 16 HEX FLANGE (13 REQ'D.)
134. SEAL — MANUAL CONTROL LEVER
135. LEVER ASSY. — MANUAL CONTROL
136. INSULATOR — GEAR SHIFT ARM
137. SWITCH ASSY. — NEUTRAL START
138. WASHER — #12 FLAT (2 REQ'D.)
139. WASHER — 6.0MM HELICAL SPG. LK. (2 REQ'D.)
140. BOLT — M6-1.0 X 40 HEX FLANGE HD. (2 REQ'D.)
141. LEVER ASSY. — THROTTLE VALVE — OUTER
142. WASHER — 8MM LOCK
143. NUT — M8 X 1.25 HEX
144. CLIP — GOV. COVER RETAINING
145. COVER — GOVERNOR
146. SEAL — 77.9MM X 3.40 RECT. SECT.
147. RETAINER — SPEEDO DRIVEN GEAR
148. PIN — 3MM X 19.9 DOWEL HRDN.
149. SEAL — 25.06MM X 2.6 O-RING
150. GEAR — SPEEDO DRIVEN
151. GOVERNOR ASSEMBLY
152. SUPPORT ASSY. — CONV. REACTOR
153. SEAL ASSY. — CONV. IMP. HUB
154. CONVERTER ASSEMBLY
155. PIN — SPEEDO RETAINING
156. SEAL ASSY. — TRANSAXLE — DIFF.
157. STRUT — LOW INTERM. BAND ANCHOR
158. SHAFT — PARKING PAWL
159. PAWL — PARKING BRAKE
160. PLUG — 12.0MM CUP
161. SPRING — PARK PAWL RETURN
162. PIN — PARKING PAWL ROLLER
163. ROLLER — PARKING BRAKE
164. SPRING ASSY. — MANUAL VALVE DETENT
165. SPRING — THROTTLE VALVE CONTROL LEVER
166. SPRING — PARK PAWL RATCHETING
167. ACTUATOR — MANUAL LEVER
168. WASHER
169. LEVER — MANUAL VALVE DETENT — INNER
170. NUT — STAMPED
171. SHAFT ASSY. — TV LEVER ACTUATING
172. NUT — M20 X 1.5 HEX
173. LEVER — PARK PAWL ACTUATING
174. SEAL — THROTTLE CONTROL LEVER SHAFT

Figure 20-2 *(continued)*.

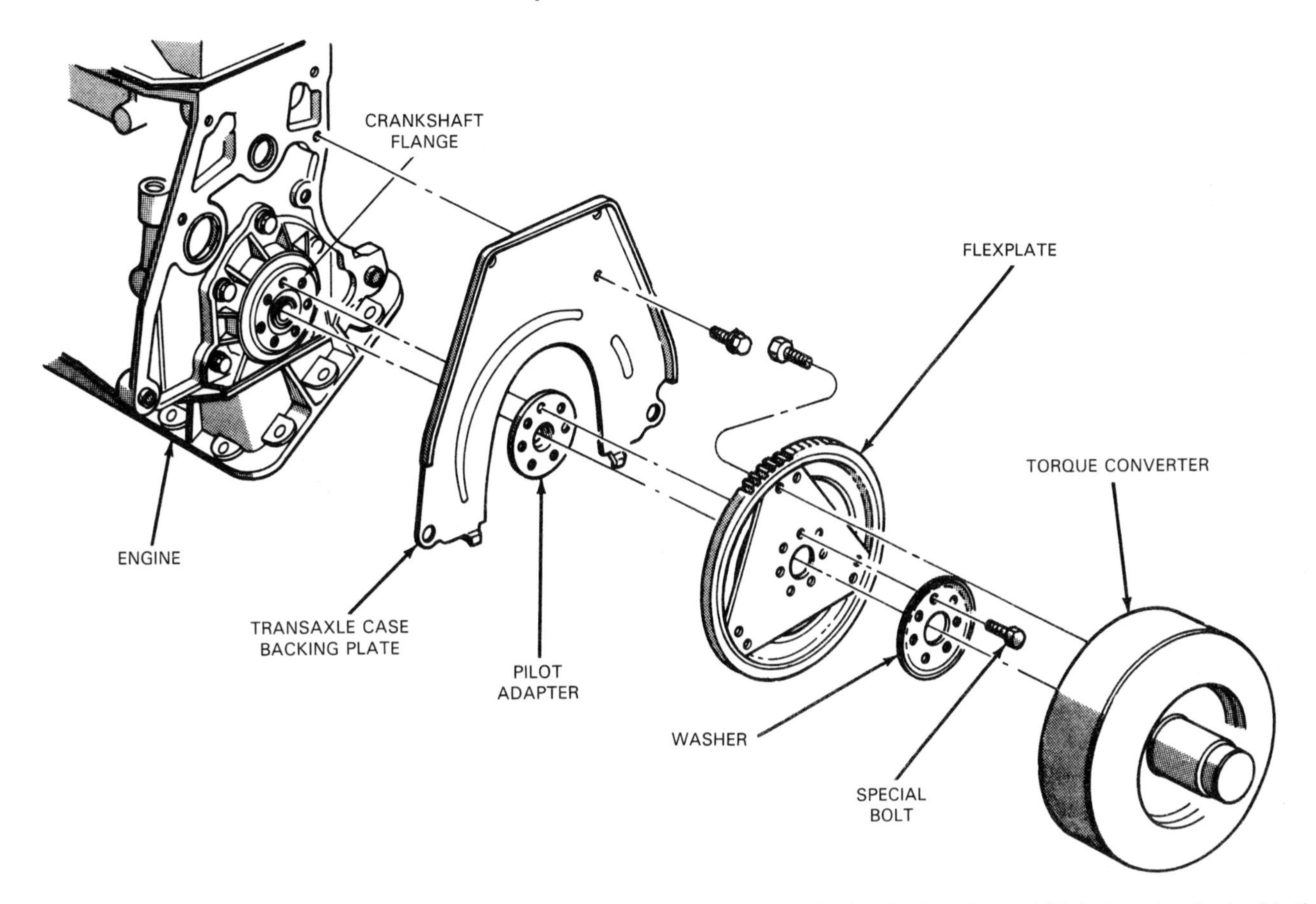

Figure 20-3. The torque converter is installed directly behind the engine. It is attached to the flexplate, which in turn, is attached to the engine crankshaft. The flexplate and torque converter turn whenever the engine turns. (Chrysler)

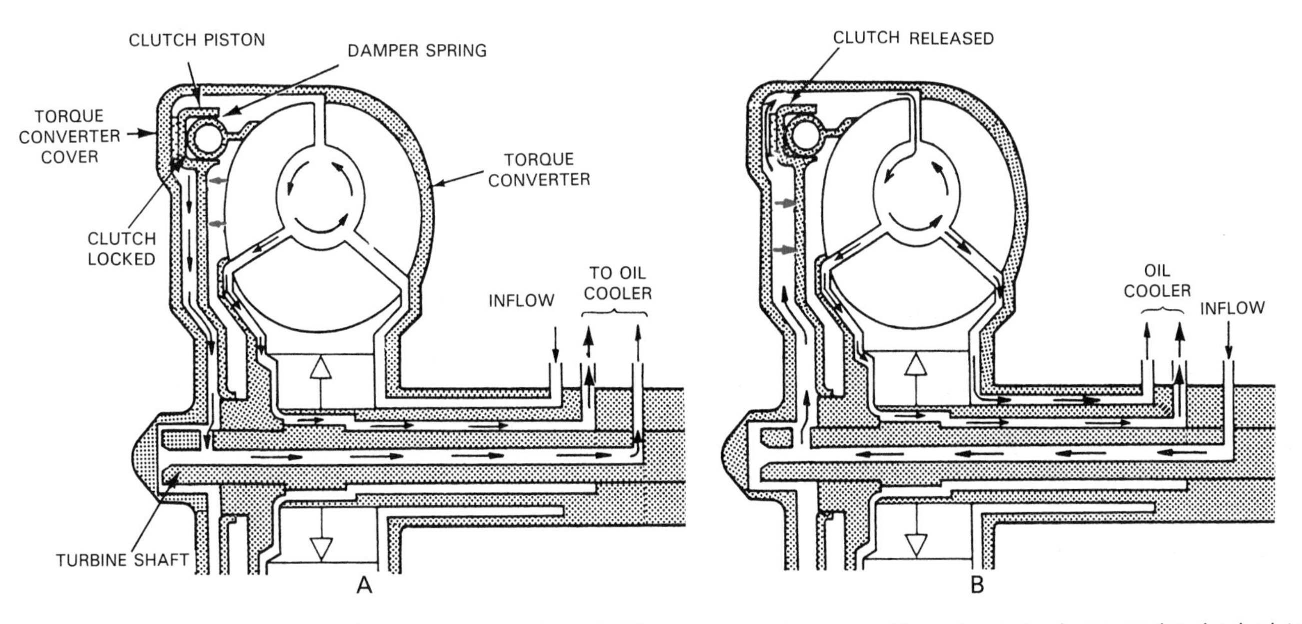

Figure 20-4. Transaxles also use lockup torque converters. A–When pressure is removed from the region between the clutch piston and the converter cover, the clutch applies, locking the turbine to the impeller and cover. B–When pressure is restored between the piston and cover, the clutch unlocks, and basic converter function resumes. (Saab)

a direct mechanical connection between the engine crankshaft and the turbine shaft. When pressure is applied to the region between the clutch piston and converter cover, the piston moves away from the cover, and the converter operates normally.

Automatic Transaxle Shafts

The automatic transaxle shafts are made of steel, hardened and tempered to withstand high loads and temperature extremes. Transaxles have several different shafts, including turbine shafts, input shafts, output shafts, and oil pump shafts. The types of shafts used and what they are called varies with the manufacturer and make of transaxle.

Automatic transaxle turbine shaft

The ***automatic transaxle turbine shaft*** is engaged with and driven by the torque converter turbine. In some designs, the turbine shaft is arranged like the input shaft of an automatic transmission on a rear-wheel drive vehicle, wherein turbine and output shaft centerlines align. This turbine shaft may be referred to as the transaxle input shaft. In other designs, there is a separate input shaft in addition to the turbine shaft, and the two shafts do not align.

Automatic transaxle input shaft

In some designs, the ***automatic transaxle input shaft*** and the turbine shaft are one and the same. Other designs have an input shaft in addition to a turbine shaft, as mentioned. The input shaft, in this case, is located below the turbine shaft and is aligned with an output shaft. In some instances, it surrounds the output shaft. The input shaft is driven by the turbine shaft through a chain drive or a set of helical gears.

Automatic transaxle output shaft

Automatic transaxle output shafts serve different functions, depending on transaxle design. In certain designs, the output shaft transfers power from the planetary gearsets to the differential assembly. In one particular transaxle, the output shaft is aligned with the turbine shaft. A gear on the end of the output shaft drives another shaft, called a *transfer shaft* (discussed shortly), which drives the differential assembly.

In other designs, the output shaft is at the output end of the differential and is driven by the differential. This shaft is situated below the turbine shaft, and it runs *through* the transaxle input shaft.

Oil pump shaft

Figure 20-5 shows a separate ***oil pump shaft*** installed in a turbine shaft. The oil pump shaft is splined to the oil pump, which would be situated at the left end (in this illustration) of the shaft. Splines on the other end of the shaft are attached to the converter hub. The torque converter drives the oil pump via the oil pump shaft. The shaft

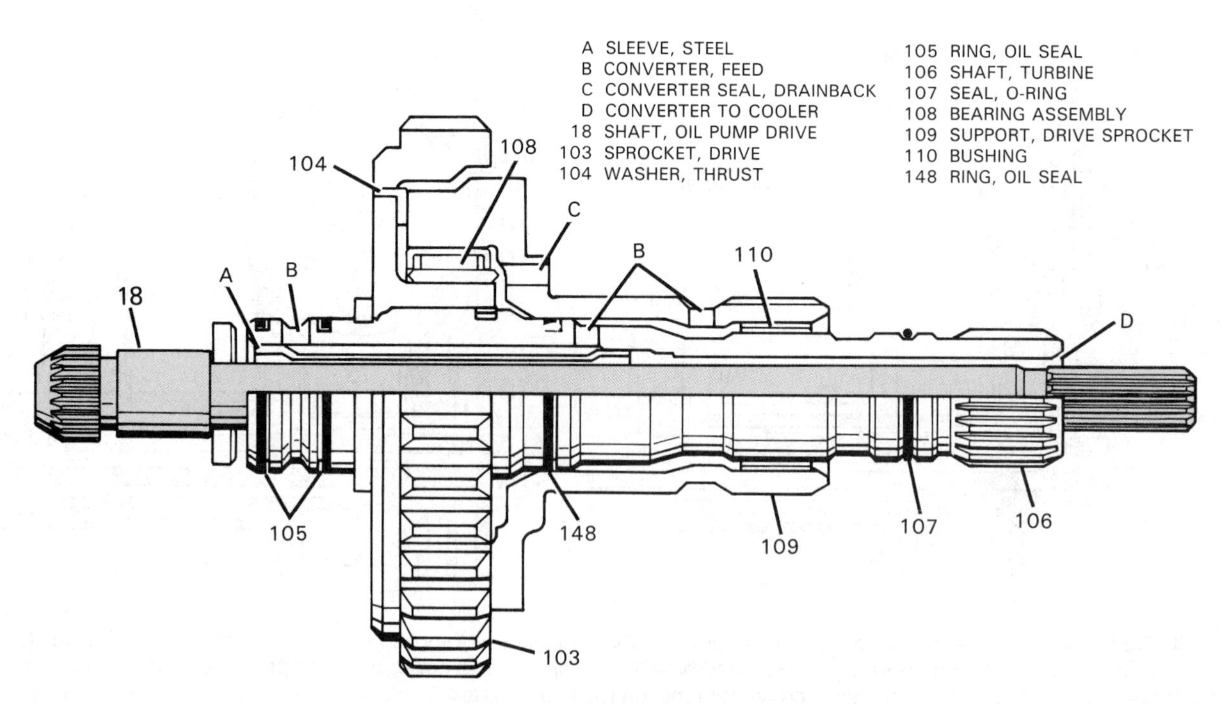

Figure 20-5. An oil pump shaft passes through the turbine shaft, as shown here. The splines on the left side of the shaft drive the oil pump. The oil passages in this shaft assembly are used to fill and control the torque converter. Note that this assembly is used with a chain drive automatic transaxle. (General Motors)

assembly contains oil passageways to deliver oil to and from the torque converter.

Stator shaft

A stator shaft is mounted on the front of some transaxle oil pumps. It provides a stationary support for the torque converter stator and overrunning clutch. The overrunning clutch must be solidly supported on a stationary part to work properly. The stator shaft has external splines that mate with internal splines on the overrunning clutch. These prevent rotation of the inner race of the clutch. The stator shaft is hollow, and the transaxle turbine shaft passes through it. The torque converter pump drive hub fits over the stator and turbine shafts.

Transfer shaft

As mentioned, some transaxles have a ***transfer shaft.*** This shaft transmits power from the transaxle output shaft to the differential assembly. The transfer shaft is positioned below the output shaft. A gear on the end of the output shaft drives a gear on the end of the transfer shaft. A helical pinion gear on the other end of the transfer shaft drives the differential ring gear.

Chain Drive

A chain drive is a power-transfer system using a chain and two sprockets. It is used in automatic transaxles to transmit motion in a small space, Figure 20-6. The drive chain transfers power from the torque converter to the transaxle transmission mechanical system. The drive sprocket is driven by the turbine shaft, and the driven sprocket drives the transaxle input shaft. The sprockets are mounted in bearing or bushing assemblies that provide support and alignment.

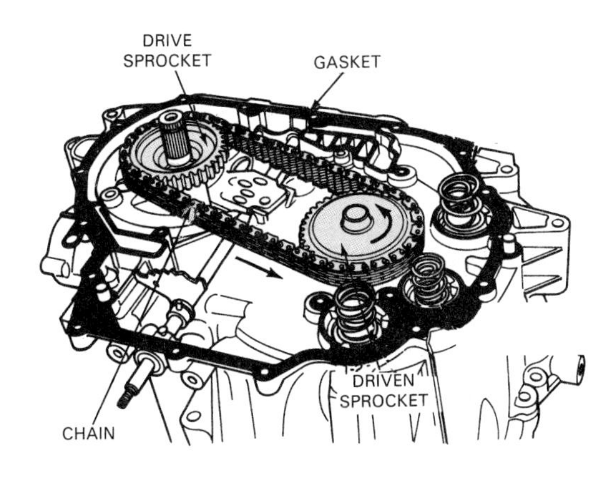

Figure 20-6. Many automatic transaxles have a chain drive assembly, enabling power transfer in a small space. (Ford)

The drive chain is a link, or silent-chain, design. This type meshes with its sprockets with a minimum of noise and friction. Figure 20-7 is an exploded view of a typical chain drive assembly. The transaxle output shaft on this design passes through the input sprocket and shaft.

Planetary Gearsets

Planetary gearsets are the means of transmitting power through the transaxle and, also, of changing gear ratios. The planetary gearsets used in automatic transaxles are the same type as found in most automatic transmissions. (Note that automatic transmissions used by one manufacturer have constant-mesh gears like those of a manual transmission. They are controlled by hydraulically operated clutch packs. Discussion here will be limited to the planetary type, since they are more common.)

A simple planetary gearset consists of a sun gear, three or four planet gears held in a carrier, and a ring gear. These gears are in constant mesh, so they seldom wear out or suffer broken teeth. A 3- or 4-speed automatic transaxle uses two or more planetary gearsets. A compound planetary gearset, Figure 20-8, is used because it can provide more forward gear ratios than a simple planetary gearset. Modified Simpson gear trains are used in most automatic transaxles. The Simpson gear train was covered in Chapter 10.

A few transaxles–4-speeds, in particular–use an older type of compound planetary gearset. This gearset, Figure 20-9, has a single ring gear and a single planet carrier with one set of long planet gears and one set of short planet gears. It also has two sun gears. The Ravigneaux gear train was covered in Chapter 10.

Planetary Holding Members

Most modern 3-speed automatic transaxles require at least two holding members be applied in any gear to move the vehicle; 4-speed transaxles require three holding members be applied. Planetary gearsets are controlled by planetary holding members. These are the components of the transaxle transmission that apply force to hold or drive other parts. They are what make the gearsets transfer power. The components that accomplish these tasks were discussed in Chapter 10. They include multiple-disc clutches, bands and servos, and overrunning clutches.

Multiple-disc clutches

Multiple-disc clutches use alternating sets of friction discs and steel plates to hold planetary gearset members stationary or lock them together for power transfer. They are engaged by a clutch apply piston that is built into the clutch assembly. Multiple-disc clutches, also called clutch packs, can be attached to the input or output shaft to drive planetary members, or they can be attached to the case to

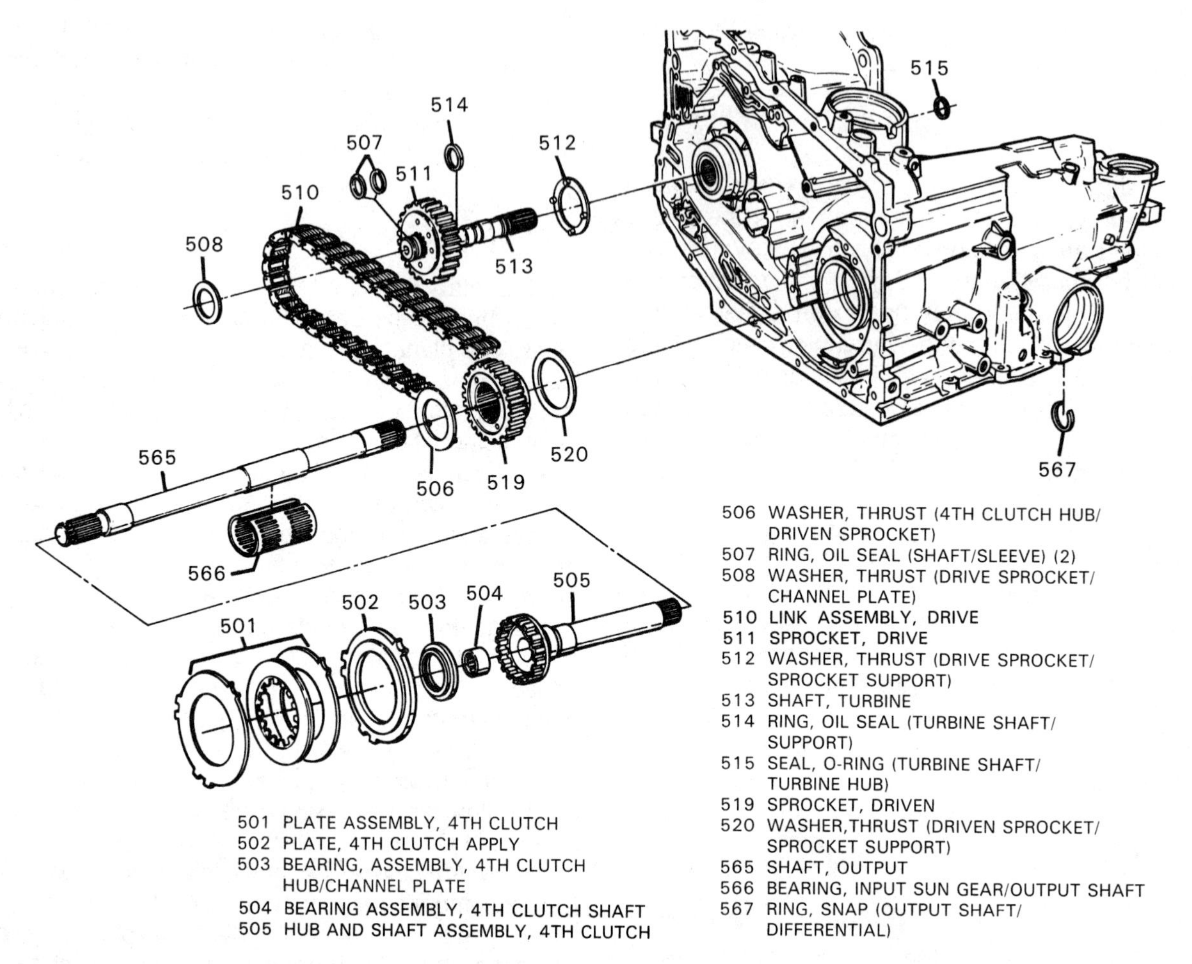

Figure 20-7. This exploded view of a chain drive assembly shows the relationship of the chain and sprockets to the automatic transaxle case. Sprocket size varies to match engine and vehicle needs. (General Motors)

hold a planetary gear stationary. A typical clutch pack is shown in Figure 20-10.

Clutch construction. The multiple-disc clutch is made up of several parts. Major components include the clutch drum, friction discs, and steel plates. Also included are the clutch apply piston, clutch pressure plate, and clutch return spring(s). In addition, the multiple-disc clutch has an associated hub that may or may not be a component of the assembly itself, and there may also be a drive shell associated with the clutch pack. Although designs vary, the parts are similar on all makes of transaxles.

A clutch drum, also called a clutch cylinder, encloses the apply piston, friction discs and steel clutch plates, pressure plate, seals, and other parts of the clutch assembly. The outer surface of the clutch drum may be a holding surface for a band. The inner surface contains channels that mate with teeth on the steel plates.

Friction discs and steel plates alternate inside of the clutch assembly for maximum holding power. The friction discs are usually covered with a lining made of a combination of paper and asbestos (or similar material). The composite material enables the discs to grip, though soaked with transmission fluid. The discs have internal teeth, or tangs, to engage the clutch hub.

The steel plates, or clutch plates, do not have any kind of lining. They have external teeth that lock into channels on the inside of the clutch drum. A friction disc fits between each steel plate. This enables the drum and hub, and parts connected to each (input shaft or planetary members), to be locked together when the clutch is engaged.

A clutch pack has an aluminum disc–the clutch apply piston–that is moved by hydraulic pressure. Movement of the apply piston clamps the friction discs and clutch plates together. Seals on the piston prevent pressure loss when the clutch is applied. Hydraulic fluid comes from the valve body through passageways in the case and shafts.

A pressure plate serves as a stop for the set of friction discs and clutch plates when the piston is applied. The clutch pressure plate is held to the drum by a large snap ring.

Return spring(s) are used to push the apply piston away from the friction discs and clutch plates after hydrau-

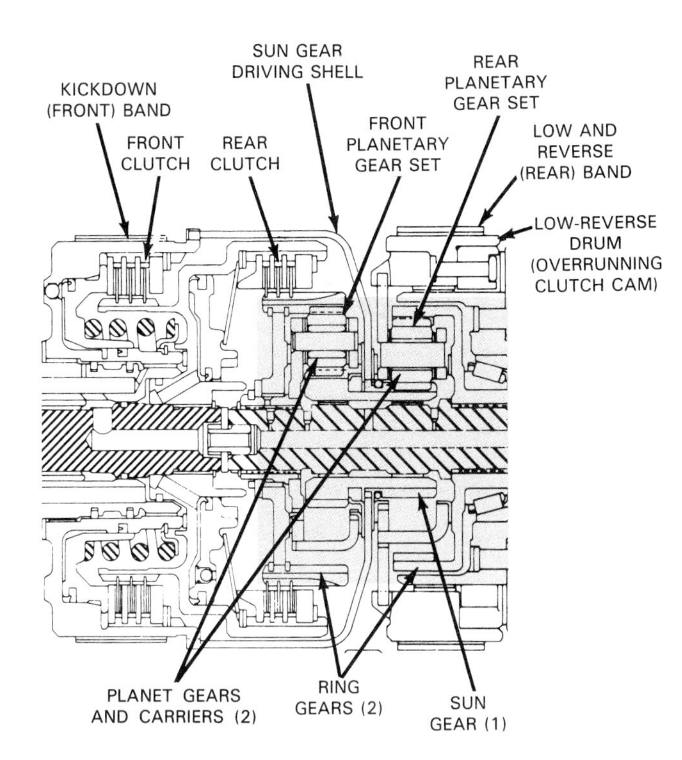

Figure 20-8. The planetary gearsets used in transaxles resemble those used in rear-wheel drive automatic transmissions. The gearset shown is the Simpson type, having two ring gears, two carriers, and a common sun gear. (Chrysler)

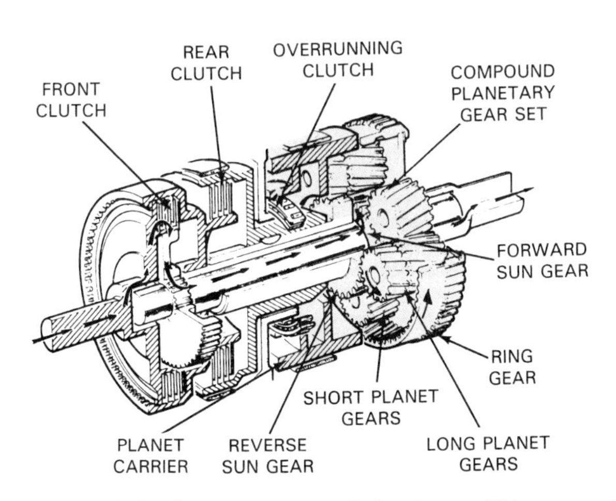

Figure 20-9. A Ravigneaux gear train is shown. This gearset is used on some automatic transaxles, especially those with four forward speeds. It has two sun gears, a single ring gear, a single planet carrier, and two sets of planet gears. (Saab)

lic pressure is released from the piston. Clutch return springs may be one of several different designs.

A hub fits inside of the clutch discs and plates and the clutch drum. External splines on the clutch hub engage the internal teeth on the friction discs. The clutch hub may also be splined to the transaxle input or output shaft or to a part of the gearset.

A drive shell is commonly used to transfer power to the planetary sun gear. It is a bell-shaped part made of sheet metal. Tangs on the shell fit into notches on the drum of the front clutch, or direct clutch. The shell surrounds the neighboring rear clutch, or forward clutch, and the front planetary gearset. The shell is splined to the sun gear, and the clutch drum, shell, and sun gear turn together. A band may be tightened around the drum to hold the sun gear stationary.

Bands and servos

Automatic transaxle transmission bands are friction devices that hold planetary gearset members. They are wrapped around clutch drums and can be tightened to stop drum rotation. Holding the drum stationary also holds the connected planetary member stationary.

The bands are anchored at one end to the case. They are applied and released by band servos, attached at the other end. The band servo is a piston and cylinder assembly. The servo is applied and released by hydraulic pressure and spring pressure.

Band and servo construction. A typical band, Figure 20-11, is a steel strap with a lining of friction material on its inner surface. This friction lining is a combination of paper and asbestos (or similar material). The composite material enables the bands to operate successfully in transmission fluid and to withstand high temperatures.

Transaxle bands attach to the band servo on one end and to the transaxle case on the other. This is shown in Figure 20-12. Many servo linkages are designed to pivot between the servo piston and the band. This increases the applying force of the servo piston. The ends of the band are usually called ears. The band ear opposite from the servo is attached to a band anchor, which holds it stationary. This band anchor may contain an adjustment screw.

A typical band servo, Figure 20-13, is composed of pistons, piston seals, and springs that are installed in a cylinder. The cylinder is usually integral with the transaxle case. Most servos have apply and release sides. Hydraulic pressures on either side of the piston cause it to apply or release the band. Figure 20-14 shows the relative positions of two servos installed in a transaxle case.

Overrunning clutches

Overrunning, or one-way, clutches are mechanical holding members used in automatic transaxles. They are used in conjunction with planetary gearset members to improve shift quality and timing. They are used in conjunction with the torque converter stator to improve efficiency.

An overrunning clutch consists of an inner and an outer race and, typically, a set of springs and rollers. The clutch performs its holding function when the rollers become jammed between the inner and outer races. A sprag clutch is another type of one-way clutch. It uses small metal

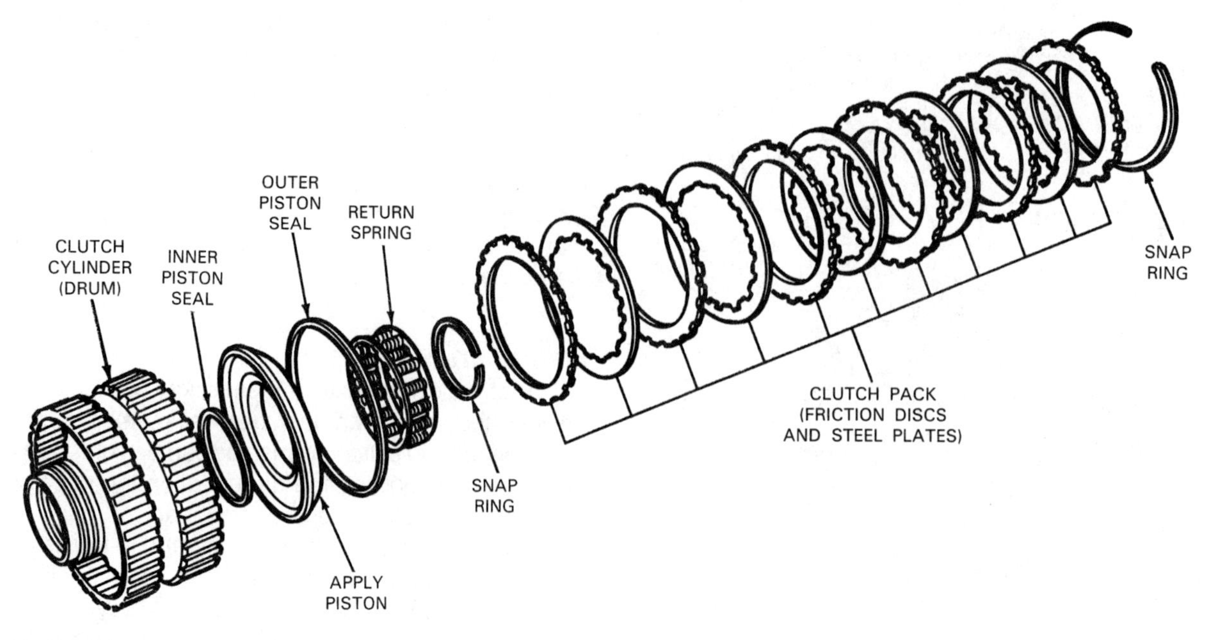

Figure 20-10. This shows an exploded view of a typical multiple-disc clutch. The discs and plates, apply piston, return spring, and drum are shown. (Ford)

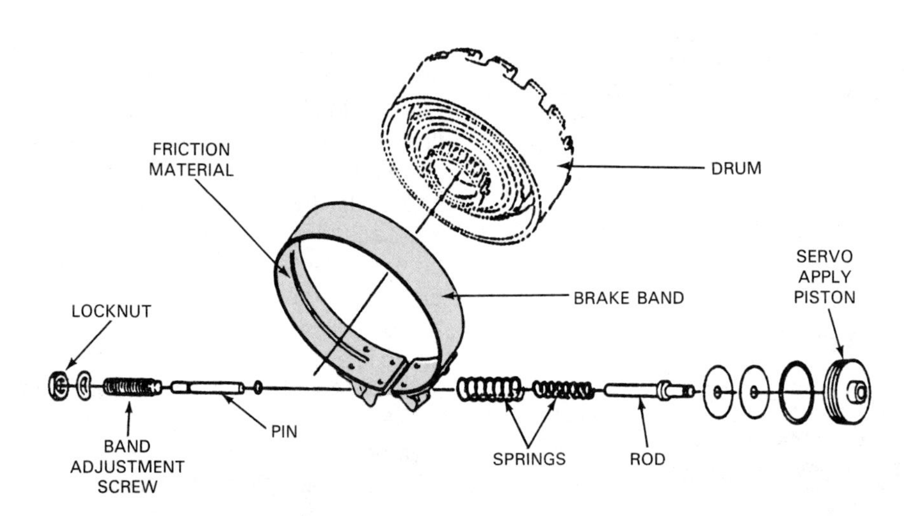

Figure 20-11. Bands are used to wrap around clutch drums. The band will stop the movement of the drum. One of the band ears is attached to the servo, and the other ear is anchored to the case. (Saab)

pieces known as sprags, instead of rollers, to perform the holding action. A sprag clutch is shown in Figure 20-15.

As related to planetary gearset operation, the overrunning clutch works by preventing backward rotation of certain planetary members during shifting. The clutch freewheels when turned in one direction, allowing the planetary member attached to it to turn. It locks up when attempt is made to turn it in the other direction. The outer race of the clutch is usually held to the transaxle case, so the planetary member becomes locked to the case when the overrunning clutch locks up. The action of an overrunning clutch is illustrated in Figure 20-16.

Accumulators

While not itself a planetary holding member, an accumulator is a hydraulic device used in the apply circuit of a band or clutch to cushion initial application. The result is a smoother shift. The accumulator essentially consists of a spring-loaded piston in a cylinder. The piston is pushed to one end of the cylinder by the spring. A hydraulic passageway connects the accumulator to the apply piston of a multiple-disc clutch or band servo.

As the clutch or servo is being applied, fluid in the passageway is received by the accumulator. The accumulator piston begins to move against spring pressure. The expanding cylinder chamber diverts some of the hydraulic fluid from the clutch or servo. As a result, pressure to the clutch or servo does not build up to line pressure immediately, and the initial application of the holding members will be soft. After the accumulator spring is fully compressed, full pressure goes to the apply piston, and the clutch or band will be held tightly.

Some accumulators have oil pressure directed to both sides of the piston. This allows the accumulator to operate during more than one shift, and it also allows the accumulator to vary its action according to overall transmission pressures. Adding this extra pressure to the accumulator is called *charging*.

Transaxle Oil Pump

The transaxle oil pump has several basic functions. It generates system pressure to operate planetary holding members and certain hydraulic valves. It keeps the moving parts of the transaxle lubricated. It keeps the torque converter filled with transmission fluid for proper operation. Finally, it circulates fluid to and from the oil cooler for heat transfer. The oil pump turns whenever the engine is running.

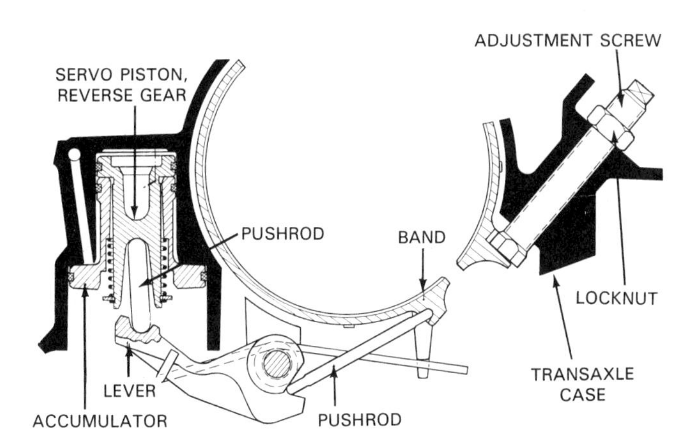

Figure 20-12. A band and servo is shown. The band is attached to the transaxle case. (Saab)

The oil pump shown in Figure 20-17 is a gear pump that is driven by an oil pump shaft, located within the turbine shaft. Other pumps are driven directly by lugs on the pump drive hub of the torque converter. The lugs fit directly into the oil pump. The oil pump turns with the torque converter. The type of drive mechanism used depends on transaxle design and parts layout.

In many transaxles, the oil pump is mounted in a location that is remote from the torque converter and valve body. When the pump is installed in this manner, special oil passages must be drilled in the case or shafts to allow the pump to send oil to other places in the hydraulic system.

Transaxle Oil Filter

Transaxle oil filters are used to remove particles from the automatic transmission fluid. Fluid being drawn into the oil pump must first pass through the filter. This keeps fluid contaminants from entering the pump or the other hydraulic components. Most filters are made of a disposable filter paper or felt. The filter is located inside the oil pan.

Transaxle Oil Cooler

Fluid in automatic transaxles is cooled in the oil cooler. This component consists of a heat exchanger built into a side tank of the engine cooling-system radiator. Cooler lines connect the transaxle to the oil cooler.

The oil cooler is immersed in cooled engine coolant. In operation, hot transmission fluid is pumped from the transaxle to the oil cooler. As the fluid passes through the cooler, its heat is transferred to the engine coolant. The cooled transmission fluid then returns to the transaxle.

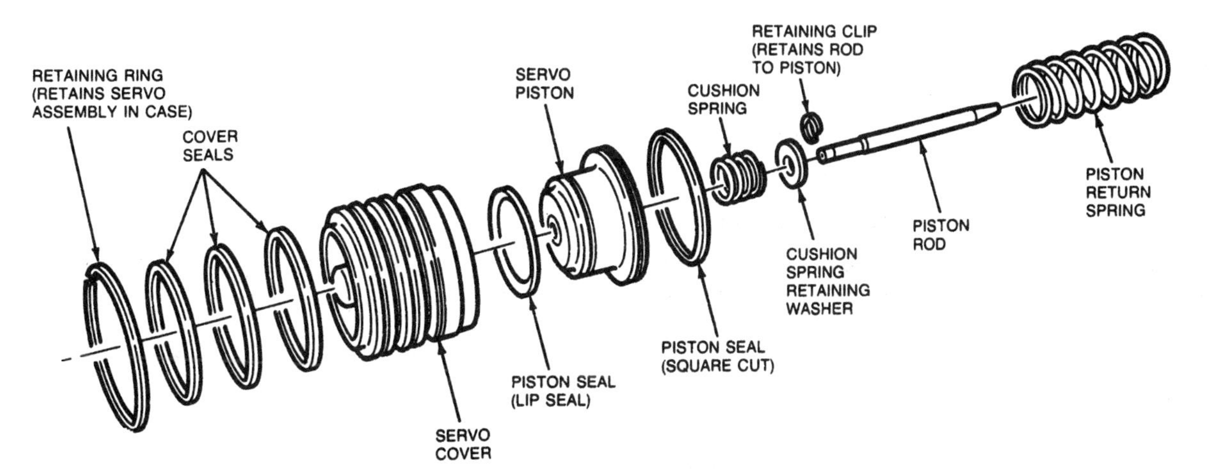

Figure 20-13. This exploded view shows a typical servo assembly removed from its specially machined cylinder. The servo ring seals prevent hydraulic fluid loss to maintain pressure. Most servos have an apply and a release side. (Ford)

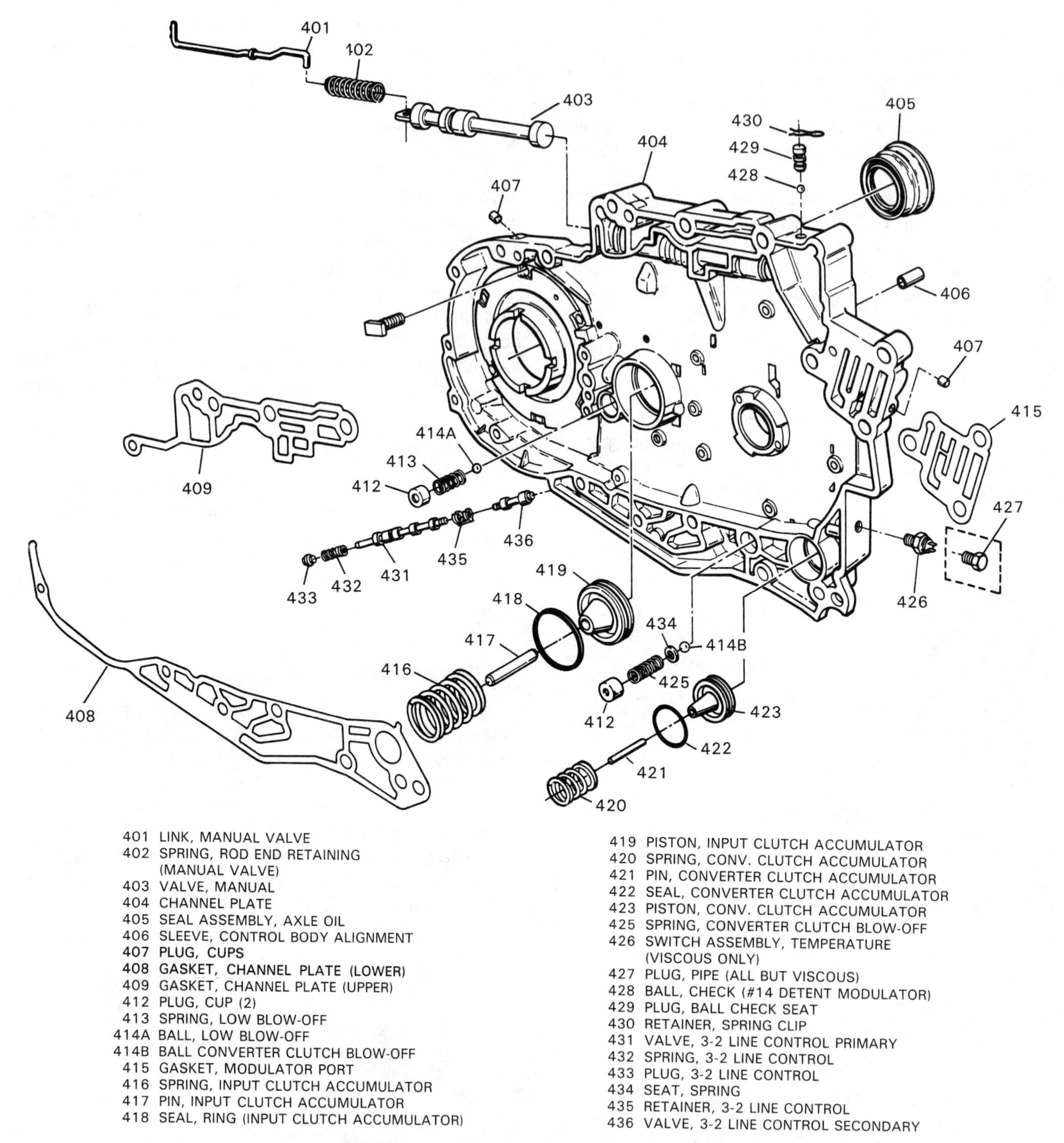

Figure 20-14. In addition to hydraulic valve and other components, this *channel plate*, which fits up against the automatic transaxle case, contains two servos. Servo springs are calibrated for each servo assembly and should not be interchanged. (General Motors)

When a vehicle is used for trailer towing or other extreme operating condition, auxiliary, or external, coolers may be installed to help reduce fluid temperature and extend the life of the fluid. These coolers consist of direct air-cooled heat exchangers that are piped to the existing fluid cooling system. They may be installed to completely replace the oil cooler in the radiator, although this is not recommended without good reason, or they may be helper units that further cool fluid that has passed through the radiator oil cooler. Auxiliary coolers are usually mounted ahead of the radiator.

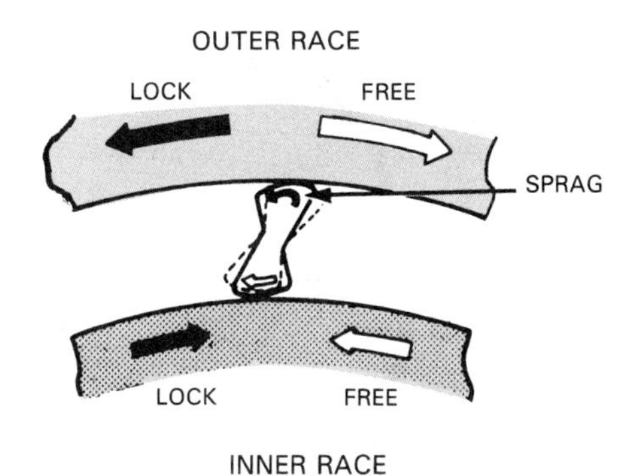

Figure 20-15. The sprag clutch is similar to the roller clutch. The sprag clutch uses sprags, instead of rollers, to perform the same basic holding function. (Honda)

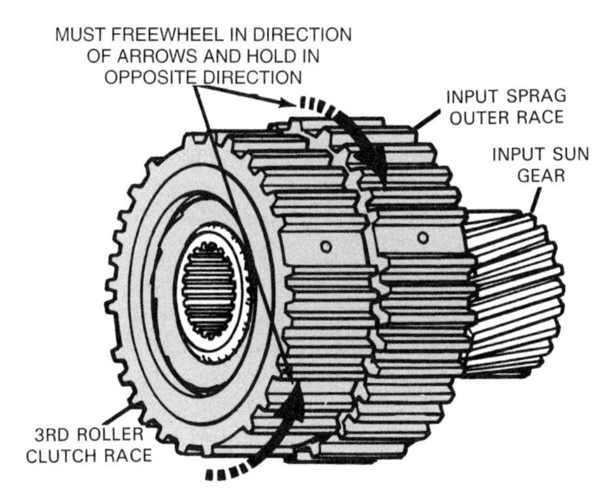

Figure 20-16. The one-way clutch locks when power is applied in one direction and unlocks when power is applied in the opposite direction. Typically, the clutch is only locked when in drive low gear; however, it will freewheel in drive low on a coastdown. (General Motors)

Valve Body

The valve body is a machined casting containing hydraulic valves and passageways. In conjunction with the hydraulic valves, it functions, in part, to redirect and modify oil flow to the transaxle holding members. The valve body may also contain accumulators and spacer plates, which restrict and help direct oil flow.

Figure 20-18 is a phantom view of a typical valve body, showing the positions of hydraulic valves and accumulators. The manual valve is operated by the driver through linkage. The position of the shift valves determines what holding members are applied; therefore, they indirectly determine the transaxle gear ratio. Other valves control the shift feel, smoothness, and timing.

To make maximum use of limited space, some transaxles contain more than one valve body. Each valve body controls a portion of the hydraulic system. Most valve bodies have gaskets and steel spacer plates installed between them and the transaxle case. Valve bodies usually contain check balls to assist in controlling oil flow. Most late model valve bodies contain solenoids that operate valves such as the converter clutch valve. Pressure sensors are installed on the valve body to tell the on-board computer what the transaxle is doing.

Hydraulic Valves

Hydraulic valves are the moving parts of the automatic transaxle that act as pressure and flow controls, directing fluid to other parts of the transaxle. Hydraulic valves are actuated manually or automatically. The same valve used in the automatic transmissions of rear-wheel drive vehicles are used in transaxles. Some valves are located in the valve body, while others lie elsewhere in the transaxle.

Valves are held in the valve body bores by snap rings, retaining pins, or retainer plates. Retainer plates are attached to the end of the valve body bores by machine screws or small bolts. In addition, springs are used in the valve body to hold valves in position when they are not in operation.

Valve springs are carefully sized to achieve a specified pressure. Specific springs are used with each valve, and they cannot be interchanged. For this reason, care should be taken when working on the valve body not to mix up springs.

There are many types of hydraulic valves used in the transaxle. These include pressure regulator, manual, shift, throttle, governor, detent, and check valves. Following is a brief discussion of each type.

Pressure regulator valve

The pressure regulator valve is a spring-loaded valve that controls overall transmission hydraulic system pressure. The primary control of transmission pressure, often called line pressure, is the spring in the pressure regulator valve. The valve is held closed by the spring. When pump output pressure becomes greater than spring pressure, the valve opens and dumps excess fluid to the oil pan or pump intake, thereby reducing pressure. The pressure regulator valve can be located on the pump or in the valve body.

Manual valve

The manual valve is located in the valve body and is operated by the driver through the shift linkage. It allows the driver to select *park, neutral, reverse,* or different drive ranges. When the shift selector lever is moved, the shift linkage moves the manual valve. The valve opens and closes passageways to route hydraulic fluid to the correct components, including other hydraulic valves, clutches, and band servos, as needed for the selected operating range.

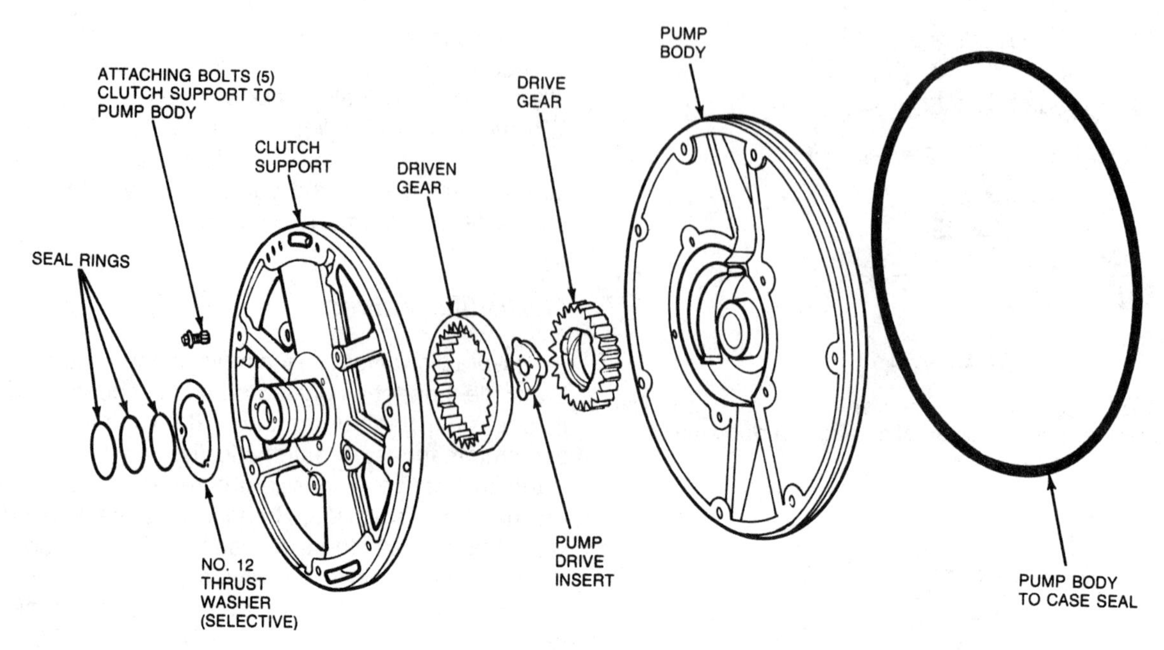

Figure 20-17. This gear pump is driven by a shaft, instead of being driven directly by the torque converter. Transaxles can use either method, depending on the manufacturer. (Ford)

Shift valves

Shift valves, located in the valve body, control transmission upshifts and downshifts. Hydraulic pressures from the other transmission valves act on the end of each shift valve. In addition, a spring, located at one end, puts a preload on the valve. Hydraulic pressure acting on one end works against hydraulic pressure *plus* spring pressure acting on the other end. The valve moves back and forth in the bore as pressures on each end change. When pressures on each end are equal, the valve is balanced and does not move.

Line pressure is directed to the center of the shift valve. It is used to apply the planetary holding members. As passageways leading to these devices are uncovered by movement of the shift valve, line pressure will pass through the valve annular groove and be routed to the holding member.

Throttle valve

The throttle valve modifies transaxle throttle pressure according to engine throttle opening. Throttle pressure is sent to transaxle valves such as the shift valves and pressure regulator valve. The throttle valve may be operated by linkage or by a vacuum modulator. In some instances, the throttle valve is computer controlled.

Most transaxle throttle valves are operated by cable linkage, as in Figure 20-19. The cable connects the throttle valve in the transaxle to the throttle lever on the vehicle throttle body or carburetor. The throttle lever, in turn, is connected to the throttle plate(s) located in the throttle body or carburetor. Engine power increases with the opening of the throttle plate and decreases with its closing.

Movement of the accelerator and, therefore, the throttle lever causes the T.V. cable to move. This movement is transferred to the throttle valve. Throttle pressure always increases as the throttle plate is opened and decreases as the throttle plate is closed. Throttle cable linkage always has some provision for adjustment.

A few transaxles have vacuum modulators. The modulator is a vacuum-operated diaphragm enclosed in a sealed container. One side of the modulator is connected to the engine intake manifold by tubing. Changes in manifold vacuum affect how much the diaphragm moves against internal modulator spring pressure.

The other side of the modulator diaphragm is connected to the throttle valve. The throttle valve is moved by the diaphragm as the engine vacuum changes, modifying transaxle throttle pressure as changes in engine load occur. If the load is high, the manifold vacuum will be low, and throttle pressure will be raised. If engine load is low, vacuum will be high, and throttle pressure will be lowered.

Governor valve

The governor valve takes line pressure at the valve and modifies it according to vehicle speed. Greater speed causes the governor to raise oil pressure at the shift valves. Lower speed causes the governor to reduce pressure at the shift valves. The governor valve is driven by the transaxle output shaft. It is mounted in the transaxle case and is typically driven through gears similar to speedometer drive and driven gears. Note, however, that a vehicle speed

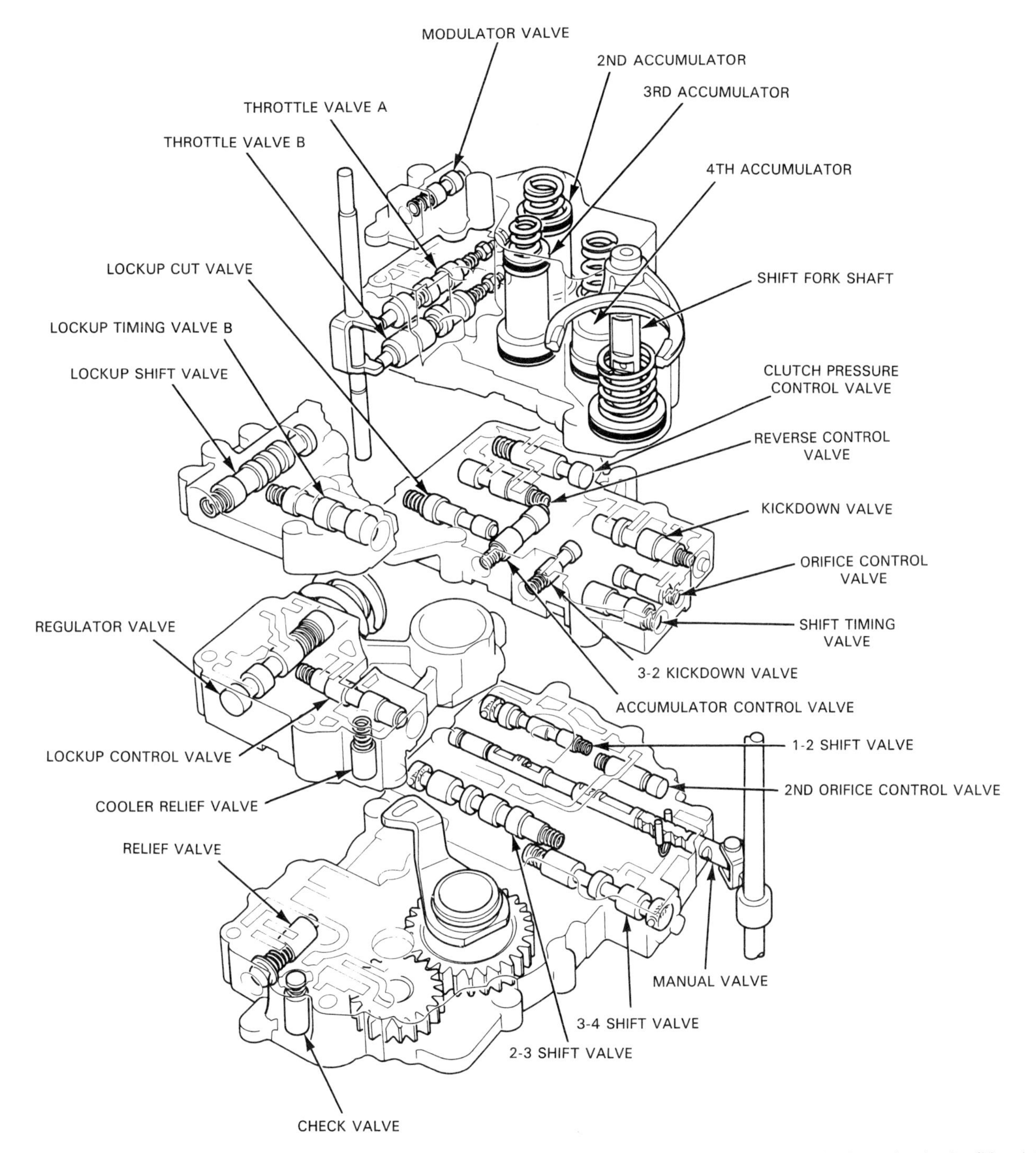

Figure 20-18. A phantom view of a valve body is shown. Note hydraulic valves and shift accumulators in the valve body. (Honda)

sensor and computer signals to a solenoid can also control the governor valve.

Two types of governors are generally found on automatic transaxles. One type contains a valve that is moved as a result of weights. As the governor rotates, the weights, acting against spring pressure, are thrown outward, causing the valve to move. This, in turn, partly opens the governor port to line pressure. At low speeds, the valve is almost completely closed, and governor pressure directed to the shift valves is very low. As vehicle speed increases, the weights open the valve. This allows increased pressure to go to the shift valves.

Note that a variation of this valve is without separate weights; the weight is an integral part of the valve itself. The motion of the governor, aided by the weight, causes the valve to move outward and open. See Figure 20-20.

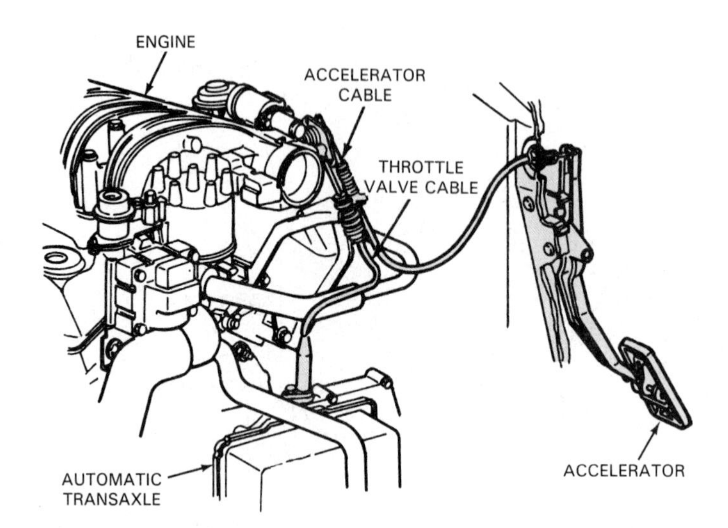

Figure 20-19. The throttle valve (T.V.) cable connects the throttle valve in the transaxle to the vehicle throttle body or carburetor. Notice the accelerator cable, or throttle linkage, shown here, which transfers movement of the accelerator to the throttle lever. (Ford)

The other, less-common kind of governor contains a valve or check balls that are closed by the motion of the governor, again, with the help of weights. This kind of governor is shown in Figure 20-21. Line pressure reaches this governor through a small opening, or orifice. The orifice limits the amount of oil that can flow in the governor circuit. At low speeds, the check balls fall open, and hydraulic fluid leaks out through the valve opening. The governor pressure is zero or close to zero. As the vehicle speeds up, the weights cause the check balls to close the valve opening. As the opening is closed off, governor pressure builds up. This type of governor is less likely to stick, but it is harder to precisely control.

Detent valve

The detent valve provides a passing gear, which is a forced downshift when the accelerator is pushed completely to the floor. The detent valve sends extra pressure to the shift valves. This extra pressure causes the shift valves to move to the downshifted position, putting the transaxle in a lower gear. The detent valve can be operated by linkage (kickdown linkage) from the engine throttle or by an electric solenoid. In some designs, there is not a separate detent valve; the throttle valve also serves this additional function.

Check valves

A check ball is a common type of check valve in an automatic transaxle. Check balls are one-way valves, which allow fluid to flow in one direction only. The steel balls are installed in chambers next to holes in the valve body spacer plate.

In operation, fluid flow in one direction will push the check ball into the spacer plate hole, sealing it. Fluid flow in the opposite direction will push the ball out of the hole allowing fluid to flow.

Automatic Transaxle Shift Linkage

The shift linkage connects the shift selector lever, which may be mounted on the floor or on the steering column, to the automatic transaxle. It consists of a series of levers and linkage rods or cables. These transfer motion to the manual valve of the transaxle valve body to select the desired operating range. The design of the shift linkage is like that on a rear-wheel drive automatic transmission, including inner and outer manual and throttle valve levers and a cockscomb on the inner manual lever.

The shift linkage also controls the operation of the parking pawl. The parking pawl mechanically locks the output shaft, preventing it from rotating. It is applied when the selector lever is moved to *park.* Figure 20-22 illustrates how a parking pawl engages the gear splined to the transaxle output shaft. When the pawl moves downward into the space between two teeth, the gear and, therefore, the output shaft are locked to the case. At the same time, the linkage moves the manual valve to neutral. Figure 20-23 shows the position of the parking gear in relation to other output shaft components.

Transaxle Differential Assembly

Differential gears in an automatic transaxle operate in the same manner as in a manual transaxle. The differential assembly allows the front wheels to turn at different speeds when the vehicle is turning a corner.

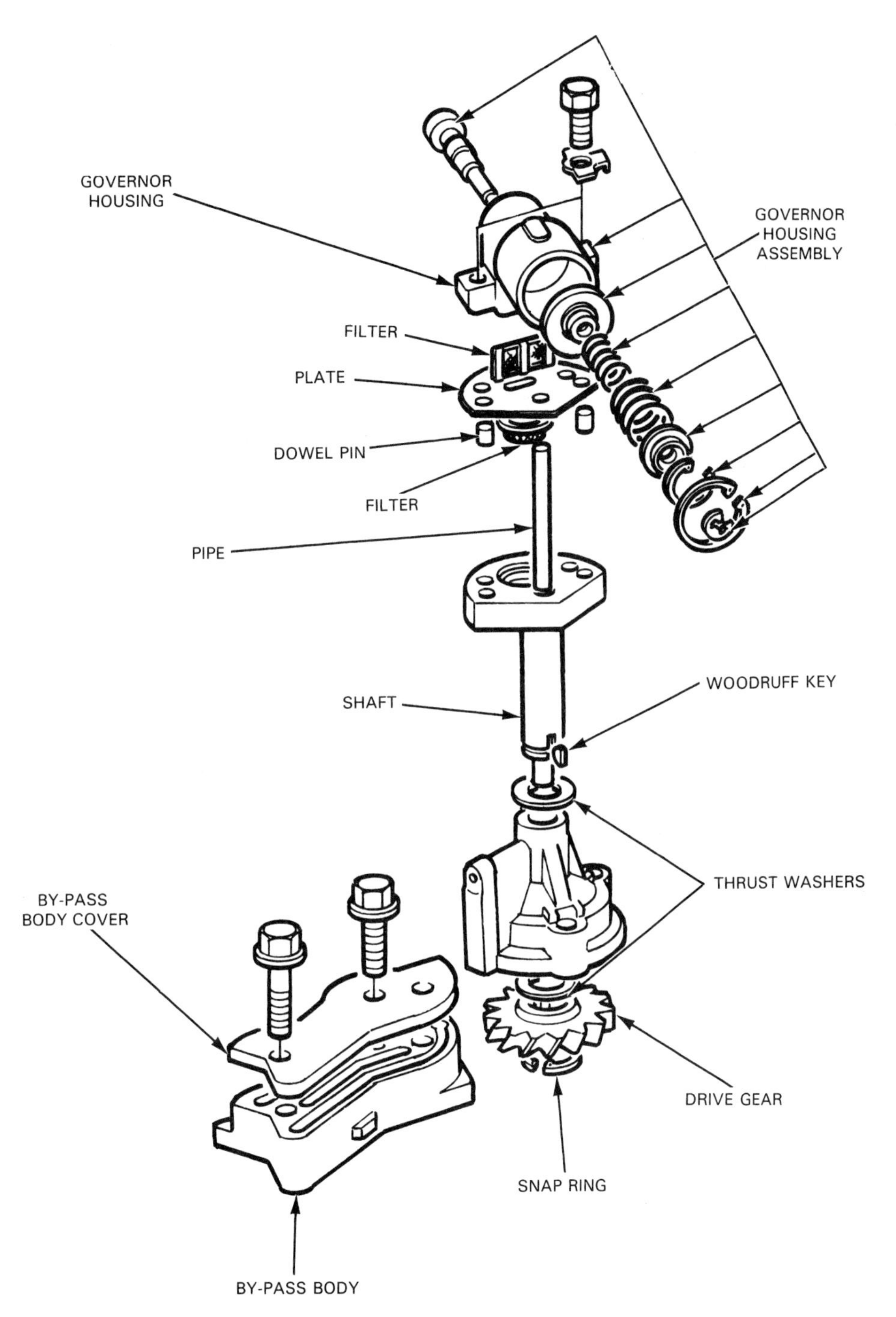

Figure 20-20. Governors are gear driven by the output shaft. The motion of the governor causes the valve to move. Governor pressure varies directly with vehicle speed. In the governor shown, the weight is integral with the valve. (Honda)

Ring and pinion gears are used in some, but not all, automatic transaxles. The drive pinion gear is installed on the output shaft. It meshes with the ring gear in the differential portion of the transaxle. Most transaxle ring and pinions are straight or helical gears, but some automatic transaxles have hypoid ring and pinion gears.

Figure 20-24 shows a differential assembly that does not use a ring and pinion. It is a combination of a planetary gearset and a differential gearset. The planetary gearset is a reduction gear that determines the final drive ratio. Placing the parts together reduces weight and complexity.

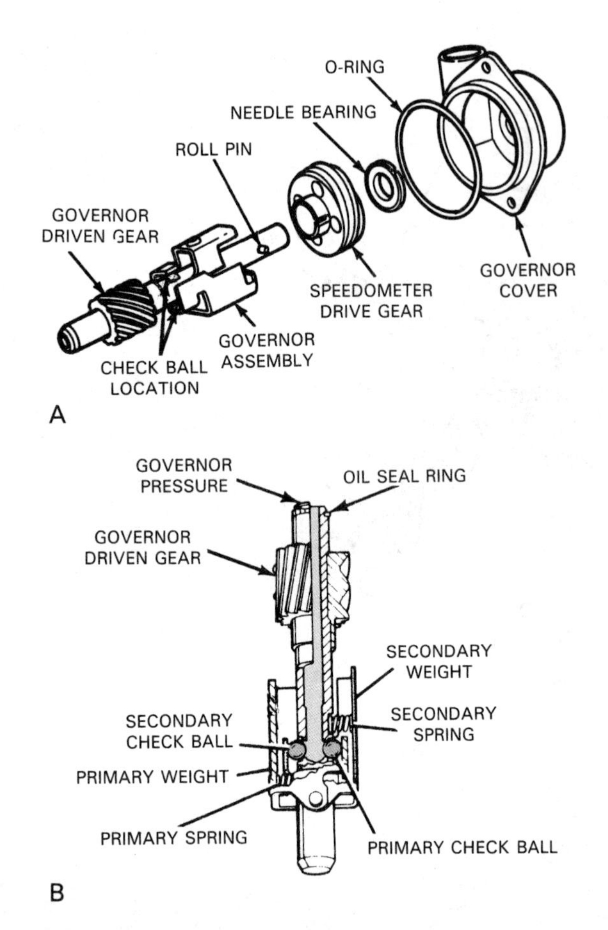

Figure 20-21. Check ball openings in this type of governor begin to close as vehicle speed increases, restricting oil flow. As flow is restricted, pressure increases. A–Exploded view of governor. B–Section view shows position of check balls, used in this type of governor. (Ford, General Motors)

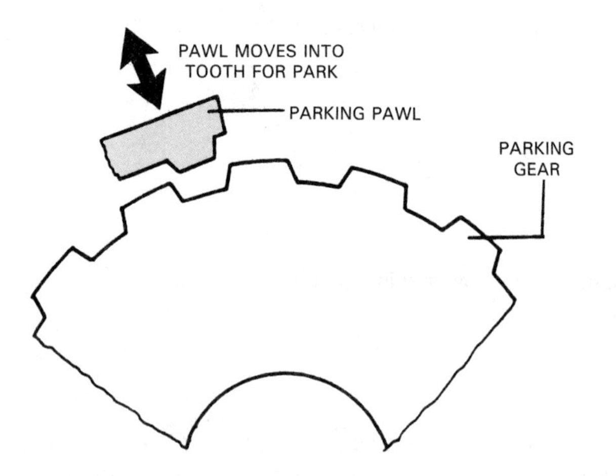

Figure 20-22. The parking pawl locks the output shaft by engaging the teeth on the parking gear, which is splined to the output shaft. The pawl locks the gear, and the output shaft cannot move. (Saab)

The housing enclosing a differential assembly is usually cast integral with the housing for the transaxle transmission. There are usually one or more sheet metal access covers, which are removed for service. Some differential housings are separate assemblies. Many of these are made of cast iron for added strength. CV axle openings on the housing have seals to retain transmission fluid.

Automatic Transaxle Case

The automatic transaxle case is normally an aluminum casting, machined to serve as a mounting and aligning surface for the moving parts of the transaxle. Most of the stationary parts of the transaxle are contained in the case.

Although most automatic transaxle cases are one-piece castings, the transaxle case can be divided into three main areas: the converter housing, the main case, and the differential housing. See Figure 20-25. The converter housing covers and protects the torque converter. The main case contains the other transmission parts. The differential housing contains the differential assembly and the inboard ends of the CV axles. Some transaxle cases have extension housings or separate sections called *channel plates,* containing some of the hydraulic passageways.

The case is carefully machined to provide an accurate mounting to the engine and to make a leakproof seal with the valve body and other hydraulic components. Cylinders are machined in the case to accept servo and accumulator pistons. Oil passageways are cast or drilled in the case to connect the hydraulic components of the transaxle. See Figure 20-26.

Sheet metal covers are installed at the end or on the sides of the case, as shown in Figure 20-26. The covers allow easier removal of transaxle components. The covers enclose certain transaxle components, such as the valve body, and many serve as oil pans. Some transaxles have a cover mounted at the top of the case.

Many external parts are installed on, or enter through, the transaxle case. Shift linkage, T.V. linkage, and oil cooler lines are examples. Many electrical wires pass through the case to operate solenoids or to connect to pressure sensors on the valve body. These wires are usually enclosed in special case connectors equipped with O-rings to prevent leaks. Speedometer assemblies are also installed on the case, as shown in Figure 20-27. The speedometer may be driven by a gear on the transaxle output shaft or differential assembly or on one of the CV axles.

Transaxle Oil Pan

The transaxle oil pan is the transmission fluid reservoir. The oil pan is installed at the bottom of the transaxle. Refer to Figure 20-26. Transaxle oil pans are wide and relatively shallow. The oil pan is attached to the transaxle case by

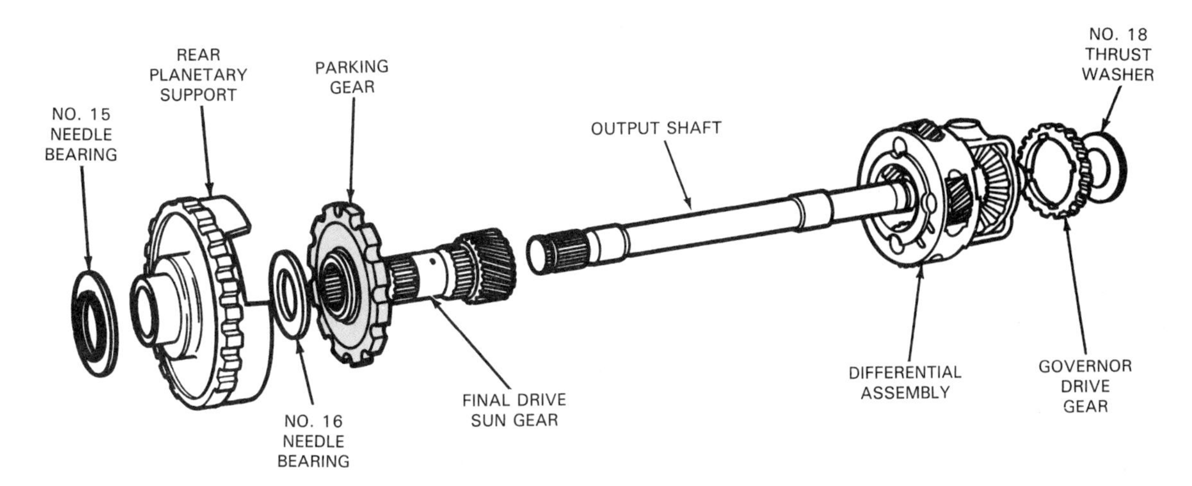

Figure 20-23. Note the position of the parking gear in this exploded view of an output shaft assembly. The parking pawl engages this gear to lock the output shaft. (Ford)

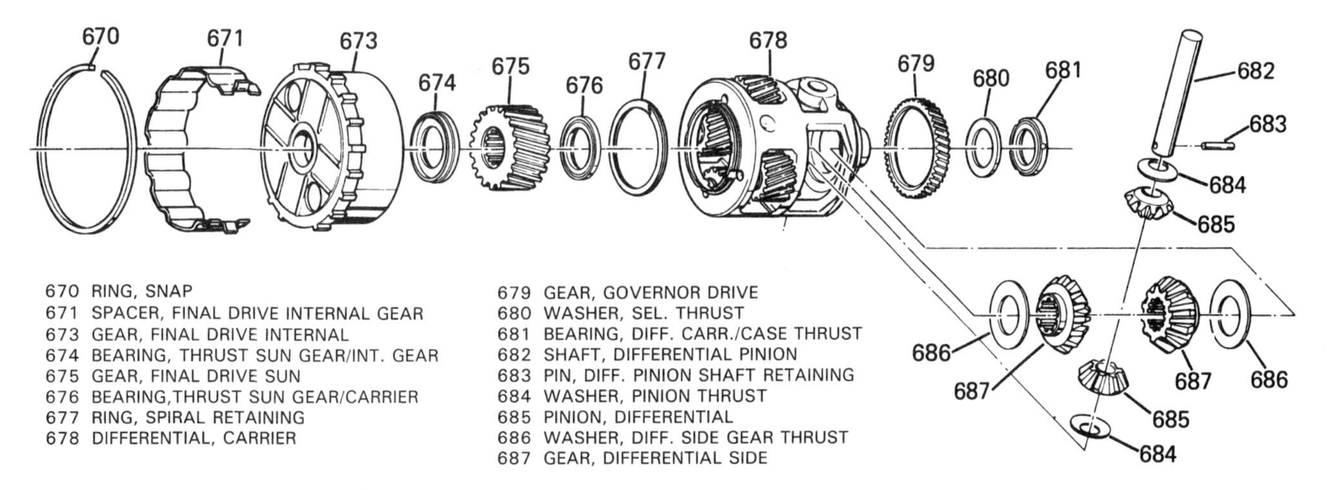

Figure 20-24. This exploded view shows planetary gears and differential gears installed into one compound carrier. The planetary gear is a final drive reduction gear. Combining parts reduces weight and complexity. (General Motors)

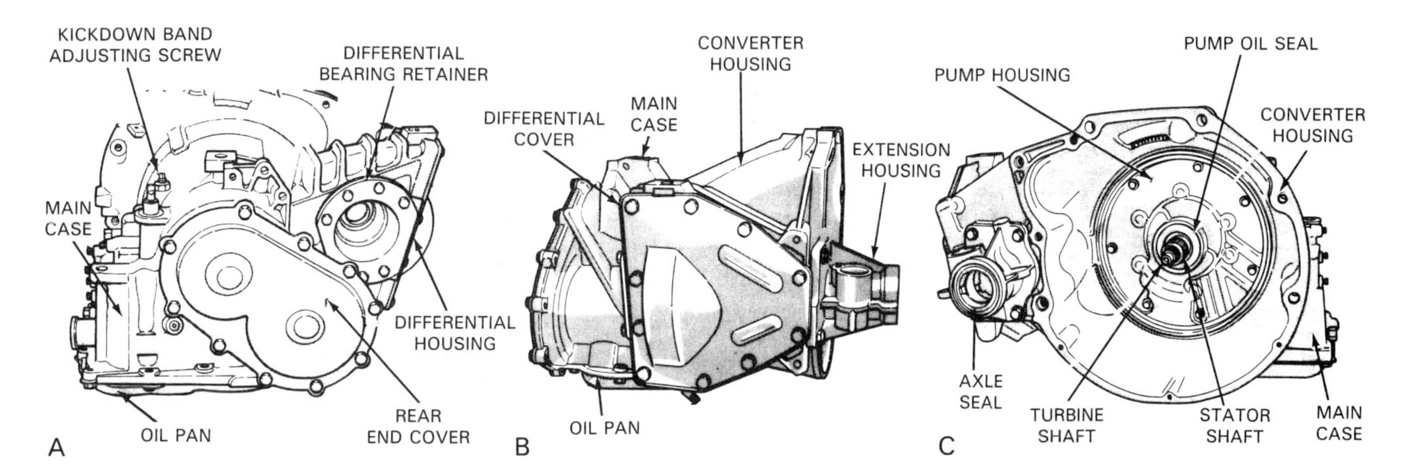

Figure 20-25. A typical automatic transaxle case is shown here. A–Note differential housing and oil pan shown in this rear view of the transaxle case. Also shown is a sheet metal cover that can be removed for transaxle service. B–Note differential cover and converter housing shown on this side view of the transaxle case. Only a few transaxles have separate extension housings at the CV axle opening. C–Notice the stator shaft and turbine shaft in this front view showing the converter housing. (Chrysler)

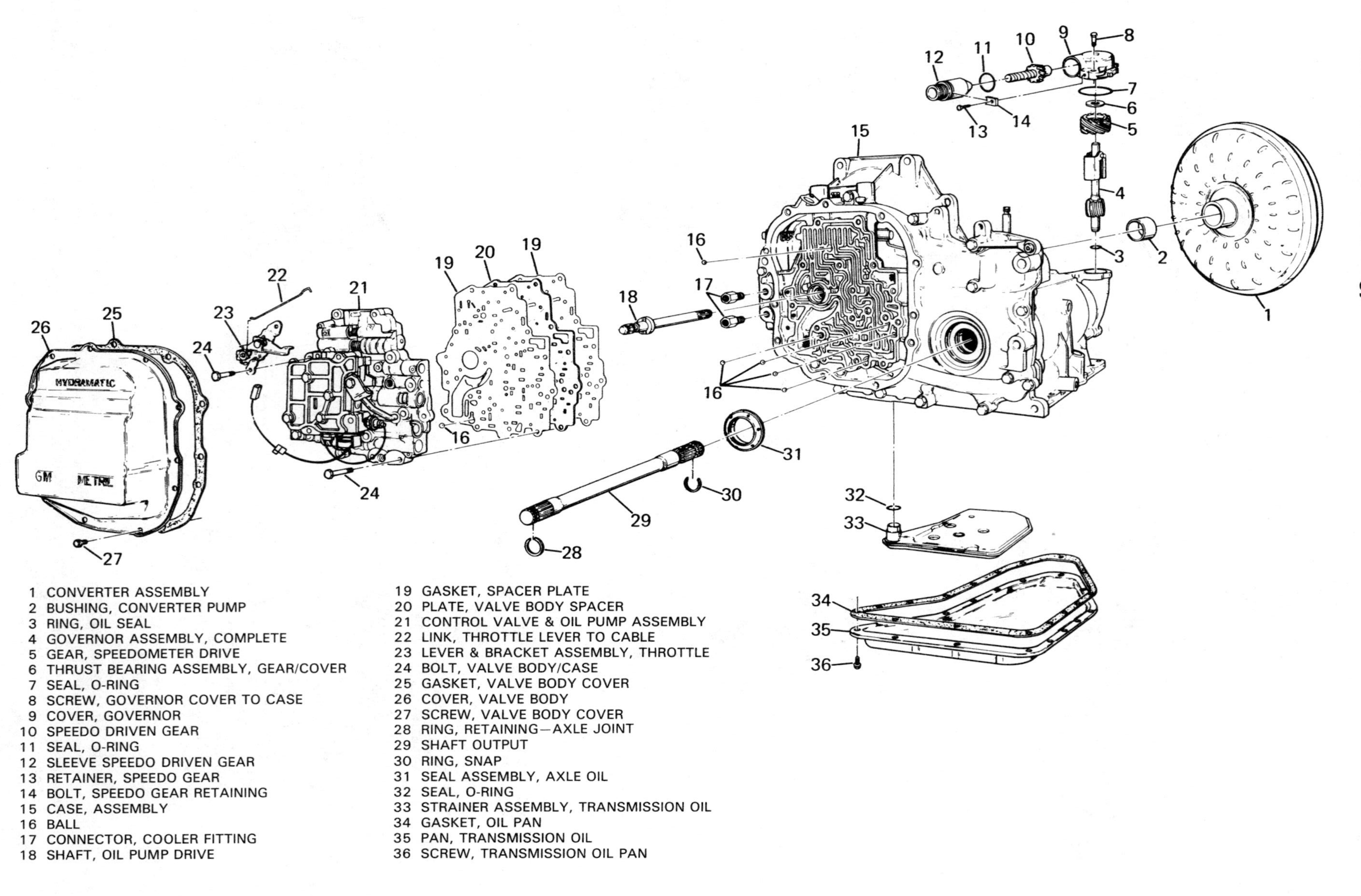

Figure 20-26. Notice oil passageways in this exploded view of a transaxle case. Sheet metal covers installed on transaxle case enclose components such as the valve body. The covers may serve as oil pans. (General Motors)

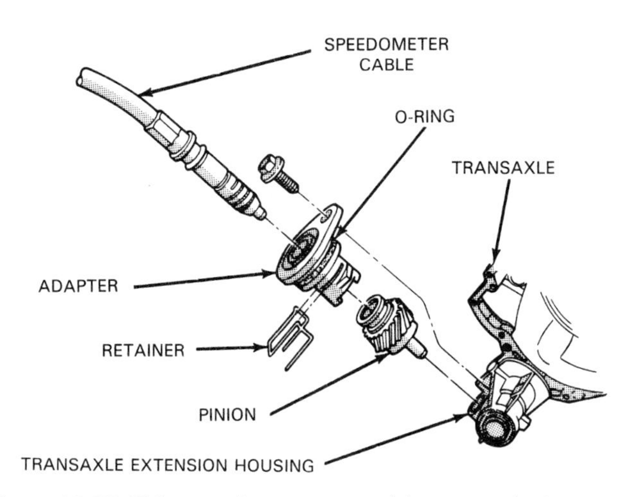

Figure 20-27. This speedometer assembly mounts in the transaxle case. The pinion gear is typically driven by the output shaft. The gear drives the speedometer cable, which gets translated by another assembly into road speed. (Chrysler)

bolts. A gasket is always used between the oil pan and case. The transaxle oil filter is installed in the oil pan.

Electronic Components

Most modern vehicles are controlled, at least in part, by an on-board computer. In addition to controlling the engine and emission systems, the on-board computer receives inputs and sends commands to the automatic transaxle. Automotive computer systems include input and output devices such as sensors and solenoids.

Through electronic hardware, the computer typically monitors engine speed, load, and throttle position, among other variables. It can then provide control for transaxle shift points, torque converter lockup, and other functions. This keeps the transaxle functioning at maximum efficiency.

Automatic Transaxle Power Flow

Power flow through transaxles is similar for all transaxle designs. The main differences depend on whether the vehicle engine is longitudinally or transversely mounted.

Figure 20-28 shows one make of transaxle used with a transverse engine. Power flows from the engine, through the torque converter and turbine (input) shaft, and on to the transaxle multiple-disc clutches. The planetary holding members match the proper gear ratio to conditions. Power flows from the clutches to the planetary gears. The planetary gears drive the output shaft. A gear on the output shaft drives a gear on the transfer shaft. The transfer shaft pinion gear, on the other end of the transfer shaft, drives the differential ring gear and case assembly. Power leaves the differential assembly and drives the CV axles and front wheels.

Figure 20-29 shows one make of transaxle used with a longitudinal engine. Engine power enters the torque converter and flows through the turbine shaft to the chain drive. From the driven sprocket, power flows through the input shaft to the planetary gears. The hydraulic system and holding members select the proper gear ratio for driving conditions. Power from the output shaft turns the pinion gear. The ring and pinion turn the power flow 90°. (They are not hypoid gears, however, because their centerlines are not offset.) Power flows through the differential assembly and drives the CV axles and front wheels.

Automatic Transaxle Hydraulic Circuits

Hydraulic circuit operation of an automatic transaxle resembles that of a rear-wheel drive vehicle. An example is shown in the hydraulic diagram of Figure 20-30. This schematic shows how oil flowing in the torque converter is controlled. The converter control valves shown are typical of lockup torque converters found on all automatic transmissions and transaxles.

Figure 20-31 illustrates oil flow to the shift valves and other hydraulic valves in an automatic transaxle valve body. Oil pressure from the pump is controlled by the pressure regulator valve. From here, oil travels to the manual valve. The manual valve sends pressurized oil to the shift valves, governor valve, and throttle valve. On this transaxle, the throttle valve is operated by a vacuum modulator.

Governor pressure and throttle pressure both act on the shift valves to control shift points. The accumulators help to cushion shifts. Other valves modify the clutch and band application to precisely control shift timing and feel.

The hydraulic diagram in Figure 20-32 illustrates oil flow in first gear. The converter is filled, and the front clutch is applied. The N-D (neutral-drive) accumulator, which functions to smooth application of the front clutch, is filled with oil. The apply side of the 1-2 shift accumulator will fill with oil upon upshift to second gear, smoothing application of the front band. Note that this transaxle does not have a lockup torque converter.

The hydraulic diagram in Figure 20-33 shows the application of a band servo in second gear. The 1-2 shift valve has moved to the upshifted position, causing line pressure to be sent to the servo piston to apply the band. Notice that oil is also going to the rear clutch, as well as to the converter and 1-2 accumulator. When the transaxle shifts to third, pressure will be sent to the release side of the servo and to the third gear clutch.

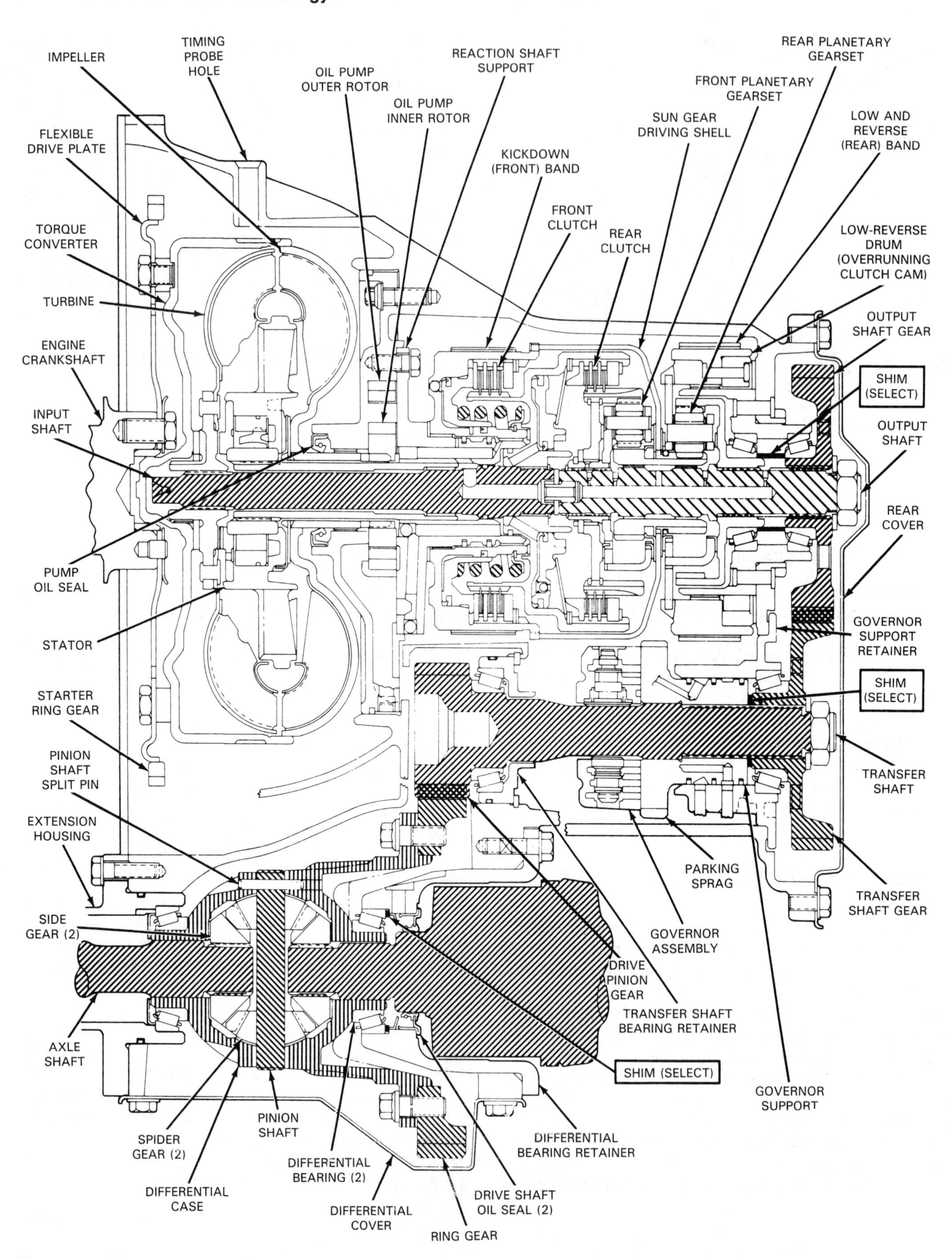

Figure 20-28. Chrysler TorqueFlite A400 series automatic transaxles are used with transverse engines. The power flow on transaxles used with transverse engines does not change angle between the engine and the drive wheels. (Chrysler)

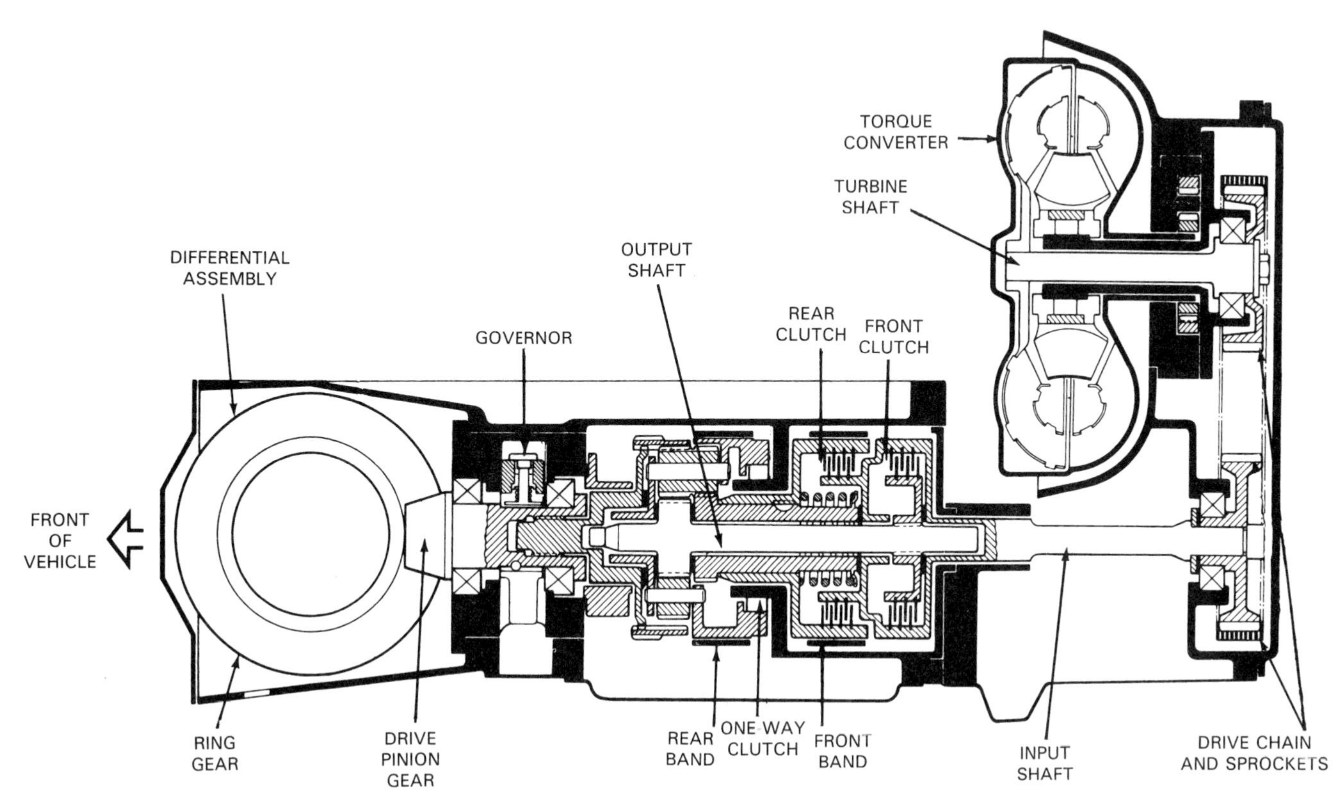

Figure 20-29. This transaxle is used with a longitudinal engine. Power takes a 90° turn through the ring and pinion. (Saab)

Automatic Transaxle Designs

The following section identifies some of the more common automatic transaxles in use today. In addition to these transaxles, refer to the Chrysler transaxle shown in Figure 20-28. This transaxle is the basis of almost all front-wheel drive Chrysler products.

Figure 20-34 shows a General Motors *Turbo Hydra-Matic 125 (THM 125).* The THM 125 is used on many General Motors' vehicles. It uses a standard torque converter. The torque converter turbine shaft drives the transaxle input shaft and gear train through a chain drive. The final drive assembly drives the differential, which drives the output shaft. The THM 125 is a 3-speed transaxle. Another version of this transaxle, the THM 125C, has a converter clutch (lockup) torque converter.

Figure 20-35 shows a Ford *AXOD.* This is an automatic transaxle with a converter clutch torque converter and chain drive. It is used on many smaller Ford products.

Figure 20-36 is a General Motors *Turbo Hydra-Matic 325-4L (THM 325-4L),* used on many General Motors vehicles with longitudinal engine placement. The THM 325 family of transaxles has been available in 3- and 4-speed versions, with and without converter clutch torque converters. The transaxle shown has an overdrive unit. The 4L designation indicates that this is a 4-speed transaxle. The differential assembly is not shown.

Figure 20-37 shows a Honda automatic transaxle. This transaxle is unique, since it does not use planetary gearsets. Helical gears are applied and released by the action of four multiple-disc clutches and a one-way clutch. The unit also has a *servo valve,* which is activated to move the *reverse selector*–a shift fork that is used to put the transaxle in reverse.

Figure 20-38 shows a transaxle used on many Ford products. It uses a unique torque converter that incorporates a planetary gearset. This gearset allows the converter to function as a lockup torque converter without the use of a lockup clutch assembly. The oil pump in this transaxle is also located at the rear of the case.

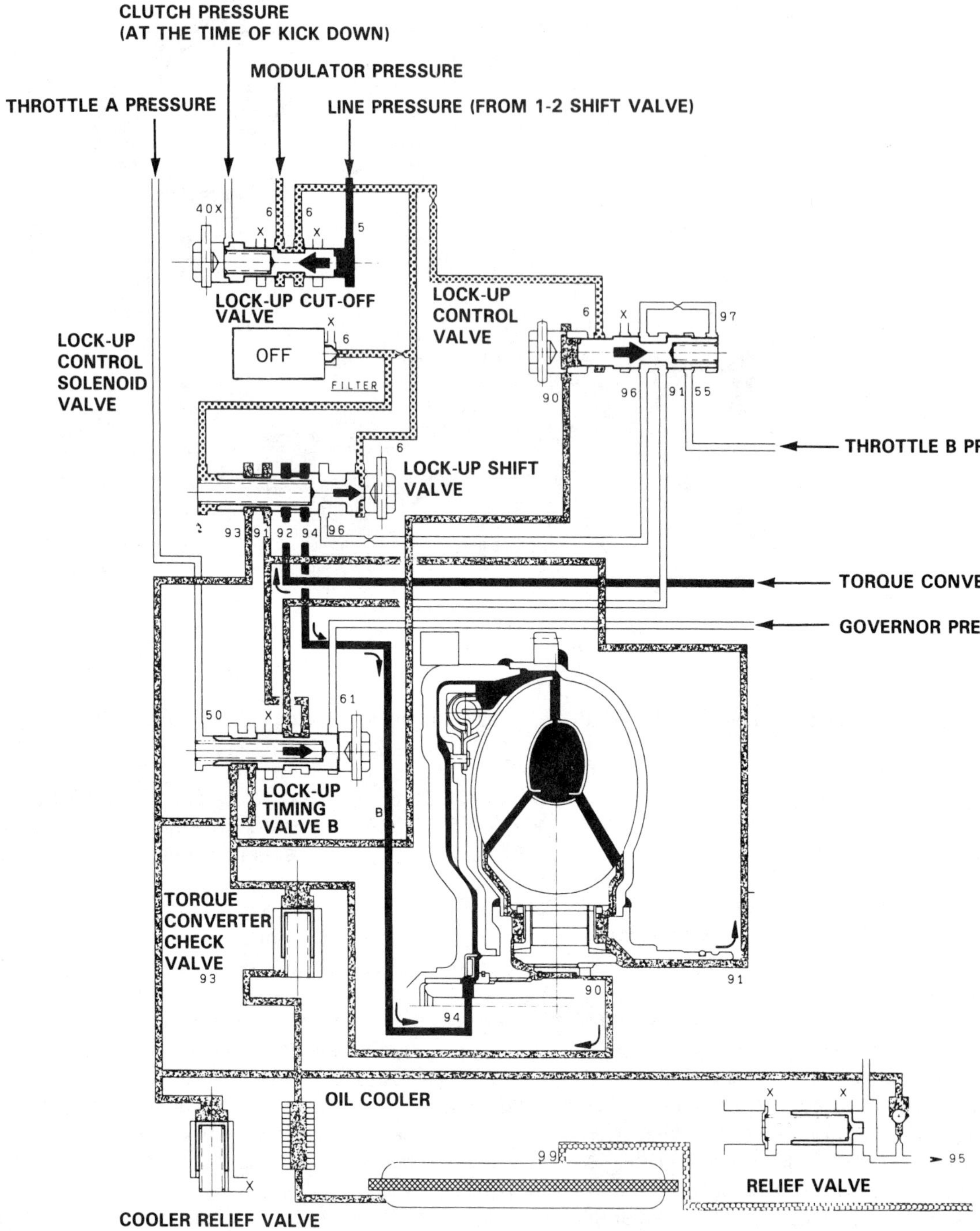

CLUTCH PRESSURE
(AT THE TIME OF KICK DOWN)
MODULATOR PRESSURE
THROTTLE A PRESSURE
LINE PRESSURE (FROM 1-2 SHIFT VALVE)
LOCK-UP CUT-OFF VALVE
LOCK-UP CONTROL VALVE
LOCK-UP CONTROL SOLENOID VALVE
OFF
FILTER
THROTTLE B PRESSURE
LOCK-UP SHIFT VALVE
TORQUE CONVERTER PRESSURE
GOVERNOR PRESSURE
LOCK-UP TIMING VALVE B
TORQUE CONVERTER CHECK VALVE
OIL COOLER
RELIEF VALVE
COOLER RELIEF VALVE

Figure 20-30. The converter clutch control system is shown in this illustration. The interaction of the hydraulic system and the clutch control valves applies the clutch at the proper time. (Honda)

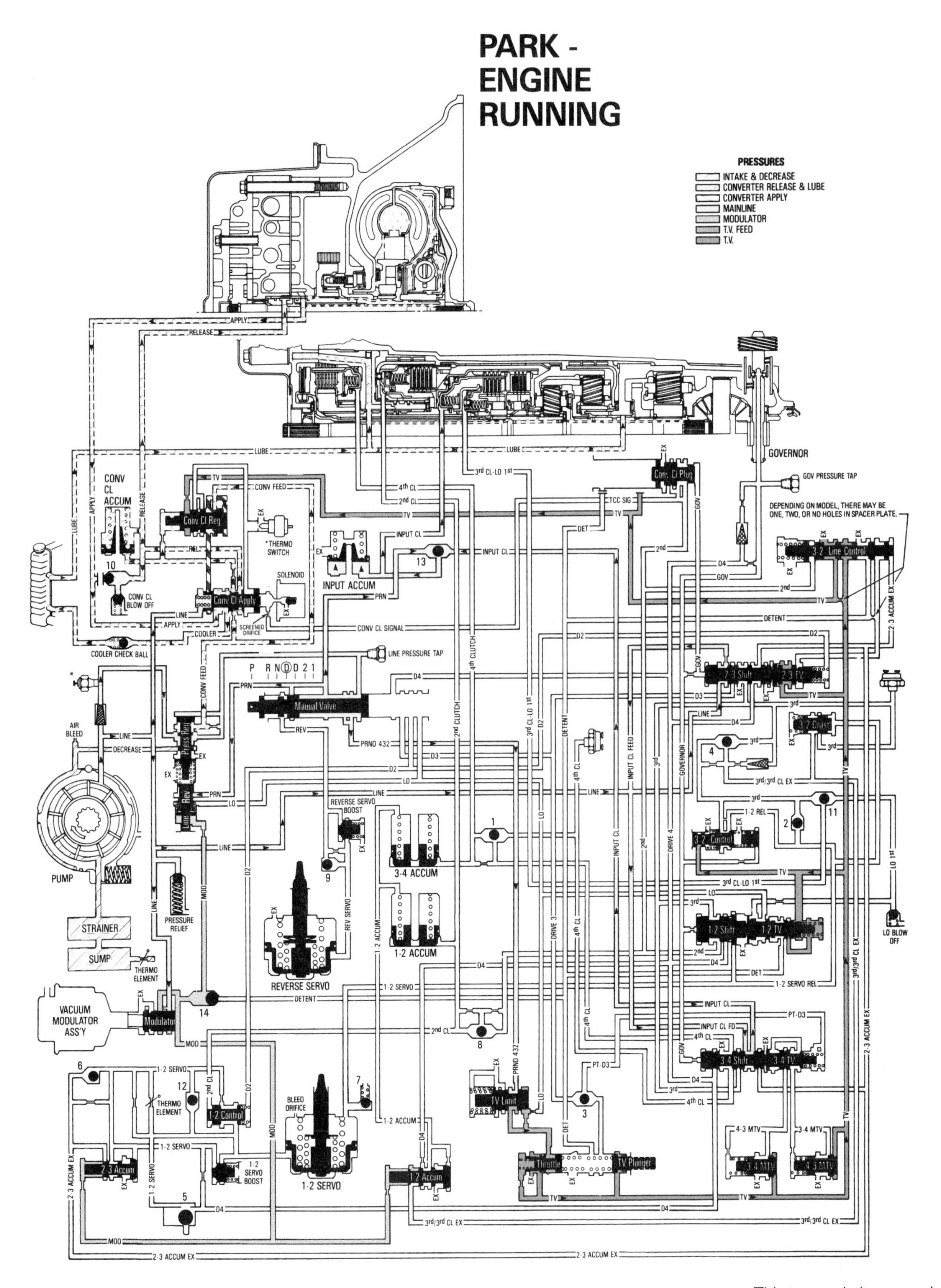

Figure 20-31. This diagram shows oil flow in a 4-speed automatic transaxle with a lockup torque converter. This transaxle is unusual in that it uses a vacuum modulator to control the throttle valve. (General Motors)

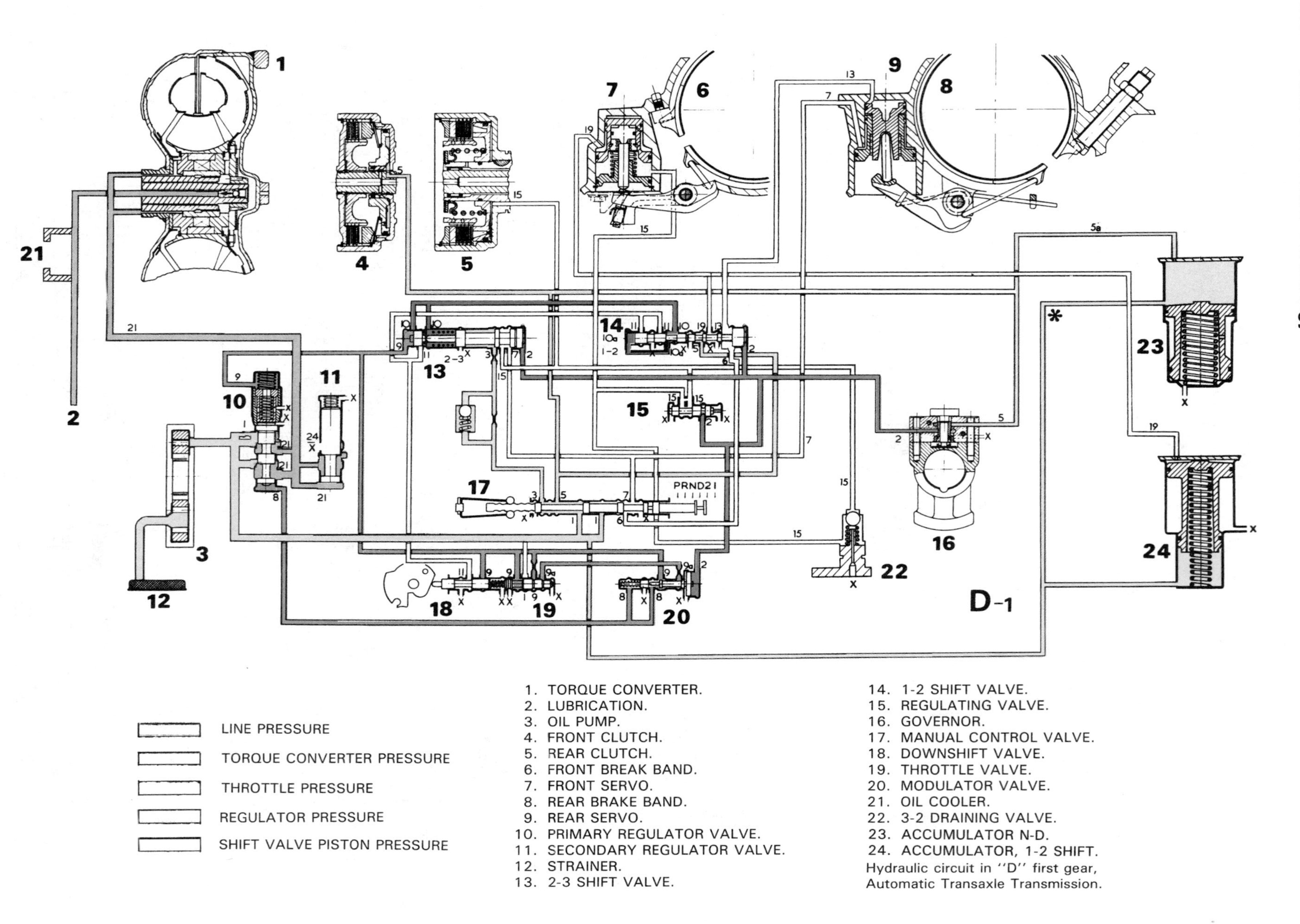

Figure 20-32. This hydraulic diagram shows a transaxle with two band servos and two clutch packs. The diagram depicts the transaxle hydraulic system in the drive first-gear position. (Saab)

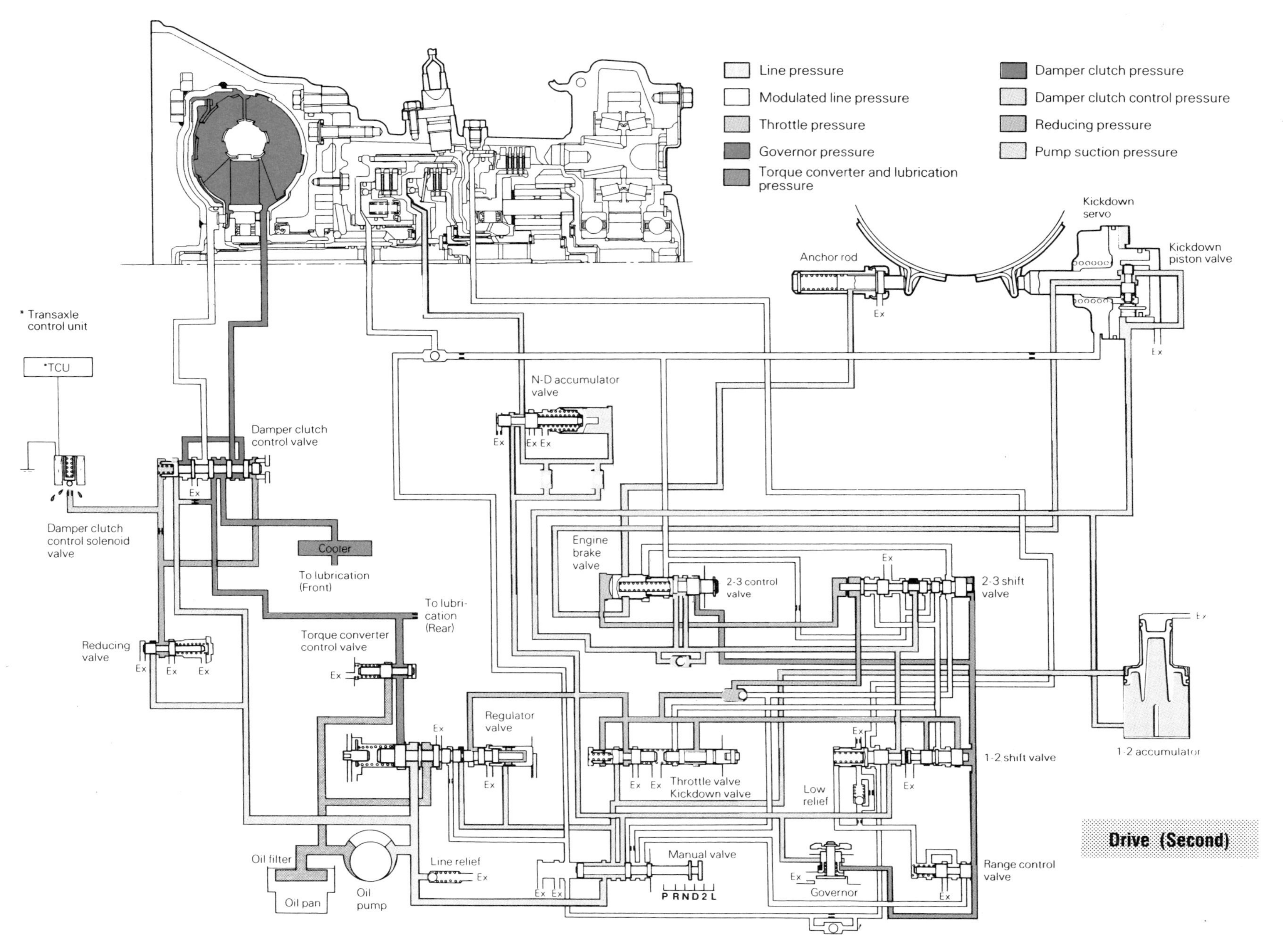

Figure 20-33. This hydraulic diagram depicts a transaxle hydraulic system in second gear. Oil is sent to the apply side of the kickdown servo, and the servo applies the band. (Ford)

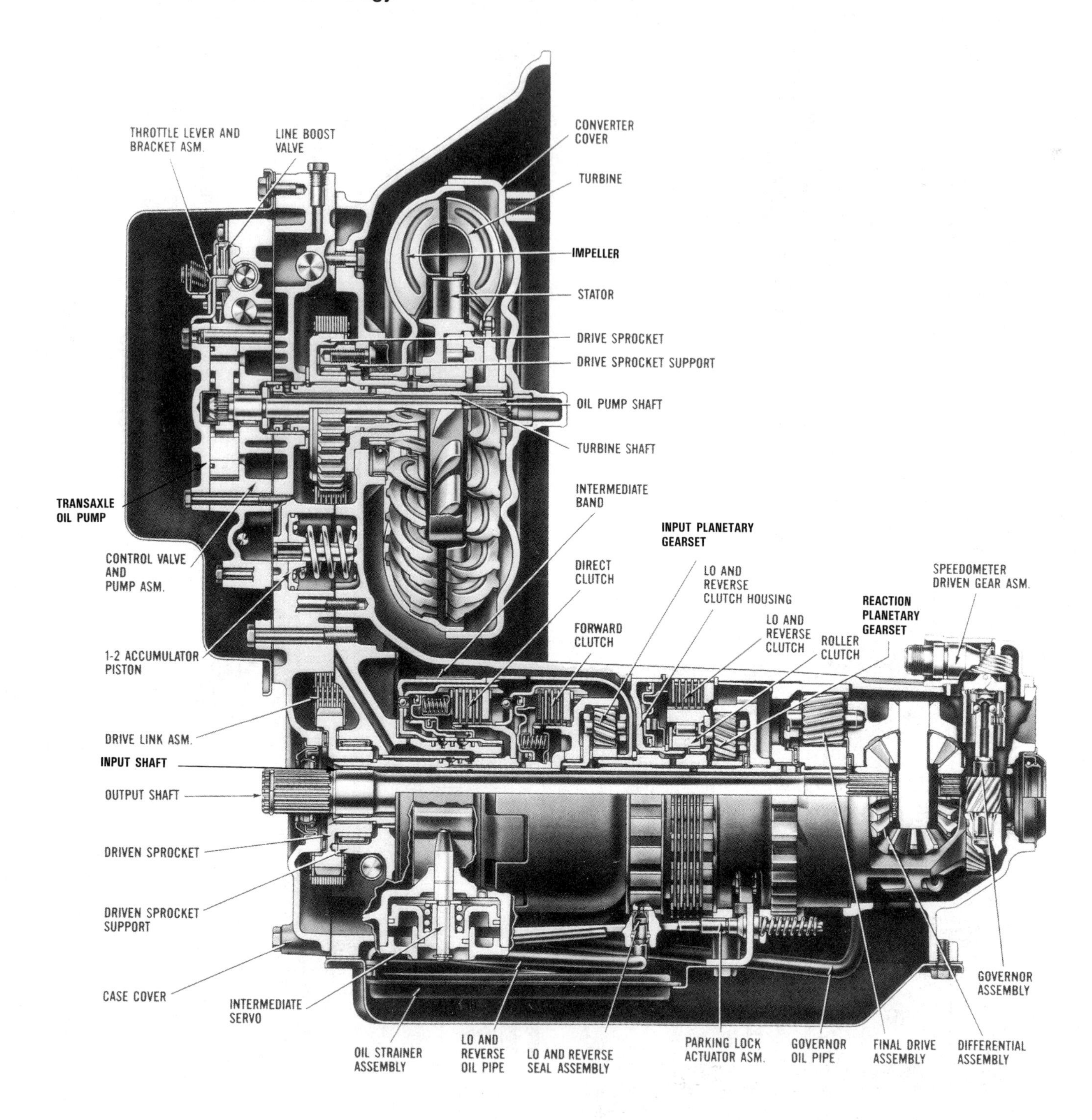

Figure 20-34. This transaxle uses a drive chain to deliver power from the torque converter to the transaxle gear train. The oil pump is remotely mounted and driven by the oil pump shaft, which lies within the turbine shaft. Note that the differential assembly of this design does not have a ring and pinion. (General Motors)

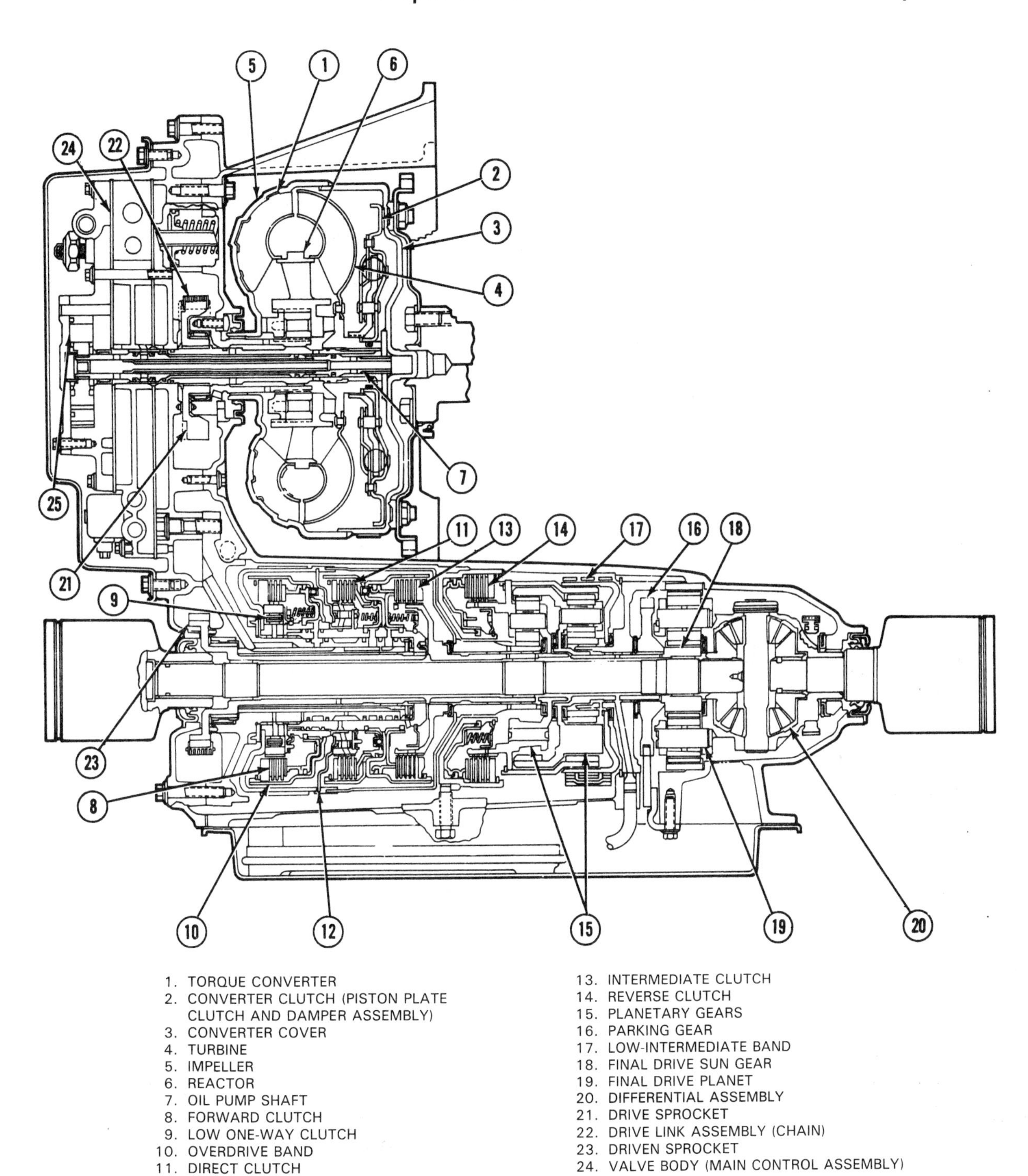

Figure 20-35. The transaxle shown here uses a chain drive and a remotely mounted oil pump. It contains a reduction planetary to obtain the final drive ratio. (Ford)

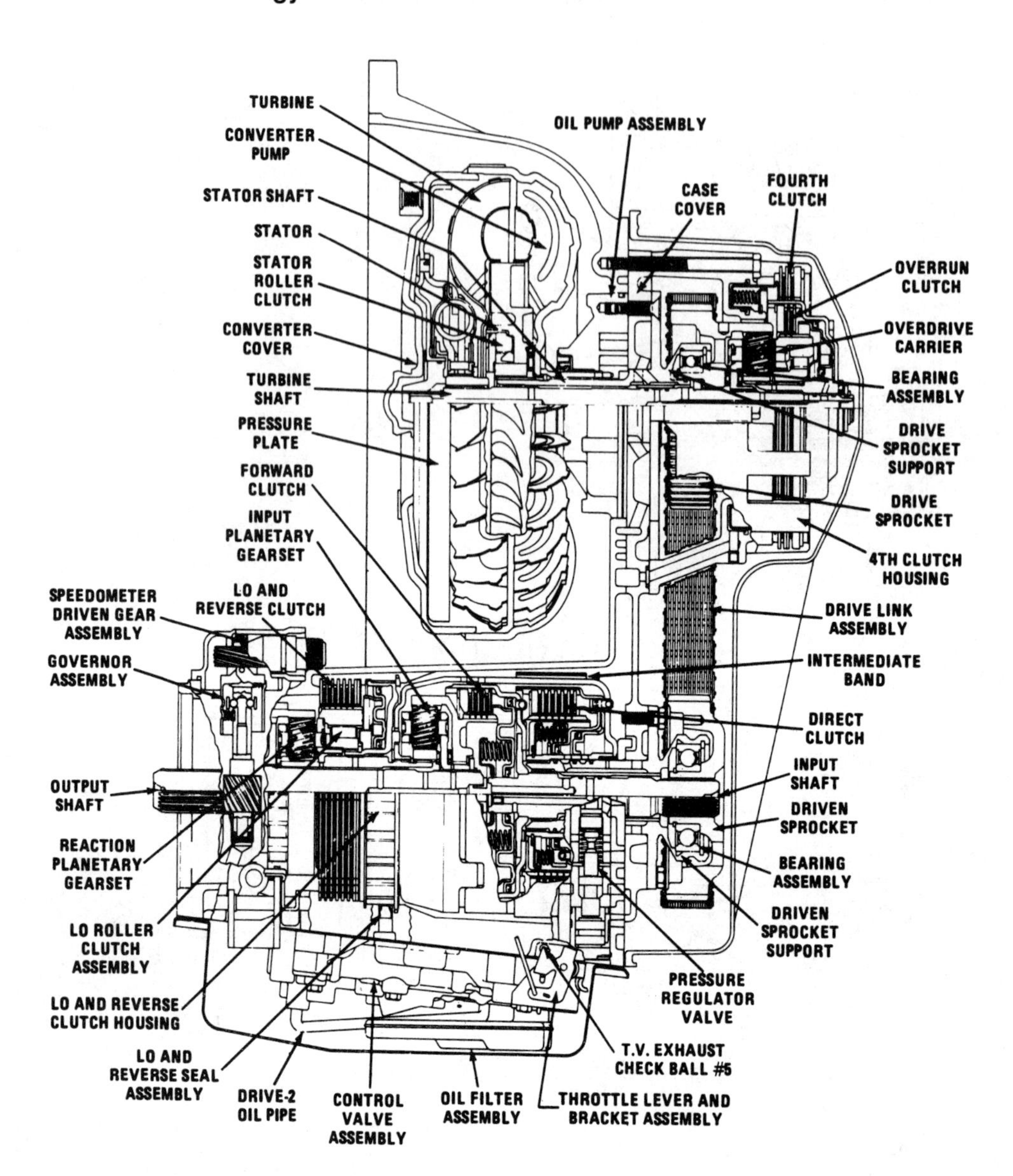

Figure 20-36. The transaxle shown here has four forward speeds and a converter clutch torque converter. It is used with a longitudinal engine. A hypoid ring and pinion gear (not shown) connects the transaxle transmission to the differential case assembly. (General Motors)

Continuously Variable Transmissions

The ***continuously variable transmission (CVT)*** has been used for many years in lawn equipment, snowmobiles, and other small engine-driven equipment. Improvements in design and materials have made the CVT practical for automotive use, particularly in front-wheel drive, low-torque applications.

The continuously variable transmission is essentially an automatic transaxle that uses a belt and two variable-diameter pulleys to transfer engine power to the drive wheels. See Figure 20-39. The variable-diameter pulleys are generally made in two halves. One-half of the pulley is fixed; the other half is adjustable, Figure 20-40. The adjustable half is operated by a hydraulic piston. The drive belt consists of approximately 300 steel segments that are held together by overlapping steel bands. See Figure 20-41. The configuration of the belt allows it to be pushed by the pulley.

Varying the effective diameter of the pulleys by moving the adjustable pulley halves in and out causes the belt to ride higher or lower in the pulley grooves. Figure 20-42 shows how changes in pulley diameter cause the belt to change position in the pulley groove. When the pulley halves are far apart, the belt rides very close to the center of the pulley (small effective diameter). When the pulley halves are close together, the belt rides farther from the center of the pulley (large effective diameter).

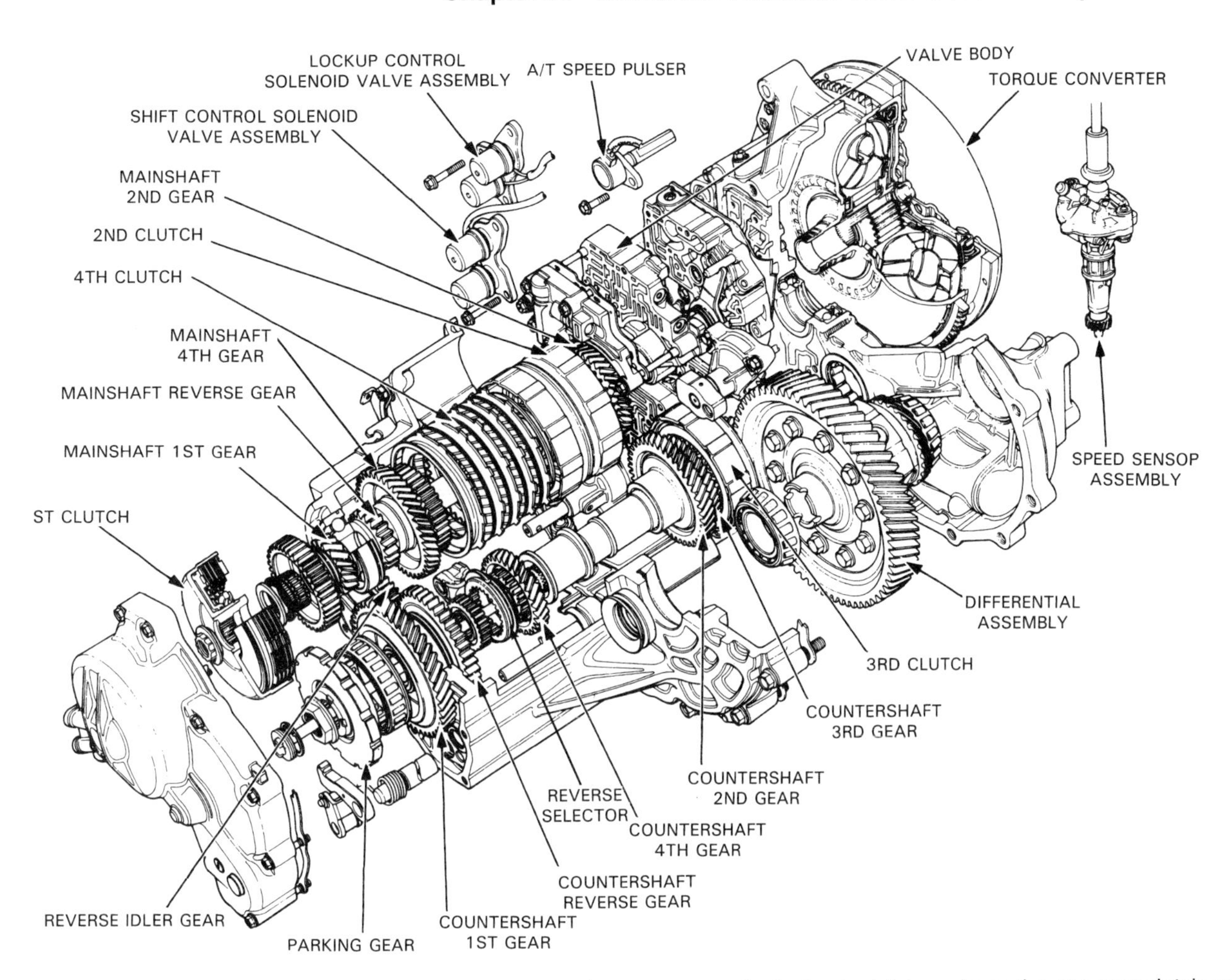

Figure 20-37. This transaxle is unusual since it does not use planetary gearsets. Instead, clutch packs and a one-way clutch apply and release meshing helical gears. (Honda)

When the effective pulley diameter is changed, the ratio between the drive pulley and the driven pulley is also changed. This ratio change is performed to match vehicle speed and engine power output. The design of the belt and the pulleys gives the CVT an unlimited number of drive ratios. Therefore, the transmission allows the engine to operate in its most efficient range at all times. This increases performance and reduces fuel consumption. Because the operation of CVT is extremely smooth, the drive ratio changes are often referred to as ***stepless shifting.***

Figure 20-43 shows a simple belt and pulley system. Engine power enters the primary pulley, passes through the belt, and exits the secondary pulley.

When a vehicle is accelerating, the primary pulley width is increased and the secondary pulley width is reduced. This causes the belt to ride close to the center of the primary pulley and near the outer part of the secondary pulley. See Figure 20-44A. Several revolutions of the small primary pulley are required to produce one revolution of the secondary pulley. This results in a reduction ratio and provides maximum power multiplication.

As a vehicle approaches cruising speed, maximum power multiplication is no longer needed. The width of the primary pulley is decreased, and the width of the secondary pulley is increased. This causes the belt to move out on the primary pulley and closer to the center on the secondary. Since the belt is approximately the same distance from the center of both pulleys, Figure 20-44B, the result is the same as driving a pulley with another pulley of the same size. A 1:1 ratio is achieved; the CVT is in direct drive.

When the vehicle is at cruising speed, the primary pulley width is at its smallest setting, causing the belt to ride on the outer portion of the pulley. The width of the secondary pulley is increased, causing the belt to ride near the center of the pulley. Since the belt is near the outside of the primary pulley and near the center of the secondary pulley, a small pulley is being driven by a large pulley. See Figure 20-44C. When the primary pulley turns through one revolution, it turns the driven secondary through more than one revolution. This is an overdrive ratio.

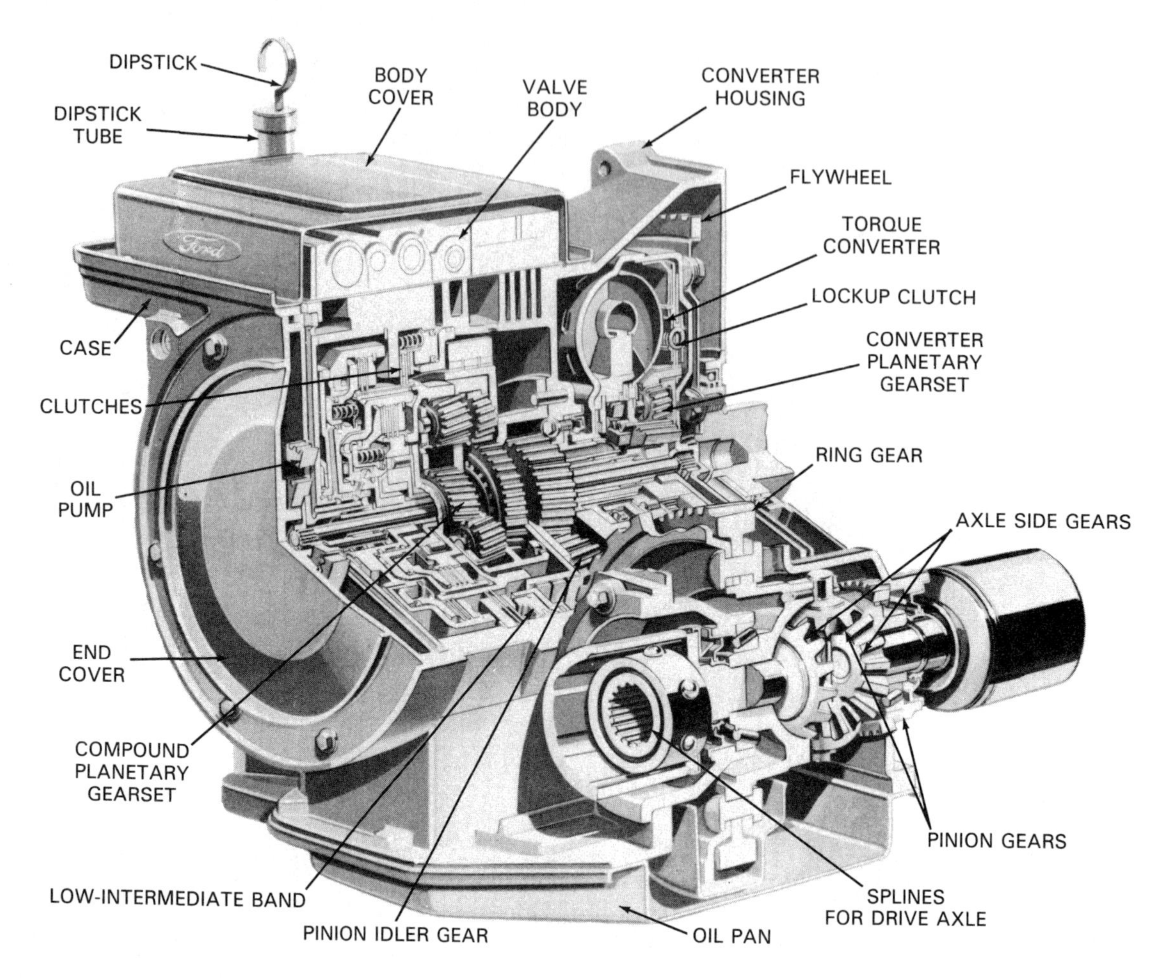

Figure 20-38. This transaxle uses a compound planetary gearset and a unique lockup torque converter clutch. Other parts are similar to those used on other kinds of transaxles. (Ford)

Summary

The automatic transaxle contains many systems and parts that are similar to the rear-wheel drive automatic transmission and differential assembly. Both sections are contained in one housing.

The torque converter is a fluid transfer device that can transmit and multiply engine power. It operates according to the same principles used to operate torque converters on rear-wheel drive vehicles. Most transaxle converters use a lockup clutch in the torque converter.

Input and output shafts transfer power and are made of hardened steel. The turbine shaft is turned by the converter. It sends power to a drive chain or directly to the gear train. The output shaft is often the connection between the transmission and differential. Some transaxles have a transfer shaft to transfer power to the differential assembly. Some transaxles have an oil pump shaft to drive the oil pump.

Some transaxles use a chain drive to transfer power between the torque converter and the gear train. Transaxles use the same planetary gears as rear-wheel drive transmissions. The gearsets are controlled with clutch packs, bands, and one-way clutches. Accumulators cushion band and clutch application by diverting hydraulic fluid. When the accumulator piston is fully retracted, full pressure is sent to the clutch or band.

The transaxle oil pump produces the fluid pressure that operates other hydraulic components and lubricates the moving parts of the transaxle. The transaxle oil filter removes particles from the automatic transmission fluid prior to circulation by the oil pump. Oil from the torque converter is pumped to an oil cooler at the front of the vehicle. The oil cooler is located in the radiator. Auxiliary coolers may be mounted ahead of the oil cooler.

The pressure regulator valve regulates hydraulic system pressure. Oil from the regulator valve goes to the valve body. Hydraulic valves in the valve body modify line pressure for proper system operation. The valves direct line pressure to the proper holding members.

The valve body contains the manual valve, shift valves, and other valves. The manual valve is moved by the shift linkage to obtain the selected operating range. The shift

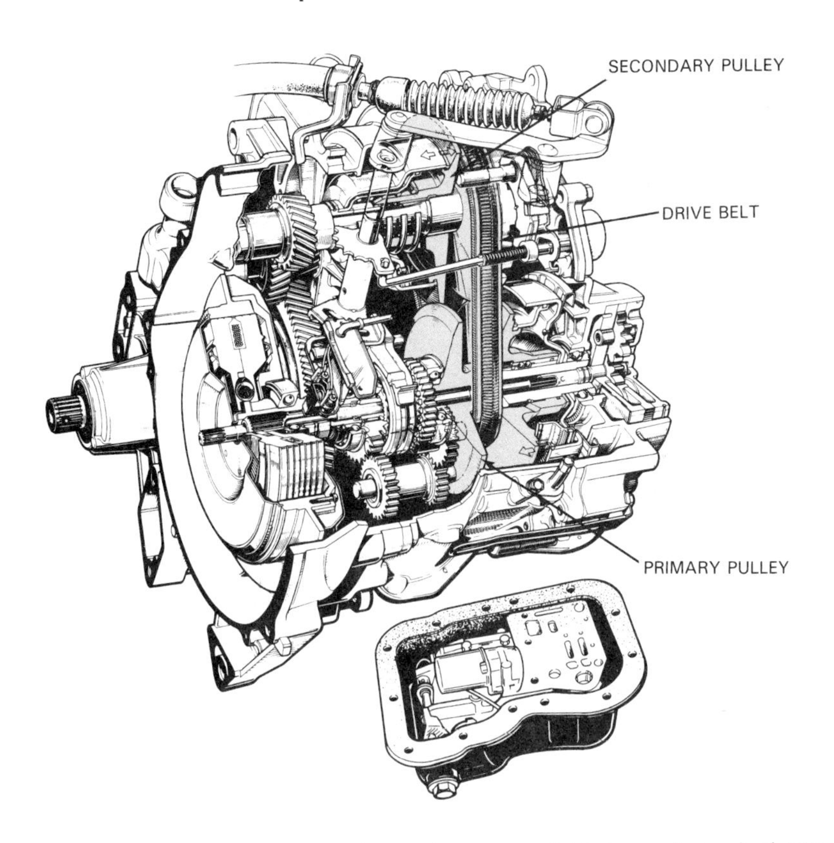

Figure 20-39. Basic design and part layout of a continuously variable transmission is shown. The most obvious difference between a CVT and a conventional transaxle is the belt and pulley system. Gears in the CVT assist in driving the belt in forward and reverse directions. An electromagnetic clutch engages and disengages the engine and transaxle. The CVT also has a differential assembly, as in a conventional transaxle. (Subaru)

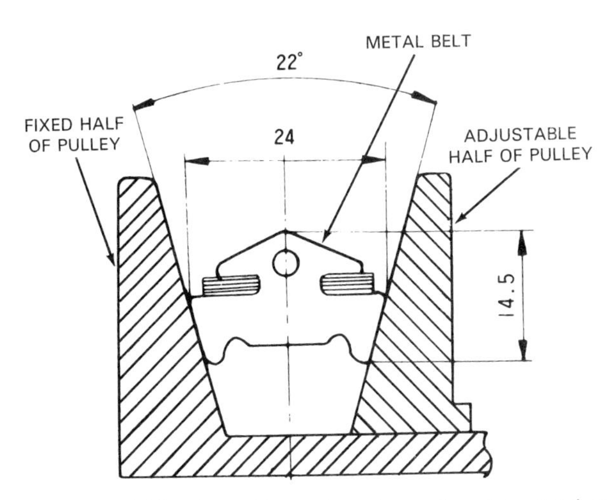

Figure 20-40. This shows the action of a split pulley used on a CVT. One half of the pulley is stationary. The other half slides on the pulley shaft. Moving the pulley halves apart decreases the effective diameter of the pulley.

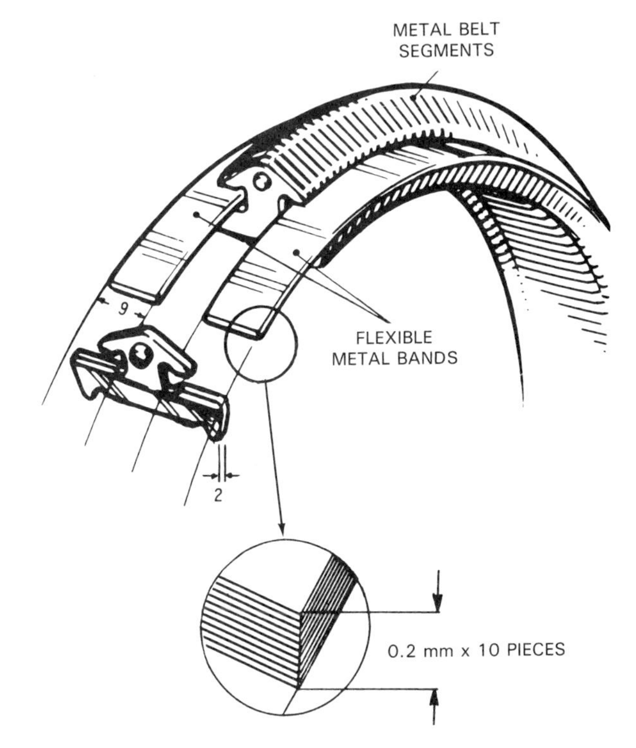

Figure 20-41. The CVT drive belt is made up of approximately 300 small metal segments, which are held together by a series of flexible steel bands. The belt is designed to be pushed, rather than pulled, by the drive pulley. (Subaru)

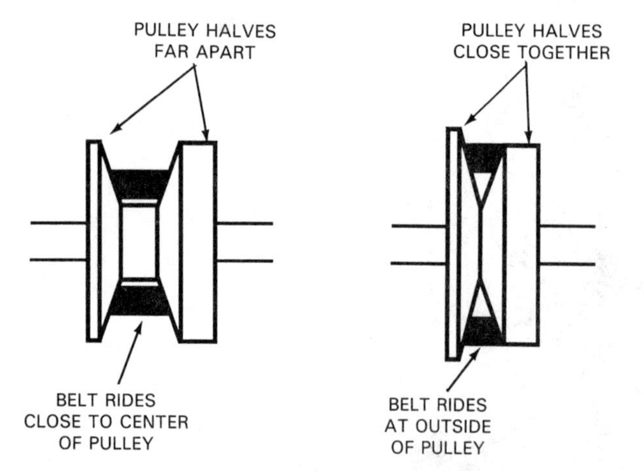

Figure 20-42. Changing the effective pulley diameter can change the position of the belt in the pulley. A–The two sides of the pulley are far apart, and the belt moves close to the center of the pulley. B–The two sides of the pulley are close together, and the belt is pushed to the outside of the pulley.

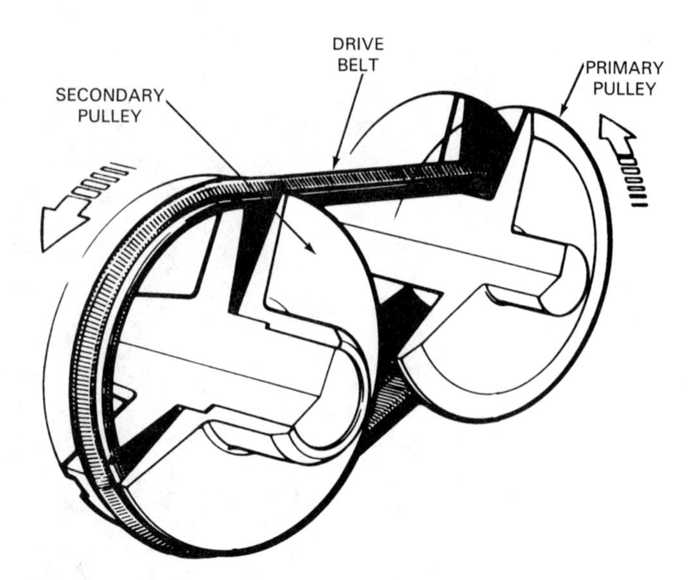

Figure 20-43. General layout of the adjustable pulleys and the drive belt is shown.

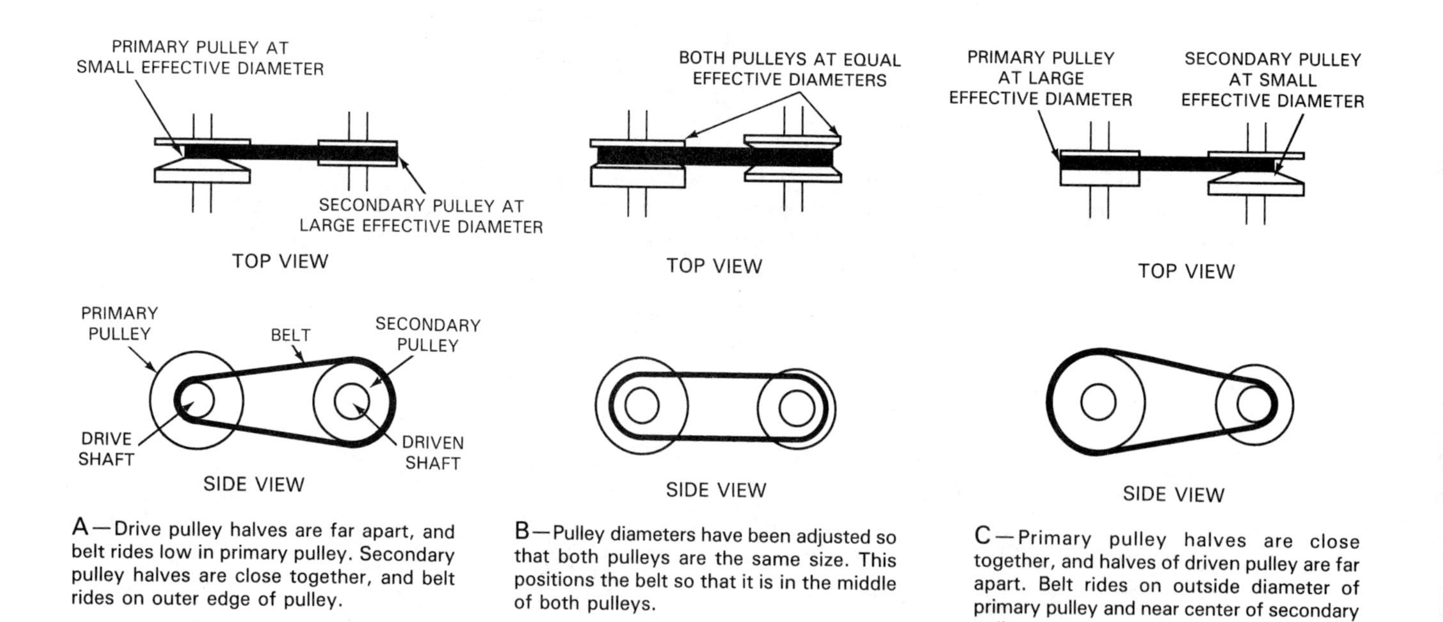

Figure 20-44. This series of illustrations shows how changing position of belt on pulleys can change pulley ratio.

valves are operated by pressure from the throttle and governor valves. The throttle pressure is raised or lowered in response to engine throttle opening. Throttle valves are operated by vacuum modulators or by a cable linkage from the throttle plate. The governor pressure varies with vehicle speed. Governor valves are driven by the output shaft through gears. The detent valve forces downshifts when the accelerator is pressed to the wide-open position.

Ring and pinion gears are sometimes used in automatic transaxles. Those used with transverse engines are straight cut or helical gears. Those used with longitudinal engines must divert power flow by 90°, just as a hypoid ring and pinion on a rear-wheel drive vehicle.

The differential unit operates in the same way as the differential in any other vehicle. The interaction of the spider and side gears allow the vehicle to make turns. Some planetary and differential gears are combined into a single unit.

The transaxle case is made of aluminum, with passages for the flow of transmission fluid between other components. Sheet metal covers are used to enclose various parts of the transaxle case. The covers make transaxle disassembly easier. An oil pan is located at the bottom of the transaxle.

Know These Terms

Automatic transaxle shafts; Transaxle oil pump; Automatic transaxle differential assembly; Automatic transaxle case; Transaxle oil pan; Automatic transaxle turbine shaft; Automatic transaxle input shaft; Automatic transaxle output shaft; Transfer shaft; Oil pump shaft; Continuously Variable Transmission (CVT); Stepless shifting.

Review Questions–Chapter 20

Please do not write in this text. Place your answers on a separate sheet of paper.

1. The _____ _____ supports and encloses the transaxle torque converter, transmission, and differential assembly.
2. A chain drive is used in an automatic transaxle mounted on a transverse engine to transfer power from the transaxle output shaft to the differential assembly. True or false?
3. A Simpson gear train may be identified by a single planet carrier and two sets of planet gears. True or false?
4. What are two different methods used for driving the transaxle oil pump?
5. Where in the automatic transaxle are the shift valves found?
6. The _____ _____ is operated by the driver through the shift linkage.
7. Some transaxles contain more than one valve body. True or false?
8. The shift valves are acted upon by _____ pressure and _____ pressure.
9. Governor valves modify pressure according to:
   a. Manifold pressure.
   b. Throttle plate position.
   c. Vehicle speed.
   d. Temperature.
10. What is the purpose of an orifice?
11. The _____ _____ mechanically locks the output shaft, preventing it from rotating.
12. An automatic transaxle mounted on a transverse engine will always have a helical ring and pinion as part of the differential assembly. True or false?
13. In a CVT, power is transmitted from drive to driven pulleys by a steel belt. True or false?
14. How does a CVT provide an infinite number of ratios?

Certification-Type Questions–Chapter 20

1. All of these can be driven by the torque converter EXCEPT:

(A) the turbine shaft.
(B) the transaxle oil pump.
(C) the transaxle input shaft.
(D) the flexplate.

2. All of these shafts are used in automatic transaxles EXCEPT:

(A) countershafts.
(B) input shafts.
(C) turbine shafts.
(D) oil pump shafts.

3. Some automatic transaxles have a transfer shaft that transmits power from the transaxle output shaft to the:

(A) input shaft.
(B) engine crankshaft.
(C) planetary gearsets.
(D) differential assembly.

4. A simple planetary gearset consists of all of these EXCEPT:

(A) a sun gear.
(B) a side gear.
(C) a ring gear.
(D) planet gears.

5. Most modern 3-speed automatic transaxles require holding members be applied in any gear to move the vehicle. Technician A says that at least two holding members are needed for this. Technician B says that only one holding member is needed. Who is right?

(A) A only
(B) B only
(C) Both A & B
(D) Neither A nor B

6. To hold planetary gearset members stationary or lock them together for power transfer, multiple-disc clutches use all of these EXCEPT:

(A) steel plates.
(B) friction discs.
(C) a transfer shaft.
(D) a clutch apply piston.

7. Technician A says that automatic transaxle transmission bands are friction devices that hold planetary gearset members. Technician B says that the bands will not work if they come in contact with transmission fluid. Who is right?

(A) A only
(B) B only
(C) Both A & B
(D) Neither A nor B

8. Overrunning clutches are used in conjunction with planetary gearset members specifically to:

(A) improve efficiency.
(B) cushion application.
(C) divert hydraulic fluid.
(D) improve shift quality and timing.

9. Technician A says that one function of the transaxle oil pump is to keep moving parts of the transaxle lubricated. Technician B says that keeping the torque converter filled with transmission fluid is one of its functions. Who is right?

(A) A only
(B) B only
(C) Both A & B
(D) Neither A nor B

10. The valve body may contain all of these EXCEPT:

(A) the oil pump.
(B) an accumulator.
(C) a spacer plate.
(D) hydraulic valves.

11. In an automatic transaxle, hydraulic valves act as all of these EXCEPT:

(A) fluid directors.
(B) actuators.
(C) flow controllers.
(D) pressure controllers.

12. All of these are hydraulic valves found in automatic transaxles EXCEPT:

(A) check balls.
(B) shift valves.
(C) body valves.
(D) detent valves.

Chapter 21

Automatic Transaxle Problems, Troubleshooting, and Service

After studying this chapter, you will be able to:

- Name and describe common types of problems that can occur in an automatic transaxle.
- Remove an automatic transaxle from a vehicle.
- Disassemble an automatic transaxle.
- Inspect the parts of an automatic transaxle.
- Assemble an automatic transaxle.
- Install an automatic transaxle in a vehicle.

This chapter is designed to familiarize you with the methods used to diagnose and repair automatic transaxles. Some procedures, especially those relating to removal and disassembly, are unique to transaxles. Many techniques have already been explained in earlier chapters on automatic transmissions, manual transaxles, and rear axle assemblies. If you have studied the procedures in these chapters carefully, you should have no difficulty in adapting them to automatic transaxle repair.

Many automatic transaxles have specific and predictable problem areas. Experienced technicians can often determine the cause of a problem based on past experience. One way to acquire some of this experience is to carefully study this chapter and to consult service publications that contain information about transaxle problems. These publications have many tips that will make transaxle diagnosis and service quicker and easier. As depicted in Figure 21-1, the automatic transaxle can have many kinds of problems.

Automatic Transaxle Problems and Troubleshooting

Automatic transaxles are complex, and it is very important that you proceed in a logical manner while troubleshooting transaxle problems. Malfunctions can be diagnosed by applying logic and following recommended troubleshooting procedures.

You should have the owner or driver describe the problem before taking any other diagnostic steps. Be aware that the driver may be unable to accurately describe the problem. Therefore, if possible, you should road test the car with the driver and have him or her point out the particular problem while it occurs. Take note as to what the operating conditions are when the problem occurs.

The most important factor in correct diagnosis is understanding the specific complaint. It may be found that the condition the vehicle owner wants corrected is, in fact, a normal condition or one that is not caused by the transaxle. Many complaints about the transaxle are caused by engine problems, restricted exhaust systems, dragging brakes, or other problems not involving the transaxle.

Before performing the road test, the transmission fluid level and condition should be checked. After the road test, check the shift linkage. Also, check the throttle and T.V. linkages or vacuum modulator, as required. Adjust linkages as necessary. Inspect all electrical connections. If these checks do not turn up any obvious problems, diagnostic checks, such as hydraulic pressure tests, can be made.

Common problems encountered in the automatic transaxle are similar to those seen in the automatic transmission and rear axle assembly. Below are some of these problems and their possible solutions.

Automatic Transaxle Slippage

Slippage occurs in the transaxle when it fails to transfer all of the engine power to the drive wheels. The engine speeds up but the vehicle does not move any faster. In extreme cases, a no-drive condition exists, wherein the vehicle cannot move at all.

Begin the diagnostic procedure by checking the fluid level and performing a road test to determine in which gear the vehicle is slipping. Then, refer to the band and clutch application charts in the factory service manual to deter-

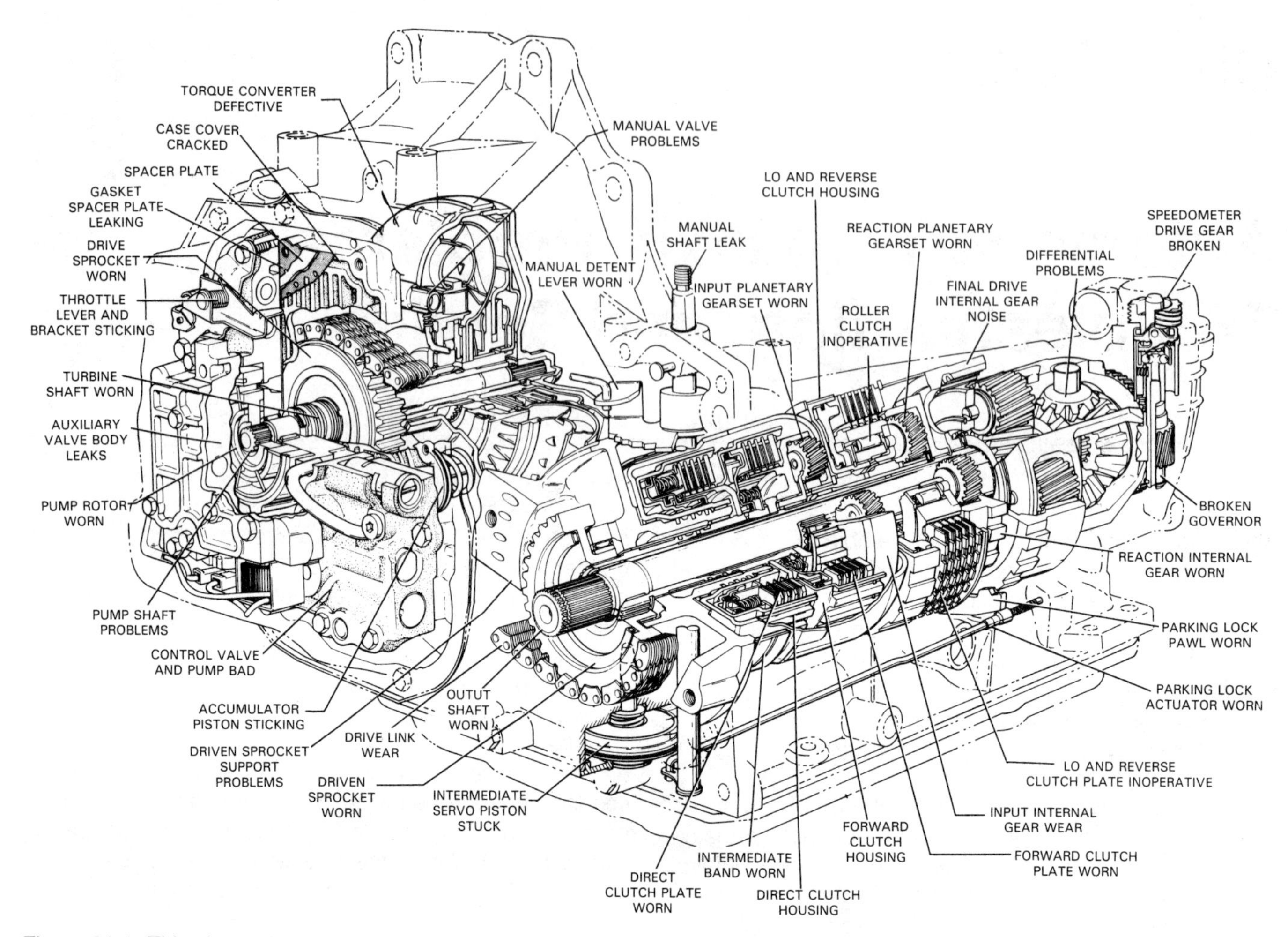

Figure 21-1. This shows the kinds of problems that can develop in an automatic transaxle. All can be diagnosed and repaired by learning correct repair procedures and following them. (General Motors)

mine which clutches and bands are applied in that gear. If the vehicle operates properly in all other gears, the hydraulic valves or the band or clutch controlling the slipping gear is probably at fault. This can usually be confirmed by presence of sludge or pieces of friction material in the transaxle oil pan, indicating burned or otherwise damaged bands or clutch friction discs and, possibly, a sticking valve.

If clearances are greater than specified, band adjustment may cure a slippage problem. However, if the band is burned, the adjustment will not help. Some servos and bands can be overhauled or replaced without removing the transaxle from the vehicle. The transaxle must be removed to overhaul clutch packs.

A vehicle slipping in every gear could be the result of an incorrect fluid level. If the level is okay, check for a clogged transaxle filter. A clogged filter may also cause a whining noise that varies with engine speed. In some cases, the transaxle operates properly at low speeds or on takeoff, but begins to slip as engine speed increases.

Next, check and adjust the T.V. linkage. Adjusting the linkage may raise shift points, but it will also raise transaxle hydraulic pressures. Higher hydraulic pressures can reduce and can sometimes eliminate slippage.

In rare cases, the torque converter may be defective. Torque converter slippage will only be noticed when starting off or during hard acceleration. If the transaxle slips in only one gear, the torque converter is probably not defective. Converter slippage is caused by a defective stator. If the stator is defective, then the entire converter is considered defective. The defective converter must be replaced in its entirety. If you suspect a defective torque converter, replace it, but not until you have investigated all other possibilities.

Shift Problems

Shift problems are those that result in improper gear changes. Examples are harsh or soft shifts, no passing gear, and incorrect shift points, including early and late upshifts and no shifts (failure to upshift or downshift at all). In some instances, shift points will vary, usually called *erratic shifting.*

Shift problems often result from improper hydraulic pressures at the shift valves. Many times, hydraulic pressure problems are caused by something external, such as a misadjusted linkage or a leaking vacuum hose. If the transaxle is not slipping, it may be possible to correct a shift problem without removing the transaxle.

If the transaxle fluid level is normal and the fluid is not burned, begin by checking the throttle valve linkage adjustment, the hose to the vacuum modulator, or sensor and actuator wiring, as applicable. Linkage should be adjusted according to specifications before making further checks.

On vehicles with a vacuum modulator, make sure that the modulator is supplied with full engine vacuum, and that the engine is properly tuned, since engine condition can affect manifold vacuum. If there is any transmission fluid in the vacuum line, or if the modulator neck is bent, the modulator should be replaced.

If the transaxle will only upshift at very high speeds, check the kickdown linkage or electric solenoid. If the detent valve is stuck in the wide-open throttle position, the transaxle hydraulic system will remain in passing gear.

If the throttle and detent systems check okay, check the governor. Modern governors are usually mounted on the transaxle case, and they can be easily removed for cleaning and inspection. After removal, check the governor valves, seal rings, and the governor housing bore. Also, check the governor screens for clogging from varnish or metal particles and look to see if something in the transaxle might be producing metal particles. These particles will cause governor sticking, sometimes over and over.

If the governor is not defective, check the valve body for valve sticking and improper placement of valves or springs. If the valve body checks out okay, check the internal passageways in the transaxle case, as well as the spacer plates and gaskets. Transaxle construction results in many areas where transmission fluid can leak away or become blocked.

Abnormal Noise

An *abnormal noise* in the automatic transaxle is an unusual sound that occurs during vehicle operation. Some transaxle noises can be heard in all gears, including neutral and park, while others are heard in only one gear.

If the noise occurs in all gears, the transaxle oil filter could be plugged. With the gear selector in neutral, the plugged filter will cause a whine at low rpm that will increase in pitch as engine speed increases. As another possibility, the noise could be a sign that something is wrong in the differential.

A noise that occurs in all gears immediately after an overhaul probably means that pump clearances are too tight. If the noise does not follow an overhaul, it is probably not the pump. It is rare for an oil pump with many miles on it to suddenly begin making noise. If you suspect the oil pump, you will have to take the transaxle apart and check pump clearances. If they are out of spec, the pump must be repaired or replaced, as necessary.

Another possible cause of a noise in every gear is the torque converter. Torque converter noise is a scraping sound that is noticed when starting off or when accelerating. The noise will stop at cruising speeds, when internal parts of the converter are turning at approximately the same speed.

If the noise seems to occur in only one gear or at certain speeds, the problem may be in a bearing or in the planetary gears. Planetary gear noise will vary with gear ratio, and it often can be heard in only one gear, usually low gear. Road test the vehicle and determine which gear is causing the noise. Check your finding with the band and clutch applications charts in the service manual to determine which planetary gearset is in use in that gear. Planetary gear noises occur when the gear train is in reduction, overdrive, neutral, or reverse. Planetaries drive as a unit in direct drive, so they will not make noise.

Lubricant Leakage

Transaxle *lubricant leakage* is an unwanted loss of fluid from inside the transaxle. Before attempting to correct the suspected transaxle leakage problem, make sure that the leaking fluid is actually coming from the transaxle and not from the engine or from the power steering, cooling, or brake systems. If you are sure that the transaxle is leaking, determine the exact source of the leak before attempting to correct it.

If the transaxle leaks with the engine running, look for leaks at the torque converter and oil pump, at the servo, accumulator, and governor covers, and at the pressure tap plugs. Also, check the oil cooler lines. Before looking for leaks, remove the front dust cover and clean the torque converter, oil pump face, and the transaxle case with solvent. Start the engine and run it for 10 minutes at 1000 rpm. Then, examine the entire transaxle for fresh fluid.

If the transaxle leaks with the engine off, try tightening the bolts on the transaxle oil pan. If this does not stop the leak, remove the oil pan and replace the gasket. Check the pan for warped sealing surfaces or punctures. Examine the CV axle seals, since they are a common source of leaks. Check seals at the manual lever/shaft assembly, T.V. cable, band adjustment screw, and electrical connectors. Many of these seals will not leak when the engine is running since oil level is lower when the transaxle hydraulic system is operating. Seals must be replaced if they leak.

If the transaxle has a vacuum modulator, it can be a source of leakage. Check this component if the transmission fluid level seems to drop without being accompanied by leakage. Look for a hole or tear in the diaphragm. A ruptured vacuum modulator diaphragm will allow transmission fluid to be drawn into the engine and burned. Also, if any transmission fluid is found on the vacuum side of the modulator, replace the modulator.

Automatic Transaxle Diagnostic Testing

The automatic transaxle can be tested by methods used for testing an automatic transmission in a rear-wheel drive vehicle. Road tests and hydraulic pressure tests are among procedures that can be performed. They are discussed in this section.

Road Test

A road test should always be performed if the car is driveable. Road testing will help determine the exact problem. Road testing can reveal problems that cannot be duplicated in the shop. One purpose of a road test is to isolate transaxle problems from engine, CV axle, exhaust, tire, and other vehicle problems. For example, roaring or rumbling noises that vary with road speed are usually caused by defective bearings or worn tires. Vibrations that vary with vehicle speed indicate a problem in the CV axles or tires instead of the transaxle.

WARNING! When road testing a vehicle, obey all traffic regulations. Try to choose uncongested roads or parking lots for testing. Do not become so engrossed in diagnosing the transaxle that you have an accident.

Hydraulic Pressure Tests

Hydraulic pressure tests can be useful to diagnose hydraulic system problems that cannot be pinpointed by any other method. The design of most transaxles forces oil to take a complicated path between components, increasing the chances of internal leaks. For this reason, many transaxles are equipped with several pressure taps, as shown in Figure 21-2. These pressure taps enable the technician to read the pressure of several hydraulic circuits while parts are operating.

Begin pressure testing by installing the pressure gauge, as shown in Figure 21-3. Then, start the engine and check pressures in neutral, all drive ranges, and reverse. Make the checks at idle, 1000 rpm, and 2000 rpm. Road test the vehicle with the gauges installed. Note the pressures in all gears and, also, as shifts occur. If necessary, change to different pressure taps and repeat the process.

Compare the gauge readings with the normal pressure ranges in the factory service manual. Figure 21-3 shows a typical set of pressure specifications. The manual will provide information on what abnormal pressure readings indicate. As a general rule, excessively high tap pressure indicates a sticking pressure regulator valve or sticking pressure relief valve. Low tap pressure indicates a worn-out oil pump, a stuck-open pressure regulator valve, an internal hydraulic system leak, or a plugged oil filter. Excessive pressure drops during shifts or a slow pressure recovery afterwards indicates internal leakage from defective piston seals, seal rings, or case gaskets. Consult the service manual band and clutch application chart to isolate the problem to a particular clutch or servo.

WARNING! Always keep safety uppermost in your mind when checking for transaxle problems. Be cautious around rotating parts when testing the transaxle with the vehicle on a lift. The CV axles and tires rotate at high speeds and can cause severe injury.

Always support the lower control arms so the front wheels are at or near their normal position in relation to the vehicle body. Operation of the CV axles at excessive angles by allowing them to hang will cause severe CV joint damage and possible injury to anyone near the vehicle. When testing a front-wheel drive vehicle on the lift, never exceed 30 mph (48 km/h).

In-Car Automatic Transaxle Service

There are a number of service procedures that can be performed without removing the transaxle from the vehicle. Obviously, checking the transaxle fluid level (and condition) is one; so is changing the fluid and filter. Making adjustments to linkages (refer to Chapter 13 for specifics), the neutral start switch, and sometimes, the transaxle transmission bands are a few other in-car service procedures.

There are still other service procedures that can be performed while the transaxle is still installed. The speedometer pinion gear, vacuum modulator, oil cooler lines, governor, valve body, and servo assemblies can be serviced in certain vehicles without removing the transaxle.

Some of the procedures mentioned are covered in the overhaul section of this chapter. More information on checking transmission fluid level and condition is given in the next several paragraphs.

Automatic Transmission Fluid Service

Like engine oil, automatic transmission fluid should be checked at specified intervals. Also, some manufacturers will recommend changing the fluid at specified intervals. The fluid can become contaminated with metal, dirt, water, and friction material from internal parts. Contaminated fluid can cause rapid part wear and premature transaxle failure.

One common cause of transaxle problems is incorrect fluid level. With a low fluid level, air will be drawn into the transaxle hydraulic system, causing erratic shifting and slipping. A high fluid level can cause foaming, aerating the fluid and causing the same problems as a low fluid level. In addition, the introduction of air into the fluid results in severe and rapid oxidizing of the fluid. Since transaxles often use less fluid than rear-wheel drive automatic transmissions, oil level is generally even more critical.

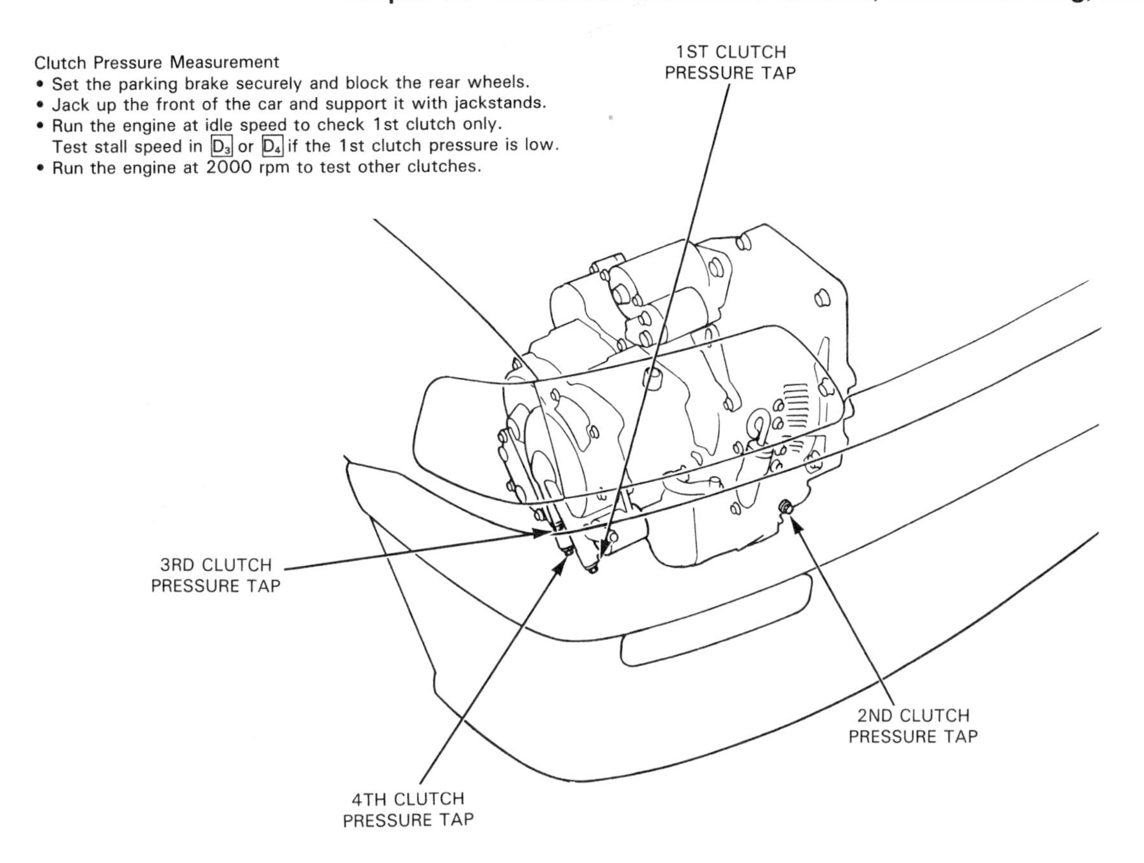

PRESSURE	SELECTOR POSITION	SYMPTOM	PROBABLE CAUSE	FLUID PRESSURE	
				Standard	Service Limit
1st Clutch	[D3] or [D4]	No or low 1st pressure	1st Clutch	785–834 kPa (8.0–8.5 kg/cm², 114–121 psi)	736 kPa (7.5 kg/cm², 107 psi)
2nd Clutch (2nd hold)	[2]	No or low 2nd pressure	2nd Clutch	785–834 kPa (8.0–8.5 kg/cm², 114–121 psi)	736 kPa (7.5 kg/cm², 107 psi)
3rd Clutch	[D3]	No or low 3rd pressure	3rd Clutch	412 kPa (4.2 kg/cm², 60 psi) (throttle control lever fully closed) 785–834 kPa (8.0–8.5 kg/cm², 114–121 psi) (throttle open more than 1/4)	363 kPa (3.7 kg/cm², 53 psi) (throttle control lever fully closed) 736 kPa (7.5 kg/cm²,107 psi) (throttle open more than 1/4)
4th Clutch	[D4]	No or low 4th pressure	4th Clutch		
	[R]		Servo valve or 4th Clutch	785–834 kPa (8.0–8.5 kg/cm², 114–121 psi)	736 kPa (7.5 kg/cm²,107 psi)

Figure 21-2. Automatic transaxle manufacturers provide pressure taps for the purpose of determining hydraulic pressures in various circuits. The number of taps varies with each manufacturer. (Honda)

Fluid levels in automatic transaxles are sometimes hard to read accurately. Many automatic transaxle dipsticks will show high readings when the fluid is cold, and after the fluid heats up, the same dipsticks will read less than full or even low. Other transaxles cannot be read accurately right after fluid is added, and these must be driven to restore the normal oil level.

The best procedure for checking fluid level in the automatic transaxle is to drive the vehicle several miles and then park on a level surface and check the fluid without

Minimum Line Pressure Check
With the parking brake and vehicle brakes applied, take the line pressure readings in the ranges and at the engine rpm indicated in the chart below.

Full Line Pressure Check
Remove the vacuum line from the modulator, and with the parking brake and vehicle brakes applied, take the line pressure readings in the ranges and at the engine rpm indicated in the chart below.

NOTICE Total running time not to exceed 2 minutes.

CAUTION Brakes must be applied at all times.

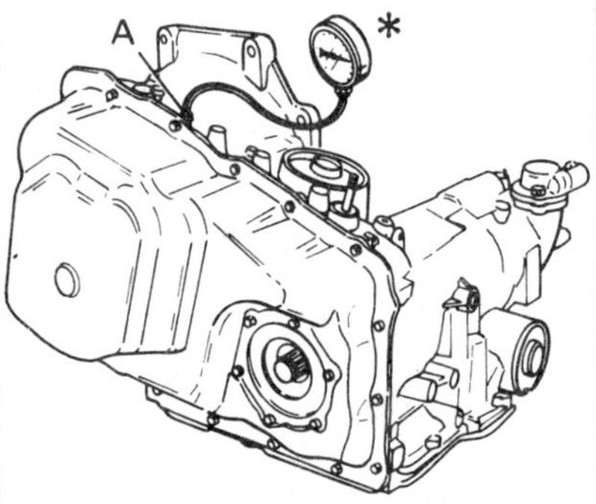

A–ATTACH OIL PRESSURE GAGE

		MODEL	7ADH, 7AFH, 7AHH, 7ALH, 7ARH, 7CAH, 7CBH		7ACH, 7BBH, 7BCH, 7BJH, 7BKH, 7BNH, 7BRH, 7BSH, 7BTH, 7BUH, 7BZH, 7FBH, 7FCH, 7FJH, 7FKH, 7FLH, 7FNH, 7FRH, 7FSH, 7FTH, 7FUH, 7FZH, 7HAH, 7HCH	
		RANGE	kPa	PSI	kPa	PSI
TRANSMISSION LINE PRESSURE	MINIMUM LINE (1250 R.P.M.)	D4,D3,D2	422 - 475	61 - 69	422 - 475	61 - 69
		D1	946 - 1324	137 - 192	998 - 1276	145 - 185
		P,R,N	422 - 475	61 - 69	422 - 4/5	61 - 69
	FULL LINE (1250 R.P.M.)	D4,D3,D2	1030 - 1266	150 - 184	1152 - 1393	167 - 202
		D1	946 - 1324	137 - 192	998 - 1276	145 - 185
		P,R,N	1436 - 1764	209 - 257	1573 - 1901	228 - 276

Line pressure is basically controlled by pump output and the pressure regulator valve. In addition, line pressure is boosted in Reverse, and Lo by the reverse boost valve.

Also, in the Neutral, Drive, and Reverse positions of the selector lever, the line pressure should increase with throttle opening because of the vacuum modulator.

Figure 21-3. An oil pressure gauge is attached to this transaxle to check line pressure. The chart is used to compare actual transaxle pressures to specifications. Incorrect readings indicate an internal hydraulic problem. (General Motors)

turning off the engine. Some transaxles will have different readings in neutral and park, so always check with the shift selector lever in the position recommended by the service manual.

NOTE! If the transaxle has a separate reservoir for differential lubricant, this level should also be checked.

While checking fluid level, the condition of the fluid should also be inspected. If the fluid smells burnt, and it is orange or dark brown, the clutch and band friction linings are probably damaged. To confirm this diagnosis, remove the transaxle oil pan. The presence of large amounts of sludge or pieces of friction material means that the holding members are burned and the transaxle must be removed for service.

Automatic Transaxle Overhaul

Do not remove the transaxle until you are sure that it must be removed. As mentioned previously, the transaxle can be serviced without removing it from the vehicle in many cases.

When your troubleshooting endeavors find major internal problems, the automatic transaxle must be removed from the vehicle. The manufacturer's service manual will give accurate instructions on how to remove, disassemble, inspect, rebuild, and install the transaxle. Special training is needed to become competent to make major internal repairs on automatic transaxles.

The remainder of this chapter presents *general* procedures explaining how to overhaul a transaxle. You should always refer to the manufacturer's service manual for specifics. Note that certain procedures included in this section–valve body service, for example–can be performed while the transaxle is still installed in the vehicle. However, they are presented here as part of a complete transaxle overhaul.

Automatic Transaxle Removal

The first step in overhauling an automatic transaxle is to remove it from the vehicle. Removal and replacement of automatic transaxles is relatively simple if you follow the steps outlined in this section.

Most automatic transaxles can be removed from underneath the vehicle without removing the engine. A few vehicles are designed so that the transaxle and engine must be removed as a unit. On others, underhood clearances are so close that it is easier to remove them this way. In this case, you must have a good engine hoist and know how to use it.

Following are some *general* procedures detailing how to remove an automatic transaxle from a vehicle. The removal process presented in this chapter assumes that the transaxle can be removed without removing the engine. Keep in mind that the procedure is much different from that used on rear-wheel drive automatic transmissions.

1. Disconnect the battery negative cable to prevent shorting wires or accidentally operating the starter.
2. While under the hood, also remove the transaxle dipstick, the speedometer cable, electrical connectors, and ground straps. Figure 21-4 illustrates some of the parts that can be disconnected before raising the vehicle. In many cases, it will be necessary to install an engine holding fixture to remove the engine and transaxle weight off the mounts.

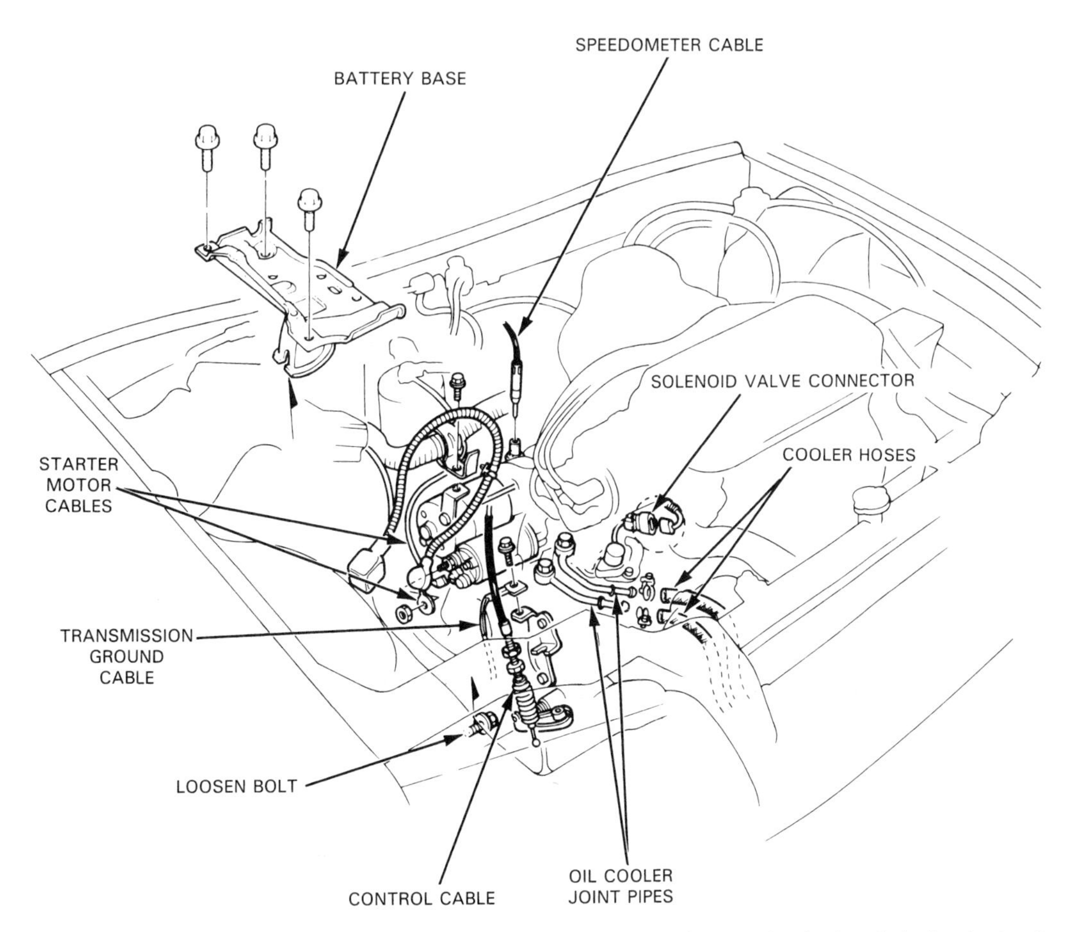

Figure 21-4. Note some of the hardware that can be disconnected from under the hood of a front-wheel drive vehicle. Exact component placement will vary, so check the service manual for procedures. (Honda)

3. Raise the vehicle to get enough clearance to remove the transaxle. If you do not have access to a hydraulic lift, always use a good hydraulic floor jack and approved jackstands.

WARNING! Never support the vehicle with wooden crates or cement blocks. Be sure the vehicle is secure before working underneath it. Do not allow the CV axles to hang.

4. Check the engine mounts to ensure that they are not broken. Broken mounts could be the cause of shifting or clutch problems.
5. Drain the transaxle fluid. While it is not absolutely necessary to drain the fluid at this point, doing it now will prevent fluid spills later. Figure 21-5 shows draining of transmission fluid and differential lubricant from one make of transaxle.
6. Remove the splash shields, strut arms or stabilizer rods, and any other hardware that will be in the way during CV axle or transaxle removal. Then, remove the CV axles. Refer to Chapter 23 for details on CV axle removal. Plug the CV axle openings to reduce oil leakage if fluid in transaxle has not yet been drained.

CAUTION! Be sure to remove the CV axles in the proper manner or the axles or other vehicle parts can be damaged.

7. Remove the exhaust pipes if necessary to gain clearance for transaxle removal. Also, on some vehicles, it may be necessary to unbolt and swing out the front frame extension or engine cradle on one side to remove the transaxle. On other vehicles, some suspension parts must be removed to gain enough clearance to lower the transaxle. Consult the specific service manual to be sure that you perform this procedure properly.
8. Disconnect the vacuum line from the vacuum modulator, if so equipped. Remove the fill tube. Disconnect the throttle valve, kickdown, and shift linkages at the transaxle. Figure 21-6 shows the location of the shift linkage in one common front-wheel drive vehicle.

   Disconnect the transaxle oil cooler lines at the transaxle. Use a backup wrench when loosening or tightening fittings to avoid twisting lines. Some cooler lines snap together and require the use of a special removal tool. Position a drip pan under the lines, as they will usually leak oil.
9. Remove the torque converter dust cover. Then, remove the converter cover attaching bolts (converter cover-to-flexplate bolts). The flexplate will have to be turned to reach all of the attaching bolts. It may be necessary to remove a cover in the inner fender to gain access to the crankshaft bolt to turn the crankshaft, as shown in Figure 21-7. You can also use the starting motor if you are careful.
10. Push the converter back away from the flexplate and remove the starter, if it is bolted to the transaxle converter housing.
11. Loosen the transaxle case-to-engine bolts.

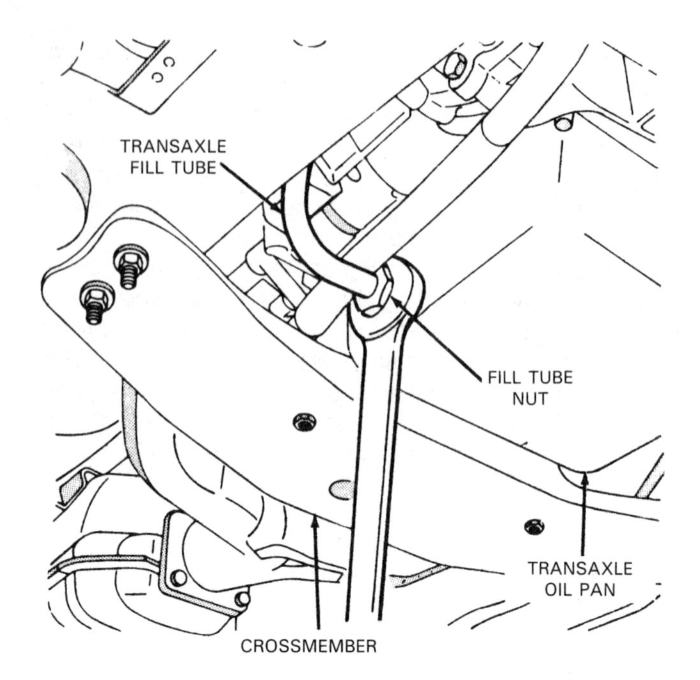

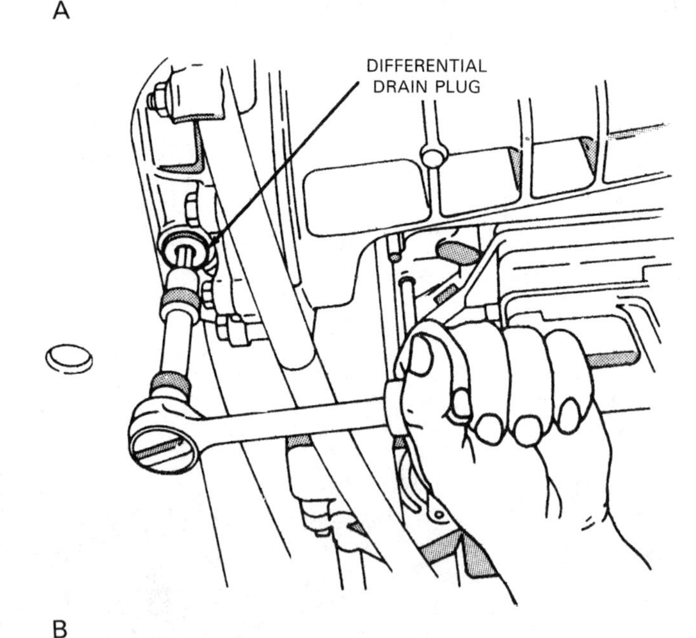

Figure 21-5. Transaxle transmission and differential are not integral in this design, and different lubricants are required for each. On most transaxles, the transaxle and differential are lubricated by the same fluid. A–Draining transmission fluid by disconnecting the fill tube at the transaxle oil pan. B–Draining the differential oil by removing the drain plug. (Chrysler)

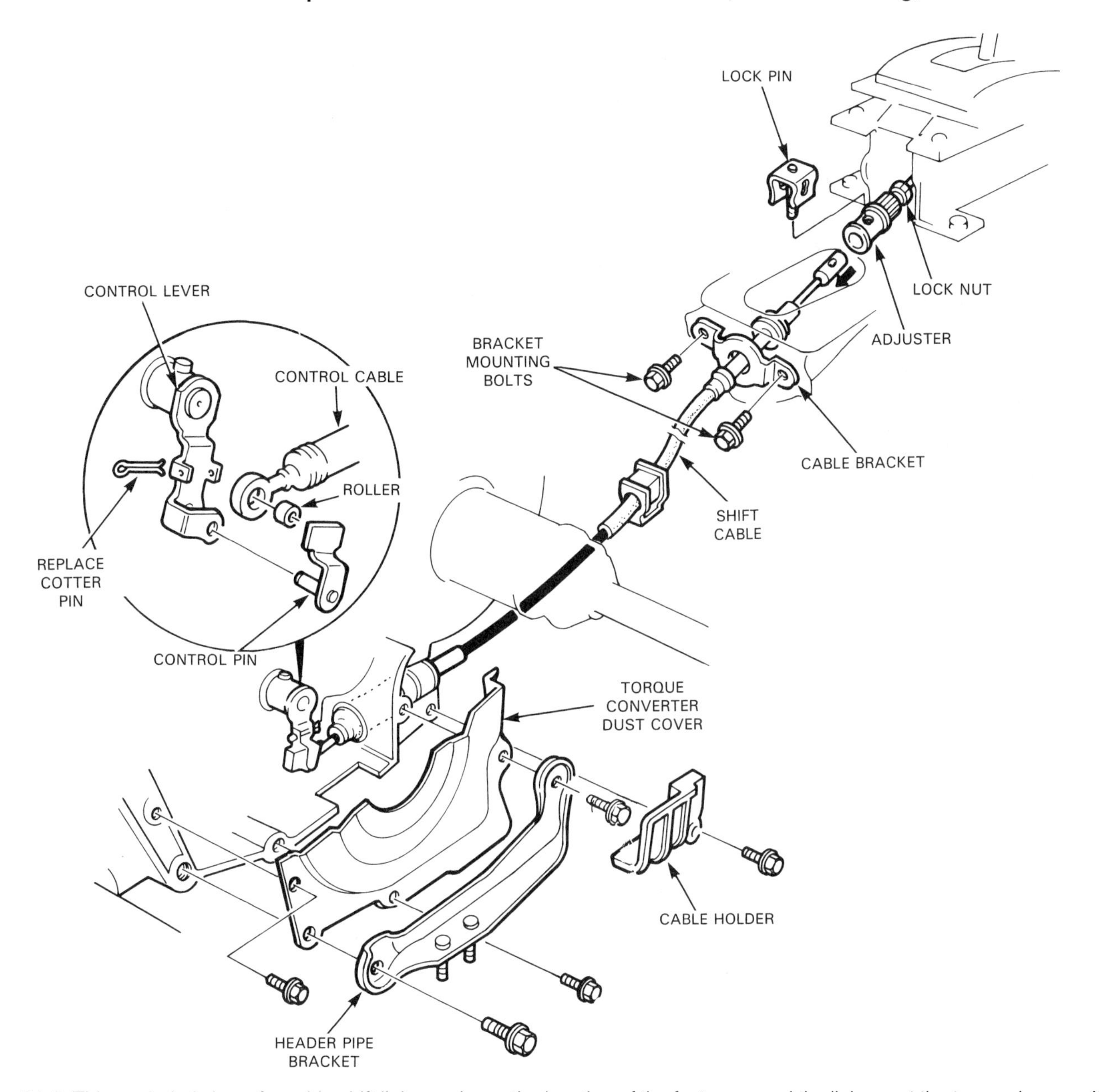

Figure 21-6. This exploded view of a cable shift linkage shows the location of the fasteners and the linkage at the transaxle case. It is only necessary to disconnect the cable at the transaxle. (Honda)

WARNING! Always leave at least two bolts holding the transaxle to the engine until you have a transmission jack securely installed under the transaxle. Although most transaxle cases are aluminum, they are still heavy. You should always use a transmission jack to support and lower the transaxle.

12. Place the transmission jack under the transaxle so that the assembly is safely supported. Raise the jack slightly to remove the weight of the transaxle from the lower transaxle mounts. Then, remove the lower transaxle mounts and any other parts holding the transaxle to the vehicle body.
13. Remove the last transaxle case attaching bolts, which hold the transaxle to the engine.
14. Slide the transaxle straight back and away from the engine, until the converter housing clears the engine flexplate. Attach a small C-clamp to edge of the converter housing or use a *torque converter holding tool* to secure the converter.
15. Lower the transaxle, watching clearances, which are usually close on front-wheel drive vehicles. Also, watch for any component that you may have forgotten to disconnect, especially any electrical connectors, ground wires, or other wiring connected to the transaxle.

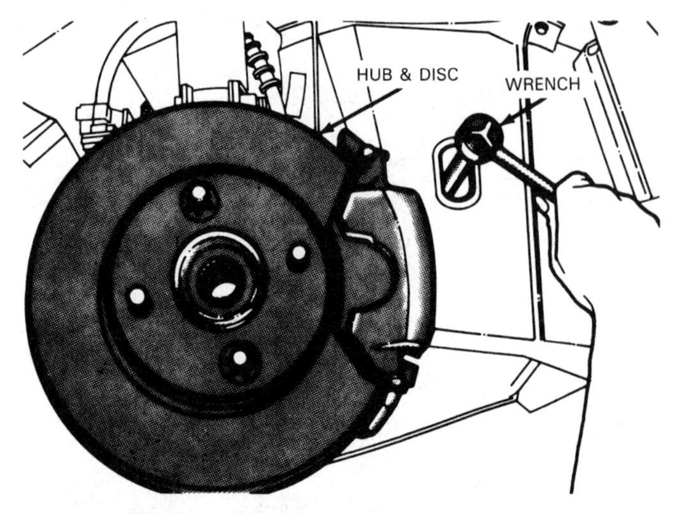

Figure 21-7. The crankshaft must be turned to gain access to the converter cover attaching bolts. To do so, it may be necessary to remove an access cover in the inner fender. With the cover removed, a wrench and extension can be used to turn the crankshaft. (Chrysler)

As soon as the transaxle is clear of the vehicle, lower it to the floor and transport it to the workbench in the lowered position. If the transaxle is very dirty, you may want to remove some of the outside dirt before beginning disassembly.

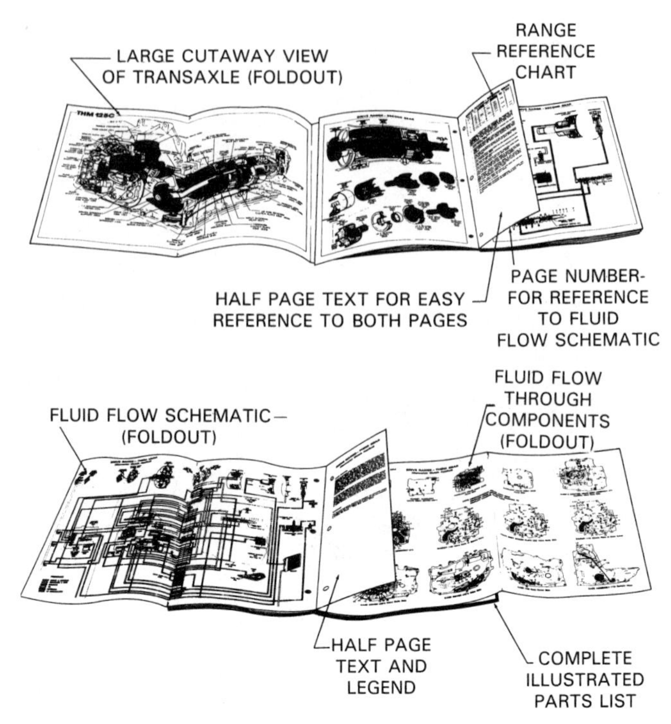

Figure 21-8. This service manual is very detailed. It covers the basic design and hydraulic circuits of one kind of automatic transaxle. This kind of service manual is a great help when rebuilding an automatic transaxle for the first time. Service manuals should be obtained before beginning the job. (General Motors)

Automatic Transaxle Disassembly

This section will cover the general principles of transaxle disassembly. Always have the proper service manual available, since transaxles vary in design. Figure 21-8 shows a typical service manual. This manual is carefully designed to be useful as a source of specific transaxle information.

Make sure that you have plenty of clean workbench space on which to perform the teardown and to place the parts. Begin disassembly by mounting the transaxle in a holding fixture.

As you remove the major parts of the transaxle, set them aside for further disassembly, as required. Work on one subassembly (pump, clutch pack, or servo, for example) at a time. During disassembly, place all related parts together on the workbench.

Remove the torque converter first. Before proceeding further, check and record the transaxle shaft end play as explained in the manufacturer's service manual. Transaxle shaft end play should be checked before transaxle disassembly and after reassembly to ensure that clearances are correct.

Remove any external transaxle components, such as vacuum modulators or case-mounted governors. Inspect all parts as they are removed. See Figure 21-9. Remove the transaxle oil pan. Remove side and end covers to gain access to the valve bodies and related linkage and solenoids. As you remove the valve bodies, carefully note the position of all valve body parts, spacer plates, oil delivery tubes, accumulators and springs, linkage parts, and check balls to aid reassembly. Removal of a valve body is shown in Figure 21-10.

The next step is to remove the drive gears or the drive chain and sprockets, as required. Figure 21-11 shows the general layout of these two drive systems. Always check for wear during disassembly.

Remove the transaxle oil pump, if it is separately mounted. Remove the front band, if the transaxle has one. Then, remove the front and rear clutch assemblies. In some transaxles, the input shaft comes out with the rear clutch. If the transaxle has a center support, remove the attaching bolts or snap rings and remove the support from the case.

CAUTION! To prevent parts damage, do not attempt to drive the center support from the transaxle case without removing the retainers.

Once the front components are removed, planetary gearsets, rear holding members, and the output shaft, as well as any case-mounted band servos or accumulators, can be removed. Then, the differential assembly can be removed from the transaxle case.

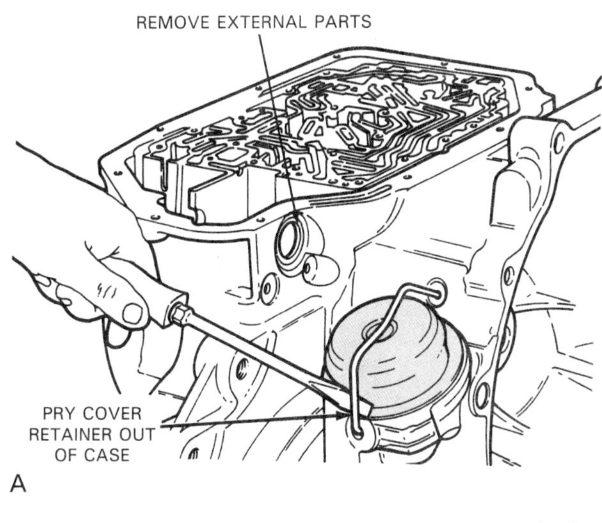

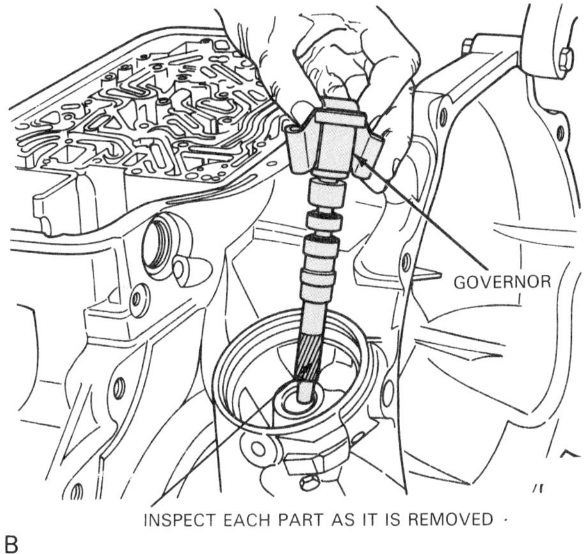

Figure 21-9. Removal of a governor is shown. Remove case-mounted transaxle components early on in disassembly, before beginning the teardown of the transaxle internals. A–Removal of the governor begins with removing the governor cover. B–Inspect all parts as you remove them. (Ford)

Automatic Transaxle Parts Cleaning

Clean all transaxle parts thoroughly, since the smallest dirt particle can cause a valve to stick or a seal to leak. Use fresh cleaning solvent. Blow parts dry with compressed air. Do not dry internal parts with rags, since this will leave lint on parts. All old gaskets should be completely removed.

As you are cleaning the parts, inspect them for varnish deposits caused by transmission fluid overheating. If there is varnish on the transaxle parts, the converter internal parts are also coated with varnish. Since the converter is a welded unit, it cannot be taken apart for repair or cleaning. It can be flushed out, however, but a heavily varnished

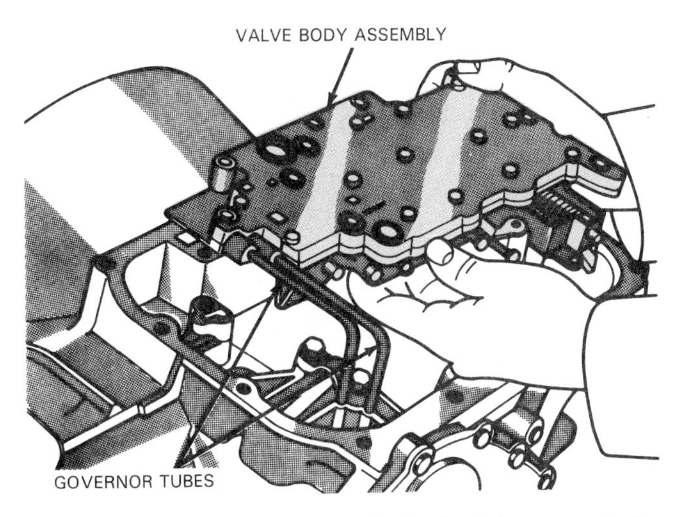

Figure 21-10. This bottom-mounted valve body is removed after the attaching bolts are removed. Governor tubes, which transport hydraulic fluid to and from governor, must be carefully removed and installed to avoid hydraulic system problems. (Chrysler)

converter should probably be replaced. If the converter is reused, it should be drained as thoroughly as possible.

If there is evidence of fluid contamination, the transmission oil cooler and lines must be flushed out with an oil cooler and line flusher. Follow up with a blast of compressed air directed into the lines.

Automatic Transaxle Parts Inspection and Repair

After the transaxle is disassembled and the internal parts are cleaned, the parts can be more closely inspected to spot defects and causes of failure. To avoid confusion, you should inspect and rebuild transaxle parts one subassembly at a time.

It is not entirely necessary to check soft parts, such as O-rings and gaskets, which will be replaced anyway. However, checking these parts can help to diagnose unusual transaxle problems. Replacement soft parts for the transaxle are usually ordered as a kit. These overhaul kits contain all of the commonly needed parts, such as gaskets, O-rings, seal rings, and new friction discs.

Transaxle hard parts, such as clutch drums, pumps, or gearsets, which are not normally replaced, must generally be ordered separately. Order replacement parts right away to eliminate lost time while waiting on parts. Carefully match all new parts against old parts to ensure that they are the correct replacements.

Automatic transaxle shaft and planetary gearset service

Carefully inspect the automatic transaxle shafts. The shafts, which are relatively small, are subject to wear.

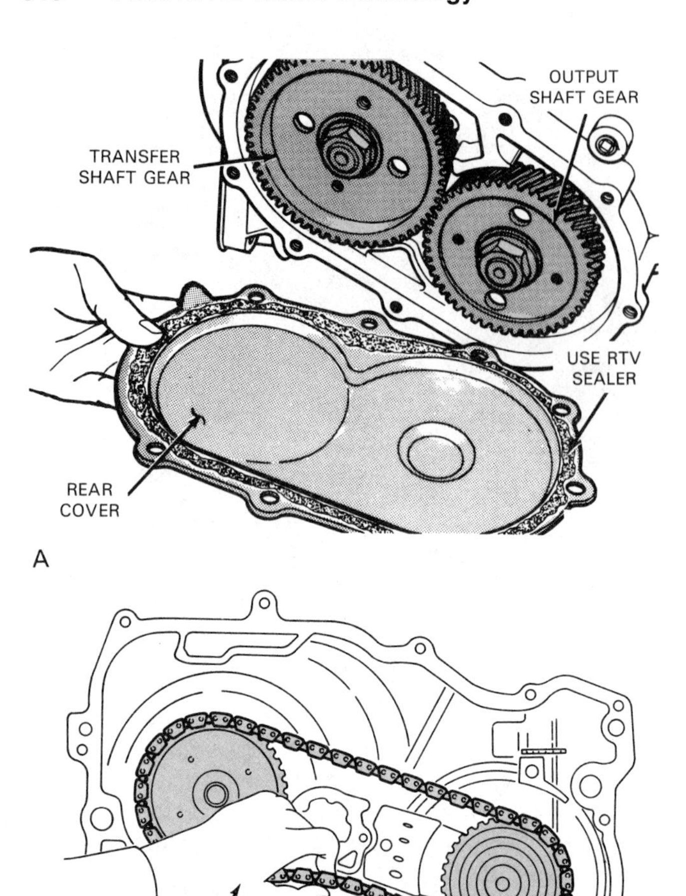

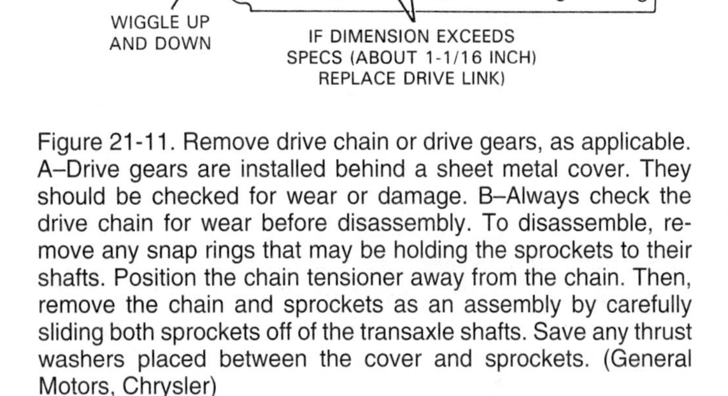

Figure 21-11. Remove drive chain or drive gears, as applicable. A–Drive gears are installed behind a sheet metal cover. They should be checked for wear or damage. B–Always check the drive chain for wear before disassembly. To disassemble, remove any snap rings that may be holding the sprockets to their shafts. Position the chain tensioner away from the chain. Then, remove the chain and sprockets as an assembly by carefully sliding both sprockets off of the transaxle shafts. Save any thrust washers placed between the cover and sprockets. (General Motors, Chrysler)

Figure 21-12 shows a typical shaft. Check the seal ring grooves and surfaces where the rings ride for wear. Also, check the splines and snap ring grooves for wear or damage, and replace the shaft if any defects are found. Many shafts contain gears and bearings, as shown in Figure 21-13.

Check the planetary gearsets for wear or damage. Typical gear damage is shown in Figure 21-14. Planet carriers seldom go bad, but they can contain bushings or needle bearings that will require attention, Figure 21-15.

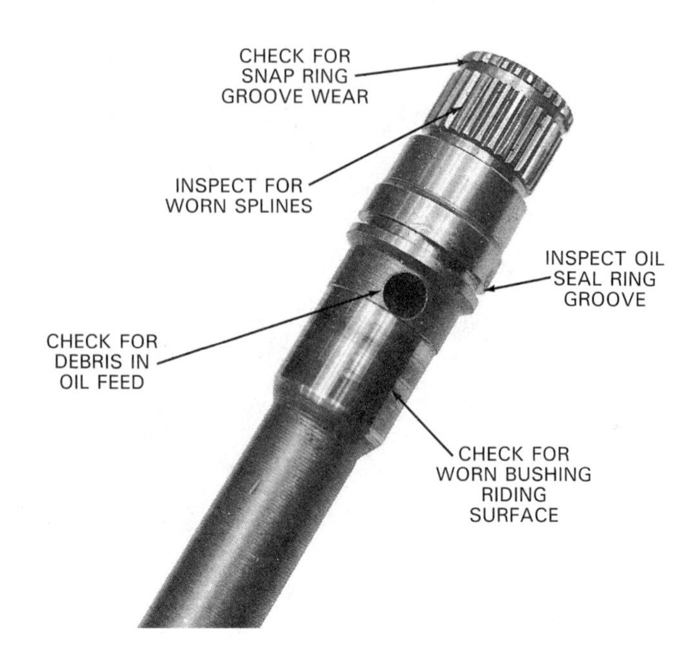

Figure 21-12. After the transaxle is disassembled, check shaft splines, bearing journals, and ring grooves for wear. (Chrysler)

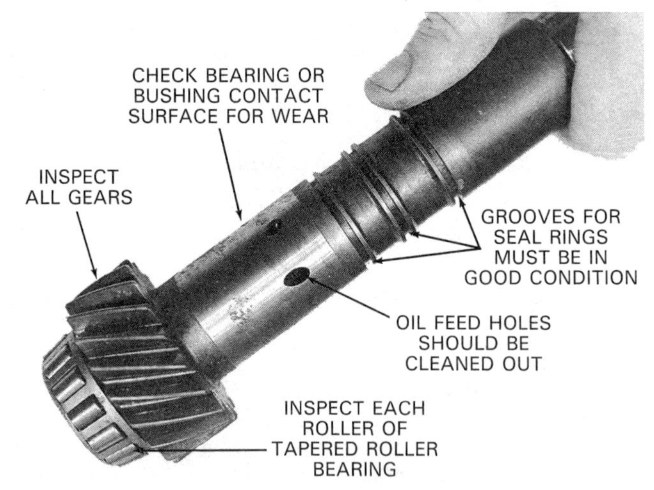

Figure 21-13. If any of the shafts contain gears and bearings, they should be closely inspected for damage. (Chrysler)

Drive chain service

Check the chain and sprockets for wear and damage. If there is any wear, the chain and both sprockets should be replaced as a unit. Refer to Chapter 3 for more information on drive chain service.

Clutch pack service

The clutch friction discs and steel plates are the most common area of transaxle failure. Examination of these can serve as a guide to general transaxle condition. Refer to Figure 21-16.

Figure 21-14. Gear teeth on all of the planetary gears should be closely inspected for chips, cracks, scoring, and wear. Also, check for worn snap ring grooves or bushings. Replace any damaged gears. (Chrysler)

To check the steel plates, take a screwdriver and remove the snap ring that holds them and the friction discs in the clutch drum; then, remove the plates and the discs. Keep all of the steel plates together. Do not interchange one set of clutch pack parts with another.

Inspect the steel plates. Look for discoloration caused by overheating, often a sign of a badly slipping clutch pack. Look for worn or scored plates, a sign of a prolonged slippage condition. Also, check the tangs for wear or damage. If there is any doubt about the condition of the plates, they should be replaced. Steel plates are sometimes included in a transaxle overhaul kit.

Friction discs are normally replaced as part of a transaxle overhaul. These parts are usually included in a transaxle overhaul kit. Worn or damaged discs may be charred, glazed, or heavily pitted. The friction lining may scrape off easily with your fingernail. In some instances, all of the friction material will be missing from the friction discs.

Use a suitable clutch pack spring compressor to remove the clutch return spring(s). Then, remove the clutch apply piston. Use a small screwdriver or ice pick to remove the piston seals, as shown in Figure 21-17. Inspect the seals. Hard or broken seals are a common cause of slippage. If the clutch friction discs were burned, expect the seals to be hard and crystallized from excessive heat. If there is also a seal around the interior of the clutch drum, replace it. Then, thoroughly clean the piston and piston bore.

Obtain new piston seals. Check them against the old ones to ensure that you are installing the proper seals on the piston. It is also a good idea to check the fit of the new

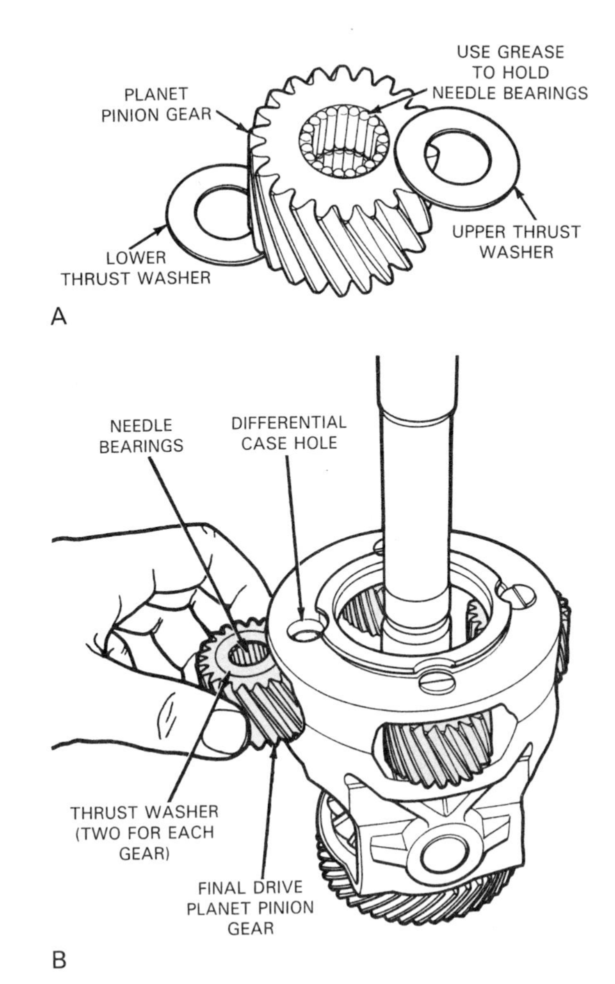

Figure 21-15. Planet carriers can be overhauled by replacing bearings and washers. This planet carrier assembly provides final drive reduction at the differential assembly. A–Typical planet-pinion gear, showing the needle bearings and thrust washers. B–Installing the planet-pinion gear in the carrier. (Ford)

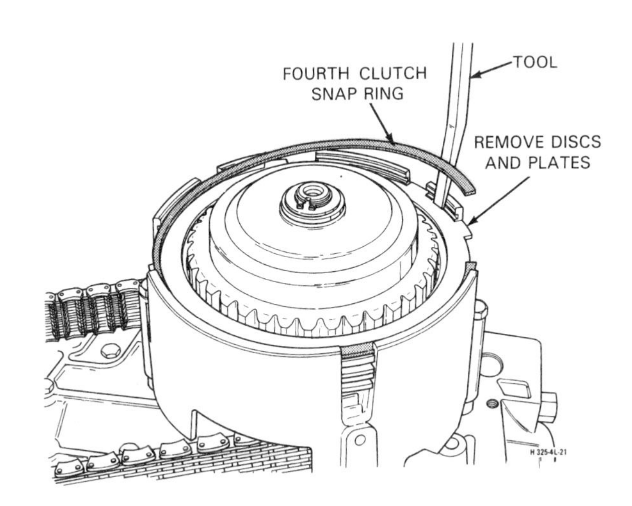

Figure 21-16. The clutch packs should be disassembled to check for burned friction discs and other problems. Most clutch packs are held together with a large snap ring, as shown here. (General Motors)

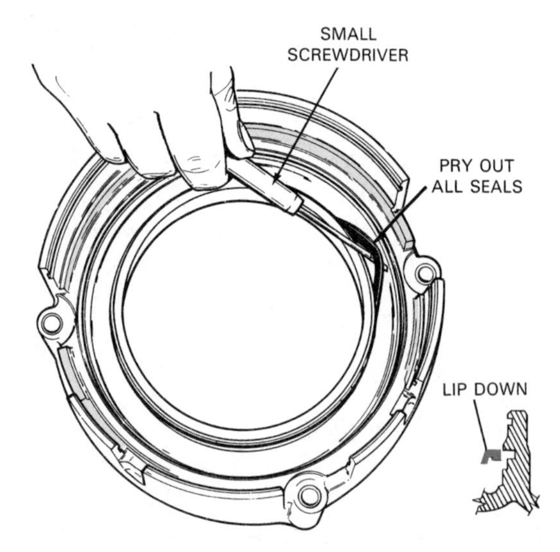

Figure 21-17. The clutch pack seals should be removed for inspection. Take a close look at them, even though they are going to be replaced. Hard or cracked seals can often cause upshift problems. (General Motors)

seals by placing them in the piston bore. They should fit snugly in the bore. Before installing the new piston seals, lubricate them with transmission fluid. Petroleum jelly can also be used, but other types of oil or grease can damage the seals. Carefully install the new piston seals on the piston. The new seals should fit snugly, but not *too* snugly.

Install the clutch piston in its bore. The type of piston seal determines the installation method. If ring seals are used, the piston can be pressed into the bore, after being thoroughly lubricated. Lip seals must be installed in the bore so that the sealing lip is directed toward the hydraulic pressure, or the back of the drum. Lip seals must be worked into place with a feeler gauge or a special seal installation tool.

Once the lip seal is in place, check the installation by trying to turn the piston. If you cannot turn it, the seal is not properly installed.

Once you are sure the piston is properly installed, put the piston return spring and spring retainer back in position on the piston. Then, use the spring compressor tool to move the return spring and retainer past the groove for the retaining snap ring. Install the snap ring and make sure that it is snug in its groove.

If the clutch pack uses a reaction plate next to the apply piston, install it now. Then, install the friction discs and steel plates, alternating them and using new parts as required. If the clutch pack uses another reaction plate at the top, install it at this time. As a final step, install the snap ring to hold the assembly together.

CAUTION! A dry friction disc will burn severely the first time the clutch is applied. To prevent this, soak new friction discs in clean transmission fluid before installing them. Allow them to soak for about 15 minutes or until they stop bubbling. Always use the correct type of automatic transmission fluid. The vehicle owner's manual should list the proper ATF. The fluid type may also be stamped on the transmission dipstick.

With the clutch pack now assembled, measure the clutch clearance, or the distance between the pressure plate and outer snap ring. For good shift characteristics, clutch pack clearances should be at the low end of the tolerances. Some transaxle manufacturers provide selective thickness snap rings or clutch plates to achieve this clearance.

Once the clutch pack has been reassembled, it is a good idea to make sure that the clutch discs and plates can turn against each other easily. Binding clutch packs will burn out quickly, Figure 21-18. Repeat the rebuilding sequence for every clutch pack in the transaxle.

Band service

Check any transaxle bands for burned facings. The facing materials can be checked in the same way the clutch friction discs were checked. If any friction material can be scraped from the band lining, the band should be replaced.

The drum that the band rides on should be inspected for scoring and wear. If found, it should be replaced. Drums that are in good condition, but having shiny band contact surfaces, can be reused. Sand the drum in the direction of

Figure 21-18. One way to check for proper clearances in the clutch pack is to partially assemble parts and try to turn them. If the parts are hard to turn, the friction discs and steel plates are binding, and the problem must be corrected. (Saab)

drum rotation to remove the shiny surface. Then, reclean the drum in solvent.

Servo and accumulator service

Disassemble the transaxle servos and accumulators. Special tools are often required to depress the servo or accumulator cover so the retaining snap rings can be removed. Do not try to remove the snap ring until the cover has been pushed in and is out of the way. With the snap ring removed, the servo parts can be removed from their cylinder. Note part positions as you disassemble. Then, check internal parts for wear and dirt.

Lubricate and install new piston seals on the servo and accumulator pistons. These are usually square-rings or teflon seals. Then, install the pistons, making sure to properly assemble all pistons and springs. Install the servo or accumulator cover using the proper seals or gaskets. The same tool used to depress the cover for disassembly can be reused to install the cover.

Note that since many modern transaxle bands are nonadjustable, band clearances must be set carefully before the transaxle is reassembled, Figure 21-19. Piston travel affects the band's holding ability. The servo is usually adjusted by using selective thickness band apply pins or shims in the servo piston.

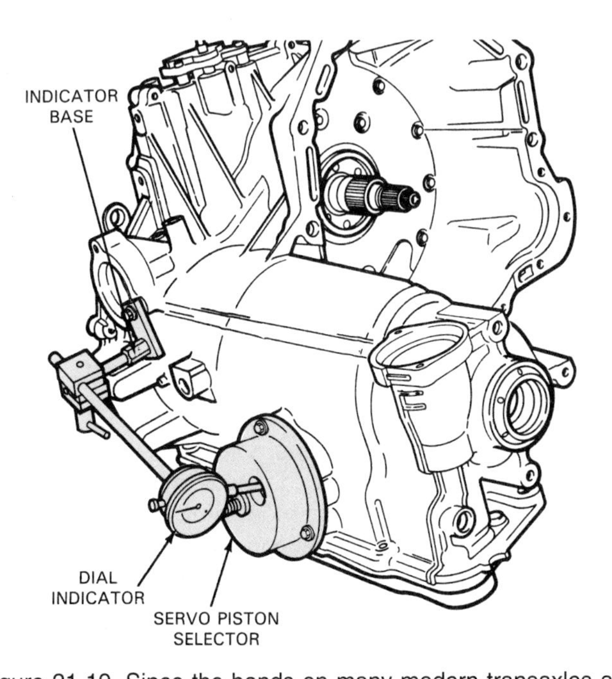

Figure 21-19. Since the bands on many modern transaxles are not adjustable, the band clearance must be carefully checked when the transaxle is torn down. The adjustment procedure usually involves using a dial indicator to check servo piston travel. The servo can be adjusted with shims or by changing the apply pin. Note that the *servo piston selector* is a special tool used to measure clearance. (Ford)

Overrunning clutch service

Check all transaxle one-way clutches to ensure that they turn in the proper direction and only in that direction. Any one-way clutch that turns in both directions, that does not turn smoothly in its intended direction, or that shows any sign of wear should be replaced. Make sure that the replacement clutch turns in the correct direction when installed.

Transaxle oil pump service

Disassemble the transaxle oil pump and inspect it for wear where the moving parts move against the pump stationary parts. Badly scored pumps can be spotted by visual examination or by feeling the suspect areas with a fingernail. Since the transaxle oil pump provides the overall hydraulic system pressure, any worn pump should be replaced.

Reinstall the pump components in the pump body, Figure 21-20. Replace any seals and replace pump bushing, if needed. Make sure that all pump parts are reinstalled in the same position that they were originally.

Hydraulic valve service

Check the hydraulic valves for free movement in their bores. Every valve should be able to drop in its bore by gravity when lightly shaken. Spring-loaded valves in the valve body should be able to return to their original positions when they are moved (carefully) with a small screwdriver. Sticking valves can sometimes be loosened by lightly polishing them or cleaning the valve bore. If proper operation cannot be restored, the valve or even the valve body assembly must be replaced.

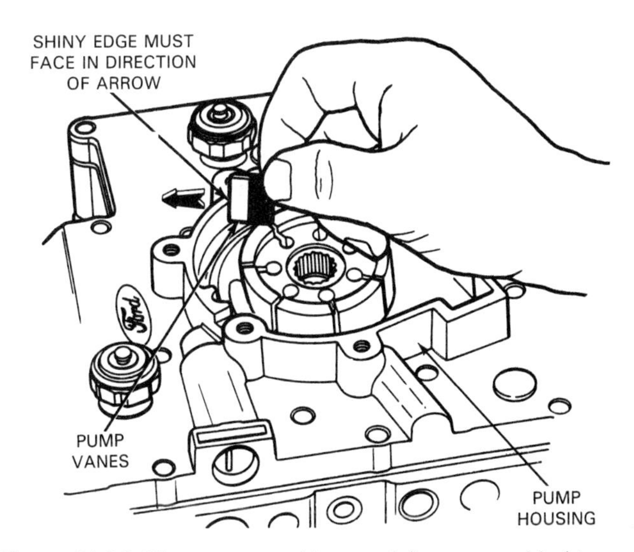

Figure 21-20. The pump must be carefully reassembled to provide proper pressures and long service. This vane pump is mounted in the side valve body of a common make of transaxle. (Ford)

CAUTION! Handle the valve body and related parts very carefully. Small nicks or dirt particles can cause valves to stick. When polishing valves, be careful not to round off the sharp edges of the valve lands. They help to keep the valve from jamming on small pieces of dirt.

Differential assembly service

The differential will require inspection when the transaxle is overhauled, unless contained in a separate housing. The differential gears should be checked for wear or damage, and the differential case should also be checked for problems. If necessary, the ring gear can be removed from the differential case. Any damaged or worn differential parts should be replaced. Adjust the differential bearings according to manufacturer's directions. Replace seals or gaskets used for separating differential lubricant from transaxle lubricant.

Bearing and thrust washer service

Finally, inspect any transaxle bearings or thrust washers that have not been checked already. Check for wear and scoring. General procedures for checking bushings and washers were discussed in Chapter 2. Be sure to use proper drivers to remove and install bushings.

If the transaxle end play was excessive, pay particular attention to the thrust washers. Selective thickness washers are available to correct end play; use thicker or thinner washers as needed.

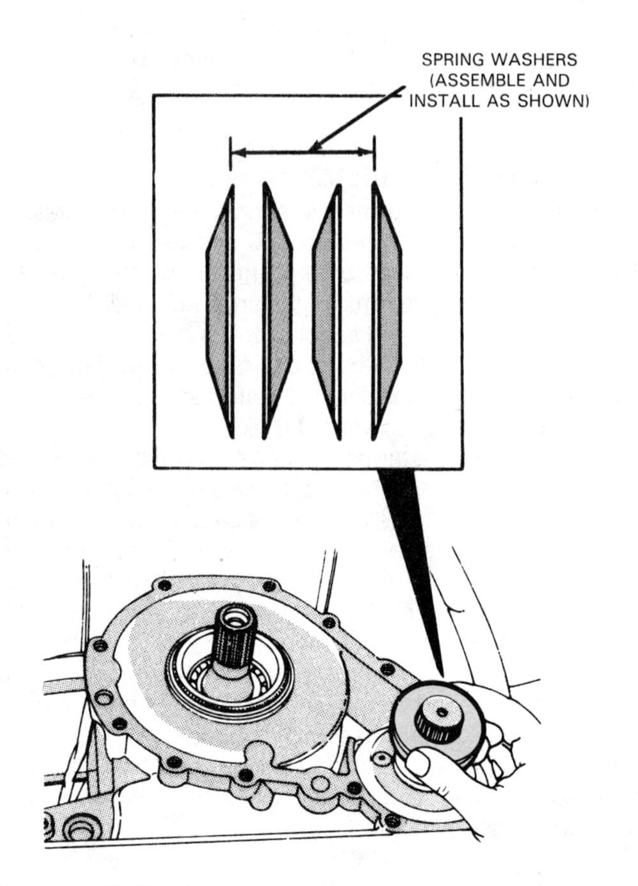

Figure 21-21. Washers are used to maintain proper clearances or to preload the drive train parts. They must always be placed in the proper position. (Chrysler)

Automatic Transaxle Assembly

Once all automatic transaxle subassemblies have been rebuilt, including clutches, planetary gearsets, shafts, and differential, install them in the transaxle case. Always refer to the manufacturer's service manual for the exact assembly procedures. It will give procedures and specifications for the exact type of transaxle.

CAUTION! Remember to soak the friction discs, bands, and all rubber parts in fresh transmission fluid before installation. Also, use plenty of lubrication and be extremely careful when aligning and seating transaxle parts.

One area to watch closely is the installation of the drive chain or gears. Make sure that all thrust washers are reinstalled in the case before installing the gears or drive chain sprockets. Some washers are spring loaded. They must be assembled properly for proper tension on the chain. See Figure 21-21.

Once thrust washers are in place, reinstall the chain and sprockets or gears, as shown in Figure 21-22. Tighten all fasteners and recheck all adjustments. Make sure that the input shaft and/or output shaft turns after tightening the fasteners. If either shaft begins to bind, you must find the cause and correct it before proceeding.

CAUTION! Never install a transmission on a vehicle if the shafts are binding. The transmission will be badly damaged.

Check the end play using a dial indicator. If the end play is not within specifications, the original selective fit thrust washer must be replaced. Install any external components, such as the vacuum modulator or governor, and adjust the bands to specifications. Some bands are not adjustable. Do not install the valve bodies yet.

Perform an air pressure test on the transaxle to ensure that clutch pistons and servos are working. Air pressure testing was covered in Chapter 12. The factory service manual will have an illustration of the necessary oil feed holes. If operation is faulty, the component may be hung up, there may be a leak or obstruction within the hydraulic passageway, or the seals, seal rings, or gaskets may not be sealing as they should.

Once air pressure checks are complete, install the spacer plate, gaskets, and any check balls in the transaxle

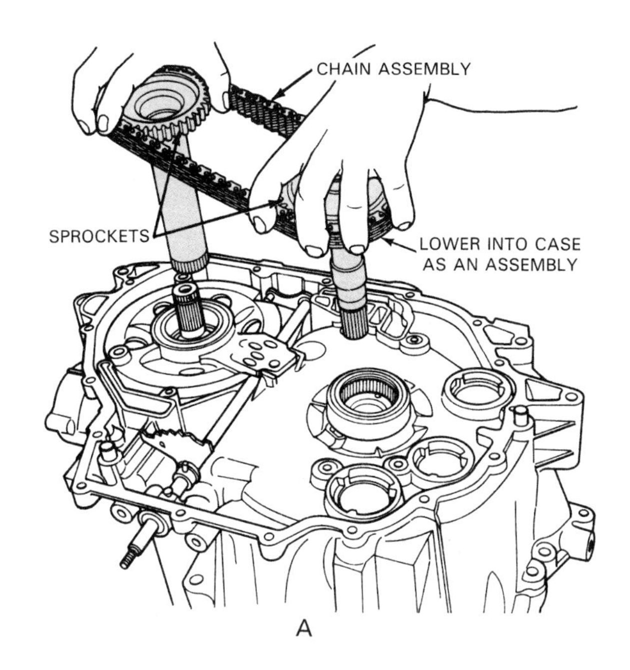

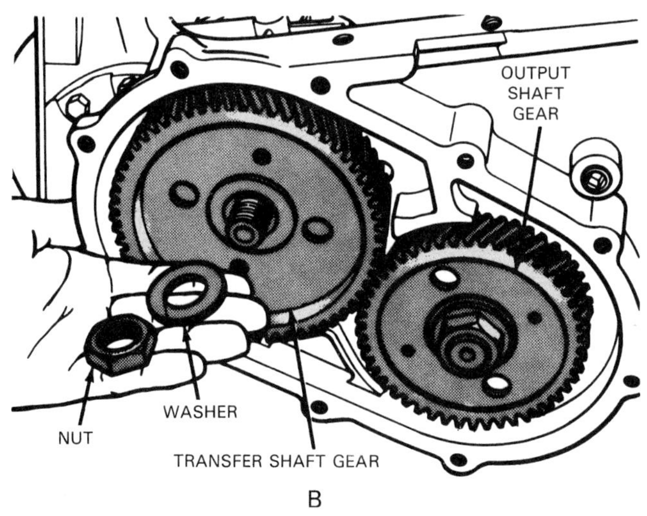

Figure 21-22. Carefully reinstall the drive chain or gears. A–The drive train and sprockets should usually be lowered into the case as an assembly. B–The drive gears can be installed separately. Be sure to install any washers and fasteners that were removed. (Ford, Chrysler)

case. Then, install the valve bodies. Carefully match the holes in the spacer plate and the replacement gaskets, paying close attention to the gasket to ensure that you have the correct replacement. Do not overtighten the bolts. Also, install any needed throttle valve and shift linkage components.

Next, install the side and/or end covers, transaxle oil pan, and any other external transaxle parts. Note that on some automatic transaxles, the output shaft passes through a side cover. The oil seal at the cover must be carefully replaced to prevent leaks. If the cover is dented or otherwise damaged, it should be replaced.

The final step in transaxle assembly is to install the torque converter. In most instances, it is easiest to turn the transaxle until the turbine shaft is pointing straight up. Then, attach two long bolts, slide hammers, or other fasteners to the converter to serve as handles. Position the converter over the turbine shaft and lower it until its internal splines mesh with splines on the turbine shaft, oil pump shaft, and stator shaft, as applicable. It may be necessary to wiggle the converter slightly to get it to engage the transaxle shafts. If the oil pump is driven by a pump drive hub, the drive lugs must fully engage the oil pump. Figure 21-23 illustrates typical converter installation.

CAUTION! Under no circumstances should you install the converter on the flexplate while it is disassembled from the rest of transmission. The splines of the transaxle shafts and converter can be damaged.

Automatic Transaxle Installation

Installing the transaxle is basically the reverse of removal. As with removal, the vehicle should be on a hoist with the engine supported. Following are some *general* procedures detailing how to install an automatic transaxle in a vehicle. Note that you should always refer to the manufacturer's service manual for specific procedures.

1. Secure the torque converter with a torque converter holding tool or a C-clamp.

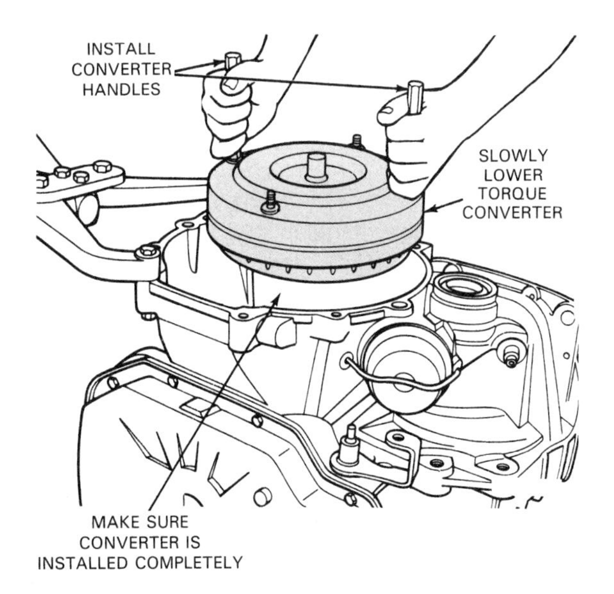

Figure 21-23. To install the torque converter, turn the transaxle until the turbine shaft faces up and lower the converter into place. Make sure that the converter is completely installed over the shaft splines, or it will bind and be damaged when the transaxle is reinstalled in the vehicle. (Ford)

2. Raise the transaxle under the vehicle with a transmission jack or floor jack until it is at the proper height to engage the engine. Then, slide the transaxle converter housing into engagement with the rear of the engine. Make sure that the converter housing fits tightly against the rear of the engine. (Remove the converter holding tool when necessary.)
3. Once the converter housing is flush with the rear of the engine, install at least two of the transaxle case-to-engine bolts.

   After the bolts are installed, make sure the torque converter still turns easily and that it is not jammed; if it is, it is not properly fitted over the transaxle shafts.

   If the converter installation is proper, install the remainder of the transaxle case-to-engine bolts. Torque all bolts to specifications.
4. Install and tighten the converter cover-to-flexplate bolts (or nuts) to specifications. Install the starter, if it was removed. Then, install the dust cover.
5. Install and tighten the engine mounts and other fasteners holding the transaxle to the vehicle frame.
6. Install the CV axles as explained in Chapter 23.
7. Install miscellaneous items, such as the oil cooler lines, modulator vacuum line, electrical connectors, and T.V., kickdown, and shift linkages. Install the exhaust pipes, if they were removed. Install the speedometer cable and dipstick tube, if they must be installed from under the vehicle.
8. Reinstall any other frame and body parts that were removed for clearance.
9. Lower the vehicle. Install the battery negative cable and other underhood components. Be sure to install all electrical components and ground wires. Check the vacuum lines at the engine. Check the various linkages for proper adjustment and movement.
10. Fill the transaxle transmission with the proper ATF. It is advisable to overfill a dry transaxle by about 2 quarts (or 2 liters), *as measured on the dipstick,* before starting the engine. If the unit has a separate reservoir for differential oil, refill it also.
11. Start the engine and immediately recheck the transmission fluid. Add fluid until the proper level is reached. Be careful not to overfill the transaxle transmission. Most transaxles will hold about 6 quarts of fluid, but check specified capacity.
12. Raise the vehicle by the front lower control arms, so that the drive wheels are raised off the ground but are near their normal position. Do not allow the wheels to hang without support. This will damage the CV joints.
13. Operate the transaxle in all gears, allowing it to upshift and downshift several times in drive gear. This will allow you to confirm that the transaxle hydraulic system is operating properly, and it will allow you to check for leaks.
14. Lower the vehicle and road test. If the transaxle shift points are incorrect, check the T.V. and throttle linkage adjustment or vacuum modulator connections, as necessary. Recheck the fluid level after the vehicle has been driven a few miles.

Summary

The fluid level on an automatic transaxle must be checked carefully, since the level will vary widely according to the temperature of the unit and whether it has been driven after adding fluid. If the transaxle has a separate differential unit, its fluid level should be checked also.

The transaxle should be removed according to a specific series of steps. The negative battery cable should be disconnected first and any hardware that can be removed from under the hood should be disconnected before raising the vehicle. Other removal steps are similar to those for a rear-wheel drive automatic transmission.

Once the transaxle is fastened in a holding fixture, it should be carefully disassembled and cleaned. Check all parts for wear, especially the clutches and band linings. The hydraulic valves should be checked for sticking and freed up if necessary. Also check the planetary gears, one-way clutches, and oil pump. Replace any defective parts.

Before beginning component reassembly, soak the clutches and bands in fresh transmission fluid for at least 15 minutes and lubricate all seals and moving parts. Carefully assemble the transaxle according to the manufacturer's directions and ensure that the transaxle shafts are free to turn. Check end play and air pressure test the clutch pistons and band servos.

Reinstall the transaxle and fill it with fluid. Start the engine and immediately recheck the fluid level. Then, lower the vehicle and road test. Make linkage adjustments if necessary.

Review Questions–Chapter 21

Please do not write in this text. Place your answers on a separate sheet of paper.

1. An automatic transaxle slips in first gear only. Technician A says that the torque converter is probably at fault. Technician B says to check the oil pan for presence of sludge and friction material, indicating a problem with a band or clutch. Who is correct?
   a. Technician A.
   b. Technician B.
   c. Both Technician A and B.
   d. Neither Technician A nor B.
2. An automatic transaxle is having shift problems. Technician A says to check the fluid level and the linkage adjustment or check for a leaking vacuum hose, if the

transaxle has a vacuum modulator. Technician B says to remove the transaxle and check the governor. Who is correct?
   a. Technician A.
   b. Technician B.
   c. Both Technician A and B.
   d. Neither Technician A nor B.
3. An automatic transaxle is noisy in all gears. Technician A says it must be the planetary gears. Technician B says it could be a plugged filter, a differential problem, or the torque converter. Who is correct?
   a. Technician A.
   b. Technician B.
   c. Both Technician A and B.
   d. Neither Technician A nor B.
4. Fluid level in an automatic transaxle seems to drop without being accompanied by leakage. Technician A says there may be a ruptured vacuum modulator. Technician B says if there was a hole or tear in the vacuum modulator, that transmission fluid would be leaking profusely onto the ground below. Who is correct?
   a. Technician A.
   b. Technician B.
   c. Both Technician A and B.
   d. Neither Technician A nor B.
5. Problems resulting from an overfilled transmission are entirely different from those resulting from an underfilled transmission. True or false?
6. What can be learned from inspection of the oil pan and its contents?
7. Which may *not* be done as an in-car service procedure?
   a. Overhaul the valve body.
   b. Service the speedometer pinion gear.
   c. Replace the oil pan gasket.
   d. Overhaul the clutch packs.
8. Before removing a transaxle for overhaul, check the ______ ______ to make sure that they are not broken, which could be causing shift problems.
9. When rebuilding an automatic transaxle, you should work on all subassemblies at the same time, so that the job goes faster. True or false?
10. What should you do when removing oil cooler lines from the transaxle?
11. The clutch friction discs and steel plates are the most common area of transaxle failure. True or false?
12. During an overhaul, gaskets, seals, and clutch friction discs should be examined for defects and replaced only if necessary. True or false?
13. Since many modern transaxle bands are nonadjustable, band clearances must be set carefully before the transaxle is reassembled. True or false?
14. When polishing a valve, you should be sure to round off any sharp edges as they may be causing the valve to jam. True or false?
15. Why should you not install the converter to the flexplate and then install the transaxle?

Certification-Type Questions–Chapter 21

1. All of these statements about automatic transaxle slippage are true EXCEPT:
   (A) a misadjusted band can cause slippage in all gears.
   (B) in some instances, slippage can be corrected without removing the transaxle from the vehicle.
   (C) the presence of sludge in the oil pan is a good indication that a slippage problem is caused by a damaged holding member.
   (D) low fluid level can cause slippage.

2. An automatic transaxle slips in second gear only. Technician A says that the problem could be caused by a defective oil pump. Technician B says that the problem could be caused by a clogged transaxle filter. Who is right?
   (A) A only
   (B) B only
   (C) Both A & B
   (D) Neither A nor B

3. All of these are examples of automatic transaxle shift problems EXCEPT:
   (A) late upshifts.
   (B) no passing gear.
   (C) noise in first gear.
   (D) harsh downshifts.

4. Technician A says that a vacuum modulator, when used, should have a leak-free connection to the intake manifold. Technician B says that transmission fluid in the modulator is a normal condition. Who is right?

(A) A only
(B) B only
(C) Both A & B
(D) Neither A nor B

5. All of these conditions can cause the automatic transaxle governor to malfunction EXCEPT:

(A) metal particles in the transmission fluid.
(B) a stuck-open pressure regulator.
(C) a worn valve or worn seal rings.
(D) a clogged governor screen.

6. Technician A says that noise in only one gear is a sign of planetary gearset damage. Technician B says that planetary gears will not make noise in overdrive. Who is right?

(A) A only
(B) B only
(C) Both A & B
(D) Neither A nor B

7. An automatic transaxle has low hydraulic pressure in one gear only. Technician A says that the oil pump may be defective. Technician B says that a servo or clutch seal may be defective. Who is right?

(A) A only
(B) B only
(C) Both A & B
(D) Neither A nor B

8. Technician A says that some automatic transaxles have a separate reservoir for the differential lubricant. Technician B says that the most accurate check of the automatic transaxle fluid level is made just after starting a vehicle which has been parked overnight. Who is right?

(A) A only
(B) B only
(C) Both A & B
(D) Neither A nor B

9. All of these may be used to gain access to the converter cover attaching bolts EXCEPT:

(A) the starting motor.
(B) a flywheel turner.
(C) turning both drive wheels in the same direction.
(D) turning crankshaft bolt with a wrench.

10. A heavily varnished torque converter should:

(A) be filled with solvent and allowed to soak.
(B) be filled with fresh transmission fluid and allowed to soak.
(C) probably be flushed out with a converter flusher.
(D) probably be replaced.

11. Replacement parts are being ordered for an automatic transaxle. Technician A says that needed parts should be determined and ordered as soon as possible to reduce waiting time. Technician B says that the new parts should be compared against the old parts to ensure that they are correct. Who is right?

(A) A only
(B) B only
(C) Both A & B
(D) Neither A nor B

12. All of these actions could cause rapid wear or burning of new clutch discs EXCEPT:

(A) setting clutch pack clearances too loose.
(B) installing the clutch piston seals upside down.
(C) not replacing the old clutch piston seals.
(D) soaking the discs in fluid before installation.

13. Technician A says that shiny band contact surfaces can be sanded and the drum reused. Technician B says that band friction linings can be sanded and the band reused. Who is right?

(A) A only
(B) B only
(C) Both A & B
(D) Neither A nor B

14. When a transaxle is reinstalled in the engine, it is discovered that the torque converter is jammed against the flywheel. Technician A says that the engine should be started to allow the converter to be positioned on the pump ears. Technician B says that the transaxle should be removed and the cause of the problem located. Who is right?

(A) A only
(B) B only
(C) Both A & B
(D) Neither A nor B

This all-wheel drive vehicle is equipped with a Torsen central differential. Torsen differentials were discussed in Chapter 16 of this text. All-wheel drive systems are discussed in Chapter 24. (Audi)

Chapter 22

Constant-Velocity Axle Construction and Operation

After studying this chapter, you will be able to:

- Explain the function of a constant-velocity axle.
- Explain the difference between equal- and unequal-length drive axle systems.
- Describe the operation of the different types of constant-velocity U-joints.
- Describe the construction of a constant-velocity axle shaft.

This chapter is designed to detail the construction and operation of constant-velocity axles. It will explain the operating principles of constant-velocity axle shafts, constant-velocity U-joints, intermediate shafts, and front wheel bearings. Studying this chapter will prepare you for the constant-velocity axle service chapter that follows.

Constant-Velocity Axles

In transaxle-equipped vehicles, the engine and transaxle are bolted together to form an assembly that is mounted at the front or rear of the vehicle. Engine power is transferred from the transaxle to the drive wheels by the ***constant-velocity axles,*** or ***CV axles.*** Figure 22-1 shows a typical front-wheel drive assembly that contains two CV axles. The term ***constant velocity*** means that the axles can transmit power through changing drive angles without variations in axle speed, or rpm. This is necessary because the drive angles of the axles must change as the drive wheels move during vehicle operation. The wheels move up and down as the vehicle travels over uneven road surfaces, and they move to the right or the left as the vehicle is turned.

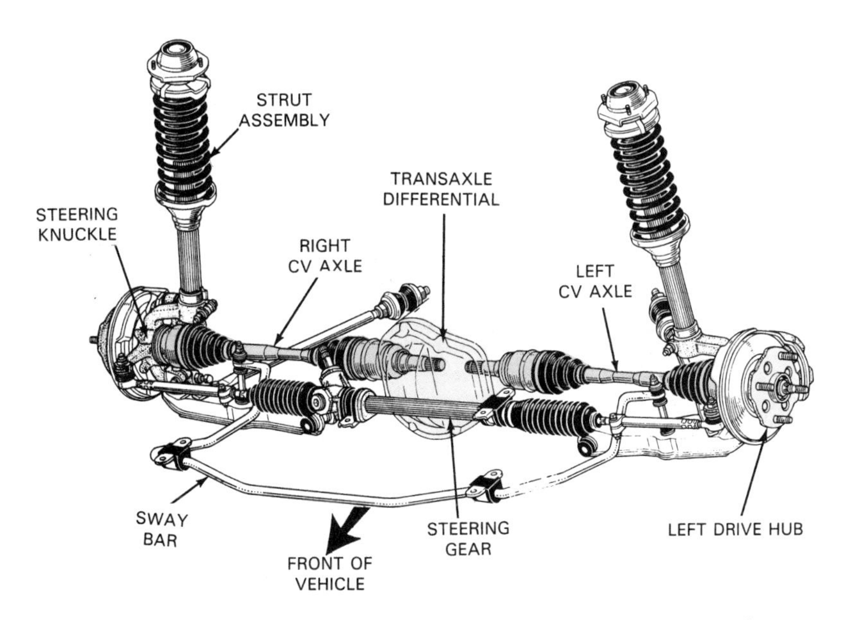

Figure 22-1. In this front-wheel drive assembly, the constant-velocity axles transfer power from the transaxle to the front wheels. (Toyota)

The inner, or inboard, ends of the CV axles are connected to the transaxle differential. The axles are driven by the differential side gears. The outer, or outboard, ends of the axles are connected to the wheel hubs. The outer ends are supported and held in alignment by steering knuckles (front-wheel drive axles) or by rear bearing supports (rear-wheel drive axles on rear- or mid-engine vehicles or that are independently suspended).

The steering knuckles are heavy mounting attachments for the wheel hubs. They house the wheel bearings and are connected to the front strut assemblies, control arms, and other front suspension parts. The steering knuckles pivot to turn the front wheels. Rear bearing supports house the rear wheel bearings and support the outer ends of the rear drive axles, but usually do not swivel for steering.

CV axles consist of *CV joints* and *CV axle shafts.* Some front-wheel drive axles and most drive axles on rear- and mid-engine vehicles are equipped with one or more non-CV flexible joints, or cross-and-roller U-joints, but these are almost always used in combination with CV joints. Properly maintained CV axles will operate with little or no vibration.

Figure 22-2 shows a typical CV axle assembly. Although there are many variations to the basic design, most CV axles contain the parts shown here. Note, however, that while most CV axles contain three axle shafts, including a *center shaft* and two *short,* or *stub,* shafts, the CV axle pictured does not have an inner stub shaft.

Equal- and Unequal-length Drive Axle Systems

In many CV axle systems, the drive axles used on each side of the transaxle are unequal in length. This is because the design of most transaxle installations places the CV axle openings in the transaxle at different distances from respective wheel hubs. This arrangement is called an ***unequal-length drive axle system,*** Figure 22-3A. The difference in axle length may cause vibrations, because each axle will be turning through a different drive angle.

The difference in drive angles in an unequal-length drive axle system may also cause one wheel to receive more torque during heavy acceleration. This produces an undesired turning effect called ***torque steer.*** Torque steer is usually more noticeable on high-powered front-wheel drive vehicles, such as those equipped with turbochargers, six-cylinder engines, or eight-cylinder engines.

To overcome torque steer, some manufacturers design their drive axle systems with a separate ***intermediate shaft*** that is mounted between the transaxle and the CV axle on the *long* side of the transaxle (side farthest from its wheel assembly). The intermediate shaft allows the use of equal-length CV shafts on each side of the vehicle. This design is called an ***equal-length drive axle system,*** Figure 22-3B. When an intermediate shaft is used, it must be supported at its outer end. A *bearing support bracket* is generally used for this purpose. An extra CV joint or another type of flexible joint is often used to join the

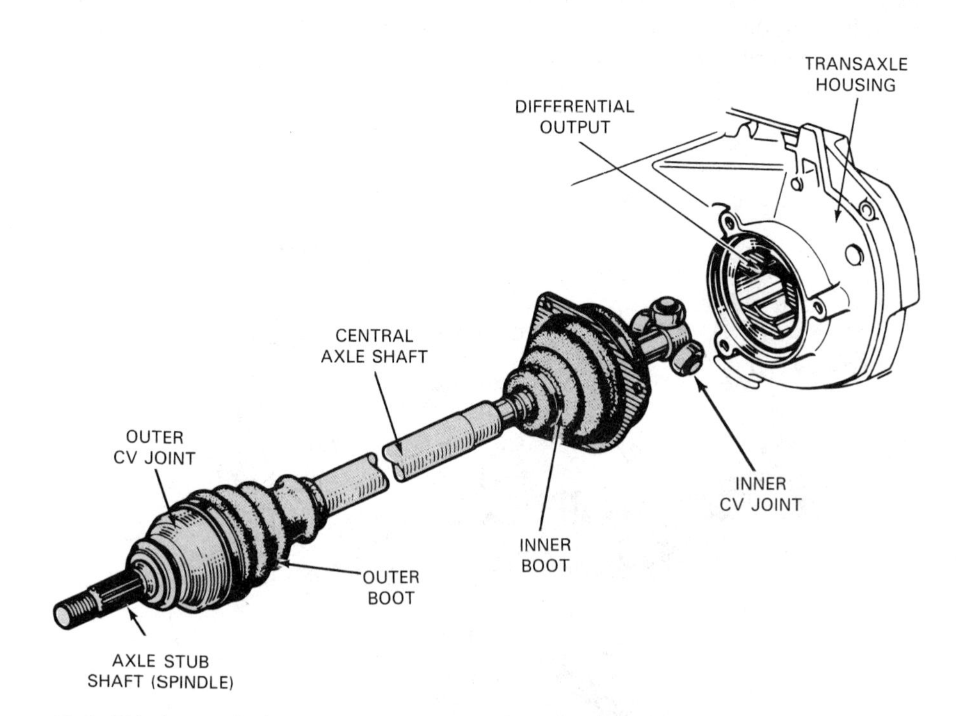

Figure 22-2. This is a typical constant-velocity axle. Notice the central axle shaft, the CV joints, and the boots. Note that this axle does not have an inner stub shaft. The inner joint fits into the transaxle housing.

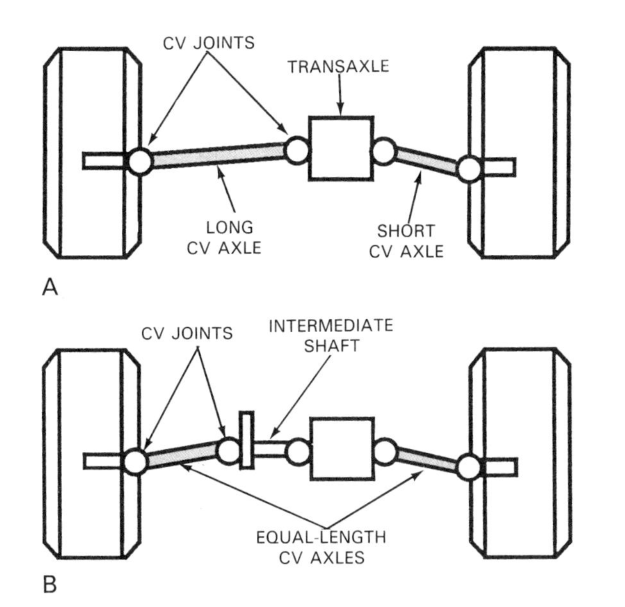

Figure 22-3. Note the differences between equal- and unequal-length drive axle systems. A–In unequal-length drive axle systems, the right and left axles are different lengths. B–In equal-length drive axle systems, an intermediate shaft is added on the long side of the transaxle. Consequently, both CV axles are the same length.

intermediate shaft to the transaxle differential. Figure 22-4 shows the differences between the equal- and unequal-length drive axle systems used by one manufacturer.

Figure 22-5 shows an intermediate shaft being used on a vehicle with front-wheel drive and a longitudinal engine. This design uses a separate differential assembly that resembles the common rear-wheel drive differential. Note that the intermediate shaft used with this drive system does not have an inboard flexible joint.

CV Axle Power Flow

Power flow through a CV axle travels out from the side gear of the transaxle differential assembly, through the inner stub shaft and CV joint, and into the center shaft. If the assembly has an intermediate shaft, the shaft transfers power from the differential to the inner CV joint and center shaft. From the center shaft, power flows through the outer CV joint, through the outer stub shaft, and into the wheel hub.

Constant-Velocity U-Joints

On all vehicles, both the front and rear axles are designed to move in relation to the chassis. This is done

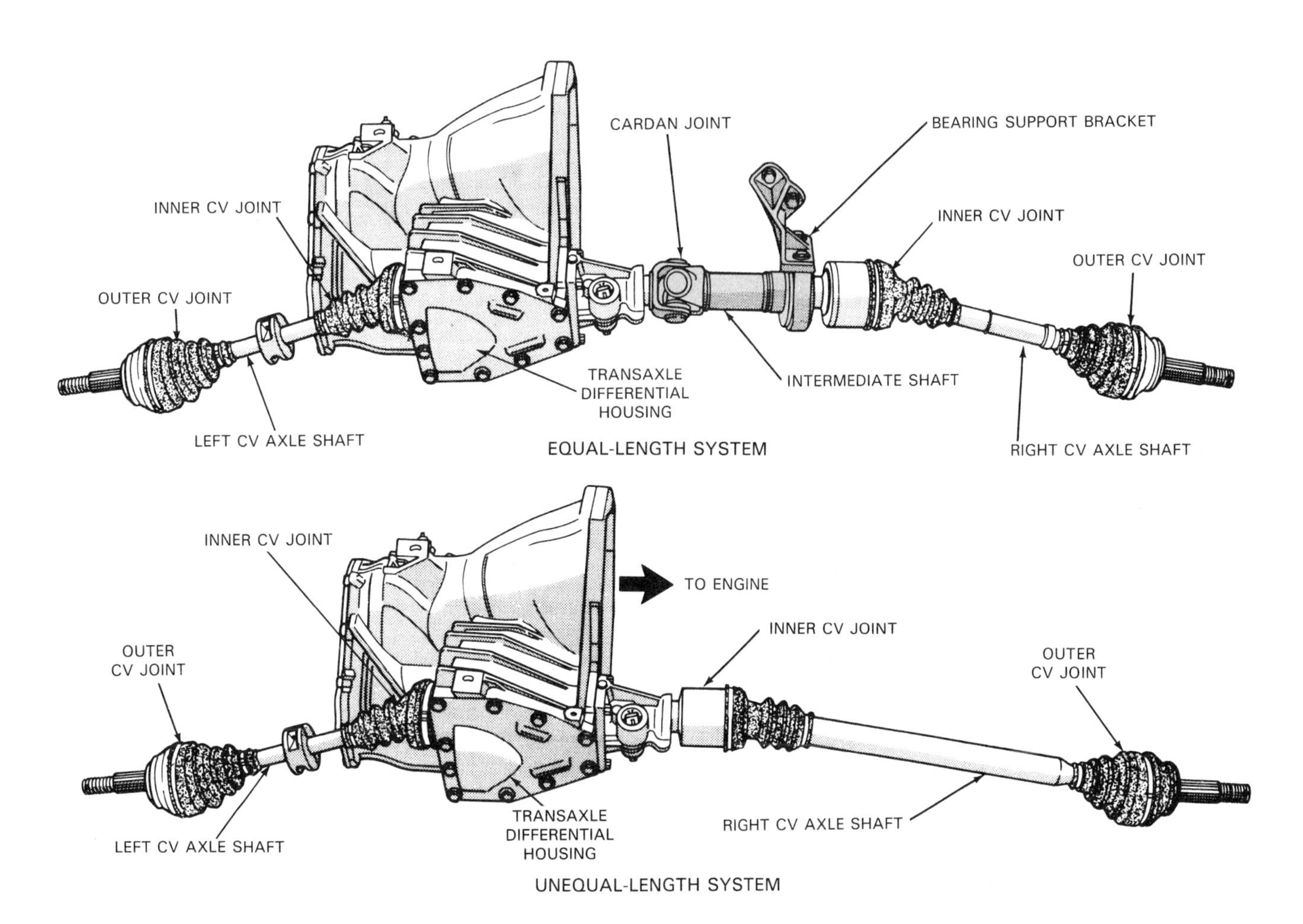

Figure 22-4. This figure shows both the equal- and the unequal-length drive axle systems used by one manufacturer. Note that the intermediate shaft requires support near its outboard end and contains a Cardan-type flexible joint on its inboard end. (Chrysler)

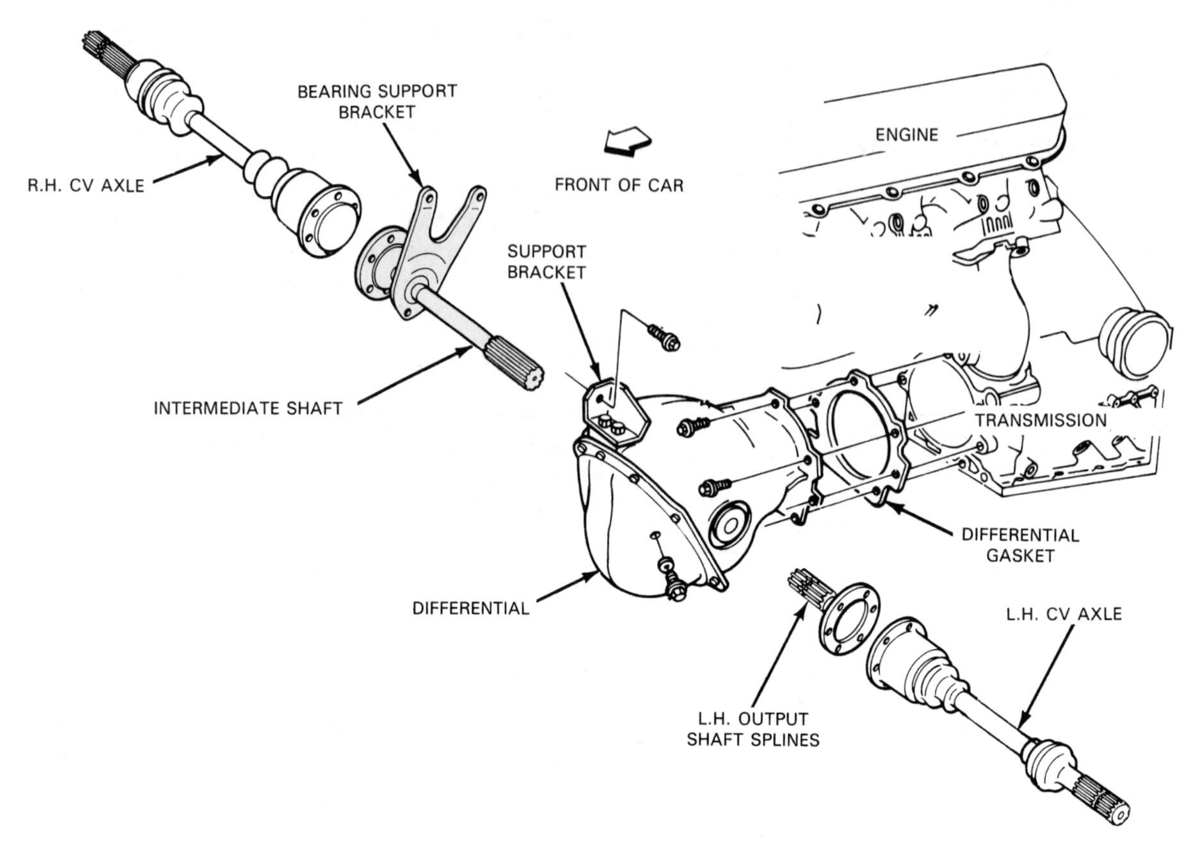

Figure 22-5. The intermediate shaft used on this front-wheel drive system has a bearing support bracket, but does not use an inboard flexible joint. (General Motors)

so that the suspension system can absorb the bumps and jolts of uneven road surfaces. Front-wheel drive vehicles have independent suspensions, which allow each wheel to move without affecting the others.

Since the transaxle is attached to the frame or chassis of the vehicle, the drive axles must be able to flex as the drive wheels move up and down or right and left. Drive wheel movement results in an almost infinite number of possible angles through which the axles must transmit power.

As mentioned, some transaxle-equipped vehicles use one or more *cross-and-roller universal joints* in the drive axle system. These joints are identical to those used on the driveline of conventional rear-wheel drive vehicles. Figure 22-6 shows a cross-and-roller universal joint installed at the outer end of an axle shaft. Another shaft equipped with cross-and-roller universal joints is shown in Figure 22-7. It resembles a small rear drive shaft assembly.

When two cross-and-roller universal joints are used on a drive axle assembly, the axle shaft usually contains internal splines to allow the shaft to change its effective length. However, some shafts do not have internal splines. Instead, a splined yoke is used at the differential. This arrangement was once common on rear-engine vehicles and was used on some older front-wheel drive systems.

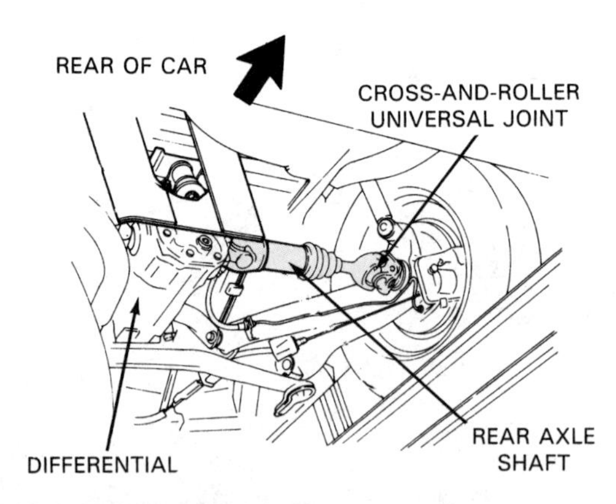

Figure 22-6. Universal joints are often used on the axle shafts of rear-wheel drive vehicles with independent suspension systems. This axle shaft uses a Cardan universal joint at the wheel.

The cross-and-roller universal joint is not a constant-velocity joint. The rigid design of this joint causes the driven yoke to speed up and slow down through its rotation when the yokes are at an angle. This produces an uneven transfer of power, resulting in vibration. For more information on cross-and-roller universal joints, refer to the related discussion in Chapter 14.

The most common universal joints used in the drive axle assemblies of modern transaxle-equipped vehicles are CV joints. These joints reduce vibration, but they are more complicated than cross-and-roller U-joints. The design of a CV joint allows it to divide the drive angle equally between the shafts that are connected through the joint.

CV joints change angles using balls or spider assemblies that roll in channels. In some designs, the balls or the spider assemblies can also slide in the channels, allowing the CV joint to change its length. This compensates for changes in the effective distance between the transaxle and the wheel hub that occur when the wheels move up and down. A joint that is designed to change its length is known as a ***plunging joint.*** Usually, only one joint in a CV axle is a plunging joint. This design eliminates the need for a slip yoke.

The two major types of CV joints used in CV axles are the *ball-and-cage joint* and the *tripod joint.* Figure 22-8 shows both types of joints installed on a single CV axle assembly. The inner joint is a tripod joint, and the outer joint is a ball-and-cage joint. When a tripod joint is used, it is almost always the inner joint and is usually a plunging joint. All CV joints are enclosed in *flexible boots.*

Ball-and-cage joints

The ***ball-and-cage joint,*** or *Rzeppa joint,* consists of a set of ball bearings, an inner race, an outer race, and a ball cage, Figure 22-9. The races form slotted channels that house the ball bearings. The ball cage holds the balls in place and allows them to rotate in the channels. The balls form a nonrigid connection between the races. Consequently, as the drive angles change, the races can change

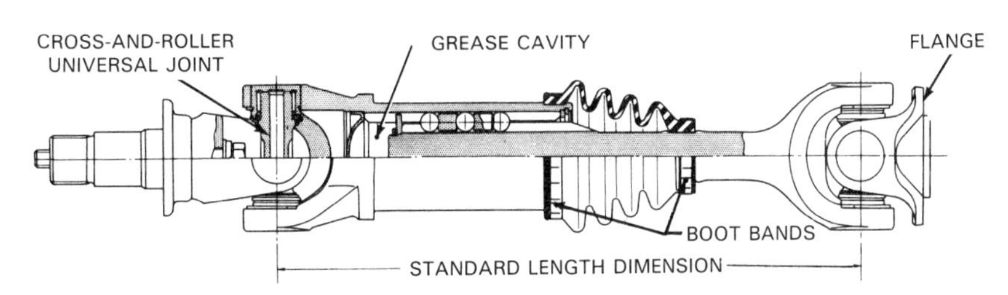

Figure 22-7. Cardan universal joints, used on this shaft, cannot compensate for changes in the effective distance between the transaxle and the wheel hubs. Therefore, an internal slip yoke is often used.

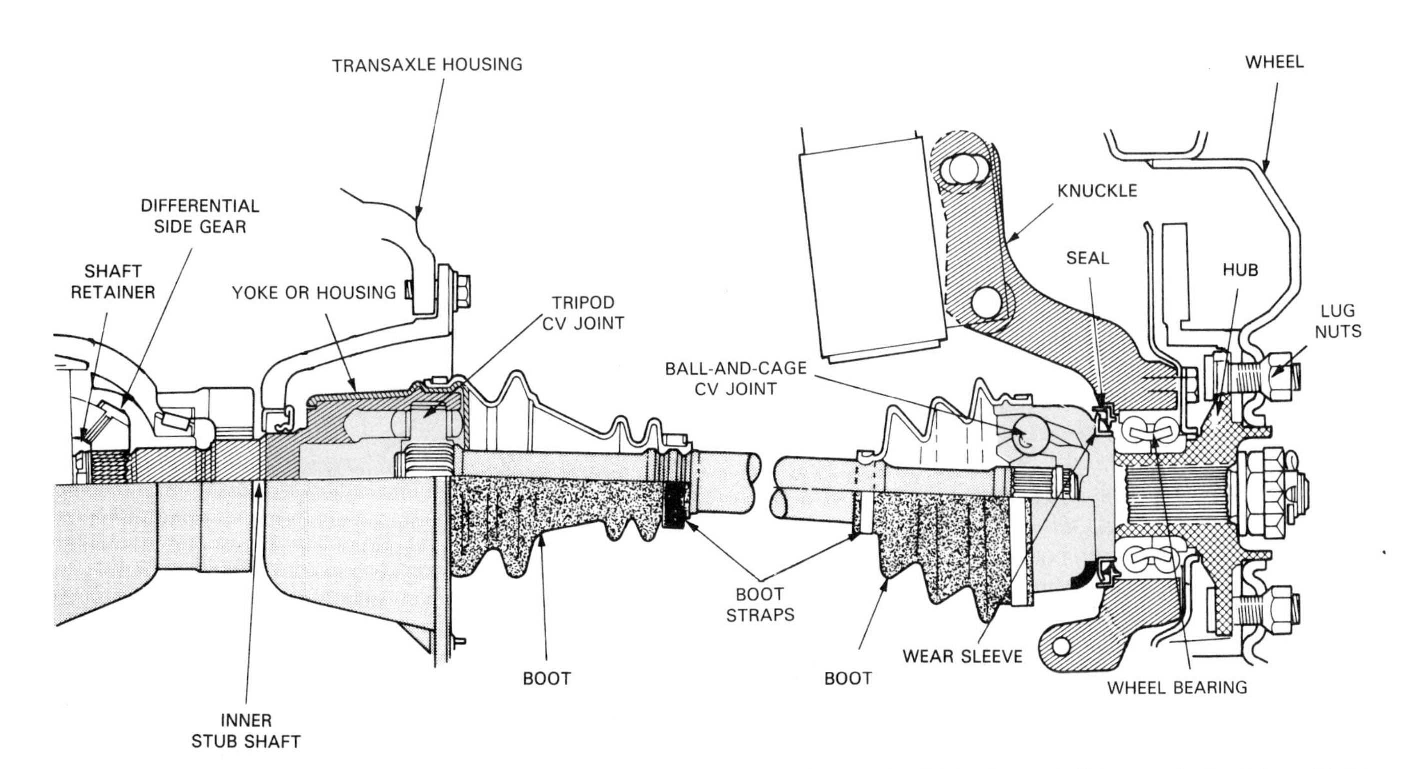

Figure 22-8. Study the two major designs of constant-velocity U-joints. When a tripod joint is used, it is usually the inboard joint. (Chrysler)

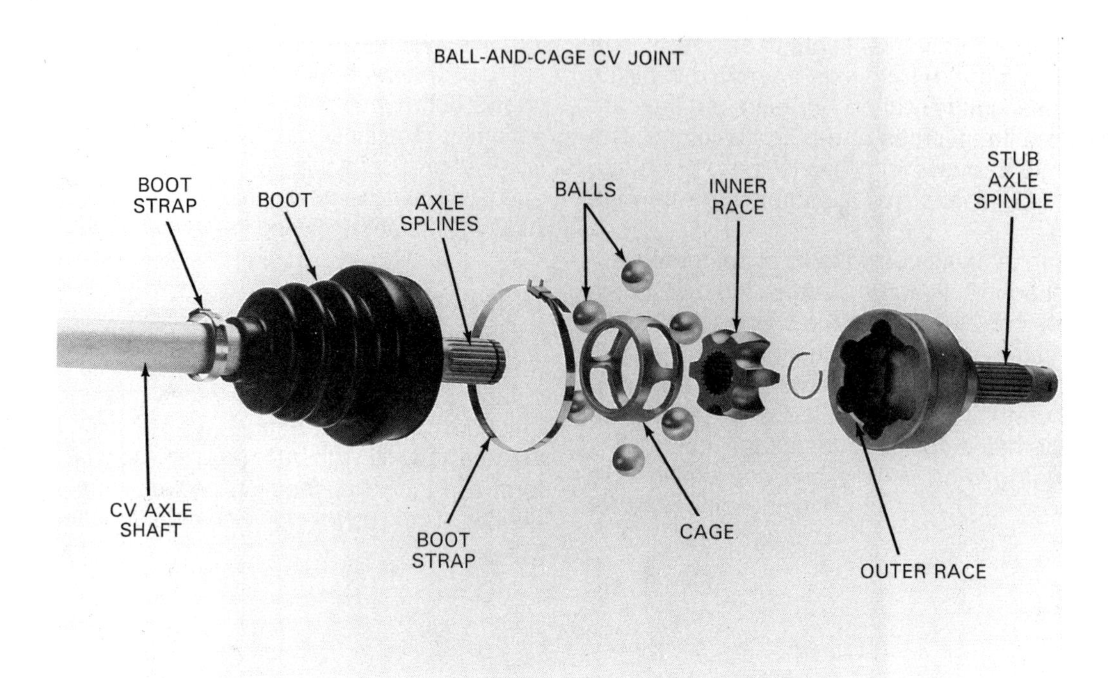

Figure 22-9. Exploded view of a ball-and-cage joint. The balls can rotate in their channels to compensate for changes in race angles. (Dana)

angles without affecting each other. Power flows from the inner race, through the balls, and into the outer race. Since the balls can roll in relation to the channel in either race, they can transmit power through changing race angles without vibration.

A cutaway of a ball-and-cage joint is shown in Figure 22-10. The inner race of this joint is splined to the axle shaft, and the outer race is part of the stub shaft that is attached to the wheel hub. Figure 22-11 is an exploded view of a plunging ball-and-cage joint. In this particular joint, the outer race is elongated, and the balls can move in a linear direction. With this design, the joint can compensate for changes in the effective distance between the transaxle and the wheel hub.

Tripod joints

The *tripod joint,* or ***tri-pot joint*** as it is sometimes called, is used as the inboard, plunging CV joint on many CV axle assemblies. This type of joint contains a three-pointed *cross,* or *spider, assembly,* which is attached to the axle shaft through a splined connection. See figure 22-12. The spider consists of three trunnions that are fitted with needle bearings and rollers. The rollers slide in channels that are cut into the outer housing of the joint assembly. Power flows into the outer housing, through the spider assembly, and into the axle shaft.

Changes in drive angle through the tripod joint are compensated for by the trunnions and rollers as they slide in the housing channels. In tripod joints, the design of the trunnions and channels allows the entire spider assembly to move back and forth inside the outer housing, changing the length of the joint. Figure 22-13 is an exploded view of a typical plunging tripod joint.

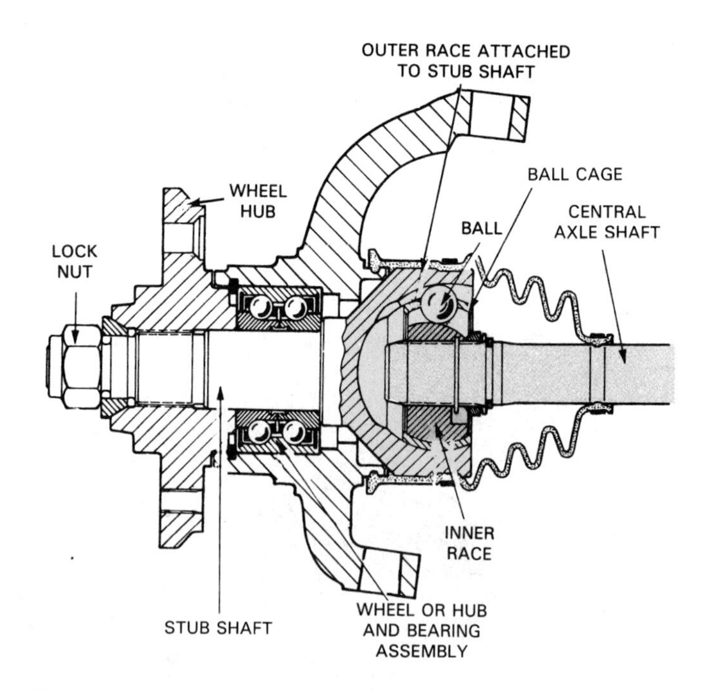

Figure 22-10. This cross section of a ball-and-cage joint shows how it is installed in relation to the steering knuckle and wheel hub. (Saab)

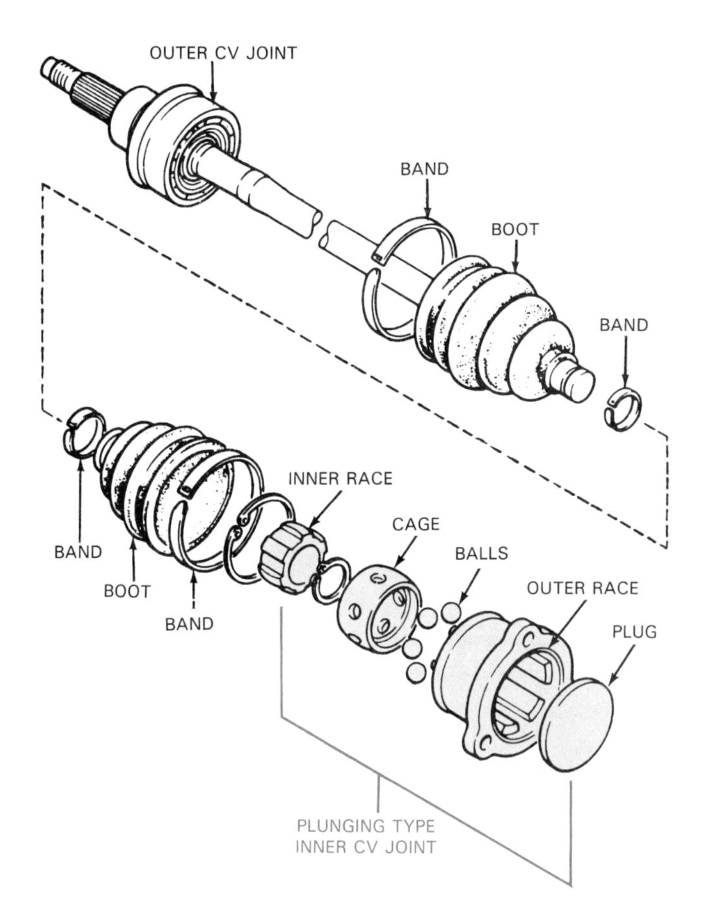

Figure 22-11. The inner CV joint shown here is a plunging ball-and-cage joint. The races in the outer housing are elongated so the balls and the cage can move back and forth. This allows for changes in axle length. (Nissan)

CV Joint Boots

As CV joints rotate, the grease that lubricates the CV joint tends to be thrown outward and away from the moving parts of the joint. Also, the placement of CV joints under the vehicle exposes them to road debris, water, and other contaminants. Therefore, flexible ***CV joint boots*** are used to protect the joints. The boots keep the lubricant in and the contaminants out.

A typical boot installed on a CV joint is shown in Figure 22-14. It completely encloses the moving parts of the joint and is retained with boot straps or clamps. The boot rotates with the axle and can flex as the drive angle changes. Boots are made of a special neoprene rubber that is not affected by the CV joint lubricant. The rubber is thick enough to resist being punctured by normal road debris. It also resists oil, water, and other contaminants.

If the boot is accidentally punctured or cut, it should be replaced. If a damaged boot is not replaced, the CV joint will be ruined by contamination and lack of lubrication. Some boots are made in two pieces and can be changed without removing the axle from the vehicle.

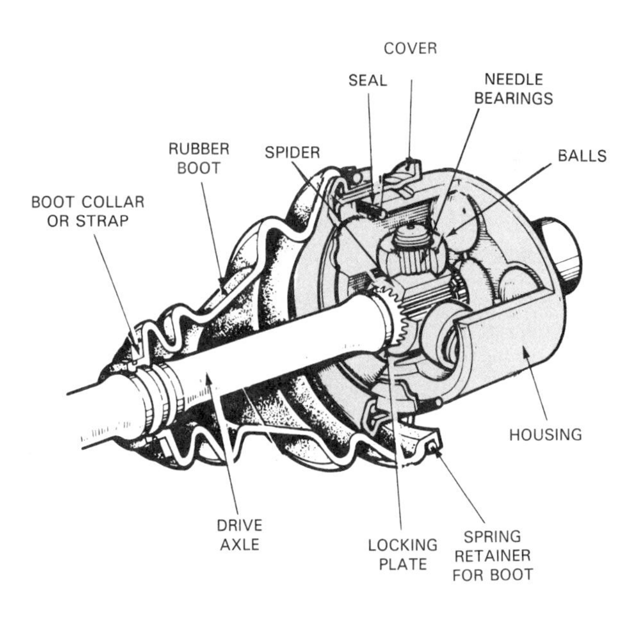

Figure 22-12. The tripod joint consists of a spider assembly and a housing with channels. The balls rotate on the spider and slide back and forth in the housing channels. (Chrysler)

Figure 22-15 is a cross section of a CV joint and boot. The boot holds enough grease to maintain proper joint lubrication. Grease that is thrown from the joint strikes the inside of the boot. When the vehicle stops moving, the grease collects at the bottom of the boot. When the vehicle begins moving again, the rotating joint picks up grease from the bottom of the boot. Special pockets are sometimes molded into the boot to help distribute the grease.

A typical boot is illustrated in Figure 22-16. Notice the grease pockets, which help ensure proper lubrication. Some boots have an alignment mark to ensure proper installation on the joint and the shaft.

CV Axle Shafts

CV axle shafts are the steel shafts that work with the CV joints to transfer power between the transaxle and the wheel hubs, Figure 22-17. Axle shafts are made of hardened steel that is machined to fit the CV joints, the differential side gears, or the side gear flanges.

Typical CV axle shafts used on a front-wheel drive vehicle include the ***central axle shaft*** and the ***stub shafts.*** The stub shafts are connected to the center shaft through the CV joints. In many systems, one or both stub shafts are part of the outer CV joint housing.

Although a few manufacturers use tubular CV axle shafts for special applications, most CV axle shafts are solid. Using small-diameter, solid axles saves space in the engine compartment. Straightness and balance of front CV axle shafts are critical because of the relatively

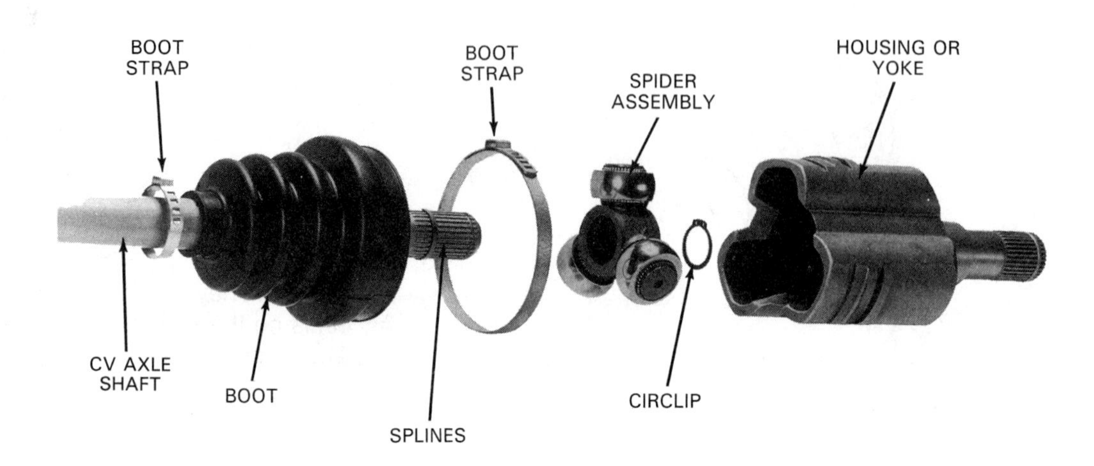

Figure 22-13. This exploded view of a tripod joint shows its major components. The spider assembly and the housing fit together to transfer power. (Dana)

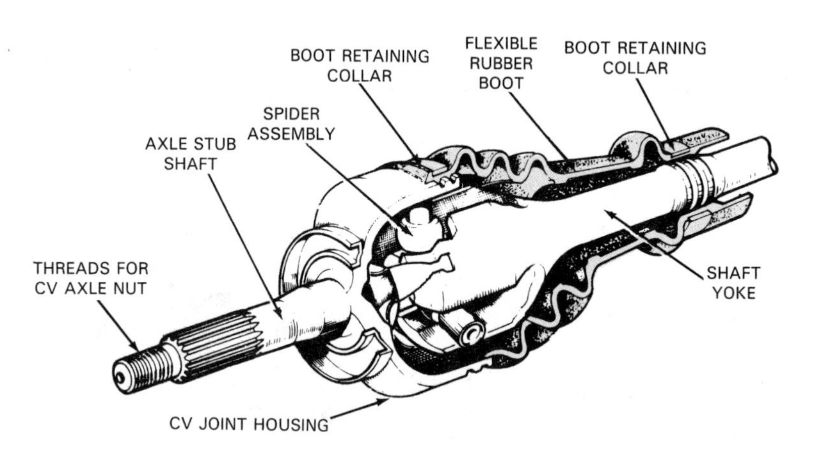

Figure 22-14. The flexible boot protects the moving parts of the CV joint and retains joint lubrication. (Chrysler)

large angles that these shafts must turn through. Additionally, an out-of-balance front CV axle shaft will more than likely cause a noticeable vibration that is felt in the passenger compartment.

CV axles are sometimes fastened to the transaxle with snap rings or roll pins, Figure 22-18. If fastened with a roll pin, the pin must be driven out before the axle can be removed from the transaxle. The snap rings used to secure some axles in the transaxle are light-tension rings. When removing these axles, the rings can be removed by lightly prying them away from the transaxle.

Occasionally, an inner stub shaft is not used, and the inner CV joint is bolted to a flange that is part of the differential side gear assembly. Removing the flange bolts will allow the axle to be easily removed from the flange. In a few systems, the axle shafts are simply held in place by the position of the transaxle and the wheel hubs.

CV Axle Wheel Hub and Bearing Assemblies

The CV axle stub shaft must be supported by the ***wheel bearings,*** which allow it to transfer power through

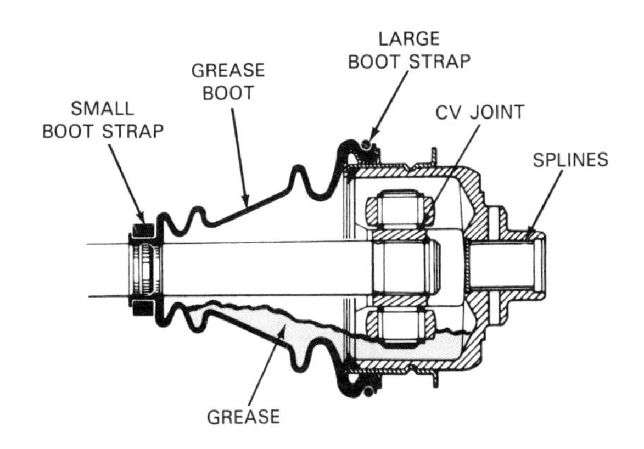

Figure 22-15. The boot should contain enough lubricant to maintain proper joint lubrication. The straps hold the boot in place and seal the boot ends. (Chrysler)

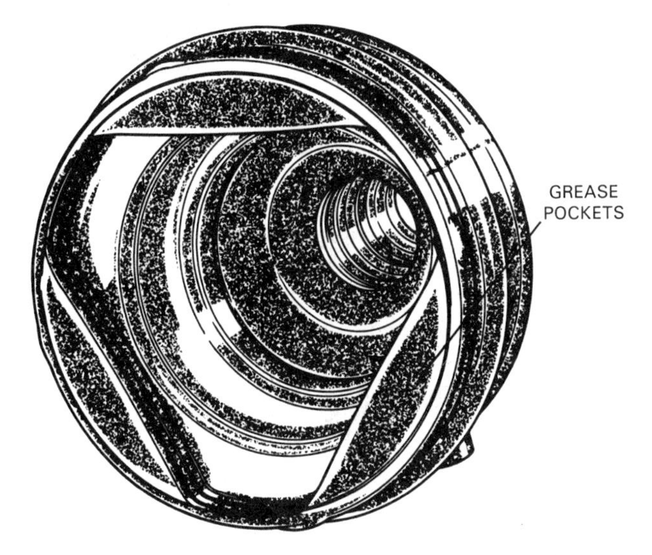

Figure 22-16. A typical CV joint boot. This boot contains special pockets to hold extra grease. Proper boot size and alignment are important. (Chrysler)

the front steering knuckle or the rear bearing support. The steering knuckle or the bearing support acts as a housing for the bearings.

In Figure 22-19, the outer stub shaft, which is sometimes called the ***spindle,*** is supported by the ***wheel hub and bearing assembly,*** or the *knuckle/hub assembly.* The shaft passes through the knuckle and rotates on the wheel bearings. The bearings support the wheel hub and maintain alignment between the hub and the knuckle.

On a vehicle equipped with an intermediate shaft, a ***bearing support bracket*** is installed at the outboard end of the shaft, Figure 22-20. The bracket houses the intermediate shaft bearing and is attached to the transaxle assembly or to a stationary part of the vehicle chassis. The transaxle differential side gears (or flanges) support the inboard end of the intermediate shaft.

An exploded view of a typical wheel hub and bearing assembly is shown in Figure 22-21. The stub shaft passes through the steering knuckle to transfer power to the hub. The stub shaft splines fit into the internal hub splines to drive the hub. The hub and the shaft are held together with a hub nut. This particular design uses two bearings that are mounted at the front and rear of the steering knuckle.

The wheel hub and bearing assembly shown in Figure 22-22 uses tapered roller bearings to support the axle shaft. The other parts of the assembly are similar to those shown in Figure 22-21. The bearing preload is adjusted by turning the hub nut.

All bearings must have an ample supply of lubricant to prevent premature wear. They must also be protected from dirt and water. Seals on the steering knuckle or the rear bearing support keep lubricant in and contaminants out. A typical lip seal is shown in Figure 22-23. The ***wear sleeve*** provides a riding surface for the sealing lip. The wear sleeve is pressed on the CV joint housing, and the seal is pressed into the steering knuckle. Many wheel hub and bearing assemblies also have a seal at the outboard end of the steering knuckle.

Figure 22-24 shows the positions of two CV axle assemblies in relation to the front suspension system, the steering system, and the front wheels. This system does not contain an intermediate shaft. For more information on bearings, see Chapter 3.

Summary

Power is transferred from the transaxle to the wheels by constant-velocity axles. The axles are driven by the side gears of the transaxle differential. The front steering knuckles or the rear bearing supports hold the axle shafts at their outer ends.

CV axle assemblies contain CV axle shafts and CV joints. CV joints transfer power more smoothly than Cardan universal joints, used in rear-wheel drive shaft assemblies.

Transaxle-equipped vehicles can have equal- or unequal-length drive axle systems. The unequal-length system has two axle shafts of different lengths. This is because the design of the transaxle places its differential outputs at different distances from respective wheel hubs. The equal-length system uses an intermediate shaft on the long side of the transaxle. The intermediate shaft makes it possible to use drive axles of equal length on both sides of the transaxle. The equal-length drive axle system reduces torque steer and traction problems. It is usually found on high-powered front-wheel drive vehicles.

Cardan joints are occasionally used on some parts of the front-wheel drive CV axle, but they are not true CV joints. These joints are used on front axles only when there is relatively little change in drive angles. There are two major types of CV joints: the ball-and-cage joint and the tripod joint.

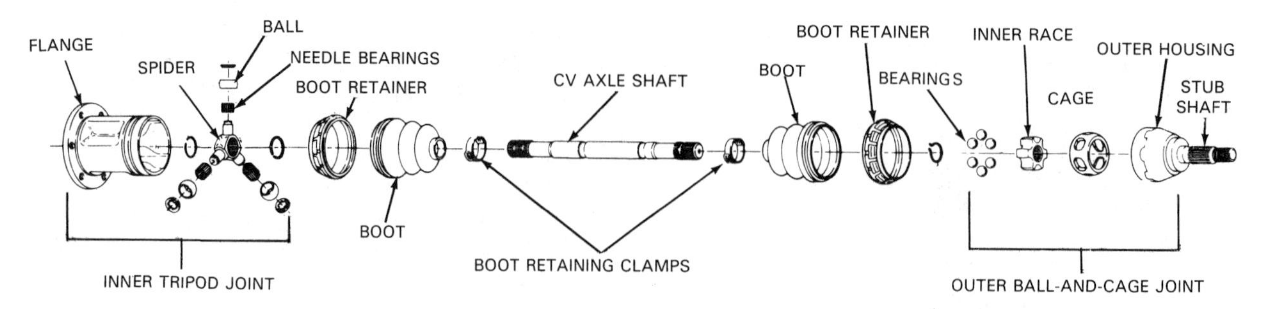

Figure 22-17. This exploded view of a CV axle assembly shows the relative positions of the axle components.

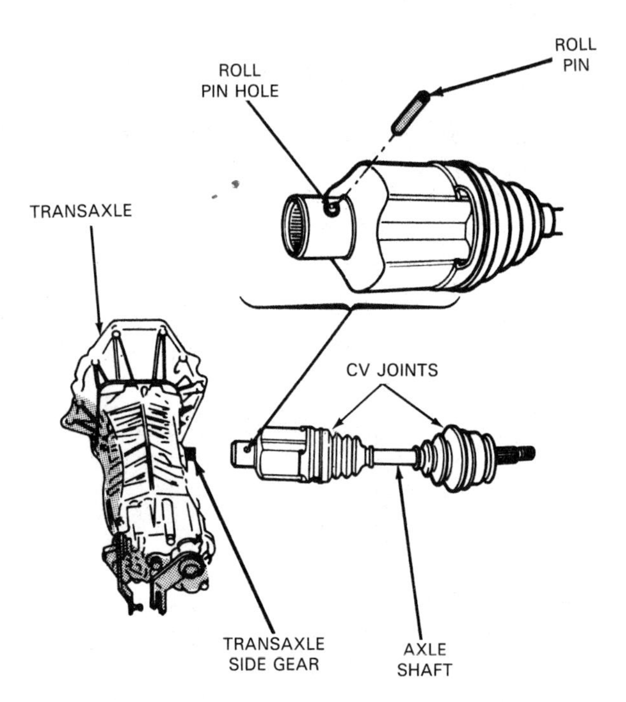

Figure 22-18. This CV axle is held to the transaxle by a roll pin. Other axles are held in place by snap rings or bolted to a transaxle flange. (Chrysler)

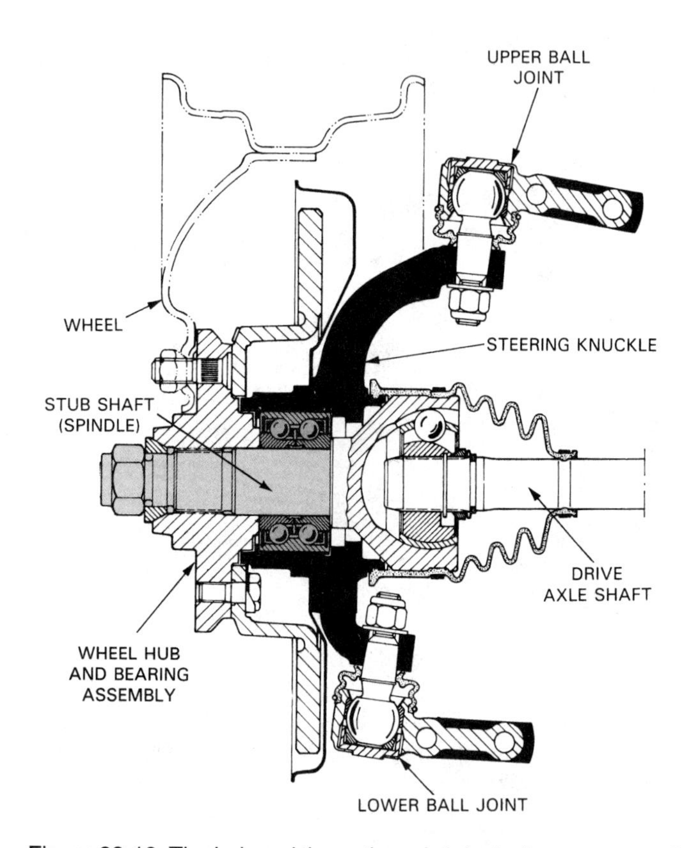

Figure 22-19. The hub and the outboard stub shaft are supported in the steering knuckle by bearings. The bearings also provide a rolling surface for the moving parts. The outboard stub shaft is often called a spindle. (Chrysler)

The ball-and-cage joint uses steel ball bearings that move between channels in the inner and outer races of the joint. Power flows from the inner race, through the balls, and into the outer race. Since the balls can roll in relation to the channel in either race, they can transmit power through changing race angles without vibration.

The tripod joint contains a spider assembly that consists of three trunnions. The trunnions are fitted with rollers that can turn in relation to the spider. The rollers can slide and rotate in the channels in the outer housing. Power flows from the outer housing channels, through the spider assembly, and into the axle shaft. The ability of the spider assembly to slide in the housing channels allows the joint to transmit power without vibration.

Some ball-and-cage joints and most tripod joints are designed to compensate for changes in the effective distance between the transaxle and the wheel hub. These joints have longer channels in their outer races, so that the balls or the spider assembly can slide as necessary. Joints with this sliding action are called plunging joints.

All CV joints must be covered by flexible boots. The boots keep lubricant in and contaminants out. CV joint boots are made of heavy neoprene rubber that resists oil, grease, and road debris. The boots are held to the CV axle shafts and the CV joint housings by straps or clamps.

The types of axle shafts found on a typical front drive vehicle include a central axle shaft and one or more stub shafts.

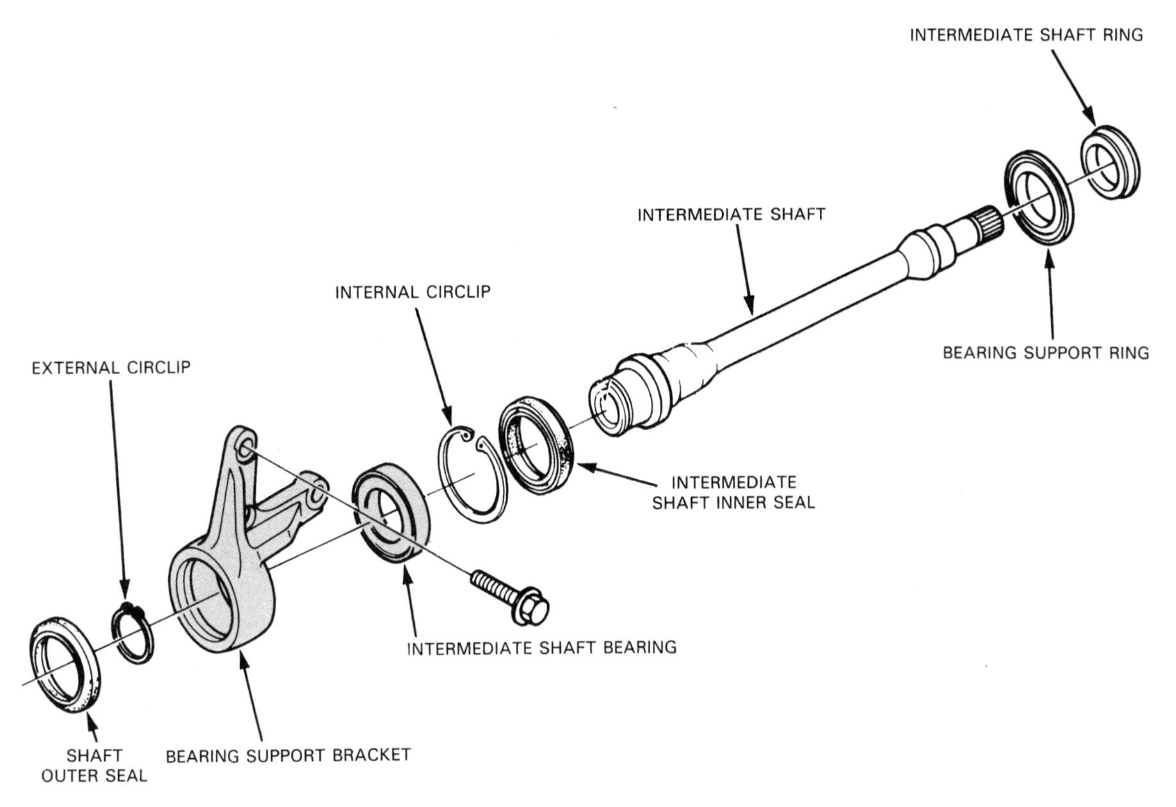

Figure 22-20. The intermediate shaft bearing is often held in a special support bracket. This bracket is attached to the transaxle or the vehicle chassis.

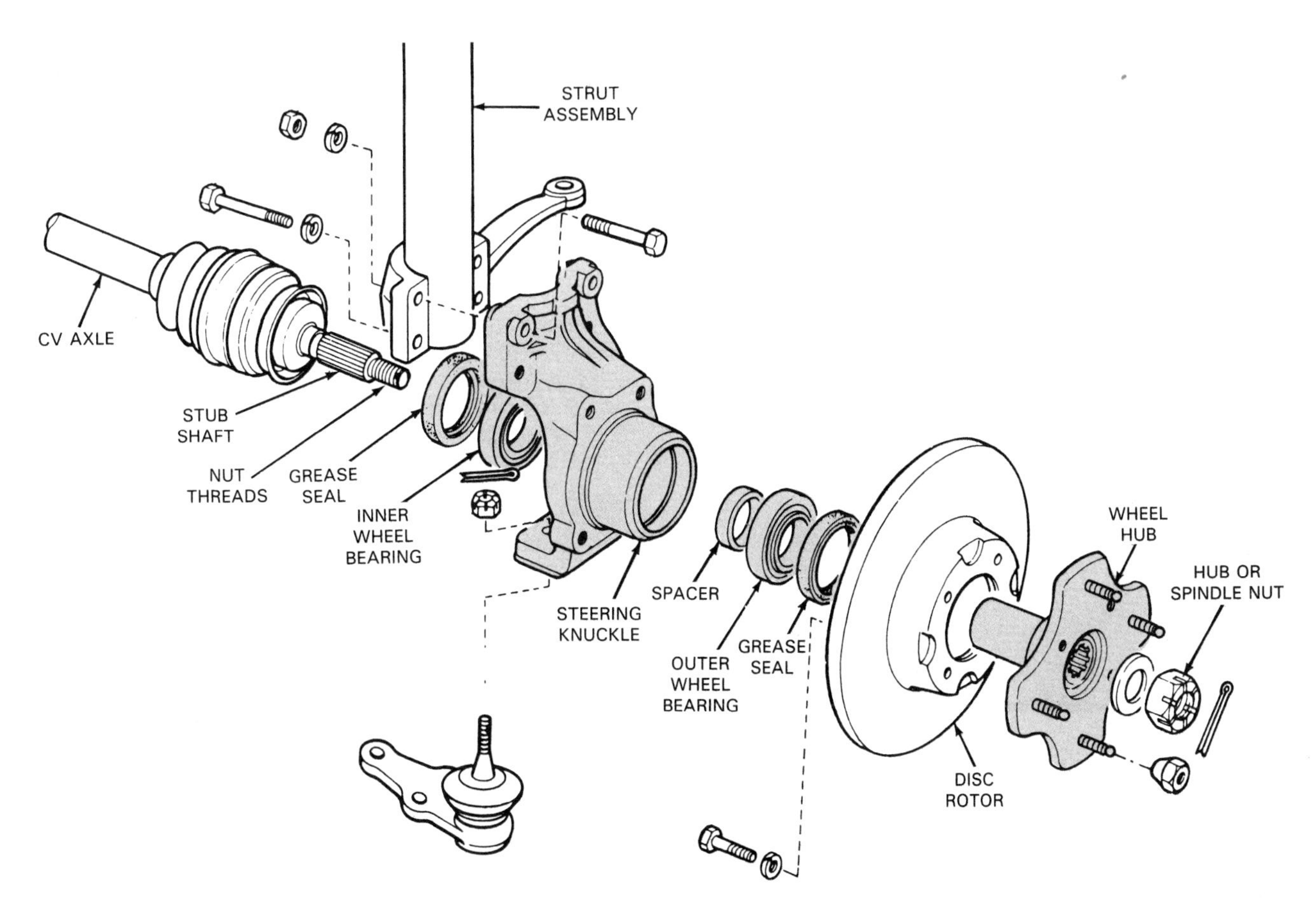

Figure 22-21. This exploded view of a spindle and a wheel hub and bearing assembly illustrates the major components on the outboard end of a CV axle. (Nissan)

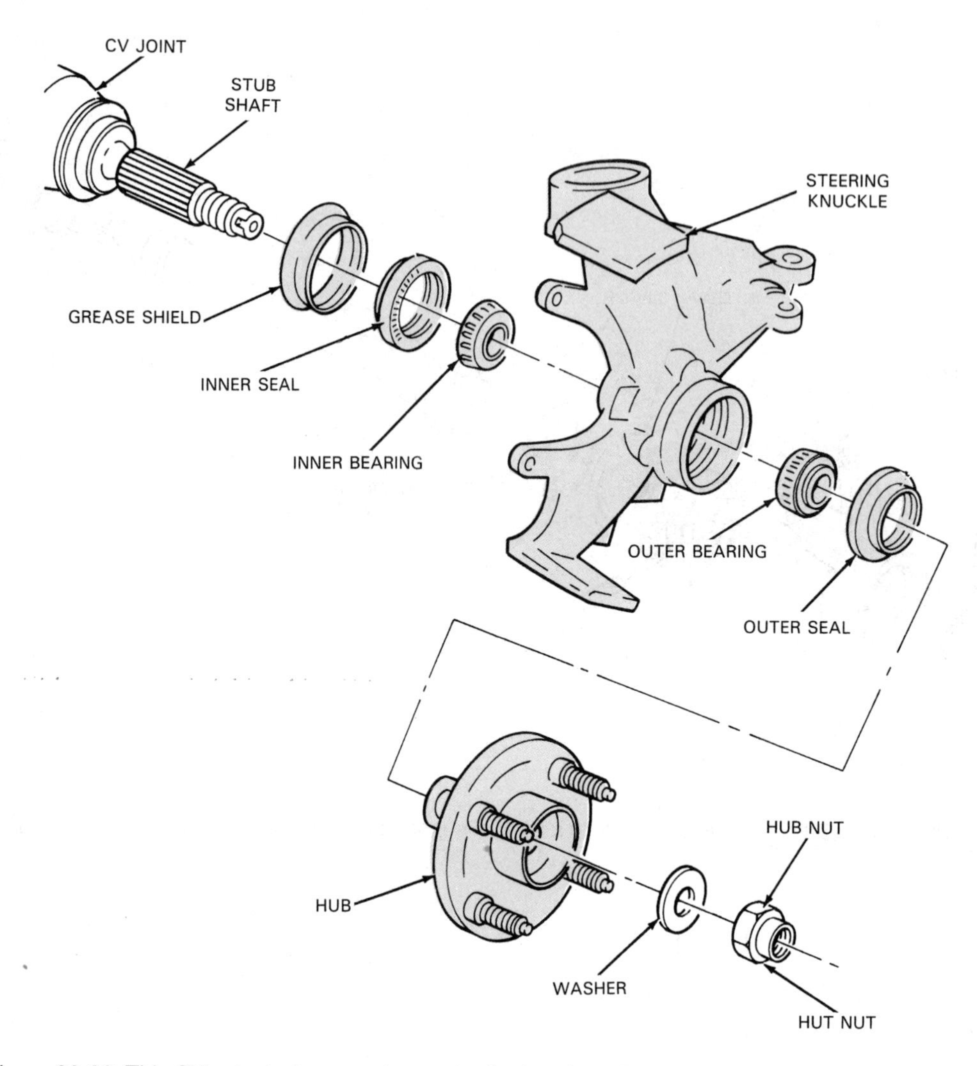

Figure 22-22. This CV axle design uses tapered roller bearings. The bearing preload is adjusted in the same manner as for front wheel bearings on a rear-wheel drive vehicle. (Ford)

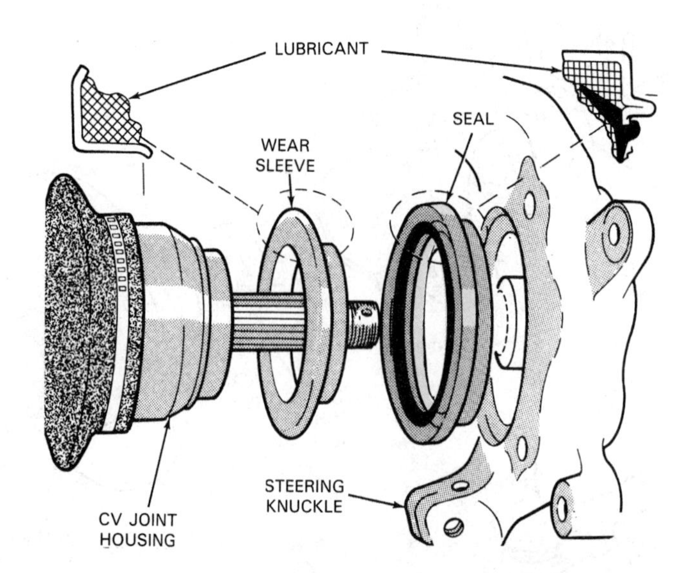

Figure 22-23. Like all moving parts, front wheel bearings require a supply of lubricant. Seals prevent the lubricant from leaking. (Chrysler)

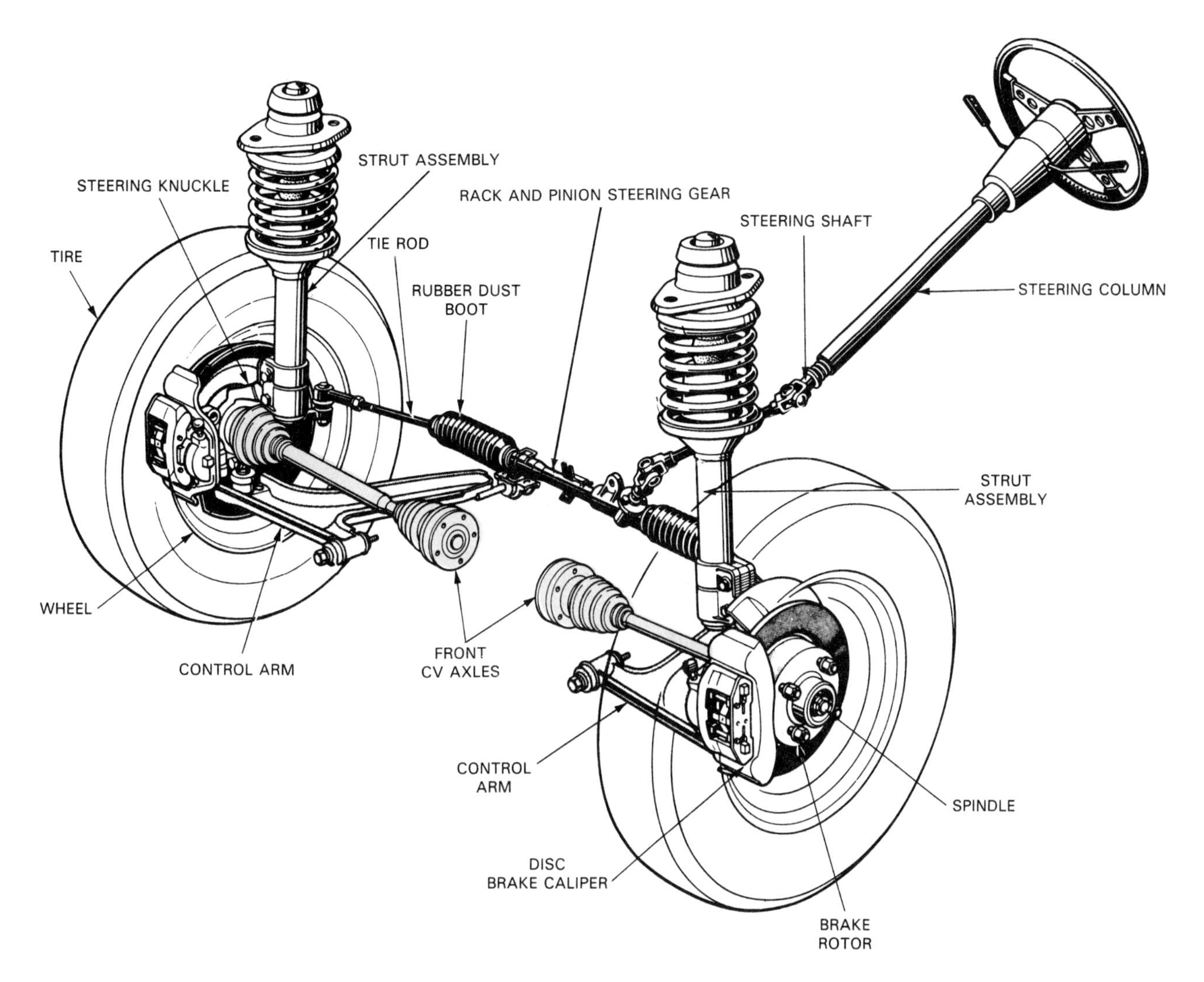

Figure 22-24. Study the position of the front CV axles in relation to the other front-end components. These axles are attached to the transaxle through bolted flanges.

Bearings in the front steering knuckle or the rear bearing support secure the outboard stub shaft, or spindle, and hold the shaft and wheel hub in alignment. Some manufacturers use ball bearings, while others use adjustable tapered roller bearings.

Know These Terms

Constant-velocity axle (CV axle); Constant velocity; Unequal-length drive axle system; Torque steer; Intermediate shaft; Equal-length drive axle system; Plunging joint; Ball-and-cage joint; Tri-pot joint; CV joint boot; CV axle shaft; Central axle shaft; Stub shaft; Wheel bearing; Spindle; Wheel hub and bearing assembly; Bearing support bracket; Wear sleeve.

Review Questions–Chapter 22

Please do not write in this text. Place your answers on a separate sheet of paper.

1. List and explain the seven major parts of a CV axle.
2. How does a CV joint operate?
3. Explain what torque steer is.
4. To prevent torque steer in unequal-length drive axle systems, _____ _____ can be used on the long side of the transaxle.
5. A cross-and-roller universal joint is a constant-velocity joint. True or false?
6. What is a plunging CV joint?
7. Explain the difference between a ball-and-cage joint and a tripod joint.
8. Why is a CV joint boot important?

9. The _____ _____ or _____ _____ acts as a housing for the wheel bearings.

10. A wear _____ can sometimes be pressed on the CV joint housing.

Certification-Type Questions–Chapter 22

1. In transaxle-equipped vehicles, engine power is transferred from the transaxle to the drive wheels by the:
- **(A)** CV axles.
- **(B)** U-joints.
- **(C)** intermediate shaft.
- **(D)** transaxle differential.

2. Difference in drive angles in an unequal-length drive axle system may cause an undesired turning effect referred to as:
- **(A)** splining.
- **(B)** drive slip.
- **(C)** torque steer.
- **(D)** wheel hop.

3. The design of the cross-and-roller universal joint causes all of these EXCEPT:
- **(A)** vibration.
- **(B)** constant velocity.
- **(C)** a slower driven yoke.
- **(D)** uneven power transfer.

4. The most common universal joints used in the drive axle assemblies of modern transaxle-equipped vehicles include all of these EXCEPT:
- **(A)** cross-and-roller U-joints.
- **(B)** tripod joints.
- **(C)** Rzeppa joints.
- **(D)** ball-and-cage joints.

5. CV joint boots are used to do all of these EXCEPT:
- **(A)** keep lubricant in.
- **(B)** protect CV joints.
- **(C)** prevent vibration.
- **(D)** keep contaminants out.

6. CV axle shafts work with CV joints to transfer power between the transaxle and the:
- **(A)** engine.
- **(B)** wheel hubs.
- **(C)** wheel bearings.
- **(D)** rear axle assembly.

7. Technician A says that straightness and balance are critical on front CV axle shafts because of the relatively large angles that they must turn through. Technician B says that an out-of-balance front CV axle shaft will more than likely cause a vibration that is felt in the passenger compartment. Who is right?
- **(A)** A only
- **(B)** B only
- **(C)** Both A & B
- **(D)** Neither A nor B

8. For it to transfer power, the CV axle stub shaft must be supported by the:
- **(A)** spindle.
- **(B)** central shaft.
- **(C)** CV axle shaft.
- **(D)** wheel bearings.

9. All of these are considered true CV joints EXCEPT:
- **(A)** tripod joints.
- **(B)** Cardan joints.
- **(C)** Rzeppa joints.
- **(D)** plunging joints.

10. The inboard end of a CV axle is connected to:
- **(A)** a wheel hub.
- **(B)** the rear bearing support.
- **(C)** the transaxle differential.
- **(D)** a stub axle.

11. All of these statements about steering knuckles are true EXCEPT:
- **(A)** they keep grease circulating.
- **(B)** they pivot to turn front wheels.
- **(C)** they connect to front suspension parts.
- **(D)** they are mounting attachments for wheel hubs.

12. CV axle shafts include all of these EXCEPT:
- **(A)** stub shafts.
- **(B)** spindles.
- **(C)** center shafts.
- **(D)** transfer shafts.

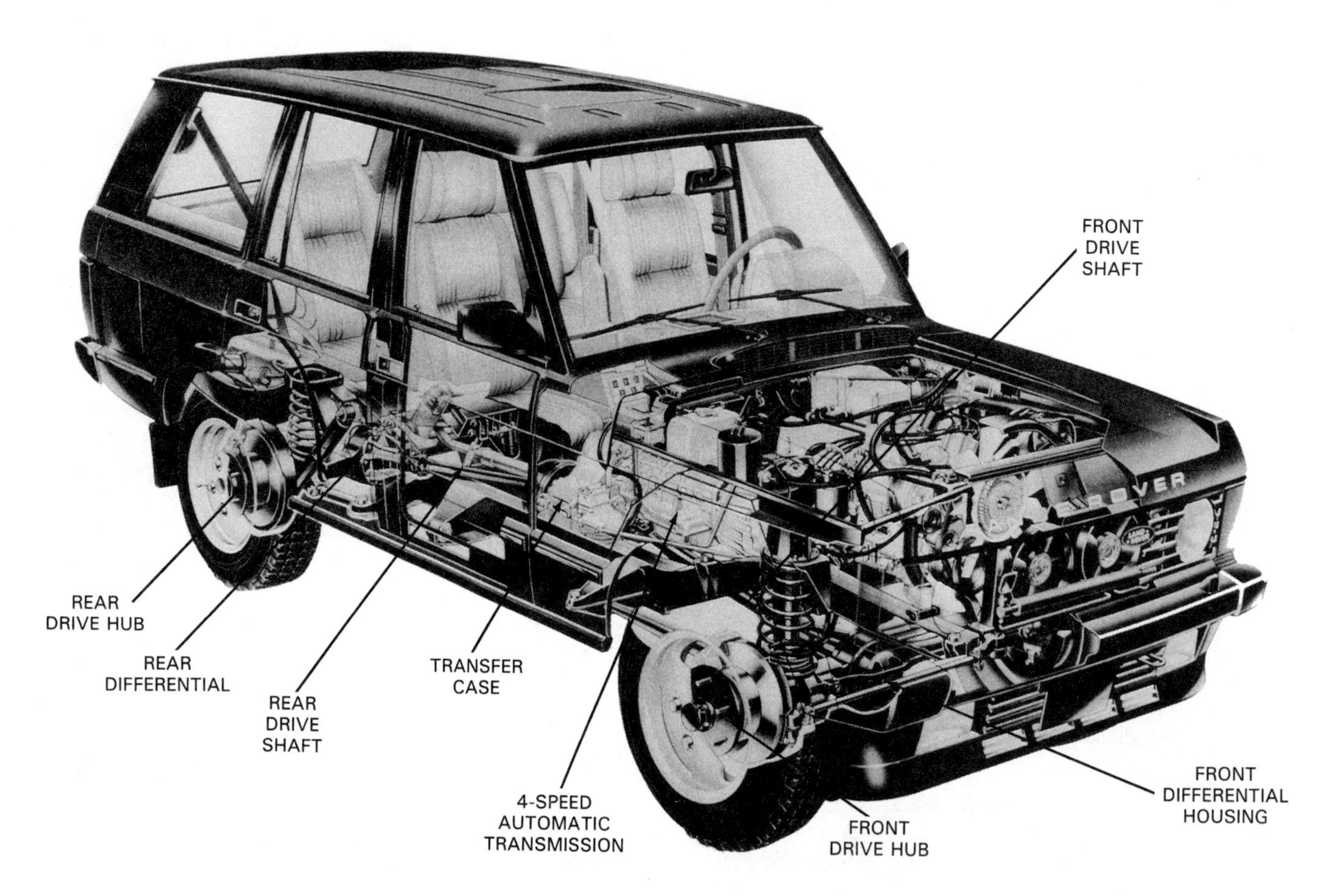

A phantom view of a sport utility vehicle showing four-wheel drive components. Four-wheel drive systems are discussed in Chapter 24 and 25 of this text. (Land Rover)

Chapter 23

Constant-Velocity Axle Problems, Troubleshooting, and Service

After studying this chapter, you will be able to:

- Troubleshoot CV axle problems.
- Remove and replace a CV axle.
- Disassemble and reassemble a CV joint.
- Install a new CV joint.
- Replace front-drive axle wheel bearings.

The importance of CV axle diagnosis and repair has grown as the number of front-wheel drive vehicles on the road has increased. This chapter will explain the proper ways to diagnose and service CV axle assemblies.

Chapter 22 contained information on the construction and operation of CV axles. If you encounter a term or concept in this chapter that you do not understand, review the material in Chapter 22.

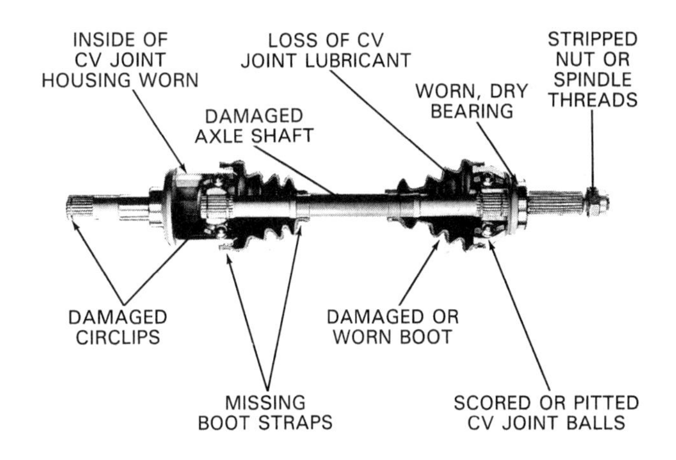

Figure 23-1. Study the common CV axle problems. (Champion Parts)

CV Axle Problems and Troubleshooting

Although the CV axle is a relatively simple device, it is subject to several problems. See Figure 23-1. CV axle problems can be grouped into three general categories: noises, vibrations, and lubricant leaks. Before deciding that a CV axle is defective, make sure that the problem is not caused by a defect in the engine, transaxle, brakes, steering and suspension components, wheel rims, or tires. Since these components are generally located in the front of the vehicle, it is easy to mistake sounds from these components for CV axle noises.

Abnormal Noise

Abnormal noises in the CV axles are unusual sounds, such as clunking, popping, or clicking noises, that are caused by defective CV axle components. Most CV axle noises occur when the wheels are turned or the brakes are applied. Dry or worn front wheel bearings, on the other hand, generally produce rumbling or roaring noises that vary with vehicle speed.

One of the most common causes of CV axle noise is a lack of lubricant in a CV joint. Lubricant may dry out after extensive use, or it may leak out through a damaged boot. A dry CV joint usually makes a popping, clicking, or snapping sound. Sometimes, regreasing the joint will solve the problem. Often, however, the joint will be worn or rusted and must be replaced or rebuilt.

Another cause of CV axle noise is a worn CV joint. Wear can occur at any point where the CV joint components move in relation to each other. Worn parts generally cause the joint to clunk or knock when the vehicle is turned or the brakes are applied. In some cases, the suspension attaching bolts or the axle bolts can become loose, allowing movement between parts.

It is sometimes difficult to determine which joint in a CV axle assembly is defective. It may be possible to see excess movement between parts on either side of a defective CV joint. To check for movement, raise the vehicle and put the transmission in park. If the vehicle has a manual transmission, it can be placed in any gear. Try to turn the wheel on the problem axle with your hand and observe the axle shafts on both sides of the suspected CV joint. Excess movement indicates a worn or damaged joint. It may help to compare movement around the joints on both sides of the vehicle. If one joint is looser than the others, it is probably defective.

Repairing a noisy CV axle can be as simple as tightening loose bolts. At the other extreme, a noisy CV axle could mean that a complete rebuild of the CV joints is necessary.

CV Axle Vibration

CV axle vibrations are pulsations or other cyclical shocks caused by worn or damaged CV axle parts. Vibrations may be caused by a bent or twisted axle shaft, worn CV joints or wheel bearings, or loose attaching bolts. Always check for worn or loose parts when vibrations are evident. If a bent CV axle shaft is suspected, it may be necessary to check the runout of the shaft.

Most vibrations will be felt through the steering wheel and, in some cases, the vehicle body. The frequency of CV axle vibrations will increase as vehicle speed increases. Since the CV axles are commonly used at the front of a vehicle, even slight vibrations will be noticeable. The usual cure for CV axle vibrations is the replacement of defective components.

Lubricant Leakage

Lubricant leakage in a CV axle generally originates from the CV joints or the wheel bearings. Most CV joint leaks are caused by torn boots or loose boot clamps. In some cases, transaxle fluid will leak from a defective seal between the transaxle housing and the CV axle shaft. Wheel bearing lubricant will often leak from a defective bearing seal.

Most leaks from the CV axle system will show up as grease or oil splattered on the underside of the vehicle near the leaking boot or seal. Grease may also be found on the boot or seal itself. Leaking seals may leave a trail of grease or oil that originates under the seal. If you squeeze a CV joint boot and hear air escaping, there is a hole in the boot or a loose clamp. Leaking boots and seals should be replaced promptly.

CV Axle Safety Precautions

CV axles rotate at high speeds and can cause severe damage. When checking for CV axle problems, always keep the following safety precautions in mind:

- If you are running the vehicle on a lift, always stay away from the spinning axles.
- When spinning the axle shafts with the vehicle on a lift, always support the lower control arms so the front wheels are at or near their normal position in relation to the vehicle body. Allowing the wheels to hang while being driven will cause the CV joints to drive through excessive angles and could cause a joint to bind, break, or fly apart violently.
- When testing a front-wheel drive vehicle on a lift, never exceed a speedometer reading of 30 miles per hour (48 kilometers per hour).
- If you road test the vehicle to check CV axle operation, always choose uncongested areas and obey all traffic regulations. This will help prevent accidents.

Road Testing

It is often necessary to road test a vehicle to determine the exact source of a problem. Often, a problem cannot be duplicated in the shop, and a road test must be conducted.

When performing a road test, try to isolate CV axle problems from engine, transaxle, tire, and other vehicle problems. Rumbling or roaring noises that increase and decrease only slightly with vehicle speed are usually caused by bad bearings. Vibrations that cycle according to vehicle speed generally indicate CV axle problems, not engine or transmission trouble.

Many clunking or popping noises in the CV axle can be diagnosed by driving the vehicle in a quiet parking lot. As the vehicle moves slowly forward, turn the wheels sharply from side to side, Figure 23-2. This will help isolate many noises. Applying the brakes with varying force may

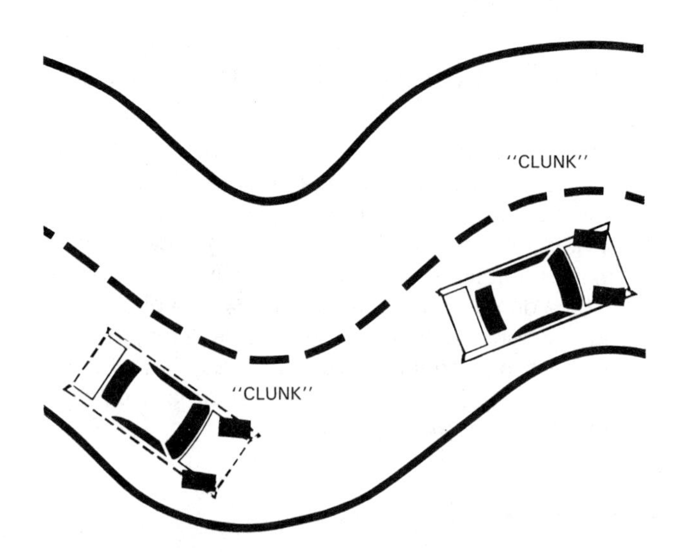

Figure 23-2. Turning the wheels from side to side will change the loads on the CV axles, isolating noises from loose or worn parts. Braking and accelerating may also help isolate the noise. (Chrysler)

also help you to isolate noises. Remember to check carefully to eliminate the steering or suspension system as the source of the noises.

CV Axle Removal

Figure 23-3 shows the parts of a typical CV axle assembly. These components must generally be removed from a vehicle when the CV joint, axle shaft, boot, or wheel bearing requires service. Almost all CV axle and wheel bearing repairs require that the axle be removed from the vehicle.

The CV axles must be removed from the transaxle before removing the transaxle from a vehicle. On many front-wheel drive vehicles, the engine and transaxle must be removed as a unit. Therefore, the CV axles must often be removed before the engine can be removed. In many cases, the CV axles can be pulled from the transaxle without removing them from the wheel hub and bearing assemblies.

There are many variations to the basic CV axle design. The procedure given below provides a general guide for removing most common CV axles. Always consult the manufacturer's service manual for the exact removal procedures.

1. Disconnect the vehicle's battery negative cable.
2. Place the vehicle transaxle in neutral.
3. Raise the vehicle and support it on approved jackstands. With some designs, you may want to loosen the stub shaft nut before raising the wheels off the floor. Remove the wheel from the wheel hub on the side to be serviced.

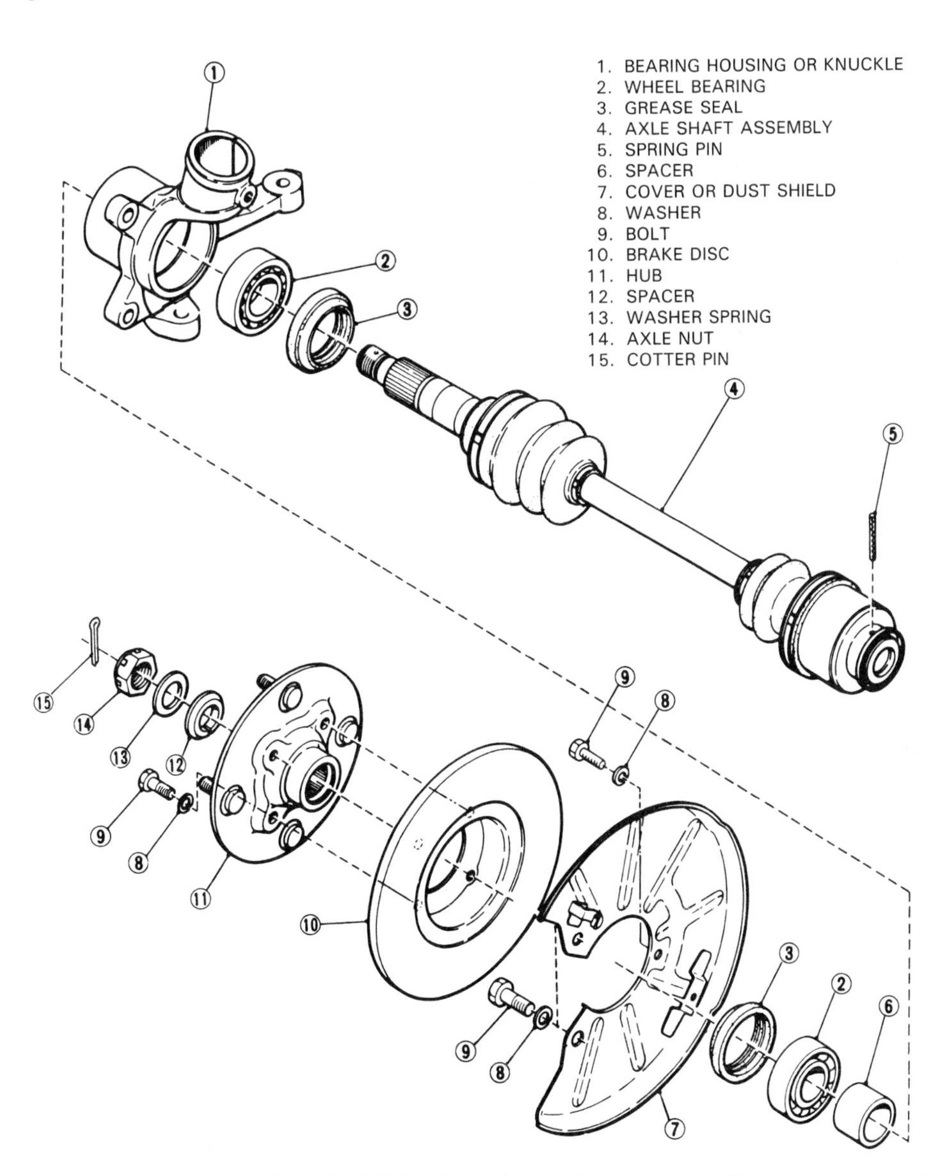

Figure 23-3. Exploded view of a typical CV axle and related components. Most transaxle-equipped vehicles have similar components. (Subaru)

4. Cover the outer CV joint boot with a shop cloth or a ***boot protector.*** See Figure 23-4. This will help prevent damage to the boot. If needed, remove the tie rod end nut and use a ball joint separator to free the tie rod end from the steering knuckle, Figure 23-5.
5. Remove the stub shaft nut cotter pin and remove the stub shaft nut, Figure 23-6. A few vehicles have left-hand threads on the right stub shaft. To remove the nut, it must be turned in a clockwise direction. Always check the service manual before trying to remove the stub shaft nut.
6. If tapered roller bearings are used for wheel bearings, remove the washer and the outer wheel bearing. Some bearing assemblies are equipped with a tensioning spring, which should also be removed at this time. Some vehicles use ball bearings, which must be pressed from the steering knuckle. The procedure for removing these bearings will be covered later in this chapter. Store all parts so they will not get damaged or dirty.

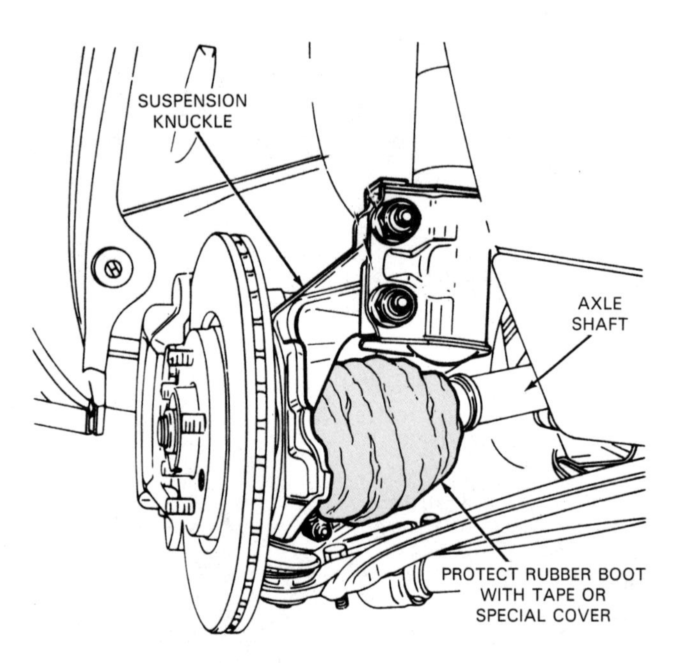

Figure 23-4. After removing the wheel and tire, cover the CV axle boot with a cloth or special protector to prevent damage during axle removal. (General Motors)

NOTE! The steering knuckle may not have to be removed when removing the CV axle. In some cases, however, the steering knuckle must be removed with the axle and the rest of the wheel hub and bearing assembly and disassembled on a workbench. Always consult the manufacturer's service manual for exact procedures.

7. Before removing the wheel hub and bearing assembly, remove the brake caliper, Figure 23-7. Most calipers can be removed without opening the hydraulic system. Pry the pads away from the rotor and remove the caliper attaching bolts. Position the caliper so that the brake hose is not strained. In many cases, the caliper can be wired to the vehicle chassis.
8. After removing the caliper, pull the wheel hub and bearing assembly from the axle stub shaft. In some cases, the bearings are pressed in the assembly, and the assembly must be removed with a special puller. See Figure 23-8.

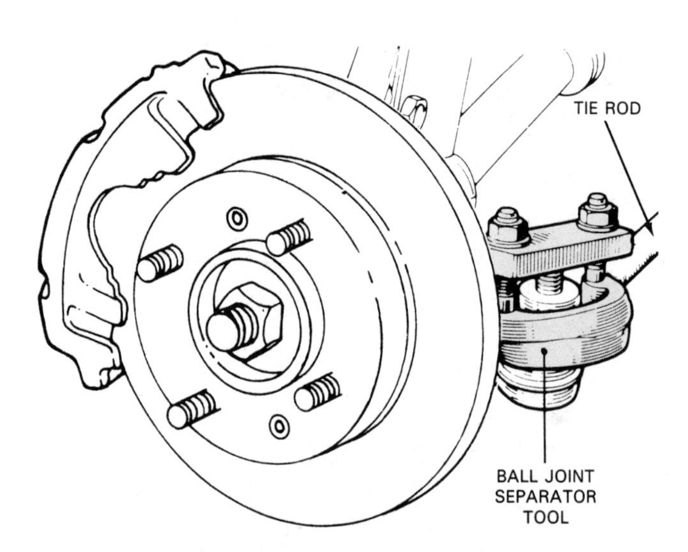

Figure 23-5. Tie rod end should be removed from the steering knuckle with a special tool. Never pound on the tie rod shaft. Pounding may damage the threads. (Chrysler)

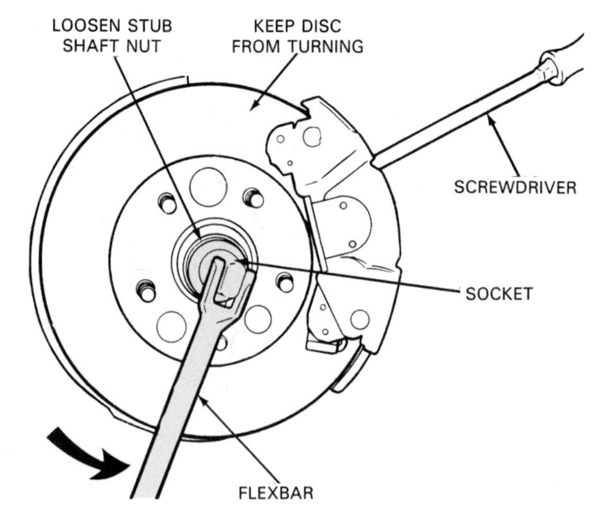

Figure 23-6. The spindle nut, or hub nut, can be removed with a large breaker bar and a socket. Leave the tire on the floor or have an assistant step on the brake to prevent the rotor from turning. If necessary, a screwdriver can be used to hold the rotor in place. (General Motors)

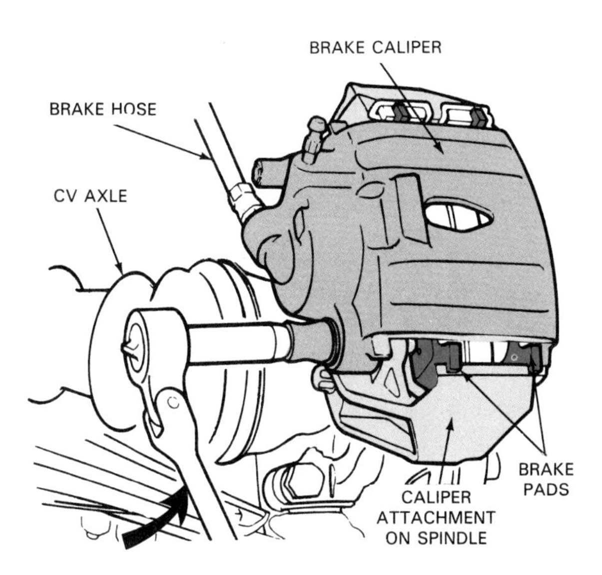

Figure 23-7. Remove the brake caliper before removing the rotor and wheel hub and bearing assembly. The caliper can be removed by prying the pads away from the rotor and removing the attaching bolts. The hydraulic system does not have to be opened when removing the caliper. (Chrysler)

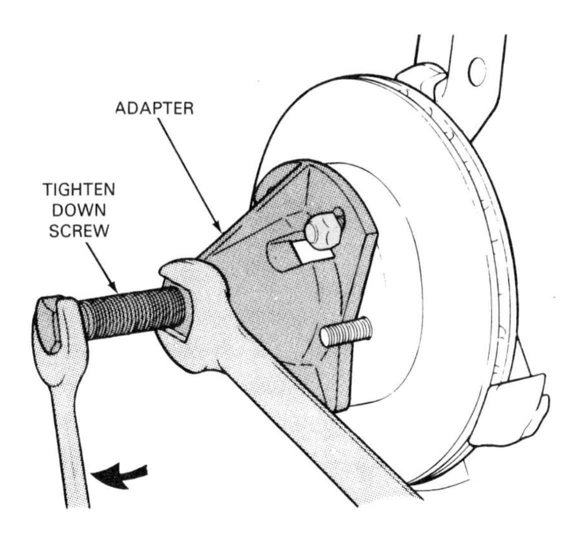

Figure 23-8. Some front-wheel drive systems use a pressed-on wheel hub and bearing assembly. A special puller tool, such as the one shown here, can be used to remove the spindle from the hub. (General Motors)

9. If necessary, remove the nut or the clamp bolt holding the ball joint stud to the steering knuckle. Next, separate the ball joint stud from the knuckle. See Figure 23-9. Free the steering knuckle by removing the bolts holding it to the lower portion of the MacPherson strut assembly, Figure 23-10.

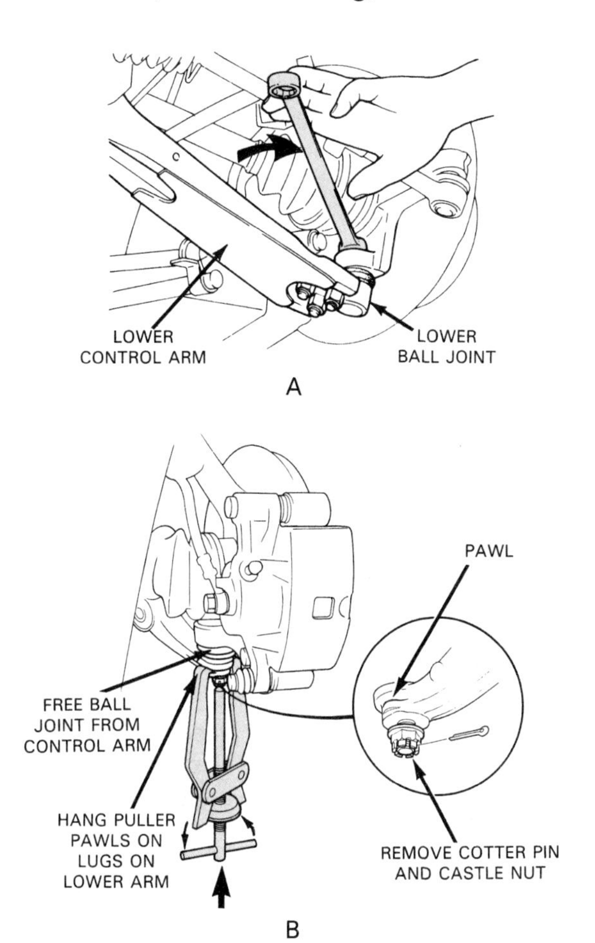

Figure 23-9. This illustration shows the removal process for a lower ball joint. Always support the lower control arm before removing the attaching nut. A–Removing the ball joint attaching nut with a hand wrench. B–Using a special puller to remove the ball joint from the steering knuckle. (Subaru, Honda)

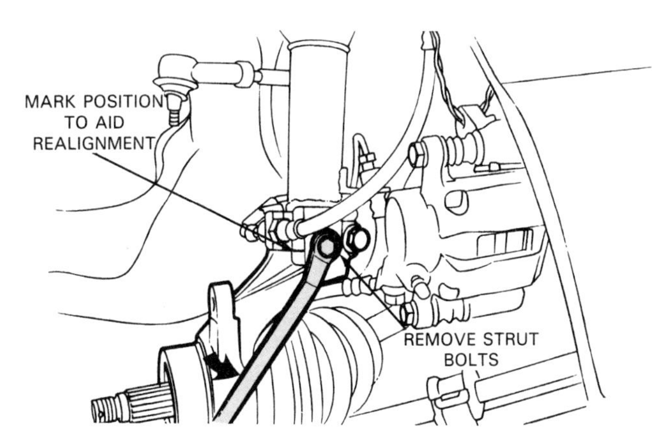

Figure 23-10. Removal of the lower bolts on a MacPherson strut suspension. Always mark the relative position of the parts to aid in front end realignment. (Chrysler)

WARNING! If the front spring is located between the lower control arm and the frame, the control arm must be supported before removing the ball joint or strut bolts. Spring tension can cause serious injury if the control arm is not supported!

10. Pull the steering knuckle assembly away from the vehicle until it clears the CV axle shaft, Figure 23-11. Then, remove the knuckle. If the steering knuckle must remain attached to other suspension parts, position it out of the way. Tie the steering knuckle to the chassis with wire if necessary.
11. If you are removing the axle that drives the speedometer drive gear, remove the gear from the transaxle.
12. You are now ready to remove the CV axle from the transaxle. The CV axles may be secured in the transaxle in one of many ways. Therefore, it is important that you consult the proper service manual for the exact removal procedure. Some CV axles are splined to the differential side gears or to the intermediate shaft on the long side of equal-length drive shaft systems. Some CV axles are held in place by a clip at the transaxle. Other axles are bolted to a flange that is attached to the transaxle. During disassembly, mark all parts so they can be reassembled in the proper order.

    Some CV axles must be removed with a special adapter that is attached to a slide hammer. See Figure 23-12. The adapter jaws are placed around the inner side of the CV joint housing. The slide hammer is then used to pull the axle from the transaxle output shaft.

    Some axles are attached with a retaining pin, which must be removed with a hammer and a punch, Figure 23-13. Once the pin is removed, the axle will slide from the transaxle output shaft. It may be necessary to pry the axle shaft loose with a pry bar placed between the inner CV joint and transaxle.

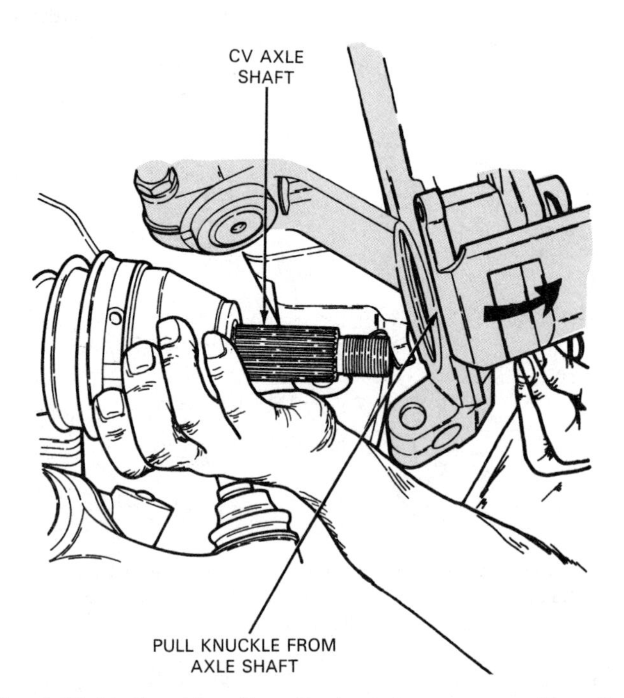

Figure 23-11. Once the other attachments are removed from the steering knuckle, it can be pulled from the CV axle shaft. (Ford)

If the axle shaft is bolted to the transaxle output stub shaft, remove the bolts, Figure 23-14. After the bolts have been removed, pull the axle shaft from the stub shaft.

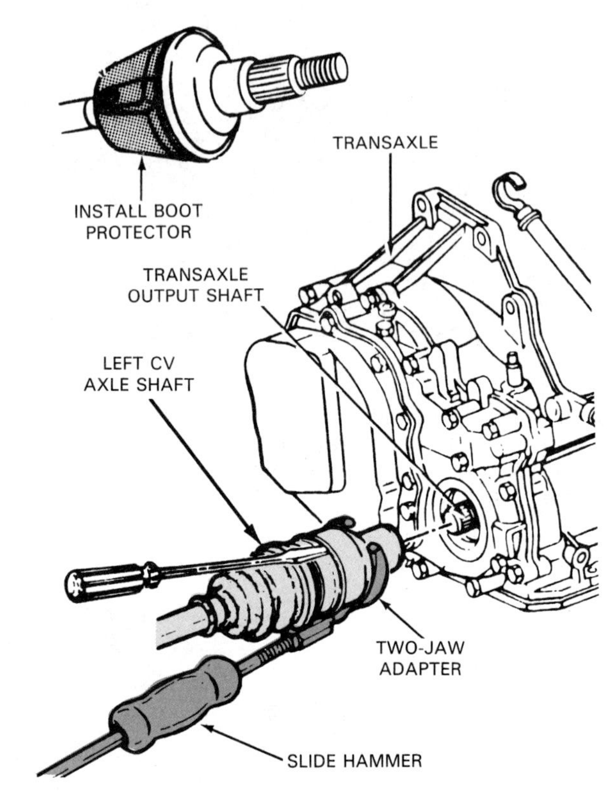

Figure 23-12. This CV axle is being removed from the transaxle with a special two-jaw adapter that is attached to a slide hammer. (General Motors)

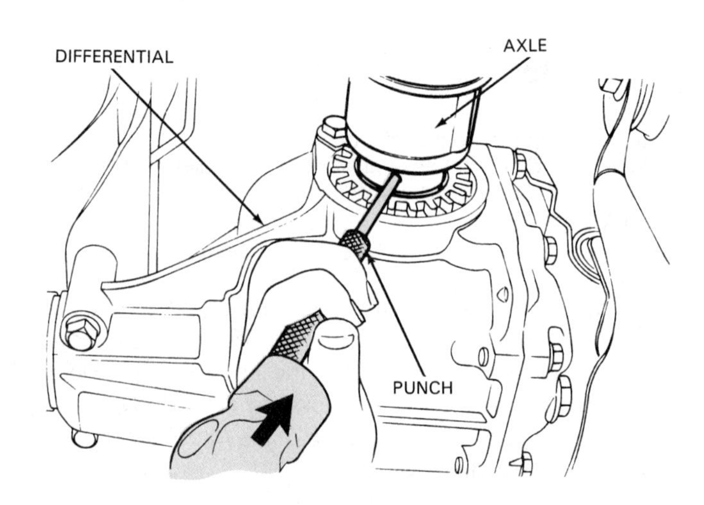

Figure 23-13. Some CV axles are held in place by a pin that is pressed through the transaxle output shaft and the CV axle shaft. After the pin is driven out, the shaft can be pried out of the transaxle.

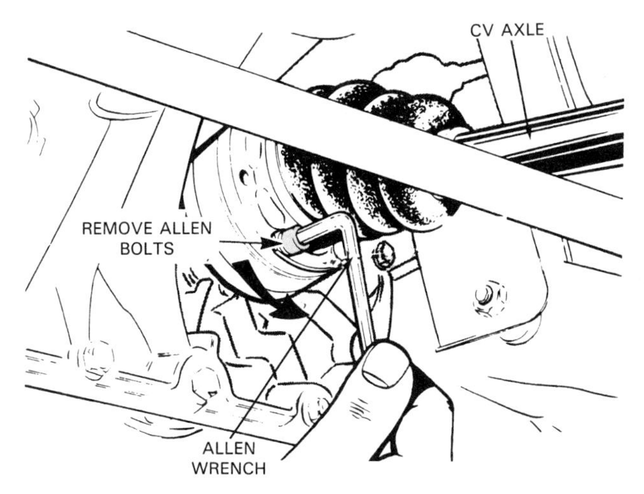

Figure 23-14. CV axles that are bolted to stub shafts at the transaxle can be unbolted and removed. Always mark the stub shafts and axle shafts before removal to ensure that they are reinstalled in their original position. (Ford)

A few CV axles are held to the transaxle by an internal circlip, or snap ring, that can only be accessed by removing the differential cover, Figure 23-15. When removing the cover, place a drip pan under the cover to catch spilled oil. After prying the clip from the axle, the shaft will slide out of the transaxle.

13. After the CV axle has been removed from the vehicle, place it on a clean workbench for disassembly. The bench should be equipped with a vise that is fitted with soft jaws to prevent damage to axle parts. If you suspect that the axle shaft is bent, check it visually first. If no damage is visible, the shaft should be mounted on V-blocks and checked with a dial indicator.

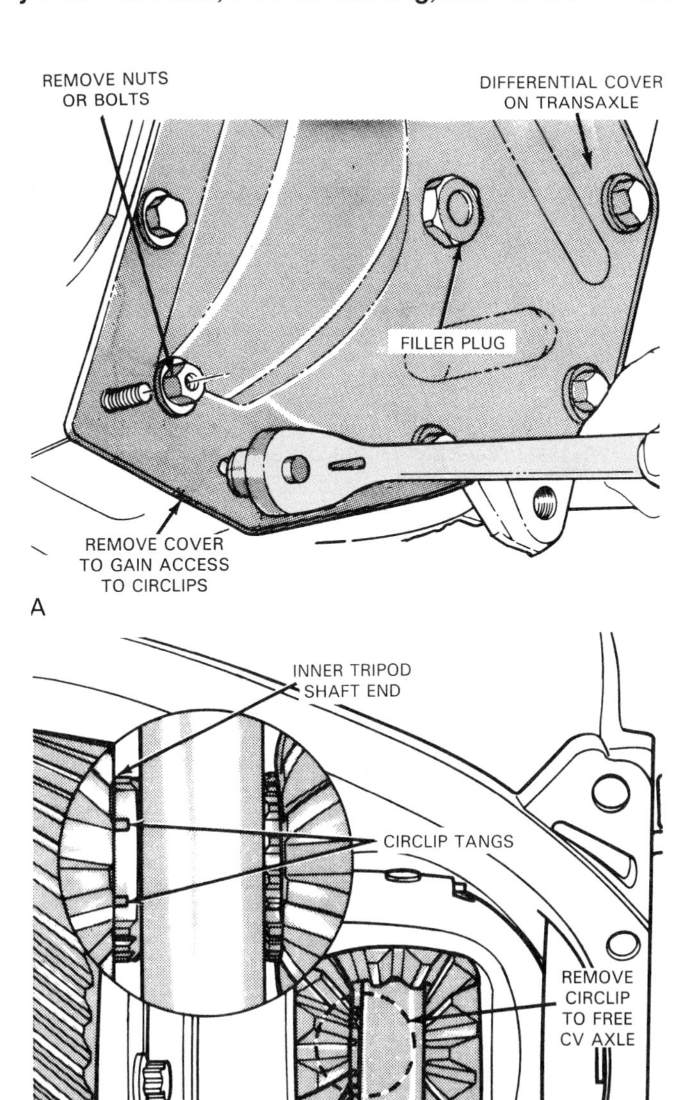

Figure 23-15. Some CV axles are held in place by a snap ring that is located inside the transaxle differential assembly. A–Removing the differential cover. B–Removing the circlip. (Chrysler)

CV Axle Parts Inspection and Repair

With the CV axles removed, each component must be carefully inspected for wear and damage. Be sure all parts are clean before inspection. Look for *any* signs of damage or wear, no matter how slight. After the defect is identified, the concerned part or parts may be repaired or replaced. Service procedures for the various CV axles and related parts are the focus of this section.

CV Axle Seal Service

After the CV axle shaft has been removed, the axle seals can be replaced, Figure 23-16. If the vehicle does not have a stub shaft at the transaxle, the first step (Figure 23-16A) is not necessary. Fluid may drip from the transaxle housing after the axle shaft is removed. Have a pan ready to catch the dripping fluid.

CV Axle Wheel Bearing Service

The wheel bearings used with CV axles are a common source of problems. Some manufacturers use ball bearings in the front wheel hub and bearing assemblies, while others use tapered roller bearings. The tapered roller bearings are similar in appearance to those used on the front wheels of rear-wheel drive vehicles. The wheel bearings

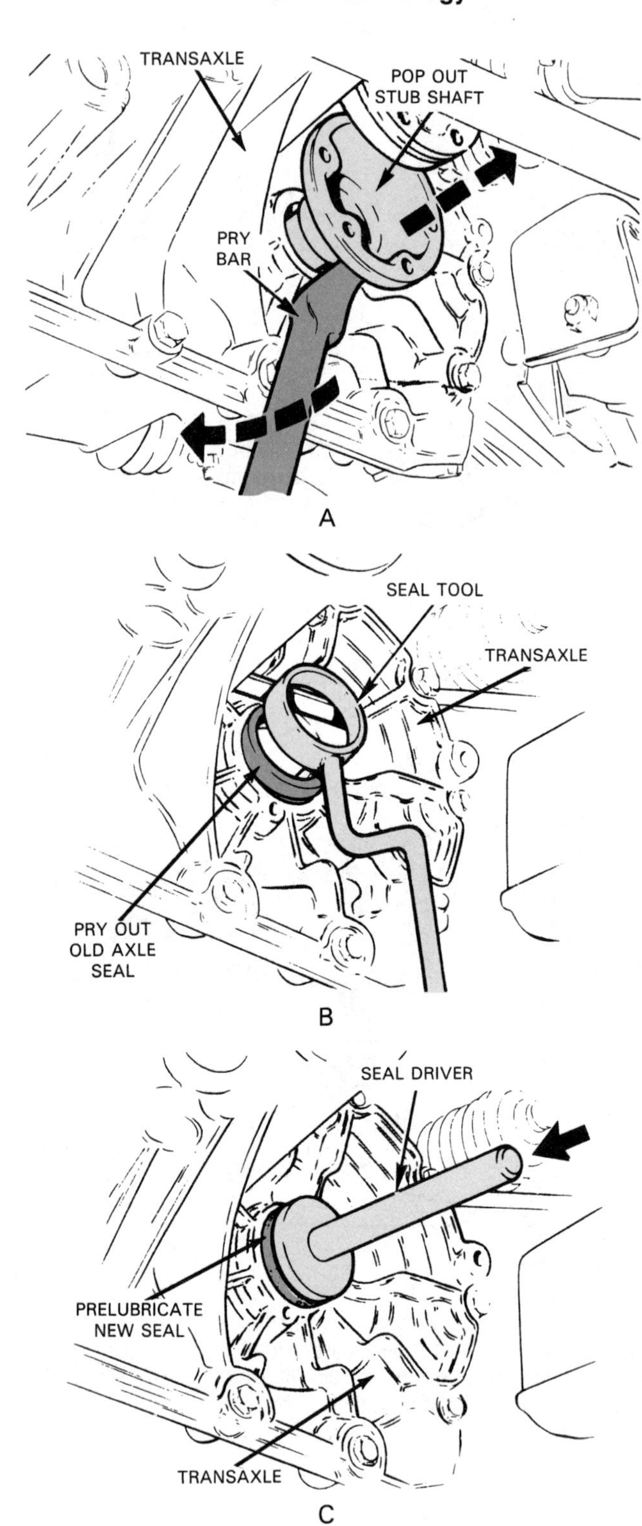

Figure 23-16. This series of steps illustrates one method of replacing a transaxle seal. A–Pry out the transaxle stub shaft. (This step is skipped if there is no stub shaft.) B–Pry out the old seal with a large screwdriver or a special removal tool. C–Install the new seal with a special installation tool. (Ford)

should be carefully inspected. Bearings should be replaced if they show signs of wear, corrosion, or damage.

Many wheel bearings are pressed into the bearing carrier or the steering knuckle, and the wheel hub is pressed into the bearing. The first step in servicing this type of assembly is to press the hub from the bearing with a hydraulic press and a proper adapter plate, Figure 23-17. The bearing can then be removed from the carrier or knuckle with a special puller.

Note that to replace some wheel hub and bearing assemblies, the steering knuckle must be removed from the vehicle. The knuckle must be removed by separating it from the upper ball joint and the strut rod assembly. After reinstalling the knuckle, the front end of the vehicle must be aligned.

In some cases, the bearing carrier or the steering knuckle uses a seal that must be pressed from the knuckle or carrier housing before the bearing can be removed. Figure 23-18 shows a seal installed in an intermediate shaft support bracket. Once this seal is removed, the bearing can be pressed out of the housing, Figure 23-19.

Some CV axle assemblies are pressed into the steering knuckle or carrier. In these cases, the shaft must be removed from the knuckle or carrier housing with a special tool. See Figure 23-20. After the shaft is removed, the outboard bearing can be removed from the housing.

If tapered roller bearings are used, the bearing races must be replaced when the bearings are replaced. Figure 23-21 shows a special driver being used to remove bearing races from a typical housing.

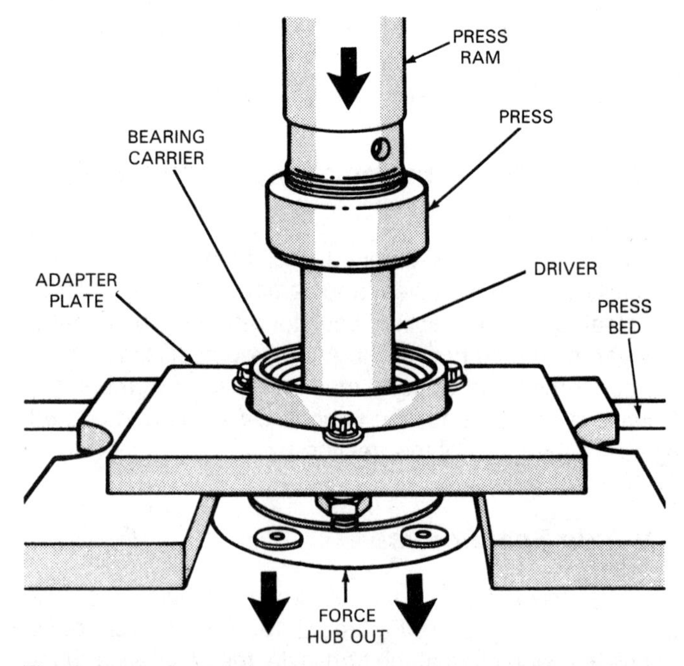

Figure 23-17. Some wheel hubs must be removed from bearings with a hydraulic press, driver, and the proper adapters. Never use excessive pressure on the housing or hub, as this can cause them to break. (Chrysler)

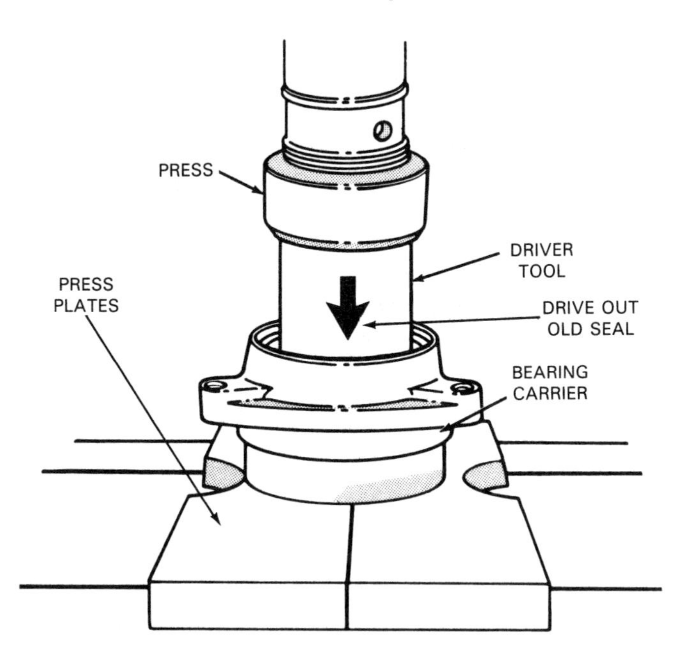

Figure 23-18. Some seals must be removed using a hydraulic press, driver, and suitable adapters. Other seals are held in place by clips. A few seals will be pressed out when the bearing is removed. (Chrysler)

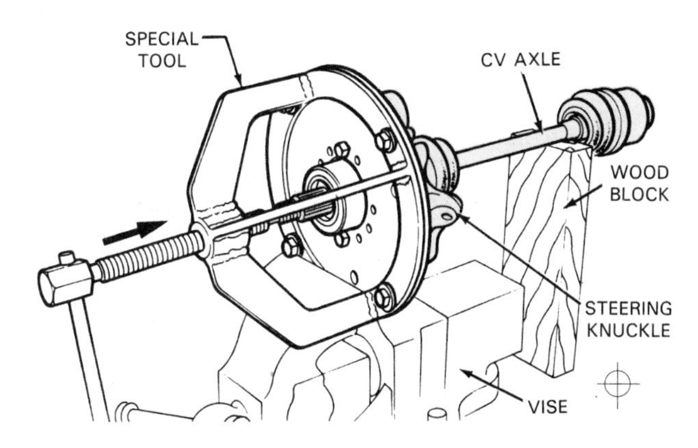

Figure 23-20. To gain access to some bearings, the CV axle and the steering knuckle assembly must be removed from the vehicle as a unit. The components are then separated with a special tool. (Subaru)

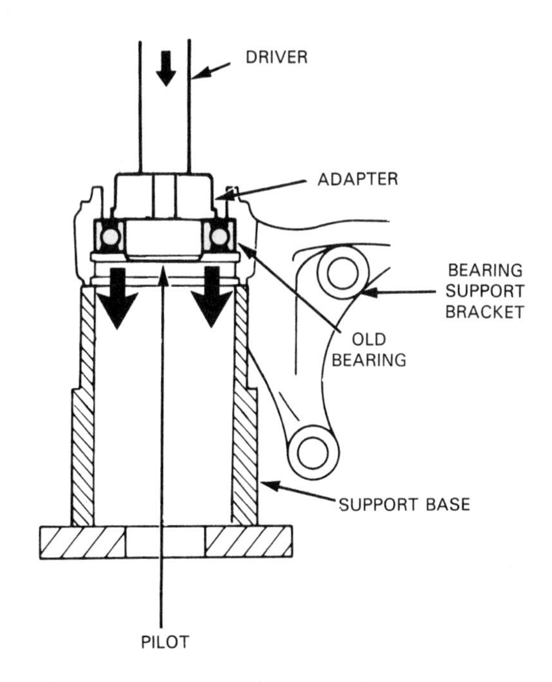

Figure 23-19. Bearings must be carefully removed. A hydraulic press, with a driver and the proper adapter, is often used to remove bearings. Always consult the service manual for exact procedures. (Honda)

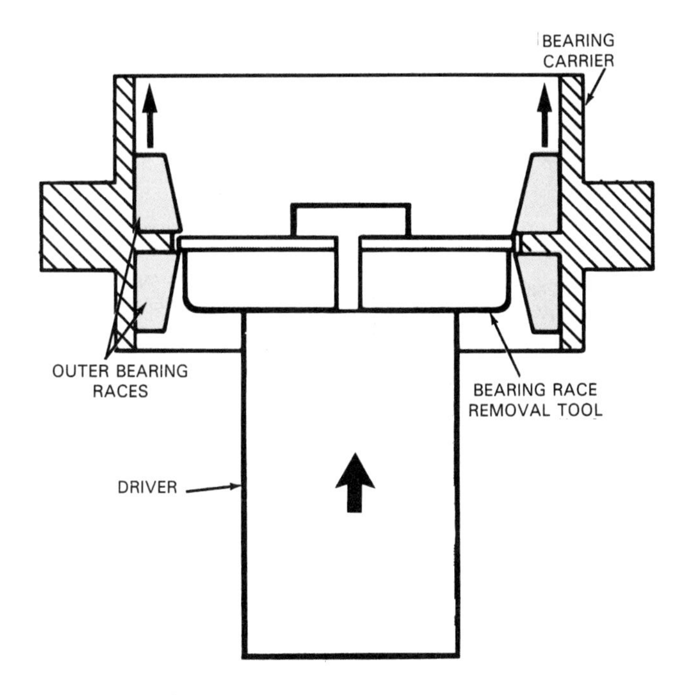

Figure 23-21. Special drivers are used to remove and install the outer bearing races used with tapered roller bearings. When using these drivers, be careful not to damage the seating surfaces in the housing. Races should not be removed unless new races are to be installed. (Chrysler)

Some bearings are pressed into the housing with a spacer between them to maintain proper bearing position. If you replace the bearings or the housing in this type of system, you must select the proper replacement spacer, or bearing problems will result. A few systems use a collapsible spacer, which is similar to the collapsible spacer used in rear-wheel drive pinions. Refer to the manufacturer's service manual for exact replacement procedures.

Install new bearings, races, and seals with the proper driver and a hydraulic press. See Chapter 3 for more information on bearing installation. Figure 23-22 shows the adapters needed to service the ball bearing and seal on a common front-wheel drive vehicle.

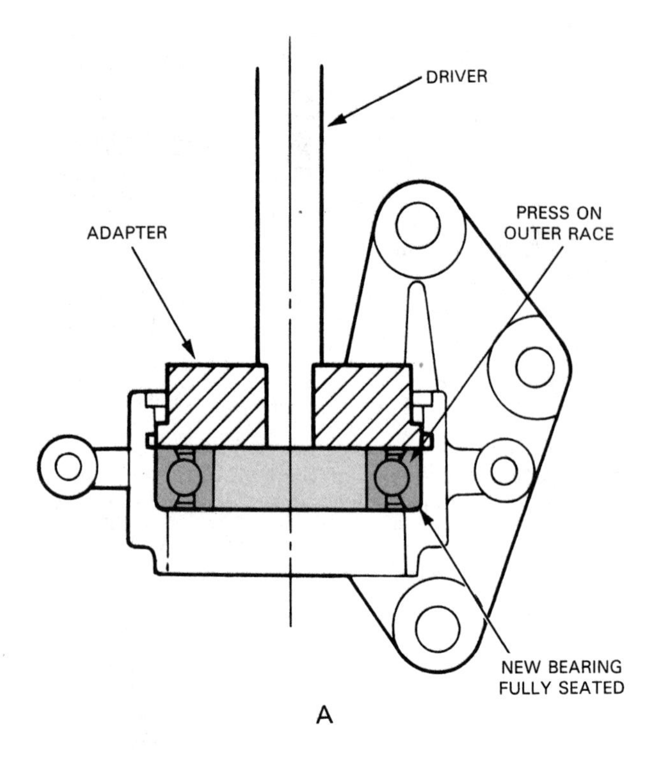

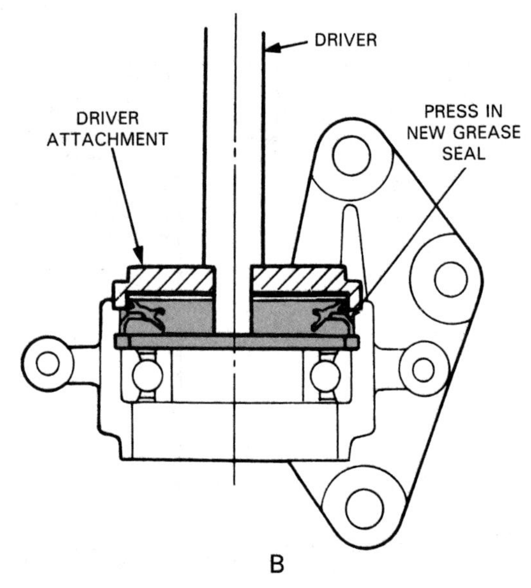

Figure 23-22. This figure illustrates the installation of a new bearing and seal in a steering knuckle housing. A–Installing the new bearing with a special adapter. B–Installing the new seal. (Honda)

Tripod Joint Service

Tripod joints, such as the one shown in Figure 23-23, are often used as inboard joints and occasionally as outboard joints in CV axle assemblies. The ability of these joints to compensate for changes in the effective distance between the transaxle and the wheel hub while allowing the axle assembly to flex makes them popular.

The various service techniques for most tripod joints all very are similar. However, you may have to choose between rebuilding or replacing the joint. See Figure 23-24. In some cases, a joint may be damaged so badly that it is easier and cheaper to replace it. Some tripod joints cannot be rebuilt.

The general procedures for rebuilding a tripod joint are presented below. Always refer to the proper service manual for specific instructions.

To begin tripod joint service, clamp the CV axle shaft in a vise. Be careful not to damage the shaft by overtightening the vise jaws. Remove the boot clamps, Figure 23-25, and pull the boot away from the tripod joint. Remove any grease that prevents a clear view of the internal parts of the joint.

Next, separate the tripod and outer housing. Remove any retaining clips or tabs as detailed in the service manual. Paint alignment marks on the tripod spider and the shaft to ease reassembly, Figure 23-26. Then, remove the spider from the shaft, Figure 23-27.

Clean the spider assembly and inspect the components for wear and damage. Pay particular attention to the needle bearings, the wear surface between the rollers and the housing, and the snap rings that hold the assembly together. Replace worn parts as needed or replace the entire assembly with a new or rebuilt unit.

To begin reassembly of the tripod joint, install a small boot clamp on the axle shaft. Then install a new boot on the axle shaft. The small end of the boot should be installed first. Boot installation is covered in detail later in this chapter.

Install the tripod spider on the CV axle shaft. Make sure that the alignment marks are positioned properly. On some tripod joints, the spider must be installed in a certain direction. On other joints, both sides of the spider are the same.

Thoroughly lubricate the spider assembly, using the type of grease recommended by the manufacturer. This grease is generally provided with the joint rebuilding kit. Most replacement joints have grease already packed into the housing. Any extra grease should be placed inside the boot to provide additional lubrication.

Install the rollers on the spider and slide the spider into the housing. Some spiders must be pressed or driven into the housing. If the spider must be installed in this way, tape the rollers to prevent them from falling off during installation. See Figure 23-28. The tape can be removed just before the rollers enter the housing. Grease all moving surfaces as recommended by the manufacturer. Make sure the retaining snap rings and tabs are in good condition and can hold the tripod in the housing.

Once the spider assembly is installed in the housing, install the boot flanges into the boot retaining grooves in the housing and clamp the boot in place. Then, check the joint for free rotation at various angles.

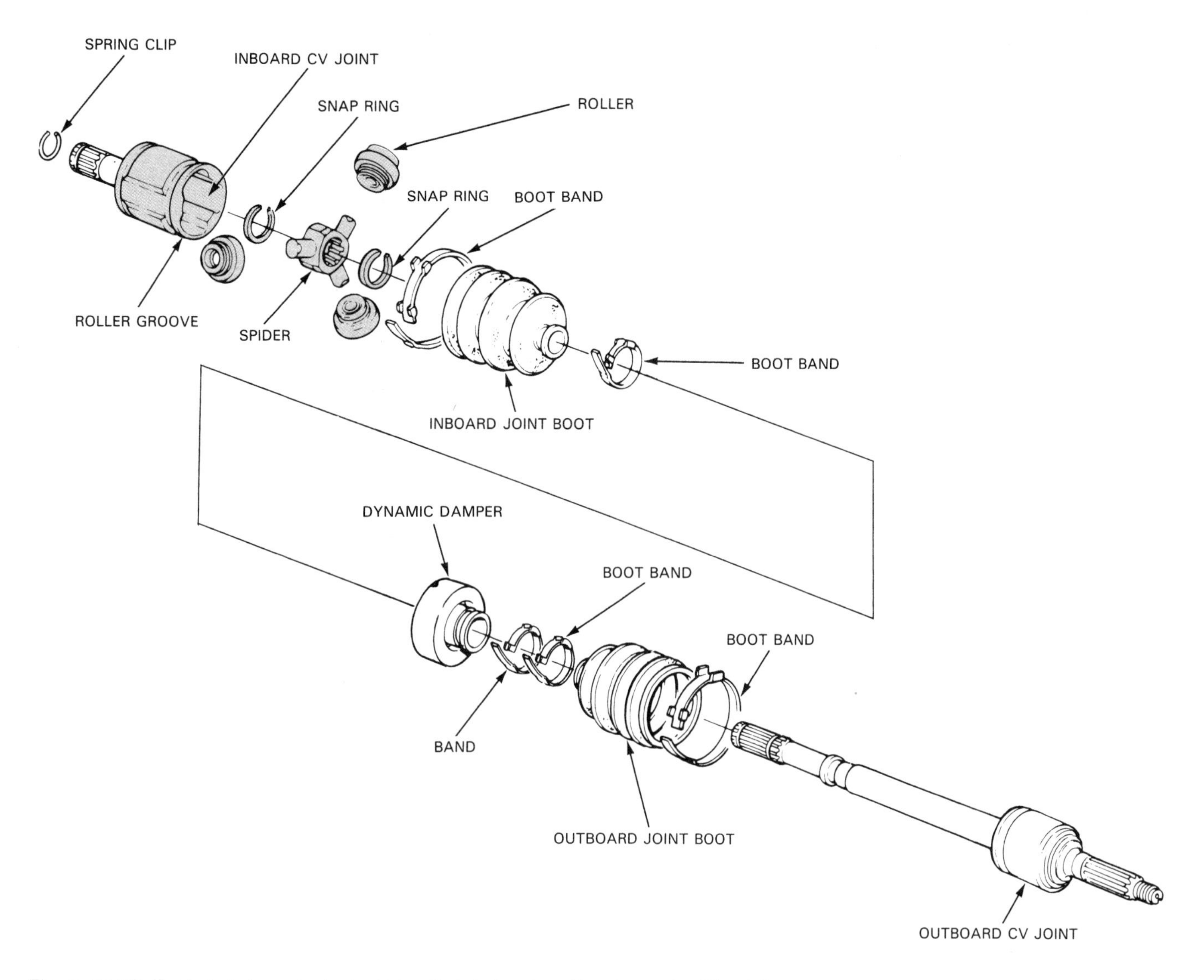

Figure 23-23. Exploded view of a typical tripod joint showing all major parts. (Honda)

Figure 23-24. This rebuilding kit contains all of the parts commonly needed to rebuild a worn CV joint. (Dana)

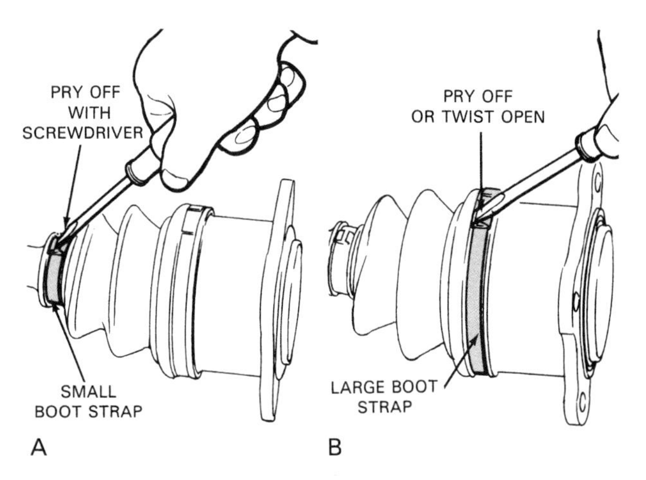

Figure 23-25. Removing boot clamps. A–Prying off the small (shaft side) clamp with a screwdriver. B–Removing the large (housing side) clamp.

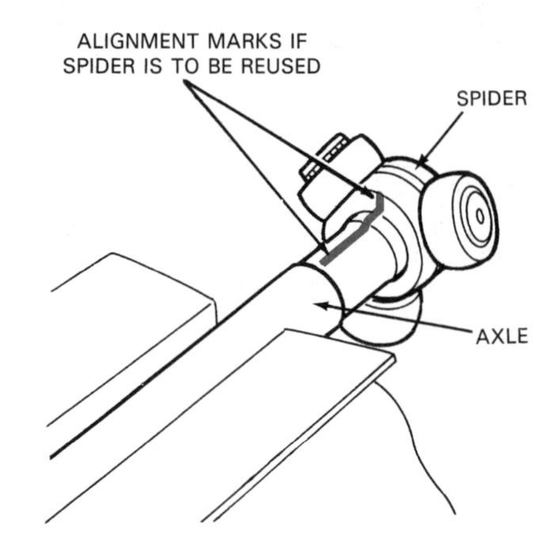

Figure 23-26. Always make alignment marks if the tripod spider will be reused. This ensures that the spider will be reinstalled correctly.

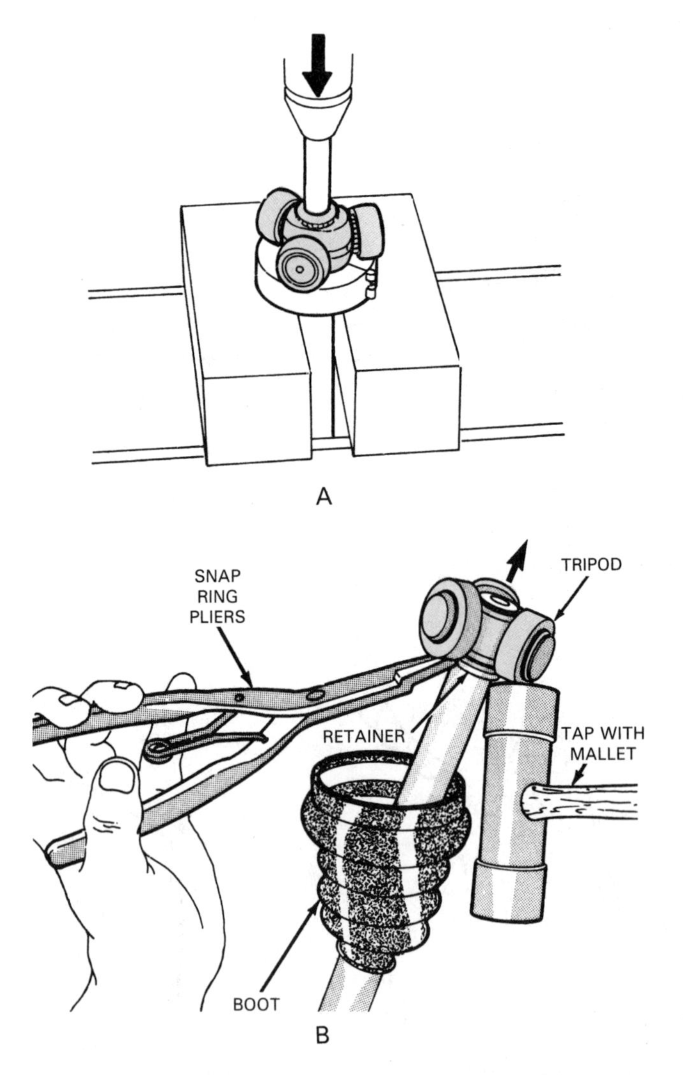

Figure 23-27. The tripod spider usually slides off the shaft, but not always. A–This tripod spider must be pressed off its shaft. B–Snap ring pliers and a mallet are used to remove this spider. (Chrysler)

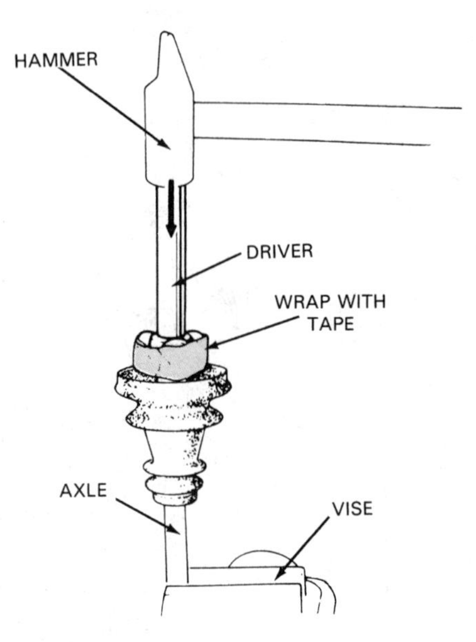

Figure 23-28. Before installing the new spider assembly in the housing, wrap the assembly with tape to keep it from coming apart. The tape can be removed just before the spider assembly enters the housing. (Chrysler)

Ball-and-Cage Joint Service

Ball-and-cage joints, such as the one shown in Figure 23-29, are the most popular type of CV joint. At least one ball-and-cage joint will be found on most CV axle shafts. Some ball-and-cage joints cannot be serviced, and the entire assembly must be replaced. If the CV joint can be serviced, consult the manufacturer's manual for exact procedures. Following is a general guide to rebuilding ball-and-cage joints:

Lightly clamp the CV axle shaft in a vise. Remove the boot clamps and pull the boot away from the joint. Clean off any grease that will prevent inspection and disassembly of the joint.

If necessary, remove the snap rings that hold the joint together and place alignment marks on the joint housing and the axle shaft to ease reassembly. See Figure 23-30. Remove the joint housing or the axle shaft from the ball-and-cage assembly. The part that is removed will vary, depending on the design of the joint.

If the ball-and-cage assembly is still inside the housing (this will be the case with most joints), carefully tilt the cage in the housing by pressing down on one side of the cage. You can then remove the ball from the opposite side of the cage. Repeat this procedure until all the balls have been removed. During disassembly, place the balls so that they can be reinstalled in the raceways from which they were removed.

Once the balls have been removed, the cage and inner cross assembly can be turned sideways and removed from the housing. After the assembly has been removed from the housing, the inner cross can be removed from the cage.

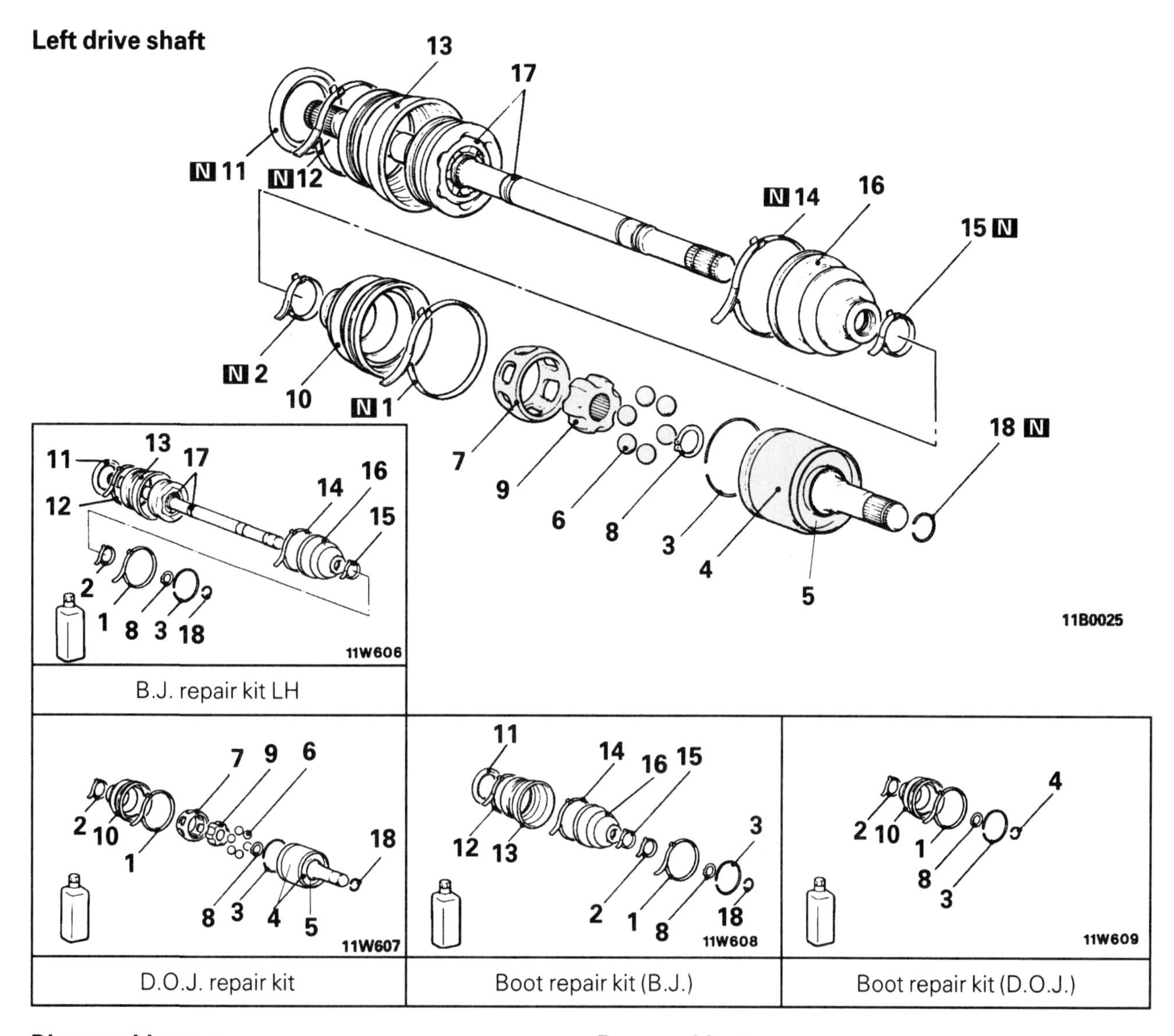

Disassembly steps

	1. Boot band A
	2. Boot band B
	3. Circlip
	4. D.O.J. outer race
←→	5. Dust cover
←→	6. Balls
←→	7. D.O.J. cage
←→	8. Snap ring
	9. D.O.J. inner race
←→	10. D.O.J. boot
←→	11. Dust cover
	12. Boot protector band
←→	13. Boot protector
	14. Boot band A
	15. Boot band B
←→	16. B.J. boot
	17. Drive shaft and B.J.
	18. Circlip

Reassembly steps

	17. Drive shaft and B.J.
→←	16. B.J. boot
→←	14. Boot band A
→←	15. Boot band B
→←	2. Boot band B
→←	10. D.O.J. boot
→←	1. Boot band A
→←	7. D.O.J. cage
→←	9. D.O.J. inner race
	8. Snap ring
→←	6. Balls
→←	5. Dust cover
→←	4. D.O.J. outer race
	3. Circlip
	18. Circlip
→←	13. Boot protector
→←	12. Boot protector band
→←	11. Dust cover

NOTE
(1) ←→ : Refer to "Service Points of Disassembly".
(2) →← : Refer to "Service Points of Reassembly".
(3) N : Non-reusable parts
(4) B.J. : Birfield Joint
(5) D.O.J. : Double Offset Joint

Figure 23-29. General procedure given here shows typical sequence followed for disassembly and reassembly of a CV axle. Ball-and-cage joint shown here is at outboard end of axle. Note repair kits that are available. (General Motors)

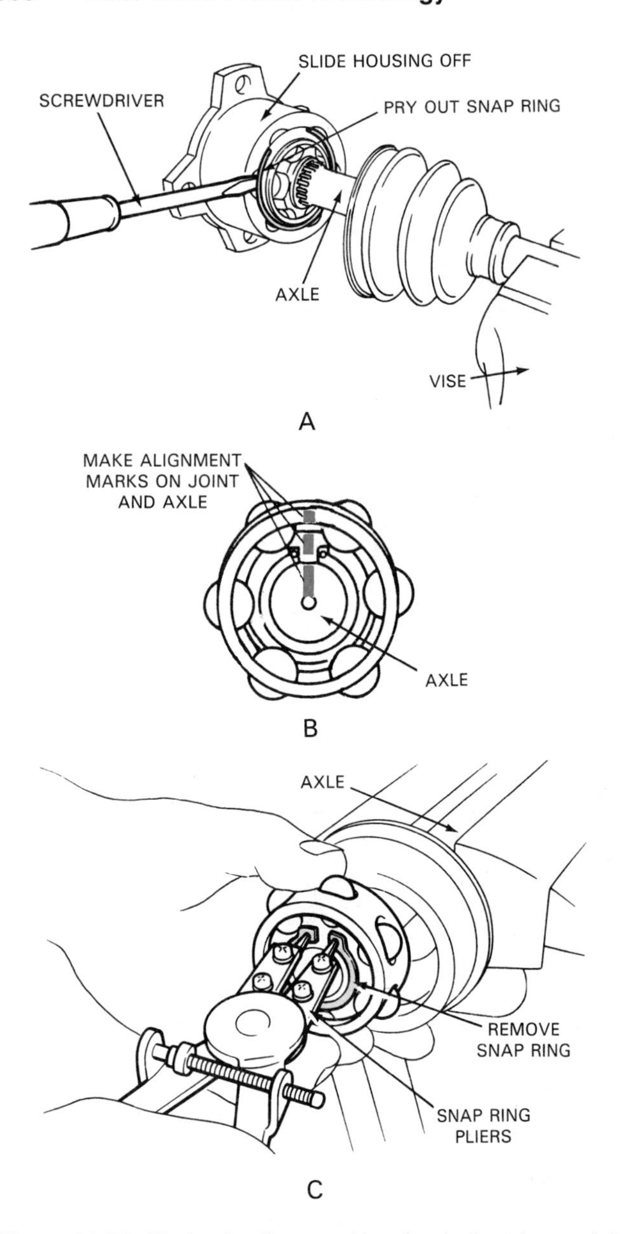

Figure 23-30. To begin disassembly of a ball-and-cage joint: A–Remove the internal snap ring that holds the cage in the housing. B–Mark the shaft and housing for proper reassembly. C–Remove the external snap ring holding the shaft to the inner cross.

The CV joint is now completely disassembled. Figure 23-31 shows the sequence of events for disassembling a typical ball-and-cage joint.

After the joint is completely disassembled, clean all the parts and inspect them for wear. The housing and inner raceways should be checked carefully for excessive wear and scoring. Check the balls for pitting, cracks, score marks, and uneven wear. Also study the ball cage for wear or cracks. Check the lubricant for contamination. Defective parts should be replaced. In some cases, the entire joint must be replaced.

To reassemble a ball-and-cage joint, align the parts and install the inner cross into the cage. Then install the cage assembly in the housing. Tilt the cage to replace the balls. Make sure that the balls are returned to the slots from which they were removed. Install the boot and the joint housing on the axle shaft. Finally, reinstall the snap rings as necessary. See Figure 23-32.

After reassembly, thoroughly grease the joint. Place extra grease in the boot to provide additional lubrication, Figure 23-33. Install the boot over the joint and tighten the attaching clamps, Figure 23-34. Check the joint for free rotation at various angles. To ensure that the joint has been properly reassembled, some manufacturers recommend checking the length of the joint between the boot clamps, Figure 23-35.

CV Joint Boot Service

The CV joint boots retain lubricant and protect the CV joint from water and dirt. The boots should be inspected whenever the technician is under the vehicle. Look for grease on the underside of the vehicle near the boot. This may indicate a punctured boot that is throwing grease as it rotates. Another cause of leakage is a clamp that is broken or loose. A leaking boot should be serviced immediately to prevent joint failure.

When working on CV axles, make sure that you do not damage the boots. When reinstalling a boot, remove any kinks from the boot bellows and position the clamps properly. Do not overtighten the clamps. To prevent damage, do not rotate the CV joints at abnormal angles when they are removed from the vehicle.

Unless you are working with a new vehicle, it is best to install new boots whenever they are removed. Although boots are inexpensive, installation is time consuming. Boot removal procedures were covered earlier in this chapter.

Some replacement CV joint boots can be installed without removing the CV axle from the vehicle. These boots are split to permit assembly over the axle shaft and the joint. Most boots, however, can only be replaced by removing the CV axle.

The boot replacement procedure presented here should be used as a general guide. Always consult the manufacturer's service manual before replacing a boot.

NOTE! The following procedure assumes that the axle shaft and the CV joint have already been disassembled.

If a boot is secured by one-piece clamps, slide the small clamp onto the axle shaft before installing the boot. Then slide the small end of the boot over the axle shaft and position it as necessary. If the shaft is splined, protect the boot by placing tape over the splines. Place the boot flange into the groove on the shaft. Position the clamp over the groove and make sure that the clamp and boot are properly aligned.

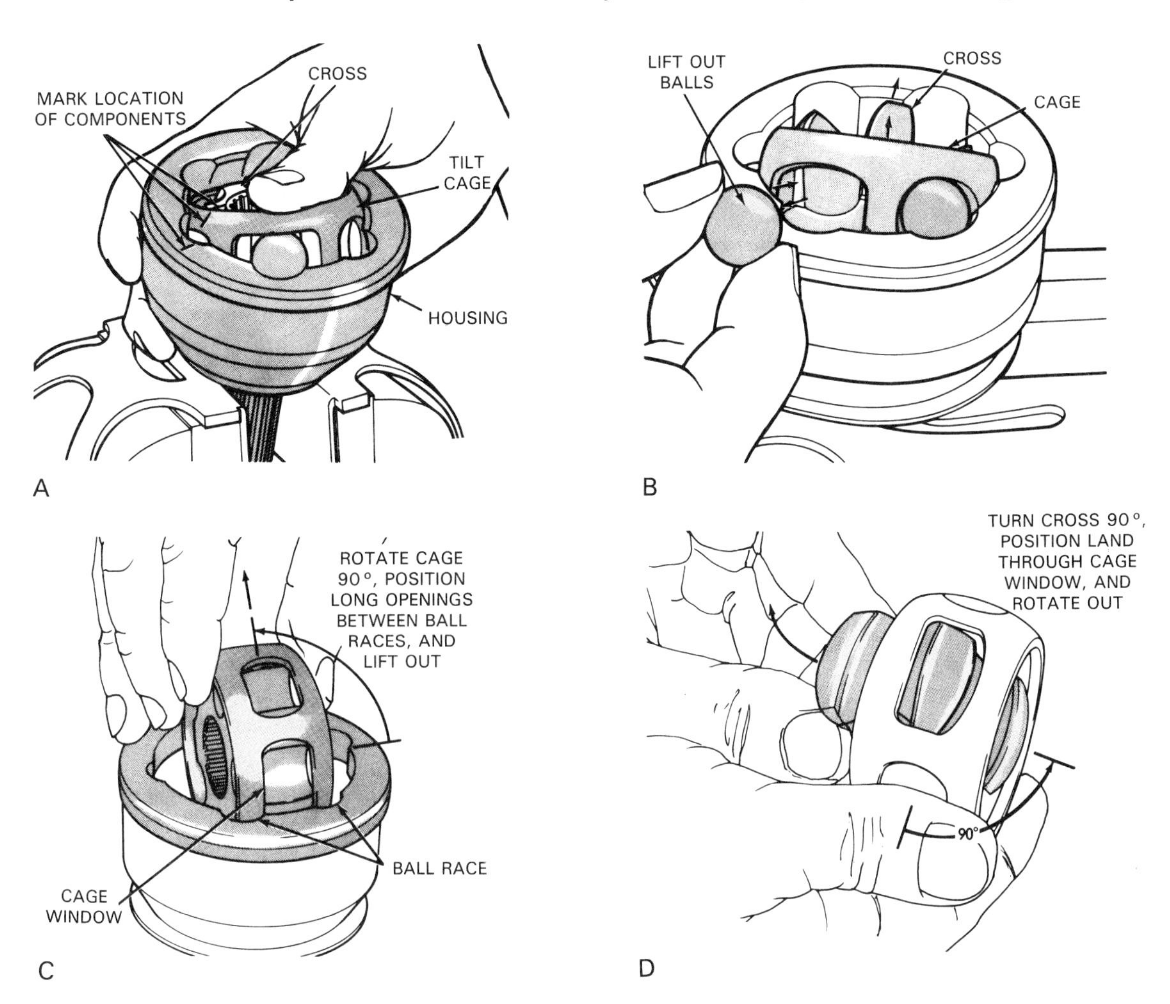

Figure 23-31. Disassembly sequence for a common type of ball-and-cage joint. A–Place the joint in a vise, mark the relative location of components, and press down on one side of the cage. This will tilt the other side of the cage upward. B–Lift out the ball on the raised side of the cage. It may be necessary to force the ball out with a small screwdriver. Repeat these steps until all of the balls are removed. C–Turn the cage at a 90° angle to the housing and lift it out of the housing. D–Remove the inner cross from the cage. The CV joint is now completely disassembled. (Chrysler)

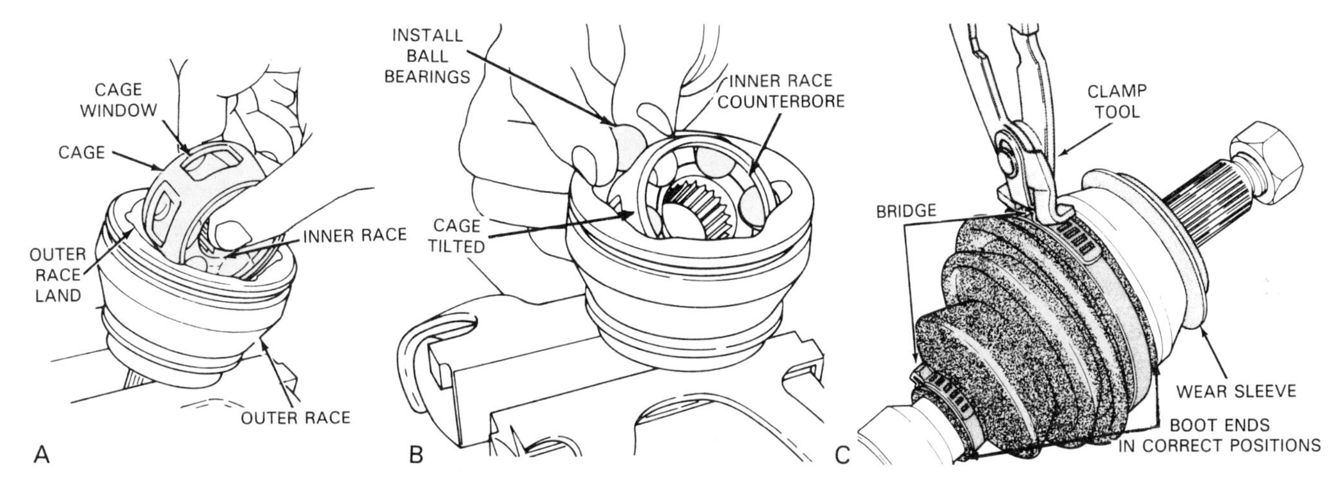

Figure 23-32. Reassembling the ball-and-cage joint. A–Place the cage and the inner cross back in the housing. B–Reinstall the balls. C–After installing snap rings and other necessary parts, fit the boot over the joint and install the boot straps. (Ford, Chrysler)

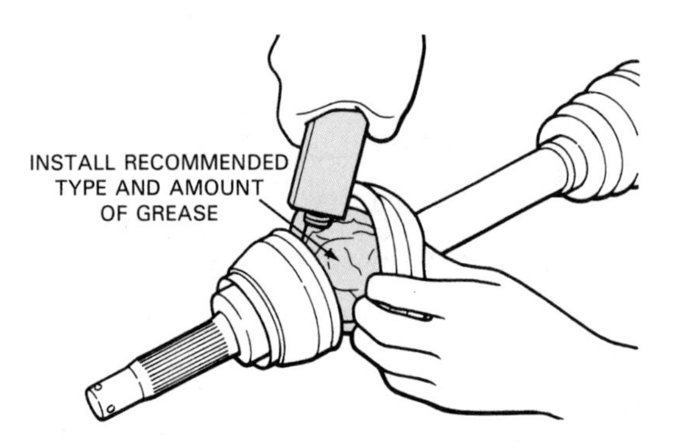

Figure 23-33. Be sure to install the proper amount of CV joint lubricant before reinstalling the boot and clamps. (General Motors)

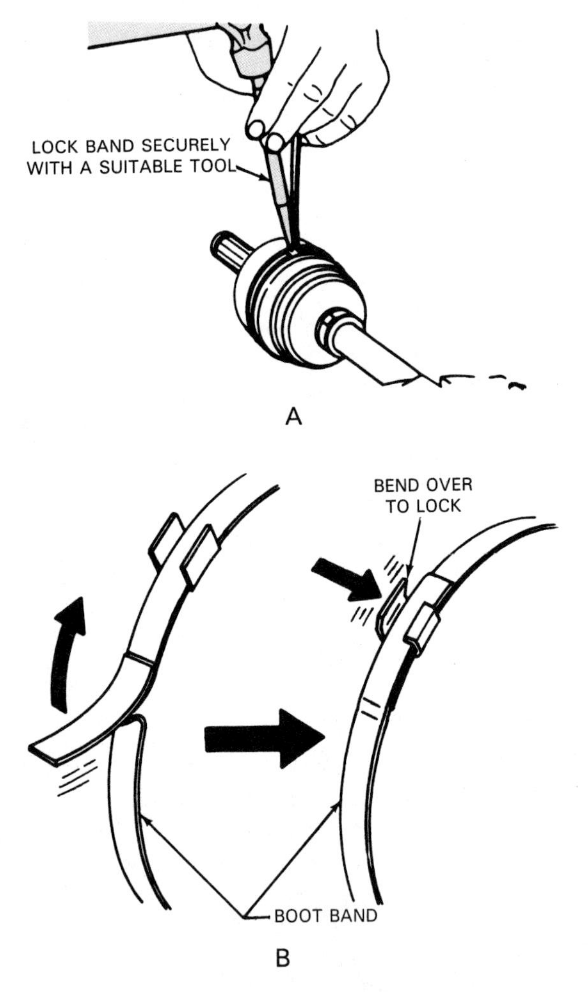

Figure 23-34. The boot clamps must be installed carefully. A–A punch and hammer are used to lock the retaining tabs over the clamp. B–This figure shows the proper method of securing the clamp. (General Motors)

Several types of clamps are used on CV joint boots. Some manufacturers use metal ladder clamps, which are similar to cooling system hose clamps. Other manufacturers use rubber rings or metal spring clamps to hold the boots in place. A common type of clamp is the strap-and-buckle clamp, Figure 23-36. A special tool must be used to install this clamp. It is important to avoid overtightening the clamps. Overtightening will damage the boot and cause a leak.

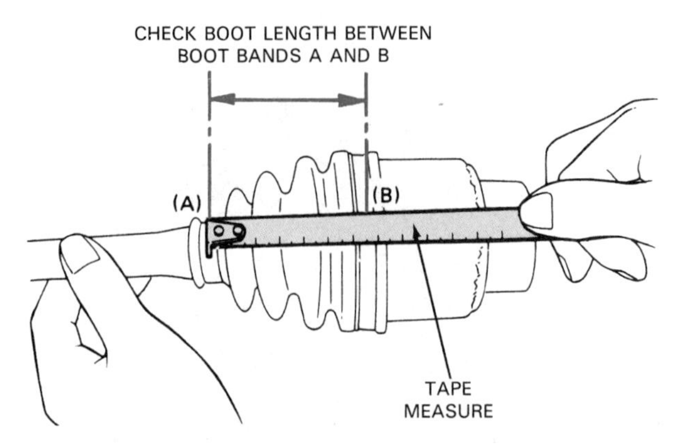

Figure 23-35. To ensure that the CV joint has been reassembled properly, some manufacturers recommend checking the installed length of the boot. (General Motors)

After installing the small clamp, fill the boot with grease as recommended by the manufacturer. Then, reassemble the CV joint.

With the joint reassembled, place the flange at the large end of the boot in the corresponding groove in the CV joint housing. Make sure that all parts are aligned properly and that the boot will not be twisted when the clamp is installed. Finally, install the large boot clamp in the same manner as the small clamp.

If the boot has wrinkled or kinked sections, smooth them out by pulling outward on both sides of the wrinkle. See Figure 23-37.

CV Axle Installation

The CV axle shaft is installed in a vehicle in the reverse order of removal. The procedure outlined here will apply to most vehicles. Always consult the factory service manual for specific installation instructions.

1. In systems where the axle shaft yoke enters the transaxle directly, lightly oil the outer surface of the shaft. Hold the inner CV joint assembly at the housing while guiding the shaft into the transaxle.
2. If the inner shaft is held in the differential by a circlip, install the clip after the axle is positioned in the transaxle. Replace the differential cover using a new gasket. If the speedometer drive gear was removed from the transaxle, it should be reinstalled at this time. If necessary, refill the transaxle with the proper type and grade of lubricant. Do not reuse fluid that has been drained from the transaxle.

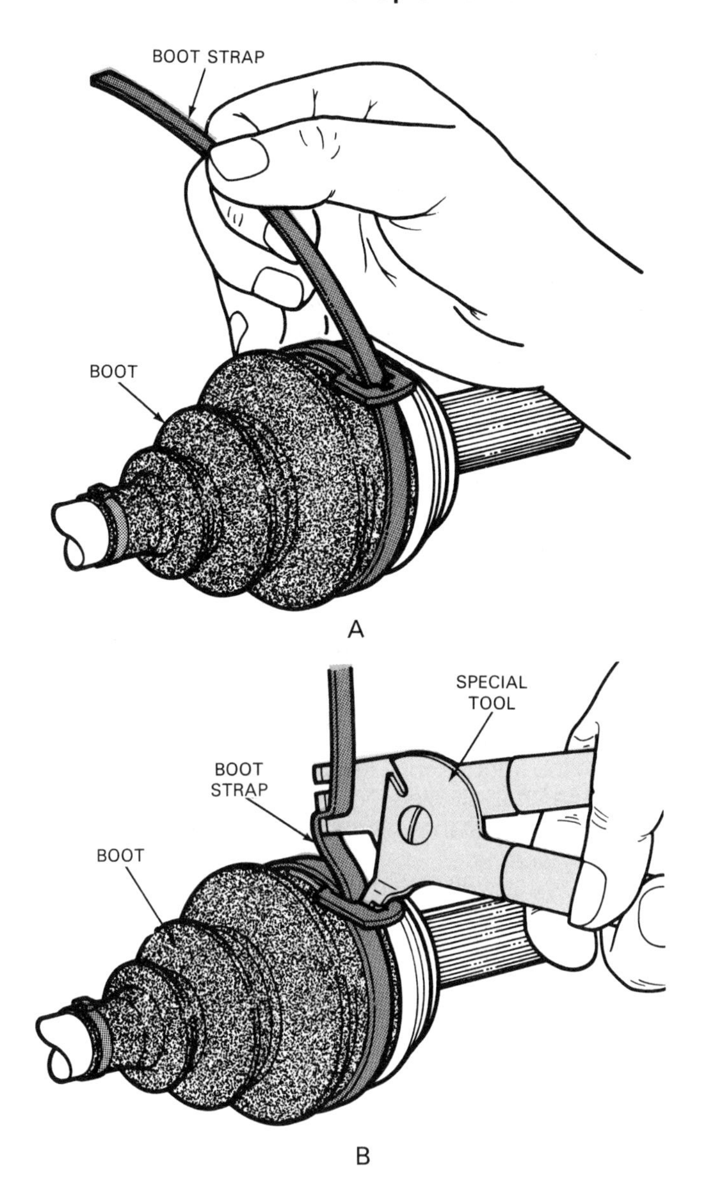

Figure 23-36. Installing a strap-and-buckle boot clamp requires the use of a special tool. A–Install strap on boot. B–Use tool to tighten the strap on the boot. (Chrysler)

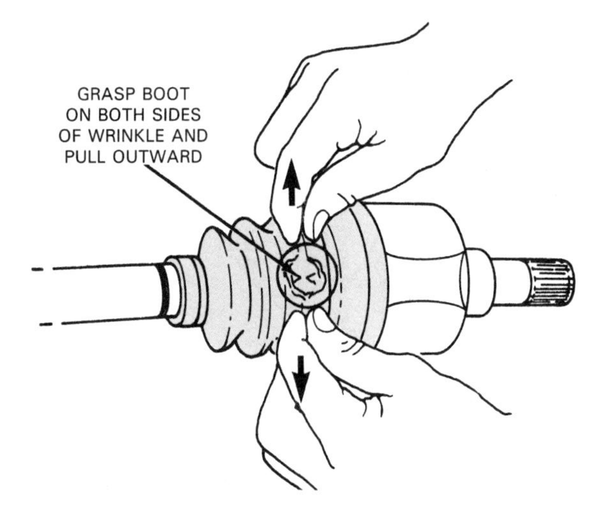

Figure 23-37. Wrinkles and kinks can be removed from the CV joint boot by pulling away from the wrinkle on both sides of the boot. Never install a CV axle with a kinked boot. A kinked boot will cause the axle to quickly wear out. (Ford)

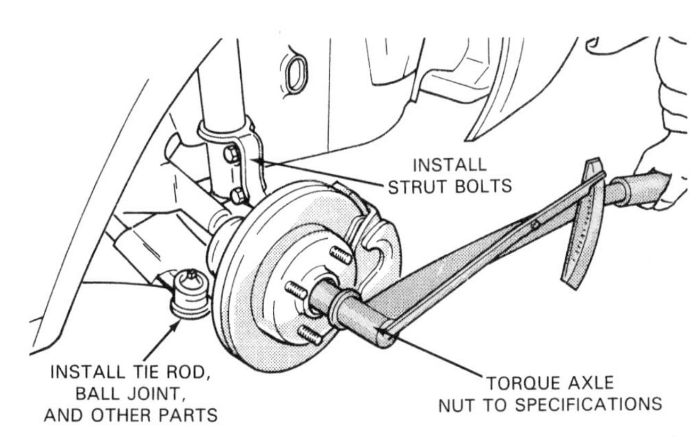

Figure 23-38. After the CV axle is installed, install the steering knuckle attachments and torque the axle nut to the specifications. (Chrysler)

If bolts are used to secure the axle to the transaxle output shaft, align the axle with the output shaft flange and install the bolts that hold the axle to the flange.

3. After securing the axle in the transaxle, install the outer CV axle shaft in the steering knuckle assembly. Then, reinstall the knuckle on the ball joint stud and the MacPherson strut assembly. Finally, reinstall the tie rod end on the knuckle and tighten the attaching nuts or bolts. Be sure to use new cotter pins where needed.
4. Reinstall the front wheel bearing, the adjusting nut, and related washers or other parts. Torque the adjusting nut to the proper specifications. See Figure 23-38. After adjusting the nut to get the proper preload, stake the nut to the stub shaft or install a new cotter pin to keep the nut from backing off. See Figure 23-39.
5. After installation, make sure that the shaft rotates freely and the CV joint boots are not kinked or collapsed. If necessary, rework the boot to remove kinks and wrinkles. In some cases, it may be necessary to loosen a boot clamp to allow air to enter a collapsed joint.
6. After the CV axle is installed properly, reinstall the wheel on the hub.
7. Lower the vehicle.
8. Reconnect the battery negative cable and conduct a road test.

NOTE! On some vehicles, the front end alignment must be checked if the steering knuckle has been disassembled.

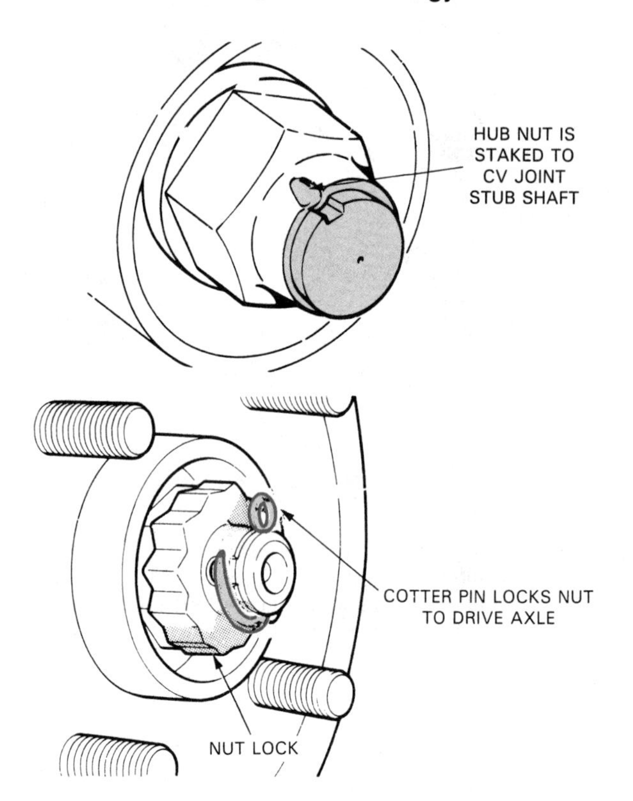

Figure 23-39. Always make sure that the spindle nut cannot work loose. Some nuts are staked to the shaft, but most use a cotter pin and a nut lock. (Ford, Chrysler)

Summary

CV axles are simple devices, but they can develop problems. CV axle problems can be grouped into three areas: noises, vibrations, and leaks. These problems are generally caused by defective CV joints, wheel bearings, or joint boots. Before working on a CV axle, always make sure that the problem is not caused by a malfunction in some other part of the vehicle. It may be necessary to put the vehicle on a lift or to perform a road test to determine the exact cause of a problem.

The CV axle must be removed from the vehicle to make most axle repairs. In many cases, the axle must be removed to repair the transaxle or remove the engine.

To begin axle removal, the vehicle must be raised and the wheel removed from the side being serviced. The front steering knuckle must be disconnected from the suspension components and pulled from the CV axle. The axle can then be removed from the transaxle. In some cases, the axle and knuckle must be removed from the vehicle as a unit.

After removal, be careful not to force the CV joints into extreme angles. Once the axle is on a clean workbench, the axle shafts should be checked for straightness, if necessary, the transaxle seals can be replaced after the axles are removed.

The wheel bearings should be checked and, if necessary, replaced. There are many bearing types and attachment methods. The technician must follow factory procedures when replacing wheel bearings.

Tripod joints are often used on CV axle shafts. To disassemble the joints, remove the boot and the retaining clips. The joint will then come apart easily. The tripod spider assembly should be checked for wear and replaced if needed. After reassembly, check the joint for proper flexibility and rotation. If the tripod joint is badly worn, it may be easier to replace the joint than to rebuild it.

The ball-and-cage joint is commonly used on CV axles. After the boot is removed, remove the retainer clips and disassemble the joint. The balls, raceways, and cage should be carefully checked for wear or corrosion. After the necessary new parts are installed, the ball-and-cage joint can be reassembled.

The CV joint boots are critical to the durability of the CV joints. If a boot is damaged, lubricant will be lost. In addition, dirt and water can enter the joint through a damaged boot. Damaged boots must be replaced promptly.

A CV axle is reinstalled in the reverse order of removal. The axle is installed in the transaxle assembly and reattached to the front steering knuckle and bearing assembly. Make sure that all steering and suspension parts are reattached and that the spindle nut is properly torqued and locked in place.

Know These Terms

CV axle vibration; Boot protector.

Review Questions–Chapter 23

Please do not write in this text. Place your answers on a separate sheet of paper.

1. Explain the three categories of CV axle problems.
2. Summarize four safety precautions that should be followed when working with CV axles.
3. During a test drive, a vehicle's drive train emits a rumbling or roaring noise that is only slightly affected by road speed. Sharp turns have little or no effect on the noise. Technician A says that the CV joints are probably dry and worn. Technician B says that the front wheel bearings are probably dry and worn. Who is correct?
   a. Technician A.
   b. Technician B.
   c. Both Technician A and B.
   d. Neither Technician A nor B.
4. Since there are many design variations, what should you do when in doubt about a CV axle service procedure?
5. The front steering knuckle must always be removed when removing a CV axle. True or false?

6. What can happen if you use engine power to drive the front wheels with the front suspension and drive axles unsupported?
7. Some stub shaft nuts have left-hand threads and must be turned counterclockwise for loosening. True or false?
8. What should you do if you suspect that an axle shaft is bent?
9. In your own words, explain how to service a tripod joint.
10. In your own words, explain the service procedure for a ball-and-cage joint.
11. Why should you install new boots during CV axle service?
12. Explain the procedure for installing CV axles.

Certification-Type Questions–Chapter 23

1. A CV axle is making a clicking noise. One of the CV joint boots is torn. Technician A says that the noise could be caused by a worn CV joint. Technician B says that regreasing the CV joint could eliminate the noise. Who is right?

(A) A only
(B) B only
(C) Both A & B
(D) Neither A nor B

2. All of these abnormal noises are typical of those coming from a dry CV joint EXCEPT:

(A) popping.
(B) clicking.
(C) snapping.
(D) rumbling.

3. CV axle vibrations can be caused by all of these EXCEPT:

(A) loose boot clamps.
(B) worn joints.
(C) a bent axle.
(D) loose axle attaching bolts.

4. When a CV joint boot is squeezed, air is heard escaping. Technician A says that the boot is torn or a boot clamp is defective. Technician B says that the boot vent is working normally. Who is right?

(A) A only
(B) B only
(C) Both A & B
(D) Neither A nor B

5. Technician A says that the CV axles must be removed from the transaxle before removing the transaxle from the vehicle. Technician B says that CV axles must always be removed from the wheel hub and bearing assemblies before removing the transaxle. Who is right?

(A) A only
(B) B only
(C) Both A & B
(D) Neither A nor B

6. A new CV joint boot is being installed on a CV axle. Technician A says that some replacement boots can be installed without removing the CV axle. Technician B says that the boot clamps can be removed with a screwdriver. Who is right?

(A) A only
(B) B only
(C) Both A & B
(D) Neither A nor B

7. All of these statements about servicing pressed-in wheel bearings are true EXCEPT:

(A) the CV axle shaft must be pressed from the wheel hub.
(B) many wheel bearings are pressed into the bearing carrier or the steering knuckle, and the wheel hub is pressed into the bearing.
(C) if tapered roller bearings are used, the bearing races must be replaced when the bearings are replaced.
(D) some bearing assemblies require installation of a collapsible spacer.

8. Technician A says that tripod joints are the most common type of universal joint found on CV axles. Technician B says that tripod joints, when used, almost always serve as the inner CV joint. Who is right?

(A) A only
(B) B only
(C) Both A & B
(D) Neither A nor B

9. When replacing a CV joint boot, which of these should be placed on the CV axle shaft first?

(A) CV joint.
(B) Large boot clamp.
(C) Small boot clamp.
(D) Lubricating grease.

10. All of these could occur if a CV joint boot is improperly installed EXCEPT:

(A) rapid wear of the CV joint.
(B) CV joint noises.
(C) loss of lubricant.
(D) transaxle failure.

11. Technician A says that on some vehicles, the front end alignment must be checked after the wheel bearings are serviced. Technician B says that some stub shaft nuts are held in place by a cotter pin. Who is right?

(A) A only
(B) B only
(C) Both A & B
(D) Neither A nor B

Chapter 24

Four-Wheel Drive Component Construction and Operation

After studying this chapter, you will be able to:

- List the basic parts of a four-wheel drive system.
- Describe the operation of a four-wheel drive system.
- Explain how power flows through a four-wheel drive system.
- Describe four-wheel drive variations.
- Explain the construction and operation of a transfer case.
- Compare transfer case design variations.
- Summarize the construction and operation of locking hubs on four-wheel drive systems.

Four-wheel drive vehicles are much more common today than they were in the past. This chapter presents the basic operating principles and construction details of typical four-wheel drive systems. The information presented here will enable you to understand the service information in the following chapter. Many of the components discussed in this chapter are similar to those studied in earlier chapters. You should have learned about basic gearing, shift mechanisms, and drive shafts in previous chapters.

Four-Wheel Drive Systems

A *four-wheel drive system* is a drive train system installed on some vehicles to drive *all four wheels* instead of just two. Four-wheel drive systems provide for better traction on slippery road surfaces than two-wheel drive systems. Not only is this helpful for off-road operation, but it increases vehicle control and safety when driving on icy or wet pavement.

In addition to the conventional drive train elements, four-wheel drive systems contain several special components. Unfortunately, the extra components add to the weight and complexity of the drive train. Also, the drag created by the additional components reduces gas mileage.

There are two major types of four-wheel drive systems. In the original four-wheel drive system, the vehicle operator directly controls the drive train parts that send power to two or four wheels. This system, which is called a ***part-time four-wheel drive system,*** is still used on many vehicles, including trucks and off-road vehicles. The newer four-wheel drive systems used on many late-model passenger cars and light trucks drive all four wheels at *all times.* These systems are called ***full-time four-wheel drive systems*** or *all-wheel drive systems.* Special drive train components are used in these systems to compensate for differences in speed and traction between the front and rear wheels. The driver has no control over the operation of full-time four-wheel drive systems.

Four-Wheel Drive Components and Designs

The drive train components of a typical four-wheel drive vehicle are shown in Figure 24-1. The major components of a four-wheel drive system include the following:

- Clutch.
- Transmission or transaxle.
- Transfer case.
- Front drive shaft assembly.
- Front differential.
- Front drive axles.
- Rear drive shaft assembly.
- Rear differential.
- Rear drive axles.

As mentioned in Chapter 6, the *clutch* is used to engage and disengage the engine and the manual transmission or transaxle. The clutch in a four-wheel drive vehicle operates in the same manner as the clutch in a two-wheel drive vehicle.

The *transmission* or *transaxle* provides different gear ratios for varying vehicle and road conditions. Either a

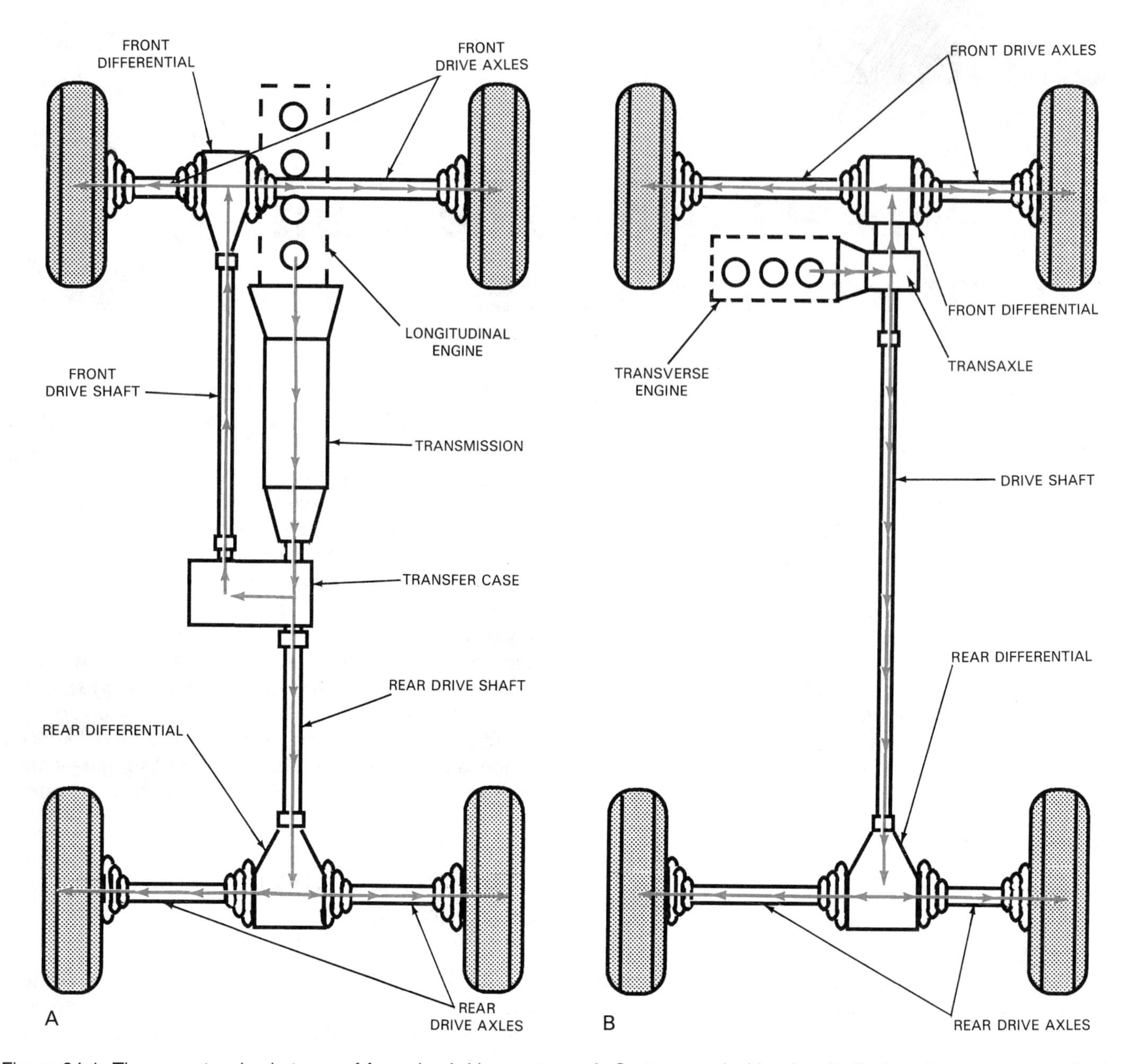

Figure 24-1. There are two basic types of four-wheel drive systems. A–System used with a longitudinal engine and a rear-wheel transmission. B–System used with a transverse engine and a front-wheel transaxle. Note that this system does not contain a separate transfer case.

manual or an automatic model can be used with four-wheel drive vehicles. The transmission or transaxle used in four-wheel drive systems operates in the same manner as those used in two-wheel drive systems.

The *transfer case* is the mechanical unit that splits power between the front and rear wheels (axles). It is attached to the rear of the transmission or transaxle and receives power from the transmission or transaxle output shaft. Power leaves the transfer case through two output shafts. In addition to various drive gears, transfer cases may contain drive chains, a differential assembly, or a *viscous coupling.* These items will be covered in detail later in this chapter.

The ***front drive shaft assembly*** extends from the front of the transfer case to the front differential. It transfers power from the transfer case to the front differential. The front drive shaft assembly has a slip yoke and universal joints. Four-wheel drive vehicles with a front-wheel drive transaxle may not have a front drive shaft assembly.

The ***front differential*** is a unit designed to transfer power from the front drive shaft assembly to the front wheels. It resembles the rear axle assembly used in a rear-wheel drive vehicle and operates in the same manner. The front differential may be a solid (one-piece) unit, or it may be *independently sprung.* On some vehicles, the differential is an integral part of the transaxle.

The ***front drive axles*** are connected to the front differential assembly and transfer power from the differential to the front wheels. The front drive axles are straight steel shafts when used with a *solid differential assembly.* When used with an independently sprung differential, the axles are equipped with cross-and-roller U-joints or CV joints for flexibility. The axle shafts used as independent front axles are sometimes called ***halfshafts.***

The ***rear drive shaft assembly*** extends from the transfer case to the rear axle assembly. It is identical in design and operation to the driveline used in two-wheel drive vehicles. Most drive shaft assemblies have at least two universal joints to provide flexibility. They also have a slip yoke to allow for changes in shaft length.

The ***rear differential*** is identical to that of a rear axle assembly used on a two-wheel drive vehicle. It contains the ring and pinion and the differential case gears. The rear differential used on trucks is usually the solid-axle type. Rear differentials used on cars may be solid-axle or, more often, independent designs.

The ***rear drive axles*** are connected to the rear differential. The axles may be straight shafts, or they may be equipped with cross-and-roller U-joints or CV joints.

Four-Wheel Drive Designs

The design of the four-wheel drive system used with a longitudinal (conventional) engine is shown in Figure 24-1A. The engine, transmission, drive shaft assembly, and rear axle assembly are all placed as they would be in a two-wheel drive vehicle. However, a transfer case is installed between the transmission and the rear drive shaft. A front drive shaft assembly, with U-joints and a slip yoke, connects the transfer case to the front axle assembly.

The front axle assembly contains a differential unit that is similar to the one used in the rear axle assembly. The front axle shafts may be solid or independent.

Figure 24-1B shows the arrangement of drive train parts on a vehicle with a transverse engine. The engine and transmission are mounted in the same way as they would be in a two-wheel drive vehicle. The transfer case is mounted on the transaxle. In some cases, the transfer case is an integral part of the transaxle. The front and rear outputs from the transfer case go to the front and rear axles. A long drive shaft connects the transfer case to the rear axle. The front differential unit and the transfer case are often installed into a single split housing. This arrangement eliminates the need for a front drive shaft.

In some four-wheel drive systems, some method of selecting two- or four-wheel drive must be provided. The standard part-time four-wheel drive system is controlled by the driver, either by turning special front-wheel hub locks or by moving a lever in the passenger compartment. This type of control system is usually found on vehicles designed for off-road use.

As previously mentioned, vehicles designed primarily for on-road use have a more sophisticated four-wheel drive system that constantly drives all four wheels. Figure 24-2 illustrates the difference between full-time and part-time four-wheel drive systems.

A full-time four-wheel drive system that is equipped with a *central differential,* to compensate for differences in speed between the front and rear axles, is shown in Figure 24-3. If one drive shaft is turning faster than the other, the differential action allows the vehicle to operate normally. The operation of differential assemblies was discussed in Chapter 16.

To prevent wheel spin at the axle with the least amount of traction, full-time four-wheel drive systems may be equipped with a limited-slip or locking differential, a centrally mounted viscous coupling (instead of a differential unit), or a combination of a coupling and a differential.

Transfer Case Details

The transfer case, like the transmission or transaxle, is a metal housing that contains shafts, gears, bearings, and other parts. The following sections identify the major transfer case components.

Transfer case housing and external components

The ***transfer case housing*** holds and aligns the transfer case shafts, gears, chains, and bearings. The housing also serves as a reservoir for the lubricating oil. Transfer case housings are usually made of cast iron or aluminum and are strengthened with cast ribs or external bracing rods. Manufacturers attempt to make the housing as compact as possible. This reduces stresses and allows the chassis of the four-wheel drive vehicle to be almost as close to the ground as the chassis of a two-wheel drive vehicle.

The transfer case housing used with a longitudinal engine is usually a two-piece design, so that it can be taken apart for service. A typical housing is shown in Figure 24-4A. One end of the housing mounts to the rear of the transmission. Gaskets at the case halves and between the case and transmission prevent oil leaks. The front and rear output yokes have seals to prevent oil loss.

Figure 24-4B is a cutaway view of a typical transmission and transfer case used with a longitudinal engine. When a shift lever is used, it extends into the transfer case to operate the shift forks. Other components, such as speedometer gears or electrical switches, may be installed in or mounted on the transfer case.

One particular transfer case that is used with a transaxle is illustrated in Figure 24-5A. It is actually a small

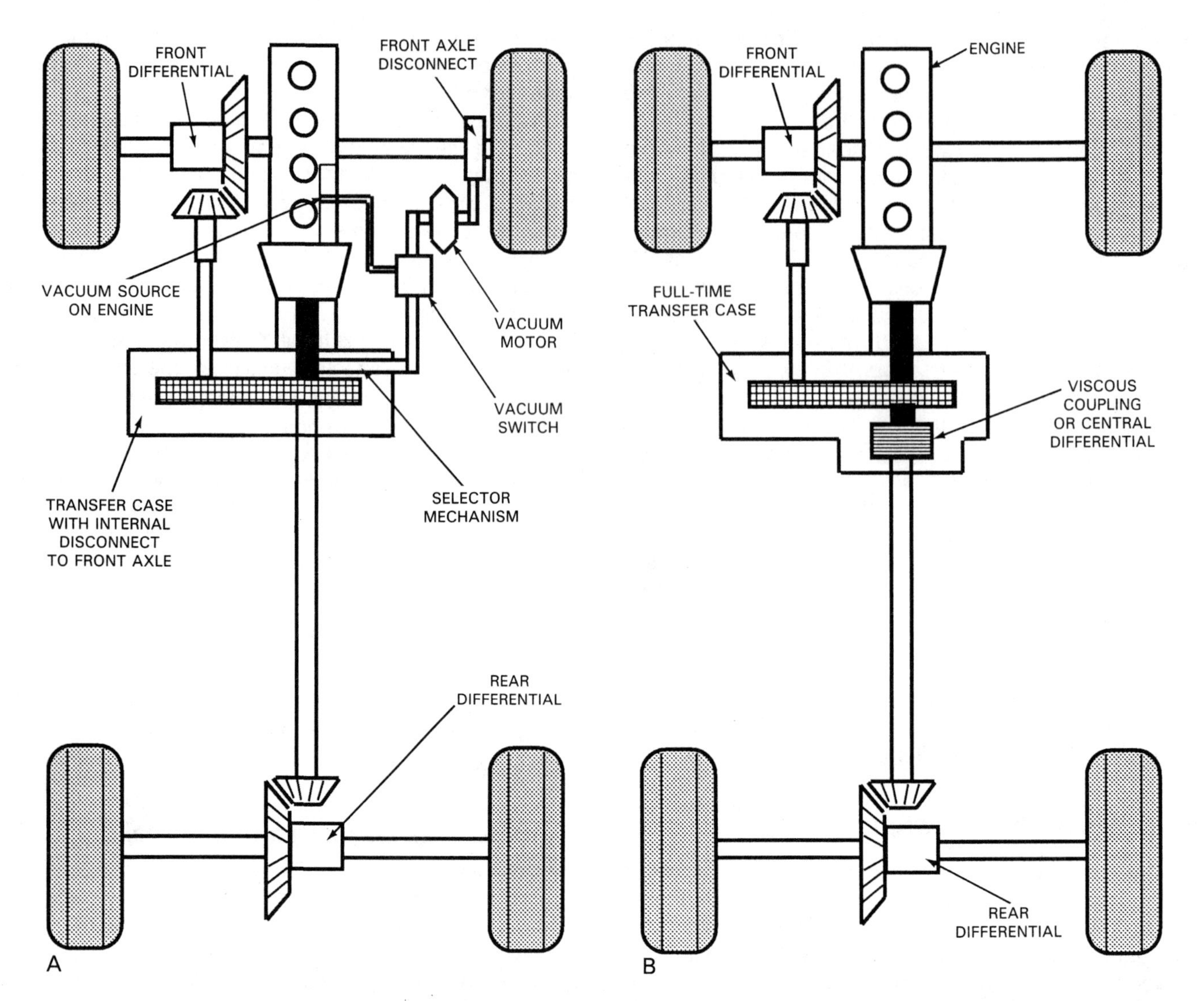

Figure 24-2. Four-wheel drive vehicles are available with either part-time or full-time systems. A–The part-time system has a selector mechanism to engage the front wheels and to provide high and low gear ratios as needed. The mechanism shown here is vacuum operated. B–The full-time system is not controlled by the driver. All wheels are driven whenever the vehicle is moving.

differential unit with bevel gears that turn the power in a 90° angle. Figure 24-5B shows this type of transfer case installed on a transaxle. This transfer case drives the rear wheels only; the front wheels are driven by the transaxle. Other transfer cases used with transverse engines drive both the front and rear wheels.

Internal transfer case parts

There are many variations of the drive mechanisms used in a basic transfer case. Figure 24-6 shows a complex transfer case, which contains many parts. This transfer case can provide two- or four-wheel drive and reduction gear drive or direct drive. Other transfer cases are simple, with only three spur (straight-cut) gears. All-wheel or full-time four-wheel drive systems usually have a single-speed transfer case. The major internal components of a typical transfer case include the mainshaft, oil pump, gears, and drive chain (if used).

The ***transfer case mainshaft,*** shown in Figure 24-7, receives power from the transmission or transaxle and transfers it to the other transfer case components. The ***transfer case oil pump,*** shown in Figure 24-8, is used to provide lubricating oil to various transfer case components. It is not used to produce hydraulic pressure, since the transfer case components operate without hydraulic pressure.

Many transfer cases do not have oil pumps and depend on oil splash for lubrication. To reduce lubrication demands, most transfer case bearings are ball or roller bearings. Tapered roller bearings and thrust bearings (Torrington bearings) are also used.

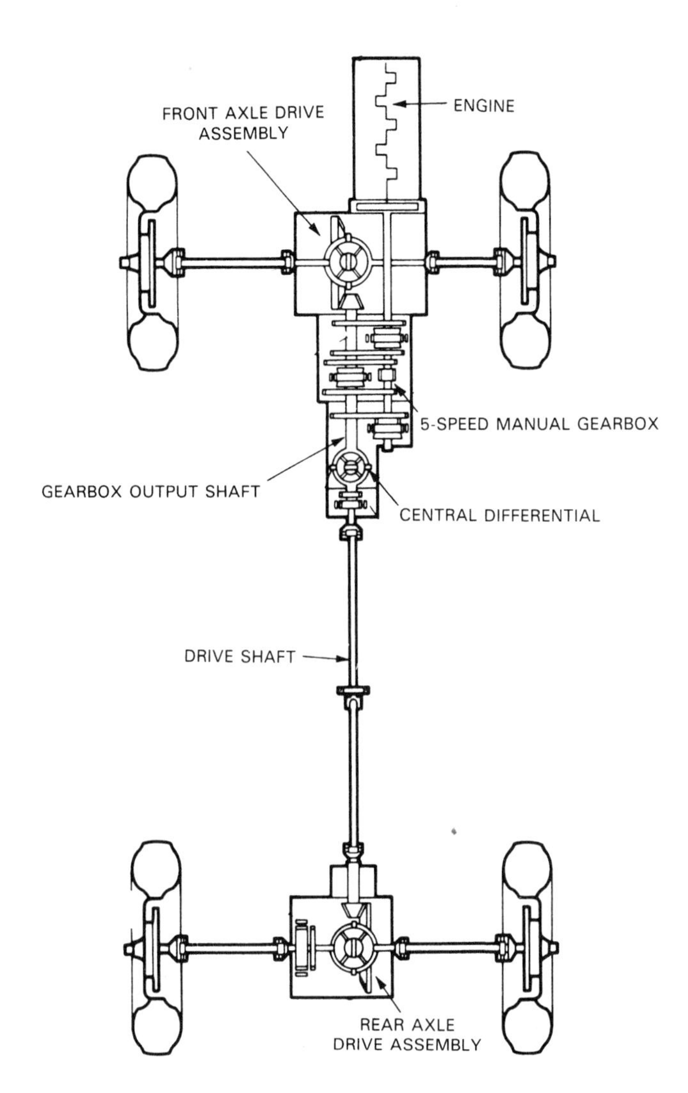

Figure 24-3. The full-time four-wheel drive system constantly drives all four wheels. The central differential unit compensates for different speeds of the front and rear wheels. (Audi)

Many types of transfer case drive mechanisms have been used over the years. Most modern transfer cases use a chain and sprockets in combination with either spur gears, helical gears, or planetary gearsets.

A typical helical gear is shown in Figure 24-9. Helical gears must be engaged and disengaged to select different gear ratios. Generally, the gear itself is not moved in and out of engagement, but is controlled by the movement of a *sliding clutch assembly,* which is similar in operation to a synchronizer assembly found in a manual transmission. Some transfer case sliding clutch assemblies have blocking rings, to eliminate gear clash.

A transfer case containing a planetary gearset is shown in Figure 24-10. The advantage of using planetary gears in the transfer case is that the gears provide a compact way to transfer power. Unlike the planetary gears used in an automatic transmission, the planetaries used in the transfer case are mechanically selected by moving the ***sliding clutch sleeve,*** or ***shift collar,*** as it is sometimes called, in and out of engagement. For more information on planetary gear basic principles, see Chapter 3.

Figure 24-11 shows a transfer case that uses gears to transfer power. The central idler shaft is used to ensure that both output shafts rotate in the same direction. In Figure 24-11A, the transfer case is in the two-wheel drive position. Power enters the transfer case from the transmission and turns the rear drive shaft assembly. The central idler shaft is also turning, but the top sliding clutch sleeve is not engaged. In Figure 24-11B, the top clutch sleeve has been engaged, and power is delivered to both the front and rear drive shaft assemblies. This transfer case can produce reduction gear drive by engaging the clutch sleeves with the idler shaft gear on the right side of the case. It can also provide a neutral by placing both sleeves in the center position. Transfer case power flow will be detailed later in this chapter.

Figure 24-12 illustrates a simple chain drive. The advantages of a chain drive are compactness and the elimination of the need for an idler gear to maintain proper output shaft rotation. The chain drive assembly consists of a chain and two sprockets. The sprocket that delivers power to the chain is called the *drive sprocket.* The other sprocket, which receives power from the chain, is called the *driven sprocket.* See Chapter 3 for additional information on chain drives.

Most drive chains used in transfer cases are multiple leaf chains, which are commonly called *silent chains.* Increasing the width of the chain increases its load-carrying capacity. See Figure 24-13.

The front and rear output shafts are connected to the transfer case gearing or to the driven sprocket, depending on the design of the transfer case. The output shaft to the rear of the vehicle is often an extension of the transmission output shaft. The output shafts are attached to the drive yokes. The shafts and yokes have matching splines. A large nut is used to secure the yoke to the output shaft.

Central differential and viscous coupling

The central differential and the viscous coupling are designed to compensate for differences in front and rear wheel speeds. They are almost always used on full-time four-wheel drive vehicles.

The ***central differential,*** Figure 24-14A, compensates for changes in speed as the spider gears walk around the side gears. Because the central differential can spin and drive the axle with the *least amount of traction,* the differential gears must be locked in some way when necessary. A few manufacturers use a driver-controlled lockout, which is often vacuum operated. However, this type of system is seldom used on cars or light-duty trucks.

To eliminate slippage, some manufacturers use a locking or limited-slip central differential. Figure 24-14B shows

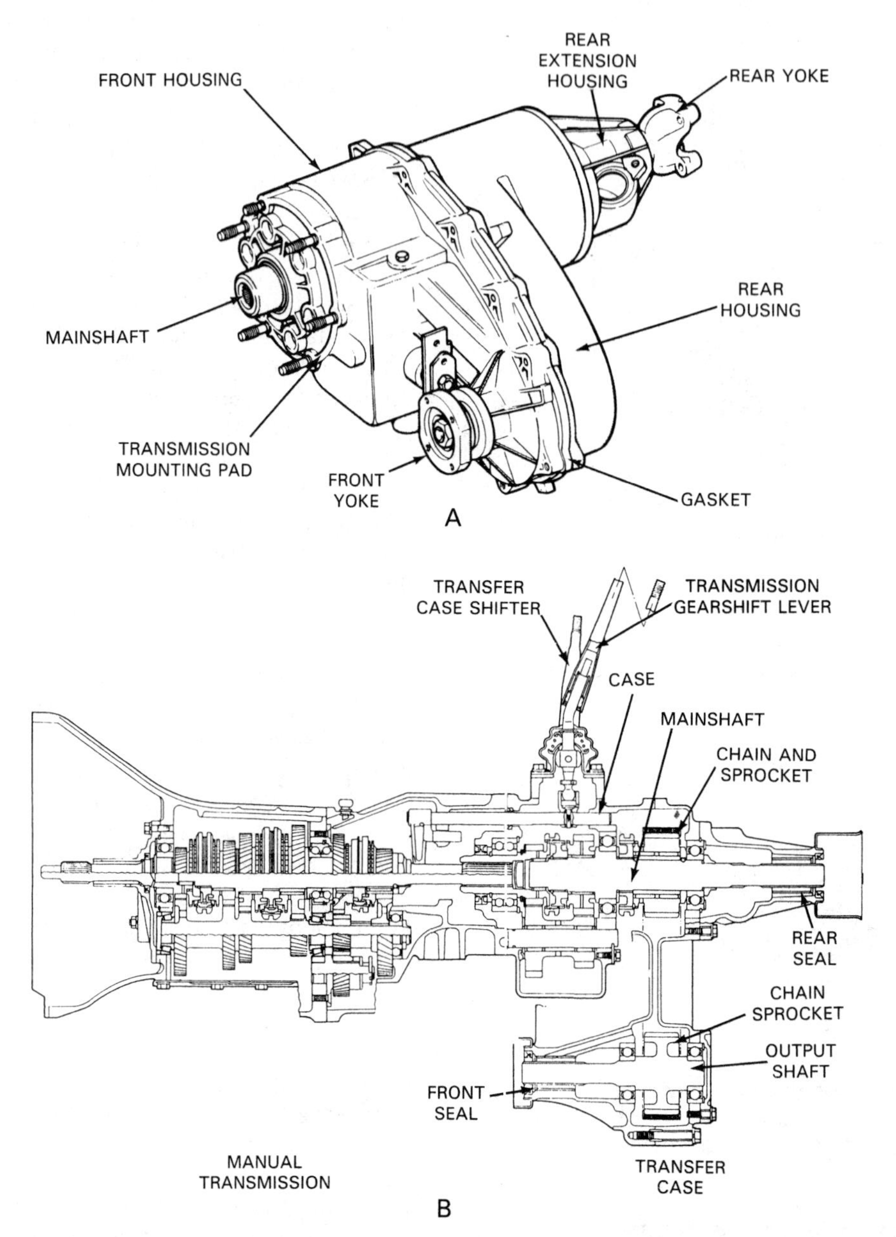

Figure 24-4. The transfer case is the "heart" of the four-wheel drive system. It divides engine power between the front and rear wheels. A–Transfer case used with longitudinal engine and rear-drive transmission. B–This cutaway shows the arrangement of the transfer case and the transmission. (Chrysler, General Motors)

a Torsen® central differential. The design of this differential's worm-and-wheel gears prevents spinning and sends power to the axle with the most traction. The gears can turn to allow slight differences in axle speeds. This type of differential was discussed in Chapter 16.

The ***viscous coupling*** is a series of alternately splined plates, which often have small ridges or other indentations. See Figure 24-14A. A few viscous couplings use sets of vanes. Half of the plates or vanes are splined to the input side of the coupling, and half are splined to the output side. The plates are close together, but they do not touch. Therefore, there is no mechanical connection between the input and the output sides of the coupling. The space between the plates is filled with a thick liquid, such as silicone. The liquid is thick enough to allow the input plates to drive the output plates. Seals keep the liquid from leaking out of the coupling and mixing with the transfer case oil.

Since the input and output plates are not locked together, the liquid allows them to slip slightly when there is

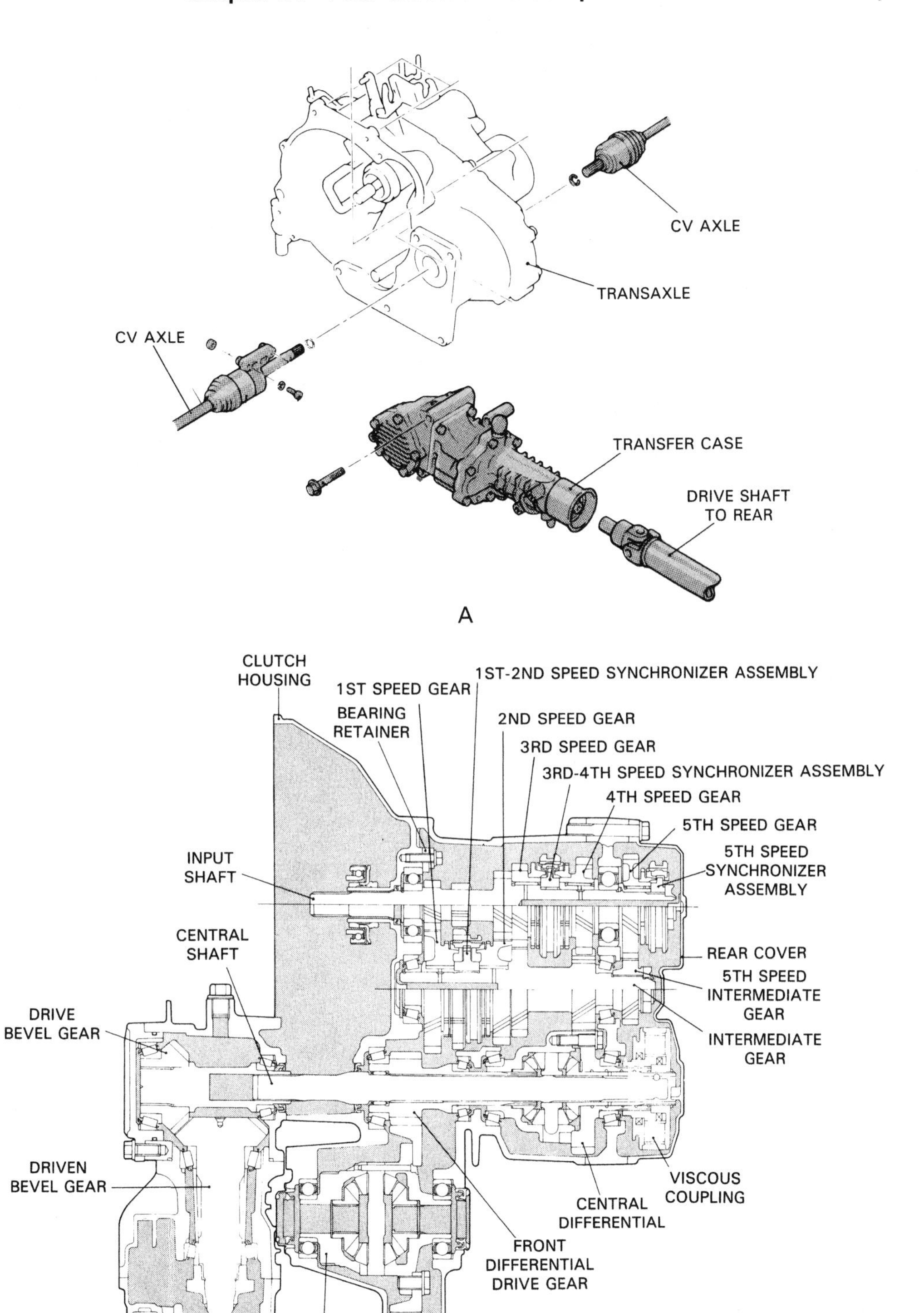

Figure 24-5. A–This transfer case is used with a front-wheel drive transaxle. It connects the transaxle output with the rear wheels. The transaxle differential assembly delivers power to the front wheels through the CV axles. B–Cutaway view of a transfer case used with a front-wheel drive transaxle showing the relative positions of the parts. The bevel gears rotate engine power 90° to turn the rear drive shaft. (Chrysler)

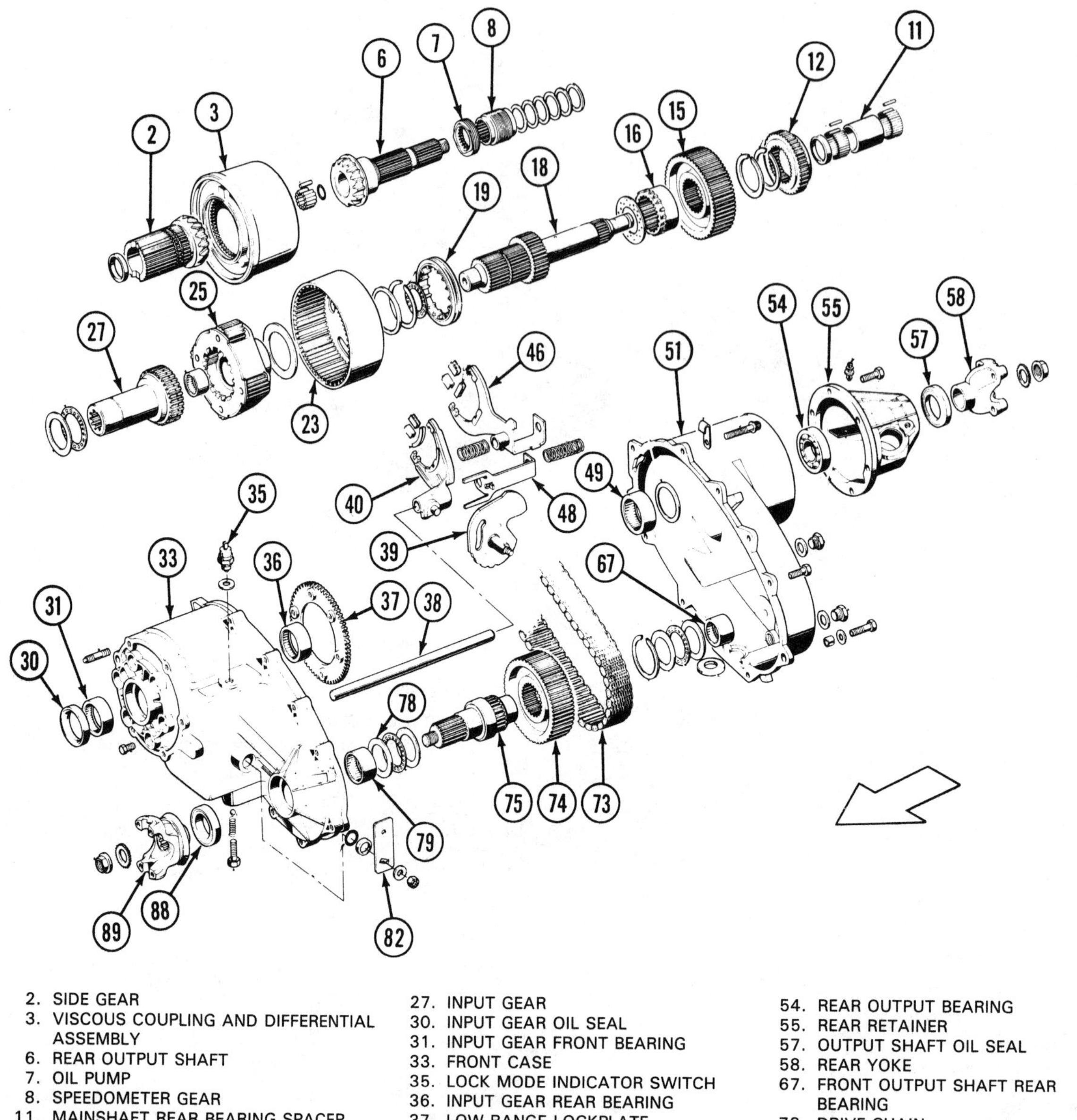

2. SIDE GEAR
3. VISCOUS COUPLING AND DIFFERENTIAL ASSEMBLY
6. REAR OUTPUT SHAFT
7. OIL PUMP
8. SPEEDOMETER GEAR
11. MAINSHAFT REAR BEARING SPACER
12. CLUTCH GEAR
15. DRIVE SPROCKET
16. SIDE GEAR CLUTCH
18. MAINSHAFT
19. CLUTCH SLEEVE
23. ANNULUS GEAR
25. PLANETARY ASSEMBLY
27. INPUT GEAR
30. INPUT GEAR OIL SEAL
31. INPUT GEAR FRONT BEARING
33. FRONT CASE
35. LOCK MODE INDICATOR SWITCH
36. INPUT GEAR REAR BEARING
37. LOW RANGE LOCKPLATE
38. SHIFT RAIL
39. RANGE SECTOR
40. RANGE FORK
46. MODE FORK
48. MODE FORK BRACKET
49. REAR OUTPUT SHAFT BEARING
51. REAR CASE
54. REAR OUTPUT BEARING
55. REAR RETAINER
57. OUTPUT SHAFT OIL SEAL
58. REAR YOKE
67. FRONT OUTPUT SHAFT REAR BEARING
73. DRIVE CHAIN
74. DRIVEN SPROCKET
75. FRONT OUTPUT SHAFT
79. FRONT OUTPUT SHAFT FRONT BEARING
82. OPERATING LEVER
88. FRONT OUTPUT SHAFT SEAL
89. FRONT YOKE

Figure 24-6. The planetary gears and the viscous coupling identify this transfer case as one designed for full-time operation. Compare this illustration with Figure 24-2B. (Chrysler)

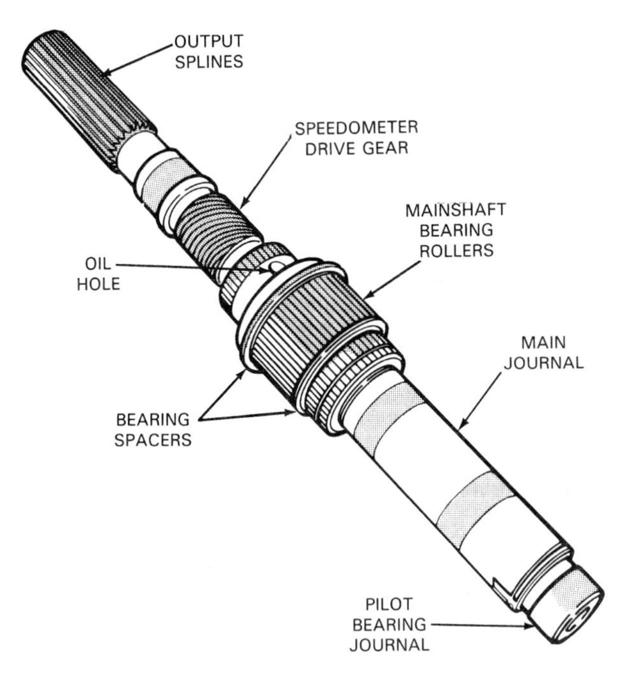

Figure 24-7. The transfer case input shaft resembles the input shaft used in transmissions and transaxles. It contains splines that mesh with the other transfer case components. (Chrysler)

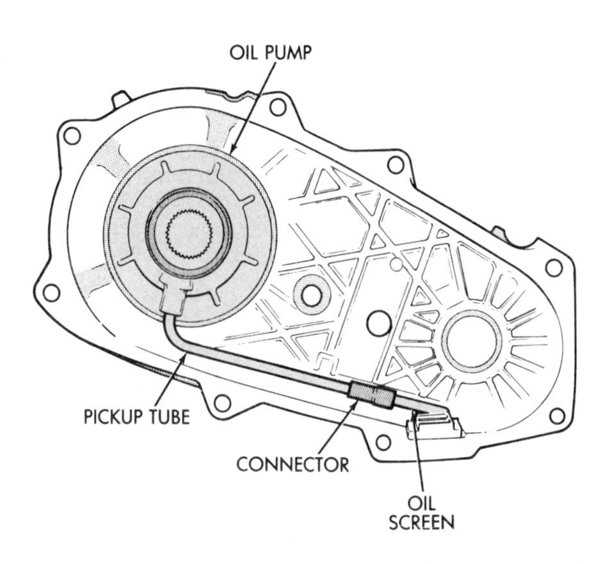

Figure 24-8. Some transfer cases use an oil pump to lubricate moving parts. An oil pump with a sump pickup tube is shown here. The pickup tube allows the pump to draw oil from the bottom of the case. Transfer cases that do not have an oil pump depend on oil splash for lubrication. (Chrysler)

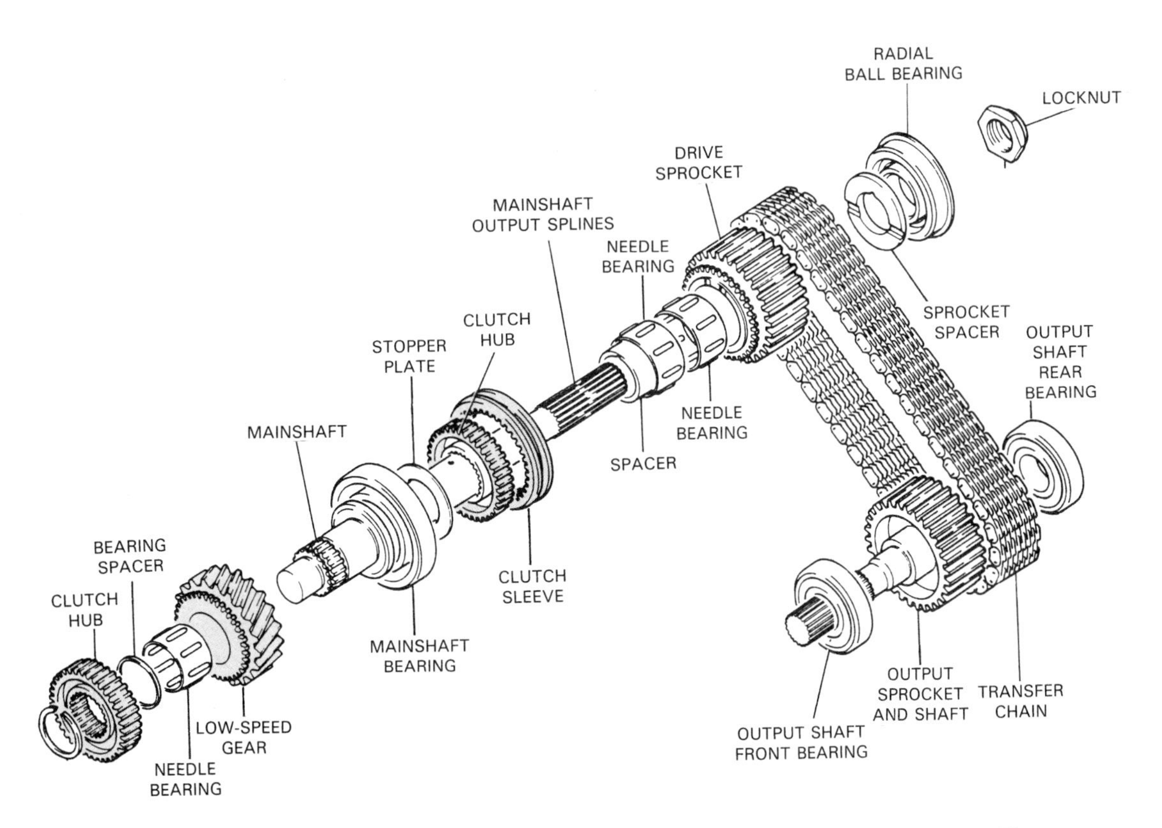

Figure 24-9. The gear-type transfer case usually contains one or more gears for selecting different gear ratios. The gears are controlled by the sliding clutch sleeve.

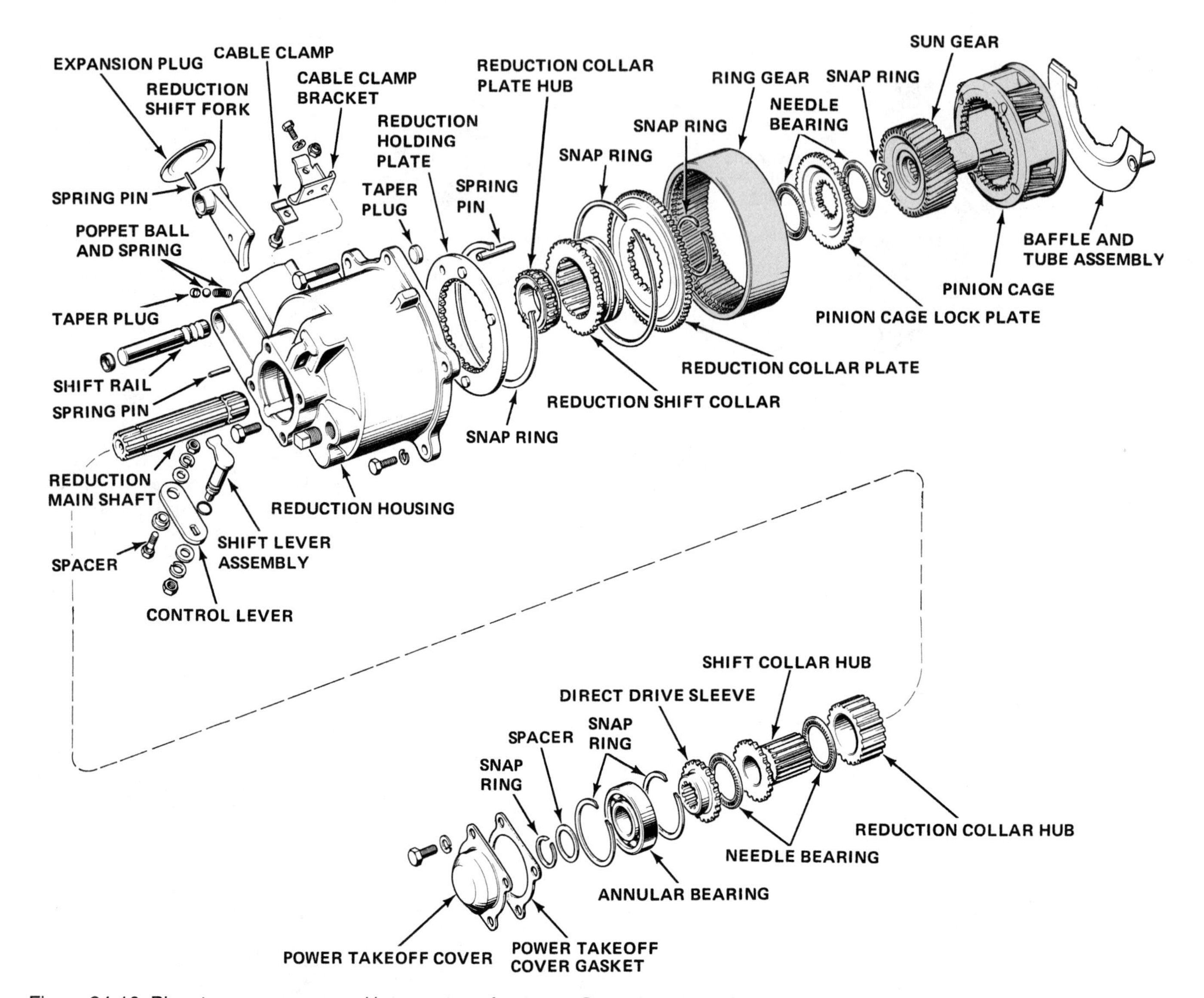

Figure 24-10. Planetary gears are used in many transfer cases. One advantage of planetary gears is that they allow the transfer case to be shifted when the vehicle is moving. Planetary gear operation was discussed in Chapter 3. (Chrysler)

a difference between front and rear axle speeds, such as when the vehicle makes a turn, allowing the vehicle to operate normally. Also, if one axle loses traction and begins spinning, the viscous coupling will continue to send power to the axle with traction.

Figure 24-15 shows a typical viscous coupling attached to the rear output shaft. Although central differentials and viscous couplings are often used together, they can be used separately.

Drive Shafts and Drive Axles

The drive shaft assemblies and drive axles used in four-wheel drive vehicles are identical to those used in two-wheel drive vehicles. Drive shaft assemblies are used to connect the transfer case to the front and rear differential units. Drive axles are used to connect the differential units to the drive wheels.

On some vehicles, the drive shafts are almost equal in length, Figure 24-16A. On other vehicles, the front drive shaft is shorter than the rear drive shaft. In some cases, a front drive shaft is not needed. Figure 24-16B shows a front drive shaft assembly, which includes conventional U-joints to provide flexibility and a slip yoke to compensate for changes in length. The rear drive shaft assembly will also have a slip yoke and U-joints.

Solid axles are usually found on trucks with four-wheel drive systems. These shafts are identical to those used for the drive axles of rear-wheel drive vehicles. Constant-velocity axles (CV axles) are utilized on many four-wheel

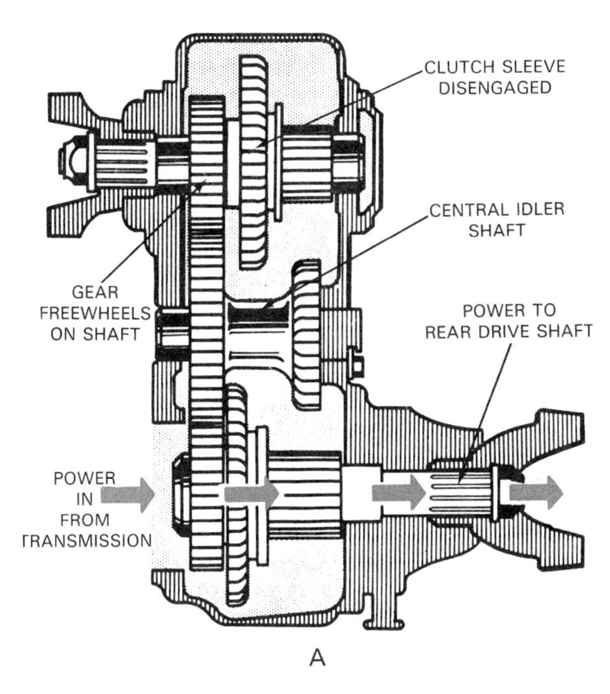

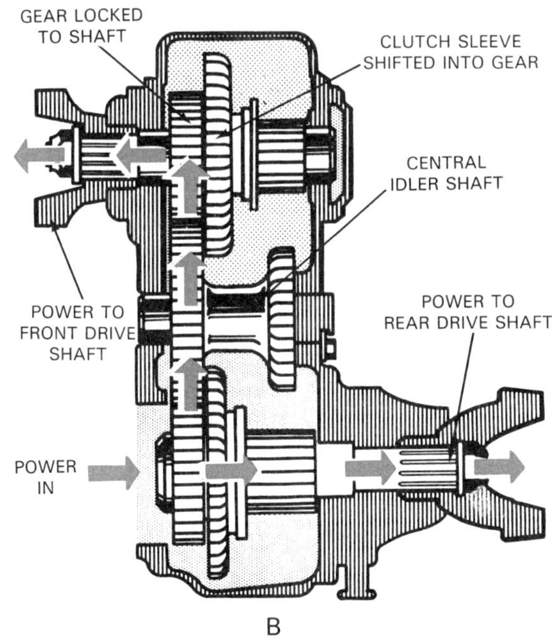

Figure 24-11. The arrows in this figure show transfer case power flow in two- and four-wheel drive positions. A–In two-wheel drive, power flows through the bottom gear of the transfer case to drive only the rear axle. B–In four-wheel drive, power flows through the bottom gear, center idler gear, and top gear to drive both axles. The idler gear ensures that the top and bottom shafts turn in the same direction. (Chrysler)

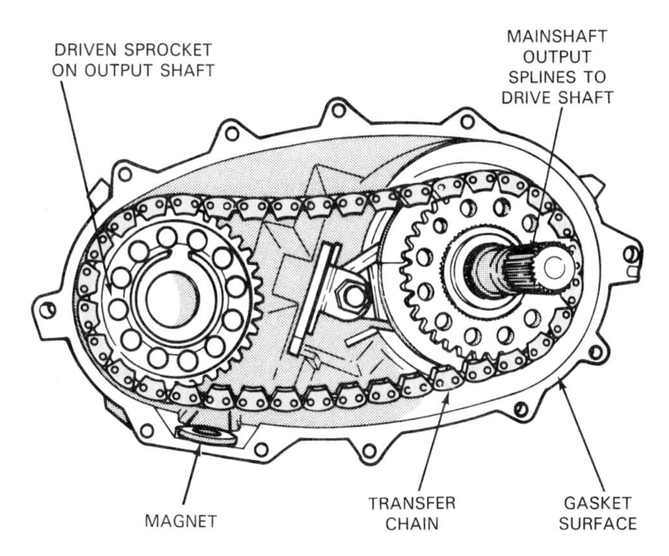

Figure 24-12. A drive chain and sprockets are often used in transfer cases to transfer power. Chain drives are compact and transfer rotation from the drive sprocket to the driven sprocket without reversing it. The drive sprocket is driven by the same shaft that turns the rear drive shaft, and the driven sprocket turns the front drive shaft. (Chrysler)

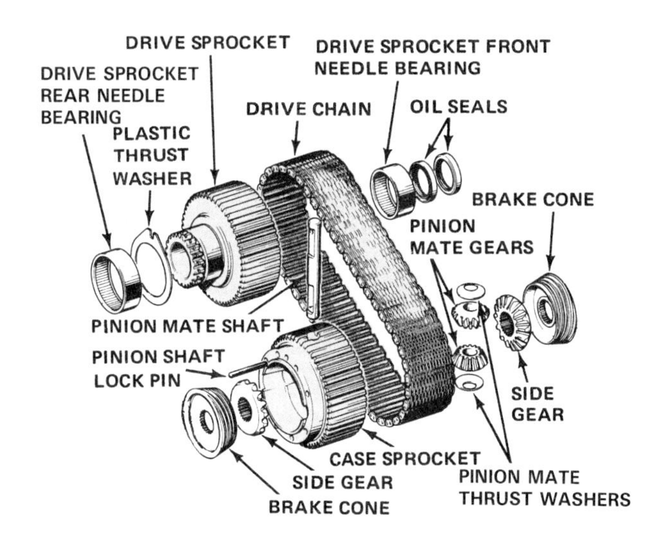

Figure 24-13. Wide, multiple-leaf chains are used to transfer power in many transfer cases. The large chains are able to handle high engine horsepower. (Chrysler)

drive automobiles. These vehicles may have CV axles at the front and a solid axle at the rear, or they may have CV axles at all four wheels. Figure 24-16C shows one type of CV joint used with a front drive axle.

Other Four-Wheel Drive Parts

Various parts may be connected to the transfer case, including the four-wheel drive indicator light switch, shifter, body ground cable, and speedometer connector. See Figure 24-17.

The axle and differential assemblies used in four-wheel drive vehicles resemble the parts used in two-wheel drive vehicles. The ring-and-pinion gears are conventional

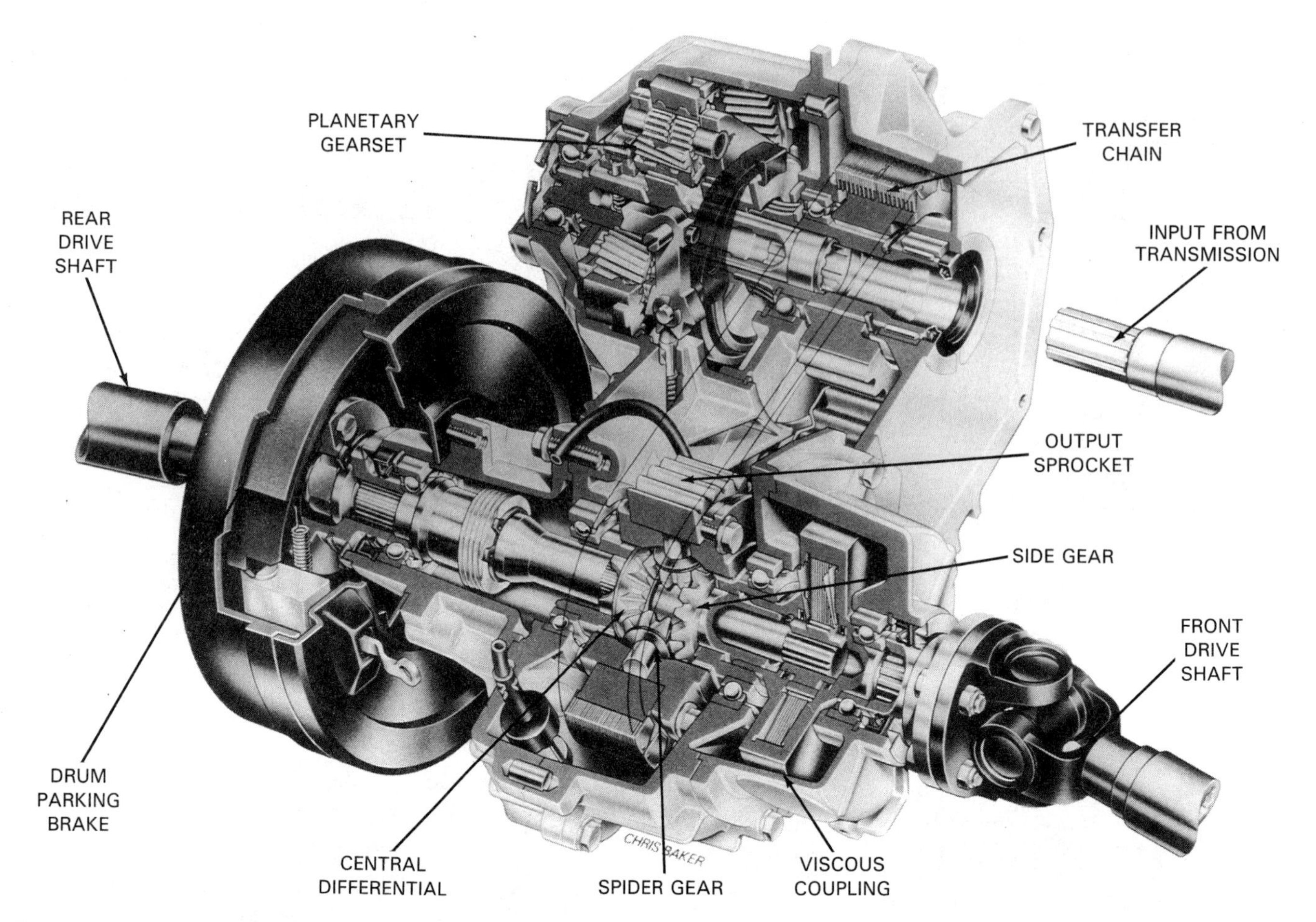

Figure 24-14. The central differential and the viscous coupling provide smooth power transfer at all times. A–The system shown uses a combination of a differential and a viscous coupling. The differential allows for changes in axle speeds, and the coupling prevents wheel slippage.

(continued)

hypoid-type gears. In most cases, the differentials used at the axles are not limited-slip models.

The steering system parts used in a four-wheel drive vehicle are also similar to those used in two-wheel drive vehicles. If the vehicle uses a front transaxle, parts will be identical to those used on the two-wheel drive version of the same vehicle (if available).

On vehicles equipped with a transmission originally designed for rear-wheel operation, the front spindle assemblies will be designed to allow the drive axles to pass through to the front hub. Other steering system parts will be similar to those used on two-wheel drive versions of the vehicle.

Transfer Case Power Flow

Because they contain many of the same parts, the transfer case can be thought of as a compact transmission. However, the transfer case parts are usually arranged for a lateral power flow. The conventional transmission is arranged for linear power flow.

Many vehicles with full-time four-wheel drive employ a single-speed transfer case, which has only one power flow path. Part-time transfer cases used in off-road vehicles usually have at least three drive combinations: two-wheel drive, high-gear position; four-wheel drive, high-gear position; and four-wheel drive, low-gear position. Figure 24-18 illustrates the power flow through a typical transfer case that uses gears to transfer power.

Two-wheel drive, high-gear position

When the shifter is in the two-wheel drive, high-gear position, power flows into the input shaft at the bottom of the gearset. See Figure 24-18A. The bottom sliding clutch sleeve moves to the left and engages with the inner teeth on the input gear, which is solidly attached to the input shaft. Because the sleeve is splined to the rear output shaft, this engagement locks the input shaft to the rear output shaft. Since the clutch sleeve locks the two shafts together, there is no gear reduction, and the transfer case is in high gear. The top sleeve is in the neutral position; the

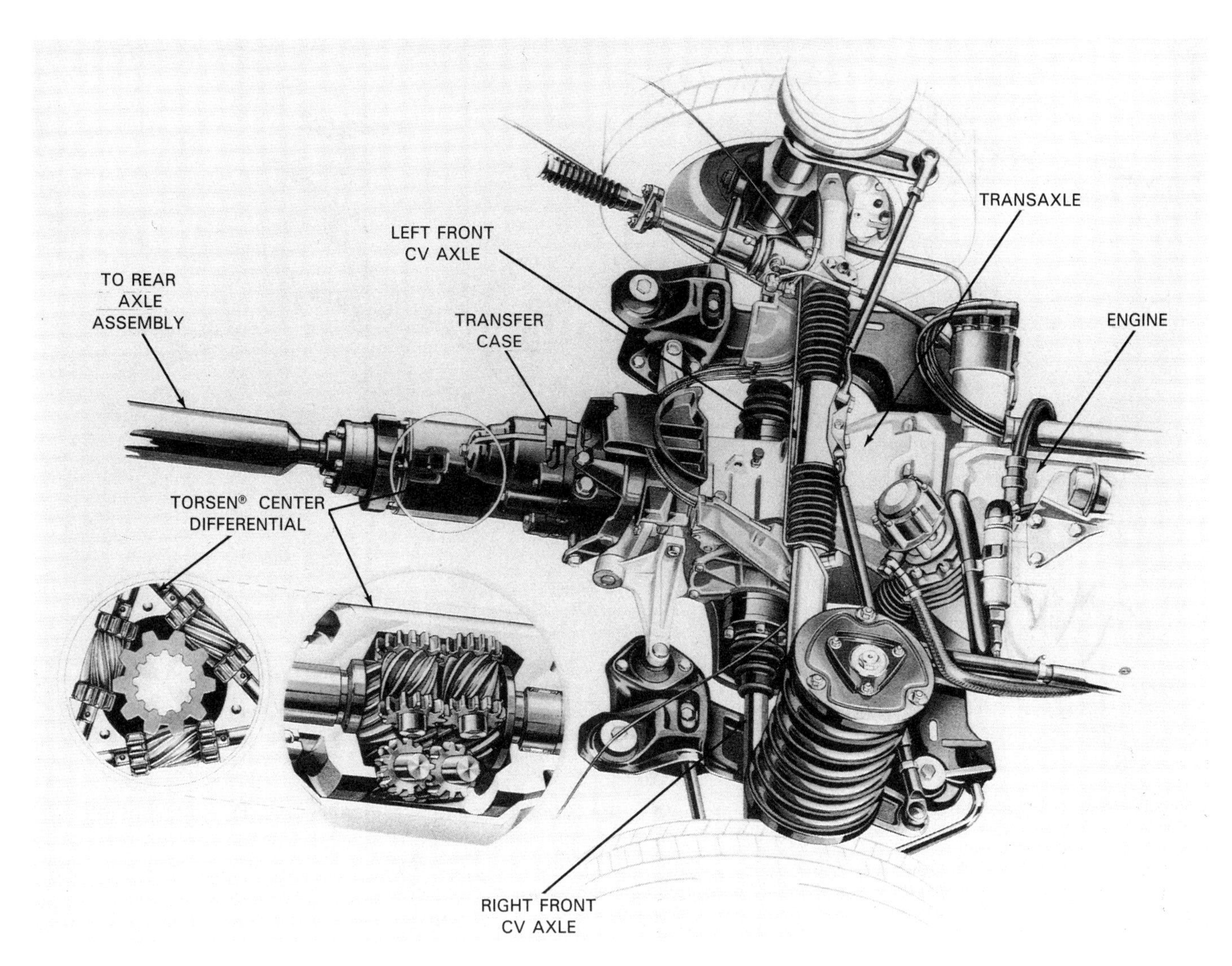

Figure 24-14 *(continued)*. B–The Torsen differential is a locking differential that takes the place of the standard differential and viscous coupling. It prevents slippage through the design of its gears. (Land Rover, Audi)

top gear freewheels, and the front output shaft does not turn. No power is transferred to the front axle.

Four-wheel drive, high-gear position

Figure 24-18B illustrates power flow when the shifter is moved to the four-wheel drive, high-gear position. The bottom sliding clutch sleeve remains in the same position it was in for the two-wheel drive, high-gear position, shown in Figure 24-18A. However, the top sleeve moves to the left and engages with the inner teeth on the top gear. Since the top clutch sleeve is splined to the front output shaft, power now flows from the top gear to the sleeve and front output shaft. Power also flows to the rear output shaft from the input shaft and bottom clutch sleeve. The clutch sleeves lock the input shaft to both output shafts; however, there is no gear reduction. Although the transfer case is in the four-wheel drive position, it remains in high gear.

Four-wheel drive, low-gear position

Figure 24-18C shows the power flow when the shifter is moved to the four-wheel drive, low-gear position. Both sliding clutch sleeves are moved to the right and their external gear teeth engage the right side gear on the idler shaft, which is the middle shaft. The right side idler gear is smaller than the two output shaft gears. Therefore, the output shaft gears turn more slowly than the idler gear, and there is torque multiplication through this gearset. Power flow is from the input gear, to the left side idler gear, to the right side idler gear, and to the two clutch sleeves and their output shafts. All four wheels are driven, and the transfer case is in low gear.

Power flow variations

Systems that provide power through planetary gearsets and drive chains work in a similar manner. How-

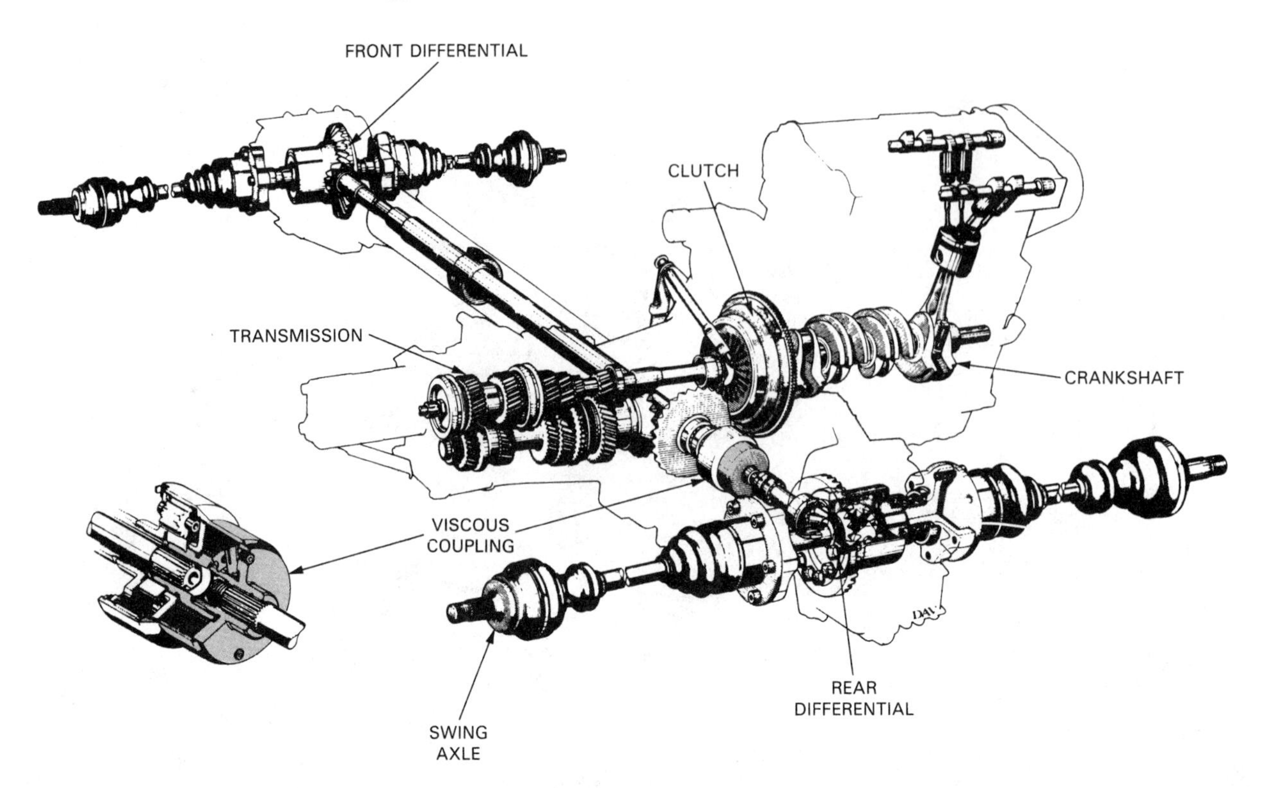

Figure 24-15. This four-wheel drive system used with a transverse mid-engine uses a viscous coupling without a central differential. The fluid in the viscous coupling compensates for differences in axle speeds and prevents excess slippage. (Peugeot)

ever, the clutch sleeves in these systems are used to hold or drive different parts of the planetary gearset. Figure 24-19 illustrates the design of a planetary type transfer case. As a general rule, transfer cases having planetary gears will use a drive chain to transmit power to the front output shaft. Transfer cases that use gears to obtain two- or four-wheel drive and to alter gear ratios use the gears themselves to transmit power.

Transfer Case Shift Mechanisms

Part-time four-wheel drive systems must have some kind of shift mechanism to allow the driver to select various gears and driving modes. Shift mechanisms can be mechanical or vacuum operated. A few late-model vehicles are equipped with transfer cases that are shifted by electric solenoids or motors. The solenoids or motors are controlled by the driver or by an on-board computer. Most all-wheel or full-time four-wheel drive systems do not have a shift linkage.

Many vehicles have simple ***mechanical shift systems*** that resemble the shift mechanisms used on manual transmissions. A shift lever that operates the shift rails of a transfer case is shown in Figure 24-20A. Because the transfer case is often mounted directly under the shift lever, the linkage can be simple and direct.

Figure 24-20B shows typical shift rails and forks used in transfer cases. Note the resemblance to the rails and forks used in manual transmissions. The forks operate the sliding clutch sleeves that select various gear ratios.

Some vehicles are equipped with a ***vacuum shift system,*** which disconnects the front axle from the drive shaft when the vehicle is in the two-wheel drive mode. See Figure 24-21A. Operation of a vacuum diaphragm moves the shift fork and shift collar, connecting or disconnecting the shafts. The vacuum diaphragm can be operated by the movement of the shift lever or by a dash-mounted switch. Sometimes, vacuum diaphragms are mounted on the transfer case to operate the shift collars inside the case. Figure 24-21B is an exploded view of a typical vacuum shift system with a dashboard switch and an indicator light. The engine intake manifold provides the vacuum to operate the diaphragm. A reservoir holds vacuum in case the engine vacuum is low when the diaphragm is activated.

The transfer case shift lever is usually mounted near the transmission gearshift lever, as shown in Figure 24-22. Some shift levers are mounted on the dashboard. Pushbuttons can also be used to control transfer case shifting.

Locking Hubs

Some vehicles use ***locking hubs*** to engage and disengage the front wheels from the front drive axles. Typical front locking hubs are shown in Figure 24-23. Although the operation of locking hubs is less complicated than other

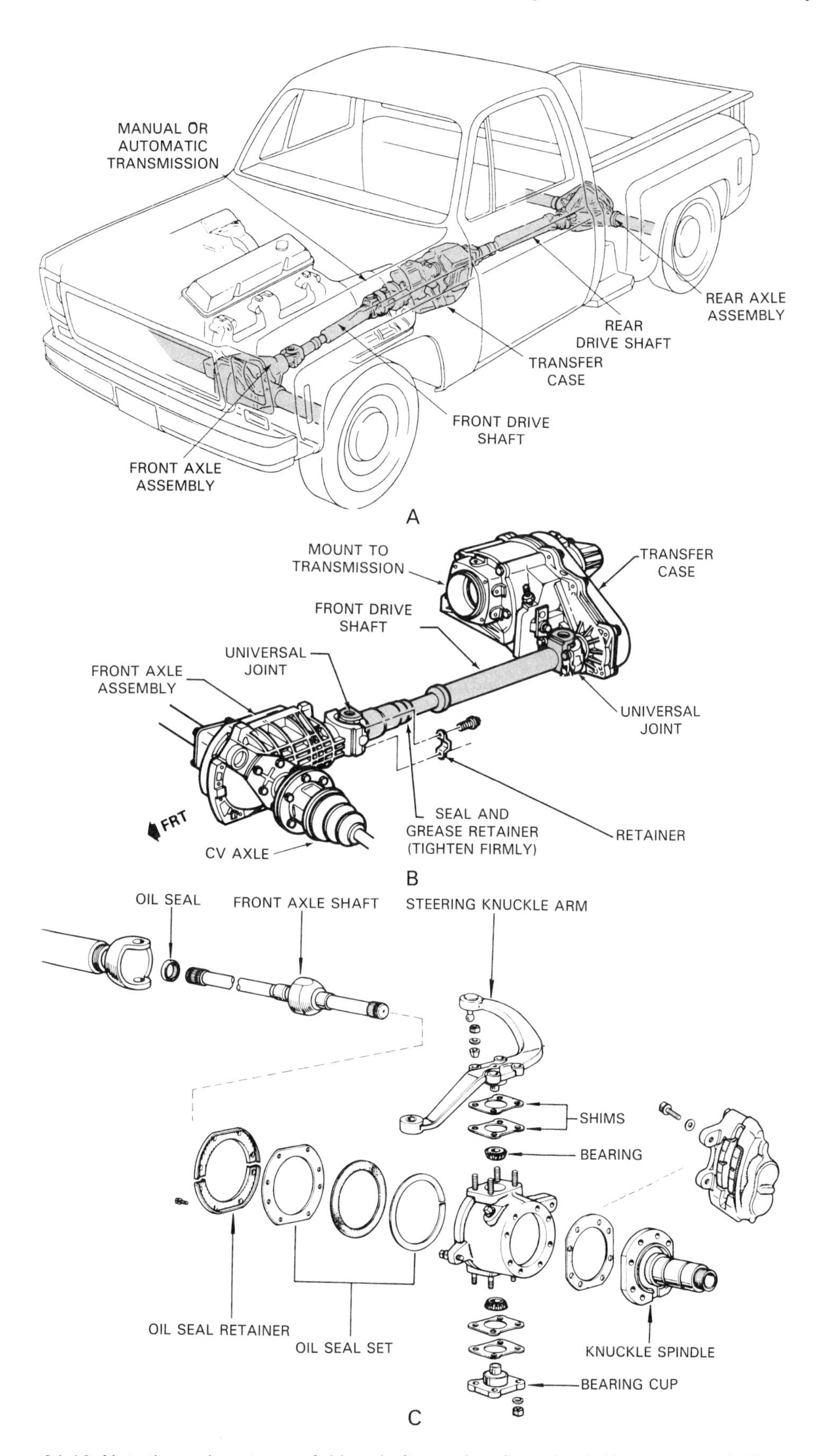

Figure 24-16. Note the various types of drive shafts used on four-wheel drive systems. A–The drive train of a longitudinal engine and transmission system includes two drive shafts with universal joints and slip yokes. B–The front drive shaft used with a longitudinal engine-transmission system is shown. The universal joints are attached to yokes at the transfer case and the front axle assembly. The shaft also has a slip yoke and grease fittings. C–The front axle shafts may be solid or independently sprung. The front shaft is equipped with a CV joint at the steering spindle. This allows the suspension to move as the wheels are turned. (General Motors)

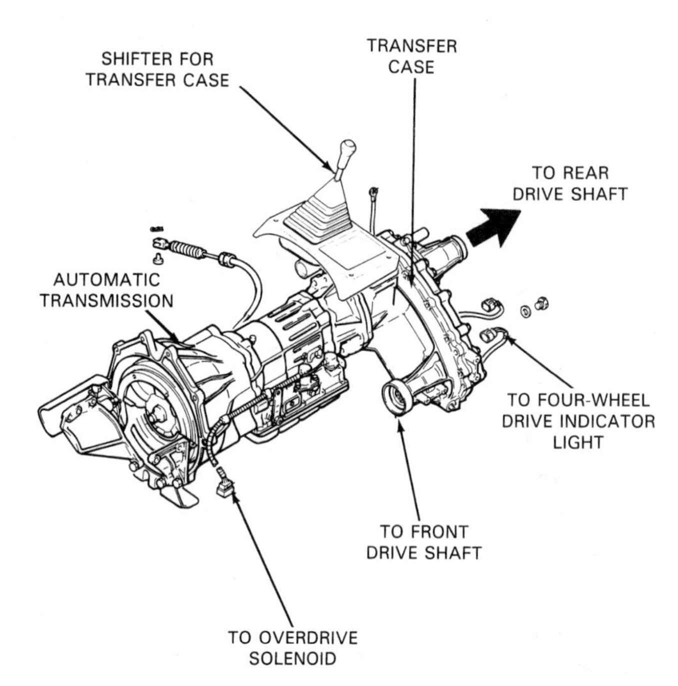

Figure 24-17. Study the parts installed on this transfer case. (Chrysler)

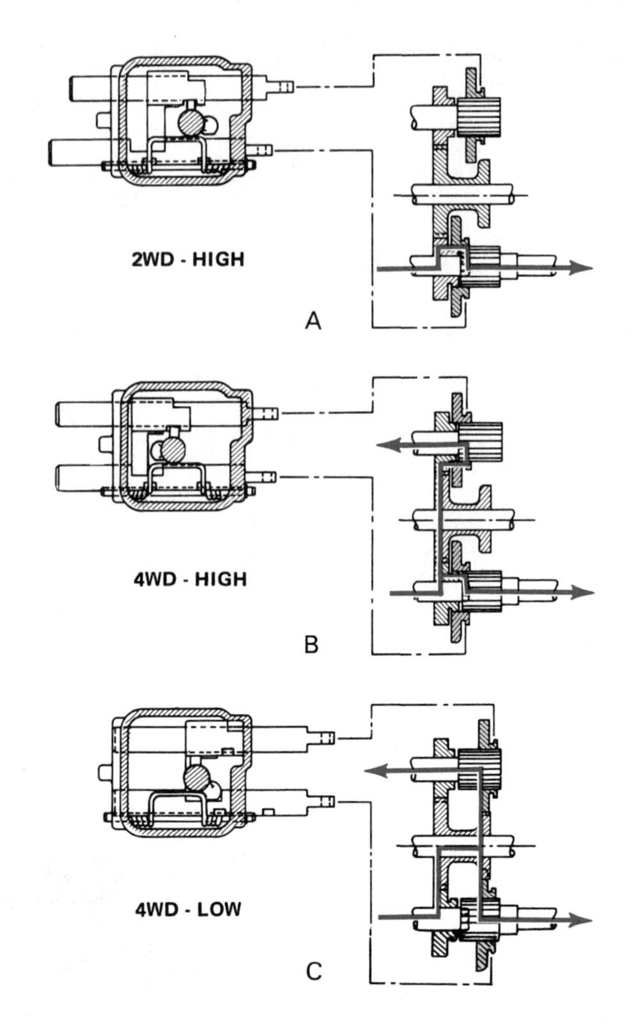

Figure 24-18. Power flow through a gear-type transfer case in various positions. Shift lever control case (left) shows position of shift lever and rails. Position of gears in transfer case is shown to right. A–Two-wheel drive, high-gear position. B–Four-wheel drive, high-gear position. C–Four-wheel drive, low-gear position. (Chrysler)

drive selection systems, they do have some disadvantages. For example, operating locking hubs requires the driver to get out of the vehicle, often in bad weather or in mud. Also, the front-wheel drive components in a locking hub system turn constantly, even when the front wheels are disengaged.

Locking hubs are sometimes used on vehicles that have provisions for disengaging the front drive components. Unlocking the hubs in this type of system prevents the front wheels from driving the axles and the differential. This reduces wear on the front-drive parts. Figure 24-24 shows an exploded view of the front locking hub mechanism. The splines of the hub are attached to the axle shaft and are engaged by turning the recessed handle at the center of the hub. This moves the hub outward for engagement or inward for disengagement.

Some front locking hubs engage automatically. They are designed to release the front drive components whenever the transfer case is in the two-wheel drive position. The action of some of these hubs is similar to that of the one-way, or overrunning, clutch used in automatic transmissions. When the transfer case is in four-wheel drive, the engine is driving the front wheels. The one-way action of the hubs causes them to lock up to deliver power. When the engine slows down or the transfer case is taken out of four-wheel drive, the front wheels begin to drive the axles. This reverses the power flow through the hubs, causing them to unlock. The front drive parts stop moving to reduce wear and increase gas mileage.

Other self-locking hubs, such as the one shown in Figure 24-25, are engaged by moving the shift lever into the four-wheel drive position and disengaged by reversing the vehicle slightly after moving the shift lever into the two-wheel drive position.

Summary

The main purpose of a four-wheel drive system is to increase traction by dividing engine power among all four wheels. There are two major four-wheel drive systems: the part-time system, which permits the operator to choose between two-wheel drive and four-wheel drive, and the full-time system, which drives all of the wheels whenever the vehicle is moving. The part-time system is used on most off-road vehicles. The full-time system is designed primarily for on-road use.

The transfer case is the "heart" of the four-wheel drive system. The transfer case housing supports and protects the moving parts and acts as a reservoir for lubricant. The design of the transfer case housing varies with the size and complexity of its internal parts and the type of transmission to which it is attached.

Power in a transfer case is controlled by the actions of meshing helical gears or planetary gearsets. Many transfer cases use a drive chain and sprockets to transfer power. Transfer case gears are selected by sliding clutch sleeves. Some transfer cases have an oil pump to provide lubricating oil at low pressures. The front and rear drive shafts extend from the transfer case to the axle assemblies.

To compensate for speed differences between the front and rear axles, many four-wheel drive systems are equipped with a central differential or a viscous coupling. The central differential, which operates in the same manner as a rear axle differential, compensates for axle speed variations.

To prevent wheel spin on slippery surfaces, locking or limited-slip differentials are often used. In some cases, viscous couplings are used to prevent slippage by the action of the thick fluid between drive plates and driven plates. These couplings can also compensate for speed variations. Some vehicles have both a differential and a viscous coupling.

The drive shafts used in four-wheel drive systems are identical to those used in two-wheel drive systems. They may be equipped with universal joints, which provide flexibility, and a slip yoke, which compensates for variations in length. Axle shafts may be solid or flexible, depending on the vehicle design and whether they are used on the front or rear axles. Flexible shafts usually have CV joints at their outboard ends.

Power flow through a transfer case varies according to the design of the unit. Helical gear transfer cases will have a central idler gear to ensure that both output shafts are driven in the same direction. Planetary gear transfer cases usually operate in combination with a drive chain.

Most full-time four-wheel drive systems do not have a shift mechanism. Transfer cases that require operator gear changes are shifted by manual linkage, vacuum diaphragms, electric solenoids, or motors. The transfer case shift lever is usually mounted near the transmission gearshift lever.

Some vehicles have locking hubs to engage and disengage the front wheels. Locking hubs are usually used when the transfer case does not have any provision for unlocking the front wheels. Locking hubs are sometimes used in combination with the transfer case to ensure that the front drive parts are not driven by the engine or the road. Although locking hubs are usually manually operated, some hubs lock and unlock automatically.

Know These Terms

Part-time four-wheel drive system; Full-time four-wheel drive system; Front drive shaft assembly; Front differential; Front drive axle; Halfshaft; Rear drive shaft assembly; Rear differential; Rear drive axle; Transfer case housing; Transfer case mainshaft; Transfer case oil pump; Sliding clutch sleeve (Shift collar); Central differential; Viscous coupling; Mechanical shift system; Vacuum shift system; Locking hub.

Review Questions–Chapter 24

Please do not write in this text. Place your answers on a separate sheet of paper.

1. Explain the difference between part-time four-wheel drive systems and all-wheel drive systems.
2. List and explain the basic components of a typical four-wheel drive system.
3. The transfer case housing holds and aligns _____, _____, _____, and other components.
4. Some transfer cases have blocking rings to eliminate gear clash. True or false?
5. What is the purpose of a viscous coupling?
6. The central differential compensates for changes in speed as the _____ gears walk around the _____ gears.
7. How does a viscous coupling work?
8. The axle shafts used on four-wheel drive vehicles differ from those used on two-wheel drive vehicles. True or false?
9. Explain power flow through a transfer case in the two-wheel drive, high-gear position.
10. Sketch the power flow through a typical transfer case in the four-wheel drive, low-gear position.

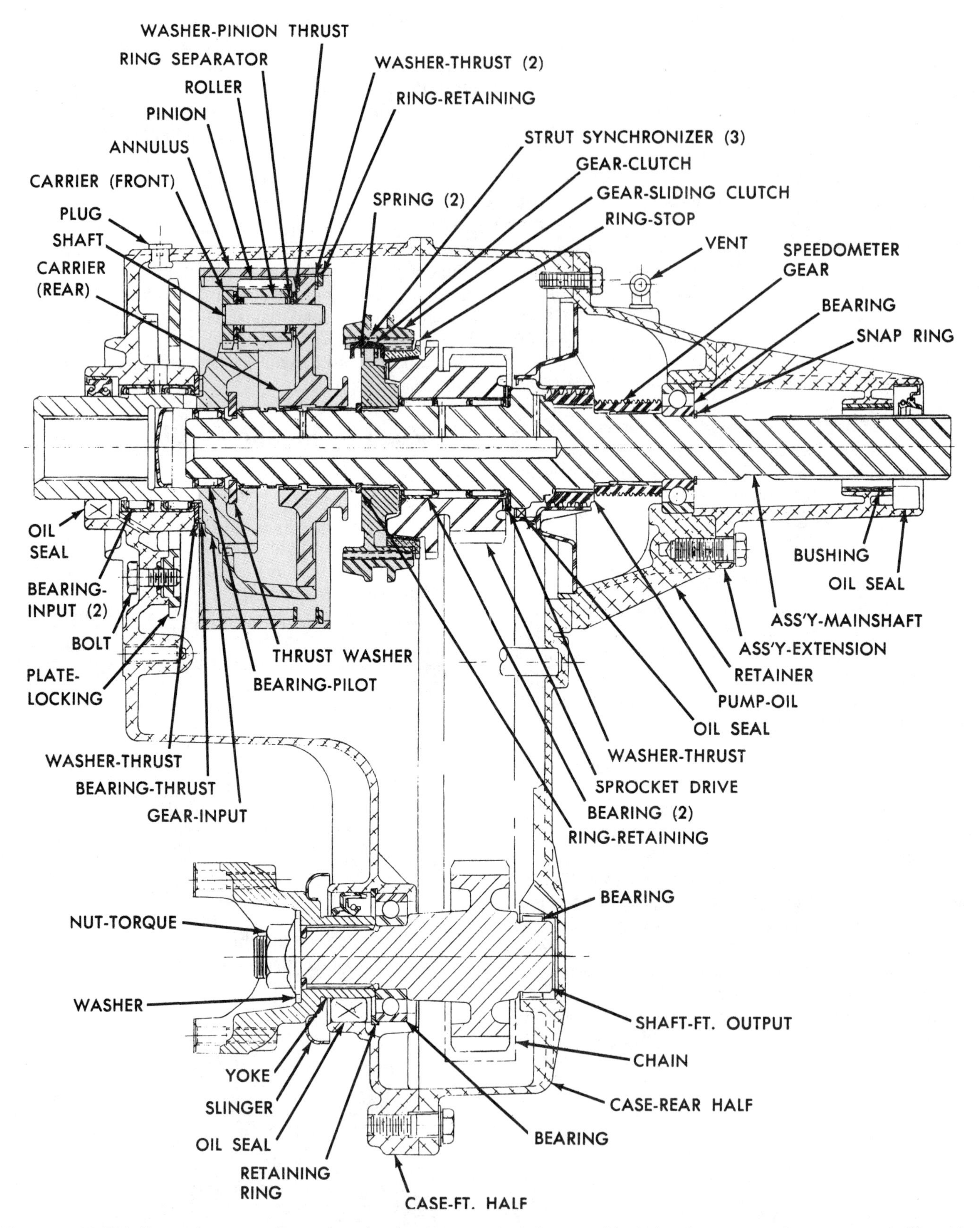

Figure 24-19. This figure shows a cross section of a planetary gear transfer case. Most planetary gear systems are used with a drive chain. The planetary system is operated by moving a sliding clutch sleeve in and out of engagement with different parts of the gearset. This transfer case is relatively simple, with no differential unit or viscous coupling. (General Motors)

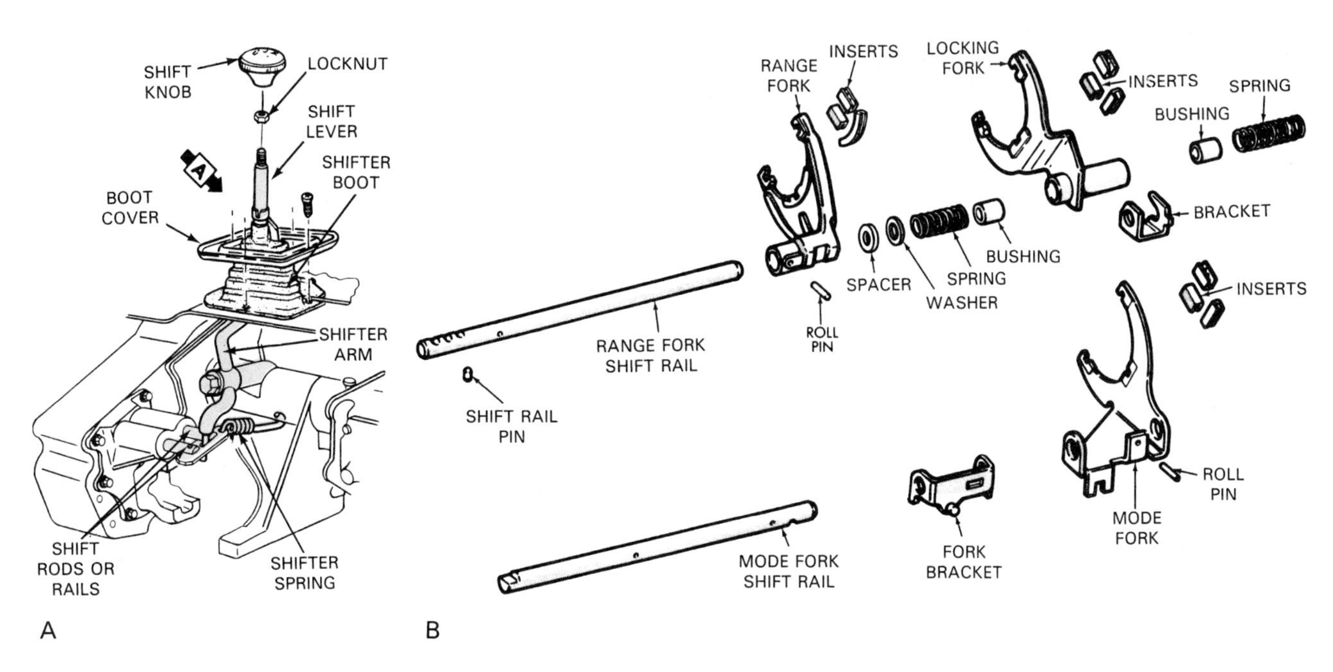

Figure 24-20. The shift linkage for most transfer cases is simple and direct. A–The shift lever and linkage are usually located on top of the transfer case. The lever is sealed at the vehicle floor by a flexible boot. B–The shift forks are attached to the shift rail and operate the sliding clutch sleeves in the transfer case. (General Motors)

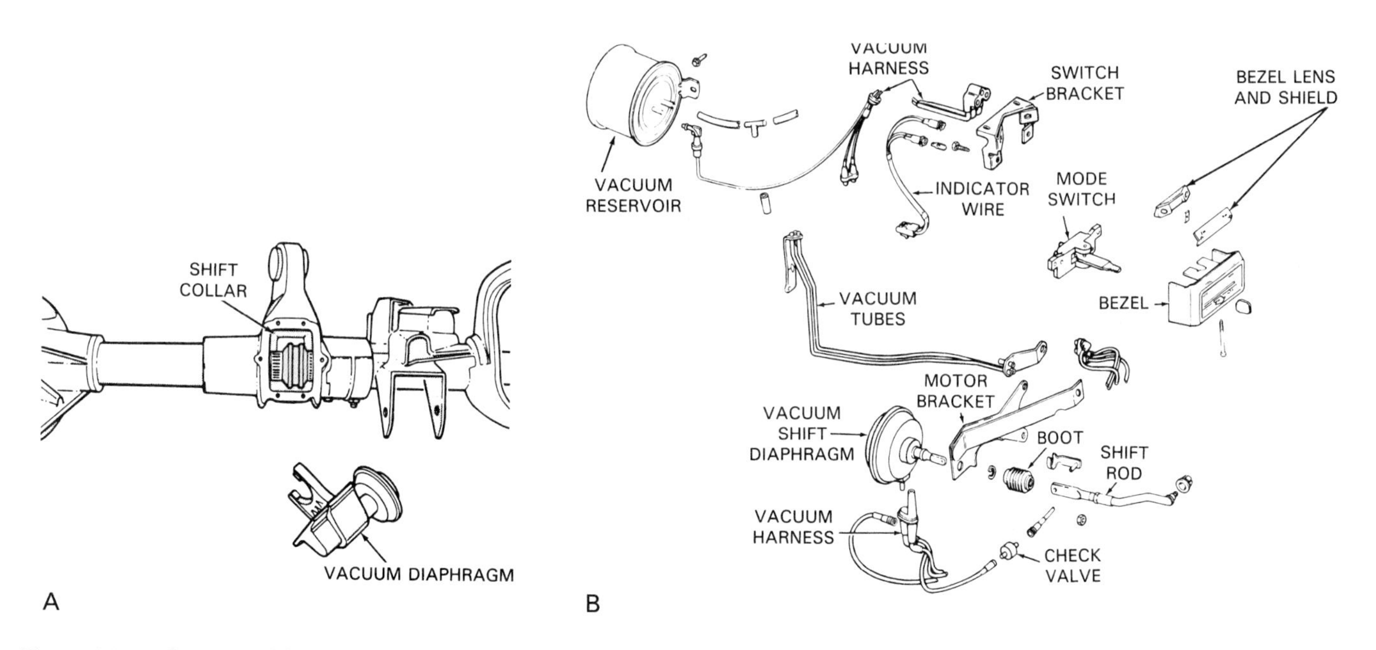

Figure 24-21. Some vehicles have vacuum-operated shift systems that use a vacuum diaphragm to operate one or more shift forks. A–This vacuum diaphragm operates the shift collar that connects and disconnects the front axles from the transfer case. B–This vacuum shift system consists of the vacuum diaphragm, vacuum source (on the intake manifold), control switch, indicator light, and a vacuum tank that acts as a reservoir. (Chrysler)

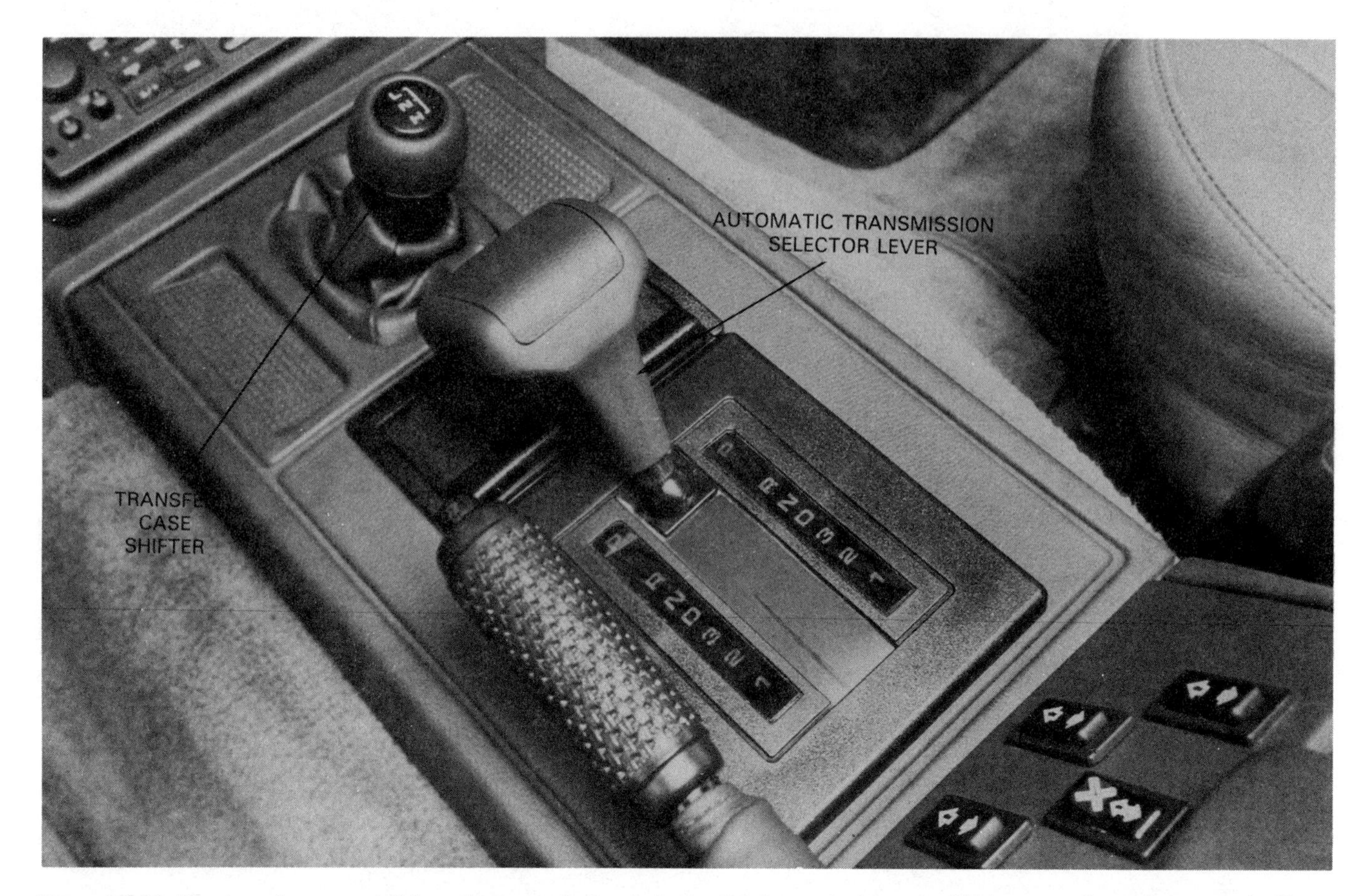

Figure 24-22. The transfer case shift lever is generally located near the transmission gearshift lever on the shifter console. Some modern systems use dash-mounted switches to activate a shifting mechanism. Most full-time systems do not have driver-operated controls. (Land Rover)

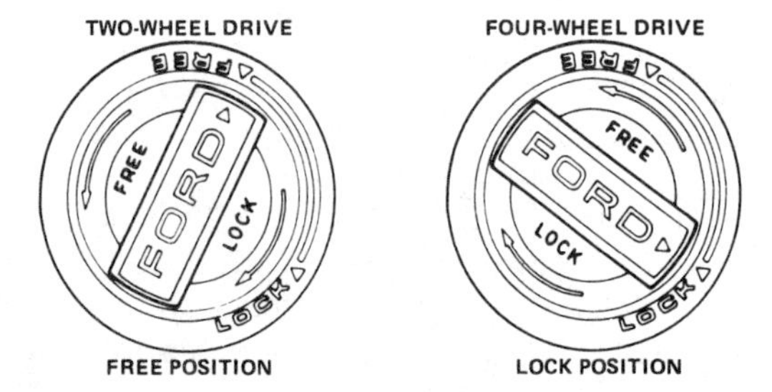

Figure 24-23. Many transfer cases are not equipped with any way to disengage the front-wheel drive components. In systems equipped with this type of transfer case, locking hubs must be used to disconnect the front wheels from the drive components. Locking hubs are often used even when the drive shaft or front axle can be disconnected from the remainder of the drive train. The hubs in these systems are unlocked to prevent the front wheels from turning the front drive components. (Ford)

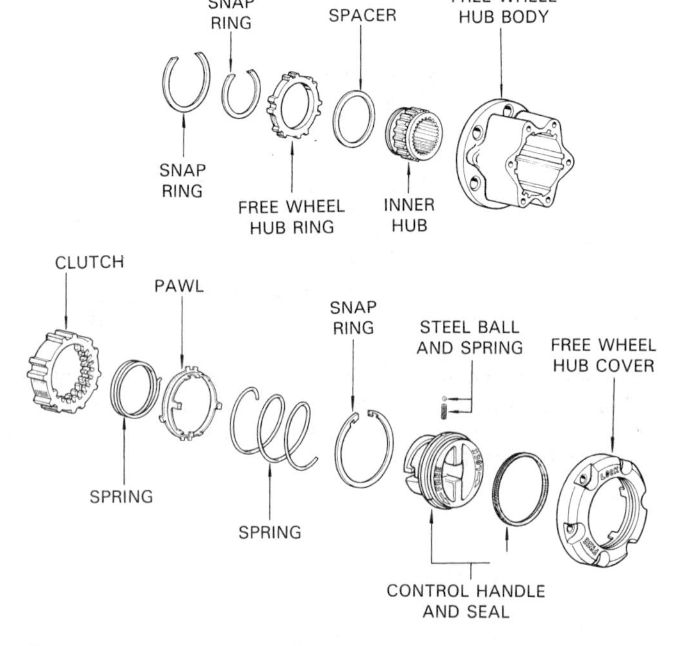

Figure 24-24. Exploded view of a front hub assembly. Note the engagement mechanism. When the splines on the axle, drive gear, and hub are engaged, the wheel is locked to the axle. When the splines are disengaged, the wheel is unlocked. (Ford)

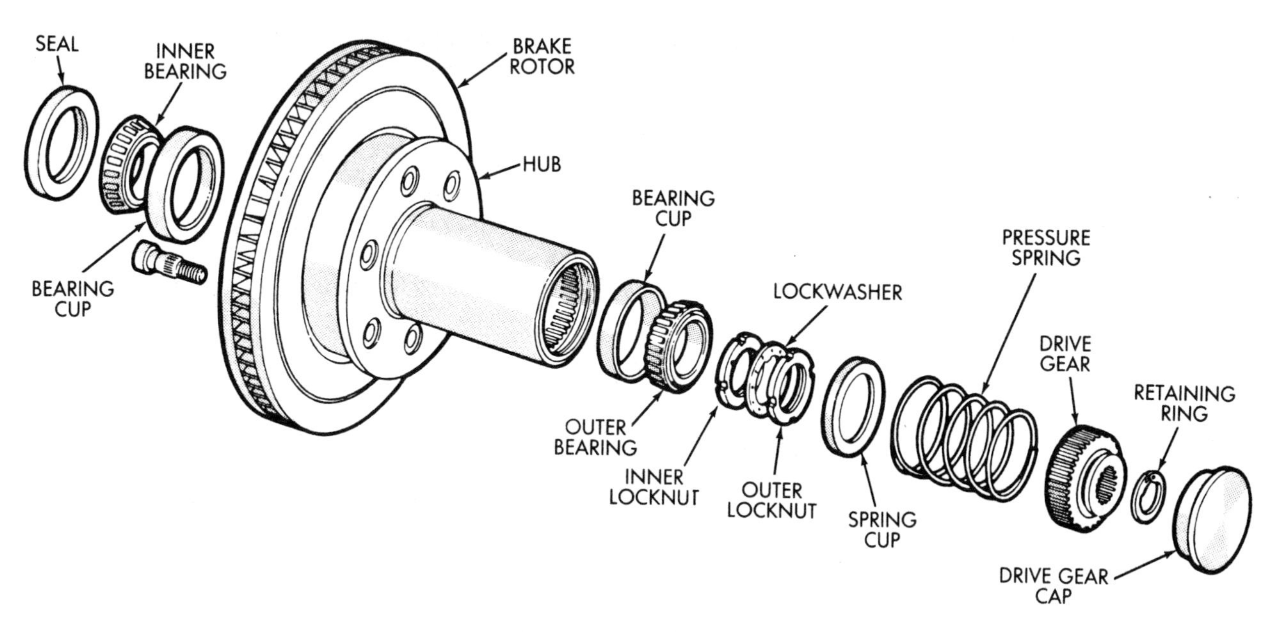

Figure 24-25. Study the parts of this automatic front locking hub. The locking mechanism in this type of hub may be operated by one-way clutch action or by reversing the vehicle after the transfer case is engaged. (Toyota)

Certification-Type Questions–Chapter 24

1. **All of these are major types of four-wheel drive systems EXCEPT:**
   (A) full-time four-wheel drive.
   (B) all-wheel drive.
   (C) part-time four-wheel drive.
   (D) 4 X 4 drive.

2. **The transfer case splits power between:**
   (A) front and rear drive chains.
   (B) a viscous coupling.
   (C) front and rear wheels.
   (D) front and rear differentials.

3. **A transfer case housing serves, in part, as a reservoir for lubricating oil. Technician A says that gaskets are used at case halves and between the case and transmission to prevent oil leaks. Technician B says that the front and rear output yokes have seals to prevent oil loss. Who is right?**
   (A) A only
   (B) B only
   (C) Both A & B
   (D) Neither A nor B

4. **Most modern transfer cases use a chain and sprockets in combination with all of these as drive mechanisms EXCEPT:**
   (A) spur gears.
   (B) helical gears.
   (C) planetary gears.
   (D) worm gears.

5. **Technician A says that a viscous coupling is designed to compensate for differences in front and rear wheel speeds. Technician B says that a central differential is designed to compensate for differences in front and rear wheel speeds. Who is right?**
   (A) A only
   (B) B only
   (C) Both A & B
   (D) Neither A nor B

6. All of these are usually found on trucks with four-wheel drive EXCEPT:

(A) solid axles.
(B) rear drive axles.
(C) front drive axles.
(D) CV axles.

7. All of these parts may be connected to the transfer case EXCEPT:

(A) a shifter.
(B) locking hubs.
(C) a body ground cable.
(D) a speedometer connector.

8. Part-time transfer cases used in off-road vehicles usually have all of these drive combinations EXCEPT:

(A) two-wheel drive, low-gear position.
(B) two-wheel drive, high-gear position.
(C) four-wheel drive, low-gear position.
(D) four-wheel drive, high-gear position.

9. Shift systems use all of these for selecting various gears and driving modes EXCEPT:

(A) locking hubs.
(B) vacuum diaphragms.
(C) mechanical linkages.
(D) electric solenoids.

10. Technician A says that the main purpose of a four-wheel drive system is to increase traction. Technician B says that the components required by four-wheel drive vehicles also help to improve gas mileage over the average two-wheel drive vehicle. Who is right?

(A) A only
(B) B only
(C) Both A & B
(D) Neither A nor B

Chapter 25

Four-Wheel Drive Problems, Troubleshooting, and Service

After studying this chapter, you will be able to:
- Remove a transfer case from a vehicle.
- Disassemble a transfer case.
- Inspect transfer case parts for wear or damage.
- Describe the parts typically replaced during a transfer case overhaul.
- Reassemble and install a transfer case.
- Disassemble, service, and reassemble a locking hub.

As a drive train technician, you must be able to service the transfer case and other components in a modern four-wheel drive system. This chapter will summarize the diagnostic and repair techniques used when servicing four-wheel drive vehicles. It draws on the material presented in previous chapters, such as information on manual transmissions and rear axle assemblies. If necessary, return to the appropriate chapters and review the material as needed. Drive train problems commonly encountered in four-wheel drive systems are illustrated in Figure 25-1.

Transfer Case Problems and Troubleshooting

Although it is relatively small, the transfer case can develop problems that affect the entire drive train. Accurately determining the source of a problem is the first step in repairing a transfer case. Figure 25-2 illustrates some of the problems that can develop in a typical transfer case.

Lubricant Leakage

Lubricant leakage in a transfer case is a loss of lubricating fluid from the housing, gasket, or seals. Leaks may be noticed as oil drips on the pavement. Always begin troubleshooting an oil leak by making sure that the leak is from the transfer case. Engine oil, transmission fluid, or power steering fluid can be mistaken for transfer case lubricant. As a vehicle is driven, oil leaking from the front components is sometimes blown onto the rear components. This makes it difficult to determine the origin of the leak.

One way to verify a transfer case leak is to check the oil level in the case. In most transfer cases, the oil level is correct if the oil just wets the bottom of the fill plug threads, Figure 25-3. If the oil level is normal, the leak is probably from some other part of the vehicle. If the oil level is low, there is generally a leak somewhere in the transfer case. If the oil level is too high, overfilling may be the cause of the leak. Overfilling may cause oil to be forced out of a seal or out through the case vent.

When looking for a transfer case leak, remember that oil will drip downward and is often blown to the rear of the vehicle. Look for leaks from loose or broken bearing retainers or from retainer gaskets. Check the seals in the area where the drive shaft yokes enter the transfer case. Leaks may also originate from loose case housing bolts and defective housing gaskets. Check for leaks from the drain and fill plugs and from bolts at the bottom of the transfer case. Some bolts must be installed with a thread sealant to prevent leaks.

Improper Engagement

When a transfer case *fails to engage properly,* no power will be transmitted to one or both sets of wheels. However, a transfer case rarely fails to drive at least one set of wheels. If the vehicle will not move in any transfer case mode, the problem is usually caused by a defective transmission or by a binding axle. If the transfer case fails in only one gear, the problem may be caused by defective internal parts or linkage.

Transfer case failure causing a *no-drive condition* in *all* gears is usually caused by stripped gears or splines, a broken drive chain, a broken shaft, or a viscous coupling that has lost its fluid. Generally, a transfer case must be removed from the vehicle to repair these defects.

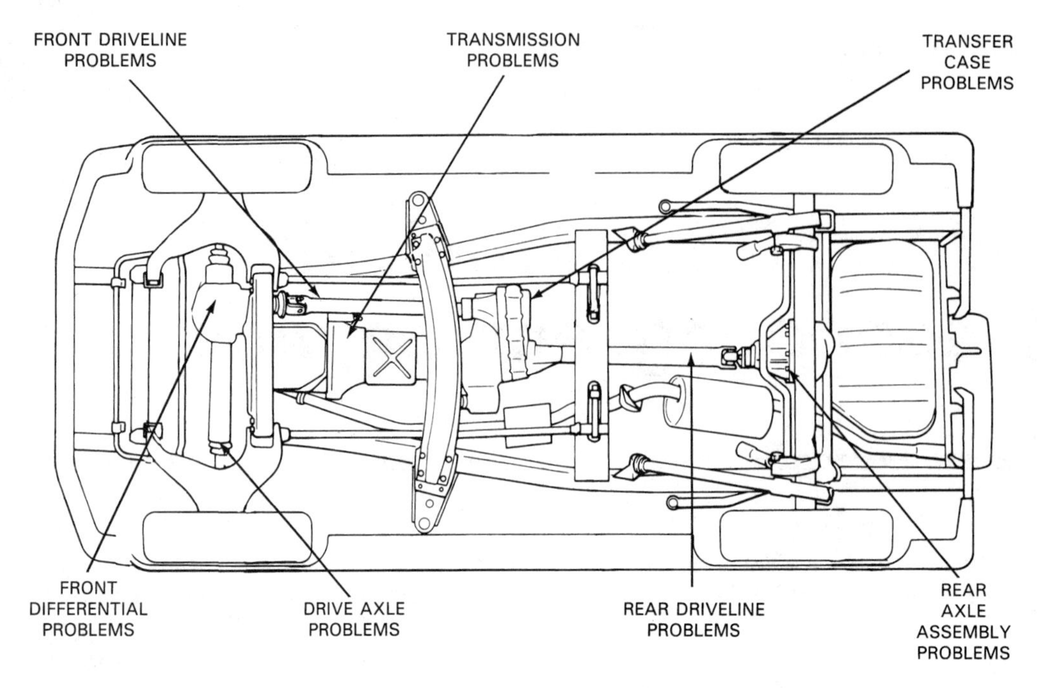

Figure 25-1. Various problems can occur in a four-wheel drive system. The transfer case, drive shafts, differentials, and locking hubs can all be sources of trouble. Be sure to pinpoint the problem before beginning repairs. (Chrysler)

Abnormal Noise

Abnormal noises in the transfer case include whines and rumbles that occur during operation. Other abnormal noises include grinding, knocking, popping, or snapping sounds. Since the transfer case is mounted directly behind the transmission, it is easy to mistake transmission or drive shaft noise for transfer case noise. Always make sure the noise is really coming from the transfer case and not some other part of the vehicle.

Transfer case noises usually vary with driving conditions. If the transfer case is the part-time type, road test the vehicle to determine the mode in which the problem occurs. If the transfer case is the full-time type, the vehicle should be placed on a lift to listen for abnormal noises, such as grinding, knocking, or excessive whining. Remember that normal drive train noises will seem louder when you are under the vehicle.

Typical causes of whining or rumbling noises in the transfer case include a low lubricant level, worn or damaged input gear, worn or damaged driven gear, worn bearings, worn planetary gears (where used), and damaged input or output shaft bearings. If the speedometer gears are installed on the transfer case, check them for damage. A worn drive chain can cause popping or snapping noises as the vehicle is accelerated. Sometimes, the transfer case will be noisy because the front and rear axle ratios do not match due to modifications. This can cause friction in the transfer case planetary gears or coupling. To repair damaged or worn transfer case parts, the unit must be removed and overhauled.

Gear Disengagement

When the transfer case *jumps out of gear,* its shift mechanism moves into neutral at undesirable times. The first step in troubleshooting this type of problem is to determine the gear in which the problem occurs. If the problem occurs in all gears, it may be caused by loose bolts on the transmission or transfer case, loose or misaligned linkage, interference between the linkage and other parts of the vehicle, or a lack of lubricant in the linkage.

If the problem occurs in only one gear, the teeth on that particular gear may be chipped or worn. Further, a blocking ring of a sliding clutch assembly may be worn, or the splines on or for the sliding clutch of the affected gear may be worn.

Hard Shifting

In some cases, it may become difficult to change transfer case gears. *Hard shifting* may be caused by a shift linkage that is bent or worn. In some cases, the linkage may simply need lubrication. Figure 25-4 shows several shift linkage adjustment and lubrication points. If the linkage is okay, the problem may be caused by bent or damaged shift forks inside the transfer case.

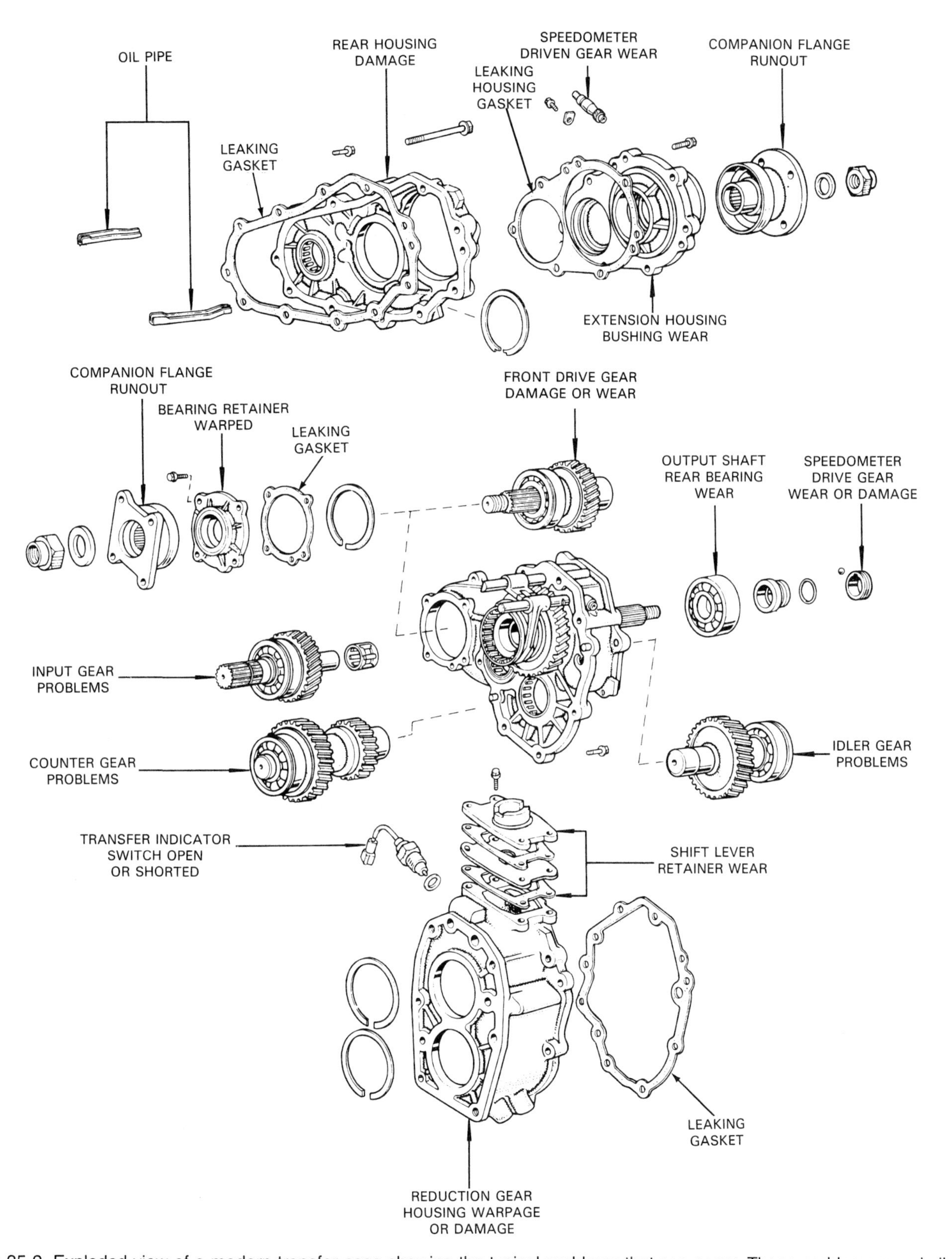

Figure 25-2. Exploded view of a modern transfer case showing the typical problems that can occur. These problems are similar to those encountered in a manual transmission. (Toyota)

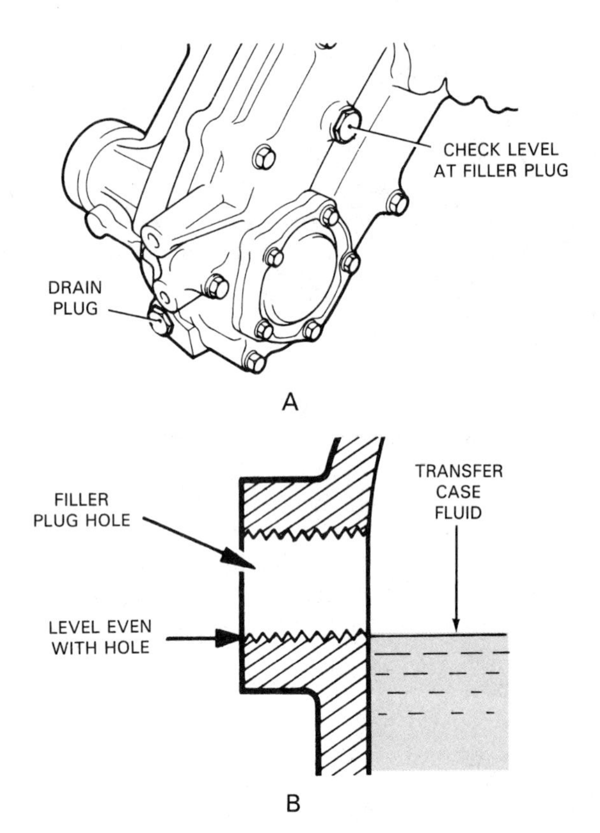

Figure 25-3. Oil level should be checked when troubleshooting transfer case problems. Some transfer cases are lubricated by the transmission fluid. In these systems, the transmission fluid level should be checked. A–Many transfer cases have a drain plug and a filler plug. Always be sure that you remove the filler plug when checking the oil level. B–In most cases, normal oil level is at the bottom of the filler plug threads. (Toyota)

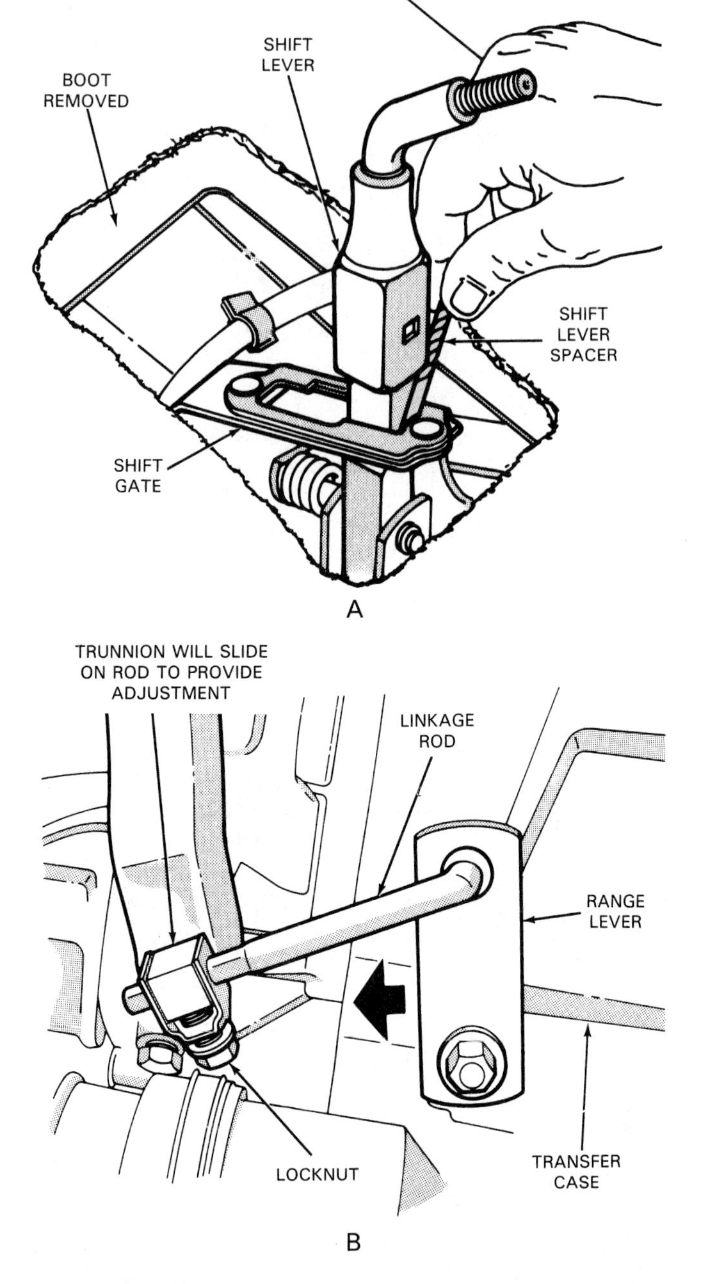

Figure 25-4. Shift linkage adjustments are critical to proper transfer case operation. Always check the linkage adjustment as part of the troubleshooting process. A–The linkage is sometimes adjusted with the help of a special tool, such as a spacer. B–Sometimes, the linkage adjustment is made with a threaded rod and trunnion that are similar to those used on manual transmission linkage. (Chrysler)

Transfer Case Shudder

Transfer case shudder is a jerking motion that typically occurs during acceleration. A shudder may also occur at low speeds on vehicles with a viscous coupling, caused by a defect in the coupling.

When diagnosing this problem, make sure that the shudder is not caused by the engine, clutch, transmission, limited-slip differential, or universal joints. Also, if the vehicle has a viscous coupling, check for silicone in the transfer case lubricant. Because silicone does not mix with oil, it will appear as globules in the lubricant. This will indicate that the silicone has leaked from the viscous coupling and that the coupling requires attention. If the viscious coupling has failed, check for different front and rear axle ratios, which may have caused it to fail.

Transfer Case Overhaul

Many of the problems encountered in four-wheel drive systems result from improper transfer case operation. Worn or damaged transfer cases must often be removed from the vehicle for service. The following sections outline the procedures for removing, servicing, and reinstalling a typical transfer case.

Transfer Case Removal

Transfer case removal is similar to manual transmission removal. Nevertheless, recommended removal procedures may vary among manufacturers. See Figure 25-5. Therefore, this chapter covers only the basics of transfer case removal. Always obtain the proper factory service manual for the particular model that you are servicing. The manual will contain important details. For example, it will tell you if the transmission and transfer case should be removed as a unit.

1. Disconnect the negative battery cable. This will prevent accidental operation of the starter when you are working on the vehicle.
2. Raise the vehicle with an approved hoist or hydraulic jack. If using a hydraulic jack, be sure to install good quality jackstands before getting under the vehicle. If the vehicle has an off-road ***skid plate*** (protective undercarriage panel), remove it. Drain the oil from the transfer case. Sometimes the transmission and transfer case share the same oil. In these systems, it is sometimes necessary to drain the oil at the transmission.
3. Place a transmission jack or stand under the transmission. This will help support the transmission and transfer case during removal.
4. Remove any ground straps and electrical connectors from the transfer case and remove the speedometer cable assembly. Disconnect the transfer case shift linkage and, if necessary, the transmission shift linkage. Vehicles with full-time four-wheel drive usually do not have linkage on the transfer case.
5. Mark the drive shafts or yokes so that they can be correctly reattached to the transfer case. Then, remove the drive shafts from the transfer case. It may not be necessary to remove the drive shaft assemblies from the vehicle. If possible, leave them attached at the differentials and secure them to the underside of the chassis with wire.
6. Remove the rear engine mount and, if necessary, the crossmember.
7. Remove any brackets or fasteners holding the transfer case to the vehicle frame.
8. Place a transmission jack (or floor jack) under the transfer case. Then, remove the bolts or nuts holding the transfer case to the transmission.

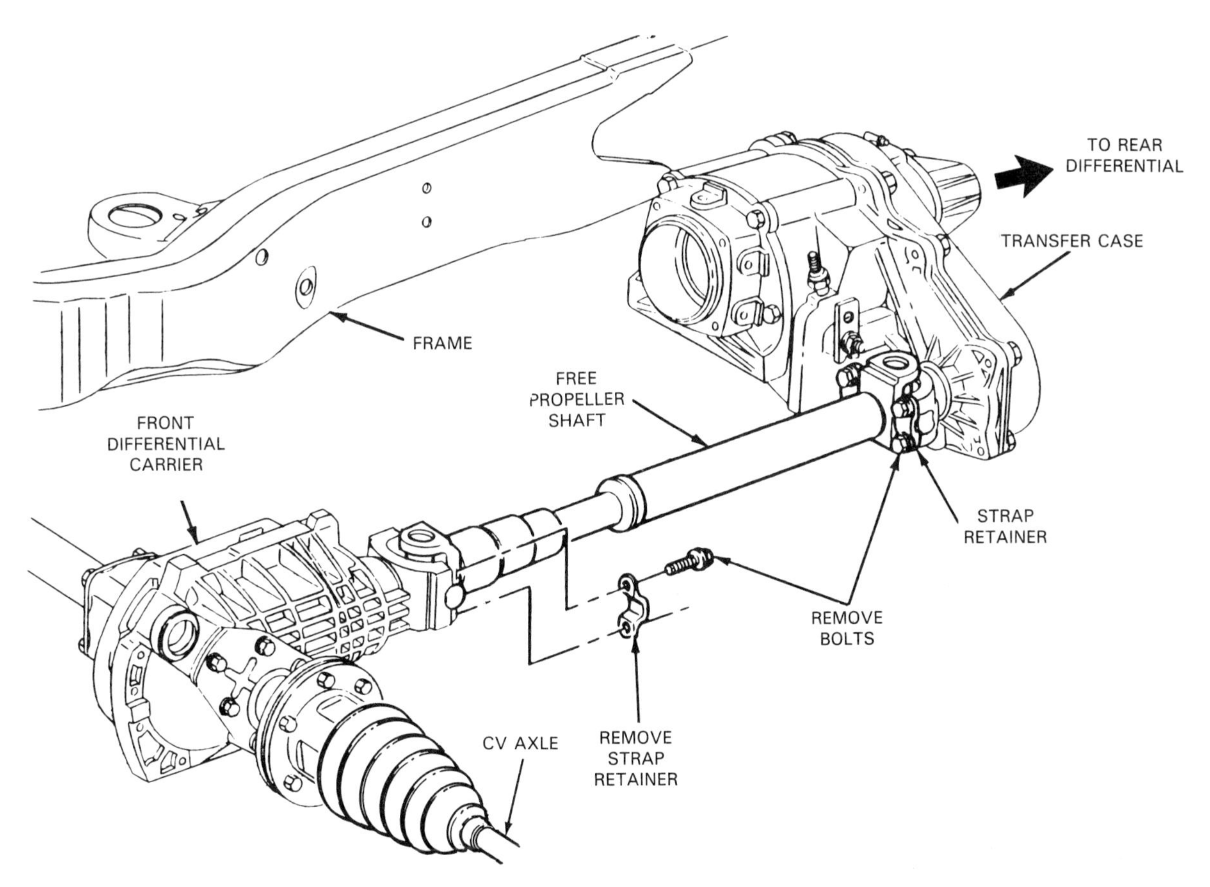

Figure 25-5. This figure shows the relationship of the transfer case to the front axle and the other parts. The removal procedure for this transfer case is similar to that for a manual transmission. Most transfer cases used with a conventional transmission are designed this way. (General Motors)

WARNING! If the transfer case is not properly supported, it will drop when the bolts holding it to the transmission are removed, causing injury or damage.

9. Slide the transfer case away from the transmission and lower the case. Note that the transmission output shaft extends into the transfer case. During removal, the transfer case must be moved away from the transmission until it clears the output shaft. As you lower the transfer case, make sure that it does not catch on adjacent parts. Also, make sure that all straps, wires, and other attachments have been disconnected.

Transfer Case Disassembly

After removing the transfer case, drain any remaining oil, clean the outside of the housing, and mount the unit securely on a clean workbench. Make sure the transfer case will not fall off the bench during the disassembly procedure, causing injury or damage.

If not removed previously, start transfer case disassembly by removing the drive shaft yokes. Most drive shaft yokes are secured to the output shafts with large nuts. The nuts can be removed with the proper wrench. The yoke should be held securely with a special tool or a large pipe wrench as the nut is loosened. Mark the front and back yokes to ease reassembly.

Remove the speedometer gear and housing from the transfer case. Also, remove electrical switches and other external transfer case parts as outlined in the service manual.

If the transfer case housing is the common two-piece type, Figure 25-6, remove the bolts holding the front and rear halves together. The halves can then be split apart. If the gasket sticks, pry the halves apart with pry bars, Figure 25-7. If the housing is a one-piece type, the front and rear retainers that hold the input shaft bearings should be removed to gain access to the internal parts. See Figure 25-8.

After accessing the internal parts, remove any snap rings that hold the transfer case gears to their shaft or secure the shafts to the housing. Then, remove the internal parts as necessary. All shafts, sliding gears, and planetary gears, if used, should be removed at this time. An exploded view of a typical transfer case is shown in Figure 25-9. Refer to the appropriate service manual for exact disassembly procedures.

Figure 25-10 shows a technician removing a mainshaft, complete with gears, shift fork, and rail. Removal procedures for other parts are similar. Some transfer cases do not have an input shaft. Instead, the transmission output shaft is splined to the inside of the transfer case input gear. If used, the differential unit and/or the viscous coupling should also be removed.

If the transfer case is equipped with a drive chain, lift the front output shaft, sprocket, and chain out of the case. In many transfer cases, the chain and sprockets are removed as an assembly. In some designs, however, it is necessary to slide the chain off the mainshaft drive sprocket during the removal process. After removing the

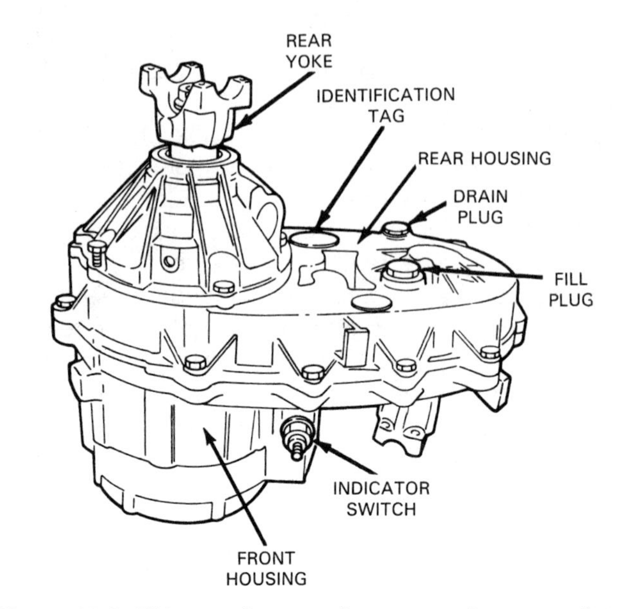

Figure 25-6. This transfer case has a two-piece or split-type housing. The identification tag should be used to order parts and to determine which service specifications to use. (General Motors)

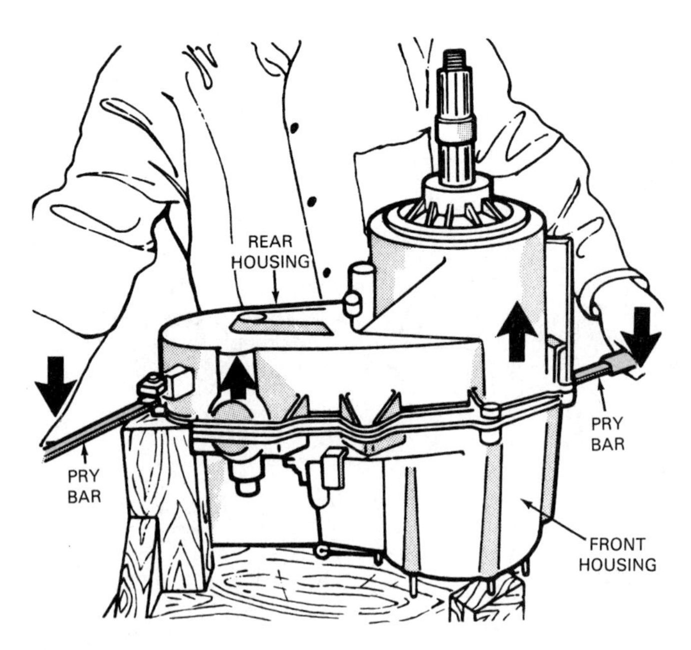

Figure 25-7. If the two halves of the transfer housing stick together, they can be pried apart with two pry bars. Pry at opposite ends of the case to avoid cracking or jamming the case halves. (Chrysler)

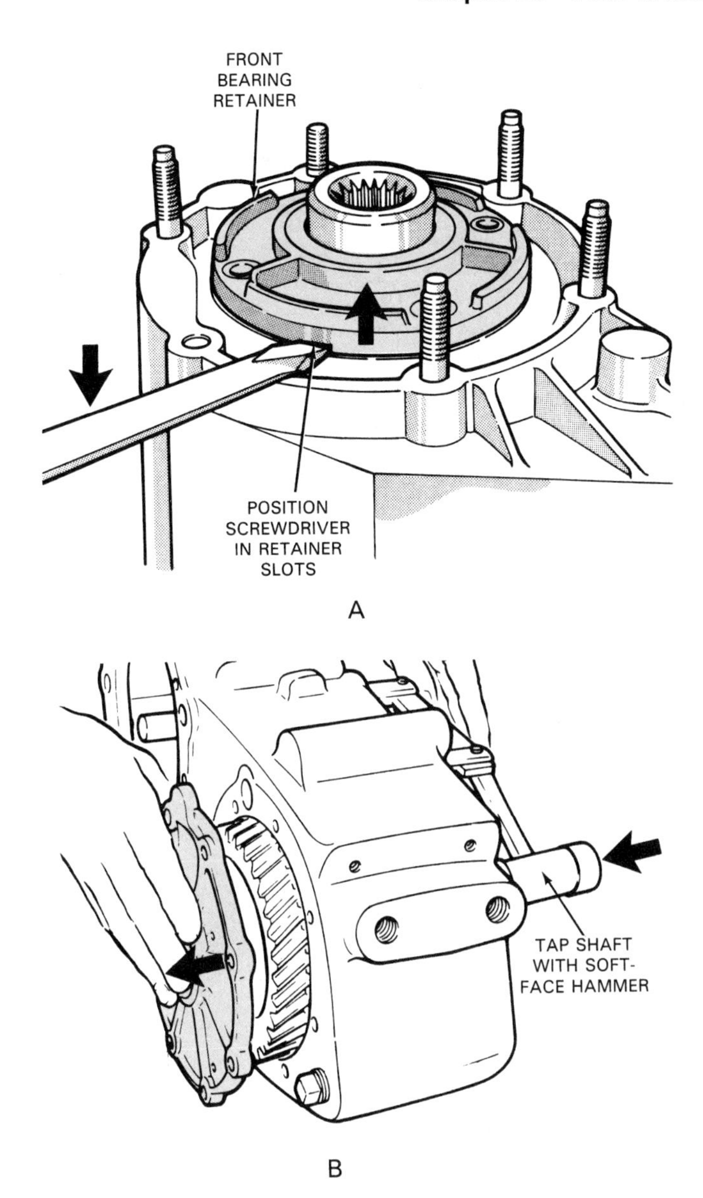

Figure 25-8. Begin disassembly of a one-piece transfer case by removing the bearing retainers. If the drive yokes were not removed during transfer case removal, they should be removed at this time. A–Most bearing retainers can be unbolted from the case and then lightly pried up. B–Basic procedure for removing the bearing retainer and shaft on a one-piece transfer case. Always use a soft hammer to avoid damaging the shaft or gears. (Chrysler, General Motors)

chain and sprockets, carefully remove any thrust washers that were located under the sprockets.

While disassembling the transfer case, note the relationship of all parts so that they can be reinstalled properly. If necessary, mark the parts with a punch or a scribe to ensure proper reassembly. This can prevent a "serious headache" when trying to put the unit back together!

Transfer Case Parts Inspection

Before inspecting the transfer case parts, scrape all gasket material from the transfer case housing, being careful not to damage the sealing surfaces. Check the bottom of the housing for needle bearings or other small parts. Clean the inside of the housing and all internal parts. Make sure that all sludge and metal particles are removed.

After cleaning, the transfer case housing should be thoroughly inspected for cracks or other damage, as should bearing retainers and extension housings. Additionally, all bushings and seals should be checked for wear.

All internal transfer case parts should be inspected for wear and damage. Examine all the shaft bearings, needle bearings, shift forks, sliding clutch sleeves, and washers. Replace worn components as necessary.

Figure 25-11 illustrates various checks that should be made to a transfer case planetary gear assembly. Examine the gears for worn, cracked, or chipped teeth. Damaged gears should be replaced. Check for wear between gear bushings and the shafts. Look for wear or damage to the splines on the shafts and to the planetary housings. Splines can strip off so cleanly that splined surfaces appear to be machined smooth. This makes it very important to compare these parts to service manual specifications.

If used, check the transfer case blocking rings of the sliding clutch assembly for wear on the outer teeth. Also, inspect the areas where the inner cone ridges contact the gear cone. (Refer to Chapter 9 for information on checking synchronizer assemblies.) Replace blocking rings showing signs of wear or damage.

It may be necessary to disassemble the mainshaft to inspect the shaft-mounted components. You should use a service manual to assist in disassembling and reassembling the mainshaft. See Figure 25-12.

Check the sliding clutch sleeves where the shift forks ride. If the shift forks are equipped with separate pads at the riding surfaces, Figure 25-13, the pads should be replaced whenever the transfer case is disassembled.

If the transfer case uses a differential assembly or a viscous coupling, it should be checked for wear and damage. Always follow the inspection procedures outlined in the factory service manual.

NOTE! Refer to the index in this textbook for more information on precision measurement, inspection, and other topics that relate to transfer case overhaul.

Transfer Case Assembly

The transfer case should be reassembled in the reverse order of disassembly. Before beginning, replace worn bushings and seals as shown in Figure 25-14.

Using the factory service manual as a guide, begin reinstalling the parts in the transfer case. See Figures 25-15 and 25-16. Remember that internal parts must be

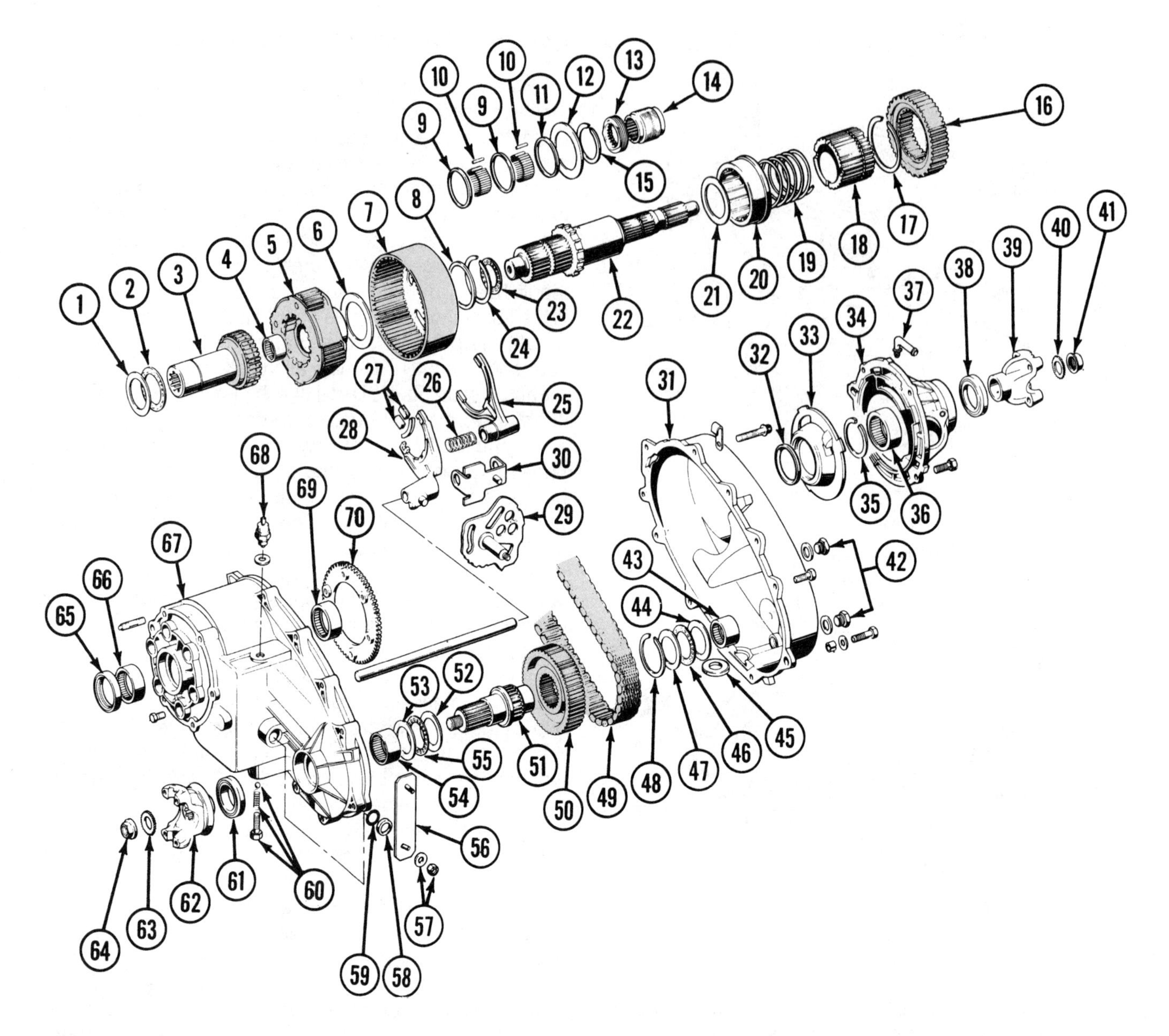

1. Input gear thrust washer
2. Input gear thrust bearing
3. Input gear
4. Mainshaft pilot bearing
5. Planetary assembly
6. Planetary thrust washer
7. Annulus gear
8. Annulus gear thrust washer
9. Needle bearing spacers
10. Mainshaft needle bearings (120)
11. Needle bearing spacer
12. Thrust washer
13. Oil pump
14. Speedometer gear
15. Drive sprocket retaining ring
16. Drive sprocket
17. Sprocket carrier stop ring
18. Sprocket carrier
19. Clutch spring
20. Sliding clutch
21. Thrust washer
22. Mainshaft
23. Mainshaft thrust bearing
24. Annulus gear retaining ring
25. Mode fork
26. Mode fork spring
27. Range fork inserts
28. Range fork
29. Range sector
30. Mode fork bracket
31. Rear case
32. Seal
33. Pump housing
34. Rear retainer
35. Rear output bearing
36. Bearing snap ring
37. Vent tube
38. Rear seal
39. Rear yoke
40. Yoke seal washer
41. Yoke nut
42. Drain and fill plugs
43. Front output shaft rear bearing
44. Front output shaft rear thrust bearing race (thick)
45. Case magnet
46. Front output shaft rear thrust bearing
47. Front output shaft rear thrust bearing race (thin)
48. Driven sprocket retaining ring
49. Drive chain
50. Driven sprocket
51. Front output shaft
52. Front output shaft front thrust bearing race (thin)
53. Front output shaft front thrust bearing race (thick)
54. Front output shaft front thrust bearing
55. Front output shaft front bearing
56. Operating lever
57. Washer and locknut
58. Range sector shaft seal retainer
59. Range sector shaft seal
60. Detent ball, spring, and retainer bolt
61. Front seal
62. Front yoke
63. Yoke seal washer
64. Yoke nut
65. Input gear oil seal
66. Input gear front bearing
67. Front case
68. Lock mode indicator switch and washer
69. Input gear rear bearing
70. Lockplate

Figure 25-9. Study this exploded view of a typical modern transfer case. The transfer cases used on many late-model vehicles have planetary gears and a drive chain to transfer power. (Chrysler)

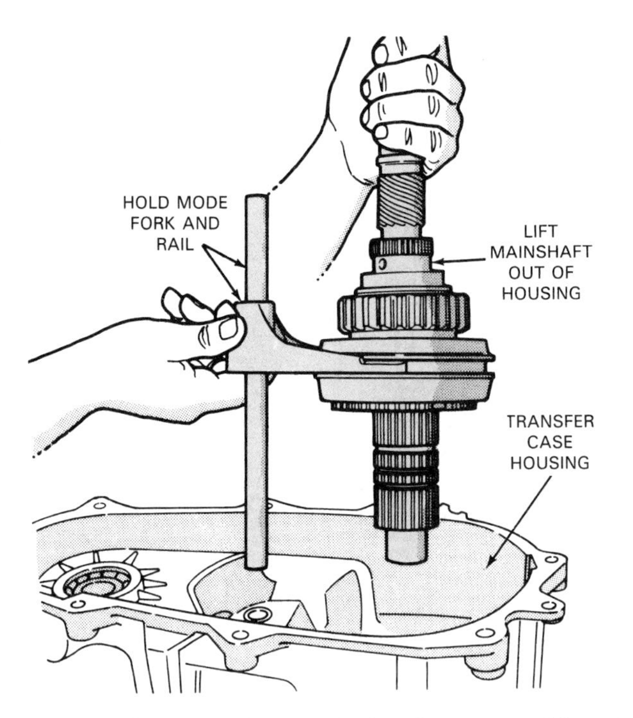

Figure 25-10. Some transfer case shaft and gear assemblies can be removed as shown. In other designs, the gears and the shaft must be removed separately. Always consult the manufacturer's service manual for exact disassembly procedures. (Chrysler)

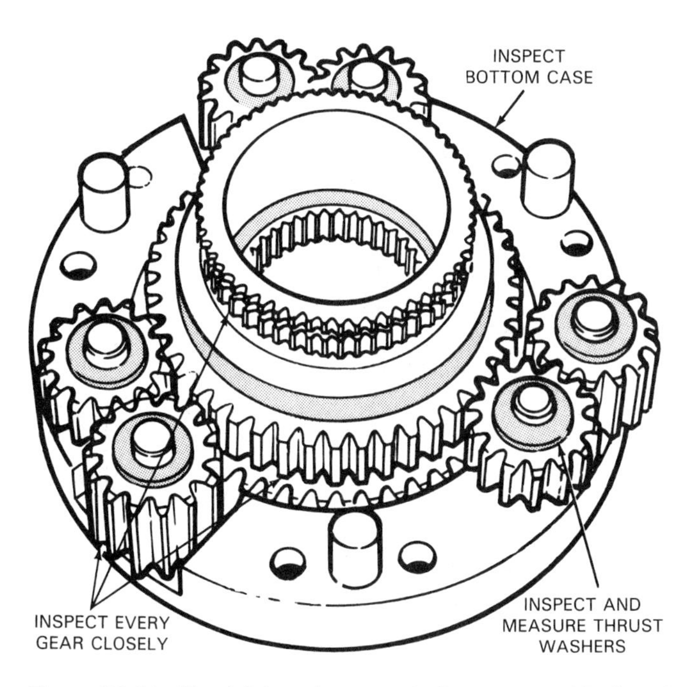

Figure 25-11. Check internal gear parts for wear or chipping at the teeth, wear or damage to internal and external splines, and wear of the thrust bearing surfaces.

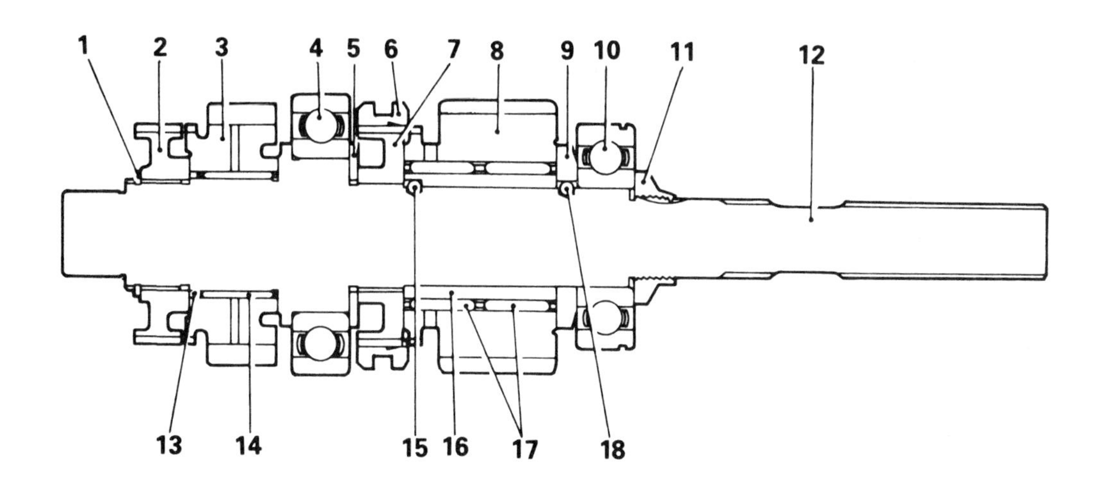

1. Snap ring
2. H-L clutch hub
3. Low speed gear
4. Ball bearing
5. Stop plate
6. 2-4WD clutch sleeve
7. 2-4WD clutch hub
8. Drive sprocket
9. Sprocket spacer
10. Ball bearing
11. Locknut
12. Rear output shaft
13. Thrust washer
14. Needle bearing
15. Steel ball
16. Sprocket sleeve
17. Needle bearing
18. Steel ball

Figure 25-12. Assembly details of a transfer case mainshaft. When disassembling or reassembling a mainshaft, consult an appropriate service manual. Referring to the manual will help prevent part damage during disassembly or reassembly. (Toyota)

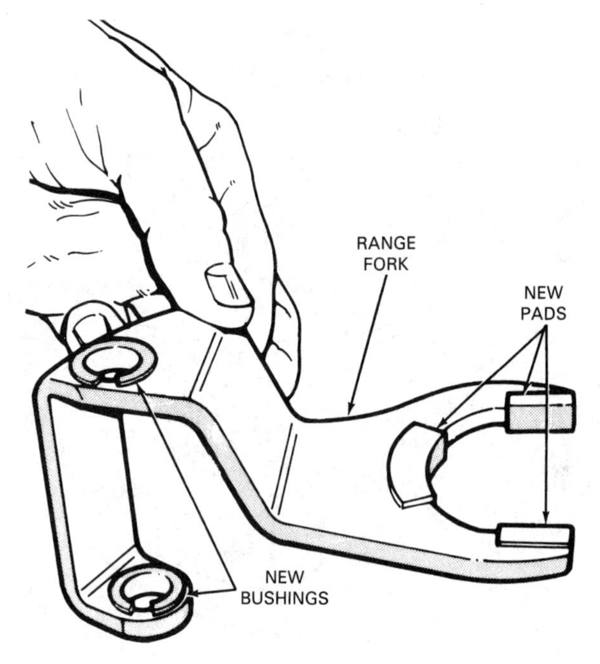

Figure 25-13. Check shift forks for wear and replace bushings or hub contact pads. Replace the entire shift fork if it is bent or worn. Seals used in areas where the shift forks or levers pass through the case should be replaced.

placed into the case in the proper order. If not, you may have to take the transfer case apart again to install forgotten parts. Reassembly procedures will vary with transfer case design. Nevertheless, the following procedure will serve as a general guide to reassembly.

Assemble the mainshaft gears and hubs onto the mainshaft. Use new parts as needed. Install all internal parts in the transfer case housing. See Figure 25-17. Note that the shift forks are usually installed at the same time as the gears that they operate. Make sure that the shafts are properly seated in the bottom of the housing to avoid binding.

Install the drive chain, making sure that the sprockets and shafts are properly aligned with the chain. Refer to Figure 25-18. After installing the chain, make sure that all thrust washers are installed. See Figure 25-19. Next, install the snap rings that hold the chain and shafts in place.

On two-piece housings, place a new gasket and/or an appropriate sealer on the mating surface of the front housing. Make one last check of all internal parts and install the rear housing on the front housing, Figure 25-20. Install and tighten the housing bolts. Then, make sure that both shafts turn. If either shaft is binding, the transfer case must be disassembled to determine the cause.

On one-piece housings, install the bearing retainers, using new gaskets. After installing the retainers, make sure that the shafts turn. Do not reinstall the transfer case if the shafts are binding; correct the problem first. If the service

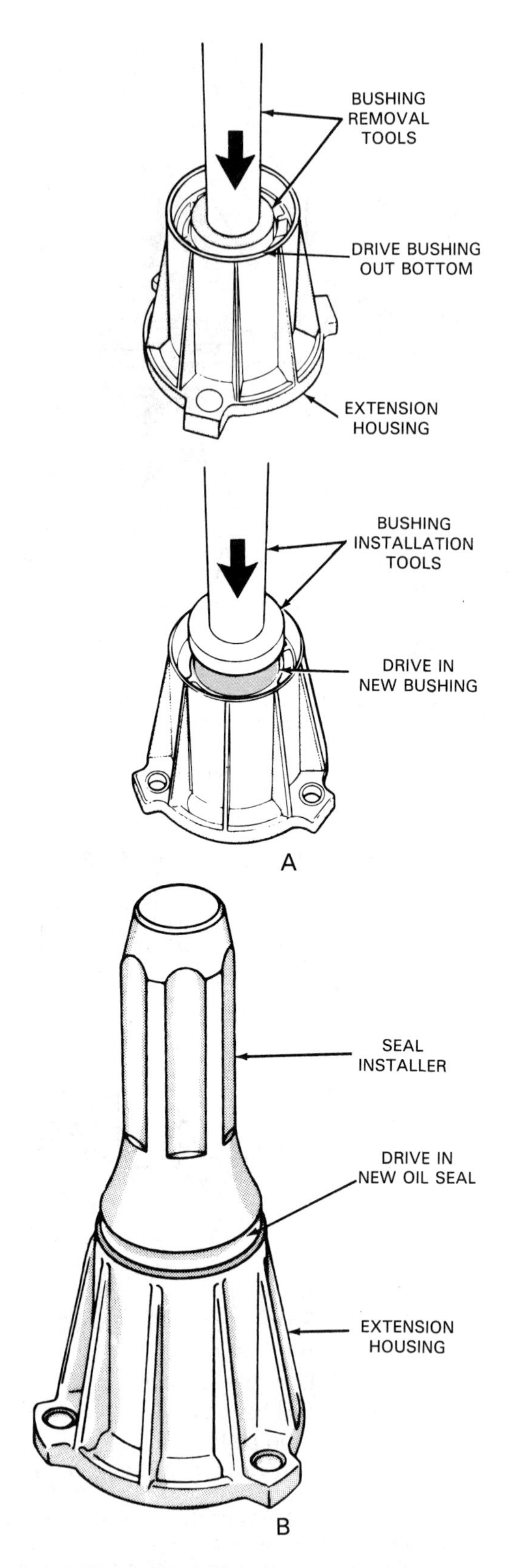

Figure 25-14. Before reassembling the transfer case, replace all worn bushings and seals. A–Removing and replacing an extension housing bushing. B–Installing a new seal in the extension housing.

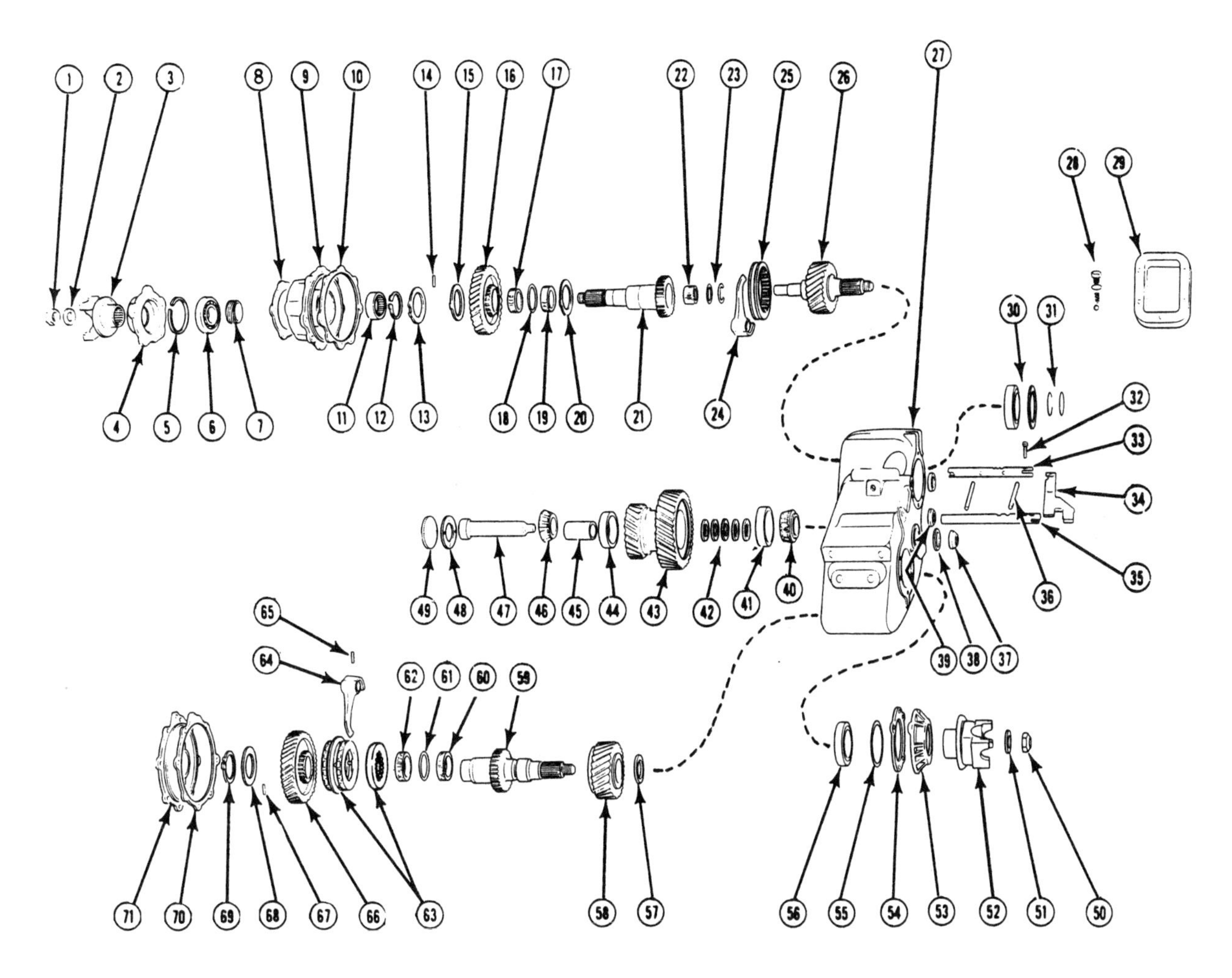

1. Rear Output Shaft Locknut
2. Washer
3. Yoke
4. Bearing Retainer and Seal Assembly
5. Snap Ring
6. Bearing
7. Speedometer Gear
8. Gasket
9. Housing
10. Gasket
11. Bearing
12. Snap Ring
13. Thrust Washer
14. Thrust Washer Lock Pin
15. Thrust Washer (Tanged)
16. Low Speed Gear
17. Needle Bearings
18. Spacer
19. Needle Bearings
20. Tanged Washer
21. Rear Output Shaft
22. Needle Bearings
23. Washer and Retainer
24. Shift Fork
25. Sliding Clutch
26. Input Shaft
27. Transfer Case
28. Poppet Spring & Ball, Light Switch Spring & Ball
29. P.T.O. Gasket and Cover
30. Input Shaft Bearing and Snap Ring
31. Snap Ring and Rubber Ring
32. Shift Link Clevis Pin
33. Range Shift Rail
34. Shift Rail Connector Link
35. Front Wheel Drive Shift Rail
36. Interlock Pins
37. Rear Idler Lock Nut
38. Washer
39. Shift Rail Seals
40. Idler Shaft Bearing
41. Bearing Cup
42. Shims
43. Idler Gear
44. Bearing Cup
45. Spacer
46. Idler Shaft Bearing
47. Idler Shaft
48. Cover Gasket
49. Rear Cover
50. Front Output Shaft Locknut
51. Washer
52. Yoke
53. Bearing Retainer and Seat
54. Gasket
55. Snap Ring
56. Front Bearing
57. Thrust Washer
58. Front Wheel High Gear
59. Front Output Shaft
60. Needle Bearings
61. Spacer
62. Needle Bearing
63. Synchronizer
64. Shift Fork
65. Roll Pin
66. Front Output Low Gear
67. Thrust Washer Lock Pin
68. Thrust Washer
69. Snap Ring
70. Rear Cover Gasket
71. Rear Cover and Bearing

Figure 25-15. Exploded view shows the arrangement of parts in a typical one-piece transfer case. One-piece transfer cases are usually found on older four-wheel drive vehicles. (General Motors)

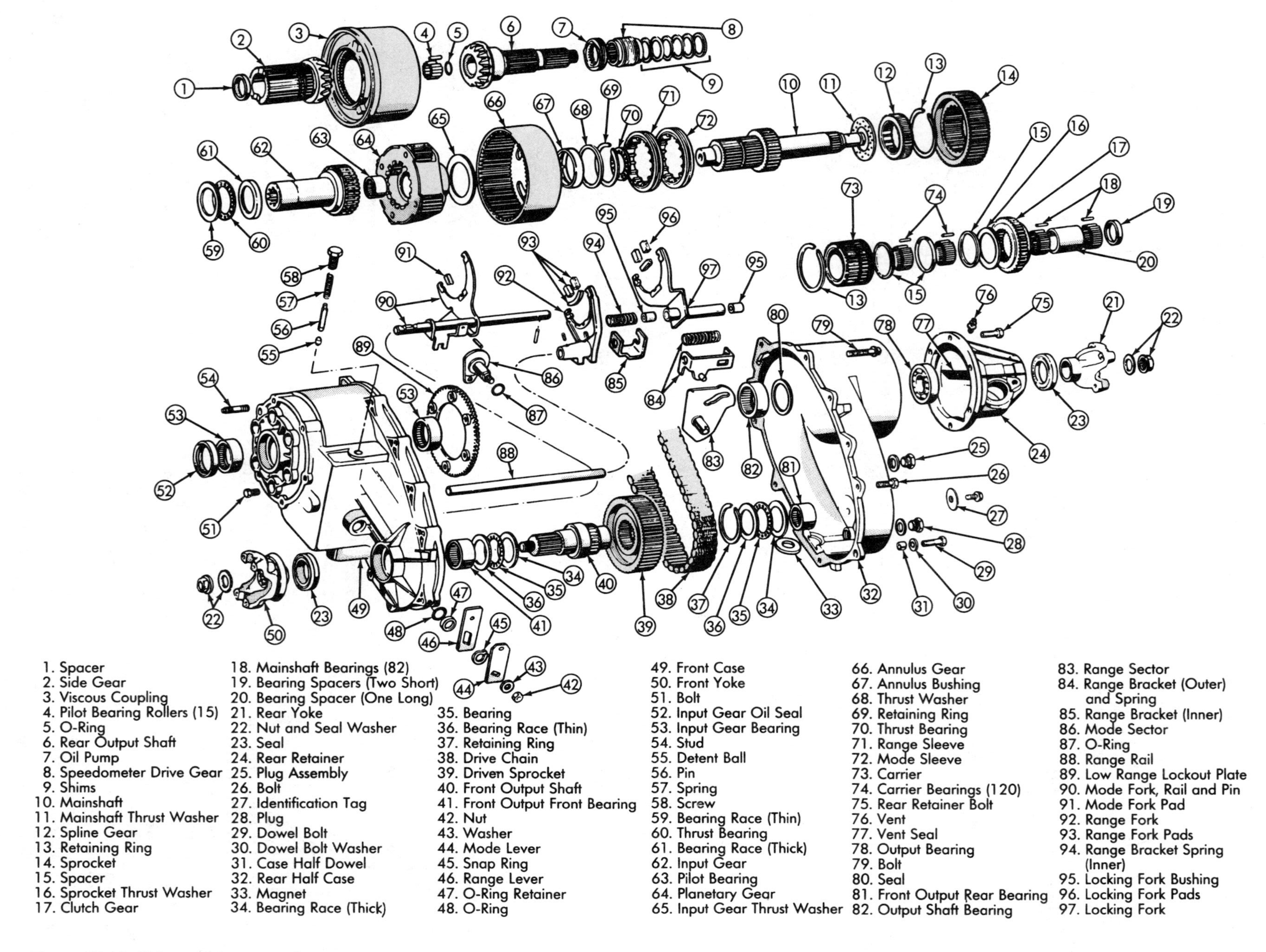

Figure 25-16. This two-piece transfer case is typical of those found on modern four-wheel drive vehicles. Notice the planetary gears, viscous coupling, and drive chain. In addition to checking the other parts, the viscous coupling should be carefully inspected for wear or leakage. Most viscous couplings cannot be disassembled

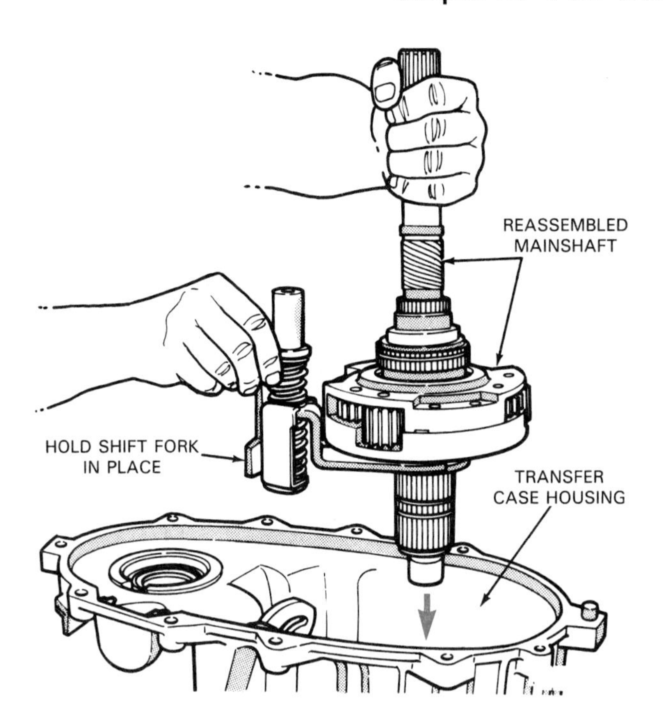

Figure 25-17. The reassembly and installation of the internal parts should be done carefully to prevent damaging parts or causing moving parts to bind. Lubricate all parts before assembly, using the same type of lubricant used in the transfer case. (Chrysler)

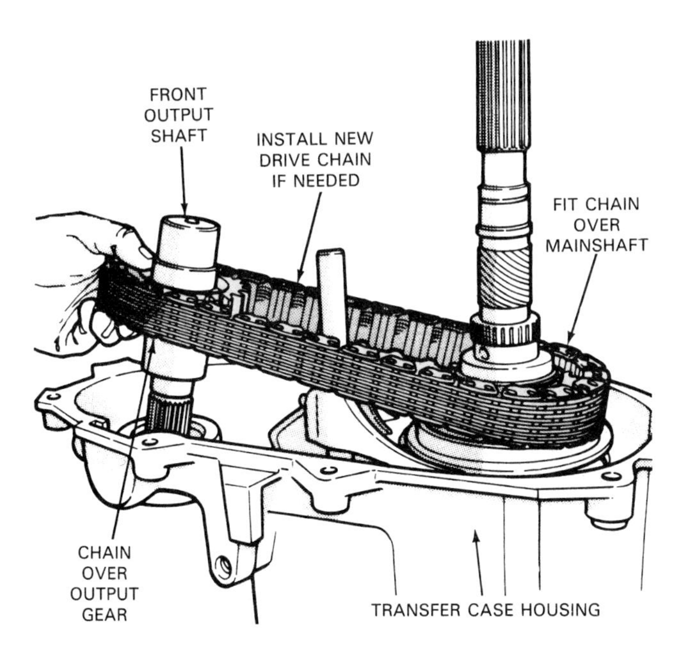

Figure 25-18. Carefully install the drive chain and sprockets. Make sure that the chain is aligned with the drive sprocket and the driven sprocket and that the sprockets are correctly installed onto the shafts. (Chrysler)

manual recommends adjusting the bearing preload, it should also be done at this time.

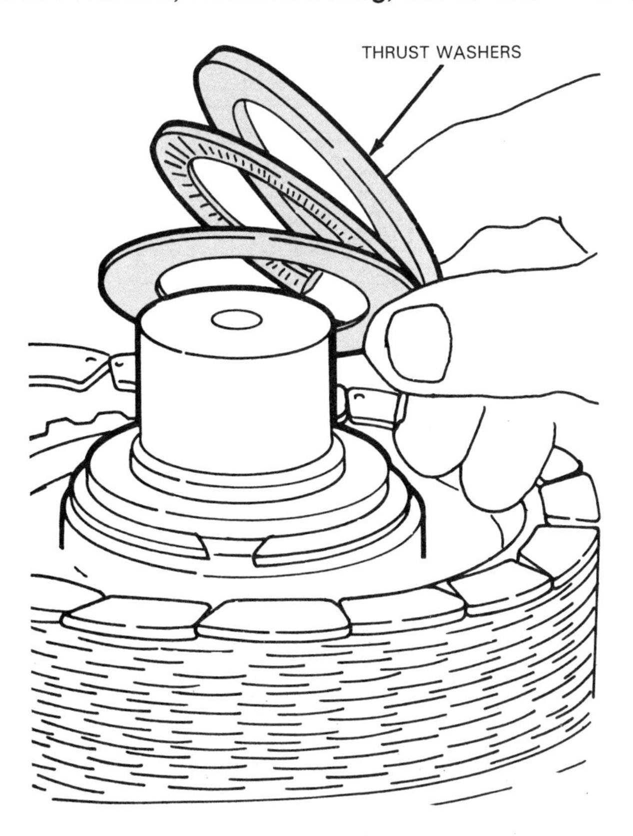

Figure 25-19. To ensure the proper alignment of the chain or other parts, always reinstall the thrust washers in the appropriate positions. Make sure the washers are installed facing the way they were originally. (General Motors)

A few transfer case manufacturers provide specifications for shaft end play. If specifications are provided, the end play should be measured. The procedure for checking end play was explained in the transmission overhaul chapters. If end play is incorrect, the rear half of the housing or the bearing retainer must be removed and thicker or thinner thrust washers must be installed.

If helical drive and driven gears are used in the transfer case, they may require adjustment. The adjustment procedure is similar to that used for a ring and pinion. See Figure 25-21. Consult the appropriate service manual for specific instructions.

Finally, install the external housing components, such as the speedometer drive gear, front and rear drive shaft yokes, electrical components, and drain and fill plugs.

Transfer Case Installation

The transfer case is installed in the reverse order of removal. Because the transfer case is heavy, you should exercise caution when lifting it or working around it.

1. Use a transmission jack or a stand jack to raise the transfer case to the level of the transmission (or transaxle). Slide the case into engagement with the trans-

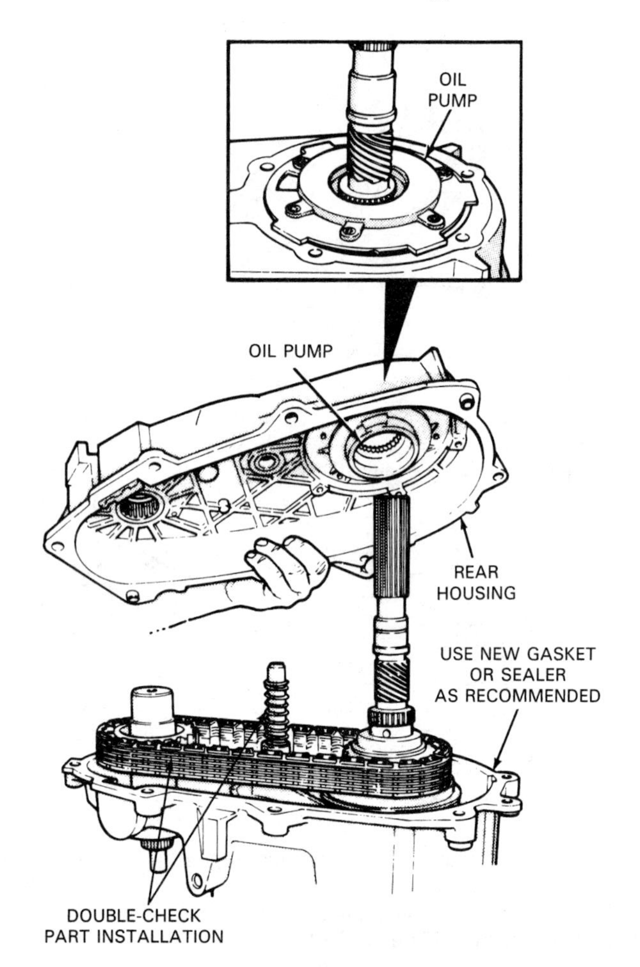

Figure 25-20. Before installing the rear half of a two-piece transfer case housing, make sure that you have installed all of the internal parts. Then, install a new gasket or sealer and carefully position the rear housing over the front housing. (Chrysler)

mission. Work carefully to prevent damaging the splines of the transmission or the transfer case. The transmission case may have alignment lugs to make installation easier.

2. Install and tighten the bolts that secure the transfer case to the transmission. Make sure that the drive shaft yokes will still turn after the transfer case is tightened.
3. Install the crossmember and rear engine mount. Remove the transmission jack or support stand.
4. Reconnect both drive shafts to the transfer case output yokes.
5. Install the transfer case shift linkage, the speedometer cable assembly, and any electrical connectors or ground straps. Install the transmission shift linkage, if removed.
6. Fill the transfer case with the proper type and amount of lubricant. Some manufacturers recommend gear oil, while others specify automatic transmission fluid. If the transmission and transfer case use the same oil, check the oil level at the transmission.
7. If necessary, install the skid plate. Then, lower the vehicle.
8. Reconnect the negative battery cable and start the engine. Road test the vehicle to make sure that the linkage adjustments are correct and that the transfer case operates properly. Recheck the oil level after the road test and inspect the transfer case for leaks.

Servicing Locking Hubs

Many four-wheel drive vehicles continue to use locking hubs at the front wheels. If you are careful and refer to the proper service manual, locking hubs will be relatively simple to repair.

Before disassembling a locking hub, raise the affected wheel. If applicable, shift the transfer case into two-wheel drive and place the hub in the lock position (manual model) to prevent it from turning. Remove the outer retainer cover screws. Then, remove the outer retainer and the shift cam from the hub.

Since hub space is limited, most of the other hub parts are held in place by one or more snap rings or retainer rings. Remove the snap rings and slide the retainer ring, cam follower, clutch gears, and other hub parts from the front axle housing. Some hub designs require you to remove the snap ring and then slightly turn the shaft to disengage the internal components. In certain designs, the front wheel bearing will come out with the other hub parts.

CAUTION! The hub parts may be heavily greased, and the snap rings may be difficult to locate. Never attempt to remove parts until you are sure that all snap rings have been removed.

Place all of the hub parts in order on a clean workbench. Carefully clean and inspect the parts for wear, damage, rust, and dirt. Replace all worn or damaged components. Springs should be carefully checked, since they can affect hub operation if they lose tension. Snap rings that are distorted or bent should be replaced. Always replace the seals and gaskets whenever a hub assembly is disassembled.

Reassemble the front axle parts is the reverse order of removal. Always refer to the manufacturer's manual for specific instructions. Thoroughly grease the hub components with a proper lubricant. Reinstall the parts in the housing in the proper order, being careful to insert all snap rings in their grooves. After all parts have been reinstalled, replace the outer retainer cover and tighten the attaching screws.

After reinstalling the hub components, it may be necessary to adjust the axle preload and end play. Typical adjustment procedures are shown in Figure 25-22. This

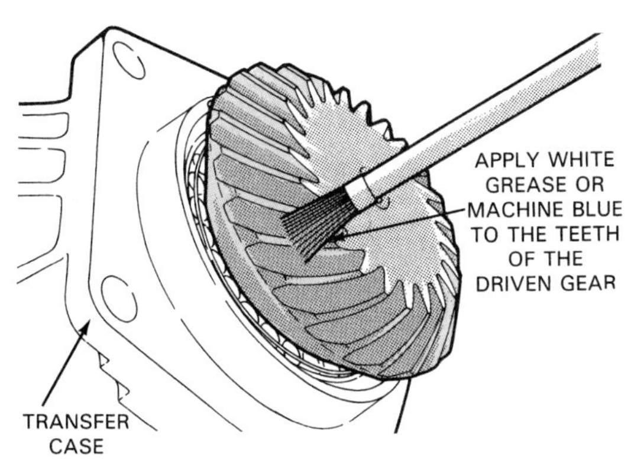

A—Paint the driven gear teeth with a light coat of machine blue or white grease.

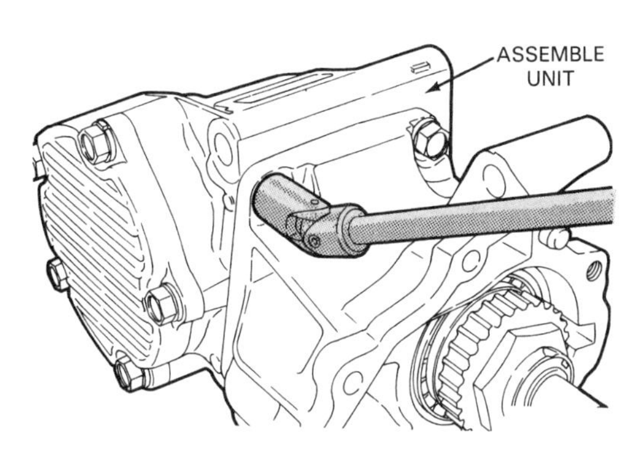

B—Install the transer case on the transaxle or the case subassembly and torque the bolts to specifications.

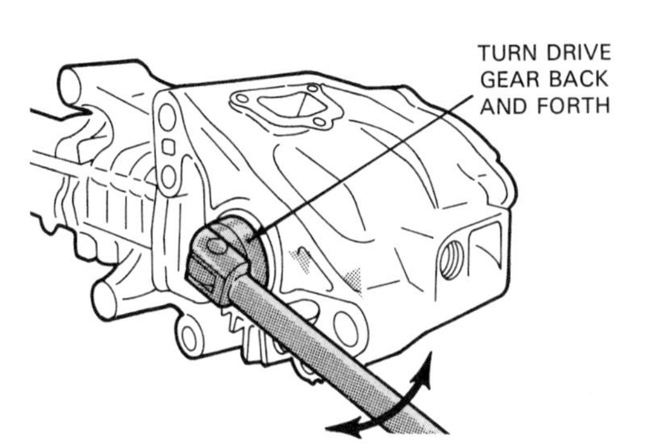

C—Turn the drive gear one revolution in each direction. Do not turn it more than one revolution.

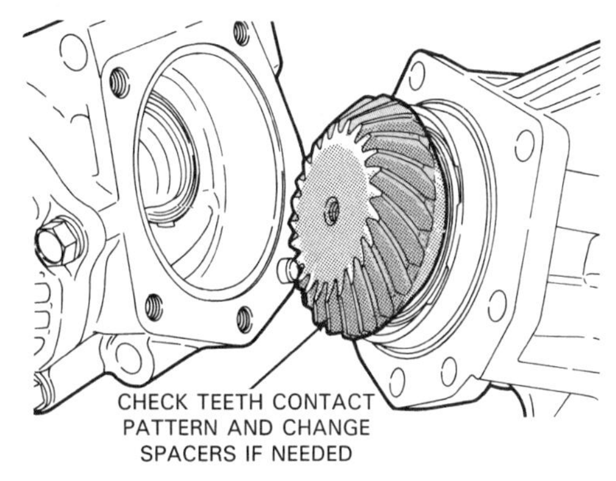

D—Remove the transfer case from the transaxle or the subassembly and check the contact pattern. If the pattern is not correct, adjust the gear positions according to factory procedures.

Figure 25-21. Study the procedure for checking the adjustment of helical gears in a transfer case. This procedure is similar to that used to check the adjustment of hypoid ring and pinion gears.

adjustment is not necessary if the front wheel bearings are separate from the hub components. Add grease if the hub is equipped with a grease fitting. Road test the vehicle in the locked and unlocked positions to ensure that the hub operates properly.

Summary

The automotive technician must be able to diagnose and repair the drive train components of four-wheel drive vehicles. Many transfer case repair techniques will be similar to the procedures used when repairing manual transmissions and rear axles.

Transfer case problems can affect the operation of the entire drive train. Always make sure that the problem is in the transfer case before beginning repairs. Transfer case oil leaks can be caused by defective seals, loose drain plugs, or loose attaching bolts. Overfilling can also cause transfer case leaks.

If the transfer case fails to drive the vehicle in any gear, the problem is usually caused by a broken or stripped internal part. If the transfer case fails to transfer power in only one gear, the problem may be in the linkage. Noises can be caused by defective internal parts or by a low oil level. Note what range the noise occurs in before attempting repairs.

If the transfer case jumps out of gear, the problem may be caused by internal part wear. In many cases, loose and maladjusted linkage can cause the transfer case to jump out of gear. A hard shifting transfer case may be caused by internal problems or by a worn or binding linkage. A defective viscous coupling may cause the transfer case to shudder.

Transfer case removal is similar to manual transmission removal. Always disconnect the negative battery cable before beginning. Then, raise the vehicle and drain the oil from the transfer case.

Support the transmission and transfer case with a transmission jack. Remove all transfer case attachments, such as linkage, wiring, and braces. Then, remove the crossmember and the bolts holding the transfer case to the transmission. Slide the transfer case back from the trans-

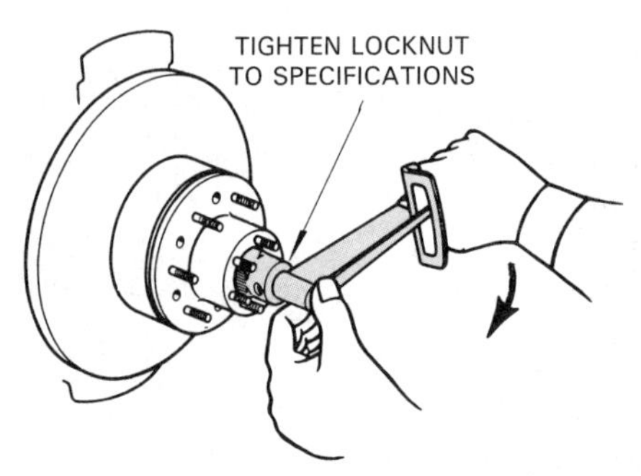

A—Start by tightening the bearing attaching nut to specifications.

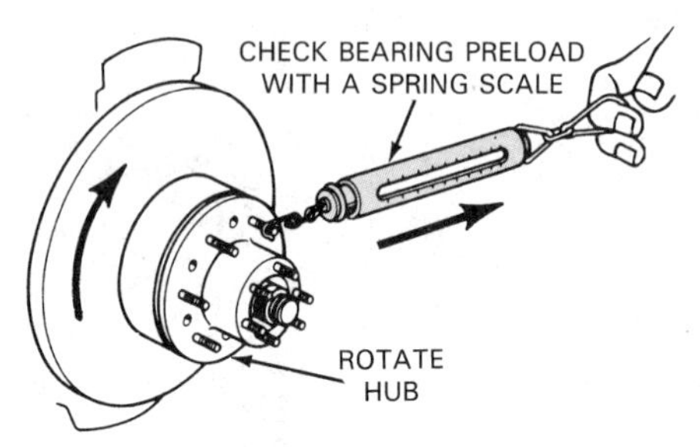

B—Check the bearing preload with a spring scale or torque wrench as called for in the service manual. Readjust as necessary.

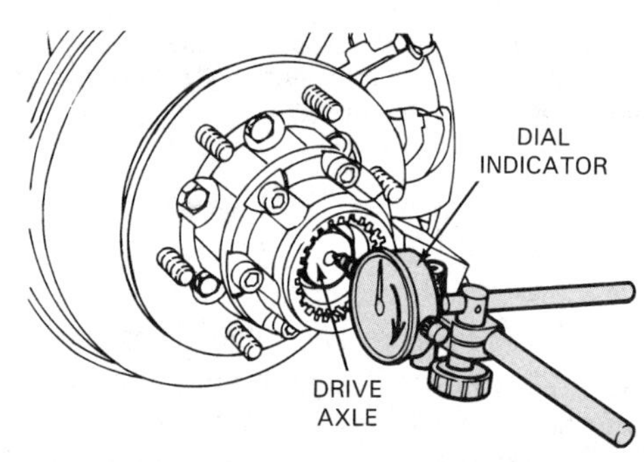

C—Check and adjust the drive axle end play. End play can be adjusted with shims or adjusting nuts, depending on the manufacturer. This step may not be necessary on some vehicles.

Figure 25-22. The adjustment procedure for front wheel bearings used with front locking hubs is similar to that used to adjust front wheel bearings on two-wheel drive vehicles. Refer to the specific manual to obtain the exact procedure for the vehicle you are servicing. (Toyota)

mission and lower it to the floor. Clean the outside of the housing before taking the transfer case apart.

Securely mount the transfer case on a workbench before beginning disassembly. Then, remove the drive shaft yokes. Remove any linkage, electrical switches, or other external transfer case parts.

Remove the internal parts of the transfer case according to procedures outlined in the factory service manual. All gears, shafts, thrust washers, bearings, chains, and sprockets should be removed. If necessary, the differential assembly and the viscous coupling should also be removed.

After removal, clean all parts and check them for wear. Pay particular attention to gears, chains, sprockets, thrust washers, bearings, and bushings. If any parts show signs of wear, replace them. Also, replace all seals and gaskets.

Reassemble the transfer case in reverse order of disassembly. Make sure that all parts are lubricated before installation. All parts should be carefully aligned during installation. When reassembly is complete, all shafts should turn easily.

If helical drive and driven gears are used, they may require adjustment. The procedures for checking and adjusting helical gears are similar to those for the hypoid ring and pinion used on a conventional rear axle vehicle.

Position the transfer case on the transmission and reinstall all attaching bolts, linkage, and other accessories. After the transfer case is installed, fill it with the appropriate lubricant and adjust the linkages. Then, road test the vehicle to ensure that the transfer case operates properly in all ranges and that the shift linkage is properly adjusted.

In most cases, locking hubs can be disassembled and repaired easily. Most of the internal components are held in place by snap rings. Removing these rings will enable the internal parts to be slid off the front axle.

After disassembling the hub, check all parts for wear, damage, or corrosion. Components prone to wear include splined parts and locking cams. Replace any defective parts. Before reassembling the hub, coat all parts with the proper lubricant. After reassembling the hub, check and, if necessary, adjust the front wheel bearings and the front axle end play.

Know These Terms

Transfer case shudder; Skid plate.

Review Questions–Chapter 25

Please do not write in this text. Place your answers on a separate sheet of paper.

1. It is sometimes hard to determine the origin of transfer case oil leaks. True or false?
2. When a transfer case fails to _____ properly, no power is transmitted to one or both sets of wheels.
3. If a transfer case fails to transfer power in any mode, the problem may be caused by a _____ _____ or _____ _____.
4. What can cause a transfer case to jump out of gear?
5. How do you remove a transfer case from a vehicle?
6. Blocking rings of the sliding clutch assemblies should be checked for wear on their outer _____ and inner _____.

7. The sliding clutch sleeve surfaces should be checked in areas where the _____ _____ rides.
8. In your own words, explain how to install a transfer case.
9. When disassembling a locking hub, the hub should be in the unlocked position. True or false?
10. After reinstalling hub components, you must often adjust axle end play. True or false?

Certification-Type Questions–Chapter 25

1. Technician A says that low oil level in the transfer case is a sign of a transfer case leak. Technician B says that overfilling the transfer case can cause lubricant leaks. Who is right?

(A) A only
(B) B only
(C) Both A & B
(D) Neither A nor B

2. All of these statements about transfer case problems are true EXCEPT:

(A) in general, a four-wheel drive vehicle that cannot move in any mode probably has a faulty transfer case.
(B) the transfer case must usually be removed from the vehicle to correct a no-drive condition.
(C) leaks from the engine or transmission are often misdiagnosed as transfer case leaks.
(D) normal drive train noises will seem louder when under the vehicle.

3. Transfer case shudder that occurs at low speed could likely be caused by:

(A) damage to the planetary gearset.
(B) a loose drive chain.
(C) a defective viscous coupling (if present).
(D) worn bearings.

4. A transfer case is being removed from a vehicle. Technician A says that marking the drive shaft yokes is not necessary unless the drive shaft is being disassembled. Technician B says that the transaxle drive shafts should be left attached to the front and rear differentials if possible. Who is right?

(A) A only
(B) B only
(C) Both A & B
(D) Neither A nor B

5. Which of these service operations can be performed without removing the transfer case from the vehicle?

(A) Viscous clutch replacement.
(B) Chain drive inspection.
(C) Linkage adjustment.
(D) Input gear replacement.

6. Technician A says that the interior of the transfer case should be thoroughly cleaned to remove sludge and metal particles. Technician B says that the interior of the transfer case should be carefully checked for small parts after disassembly is complete. Who is right?

(A) A only
(B) B only
(C) Both A & B
(D) Neither A nor B

7. Globules of silicone have been found in the oil drained from a transfer case. Technician A says that this is a sign of severe oil breakdown. Technician B says that the viscous coupling could be defective. Who is right?

(A) A only
(B) B only
(C) Both A & B
(D) Neither A nor B

8. Technician A says that some manufacturers specify gear oil for the transaxle lubricant. Technician B says that some manufacturers specify using automatic transmission fluid in the transaxle. Who is right?

(A) A only
(B) B only
(C) Both A & B
(D) Neither A nor B

9. Locking hubs on the front wheels are held in place by:

(A) a central nut and cotter pin.
(B) snap rings.
(C) a press fit on the spindle.
(D) case-hardened machine screws.

10. To prevent difficulty when reassembling the transfer case, the technician should:

(A) order parts before disassembly.
(B) replace the transfer case as an assembly.
(C) note the location of parts as the transfer case is disassembled.
(D) apply grease profusely.

Chapter 26

Drive Train Electronics

After studying this chapter, you will be able to:

- Explain how a computer can be used to control drive train operation.
- Describe the various sensors and actuators found in computer controlled drive trains.
- Use computer self-diagnosis to find problems in a drive train system.
- Test and service drive train sensors and actuators.
- Replace a computer and its PROM.

Almost every modern vehicle has some connection between its on-board computer and its drive train. The computer may simply receive information from the drive train components, or it may actually control the operation of the transmission, torque converter, transfer case, or traction control system. Increasingly, the modern vehicle will use an on-board computer to precisely coordinate the operation of the engine, transmission, steering, suspension, and other systems.

This chapter is designed to introduce the basic operating principles of automotive computers and outline the ways in which computers are used to control drive train components. It will also explain the operation of the most common input sensors, output devices, and computer configurations that influence drive train operation.

After studying this chapter, you will be familiar with the basic procedures for diagnosing and servicing computerized drive train components. Particular emphasis is placed on using the computer's built-in diagnostic capability.

Computerized Drive Trains

In vehicles with a ***computerized drive train,*** an on-board computer makes decisions concerning transmission shift points and pressures, torque converter lockup, and other functions.

Figure 26-1 is a schematic of an automatic transmission that is almost completely controlled by the on-board computer. The computer makes all of the decisions concerning shift patterns, operation of the converter clutch, and overdrive selection.

There are three drive train areas where computer controls have gained a foothold. Computer controls are used extensively in the operation of automatic transmissions (transaxles), traction control systems, and four-wheel drive systems.

Computer System Components

The basic components found in most automotive computer systems include sensors, computers, and actuators. In a typical system, the computer receives input signals from the sensors. Inputs from sensors provide the computer with information about engine speed, vehicle speed, throttle position, and various other conditions. Based on the inputs received, the computer makes decisions and produces signals that reflect these decisions. The computer signals are called outputs and are sent to output devices called actuators. The actuators can change operating conditions if necessary.

Drive Train Sensors

Drive train sensors provide the electrical input signals to the computer. Sensors, such as the one shown in Figure 26-2A, are attached to the computer through a wiring harness. They provide the computer with information about engine speed, vehicle speed, transmission pressures, and shift selector lever position. When the sensor is located inside the transmission, the wiring must enter the transmission through a special electrical case connector, Figure 26-2B. Case connectors are usually made of plastic to help insulate the wiring and are equipped with an O-ring seal to prevent oil leaks.

There are many sensors used in computer control systems. The number and type of sensors used will vary

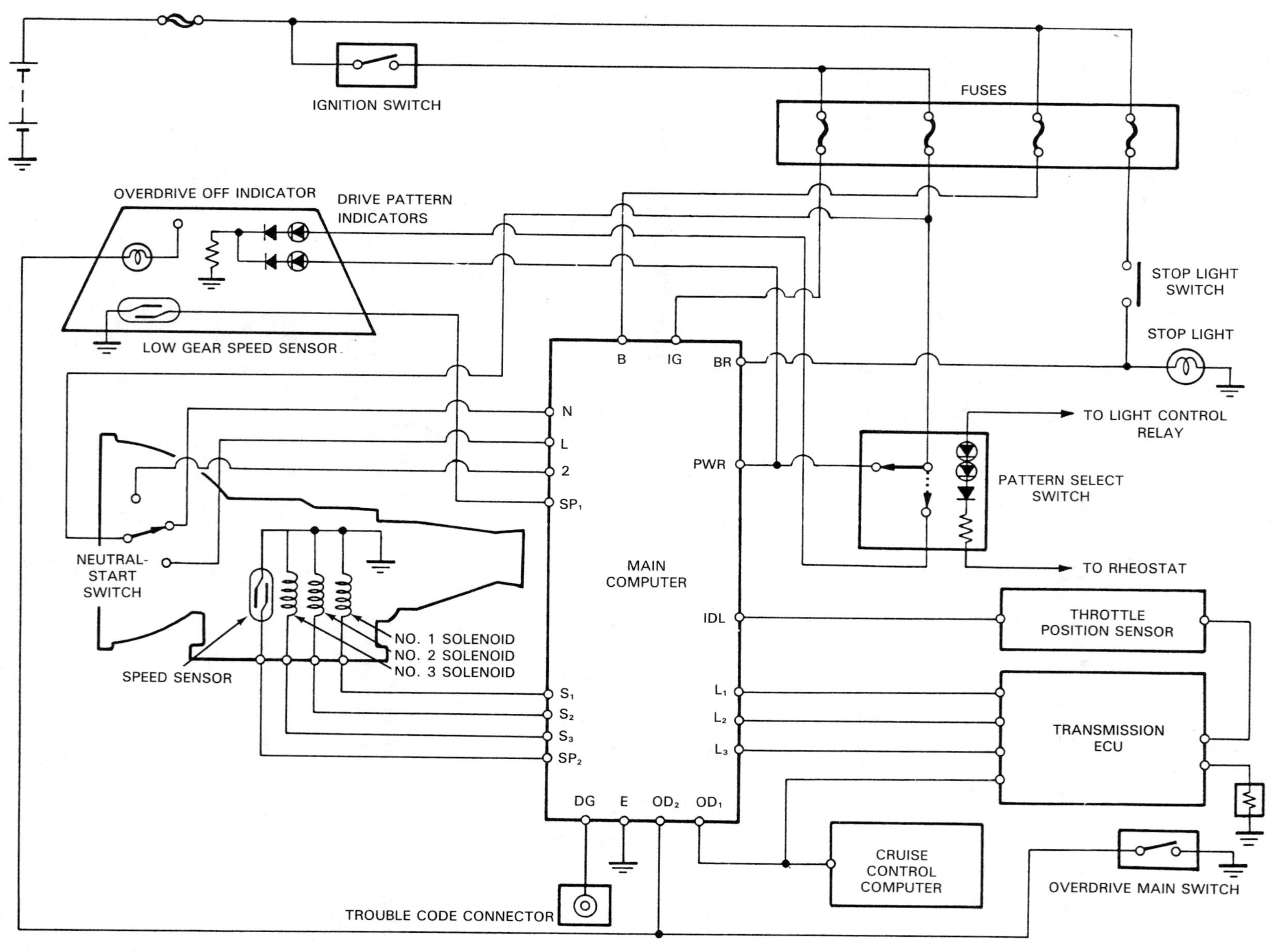

Figure 26-1. Schematic diagram of a computer controlled automatic transmission. The transmission was one of the first drive train components to be controlled by a computer.

with drive train designs. The sensors identified in the following sections are used in many common drive train computer systems. In some cases, the drive train sensors deliver information through another computer, such as the engine control computer or the cruise control computer. These computers relay pertinent information to the drive train computer.

NOTE! Although many types of sensors are installed on modern vehicles, not all of them provide inputs that are used to control the drive train. Many sensors are designed to help control the engine or other vehicle systems. Always check the manufacturer's service manuals to determine which sensors affect drive train operation.

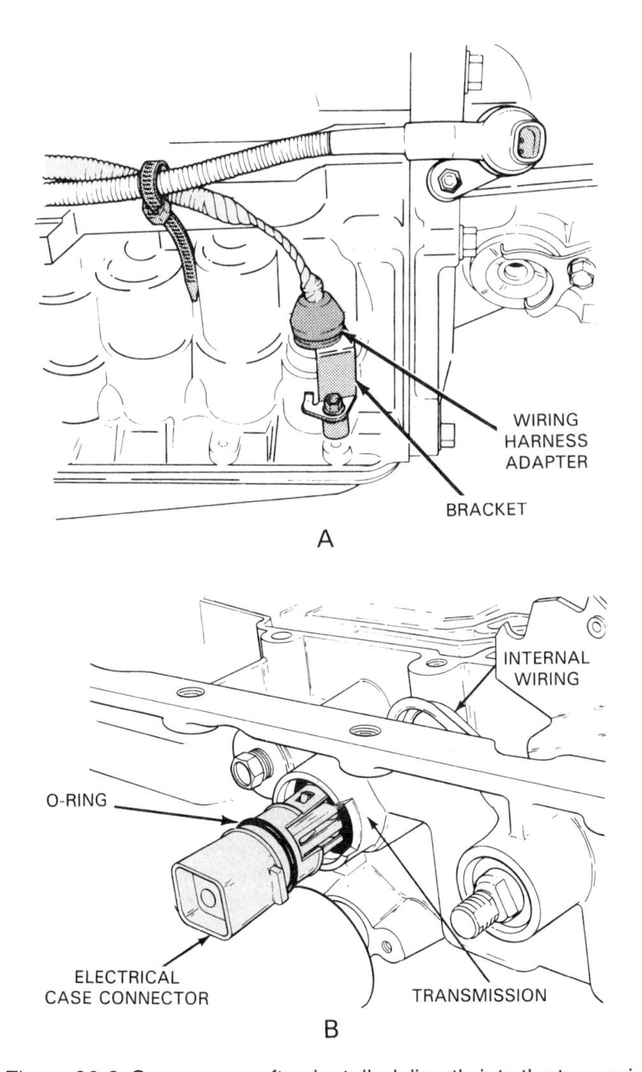

Figure 26-2. Sensors are often installed directly into the transmission, transaxle, or transfer case. A–The transmission sensor shown here is connected to the on-board computer by a weatherproof wiring harness. B–The transmission case connector allows the wiring to enter the transmission and prevents the oil from leaking out. (Chrysler, General Motors)

Categories of sensors

Sensors can be divided into two general categories. These include *active sensors* and *passive sensors.*

Active sensors create their own voltage signal by generating electricity. Some active sensors create electricity from movement. For example, a turning drive train part creates a fluctuating magnetic field in most speed sensors. The fluctuating magnetic field induces a voltage in the sensor. These sensors are sometimes called ***pulse-generating sensors.*** Other active sensors, such as oxygen sensors, generate electricity from chemical activity.

Passive sensors depend on an outside source of electricity for proper operation. In most cases, the computer supplies a reference voltage of about 5 volts to the sensor. The passive sensor modifies the reference voltage and sends it back to the computer. Throttle position sensors and manifold pressure sensors are examples of common passive sensors.

Some passive sensors simply send an *on* or *off* signal to the computer. The vehicle brake light switch and the air conditioner compressor switch are typical examples of this type of sensor.

Backup/neutral safety switches

Backup switches consist of electrical contacts that are activated when the vehicle is put into reverse gear. These switches are used to operate the backup lights. The *neutral start,* or *neutral safety, switch* prevents a vehicle from being started if it is not in park or neutral. This switch is connected in series with the starter solenoid. When the gear selector is not in park or neutral, the switch is open and the starter will not operate.

On most modern vehicles, the backup switch and the neutral safety switch are incorporated into one assembly, which is called a ***combination switch.*** Many combination switches are installed inside the transmission case, Figure 26-3A. Others are installed on the outside of the transmission, Figure 26-3B. Some switches are installed on the shift linkage inside the passenger compartment, Figure 26-3C. On floor-shift models, the switch is usually installed in the console, and on column-shift models, it is installed on the steering column near the firewall.

Throttle position sensors and transducers

Throttle position sensors and ***throttle transducers*** are sensing devices installed on or near the throttle plate to sense changes in throttle plate opening. A computer reference voltage flows to the throttle position sensor.

Throttle position sensors contain a variable resistor or a series of contacts. Throttle transducers contain a wire coil and metal plunger. The basic operation of the sensor and the transducer is the same. Changes in throttle opening cause the amount of current flowing through the sensor or transducer to change. This change in current affects the

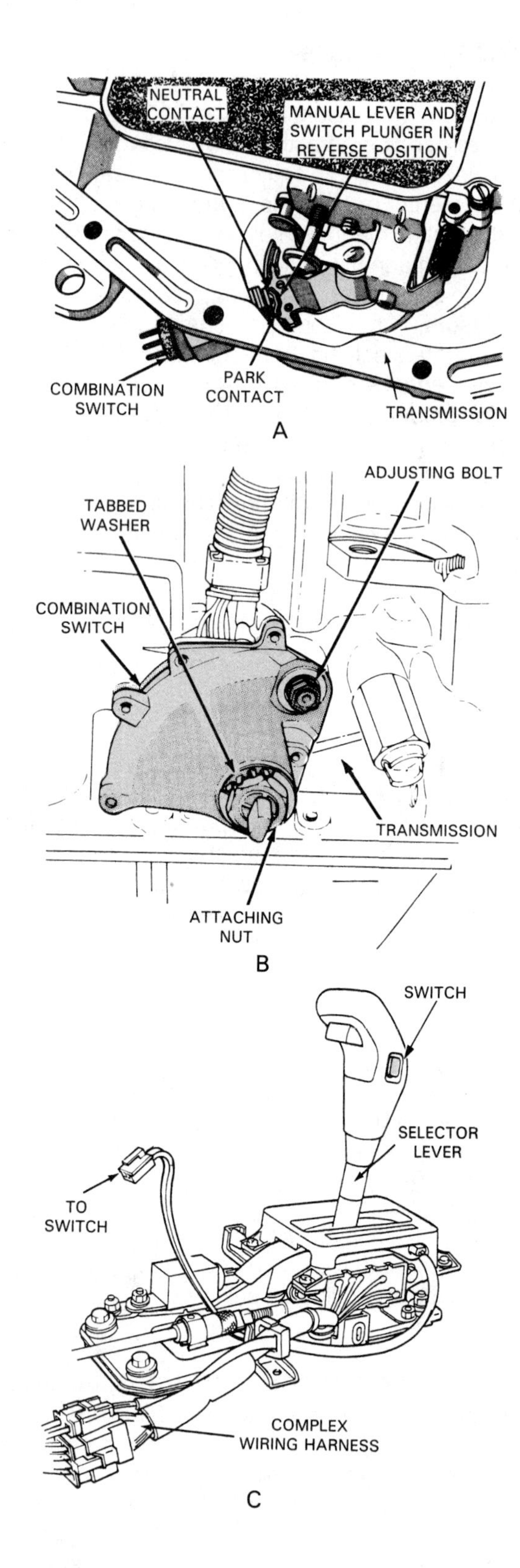

Figure 26-3. Neutral safety switches and backup light switches can be installed in several locations on the vehicle. A–Switch installed inside the transmission case. B–Switch installed on the transmission case. C–Switch installed on the shift selector lever inside the passenger compartment. (Chrysler, Honda)

voltage that the computer receives. The computer interprets the change in voltage as a change in throttle position.

Figure 26-4A shows a typical throttle position sensor. Most modern throttle position sensors can be adjusted by loosening the attaching screws and moving the sensor body until a specified output voltage is obtained.

A cutaway view of a throttle position sensor is shown in Figure 26-4B. Notice the electrical contacts. As the throttle position is changed, the contact plate rotates and sends current to the stationary points. Voltage signals from the stationary points inform the computer when the throttle opening changes.

A variation of the throttle position sensor is the *kickdown switch,* Figure 26-5. In early model vehicles, its function was primarily to shift the transmission to a lower gear when the throttle was pressed to the wide-open position. In most modern vehicles, however, the kickdown

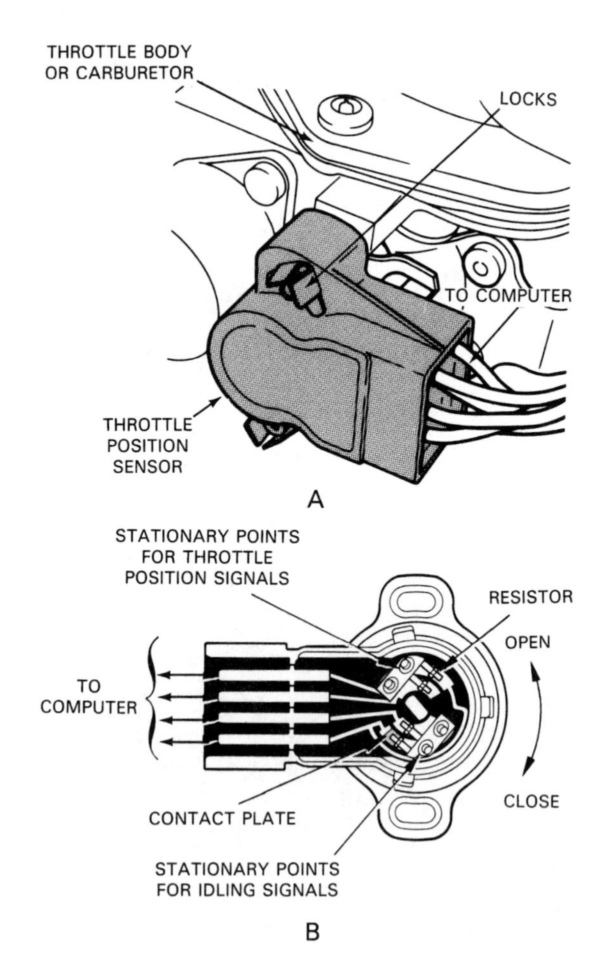

Figure 26-4. The throttle position sensor is designed to sense changes in throttle opening. A–This sensor is mounted on the throttle body assembly. B–This cutaway view of a throttle position sensor shows the internal contacts. Movement of the throttle causes the electrical path to shift along the stationary points, modifying the signal that is sent to the computer.
(Chrysler, Lexus)

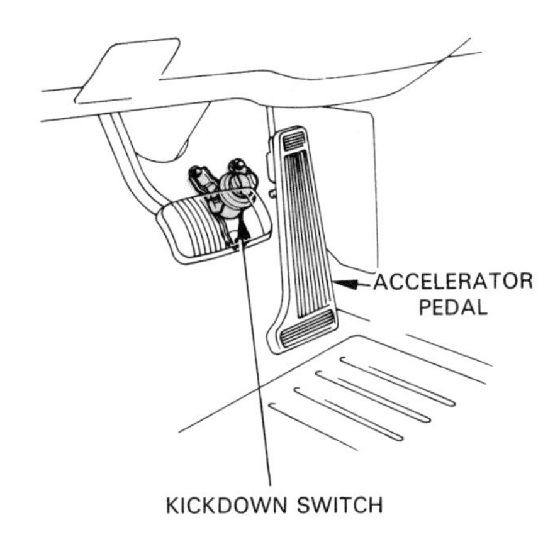

Figure 26-5. The kickdown switch is used to send a wide-open throttle signal to the computer. The signal can be used to shift the transmission into a lower gear (passing gear) or to modify the shift pattern and system pressures to match the increase in engine output. (Lexus)

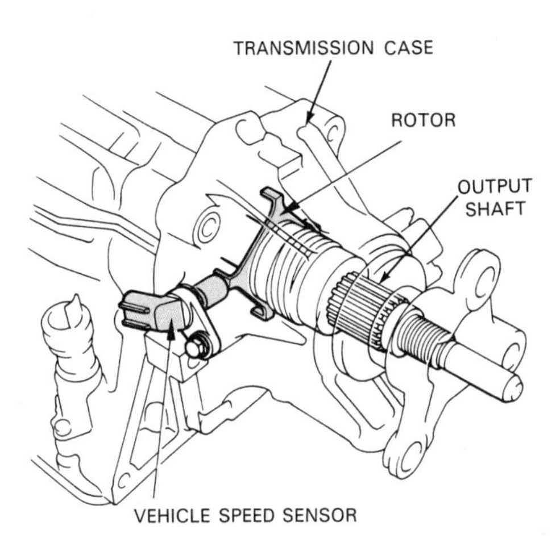

Figure 26-6. Vehicle speed sensors are used by the computer to determine how fast the vehicle is moving. The speed sensor rotor is generally installed on one of the output members of the final drive. (Lexus)

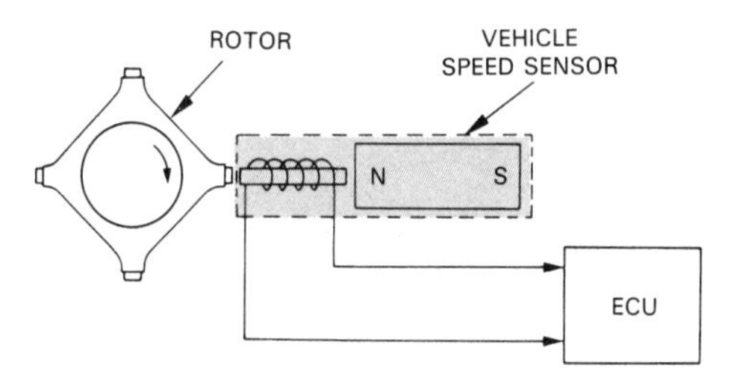

Figure 26-7. Speed sensor operation. The speed sensor, activated by a rotor or toothed wheel, is comprised of a magnet and coil assembly. (Lexus)

switch sends a voltage signal to the computer. The computer uses this input signal to control engine and transmission functions.

The ***idle switch*** is also a variation of the throttle position sensor. It is used to tell the computer when the engine is at idle and when it is operating at speeds above idle. Most idle switches are simple ON/OFF switches. The throttle position sensor shown in Figure 26-4B contains a built-in idle switch. A few idle switches consist of a single positive electrical contact. The contact is grounded to a piece of the throttle linkage when the engine is idling. The grounded connection is broken when the engine is operating above idle.

Vehicle speed sensors

The ***vehicle speed sensor,*** also called the ***road speed sensor,*** is a magnetic sensor used to measure vehicle speed. This sensor is installed on some part of the final drive or after the final transmission gears. This mounting ensures that the input signal will always be matched to road speed, not engine speed. A vehicle speed sensor installed on a transmission extension housing is shown in Figure 26-6.

The vehicle speed sensor contains a small magnet and a wire coil, Figure 26-7. A toothed wheel or rotor is installed on the rotating shaft. When the rotor or wheel rotates next to the sensor, the magnetic field is made to fluctuate. This fluctuating magnetic field induces a small pulsating dc current in the sensor. The current is sent to the computer as a speed signal. As shaft speed increases, the rate of the changing magnetic field increases. Consequently, the magnitude of the current increases.

One version of the vehicle speed sensor consists of a light-emitting diode (LED), a rotating mirror, and a light-sensitive electronic device, or *photocell.* This type of sensor is usually part of the speedometer cable assembly. As the mirror is turned by the speedometer cable, it varies the amount of light reaching the photocell. The photocell is capable of developing a small voltage from the light that it receives from the LED.

The photocell sends the voltage, which is proportional to the vehicle speed, to the computer. This type of sensor is commonly used to provide a vehicle speed signal to the cruise control unit.

Engine speed sensors

The ***engine speed sensor*** is used to measure engine rpm. The sensor is generally installed ahead of the transmission gears. See Figure 26-8. The engine speed sensor operates in the same manner as a vehicle speed sensor. A toothed wheel or rotor rotates next to the sensor, causing a fluctuating magnetic field. This fluctating magnetic field induces a pulsating dc current in the sensor. The current varies with engine speed and is sent to the computer as an input.

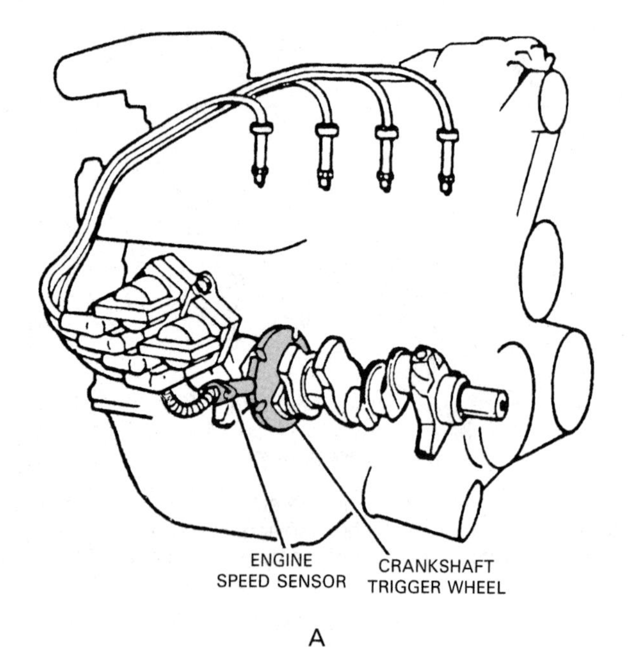

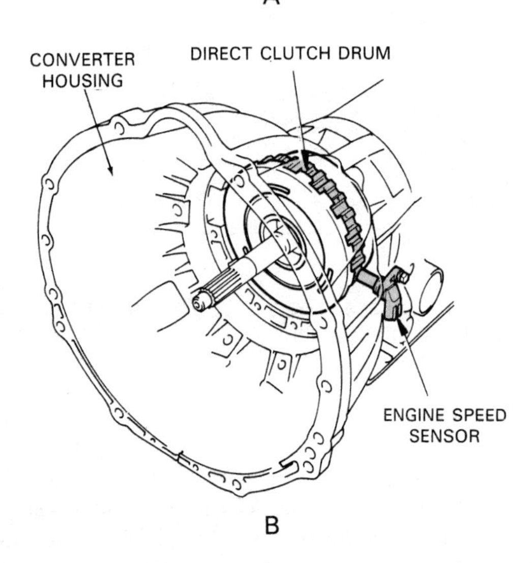

Figure 26-8. A–Engine speed sensor mounted near crankshaft trigger wheel. This type of sensor is often called a crankshaft position sensor. B–Speed sensor located in the transmission case. This sensor monitors the speed of input shaft rotation. (General Motors, Lexus)

Some computer systems use three speed sensors. One sensor measures road speed, one measures engine speed, and the third is positioned to measure the speed of the torque converter and transmission input shaft.

Temperature sensors

The ***engine temperature sensor*** converts engine temperature to an electrical signal. Temperature sensors are installed in an engine coolant passage to measure the coolant temperature. Coolant temperature is a reliable indicator of overall engine temperature. The temperature sensor can be an ON/OFF switch or a variable resistor. Figure 26-9 illustrates a temperature sensor.

The switch-type temperature sensor provides a signal to the computer indicating that the engine has reached a certain operating temperature. The switch-type temperature sensor was widely used with early automotive computers to ensure that the computer would not begin controlling engine operation until the engine was thoroughly warmed up.

The resistor-type temperature sensor can tell the computer the exact engine temperature. This allows the modern automotive computer to make exact adjustments based on engine temperature.

Some temperature sensors installed on newer vehicles measure the temperature of the air entering the intake manifold. This temperature reading, which is of more use to the engine control computer than the drive train computer, is used to match air/fuel ratios to outside air temperatures. Nevertheless, this information is sometimes sent to the drive train computer to more precisely control shift patterns or hydraulic pressure.

Pressure switches

Pressure switches are pressure-sensing devices. In a computer controlled drive train, they are installed in the valve body and send the computer information about the transmission's hydraulic system. Pressure switches are simple ON/OFF switches. See Figure 26-10. The typical pressure switch used in a computer controlled drive train is threaded into an oil passage leading to a hydraulic holding member (clutch or band servo).

Voltage is sent to the pressure switch from the computer. The switch contacts are normally open when the switch is not under pressure. When pressure is applied to the switch, the contacts close, and current flows back to the computer. For example, when hydraulic pressure is sent into a passage to operate a holding member apply piston, it closes the contacts in the pressure switch. Volt-

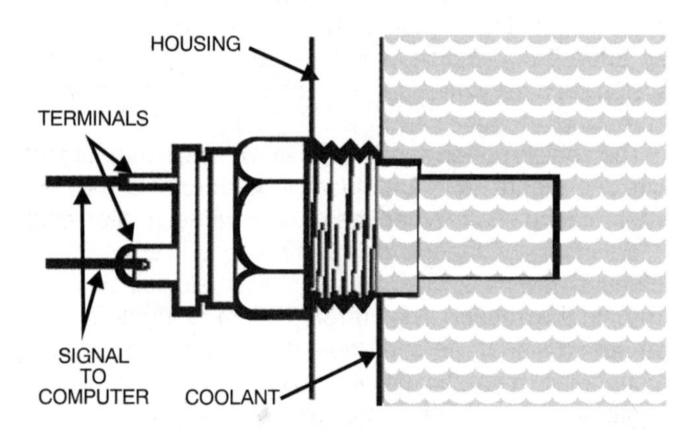

Figure 26-9. The engine temperature sensor is always installed in a coolant passage. It converts temperature changes into electrical signals.

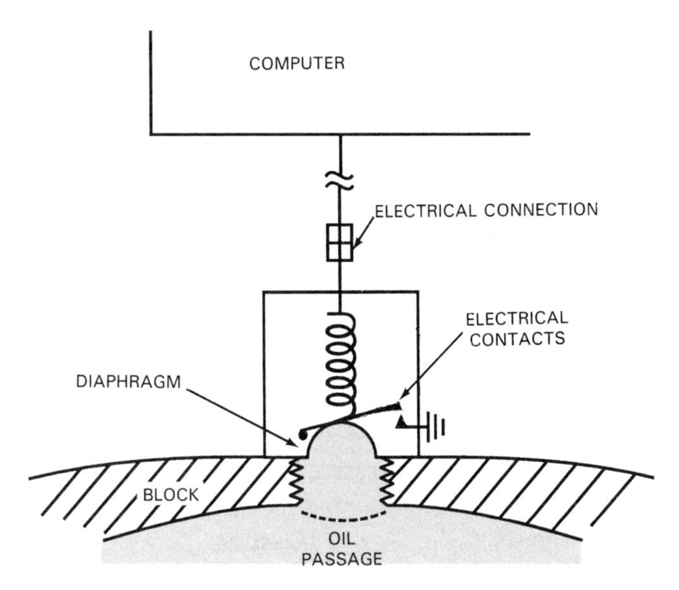

Figure 26-10. The typical hydraulic pressure sensor is a switch that tells the computer if a hydraulic circuit is pressurized. Oil pressure acts on the switch diaphragm to open or close the electrical contacts.

age that flows through the switch informs the computer that the particular holding member is applied. The computer uses this signal to determine which gear the transmission is in and to monitor other operating conditions.

Air pressure sensors

Air pressure sensors detect air pressure changes and convert these changes into an electrical signal. Drive train computers utilize inputs from air pressure sensors to help control drive train operation. The two major types of air pressure sensors are ***barometric (outside) pressure sensors,*** which are often called BARO, or altitude, sensors, and ***intake manifold vacuum sensors,*** which are sometimes called manifold absolute pressure sensors, or MAP sensors.

Many barometric pressure sensors and intake manifold vacuum sensors are combined into a single unit. In most cases, the sensor is a dual-purpose unit that also sends signals to the engine control computer.

Drive Train Computers

The computer used to control the drive train may be the same computer that controls the engine and other vehicle systems. Frequently, however, the drive train computer is one of several on-board computers used to control vehicle operation. It is not unusual for a vehicle to be equipped with separate computers to control the engine, transmission, suspension, steering, cruise control, dashboard indicators, etc. The ***drive train computer*** can operate as an independent unit, or it can be connected with other computers that control various systems.

Drive train computers can be located in many parts of the car. Typical locations are under the dashboard, inside the right kick panel, inside the left or right front fender, under one of the front seats, or on the radiator support. In a few cases, the computers are found under the rear seat or in the trunk.

All on-board computers are solid-state microprocessors. They are collections of computer chips, power transistors, and printed circuits. The solid-state design of the computer allows it to function under the extremes of heat, cold, vibration, and moisture encountered in automotive service.

The names used for the drive train computer vary among manufacturers. Commonly used names include computer, controller, electronic transmission controller, electronic control unit (ECU), and microprocessor. Whatever the name, every on-board computer performs the same job: processing inputs and delivering output commands.

The computer receives inputs from sensors and electrical switches. Many of these inputs are varying low-voltage signals. These signals are called analog signals. The first step in processing analog input signals in a digital computer is to convert them to digital signals. Digital signals are number readings that vary in distinct steps. The digital computer contains internal conversion circuits that turn the incoming analog signals into digital signals. For more information on analog and digital signals, refer to Chapter 5.

Once the inputs are converted to digital form, they are processed by the microchips in the computer. The computer compares the inputs with information stored in its memory. If the inputs do not match the memory settings, the computer changes its output signals. The output signals are sent to output devices, which alter operating conditions until the proper input readings are produced. This process of reading inputs and delivering outputs occurs many times per second.

Computer control systems are often called ***feedback systems,*** because they rely on the sensors to provide information, or feedback, on the changes that the computer is making. The computer can make changes to compensate for variations in vehicle operation and changes in outside factors, such as air temperature and road grade. They can also compensate for part wear. Many computers have a learning capability. These computers can react to long-term changes in the input signals and incorporate these changes into their control patterns, which are used to govern the output devices.

Drive Train Actuators

The output devices that the computer uses to change operating conditions are called ***actuators.*** The computer controlled drive train uses two general types of actuators: servo motors and electrical solenoids. ***Servo motors*** are electric motors that are commonly used to control the operation of engine throttle plates. Servo motors are used on several vehicles to provide hydraulic pressure to the traction control system. Servo motors may also be used to operate transfer case linkages and to provide lockout protection to the gear selectors.

Electrical solenoids are the most commonly used actuators in drive train control systems. They can be used to control the flow of oil to hydraulic valves and are sometimes designed to operate the valves directly. In a few cases, a solenoid is used to control vacuum to a diaphragm that operates a drive train selector linkage.

A typical transmission solenoid is shown in Figure 26-11A. This particular sensor is used to control the flow of oil to the torque converter lockup clutch. Figure 26-11B shows several solenoids installed on the case of a common transmission.

An electrical schematic of a transmission control solenoid is shown in Figure 26-12. Current from the computer energizes the solenoid. Notice that one of the solenoid's electrical connections is completed through a switch that is connected to the brake pedal. This ensures that the solenoid will not be energized when the brakes are applied.

In the past, solenoids were used primarily to control the operation of the lockup converter. In addition to this application, most modern transmissions use solenoids to control hydraulic valves in the shift control system.

Figure 26-13 shows a typical solenoid and its related hydraulic circuit. In Figure 26-13A, the solenoid is not energized. The hydraulic pressure in the circuit holds the valve to the right. In Figure 26-13B, the solenoid is energized and oil flows out of the solenoid passage. Since oil is flowing by the solenoid, the pressure in the circuit drops and the valve spring moves the valve to the left. If you look closely at Figure 26-13, you will notice that moving the valve connects some of the oil passages that were initially closed off. This solenoid and valve combination can be used to cause the transmission to shift by redirecting oil in the hydraulic system.

In some transmissions, the solenoids operate the valves directly. The solenoid may be attached to the valve with a small pushrod. In some cases, the valve is designed with an extension that is attached to the solenoid. The valve is usually spring loaded, so it quickly returns to the off position when the solenoid is de-energized.

Figure 26-14 shows three types of transmission control solenoid configurations used on one modern vehicle. Figure 26-14A shows a solenoid with a simple plunger-type valve that is used to block a hydraulic passage. When the valve is closed, oil cannot escape and full pressure is maintained in the circuit. When the valve is open, oil drains and pressure drops.

Figure 26-14B shows a solenoid and valve assembly that controls flow in the same manner that a conventional shift valve controls holding member application. The valve

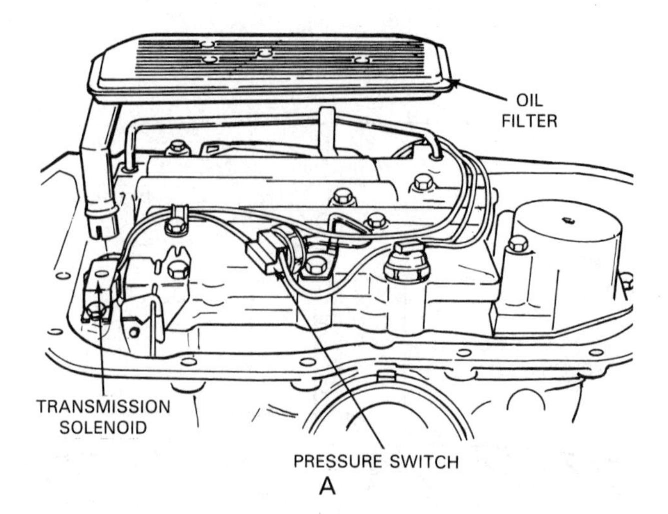

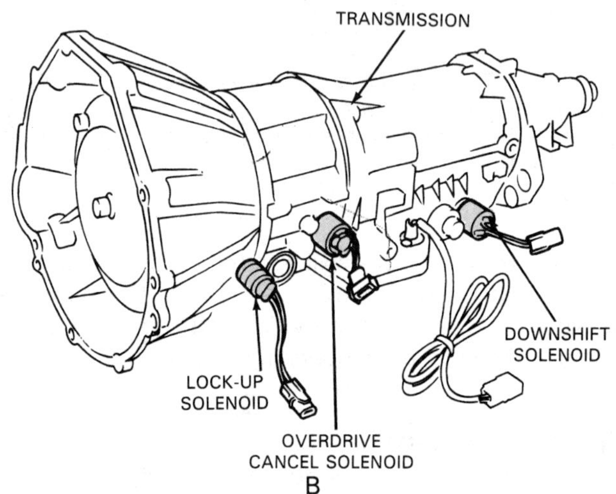

Figure 26-11. Transmission output solenoids can be installed at any spot on the transmission, depending on the type of solenoid and the routing of the internal oil passages. A–Solenoid located inside the transmission oil pan. B–Solenoids threaded into the side of the transmission case. (General Motors, Toyota)

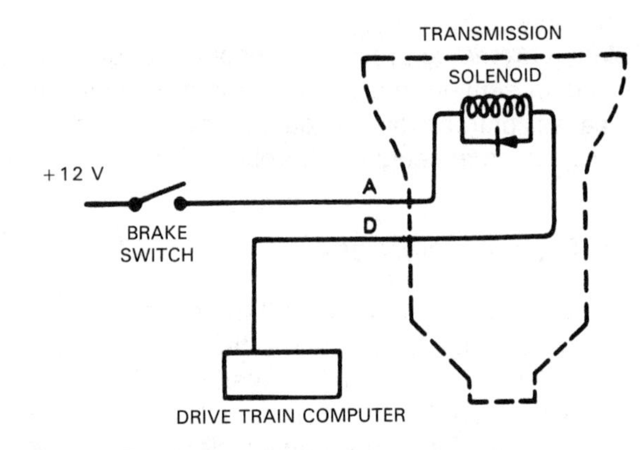

Figure 26-12. This figure shows a typical solenoid wiring diagram. Current from the computer (ECM) enters the solenoid and travels to ground. (General Motors)

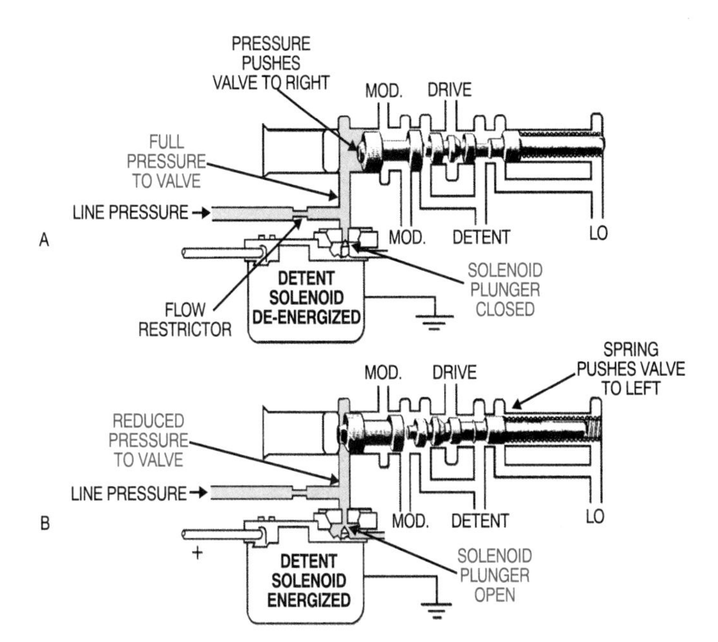

Figure 26-13. Solenoid operation of a hydraulic valve. A–When the solenoid is de-energized, oil is stopped at the solenoid plunger and full oil pressure is sent to the valve. This causes the valve to move to the right. B–When the solenoid is energized, the plunger is pulled off its seat and oil flows out the opening. Oil pressure drops, and the valve spring can move the valve to the left, connecting various oil passages. (General Motors)

is operated directly by a solenoid, which receives an electrical signal from the computer. The solenoid has only *on* (energized) and *off* (de-energized) positions.

Figure 26-14C shows the most advanced type of a pressure control solenoid. This solenoid is able to position the valve according to the strength of the electrical signal from the computer. Variations in current increase or decrease the magnetic field of the solenoid and, therefore, alter the distance that the solenoid can move the valve against return spring pressure. The computer operates this solenoid to control shift feel and overall transmission pressures. A servo motor would be ideal for this actuator application.

Figure 26-15 is a schematic of a drive train control computer with output solenoids. Notice that there is one solenoid to control the torque converter, one to control downshifting, and another to lock out the overdrive system when the driver selects the no-overdrive position. Selecting the no-overdrive position also turns on a light in the driver's compartment to remind the driver that the overdrive has been locked out.

Typical Computer Controlled Drive Train Systems

As previously mentioned, the sensors, actuators, and computers are combined into a complete control system that controls drive train operation. Computer control sys-

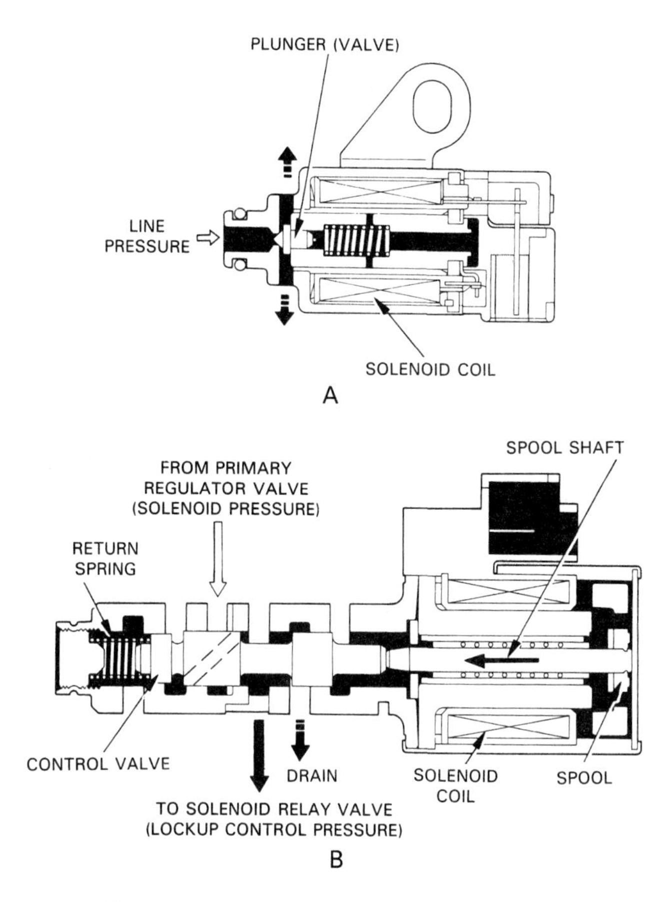

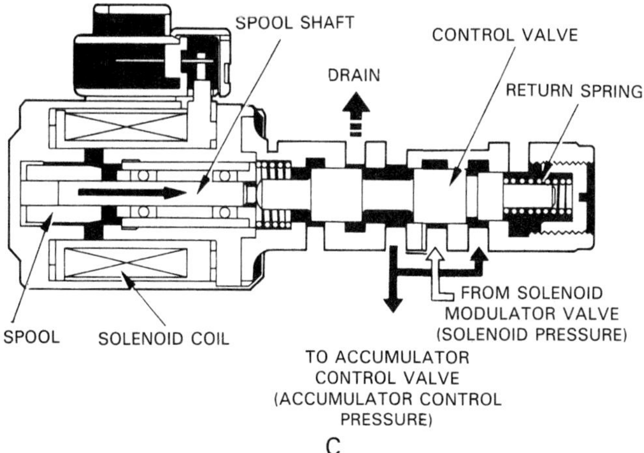

Figure 26-14. Three types of modern solenoid configurations. A–Solenoid and plunger valve. This design relieves pressure to move a hydraulic valve located in a valve body passage. B–Solenoid-operated hydraulic valve. This is a two-position valve that is operated directly by the solenoid. C–Solenoid-positioned hydraulic valve. This valve can be moved to an infinite number of positions, depending on the amount of current sent to the solenoid by the computer. (Lexus)

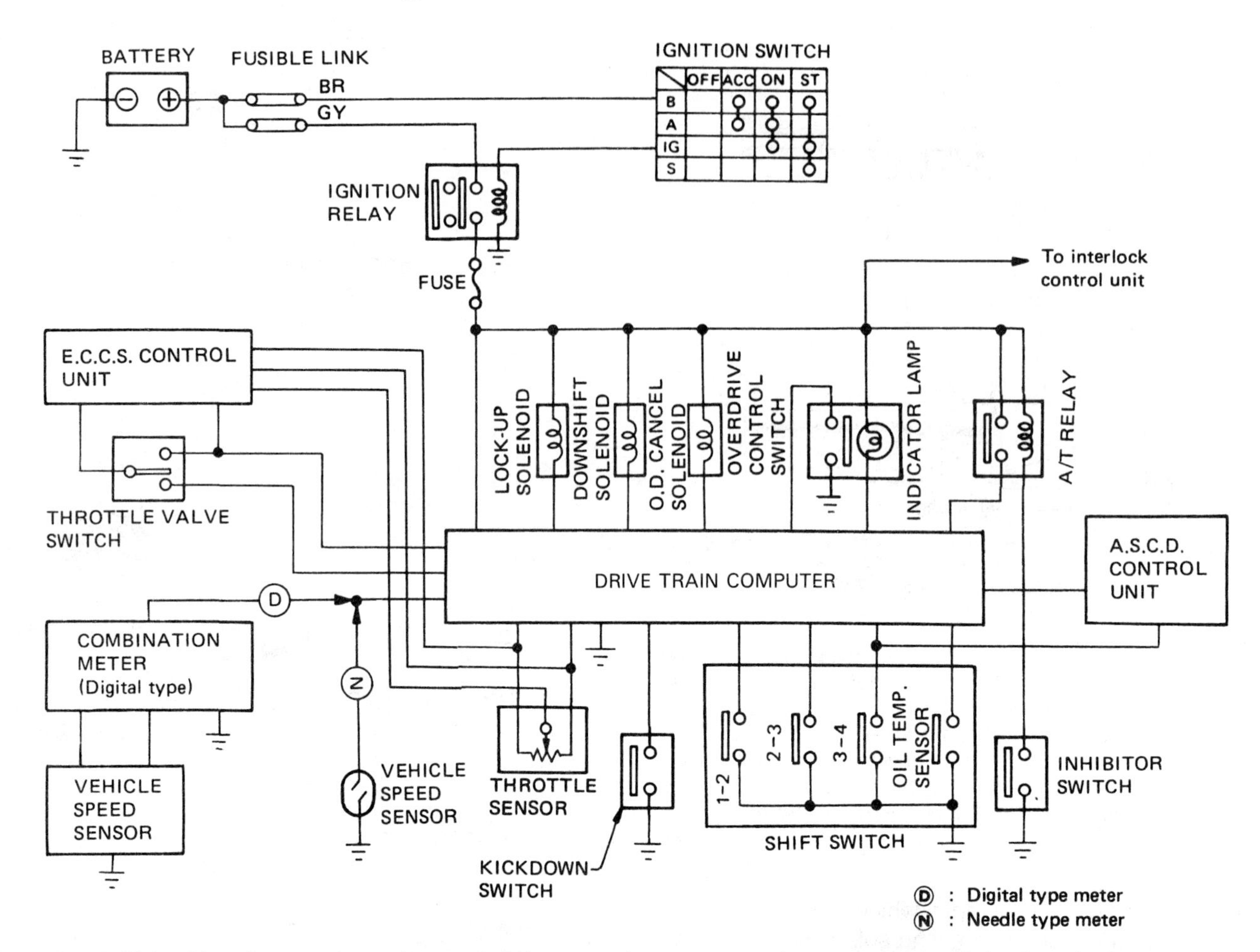

Figure 26-15. This wiring diagram shows the transmission control computer used on one vehicle. Notice how the input sensors and output solenoids are connected to the main computer. The computer is connected to the engine control computer and other computers in the vehicle. (Toyota)

tems are commonly called ***networks.*** Networks are represented by electrical schematics (discussed shortly) and hydraulic schematics.

Figure 26-16 is an electrical schematic of an electronically controlled 4-speed transaxle that is used on many modern vehicles. It consists of a drive train computer, input devices, and output devices. The drive train computer is connected to the engine control computer by a group of bundled wires called a ***bus.*** The engine and transaxle controllers exchange information to control the operation of both the engine and transaxle. The engine controller informs the transaxle computer of engine speed and temperature. It also signals the transaxle computer when the engine is idling.

In addition to the inputs from the engine computer, the transaxle computer receives inputs from the throttle position sensor, the output speed sensor, the turbine speed sensor, and the transaxle hydraulic pressure switches. The output speed sensor tells the computer how fast the vehicle is moving, and the turbine speed sensor tells the computer how fast the transmission input shaft is turning. Internal transaxle pressure switches provide the computer with information on the operation of the hydraulic control system.

The transaxle computer operates the transaxle by energizing and de-energizing the solenoids. The solenoids are installed in a solenoid pack that is mounted to the side of the transaxle case. The solenoid pack also contains the hydraulic pressure switches.

Figure 26-17 shows the hydraulic system of the transmission controlled by the computer system illustrated in Figure 26-16. Note that there are no shift valves, no governor valve, and no throttle valve. All shift decisions are made by the controller through the solenoids.

Two of the solenoid control valves are normally open, which means that they are closed only when their solenoid is energized. The other two solenoid control valves are normally closed. These valves open only when their related solenoid is energized. When the solenoids operate, hydraulic pressures are directed to the proper holding members to place the transaxle in the appropriate gear.

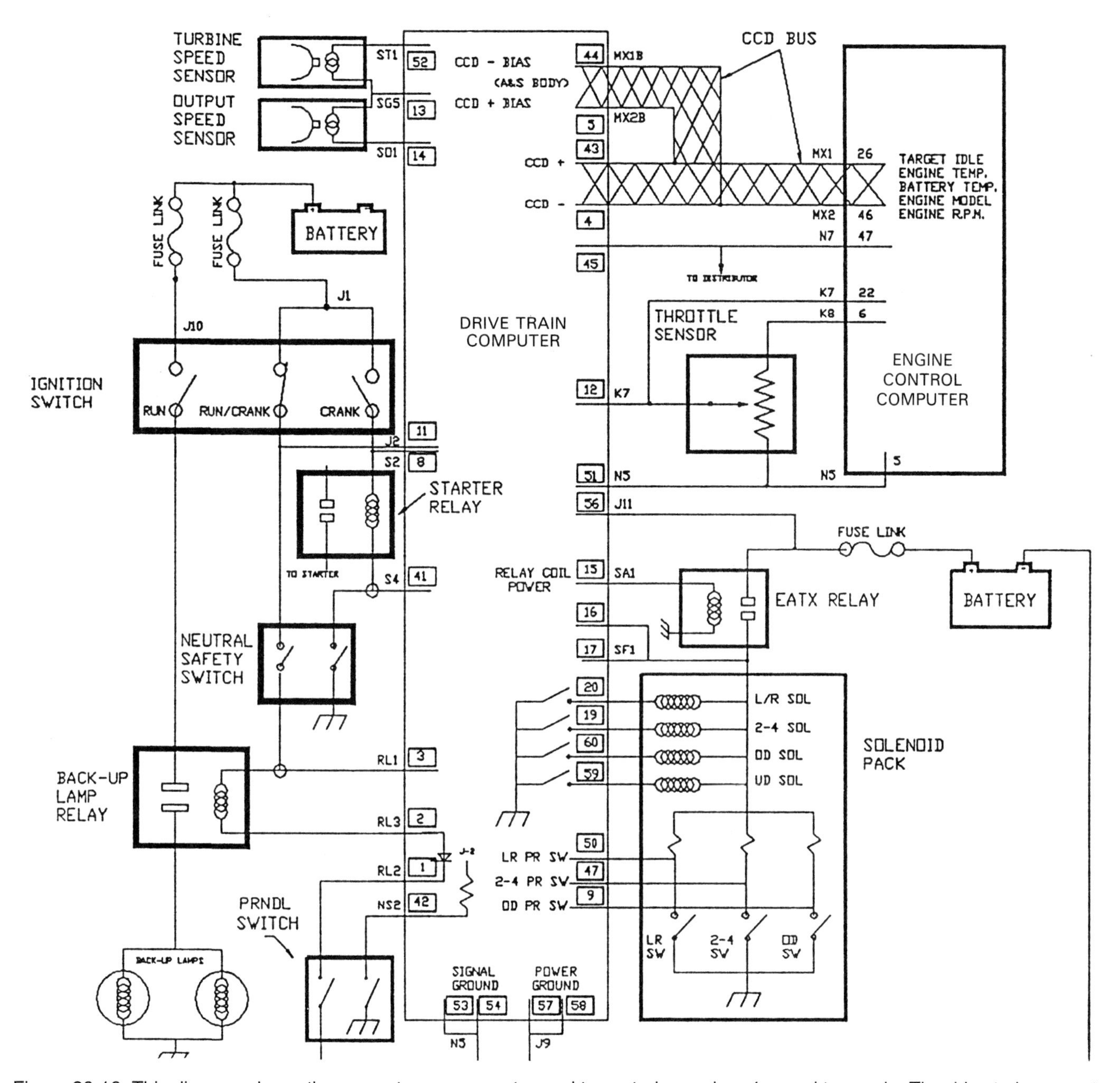

Figure 26-16. This diagram shows the computer components used to control a modern 4-speed transaxle. The drive train computer receives inputs from the sensors, the ignition switch, the shift selector lever, and the engine control computer. Outputs are delivered to a separate solenoid pack, which is installed on the transaxle case. (Chrysler)

The transaxle contains a conventional oil pump, filter, pressure regulator valve, and manual valve. Check valves and accumulators are also used to control shift feel. The torque converter lockup clutch valves are controlled by the operation of the shift solenoids.

Figure 26-18 shows the computer system used on a continuously variable transmission. The computer controls an electromagnetic clutch, which replaces the torque converter. It also operates a solenoid that controls the line pressure. Inputs to the computer include the shift position indicator, stoplight switch, vehicle speed sensor, air conditioner clutch switch, and two ON/OFF throttle position switches. This transmission computer also receives information from the engine control computer. The engine computer sends information about engine temperature, speed, and torque to the drive train computer.

Figure 26-19 is a diagram of a computer control system that controls both the engine and the transmission of a particular vehicle. A single on-board computer controls the engine fuel and ignition systems and the transmission shift

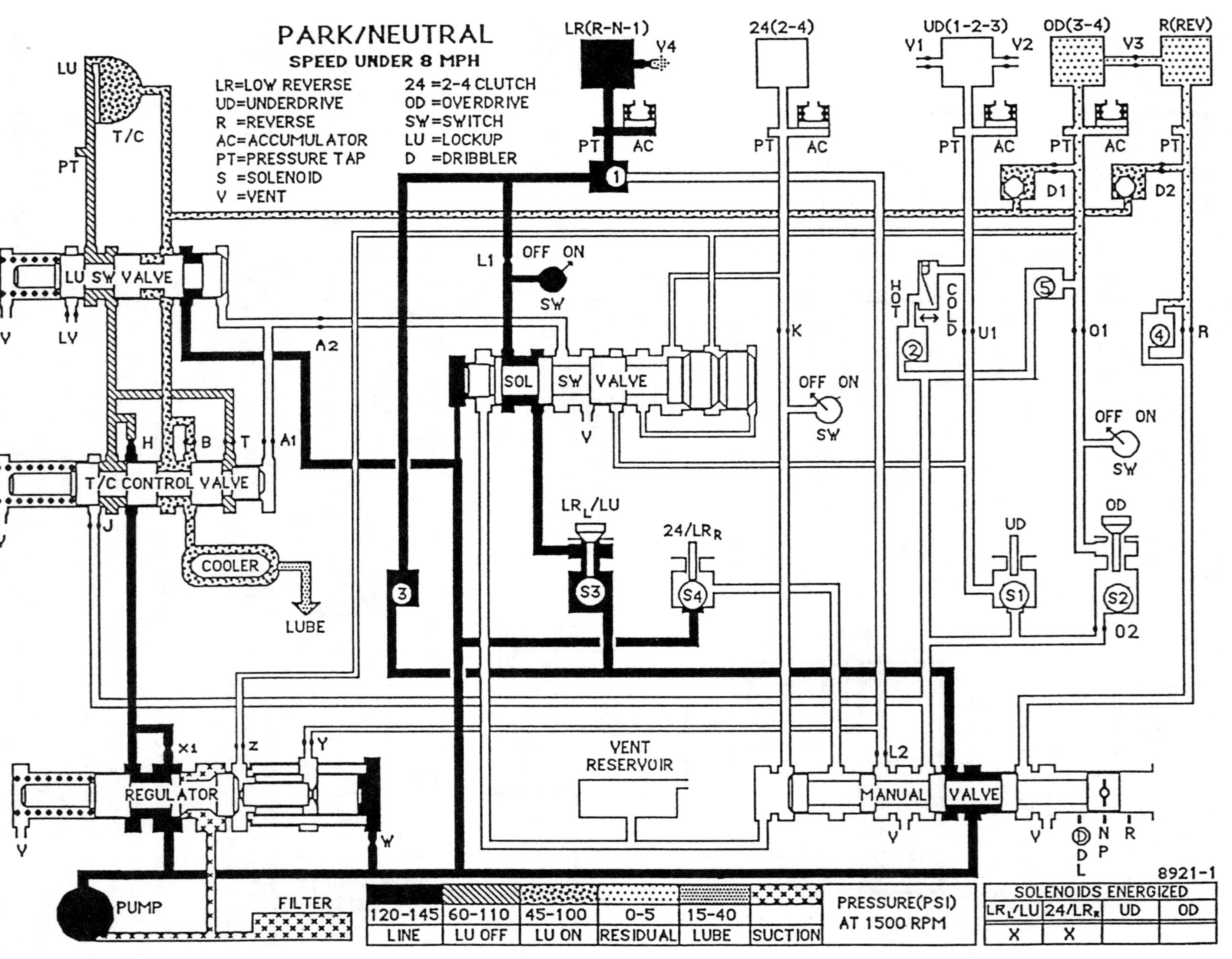

LR$_L$/LU	24/LR$_R$	UD	OD
X	X		

Figure 26-17. Schematic of the hydraulic system of a computer controlled 4-speed transaxle. Notice that there are no shift valves in this system. All transmission shifts are controlled by the four solenoids (S1, S2, S3, and S4). (Chrysler)

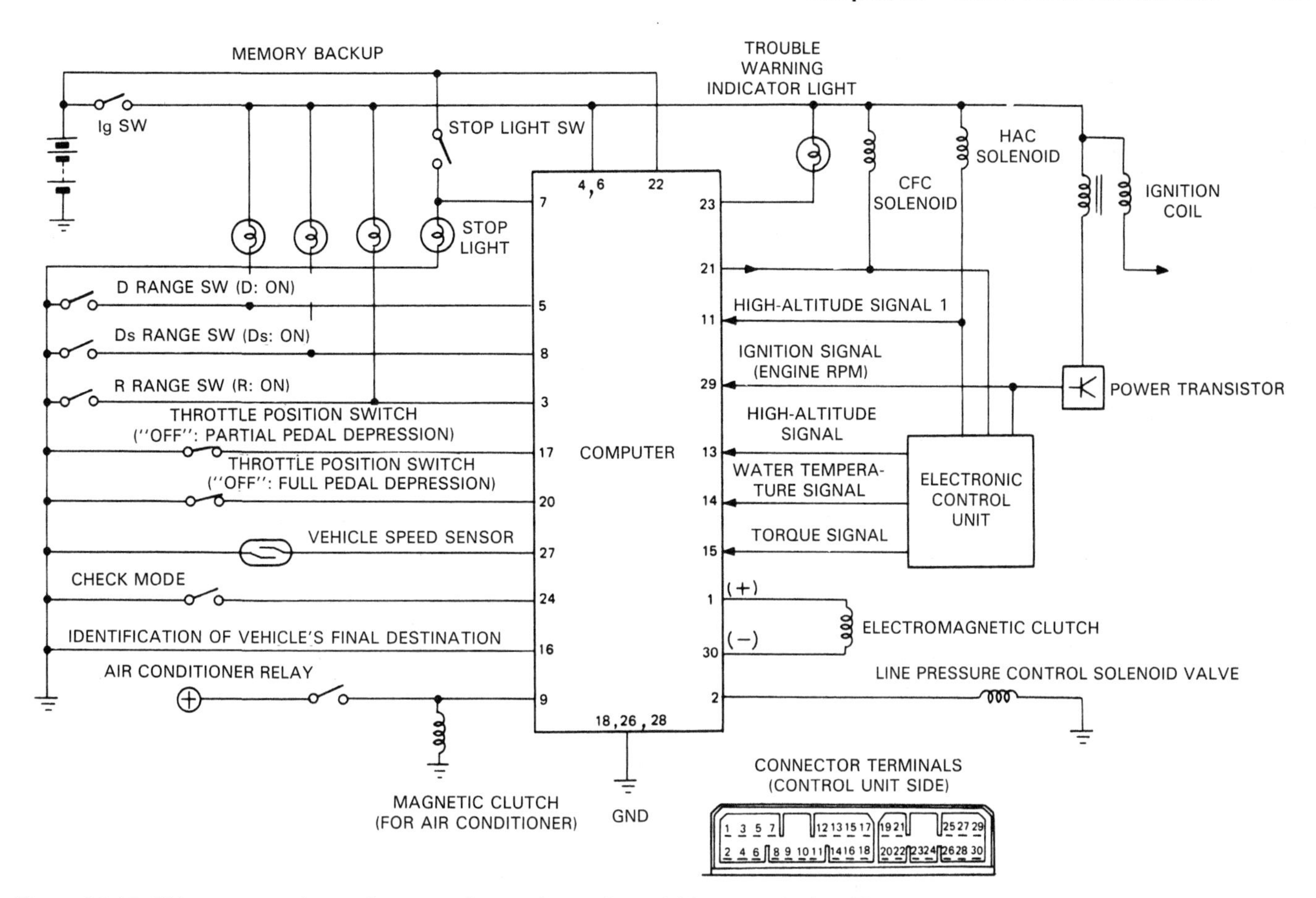

Figure 26-18. This computer is used to control a continuously variable transmission. The computer inputs are similar to those used on other computer controlled transmissions. However, the outputs are sent to the electromagnetic clutch and to a solenoid that controls the transmission's hydraulic pressure. (Subaru)

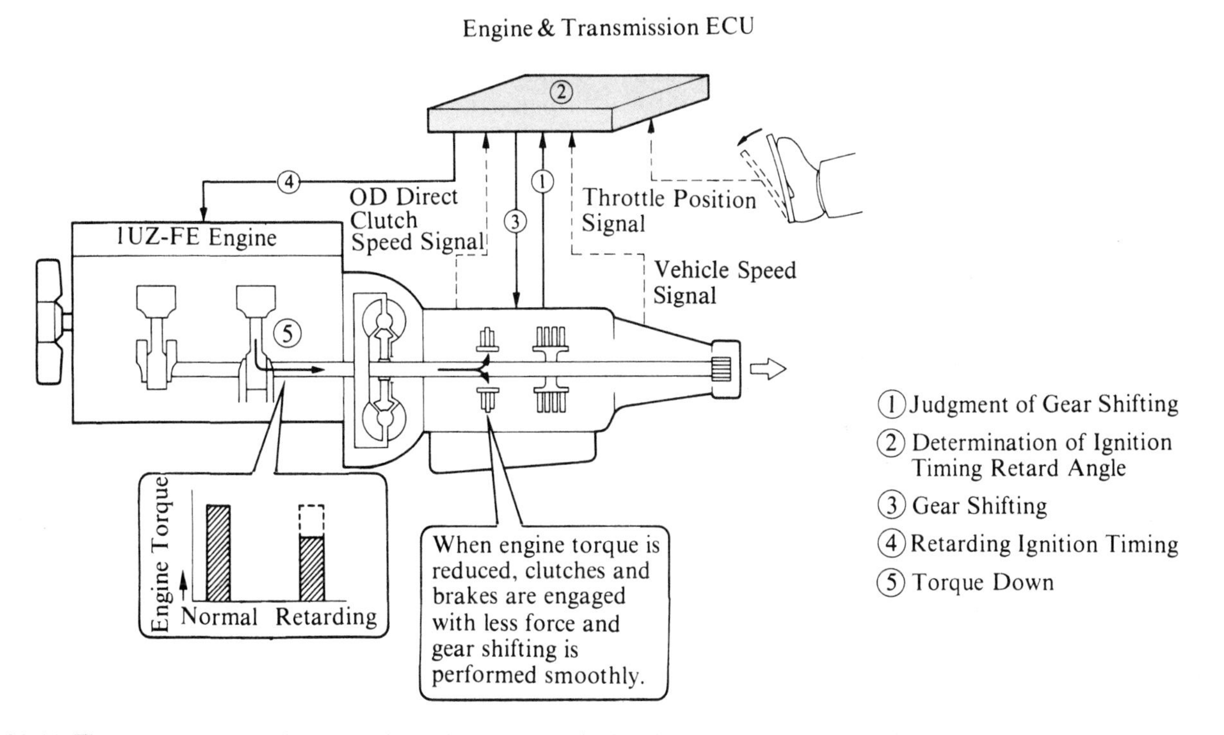

Figure 26-19. The computer control system shown here uses a single microprocessor to control both the engine and the transmission. (Lexus)

points and hydraulic pressures. A unique part of this system is an engine torque reducing feature. As the transmission upshifts, the computer retards engine timing to slightly reduce torque. This allows the transmission to shift more smoothly and reduces engine and transmission wear. Once the transmission is firmly in the next gear, engine timing is advanced to increase torque.

Figure 26-20A is a schematic of a traction control system used on one particular vehicle. It controls vehicle traction during turns and on slippery road surfaces by processing inputs from all four wheels and the steering wheel. The position of the steering wheel and differences in wheel speeds inform the computer when a wheel loses traction. The computer outputs modify the operation of the ignition and fuel systems to reduce power until full traction is restored.

Figure 26-20B shows the location of the parts of another traction control system. In addition to controlling the engine and drive train, this system controls part of the brake system. In operation, it turns on a small electric motor to pressurize the brake system. The system can apply the brakes at any wheel that is losing traction.

Servicing Drive Train Electronics

The computerized parts of the drive train do not require maintenance. The computer and most sensors and output devices are sealed solid-state units that cannot be adjusted. The majority of computer system parts are *replaced,* and then, only when they develop problems. Sometimes, these problems are common to a particular system. At other times, the problems are unusual and are difficult to pinpoint.

Troubleshooting is an important part of servicing a drive train computer system. Keep in mind that locating a problem can be complicated by the fact that the sensors and the computer generally fail internally and cannot be checked visually for problems. Fortunately, the drive train computer system on most late-model vehicles has the ability to diagnose itself.

Drive Train Computer Self-Diagnosis

The most important technique to remember when troubleshooting a computer controlled drive train system is to take logical, systematic steps. Before beginning computer system diagnosis, always look for conventional drive train problems. Check for low fluid levels, maladjusted or disconnected linkage, worn parts, loose bolts, disconnected or corroded wiring, and other obvious problems before investigating the computer system.

To begin diagnosing the drive train computer system, you must place the computer in the self-diagnostic mode. Methods for triggering the self-diagnostic mode vary, but in most cases, it is accessed by making a wiring connection. The connection is generally made between two computer self-test terminals or between the computer self-test terminal and ground.

Figure 26-21 shows a computer and self-test connectors that are located under the dashboard of one particular vehicle. On a few vehicles, the technician can trigger computer self-diagnosis by performing a certain sequence of operations with the dashboard controls.

NOTE! Always check the correct service manual for the self-diagnosis triggering procedure. The improper procedure can damage the computer system.

Once the system is in the self-diagnostic mode, the computer produces a sequence of voltage pulses that occur at certain intervals. The sequence of pulses represents one or more ***trouble codes.*** Trouble codes are numbered codes that represent specific areas in the computer system. The codes are used to pinpoint problem areas. The pulses are communicated to the technician by a series of flashes of the dashboard computer light or computer-mounted LEDs. In some cases, an analog meter is used to monitor the voltage pulses. The method for interpreting pulses varies among manufacturers. Consult the appropriate service manual for specific instructions. After the trouble codes have been identified, they can be compared with the *trouble code chart* in the service manual to determine the problem area.

Figure 26-22 shows the trouble code chart for a specific computerized drive train system. Notice that it lists the trouble code numbers, the system with the defect, the probable cause, and the parts that require further checking.

As previously mentioned, the trouble codes are used to find the general problem area in a computer system. In most cases, however, further checking will be needed to pinpoint the problem. If trouble codes are not displayed when the self-diagnostic mode is triggered, do not assume that the computer system is operating correctly. Some areas or components are not connected to the computer diagnostic system. These areas can often be checked with an ohmmeter or a voltmeter. In some cases, components in these areas must be checked by replacing them with a part that is known to be good and rechecking operation.

Many problems in the computer system are not caused by defective components. Because the voltages produced by sensors are very low, small changes in the resistance of wires or connections can cause incorrect sensor readings. Shorted or disconnected wires can cause system malfunctions and may damage the computer or other devices. Before checking system components, check all wires and connections. Also check for incorrect linkage adjustments and disconnected vacuum lines. Often, a computer system problem can be solved by performing these simple checks.

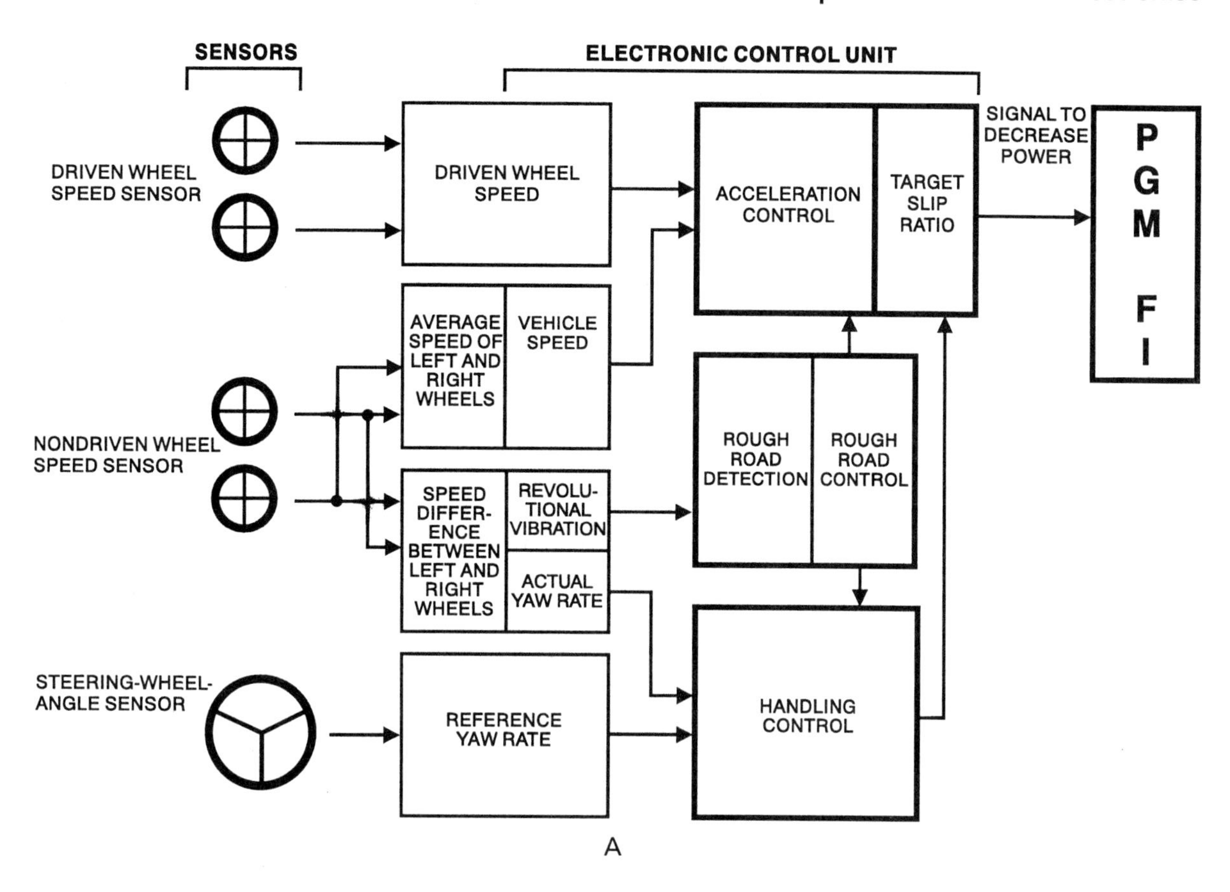

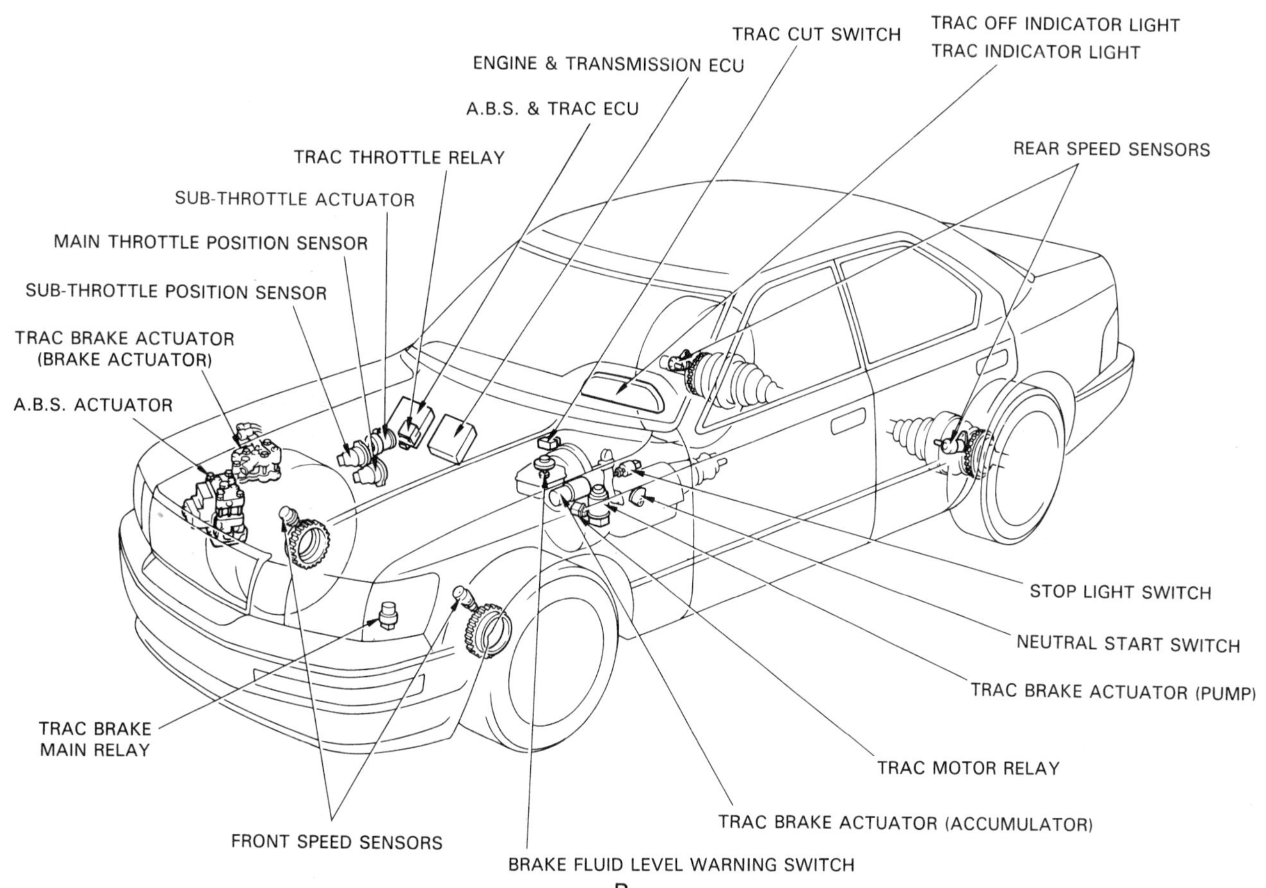

Figure 26-20. On-board computers can be used to control traction by reducing wheel spin. A–The system shown here controls traction by reducing engine output when wheel slippage begins. B–This system prevents wheel spin by applying the brake on the wheel that is slipping. It is connected to the anti-lock braking system (ABS). Both systems use wheel-mounted sensors at all four wheels to determine when slippage begins. (Toyota, Lexus)

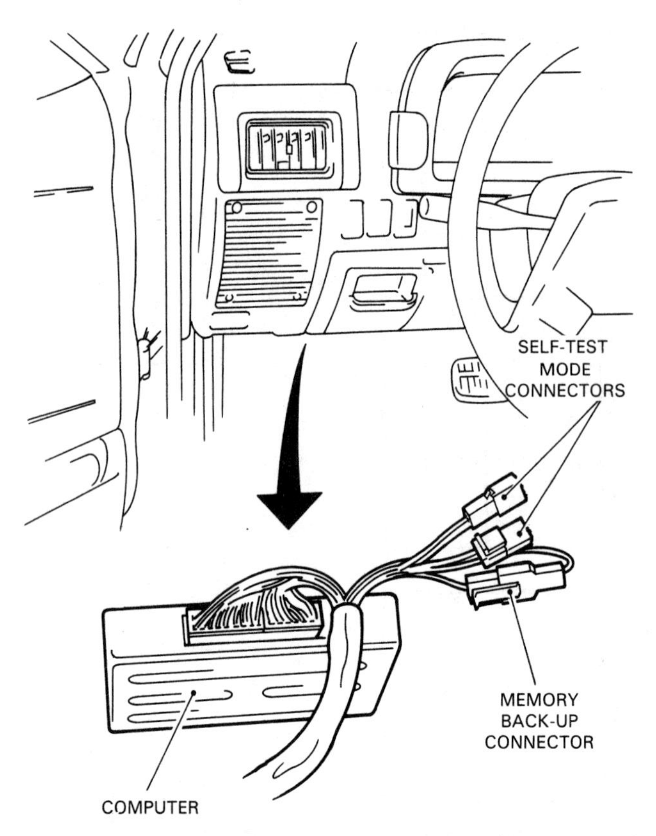

Figure 26-21. The location of the self-diagnostic connector is shown for one particular vehicle. The connector can be located in one of several places. Consequently, the manufacturer's service manual should be consulted to determine the connector location on a specific vehicle. (Subaru)

Using Wiring Diagrams

In many cases, the exact design of the computer system electrical circuits is not obvious. Therefore, the technician must often use a wiring diagram to determine how circuits are arranged when troubleshooting electrical systems.

A ***wiring diagram*** is a set of lines and symbols that represents electrical circuits. The wiring diagram can be thought of as a map of the electrical system. Tracing the circuit will enable the technician to determine which electrical devices are used in a particular system and how they are connected.

A typical wiring diagram is shown in Figure 26-23. This diagram illustrates how various components are connected in one particular computer controlled drive train system. To trace a circuit in a wiring diagram, you must follow the circuit from the part that is not operating properly until you reach the components or wires that could be the source of the problem.

For example, if a vehicle equipped with the system illustrated in Figure 26-23 will not go into overdrive, the place to start when tracing the overdrive circuit is at the overdrive solenoid. The solenoid itself could be bad, but there is no way to tell without removing the transmission oil pan. Tracing the wiring back along the diagram will result in reaching the overdrive relay, which also could be the problem. However, tracing the circuit further shows that the relay is controlled by the overdrive switch and an engine coolant temperature switch. Both of these units get their electricity through a single fuse. With the help of the

Trouble code	System in trouble	Probable cause	Parts to check
13	D-range switch signal system	D-range switch signal circuit open or shorted.	① Wire harness and connector ② D-range inhibitor switch ③ Control unit
14	Ds-range switch signal system	Ds-range switch signal circuit open or shorted.	① Wire harness and connector ② Ds-range inhibitor switch ③ Control unit
15	R-range switch signal system	R-range switch signal circuit open or shorted.	① Wire harness and connector ② R-range inhibitor switch ③ Control unit
21 (*1)	Torque signal system	Torque signal remains "ON" or "OFF".	① Wire harness and connector ② Control unit ③ EFC control unit

Figure 26-22. Troubleshooting chart for one type of on-board computer. After determining the problem areas, problems can be pinpointed by further checking. (Subaru)

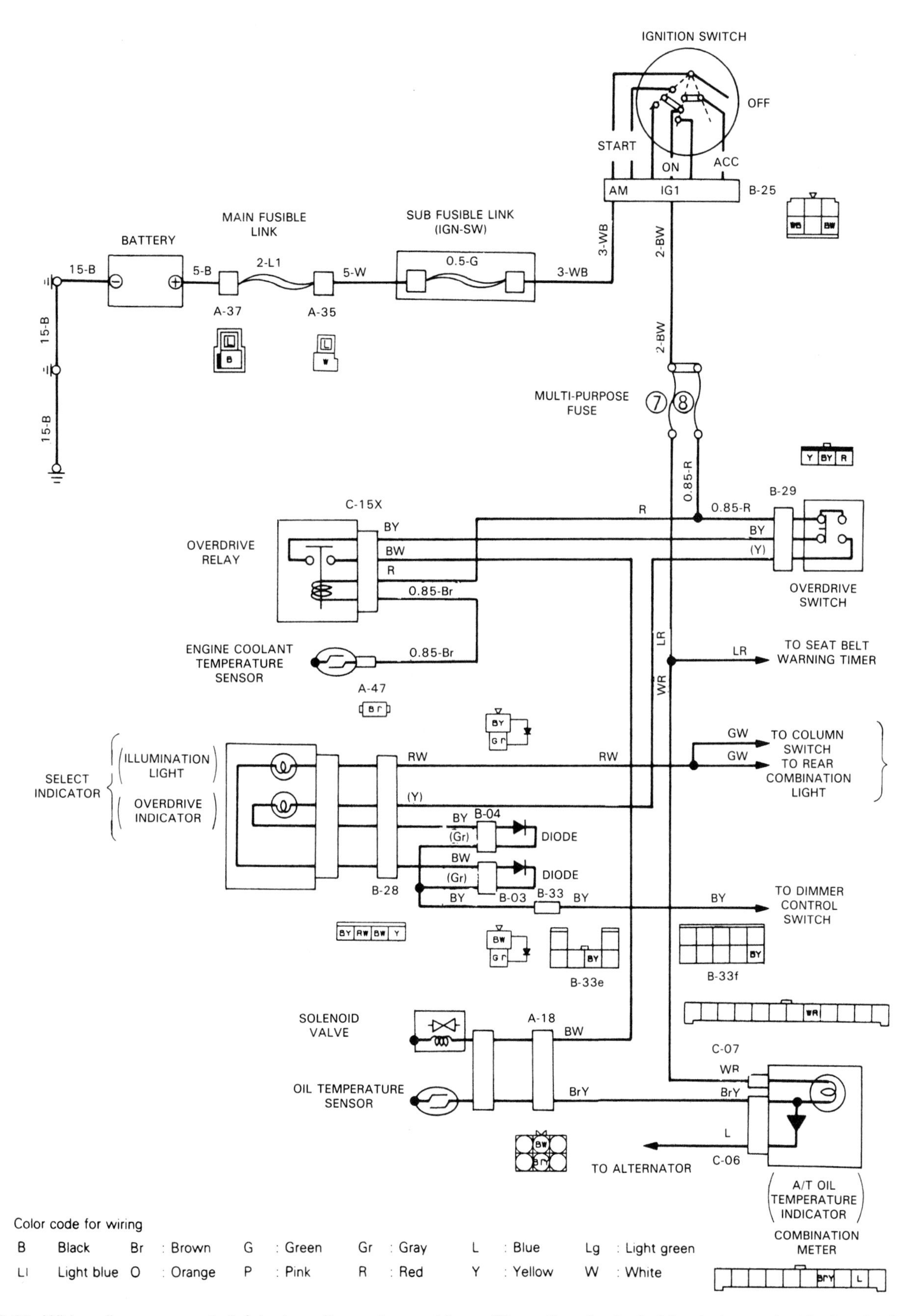

Figure 26-23. Wiring diagrams are helpful when diagnosing problems. They allow the technician to trace electrical paths from the problem area to the probable causes. Always obtain the proper diagram for the vehicle being serviced.

diagram, the technician can bypass or replace any of the components in the circuit to find the problem. In this case, the fuse is the part most likely to fail and is relatively easy to replace.

If replacing the fuse does not solve the problem, but bypassing the relay causes the solenoid to activate, the technician knows the solenoid is good, and the problem is in the relay or the switches. If bypassing one of the switches causes the relay to operate, the bypassed switch is defective. Most common electrical system problems can be located by using a wiring diagram and by following logical troubleshooting procedures.

Types of diagrams

There are essentially two kinds of wiring diagrams used with computer systems, the *direct wiring diagram,* which shows exact wiring connections, and the *logic diagram,* or *flow diagram,* which shows the major computer inputs and outputs in a general way. Sometimes, hydraulic system flows are shown in logic diagrams. In most cases, the logic diagram is used to obtain general information on system operation, while the wiring diagram is used for circuit tracing.

Figure 26-24 is a logic diagram. Compare it with the wiring diagram in Figure 26-23. The logic diagram shows the general path of the electrical and the hydraulic circuits. While this type of diagram is useful for gaining an understanding of the system, it must be supplemented by the proper direct wiring diagram when tracing a circuit.

When you must trace wiring, use the wiring diagram that was produced specifically for the vehicle at hand. Improvements are continually being made in computer systems, and an older or newer wiring diagram is likely to be very different from the actual wiring system.

Drive Train Sensor Service

General procedures for checking the major classes of sensors are explained in the following sections. There are many sensors used in vehicle computer systems. The sensors covered in this chapter apply most directly to the drive train. There may be other sensors that affect the drive

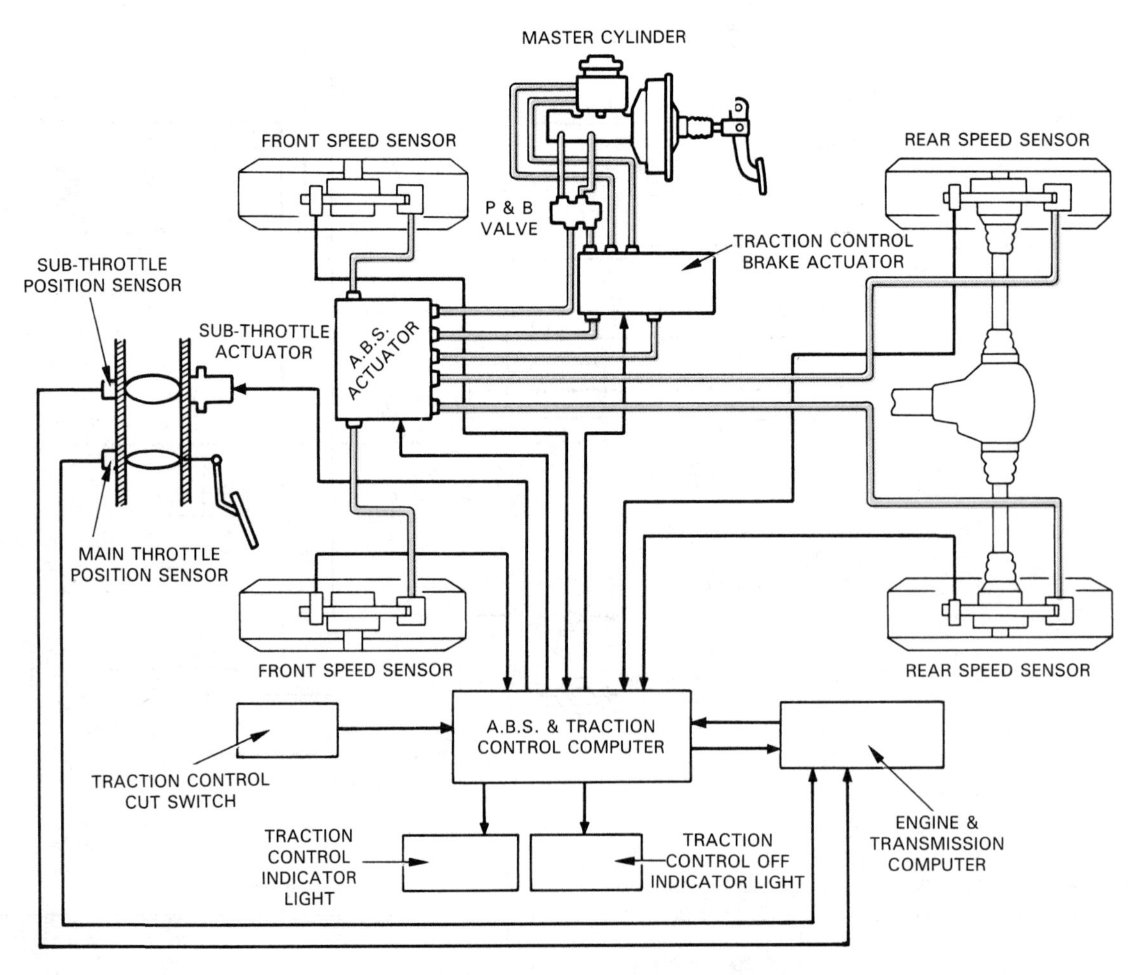

Figure 26-24. Logic diagram for a typical traction control system. It shows the overall electrical flows of the traction control system and the anti-lock braking system. It also illustrates the related hydraulic connections. (Lexus)

train control system, depending on the system design. Refer to the proper factory service manual for exact service procedures.

Switching sensor service

Switching sensors are ON/OFF switches used to indicate pressure, vacuum, or the operation of various electrical devices. Switches were used on vehicles long before the introduction of computers, and many of the testing procedures for switching sensors resemble those used for conventional switches.

Some switching sensors, such as the vacuum switch illustrated in Figure 26-25, can be adjusted. Begin the adjustment procedure by removing the wiring connector and connecting a self-powered test light. Start the engine and run it at a high idle to develop high manifold vacuum. Adjust the vacuum switch by turning the adjustment screw with a hex wrench. In the adjustment procedure shown in Figure 26-25, the test light will turn off when the proper adjustment setting is reached.

The backup light/neutral safety switch must be adjusted when: the vehicle will not start in park or neutral; the vehicle will start in any other gear; the switch seems to be the source of a computer problem; or the backup lights do not come on when the vehicle is in reverse.

Some switches are located on or inside the transmission, while others are attached to the shift linkage. A typical backup light/neutral safety switch is illustrated in Figure 26-3.

To adjust the backup light/neutral safety switch, loosen the switch adjustment mechanism. Place the shift selector lever in neutral, and try to start the engine. If the starter will not crank the engine, move the neutral safety switch in small steps and recheck starter operation. It is often easier to have a helper hold the ignition switch in the start position as you move the backup light/neutral safety switch. Once the engine starts in neutral, tighten down the switch adjustment mechanism and check switch operation in all other gears. The engine should crank only when the shift selector is in neutral or park. The backup lights should come on in reverse gear only. Readjust the switch as needed. In most cases, adjusting the switch for proper starter operation will set its contacts for the other shift selector lever positions.

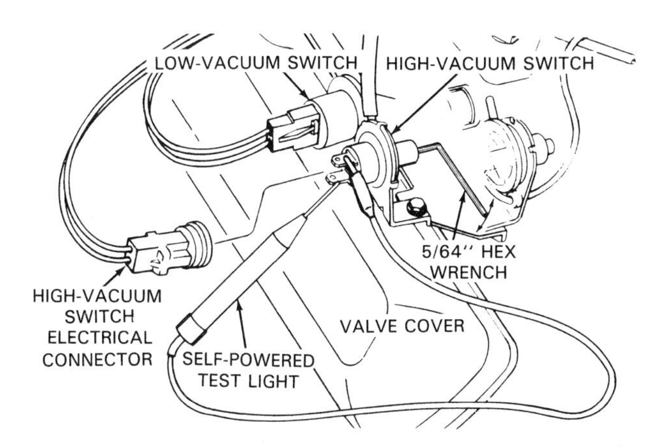

Figure 26-25. This vacuum switch can be adjusted to provide the proper signal to the computer. Many sensors are not adjustable and must be replaced if they do not provide the proper inputs. (General Motors)

NOTE! Some neutral safety switches are adjusted using a special alignment pin. If the pin can slide through holes in the movable and stationary switch parts when the shift selector lever is in park, the switch is adjusted properly.

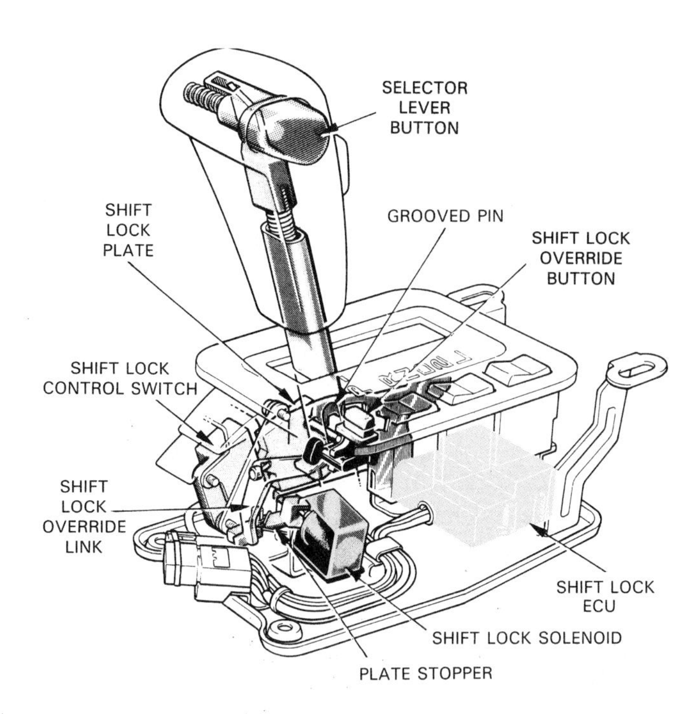

Figure 26-26. This variation of a neutral safety switch uses a small computer and a solenoid to prevent damage to the drive train and engine. The shift selector lever is connected to a small computer. The computer triggers the solenoid, which will not allow the lever to be moved out of gear unless the brake is applied. (Lexus)

A unique version of the neutral safety switch is shown in Figure 26-26. A control unit, called a *shift lock ECU,* and a solenoid are used to prevent the shift selector lever from moving out of the park position until the brake pedal is depressed. This keeps the selector lever from being moved accidentally. It also prevents shock loads on the transmission that occur when it is shifted into gear at high engine rpm. Such shock loads can damage the engine or the drive train.

The shift lock ECU also operates a solenoid in the steering column, Figure 26-27. The solenoid is not energized unless the shift selector lever is in park. This prevents the vehicle from being started when the shift

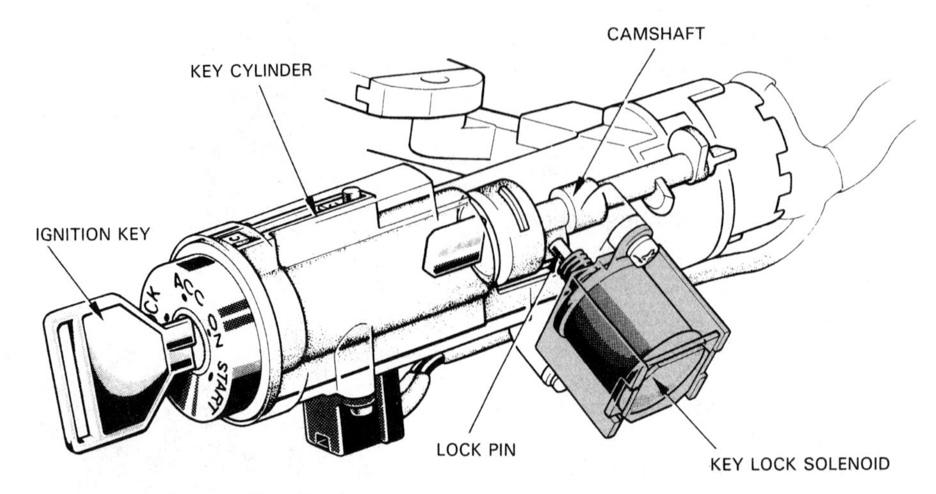

Figure 26-27. This solenoid is installed on the ignition key assembly and is operated by the computer shown in Figure 26-26. It prevents accidental starts in gear and reduces the chance of vehicle theft. (Lexus)

selector lever is not in park. The ECU and solenoid take the place of linkage devices that prevent a vehicle with an automatic transmission from starting when the shift selector lever is in the wrong gear.

Pulse-generating sensor service

Pulse-generating sensors are used to measure the rotating speed of the engine crankshaft, the transmission input shaft, the output shaft, or other components. Many pulse-generating speed sensors can be checked with an ohmmeter, Figure 26-28A, or an AC voltmeter without being removed from the vehicle. Compare the meter reading with factory specifications. Some pulse-generating sensors can be checked with a magnet, Figure 26-28B. After removing the sensor from the vehicle, use a magnet to simulate the fluctuating magnetic field created by the turning shaft. The sensor leads should be attached to a voltmeter that can register small voltage pulses. Moving the sensor near the magnet should produce a voltage reading. If the pulse-generating sensor fails the ohmmeter or voltmeter tests, it should be replaced.

Throttle position sensor service

Throttle position sensors are often checked with an ohmmeter. Remove the wiring harness connector and attach the ohmmeter to the sensor leads. See Figure 26-29. Resistance should vary as the throttle is opened and closed. If it does not, replace the sensor. A variation of this procedure involves using a voltmeter to measure voltage drops at specified throttle positions. If the voltage readings are not within specifications, the sensor is defective. In many cases, a scan tool can be used to check the operation of the throttle position sensor.

Some throttle position sensors can be adjusted. Attach a voltmeter or an ohmmeter as required and loosen the adjustment lock screws. Adjust the sensor to obtain the proper reading and retighten the lock screws. Recheck the reading before removing the meter.

NOTE! In many cases, throttle position sensors can only be adjusted with the help of a special computer system tester. Always check the factory service manual for exact adjustment procedures.

Temperature sensor service

Most temperature sensors will register a low resistance when cold. Resistance will increase as the sensors warm up. The easiest way to check a temperature sensor is to leave it installed and make ohmmeter tests as the engine temperature changes. A sensor that does not change its resistance as temperature increases should be replaced. If necessary, digital thermometer or pyrometer can be used to measure sensor temperature. Resistance can then be measured at specific temperatures and compared to specifications. See Figure 26-30.

Hydraulic pressure sensor service

The pressure sensors used in drive train applications are almost always simple ON/OFF switches. These sensors are used to indicate which gear the automatic transmission is in and which hydraulic circuits are pressurized. Pressure sensors can be checked with an ohmmeter without removing them from the vehicle. The electrical connector to the sensor should be disconnected. One lead of the ohmmeter should be connected to the sensor lead, and the other lead should be connected to ground.

The vehicle can be operated on a lift until the hydraulic system activates the sensor. If the ohmmeter reading does not change when the sensor is activated, the sensor is defective.

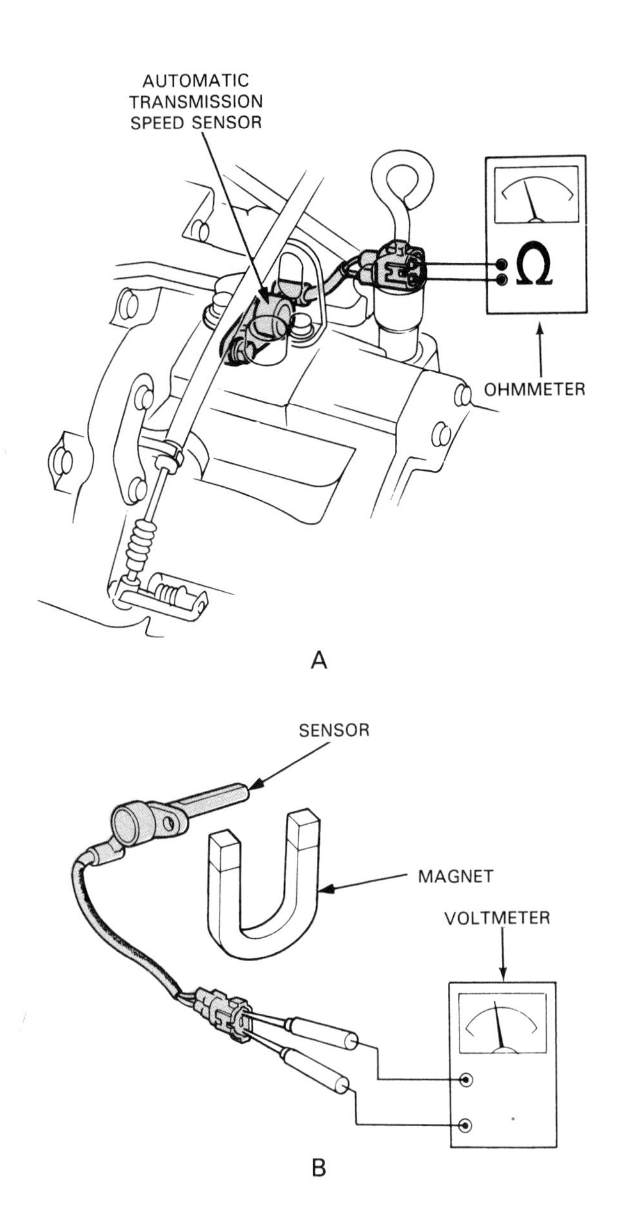

Figure 26-28. There are two ways to check a speed sensor. A–Many speed sensors can be checked with an ohmmeter without removing them from the vehicle. B–If the sensor is removed from the vehicle, it can be checked with a magnet and a voltmeter. (Honda)

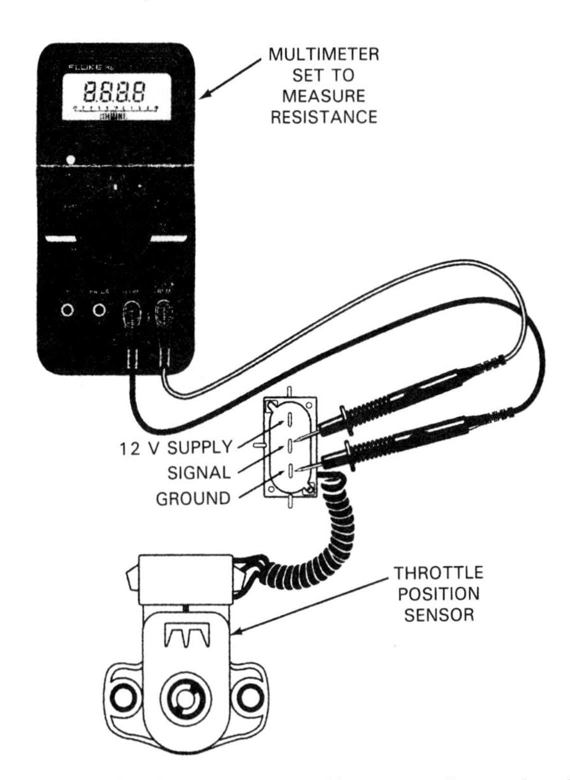

Figure 26-29. An ohmmeter or a multimeter can be used to check the throttle position sensor for proper operation. (Fluke)

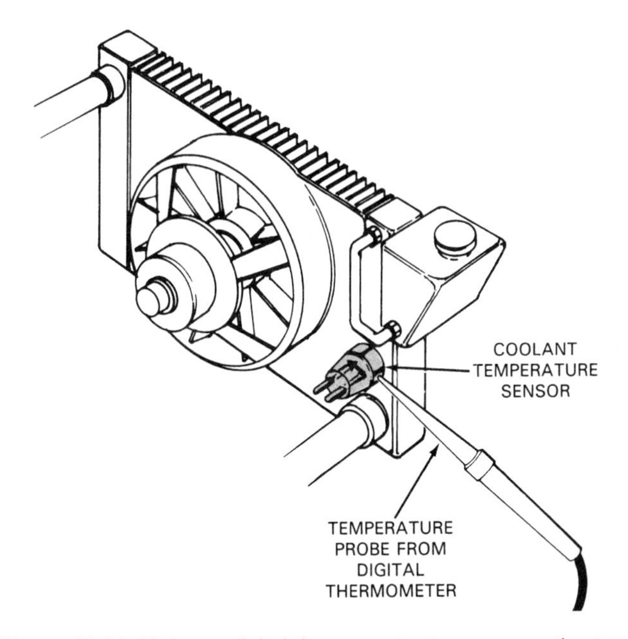

Figure 26-30. Using a digital thermometer to measure the temperature at the base of the sensor.

CAUTION! Never activate a hydraulic pressure sensor with air pressure. The sensor diaphragm may be destroyed. Correct test results are not possible when the sensor is checked with air pressure.

Air pressure sensor service

Many air pressure sensors can be checked and adjusted. In some cases, however, proper operation can only be verified by substitution. The factory service manual should be consulted before attempting to test or adjust an air pressure sensor. The most common air pressure sensor problems are cracked, plugged, or loose hoses to the sensor. Since the sensor operates on very low voltages, wire connections should be closely inspected when a pressure sensor problem is suspected.

General service rules

Sensor replacement is relatively easy. Most sensors simply unscrew from the part in which they are mounted. Remember to disconnect the negative battery cable before disconnecting sensor wiring. The slightest voltage surge can damage the computer or the sensor.

When replacing any sensor, make sure that you replace the related O-rings, gaskets, and/or other sealing components, Figure 26-31. If you replace a threaded sensor that installs without a gasket or seal, lightly coat the threads with sealant before installation.

Drive Train Actuator Service

The actuators used on drive train components can be checked by using jumper wires to actuate them. When testing solenoids, connect one wire between the solenoid case and ground. Then connect another wire to the solenoid positive lead and the battery positive terminal. Solenoids that are energized will produce a click or snap to indicate that they are working. Solenoids can be checked on or off the vehicle.

If a solenoid has been determined to be defective, it should be replaced. Typical solenoid testing and replacement procedures are shown in Figure 26-32. The solenoid shown is equipped with a filter, which should be cleaned or replaced when the solenoid is removed.

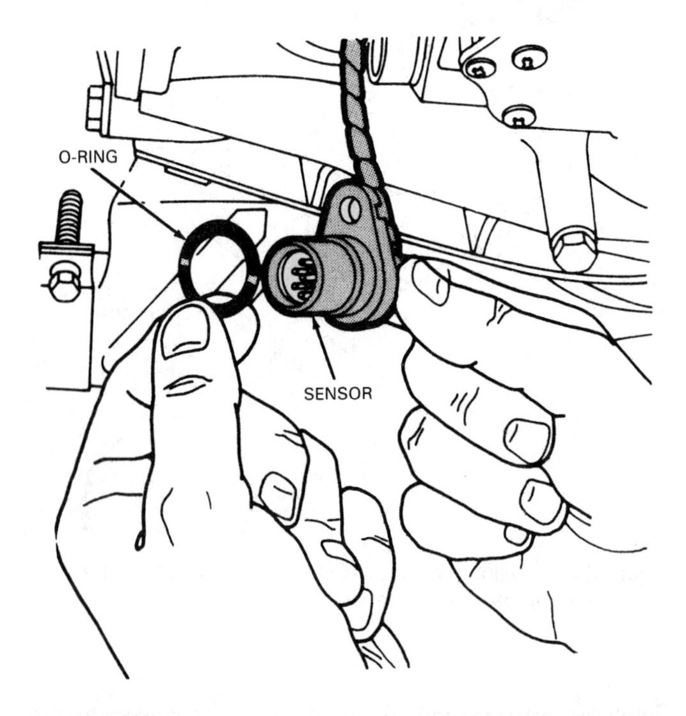

Figure 26-31. When a transmission-mounted sensor is removed or replaced, always replace the seal. This particular sensor is sealed with an O-ring. In some cases, manufacturers recommend the use of sealer on threaded assemblies. (Chrysler)

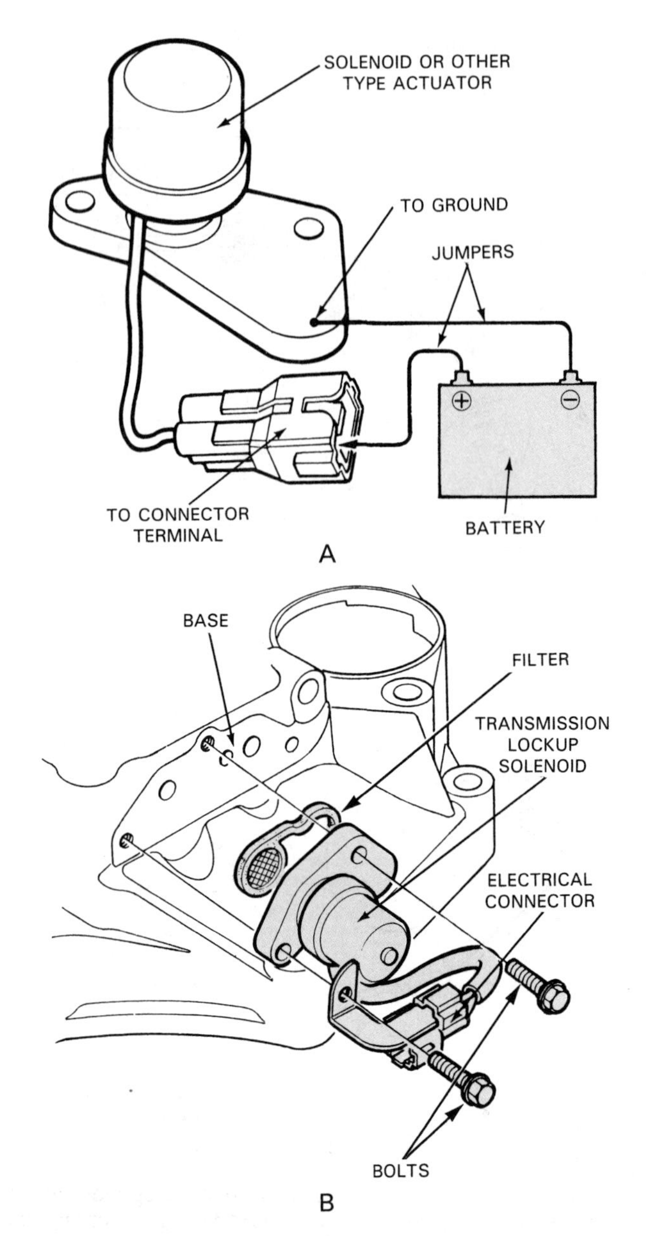

Figure 26-32. Solenoid service. A–Solenoids can be checked by operating them with battery current. B–Most solenoids are easy to remove and replace. Always change any related seals or filters. (Toyota)

Some solenoids will be destroyed by battery voltage. These particular solenoids cannot be checked with jumper wires. Instead, they should be checked for damaged insulation, open windings, and grounded windings. Connect an ohmmeter as shown in Figure 26-33. If the ohmmeter reads infinity or no resistance when the leads are placed on the positive terminal and the case, as shown, replace the solenoid.

Like solenoids, electronic servo motors can be checked by operating them with battery current. Motors can also be checked with an ohmmeter to isolate broken or grounded motor wiring.

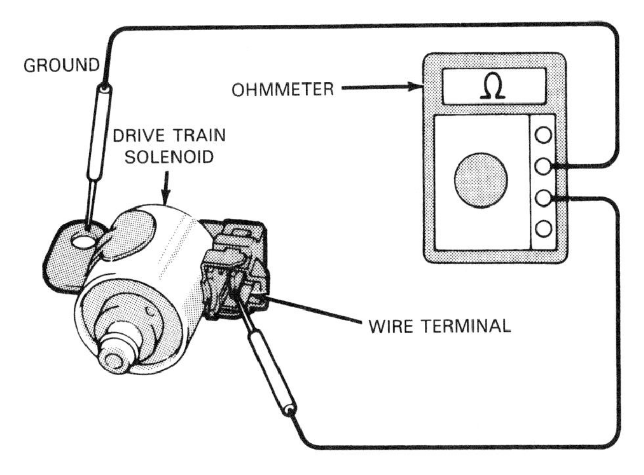

Figure 26-33. Solenoids can be checked with an ohmmeter. Place one lead on the case and one lead on the input terminal. If the solenoid resistance is zero or infinity, the solenoid is defective. (Toyota)

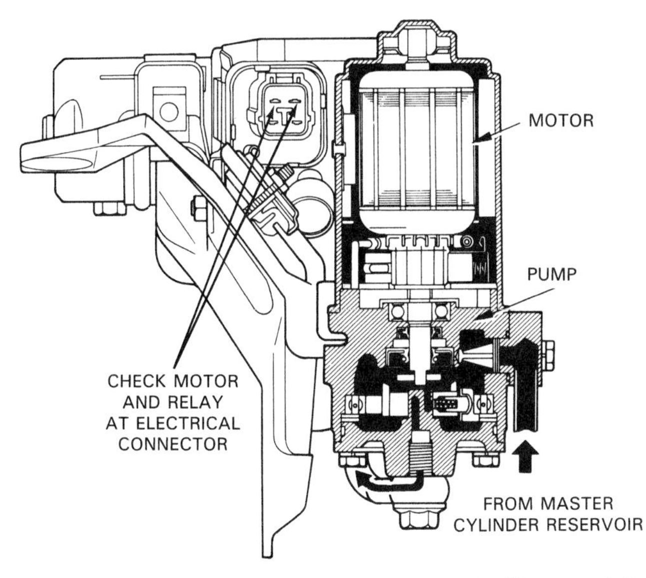

Figure 26-34. This electric motor can be checked by energizing the motor relay with battery current. If the motor does not turn, it is defective. The motor can also be checked with an ohmmeter. (Lexus)

Figure 26-34 shows a motor that is used to develop the hydraulic pressures in a traction control system. This motor is controlled by the on-board computer. During testing procedures, the motor can be operated with jumper wires. If the motor operates when battery voltage is applied, some other part of the system is defective. Check the control relay and the computer system for problems.

Drive Train Computer Service

The computer itself is usually the last part to be tested because it is the part least likely to fail. The computer's solid-state construction makes it less prone to damage from vibration, and it is usually mounted in a location that is protected from extreme temperatures and moisture. In addition, the computer has self-diagnostic capability, so it can often pinpoint its own problems.

Figure 26-35 shows two ways in which a computer can display trouble codes. Techniques for accessing and interpreting computer trouble codes were discussed earlier in this chapter.

In some cases, the computer can only be checked by replacing it with a unit that is known to be working properly and rechecking system operation. In many cases, however, it is possible to make ohmmeter or voltmeter checks of the computer. Figure 26-36 shows the procedure for checking voltages on one specific computer. On a few computers, the operating settings can be changed with adjustment screws. Consult the proper service manual for exact checking and adjustment procedures.

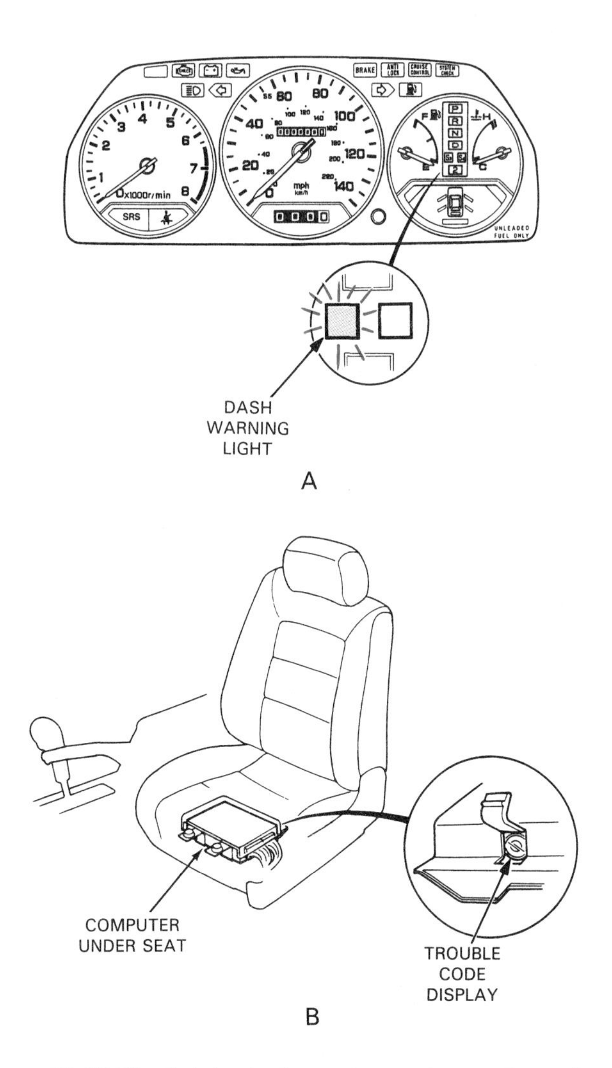

Figure 26-35. The first step in diagnosing most computer systems is to locate the self-diagnostic readout light. A–A dashboard-mounted warning light. B–An LED (light-emitting diode) trouble code display. (Honda)

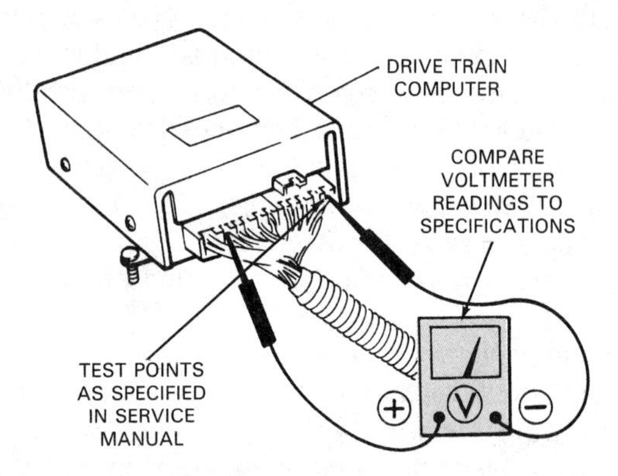

Figure 26-36. This figure shows a method of checking computer inputs and outputs with a voltmeter. This procedure should only be used if it is recommended by the manufacturer. (Toyota)

The computer usually is replaced as a unit. In some computers, however, the ***programmable read only memory (PROM),*** which is the part of the computer that stores the preset adjustment specifications, can be removed from a computer. There are two occasions when a PROM should be removed from a computer:

- When the PROM is defective.
- When the computer is replaced and the original PROM is to be reused.

Note that since the PROM information instructs the computer to operate correctly with a certain engine, transmission, axle ratio, body weight, etc., it is important that the PROM match the vehicle.

Figure 26-37 shows the procedure for removing and installing the PROM in a typical computer assembly. Begin by disconnecting the negative battery cable. Then remove the computer wiring harnesses, Figure 26-37A, and remove the PROM cover, which is generally held in place with two small sheet metal screws.

Remove the PROM from the computer, Figure 26-37B. Some manufacturers recommend using a special PROM tool to grasp the PROM and pull it out of its socket.

CAUTION! Avoid touching the PROM with your fingers. Static electricity or oil from your hands can affect PROM operation.

If an old PROM is being installed in a new computer, check that the PROM pins (terminals) are straight. Straighten the pins if needed. Also, check the installation reference mark on the PROM. A PROM usually has an indentation or other mark to indicate the proper direction of installation. Make sure the PROM part number is correct for the vehicle body and engine. Engage the PROM pins with the sockets in the computer. Make sure that any reference marks are correctly positioned.

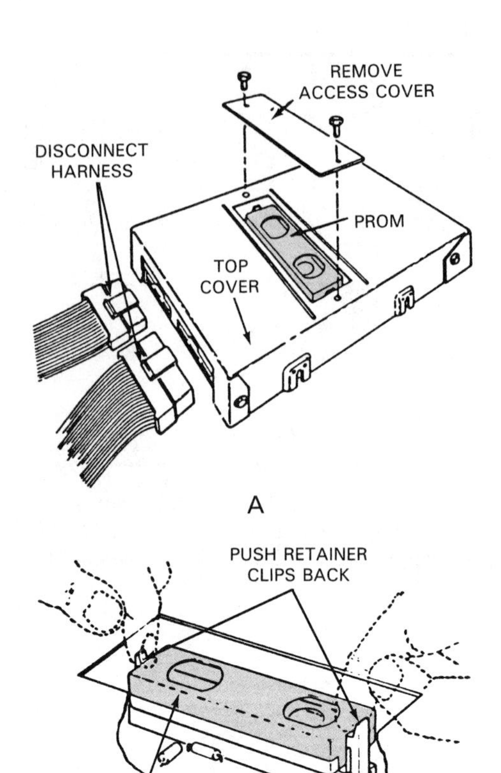

Figure 26-37. If a PROM must be replaced, it should be handled carefully. Before servicing the computer, remove the negative battery cable. A–Disconnect the wiring harness and remove the PROM access cover. B–Push the retainer clips back and remove the PROM, pulling it straight up from the computer. Reverse these procedures when installing a PROM. (General Motors)

CAUTION! If the PROM is installed backwards, it will usually be damaged and require replacement.

Using a blunt tool or a small piece of wood, press the PROM into the computer. Push on each corner to make sure that the PROM pins are fully seated in their sockets. If the PROM is held in place by clips, make sure the clips are fully engaged.

Some PROMs use a separate carrier. The ***carrier*** is a plastic part that surrounds the outside of the chip. If a carrier is used, you must use a blunt tool to engage the PROM in the carrier. Before installing the PROM in the computer, make sure the top of the PROM chip is flush with the top of the carrier. Install the carrier and PROM in the computer with the reference mark positioned properly. First, press down on the carrier only. Then, press on the corners of the PROM until it is fully seated in the computer.

Replace the PROM access cover and reinstall the computer in the vehicle. Reconnect the computer wiring harnesses and the negative battery cable. Then turn on the ignition and access the computer's self-diagnostic system. If trouble codes have been set, the PROM may not be fully seated or may have a bent pin. Recheck the installation if necessary.

Summary

An on-board computer is used to control the drive train on many vehicles. Drive train technicians will often be called on to work on computer systems as part of their job.

The drive train computer is most often used to control the operation of the automatic transmission or transaxle. Some late-model vehicles have a computer to control the operation of the four-wheel drive system or to provide traction control on turns or slippery surfaces.

Drive train sensors provide the inputs to the computer. Sensors fall into two categories: active sensors, which create the electrical signals that are sent to the computer; and passive sensors, which depend on an outside voltage source to produce an electrical signal.

Drive train sensors monitor gear selector position, throttle position, road speed, engine speed, engine temperature, hydraulic pressure, and air pressure. The sensors convert these factors into electrical inputs that are delivered to the computer.

Drive train actuators fall into two broad categories: solenoids and motors. The majority of actuators are electric solenoids. Solenoids are used to control the flow of oil in hydraulic circuits. Some solenoids open oil pressure passages, allowing oil pressure to drop and the valves to open. Solenoids can also operate a valve directly.

A computer control system typically contains several sensors and actuators. The drive train computer may communicate with other vehicle computers to receive and provide inputs. In some cases, the drive train computer makes all of the shifting decisions, while in others, it controls only part of the hydraulic system.

The computer itself is a solid-state microprocessor that converts analog input signals to digital signals, compares them to preset information in its memory, and sends output commands to actuators. The actuators modify drive train operation to correct any improper input signals. The computer can be found in one of several places on the vehicle. It is usually located in a spot that provides maximum protection from heat, cold, dirt, and vibration.

Before deciding that the computer system is the source of a problem, check all other components. Also check for faulty wiring connections and corroded grounds. Minor adjustments and parts checking may solve the problem.

Most modern computers have a self-diagnostic feature. The computer self-diagnosis can be accessed easily, usually by connecting two wires or grounding a special connector. The computer will produce a series of voltage pulses that are communicated by a series of flashes of the dashboard light or of a computer-mounted LED. In some systems, an analog meter must be used to monitor voltage pulses. The pulses must be interpreted and compared to a factory trouble code chart to determine the problem area. The technician can then proceed to make detailed checks of the problem area.

Although some computer systems can only be diagnosed with special testers, most can be at least partially diagnosed by reading the trouble light codes and by using simple test equipment.

Wiring diagrams are a map of the computer electrical system. The diagrams can be traced back from a problem area to isolate defective components or wiring. The two types of wiring diagrams are logic diagrams and direct wiring diagrams. Logic diagrams give an overall picture of system operation. Direct wiring diagrams are used when tracing electrical problems and performing electrical tests. Always obtain the correct wiring diagram for the system that you are troubleshooting.

Most drive train sensors can be checked with standard equipment. In some cases, sensors can be adjusted. Switch-type sensors can be checked with a test light or an ohmmeter. Neutral safety switches and backup light switches can be adjusted by using the same procedures that apply to the switches used on older vehicles. Pulse-generating (speed) sensors, throttle position sensors, and pressure sensors can be checked with ohmmeters or voltmeters. See factory specifications for specific instructions.

Many actuators can be checked by activating them with battery voltage. If a solenoid will not operate on battery voltage, it should be replaced. Most solenoids can also be checked with an ohmmeter. Motors can be checked by trying to operate them with battery voltage or by checking them with an ohmmeter.

In most systems, the computer is the part least likely to fail. In addition, it can generally pinpoint its own problems through its self-diagnostic system. However, proper computer operation can only be verified by replacing the unit with a computer that is known to be working correctly. If the computer is faulty, it must be replaced.

Some computers have a separate memory unit called a PROM. The PROM can be replaced if it is defective. It is important to replace the PROM with a exact duplicate, because each PROM is calibrated to the particular vehicle in which it is installed. Be careful when replacing a PROM. Slight contamination or static electricity can damage the unit. Always recheck computer operation after replacing the PROM. When a new computer is installed in a vehicle, the original PROM is often removed from the old computer and inserted in the new unit.

Know These Terms

Computerized drive train; Active sensor; Pulse-generating sensor; Passive sensor; Backup switch; Combination switch; Throttle position sensor; Throttle transducer; Idle switch; Vehicle speed sensor; Road speed sensor; Engine speed sensor; Engine temperature sensor; Pressure switch; Air pressure sènsor; Barometric pressure sensor; Intake manifold vacuum sensor; Drive train computer; Feedback system; Actuator; Servo motor; Network; Bus; Trouble code; Wiring diagram; Switching sensor; Programmable read only memory (PROM); Carrier.

Review Questions–Chapter 26

Please do not write in this text. Place your answers on a separate sheet of paper.

1. What is a computerized drive train?
2. Describe computer inputs and outputs.
3. _____ sensors produce their own voltage, while _____ sensors require a reference voltage from the computer.
4. What is the value of a typical reference voltage?
5. Name three types of sensors that can be used to detect engine throttle movement.
6. Explain the operation of a speed sensor.
7. The most common type of actuator is a:
   a. Relay.
   b. Servo motor.
   c. Lightbulb.
   d. Solenoid.
8. Explain how actuators are used to control the operation of an automatic transmission or transaxle.
9. Late-model vehicles generally have only one on-board computer. True or false?
10. Explain the difference between analog signals and digital signals.
11. How do you use a computer's self-diagnostic capabilities to locate problems in a computer controlled drive train?
12. How do you check solenoid-type actuators for proper operation?

Certification-Type Questions–Chapter 26

1. In vehicles with a computerized drive train, an on-board computer makes decisions concerning all of these EXCEPT:

(A) torque converter lockup.
(B) transmission shift points.
(C) transmission shift pressures.
(D) drive axle speed.

2. Computer controls are used extensively in the operation of all of these EXCEPT:

(A) automatic transmissions.
(B) traction control systems.
(C) manual transmissions.
(D) four-wheel drive systems.

3. All of these basic components are found in most automotive computer systems EXCEPT:

(A) plates.
(B) sensors.
(C) actuators.
(D) computers.

4. Based upon inputs received about engine and vehicle speed, computer outputs are made. Technician A says that sensors receive these outputs and change operating conditions. Technician B says that actuators receive outputs and change operating conditions. Who is right?

(A) A only
(B) B only
(C) Both A & B
(D) Neither A nor B

5. Which of these general categories can sensors be divided into?

(A) Active and passive.
(B) Active and nonactive
(C) Actuating and sensitive.
(D) Passive and nonpassive.

6. Technician A says that a combination switch is used to operate a vehicle's backup lights. Technician B says that the switch prevents the vehicle from being started unless it is in park or neutral. Who is right?

(A) A only
(B) B only
(C) Both A & B
(D) Neither A nor B

7. Technician A says that the solid-state design of an on-board computer allows it to function under conditions of extreme heat and extreme cold. Technician B says that the solid-state components are seriously affected by vibration and moisture. Who is right?

(A) A only
(B) B only
(C) Both A & B
(D) Neither A nor B

8. Technician A says that signals that vary continuously, or that may assume any value within a defined range, are called digital signals. Technician B says that analog signals vary in distinct steps. Who is right?

(A) A only
(B) B only
(C) Both A & B
(D) Neither A nor B

9. All of these are output devices that the on-board computer uses to change operating conditions EXCEPT:

(A) sensors.
(B) actuators.
(C) servo motors.
(D) electrical solenoids.

10. Technician A says that when troubleshooting a computer controlled drive train system, a computer system diagnosis should first be conducted. Technician B says that looking for conventional drive train problems should be the first step. Who is right?

(A) A only
(B) B only
(C) Both A & B
(D) Neither A nor B

11. All of these are wiring diagrams used with computer systems EXCEPT:

(A) flow diagrams.
(B) logic diagrams.
(C) exact diagrams.
(D) direct wiring diagrams.

12. Preset adjustment specifications are stored in a solid-state device called a:

(A) passive-ready option maker (PROM).
(B) programmable read only memory (PROM).
(C) pulsating responsive output motor (PROM).
(D) programmable road option memory (PROM).

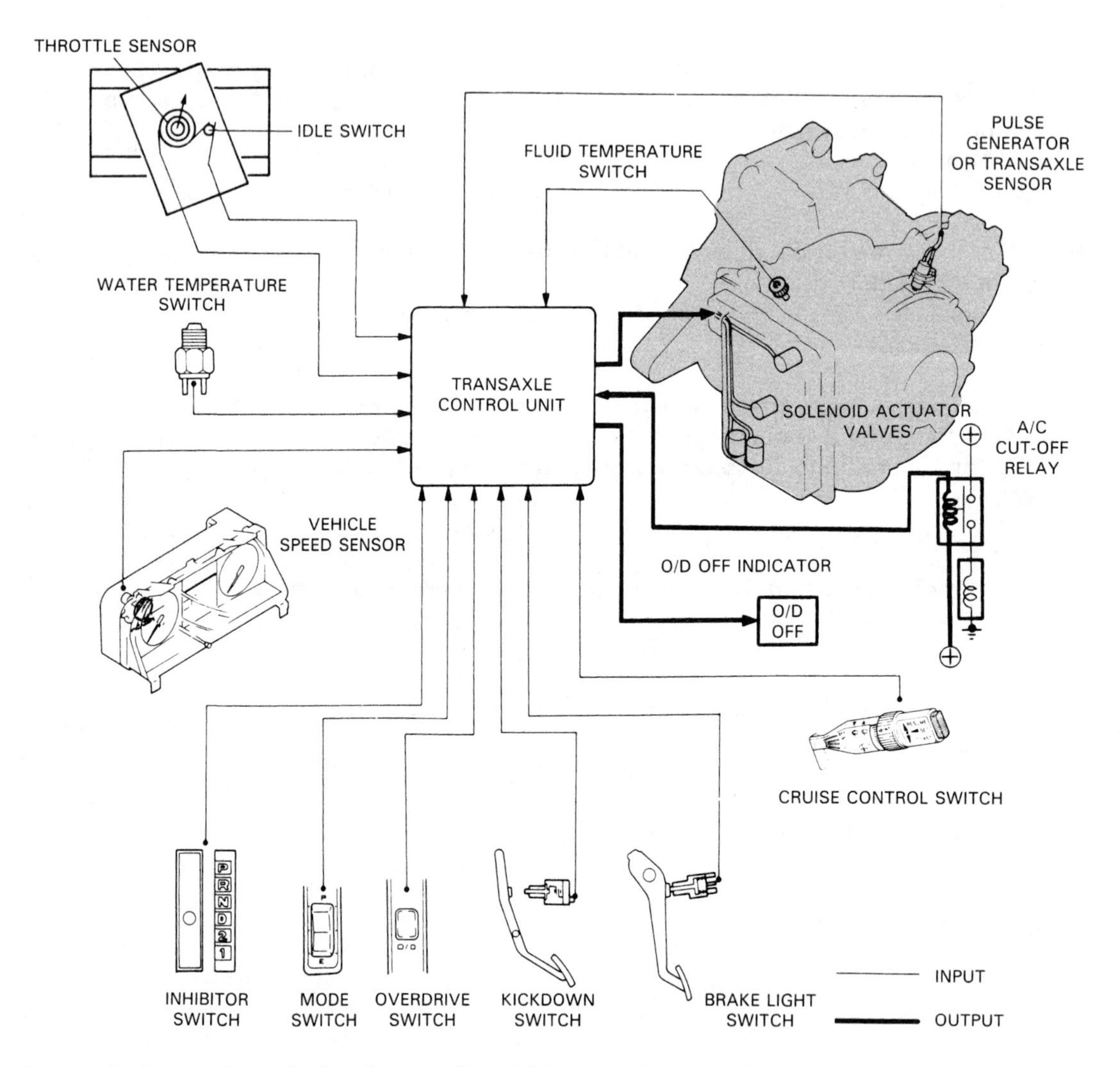

A computer is used for controlling the operation of this automatic transaxle.

Chapter 27

Career Success

After studying this chapter, you will be able to:

- Discuss the advantages and disadvantages of working in the automotive service industry.
- Explain the importance of being dedicated to your profession.
- Identify the major sources of employment in the automotive industry.
- Identify three distinct levels of automotive service professionals.
- Fill out a typical job application.
- Properly conduct yourself during a job interview.

This chapter is designed to provide an overview of career opportunities in the automotive service industry. It also details the qualities needed to become a successful automotive technician. Studying this chapter will help you to get a job in the automotive industry.

The Trials and Rewards of Automotive Service Careers

The automotive service industry has provided steady, well-paying jobs for many millions of people for many years. It will continue to provide excellent employment opportunities for years to come. Like any career, automotive service has its drawbacks, but it also has its rewards.

There are many advantages to working in the automotive service industry, including the opportunity to work with your hands and the enjoyment of fixing something that is broken. Auto repair salaries are usually competitive with those for similar jobs, and it is a secure profession in which a good technician can always find work.

Unfortunately, most people in the auto service business work long hours. The diagnosis and repair procedures can be mentally taxing and physically demanding. Working conditions are often hot and dirty. The technician often has to deal with difficult, condescending, or dishonest customers. Automotive service has never been a prestigious career, although this is changing as vehicles become more complex and technicians become highly skilled.

The Importance of Dedication

Although it is important to diagnose and repair vehicles properly, you must have the right attitude to succeed in the automotive repair industry. You must have the ***dedication*** to perform all aspects of the job to the best of your ability. Sometimes, maintaining the right attitude is difficult when you are assigned a job that you dislike. Unfortunately, automotive technicians are often called on to work outside their area of interest. Dedication will give you the attitude to accept the occasional unpleasant job. Some of the traits that indicate dedication include dependability, quality workmanship, the ability to work as part of a team, and the willingness to learn new things.

Dependability

Shop owners will find it hard to schedule work or guarantee that a repair will be completed on time if they cannot count on their employees to show up for work. ***Dependability*** involves reporting to work when you are scheduled and performing the tasks that are expected of you. Occasionally, things will happen that are beyond your control, but a pattern of arriving late, leaving early, or not showing up at all tells your employers that you are not serious about working for their company.

Quality Workmanship

Customers and employers expect each job to be done properly. Sloppy repairs will often lead to angry customers. In many cases, time is wasted because the improper repair work must be corrected. Careless work may even be the cause of an accident, resulting in property damage, injury, or death. Always try to do every job properly the first time.

Ability to Work as Part of a Team

The people in a shop must often pull together to complete a project or diagnose a problem. It is important to work together for the good of the company, even when your own immediate advantage is not apparent. Taking a me-against-them attitude will backfire in the long run.

Willingness to Learn

Unfortunately, many people quit using their brain when they leave school. Do not let this happen to you! To ensure that you remain an asset to your employer, always keep up with the latest developments in the automotive field and become certified in as many areas as possible. Certification is covered in Chapter 28.

Technician Payment Methods

Technicians are paid in one of several ways. Many technicians receive a ***commission*** for the work they perform. Commission simply means that, instead of being paid by the hour, the technicians are given a percentage of the total charge for the work they perform.

Some technicians are paid a flat rate for each job they complete. ***Flat rates*** are based on the number of hours a job should take. Flat rate hours are published by vehicle manufacturers and by independent manual publishers, such as Motor and Chilton. Jobs are generally broken into tenths of an hour (6 minute intervals). If, for example, a transmission overhaul is listed at 12.3 hours, the job can be performed in 12 hours and 18 minutes. The flat rate hours listed in a *flat rate manual* are multiplied by the technician's ***hourly pay rate*** (amount the technician makes per flat rate hour) to determine how much the technician will be paid for performing the job. For example, if the flat rate for replacing a clutch is 2 hours and the hourly pay rate is $15.00 per hour, the technician will make $30.00 for performing the job. Regardless of the amount of time it takes the technician to complete the job, he or she will be paid for 2 hours work.

Commission and flat rate pay systems can be very profitable. The disadvantage of these systems is that lack of work and unforeseen problems, such as stripped bolts or problems getting parts, can reduce your income. However, your pay rate cannot be less than the minimum wage set by the U.S. Department of Labor multiplied by the number of hours that you work.

The other major pay methods are ***hourly wage*** and ***salary***. Technicians who are paid an hourly wage receive a specific amount of money for each hour they work. When these technicians work more than 40 hours in a week, they often receive overtime pay. Overtime pay is usually one and a half times an employee's regular hourly wage.

Many management employees and some technicians are paid a salary that does not directly correspond to the number of hours worked. Salaries are usually computed on a weekly or monthly basis, and there is no provision for overtime pay. This pay method provides a cushion for slow workdays or problem repair jobs, but does not allow the technician to make more money when there is a lot of work.

Many shops use a system that combines hourly wages or salary with commission. The employee is paid an hourly wage (or a salary) and also receives a commission on the work that he or she performs. In some cases, commission is based on the value of parts that the technician installs.

Types of Automotive Service Facilities

There are many types of automotive service facilities in which a technician can work. Traditionally, the place to get started in the automotive repair business was the corner service station. Unfortunately, most small service stations have been replaced by self-serve facilities. Nevertheless, many employment opportunities in vehicle repair still exist. Even the smallest community has at least one automotive repair facility. Typical service facilities are discussed in the following sections.

New Vehicle Dealerships

All ***new vehicle dealerships*** must have large, well-equipped service departments to meet the warranty service requirements of the vehicle manufacturer. These service departments are usually equipped with the special testers, tools, and service literature needed to service the vehicles sold by the dealer. Dealership service departments are also equipped with lifts, parts cleaners, hydraulic presses, brake lathes, electronic test equipment, and other equipment for efficiently servicing vehicles. In most cases, however, the technician has to provide his or her own hand and air tools. Dealers generally stock common parts and are tied into a factory parts network that allows them to quickly obtain any part that is not in stock. See Figure 27-1.

Pay scales at most dealerships are based on flat rate hours. The rate per hour is competitive between dealerships in the same area. The number of hours the technician is paid for depends on how much work comes in and how fast the technician can complete it. If you work fast and ample work is available, the pay can be excellent. Additionally, most modern dealerships offer some type of benefits package, which may include health insurance and a retirement plan.

Dealership working conditions are relatively good, and most of the vehicles serviced are new or well-maintained older models. Since the dealer must fix all vehicle systems, the technician is exposed to a variety of work. Although

Figure 27-1. New vehicle dealerships offer a variety of repair tasks for the automotive technician.

Figure 27-2. Chain store automotive service centers employ many technicians.

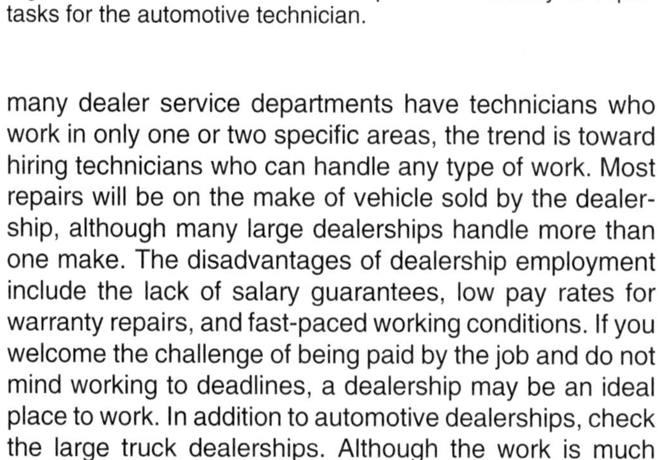

many dealer service departments have technicians who work in only one or two specific areas, the trend is toward hiring technicians who can handle any type of work. Most repairs will be on the make of vehicle sold by the dealership, although many large dealerships handle more than one make. The disadvantages of dealership employment include the lack of salary guarantees, low pay rates for warranty repairs, and fast-paced working conditions. If you welcome the challenge of being paid by the job and do not mind working to deadlines, a dealership may be an ideal place to work. In addition to automotive dealerships, check the large truck dealerships. Although the work is much heavier, the pay is usually somewhat higher and working conditions are not as hectic.

Chain Store Auto Service Centers

Many national chain stores and department stores have ***auto service centers,*** which perform various types of automotive repairs. Companies with these types of service centers include Sears, Montgomery Ward, and K-Mart, Figure 27-2. Automotive service centers often hire technicians for entry-level jobs, but they may also employ experienced technicians. In most service centers, technicians are paid a salary and also receive a commission for the work they perform. Pay scales for the various types of work are competitive, and most companies of this type offer generous benefit packages. One advantage of working for chain store service centers is the chance for advancement into other areas, such as sales or management.

One disadvantage of working at the average auto center is the lack of variety. Most auto centers concentrate on only a few repairs, such as alignment and brake work, and turn down most other repairs. The work can become monotonous. Although the job pressure is usually less than that at dealerships, customers still expect their vehicles to be repaired in a reasonable period of time. If you enjoy working in only one or two areas of automotive repair, this type of facility may be ideal for you.

Specialty Shops

Brake, muffler, and other ***specialty shops*** generally offer good working conditions and excellent pay. Most of these shops concentrate on their major specialty, but may perform a few other types of repair. Typical examples of specialty shops include Midas muffler and brake shops, AAMCO transmission shops, and Western Auto Service Centers. Technicians at these shops are usually paid a salary plus commission. Pay scales are generally competitive with other types of service facilities. A disadvantage of working in a specialty shop is the lack of variety. Since these shops concentrate on a few repair areas, the work can become monotonous.

Many specialty shops are franchise operations, and the demands of the franchise can create problems. If the prime purpose of the shop is to sell tires, for instance, the technician who was hired to do brake repairs may be forced to spend time installing tires. This can be annoying to technicians who want to be doing the job for which they were hired.

General Repair Facilities

There are millions of ***general repair facilities*** in operation today. Many of these facilities are independently owned. As places to work, they can range from excellent to terrible. Many shops are run by competent, fair-minded managers, have first-rate equipment, and provide good working conditions. Other shops may have very little equipment, low pay rates, and extremely poor working conditions. As a prospective employee, you should carefully check all aspects of a shop's environment before agreeing to work there. Technicians at general repair shops are usually paid salary plus commission.

General repair shops perform a variety of repair work on different types of vehicles. However, many of these shops avoid repair tasks that require special equipment, such as automatic transmission service, front end alignment, and air conditioning service. The general repair shop can be a good place to work if you like to be involved in the diagnosis and repair of many types of vehicles.

Government Agencies

Many local, state, and federal ***government agencies*** maintain their own vehicles. Government-operated repair shops, such as the one in Figure 27-3, can be good places to work. Pay is straight salary and is usually set by law. Although government pay scales are generally lower than those for private industry, the benefits are excellent. Pay raises, while relatively small, are regular. Most government shop employees work a 35- or 40-hour week and have the same holidays as other government employees.

The working conditions in most government-operated shops are good, without the stress of deadlines or the hassles of dealing with customers. Hiring procedures are more complicated than those for other auto repair shops. Prospective employees must take civil service examinations, which often have little to do with automotive subjects. Some government agencies require a certain level of education, a thorough background check, and a list of former employers or other references.

If you would be interested in working for a government-operated repair shop, contact your state employment agency for the addresses of state and federal employment offices in your area.

Figure 27-3. Government-operated repair shops offer the technician job security and excellent benefits.

Large Companies and Corporations

Large companies and corporations often own many vehicles and have repair facilities to perform preventative maintenance and repairs. Organizations of this type include trucking companies, utility companies, and car rental agencies.

Pay methods and working conditions at these facilities are similar to those at government agencies. Most technicians are paid an hourly wage, and pay scales are competitive with local repair shops. Most large companies offer good benefits packages. Because having a vehicle out of service often costs a company more than the repair, emphasis is placed on completing repairs quickly and efficiently. One disadvantage of this type of work is lack of variety in the type of vehicles encountered, although the nature of problems and the types of repairs performed will vary widely.

Entrepreneurship

Many people dream of going into business for themselves. Starting a business can be a profitable option for the good technician. However, in addition to mechanical and diagnostic ability, business owners must have a certain type of personality to be successful. They must be able to shoulder responsibilities, handle problems, and look for practical ways to increase business and make a profit. They must also maintain a clear idea of the long- and short-term plans that must be made. A person who starts a business is often called a ***entrepreneur.***

When you own an automotive service and repair business, you are responsible for repairs, inventory, bookkeeping, debt collection, and many other business tasks. Starting your own shop requires a large investment in tools, equipment, and working space. If money must be borrowed to start the business, you will be responsible for paying it back. However, many people enjoy the feeling of independence. If you have the personality to deal with the challenges of owning a business, you may enjoy being your own boss.

Another possible option for self-employment is to obtain a franchise from a national chain. A ***franchise*** is a shop that is part of a national chain, but is owned and operated by a individual. Owning a franchise operation removes some of the headaches of starting a business. Many nationally recognized muffler, tire, transmission, and tune-up businesses are franchise shops that have local owners. The owners enjoy the advantages of the franchise affiliation, including national advertising, reliable and reasonable parts supplies, and employee benefit programs. Disadvantages of a franchise arrangement include high franchise fees and startup costs, lack of local advertising, and loss of control of shop operations to the franchise headquarters.

Levels of Automotive Service Employees

The public tends to classify all people who service automobiles as "mechanics." However, there are three distinct levels of automotive service professionals: helpers, installers, and certified technicians. Although the levels are not official, they tend to be used throughout the automotive repair industry. It would be possible to break these levels down into sublevels, but the general skill classifications can be adequately covered by the three levels.

Helpers

The ***helper*** performs simple service and maintenance tasks, such as installing and balancing tires, changing engine oil and filters, and installing batteries. The skills required of the helper are low, and the helper is paid less than other automotive service employees. However, starting as a helper is a good way for many people to advance into other service areas.

Installers

The ***installer*** is the service person who installs parts, such as suspension components and exhaust system components. In some cases, these employees install brake master cylinders, alternators, and voltage regulators. Installers seldom perform complicated repair work and generally do not diagnose vehicle problems. Installers are paid more than helpers, but less than certified technicians. As they gain experience, many installers become certified automotive technicians.

Certified Technicians

Certified technicians are the most experienced and well-trained of all automotive service professionals. Most modern technicians have received ASE certification in at least one automotive area, and many are certified in all car or truck areas. The certified technician is able to successfully diagnose and repair vehicle systems in every area in which he or she is certified and can perform many other service jobs. The certified technician is paid more than helpers and installers.

Other Opportunities in the Auto Service Industry

There are many other opportunities available to the automotive technician. Most of these opportunities still involve the servicing of vehicles, but may not require the physical work. If you like cars and trucks, but are unsure whether you want to make a career of repairing them, one of these jobs may be for you.

Shop Foreman or Service Manager

As a technician, the promotion that you will most likely be offered is to the position of ***shop foreman.*** Many repair facilities are large enough to require one or more foremen or service managers. If you move into management from the shop floor, your salary will increase, and you will be in a cleaner, less physically demanding position. Many technicians enjoy management positions because they allow them to troubleshoot vehicles without the drudgery of making the repairs.

The disadvantage of taking a management position is that you will no longer be dealing exclusively with the principles of machines. Instead, you will be dealing with people. Both the customers and the technicians will have problems and complaints that you will have to contend with. Unlike a mechanical problem, these problems require considerable personality and tact.

A manager is generally responsible for a great deal of paperwork. Record keeping requires a considerable amount of desk time. If you do not mind the paperwork and enjoy working with people, a career in management may be for you.

Sales Representative or Service Adviser

The ***sales representative*** or the ***service adviser*** performs a vital service, since most repairs will not be performed unless the owner is sold on the necessity of having them done. Sales representatives are not necessary in many small shops, but they are often an important part of dealership service departments, chain store service centers, and specialty repair shops.

The sales representative may enjoy a large income and may be directly responsible for the majority of the business in the shop. However, selling takes a lot of persuasive ability and diplomacy. If you are interested in dealing with the public and welcome the challenge of selling a product or service, a sales job may be for you.

Parts Person

One vital automotive service position that is often overlooked is the ***parts person.*** The parts person generally works in a parts outlet, such as a dealership parts department, an independent parts store, a retail parts department, or a combination parts and service outlet. These facilities must meet the needs of technicians and do-it-yourselfers.

Parts persons are trained to keep the supply of parts flowing through the system until they reach their ultimate destination–the vehicle. Parts must be carefully checked into the parts department and stored so that they can be found when needed. When a specific part is required, it must be located and delivered to the person requesting it. When inventory is low, the parts person must order additional parts to increase the inventory. If a part is required that is not carried in stock, it must be ordered.

Although the job may not pay as well as other areas of automotive service, the rates are comparable to similar jobs. The position of parts person is a good job choice for many people.

Getting the Right Job

There are essentially two obstacles to getting a job: finding the right job opening and successfully applying for the job. The following sections contain hints for overcoming these obstacles.

Finding Job Openings

Before applying for a job, you must know about a suitable opening. A good place to begin your job search is at your school. Discuss your career plans with your instructors. They often have contacts in the local automobile industry and may be able to recommend you to a company. Another good place to begin your search is the classified section of your local newspaper. Automotive jobs, especially those at automotive dealerships, specialty and franchise shops, and independent repair shops, are often advertised in newspapers.

Visit your state employment agency or a local job service agency. Most of these organizations keep records of job openings throughout the state and may be able to connect to a data bank of nationwide job openings. Some private employment agencies specialize in automotive placement. If there is an agency of this type in your area, arrange for an interview.

Visit local repair shops that you are interested in working for. Sometimes, these shops have an opening that has not been advertised. If you are interested in working for a service center in a chain store or department store, most of these facilities have a personnel department that you can visit to fill out an application. Even if no jobs are available at the time you apply, your application will be placed on file in the event that a job becomes available in the future.

Items to think about when considering applying for a particular job include work hours, working conditions, salary, and the general attitude of the employer. If you are not satisfied with what you discover, you may want to look elsewhere for employment.

Applying For the Job and Interviewing

No matter how well qualified you may be, you will not get a job if you make a poor impression either on the application or during the job interview. It is important to put your best foot forward in this critical phase of the process.

Create a good impression when you fill out the employment application. Type or neatly print the application. Fill in all blanks and completely explain voids in your education or employment history. List all of your educational qualifications, including those that may not apply directly to the automotive industry. Figure 27-4 shows a typical employment application.

If you are called for an interview, try to arrange a morning appointment since people are most likely to be in a positive mood in the morning. Dress neatly and arrive on time. When introduced to the interviewer, make an effort to repeat and remember his or her name. Speak clearly when answering the interviewer's questions. Do not smoke or chew gum during the interview. State your qualifications for the job without bragging, but be careful not to belittle your accomplishments. Make sure there are no misunderstandings about salary, benefits, or work hours. At the conclusion of the interview, thank the interviewer for taking the time to meet with you. If you do not hear from the interviewer within a few days, it is permissible to make a brief and polite follow-up call.

Summary

The automotive service industry provides employment for many people and will continue to do so. Advantages of an automotive service career include interesting work, job security, and the enjoyment of diagnosing and correcting problems. Automotive service has some disadvantages, such as long hours, hard work, lack of status, and difficulties in dealing with the public.

There are many places to work as an automotive technician, including vehicle dealerships, chain store auto centers, specialty shops, independent repair shops, and government agencies. Some people prefer to start their own business, either as an independent owner or as part of a franchise system. People who start a business are commonly called entrepreneurs.

There are three general levels of automotive service employees: the helper, the installer, and the certified technician. The helper does simple tasks, such as mounting tires and changing oil. Many helpers move up into the other levels after a short time. The installer installs new parts, such as suspension system components and exhaust system components. The certified technician performs the most complex diagnosis and repair jobs.

Other opportunities in the automotive service field include careers in management or sales. Another employment possibility is in the automotive parts business.

The first step in obtaining a job in automotive service is to locate job openings. Try the local newspaper and the state job service. Visit repair shops in your area. Most chain store service centers have personnel departments where you can fill out a job application. To get a job, you must make a good impression. Fill out job applications carefully

APPLICATION FOR EMPLOYMENT				
PERSONAL INFORMATION			Date	
Social Security Number				
Name				
Present Address		City	State	Zip
Permanent Address		City	State	Zip
Phone Number		If hired, in case of emergency contact, Name	Phone	Relationship
Are you related to anyone in our company?	If so, whom? Name Department			Referred By
EMPLOYMENT DESIRED				
Position			Date you can start	Expected Salary
Are you presently employed?	If employed, may we contact your present employer?		Have you ever applied at our company before? If so, when?	
EDUCATION	Name and Location of School	Years Attended	Date Graduated	Courses
GRAMMAR SCHOOL				
HIGH SCHOOL				
COLLEGE				
TRADE SCHOOL				
EXPERIENCE				
Employed by Address		Title	Phone	From / To
Duties				
Employed by Address		Title	Phone	From / To
Duties				
Employed by Address		Title	Phone	From / To
Duties				
What foreign languages do you speak?			Typing wpm	
U.S. Military or Naval Service		Rank	Do you presently serve in the National Guard or Reserve?	
References				
References				
References				

Figure 27-4. Always fill out employment applications neatly and carefully. The application is often the first chance to make a good impression on a potential employer.

and neatly, listing your qualifications honestly. When invited to a job interview, dress neatly, arrive on time, and be courteous. Answer all questions without overstating or understating your abilities and experience. If you do not hear from the interviewer within a few days, it is acceptable to follow up with a brief phone call.

Know These Terms

Dedication, Dependability, Commission, Flat rates, Hourly pay rate, Hourly wage, Salary, New vehicle dealerships, Auto service centers, Specialty shops, General repair facilities, Government agencies, Entrepreneur, Franchise, Helper, Installer, Certified technicians, Shop foreman, Sales representative, Service adviser, Parts person.

Review Questions–Chapter 27

Please do not write in this text. Place your answers on a separate sheet of paper.

1. List some of the advantages of working in the automotive service field.
2. List the disadvantages of working in the automotive service field.
3. The common automotive repair facilities in which an automotive technician can get started include:
   a. Specialty shops.
   b. Chain store service centers.
   c. New vehicle dealers.
   d. All of the above.
4. When a technician is paid by the flat rate method, he or she is being paid:
   a. For the amount of repair work that he/she does.
   b. For the number of hours spent in the shop.
   c. A salary based on 35 or 40 hours per week.
   d. A percentage of the profit that the shop makes.
5. Technicians in department store automotive centers are usually paid by what type of system?
   a. Salary only.
   b. Salary plus commission.
   c. Commission for work performed only.
   d. Both b and c.
6. A general repair shop can be a good place to work if you like:
   a. Working on only one type of vehicle.
   b. One kind of repair work.
   c. A variety of repair work.
   d. None of the above.
7. An advantage of working for a government-operated repair shop is:
   a. Steady pay.
   b. Good benefits.
   c. High wages.
   d. Both a and b.
8. Describe the activities of the three general classes of automotive service employees: *helper, installer,* and *certified technician.*
9. There are two major obstacles to getting a job. Name them.
10. The time to start making a good impression on a possible employer is:
    a. When you fill out the employment application.
    b. During the job interview.
    c. In a follow-up letter or call.
    d. As soon as you get the job.

Certification-Type Questions–Chapter 27

1. **All of these are traits that indicate dedication EXCEPT:**
   (A) dependability.
   (B) competitive attitude.
   (C) quality workmanship.
   (D) willingness to learn.

2. **In comparison to similar jobs, auto repair salaries are usually:**
   (A) equal.
   (B) lower.
   (C) higher.
   (D) competitive.

3. All of these are typical methods of payment for automotive technicians EXCEPT:
(A) salary.
(B) commission.
(C) hourly wage.
(D) monthly stipend.

4. Flat rates are based on:
(A) a technician's abilities.
(B) the number of hours a job takes.
(C) the number of hours a job should take.
(D) a set percentage of a job's total cost.

5. All of these are facilities that employ people as automotive service technicians EXCEPT:
(A) specialty shops.
(B) self-serve gas stations.
(C) new vehicle dealerships.
(D) chain store auto service centers.

6. A disadvantage of dealership employment is:
(A) low pay rates for warranty repairs.
(B) the absence of any type of benefits package.
(C) a lack of service equipment.
(D) slow-paced working conditions.

7. An entrepreneur is:
(A) a technician in training.
(B) a technician who specializes in a particular type of repair.
(C) someone who starts his or her own business.
(D) the top technician at an establishment.

8. All of these are different levels of automotive service employees EXCEPT:
(A) clerk.
(B) helper.
(C) installer.
(D) certified technician.

9. Automotive service jobs that generally do not involve hands-on repairwork include all of these EXCEPT:
(A) parts person.
(B) shop foreman.
(C) service regulator.
(D) sales representative.

10. All of these are recommended strategies for a job-search campaign EXCEPT:
(A) discussing your career plans with your instructors.
(B) reading the classified section of your local newspaper.
(C) visiting a local job service agency.
(D) idly waiting for potential employers to contact you.

11. All of these are a necessary part of getting a job EXCEPT:
(A) applying for the job.
(B) making a favorable first impression.
(C) finding the right job opening.
(D) understating your qualifications.

This ASE certified technician is using a handheld scanner to help diagnose an operating problem. (Owatonna Tool)

Chapter 28

Certification in Drive Train Service

After studying this chapter, you will be able to:

- Explain why technician certification is beneficial.
- Describe the process of registering for ASE tests.
- Explain what is done with ASE test results.
- Explain how to take the ASE tests.

This chapter will explain the benefits of *National Institute for Automotive Service Excellence (ASE)* certification. It will also describe how to apply for and take the ASE tests. The ASE tests that are directly applicable to drive trains are the *Automatic Transmission/Transaxle* and *Manual Drive Train and Axles* tests. In addition, the ASE test covering *Electrical Systems* is relevant to drive trains, among other mechanical systems. You should also consider taking the *Engine Performance* test, since the condition of the engine and its related systems has a big effect on the performance of the drive train components. Becoming certified is voluntary. However, most repair shops prefer to hire certified technicians because these individuals have demostrated a desired level of expertise.

Purpose of ASE

The ***National Institute for Automotive Service Excellence (ASE)*** is a nonprofit organization that was formed to promote high standards of automotive service and repair. To accomplish this, the ASE administers a series of written tests on various aspects of automotive theory and repair. Tests are also available in the truck repair, auto body repair, and engine machining areas. The ASE tests are called ***standardized tests*** because everyone who takes them is tested on the same material. Technicians who pass one or more of the ASE tests and meet certain experience requirements are certified in the repair areas covered by the tests passed. If a technician can pass all of the tests in the automotive area or the heavy truck area, he or she is considered a ***certified master technician.***

ASE Tests

The concept of setting standards of excellence for skilled jobs is not new. In ancient times, metalworkers, weavers, potters, and other artisans were expected to conform to certain standards of work quality. In many cases, the need for quality standards resulted in the establishment of associations of skilled workers who set the standards and enforced rules of conduct. Many modern American labor unions have descended from these early associations.

Certification processes for aircraft technicians, aerospace workers, and electronics technicians have existed since the beginning of these industries. However, this is not the case in the automotive industry. Automobile manufacturing and repair began as a fragmented industry, made up of many small vehicle manufacturers and thousands of small repair shops. Although the number of vehicle manufacturers decreased, repair facilities continued to increase in number and variety. Due to the decentralized nature of the automotive repair industry, standards are difficult to establish. For over 50 years, there was no unified set of standards for measuring a technician's knowledge. All people who worked on cars could call themselves automotive mechanics, even if they were unqualified. This situation resulted in unneeded or unsatisfactory repair work. Consequently, a large segment of the public came to regard mechanics as unintelligent and dishonest.

This situation changed in 1975, when the National Institute for Automotive Service Excellence (ASE) was established to develop a certification program for automobile technicians. The purpose of the ***ASE certification program*** is to identify and reward skilled and knowledgeable technicians. Periodic recertification requirements provide technicians with an incentive for updating their skills and also provide guidelines for keeping up with current technology. The certification program allows potential employers and customers to identify good technicians. It also helps the technician advance his or her career. The program is not mandatory on national or state levels, but many repair shops now hire only ASE certified technicians. Approximately 500,000 persons are now ASE certified in one or more areas.

Other activities in which ASE is involved include encouraging the development of effective training programs, conducting research on instruction methods, and publicizing the advantages of technician certification. ASE is managed by a board, which is made up of people from the automotive and truck service industries, motor vehicle manufacturing industries, state and federal government agencies, schools and educational groups, and consumer associations.

One benefit that the ASE certification program has brought to the automotive industry is increased respect for automotive technicians. This has resulted in better pay, improved working conditions, and increased standing in the community.

Applying for the ASE Tests

Anyone may take the ASE tests; however, a passing grade does not mean automatic certification. To receive certification, the applicant must also have at least two years of experience as an automobile or truck technician. This experience does not have to be in any specific area of automotive service. In some cases, training programs, apprenticeship programs, or time spent performing similar work can be substituted for all or part of the work experience.

ASE tests are given in the spring and the fall of each year. The tests are usually held during a two-week period and are given on weeknights and Saturdays. The actual test administration procedures are performed by ***ACT,*** a nonprofit organization experienced in administering standardized tests. The tests are given at designated test centers at over 300 locations in the United States. If necessary, special test centers can be set up in remote locations. However, there must be enough potential applicants for the establishment of a special test center to be practical.

To apply to take ASE tests, begin by acquiring a registration form, Figure 28-1. To obtain the most current form, contact ASE at the following address:

National Institute for Automotive Service Excellence
13505 Dulles Technology Drive
Herndon, VA 22071-3415

ASE will send you the proper form and an ***ASE Information Bulletin,*** which explains the procedure for completing the form. When you receive the form, fill it out carefully, recording all necessary information. Work experience should be listed on the application form. If there is any doubt about what should be placed in a particular

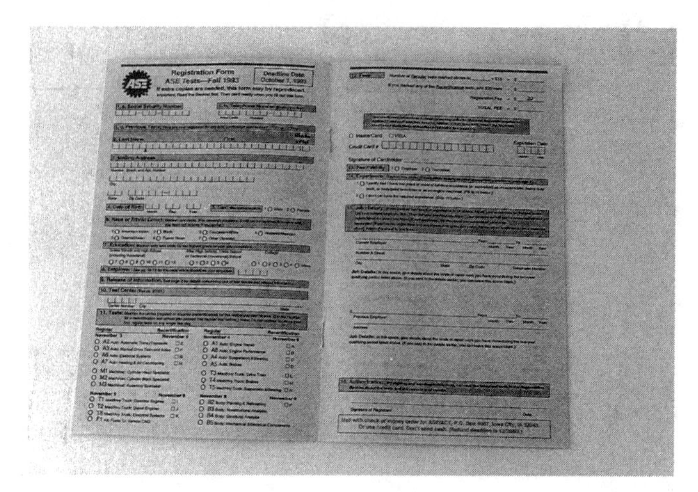

Figure 28-1. Applicants must complete a registration form before taking ASE tests.

space, consult the ASE Information Bulletin. Be sure to determine the closest test center, and record its number in the appropriate space. Most test centers are located at local colleges, high schools, or vocational schools. Although you may apply to take all the tests being given, fewer tests can be taken if desired. In some cases, you may wish to take only one test. In addition to the application, you must include a check or money order to cover all necessary fees. A fee is charged to register for the test series, and a separate fee is charged for each test to be taken. Refer to the latest ASE Information Bulletin for the current fee structure. In some cases, employers pay the registration and test fees. Check with your employer before submitting your application.

To be accepted for either the spring or fall ASE tests, your application and payment must arrive at ASE headquarters at least one month before the test date(s). To ensure that you can take the test at the test center of your choice, send in the application as early as possible.

After receiving your application and fees, ASE will send you an ***admission ticket*** to the test center. You should receive the ticket by mail about two weeks after submitting your application. See Figure 28-2. If the test dates are less than two weeks away and your admission ticket has not arrived, contact ASE using the phone number given in the ASE Information Bulletin. If the desired test center is filled when ASE receives your application, you will be instructed to report to the nearest center that has an opening. If it is not possible to go to the alternate test center, contact ACT immediately. The phone number for ACT can also be found in the ASE Information Bulletin.

Test Results

After taking the ASE tests, be prepared to wait six to eight weeks for the tests to be graded and the results to be sent out. You will receive a *confidential* report of your

National Institute for Automotive Service Excellence

ACT, P.O. Box 4007, Iowa City, Iowa 52243, Phone: (319) 337-1433

Admission Ticket

Test Center to which you are assigned:

GREENVILLE TECHNICAL COLLEGE
CONTINUING EDUCATION BLDG-RM 136
SOUTH PLEASANTBURG DRIVE
GREENVILLE, SOUTH CAROLINA 29607

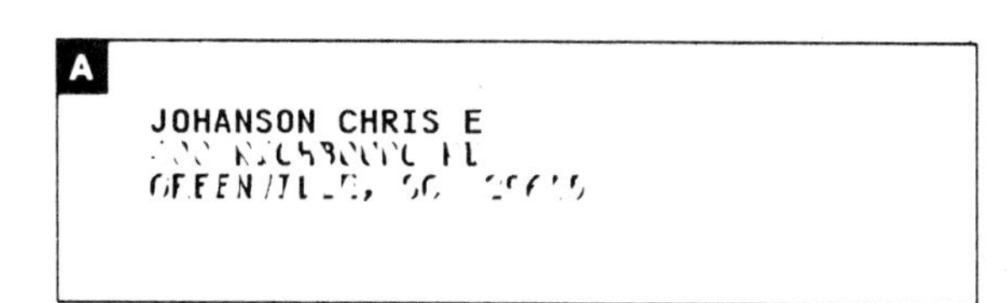

A

JOHANSON CHRIS E

REGULAR TESTS (Late arrivals will not be admitted.)		
DATE	REPORTING TIME	TEST(S)

RECERTIFICATION TESTS (Late arrivals will not be admitted.)
TESTED ONLY ON MAY 12 AT 7:00 PM
A1*, A2*, A3*, A4*, A5*, A6* A7*, A8*

TEST CODE KEY

A1 Auto: Engine Repair	A6 Auto: Electrical Systems	M3 Machinist: Assembly Specialist	T5 Med/Hvy Truck: Suspension & Steering
A2 Auto: Automatic Trans/Transaxle	A7 Auto: Heating & Air Conditioning	T1 Med/Hvy Truck: Gasoline Engines	T6 Med/Hvy Truck: Electrical Systems
A3 Auto: Manual Drive Train & Axles	A8 Auto: Engine Performance	T2 Med/Hvy Truck: Diesel Engines	B1 Body: Body Repair
A4 Auto: Suspension & Steering	M1 Machinist: Cylinder Head Specialist	T3 Med/Hvy Truck: Drive Train	B2 Body: Painting & Refinishing
A5 Auto: Brakes	M2 Machinist: Cylinder Block Specialist	T4 Med/Hvy Truck: Brakes	

See Notes and Ticketing Rules on reverse side. An asterisk (*) indicates your certificate in these areas is expiring.

SPECIAL MESSAGES

-CHECK YOUR TESTS AND TEST CENTER TO BE SURE THEY ARE WHAT YOU REQUESTED. IF EITHER IS INCORRECT, CALL 319/337-1433 IMMEDIATELY. TESTS CANNOT BE CHANGED AT THE TEST CENTER.
-IF YOU MISS ANY EXAMS FOR WHICH YOU ARE REGISTERED, YOU MAY OBTAIN A REFUND OF YOUR TEST FEES BY FOLLOWING THE INSTRUCTIONS ON THE BACK OF THIS SHEET. THE REFUND DEADLINE IS JUNE 22, 1992. REFUND CHECKS WILL BE ISSUED IN AUGUST.

MATCHING INFORMATION: The information printed in blocks B and C at the right was obtained from your registration form. It will be used to match your registration information and your test information. Therefore, the information at the right must be copied EXACTLY (even if it is in error) onto your answer folder on the day of the test. If the information is not copied exactly as shown, it may cause a delay in reporting your test results to you.

IF THERE ARE ERRORS: If there are any errors or if any information is missing in block A above or in blocks B and C at the right, you must contact ACT immediately. DO NOT SEND THIS ADMISSION TICKET TO ACT TO MAKE SUCH CORRECTIONS.

Check your tests and test center to be sure they are what you requested. If either is incorrect, call 319/337-1433 immediately. Tests cannot be changed at the test center. **ON THE DAY OF THE TEST,** be sure to bring this admission ticket, positive identification, several sharpened No. 2 pencils, and a watch if you wish to pace yourself.

B FIRST FIVE LETTERS OF LAST NAME

J	O	H	A	N

C SOCIAL SECURITY NUMBER OR ACT IDENTIFICATION NUMBER

SIDE 1

Figure 28-2. The admission ticket must be presented on the day of the test.

performance. The report will show your score on the test and indicate whether this score is sufficient for certification. The test questions are subdivided into general sections to help you determine the areas that require more study. For instance, the engine performance test questions will be divided into such subsections as ignition, fuel, starting and charging systems, engine mechanical systems, and computer control systems. Also included with the report is a certification document for the tests that you passed, an ASE shoulder patch, and a pocket card listing all of the areas in which you are certified. See Figure 28-3.

As previously mentioned, ASE test results are confidential. To protect your privacy, test results will be mailed to your home address. The only information that ASE will release to an employer is a confirmation of certification in a particular area. This is true even if your employer has paid the test fees. If you want your employer to know exactly how you performed on the ASE tests, you must provide him or her with a copy of your test results.

If you fail a certification test, you can retake it as many times as desired. However, you (or your employer) must pay all of the applicable registration and test fees again. You should study all available information that pertains to the areas in which you did poorly. A copy of the ***ASE Preparation Guide*** may help you sharpen your skills in these areas. The ASE Preparation Guide is free and can be obtained by filling out the coupon at the back of the ASE Information Bulletin.

Recertification Tests

Once you are certified in any area, you must take a ***recertification test*** every five years to maintain your certification. Recertification test questions concentrate on recent developments. This ensures that certified technicians will keep up with technology changes. The process for applying to take recertification tests is similar to that for original certification tests. Use the same form and enclose the proper recertification test fees. If you allow your certification to lapse, you must take the regular certification test(s) to regain your certification.

Figure 28-3. This shoulder patch identifies the person wearing it as a certified master technician.

Tips on Taking ASE Tests

Be sure to bring your admission ticket when you report to the test center. When you arrive at the center, you will be asked to produce the admission ticket and a driver's license or other photographic identification. In addition to these items, bring several Number 2 pencils. Although pencils may be made available at the test center, extra pencils may save time if your original pencil breaks.

Follow all instructions given by the test administrators. During the test, read each question carefully before deciding on a proper answer. After completing all the questions in a particular test, recheck your answers to ensure that you did not make a careless error. In most cases, rechecking your answers more than once is unnecessary and may lead you to change correct answers to incorrect ones. The time allowed for each test is usually about four hours. However, you may leave after completing your last scheduled test and handing in all test material.

Types of ASE Test Questions

ASE tests are designed to measure your knowledge of three things:

- The operation of various automotive systems and components.
- The diagnosis and testing of various automotive systems and components.
- The repair of automotive systems and components.

Each ASE test contains between 40 and 80 questions, depending on the subject matter. All test questions are multiple choice and contain four possible answers. These questions are similar to the multiple-choice questions used in this chapter. Sample questions are given below. The answer to each question is explained in detail.

One-part question

1. The clutch is released by which one of these?

(A) Pressure plate springs.
(B) Clutch fork spring.
(C) Foot pressure on the clutch pedal.
(D) Weights on the release lever.

In this question, you must choose the best answer out of all of the possibilities. The only way that the clutch can be released is by pushing on the clutch pedal. Therefore, "(C) Foot pressure on the clutch pedal." is the correct answer.

Two-part question

1. Technician A says that a locking rear differential assembly can be refilled with regular gear oil. Technician B says that the differential assembly allows the vehicle to turn corners without wheel hop. Who is right?

(A) A only **(C)** Both A & B
(B) B only **(D)** Neither A nor B

This question requires you to read two statements and decide if they are true. Both statements can be true, or both can be false. In some cases, only one of the statements is true. In this question, the statement made by Technician A is wrong, since a locking rear wheel differential assembly must be refilled with special non-slip oil. Technician B's statement is correct, since the purpose of the differential assembly is to allow the vehicle to turn corners without wheel hop. Therefore, the correct answer is "(B) B only." Note that Technician A and Technician B appear in many ASE test questions. You must carefully evaluate the statements of each technician before deciding which answer is correct.

Negative questions

Some questions are called negative questions. These questions require you to identify the *incorrect* answer. Negative questions will usually contain the word "except."

1. A manual transmission contains all of these bearings EXCEPT:

(A) an input shaft bearing.
(B) countershaft bearings.
(C) a pilot bearing.
(D) an output shaft bearing.

Since the pilot bearing is installed in the flywheel and there is no bearing with that name used in the transmission, the correct answer is "(C) a pilot bearing."

A variation of the negative question contains the word "least."

1. An automatic transmission installed in a late-model car slips during acceleration. Which of these defects is the *least* likely cause?

(A) Clogged transmission oil filter.
(B) Defective oil pump.
(C) Maladjusted throttle linkage.
(D) Low oil level.

In this case, the least likely cause of transmission slipping is maladjusted throttle linkage, which is much more likely to cause shifting problems than slipping. Therefore, the correct answer is "(C) Maladjusted throttle linkage."

Completion questions

Some test questions are simply sentences that must be completed. One of the four possible answers correctly completes the sentence. An example of a completion question follows.

1. The inclinometer is used to measure:

(A) rear axle angles.
(B) driveline angles.
(C) front drive axle angles.
(D) transmission slip yoke angles.

Once again the question calls for the best answer. The inclinometer is used to measure front and rear U-joint angles, so "(B) driveline angles." is correct.

Summary

At one time, the automotive industry was one of the few major industries that did not have a testing and certification program. This resulted in decreased professionalism in the automobile industry, often leading to poor or unneeded repairs and decreased status for automobile technicians. The National Institute for Automotive Service Excellence (ASE) was established in 1975 to promote high standards and to help overcome the negative image and low status of the automotive repair industry. ASE tests and certifies automotive technicians in major areas of automotive repair. This has increased the skill level of technicians, resulting in better service for customers and increased benefits for technicians.

ASE tests are given in the spring and the fall of each year. Anyone can register to take the tests by filling out the registration form and paying the registration and test fees. The registrant must select the test center that he or she would like to attend. To be considered for certification, the registrant must have two years of hands-on experience as an automotive technician. Proof of this experience should also be included with the registration form. About three weeks after applying for the test, the technician will receive an admission ticket, which he or she must bring to the test center on the day of the test.

Test results will arrive six to eight weeks after the test session. Results are confidential and will be sent only to the home address of the person who took the test. If a test is passed and the experience requirement has been met, the technician will be certified in the specific test area for five years. Anyone who fails a test can take it again during the next session. Tests can be taken as many times as necessary. Recertification tests can be taken at the end of the five-year certification period.

The ASE test questions will test your knowledge of general system operation, problem diagnosis, and repair techniques. All of the questions are multiple choice and usually contain four possible answers. The questions must be read carefully. The entire test should be rechecked one time to catch careless mistakes.

Know These Terms

National Institute for Automotive Service Excellence (ASE); Standardized test; Certified master technician; ASE certification program; ACT; ASE Information Bulletin; Admission ticket; ASE Preparation Guide; Recertification tests.

Review Questions–Chapter 28

Please do not write in this text. Place your answers on a separate sheet of paper.

1. The automobile repair industry was one of the first American industries to have a certification program. True or false?
2. The National Institute for Automotive Service Excellence was founded in 1975. True or false?
3. ASE tests are given four times each year. True or false?
4. One ASE test can be taken a total of three times. True or false?
5. Recertification tests must be taken every three years. True or false?
6. To enter the test center, what must the technician present?
7. What three things are measured by ASE tests?
8. A negative question will usually contain the word ______.
9. It is recommended that you do not recheck your answers more than:
   a. Once.
   b. Twice.
   c. Three times.
   d. There is no limit.

Certification-Type Questions–Chapter 28

1. ASE encourages high standards of automotive service and repair by providing technicians with the opportunity to take a series of:

(A) written tests.
(B) hands-on tests.
(C) oral tests.
(D) essay tests.

2. A standardized test is:

(A) a comprehensive test covering all aspects of a subject.
(B) used to check all people taking it on the same material.
(C) a test that is printed on a standard type and size of paper.
(D) the same thing as an IQ test.

3. Technician A says that an ASE certified master technician is one who has passed all of the ASE tests in the automotive or heavy truck areas and who has at least two years of hands-on experience. Technician B says that it is one who has passed both written and hands-on tests and who has demonstrated that he or she can change oil, balance tires, and install batteries. Who is right?

(A) A only
(B) B only
(C) Both A & B
(D) Neither A nor B

4. The advantages that ASE certification has brought to automotive technicians include all of these EXCEPT:

(A) increased respect.
(B) better working conditions.
(C) lower pay scales.
(D) increased standing in the community.

5. ASE tests are offered:

(A) once each year.
(B) twice each year.
(C) 4 times each year.
(D) 12 times each year.

6. Which of these are held at night and on weekends?

(A) ASE tests.
(B) ASE training sessions.
(C) ASE registrations.
(D) ASE review sessions.

7. **ASE provides test results to the person who took the test and to:**
   (A) whoever paid for the test.
   (B) the technician's employer.
   (C) the last educational institution the technician attended.
   (D) nobody else.

8. **A technician can retake certification tests:**
   (A) 2 times only.
   (B) 4 times only.
   (C) 5 times only.
   (D) as many times as desired.

9. **To maintain their certification, technicians must take a recertification test every:**
   (A) 2 years.
   (B) 5 years.
   (C) 10 years.
   (D) 15 years.

10. **ASE test questions resemble the questions:**
   (A) in college-level courses.
   (B) in essay-type tests.
   (C) in verbal examinations.
   (D) at the end of each chapter of this text.

Useful Tables

CONVERSION CHART

METRIC/U.S. CUSTOMARY UNIT EQUIVALENTS

Multiply:	by:	to get:	Multiply: by:	to get:
ACCELERATION				
feet/sec^2	X 0.3048	= metres/sec^2 (m/s^2)	X 3.281	= feet/sec^2
inches/sec^2	X 0.0254	= metres/sec^2 (m/s^2)	X 39.37	= inches/sec^2
ENERGY OR WORK (watt-second = joule = newton-metre)				
foot-pounds	X 1.3558	= joules (J)	X 0.7376	= foot-pounds
calories	X 4.187	= joules (J)	X 0.2388	= calories
Btu	X 1055	= joules (J)	X 0.000948	= Btu
watt-hours	X 3600	= joules (J)	X 0.0002778	= watt-hours
kilowatt - hrs	X 3.600	= megajoules (MJ)	X 0.2778	= kilowatt - hrs
FUEL ECONOMY AND FUEL CONSUMPTION				
miles/gal	X 0.42514	= kilometres/litre (km/L)	X 2.3522	= miles/gal

Note:
235.2/(mi/gal) = litres/100km
235.2/(litres/100 km) = mi/gal

Multiply:	by:	to get:	Multiply: by:	to get:
LIGHT				
footcandles	X 10.76	= lumens/metre^2 (lm/m^2)	X 0.0929	= footcandles
PRESSURE OR STRESS (newton/sq metre = pascal)				
inches Hg(60°F)	X 3.377	= kilopascals (kPa)	X 0.2961	= inches Hg
pounds/sq in	X 6.895	= kilopascals (kPa)	X 0.145	= pounds/sq in
inches H_2O(60°F)	X 0.2488	= kilopascals (kPa)	X 4.0193	= inches H_2O
bars	X 100	= kilopascals (kPa)	X 0.01	= bars
pounds/sq ft	X 47.88	= pascals (Pa)	X 0.02088	= pounds/sq ft
POWER				
horsepower	X 0.746	= kilowatts (kW)	X 1.34	= horsepower
ft-lbf/min	X 0.0226	= watts (W)	X 44.25	= ft-lbf/min
TORQUE				
pound-inches	X 0.11298	= newton-metres (N·m)	X 8.851	= pound-inches
pound-feet	X 1.3558	= newton-metres (N·m)	X 0.7376	= pound-feet
VELOCITY				
miles/hour	X 1.6093	= kilometres/hour (km/h)	X 0.6214	= miles/hour
feet/sec	X 0.3048	= metres/sec (m/s)	X 3.281	= feet/sec
kilometres/hr	X 0.27778	= metres/sec (m/s)	X 3.600	= kilometres/hr
miles/hour	X 0.4470	= metres/sec (m/s)	X 2.237	= miles/hour

COMMON METRIC PREFIXES

mega	(M) = 1 000 000 or 10^6	centi	(c) = 0.01	or 10^{-2}
kilo	(k) = 1 000 or 10^3	milli	(m) = 0.001	or 10^{-3}
hecto	(h) = 100 or 10^2	micro	(μ) = 0.000 001	or 10^{-6}

METRIC/U.S. CUSTOMARY UNIT EQUIVALENTS

Multiply:	by:	to get:	Multiply: by:	to get:
LINEAR				
inches	X 25.4	= millimetres (mm)	X 0.03937	= inches
feet	X 0.3048	= metres (m)	X 3.281	= feet
yards	X 0.9144	= metres (m)	X 1.0936	= yards
miles	X 1.6093	= kilometres (km)	X 0.6214	= miles
inches	X 2.54	= centimetres (cm)	X 0.3937	= inches
microinches	X 0.0254	= micrometres (μm)	X 39.37	= microinches
AREA				
inches^2	X 645.16	= millimetres^2 (mm^2)	X 0.00155	= inches^2
inches^2	X 6.452	= centimetres^2 (cm^2)	X 0.155	= inches^2
feet^2	X 0.0929	= metres^2 (m^2)	X 10.764	= feet^2
yards^2	X 0.8361	= metres^2 (m^2)	X 1.196	= yards^2
acres	X 0.4047	= hectares (10^4m^2) (ha)	X 2.471	= acres
miles^2	X 2.590	= kilometres^2 (km^2)	X 0.3861	= miles^2
VOLUME				
inches^3	X 16387	= millimetres^3 (mm^3)	X 0.000061	= inches^3
inches^3	X 16.387	= centimetres^3 (cm^3)	X 0.06102	= inches^3
inches^3	X 0.01639	= litres (L)	X 61.024	= inches^3
quarts	X 0.94635	= litres (L)	X 1.0567	= quarts
gallons	X 3.7854	= litres (L)	X 0.2642	= gallons
feet^3	X 28.317	= litres (L)	X 0.03531	= feet^3
feet^3	X 0.02832	= metres^3 (m^3)	X 35.315	= feet^3
fluid oz	X 29.57	= millilitres (mL)	X 0.03381	= fluid oz
yards^3	X 0.7646	= metres^3 (m^3)	X 1.3080	= yards^3
teaspoons	X 4.929	= millilitres (mL)	X 0.2029	= teaspoons
cups	X 0.2366	= litres (L)	X 4.227	= cups
MASS				
ounces (av)	X 28.35	= grams (g)	X 0.03527	= ounces (av)
pounds (av)	X 0.4536	= kilograms (kg)	X 2.2046	= pounds (av)
tons (2000 lb)	X 907.18	= kilograms (kg)	X 0.001102	= tons (2000 lb)
tons (2000 lb)	X 0.90718	= metric tons (t)	X 1.1023	= tons (2000 lb)
FORCE				
ounces — f (av)	X 0.278	= newtons (N)	X 3.597	= ounces — f (av)
pounds — f (av)	X 4.448	= newtons (N)	X 0.2248	= pounds — f (av)
kilograms — f	X 9.807	= newtons (N)	X 0.10197	= kilograms — f

TEMPERATURE

°F -40 0 32 40 80 98.6 120 160 200 212 240 280 320 °F
°C -40 -20 0 20 40 60 80 100 120 140 160 °C

°Celsius = 0.556 (°F — 32)   °F = (1.8°C) + 32

TAP/DRILL CHART

COARSE STANDARD THREAD (N. C.) Formerly U. S. Standard Thread					FINE STANDARD THREAD (N. F.) Formerly S.A.E. Thread				
Sizes	Threads Per Inch	Outside Diameter at Screw	Tap Drill Sizes	Decimal Equivalent of Drill	Sizes	Threads Per Inch	Outside Diameter at Screw	Tap Drill Sizes	Decimal Equivalent of Drill
1	64	.073	53	0.0595	0	80	.060	3/64	0.0469
2	56	.086	50	0.0700	1	72	.073	53	0.0595
3	48	.099	47	0.0785	2	64	.086	50	0.0700
4	40	.112	43	0.0890	3	56	.099	45	0.0820
5	40	.125	38	0.1015	4	48	.112	42	0.0935
6	32	.138	36	0.1065	5	44	.125	37	0.1040
8	32	.164	29	0.1360	6	40	.138	33	0.1130
10	24	.190	25	0.1495	8	36	.164	29	0.1360
12	24	.216	16	0.1770	10	32	.190	21	0.1590
1/4	20	.250	7	0.2010	12	28	.216	14	0.1820
5/16	18	.3125	F	0.2570	1/4	28	.250	3	0.2130
3/8	16	.375	5/16	0.3125	5/16	24	.3125	I	0.2720
7/16	14	.4375	U	0.3680	3/8	24	.375	Q	0.3320
1/2	13	.500	27/64	0.4219	7/16	20	.4375	25/64	0.3906
9/16	12	.5625	31/64	0.4843	1/2	20	.500	29/64	0.4531
5/8	11	.625	17/32	0.5312	9/16	18	.5625	0.5062	0.5062
3/4	10	.750	21/32	0.6562	5/8	18	.625	0.5687	0.5687
7/8	9	.875	49/64	0.7656	3/4	16	.750	11/16	0.6875
1	8	1.000	7/8	0.875	7/8	14	.875	0.8020	0.8020
1 1/8	7	1.125	63/64	0.9843	1	14	1.000	0.9274	0.9274
1 1/4	7	1.250	1 7/64	1.1093	1 1/8	12	1.125	1 3/64	1.0468
					1 1/4	12	1.250	1 11/64	1.1718

BOLT TORQUING CHART

METRIC STANDARD						
GRADE OF BOLT		5D	.8G	10K	12K	
MIN. TENSILE STRENGTH		71,160 P.S.I	113,800 P.S.I.	142,200 P.S.I.	170,679 P.S.I.	
GRADE MARKINGS ON HEAD		5D	8G	10K	12K	SIZE OF SOCKET OR WRENCH OPENING
METRIC		FOOT POUNDS				METRIC
BOLT DIA.	U.S. DEC EQUIV.					BOLT HEAD
6mm	.2362	5	6	8	10	10mm
8mm	.3150	10	16	22	27	14mm
10mm	.3937	19	31	40	49	17mm
12mm	.4720	34	54	70	86	19mm
14mm	.5512	55	89	117	137	22mm
16mm	.6299	83	132	175	208	24mm
18mm	.709	111	182	236	283	27mm
22mm	.8661	182	284	394	464	32mm

SAE STANDARD / FOOT POUNDS						
GRADE OF BOLT	SAE 1 & 2	SAE 5	SAE 6	SAE 8		
MIN. TEN STRENGTH	64,000 P.S.I.	105,000 P.S.I.	133,000 P.S.I.	150,000 P.S.I.		
MARKINGS ON HEAD					SIZE OF SOCKET OR WRENCH OPENING	
U.S. STANDARD	FOOT POUNDS				U.S. REGULAR	
BOLT DIA.					BOLT HEAD	NUT
1/4	5	7	10	10.5	3/8	7/16
5/16	9	14	19	22	1/2	9/16
3/8	15	25	34	37	9/16	5/8
7/16	24	40	55	60	5/8	3/4
1/2	37	60	85	92	3/4	13/16
9/16	53	88	120	132	7/8	7/8
5/8	74	120	167	180	15/16	1.
3/4	120	200	280	296	1-1/8	1-1/8

DECIMAL CONVERSION CHART

FRACTION	INCHES	M/M
1/64	.01563	.397
1/32	.03125	.794
3/64	.04688	1.191
1/16	.06250	1.588
5/64	.07813	1.984
3/32	.09375	2.381
7/64	.10938	2.778
1/8	.12500	3.175
9/64	.14063	3.572
5/32	.15625	3.969
11/64	.17188	4.366
3/16	.18750	4.763
13/64	.20313	5.159
7/32	.21875	5.556
15/64	.23438	5.953
1/4	.25000	6.350
17/64	.26563	6.747
9/32	.28125	7.144
19/64	.29688	7.541
5/16	.31250	7.938
21/64	.32813	8.334
11/32	.34375	8.731
23/64	.35938	9.128
3/8	.37500	9.525
25/64	.39063	9.922
13/32	.40625	10.319
27/64	.42188	10.716
7/16	.43750	11.113
29/64	.45313	11.509
15/32	.46875	11.906
31/64	.48438	12.303
1/2	.50000	12.700

FRACTION	INCHES	M/M
33/64	.51563	13.097
17/32	.53125	13.494
35/64	.54688	13.891
9/16	.56250	14.288
37/64	.57813	14.684
19/32	.59375	15.081
39/64	.60938	15.478
5/8	.62500	15.875
41/64	.64063	16.272
21/32	.65625	16.669
43/64	.67188	17.066
11/16	.68750	17.463
45/64	.70313	17.859
23/32	.71875	18.256
47/64	.73438	18.653
3/4	.75000	19.050
49/64	.76563	19.447
25/32	.78125	19.844
51/64	.79688	20.241
13/16	.81250	20.638
53/64	.82813	21.034
27/32	.84375	21.431
55/64	.85938	21.828
7/8	.87500	22.225
57/64	.89063	22.622
29/32	.90625	23.019
59/64	.92188	23.416
15/16	.93750	23.813
61/64	.95313	24.209
31/32	.96875	24.606
63/64	.98438	25.003
1	1.00000	25.400

INDEX–GLOSSARY REFERENCE

The Index-Glossary Reference serves two functions: it serves as an index and as a technical dictionary. It can be used like a conventional index for finding topics in the body of the book. It also provides a method of quickly finding definitions of technical terms, like a conventional glossary.

Boldface type is used to give page numbers for definitions. If you need a technical word explained in one or two sentences, simply turn to the page number printed in darker type; there you will find the defined term printed in ***boldface italics.***

The Index-Glossary Reference is educationally superior to a conventional glossary because it allows you to obtain more information about the term being questioned. You can read the definition in the context of the book and also refer to the illustrations that "paint a picture" about the new term.

While you are looking up the meaning of a particular term, you might also want to read any definitions of other related words to build a background for more fully comprehending the new technical term. This will help you learn the essential "language of an automotive technician."

A

B

C

D

G

H

I

J

K

L

M

N

O

P

R

S